Textbook of Children's Environmental Health

T0073877

Textbook of Children's Environmental Health

Second Edition

Edited by

RUTH A. ETZEL

and

PHILIP J. LANDRIGAN

OXFORD
UNIVERSITY PRESS

Oxford University Press is a department of the University of Oxford. It furthers
the University's objective of excellence in research, scholarship, and education
by publishing worldwide. Oxford is a registered trade mark of Oxford University
Press in the UK and certain other countries.

Published in the United States of America by Oxford University Press
198 Madison Avenue, New York, NY 10016, United States of America.

Library of Congress Cataloging-in-Publication Data
Names: Landrigan, Philip J., editor of compilation. |
Etzel, Ruth Ann, editor of compilation.
Title: Textbook of children's environmental health /
Edited by Ruth A. Etzel and Philip J. Landrigan.
Description: Second edition. | New York, NY, United States of America :
Oxford University Press, [2024] |
Includes bibliographical references and index.
Identifiers: LCCN 2023035497 (print) | LCCN 2023035498 (ebook) |
ISBN 9780197662533 (pb) | ISBN 9780197662526 (hb) |
ISBN 9780197662557 (epub) | ISBN 9780197662564
Subjects: MESH: Child—United States. | Environmental Health—United States. |
Environmental Exposure—adverse effects—United States. |
Environmental Pollutants—adverse effects—United States.
Classification: LCC RJ102 .T49 2024 (print) | LCC RJ102 (ebook) |
DDC 362.19892—dc23/eng/20231025
LC record available at https://lccn.loc.gov/2023035497
LC ebook record available at https://lccn.loc.gov/2023035498

DOI: 10.1093/oso/9780197662526.001.0001

Paperback printed by Marquis Book Printing, Canada
Hardback printed by Bridgeport National Bindery, Inc., United States of America

Contents

PART II ENVIRONMENTS

PART III ENVIRONMENTAL HAZARDS

PART IV THE ENVIRONMENT AND DISEASE IN CHILDREN

PART V PREVENTION AND CONTROL OF DISEASES OF ENVIRONMENTAL ORIGIN IN CHILDREN

Acknowledgments

This volume builds on the work of many people over many decades. They include pediatricians, toxicologists, lawyers, and leaders of government agencies: persons of vision, courage, integrity, and altruism who have worked—sometimes in the face of enormous adversity—to help mend this world and make it a better place for all our children.

In particular, we are grateful to the authors of the 60 chapters in the first edition because their work laid the foundation for this edition. We owe a debt of gratitude to Robert W. Miller, MD, DrPH, the visionary pediatrician who founded the discipline of children's environmental health. We recognize the pioneering work of our sister and brother pediatricians and scientists who have patiently worked to build the discipline of children's environmental health over many years—Sophie J. Balk, MD, FAAP; Henry Falk, MD, MPH; Eunhee Ha, MD, PhD; Richard J. Jackson, MD, FAAP; Amalia LaBorde, MD, PhD; Bruce P. Lanphear, MD, MPH; Jerome Paulson, MD, FAAP; and Peter D. Sly, MD, who have generously contributed to this volume. We thank the former agency directors and US government officials who have supported the growth of children's environmental health: David P. Rall, PhD, MD; Kenneth Olden, PhD; and Linda Birnbaum, PhD, of the National Institute for Environmental Health Sciences; Barry L. Johnson, PhD, of the Agency for Toxic Substances and Disease Registry; and Vernon Houk, MD, of the Centers for Disease Control and Prevention. Deep thanks to the American Academy of Pediatrics for nurturing the discipline of children's environmental health. And finally, special thanks to our families, without whose love and unwavering support we could never have produced this volume—RAE thanks Raymond and Marian and Richard Hofrichter. PJL thanks Mary, Mary Frances, Christopher, Elizabeth, Jack, Ryan, Mary Katya, Sara, Gabriel, Aelish, and Isaac.

About the Editors

Ruth A. Etzel, MD, PhD, is an internationally recognized pediatrician, environmental epidemiologist, and preventive medicine specialist. She performed the first study documenting that children with secondhand exposure to tobacco smoke had measurable exposure to nicotine. Her pioneering work led to nationwide efforts to reduce indoor exposure to tobacco, including the ban on smoking in US airliners. She also produced the first research to show that exposure to toxigenic molds in the home could be dangerous to infants' health. From 2009 to 2012, she led the World Health Organization's activities to protect children from environmental hazards. She was the first pediatrician to serve as Director of the Office of Children's Health Protection at the US Environmental Protection Agency. She led the President's Task Force on Environmental Health Risks and Safety Risks to Children in developing the Federal Action Plan to Reduce Childhood Lead Exposures and Associated Health Impacts. She is the founding editor of *Pediatric Environmental Health*, now in its fourth edition.

Philip J. Landrigan, MD, MSc, is a pediatrician and epidemiologist. His research examines the connections between toxic chemicals and children's health. His studies of lead poisoning demonstrated that lead is toxic to children even at very low levels, and his work contributed to the US government's 1975 decision to remove lead from paint and gasoline, actions that reduced blood lead levels in the United States by 95% and increased children's average IQ by 5 points. A study he led in the 1990s at the National Academy of Sciences defined children's unique susceptibilities to pesticides and catalyzed fundamental revamping of US pesticide policy. From 2015 to 2017, Dr. Landrigan co-chaired the *Lancet* Commission on Pollution and Health, which reported that pollution causes 9 million deaths annually and is an existential threat to planetary health. In 2022–2023, he led the Minderoo-Monaco Commission on Plastics and Human Health, which analyzed plastics' negative impacts on human health across the plastic life cycle. Dr. Landrigan directs the Program for Global Public Health and the Common Good at Boston College.

Contributors

A. Kofi Amegah, PhD
Public Health Research Group
Department of Biomedical Sciences
University of Cape Coast
Cape Coast, Ghana

Katherine Arnold
Private Consultant
Tucson, AZ, USA

Robert G. Arnold, PhD
Professor Emeritus
Department of Chemical and Environmental
   Engineering
University of Arizona
Tucson, AZ, USA

Hilal Yildiz Atar, MD
Fellow
Neonatal-Perinatal Medicine
Case Western Reserve University School of
   Medicine
Cleveland, OH, USA

Somaiyeh Azmoun
Research Assistant
PhD candidate in Public Health
Department of Environmental Health
   Sciences
Robert Stempel College of Public Health and
   Social Work
Florida International University
Miami, FL, USA

Dean Baker, MD, MPH
Professor Emeritus of Medicine, Pediatrics,
   and Epidemiology
University of California
Irvine, CA, USA

Sophie J. Balk, MD
Attending Pediatrician
Children's Hospital at Montefiore
Professor of Clinical Pediatrics
Albert Einstein College
   of Medicine
New York, NY, USA

Dana Boyd Barr, PhD
Professor, Exposure Science and
   Environmental Health
Gangarosa Department of
   Environmental Health
Rollins School of Public Health
Emory University
Atlanta, GA, USA

Cynthia F. Bearer, MD, PhD
William and Lois Briggs Chair in Neonatology
Chief, Division of Neonatology
UH Rainbow Babies and Children's Hospital
Professor, Pediatrics
Case Western Reserve University School of
   Medicine
Cleveland, OH, USA

David C. Bellinger, PhD
Professor of Pediatrics
Harvard Medical School
Professor, Harvard School of Public Health
Children's Hospital Boston
Boston, MA, USA

Fiorella Belpoggi, PhD
Scientific Director
Ramazzini Institute
Cesare Maltoni Cancer Research Centre
Bologna, Italy

Aparna Bole, MD
Associate Professor
Department of Pediatrics
Case Western Reserve University School of
   Medicine
Cleveland, OH, USA

Meghan Buran, MPH
Senior Professional Research Assistant
Colorado School of Public Health
Aurora, CO, USA

Miranda Brazeal, PhD, MPH
Site Coordinator—Researcher
University of Illinois at Chicago
Chicago, IL, USA

Annemarie Charlesworth, MA
Director, Community Engagement Core
UCSF EaRTH Center
University of California
San Francisco, CA, USA

Leda Chatzi, MD, PhD
Professor
Department of Environmental Toxicology
University of California at Davis
Davis, CA, USA

Yu Chen, PhD, MPH
Professor of Epidemiology
Departments of Population Health and
  Environmental Medicine
New York University School of Medicine
New York, NY, USA

Alicia Cousins, MD
Fellow
Neonatal-Perinatal Medicine
UH Rainbow Babies and Children's Hospital
Case Western Reserve University School of
  Medicine
Cleveland, OH, USA

Yenny Fariñas Diaz, MS
PhD candidate in Public Health
Department of Environmental Health
  Sciences
Robert Stempel College of Public Health and
  Social Work
Florida International University
Miami, FL, USA

Kristie L. Ebi, PhD, MPH
Professor, Center for Health and the Global
  Environment
University of Washington
Seattle, WA, USA

Ruth A. Etzel, MD, PhD
Environmental and Occupational Health
Milken Institute School of Public Health
The George Washington University
Washington, DC, USA

Sarah Evans, PhD, MPH
Department of Environmental Medicine and
  Public Health
Icahn School of Medicine at Mount Sinai
New York, NY, USA

Jeffrey J. Fadrowski, MD, MHS
Associate Professor of Pediatrics
Johns Hopkins University School of Medicine
Baltimore, MD, USA

Henry Falk, MD, MPH
Adjunct Professor of Environmental Health
Rollins School of Public Health
Emory University
Atlanta, GA, USA

Winnie Fan
Medical student
University of California
San Francisco, CA, USA

Shohreh F. Farzan, PhD
Assistant Professor of Population and Public
  Health Sciences
Keck School of Medicine of University of
  Southern California
Los Angeles, CA, USA

Stephanie Ford, MD
Neonatologist
UH Rainbow Babies and Children's Hospital
Assistant Professor
Case Western Reserve University School of
  Medicine
Cleveland, OH, USA

Rebecca C. Fry, PhD
Carol Remmer Angle Distinguished Professor
  and Associate Chair
Department of Environmental Sciences and
  Engineering
Director, Institute for Environmental Health
  Solutions
Director, Graduate Studies, Curriculum in
  Toxicology
University of North Carolina
Chapel Hill, NC, USA

Jennifer Fuller
Department of Industrial and Management
  Systems Engineering
West Virginia University
Morgantown, WV, USA

Richard Fuller
President
Pure Earth
New York, NY, USA

Maida P. Galvez, MD, MPH
Professor
Department of Environmental Medicine and
    Public Health
Icahn School of Medicine at Mount Sinai
New York, NY, USA

Panos G. Georgopoulos, PhD
Professor of Environmental and Occupational
    Medicine, Exposure Measurement and
    Assessment
Robert Wood Johnson Medical School
Piscataway, NJ, USA

Philippe Grandjean, MD, DMSc
Professor of Environmental Medicine
University of Southern Denmark
Odense, Denmark

Eric J. Grant, PhD
Associate Chief of Research (retired)
Radiation Effects Research Foundation
Hiroshima/Nagasaki, Japan

Kristina K. Gustafson, MD
Associate Professor
Department of Pediatrics
Medical University of South Carolina
Charleston, SC, USA

Eunhee Ha, MD, PhD
Dean and Professor
Ewha Womans University College of Medicine
Seoul, Republic of Korea

Cara J. Hamann, MPH, PhD
Assistant Professor
University of Iowa College of Public Health
Core Director for Training and Education
University of Iowa Injury Prevention
    Research Center
Iowa City, IA, USA

Summer Sherburne Hawkins, PhD, MS
Associate Professor
School of Social Work
Boston College
Boston, MA, USA

Denis L. Henshaw, PhD
Emeritus Professor of Human Radiation
    Effects
School of Chemistry
University of Bristol
Bristol, UK

David G. Hoel, PhD
Distinguished University Professor
Department of Public Health Sciences
College of Medicine
Medical University of South Carolina
Charleston, SC, USA

Amy A. Hunter, MPH, PhD
Assistant Professor
Department of Public Health Sciences,
    UConn Health
Department of Pediatrics, UConn Health
Farmington, CT, USA

Richard J. Jackson, MD, MPH
Professor Emeritus
UCLA Fielding School of Public Health
Berkeley, CA, USA

David E. Jacobs, PhD, CIH
Chief Scientist
National Center for Healthy Housing
Adjunct Associate Professor
University of Illinois Chicago School of
    Public Health
Chicago, IL, USA

Sandra H. Jee, MD, MPH
Professor of Pediatrics
Co-Director, Finger Lakes Children's
    Environmental Health Center
University of Rochester Medical Center,
    General Pediatrics
Rochester, NY, USA

Margaret R. Karagas, MS, PhD
James W. Squires Chair and Professor of
    Epidemiology
Geisel School of Medicine at Dartmouth
Hanover, NH, USA

Catherine J. Karr, MD, PhD
Professor, Pediatrics/Environmental/
    Occupational Health Sciences
Adjunct Professor, Epidemiology
Director, NW Pediatric Environmental Health
    Specialty Unit
University of Washington
Seattle, WA, USA

Matthew C. Keifer, MD, MPH
Professor Emeritus of Medicine and
    Public Health
University of Washington
Seattle, WA, USA

Margaret Kuper-Sassé, MD
Assistant Professor, Pediatrics
Case Western Reserve School of Medicine
Cleveland, OH, USA

Amalia Laborde, MD, PhD
Professor
Toxicology Department
Centro de Información y Asesoramiento
    Toxicológico
Environmental Pediatric Unit
Faculty of Medicine
University of the Republic
Montevideo, Uruguay

Michele La Merrill, PhD
Associate Professor
Department of Environmental
    Toxicology
University of California at Davis
Davis, CA, USA

Philip J. Landrigan, MD, MSc
Director, Program for Global Public Health
    and the Common Good
Boston College
Boston, MA, USA

Bruce P. Lanphear, MD, MPH
Professor
Child and Family Research Institute, BC
    Children's Hospital
Faculty of Health Sciences, Simon Fraser
    University
Vancouver, BC, Canada

Danielle Laraque-Arena, MD
Professor
Departments of Epidemiology
    and Pediatrics and Columbia
    Center for Injury Science
Columbia University
Senior Research Scientist/Senior Scholar
New York Academy of Medicine
New York, NY, USA

Barbara C. Lee, PhD
Senior Research Scientist
Director, National Children's Center
    for Rural and Agricultural Health
    and Safety
Marshfield Clinic Research Institute
Marshfield Clinic
Marshfield, WI, USA

Edward D. Levin, PhD
Professor of Psychiatry and Behavioral
    Sciences
Department of Psychiatry and Behavioral
    Sciences
Duke University Medical Center
Durham, NC, USA

Barry S. Levy, MD, MPH
Adjunct Professor of Public Health
Tufts University School of Medicine
Boston, MA, USA

Roberto G. Lucchini, MD
Professor, Department of Environmental
    Health Sciences
Robert Stempel College of Public Health and
    Social Work
Florida International University
Miami, FL, USA

Daniele Mandrioli, MD, PhD
Director
Cesare Maltoni Cancer Research Center
Ramazzini Institute
Bologna, Italy

Mana Mann, MD, MPH
Department of Pediatrics
Flushing Hospital
New York, NY, USA

Keith Martin, MD
Executive Director
Consortium of Universities for Global Health
Washington, DC, USA

James T. McElligott, MD
Associate Professor
Department of Pediatrics
Medical University of South Carolina
Charleston, SC, USA

J. David Miller, PhD
Department of Chemistry
Carleton University
Ottawa, Ontario, CAN

Abby Nerlinger MD, MPH
Assistant Professor of Pediatrics
Nemours Children's Hospital, Delaware/
    Sidney Kimmel Medical College at Thomas
    Jefferson University
Wilmington, DE, USA

Nicholas C. Newman, DO, MS
Assistant Professor of Pediatrics and
    Environmental & Public Health Sciences
University of Cincinnati College of Medicine
Cincinnati, OH, USA

Parinya Panuwet, PhD, MS, MSPH
Research Assistant Professor
Gangarosa Department of
    Environmental Health
Rollins School of Public Health
Emory University
Atlanta, GA, USA

Archana Patel, MD, PhD
Program Director
Lata Medical Research Foundation
Nagpur, India

Hester Paul, MS
National Director for Eco-Healthy Child Care
Children's Environmental Health Network
Washington, DC, USA

Jerome A. Paulson, MD
Professor Emeritus of Pediatrics and of
    Environmental and Occupational Health
George Washington University School of
    Medicine and Health Sciences and George
    Washington University Milken Institute
    School of Public Health
Washington, DC, USA

Frederica Perera, DrPH, PhD
Professor of Public Health
Director Translational Research and Founding
    Director
Columbia Center for Children's
    Environmental Health
Mailman School of Public Health
Columbia University
New York, NY, USA

Alasdair Philips, DAgE
Trustee
Children with Cancer UK
London, UK

Kelsey Phinney, MPH
Project Coordinator
Colorado School of Public Health
Aurora, CO, USA

Margaret Pinder, BSPH
University of North Carolina
Chapel Hill, NC, USA

Virginia A. Rauh, ScD
Vice Chair, Heilbrunn Department of
    Population and Family Health
Mailman School of Public Health
Columbia University
New York, NY, USA

Kimberly Rauscher, ScD, MA
Professor
School of Public and Population Health
Boise State University
Boise, ID, USA

Nicholas Rickman, MS
Case Western Reserve University School of
    Medicine
Cleveland, OH, USA

James R. Roberts, MD, MPH
Professor
Department of Pediatrics
Medical University of South Carolina
Charleston, SC, USA

P. Barry Ryan, PhD
Professor
Exposure Science and Environmental
    Health
Gangarosa Department of
    Environmental Health
Rollins School of Public Health
Emory University
Atlanta, GA, USA

Jonathan M. Samet, MD, MS
Dean and Professor
Colorado School of Public Health
Aurora, CO, USA

Jennifer Sample, MD
Pediatrician and Toxicologist
PediaTox, LLC
Platte City, MO, USA

Lucy Schultz, MA
Medical Student
Icahn School of Medicine at
    Mount Sinai
New York, NY, USA

Laura Schwab-Reese, MA, PhD
Assistant Professor
Department of Public Health
Purdue University
West Lafayette, IN, USA

Christian Sewor, PhD
Public Health Research Group
Department of Biomedical Sciences
University of Cape Coast
Cape Coast, Ghana

Perry E. Sheffield, MD
Associate Professor
Departments of Pediatrics and Preventive
    Medicine
Icahn School of Medicine at Mount Sinai
New York, NY, USA

Antonio J. Signes-Pastor, PhD
GenT Distinguished Researcher
Nutrition Epidemiology Group, The Miguel
    Hernández University, Alicante, Spain
Center for Biomedical Research in
    Epidemiology and Public Health
    Network, The Carlos III Health Institute,
    Madrid, Spain
Alicante Health and Biomedical Research
    Institute
Spain
Department of Epidemiology
Geisel School of Medicine
Dartmouth College
Lebanon, NH, USA

Peter D. Sly, MD, DSc
Director
Children's Health and Environment
    Program
Child Health Research Centre
University of Queensland
St. Lucia, Queensland, Australia

Annemarie Stroustrup, MD, MPH
Assistant Professor of Pediatrics and
    Preventive Medicine
Icahn School of Medicine at Mount Sinai
New York, NY, USA

Frederick J. Suchy, MD
Senior Research Strategist
University of Colorado School of Medicine
Aurora, CO, USA

Kam Sripada, PhD, EdM
Neuroscientist
Norwegian University of Science and
    Technology
Trondheim, Norway

Kurt Straif, MD, MPH, PhD
Research Professor
Program for Global Public Health and the
    Common Good
Schiller Institute for Integrated Science and
    Society
Boston College
Boston, MA, USA

Shanna H. Swan, PhD
Professor
Department of Environmental Medicine and
    Public Health
Icahn School of Medicine at
    Mount Sinai
New York, NY, USA

Pooja Sarin Tandon, MD, MPH
Associate Professor
University of Washington and Seattle
    Children's Hospital
Seattle, WA, USA

Cansu Tokat, MD
Case Western Reserve University School of
    Medicine
Cleveland, OH, USA

Peter van den Hazel, MD, PhD
Environmental health consultant
International Network for Children's Health,
    Environment and Safety
Ellecom The Netherlands

Dwan Vilcins, PhD
Group Leader, Environmental
    Epidemiology
Children's Health and Environment
    Program
The University of Queensland
Brisbane, Australia

Roshan Wathore
Scientist
CSIR-National Environmental Engineering
    Research Institute,
Nagpur, India

Virginia M. Weaver, MD, MPH
Adjunct Associate Professor of Environmental
 Health and Engineering
Collaborating Faculty, Johns Hopkins
 Education and Research Center for
 Occupational Safety and Health
Johns Hopkins Bloomberg School of
 Public Health
Baltimore, MD, USA

Darcy K. Weidemann, MD, MHS
Associate Professor of Pediatrics
Children's Mercy Kansas City
University of Missouri—Kansas City School
 of Medicine
Kansas City, MO, USA

Clifford P. Weisel, PhD
Professor
Department of Environmental and
 Occupational Health
UMDNJ-Robert Wood Johnson
 Medical School
New Brunswick, NJ, USA

Tracey Woodruff, PhD
Director
Program on Reproductive Health and the
 Environment
University of California—San Francisco
Oakland, CA, USA

Fen Wu, PhD
Senior Research Scientist
Division of Epidemiology
Department of Population Health
NYU Grossman School of Medicine
New York, NY, USA

Marya G. Zlatnik, MD, MMS
Professor
Obstetrics, Gynecology, & Reproductive
 Sciences
University of California—San Francisco
San Francisco, CA, USA

PART I
INTRODUCTORY/OVERVIEW CHAPTERS

1

Children's Environmental Health

A New Branch of Pediatrics

Philip J. Landrigan and Ruth A. Etzel

Children's environmental health is an academic discipline, a branch of pediatrics, that studies how environmental exposures in early life—chemical, physical, nutritional, and social exposures—influence health, development, and risk of disease in infancy, childhood, and across the entire human life span.

The focus of children's environmental health is on the discovery, diagnosis, treatment, and prevention of diseases in children that are associated with harmful exposures in the environment. Children's environmental health also studies how positive environments protect children's health and nurture healthy growth and development. Children's environmental health is sometimes termed "environmental pediatrics."

Children's environmental health is based on a very inclusive definition of childhood. It is concerned with environmental exposures that occur during pregnancy as well as in infancy, childhood, and adolescence. It considers parental exposures prior to conception that may influence the health of children. It traces the influence of environmental exposures in early life on health and development across the entire human life span—from conception, through the embryonic and fetal periods, into infancy, childhood, and adolescence and on into adulthood and even to extreme old age.[1]

Children's environmental health considers the environment broadly. It recognizes that children's environments are complex, are comprised of many layers, and change over time. It therefore studies the influences on children's health of chemical exposures in early life,[2] the nutritional environment in the mother's womb,[1] the built environment,[3] stress,[4] and the social environment.[5] It studies interactions among these multiple environments at different life stages.[6] It examines interactions between environmental exposures, poverty, and social injustice.[7] It examines the influences of the environment on the human genome and epigenome.[8]

Children's environmental health is highly interdisciplinary. It spans and brings together general pediatrics and numerous pediatric subspecialties as well as epidemiology, occupational and environmental medicine, medical toxicology, industrial hygiene, and exposure science. Beyond medicine, children's environmental health links to architecture, urban planning, social work, education, ecology, economics, and political science.

Research in children's environmental health seeks to discover the environmental causes of disease and dysfunction in children. It evaluates the benefits to children's health of positive changes in the environment. And in the realm of advocacy and practice, children's environmental health translates the findings of research into evidence-based blueprints for the prevention of disease and the protection of children's health.

Philip J. Landrigan and Ruth A. Etzel, *Children's Environmental Health* In: *Textbook of Children's Environmental Health,* Second Edition. Edited by: Ruth A. Etzel and Philip J. Landrigan, Oxford University Press. © Oxford University Press 2024.
DOI: 10.1093/oso/9780197662526.003.0001

The ultimate goals of children's environmental health are to safeguard children's health; improve the environments where children live, learn, and play; and thus enable children to thrive and prosper. Children's environmental health honors children.[9] It advocates for the creation of healthy environments where children can live happily and achieve their highest potential.[3]

Interest in children's environmental health has grown rapidly in recent years. This growth has been catalyzed by rapidly rising rates of noncommunicable diseases in children—asthma, cancer, autism, attention-deficit/hyperactivity disorder (ADHD), birth defects, obesity, and diabetes. It has been further stimulated by the growing recognition that harmful environmental exposures in early life can cause disease in childhood and that these exposures can have consequences that reverberate across the human life span and influence health and development even into extreme old age. Indeed, it is now widely recognized that many diseases of adult life have their origins in events and exposures that occur in utero and during early childhood. This interconnectedness of early-life exposure with later health outcomes is known as the *developmental origins of adult disease*.[1]

The rapid, highly complex, and tightly choreographed growth and development that is the defining feature of fetal life creates unique windows of vulnerability to environmental exposures. These windows of vulnerability have no counterpart in adult life. Even very low-dose exposures to harmful environmental influences during these sensitive periods can program the body toward diseases that may appear early in life or not become manifest until years or decades later.

Changing Environment and Changing Diseases

The environments in which children live have changed profoundly in the past 200 years. During the same time, patterns of disease in children have changed greatly.[10] The field of children's environmental health has developed against the background of these changes.

Two centuries ago, average life expectancy in the United States, Western Europe, and Japan was about 40–45 years. The infant mortality rate ranged between 220 and 340 per 1,000 births, and 30% of children died before their first birthday. The major illnesses of children were almost all infectious diseases: pneumonia, dysentery, cholera, smallpox, typhoid fever, pertussis, and measles.[11]

But, since that time, the infectious diseases have largely been controlled in the industrially developed counties, and child health has improved remarkably. Despite the very large impacts of emerging infections such as HIV/AIDS, Ebola virus disease, Zika virus disease, and COVID-19, infectious diseases have a much smaller impact on patterns of disease and death in children than they did a century ago. Thanks to vaccines and antibiotics, the ancient scourges of tuberculosis, measles, whooping cough, and polio no longer decimate children. Infant mortality has fallen by 90%.[10] Death rates across all ages have decreased by more than 50%. Life expectancy has nearly doubled. Today the major pediatric health challenges are the noncommunicable diseases. Incidence and prevalence rates of these diseases are on the rise.

The change in patterns of disease and death from infectious to noncommunicable diseases that parallels industrial development is termed the *epidemiological transition*.[10]

The principal drivers of the epidemiological transition were public health interventions and broad-scale improvements in the environment engineered in the 19th and early 20th centuries.[12] These included construction of reservoirs and aqueducts that for

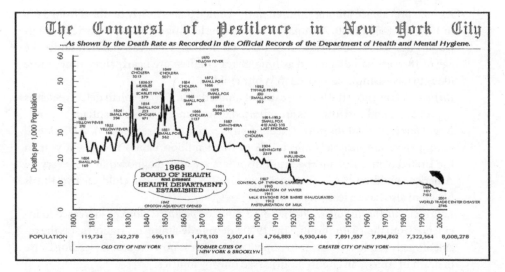

Figure 1.1. The epidemiological transition in New York City, 1800–2007

the first time brought safe drinking water into cities, construction of sewage systems to remove waste, provision of wholesome food, the control of insect vectors, and the passage of laws and regulations that set legal standards for decent housing. These environmental changes contributed substantially to the control of cholera, typhoid fever, yellow fever, tuberculosis, and the other ancient infectious diseases.

Figure 1.1 charts the epidemiological transition in New York City. The line across the graph traces the mortality rate (the number of deaths per 1,000 inhabitants) in New York from 1800 to the present. It is noteworthy that the great decline in mortality that marked the start of the epidemiological transition began in the 1860s, soon after construction of New York City's reservoir and aqueduct system and nearly 80 years before the discovery of penicillin.

The epidemiological transition provides dramatic historical documentation of the power of the environment to shape patterns of health and disease in children. Similar transitions have been observed in country after country over the past two centuries as countries have moved through the stages of industrial development. They continue to the present.

The Environment and Disease in Children Today

Today, in the aftermath of the epidemiological transition and despite the ever present threats of HIV/AIDS, tuberculosis and newly emerging infections such as Zika, Ebola, and COVID-19, the principal causes of illness, disability, and death among children in high-income countries such as the United States, Japan, and Western Europe and are no longer the infectious diseases. Instead, the major causes of morbidity and mortality among children in the world's industrially developed countries are noncommunicable diseases—the "new pediatric morbidity."[13] Incidence and prevalence rates of these diseases are high, and many are increasing. Evidence is strong and growing that harmful exposures in the environment are important causes of these diseases.

- *Asthma* prevalence among children in the United States nearly doubled from 3.6% in 1980 to 5.8% in 2020. Asthma has become the leading cause of pediatric hospitalization and school absenteeism. In 2020, 4.2 million American children under the age of 18 years had diagnosed asthma. Black children are nearly three times more likely to have asthma compared to White children[14];
- *Birth defects* are now the leading cause of infant death. Certain birth defects, such as hypospadias and gastroschisis are reported to have increased substantially[15];
- *Neurodevelopmental disorders, including dyslexia, mental retardation, ADHD, and autism spectrum disorders* affect 17% of the 4 million babies born each year in the United States. Autism spectrum disorders are now diagnosed in 1 of every 44 American children.[16] ADHD is diagnosed in 10% of American children, two-thirds of whom also have a learning disability[17];
- *Leukemia and brain cancer, the two most common pediatric cancers,* increased in incidence from the 1970s through the 1990s, despite declining mortality that is the result of vastly improved cancer treatments.[18] Cancer is now the second leading cause of death in American children, surpassed only by injuries;
- *Testicular cancer* in young men ages 15–30 years, has more than doubled in incidence and is occurring at younger ages[18];
- *Childhood obesity* has almost quadrupled in prevalence—from 5% in the 1970s to nearly 20% today.[19] *Type 2 diabetes,* known formerly as "adult-onset diabetes" and almost never previously diagnosed in childhood, has increasingly begun to be seen among children and at ever younger ages and is the direct consequence of rising rates of childhood obesity.

The Historical Origins of Children's Environmental Health

Parents have understood from time immemorial that children are highly sensitive to hazards in the world around them. Pediatricians have studied children's unique susceptibilities to toxic chemicals and other environmental exposures since the specialty of pediatrics was established in the 1860s by Dr. Abraham Jacobi at the Mount Sinai Hospital in New York City.

But it was not until the 1990s that three areas of research and policy investigation came together to form the modern academic discipline of children's environmental health.* These three convergent areas of scientific inquiry that gave rise to children's environmental health are

*1. Pediatric toxicology;
2. Nutritional epidemiology; and
3. Social science research.

* We are deeply indebted to Fiona Stanley, MD, MSc, FFPHM(U.K.), FAFPHM(Aust.), Distinguished Research Professor in the School of Paediatrics and Child Health at the University of Western Australia, for having first developed this historical paradigm.

Pediatric Toxicology and Children's Environmental Health

Pediatric toxicology is the oldest of these three disciplines. The earliest beginnings of children's environmental health may be traced to a series of astute clinical observations and epidemiological studies of children exposed to toxic chemicals. These reports first began to appear in the published literature in the late 19th and early 20th centuries. They include

- Reports from Queensland, Australia, in 1897 and 1904, describing an epidemic of acute lead poisoning in young children who had ingested lead-based paint[20];
- An epidemic of leukemia and other cancers among children in Hiroshima and Nagasaki exposed to ionizing radiation in the atomic bombings[21];
- An increased prevalence of microcephaly among infants in Hiroshima and Nagasaki exposed in the first trimester of pregnancy to ionizing radiation[22];
- An epidemic of microcephaly, cerebral palsy, mental retardation, convulsions, and blindness among children born in the 1950s and 1960s in Minamata, Japan. The epidemic was traced to consumption by pregnant women of local fish contaminated with methylmercury released into Minamata bay by a nearby chemical manufacturing plant. Exposed mothers were only minimally affected[23];
- An epidemic of phocomelia that followed ingestion of thalidomide as an antiemetic in early pregnancy.[24] More than 15,000 cases were reported worldwide;
- Cases of adenocarcinoma of the vagina in young women exposed in utero to the synthetic hormone diethylstilbestrol (DES) taken by their mothers to prevent miscarriage.[25]

These early studies established that toxic chemicals in the environment can have devastating effects on the health of children and that these effects can persist for a lifetime. They established that toxic chemicals can cross the placenta from mother to baby to cause injury to the fetus in utero. The stark disparities seen in the Minamata, thalidomide, and DES tragedies between disease risk in exposed mothers and their infants established that infants and young children have unique patterns of exposure and exquisite vulnerabilities that have no counterparts in adult life.

Four organizations that played key roles in further advancing the study of pediatric environmental health were the American Academy of Pediatrics Council on Environmental Health, the Children's Environmental Health Network, the Academic Pediatric Association, and the World Health Organization (WHO).

The American Academy of Pediatrics Council on Environmental Health and Climate Change, formerly the AAP Committee on Environmental Hazards, played a key role in educating pediatricians about the hazards to children's health of toxic exposures in the environment. This vibrant Committee, championed by Dr. Robert Miller (see Box 1.1), has played the additional very important role of serving as an incubator for many of today's US leaders in the field of children's environmental health, including the two editors of this textbook and many of the chapter authors.

The Kids and Environment Project, now the Children's Environmental Health Network (CEHN) (www.cehn.org), a nongovernmental organization based originally in California and now in Washington, DC, has played a key role since the late 1980s in educating pediatricians, nurses, and other health professionals about environmental threats to children's health. Under the visionary leadership of Joy Carlson, a dynamic health

Box 1.1 Robert W. Miller, MD, DrPH

Robert Miller (Figure 1.2), a pediatrician who served for many years as an epidemiologist with the US National Cancer Institute, is considered the father of children's environmental health. He deserves great recognition as the first scientist to move beyond study of specific outbreaks of environmental disease in children to the realization that fetuses, infants, and children are unusually sensitive to a wide range of environmental hazards.

Figure 1.2. Robert W. Miller, MD

Dr. Miller began his work as a radiation epidemiologist in Hiroshima and Nagasaki, and he was one of the first scientists to recognize the epidemic of leukemia in children who had survived the acute effects of the atomic bombings.[21] He returned to the United States, and, in the late 1960s, championed the American Academy of Pediatrics Committee on Environmental Hazards (now the Council on Environmental Health and Climate Change), which for six decades has been a key incubator of leaders in children's environmental health.

Dr. Miller taught that every pediatrician must be an "alert clinician," always open to the possibility of discovering new diseases in children caused by toxic exposures in the environment.[26] He considered diagnostic vigilance an essential component of primary care pediatrics.

Examples of diseases of environmental origin first documented by alert clinicians and cited by Dr. Miller include phocomelia in babies exposed thalidomide,[24] adenocarcinoma of the vagina in girls exposed in utero to DES,[25] and microcephaly in babies exposed in utero to ionizing radiation.[22] Recognition of these sentinel events by astute clinicians led in every case to disease prevention.

Figure 1.3. Ellen Crain, MD, PhD (center) flanked by John Balbus, MD, MPH and Ruth Etzel, MD, PhD

educator, and with financial support from the US National Institute of Environmental Health Sciences (NIEHS), CEHN convened the first national US policy conference in children's environmental health in Washington in 1994. This was a landmark event that contributed importantly to raising public and professional awareness of environmental threats to children's health.[27,28] Today, the inspirational leader of CEHN is Nse Obot Witherspoon.

The Ambulatory Pediatric Association, now the Academic Pediatric Association (APA), was very important in solidifying the position of children's environmental health within academic medicine in the United States. Ellen Crain, MD, PhD, Professor Emeritus of Pediatric Emergency Medicine at Albert Einstein College of Medicine and past-president of APA, spearheaded this effort (Figure 1.3).

Under Dr. Crain's leadership, the APA established fellowship training programs to educate future leaders in children's environmental health in academic health centers across the United States and Canada. Starting in 2002, the APA regularly convened retreats for pediatric environmental health scholars that bring these fellows together. These retreats enabled fellows in pediatric environmental health training programs across North America to meet each other, learn of each other's work, and form the professional bonds that are so important for a cadre of leaders in a newly emerging field.

The WHO set up a Task Force for the Protection of Children's Environmental Health in 1999. The organization was deeply engaged in children's environmental health and played a critical role in making public health leaders in countries around the world aware of environmental threats to children's health. The WHO Regional Office for Europe and the Pan American Health Organization also formed regional programs in children's environmental health. The United Nations Children's Fund (UNICEF) and the United Nations Environment Programme also have developed programs in children's environmental health and are collaborating with the WHO. Jenny Pronczuk de Garbino, MD, a visionary Uruguayan clinician and toxicologist, created and directed the WHO activities on children's environmental health until her untimely death in 2011.

Nutritional Epidemiology and Children's Environmental Health

Classic studies begun in the late 1980s by Professor David Barker and colleagues at the University of Southampton in southern England have shown that the nutritional environment in early life can exert powerful influence on health in childhood and across the life span. Barker et al. reported that babies who were undernourished in utero have significantly increased risks *decades* later for pathological weight gain, diabetes, hypertension, and cardiovascular disease.[1] These studies gave powerful insight into the unique vulnerabilities of early development. They gave rise to the "developmental origins of adult disease" hypothesis and to life stage research that examines developmental trajectories over the entire human life span.

Recent research suggests that epigenetic modification of fetal genes based on metabolic cues received in pregnancy from the mother may be a mechanism that accounts for many of these observations.[29] It is suggested that toxic chemicals in the environment may influence early growth and development by causing similar epigenetic changes in gene expression.

Social Science Research and Children's Environmental Health

A seminal report from the US National Academy of Sciences, *From Neurons to Neighborhoods*, documented the power of social environments and maternal stress during pregnancy to influence children's health and development.[5] A positive social environment can produce beneficial impacts that persist across the life span. Healthy cities, healthy communities, healthy schools, and early intervention programs such as Head Start have all been documented to enhance the likelihood that children will lead productive lives with reduced risk of illness. By contrast, early stress can have deleterious effects and increase the risk of asthma, allergy, and obesity.[4] Epigenetic modification of fetal gene expression may be a mechanism that accounts for the impacts of stress on health.[29]

Birth of the Discipline Children's Environmental Health

A catalytic event that drew together these three lines of scientific enquiry that were each separately examining the impacts of early-life exposures on children's health and development was the publication, in 1993, of a report from the US National Academy of Sciences (NAS), *Pesticides in the Diets of Infants and Children*.

This seminal report documented children's unique patterns of exposure and their exquisite vulnerability to pesticides and other toxic chemicals in the environment. It directed the attention of elected officials and national policymakers in the United States to children's unique vulnerability and made the consideration of children's vulnerabilities an urgent matter of national policy. The NAS report identified four differences between children and adults that contribute to children's heightened susceptibility to pesticides and other toxic chemicals:

- Children have proportionately greater exposures than adults to toxic chemicals on a body-weight basis;
- Children's metabolic pathways are immature;
- Children's extremely rapid, but exquisitely delicate developmental processes are easily disrupted; and
- Children have more future years of life than adults to develop chronic diseases that may be triggered by environmental exposures in early life.

The detailed findings of the NAS report are presented in Chapter 2.

The 1993 NAS report produced a paradigm shift in US children's environmental health policy. It led to new legislative and administrative initiatives to better protect infants and children against environmental health threats, most notably the Food Quality Protection Act of 1996, the US law governing use of pesticides and the only federal environmental statute in the US to contain explicit provisions for the protection of children. It catalyzed a Presidential Executive Order of 1997 on Children's Health and the Environment that directed all agencies of the US government to consider impacts on children's health and safety of any major new policy initiative. It has had profound impacts on exposure science, toxicology, risk assessment, and environmental regulation. It was a powerful catalyst that greatly increased federal research investment in children's environmental health. These ramifications for children's environmental health of the NAS report are explored more fully in Chapter 58.

Over the past three decades children's environmental health has grown exponentially and become an increasingly visible and important area of pediatrics. A series of advances have helped to establish the credibility of children's environmental health as an academic discipline.

- National and international conferences have been held on the impact of the environment on the children's health.
- National networks of research centers in children's environmental health have been established in the United States and the Republic of Korea.
- A network of clinically oriented Pediatric Environmental Health Specialty Units has been established in the United States. This network has extended internationally to include units in Canada, Mexico, Argentina, Uruguay, and Spain.

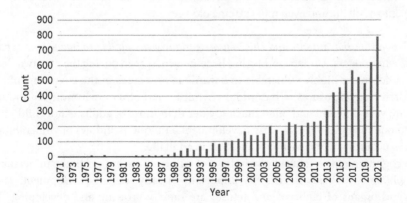

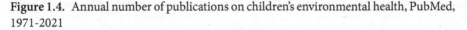

Figure 1.4. Annual number of publications on children's environmental health, PubMed, 1971-2021

- The number of papers on topics in children's environmental health published in the peer-reviewed literature has increased greatly (Figure 1.4).
- The American Academy of Pediatrics has published *Pediatric Environmental Health*,[30] the "Green Book," now in its fourth edition.
- Training programs have been launched for the education of pediatricians and research scientists in children's environmental health. The first was launched in the United States, in 2001, by the Ambulatory Pediatric Association with support from the *Eunice Kennedy Shriver* National Institute of Child Health and Human Development (NICHD) and the NIEHS.
- Prospective epidemiological studies are under way to discover new associations between environmental exposures in early life and the health of children. These include the Japan Environment and Children Study (JECS)[31] and the Shanghai Children's Study.
- International interest in children's environmental health has grown. The first international meeting on children's environmental health was held in Amsterdam, in 1996.[28] Out of that meeting the International Network on Children's Health, Environment and Safety (INCHES) was created.
- In 1997, the eight leading economic nations, the G-8, issued a declaration at a meeting in Miami in support of children's environmental health. All eight countries—the United States, Canada, Great Britain, France, Germany, Italy, Japan, and Russia—agreed to make the protection of children's health against environmental threats a national priority.
- Children's environmental health has been featured at the annual conferences of the International Society for Environmental Epidemiology.

Three powerful new insights have catalyzed recent growth of the discipline of children's environmental health:

1. Recognition that children at all ages, and especially in the embryonic and fetal periods, are uniquely vulnerable and exquisitely sensitive to hazardous exposures in the environment;
2. Growing understanding that changes in the environment—positive as well as negative—are powerful drivers of health and disease in children; and
3. Realization that environmental exposures in early life can influence health and disease not only in infancy and childhood, but also across the entire human life span, into adult life and even into extreme old age.

Recognition that infants and children are uniquely susceptible to hazardous exposures in the environment. Understanding has become widespread during the past two decades that children are highly vulnerable to hazardous exposures in the environment.[13–25,32] Children have exposures to hazardous chemicals and other environmental influences that are very different and often much greater than those of adults. And, children have enormous susceptibilities in early development—unique "windows of vulnerability" to toxic exposures—that have no counterpart in adult life.

Recognition of the concept of windows of vulnerability has led to the understanding that *timing of exposure* is critically important in early human development. The tissues and organs of embryos and fetuses are rapidly growing and developing. These complex and delicate developmental processes are uniquely sensitive to disruption by

---

Box 1.2 Subclinical Toxicity: The Case of Lead

The first description of lead poisoning in children was a report of a cluster of cases of acute intoxication published from Australia more than 100 years ago.[20] Prior to that time, lead poisoning had been known only as an occupational disease affecting adult workers. Lead moved beyond the workplace to affect children in communities following the commercial introduction and wide-scale marketing of lead-based paint. This was an early example of the dissemination of toxic chemicals into the environment without advance consideration of their possible effects on children's health, a recurrent pattern of behavior discussed in detail in Chapter 3.

The affected children in Australia became sleepy and disoriented and then developed coma and convulsions. These symptoms occurred in the absence of fever. Some died. The illnesses were thought initially to be a form of tropical encephalitis. However, the absence of fever argued against infection. Finally, after several years of detective work, the children's illnesses were diagnosed as lead poisoning.[20] The source was traced to peeling lead-based paint on the verandas of the children's homes. While playing on their verandas, the children had ingested chips and dust that eroded from this paint.

The subclinical toxicity of lead was first recognized in the 1970s. Clinical and epidemiological studies of asymptomatic children with elevated blood lead levels showed that they had statistically significant decreases in intelligence (IQ score).[36,37] They also had shortening of attention span, alteration of behavior, and problems in memory. Long-term follow-up found that these children went on to have elevated rates of reading difficulties, school failure, and incarceration. These studies made clear that subclinical toxicity, while clinically silent, can cause significant injury that lasts a lifetime.

---

environmental influences. Exposures sustained during windows of early vulnerability, even to extremely low levels of toxic materials, can cause lasting damage.

The unique vulnerabilities of infants and children to hazards in the environment is discussed in detail in Chapter 2.

Understanding that exposures in the environment—positive as well as negative—are powerful drivers of health and disease in children. Today it is widely understood that the environment exerts powerful influence over patterns of health and disease in children. The WHO estimates that 24% of the global burden of disease today and 36% of all disease among children 0 to 14 years of age is attributable to harmful exposures in the environment.[33] In the United States, nearly half of the population lives in a county where air pollution exceeds federal standards, nearly 50 million people drink water with illegal concentrations of toxic chemicals or bacteria, and one-fourth of the population lives within 1 mile of a hazardous waste site.

Initial recognition that environmental exposures can cause disease in children typically involved clinically obvious poisonings in high-dose, acute exposure scenarios. Subsequent, more sophisticated research has then documented less striking but nonetheless serious adverse effects at lower levels of exposure.[34]

This sequence of discovery has led to the recognition that toxic chemicals and other hazards in the environment exert a range of adverse effects—some are clinically evident, but others can be discerned only through special testing and are not evident on the

Box 1.3 Subclinical Toxicity: The Case of Methylmercury

Like lead, methylmercury was first established as neurotoxic in workers exposed occupationally. The toxicity of methylmercury to children was first recognized in the
1960s, in Minamata, a fishing village in southwestern Japan. An epidemic of spasticity, blindness, and profound mental retardation occurred there among infants born
to mothers who had consumed methylmercury-contaminated fish from Minamata
Bay during pregnancy.[23] Figure 1.5 is a haunting picture of a child damaged in utero
by methylmercury in Minamata. The child's mother is physically unaffected.

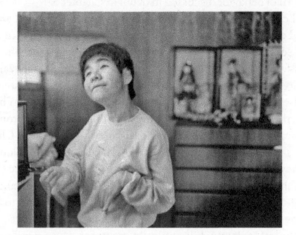

Figure 1.5. A patient born with Minamata Disease
Source: Photo by Chris Steele-Perkins.

  The source of the mercury in Minamata was traced to a chemical manufacturing
plant that had discharged metallic mercury into Minamata Bay. This metallic mercury was converted by marine microorganisms in the bay to methylmercury. The
methylmercury then accumulated and reached high levels in locally caught fish.
  More recent studies have documented that methylmercury can cause subclinical
toxicity to infants exposed prenatally to levels insufficient to produce obvious symptoms. A prospective study of New Zealand children exposed in utero found a 3-point
decrement in IQ as well as alterations in behavior.[38] A large prospective study of prenatally exposed children in the Faroe Islands found evidence for impairments in
memory, attention, language, and visuo-spatial perception.[39] A prospective cohort
study in the Seychelles provided limited support for prenatal neurotoxicity.[40] The US
NAS reviewed these studies and concluded that there was strong evidence for the fetal
neurotoxicity of methylmercury, even at low levels of exposure.[41]

standard examination. Hence they are termed "subclinical toxicity." The underlying concept is that there exists a dose-dependent continuum of toxic effects in which clinically
obvious effects have their subclinical counterparts.[35] Lead and methylmercury provide
examples (see Box 1.2 and Box 1.3).

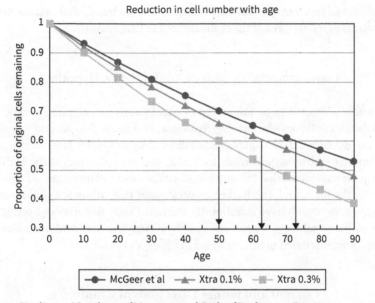

Figure 1.6. Declining Numbers of Neurons and Risk of Parkinson Disease

Cell populations in the substantia nigra, the area of the brain affected by Parkinson Disease (PD), normally diminish with age and do not regenerate. By about age 72, about 40% of these cells have normally been lost. If this decline is accelerated by toxic exposures in the environment by as little as 0.1% per year, loss of 40% of the cells will occur by about age 64. If cell loss is accelerated by 0.3% per year, 40% of cells will be gone by age 50 (45). A smaller population of cells in the substantia nigra increases risk for PD.

Subclinical Toxicity: A Widespread Phenomenon

The concept of subclinical toxicity has now extended far beyond lead and methylmercury.[35] It has become a key foundation of the field of children's environmental health. It includes the effects in numerous organ systems of lower-dose exposures to a wide range of toxic chemicals. And it has come to be understood that wide-scale subclinical toxicity caused by toxic chemicals that are widely disseminated in the modern environment can adversely affect the health and well-being of entire societies. These concepts are discussed in detail in Chapter 3.

Realization that environmental exposures in early life may influence health and disease across the entire human life span. Incidence rates of certain cancers and also of Parkinson disease and Alzheimer disease, the two most common neurodegenerative disorders of later life, have increased in recent years.

These trends are too rapid to be of genetic origin. They therefore raise the possibility that environmental exposures—including exposures in fetal and early postnatal life—may be contributing to chronic diseases in adult life, even into extreme old age.[1,42]

A possible mechanism connecting early exposures with these late effects is that early exposures may initiate cascades of changes within cells that can lead ultimately to malignancy or that can reduce the numbers of neurons in critical areas of the brain to levels below those needed to maintain function in the face of advancing age.[43] Some of these changes may be mediated by epigenetic modulation of gene expression by toxic chemicals or other environmental exposures.[29] Figure 1.6 shows the possible long-term consequences of exposures in early life to neurotoxic chemicals that destroy neurons in the substantia nigra, the area of the brain affected in Parkinson disease.

Developmental Origins of Adult Disease Hypothesis

Diseases now suspected to be influenced in part by early environmental exposures include diabetes, cardiovascular disease, dementia, Parkinson disease, and cancer. This is termed the "developmental origins of adult disease" hypothesis.[42] It represents an extension of landmark observations made by Barker and his colleagues at the University of Southampton who found that the nutritional environment in utero can influence health throughout the life span.[1] The Barker group showed that babies who are insufficiently nourished in the womb have significantly increased risks five, six, and seven decades later for pathological weight gain, diabetes, hypertension, and cardiovascular disease. These topics are discussed in further detail in Chapters 14 and 23.

Current and Future Directions of Children's Environmental Health

Today children's environmental health is a vibrant, highly collaborative, and intensely interdisciplinary field. Thanks to recent increases in governmental and philanthropic support, research, and practice in children's environmental health are moving forward on many fronts. Among these are

- Epidemiology
- Genetics and epigenetics
- Economics
- Evidence-based disease prevention

Epidemiology

Epidemiology—in particular, prospective birth cohort epidemiological studies that enroll women before and during pregnancy and measure environmental exposures as they occur during pregnancy and then follow the children longitudinally through infancy, childhood, and beyond—have become powerful engines for the scientific discovery of associations between early-life exposures and disease. A great strength of the prospective study design is that it permits unbiased assessment of exposures as they actually are occurring during pregnancy or early infancy, months or years before the onset of clinically evident disease or dysfunction. The prospective design thus reduces recall bias and is crucial for studies that require accurate assessments of environmental exposures in early life.

Prospective epidemiologic studies are especially powerful when they incorporate measurements of exposures in the environment[44] as well as biomarkers of exposure, individual susceptibility, and the precursor states of disease—a cutting-edge approach known as *molecular epidemiology*.[45] With increasing deployment of prospective biomarker-based epidemiological studies, the pace of scientific discovery in environmental pediatrics has accelerated.

The contributions of epidemiology to children's environmental health are discussed in detail in Chapter 6. The contributions of exposure science are covered in Chapters 7 and 8.

Genetics and Epigenetics

Children's environmental health considers the multiple ways in which children's environments interact with the human genome to influence health and disease. Research in children's environmental health studies gene–environment interactions, gene–gene interactions, and epigenetic modifications of gene expression that may be triggered by environmental exposures.[29,46]

Only a very few human diseases—most of them single-gene disorders such as Tay-Sachs disease or sickle cell disease—are of purely genetic origin. Likewise, very few diseases are of purely environmental origin—most of them are severe, acute, high-dose intoxications such as acute, high-dose lead poisoning or Minamata disease.[23] The discipline of children's environmental health is thus based on the understanding that most human disease arises from interactions between the genome and the environment.[8] Dr. Kenneth Olden, former Director of the US NIEHS, famously observed that "Genetics loads the gun, and the environment pulls the trigger."

The topic of epigenetics and children's environmental health are discussed in detail in Chapter 10.

Economics

Disease and dysfunction of environmental origin in children is very costly. Understanding of these costs can catalyze disease prevention efforts. A recent analysis of the medical and societal costs associated with illness of environmental origin among American children found these costs to amount to US$76.6 billion annually.[47] The principal contributor to these great costs is lifelong reduction in intelligence caused by exposures to neurotoxic chemicals such as lead, methylmercury, and PCBs. Widespread loss of intelligence imposes great economic burdens on society and leads to lifelong reductions in economic productivity.

The contributions of economics to the discipline of children's environmental health are discussed in Chapter 12.

Evidence-Based Prevention of Environmental Disease

Diseases caused by toxic chemicals and other environmental hazards can be prevented. Because the exposures that cause these diseases are the consequence of human activity, they can be prevented by modification or cessation of that activity. Prevention of diseases of environmental origin will require deliberate and concerted action across many sectors of society to reduce hazards and improve the quality of the many environments where children live: the womb (the first environment), the home, communities, and schools.

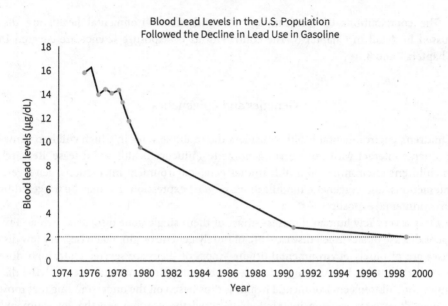

Figure 1.7. Blood lead levels in the U.S. population followed the decline in lead use in gasoline

Source: CDC National Report on Human Exposure to Environmental Chemicals, March 2001.

Disease prevention saves lives, enhances the quality of children's lives, reduces health care and education costs, and increases national productivity. Disease prevention is the ultimate goal of children's environmental health. Effective prevention of environmental disease and dysfunction in children requires the translation of research findings into evidence-based advocacy.

An extraordinarily successful example of evidence-based prevention of environmental disease is the United States' experience in removing lead from gasoline. This single action, catalyzed in large part by discovery of the subclinical toxicity of lead to young children, reduced childhood lead poisoning by more than 90% and raised the average IQ of American children by 2–4 points. Moreover it has produced an economic benefit estimated to be approximately $200 billion in each US birth cohort born since the 1980s.[48] This economic benefit resulted largely from increases in productivity that followed population-wide increases in intelligence. This success has now been repeated in countries around the world (Figure 1.7).

Organization and Use of This Textbook

The purpose of this textbook is to present a comprehensive summary of children's environmental health. We intend it to serve as an authoritative guide and an intellectual foundation for this new and rapidly expanding academic discipline.

We start with Part I, "Introductory/Overview Chapters"—chapters covering what we feel are the basic concepts that a student needs to understand in order to function in the field of children's environmental health. We have tried to weave a global perspective through these chapters and indeed throughout the entire volume. Children's environmental health has become very much an international discipline, and we have endeavored to make the volume reflect that reality.

In Part II, we present a series of chapters that describe the various "Environments" in which children live. We start with a child's first environment—the womb. We then proceed chronologically and outward to finish this section with a chapter on the changing global environment and its implications for the health of children.

Part III presents a series of chapters that review the major "Environmental Hazards" that confront children in today's world. Each chapter begins with a description and definition of the hazard. Then come data describing the temporal and geographic distribution of the hazard. Next comes a description of the routes through which children are exposed to each hazard. Then follow descriptions of the diseases associated with each hazard.

Part IV presents "The Environment and Disease in Children." This includes a series of chapters that describe the major categories of illness in children that have been linked to hazardous exposures in the environment. These chapters describe the clinical and epidemiological aspects of each disease and present up-to-date information on what is known of environmental causation. As appropriate, gene–environment and epigenetic aspects of causation are discussed, as are questions of diagnosis, clinical management, treatment, and prevention.

The volume concludes with Part V on the "Prevention and Control of Diseases of Environmental Origin in Children," beginning with chapters on the clinical practice of children's environmental health. Next come a series of policy-oriented chapters that describe relevant policies in key countries and globally. This section and the volume conclude with chapters on poverty, disparity and injustice; war, terrorism, and children's health; natural disasters and children's health; and new frontiers in children's environmental health.

The anticipated audiences for this volume are many because the subject matter cuts across many fields and the volume is global in its scope. The volume may be useful for courses in environmental health, especially in schools of public health, nursing schools, and in undergraduate curricula in public health. Potentially interested readers may include

- Pediatricians
- Pediatric residents and trainees
- Family physicians
- Obstetricians, especially specialists in maternal-fetal medicine
- Nurses
- Medical and nursing students
- Developmental psychologists
- Graduate students in the life sciences, especially in developmental biology and the neurosciences
- Public health and environmental health researchers and practicing professionals
- Pediatric and clinical toxicologists
- Health economists
- Health and environmental decision-makers and risk assessors in ministries of the environment and public health

- Civil society organizations concerned with providing information and advice on protecting children against environmental threats to health, and
- Elected officials at all levels of government who seek information to help them better protect children.

We hope that you will find the volume interesting and informative. We encourage you to write us with suggestions for improvement and expansion of the text and organization of the volume.

References

1. Barker DJ. The developmental origins of adult disease. *J Am Coll Nutr.* 2004;23:588S–95S.

2. Landrigan PJ, Fuller R, Fisher S, Suk WA, Sly P, Chiles TC, Bose-O'Reilly S. Pollution and children's health. *Sci Total Environ.* 2019 Feb 10;650(Pt 2):2389–94.

3. Jackson RJ, Sinclair S. *Designing Healthy Communities.* San Francisco: Jossey-Bass; 2011.

4. Oh DL, Jerman P, Silvério Marques S, Koita K, Purewal Boparai SK, Burke Harris N, Bucci M. Systematic review of pediatric health outcomes associated with childhood adversity. *BMC Pediatr.* 2018 Feb 23;18(1):83.

5. National Academy of Sciences. *From Neurons to Neighborhoods: The Science of Early Childhood Development.* Washington DC: National Academies Press; 2000.

6. Wild CP. Complementing the genome with an "exposome": the outstanding challenge of environmental exposure measurement in molecular epidemiology. *Cancer Epidemiol Biomarkers Prev.* 2005;14:1847–50.

7. Chesney ML, Duderstadt K. Children's rights, environmental justice, and environmental health policy in the United States. *J Pediatr Health Care.* 2022 Jan-Feb;36(1):3–11.

8. Schwartz D, Collins F. Environmental biology and human disease. *Science.* 2007;316:695–6.

9. Cavoukian R, Olfman S. *Child Honouring: How to Turn This World Around.* Salt Spring Island, BC, Canada: Homeland Press; 2006.

10. Omran AR. The epidemiologic transition: a theory of the epidemiology of population change. *Milbank Mem Fund Q.* 1971;49:509–38.

11. US Burden of Disease Collaborators. The state of US health, 1990–2016: burden of diseases, injuries, and risk factors among US states. *JAMA.* 2018 Apr 10;319(14):1444–72.

12. Halliday S. *The Great Stink of London: Sir Joseph Bazalgette and the Cleansing of the Victorian Metropolis.* London: Sutton Publishing; 1999.

13. Haggerty R, Rothman J. *Child Health and the Community.* New York: John Wiley & Sons; 1975.

14. Centers for Disease Control and Prevention. Most recent national asthma data. 2021. Available at: www.cdc.gov/asthma/most_recent_national_asthma_data.htm.

15. Kirby RS. The prevalence of selected major birth defects in the United States. *Semin Perinatol.* 2017 Oct;41(6):338–44.

16. Centers for Disease Control and Prevention. 2012. Data and statistics on autism spectrum disorders (ASDs). Available at: https://www.cdc.gov/ncbddd/autism/signs.html

17. Centers for Disease Control and Prevention. 2012. Data and statistics about ADHD. Available at: https://www.cdc.gov/ncbddd/adhd/data.html.

18. National Cancer Institute. SEER database. 2023. Available at: http://seer.cancer.gov.

19. Centers for Disease Control and Prevention. Prevalence of childhood obesity. Available at: https://www.cdc.gov/obesity/data/childhood.html.

20. Turner AJ. Lead poisoning among Queensland children. *Aust Med Gazette*. 1897;16:475–79.

21. Miller RW. Delayed effects occurring within the first decade after exposure of young individuals to the Hiroshima atomic bomb. *Pediatrics*. 1956;18:1–18.

22. Miller RW, Blot WJ. Small head size after exposure to the atomic bomb. *Lancet*. 1972;2:784–7.

23. Harada H. Congenital Minamata disease: intrauterine methylmercury poisoning. *Teratology*. 1978;18:285–8.

24. Lenz W. Chemicals and malformations in man. In: Fishbein M, ed. *Second International Conference on Congenital Malformation*. New York: International Medical Congress; 1963:263–71.

25. Herbst AL, Hubby MM, Azizi F, Makii MM. Reproductive and gynecologic surgical experience in diethylstilbestrol-exposed daughters. *Am J Obstet Gynecol*. 1981;141:1019–28.

26. Miller RW. How environmental hazards in childhood have been discovered: carcinogens, teratogens, neurotoxicants, and others. *Pediatrics*. 2004;113:945–51.

27. Etzel RA. Developmental milestones in children's environmental health. *Environ Health Perspect*. 2010;118:A420–1.

28. Landrigan PJ. Children's environmental health: a brief history. *Acad Pediatr*. 2016;16:1–9.

29. Bollati V, Baccarelli A. Environmental epigenetics. *Heredity (Edinb)*. 2010 Jul;105(1):105–12.

30. American Academy of Pediatrics Council on Environmental Health. *Pediatric Environmental Health*, 4th ed. Etzel RA, Balk SJ, eds. Itasca, IL: American Academy of Pediatrics; 2019.

31. Government of Japan. Japan environment and children's study. Ministry of the Environment. 2023. https://www.env.go.jp/chemi/ceh/en/

32. National Academy of Sciences. *Pesticides in the Diets of Infants and Children* (Chair: Landrigan PJ). Washington, DC: National Academies Press; 1993.

33. World Health Organization. Children's environmental health. 2012. Available at: https://www.who.int/health-topics/children-environmental-health#tab=tab_1.

34. Grandjean P, Landrigan PJ. Neurobehavioural effects of developmental toxicity. *Lancet Neurol*. 2014;13:330–8.

35. Landrigan PJ. The toxicity of lead at low dose. *Br J Ind Med*. 1989;46:593–6.

36. Landrigan PJ, Whitworth RH, Baloh RW, Barthel WF, Staehling NW, Rosenblum BF. Neuropsychological dysfunction in children with chronic low-level lead absorption. *Lancet*. 1975;1:708–12.

37. Needleman HL, Gunnoe C, Leviton A, et al. Deficits in psychologic and classroom performance of children with elevated dentine lead levels. *N Engl J Med*. 1979;300:689–95.

38. Kjellström T, Kennedy P, Wallis S, et al. Physical and mental development of children with prenatal exposure to mercury from fish. Stage 2, interviews and psychological tests at age 6 (Report 3642). Stockholm: National Swedish Environmental Protection Board; 1989.

39. Grandjean P, Weihe P, White RF, et al. Cognitive deficit in 7-year-old children with prenatal exposure to methylmercury. *Neurotoxicol Teratol*. 1997;19:417–28.

40. Myers GJ, Davidson PW, Cox C, et al. Prenatal methylmercury exposure from ocean fish consumption in the Seychelles child development study. *Lancet*. 2003;361:1686–92.

41. National Research Council. *Toxicological Effects of Methylmercury*. Washington, DC: National Academies Press; 2000.

42. Barouki R, Gluckman PD, Grandjean P, Hanson M, Heindel JJ. Developmental origins of non-communicable disease: implications for research and public health. *Environ Health*. 2012;11:42.

43. Weiss B. Lead, manganese, and methylmercury as risk factors for neurobehavioral impairment in advanced age. *Int J Alzheimers Dis*. 2010;2011:607543.

44. National Academy of Sciences. *Exposure Science in the 21st Century*. Washington, DC: National Academies Press; 2012.

45. Perera FP, Weinstein IB. Molecular epidemiology: recent advances and future directions. *Carcinogenesis*. 2000;21:517–24.

46. Skinner MK. Role of epigenetics in developmental biology and transgenerational inheritance Birth Defects Res C 2011;93:51–5.

47. Trasande L, Liu Y. Reducing the staggering costs of environmental disease in children, estimated at $76.6 billion in 2008. *Health Affairs*. 2011;30:863–70.

48. Grosse SD, Matte TD, Schwartz J, Jackson RJ. Economic gains resulting from the reduction in children's exposure to lead in the United States. *Environ Health Perspect*. 2002;110:563–9.

2

Children's Exquisite Vulnerability to Environmental Exposures

Ruth A. Etzel and Philip J. Landrigan

A fundamental maxim of pediatrics and also of children's environmental health is that *children are not "little adults."*[1] Children are far more sensitive than adults to toxic chemicals and other harmful exposures in the environment. This sensitivity reflects the combination of children's disproportionately greater exposures to toxic materials in the environment plus their exquisite biological sensitivity.

Children's susceptibility is especially great during periods in early development—unique "windows of vulnerability"—when their vital organs are forming and rapidly developing.[2] Most of these windows occur in embryonic and fetal life and some in early childhood. Each organ of the body develops at its own pace, with specific sensitive periods of rapid development.[3-10] See Table 2.1. Exposures to even minute quantities of toxic chemicals during these sensitive periods can lead to permanent and irreversible injury to the organs. These windows of vulnerability have no counterpart in adult life.

Recognition of the vulnerability of children, infants, and fetuses to toxic chemicals in the environment was a watershed event. This discovery catalyzed development of the academic discipline of children's environmental health. It also triggered two further insights that have now themselves become cornerstones of children's environmental health:

- Exposures to toxic chemicals in early life are important causes of disease and dysfunction in children and also in adults, and
- Diseases caused by chemicals can successfully be prevented, thus saving lives, enhancing the quality of life, reducing healthcare and education costs, and increasing national productivity (see Chapter 1 for further discussion).

Historical Background

Current understanding of children's vulnerability to environmental hazards had its origins in a series of seminal clinical observations and epidemiological studies.[11] Among these early reports were

- *Thalidomide and phocomelia.* In the 1950s and 1960s, pediatricians in Europe began to see large numbers of newborn babies with a previously rare birth

Ruth A. Etzel and Philip J. Landrigan, *Children's Exquisite Vulnerability to Environmental Exposures* In: *Textbook of Children's Environmental Health*, Second Edition. Edited by: Ruth A. Etzel and Philip J. Landrigan, Oxford University Press.
© Oxford University Press 2024. DOI: 10.1093/oso/9780197662526.003.0002

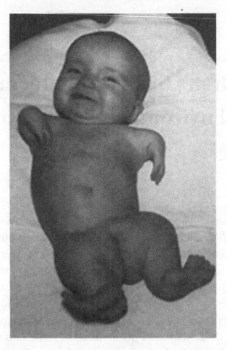

Figure 2.1. Infant with phocomelia following exposure *in utero* to thalidomide

defect of the limbs termed "phocomelia" (Figure 2.1). (The term comes from Greek: *phokos* = seal [the marine mammal] and *melia* = limb; these babies' vestigial limbs were thought to resemble a seal's flippers.) Clinical and epidemiological studies found that virtually all of these babies had been exposed in the womb to thalidomide, a newly invented sedative that had been prescribed to women during the first trimester of pregnancy to alleviate morning sickness. Thalidomide was most harmful when it was taken between days 34 and 50 of pregnancy, which is precisely the time when the limbs are forming.[12] In addition to interfering with limb formation, thalidomide was subsequently found to be associated also with deformed eyes, ears, hearts, alimentary and urinary tracts, blindness, deafness, and increased risk of autism.[13] The drug was never licensed in the United States, but more than 10,000 cases of phocomelia were reported worldwide (8,000 in Germany alone) before thalidomide was removed from the market and the epidemic halted.[14] The mothers who took the medication were physically unaffected. In the 2010s, a new generation of thalidomide survivors was born in Brazil, where thalidomide is used to treat the complications of leprosy.[15] Thalidomide also was used to treat COVID.[16]

- *Leukemia and microcephaly in atomic bomb survivors.* An epidemic of leukemia was reported in the 1940s and 1950s among children in Hiroshima and Nagasaki who had been exposed to ionizing radiation in the atomic bombings.[17] Cases of leukemia began to be seen in these children in the first 2–3 years after the attacks. Incidence peaked approximately 7 years after the bombing and then declined. Risk of leukemia was highest in the children who had been most heavily irradiated. Risk was much greater in children than in adults with similar radiation doses. This study and subsequent studies of leukemia in children exposed to X-rays in utero[18] established that infants and children are much more sensitive than adults to ionizing radiation.

An additional finding in these studies was that infants who were exposed in utero to ionizing radiation had an increased incidence of microcephaly. This presumably reflected radiation injury to their developing brains. There was no comparable neurological damage observed in adults.

- *Minamata disease (congenital methylmercury poisoning).* An epidemic of microcephaly, cerebral palsy, mental retardation, blindness, spasticity, and convulsions was reported in the 1960s among children living in the fishing village of Minamata, Japan.[19] Investigation revealed that the epidemic was caused by the ingestion by pregnant women of fish and shellfish contaminated with methylmercury. The mothers were only minimally affected.[20] Minamata disease is described in detail in Chapter 1.

- *The DES episode.* Diethylstilbestrol (DES), a synthetic estrogen, was prescribed to as many as 5 million pregnant women in the United States in the 1960s and early 1970s to block spontaneous abortion and promote fetal growth. A decade later, gynecologists began observing cases of a rare malignancy, adenocarcinoma of the vagina, in young women. Peak incidence was in the years immediately after puberty. Epidemiologic analysis found that the great majority of the young women with vaginal cancer had been exposed in utero to DES. Their mothers were physically unaffected. Further long-term follow-up studies have shown that, after age 40, DES daughters have a 2.5-fold increased incidence of breast cancer[21] as well as an increased risk of pancreatic cancer.[22] Studies of the health of the third generation suggest the intergenerational transmission of DES health effects.[23]

These tragic episodes marked the beginning of the academic discipline of children's environmental health. They showed clearly that infants are far more sensitive to toxic chemicals than are their mothers. They demonstrated that toxic chemicals from the environment can enter women's bodies, cross the placenta from mother to fetus, and cause disease in the child. They destroyed forever the myth that the placenta is an impenetrable barrier that protects infants in the womb against toxic chemicals.

These episodes occurred in the era when concern about the environment and about links between the environment and human health was just beginning to emerge in America. They occurred during the time in which Rachel Carson was writing her transformative book, *Silent Spring.* The publication of *Silent Spring* in 1962 is widely considered to mark the birth of the American environmental movement.[24]

In response to these episodes, the American Academy of Pediatrics (AAP) established a Committee on Radiation Hazards and Epidemiology of Malformations. The purpose of this Committee, formed in 1957, was to advise pediatricians about the diagnosis, treatment, and prevention of environmental hazards. Robert W. Miller, an American pediatrician who had led epidemiologic studies of children's health in Hiroshima and Nagasaki after the atomic bombings, was instrumental in establishing this Committee. Dr. Miller is profiled in Chapter 1. This Committee still exists today; it has grown and been renamed the AAP Council on Environmental Health and Climate Change. This Council has been an important source of authoritative reports and important recommendations on a wide range of environmental hazards to children.[25] It has been central to the establishment of the academic discipline of children's environmental health, and it has been an incubator of leaders in children's environmental health.

Report on Pesticides in the Diet by the US National Academy of Sciences

A key event that crystallized findings from earlier studies built on the growing environmental movement in the United States and greatly accelerated understanding of the vulnerability of infants and children to toxic hazards in the environment was the publication, in 1993, of a report by the US National Academy of Sciences (NAS) on *Pesticides in the Diets of Infants and Children*.[1]

The NAS report found that children have unique patterns of exposure and unique susceptibilities to hazardous exposures in the environment that have no counterparts in adult life. The report recommended that children must have special protections against environmental threats to health in law, regulation, and risk assessment. The central conclusion of this report was that "Children Are Not Little Adults."

This report was commissioned by the Committee on Agriculture of the US Senate. It was catalyzed by growing concern about potential risks to children's health of pesticides in fruits and vegetables, concern that stemmed from the recognition that nearly all fruits and vegetables sold in commercial markets in industrialized countries, except those certified "organic," contain measurable levels of one of more pesticides.[1]

Prior to publication of the NAS report, virtually all research in toxicology and all risk assessment and policy formulation in environmental health had focused on protection of the "average adult." That research took little cognizance of the unique exposures or the special susceptibilities of fetuses, infants, and children.

The old approach to regulation of pesticides in food provides an example. Prior to publication of the NAS report, the levels of pesticides, termed "tolerance levels," that were permitted on fruits and vegetables sold in markets were set at levels considered to be safe for adults. However, those older tolerances had two fundamental shortcomings. A first problem was that they were not health-based. Instead the tolerance-setting process weighed the protection of human health against the costs of regulation to agricultural producers and tried to strike a balance between the two, often to the detriment of public health. A second even more serious shortcoming was that the older pesticide tolerances paid no attention to the unique exposures or special susceptibilities of infants and children. They assumed that the population was comprised solely of adults and that a single tolerance level would protect people of all ages against pesticides in agricultural products.

The NAS report fundamentally changed that paradigm. It built on parents' age-old understanding that children are exquisitely sensitive to hazards in the world around them. It built on research into children's vulnerabilities that had begun with the founding of the specialty of pediatrics in the 1850s by Dr. Abraham Jacobi. The report's main finding was that children are uniquely vulnerable to pesticides and other toxic chemicals in the environment.[1]

The findings of the NAS report produced a profound shift in public policy. For the first time ever, it brought the issue of children's sensitivity to toxic chemicals to the attention of national policymakers and elected officials in the United States and other countries. This increase in policymakers' attention to children catalyzed the recent exponential growth of children's environmental health that is discussed in Chapter 1.

The NAS report identified four differences between children and adults that contribute to children's heightened susceptibility to pesticides and other toxic chemicals.

1. *Children have proportionately greater exposures than adults to toxic chemicals on a body-weight basis.* Children's increased exposures are due in part to their

disproportionately large intakes of air, food, and water. Infants, for example, have respiratory rates that are twice as great as those of adults, and, on a body weight basis, they inhale twice as much air. Thus they are at increased risk of absorbing airborne toxins. Likewise, children eat more food and drink much more water per pound of body weight than do adults. For example, children take in three to four times more calories per kilogram than adults, and a 6-month-old infant drinks seven times more water per kilogram body weight than an adult. Children's diets are very different from those of adults. Children also consume more milk, more fruits, and often more vegetables. Moreover children have unique food preferences and eat a much less varied diet than most adults. For example, an average 1-year-old drinks 21 times more apple juice and 11 times more grape juice than an adult. The consequence of all these differences is that they place children at much greater risk of ingesting any pesticides or toxic chemicals that may be present in those favored foods.

The differences between children and adults in dietary exposure to toxic chemicals become especially striking when one moves beyond consideration of average exposures to examination of the full range of exposures to infants and children. Some children consume extraordinary amounts of certain foods such as milk, juices, and fruits for certain periods of time in early infancy and childhood. If those foods happen to be contaminated with relatively high levels of pesticides—and worse yet if they are contaminated by multiple pesticides—the cumulative dose delivered to certain infants in the population can be quite significant. The NAS report presented quantitative analyses of the full range of children's exposures to benomyl, aldicarb, and combinations of organophosphate pesticides.[1] These analyses demonstrated that several thousand children in the United States are exposed each day to levels of these pesticides sufficiently high to cause toxic injury.

Another physiologic difference between children and adults is that children have a larger surface-to-volume ratio and more permeable skin, two factors that lead to greater dermal absorption of toxic chemicals (see Box 2.1 and Box 2.2).

Children's age-related behaviors further magnify their intake of toxic chemicals from the environment. Children behave differently from adults, and their behaviors change as they develop. Most children actively explore their environments, and young children engage in frequent hand-to-mouth and object-to-mouth behavior. This normal oral exploratory behavior can lead to significant ingestion of toxic substances.

Children spend their time in different physical locations than adults, and these differences magnify their exposures. Infants and young children, for example, spend much of their time on the floor. They are therefore at disproportionate risk of exposure to house

---

Box 2.1 Case Study: Scrotal Cancer in Chimney Sweeps

A classic example is an epidemic of skin cancer of the scrotum among preteen and adolescent boys described in 1775 by the eminent London surgeon, Sir Percival Pott. These boys were employed as chimney sweeps in Victorian England.[26] They were placed in rope harnesses and lowered naked into chimneys that were too narrow for adults. This work led to heavy dermal exposure to soot containing carcinogenic polycyclic aromatic hydrocarbons. Soot was trapped in the skin folds of the scrotum and absorbed. Repeated exposures over many years led to the formation of squamous cell carcinomas.

Box 2.2 Case Study: Children's Disproportionate Exposure to Airborne Mercury

---

Mercury vapor is heavier than air, and thus the highest concentrations of airborne mercury vapor occur near the floor. Before 1991, many brands of interior latex paint sold in the United States contained mercury as a preservative. Also, paint stores sold mercury salts that could be added to paint for the control of mildew. During the first several months after this mercury-containing paint was applied to a wall, mercury vapor was released from the paint into the indoor air, sometimes exposing people to high levels of mercury. Acrodynia, a form of pediatric mercury poisoning, was a result.[28] In one case, a 4-year-old boy became poisoned after the entire interior of his fire-damaged home had been painted with 17 gallons of paint containing mercury.[29] Remarkably, four adult family members living in the same house under the same conditions remained unaffected, although urine tests documented that they, too, had been exposed to elevated levels of mercury.[29]

---

dust that may be contaminated by lead or pesticides. And toddlers, because of their short stature, breathe air that is much closer to the ground than that inhaled by adults. They are therefore at increased risk of inhaling vapors of solvents or pesticides that may form layers near the floor.[27] In addition, children are exposed to preschool or school classroom environments and playgrounds. Schools and playgrounds are too often built on relatively undesirable lands, and the facilities may be old, poorly maintained and poorly ventilated.

2. *Children's metabolic pathways are immature.* Children's ability to metabolize and excrete toxic chemicals is different from that of adults. In some instances, infants are actually at lower risk than adults because they cannot convert chemicals to their toxicologically active forms. But, in many other cases, they are more vulnerable because children are less able than adults to detoxify and excrete toxic compounds. As a result, many toxic chemicals have prolonged half-lives in children's bodies. Organophosphate pesticides provide an example. The half-life of the widely used organophosphate chlorpyrifos in the bloodstream of an adult is about 6 hours. But in an infant the half-life of chlorpyrifos is 36 hours, which means that, in an infant, this biologically active molecule has much more time to cause cellular injury.[25]

Another manifestation of children's immaturity is the rapidity of their metabolism. This can result in increased risk following exposure to certain toxic materials, as illustrated by the disproportionate occurrence of carbon monoxide poisoning in children as compared to adults. Children are more susceptible to carbon monoxide because their developing organ systems have high metabolic rates and high oxygen demand. They are therefore severely affected by the oxygen deprivation that results when carbon monoxide combines with hemoglobin and blocks oxygen transport in the bloodstream.[25] There have been cases reported in which a snowbound automobile containing high levels of carbon monoxide was found with adults in the front seat, unconscious, but alive, while children in the back seat were dead.[25] A fetus is also more vulnerable than an adult to carbon monoxide poisoning. Fetal blood has a higher affinity for carbon monoxide than does adult blood, and a fetus eliminates carboxyhemoglobin more slowly than an adult.

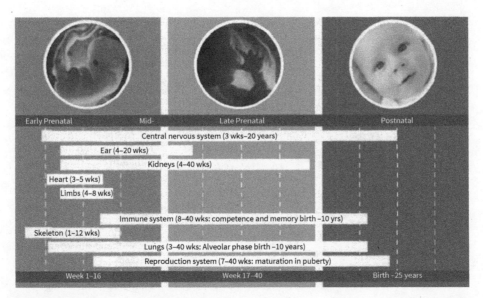

Figure 2.2. Stages of human development
Source: Courtesy of Dr Jerrold Heindel, US National Institute of Environmental Health Sciences.

3. *Children's extremely rapid but exquisitely delicate developmental processes are easily disrupted.* Rapid, complex, and highly choreographed growth and development take place in embryonic and fetal life as well as in the first years after birth, as illustrated in Figure 2.2. This great complexity creates *windows of vulnerability*, periods of heightened sensitivity to toxic chemicals that exist only in early development and have no counterpart in adult life.[2]

The developing brain, for example, is very vulnerable to radiation injury during pregnancy. Mental retardation and microcephaly occurred in children born to women who were pregnant when the atomic bomb was dropped on Hiroshima.[30] The increased vulnerability of the infant respiratory tract is largely due to the prolonged development of the infant lungs. The lungs are growing rapidly during the first year of life, [31] and inhalation of toxigenic molds during early infancy can result in acute pulmonary hemorrhage.[32] Additional alveoli continue to form up until the fourth year of life.[31] Exposure to secondhand smoke during this time has harmful effects on the developing lungs. Because the adult lung and respiratory tract are mature, secondhand smoke does not have these same effects. Similarly, outdoor pollutants such as airborne particulates and nitrogen dioxide have been linked to significant deficits in lung growth in fourth-graders, but less significant deficits were noted in seventh- and tenth-graders living in the same polluted area.[33]

4. *Children have more time than adults to develop chronic diseases that may be triggered by environmental exposures in early life.* Many diseases triggered by toxic chemicals, such as cancer and neurodegenerative diseases, are now understood to evolve through multistage, multiyear processes over the course of many years or even decades. Because children have more years of future life than most adults, they have much more time to develop chronic diseases that may be initiated by early exposures.[34] This understanding builds on the observation that the nutritional environment in utero can influence health across the entire human life span even into extreme old age.[35]

Table 2.1 Developmental stages and special environmental health risks during each stage

Developmental stage	Time period	Developmental milestones	Special environmental health risks
Embryonic	8 days to 9 weeks of pregnancy	Organogenesis at approx. days 20 to 60 days of gestation	Thalidomide and phocomelia (day 34–50)
Fetal	9 weeks of pregnancy to birth	Control of autonomic nervous system at approx 24 weeks	Microcephaly and mental retardation with in utero radiation exposure at 8–15 weeks
Infancy	Birth to 12 months	Rolling over at 2–3 months, sitting at 3 months, standing with support at 6 months, walking begins at 10–12 months	Mercury vapor and acrodynia Nitrates and methemoglobinemia Secondhand smoke and lung diseases Toxigenic molds and pulmonary hemorrhage
Young toddlers Older toddlers Preschoolers	1 to 2 years 2 to 3 years 3 to 5 years	Self-feeding at 1 Toilet-trained at 2 Motor skills at 5	Radiation and thyroid cancer—Chernobyl
School-aged	5 to 12 years	Specific synapse formation in brain	Nicotine and addiction Soot exposure and cancer of scrotum
Adolescence	12 to 19 years	Maturation of organs	Soot exposure and cancer of scrotum

Case Studies in Vulnerability

The unique vulnerability of infants and children to toxic chemicals in the environment is illustrated by three case studies:

1. Examination of the impacts on brain development of early exposures to neurotoxic chemicals;
2. Examination of the effects on multiple organ systems of early exposures to endocrine-disrupting chemicals; and
3. Examination of the effects on the developing lung of exposures to toxigenic molds.

Neurotoxic Exposures in Early Development

Exposures to even minute quantities of neurotoxic chemicals during early brain development can cause devastating damage to the brain and nervous system that have no counterpart in adult life (see also Chapters 9 and 49). This vulnerability of the developing human brain to toxic insult is a direct consequence of the extraordinary complexity of early brain development.[36] In the 9 months of pregnancy, the human brain and spinal cord must develop from a thin strip of cells along the dorsal surface of the embryo into a complex organ comprised of billions of precisely located, highly interconnected and specialized cells. Brain development requires that neurons move along precise pathways from their points of origin to their assigned locations, that they establish connections with other cells near and distant, and that they learn to intercommunicate. Each connection between and among neurons must be precisely established at a particular point in development. Redundant connections need to be pruned away through programmed cell death, or *apoptosis*. All of these processes must take place within a tightly controlled time frame, in which each developmental stage must be reached on schedule and in the correct sequence.

Because of its extraordinary complexity, brain development is highly sensitive to toxic environmental exposures. Any toxic or other environmental exposure that interferes with the tightly orchestrated sequence of events involved in brain formation is likely to have profound effects on intellect, behavior, and other function.[5] If a developmental process in the brain is halted or inhibited, if cells fail to migrate in the proper sequence to their assigned locations, if synapses fail to form, or if pathways are not established, there is only limited potential for later repair and the consequences can be permanent.

Postnatally the human brain continues to develop, and the period of heightened vulnerability therefore extends throughout infancy and into childhood. While most neurons have been formed by the time of birth, growth of support cells and development and myelinization of axons continue through adolescence[36]; see Box 2.3.

Box 2.3 Early Life Exposures to Organophosphate Pesticides Can Cause Developmental Retardation

---

Organophosphate (OP) insecticides are the most commonly utilized class of pesticides in the world (see also Chapter 35). They are well known to be acutely toxic to the brain and nervous system at high doses. They kill insects and also cause acute human poisoning by inhibiting acetylcholinesterase, an enzyme found in the nervous systems of both insects and mammals. Acetylcholinesterase inhibition leads to accumulation of acetylcholine in the brain and nerves, which in turn causes nausea, vomiting, diarrhea, excessive salivation, meiosis, coma, convulsions, and, in extreme cases, death by respiratory failure. The "nerve gas," sarin, is a member of the OP family.

The developmental toxicity of the OP pesticides has come to be recognized only in recent years. It appears to be quite distinct from acute OP toxicity and to be mediated via different cellular mechanisms.[37] Studies of the developmental toxicity of OP pesticides have focused especially on chlorpyrifos, a member of the OP family extensively used until a few years ago to control insects in schools and homes in the United States and still used in agriculture. Studies of the developmental toxicity of chlorpyrifos in newborn baby rodents (whose developmental stage is roughly equivalent to that of a 7-month human fetus) have shown that even very low doses of chlorpyrifos can disrupt the basic cellular machinery in the developing brain that controls neural cell maturation and synapse formation.[38] The consequences are reduced numbers of neurons in the brains of newborn rodents followed by learning deficits and behavioral abnormalities.

Prospective epidemiological studies have found that chlorpyrifos also is linked to developmental neurotoxicity in human infants. Infants exposed in utero to chlorpyrifos have been found to have smaller head circumference at birth than unexposed babies. Reduced head circumference at birth is an indicator of delayed brain growth during pregnancy and a predictor of delayed development and learning disabilities.

This effect is most pronounced among infants born to mothers with low expression levels of the enzyme paraoxonase, an enzyme critical to the metabolic breakdown of OP pesticides in the body.[39] This effect disappeared after a ban on residential use of chlorpyrifos was imposed in the United States in 2001.[40]

Follow-up studies of children with biochemically documented exposures to chlorpyrifos in utero have found evidence for developmental delays. These children also had increased prevalence of attention deficit hyperactivity disorder (ADHD).[41–43] A prospective epidemiological follow-up study suggested that prenatal exposure to chlorpyrifos may be associated with pervasive developmental disorder—not otherwise specified (PDD-NOS), a form of autism spectrum disorder.[44]

Although residential use of chlorpyrifos has now been banned in the United States, dozens more OPs are approved for home and school use, and many more OPs are used extensively in agriculture. Their potential impacts on brain development in early childhood are largely unexplored.

Endocrine Disruption and Early Development

Endocrine disruptors are synthetic chemicals that can mimic, alter, magnify, and block the effects of naturally occurring hormones, such as estrogen, testosterone, growth hormone, insulin, and thyroid hormone (see also Chapters 40 and 41). Synthetic endocrine disruptors are manufactured in volumes of millions of pounds per year. They include phthalates, bisphenol A, perchlorate, certain pesticides, brominated flame retardants, certain metals, and dioxins. These chemicals are widespread today in consumer products such as soaps, shampoos, perfumes, and plastics. They are common contaminants in air, food, and drinking water.

Exposures to even minute quantities of endocrine disruptors during windows of vulnerability in utero and in early childhood have been shown to be capable of producing serious effects on health (Box 2.4).

Phthalates appear to be toxic to the developing brain and nervous system. Childhood exposure to phthalates appears to be associated with lower IQ scores.[46] A systematic review found that most studies demonstrated statistically significant inverse relationships between maternal urinary phthalate concentrations during pregnancy and subsequent outcomes in children's cognitive and motor scales, especially in boys rather than girls.[47]

Box 2.4 The Effects of Phthalates on Reproductive Function and Brain Development.

Phthalates are a widely used family of chemicals used as plasticizers to confer flexibility to rigid plastics and also used in personal-care products, lacquers, varnishes, and timed-release coatings for some medications.

Phthalates are endocrine disruptors, and several phthalates possess anti-androgenic activity and reduce testosterone levels. In animal studies, evidence of anti-androgenic effects associated with phthalate exposure in early life include impaired Leydig cell function, hypospadias, and undescended testicles. In humans, prenatal exposure to phthalates has been linked to lower serum testosterone levels in newborn and adult males and adverse effects on adult sperm. Prenatal exposure to phthalates has also been linked to shortening of the ano-genital distance in baby boys, a finding indicative of in utero feminization.[45]

Pulmonary Toxicity from Exposures to Molds
during Early Infant Development

Homes that have been flooded and remain water-damaged and chronically damp are an ideal environment for the growth of a variety of fungi. Some of these fungi are toxigenic, i.e., they produce very potent toxins. Mold toxins such as aflatoxins are well known to cause acute human poisoning; some mycotoxins have been suspected of use in biological warfare.[48] The pulmonary toxicity of the mycotoxins for infants has only been recognized recently. Some mycotoxins preferentially affect rapidly dividing cells and lung cells are dividing very rapidly during the first months of life, making infants most vulnerable (Box 2.5).

Box 2.5 Early Life Exposures to Toxigenic Molds Can Cause Infant Pulmonary Hemorrhage

---

A cluster of 8 cases of pulmonary hemorrhage among infants in Cleveland in the 1990s prompted a case-control study that discovered a new association between exposure to fungi, including the toxigenic fungus *Stachybotrys chartarum*, in the indoor environment and infant pulmonary hemorrhage.[32] The isolates of *Stachybotrys* from the case infants' homes were documented to produce toxins.[49] Also, *S. chartarum* was found by quantitative polymerase chain reaction (qPCR) in the respiratory secretions of four of the cases.[50] A reanalysis of the data by a second group of independent analysts confirmed the statistically significant association between exposure to fungi and infant pulmonary hemorrhage.[51] Several case reports that lend support to this association have appeared in the literature. Exposure to household fungi has been linked with acute pulmonary hemorrhage among young infants in Kansas City, Missouri and Delaware.[52–54] A case of acute pulmonary hemorrhage in a North Carolina infant was associated with exposure to a moldy home from which *Trichoderma*, a toxin-producing fungus was isolated.[55] Cases also have been reported from New Zealand[56] and Oman.[57] *Stachybotrys* has been isolated from the bronchoalveolar lavage fluid of a 7-year old child with lung bleeding and pulmonary hemosiderosis in Houston.[58] Mold and mycotoxins were identified in the home and the lungs, liver and brain of a 16-month-old male who died from pulmonary hemorrhage in California.[59] Though infants are most vulnerable, in one study from Hungary nose bleeds were documented among adult workers exposed to aerosols of *Stachybotrys* in contaminated hay.[60]

A New Paradigm for Toxicology

Understanding has become widespread in the past two decades that children are highly sensitive to toxic chemicals and other hazardous exposures in the environment. Exposures in early life, even to extremely low levels of toxic materials can cause lasting damage to embryos, fetuses, and young children. It is now understood that children have susceptibilities in early development—unique "windows of vulnerability"—that have no counterpart in adult life.[61]

It also has come to be understood that *timing of exposure* is critically important in early development. The level of lead or methylmercury that can injure the developing brain of a fetus is far lower than the level that causes injury to a 5-year-old, and that level in turn is much lower than the level that can injure an adult. Careful studies of birth defects in children exposed prenatally to thalidomide have shown that exposures between days 34 and 27 of pregnancy can cause defects of the ears, while exposures between days 34 and 50 cause phocomelia, and exposures between days 20 and 24 are associated with increased risk of autism. These effects have no parallel in toxicological studies in adults.

In light of these new findings, it is necessary to reconsider the ancient toxicological principle that "the dose makes the poison." This principle, which is attributed to Paracelsus, the "Father of Toxicology," states that the greater the level of exposure to a toxic chemical (the "dose") the more severe will be its effects on health.[62]

Paracelsus's principle has been a powerful organizing theorem in toxicology for five centuries. It still has enormous validity today. But it fails to explain how very small exposures to toxic chemicals in early development can have profound and lasting impacts on health. These findings therefore suggest the need for a new corollary principle in toxicology that "in early development, the timing makes the poison."

In the years ahead, this new understanding of the vulnerability of infants and children to environmental hazards will need to be translated into pediatric and public health practice, risk assessment, regulation, and legislation (see Chapters 58 through 61). The protection of children's health demands no less.

References

1. National Academy of Sciences. *Pesticides in the Diets of Infants and Children*. Washington, DC: National Academies Press; 1993.

2. Rice D, Barone S Jr. Critical periods of vulnerability for the developing nervous system: evidence from humans and animal models. *Environ Health Perspect*. 2000;108:511–33.

3. Piñeiro-Carrero VM, Piñeiro EO. Liver. *Pediatrics*. 2004;113(4 Suppl):1097–106.

4. Solhaug MJ, Bolger PM, Jose PA. The developing kidney and environmental toxins. *Pediatrics*. 2004;113(4 Suppl):1084–91.

5. Rodier PM. Environmental causes of central nervous system maldevelopment. *Pediatrics.* 2004;113(4 Suppl):1076–83.

6. Greim HA. The endocrine and reproductive system: adverse effects of hormonally active substances? *Pediatrics.* 2004;113(4 Suppl):1070–75.

7. Sreedharan R, Mehta DI. Gastrointestinal tract. *Pediatrics.* 2004;113(4 Suppl):1044–50.

8. Finkelstein JN, Johnston CJ. Enhanced sensitivity of the postnatal lung to environmental insults and oxidant stress. *Pediatrics.* 2004;113(4 Suppl):1092–96.

9. Kajekar R. Environmental factors and developmental outcomes in the lung. *Pharmacol Ther.* 2007;114(2):129–45. doi:10.1016/j.pharmthera.2007.01.011.

10. Mancini AJ. Skin. *Pediatrics.* 2004;113(4 Suppl):1114–9.

11. Miller RW. How environmental hazards in childhood have been discovered: carcinogens, teratogens, neurotoxicants, and others. *Pediatrics.* 2004;113:945–51.

12. Lenz W. Chemicals and malformations in man. In: Fishbein M, ed. *Second International Conference on Congenital Malformation.* International Medical Congress: New York; 1963:263–71.

13. Arndt TL, Stodgell CJ, Rodier PM. The teratology of autism. *Int J Dev Neurosci.* 2005;23:189–99.

14. Vargesson N. Thalidomide-induced teratogenesis: history and mechanisms. *Birth Defects Res C Embryo Today.* 2015;105(2):140–56. doi:10.1002/bdrc.21096.

15. Vianna FSL, Schüler-Faccini L, Leite JCL, de Sousa SHC, da Costa LMM, Dias MF, Morelo EF, Doriqui MJR, Maximino CM, Sanseverino MTV. Recognition of the phenotype of thalidomide embryopathy in countries endemic for leprosy: new cases and review of the main dysmorphological findings. *Clin Dysmorphol.* 2013;22(2):59–63. doi:10.1097/MCD.0b013e32835ffc58.

16. Li Y, Shi K, Qi F, et al. Thalidomide combined with short-term low-dose glucocorticoid therapy for the treatment of severe COVID-19: a case-series study. *Int J Infect Dis.* 2021;103:507–13. doi:10.1016/j.ijid.2020.12.023.

17. Miller RW. Delayed effects occurring within the first decade after exposure of young individuals to the Hiroshima atomic bomb. *Pediatrics.* 1956;18:1–18.

18. Stewart AM. Leukemia and other neoplasms in childhood following radiation exposure in utero: a general survey of present knowledge. *Br J Radiol.* 1968;41:718–19.

19. Tsuda T, Yorifuji T, Takao S, Miyai M, Babazono A. Minamata disease: catastrophic poisoning due to a failed public health response. *J Public Health Policy.* 2009;30(1):54–67. doi:10.1057/jphp.2008.30.

20. Harada H. Congenital Minamata disease: intrauterine methylmercury poisoning. *Teratology.* 1978;18:285–8.

21. Herbst AL, Hubby MM, Azizi F, Makii MM. Reproductive and gynecologic surgical experience in diethylstilbestrol-exposed daughters. *Am J Obstet Gynecol.* 1981;141:1019–28.

22. Troisi R, Hatch EE, Titus L, et al. Prenatal diethylstilbestrol exposure and cancer risk in women. *Environ Mol Mutagen.* 2019;60(5):395–403. doi:10.1002/em.22155.

23. Titus L. Evidence of intergenerational transmission of diethylstilbestrol health effects: hindsight and insight. *Biol Reprod.* 2021;105(3):681–6. doi:10.1093/biolre/ioab153.

24. Carson R. *Silent Spring.* Cambridge, MA: Riverside Press, 1962.

25. American Academy of Pediatrics Council on Environmental Health. In: Etzel RA, Balk SJ, eds. *Pediatric Environmental Health,* 4th ed. Itasca, IL: American Academy of Pediatrics; 2019.

26. Pott P. *Chirurgical Observations Relative to the Cataract, the Polypus of the Nose, the Cancer of the Scrotum, and Different Kinds of Ruptures, and Mortification of the Toes and Feet.* London: Hawes, Clark, Collins, 1775.

27. Etzel RA. The special vulnerability of children. *Int J Hyg Environ Health.* 2020;227:113516. doi.org/10.1016/j.ijheh.2020.113516

28. Hirschmann SZ, Feingold M, Boylen G. Mercury in house paint as a cause of acrodynia: effect of therapy with N-acetyl-D,L-penicillamine. *N Engl J Med.* 1963;269:889–93.

29. Agocs MM, Etzel RA, Parrish RG, et al. Mercury exposure from interior latex paint. *N Engl J Med.* 1990;323:1096–101.

30. Schull WJ, Otake M. Cognitive function and prenatal exposure to ionizing radiation. *Teratology.* 1999;59:222–6.

31. Kajekar R. Environmental factors and developmental outcomes in the lung. *Pharmacol Ther.* 2007;114(2):129–45. doi:10.1016/j.pharmthera.2007.01.011

32. Etzel RA, Montaña E, Sorenson WG, et al. Acute pulmonary hemorrhage in infants associated with exposure to *Stachybotrys atra* and other fungi. *Arch Pediatr Adolesc Med.* 1998;152(8):757–62. doi:10.1001/archpedi.152.8.757

33. Gauderman WJ, Gilliland GF, Vora H, et al. Association between air pollution and lung function growth in southern California children: results from a second cohort. *Am J Respir Crit Care Med.* 2002;166:76–84.

34. Landrigan PJ, Sonawane B, Butler RN, Trasande L, Callan R, Droller D. Early environmental origins of neurodegenerative disease in later life. *Environ Health Perspect.* 2005;113:1230–3.

35. Barker DJ. The developmental origins of adult disease. *J Am Coll Nutr.* 2004;23:588S–95S.

36. Grandjean P, Landrigan PJ. Developmental neurotoxicity of industrial chemicals. *Lancet.* 2006;368:2167–78.

37. Slotkin TA, Levin ED, Seidler FJ. Comparative developmental neurotoxicity of organophosphate insecticides: effects on brain development are separable from systemic toxicity. *Environ Health Perspect.* 2006;114:746–51.

38. Slikker W, Xu ZA, Levin ED, Slotkin TA. Mode of action: disruption of brain cell replication, second messenger, and neurotransmitter systems during development leading to cognitive dysfunction: developmental neurotoxicity of nicotine. *Crit Rev Toxicology.* 2005;35:703–11.

39. Berkowitz GS, Wetmur JG, Birman-Deych E, et al. In utero pesticide exposure, maternal paraoxonase activity, and head circumference. *Environ Health Perspect.* 2004;112:388–91.

40. Lovasi GS, Quinn JW, Rauh VA, et al. Chlorpyrifos exposure and urban residential environment characteristics as determinants of early childhood neurodevelopment. *Am J Public Health.* 2011;101:63–70.

41. Engel SM, Wetmur J, Chen J, et al. Prenatal exposure to organophosphates, paraoxonase 1, and cognitive development in childhood. *Environ Health Perspect.* 2011;119:1182–8.

42. Rauh V, Arunajadai S, Horton M, et al. 7-year neurodevelopmental scores and prenatal exposure to chlorpyrifos, a common agricultural pesticide. *Environ Health Perspect.* 2011;119:1196–201.

43. Bouchard MF, Chevrier J, Harley KG, et al. Prenatal exposure to organophosphate pesticides and IQ in 7-year old children. *Environ Health Perspect.* 2011;119:1189–95.

44. Rauh VA, Garfinkel R, Perera FP. Impact of prenatal chlorpyrifos exposure on neurodevelopment in the first 3 years of life among inner-city children. *Pediatrics.* 2006;118:e1845–59.

45. Swan SH. Environmental phthalate exposure in relation to reproductive outcomes and other health endpoints in humans. *Environ Res.* 2008;108:177–84.

46. Cho SC, Bhang SY, Hong YC, et al. Relationship between environmental phthalate exposure and the intelligence of school-age children. *Environ Health Perspect.* 2010;118:1027–32.

47. Martínez-Martínez MI, Alegre-Martínez A, Cauli O. Prenatal exposure to phthalates and its effects upon cognitive and motor functions: a systematic review. *Toxicology.* 2021;463:152980. doi:10.1016/j.tox.2021.152980.

48 Mirocha CJ, Pawlosky RA, Chatterjee K, Watson S, Hayes W. Analysis for *Fusarium* toxins in various samples implicated in biological warfare in Southeast Asia. *J Assoc Off Anal Chem.* 1983 Nov;66(6):1485–99.

49. Jarvis BB, Sorenson WG, Hintikka EL, et al. Study of toxin production by isolates of *Stachybotrys chartarum* and *Memnoniella echinata* isolated during a study of pulmonary hemosiderosis in infants. *Appl Environ Microbiol.* 1998;64(10):3620–5.

50. Dearborn DG. Mold. In: Landrigan PJ, Etzel RA, eds. *Textbook of Children's Environmental Health.* New York: Oxford University Press; 2014:352–61.

51. Etzel RA. Stachybotrys. *Curr Opin Pediatr.* 2003;15:103–6.

52. Knapp JF, Michael JG, Hegenbarth MA, et al. Case records of the Children's Mercy Hospital, Case 02-1999: a 1-month-old infant with respiratory distress and shock. *Pediatr Emerg Care.* 1999;15:288–93.

53. Flappan SM, Portnoy J, Jones P, et al. Infant pulmonary hemorrhage in a suburban home with water damage and mold (*Stachybotrys atra*). *Environ Health Perspect.* 1999;107:927–30.

54. Weiss A, Chidekel AS. Acute pulmonary hemorrhage in a Delaware infant after exposure to *Stachybotrys atra*. *Del Med J.* 2002;74:363–8.

55. Novotny WE, Dixit A. Pulmonary hemorrhage in an infant following 2 weeks of fungal exposure. *Arch Pediatr Adolesc Med.* 2000,154:271–5.

56. Habiba A. Acute idiopathic pulmonary haemorrhage in infancy: case report and review of the literature. *J Paediatr Child Health.* 2005;41(9-10):532–3.

57. Al-Tamemi S, Al-Kindi H. Acute idiopathic pulmonary haemorrhage in a 2 month old infant. Case report and review of the literature. *Sultan Qaboos Univ Med J.* 2009;9(2):170–4.

58. Elidemir O, Colasurdo GN, Rossmann SN, et al. Isolation of *Stachybotrys* from the lung of a child with pulmonary hemosiderosis. *Pediatrics.* 1999;104:964–6.

59. Thrasher JD, Hooper DH, Taber J. Family of 6, their health and the death of a 16 month old male from pulmonary hemorrhage: Identification of mycotoxins and mold in the home and lungs, liver and brain of deceased infant. *J Clin Toxicol.* 2014;2:1–9. https://globaljournals.org/GJMR_Volume14/1-Family-of-Six-their-Health.pdf

60. Forgacs J, Caril WT. Mycotoxicoses. *Adv. Vet. Med.* 1962: 7273–293.

61. Barouki R, Gluckman PD, Grandjean P, Hanson M, Heindel JJ. Developmental origins of non-communicable disease: implications for research and public health. *Environ Health.* 2012;11:42.

62. Binswanger HC, Smith KR. Paracelsus and Goethe: founding fathers of environmental health. *Bull World Health Organ.* 2000;78:1162–4.

3

The Chemical Environment and Children's Health

Philip J. Landrigan and Ruth A. Etzel

Children's environments have changed profoundly in the past 75 years. A critically important component of this change has been the synthesis, manufacture, and wide dissemination into the environment of thousands of new synthetic chemicals (Figure 3.1).

An estimated 325,000 synthetic chemicals and chemical mixtures have been invented since 1950.[1] The great majority did not exist in nature 75 years ago. These chemicals are used in millions of consumer products ranging from food packaging to clothing, building materials, motor fuels, cleaning products, cosmetics, medicinal products, toys, and baby bottles. They are widely disseminated in the environment. Children's exposures to synthetic chemicals have become the focus of great concern in children's environmental health.[2]

Some synthetic chemicals have benefitted children's health. Antibiotics have helped reduce mortality from the major communicable diseases. Chemical disinfectants have reduced deaths from cholera and other waterborne diseases. Chemotherapeutic agents have made possible the treatment of many childhood cancers.

But new chemicals have also been responsible for tragic episodes of disease, death, and environmental degradation. Many of these episodes have resulted in severe injury to children.

A recurrent theme in far too many of these episodes is that new chemicals have been brought to market with great enthusiasm and no testing, came into wide use, were extensively disseminated into the environment, and then, belatedly, after years and even decades of use, were found to have harmed children's health and damaged the environment.[2] In almost all of these instances, commercial introduction of the chemical preceded any systematic effort to assess potential toxicity. Especially absent were any efforts to examine possible impacts on children's health or potential to disrupt early development. Examples of inadequately tested chemicals that resulted in tragedy include thalidomide, asbestos, DDT, tetraethyl lead added to gasoline, polychlorinated biphenyls (PCBs), diethylstilbestrol (DES), and the ozone-destroying chlorofluorocarbons (CFCs).

A second theme in many of these episodes is that early warnings of danger have been ignored. As a result, efforts to control exposures and prevent injury were delayed, sometimes for decades.[3] In some instances, industries with deeply vested commercial interests in protecting markets for hazardous technologies, such as the tobacco and asbestos industries, have actively opposed efforts to understand and control children's exposures

Philip J. Landrigan and Ruth A. Etzel, *The Chemical Environment and Children's Health* In: *Textbook of Children's Environmental Health*, Second Edition. Edited by: Ruth A. Etzel and Philip J. Landrigan, Oxford University Press. © Oxford University Press 2024.
DOI: 10.1093/oso/9780197662526.003.0003

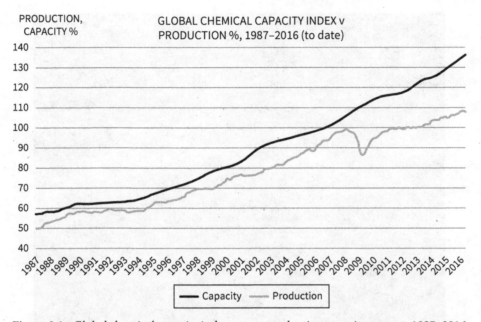

Figure 3.1. Global chemical capacity index versus production capacity percent, 1987–2016
Source: IeC Analysis: American Chemistry Council Data.

to these materials. These industries have used highly sophisticated disinformation campaigns to confuse the public and discredit science. They have attacked heroic pediatricians and environmental scientists who called attention to the risks of emerging technologies and new chemicals.[4] Such was the case with lead and mercury. It is happening today with the organophosphate and neonicotinoid insecticides and with the herbicide glyphosate. See Figure 3.2.

High Production Volume Chemicals

High production volume (HPV) chemicals are defined by the US Environmental Protection Agency as chemicals produced or imported into the United States in quantities of more than 1 million pounds or 500 tons per year.[1] HPV chemicals are those synthetic chemicals that are in widest use, that are most widely dispersed in the modern environment, and that therefore have the greatest potential for human exposure. HPV chemicals are found in a great array of consumer goods, cosmetics, medications, motor fuels, and building materials. They are detectable in much of the United States in air, food, and drinking water. Children are at high risk of exposure.

In national surveys conducted by the Centers for Disease Control and Prevention (CDC), measurable quantities of more than 200 HPV chemicals are routinely detected today in the blood and urine of virtually all Americans, including pregnant women.[5] Elevated levels of HPV chemicals are seen also in the breast milk of nursing mothers and the cord blood of newborn infants.[6]

Figure 3.2. Chemical plant

Widespread Failure to Test Chemicals for Possible Toxicity

Widespread failure to test new chemicals for toxicity before they come to market is the single most worrisome aspect of recent rapid increases in chemical production. Failure to test chemicals for safety or toxicity represents a grave lapse of stewardship. It puts children at daily risk of exposure to materials whose potential harms are virtually unknown. This failure reflects a combination of industry's unwillingness to take responsibility for the chemicals they produce coupled with failure of government.[2] The legislative and regulatory origins of this failure are discussed in Chapter 58.

The consequences of systematic failure to test chemicals are the following:

- No information on potential toxicity is publicly available for more than half of the HPV chemicals currently in widest commercial use.
- No information on developmental toxicity or capacity to harm infants and children is available for about 80% of HPV chemicals.[2]

Even less is known about the potential effects on children's health of simultaneous exposures to combinations of synthetic chemicals. Almost nothing is known about how these chemicals may interact with one another in the human body to possibly cause synergistic effects on children's health.

Examples of synthetic chemicals that are currently on the market and caused widespread human exposure and environmental contamination before any assessment of their potential hazards include phthalates, bisphenols, per- and polyfluoroalkyl substances (PFAS), brominated flame retardants, neonicotinoid insecticides, and glyphosate, the world's most widely used herbicide. All of these materials are produced in extremely high volumes (Table 3.1).

Table 3.1 Production volumes of major synthetic chemicals

Chemical	Annual global production
Bisphenol A	2.5–3 million[7]
Brominated flame retardants	~400,000[8]
Di-ethylhexyl phthalate	4 million[9]
Per- and polyfluoroalkyl substances (PFAS)	>1 million[10]
Neonicotinoid insecticides (imidacloprid)	20,000[11]
Glyphosate	1.065 million[12]

Toxic Chemicals and Disease in Children

Evidence is strong and continuing to accumulate that toxic chemicals released into the environment are important causes of disease, disability, and death in children. The pace of scientific discovery in this area has quickened in recent years. New discoveries of associations between childhood diseases and chemical exposures in early life are being made with increasing rapidity.

This scientific progress has been driven by a series of technical and methodological advances. These include the development of extraordinary new techniques in analytical chemistry that make possible the accurate measurement of several hundred synthetic chemicals in 10–20 mL of blood or urine,[5] the decoding of the human genome, increased understanding of the epigenome, and advances in information technology that can support the enormous volumes of data generated by current studies.

An especially important methodological advance has been the development of *prospective birth cohort epidemiologic studies*. These studies enroll women in pregnancy, measure prenatal environmental exposures through maternal assessment during pregnancy as they are actually occurring, and then follow the children longitudinally, sometimes over many years.[13]They have become powerful engines for the scientific discovery of associations between early-life exposures and disease. A great strength of the prospective study design is that it permits unbiased assessment of exposures months or years before the onset of disease or dysfunction. This design thus reduces recall bias and supports accurate assessments of exposures that occurred in pregnancy or during early childhood. A second great advantage of prospective studies is that they enable the linkage of individual exposures in children with individual health outcomes.

The increasing use of prospective birth cohort studies has generated information on a series of previously unrecognized associations between environmental exposures and disease in children. These include

- Prenatal exposure to PCBs is linked to reduction in children's intelligence.[14] PCBs are an environmentally persistent class of synthetic chemicals that accumulate to high levels in certain species of fish. Human exposure is principally the consequence of maternal consumption of contaminated fish before and during pregnancy. Although PCBs are no longer manufactured in the United States, they were used extensively for many years in the manufacture of electrical equipment such as transformers, and they continue to be important contaminants today because they are highly persistent in the environment.

- Prenatal exposure to the organophosphate insecticide chlorpyrifos is associated with reduced head circumference at birth and developmental delays.[15] Small head circumference at birth is an indicator of delayed brain growth during pregnancy.
- Prenatal organophosphate exposures are associated with cognitive impairments and with increased incidence of autism spectrum disorder.[16-20]
- Baby boys exposed in utero to phthalates, a widely used class of chemicals found in plastics, cosmetics, and many common household products that act as endocrine disruptors, are at increased risk of shortened ano-genital distance, an indicator of decreased masculinization of the reproductive organs.[21] A shortened ano-genital distance in infancy is predictive of decreased fertility in adult life.[22]
- Children exposed in utero to phthalates appear to be at increased risk of behavioral abnormalities, including abnormalities in gender-specific behaviors.[23,24]
- Prenatal exposure to bisphenol A, a synthetic chemical used to manufacture polycarbonate plastics and an endocrine disruptor, is linked to behavioral abnormalities.[25]
- Prenatal exposures to the metals arsenic and manganese are linked to neurodevelopmental impairment and reduced IQ in children.[26,27]

Are There Additional Still Undiscovered Toxic Chemicals?

The answer to this question is not known. Because so few of the thousands of synthetic chemicals currently on the market have been tested for toxicity, it is not whether there are chemicals in wide use today—hazards lurking in plain sight—that may be causing still undiscovered silent injury to children's developing brains and other organ systems.

Until toxicological and epidemiological studies are conducted to specifically seek ill effects associated with early-life exposures to untested chemicals, disease and dysfunction caused by them can go unrecognized for years or even decades.

Case Study: Tetraethyl Lead The "silent epidemic" of childhood lead poisoning that affected millions of children in the United States following the addition of tetraethyl lead to gasoline is a sobering example of a widely used chemical whose hazards to children's health went unrecognized for decades.[28]

Tetraethyl lead is an organic lead compound. It was added to automotive gasoline from the 1940s to the early 1980s to increase engine performance. Millions of American children were exposed to lead from gasoline over a period of nearly five decades. At peak use in the 1970s, annual consumption of tetraethyl lead in gasoline in the United States was nearly 100,000 tons. Virtually all of this lead was released into the environment through the exhaust pipes of cars and trucks. It caused extensive environmental contamination of soil and dust, especially within cities and along roadways. The average blood lead level of US children in that era was close to 20 µg/dL. Many children suffered unrecognized brain injury as the result of their exposure to lead from gasoline.[29]

It is estimated that the epidemic of subclinical lead poisoning may have reduced the number of children with truly superior intelligence (IQ scores above 130 points) by more than 50% and likewise caused a more than 50% increase in the number with IQ scores below 70.[30] In the United States alone, the aggregate number of children at risk of

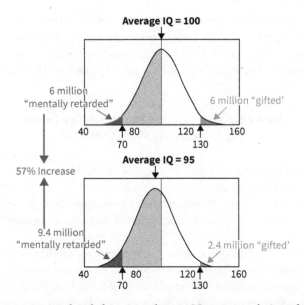

Figure 3.3. Losses associated with five-point drop in IQ on a population of 100 million
Sources: Weiss B. Neurobehavioral toxicity as a basis for risk assessment. Trends Pharmacol Sci. 1988;9(2):59–62. Gilbert SG, Weiss B. A rationale for lowering the blood lead action level from 10 to 2 microg/dL. Neurotoxicology. 2006;27(5):693–701.

exposure to airborne lead during the 40 years of its use was approximately 100 million. See Figure 3.3.

As a result of this widespread exposure to lead, there was an increase in the number of children who did poorly in school, required special education and other remedial programs, and could not contribute fully to society when they became adults.[31] At the same time, there was a reduction in the number of children with truly superior intelligence.

For synthetic chemicals in wide use such as tetraethyl lead, the population at risk of subclinical toxicity can be very large. While subclinical effects may be small at the individual level, the aggregate effects at a population level can have far-reaching consequences.[30] And until this toxicity is recognized, the damage can proceed unchecked for many years.

Widespread exposures to neurotoxic chemicals threaten societal sustainability, undermine national security, and decimate a country's future leadership. It has been speculated that exposure of the Roman ruling classes to lead with subsequent widespread brain injury and reduced fertility accelerated the fall of Rome.[32]

The hypothesis is credible that among the hundreds of still untested synthetic chemicals to which children in industrialized countries are exposed routinely today there may be chemicals in wide use whose adverse effects on children's health have not yet been recognized.[33] Epidemiological studies, especially prospective birth cohort studies that incorporate the tools of molecular epidemiology, have great potential to generate new discoveries of previously unrecognized associations between early exposures to these chemicals and pediatric disease.

David P. Rall, PhD, MD, a former Director of the National Institute of Environmental Health Sciences, once stated that "If thalidomide [a drug widely used in the 1950s and 1960s to treat morning sickness in early pregnancy] had caused a ten-point loss of IQ instead of obvious birth defects of the limbs, it would probably still be on the market."[34]

How Many Synthetic Chemicals Can Injure the Developing Human Brain?

The relatively small number of chemicals that have been documented to cause injury to the developing human brain should be viewed as a tip of an iceberg that could be very large. Figure 3.4 relates the few chemicals with known capacity to injure the developing brain to the larger universe of inadequately tested synthetic chemicals that surrounds them.[33]

To try to answer the question of how many chemicals might be capable of causing damage to the developing human brain, a systematic review was conducted of the world's literature. This review examined the Hazardous Substances Data Bank of the US National Library of Medicine and other relevant data sources.[33] The search produced a list of approximately 200 industrial chemicals that are documented to be neurotoxic in adult humans. These are chemicals that have caused serious, clinically obvious, acute effects in workers industrially exposed to them. Additionally, this search produced a second list of approximately 1,000 chemicals that have been found to be toxic to the nervous systems of animal species. None of the 1,200 neurotoxic chemicals identified through this search

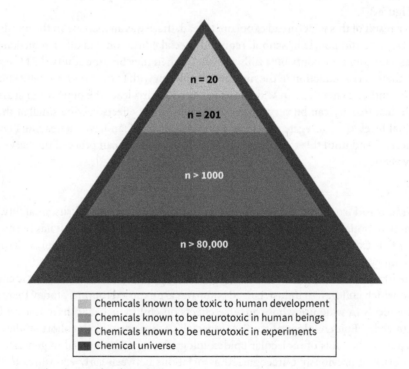

n = 20

n = 201

n > 1000

n > 80,000

Chemicals known to be toxic to human development
Chemicals known to be neurotoxic in human beings
Chemicals known to be neurotoxic in experiments
Chemical universe

Figure 3.4. The chemical iceberg (33)

has been tested to determine whether it has potential to cause damage to the developing human brain and nervous system.

Prevention of Disease Caused by Toxic Chemicals

When information is generated about the toxicity of synthetic chemicals to the health of children, the findings of this research can be translated into evidence-based programs of disease prevention. Such translation is, however, seldom easy or automatic. It requires that pediatricians and other health scientists make their findings known to policymakers and the public and that policymakers listen to the science.

Examples of scientific discoveries of disease caused by toxic chemicals that have successfully translated into prevention programs and yielded great gains for children's health include

- A Congressionally mandated ban on production of PCBs, which led to a reduction in the number of children with PCB-induced losses in intelligence[14];
- Elimination of residential uses of neurotoxic organophosphate pesticides, which led to reductions in the number of babies with low birth weight and small head circumference[35];
- Reduction in exposures to arsenic in well water, which led to reductions in skin and liver cancer[36]; and
- Reductions in air pollution emissions in the United States by more than 70% since 1970, driven by implementation of the Clean Air Act. These improvements in air quality have led to substantial reductions in infant mortality in American children and to striking reductions in morbidity and mortality from cardiorespiratory disease in American adults.[37] They have also produced great benefits for the economy by reducing healthcare costs and increasing the productivity of a healthier, more highly intelligent, longer-lived population. Every dollar invested in air pollution control in the United States since 1970 is estimated to have yielded $30 in economic benefit.[38]

The Alert Clinician

The importance of the pediatric practitioner as a critical first line of defense against the unanticipated hazards of new chemicals cannot be overemphasized.[39] Regardless of how extensively a new chemical is tested for possible hazard, the possibility will always remain that unforeseen, previously undetected toxic effects will become evident only after the chemical has been widely disseminated and millions of children exposed. It therefore falls to the astute pediatric practitioner to discover these novel outcomes, recognize their significance as "sentinel health events,"[40] and act on that information.

The environmental history is a critically important tool for clinical recognition of new associations between toxic environmental exposures and disease in children. The importance of the environmental history as a key tool for the discovery of disease of environmental origin in children is discussed in Chapter 56.

References

1. Wang Z, Walker GW, Muir DCG, Nagatani-Yoshida K. Toward a global understanding of chemical pollution: a first comprehensive analysis of national and regional chemical inventories. *Environ Sci Technol*. 2020 Mar 3;54(5):2575–2584.

2. Landrigan PJ, Goldman LR. Children's vulnerability to toxic chemicals: a challenge and opportunity for health and environmental policy. *Health Affairs*. 2011;30:842–850.

3. European Environment Agency. *Late Lessons from Early Warnings: The Precautionary Principle 1896–2000*. Copenhagen: European Environmental Agency; 2002.

4. Michaels D. *The Triumph of Doubt: Dark Money and the Science of Deception*. London: Oxford University Press; 2020.

5. Centers for Disease Control and Prevention. National report on human exposure to environmental chemicals. 2023. Available at: https://www.cdc.gov/exposurereport/index.html.

6. Woodruff TJ, Zota AR, Schwartz JM. Environmental chemicals in pregnant women in the US: NHANES 2003–2004. *Environ Health Perspect*. 2011;119:878–885.

7. CDC. National Biomonitoring Summary https://www.cdc.gov/biomonitoring/BisphenolA_BiomonitoringSummary.html

8. Boer J de, Stapleton HM. Toward fire safety without chemical risk. https://www.sciencemagazinedigital.org/sciencemagazine/19_april_2019/MobilePagedArticle.action?articleId=1482048

9. Stamatelatou K, Pakou C, Lyberatos G. Occurrence, toxicity, and biodegradation of selected emerging priority pollutants in municipal sewage sludge. In: Moo-Young M, ed. *Comprehensive Biotechnology* (2nd ed.). Academic Press; 2011:473–484. doi:10.1016/B978-0-08-088504-9.00496-7

10. Evich MG, Davis MJB, McCord JP, et al. Per- and polyfluoroalkyl substances in the environment. *Science*. 2022;375 (6580):eabg9065. doi:10.1126/science.abg9065

11. Simon-Delso N, Amaral-Rogers V, Belzunces LP, et al. Systemic insecticides (neonicotinoids and fipronil): trends, uses, mode of action and metabolites. *Environ Sci Pollut Res Int*. 2015;22(1):5–34. doi:10.1007/s11356-014-3470-y

12. Research report on China's glyphosate industry (2018–2022). https://www.prnewswire.com/news-releases/research-report-on-chinas-glyphosate-industry-2018-2022-300689149.html

13. Landrigan PJ, Trasande L, Thorpe LE, et al. The National Children's Study: a 21-year prospective study of 100,000 American children. *Pediatrics*. 2006 Nov;118(5):2173–2186.

14. Jacobson JL, Jacobson SW. Intellectual impairment in children exposed to polychlorinated biphenyls in utero. *N Engl J Med*. 1996;335:783–789.

15. Berkowitz GS, Wetmur JG, Birman-Deych E, et al. In utero pesticide exposure, maternal paraoxonase activity, and head circumference. *Environ Health Perspect*. 2004;112:388–391.

16. Rauh V, Arunajadai S, Horton M, et al. 7-year neurodevelopmental scores and prenatal exposure to chlorpyrifos, a common agricultural pesticide. *Environ Health Perspect*. 2011;119:1196–1201.

17. Hertz-Picciotto I, Sass JB, Engel S, et al. Organophosphate exposures during pregnancy and child neurodevelopment: Recommendations for essential policy reforms. *PLoS Med*. 2018;15(10):e1002671.

18. Bouchard MF, Chevrier J, Harley KG, et al. Prenatal exposure to organophosphate pesticides and IQ in 7-year old children. *Environ Health Perspect*. 2011;119:1189–1195.

19. Gunier RB, Bradman A, Harley KG, et al. Prenatal residential proximity to agricultural pesticide use and IQ in 7-year-old children. *Environ Health Perspect.* 2017;125(5):057002.

20. Rauh VA, Garfinkel R, Perera FP, et al. Impact of prenatal chlorpyrifos exposure on neurodevelopment in the first 3 years of life among inner-city children. *Pediatrics.* 2006;118:e1845–1859.

21. Swan SH, Sathyanarayana S, Barrett ES, et al.; TIDES Study Team. First trimester phthalate exposure and anogenital distance in newborns. *Hum Reprod.* 2015;30(4):963–972.

22. Parra MD, Mendiola J, Jørgensen N, et al. Anogenital distance and reproductive parameters in young men. *Andrologia.* 2016;48(1):3–10.

23. Engel SM, Miodovnik A, Canfield RL, et al. Prenatal phthalate exposure is associated with childhood behavior and executive functioning. *Environ Health Perspect.* 2010;118:565–571.

24. Evans SF, Raymond S, Sethuram S, et al. Associations between prenatal phthalate exposure and sex-typed play behavior in preschool age boys and girls. *Environ Res.* 2021;192:110264.

25. Braun JM. Early-life exposure to EDCs: role in childhood obesity and neurodevelopment. *Nat Rev Endocrinol.* 2017;13(3):161–173.

26. Wasserman GA, Liu X, Parvez F, et al. Water arsenic exposure and intellectual function in 6-year-old children in Araihazar, Bangladesh. *Environ Health Perspect.* 2007;115:285–289.

27. Wasserman GA, Liu X, Parvez F, et al. Water manganese exposure and children's intellectual function in Araihazar, Bangladesh. *Environ Health Perspect.* 2006;114:124–129.

28. Rosner D, Markowitz G. A "gift of God"?: The public health controversy over leaded gasoline during the 1920s. *Am J Public Health.* 1985;75(4):344–352.

29. Needleman HL. The removal of lead from gasoline: historical and personal reflections. *Environ Res.* 2000;84(1):20–35.

30. Gilbert SG, Weiss B. A rationale for lowering the blood lead action level from 10 to 2 microg/dL. *Neurotoxicology.* 2006;27(5):693–701.

31. Needleman HL, Schell A, Bellinger D, et al. The long-term effects of exposure to low doses of lead in childhood: an 11-year follow-up report. *N Engl J Med.* 1990;322(2):83–88.

32. Gilfillan SC. Lead poisoning and the fall of Rome. *J Occup Med.* 1965;7:53–60.

33. Grandjean P, Landrigan PJ. Developmental neurotoxicity of industrial chemicals: a silent pandemic. *Lancet* 2006;368:2167–2178.

34. Weiss B. Food additives and environmental chemicals as sources of childhood behavior disorders. *J Am Acad Child Psychiatry.* 1982;21:144–152.

35. Whyatt RM, Rauh V, Barr DB, et al. Prenatal insecticide exposures and birth weight and length among an urban minority cohort. *Environ Health Perspect.* 2004;112:1125–1132.

36. Graziano JH, van Geen A. Reducing arsenic exposure from drinking water: different settings call for different approaches. *Environ Health Perspect.* 2005;113:A360–361.

37. Rajagopalan S, Landrigan PJ. Pollution and the heart. *N Engl J Med.* 2021;385(20):1881–1892.

38. US Environmental Protection Agency. Benefits and costs of the Clean Air Act from 1990 to 2020. 2011. Available at: https://www.epa.gov/sites/default/files/2015-07/documents/fullreport_rev_a.pdf

39. Rutstein DD, Berenberg W, Chalmers TC, et al. Measuring the quality of medical care: a clinical method. *N Engl J Med.* 1976;294:582–588.

40. Miller RW. How environmental hazards in childhood have been discovered: carcinogens, teratogens, neurotoxicants, and others. *Pediatrics.* 2004;113:945–951.

4

The Changing Global Environment and Children's Health

Perry E. Sheffield, Lucy Schultz, and Kristie L. Ebi

The global environment is changing on a massive, rapid, and historically unprecedented scale. Extraordinary growth in human populations around the world—from almost 8 billion persons today to more than 9 billion anticipated by 2050—is one of the principal drivers of this change (Figure 4.1).[1] Per capita consumption has also increased and varies widely based on access and lifestyle differences. Humans extract 60 billion tons of Earth's resources each year.[2]

The ways in which developing and developed nations, at times irrespective of a country's population growth rate, are using natural resources, in turn, affects the global environmental processes that regulate climate, air quality, water availability, and crop yields. These environmental processes are fundamental support systems for society, culture, and human life on Earth. Nature's contribution to quality of life varies based on social and geographic group, exacerbating the disadvantages of some groups and the power of others.[2]

Changes to global biological, ecological, and chemical cycles have impacts on regional and local environments. These regional and local changes can directly affect human health and can especially affect the health and well-being of children. Three factors determine the health risks of global environmental change: (a) the hazard itself, such as climate change, increasing the frequency and intensity of extreme weather events or biodiversity loss; (b) the populations and regions exposed to the hazard; and (c) the underlying vulnerability of the exposed populations, where vulnerability is defined as their propensity or predisposition to be adversely affected and the capacity of health systems to prepare for changing disease patterns.[3] A wide range of factors, including anthropogenic environmental change, socioeconomic development, structural barriers faced by the affected populations, and global trade influence hazards, exposure, and vulnerability. For example, demand for palm oil, a vegetable oil used in processed foods and cosmetics, has driven slash-and-burn forest clearing in tropical developing countries. This practice has led vulnerable neighboring populations to experience increased health risks, such as respiratory disease, cardiovascular disease, neurodevelopmental impacts, and mortality among children.[4] This chapter provides an overview of the myriad connections discovered in recent decades between the global environment and the health and well-being of human populations. Actions that promote environmental conservation and sustainability can benefit children's health by anticipating, avoiding, and reducing the magnitude and extent of global environmental change. The three major areas of global change covered in this chapter are (a) biodiversity loss; (b) deforestation,

Perry E. Sheffield, Lucy Schultz, and Kristie L. Ebi, *The Changing Global Environment and Children's Health*, Second Edition. In: *Textbook of Children's Environmental Health*. Edited by: Ruth A. Etzel and Philip J. Landrigan, Oxford University Press.
© Oxford University Press 2024. DOI: 10.1093/oso/9780197662526.003.0004

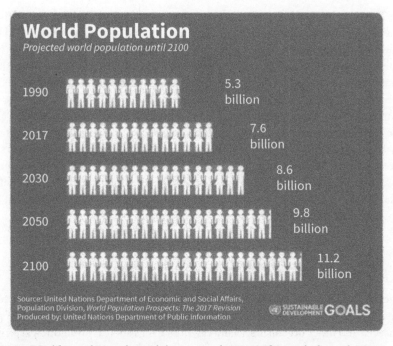

Figure 4.1. World population through history and projected growth through 2100

Source: World population prospects—2017 revision: Global population | multimedia library—United Nations department of economic and social affairs. United Nations. June 21, 2017. https://www.un.org/en/desa/world-population-prospects-2017-revision-global-population.

desertification, and ocean degradation; and (c) climate change with resulting sea level rise. These global changes have both direct and indirect consequences for children's health with vastly higher tolls on Black and brown and children and those from lower income communities.

Scope and Nature of the Problem

Loss of Biodiversity

Biodiversity, shorthand for "biological diversity," is a measure of the concentration and the degree of variety among the forms of life in individual ecosystems and on planet Earth.[2] An ecosystem such as the Spiny Forest of southern Madagascar, the giant sequoia ecosystem of the western United States, or the coral reefs in many of the world's oceans is a complex, interdependent unit that functions as a whole. An ecosystem typically includes not only plants and animals but also microorganisms and the nonliving environment. Biodiversity is a measure of the health of an ecosystem.

Since life began on Earth, five major mass extinctions have occurred—defined as the dying out of an abnormally large number of species within a short period of time. Each led to large and sudden drops in biodiversity. These extinctions were triggered by major global disruptions such as wide-scale Siberian volcanic eruptions 250 million years ago and an asteroid impact in the Gulf of Mexico 65 million years ago.[5]

The Earth is currently experiencing another period of mass extinction driven by human activity. Approximately 25% of plant and animal species are threatened by extinction.[2] Biodiversity loss, locally and globally, is the consequence of this current extinction. Because individual species can play essential roles in an ecosystem, extinction of specific species can impact the functioning of entire ecosystems. Loss of biodiversity is an important component of the changing global environment.

Contributors to the current loss of biodiversity include habitat destruction, climate change (discussed later in this chapter), invasive species, overexploitation (as is occurring with the majority of the world's fisheries), and nutrient pollution (as occurs when fertilizer runoff from agricultural fields causes overgrowth of phytoplankton in nearby waterways or oxygen depletion in river deltas). These factors sometimes affect only individual species, but more commonly they disrupt the food sources or the habitat necessary to support entire ecosystems.

Biodiversity Loss and Drug Discovery

Biodiversity loss impacts medical science and public health by reducing opportunities for the discovery of new drugs and diagnostics that are the natural products of plants, fish, insects, and other organisms. Many essential drugs such as antibiotics, aspirin, and cancer medications were found in nature.[2] Two examples of past discoveries that fundamentally changed our ability to treat childhood illnesses are

- The discovery and isolation post World War II of vinca alkaloid compounds from the Madagascar rosy periwinkle flower (*Catharanthus roseus*, formerly *Vinca rosea*). These unique molecules are the source of vincristine and vinblastine. In combination with other drugs, the vinca alkaloids transformed the treatment of acute childhood leukemia and turned it from a fatal to an often-curable disease.
- The discovery in cone snails of unique peptide toxins that could be transformed into new classes of nonopioid, nonaddictive pain medications. These molecules could prevent in utero exposure to opiates and thus curb the epidemic of opiate addiction among newborn babies, a problem that increased threefold in frequency over the past decades.[6] There are about 700 species of cone snails, each producing up to 200 distinct peptide toxins. However, many species of cone snails are threatened by the destruction of the tropical oceans and coral reefs that are their primary habitats.[7] If these species become extinct, their molecules will be lost forever.

Biodiversity Loss and Human Health

Loss of biodiversity can contribute to the emergence of new or spread of existing infectious diseases, food insecurity, loss of potential for a balanced diet (particularly for some Indigenous and traditional communities), loss of the flood protection provided by ecosystems such as mangroves and coral reefs, and loss of ecotourism, one of the fastest growing areas of tourism globally and an important source of economic livelihood in many low- and middle-income countries.

The concept of biodiversity first gained wide recognition at the 1992 Earth Summit in Rio de Janeiro. In the past 10–15 years, an understanding of biodiversity and its role in quality of life has grown significantly due to advances in modeling and informatics from remote observation, but global policy has not yet caught up to assure human population protections.[2]

Deforestation, Desertification, and Ocean Degradation

Human activity has degraded ecosystems through deforestation, desertification, and ocean degradation. Each of these contributes to health risks through loss of biodiversity, increasing incidence of droughts and floods, and heightening food insecurity.

Deforestation is a reduction in the amount of land covered by forests. From 2015 to 2020, 10 million hectares of forest were cleared per year, most often due to agricultural expansion.[8] Deforestation occurs unequally across regions, with forest area stabilizing in high-income nations and decreasing by 30% in low-income nations since 1990.[2] More than 90% of deforestation occurs in the tropics, with Africa experiencing the largest amount of deforestation worldwide.[9] Globally, replanting efforts have led to tree plantations—often pine and *Eucalyptus*,[8] though these do not match natural forests in biodiversity[10] and carbon sequestration capacity.[11]

Forests provide numerous benefits to local and distant human populations, including specific products such as timber. Forests are home to half of species on land[9] and provide broader benefits such as climate regulation, carbon sequestration, and flood control. Trees have microclimate effects via shading and surface cooling and provide 75% of the world's fresh water.[9] Thus, deforestation can adversely affect human health via economic impacts and loss of the broader supports that forests provide.[2]

Desertification is defined as the degradation of drylands in arid, semiarid, and dry subhumid areas resulting from various factors such as climatic changes and human activities.[12] Desertification is the process by which land that was already moderately dry becomes even drier and less usable. One-quarter of the world's land and one-fifth of the population experience desertification,[2] with Asia and Africa experiencing the majority.[12] Unlike naturally occurring deserts that can support ecosystems, desertification results in land that is bare soil without vegetation, thus limiting biological and economic potential.[2]

Ocean degradation refers to overexploitation of marine resources such as fish or other organisms; destruction and pollution of coastal areas, specifically mangrove ecosystems; and changes to the larger ocean through such processes as ocean acidification, which is driven by increasing global emissions of carbon dioxide. These factors have led to the loss of a staggering percentage of the world's ocean and coastal habitats. Since 1870, half of all coral reefs have been lost and continue to decrease by 4% each decade. One-third of the world's fish stocks are currently overexploited, and mangrove populations continue to decline,[2] being cleared for wood, development of coastal areas, or aquaculture.

The health of ecosystems is affected by desertification, deforestation, and ocean degradation, thus affecting human health as well. Change in the kinds and relative proportions of insect vectors is a consequence of ecosystem degradation. As biodiversity dilutes vector populations, reducing diversity can concentrate disease-carrying vectors.[2] For example, mosquito species diversity declines following deforestation; the species of mosquitoes that increase in number are those effective at transmitting malaria, a major contributor to childhood morbidity and mortality. By disrupting the complexity and interactions of ecosystems, land degradation practices drive the risk of disease and future pandemics.[13] In addition, land clearing practices like deforestation are associated with decreased microbial diversity within the human body that plays an important role in the human immune system. Loss of microbiota diversity is associated with a rise in in-flammatory bowel disease, asthma, and other autoimmune diseases.[2]

Deforestation, desertification, and ocean degradation also play a role in the incidence of droughts and flooding. Forests affect local climate by increasing regional rainfall due to evapotranspiration from the forest.[9] Deforestation can thus lead to less rainfall for nearby agricultural crops and to periods of decreased availability of fresh water and prolonged drought. Since 1990, the number of areas affected by extreme droughts has consistently increased.[14] Drier land with less vegetative cover due to desertification can compound the harm of droughts, leading to worsening dust storms that increase respiratory and eye problems. (See Chapter 25 on air pollution and Chapter 48 on asthma and allergy.) Because forests have the capacity to absorb and store large quantities of rainfall, deforestation and desertification increase surface water runoff and thus the risk of flooding in downstream areas. The degradation of ocean ecosystems like mangroves reduces coastal area protection from flooding.[2] Reduction in crop yields from currently arable land due to deforestation and desertification, along with a collapse of global fisheries, could lead to greater food insecurity in areas of the world already contending with significant child malnutrition (discussed more later in this chapter).

Climate Change

Burning of fossil fuels and deforestation release greenhouse gases such as carbon dioxide and methane to the atmosphere. Greenhouse gas emissions have doubled since 1980.[2] These gases are heat-trapping and therefore increase the level of energy in the climate system. This increased energy has multiple effects, including increasing global surface temperatures; changing patterns of precipitation (rain as well as snowfall); increased frequency, intensity, duration, and spatial extent of extreme weather and climate events; and rising sea level. Rising sea level is the consequence of ocean expansion and melting ice. Climate change is intricately related to each of the global environmental change processes discussed in this chapter.

Climate change produces multiple health effects. More direct effects include disease or injury directly associated with extreme weather and climate events, such as dehydration from extreme heat exposure, fear and anxiety due to extreme weather events, or drowning secondary to flooding following heavy precipitation. Injuries, illnesses, and deaths result from changes to ecosystems, air quality, and other factors that were themselves altered by climate change. Malnutrition results from reduced crop yields that are the consequence of changing precipitation or increased incidence of vector-borne disease resulting from changes in the seasonality and geographic range of the disease vector and other transmission dynamics.[15]

Children are disproportionately impacted by a changing climate. In addition to greater impacts from all the previously noted exposures related to global environmental degradation, children born in 2020 are projected to endure two to seven times more extreme weather events during their lifetimes than people born in 1960.[16] This will mean prolonged exposure to environmental hazards, the effects of which will accumulate across their lifetimes. Children, particularly those from low- and middle-income countries, are and will continue to be the most affected by climate change. Children are more likely to experience an adverse health outcome after an extreme weather event. Undernutrition, diarrheal diseases, and malaria, which have increased in incidence with climate change, are leading causes of mortality and morbidity for young children. Treating adverse health outcomes due to climate change is estimated to cost the United States billions of dollars each year. Assuming a constant world except for higher temperatures, by 2030, the global cost of treating the additional cases of malnutrition, diarrheal disease, and malaria could amount to between US$5 billion and 16 billion per year, or almost as much as the current annual US overseas development assistance for health.[17]

The effects of climate change on child health also highlight global injustices and systemic barriers and oppression. While wealthier nations have been able to adapt to global environmental change, less-developed nations (and lower income communities in wealthy nations) lack the resources needed to prepare.[14] Within the United States, 94% percent of urban neighborhoods historically subject to housing discrimination, known as "redlining," have a surface temperature up to 7°C higher than neighboring non-redlined areas.[18] In addition, an unjust discrepancy exists between contributions to and effects of climate change: the majority of the climate change–related impacts on children occur in Africa, yet the sources of greenhouse gases are principally the countries of North America, Western Europe, and Asia, as illustrated in Figure 4.2.[19]

Numerous additional climate-sensitive health outcomes affect children's health.[20] These include respiratory illnesses from changing patterns of air pollution (see Chapter 25 on air pollution and Chapter 48 on asthma and allergy) as more than 90% percent of children worldwide are exposed to levels of fine particulate matter higher than the World Health Organization (WHO) recommendation. Air pollution and fine particulate matter due to the burning of fossil fuels is the largest global environmental risk factor for premature death, contributing to 7 million deaths worldwide.[21] High temperatures due to climate change have also impacted aeroallergens such as pollen and mold, correlating with an increase in allergic rhinitis.[22]

Vector-borne infectious disease such as dengue fever and chikungunya, and water borne illnesses pose another risk to children due to climate change (see Chapters 28 and

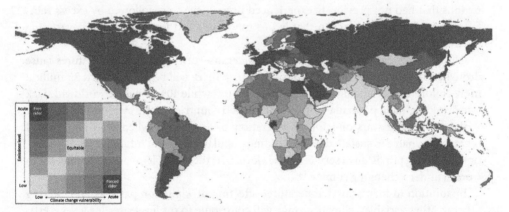

Figure 4.2. Global inequity in contribution to and burden of impact from climate change. Map depicts 2030 projections. Countries with highest emissions and lowest climate change vulnerability appear as dark red ("free riders") and those with lowest emissions and highest vulnerability appear as dark green ("forced riders"). Countries in yellow have GHG emissions concomitant with their vulnerability quintile. Countries with insufficient data appear as grey.

Source: Althor, G., Watson, J. & Fuller, R. Global mismatch between greenhouse gas emissions and the burden of climate change. Sci Rep 6, 20281 (2016). https://doi.org/10.1038/srep20281

29 on water pollution and sanitation and hygiene), as 9 of the 10 highest years of dengue fever transmission occurred since 2000. Children experience the most severe consequences of dengue fever and are more susceptible to diarrheal diseases.[21]

With record high temperatures in 2020,[14] children are experiencing more frequent heat waves than ever before. The impact of heat exposure is of particular concern in an increasingly urbanized world (see Chapter 18 for other health effects of the built environment). Young children are particularly susceptible to electrolyte imbalance, fever, respiratory disease, and kidney disease due to extreme heat events.[21] During fetal development, higher temperatures and heat waves correlate to increase of adverse pregnancy outcomes, such as stillbirth and preterm births.[23] Heat waves also have an economic impact on families due to a loss of work hours.[21]

Extreme storms or population displacement because of sea level rise, increased conflict, or expansion of nonarable land have led to more incidences of posttraumatic stress disorder (PTSD), depression, anxiety, and substance abuse among children (see Chapter 64 on natural disasters). The stress of climate change begins during fetal development, impacting behavior and motor and intellectual function.[24] Chronic worry and anxiety in children lead to alter cortisol signaling that may have adverse impacts on organ systems and predispose children to disease later in life.[25] Social stressors also make important contributions to a wide range of non–mental health-related illnesses, such as asthma. (See Chapter 5 for more on the social and behavioral influence on child health and development.)

Global change processes are increasing the probability that these and other health impacts may occur jointly. For example, hotter and drier climates have led to more wildfire exposures, which increased in 72% of countries since 2001–2004.[14] Wildfire smoke has both local as well as distant air quality effects. Wildfires lead not only to greater air pollution exposures but also to secondary impacts through social stress from population displacement or lifestyle disruption. Another example of concomitant hazards due to climate change occurred in British Columbia in 2021. After a period of extreme drought and heat waves, the area faced an unprecedented season of wildfires occurring in coastal regions that had not previously experienced fires. These were followed by excess rainfall on recently burned terrain that led to mudslides and flooding, again devastating thearea.[26]

Climate change is exacerbating food and water insecurity as rising temperatures cause drought and the reduction of crop production.[14] There will be an estimated 25 million more children experiencing malnutrition by 2050,[25] while 40% of people worldwide lack safe drinking water, particularly in less-developed countries such as within sub-Saharan Africa.[2] Resource wars and forced migration have resulted when land and coastal regions are unable to sustain the surrounding populations due to inhibited food production, declining provision of services, or sea level rise: these problems are also projected to worsen under a changing climate.[14]

In addition to increasing temperatures affecting precipitation patterns and altering other weather variables, climate change will contribute to sea level rise. Sea level is estimated to reach up to 2 meters above current levels within the next 80 years under high greenhouse gas emissions.[14,27] This change, coupled with ongoing coastal development, is expected to increase the size of the population at risk of storm surge, flooding, erosion, and infrastructure impacts, all of which are expected to be particularly severe in small island nations.[3] Three-quarters of the world's 20 megacities (defined as cities with greater than 10 million inhabitants) will also potentially be impacted by rising sea levels and increasing storm surge. Saltwater intrusion from changes in sea level or storm surge could affect water quality in small island nations and in these global megacities.[28]

Prevention

Countries have fallen short in responding to the growing health risks associated with climate change. In 2020, 63% of nations lacked an emergency framework necessary to prepare for the COVID-19 pandemic and other climate-related health emergencies. Although many public health agencies and organizations have impressive records of controlling the burden of health outcomes associated with global environmental changes, current and planned programs and activities need to be modified and expanded to address the additional risks posed by ongoing changes in biodiversity, deforestation, desertification, ocean degradation, and climate change. Public health organizations and institutions need to incorporate approaches that address risks that are changing over time and space while simultaneously integrating national- and community-level perspectives. Taking an iterative management approach—one that includes constant monitoring and reevaluation, a systems-based approach, models of future impacts, and significant stakeholder engagement—can help reduce current levels of health burdens as diseases change their geographic range and incidence in response to environmental conditions.[29,30]

Effectively addressing the health risks of global change in the short term focuses on reducing underlying vulnerabilities and root causes. Many of the health risks of global change for children are intimately connected with poverty, including lack of access to safe water, and lack of sanitation, nutritious food, and healthcare. Achieving the United Nations' Sustainable Development Goals (SDGs)[31] would increase the resilience of children to risks today and in the future. In some situations, reducing the number of people moving into harm's way, as in coastal areas subject to storm surges, may be the most effective approach to reducing risks.

These complex environmental changes cannot be solved in the short term but only made better or worse. Short-term prevention activities and strengthening health systems can address current deficits in managing the risks. In the longer term, evidence-based programs of prevention are needed that address the causes of global change and the underlying drivers of vulnerability.

Short-Term Strategies

Short-term prevention strategies aim to increase resilience to the health risks of global change. The majority of health impacts expected to result from global environmental change are exacerbations of existing health problems, such as worsening of child malnutrition or malaria, rather than entirely new problems. Public health departments and healthcare delivery systems typically have some of the programs needed to combat these risks. For example, disease surveillance programs can provide important first steps in building an understanding of trends and human–environment interactions.

Projected local changes in the environment and other drivers of health outcomes need to inform the implementation of new or the modification of current programs to enhance individual and community preparedness. Such preparation would mean, for example, that clinicians in areas at increasing risk of dengue fever include dengue in their differential diagnosis, be prepared to treat affected patients quickly and appropriately, and participate in public health surveillance of reportable diseases. Similarly, emergency management offices in low-lying coastal zones need to take seriously the threat of a severe storm surge from increasingly violent storms and prepare appropriate contingency plans. It is important that programs incorporate strategies for regular monitoring, evaluation, and modification to evaluate their current and likely future effectiveness under a changed climate. Awareness of global change is a critical first step in the process.

To further improve the ability of public health systems to protect children from global and local environmental change, child-specific vulnerabilities need to be central to large-scale adaptation efforts. For example, growing evidence suggests that children of many ages, not only infants, are impacted by heat and through many disease pathways, not only heat-related illness.[32] Short-term strategies should aim to identify and target such modifiable risk factors for vulnerable subgroups.

Roles and Responsibilities for Adaptation

Effective adaptation to a changing global climate requires action at individual, community, and national and multinational scales. The first step is for all decision-makers to identify, for their community and region, their vulnerabilities to climate change impacts and the shorter-term adaptation options that could increase resilience to current and projected impacts. Examples of possible actions at various scales include

- *Individual.* Individuals can take action by increasing awareness about global change issues and reducing their ecological footprint (the land area used to produce the resources consumed and absorb the waste produced) and their carbon footprint (the total greenhouse gases created by an individual's lifestyle, measured in carbon dioxide equivalents). Some reduction measures include (a) eating locally to reduce the energy required for the production and delivery of foodstuffs to one's plate and (b) using human-powered transportation that has short-term health benefits by increasing physical activity and global longer-term benefits.
- *Community.* Many risks are best managed locally, such as heat wave early warning systems. Much can be learned from experiences with managing changes in climate variability that have recurred within a community's collective memory. Furthermore, people can encourage practices that make individual actions more feasible, such as policies to increase the affordability of local food and human-powered transport to decrease the barriers for adoption of such changes by others. It is also essential to educate the public and decision-makers about the interconnectedness of environmentally sustainable practices and protection of human health. Only with wider-scale appreciation of such connections can the political and social will be developed to create wide-scale change. For example, even in the absence of sufficient national action to reduce carbon pollution, many cities are making strides to increase sustainability and reduce the health and other risks of global change.[33]
- *National and multinational.* Through laws, regulations, and conventions, national governments provide the enabling conditions for implementing national adaptation strategies that facilitate subnational governments to take actions to increase resilience to global changes. National governments often fund the research and

development that underpin adaptation initiatives. Furthermore, national governments provide the policies that encourage sustainable development, including encouraging energy conservation, sustainable agricultural practices, and technology transfer to decrease current and future health risks. Collaboration across nations, as outlined in the Paris Agreement, is essential, but, to meet the goal of greenhouse gas emission reduction by half within a decade, the current pace of a transition to a low carbon economy must significantly increase.[14]

Adaptation can begin today at all levels: individual to regional to multinational. Sufficient knowledge exists of the current and projected risks of global environmental changes and of effective and efficient options to initiate risk-reduction actions now.

Long-Term Strategies

Prevention strategies need to consider how ongoing environmental and societal changes could alter the magnitude and pattern of climate-sensitive health outcomes and sustainable development pathways. As acknowledged by the United Nations' SDGs, lasting advances in global health will be achieved only by ensuring long-term environmental sustainability through adaptation and mitigation.[31] Importantly, many mitigation policies also have many short-term health benefits (called *co-benefits*) whose economic value meets or exceeds the cost of these policies.[34] Benefits for children include improved lung development from better air quality and increased health from more physical activity from pedestrian-powered transport.

Long-term strategies require a sustained commitment and an increased appreciation by decision-makers of global interconnectedness between humans and the environment and between different human populations and nations, including how their interactions affect the health of children.

Notes

1. United Nations, Department of Economic and Social Affairs, Population Division (2017). World Population Prospects: The 2017 Revision, Key Findings and Advance Tables. Working Paper No. ESA/P/WP/248. Available at: https://esa.un.org/unpd/wpp/publications/files/wpp2017_keyfindings.pdf (Figure 2 on page 2).
2. Brondízio ES, Settele J, Díaz S, Ngo HT, eds. IPBES (2019): Global assessment report of the Intergovernmental Science-Policy Platform on Biodiversity and Ecosystem Services. Bonn: IPBES Secretariat; 2019.
3. Field CB, Barros V, Stocker TF, et al, eds. *Managing the Risks of Extreme Events and Disasters to Advance Climate Change Adaptation. A Special Report of Working Groups I and II of the Intergovernmental Panel on Climate Change.* Cambridge: Cambridge University Press; 2012.
4. Kadandale S, Marten R, Smith R. The palm oil industry and noncommunicable diseases. *Bull WHO.* 2018;97(2):118–28. doi:10.2471/blt.18.220434
5. Barnosky AD, Matzke N, Tomiya S, et al. Has the Earth's sixth mass extinction already arrived? *Nature.* 2011;471:51–7.
6. Patrick SW, Schumacher RE, Benneyworth BD, Krans EE, McAllister JM, Davis MM. Neonatal abstinence syndrome and associated health care expenditures: United States, 2000–2009. *JAMA.* 2012;307:1934–40.

7. Chivian E, Bernstein AS. *Sustaining Life: How Human Health Depends on Biodiversity.* New York: Oxford University Press; 2008.

8. Food and Agricultural Organization of the United Nations. Global Forest Resources Assessment 2020: Main report. Rome 2020. https://doi.org/10.4060/ca9825en

9. Pacheco P, Mo K, Dudley N, et al. Deforestation *Fronts: Drivers and Responses in a Changing World.* Gland, Switz: WWF; 2021.

10. Altamirano A, Miranda A, Aplin P, et al. Natural forests loss and tree plantations: large-scale tree cover loss differentiation in a threatened biodiversity hotspot. *Environ Res Lett.* 2020;15(12):124055. doi:10.1088/1748-9326/abca64

11. Lewis SL, Wheeler CE, Mitchard ETA, Koch A. Regenerate natural forests to store carbon. *Nature.* 2019;568:25–8.

12. Mirzabaev AJ. Wu J, Evans F, et al. Desertification. In: Shukla PR, Skea J, Calvo Buendia E, et al. eds., *Climate Change and Land: An IPCC Special Report on Climate Change, Desertification, Land Degradation, Sustainable Land Management, Food Security, and Greenhouse Gas Fluxes in Terrestrial Ecosystems.* Geneva, Switzerland:Intergovernmental Panel on Climate Change 2019.

13. Daszak P, Amuasi J, das Neves CG, et al.; IPBES. *Workshop Report on Biodiversity and Pandemics of the Intergovernmental Platform on Biodiversity and Ecosystem Services.* Bonn: IPBES Secretariat; 2020. doi:10.5281/zenodo.4147317

14. Romanello M, McGushin A, Di Napoli C, et al. The 2021 report of the Lancet Countdown on health and climate change: code red for a healthy future [published correction appears in *Lancet* 2021 Dec 11;398(10317):2148]. *Lancet.* 2021;398(10311):1619–62. doi:10.1016/S0140-6736(21)01787-6

15. IPCC. Summary for policymakers. In: Pörtner HO, Roberts DC, Poloczanska ES, et al. eds., *Climate Change 2022: Impacts, Adaptation, and Vulnerability.* Contribution of Working Group II to the Sixth Assessment Report of the Intergovernmental Panel on Climate Change Cambridge: Cambridge University Press; 2022.

16. Thiery W, Lange S, Rogelj J, et al. Intergenerational inequities in exposure to climate extremes. *Science.* 2021;374(6564):158–60. doi:10.1126/science.abi7339

17. Mock CNR, Kobusingye NO, Smith KR, eds. 2017. *Injury Prevention and Environmental Health. Disease Control Priorities.* 3rd ed. Volume 7. Washington, DC: World Bank; 2017. doi:10.1596/978-1-4648-0522-6

18. Hoffman JS, Shandas V, Pendleton N. The effects of historical housing policies on resident exposure to intra-urban heat: a study of 108 US urban areas. *Climate.* 2020;8(1):12. doi:10.3390/cli8010012

19. Althor G, Watson J, Fuller R. Global mismatch between greenhouse gas emissions and the burden of climate change. *Sci Rep.* 2016;6:20281. https://doi.org/10.1038/srep20281

20. Sheffield PE, Landrigan PJ. Global climate change and children's health: threats and strategies for prevention. *Environ Health Perspect.* 2011;119:291–8

21. Watts N, Amann M, Arnell N, et al. The 2019 report of The Lancet Countdown on health and climate change: ensuring that the health of a child born today is not defined by a changing climate. *Lancet.* 2019;394(10211):1836–78. doi:10.1016/S0140-6736(19)32596-6

22. Fuertes E, Butland BK, Ross Anderson H, et al. Childhood intermittent and persistent rhinitis prevalence and climate and vegetation: a global ecologic analysis. *Ann Allergy Asthma Immunol.* 2014;113(4):386–92.e9. doi:10.1016/j.anai.2014.06.021

23. Chersich MF, Pham MD, Areal A, et al. Associations between high temperatures in pregnancy and risk of preterm birth, low birth weight, and stillbirths: systematic review and meta-analysis. *BMJ.* 2020;371:m3811. doi:10.1136/bmj.m3811

24. Burke SEL, Sanson AV, Van Hoorn J. The psychological effects of climate change on children. *Curr Psychiatry Rep.* 2018;20(5):35. doi:10.1007/s11920-018-0896-9

25. McMichael AJ. Climate change and children: health risks of abatement inaction, health gains from action. *Children (Basel)*. 2014;1(2):99–106. doi:10.3390/children1020099

26. Palmer J. The one-two punch of fires and mudslides. *Nature*. 2022;601:184–6.

27. IPCC. Summary for policymakers. In: Masson-Delmotte VP, Zhai A, Pirani SL, et al., eds. *Climate Change 2021: The Physical Science Basis*. Contribution of Working Group I to the Sixth Assessment Report of the Intergovernmental Panel on Climate Change. Cambridge: Cambridge University Press; 2021.

28. Environmental Protection Agency. Climate change: international impacts and adaptation, 2012 Available at: http://www.epa.gov/climatechange/impacts-adaptation/international.html

29. Ebi K. Climate change and health risks: assessing and responding to them through "adaptive management." *Health Aff (Millwood)*. 2011;30:924–30.

30. Williams BK, Szaro RC, Shapiro CD. *Adaptive Management: The U.S. Department of the Interior Technical Guide*. Washington, DC: US Department of the Interior; 2009.

31. United Nations. The 17 goals of sustainable development. https://sdgs.un.org/goals

32. Bernstein AS, Sun S, Weinberger KR, Spangler KR, Sheffield PE, Wellenius GA. Warm season and emergency department visits to U.S. children's hospitals. *Environ Health Perspect*. 2015 Jan;130(1):17001. doi:10.1289/EHP8083. PMID: 35044241; PMCID: PMC8767980

33. ICLEI. Local governments for sustainability. 2012. Available at: http://www.iclei.org/our-members.html

34. Hess JJ, Ranadive N, Boyer C, et al. Guidelines for modeling and reporting health effects of climate change mitigation actions. *Environ Health Perspect*. 2020;128(11):115001. PMID: 33170741. https://doi.org/10.1289/EHP6745

5

Social and Behavioral Influences on Child Health and Development

Summer Sherburne Hawkins

Introduction

While medical care has long been considered the main determinant of population health, even in countries with universal healthcare, disparities persist.[1] The discourse has turned to consider the broader context in which individuals reside. How can children be in good health if their schools, communities, and local and state policies do not protect or support healthy behaviors and outcomes? Across the factors that determine population health, genetics, biology, and health behaviors are estimated to account for approximately only one-quarter.[2] The remaining determinants—the social, physical, and economic environments and medical care—are the major drivers of population health.

The social determinants of health, as defined by the US Centers for Disease Control and Prevention, are the "conditions in the environments where people are born, live, learn, work, play, worship, and age that affect a wide range of health, functioning, and quality-of-life outcomes and risks."[3] In essence, the social determinants impact population health by shaping where and how people live. The five domains of the social determinants of health are economic stability, education access and quality, healthcare access and quality, neighborhood and built environment, and social and community context.[3] Foreshadowing the discussion that follows, examples include available and affordable housing, neighborhood safety, access to green spaces, discrimination, and social capital.

While context matters, it is widely recognized that the unequal environments created by the social determinants contribute to health disparities. Data from the 2019–2020 National Survey of Children's Health revealed a gradient in children's parent-reported health by household income. Parents in households at less than 100% of the federal poverty level were four times more likely to report their children were in good, fair, or poor health (16.7%) than parents of children in households at 400+% (4.6%) (Figure 5.1). The varying social, physical, and economic exposures shape children's environments and subsequently influence their health behaviors and outcomes.[1] For example, children from low-income households and Black and Hispanic children are more likely to have asthma.[4] Because poor housing conditions are a major risk factor, this suggests that one solution may be to improve housing quality. However, poor housing conditions are concentrated in low-income communities and communities of color, which are also more likely to experience other environmental injustices.[5] Further upstream, racially segregated housing is a result of institutionally racist policies, such as redlining—a discriminatory practice of denying financial services to communities based on race, and the

Summer Sherburne Hawkins, *Social and Behavioral Influences on Child Health and Development*, Second Edition. In: *Textbook of Children's Environmental Health*. Edited by: Ruth A. Etzel and Philip J. Landrigan, Oxford University Press.
© Oxford University Press 2024. DOI: 10.1093/oso/9780197662526.003.0005

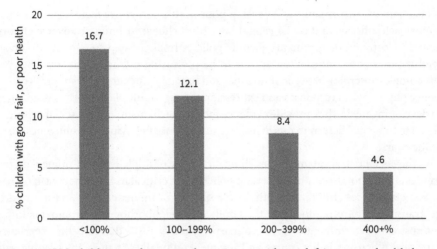

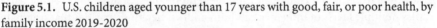

Figure 5.1. U.S. children aged younger than 17 years with good, fair, or poor health, by family income 2019-2020

Source: National Survey of Children's Health 2019-2020. Data query from the Child and Adolescent Health Measurement Initiative, Data Resource Center for Child and Adolescent Health. Available from: https://www.childhealthdata.org/browse/survey/results?q=8446&r=1&g=900

ramifications are still evident.[5] Interventions are needed across this cascade, including measures to improve housing conditions and clinically treating children; however, without addressing the fundamental causes of institutional racism and discriminatory housing policies, disparities in asthma will likely continue.[6]

The mechanisms for how the social determinants of health impact children are unequivocally complex. However, it is critical to consider how upstream determinants shape and influence population health both directly, such as through exposures, but also indirectly, by influencing health-related behaviors. It is well-established that early exposure to poverty is associated with adverse health behaviors and outcomes during childhood and across the life course.[6] Children living in more disadvantaged circumstances at home or in their communities have a higher risk of exposures, for example, to poorer nutrition, environmental pollutants, and violence.[5,7,8] Children's experiences of poverty also elicit stress responses as a result of inadequate resources as well as family conflict related to this instability. Elevated risks of poor health go beyond the physical environment as social exposures indirectly influence health behaviors and outcomes.[9] Exposure to family and neighborhood violence heightens children's stress responses, can increase psychological distress, raise blood pressure, and disturb sleep.[10-12] Furthermore, family and community exposure to risky health behaviors has also been shown to increase children's uptake of related behaviors through modeling and normalizing of behaviors that are negatively associated with long-term health.[13-15]

Whether a result of stress related to poverty or experiences of discrimination, there is increasing evidence of how social experiences get "under the skin" to affect health. Chronic exposure to these onslaughts has been shown to induce physiological changes through inflammatory responses.[6,9] There is also evidence demonstrating that socioeconomic and social stressors can cause epigenetic changes, which can contribute to the intergenerational transmission of stress that influences health and disease.[6,9] It is important to note that not all children who experience disadvantaged social, physical, and economic environments will develop poor health. Protective factors such as social support, self-efficacy, and social capital have been shown to mitigate adverse circumstances.[9] Upstream determinants such as policies implemented to support disadvantaged families and communities appear to be promising public health strategies.[16-18] This complex interplay between individual and broader exposures and supports can influence physiological responses and expression through multiple pathways. While these are developing fields, they provide evidence for the mechanisms of how social exposures related to the social determinants of health can influence behaviors and outcomes across the life course.

As the social determinants of health occur across multiple levels of influence, from families and schools to communities and policies, it is critical to assess how each of these contexts influences children's health. These spheres of influence also vary as children age, transitioning from a focus on the family and neighborhood environments to further upstream determinants.[19] A life course perspective posits that early life experiences shape health across the life course and intergenerationally. As such, intervening early and across multiple levels of influence can both address health disparities that originate in childhood and change the contexts in which children experience them over time.[19] Using a socioecological perspective, this chapter presents the social determinants of children's health across the levels of the family, community, and social conditions, schools, and policy.[1,20] Dr. Tom Frieden proposes in the health impact pyramid that addressing upstream determinants takes collective action and the most effort for structural change but will have the largest impact on population health.[21] Through the lens of the health impact pyramid, this chapter focuses on determinants with the potential for the largest effects by changing contexts.[21]

Family

The family environment shapes children's health through both the physical environment of where children live as well as exposures through the social environment. Housing itself is a critical social determinant. Green and colleagues outline four components of housing that influence children's health: (1) quality, specifically housing that is free of environmental exposures; (2) stability, including frequent moves, evictions, or homelessness; (3) affordability, defined as the housing cost burden; and (4) neighborhood, including outdoor green spaces, schools, job opportunities, and transportation.[5] Children from low-income communities and communities of color are more likely to experience low-quality housing, housing instability, and a high housing cost burden, and live in neighborhoods with fewer resources.[5] Housing solutions with the largest impact will target the upstream factors of policy development to increase access to safe and affordable housing and investment in neighborhoods rather than rely on individual-level solutions. For example, despite decreasing blood lead levels in children over recent decades, lead exposure still persists. Hauptman and colleagues found that more than half of children in their study had detectable blood lead levels, and 1.9% had elevated levels.[22] These children were more likely to be on Medicaid and live in communities with pre-1950 housing and greater poverty. Lead exposure, as an example of other housing-related exposures, intersects with the areas of housing quality, affordability, and the neighborhood, suggesting that solutions need to be multisectoral, with multiple stakeholder involvement.

Families that experience housing insecurity often face food insecurity. The components of quality, stability, affordability, and neighborhood are also applicable.[5] Families cycle in and out of food insecurity, often experiencing more shortages just prior to government assistance payments. The burdens of the higher costs of healthy foods and lower accessibility of fresh whole foods are disproportionately experienced by low-income families.[7] Laraia and colleagues posit that low-income families face additional stressors of living in poverty and the accompanying uncertainty, which translate into psychological and biological responses that influence diet quality.[7] Job insecurity and housing insecurity, also more likely to be experienced by low-income families, can influence diet quality through lower wages, financial instability, and time constraints in food preparation and the appropriate storage of foods.[7] Taken together, these findings suggest that policies across sectors which help provide stable housing, food, and employment will provide a broader range of support for children and families with wider-reaching effects on health.

The social environment of families in response to the social determinants indirectly influence children's health. Increasing evidence suggests that adverse childhood experiences (ACEs) early in life are associated with adverse physical and mental health problems across the life course.[23] ACEs can affect children directly through experiences of abuse and neglect or indirectly through exposure to parental conflict, financial problems, substance abuse, or mental illness.[23] Hughes and colleagues found that individuals who experienced four or more ACEs were at higher risk for noncommunicable and communicable diseases compared to those who experienced none, with the highest risk for sexual risk-taking, poor mental health, problematic alcohol use, drug use, and interpersonal and self-directed violence.[23]

Just as the social determinants of health shape the environmental context and engagement in health behaviors by parents and family members, the modeling of these behaviors has been associated with children's own health behaviors. Research has shown that family members' and friends' smoking increases uptake of youth smoking,[13] parents' physical activity is associated with children's physical activity,[14] and parental food habits and sharing family meals influences children's eating behaviors.[15] In sum, this body of evidence highlights the role of the family and the need for a life course perspective in addressing the social determinants to prevent the intergenerational transmission of poor health.

Community and Social Conditions

Children's family environments are inextricably linked with the community and social conditions in which they are embedded. This section focuses on those components of the built and social environments influenced by the social determinants that help shape children's experiences and health outcomes. It is necessary to recognize that the unequal exposure of low-income communities and communities of color to environmental hazards is a result of environmental injustices and unequal protections for these communities. This historical context is essential for understanding how the social determinants influence environmental exposures and contribute to disparities in children's health. For example, children who live in low-income communities and communities of color are more likely to be exposed to outdoor and indoor air pollution,[8] and there is evidence that in utero exposure to air pollution ($PM_{2.5}$) increases the risk of infants born with low birth weight and preterm as well as the risk of longer-term effects on children's respiratory, immune, neurological, and cardiovascular systems.[24]

While the scope of the built environment is broad, there is consensus about the importance of providing children with opportunities to be outside, access to green spaces and healthy whole foods, and safe neighborhoods.[7,25] A review of the built environment by Ortegon-Sanchez and colleagues found evidence that children's physical activity was positively associated with residential connectivity, density and walkability; accessibility and proximity to activity-promoting locations; availability of public spaces and parks; and safety.[25] Many of these same features were also associated with children's lower body mass index (BMI).[25] Conversely, increased sedentary time was associated with decreased walkability and street connectivity and concerns about safety. As noted previously, the location and availability of affordable and nutritious whole foods within communities influence children's diet quality and risk of food insecurity.[7] Considering the broader implications of the built environment and food access, these aspects not only influence health behaviors, but likely also contribute to disparities in noncommunicable diseases such as obesity, with the highest levels among children from low-income households.[26]

An accumulating body of research is demonstrating that the social environment within the broader community context impacts children's health. Systemic racism, which manifests as discriminatory policies and practices at the systems level, is evident in sectors within the community including housing, employment, schools and education, and healthcare. Cave and colleagues undertook a comprehensive review of the impact of interpersonal and systemic racism on children's health as reported directly by children themselves or their caregivers.[27] They found that children's exposure to racial discrimination was associated with poor mental health, including depression and low self-esteem; behavior problems such as delinquency; and substance use.[27] The risk of poor health varied by both the duration and timing of exposure to racial discrimination across childhood.

Additional stressors experienced by children are exposures to community violence and perceptions of safety. There is extensive evidence that community violence exposure increases children's psychological distress, including internalizing and externalizing behaviors.[11] A systematic review by Wright and colleagues found consistent associations between community violence exposure with elevated blood pressure and sleep problems in children,[12] but highlight that a limited range of health outcomes have been studied. Mayne and colleagues also found evidence that living in neighborhoods with high crime and violence or lower perceived safety was associated with greater perceived sleep problems in children.[10]

Not all children who experience adversity have poor health, and it is important to consider protective factors that allow children to thrive. For example, a review by Ozer and colleagues found that close family relationships reduced internalizing and externalizing behaviors among youth exposed to community violence, but few studies examined school- or community-level factors.[28] *Social capital* is the positive social networks and relationships that support social, economic, and health outcomes, and it has been identified as an important component of children's health. McPherson and colleagues considered the role of family social capital (often referred to as family characteristics) and community social capital in reducing children's engagement in risky healthy behaviors.[29] At the family level, close parent–child relationships were associated with decreased use of tobacco, alcohol, drugs and decreased engagement in risky sexual behaviors, while parental monitoring also reduced risky sexual behaviors.[29] At the community level, peer networks were associated with an increased risk of tobacco use. High-quality school environments were found to reduce substance use, and the social (rather than personal) aspects of religiosity were protective across health behaviors.[29] Taken together, this growing body of research demonstrates that children's health and related behaviors are influenced by both their location and related environmental exposures, attributes of the environment within their communities, and the social milieu.

School

The social determinants of health shape the school environment and influence children's academic, health-related, and social experiences. As many school districts across the US are funded by local property taxes, school quality varies, and, consequently, children's experiences and related health effects vary as well. The physical school environment, including school buildings, can affect child health. For example, child asthma is socially patterned, and environmental exposures are risk factors for the development of asthma.[4] A review by Esty and Phipatanakul found that the indoor school environment harbors allergens, molds, pollutants, and endotoxins that have been associated with increased pediatric morbidities.[30] As the levels of influence go further upstream, it is also important to consider the potential compounding effects of exposure to multiple social determinants, such as housing and school conditions, on children's health.

How children travel to school, opportunities for physical activity, and food policies have all been shown to influence children's health and health behaviors. Studies have found that active travel to school was associated with street connectivity, residential density, walkability (e.g., sidewalks), and safety.[25] A review by Neil-Sztramko and colleagues found that within-school physical activity, physical education programs, and before- and after-school programs were associated with increases in children's physical activity and fitness but had little effect on reducing BMI.[17] Food policies also shape food access and availability in schools, which subsequently influence children's food intake, but there is limited evidence for their impact on overweight and obesity or BMI.[18] In addition, policies which increase provisions of fruits and vegetables in schools have been shown to increase children's consumption, while policies that reduce competitive foods and beverages reduce consumption of sugar-sweetened beverages and unhealthy snacks.[18] School meal standards have also been found to increase fruit intake and reduce fat and sodium. Over the past two decades, federal and local school food policy changes have increased access to healthy foods. A review by Mansfield and Savaiano evaluated school wellness policies and the federal Healthy, Hunger-Free Kids Act and found that increasing access to healthy foods during school lunch improved dietary intake, including consumption of fruits and vegetables, but few schools met the target nutrition standards.[31] It is also critical to recognize that these changes in children's physical activity and food intake only occur during school hours and throughout the academic year, leaving some children vulnerable to differential access based on their family circumstances and community.

The healthcare system is a social determinant related to health coverage, provider availability, and quality of care, which all influence children's health. However, more focus is needed on the roles of school-based services in supporting children's health, particularly children without conventional and consistent sources of medical care. Knopf and colleagues found that most studies have evaluated in-school clinics serving low-income and racial/ethnic minority high school students.[32] The presence and use of school-based health centers were associated with improvements in health-related outcomes, including vaccination, asthma morbidity, contraceptive use among females, illegal drug use, and alcohol use. Sanchez and colleagues reviewed school-based mental health services for elementary-school children.[33] The largest effects were seen for targeted interventions, with some evidence for reductions in mental health problems and externalizing behaviors.[33] Ultimately, if school-based health and mental health services are based in the highest-risk schools, then the evidence suggests that these services could help reduce health disparities.

Although many features of the social environment overlap with those experienced by children in the community, bullying and cyberbullying are often unique to school settings. While traditional bullying typically includes physical contact, verbal harassment, and rumor spreading,[34] cyberbullying consists of intentional acts or behaviors using electronic forms of contact.[35] While both types are often conducted repeatedly and over time, the degree of anonymity online often causes the intensity of cyberbullying to be greater than for traditional bullying. Moore and colleagues found that bullying victimization was strongly associated with mental health problems, including anxiety, depression, self-injury, and suicidal ideation and attempts, as well as smoking and illicit drug use.[34] Hamm and colleagues focused on the effects of cyberbullying victimization on mental health and found consistent evidence for a positive association between cyberbullying and depression.[35] With the increasing availability of social media and emergence of new platforms, more research is needed on the effects of the digital environment on children's mental and physical health and health behaviors in this ever-changing field.

Policy

The policy environment influences the most upstream determinants of children's health. As such, policies and programs that intervene at the structural level and address the fundamental causes are likely to have the largest effects on population health.[9,21] For example, Thornton and colleagues identified promising interventions and policies addressing the social determinants that have been shown to improve the lives of those at the highest risk, such as high-quality preschool education and income supplements.[36] This section reviews examples of policies and programs that govern many of the downstream determinants that were raised previously.

Historical housing policies, such as redlining, continue to shape communities and children's health.[5] For example, Nardone and colleagues found that age-adjusted rates for asthma-related emergency department visits, including children's visits, were more than two times higher in historically redlined census tracts than in those not redlined in the 1930s.[37] The public health community needs to ask how many more generations will continue to experience the ramifications of discriminatory policies? The solution is not to only improve housing, but also to address the upstream policies and practices that created residential segregation. In addition, Vasquez-Vera and colleagues found that the threat of eviction increased experiences of negative mental (depression, anxiety, psychological distress, suicides) and physical (poor self-reported health, high blood pressure, child maltreatment) health outcomes.[38] These findings suggest that interventions in the housing sector and with financial institutions are needed to protect housing security and reduce the threat of evictions, particularly for low-income communities and communities of color who are at the highest risk.[5]

Federal and local governments have created policies and programs to address the social determinants of health, which may directly affect children's health or indirectly affect them by changing the contexts experienced by families and communities. Federal food assistance programs such as the Special Supplemental Nutrition Program for Women, Infants, and Children, Supplemental Nutrition Assistance Program, and the National Food Lunch program have all been shown to reduce children's food insecurity.[39] However, there continue to be households with children eligible for benefits that do not register and households who participate in these programs but still experience food insecurity.[39] Recognizing that the costs of healthy whole foods is a known barrier for diet quality, Afshin and colleagues examined the effects of changes in food prices on dietary intake in adults and children.[40] They found that decreases in price increased consumption of fruits and vegetables without significant effects on more healthful beverages. In contract, price increases, through a tax, decreased consumption of sugar-sweetened beverages and fast food. There was some evidence that price decreases reduced BMI, but no evidence for price increases on sugar-sweetened beverages or fast-food consumption. Taxes versus subsidies are important policy tools that localities can implement to influence dietary intake.[40,41] Fiscal policies also are some of the most effective means by which to influence substance use. A substantial body of evidence has demonstrated that cigarette and alcohol taxes reduce consumption, with the largest effects among low-income populations who are the most price sensitive.[41] My own research has highlighted the importance of assessing the downstream effects of fiscal policies on children's health, with evidence that cigarette tax increases decrease smoking among adolescents[42] and pregnant women,[43,44] with subsequent improvements in birth outcomes.[44,45]

Although the Affordable Care Act (ACA) was intended to benefit adults through increased health insurance coverage, there is evidence of positive spillover effects on children. Prior to the ACA, most low-income children were eligible for health insurance through Medicaid or the Children's Health Insurance Program, but not all were registered. An innovative study by Hudson and Moriya found that Medicaid expansion also increased the number of eligible children enrolled in public insurance.[46] While subsequent health effects were not examined, increasing insurance coverage has important implications for previously uninsured children.

Finally, it is important to consider the role of policies in influencing the social environment and subsequent effects on children's health. For example, Austin and colleagues reviewed the literature at each level of the socioecological model to identify risk and protective factors for child maltreatment.[16] Focusing on more upstream determinants in communities, they identified that community disadvantage (measured by increasing levels of poverty and unemployment) and neighborhood crime and violence increased risk for child maltreatment.[16] However, community access to health, social, and educational services as well as social cohesion were found to be protective. At the policy level, they found that economic policies that reduced household income and generated financial instability, such as recessions and regressive taxes, increased risk for child maltreatment.[16] In contrast, social and economic policies that reduced child-related costs and increased household income, such as minimum wage increases, were found to be protective. Taken together, expanding our view of how to address pressing child health issues through a multilevel approach that also targets the fundamental causes has the potential to reduce disparities in children's health.

Conclusion

The COVID-19 pandemic affected the social determinants of health and magnified the existing health disparities experienced by low-income communities and communities of color.[47] Suarez-Lopez and colleagues explain how the pandemic threatened children's health and development in five areas: (1) changes in environmental exposures, including indoor air pollution, lead exposure, and reduced interactions with the built environment; (2) changes in the food environment, with increases in benefits and safety net programs but reduced access to school meals; (3) reductions in educational resources and limitations of remote schooling; (4) changes in the social environments related to family stressors and exposures; and (5) increased social injustice and discrimination during a time of racial reckoning.[48] The direct and indirect effects of COVID-19 on children's physical and mental health are becoming increasingly evident, with worse outcomes for those children who were already at higher risk prior to the pandemic. While one success of the pandemic was the increase in telehealth to deliver healthcare, including specialist and mental health services, limitations are evident.[49] Reduced access to devices, internet access, interpretation services, and privacy have all disproportionately affected those at highest risk. If only the overall effects are considered, which may appear beneficial on face value, disparities can increase if the most advantaged populations benefit and those with fewer resources cannot access these resources.

Increased recognition of the critical role of social determinants is evidenced by one of Healthy People 2030's five overarching goals related to creating "social, physical, and economic environments that promote attaining the full potential for health and well-being for all."[3] The social determinants intersect with the family, community, school, and policy contexts to shape children's health. Addressing health disparities using a health equity approach will provide resources to those populations disproportionately affected by poor health. There is no reason why children from low-income families should have poorer health than those from higher income levels (Figure 5.1). However, it is essential to note that it is not just children with the lowest incomes who have poor health, but children at each step in the income gradient.

The conversation is changing, and the social determinants of health are receiving increasing attention in clinical work, community groups, schools, and policy circles. Context matters not only for children's health in the short-term but also throughout the life course. Although altering the fundamental causes may be the greatest challenge,[21] improving the context in which children live will create system-level change that both mitigates health disparities and improves population health.

References

1. Braveman P, Egerter S, Williams DR. The social determinants of health: coming of age. *Annu Rev Public Health*. 2011;32:381–98.

2. Centers for Disease Control and Prevention. NCHHSTP Social Determinants of Health (SDH) FAQ. 2022. Available at: https://www.cdc.gov/nchhstp/socialdeterminants/faq.html

3. US Department of Health and Human Services. Healthy People 2030: social determinants of health. 20233. Available at: https://health.gov/healthypeople/objectives-and-data/social-determinants-health

4. Zahran HS, Bailey CM, Damon SA, Garbe PL, Breysse PN. Vital signs: asthma in children: United States, 2001–2016. *MMWR Morb Mortal Wkly Rep*. 2018;67(5):149–55.

5. Green KA, Bovell-Ammon A, Sandel M. Housing and neighborhoods as root causes of child poverty. *Acad Pediatr*. 2021;21(8S):S194–S9.

6. Shonkoff JP, Slopen N, Williams DR. Early childhood adversity, toxic stress, and the impacts of racism on the foundations of health. *Annu Rev Public Health*. 2021;42:115–34.

7. Laraia BA, Leak TM, Tester JM, Leung CW. Biobehavioral factors that shape nutrition in low-income populations: a narrative review. *Am J Prev Med*. 2017;52(2S2):S118–S26.

8. Mathiarasan S, Huls A. Impact of environmental injustice on children's health-interaction between air pollution and socioeconomic status. *Int J Environ Res Public Health*. 2021;18(2):795.

9. Braveman P, Gottlieb L. The social determinants of health: it's time to consider the causes of the causes. *Public Health Rep*. 2014;129(Suppl 2):19–31.

10. Mayne SL, Mitchell JA, Virudachalam S, Fiks AG, Williamson AA. Neighborhood environments and sleep among children and adolescents: a systematic review. *Sleep Med Rev*. 2021;57:101465.

11. Fowler PJ, Tompsett CJ, Braciszewski JM, Jacques-Tiura AJ, Baltes BB. Community violence: a meta-analysis on the effect of exposure and mental health outcomes of children and adolescents. *Dev Psychopathol*. 2009;21(1):227–59.

12. Wright AW, Austin M, Booth C, Kliewer W. Systematic review: exposure to community violence and physical health outcomes in youth. *J Pediatr Psychol*. 2017;42(4):364–78.

13. Wellman RJ, Dugas EN, Dutczak H, et al. Predictors of the onset of cigarette smoking: a systematic review of longitudinal population-based studies in youth. *Am J Prev Med*. 2016;51(5):767–78.

14. Matos R, Monteiro D, Amaro N, et al. Parents' and children's (6–12 years old) physical activity association: a systematic review from 2001 to 2020. *Int J Environ Res Public Health*. 2021;18(23):359–66.

15. Scaglioni S, De Cosmi V, Ciappolino V, Parazzini F, Brambilla P, Agostoni C. Factors influencing children's eating behaviours. *Nutrients*. 2018;10(6):706.

16. Austin AE, Lesak AM, Shanahan ME. Risk and protective factors for child maltreatment: a review. *Curr Epidemiol Rep*. 2020;7(4):334–42.

17. Neil-Sztramko SE, Caldwell H, Dobbins M. School-based physical activity programs for promoting physical activity and fitness in children and adolescents aged 6 to 18. *Cochrane Database Syst Rev*. 2021;9:CD007651.

18. Micha R, Karageorgou D, Bakogianni I, et al. Effectiveness of school food environment policies on children's dietary behaviors: a systematic review and meta-analysis. *PLoS One*. 2018;13(3):e0194555.

19. Braveman P, Barclay C. Health disparities beginning in childhood: a life-course perspective. *Pediatrics*. 2009;124(Suppl 3):S163–75.

20. Centers for Disease Control and Prevention. The social-ecological model: a framework for prevention. Available at: https://www.cdc.gov/violenceprevention/about/social-ecologic almodel.html?CDC_AA_refVal=https%3A%2F%2Fwww.cdc.gov%2Fviolenceprevent ion%2Fpublichealthissue%2Fsocial-ecologicalmodel.html.

21. Frieden TR. A framework for public health action: the health impact pyramid. *Am J Public Health*. 2010;100(4):590–5.

22. Hauptman M, Niles JK, Gudin J, Kaufman HW. Individual- and community-level factors associated with detectable and elevated blood lead levels in US children: results from national clinical laboratory. *JAMA Pediatrics*. 2021;175(12):1252–60.

23. Hughes K, Bellis MA, Hardcastle KA, et al. The effect of multiple adverse childhood experiences on health: a systematic review and meta-analysis. *Lancet Public Health*. 2017;2(8):e356–e66.

24. Johnson NM, Hoffmann AR, Behlen JC, et al. Air pollution and children's health: a review of adverse effects associated with prenatal exposure from fine to ultrafine particulate matter. *Environ Health Prev Med*. 2021;26(1):72.

25. Ortegon-Sanchez A, McEachan RRC, Albert A, et al. Measuring the built environment in studies of child health-A meta-narrative review of associations. *Int J Environ Res Public Health*. 2021;18(20):10741.

26. Ogden CL, Carroll MD, Fakhouri TH, et al. Prevalence of obesity among youths by household income and education level of head of household: United States 2011–2014. *MMWR Morb Mortal Wkly Rep*. 2018;67(6):186–9.

27. Cave L, Cooper MN, Zubrick SR, Shepherd CCJ. Racial discrimination and child and adolescent health in longitudinal studies: a systematic review. *Soc Sci Med*. 2020;250:112864.

28. Ozer EJ, Lavi I, Douglas L, Wolf JP. Protective factors for youth exposed to violence in their communities: a review of family, school, and community moderators. *J Clin Child Adolesc Psychol*. 2017;46(3):353–78.

29. McPherson KE, Kerr S, Morgan A, et al. The association between family and community social capital and health risk behaviours in young people: an integrative review. *BMC Public Health*. 2013;13:971.

30. Esty B, Phipatanakul W. School exposure and asthma. *Ann Allergy Asthma Immunol*. 2018;120(5):482–7.

31. Mansfield JL, Savaiano DA. Effect of school wellness policies and the Healthy, Hunger-Free Kids Act on food-consumption behaviors of students, 2006–2016: a systematic review. *Nutrition Rev*. 2017;75(7):533–52.

32. Knopf JA, Finnie RK, Peng Y, et al. School-based health centers to advance health equity: a community guide systematic review. *Am J Prev Med*. 2016;51(1):114–26.

33. Sanchez AL, Cornacchio D, Poznanski B, Golik AM, Chou T, Comer JS. The effectiveness of school-based mental health services for elementary-aged children: a meta-analysis. *J Am Acad Child Adolesc Psychiatry*. 2018;57(3):153–65.

34. Moore SE, Norman RE, Suetani S, Thomas HJ, Sly PD, Scott JG. Consequences of bullying victimization in childhood and adolescence: a systematic review and meta-analysis. *World J Psychiatry*. 2017;7(1):60–76.

35. Hamm MP, Newton AS, Chisholm A, et al. Prevalence and effect of cyberbullying on children and young people: a scoping review of social media studies. *JAMA Pediatrics*. 2015;169(8):770–7.

36. Thornton RL, Glover CM, Cene CW, Glik DC, Henderson JA, Williams DR. Evaluating strategies for reducing health disparities by addressing the social determinants of health. *Health Aff (Millwood)*. 2016;35(8):1416–23.

37. Nardone A, Casey JA, Morello-Frosch R, Mujahid M, Balmes JR, Thakur N. Associations between historical residential redlining and current age-adjusted rates of emergency department visits due to asthma across eight cities in California: an ecological study. *Lancet Planet Health*. 2020;4(1):e24–e31.

38. Vasquez-Vera H, Palencia L, Magna I, Mena C, Neira J, Borrell C. The threat of home eviction and its effects on health through the equity lens: a systematic review. *Soc Sci Med.* 2017;175:199–208.

39. Gundersen C, Ziliak JP. Childhood food insecurity in the U.S.: trends, causes, and policy options. *Future of Children.* 2014;24:1–19.

40. Afshin A, Penalvo JL, Del Gobbo L, et al. The prospective impact of food pricing on improving dietary consumption: a systematic review and meta-analysis. *PLoS One.* 2017;12(3):e0172277.

41. Chaloupka FJ, Powell LM, Warner KE. The use of excise taxes to reduce tobacco, alcohol, and sugary beverage consumption. *Annu Rev Public Health.* 2019;40:187–201.

42. Hawkins SS, Bach N, Baum CF. Impact of tobacco control policies on adolescent smoking. *J Adolesc Health.* 2016;58(6):679–85.

43. Hawkins SS, Baum CF. Impact of state cigarette taxes on disparities in maternal smoking during pregnancy. *Am J Public Health.* 2014;104(8):1464–70.

44. Hawkins SS, Baum CF. The downstream effects of state tobacco control policies on maternal smoking during pregnancy and birth outcomes. *Drug Alcohol Dependence.* 2019;205:107634.

45. Hawkins SS, Baum CF, Oken E, Gillman MW. Associations of tobacco control policies with birth outcomes. *JAMA Pediatrics.* 2014;168(11):e142365.

46. Hudson JL, Moriya AS. Medicaid expansion for adults had measurable "welcome mat" effects on their children. *Health Aff (Millwood).* 2017;36(9):1643–51.

47. Williams DR, Cooper LA. COVID-19 and health equity: a new kind of "herd immunity." *JAMA.* 2020;323(24):2478–80.

48. Suarez-Lopez JR, Cairns MR, Sripada K, et al. COVID-19 and children's health in the United States: consideration of physical and social environments during the pandemic. *Environmental Research.* 2021;197:111160.

49. Curfman A, McSwain SD, Chuo J, et al. Pediatric telehealth in the COVID-19 pandemic era and beyond. *Pediatrics.* 2021;148(3):e2020047795.

6

Epidemiology

A Tool for Studying Environmental Influences on Children's Health

Dean Baker

Epidemiology is defined as the study of the distribution and determinants of health and disease in populations. Epidemiological studies have made enormously important contributions to understanding how environmental factors influence health and disease among children. For example, worldwide increases in the prevalence of asthma and atopic diseases were documented in descriptive epidemiologic studies.[1] Asthma risk factors were then identified in cross-sectional, case-control, and panel studies. Prospective cohort studies evaluated how environmental factors during the intrauterine and early-life periods modify the risk of asthma.[2-4] As the number of epidemiological studies grew, investigators combined findings using methods such as meta-analysis and pooling to evaluate consistency across studies and increase statistical power to identify important but subtle risk factors.[5,6] When ethically possible, intervention studies have been conducted to identify strategies to prevent or reduce asthma morbidity.

This progression of epidemiological studies has occurred for many environmental factors, such as heavy metals, pesticides, air pollution, synthetic organic compounds, and, more recently, the urban environment and climate change. In parallel, studies have focused on environmental factors that could impact on diseases and development, including low birth weight and birth defects, neurodevelopment, lung growth, obesity, and cancers.

The purpose of this chapter is to provide a review of epidemiological concepts and basic study designs as context for understanding the epidemiological studies described in this volume and epidemiological studies being published in journals.

A fundamental premise of epidemiology is that diseases are not distributed evenly across populations. Disease rates are different in different groups of people—in males versus females and in people of different races and socioeconomic status. Rates of disease vary from place to place and over time. By using the tools of epidemiology to study patterns of disease in different populations, some of whom may be exposed to environmental factors while others are not, it is possible to identify environmental causes of disease.

Populations and Population Samples

A *population at risk* is defined in epidemiology as those persons within a population who are at risk of developing a particular disease. It is seldom feasible and not

Dean Baker, *Epidemiology* In: *Textbook of Children's Environmental Health*, Second Edition. Edited by: Ruth A. Etzel and Philip J. Landrigan, Oxford University Press. © Oxford University Press 2024. DOI: 10.1093/oso/9780197662526.003.0006

necessary to study an entire population in order to evaluate associations between exposure and disease. Epidemiological studies therefore draw samples from populations to examine associations between exposure and disease within the population samples.

The *source population* is defined as the population at risk from which a sample population is drawn. A source population might consist, for example, of all children of a certain age living in a specific region. It is important to have a clear definition of the source population because it is the population to which the findings of a study are most immediately applicable.

A *study population* consists of the individuals from the source population who are chosen for inclusion in an epidemiological study. Different types of epidemiological studies use different strategies to select study populations. In a *cohort study*, a sample of apparently healthy individuals is drawn from the source population. Such a study population might consist, for example, of children of a certain age who live in communities that have different levels of air pollution. This study population is then followed over time to assess the occurrence of disease, such as asthma, or lung growth in relation to air pollution.[4,7]

In a *case-control study*, by contrast, the study population consists of a sample of cases of disease drawn from the source population over a particular time period. These persons with disease are compared to people without disease, termed "controls," who are selected from the same source population over the same time period. The goal of comparing cases and controls in a case-control study is to identify risk factors for disease to answer the question, "Why do some people in a population develop the disease, while others stay healthy?"

A key methodological issue in any epidemiological study is to understand clearly the relationship between the source population and the study population. An important step in drawing a sample for a study population is to specify a *sampling frame*, which is a list of the sampling units (e.g., individuals, prenatal clinics, households, schools) in the source population. The study population is then sampled from the sampling frame using established sampling methods, such as a simple or stratified random sample, cluster sample, or multistage sample. The investigators should monitor whether the study is successful in achieving a representative sample in which the study population is an accurate reflection of the source population. If the study population is not representative of the source population, the findings may not be valid.

Validity, Error, and Bias

The goal of an epidemiological study is to obtain an accurate estimate of the association between an environmental exposure or other risk factor and a disease. The strategy for accomplishing this goal is to develop a design for the study that minimizes the potential for error.[8]

In epidemiological studies, error can be random or systematic. *Random error* can occur during the process of sampling a study population from the source population. Random error can also arise from the inherent variability of biological data (e.g., random fluctuation in blood pressure or inaccuracy in measuring children's height). Random error is typically greatest in studies with small numbers of participants. It can be reduced by doing a larger study.

Systematic error or *bias* is the consequence of flaws in the study design or in the procedures used to select the study population or to measure risk factors and health outcomes in the study population. Systematic bias cannot be reduced by doing a larger study. *Validity* is defined in epidemiology as the extent to which systematic errors are controlled.

Three principal forms of systematic error have been identified in epidemiology: selection bias, information bias, and confounding.

Selection bias arises when the procedures used to recruit study participants from the source population are flawed and the study population is therefore systematically different from the source population. This bias can occur, for example, if a study does not define a sampling frame, if the sampling frame is flawed, or if the investigators use a nonprobability sampling method such as asking for volunteers from the sampling frame. Selection bias is not usually a problem in a cohort study involving complete follow-up. However, bias can occur in a cohort study if the degree of exposure influences participation in the study or if a selected subset of participants is differentially lost to follow-up. For example, in a cohort study of air pollution and asthma, selection bias could occur if families in polluted areas migrate away from the study area and are lost to follow-up when their children develop respiratory symptoms.

Selection bias is of greater concern in case-control studies because these studies entail sampling from a source population. Selection bias can occur in a case-control study if controls are chosen in a nonrepresentative manner; for example, if exposed individuals were more likely to be selected as controls than were nonexposed individuals.

The systematic error of selection bias is different from random sampling error. Selection bias cannot be avoided just by sampling a large study population. It requires paying careful attention to factors, such as the sampling frame and sampling methods, which may distort the sampling process or cause the study population not to be representative of the source population.

Information bias results from misclassification of the study participants with respect to either exposure or health outcome status. Information bias is also called *measurement bias* or *misclassification bias* because information errors commonly arise while measuring or classifying study variables. Two general forms of information bias are recognized.

Nondifferential information bias occurs when the misclassification of exposure is independent of the disease outcome or, conversely, when misclassification of disease outcome is independent of the true exposure. An example of nondifferential information bias is the use of a child's residence as a surrogate measure of exposure to air pollution from automobile traffic since the actual air pollution at the child's residence is typically not measured and because the child spends time away from that residence at locations with different levels of air pollution. Nondifferential information bias usually reduces the likelihood of discerning a link between air pollution and asthma because it results in collection of suboptimal information on environmental exposure.

Differential information bias occurs when the likelihood of exposure misclassification differs between diseased and nondiseased persons, or the likelihood of disease misclassification differs between exposed and nonexposed persons. *Recall bias* is a specific form of differential information bias. For example, in a case-control study, recall of a past exposure such as maternal exposure to pesticides during pregnancy might be different among mothers of children with developmental delays than among mothers whose children have normal development. Similarly, families living near an industrial site may report more nose bleeds and respiratory symptoms in their children than families living some distance from the site because they are more concerned about a possible link between their symptoms and the site's emissions.

Measurement error refers to errors made in measuring study variables. This term refers to the validity of a particular measure of exposure or disease, while information bias refers to the validity of a study's effect estimate. Measurement errors can be systematic (e.g., a spirometer that is always incorrectly calibrated) or random (e.g., because of test-to-test variability). Systematic error also can occur if the study personnel conduct study procedures, such as asking interview questions, measuring lung function, or doing examinations, differently because of a participant's exposure status. Strategies such as making sure that study personnel who collect health outcome data do not know participants' exposure status, called *blinding*, can reduce the likelihood of this bias. Systematic measurement errors reduce a measurement's validity, while random measurement errors reduce precision. In both instances, these measurement errors will result in information bias and reduce the validity of the effect estimate.

A major decision in planning an epidemiologic study of environmental factors is to evaluate the feasibility and cost of alternative methods to measure the potential environmental exposures. It can be very expensive to obtain detailed, personal environmental measurements. Investigators therefore often have to use less costly, but less accurate measures. Chapters 7 and 8 present more information about exposure science in children's environmental health.

Confounding occurs when the exposed and nonexposed groups in a study are not comparable even when there is no selection bias.[9] Confounding typically arises when the exposed and nonexposed groups have differences in such characteristics as age, gender, race, or socioeconomic status or have differences in their exposure to risk factors other than the risk factors under study. Confounding introduces systematic error into a study because it influences apparent risk of disease associated with the exposure being studied. Depending on the relationships between the exposure, the confounder, and the health outcome under study, confounding can either increase or decrease the apparent strength of an association between an exposure and a health outcome.

Multiple confounding factors can occur simultaneously and may act in opposite directions, making clarification of the net bias difficult. For example, multiple confounding factors have been identified in epidemiological studies of lead exposure and neurodevelopment, including maternal IQ, quality of the home environment (i.e., HOME Inventory), birth weight, and maternal education.[10,11]

In epidemiological studies, confounding can be controlled in the study design, during the analysis, or both.[9] Control of confounding at the design stage is accomplished by randomization, restriction, or matching. *Randomization* can occur in experimental studies when it is ethical and practical for investigators to assign participants to exposure or treatment status because it could be expected that the randomly assigned groups of participants would be comparable on other factors. However, it is rarely possible to study environmental hazards using experimental study designs. *Restriction* entails narrowing the ranges of values of the potential confounders, for example, by restricting the study to white females in a particular age group. However, this approach may limit the amount of information provided by the study. Another strategy is to match study participants on potential confounders (e.g., gender and race). Technically *matching* does not control for confounding; indeed, it causes confounding unless analyzed appropriately during data analysis. But matching can substantially increase efficiency by selecting controls that are very similar to the cases in one or more known confounding factors. Matching is most commonly used in case-control studies.

The most common approach to controlling confounding during data analysis is to use multivariate analytical methods to model the effect of exposure while adjusting for confounders. The conceptualization of confounding and methods to control for confounding in data analysis have become much more sophisticated, including use of causal models to identify potential confounders (e.g., directed acyclic graphs), to clarify criteria to consider a covariate to be a confounder, and to use appropriate multivariate models.[9,11]

Epidemiological Study Designs

All of the traditional epidemiological study designs have been used to study environmental influences on children's health. The most powerful of these is the *prospective birth cohort design*, which is discussed in the next section. This section reviews the traditional epidemiological study designs, progressing generally from the least to the most definitive of the study designs.

Descriptive studies examine the distribution of environmental factors or health conditions in defined populations. Descriptive studies do not formally evaluate associations between exposures and health outcomes. Descriptive data are commonly used to examine patterns of disease by place, time, and person. Most descriptive studies are based on existing mortality or morbidity statistics, such as hospital discharge data, or population surveys, such as the US National Health and Nutrition Examination Survey (NHANES), which collect information on health status, health behaviors, home characteristics, and biomarkers of environmental exposures.[12,13] They examine patterns of disease by age, gender, or ethnicity for specified time periods or geographical areas. Examples of descriptive studies include those that have reported on the concentrations of chemicals in the serum of children or on the changing prevalence of diseases such as asthma or autism spectrum disorders. Geographical patterns or temporal trends identified in descriptive studies may lead to hypotheses about environmental factors and disease that can be evaluated in analytical epidemiology studies.

Ecological studies examine associations between environmental exposures and health outcomes using groups of people, rather than individuals, as the unit of analysis. An ecological study compares aggregate measures of exposure, such as average exposure or proportion of population exposed, with aggregate measures of health outcome rates, for the same population. Ecological studies have provided important evidence on the role of environmental factors in human health, considering that many exposure occur on a population level.[14]

Ecological analyses often use *geographical areas*, such as counties or states, as the basis for defining the groups. Geographical ecological studies have, for example, been conducted to evaluate the association between risk of cancer and concentration of per- and poly-fluoroalkyl substances (PFAS) in water supplies across communities[15] or environmental factors associated with spatial heterogeneity of COVID-19 indices in France.[16] Although comparison of geographical units is conceptually straightforward, determining exposure levels precisely can be difficult if the geographical areas providing the health outcome data do not correspond geographically to the distribution of the environmental factors being studied. Water monitoring data, for example, may not be collected for an administrative unit such as a township, yet this may be the unit of analysis for which the health outcome rates are estimated.

Controlling for risk factors that may cause bias and confounding is difficult in ecological studies. Groups of people from different regions often live in different socioeconomic conditions, have different lifestyles, and have different exposures. The result is
that exposures to multiple factors and not only to the exposure under study may vary
between groups. Therefore, a study that attempts to examine the association between an
exposure of interest and a health outcome may be confounded by different rates of other
risk factors in the groups.

The groups studied in an ecological study also may be defined by *time period*.
Temporal comparisons or time-trend studies examine associations between changes in
exposure to environmental factors and changes in health outcome rates over time within
the same source population. One of the advantages of making a temporal comparison is
that population characteristics that could bias group-level associations may be relatively
constant over time in the same geographical area. However, the study of trends over long
periods of time may be complicated by substantial changes in demographics, such as
increasing population density and industrialization. Furthermore, available data on exposure and health outcome rates may not be comparable over time.

Cross-sectional studies examine associations between environmental exposures and
health status or disease prevalence at a particular point in time. For example, investigators may test lung function in a study population and at the same time do personal air
monitoring or ask questions about exposure to indoor air pollution. In environmental
epidemiology, cross-sectional studies have been used to assess associations between environmental exposures and subclinical measures of lung, neurological, immune, and reproductive functions.

Cross-sectional studies may be based on national population studies or make use of
existing data. The NHANES databases, for example, have been used by investigators to
characterize the concentrations of chemicals in children's blood and to examine associations between markers of exposure, such as blood lead concentration or urinary
phthalate metabolites, and health outcomes such as reported asthma and attention deficit hyperactivity disorder (ADHD). Many studies in communities are cross-sectional
in design, although they may use questionnaires, biomarkers of past exposure, or other
sources of data to estimate past exposures. Such studies may be called *prevalence* studies
because the health status is associated with earlier exposures, although many articles do
not distinguish between "cross-sectional" and "prevalence" study designs.

Selective survival is a major potential challenge for cross-sectional studies. Selective
survival is a type of selection bias because the study population may not include people
who have left the source population because of their earlier exposures or because they
developed disease or died. Selective survival may create difficulty in interpreting the "di-
rectionality" of an observed association.

Measurement error that produces *reverse causality* is a second potential problem in
cross-sectional studies. Reverse causality can occur if a toxic substance changes the biological systems involved in the toxicokinetics of metabolism, distribution, storage, or
excretion. For example, studies have suggested that perfluorinated compounds such as
perfluorooctanoic acid (PFOA) can change kidney glomerular filtration rate and thus
influence the serum level of the chemical. Such an effect would reduce the validity of
serum PFOA concentration as a biological indicator of past exposure.[17]

Case-control studies examine associations between exposures and a disease by comparing cases (individuals who have developed a disease) and unaffected controls who
are a sample of the source population from which the cases were identified. Controls

are usually individuals who are similar to the cases in terms of risk characteristics but who have not developed the disease in question. Having selected cases and controls, the investigators then determine the prior exposure status of each by examining exposure records, obtaining information through questionnaires, or measuring biomarkers of past exposure.

An efficiency of the case-control design is that exposure is measured only in the cases and controls and not for the entire source population, thus decreasing the study cost. The case-control study design is especially efficient for studying rare diseases. Thus, case-control studies have been used extensively to study the role of environmental factors in the development of cancer because most types of cancer have long induction periods and are rare in the general population.

Two major disadvantages of the case-control study design are, first, that identifying an appropriate control population can be difficult, and, second, there is higher risk of bias in classifying exposure because exposure status is determined after the disease has developed. An example of potential bias in sampling a control population is seen in case-control studies of childhood leukemia in relation to residential exposure to low-frequency electromagnetic fields. Initial studies of this association were problematic because they did not match sufficiently on socioeconomic status, which was associated with residential proximity to high-power transmission lines. Similarly, recall bias has been noted in case-control studies of children with autism spectrum disorders in which parents were asked to report on their child's gestational or early-life exposures years after the cases had been diagnosed. Some of the limitations with assessing past exposures are now being addressed by increasing use of biomarkers of past exposures, such as measuring persistent chemicals or metabolites in body tissue, or biomarkers of early effects such as epigenetic changes in microRNA or DNA methylation (see Chapter 10).

Time-series and *panel studies* have been used effectively to study associations between exposures and health outcomes that change over time. For example, time-series studies have examined temporal associations between air pollution levels and emergency room visits for asthma. Panel studies have been used to follow small cohorts of asthmatic children to make repeated measurements of concentrations of air pollutants and lung function or daily medication use.[18]

In *cohort studies*, the study population consists of a sample of individuals from the source population who are at risk of developing particular health outcomes. The cohort members are followed over time to determine the incidence of health outcomes among participants with different personal characteristics and exposure levels. Cohort studies have been very important for evaluating the role of environmental exposures on asthma, respiratory illnesses, lung development, and neurological outcomes in children and adolescents. The sampling frame can be defined without regard to a specific environmental factor, such as a population-based sample that could be used to assess the effects of multiple personal and environmental risk factors, or it could be defined to assess the effects of exposures to specific environmental factors or in specific populations. An example is a cohort study of children sampled from schools in different regions of Southern California with different levels of ozone and other air pollutants. The children were followed for years to assess incident asthma, lung growth, and other health outcomes.[4]

The prospective cohort study design is considered the most definitive of the observational study designs because the investigators identify the study population and then follow the causal events longitudinally as they unfold, from exposure until the development of the health outcome. The likelihood of information bias is minimized and the temporal relationship between exposure and health outcome is clearly determined, thus eliminating any possibility of reverse causality. The cohort study design also allows for the analysis of multiple health outcomes in relation to an exposure. Cohort studies can, however, be expensive if the study population is large or the follow-up period is long, and they can be inefficient if the health outcome is rare.

Prospective Birth Cohort Studies

Prospective birth cohort studies have been used around the world to examine the contributions of environmental exposures to diseases in children. In this study design, investigators enroll pregnant women—often identified through sources of prenatal care—or mother–child pairs at the time of the child's delivery and then follow the family over time. A well-known example is the birth cohort study of children in the Faroe Islands, in which the investigators identified infants at birth; used cord blood, cord tissue, and maternal hair to estimate prenatal exposures to methylmercury; and then followed the children through adolescence with periodic examinations to assess the effect of the methylmercury exposure on neurobehavioral function. The investigators showed that methylmercury exposure in early life is significantly associated with deficits in motor, attention, and verbal neurobehavioral tests that persisted until at least 14 years of age.[19] This study was very influential in shaping public health guidelines for consumption of fish that may contain methylmercury. This cohort study has been extended with multiple waves of participants to contribute to our understanding of other toxicants (e.g., PFAS) and health outcomes, such as immune dysfunction indicated by altered antibody response to vaccinations.[20,21] A well-known example of an early birth cohort study of lead exposure is the Port Pirie cohort study in Australia that has followed the birth cohort for more than three decades into adulthood.[22]

The number of prospective birth cohort studies has increased dramatically during the past two decades. Many studies are relatively small, with only a few hundred participants, and have focused on specific populations or environmental hazards such as the study of methylmercury in the Faroe Islands or pesticides in farmworker communities. However, some studies are based on much larger and more representative population samples. These studies have more statistical power to evaluate multiple environmental factors and interactions between environmental and individual factors during specific periods of gestation and early life. Some of the better known large birth cohort studies include the Avon Longitudinal Study of Pregnancy and Childhood (England, $n = 14{,}541$), the Generation R Study (Netherlands, $n = 9{,}778$), the Danish National Birth Cohort (Denmark, $n = 101{,}042$), and the Norwegian Mother and Child Cohort Study (Norway, $n = 100{,}000$). Large national birth cohort studies of 100,000 mother–child pairs have also been initiated in several countries in Asia, including the Japan Environment and Children's Study and the Korean Children's Environmental Health Study. National birth cohort studies were initiated in the United States and the United Kingdom but were terminated due to cost and complexity, although these countries have many smaller prospective cohort studies with recruitment during childhood and birth cohort studies which involve collaboration among the investigators.

The reason prospective studies are widely used is twofold. First, they can measure environmental exposures—chemical, nutritional, or psychological—at specific points in time during early human development as those exposures are occurring, thus increasing measurement accuracy and minimizing information bias. Second, prospective birth co-hort studies that measure individual characteristics and environmental exposures re-peatedly over time are uniquely well suited for discovering associations between early exposures and later disease. Early exposures may cause immediately obvious disease, such as congenital malformations or low birth weight. But environmental exposures can also act via epigenetic mechanisms or by modifying neurological development during critical developmental periods to produce delayed effects that become evident years or decades later (see Chapter 10). A further advantage is that these studies permit collection of exposure data over time, so they may be able to pinpoint vulnerable periods during gestation and early postnatal life for the exposures. Large cohort studies also can evaluate environmental factors within a multilevel framework, which considers exposures at the individual, family, neighborhood, and societal levels.

Some challenges to conducting birth cohort studies are the complex logistics, high costs of recruiting and following participants and collecting data at multiple time points over many years, and high participant burden because of repeated data collection.[23] There is concern that high participant burden will lead to selection bias during cohort recruitment and to selective retention during follow-up. A common strategy to reduce participant burden and study costs is to collect environmental and biological samples and then archive most of the samples for later analysis. Virtually all birth cohort studies collect and store biological specimens. This approach means that final use of the envi-ronmental and biological samples does not have to be determined in advance. By using a *nested case-control analysis* that compares cohort members who develop particular health outcomes with a random sample of similar participants who do not develop the outcome, a study can use stored samples to address multiple study objectives and can re-duce costs by analyzing specimens only for the cases and controls rather than for the full cohort. There are some limitations to this strategy. It can be expensive to maintain large repositories of stored samples. Multisite studies can have complicated logistics if the field sites have to collect, process, and ship specimens to repositories for further process-ing and long-term storage. Furthermore, despite this approach being used commonly in birth cohort studies, measurement validity, such as stability of specimens over many years, has not been adequately evaluated.

Another common strategy to reduce participant burden and costs in prospective co-hort studies is to make use of existing environmental data and assess health outcomes by accessing medical and school records or linking participants to national healthcare records rather than collecting all of these data specifically for the study. Some of the larg-est birth cohort studies, such as the Danish and Norwegian birth cohort studies, have relied primarily on linkage with national healthcare records to assess health outcomes. This approach reduces costs, but it also limits the range of environmental factors and health outcomes that can be assessed. This approach is not feasible in many countries that do not have an infrastructure of environmental monitoring or health systems that allow the collection and linkage of health status data.

Collaboration and Harmonization Among Birth Cohort Studies

International efforts to summarize, evaluate, and plan for collaboration among the many birth cohort studies taking place in different regions of the world have been an important development in the implementation of studies. For example, several programs have been funded to facilitate collaboration among European birth cohorts, including the European Child Cohort Network (eucconet.site.ined.fr) and the Environmental Health Risks in European Birth Cohorts (ENRIECO) program (isglobal.org).[24,25] A Birth Cohort Consortium of Asia was established and, as of 2022, includes 31 cohorts in 16 Asian countries (bicca.org).[26] This approach is important because the high cost and complex logistics of implementing very large, multicenter studies becomes increasingly difficult to be undertaken by one institute or agency even for large, wealthy countries. At the same time, efforts to combine data from multiple studies have been difficult because earlier studies did not use consistent measures of exposure or disease outcomes, time periods of measurement, or methods for measuring biomarkers and chemical contaminants in environmental media. Approaches for collaboration among birth cohort studies have been for investigators with national or international agencies such as the World Health Organization to gather and summarize study methods and findings of studies.[25,27] Collaboration provides opportunity to harness a wider range of expertise to address research questions of public health interest. By sharing ideas and expertise, scientists involved with birth cohorts may improve the efficiency in data collection and analyses. To discuss a path toward data harmonization, representatives from several large birth cohort studies and biomonitoring programs formed a collaborative group, the Environment and Child Health International Birth Cohort Group.[27] For studies in which the objectives and methods are sufficiently compatible, investigators have used meta-analysis and pooling to make use of the studies' combined data. Pooling of data from different studies allows the opportunity to consider hypotheses beyond the scope achievable by individual studies or single groups of investigators.[28] Combining or pooling data can not only improve statistical power to address rare outcomes and complex associations but can also improve causal inference because it permits cross-cohort comparisons and replications. Indeed, heterogeneity of findings across studies can provide insights about underlying causes of inequalities in disease and health-related behaviors among different populations of the world.

References

1. ISAAC Steering Committee. Worldwide variations in the prevalence of asthma symptoms: the International Study of Asthma and Allergies in Childhood (ISAAC). *Eur Resp J* 1998;12:315–35.

2. Keil T, Kulig M, Simpson A, et al. European birth cohort studies on asthma and atopic diseases: II. Comparison of outcomes and exposures: a GA^2 LEN initiative. *Allergy* 2006;61:1004–11.

3. Lodge CJ, Dharmage SC. Breastfeeding and perinatal exposure, and the risk of asthma and allergies. *Curr Opin Allergy Clin Immunol*. 2016;16:231–6.

4. Garcia E, Berhane KT, Islam T, McConnell R, Urman R, Chen Z, et al. Association of changes in air quality with incident asthma in children in California, 1993–2014. *JAMA*. 2019;321:1906–15.

5. Carlsen KCL, Roll S, Carlsen K-H, et al. Does pet ownership in infancy lead to asthma or allergy at school age? Pooled analysis of individual participant data from 11 European birth cohorts. *PloS ONE*. 2012;7:e43214.

6. Khreis H, Kelly C, Tate J, Parslow R, Lucas K, Nieuwenhuijsen M. Exposure to traffic-related air pollution and risk of development of childhood asthma: a systematic review and meta-analysis. *Environ Int.* 2017;100:1–31.

7. Urman R, McConnell R, Islam T, Avol EL, Lurmann FW, Vora H, et al. Associations of children's lung function with ambient air pollution: joint effects of regional and near-roadway pollutants. *Thorax.* 2014;69:540–7.

8. Rothman K, Lash T. Epidemiology study design with validity and efficiency considerations. In: Lash T, VanderWeele T, Haneuse S, Rothman K, eds., *Modern Epidemiology.* Philadelphia: Wolters Kluwer; 2021.

9. VanderWeele T, Rothman K, Lash T. Confounding and confounders. In: Lash T, VanderWeele T, Haneuse S, Rothman K, eds., *Modern Epidemiology.* Philadelphia: Wolters Kluwer; 2021.

10. Lanphear BP, Hornung R, Khoury J, Yolton K, Baghurst P, Bellinger DC, et al. Low-level environmental lead exposure and children's intellectual function: an international pooled analysis. *Environ Health Perspect.* 2005;113:894–9.

11. Van Landingham C, Fuller WG, Schoof RA. The effect of confounding variables in studies of lead exposure and IQ. *Crit Rev Toxicol.* 2020;50:815–825.

12. CDC. National Health and Nutrition Examination Survey, 2013–14, Overview. 2022 Mar. CDC: National Center for Health Statistics (publication CS239212). http://www.cdc.gove/nhanes

13. Sobus JR, DeWoskin RS, Tan YM, Pleil JD, Phillips MB, George BJ, et al. Uses of NHANES biomarker data for chemical risk assessment: trends, challenges, and opportunities. *Environ Health Perspect.* 2015;123:919–27.

14. Blanco-Becerra LC, Pinzón-Flórez CE, Idrovo ÁJ. Estudios ecológicos en salud Ambiental: más allá de la epidemiología [Ecological studies in environmental health: beyond epidemiology]. *Biomedica.* 2015;35:191–206.

15. Messmer MF, Salloway J, Shara N, Locwin B, Harvey MW, Traviss N. Risk of cancer in a community exposed to per- and poly-fluoroalkyl substances. *Environ Health Insights.* 2022;16:1–16.

16. Gaudart J, Landier J, Huiart L, Legendre E, Lehot L, Bendiane MK, et al. Factors associated with the spatial heterogeneity of the first wave of COVID-19 in France: a nationwide geo-epidemiological study. *Lancet Public Health.* 2021;6:e222–31.

17. Zhu Y, Shin HM, Jiang L, Bartell SM. Retrospective exposure reconstruction using approximate Bayesian computation: a case study on perfluorooctanoic acid and preeclampsia. *Environ Res.* 2022;209:112892.

18. Delfino RJ, Staimer N, Tjoa T, et al. Personal and ambient air pollution exposures and lung function decrements in children with asthma. *Environ Health Perspect.* 2008;116:550–8.

19. Debes F, Budtz-Jørgensen E, Weihe P, White RF, Grandjean P. Impact of prenatal methylmercury exposure on neurobehavioral function at 14 years. *Neurotoxicol Teratol.* 2006;28:536–47.

20. Grandjean P, Heilmann C, Weihe P, Nielsen F, Mogensen UB, Budtz-Jørgensen E. Serum vaccine antibody concentrations in adolescents exposed to perfluorinated compounds. *Environ Health Perspect.* 2017;125:077018.

21. Shih YH, Blomberg AJ, Bind MA, Holm D, Nielsen F, Heilmann C, et al. Serum vaccine antibody concentrations in adults exposed to per- and polyfluoroalkyl substances: a birth cohort in the Faroe Islands. *J Immunotoxicol.* 2021;18:85–92.

22. Searle AK, Baghurst PA, van Hooff M, Sawyer MG, Sim MR, Galletly C, et al. Tracing the long-term legacy of childhood lead exposure: a review of three decades of the Port Pirie cohort study. *Neurotoxicology*. 2014;46–56.

23. Luo ZC, Liu JM, Fraser WD. Large prospective birth cohort studies on environmental contaminants and child health–goals, challenges, limitations and needs. *Med Hypotheses*. 2010;74:318–24.

24. Vrijheid M, Casas M, Bergström A, et al. European birth cohorts for environmental health research. *Environ Health Perspect*. 2012;120:29–37.

25. Gehring U, Casas M, Brunekreef B, Bergström A, Bonde JP, Bottom J, et al. Environmental exposure assessment in European birth cohorts: results from the ENRIECO project. *Environ Health*. 2013;12, 8.

26. Kishi R, Zhang JJ, Ha E-H, Chen P-C, Tian Y, Xia Y, et al. Birth Cohort Consortium of Asia. *Epidemiology*. 2017;28:S19–S34.

27. Nakayama SF, Espina C, Kamijima M, Magnus P, Charles MA, Zhang J, et al. Benefits of cooperation among large-scale cohort studies and human biomonitoring projects in environmental health research: an exercise in blood lead analysis of the Environment and Child Health International Birth Cohort Group. *Int J Hyg Environ Health*. 2019;222:1059–67.

28. Etzel R, Charles M-A, Dellarco M, Gajeski K, Jöckel K-H, Hirschfeld S, et al. Harmonizing biomarker measurements in longitudinal studies of children's health and the environment. *Biomonitoring*. 2014;1:10.2478.

7

Exposure Science to Protect Children's Health

Clifford P. Weisel

Overview

Over the previous several decades, understanding the role of exposure science in elucidating how environmental contaminants affect children's health has been recognized as pivotal to mitigate/prevent contacts with contaminants that lead to adverse health outcomes.[1-3] This includes contacts and then exposures by children and/or the fetus. Exposure science links environmental science and disciplines within environmental health sciences that include toxicology, epidemiology, and clinical evaluations (Figure 7.1).[4,5] Although emission sources and processes that release contaminants into and transport them though one or more media (air, water, soil, or food) relate to the presence of environmental hazards, it is how, to whom, and the duration of contact with contaminants (i.e., exposure) that causes health effects. To reduce or eliminate adverse health effects requires identifying whether and how children contact agents released into the environment because, while the toxicity of substance is an inherent property, the levels of exposures can be altered. Exposure science also informs how toxicants enter the human body following a child's contact with a toxicant by inhalation, ingestion, and/or dermal routes. To reduce the incidence or persistence of environmentally mediated childhood diseases, we need to mitigate toxicant exposures by eliminating or reducing their emissions and environmental levels or by altering children's activities and behaviors that lead to contact.

Definitions and Foundations

Exposure science is[6] the study of human contact with chemical, physical, or biological agents occurring in their environments, and it advances knowledge of the mechanisms and dynamics of events either causing or preventing adverse health outcomes. *External exposure* is contact with a material at a boundary (nose, skin, or mouth) between the human (i.e., a child) and the "environment" at a concentration present in the environment over an interval of time.

Applications of exposure science start by determining whether there is "contact." Without contact there is no exposure; therefore, no adverse health outcome associated with environmental contaminants would be experienced by a child. Contact with environmental agents can be either good or bad. For example, our ability to live requires continuous contact and exposure to oxygen by the respiratory system through inhalation. In contrast, a short-term inhalation contact by a child with a toxic chemical (e.g., a cleaning agent) can lead to an exposure which causes an acute health effect. Exposure is also tightly bound to the time course of contact with an agent(s) (acute, subchronic, chronic)

Clifford P. Weisel, *Exposure Science to Protect Children's Health* In: *Textbook of Children's Environmental Health,*
Second Edition. Edited by: Ruth A. Etzel and Philip J. Landrigan, Oxford University Press. © Oxford University Press 2024.
DOI: 10.1093/oso/9780197662526.003.0007

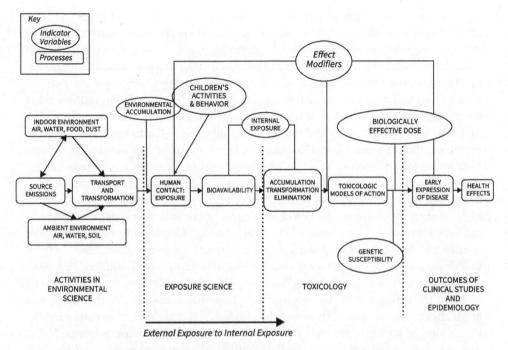

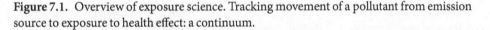

External Exposure to Internal Exposure

(Adapted from Lioy and Weisel 2014[5], modified January 2022, Previously adapted from Lioy 1990 and Lioy 2010)

Figure 7.1. Overview of exposure science. Tracking movement of a pollutant from emission source to exposure to health effect: a continuum.

that could result in adverse health effects. Throughout the 20th century the approach to defining cause-and-effect relationships was driven by toxicological research and not exposure. Furthermore, there was a lack of attention to children's or pregnant women's environmental health issues even though these vulnerable individuals could experience adverse effects at lower exposures than healthy, non-pregnant adults. Environmental regulations have focused on levels in the outside environment, even though children spend in excess of 85% of their time indoors, primarily in their residences, with the percentage of time indoors being higher for younger children.[7] While regulations to control emissions of toxicants to directly or indirectly protect the public's health consider the most vulnerable individuals, which often are pregnant women and children, based on vulnerabilities of their biological systems and stage of development, there has been less emphasis placed on the activities that they are engaged and that lead to exposures[8,9] Within the United States, the National Ambient Air Quality Standards (NAAQS) are designed to protect the public's health—including children's—from outdoor air pollutants. However, the NAAQS are inadequate for addressing total human inhalation exposures because there are only outdoor standards for six criteria pollutants. While the US Environmental Protection Agency (EPA) recognizes other outdoor pollutants, referred to as *hazardous air pollutants* (HAPs), EPA federal regulations do not address indoor air even though many of the HAPs and several of the criteria pollutants (nitrogen oxides, particulate matter, carbon monoxide, lead) can have indoor sources that result in indoor levels—where children spend the vast majority of their time—exceeding outdoor levels.[10] Indoor air quality (IAQ) became a focus of research following the energy crises of the 1970s, when guidelines for construction of new buildings and retrofitting of older buildings shifted to more airtight homes to reduce energy use. These guidelines resulted in less air exchange with outdoor air and chemicals, particulates, and biologics emitted or used inside adversely affecting IAQ and leading to increased inhalation exposure. While governmental, industrial, nonprofit, and academic researchers are now devoting more attention to indoor air pollution and chemistry, only limited regulation of toxicants indoors exists for nonoccupational settings. Rather, the contents of consumer products and building materials are regulated to try to limit what might be released into the indoor air. One example is the compound formaldehyde, a respiratory irritant and carcinogen that has been linked to childhood asthma.[11] While formaldehyde is emitted into the outdoor air from combustion processes and formed by reactions of chemicals, formaldehyde also can be released from the insulation commonly used in homes and furnishings, thus causing high indoor air levels. To minimize a child's formaldehyde and other hazardous chemical inhalation exposure, considerations beyond just ambient air levels are needed. The home is where children and pregnant women have the greatest exposures, often resulting from indoor sources, and this exposure is underestimated when using only outdoor air levels.

Exposure science evolved during the 1980s with the first studies focused on exposure and the emergence of risk assessment.[1,4] Exposure assessment methodology was originally developed as an analysis tool. It provided "screening" results to determine the significance of toxicants present at waste sites in causing human health effects. Initial exposure assessment and hazard characterizations of an agent were highly dependent on the type of associated health outcome, with cancer and non-cancer endpoints having different exposure time frames of importance. Furthermore, the same agent can have multiple exposure routes: inhalation, dermal, ingestion, and occasionally injection. To compare exposures across routes, an internal exposure (or internal dose) is calculated, frequently in units of mg/unit body weight/day.[12] Normalizing the dose to body weight is important for evaluating potential impacts of an exposure on children.

With the 1996 passage of the Food Quality Protection Act (FQPA), the field began to characterize cumulative (multiple compounds) and aggregate (multiple routes) exposures.[13] The FQPA required federal agencies to examine children's exposure to pesticides. It was recognized that children's pesticide exposures could result from nondietary ingestion, in a similar manner to lead dust exposures in homes. However, exposures derived from all routes of entry into the body had a twist for pesticides. Traditional public health practice had always considered pesticides as residues after deposition on to surfaces. However, many pesticides are semi-volatile and therefore prone to emission into the indoor air at room temperature, followed by redeposition throughout a residence or other buildings. Thus, semi-volatile compounds like pesticides have dynamic, not static properties.[14] However, the most important feature of the dynamic process, especially for pesticides, was that some classes of pesticides have an affinity for the polyurethane foam in plush materials. The volatilized pesticides not only redeposited on toys or other plush objects, they accumulated there; thus, those toys acted as a reservoir and new exposure source, as observed for the pesticide chlorpyrifos.[15] Because of this non-dietary exposure route and chlorpyrifos's neurological effects in children, the US EPA banned spraying of chlorpyrifos in residential settings in 2000.[16] However, chlorpyrifos was still permitted for agricultural uses. The 2000 US ban decreased chlorpyrifos's loading in carpets—an exposure source for children—and resulted in lower blood levels. However, political and economic considerations affected this pesticide's continued use in the United States and Europe, first allowing its continued use for agricultural purposes in the face of lawsuits and then subsequently reducing its availability to the market for economic reasons.[16]

Here, reductions in childhood exposure were not solely based on scientific exposure and risk considerations, though those were the basis for initial US EPA regulations.

Exposure: Environmental and Health Sciences

Exposure studies provide information on the processes and intensity of single or multiroute exposures of children. The activities associated with the left side of Figure 7.1 provide information on sources of pollutants, their movement and accumulation in the environment, and the chemical, biological, and physical transformations that can occur and lead to contact with a child. The right side includes the results from toxicological and human health-based research.

Risk management benefits from modern exposure characterization since the results can be used to direct control strategies for specific situations or the removal of harmful chemicals from children's toys and personal products. Deriving a source-to-exposure-to-effect relationship starts by determining how a source can be linked by one or more pathways and exposure routes. Currently, this is difficult to measure for very young children (i.e., infants and toddlers) since they cannot wear personal monitors or take representative samples (e.g., water they drink).

Routes of Exposure

Exposure routes of concern for children are the same portals of entry for a toxicant into any human body.[12] As shown in Figure 7.1, once the toxicant has entered the body the focus of attention goes from external exposure to internal exposure (i.e., biomarkers) and, finally, the biologically effective dose. The movement of material from the source to people is the *exposure pathway* (air, water, soil, dust, or food) with four actual routes of exposure: inhalation, dermal, ingestion, and injection.

Inhalation involves breathing in the toxicant (whether chemical, physical, or biological) and then direct contact in one or more of the three major regions of the respiratory system: the nasal-pharynx (nose, mouth, and pharynx), the thoracic (the bronchial tree), and air exchange region (bronchioles and the alveoli sacks). Deposition or absorption of species occurs differently in these regions based on the solubility of gaseous species and the size and shape of particulate species. In the extrathoracic region, highly soluble gaseous species are absorbed and large particles (>10 µm) are deposited. In the tracheo-bronchial region, the airways go through a series of dichotomous branching which result in a higher linear velocity of each breath and the deposition of smaller particles. The bronchiole is the gas exchange region of the respiratory tract and the final location for particle deposition until the breath is expired. The mass deposited in the respiratory tract also depends on breathing rate, the extent of solubilization of different particles, and their interaction with or absorption through the respiratory lining.

Ingestion involves the consumption of a toxicant. Usually, ingestion exposure occurs because the toxicant is part of the material (such as food and liquids) being consumed by a child. Chemical food contamination can be caused by multiple processes, with the most prevalent being uptake by growing fruits and vegetables in contaminated soil, accumulation of toxicants in animal feed, the deposition and retention of pesticides sprayed on food, and transference of contaminants to food from surfaces during preparation or handling (e.g., dragging a lollipop on a rug). Biological contamination of food by bacteria can occur during handling by individuals who are ill and not practicing appropriate hygienic procedures and from improper conditions during the food's storage or transport. Contaminated drinking water consumed or used in food preparation can also be a source for exposure to chemical and biological agents. Physical contamination can occur when a foreign object inadvertently is incorporated into food products or natural non-food materials are left in the food. This can happen at multiple points during the production and manufacturing process.

"Incidental" ingestion occurs when children ingest contaminated soil or dust and chew or suck on contaminated objects. The latter is a common pathway for pesticide and lead exposures. Transference of contaminants onto food from unwashed hands after touching contaminated dust on floors, tables, and multiple other surfaces or objects leads to incidental ingestion. Characterization of ingestion exposure from food sources can be evaluated based on the contaminant level in food, including contributions from food preparation and cooking, combined with an evaluation of which foods are actually consumed by children. It can be difficult to determine average food contaminant levels when multiple food types need to be traced through the food distribution system, along with predicting contributions from a variety of food preparation/handling methods. The food consumed often differs from the meal as prepared since children (as well as adults) have preferences about what they eat. It is also important to consider food consumed as snacks between meals, which may be eaten away from a table and potentially contaminated by house dust when children consume snacks while seated or playing on the floor.

Dermal exposure involves the deposition of contaminants on the skin surface causing irritation and/or the absorption into the body. The hands are the most prominent section of a child's body that contacts their surroundings, outside of showering, bathing, or swimming, since clothing is often worn on other portions of the body. However, if clothing becomes wet it can facilitate dermal exposure by being a repository for contaminants. Chemical penetration through the skin varies across different parts of the body due to differences in the thickness of the epidermis. The lipophilic nature of the contaminant is a primary determinant for the amount of penetration of a compound through the skin. When the skin is damaged, as when a child has an abrasion or cut, additional dermal absorption of chemical and biological agents can occur; the latter can result in infections. Dermal absorption of vapors, while typically smaller than from liquids when both contact the skin, has been estimated to be an important exposure route for many semi-volatile organic compounds.[17] Children continually touch surfaces, particularly floors and dirt, without washing their hands, potentially enhancing dermal absorption.

Injection exposure of a toxicant into the body can occur during vaccination or other activity, legal or illegal. Occasionally penetration can occur from particles or materials that are emitted at high speed and contact the skin. This can cause material to be embedded within the skin that can lead to a rash or continued systemic release of agents.

Defining Equations for Exposure Calculations

Exposure (E) is represented by:

$$E = \int_{t_0}^{t_1} C(t)dt \qquad (7.1)$$

and

$$E = \sum_{i=1}^{n} \Delta C_i \times \Delta t_i \qquad\qquad (7.2)$$

where for Equation 7.1 $C(t)$ is the concentration at time t, from t_0 to t_1, and for Equation 7.2, the summation is over n locations that a person travels through in a day. ΔC_i is the average concentration, and Δt_i the time spent in each location.

Equation 7.1 represents the total (continuous) exposure for an interval of time. The second equation represents the summation of all the "contacts" that lead to exposures within microenvironments, the locations with a "uniform" exposure concentration, through which a person passes during a day. For inhalation exposures examples of microenvironments could be the home; outdoors in a city, suburban or rural setting; automobile or other transportation modes; and a daycare center, school, or workplace. Equation 7.2 is more commonly used because there are very few instances in which total personal exposure is directly measured, while concentration data in locations encountered during an individual's daily activities may be available or estimable from other information and can serve as inputs for exposure models. For ingestion and dermal exposures, the contact is usually not continuous but occurs in discrete episodes since, unlike breathing, which is a continuous activity, eating and touching surfaces are distinct events. To estimate their exposures, the duration and frequency of the contact are needed.

Exposure Science Measurement Designs

Two approaches, direct and the indirect, are used to measure exposure. Collecting samples for characterization of exposure using the direct measurement method requires coordination with the individuals being evaluated so that samples can be collected in all locations, media encountered, and items consumed relevant to the contaminant(s) being studied. The direct method of exposure measurement uses personal monitoring (external exposures) or biological monitoring (internal exposures) which requires specific consideration of whether a child can wear/utilize a personal monitor or provide the biological sample (blood, urine) on the schedule required, respectively. This can be particularly problematic for babies and young children.

The indirect methods infer exposures from determining concentrations in each media contacted, combined with knowing the contact times and frequencies. For example, measuring air pollutants in the locations the participant visits (i.e., home, school, work, etc.) combined with activity patterns of participants for inhalation exposure; or for waterborne contaminants, measuring the amounts in tap water, bottled water, and other fluid concentrations, coupled with the daily amount of each fluid consumed for ingestion exposure and the shower or bathing duration and frequency and other water contacts for dermal exposure. This method usually develops sampling strategies that are linked to locations of interest, populations at risk, and duration of activities.

Human Activities and Behavior

A major component of exposure science is the collection of human behavior and time activity information. Data for children cannot be readily extrapolated from adults because their behaviors and activities differ, resulting in different contacts with a toxicant even when individuals are present in the same locations. Behavior and activity data have been primarily obtained using survey research, with the questions primarily focused on locations, time spent in various locations, and the activities participated in while in each location.[7,18] Questionnaires also have identified behaviors that affect exposure. For example, parental surveys have reported the number of times a child touches surfaces and then put his or her hands in his or her mouth per hour. Those initial estimates were well below 10 per hour, but videotaping children over the course of a day in their homes subsequently documented that young children have a much higher rate of hand-to-mouth contacts.[19] This observational method provided more realistic information on the frequency of hands touching surfaces and therefore improved estimates of a child's contact with pesticides present on home surfaces than those obtained from recall questionnaires. It also opened a window into the world of children's behaviors and changed the evaluation of exposures. For instance, children were observed sitting on containers that may contain hazardous substances, playing in open crop fields, and chewing on many types of surfaces and objects, thereby potentially having high exposures. These tools have helped validate the range of values ascribed to variables used in exposure models and identified that exposures vary with age, gender, and culture/ethnicity. New approaches to obtain behavioral and activity data now use global positioning system (GPS) and phone video cameras.[20–22] Ethical and privacy issues must be addressed, however, when using these methods. Data tracking methods show differences from self-reported questionnaires.[23] Thus, the potential exists to improve the validity of each input value used in exposure models, providing better evaluations of environmental health problems and risk management.

Application of the Continuum

Determining whether a child has been exposed to an environmental agent should start with a continuum (Figure 7.1). The words "pollution source" usually conjure up images of an industrial stack, effluent pipe, a large tract of land surrounded by a fence, or the emissions from a motor vehicle. However, many smaller emission sources (e.g., gas stoves, tobacco smoke, dry-cleaned clothing, household cleaning products) used in the home can be more important contributors to a child's total exposure because of the proximity of a source to the child and the duration of the contact. Some sources are obvious while others are subtle. *Proximate* or *near-field* sources become an important part of the solution when attempting to determine what is leading to an adverse exposure to children in their homes.[24] The EPA has developed databases and exposure models that specifically incorporate chemicals within consumer products because they often are the largest sources contributing to exposures.[25,26]

Rugs and carpets in the home are a proximate source that can lead to a child's exposure to many toxicants.[27] We use carpets to keep a space warm, reduce noise, provide an esthetically pleasing space, prevent injury, etc. However, an unintended consequence of carpets is an accumulation of microscopic materials over time—some toxic—which may or may not degrade over time, thus resulting in a reservoir of toxicants that a child play-ing on a carpet can contact and be exposed to.

Exposure characterization needs to incorporate the concept of multiple sources and media for many pollutants. To eventually reduce an exposure, the largest sources must be identified and linked to routes of exposure. One example is pesticides sprayed in the home during pest control applications, transported into the home from the outside air, and tracked indoors after walking on areas treated with pesticides if shoes are not removed or cleaned prior to entering the living space. Other toxicants that accumulate in carpets include lead-contaminated dirt tracked in by people and animals from the yard, infill from athletic field turf, and polycyclic aromatic hydrocarbons deposited from to-bacco smoking, fireplaces, or wood- and coal-burning stoves. Children playing on rugs whose hands are sticky from saliva or food residue can contact dust containing these toxins, resulting in potentially significant exposures that confound attempts to link a specific chemical or other toxic agent to a disease or a particular emission source as the primary exposure pathway. Rugs can also be a media for mold growth in humid environ-ments and accumulate allergens, including pollen, pet dander, and dust mite dropping. These can also become resuspended in the air during typical home activities, leading to inhalation exposure.[14]

Lead is a toxicant with multiple sources and exposure routes that still plagues children in developed countries and globally.[28] Lead can be found in the following places:

- *Air*: From point industrial and mining sources, which vary over time depending on the strength of the emissions and local meteorology.
- *Drinking water*: Predominantly from leaching out of pipes and fittings in a home. Variations in controls on water corrosivity and plumbing fixtures alter the amount of lead delivered to the tap and that leads to lead ingestion exposure.
- *Food*: While typically low and very sporadic in the food supply, lead could be present from food grown in contaminated soil, prepared with contaminated drink-ing water, or if the preparation or handling is done on dusty surfaces or using con-tainers with lead glaze.
- *Soil*: In urban high-traffic settings prior to the ban on leaded gasoline (1996 in the United States) or near facilities that emit lead particulates. Deposited lead will re-main in soil until all topsoil is cleaned up or remediated and covered with clean soil. Lead from exterior paint can crumble, peel, or chalk contributing to lead in the soil near homes and presenting an exposure pathway to children playing in their yards or if tracked indoors.
- *Homes*: Where leaded paint was used (prior to 1978 in the United States, earlier for other countries). Peeling lead-based paint chips present a potential high exposure due to direct ingestion or becoming part of house dust. Lead occasionally is present in glazes and paints on pottery, in some cosmetics, as a dye or pigment in some cul-tural activities, and in brightly colored toys and jewelry. Lead can be brought home on the clothing of workers and on the shoes or feet of all occupants or pets.

Exposure Measurements

Exposure measurements for children are optimally conducted by measuring each media that the child contacts in each relevant location containing the contaminant of interest,[29] although for very young children this can be very challenging. Alternately, biomarker specimens can be taken that reflect total exposures.

Microenvironments

The concept of microenvironments is a critical component of exposure modeling for risk assessment. Microenvironments are individual or aggregate locations or activities within a location that have a homogeneous concentration of the pollutant of interest. Thus, a microenvironment is where all individuals would potentially have the same exposure. If an individual moves to a different location or engages in a different activity, that would be a different microenvironment and therefore a different exposure. In calculating exposure, the concentration associated with each microenvironment is multiplied by the time spent in that microenvironment and is then summed across all microenvironments encountered over time (Equation 7.2). However, this does not account for variability in the intensity of exposure caused by different human behaviors. A simple listing of typical microenvironments might include outdoors; indoors at a residence, daycare, school, work, or entertainment venue; and in transit. The selection of unique microenvironments will vary with the contaminant being considered and what people do and where. For example, to determine exposure to water disinfection by-products, the shower stall and bathroom should be added to the residential indoor environment because these are important sources in terms of greatly elevated air concentrations and dermal contact compared to other rooms in a home.

Biological Monitoring

A biological marker of exposure is a physiological response or measured concentration of a toxicant, its metabolite, or adduct within body fluids or tissues that is proportional to an external exposure.[30] Biomarkers can be measured in multiple body tissues or fluids. When the biomarker is specific to the agent a child was exposed to, its presence confirms that an exposure has occurred. A common biomarker is blood lead level. However, some biomarkers, particularly metabolites and physiological responses, are not specific to a single chemical and thus do not confirm an exposure to a single agent. Furthermore, biomarker concentrations typically change in the body with time as the biomarker is metabolized or excreted. The concentration change often follows an exponential time curve, although most measurements are made in samples collected at a single point in time or at a few time points. Thus, the lack of a measurable biomarker level does not mean an exposure has not occurred since, after several biological half-lives, its concentrations could decrease to below a detectable level. For example, many volatile organic compounds have short half-lives so biomarker measurement days after a single exposure will not be informative of the exposure.[31] Compounds with long biological residence times (years) are the most easily interpreted.[30] Current examples are the family of compounds known as perfluorinated alkyl substance (PFAS), which have ubiquitous exposures across developed countries, and polybrominated diphenyl ethers (PBDE) and polybrominated biphenyls (PBBs), which have been used as flame retardants, including in children's sleepwear and clothing.

Since the mid-1990s the US Centers for Disease Control and Prevention (CDC) and similar organizations in other countries have routinely measured hundreds of organic compounds and multiple metals in the blood and/or urine of adults and children older than 6 years.[32] These measurements provide trends for nationwide distribution of biomarker levels and can be used to determine whether an individual's levels are within the national norms or very high or very low. Biomarker levels also can be altered by genetic polymorphisms in metabolic enzymes or phenotypes that induce or suppress metabolism, thereby altering their metabolic rate. This results in potentially large interindividual differences in biomarker levels from the same exposure. Thus, biomarker measurements removed from information about exposure(s) do not provide complete information for determining when or how an exposure occurred or its duration and magnitude.

Personal Monitoring

Inhalation exposure: Personal measurements characterizing air pollutant levels in the breathing zone of a child are important direct measures of exposure.[33] These are collected as active samples, using pumps to draw air though adsorbents or denuders for gaseous species and filters for particulate species, or as passive samples, whose sampling rates are based on the diffusion of chemicals or particles to the sampling surface. Generally, these provide average or integrated air concentrations, and the sample is returned to a laboratory for analysis. Sensors combined with data loggers are used to continuously measure and store the concentration for subsequent transfer to an electronic database. Sensors provide information on acute or peak and average exposures and can give feedback to the wearer to potentially alter behavior to reduce exposures.[34,35] Some devices are small and light so can be worn by children and pregnant women. Alternately robotic platforms that simulate movement of infants and toddlers have been used as surrogates to estimate their inhalation (and dermal) exposures.[36–39]

Dermal exposure: Hand wipes and patches worn on clothing are used to determine potential dermal exposure. These approaches have been used to evaluate children's exposure to lead, pesticides, and semi-volatile organic compounds. Because hand wipes represent what is present on the hand at the time of sampling, they may not represent the total contaminant absorbed through the skin or ingested through hand-to-mouth contact.

Ingestion exposure: Food sampling can use the *duplicate plate technique*, where the participant prepares two identical dishes and only contaminants in the food consumed are used to determine potential ingestion exposure. Fully processed food and not just the same raw ingredients should be used to account for contributions from food handling and processing. Snack food should also be collected and analyzed. However, inadvertent ingestion exposure is rarely directly measured. Rather, contaminants on hands, in dust, and on surfaces are measured and combined with estimates of the quantity of hand-to-mouth transference of dust and dirt ingested to assess potential exposures.

Exposure and Source to Dose Models

Exposure models have been developed to define population exposures, and individual personal exposures and have been linked to *physiologically based pharmacokinetic models*

(PBPK).[40,41] Where appropriate, exposure models and modeling systems borrow tools from other fields, such as movement of pollutants through the environment, and augment the modeling framework with components that are specifically tailored to the needs of the field, such as modeling human activity and physiology. Multiple models integrated into a modeling system are important to the continuum described in Figure 7.1; namely, to establish source-to-dose relationships. Exposure models that are either individual- or population-based require that information on the child's activities, behavior, and duration of contact be linked to the concentration found in each environment encountered.

References

1. Lioy PJ. Assessing total human exposure to contaminants: a multidisciplinary approach. *Environ Science Technol.* 1990;24:938–45.

2. Lippmann M, Leikauf GD. *Environmental Toxicants: Human Exposures and Their Health Effects.* 4th ed. New York: John Wiley & Sons; 2020.

3. National Research Council. *Exposure Science in the 21st Century: A Vision and a Strategy.* Washington, DC: The National Academies Press; 2012.

4. Lioy PJ. Exposure science: a view of the past and milestones for the future. *Environ Health Perspect.* 2010;118:1081–90.

5. Lioy PJ, Weisel CP. Exposure science to protect children's health. In: Landrigan PJ, Etzel RA, eds. *Textbook of Children's Environmental Health.* 1st ed. Oxford: Oxford University Press; 2014:58–68.

6. Ott WR. Human exposure assessment: the birth of a new science. *J Exposure Analysis Environ Epidemiol.* 1995;5:449–72.

7. United States Environmental Protection Agency (EPA). *Child-Specific Exposure Factors Handbook.* Washington, DC: Office of Research and Development; 2008:687.

8. Ferguson A, Penney R, Solo-Gabriele H. A review of the field on children's exposure to environmental contaminants: a risk assessment approach. *Int J Environ Res Public Health.* 2017;14:265–90.

9. Hines RN, Sargent D, Autrup H, et al. Approaches for assessing risks to sensitive populations: lessons learned from evaluating risks in the pediatric population. *Toxicol Sci.* 2010;113:4–26.

10. Goldstein AH, Nazaroff WW, Weschler CJ, Williams J. How do indoor environments affect air pollution exposure? *Environ Sci Technol.* 2021;55:100–8.

11. McGwin G, Lienert J, Kennedy JI. Formaldehyde exposure and asthma in children: a systematic review. *Environ Health Perspect.* 2010;118:313–7.

12. Lioy PJ, Weisel CP. *Exposure Science: Basic Principles and Applications.* New York: Elsevier/Academic Press; 2014.

13. Naidenko OV. Application of the Food Quality Protection Act children's health safety factor in the U.S. EPA pesticide risk assessments. *Environ Health.* 2020;19:16.

14. Lioy PJ. Employing dynamical and chemical processes for contaminant mixtures outdoors to the indoor environment: the implications for total human exposure analysis and prevention. *J Exposure Sci Environ Epidemiol.* 2006;16:207–24.

15. Gurunathan S, Robson M, Freeman N, et al. Accumulation of chlorpyrifos on residential surfaces and toys accessible to children. *Environ Health Perspect.* 1998;106:9–16.

16. Hites RA. The rise and fall of chlorpyrifos in the United States. *Environ Sci Technol.* 2021;55:1354–8.

17. Weschler CJ, Nazaroff WW. SVOC exposure indoors: fresh look at dermal pathways. *Indoor Air.* 2012;22:356–77.

18. Glen G, Smith L, Isaacs K, McCurdy T, Langstaff J. A new method of longitudinal diary assembly for human exposure modeling. *J Exposure Sci Environ Epidemiol.* 2008;18:299–311.

19. Freeman NC, Jimenez M, Reed KJ, et al. Quantitative analysis of children's microactivity patterns: The Minnesota Children's Pesticide Exposure Study. *J Exposure Analysis Environ Epidemiol.* 2001;11:501–9.

20. Maddison R, Ni Mhurchu C. Global positioning system: a new opportunity in physical activity measurement. *Int J Behav Nutrition Physical Activity.* 2009;6:73.

21. Ma X, Longley I, Gao J, Salmond J. Assessing schoolchildren's exposure to air pollution during the daily commute: a systematic review. *Sci Total Environ.* 2020;737:140389.

22. Bagot KS, Matthews SA, Mason M, et al. Current, future and potential use of mobile and wearable technologies and social media data in the ABCD study to increase understanding of contributors to child health. *Develop Cogn Neurosci.* 2018;32:121–9.

23. Elgethun K, Yost MG, Fitzpatrick CT, Nyerges TL, Fenske RA. Comparison of global positioning system (GPS) tracking and parent-report diaries to characterize children's time-location patterns. *J Exposure Sci Environ Epidemiol.* 2007;17:196–206.

24. Pellizzari ED, Woodruff TJ, Boyles RR, et al. Identifying and prioritizing chemicals with uncertain burden of exposure: opportunities for biomonitoring and health-related research. *Environ Health Perspect.*2019;127:126001.

25. Huang L, Ernstoff A, Fantke P, Csiszar SA, Jolliet O. A review of models for near-field exposure pathways of chemicals in consumer products. *Sci Total Environ.* 2017;574:1182–208.

26. Jolliet O, Huang L, Hou P, Fantke P. High throughput risk and impact screening of chemicals in consumer products. *Risk Analysis.* 2021;41:627–44.

27. Lioy PJ, Freeman NC, Millette JR. Dust: a metric for use in residential and building exposure assessment and source characterization. *Environ Health Perspect.* 2002;110:969–83.

28. O'Connor D, Hou D, Ye J, et al. Lead-based paint remains a major public health concern: a critical review of global production, trade, use, exposure, health risk, and implications. *Environ Int.* 2018;121:85–101.

29. Cohen Hubal EA, Sheldon LS, Burke JM, et al. Children's exposure assessment: a review of factors influencing children's exposure, and the data available to characterize and assess that exposure. *Environ Health Perspect.* 2000;108:475–86.

30. Angerer J, Ewers U, Wilhelm M. Human biomonitoring: state of the art. *Int J Hygiene Environ Health.* 2007;210:201–28.

31. Weisel CP. The development of biomarkers. In: Nieuwenhuijsen MJ, ed. *Exposure Assessment in Environmental Epidemiology.* 2nd ed. Oxford: Oxford University Press; 2015:349–60.

32. CDC. *Fourth National Report on Human Exposure to Environmental Chemicals: Updated Tables.* 2021 Mar. Atlanta, GA: U.S. Department of Health and Human Services; 2021:595.

33. Weisel CP, Zhang J, Turpin BJ, et al. Relationship of Indoor, Outdoor and Personal Air (RIOPA) study: study design, methods and quality assurance/control results. *J Exposure Analysis Environ Epidemiol.* 2005;15:123–37.

34. Becker AM, Marquart H, Masson T, Helbig C, Schlink U. Impacts of personalized sensor feedback regarding exposure to environmental stressors. *Curr Pollution Rep.* 2021;7:579–93.

35. Helbig C, Ueberham M, Becker AM, Marquart H, Schlink U. Wearable sensors for human environmental exposure in urban settings. *Curr Pollution Rep.* 2021;7:417–33.

36. Sagona JA, Shalat SL, Wang Z, et al. Comparison of particulate matter exposure estimates in young children from personal sampling equipment and a robotic sampler. *J Exposure Sci Environ Epidemiol.* 2017;27:299–305.

37. Wu T, Täubel M, Holopainen R, et al. Infant and adult inhalation exposure to resuspended biological particulate matter. *Environ Sci Technol.* 2018;52:237–47.

38. Zhang L, Yao M. Walking-induced exposure of biological particles simulated by a children robot with different shoes on public floors. *Environ Int.* 2022;158:106935.

39. Zhou J, Mainelis G, Weisel CP. Pyrethroid levels in toddlers' breathing zone following a simulated indoor pesticide spray. *J Exposure Sci Environ Epidemiol.* 2019;29:389–96.

40. Georgopoulos PG, Lioy PJ. From a theoretical framework of human exposure and dose assessment to computational system implementation: the Modeling ENvironment for TOtal Risk Studies (MENTOR). *J Toxicol Environ Health Part B, Crit Rev.* 2006;9:457–83.

41. Wambaugh JF, Bare JC, Carignan CC, et al. New approach methodologies for exposure science. *Curr Opinion Toxicol.* 2019;15:76–92.

8

Geographic Information Systems in Children's Environmental Health

Panos G. Georgopoulos

The Historical Origins of Medical Geography

Two thousand five hundred years ago, in the 5th century BC, Hippocrates, the father of medicine, in his classic work *On Airs, Waters, and Places* observed that human health is affected by the combination of (a) human behavior, which includes elements such as eating and exercise habits and (b) geographic location—specifically the environmental attributes associated with location, such as its climate and the "quality" of its air, water, and soil. This was the first recorded recognition of the importance of the environment as a determinant of human health.

The formal birth of a discipline focused on human health and the environment is considered to have taken place in 1792, with the publication of L. L. Finke's treatise (in German) on Medical Geography.[1] About half a century later, in 1854, another pivotal event occurred in London, when John Snow, the "Father of Epidemiology," plotted the locations of cholera cases on a map and discovered a spatial pattern of cases that centered on Broad Street, in Soho (Figure 8.1). This observation famously led Snow to have the Broad Street pump handle removed; although the mechanisms of cholera pathogenesis and transmission were not to be learned for another 40 years, the geographical pattern of the data was sufficiently clear to guide an action of public health intervention and mitigation.

While the early use of geographic information through maps for characterizing disease focused primarily on tracking the spread of epidemics (a 1694 map of plague in Bari, Italy; 1798 and 1819 maps of yellow fever and an 1821 map of typhoid fever in New York City[1]), more complex maps with overlaid demographics and even environmental indicators (such as the 1878 map of offensive odors in Boston) also appeared. In the United States, the Board of Health produced national maps beginning in the late 1800s that examined such statistics as infant mortality rates by state.[2] These maps presented visual geographic information in a manner that shares many attributes with current maps produced by computer-based *geographic information systems* (GIS). A 1912 report on studies of infant mortality, from the then newly established "Children's Bureau" of the US Department of Health and Senior Services, contains the following statement, quoted in Curtis and Leitner[2]:

> [T]he coincidence of a high infant mortality rate with low earnings, poor housing, the employment of the mother outside the home, and large families was indicated in these studies. They all showed great variation in infant mortality rates, not only in different

Panos G. Georgopoulos, *Geographic Information Systems in Children's Environmental Health* In: *Textbook of Children's Environmental Health*, Second Edition. Edited by: Ruth A. Etzel and Philip J. Landrigan, Oxford University Press.
© Oxford University Press 2024. DOI: 10.1093/oso/9780197662526.003.0008

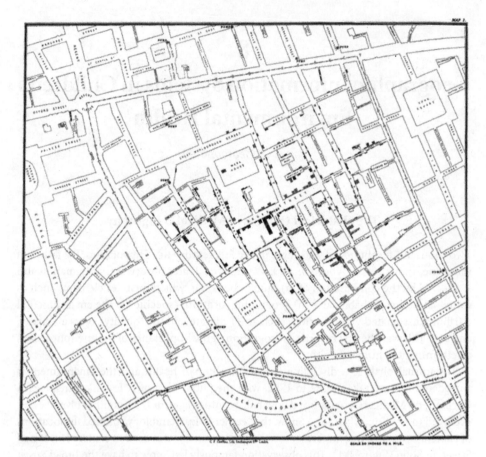

Figure 8.1. Reproduction of John Snow's seminal map of cholera cases (indicated by stacked rectangles) in the London epidemic of 1854. John Snow used geospatial information, i.e., the clustering pattern of cases around the pump located at the intersection of Broad and Cambridge Streets, for an innovative assessment of a serious public health problem.
Source: https://www.ph.ucla.edu/epi/snow/highressnowmap.html.

parts of the United States but also in different parts of the same state and in the same city, town, or rural district. These differences were found to be caused by different population elements, widely varying social and economic conditions, and differences in the appreciation of good prenatal and infant care and the facilities available for such care.

Although "environmental conditions" are not explicitly listed in this statement, the importance of geographic location and of "great variation" in infant mortality across multiple geographical scales was fully recognized.

Advances in computing power, starting in the second half of the 20th century, reduced dramatically the amount of human labor required to perform geographical analyses such as those described above, especially when these analyses were considering issues extending across large spatial domains. Furthermore, the era of personal and mobile computing and the development of the internet made hardware and software tools much more affordable and almost universally available. At the same time, accessibility of the data that are needed to drive geographical analyses evolved rapidly. Thus, while in the early 1990s GIS installations involved expensive licenses that were used almost exclusively by trained specialists with backgrounds in geography and computational science, and data manipulation required development of scripts in platform-specific languages, today, software with menu-driven graphical interfaces is available for personal computers, making its use feasible to scientists with a wide range of backgrounds.

This chapter presents a brief overview of GIS concepts and terminology along with an introductory discussion of important issues and challenges associated with developing and interpreting GIS-based visualizations and analyses of environmental health data. For the reader who wishes to learn more about this fascinating and rapidly evolving field, numerous textbooks and monographs are available, such as Longley at al.,[3] Koch,[4] Curtis and Leitner,[2] Pfeifer et al.,[5] Brown et al.,[6] Emch et al.,[7] and Cromley and McLafferty.[8]

Basic Geographic Information System
Concepts and Terminology

A GIS can be defined as a computer-based "toolbox"[9] with a variety of tools that allow the input (or automatic capture), storage, retrieval, analysis, and display/visualization of a wide range of spatial and spatiotemporal data. These tools reflect developments in a variety of fields, including database technologies, automated data recording and transmission from geo-positioning systems and wireless sensors, numerical computation methods, pattern recognition methods, multivariate statistics and geostatistics, computer visualization methods, and cartography.

The most important requirement for all of the many types of data that go into a GIS analysis is that each data point must be *georeferenced*, that is, assigned some type of geographical location, such as longitude and latitude, and in most cases given a "time stamp" (if the data change with time). The location can be an actual "point" corresponding to a specific set of geographic coordinates (or an exact "address"), or it can be an "areal unit" with boundaries that are typically defined according to an "administrative" jurisdiction (i.e., political boundaries of countries, states, counties, towns, census tracts, blocks, or postal zip codes). In Figure 8.2, which shows a three-frame record of temporal change in global under-5 mortality rates (probability of dying before age 5 per 1,000 live births) over a 25-year period (1990–2015), areal units are defined by country boundaries. In some cases, physical boundaries that are either manmade (such as major roadways) or natural (e.g., the boundaries of a watershed) can be used to define areal units.

Information is stored in the GIS in the form of "information layers," as depicted schematically in Figure 8.3 (top). These layers can include demographic, socioeconomic, environmental quality, land use, climatological, hydrogeological, and ecological data. Data in the different layers are interlinked in the GIS database through their spatial and temporal coordinates. Retrieval of multiple types of data for *receptor locations*, where individuals or members of populations of concern spend time (residences, workplaces, schools, public spaces, etc.), can be performed, using GIS functions to characterize local (ambient and "microenvironmental") conditions at each of these locations. This information, when combined with GIS-based time-location-activity records [shown

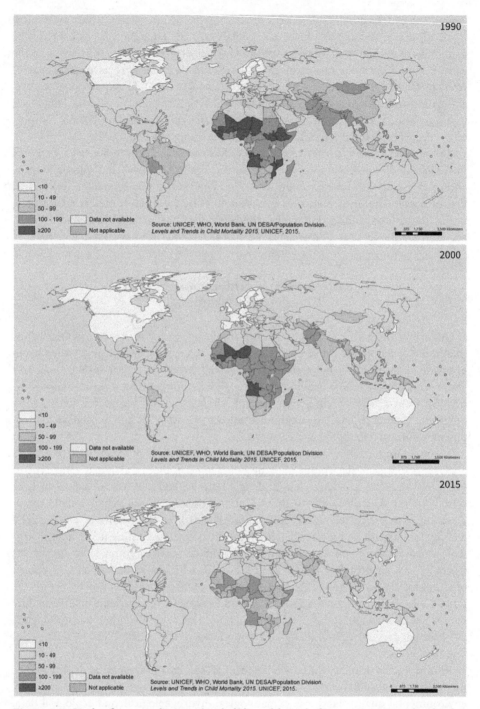

Figure 8.2. Under-five mortality rate (probability of dying before age 5 per 1000 live births) per year 1990 (top), 2000 (middle), and 2015 (bottom)

Source: World Health Organization Global Health Observatory Map Gallery, https://www.who.int/data/gho/map-gallery.

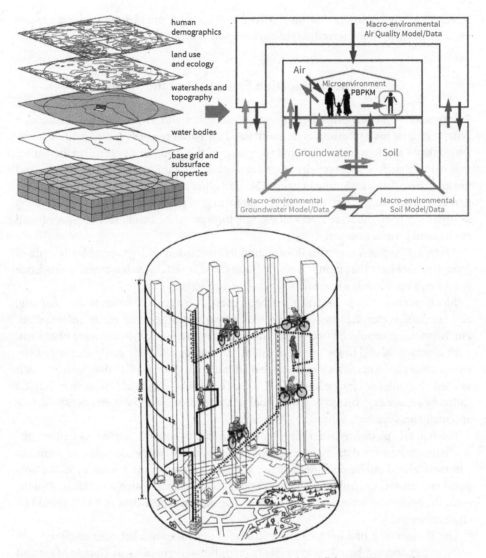

Figure 8.3. **(top):** Schematic depiction of "information layers" for a GIS map relevant to environmental health. The information can be used to characterize outdoor and indoor microenvironments and to associate human exposures with environmental stressors. **(bottom):** GIS can be used to store, manage, and analyze various aspects of information related to time–location–activity patterns that determine individual and population exposures. (Figure sources: top left—USEPA[21]; top right—Georgopoulos et al.[22], adapted with permission, Copyright 1997 American Chemical Society; bottom—Chrisman[23], reprinted with permission from John Wiley & Sons, Inc., Copyright 2001).

schematically in Figure 8.3 (bottom)]) of the individuals or populations under study, can produce detailed characterizations of environmental exposures.

Geographic Information System Types and Standards

Traditionally, GIS platforms were divided into *raster* and *vector* systems. In a *raster system*, geographical information is maintained on a uniform surface grid, and each variable takes a single value in each cell (or pixel) of this grid. This system is efficient for maintaining surface coverage data that are available at a uniform resolution (e.g., land use and cover data), including imagery data and other information from remote sensors that scan areas in a uniform manner. The uniform grid pixels are fixed in space, allowing information from multiple layers at a single location to be combined, compared, and analyzed in a direct manner.

Vector GIS systems maintain information in association with geographically defined geometrical objects (i.e., points, lines, polygons, and areas). In such systems, a residence is a point, a bus route is a line, and a zip code is a polygon.

More recently, the boundaries between these two types of systems are blurring, and modern vector GIS systems can handle various aspects of raster information. Furthermore, it should be noted that though there are many GIS software platforms, both commercial and in the public domain, data can be relatively easily exchanged between these platforms as long as they are maintained in standard GIS data formats, such as ESRI Shapefiles or Google Earth KML files. The Open Geospatial Consortium (OGC) (http://www.ogc.org) brings together public- and private-sector stakeholders to define and maintain standards across the GIS field.

Current GIS platforms generally offer built-in *spatial database engines* as well as connectivity options for data maintenance and retrieval across a wide range of common commercial and public domain database systems. As long as the necessary spatiotemporal coordinates are linked with data residing in a database (such as MySQL, Oracle, etc.), the process of accessing, analyzing, and visualizing these data in a GIS should be straightforward.

Until recently, it had been considered useful to distinguish between *mapping software*, either desktop-based, or available from online servers (such as Google Maps and OpenStreetMap) and *GIS software*, such as ArcGIS and QGIS.[3] The former traditionally produced maps within only a range of predefined specifications, whereas the latter provided functions involving data input, management, analysis, and fully customized visualization/presentation. However, as systems evolve, this distinction is becoming less clear, and platforms that were traditionally considered to be mapping software have gradually been adding substantial functionality that allows both data input and customization.[10]

Publicly Available Data for Geographic Information System Applications

An enormous amount of GIS-relevant data (socioeconomic/consumer as well as ecological/environmental) have been collected in recent years by numerous public agencies and private-sector entities across the globe. This has resulted in a virtual "information overload." Unfortunately, the amount of data collected does not necessarily imply either quality or completeness, often due to inefficiencies in planning and coordination.

In the United States, GIS installations are maintained by federal, state, and very often by county and local agencies. Many of these sources provide online access to potentially very useful GIS data that can support various analyses, including public health and environmental health studies. Valuable information and data directly relevant to children's environmental health can be retrieved from the sites of agencies such as the US Geological Survey (USGS), the Environmental Protection Agency (EPA), the Centers for Disease Control and Prevention (CDC), and the National Cancer Institute (NCI). In addition to making GIS data available to download, these agencies provide various user-friendly online "mapping tools" for interactive creation of customized maps depicting information as specified by the user. An example is the EPA's *EnviroMapper* (https://enviro.epa.gov/enviro/em4ef.home) for the creation of maps of environmental metrics (https://enviro.epa.gov/enviro/em4ef.home). A second example is the EPA's Environmental Justice Screening and Mapping Tool *EJscreen* (https://www.epa.gov/ejscreen), which uses a nationally consistent dataset and approach for producing maps and reports that combine environmental and demographic indicators. A third example are the interactive maps of cancer statistics provided by CDC/NCI through the State Cancer Profiles portal (https://statecancerprofiles.cancer.gov). Though these online tools are limited compared to a full-scale GIS, they provide easily accessible geospatial information to the public.

From the perspective of the potential data consumer, a useful development has been the creation of a US "government-wide" geospatial data portal (www.geoplatform.gov) that provides links to data collected across government agencies as well as an infrastructure for analyzing these data. International data are not as widely available. Nevertheless, many sites provide online access to various types of health-related geographic data. The World Health Organization (WHO), for example, has "pre-made" maps on global health issues available online at http://gamapserver.who.int/mapLibrary/.

Basic Geographic Information System Operations and Mapping Options

Visualization via GIS can be seen as a kind of "interactive cartography." The GIS user has many choices in mapping/plotting data and in applying various operations on them to improve the interpretation and enhance the visualization of available information. Typical basic operations available in a GIS include *transformations between coordinate system projections, layering, buffering, aggregation*, and *performing Boolean expression queries* on the datasets. Information on these operations can be found in any introductory GIS textbook and therefore will not be discussed in detail here. Only brief informal definitions follow, accompanied by words of caution, as appropriate.

Geographic projection transformation: In general, if an analysis depends on a particular spatial property (area, length, direction, shape), then an appropriate coordinate projection should preserve that property (at least locally) on the map. Most of the maps displayed in this chapter focus on a domain covering the contiguous United States (CONUS) and depict properties that require areas to be preserved across the map to avoid "information distortion." A *Lambert azimuthal equal area* projection often is used for this purpose. By contrast, local-scale studies would typically use other projections, such as *state plane* or *universal transverse Mercator* (UTM). The reader should make sure, when comparing variables presented in different maps, that the maps employ consistent and comparable geographical attributes (projection, zone, geodetic datum, scale).

Layering: This involves the combined presentation of information from different data layers. For example, environmental, demographic, and health outcomes could be plotted together (or on maps with identical attributes) to explore potential associations and formulate hypotheses for further study. An example would be a study examining links between air pollution levels, racial distribution within zip codes, and incidence of child-hood asthma.

Buffering: This refers to characterization of information from multiple layers in an area surrounding specific map elements. Buffering involves the creation of a zone (of either constant or variable width) around a point, line, or polygon. Queries determine which entities occur either within or outside the buffer.

Aggregation: This involves bringing together information from various locations or areal units to characterize (or assign variable values) to a larger area that encompasses these locations. This is a process that ranges in complexity depending on the type of available data.

Queries involving Boolean ("and/or") expression operations are used to extract specific subsets of information from the typically multiattribute datasets residing in a GIS. A user may, for example, use such queries to map the distribution of a population fraction that belongs to a particular gender, age group, and lives in a household with a specified income range.

Data Visualization Options

Several basic data visualization options are available today within typical GIS installations. These include *choropleth maps, isopleth maps, simple dot/symbol maps*, and *proportional (graduated) symbol maps*.

The *choropleth map*, which uses colors or shades corresponding to value ranges of the variable studied, is the most used data visualization technique. The approach is often treated as the "default" option by GIS users. The top panel of Figure 8.4, which shows modeled estimates of county-level annual average airborne diesel particle concentrations across the CONUS, is a typical example of a choropleth map. The reader should be aware that the choropleth map option can easily miss or misinterpret peaks and patterns in the data, if, for example, the number and values of "break points" in the value ranges that correspond to different colors or shades are not properly selected. There are various ways of selecting these break points, including use of equal steps, logarithmic steps, fixed quantiles, or, as in the case of Figure 8.4, values that correspond to a fraction or a multiple of an environmental or health benchmark.

An advantage of using an interactive GIS map rather than a static map is that the user can, and should, explore various option settings to select the most appropriate visualization. In cases of highly variable spatial values, supplementing the choropleth map with a 3D visualization that adds "column height" to color range for depicting values of the quantity of concern, as in the bottom panel of Figure 8.4, can be useful for identifying outliers and peaks that are "hidden" in the 2D map.

The choropleth option can be problematic when the areas of the geographic units vary substantially (e.g., counties or census tracts with similar population sizes but very

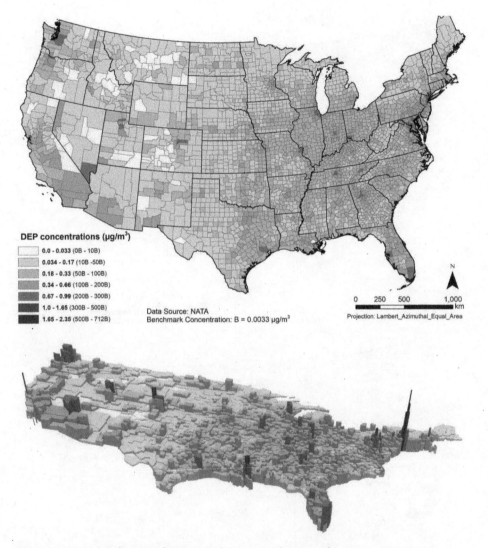

DEP concentrations (µg/m³)

	0.0 - 0.033 (0B - 10B)
	0.034 - 0.17 (10B -50B)
	0.18 - 0.33 (50B - 100B)
	0.34 - 0.66 (100B - 200B)
	0.67 - 0.99 (200B - 300B)
	1.0 - 1.65 (300B - 500B)
	1.65 - 2.35 (500B - 712B)

Data Source: NATA
Benchmark Concentration: B = 0.0033 µg/m³

N

0   250   500   1,000 km

Projection: Lambert_Azimuthal_Equal_Area

Figure 8.4. Modeled county-level annual average airborne ambient concentrations of diesel exhaust particles (µg/m³) across the contiguous U.S. counties for 2014 (Data source: USEPA 2018 National Air Toxics Assessment, http://www.epa.gov/nata). The break points for the different colors used in this choropleth map correspond to multiples of a Cancer Health Benchmark B = 0.0033 µg/m³ (see www.nj.gov/dep/airtoxics/diesemis.htm)

different areas) within the spatial domain considered. In such a case, use of a proportional symbol map may provide a visually more appropriate depiction of the variable of concern. Figure 8.5, which summarizes 2000–2004 county-level data on mortality rates of childhood leukemia (age 0–14 years per 100,000 person-years), demonstrates the utility of this approach. It shows that the choropleth map does not appropriately "resolve" counties in the eastern half of the nation (top panel), while a proportional symbol map offers a clearer representation (bottom panel).

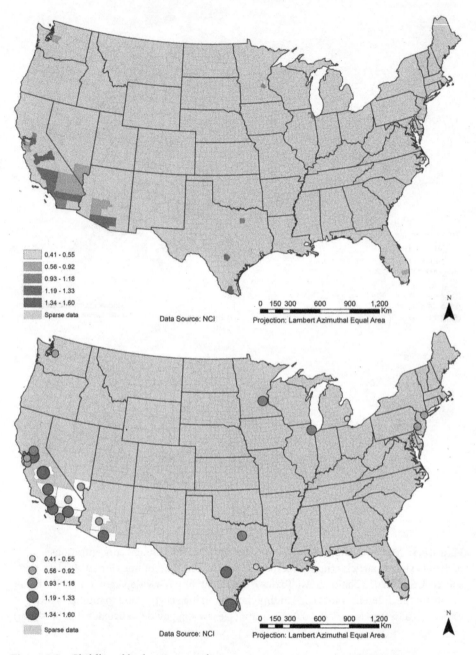

Figure 8.5. Childhood leukemia mortality rates, ages 0-14 years per 100,000 person-years by county for 2000–2004: Choropleth map (top panel) and proportional symbol map (bottom panel)

Source: National Cancer Institute, http://www.cancer.gov.

Spatial Analysis with GIS: Caveats Concerning Issues of Scale and Aggregation

The simplest methods of spatial analysis involve some form of regression performed on a domain of geographical "areal units" such as counties, municipalities, zip codes, census tracts, and possibly point locations. A metric of a "dependent" variable (such as low birth weight) for each geographical unit is regressed against metrics representing various demographic, socioeconomic, behavioral, and environmental variables (such as income, age of mother, education, consumption of particular foods, quality of drinking water, and quality of ambient air) at the same location and that are thought to influence low birth weight. "Global" (i.e., domain-wide) associations can be identified through this process for further study. Figure 8.6 presents an example of this type of analysis, showing county-level maps of the percentages of individuals living below the poverty level (top) and the percentages of low birth weight (<2.5 kg) deliveries (middle), along with a regression scatterplot (bottom) showing a positive association of these variables. Associations between geographically defined factors will in general be variable across space.

The study of *outlier* areas, where the association is either stronger or weaker than across the whole domain, can be especially informative. Outliers can provide insights on the nature of the association and on how other variables may affect the association. For example, in Dr. John Snow's studies of cholera in London described above, important information was derived from study of families living many blocks distance from the Broad Street who, despite their distance from the source of contamination, developed cholera. Further investigation revealed that several of these families had carried water from the Broad Street pump to their houses because they liked the taste.

Numerous advanced computational analysis methods for studying spatial (and spatiotemporal) multivariate associations of geographically defined variables exist. They derive from scientific fields that range from classical geostatistics to stochastic field approaches and modern Bayesian information theory. A rapidly growing body of literature that includes scientific journals, specialized monographs, and textbooks is available on the subject. Furthermore, many specialized software packages that implement newer computational methods for GIS also are available, either as stand-alone applications or as "extensions" and "add-ons" that complement popular commercial and public domain GIS platforms. The interested reader can find excellent introductions to the field in de Smith et al.,[11] Pfeifer et al.,[5] Haining,[12] Waller and Gotway,[13] and Lawson,[14] whereas more advanced or specialized overviews, including applications of various software platforms, are presented in, for example, Kanevski,[15] Carr and Pickle,[16] Lawson,[17] Cressie and Wikle,[18] Fischer and Getis,[19] and Kresse and Danko.[1]

GIS offer an essential framework for bringing together and "fusing" not only data from diverse observations and field studies, but also outcomes from mechanistic environmental and socioeconomic modeling studies, thus allowing multiple levels of cross-evaluation and hypothesis formulation and testing. For example, current versions of many widely used mechanistic models of environmental transport and fate have been adapted to operate in conjunction with GIS and associated databases, accepting inputs and producing outputs in the standard formats and structures supported by these systems. In fact, the evolving trend is toward tighter integration and "interoperability" of GIS and environmental models[20] aiming to achieve more streamlined and efficient analyses of environmental problems, optimizing data management aspects of the process and reducing the potential for human error.

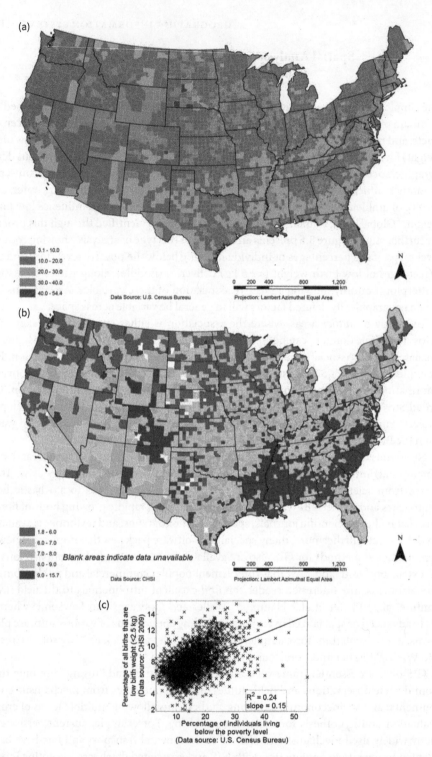

Figure 8.6. (a) Percentage of individuals living below the poverty level in all contiguous US counties for 2008; (b) Percentage of all births that are low birth weight (<2.5 kg) in all contiguous U.S. counties for years 1996–2005; (c) Regression scatterplot showing the percentage of individuals living below the poverty level vs corresponding percentage of low birth weight births by county, 1996–2005

Sources: U.S. Census Bureau, http://census.gov and Community Health Status Indicators Report. For details see Georgopoulos et al.[24]

Challenges to GIS Analysis

Two particularly important challenges to GIS analysis are (a) the spatial (and in some cases spatiotemporal) dependence of data, and (b) the effects of scale (or resolution and aggregation).

The challenge of the spatial dependency becomes evident in a GIS map where the value assigned to each variable in each geographical unit (e.g., census tract or zip code) is assumed to be independent from the values assigned to the same variable in other neighboring units. In reality, neither environmental properties (e.g., airborne concentrations of contaminants) nor social conditions (e.g., crime rates) would be expected to abruptly change at administrative boundaries such as county lines. Instead there usually is a gradual variation of values and some degree of dependence among neighboring units. But when a single metric, typically an average value, is used to characterize the entire geographical unit, the consequence can be the appearance of sharp and unrealistic gradients at the unit boundaries that can be easily misinterpreted by the public, especially when presented in map form.

The second challenge to GIS, known as the modifiable area unit problem (MAUP), reflects the fact that the results of any analysis can vary according to the level and type (resolution and configuration) of geographical aggregation (country, state, county, township, zip code, census tract, census block) employed in the study. Revisiting Figure 8.2 and considering in more detail the spatial distribution of under-5 mortality rate for the CONUS presented in Figure 8.7 with state-level data (top panel) and available county-level data (bottom panel) demonstrates this point. Though at the nationwide level (shown in Figure 8.2) the United States is assigned a rate of less than 10, post-2000, Figure 8.7 identifies areas where this rate is still actually larger than 10 for 2009.

In another example,[2] an analysis of infant deaths at the zip code level might find no significant relationship between infant mortality and birth weight if there is substantial variation of income within the zip code and affluent mothers are able to afford the care needed to cope with medical conditions arising from the low-birth-weight delivery. However, a relationship might become evident if the scale changed to the census block level with more uniform socioeconomic conditions prevailing within each such unit.

Though an increase in resolution may often address this type of problem, it should not be considered a panacea because, in fact, there may not be sufficient data at the higher resolution (i.e., smaller geographical unit) to derive reliable and representative statistics for the variable of concern. Some of the advanced analytic methods now available incorporate approaches to probabilistically quantify uncertainty distributions for variables of concern at different scales (i.e., different sizes of the geographical unit) to identify scales appropriate for the needs of specific study objectives.

The Present and Future of GIS in Environmental Health: Mobile Devices and Ubiquity

In recent years, the general public has been experiencing, though unknowingly, a dramatic increase in accessibility to GIS services. Whenever someone is using a smartphone or other wirelessly connected mobile device, such as a tablet, to quickly find out whether there is a theater within a 10-mile radius, showing a specific film with show times starting between 7 and 8 PM, she or he is in fact performing a "query" that just a few years

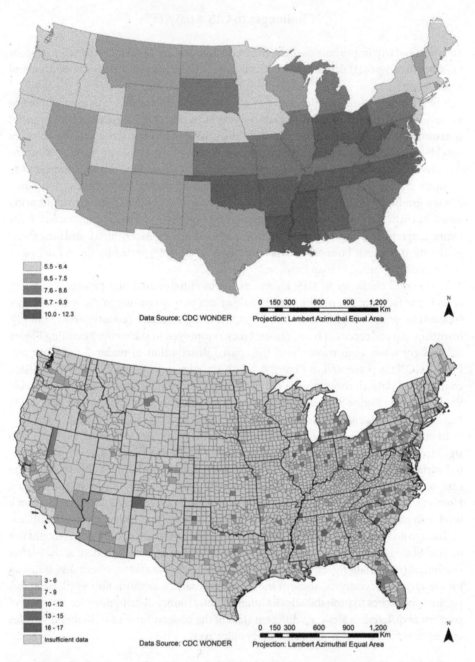

Figure 8.7. Under-5 mortality rate (probability of dying by age 5 per 1,000 live births) for the year 2009 (top) in all contiguous U.S. States; (bottom) in all contiguous U.S. counties for which there are sufficient data

Source: Centers for Disease Control and Prevention, National Center for Health Statistics, http://wonder.cdc.gov/cmf-icd10.html.

ago would require access to a GIS system via scripting commands. Now the smartphone is acting as a "client" to a remote GIS server, and the query takes place through a user-friendly screen-touch interface—or even through verbal communication via a voice-interpreting computer program. Similar queries can be used to retrieve information on nearby healthcare facilities, schools, childcare centers, etc., as well as on gas stations, industrial sites, or landfills; on toxic releases and spills that have taken place within a user-defined area of concern; or on most recently measured air quality data by monitors within a prescribed radius around the user's or any other defined location.

Demographic and socioeconomic data, directly relevant to environmental health, also are available to smartphone users: map visualization of this information takes place on the smartphone via online platforms that include Google Maps, Apple Maps, and ESRI/ArcGIS. Numerous applications (or "apps") that can perform the above-mentioned information retrieval and visualization tasks are available for different types of mobile devices and operating systems (e.g., iOS and Android). These same devices also typically have built-in geo-positioning systems (GPS), that, in conjunction with either local or remote (on server) software, can track and record the location of the user, provide location-aware environmental risk-relevant information, or determine the optimal route for getting to a healthcare facility. Though the software and data that perform and support these actions reside on remote servers and function "behind the scenes," they reflect the evolving state-of-the-art in GIS technology. These new developments present great potential not only for the retrieval and interpretation of existing information, but also for the design of future environmental health studies that will collect and analyze multiple layers of data with increased accuracy and efficiency.

Acknowledgments

This work was supported in part by the National Institute of Environmental Health Sciences (NIEHS)-sponsored Center for Environmental Exposures and Disease (Grant #: NIEHS P30ES005022). The author is grateful to Zhongyuan Mi of the Environmental and Occupational Health Sciences Institute for his assistance.

References

1. Kresse W, Danko DM. *Springer Handbook of Geographic Information.* 2nd ed. New York: Springer; 2021.

2. Curtis A, Leitner M. *Geographic Information Systems and Public Health: Eliminating Perinatal Disparity.* Hershey: IRM Press; 2006.

3. Longley PA, Goodchild MF, Maguire DJ, Rhind DW. *Geographic Information Science and Systems.* 4th ed. Hoboken, NJ: Wiley; 2015.

4. Koch T. *Cartographies of Disease: Maps, Mapping, and Medicine.* Redlands, CA: ESRI Press; 2005.

5. Pfeiffer D, Robinson TP, eds. *Spatial Analysis in Epidemiology.* New York: Oxford University Press; 2008.

6. Brown T, McLafferty S, Moon G. *A Companion to Health and Medical Geography.* Malden, MA: Wiley-Blackwell; 2010.

7. Emch M, Root ED, Carrel M. *Health and Medical Geography*. 4th ed. New York: Guilford; 2017.

8. Cromley EK, McLafferty S. *GIS and Public Health*. 2nd ed. New York: Guilford; 2012.

9. Clarke KC. *Getting Started with Geographic Information Systems*. 5th ed. Upper Saddle River, NJ.: Prentice Hall; 2010.

10. DuVander A. *Map Scripting 101: An Example-Driven Guide to Building Interactive Maps with Bing, Yahoo!, and Google Maps*. San Francisco, CA: No Starch Press; 2010.

11. de Smith MJ, Goodchild MF, Longley PA. *Geospatial Analysis: A Comprehensive Guide to Principles, Techniques, and Software Tools*. 6th ed. London: Winchelsea Press; 2018.

12. Haining RP. *Spatial Data Analysis: Theory and Practice*. Cambridge: Cambridge University Press; 2003.

13. Waller LA, Gotway CA. *Applied Spatial Statistics for Public Health Data*. Hoboken, NJ: Wiley; 2004.

14. Lawson A. *Statistical Methods in Spatial Epidemiology*. 2nd ed. Chichester, UK: Wiley; 2006.

15. Kanevski M. *Advanced Mapping of Environmental Data: Geostatistics, Machine Learning and Bayesian Maximum Entropy*. Hoboken, NJ: ISTE/Wiley; 2008.

16. Carr DB, Pickle LW. *Visualizing Data Patterns with Micromaps*. Boca Raton, FL: Chapman & Hall/CRC; 2010.

17. Lawson A. *Bayesian Disease Mapping: Hierarchical Modeling in Spatial Epidemiology*. Boca Raton: CRC Press; 2009.

18. Cressie NAC, Wikle CK. *Statistics for Spatio-Temporal Data*. Hoboken, NJ: Wiley; 2011.

19. Fischer MM, Getis A, eds. *Handbook of Applied Spatial Analysis: Software Tools, Methods and Applications*. Heidelberg: Springer; 2009.

20. Brimicombe A. *GIS, Environmental Modeling and Engineering*. 2nd ed. Boca Raton, FL: CRC Press/Taylor & Francis Group; 2010.

21. USEPA. *Multimedia, Multipathway, and Multireceptor Risk Assessment (3MRA) Modeling System. Volume I: Modeling System and Science*. Report No. EPA530-D-03-001a. Research Triangle Park, NC: U.S. Environmental Protection Agency; 2003.

22. Georgopoulos PG, Walia A, Roy A, et al. An integrated exposure and dose modeling and analysis system. 1. Formulation and testing of microenvironmental and pharmacokinetic components. *Environ Sci Technol*. 1997;31(1):17–27.

23. Chrisman NR. *Exploring Geographical Information Systems*, 2nd ed. New York: Wiley; 2002.

24. Georgopoulos PG, Brinkerhoff CJ, Isukapalli S, Dellarco M, Landrigan PJ, Lioy PJ. A tiered framework for risk-relevant characterization and ranking of chemical exposures: applications to the National Children's Study (NCS). *Risk Analysis*. 2014;34(7):1299–316.

9

Developmental Toxicology and Children's Environmental Health

Edward D. Levin

The process of development from two gametes combining to form a single-cell zygote to then proceed to develop into a fully formed and integrated organism requires a complex series of molecular, cellular, tissue, organ, and organ system physiological events to be performed in the correct sequence within specific windows of time. Early development sets the foundation for health across the entire life span.

Development from parental gametes to adult offspring is a complex and dynamic process. The essential plan is not necessarily fragile. Because correct performance of developmental events is critical to the survival of the individual and the species, numerous controls have evolved that enable developmental processes to perform sufficiently well even in the face of the ancient challenges that humans have confronted for millennia, such as variations in diet, disease, and stress. However, sufficient performance is not necessarily optimal performance, as demonstrated by studies showing that variations in diet, disease, and stress during vulnerable periods in early development can have long-lasting and deleterious effects on health.

Of greater concern are exposures to new synthetic chemicals that have been invented only in the past few generations. These exposures to relatively recent anthropogenic toxicants that were never before encountered during evolution pose potentially serious and frequently uncharted dangers to development. Because of this lack of evolutional adaptation, too frequently, there is insufficient physiological defenses against these toxic dangers.

There could be adaptations to low-level toxicant exposures during development or across generations via epigenetic effects. However, in an environment with changing environmental toxicant exposures there could be mis-adaptation, when an individual develops under the influence of one toxicant but then that individual later in life or their offspring are faced with new types of toxicant exposures.[1]

Developmental exposures to novel compounds can have particularly insidious and long-lasting effects when synthetic compounds are designed to be biologically active or by happenstance affect key biological processes. While there are a variety of compensatory mechanisms to help keep the complex processes of development on track, multiple chemical exposures or chemical exposures combined with genetic susceptibilities or other conditions that affect critical processes of development and their control mechanisms can be especially devastating. Other chapters in this book discuss in great detail the progress that has been made in understanding the variety of ways by which exposures to environmental contaminants, drugs of abuse, or iatrogenic effects of therapeutic drugs can disrupt development and produce lifelong impairment (see Chapters 14 and 47).

Edward D. Levin, *Developmental Toxicology and Children's Environmental Health* In: *Textbook of Children's Environmental Health*, Second Edition. Edited by: Ruth A. Etzel and Philip J. Landrigan, Oxford University Press. © Oxford University Press 2024. DOI: 10.1093/oso/9780197662526.003.0009

Key to finding better ways of controlling these adverse impacts is enhanced understanding of the vulnerable processes of development and of how toxic environmental exposures can perturb these processes beyond the ability of physiologic mechanisms to compensate. This is the area in which the science of developmental toxicology makes its contribution.

When dealing with development, the commonly cited quote by Paracelsus that "The dose makes the poison" must be amended to include the realization that "The timing of exposure also makes the poison" (see also Chapter 2). Development is by definition a sequence of biological processes that must occur in the proper order to construct a complete organism. Toxic effects that result from exposures in early development depend on precisely when during development an exposure occurs. Most obviously, it is the particular organ or tissue under construction at the time of the toxic exposure that is the prime target for disruption. Even when an exposure is ephemeral and takes place only in a brief window of time when a particular developmental process is under way, permanent impairment can result. Some effects of early exposure are clinically obvious, such as the phocomelia caused by exposure to thalidomide. There are severely toxic effects that impact relatively few people; there are also more subtle toxic effects that impact far greater numbers of people because so many are exposed to lower levels of toxicants. These more common types of toxicities are the less overt and may not be fully obvious until later phases of development. Such is the case for toxic impacts on genomic imprinting. Epigenetic alterations in the annotation of DNA may not become evident until much later in life when the affected portion of the genome plays a critical role in development. Thus, the incipient toxic event can long precede the disrupted developmental process.

Widespread Effects of Low-Level Toxic Effects

At very high doses of exposure to a toxicant all individuals are adversely impacted. High-level exposures have dire effects, but the incidence of such exposures is thankfully rare. Lower-level exposures are much more prevalent. They have lesser effects on each affected individual. Nonetheless, a modest effect on a population-wide scale can have as great or greater a societal impact as a highly toxic effect on a few individuals. For example, an estimated five-point deficit in IQ occurred in millions of young children in the United States and countries around the world from the 1950s through the 1970s as the consequence of population-wide, low-level exposure to lead from gasoline. A general dampening of productivity resulted with this reduction in the number of highly intelligent persons in the population and a corresponding expansion of the lower tail of the IQ distribution (see also Chapters 3 and 31).

Across the population many variables influence reactions to toxic exposures. For example, at the same dose of exposure some individuals have very adverse effects while others can tolerate the toxic exposure quite well. Discovering these determinants of vulnerability is crucial for understanding the risks of low-level exposure. This is another area where the science of developmental toxicology makes an important contribution.

There are two ways to study the mechanisms of vulnerability. One is *failure analysis*, seeing how some individuals react poorly to exposures to toxicants. This certainly accurately identifies those at risk, but only after the toxic damage is done. Their functional impairments are by definition apparent. Connecting the impairment to the exposure is the task of the scientist. However, this approach is often fraught with difficulty since there are often many possible explanations for impairment and only some of these involve low-level exposure to a toxicant. Thus, it is often very difficult to prove that a toxic exposure has caused impairment in a particular individual. In addition, determining the causative chain of events from the toxic exposure through the physiological derangement to the functional impairment can be extremely difficult. A second, potentially useful approach for discerning mechanisms of vulnerability is to study toxicant effects on individuals who do not show adverse effects. A complex set of physiological reactions have evolved that protect against adverse phenotypic effects. Instead of failure analysis, we can study how robust individuals compensate for toxic exposures to avoid impairment.

Enhanced Exposures

Children's behaviors can enhance their exposures to a variety of toxic chemicals in their environments. For example, crawling can increase proximity to toxic contaminants. During the infant and toddler stages, children spend considerable time near the floor, where much of the dust containing environmental toxicants settles.[1] Another childhood activity that enhances exposure is hand-to-mouth behavior. This is an innate response of young children that can increase their ingestion of contaminants from the environment. *Pica*, the consumption of non-food items, is another ingestive behavior to which some children are prone and that has been found to increase toxicant load. This is particularly of concern when the exposure is highly toxic, as is the case with lead. Lead pica has been shown to result in very high body burdens.[2] Lead pica is further exacerbated by the fact that lead acetate is sweet to the taste.[3]

Sole-source nutrition can increase both the exposure and the impact of toxic chemicals. Breast-feeding is the classic example. Breast-feeding is highly beneficial for infant development. If the mother carries a heavy burden of lipophilic toxicants such as poly-chlorinated biphenyls (PCBs) or polybrominated diphenyl ethers (PBDEs), these compounds will readily cross through the milk to the nursing infant. Bottle-fed infants also can be very vulnerable to high-dose exposures from sole-source nutrition. Thus soy-based formula can provide an excess of phytoestrogens, which, in light of animal models and human research, could have persisting effects on both male and female development following early-life exposure.[4,5] The net estrogenic activity of the infant formula can be further increased by bisphenol A (BPA) coming from the plastic bottles used for feeding.[6]

Dietary Constituents

Both caloric deficiency and caloric excess have been found in numerous studies to harm the health of children. Overnutrition from excessive caloric intake can result in obesity, diabetes, and, later in life, in hypertension and cardiovascular disease.

Specific toxic effects can also result from excessive intake of particular dietary components. For example, vitamin A (retinol) is essential for optimal development and especially brain development. It is involved in gene expression, cell differentiation, proliferation, migration, and programmed cell death via actions at retinoid acid receptors.[7] However, adverse effects on neurodevelopment can result both from overdoses as well as from deficiency of vitamin A during pregnancy because of the key role that retinoic acid receptor signaling plays in neural tube closure.[8-10] These effects have been observed repeatedly in experimental animal studies.[11] The consequences of excess vitamin A in pregnancy include severe malformations of the brain, increased risk for spina bifida and other neural tube defects, and a variety of craniofacial, limb, cardiac, and urogenital abnormalities. Between a fortified diet and vitamin supplements, pregnant women can easily ingest doses of vitamin A that exceed recommended levels.[8]

Disparities in Toxic Exposures Contributing to Societal Group Differences

Highly exposed groups are of special concern in developmental toxicology. Particular groups within the population may have toxic exposures that are much greater than the majority of the population. For example, blood lead levels have decreased over the past 25 years as evidence has accumulated documenting the deleterious neurobehavioral effects of low-level lead exposure. However, disproportionately greater numbers of African American children in the United States compared to Caucasian children still have elevated blood lead. The percent of African American children who have blood lead levels greater than 5 μg/dL is more than three times that of Caucasian children (17.4% vs. 5.6%).[12] Higher blood lead levels in African American children may at least partially account for the gap in scholastic achievement between these children and Caucasian children.[13]

Sex Differences in Response

There are typical sex differences in all mammals, including humans, in addition to the primary and secondary sex characteristics. Developmental toxicants can have selective effects on either males or females, and they can also cause diminution of typical sex differences. For example, early postnatal exposure of rats to the organophosphate insecticide chlorpyrifos diminishes the typical sex differences in spatial memory and locomotor activity.[14,15] Evidence for a sex-selective effect of chlorpyrifos on cognitive impairment has also recently been reported in boys, but not in girls in an epidemiological study.[16]

Long-Term Effects

Every person has gone through development. All humans are either children or former children. All adults started as zygotes, passed through gestation, and grew through infancy, childhood, and adolescence. Flaws in the developmental process have consequences therefore not only in children, but also in adults. Inherent in the study of developmental toxicology is consideration of the adult expression of toxic events in early life. Particularly severe impairments may be evident in childhood, but more subtle toxic exposures may not have their full impacts on health until later in life. Because of this lifelong evolution of developmental toxicity, it is important to be aware of the delayed consequences of older toxicants which have been identified and removed from use. Persons who were exposed to these materials decades ago when they were children will continue to be at risk of health effects from those exposures throughout the rest of their lives.

A case in point is lead. We have substantially decreased higher exposures to lead. But we also need to attend to the lifelong effects of high exposures in the past by developing adequate treatments for the persisting toxic impairments of higher developmental lead exposure as it interacts with processes of disease and aging.

Much progress has been made in developmental toxicology and in understanding the delayed consequences of toxic exposures in early life, but many challenges remain. Lead illustrates both our past gains and our future challenges. Half of the population in the United States is estimated to have had adverse blood lead levels during development.[17] Much progress has been made in reducing exposure to lead. This progress was triggered by epidemiological studies that detected behavioral impairments in children with low-level elevations of lead in blood. It was further catalyzed by experimental animal studies demonstrating the cause-and-effect relationship between low lead exposure in early development and neurobehavioral impairment. However, despite this success, further progress is necessary to prevent and manage the neurotoxicity still caused today by lead. We must not fall victim to the fallacy of small numbers. The fact that, in the United States, mean blood lead concentrations in children ages 1–5 years have fallen from more than 13 µg/dL in the late 1970s[18] to approximately 2 µg/dL in the early 2000s[12] is evidence of improvement, but we should not subscribe to the false belief that just because 2 is a small number the job is nearly done. The number 2 is merely an artifact of the units of measure: 2 µg/dL = 20 µg/L = 2,000 ng/dL. The real concern is the risk posed by a level of exposure, not the number describing it.

Another point of concern is that there is no single threshold for toxic risk. There is a spectrum of thresholds for the variety of individuals who comprise a population that depends on a number of factors that include age at exposure, genetic factors, concurrent disease states, and other concomitant exposures. Thresholds for toxicity also vary within individuals depending on the organ system or function in question. To accommodate this spectrum of thresholds and in the face of incomplete knowledge of the mechanisms of toxicity and consequences of intoxication, toxicologists have taken a conservative approach and used safety factors to decrease the permitted level of exposure to protect the population. In the case of lead, no such safety factors are in place. If safety factors were used for lead, the entire population would be included in the at-risk population even though the average blood lead value is now "only" about 2 mg/dL.

Another fallacy that must be avoided is the notion that removal of a toxicant from the environment, or at least removing children from exposure, solves the problem of toxicity. This view abdicates our responsibility for those who have already been exposed. All adult humans were once embryos, fetuses, newborns, infants, children, and adolescents. Developmental exposures have lifelong effects. Children exposed today will continue to have adverse effects expressed as their lives progress. It is also true that we children of the past continue to carry with us the toxic legacy of our exposures decades ago. Ideally, information about developmental toxicity can be used to reduce exposures in the future, but we also need to address the consequences of past exposures and develop treatments for them. We must also consider the possibility of intergenerational toxicity of past toxic exposures if toxicant-induced epigenetic changes are carried through the gametes.

A potentially useful approach for discerning mechanisms of vulnerability, which is often overlooked, is to study toxicant effects on endophenotypes in individuals who do not develop adverse effects following exposure. A complex set of physiological reactions have evolved over the millennia to protect us against adverse phenotypic effects. Instead of failure analysis it will be important to study how robust individuals compensate for toxic exposures. These are the critical mechanisms that, when deficient, open the way for susceptible individuals to fare poorly to low-dose exposures.

We must continue with vigilance to conduct epidemiological studies to detect, reduce and eliminate human exposures to toxic chemicals. But we must also recognize that epidemiological findings of toxicity are in reality admissions of failure. Epidemiology is essentially failure analysis of toxic exposures to which people have suffered. Much better, we should aim to identify toxic risks before they cause damage to humans. The goal of toxicology must be to predict toxic risk before chemicals are released into human contact.

Future Developments

Future developments in the field of developmental toxicology will take at least two forms. First, new and possibly toxic exposures will need to be monitored, studied, and ideally avoided. Mandatory premarket testing of new chemicals and processes before they enter commercial markets is a highly efficient way to accomplish this goal. Second, new discoveries about critical biological processes are needed that will enable us to more accurately gauge toxic risks during development. A few examples of discoveries of biological processes important for toxicity are oxidative stress and epigenetic, microbiome, and paternal toxicity effects on development. Unlike with maternal exposure, after conception the offspring is not directly exposed to the toxicant, but paternal exposure before conception does alter epigenetic imprints on sperm DNA[19-21] and neurobehavioral development in the offspring.[22-24]

Characterization of potential new toxicants and testing of untested potential toxicants to which people are already exposed will require more efficient means of toxicological assessment. Much enthusiasm has been expressed for the promise of high-throughput assays for quickly determining the toxic risk of great numbers of compounds and mixtures. It is true that much information can be collected in a short period of time using these methods. However, the significance of this information is often not clear. More complex in vitro screens are being developed, but none replicates the intricate signaling sequences involved in developmental processes that have evolved over millions of years for the integrated organisms of concern today. Rather than abandoning classic in vivo models of toxicity for the newer high-throughput assays, an alliance would be more beneficial. It would be advantageous to use both types of models and others in a spectrum of approaches to ascertain which toxicants pose excessive risks.

In particular, a fruitful approach may be to use the classic in vivo models to discover additional physiological processes that are key mediators of toxic effects and then to devise high-throughput in vitro assays to screen numerous chemicals for this endophenotypic mechanisms. This approach recognizes the truth that our current knowledge of biology is quite incomplete. This two-stroke engine of biological process discovery and chemical screening can efficiently evaluate the set of unknown possible toxicants in a biologically informative way. It is good that advancements in vitro models have become much cheaper and with higher throughput because not only will we need these types of models to screen the many tens of thousands of compounds and mixtures currently uncharacterized, but we will also have to rerun the full complement of potential toxicants again and again as additional pathways of toxicity are discovered.

Intermediate models using invertebrates, such as *Caenorhabditis elegans* and *Drosophila*, and aquatic models, such as zebrafish and medaka, offer some of the economy and molecular access of in vitro models with the advantages of being fully integrated organisms[25,26] that are the products of many millions of years of evolution complete with complex mechanisms of communication and adaptation. Intermediate models can be valuable in the spectrum of experimental approaches to predictive toxicology.

Conclusion

Development starting with parental gametes and continuing with gestation, infancy, childhood, and adolescence provides the foundation for health throughout the whole of life. The complex sequence of physiological events that comprise development can be disrupted by toxic exposures. Given the importance of developmental processes for health and survival, systems have evolved for regulating them under changing circumstances. Toxic exposures that impede this regulation can produce lifelong impairments.

In addition to vetting potential toxicants to learn which can affect the processes of developmental control, we need also to conduct basic research to discover additional important developmental control processes so that they, too, can be monitored. Healthy development is fundamental to lifelong well-being. It is essential that we protect our children.

Acknowledgments

The author was supported by the Duke University Superfund Research Center ES010356.

References

1. Chaudhuri N, Frechtengarten L. Where the child lives and plays. In: Pronczuk de Garbino J, ed., *Children's Health and the Environment*. World Health Organization; 2004: 29–39.

2. Chisolm JJ. Lead poisoning. *Sci Am*. 1971;224:15–23.

3. Mason DJ, Safford HR. Palatability of sugar of lead. *J Comp Physiol Psychology*. 1965;59:94–7.

4. Cimafranca MA, et al. Acute and chronic effects of oral genistein administration in neonatal mice. *Biol Reprod*. 2010;83:114–21.

5. Jefferson, WN, Patisaul, HB, Williams CJ. Reproductive consequences of developmental phytoestrogen exposure. *Reproduction*. 2012;143:247–60.

6. vom Saal FS, Hughes C. An extensive new literature concerning low-dose effects of bisphenol A shows the need for a new risk assessment. *Environ Health Perspect*. 2005;113:926–33.

7. Goodman AB. Chromosomal locations and modes of action of genes of the retinoid (vitamin A) system support their involvement in the etiology of schizophrenia. *Am J Med Genet*. 1995;60(4):335–48.

8. Goldberg JS. Monitoring maternal beta carotene and retinol consumption may decrease the incidence of neurodevelopmental disorders in offspring. *Clin Med Insights Reprod Health*. 2012;6:1–8.

9. Rothman KJ, et al. Teratogenicity of high vitamin A intake. *N Engl J Med.* 1995;333(21):1369–73.

10. Maden M, Holder N. The involvement of retinoic acid in the development of the vertebrate central nervous system. *Development.* 1991;Suppl 2:87–94.

11. Shenefelt RE. Gross congenital malformations. Animal model: treatment of various species with a large dose of vitamin A at known stages in pregnancy. *Am J Pathol.* 1972;66(3):589–92.

12. Jones RL, et al. Trends in blood lead levels and blood lead testing among US children aged 1 to 5 years, 1988–2004. *Pediatrics.* 2009;123:e376–e385.

13. Miranda ML, et al, Environmental contributors to the achievement gap. *Neurotoxicology.* 2009;30:1019–24.

14. Aldridge JE, et al. Developmental exposure of rats to chlorpyrifos leads to behavioral alterations in adulthood, involving serotonergic mechanisms and resembling animal models of depression. *Environ Health Perspect.* 2005;113(5):527–31.

15. Levin ED, et al, Persistent behavioral consequences of neonatal chlorpyrifos exposure in rats. *Develop Brain Res.* 2001;130:83–9.

16. Horton MK et al. Does the home environment and the sex of the child modify the adverse effects of prenatal exposure to chlorpyrifos on child working memory? *Neurotoxicol Teratol.* 2012;34:534–41.

17. McFarland J, Hauer ME, Reuben A. Half of US population exposed to adverse lead levels in early childhood. *Proc Natl Acad Sci.* 2022 l119(11):e2118631119.

18. Pirkle JL et al. The decline in blood lead levels in the United States: the national health and nutrition examination surveys (NHANES). *JAMA.* 1994;272:284–91.

19. Acharya KS, et al. Epigenetic alterations in cytochrome P450 oxidoreductase (Por) in sperm of rats exposed to tetrahydrocannabinol (THC). *Sci Rep.* 2020;10(1):12251.

20. Murphy SK, et al. Cannabinoid exposure and altered DNA methylation in rat and human sperm. *Epigenetics.* 2018;13(12):1208–21.

21. Schrott R, et al. Sperm DNA methylation altered by THC and nicotine: vulnerability of neurodevelopmental genes with bivalent chromatin. *Sci Rep.* 2020;10(1):16022.

22. Hawkey AB, et al. Paternal nicotine exposure in rats produces long-lasting neurobehavioral effects in the offspring. *Neurotoxicol Teratol.* 2019;74:106808.

23. Holloway ZR, et al. Paternal factors in neurodevelopmental toxicology: THC exposure of male rats causes long-lasting neurobehavioral effects in their offspring. *Neurotoxicology.* 2020;78:57–63.

24. Slotkin TA, Levin ED, Seidler FG. Paternal cannabis exposure prior to mating, but not δ9-tetrahydrocannabinol, elicits deficits in dopaminergic synaptic activity in the offspring. *Toxicol Sci.* 2021;184(2):252–64.

25. Russell RW. Essential roles for animal models in understanding human toxicities. *Neurosci Biobehav Rev.* 1991;15:7–11.

26. Russell RW. Interactions among neurotransmitters: their importance to the "Integrated Organism." In: Levin ED, Decker MW, Butcher LL, eds., *Neurotransmitter Interactions and Cognitive Function.* Bern: Berkhäuser; 1992:1–14.

10

Epigenetics and Children's Environmental Health

Rebecca C. Fry and Margaret Pinder

Epigenetics

Epigenetic modifications represent heritable changes in gene expression that are not associated with changes in the primary DNA sequence but rather represent modifications "on top" of the genome.[1] Epigenetic mechanisms include the altered methylation of DNA, altered expression of microRNAs (miRNAs), and modifications of histones. Of these three forms of modifications, DNA methylation is the best studied in relation to children's environmental health. Within mammalian genomes, DNA methylation most frequently occurs through the addition of a methyl group to the cytosine of CpG dinucleotides, which are nonrandomly distributed throughout the genome, most commonly in unmethylated CpG islands within gene promoters.[1] DNA methylation, particularly when located in the promoter regions of genes, may be associated with decreased gene expression due to the inhibition of transcription by DNA binding proteins or by modified access of promoter regions to transcriptional machinery.[1] miRNAs are noncoding RNAs 18–24 nucleotides in length that can regulate gene expression in different ways. Specifically, miRNAs can bind to their target mRNAs, resulting in degradation or transcriptional inhibition.[1] A final regulatory mechanism that alters gene expression is the modification of histones through posttranslational modifications that alter chromatin structure and subsequent transcriptional activation.[1]

DNA Methylation and Children's Health

Birth Outcomes

The relationship between prenatal exposure to toxic metals and altered DNA methylation of imprinted genes has been observed in several population-based studies around the world. These results are intriguing because imprinted genes control fetal size and development, and they have parent-of-origin dependent expression. Among the studies highlighting these relationships is a pregnancy cohort in Gómez Palacio, Mexico, where infants were exposed in utero to arsenic levels above 200 μg/L.[7] These levels far exceed the recommended limit of 10 μg/L set by the World Health Organization and US Environmental Protection Agency. Arsenic-associated differences in DNA methylation levels were analyzed in cord blood and 16 genes were identified. Highlighting the impact

Rebecca C. Fry and Margaret Pinder, *Epigenetics and Children's Environmental Health* In: *Textbook of Children's Environmental Health*, Second Edition. Edited by: Ruth A. Etzel and Philip J. Landrigan, Oxford University Press. © Oxford University Press 2024.
DOI: 10.1093/oso/9780197662526.003.0010

of these epigenetic changes on mRNA abundance, these same genes also displayed changes in transcript level. Among these genes was potassium voltage-gated channel subfamily Q member 1 (KCNQ1), a known imprinted gene. Furthermore, in relation to children's health outcomes, the DNA methylation levels of a set of seven genes, including KCNQ1, were associated with differential birth outcomes, including head circumference and gestational age.[7]

Another example of linkages between in utero exposure to toxic metals, DNA methylation of imprinted genes, and birth outcomes was provided through a pregnancy cohort located in Durham County, North Carolina.[10,11] Specifically, exposure to low levels of cadmium in utero was associated with lower birth weight. In addition, decreases in cord blood DNA methylation within the regulatory region of maternally expressed 3 (MEG3) was observed in males, while in females lower DNA methylation was observed in the regulatory region of paternally expressed 3 (PEG3).[10] These studies also highlighted the sex-specific trends in DNA methylation patterning. Within the same cohort, exposure to lead in utero was also associated with differential methylation of imprinted genes and lower birth weight. Higher maternal lead levels were associated with increased cord blood DNA methylation within the regulatory region of the MEG3 gene. Increased levels of lead exposure were not only associated with lower birth weight, but also with more rapid weight-for-height z-score gains, a measure of adiposity and a risk factor for childhood obesity.[11] A separate study in a cohort in Rhode Island investigated DNA methylation and gene expression changes in relation to in utero exposure to cadmium.[8] In contrast to the studies in Mexico and North Carolina detailed above, here the assessment of CpG methylation was carried out in placental-derived DNA. Highlighting the functionality of the epigenetic mark, methylation in the promoter region of the imprinted gene proto-cadherin alpha subfamily C, 1 (PCDHAC1) was observed and correlated with decreased expression of the gene. PCDHAC1 expression was positively associated with fetal growth, highlighting that low levels of cadmium exposure have an adverse effects on fetal size.[8] The identification of imprinted genes as a target for environmental chemicals is not limited to metals exposure. For example, there is evidence that insulin-like growth factor 2 (IGF2) is differently methylated in placental DNA following exposure to phthalates.[12] Specifically, a significant negative association between IGF2 DNA methylation and levels of phthalate exposure was observed in a cohort from Wenzhou, China. These results displayed even stronger associations between these two variables among infants with fetal growth restriction (FGR), suggesting that placental DNA methylation may mediate the known relationship between phthalate exposure and adverse birth outcomes.[12]

Apart from imprinted genes, evidence supports linkages between chemical exposure and DNA methylation within the mitochondria. The effects of maternal smoking in utero on mitochondrial DNA (mtDNA) methylation, mitochondrial function, and low birth weight was studied in a cohort in Genk, Belgium.[13] Tobacco exposure was associated with increased risk of lower birth weights compared to infants who were not exposed. Additionally, smoking during pregnancy was associated with lower relative mtDNA content and higher absolute mtDNA methylation levels of MT-RNR1, which is a ribosomal RNA molecule that helps regulate the folate cycle, metabolic homeostasis, and insulin sensitivity. Although the analysis did not reveal that any of the studied molecular factors mediated the association between tobacco smoke exposure and birth weight, this study indicates that mitochondrial DNA function and methylation are potential biomarkers of the association between tobacco exposure and low birth weight.[13] Mediation analysis was used to reach a similar conclusion in a study in Sherbrooke, Canada, that examined the relationship between maternal smoking, placental DNA methylation, and birth weight.[15] Here exposure to maternal smoking in utero was associated with lower birth weight, and this relationship was mediated by DNA methylation in seven different genes related to fetal growth. The strongest mediation relationship was observed with pre-b-cell leukemia transcription factor 1 (*PBX1*), which is related to skeletal programming and patterning.[15] Supporting the association between smoking, the epigenome, and children's health, a study in Houston, Texas, investigated the associations between tobacco smoke, global methylation, gene expression, and birth weight.[9] Differential DNA methylation in six or more CpG sites was attributable to birth weight reduction, mediated by maternal smoking.[9] Another study in rural Bangladesh measured DNA methylation in cord blood and found that cadmium-related DNA methylation changes also were associated with lower birth weight.[14]

To summarize, several human population-based studies assess the relationships among chemicals in the environment, epigenetic modifications, and birth outcomes such as size at birth. These studies span the study of exposures to organic and inorganic chemicals including arsenic, cadmium, lead, phthalates, and tobacco smoke and characterize exposures that occur in utero. Related to target tissues, the majority of these studies analyzed DNA methylation patterns in cord blood, while some focused on the placenta. Notable genes that displayed altered methylation include imprinted genes such as *IGF2*, *PEG3*, and *MEG3*. In terms of an underlying mechanism, imprinted genes are intriguing targets because they are known to control the body size of infants.[16]

Immune Function in Children

The relationship between prenatal exposure to environmental chemicals and altered DNA methylation of immune-related genes and pathways has been observed in several population studies around the world. A research study in Providence, Rhode Island, examined the association between maternal smoking, placental methylation, and health outcomes.[17] Out of 21,551 autosomal CpG loci investigated within placentas exposed to maternal tobacco smoke, several differently methylated loci were found within runt-related transcription factor 3 (*RUNX3*), which is associated with immune system development. In addition to a role in immune system development, DNA methylation at one of the loci within the *RUNX3* gene was associated with decreased gestational age, thus highlighting the cross-talk of genes and their biological functions.[17]

A more specific relationship between environmental toxicants, DNA methylation, and immune function can be seen in a study conducted in a cohort of children in Fresno, California.[18] Childhood exposure to high levels of polycyclic aromatic hydrocarbons (PAHs) was associated with increased methylation at the forkhead box P3 (*FOXP3*) locus. Highlighting the functional impact of this epigenetic mark, DNA methylation was associated with differential expression of the FOXP3 protein and immune cell dysfunction. Moreover, PAH exposure was associated with impaired Treg function, increases in total plasma IgE levels, decreases in interleukin (IL)-10 protein expression, and increases in interferon (IFN)-γ protein expression. Interestingly, these associations became stronger as the length of exposure increased, suggesting a chronic immune response.[18]

Taken together, these studies have analyzed the association between exposure to tobacco smoke and PAHs and highlight the effects of exposures occurring during the in utero period and in childhood. One of these studies analyzed methylation patterns in childhood blood samples, while one focused on the placenta as the target tissue. Notable genes and pathways that displayed altered methylation include genes such as the *FOXP3* locus and the *RUNX3* gene, as well as cellular pathways that play a role in other immuno-logical systems such as regulatory T cells (Tregs). In terms of a mechanistic link, *FOXP3* and *RUNX3* genes are known for their role in immune system development, including immunoglobin and interleukin function. Regulatory T cells are also associated with immunity, and they are vital for modulating the immune response, preventing autoimmune diseases, and maintaining immunological homeostasis and tolerance.

Respiratory and Vascular Health in Children

Several studies have investigated the associations between chemicals in the environment, epigenetic modifications, and children's respiratory health. Investigators in Southern California analyzed the relationships among in utero exposure to air pollution, including $PM_{2.5}$, PM_{10}, nitrogen dioxide (NO_2), and ozone (O_3); global DNA methylation; and cardiovascular disease risk.[19] DNA methylation and demethylation percentages varied with different levels of exposure, different chemical exposures, and different mixtures of chemical exposures. Using newborn blood spots, the study concluded that systolic blood pressure and carotid intima-media thickness, which are biomarkers of cardiovascular disease risk, were associated with first-trimester ozone exposure within the strata of DNA methyltransferase single nucleotide polymorphisms (SNPs).[19] These results suggest that epigenetic and genetic variations in early life have the potential to mediate the known associations between air pollution and cardiovascular risk.

Associations among chemical exposures, DNA methylation, and asthma have also been studied. Specifically, in a cohort in New York City, in utero exposure to PAHs through maternal inhalation was associated with increased cord blood DNA methylation at the 5′-CpG island of acyl-coa synthetase long chain family member 3 (*ACSL3*), a gene that is related to fatty acid metabolism.[20] Increased methylation was not only associated with higher levels of maternal PAH exposure (>2.41 ng/m^3), but it was also associated with parental reports of asthma symptoms prior to age 5.[20] Phthalates may also increase the risk of asthma for children. Children in Taipei, Taiwan, with exposure to phthalates in food showed differential methylation in the androgen receptor (*AR*), tumor necrosis factor-α (*TNF-α*), and *IL-4*, which are associated with tumor necrosis, growth, and immunity.[21] These methylation changes were associated with differences in gene expression, and lower methylation of the 5′-CpG island of *TNF-α* was associated with childhood asthma, suggesting that DNA methylation may mediate the effects of exposure to phthalates on children's health outcomes, including asthma.[21]

In summary, these studies have investigated the associations between chemicals in the environment, epigenetic modifications, and children's respiratory health. These studies span exposure to air pollution (including $PM_{2.5}$, PM_{10}, NO_2, and O_3), PAHs, and phthalates and focus on an exposure window within the in utero period. Notable genes and pathways that displayed altered methylation include the 5′-CpG islands of genes such as *ASCL3*, as well as genes that play a role in other biological processes such as immunity and growth. Global DNA methylation, induced by environmental chemical exposures, was also implicated as an epigenetic modification with respiratory health consequences for children. *ASCL3* is known for its role in fatty acid metabolism. Genes such as *AR*, *TNF-α*, and *IL-4* are associated with gene expression, tumor necrosis, and inflammation, respectively. The majority of these studies analyzed methylation patterns in blood spots, while one focused on cord blood as the target tissue.

Neurocognition and Neurodevelopment in Children

Several studies have investigated the relationship between environmental chemicals, epigenetic modification and children's neurocognition/neurodevelopment. For example, a cohort in Tongliang County, China, exposed to PAHs in utero displayed associations between global LINE-1 DNA methylation in cord blood and PAH-DNA adducts, which are a potential biomarker of cancer that were correlated to a local coal-fired power plant.[22] Global LINE-1 DNA methylation was also associated with IQ scores. Although mediation analysis did not indicate that LINE-1 DNA methylation was a direct mediator between IQ scores and PAH-DNA adducts, directional analysis revealed that higher levels of prenatal exposure to PAHs were associated with lower IQ scores at age 5.[22]

Other studies have focused on the impact of mercury exposure on children's neurodevelopmental health. Maternal blood samples were used to assess in utero exposure to mercury in a study in Massachusetts, and analysis of cord blood showed that exposure was associated with lower regional methylation at paraoxonase 1 (PON1) among males.[23] This gene is associated with neurodevelopment, and the observed site-specific DNA methylation persisted into early childhood. DNA methylation changes at the PON1 locus were a predictor of lower cognitive test scores during early childhood, indicating that PON1 may modulate mercury-induced health outcomes or act as a sex-specific biomarker.[23] Last, a study in Providence, Rhode Island, examined the relationship between maternal ingestion of mercury, placental DNA methylation, and neurodevelopment.[24] In utero exposure to mercury was associated with hypomethylation in six loci within collagen type XXVI alpha 1 chain (EMID2), and methylation at these sites was correlated with EMID2 gene expression. Furthermore, methylation within these six loci was significantly associated with a high-risk neurobehavioral profile using the Neonatal Intensive Care Unit Network Neurobehavioral Scales.[24] Although mediation analysis was not used to determine the exact relationship between mercury exposure, DNA methylation, and children's health outcomes, these results suggest that EMID2 hypomethylation could provide a mechanistic link between mercury exposure and adverse neurological outcomes in children.

In summary, these studies have analyzed the associations between chemicals in the environment, epigenetic modifications, and neurocognition in children. These studies include exposure to PAHs and mercury. All focus on an exposure window of the in utero period. The majority of these studies analyzed methylation patterns in cord blood, while one focused on the placenta as the target tissue. These results demonstrate that the region around the PON1 and EMID2 genes are examples of notable genes and pathways that displayed altered methylation. Global DNA methylation, induced by exposure to environmental chemicals, was also implicated as an epigenetic modification with neurological health consequences for children. PON1 is known for its role in oxidative protection and metabolization of toxic oxidized lipids; EMID2 is a known proton coding gene associated collagen formation, but its specific function is unknown.

Noncoding RNAs and Children's Health

While DNA methylation is the most well-studied epigenetic mark assessed in relation to children's environmental health, several studies have explored the associations between chemicals in the environment, children's health, and other epigenetic modifications, including noncoding RNAs. For example, exposure differences in $PM_{2.5}$ and formaldehyde were observed between asthmatic and nonasthmatic children in China and tied to altered expression levels of miR-155 in the group with asthma.[25] In a separate study in adolescents from Hong Kong, urinary levels of arsenic and lead were associated with decreased levels of miR-21.[26] MiR-21 levels were also associated with microalbuminuria, an early biomarker of vascular or kidney disease. Although toxic metals were not directly associated with microalbuminuria among the participants of this study, the associations between miRNA levels and both variables might imply that microRNAs are involved in the biological pathways connecting exposure to environmental chemicals to organ damage.[26] Finally, in the Rhode Island Child Health Study, placental cadmium concentrations were associated with altered expression of four long noncoding RNAs (lncRNAs) as well as birth weight.[27] These results provide evidence that disruptions of placental lncRNAs may be a possible mechanism for cadmium's toxic health effects.

Mechanisms by Which Environmental Chemicals Modify the Epigenome

As detailed above, epigenetic machinery is subject to modification following exposure to a wide range of environmental chemicals. There are various mechanisms by which environmental contaminants can result in either gene-specific or global methylation, as well as modifications to histones. For example, chemicals can alter DNA methyltransferase (DNMT) activity, the enzyme responsible for adding methyl groups to cytosine. In addition, there is evidence that environmental chemicals can directly alter the functionality of ten-eleven translocation protein (TET), which is an enzyme family involved in demethylation.[2] In addition, chemical exposure can alter the availability of S-adenosylmethionine, a cofactor necessary for methylation, which would similarly affect global DNA methylation patterns.[2]

A hypothesis for gene-specific DNA methylation is the *transcription factor occupancy theory*, which proposes that gene-specific methylation is the result of the presence or absence of transcription factors and their response to specific environmental contaminants. For example, transcription factor activation in response to a chemical contaminant may result in cellular adaptations that prevent DNA methyltransferase from accessing a specific DNA locus, resulting in site-specific hypomethylation.[3] Conversely, hypermethylation may result from the repression of transcription factors following exposure to environmental chemicals, which would allow DNMT to access specific genomic loci.[3]

MicroRNAs are often differentially expressed in association with exposure to environmental toxicants. A proposed mechanism underlying this alteration is that transcription factors bind to specific miRNAs and alter their expression. For example, there is evidence that placental miRNAs are not only susceptible to altered expression in relation to pregnancy complications such as obesity and preeclampsia, but they are differentially expressed in placentas that have been exposed to environmental chemicals.[4] There is evidence that certain environmentally responsive transcription factors are enriched for promoter binding among environmentally responsive miRNAs, which indicates that the identified transcription factors are likely transcriptional regulators of miRNAs with altered functionality following exposure to environmental toxicants.[5]

Finally, histone modifications, such as histone methylation and histone acetylation, are orchestrated by mechanisms nearly identical to those that regulate those epigenetic modifications on DNA strands. Histone methylation is catalyzed by histone methyltransferase and occurs in the presence of S-adenosylmethionine, which is the same cofactor required for DNA methylation and whose abundance can be regulated by the presence or absence of environmental chemicals.[6] As demonstrated in the previous discussion about DNA methylation and chemical modifiers, changes in the chemical conditions necessary for methylation, demethylation, and other epigenetic modifications can affect the frequency and location of these alterations.

Future Directions

There are four important research themes that emerge in relation to children's environmental health and the epigenome. First, many of the current studies focus on chemical exposures that occur in utero. A growing body of evidence suggests that preconception exposure to environmental chemicals may have epigenetic effects.[28] Therefore, expanding children's studies to include the preconception period will help illuminate the importance of this time as a development window of susceptibility. Second, while maternal chemical exposures and their associated epigenetic modifications are the focus of many studies to date, it is necessary to explore the roles that paternal exposures have on the epigenomes of their offspring. Third, another vital aspect of future studies is consideration of the target tissue of study and the potential biological impact of such modifications as well as the inclusion of a measured function of the epigenetic mark. Finally, there is a need for studies that focus on the health effects associated with chemical mixtures. Environmental chemical exposures occur in combinations and thus simultaneous exposure to more than one of the chemicals will likely modulate the effects of each toxicant on children's health. Taken together, this chapter highlights studies that support the link between environmental chemical exposures, the epigenome, and health outcomes in children.

References

1. Feng S, Jacobsen SE, Reik W. Epigenetic reprogramming in plant and animal development. *Science*. 2010;330(6004):622–7.

2. Martin EM, Fry RC. Environmental influences on the epigenome: Exposure-associated DNA methylation in human populations. *Ann Rev Public Health*. 2018;39(1):309–33.

3. Martin EM, Fry RC. A cross-study analysis of prenatal exposures to environmental contaminants and the epigenome: Support for stress-responsive transcription factor occupancy as a mediator of gene-specific CPG methylation patterning. *Environmental Epigenetics*. 2016 Jan;2(1):1–9.

4. Addo KA, Palakodety N, Hartwell HJ, Tingare A, Fry RC. Placental microRNAs: Responders to environmental chemicals and mediators of pathophysiology of the human placenta. *Toxicol Rep*. 2020 Aug 15;7:1046–56.

5. Sollome J, Martin E, Sethupathy P, Fry RC. Environmental contaminants and microRNA regulation: Transcription factors as regulators of toxicant-altered microRNA expression. *Toxicol Appl Pharmacol*. 2016 Dec 1;312:61–6.

6. Zhang Y, Sun Z, Jia J, Du T, Zhang N, Tang Y, et al. Overview of histone modification. In: Fang D and Han J, eds. *Histone Mutations and Cancer*. Springer; 2020:1–16.

7. Rojas D, Rager JE, Smeester L, et al. Prenatal arsenic exposure and the epigenome: Identifying sites of 5-methylcytosine alterations that predict functional changes in gene expression in newborn cord blood and subsequent birth outcomes. *Toxicol Sci*. 2014;143(1):97–106.

8. Everson TM, Armstrong DA, Jackson BP, Greem BB, Karagas MR, Marsit CJ. Maternal cadmium, placental PCDHAC1, and fetal development. *Reprod Toxicol*. 2016;65:263–71.

9. Suter M, Ma J, Harris AS, et al. Maternal tobacco use modestly alters correlated epigenome-wide placental DNA methylation and gene expression. *Epigenetics*. 2011;6(11):1284–94.

10. Vidal AC, Semenova V, Darrah T, et al. Maternal cadmium, iron and zinc levels, DNA methylation and birth weight. *BMC Pharmacol Toxicol*. 2015;16(1):20.

11. Nye MD, King KE, Darrah TH, et al. Maternal blood lead concentrations, DNA methylation of MEG3 DMR regulating the DLK1/meg3 imprinted domain and early growth in a multi-ethnic cohort. *Environ Epigenet*. 2016;2(1):dvv009.

12. Zhao Y, Chen J, Wang X, Song Q, Xu H-H, Zhang Y-H. Third trimester phthalate exposure is associated with DNA methylation of growth-related genes in human placenta. *Sci Rep*. 2016;6(1):33449.

13. Janssen BG, Gyselaers W, Byun H-M, et al. Placental mitochondrial DNA and CYP1A1 gene methylation as molecular signatures for tobacco smoke exposure in pregnant women and the relevance for birth weight. *J Transl Med*. 2017;15(1):5.

14. Kippler M, Engström K, Mlakar SJ, et al. Sex-specific effects of early life cadmium exposure on DNA methylation and implications for birth weight. *Epigenetics*. 2013;8(5):494–503.

15. Cardenas A, Lutz SM, Everson TM, Perron P, Bouchard L, Hivert MF. Mediation by placental DNA methylation of the association of prenatal maternal smoking and birth weight. *Am J Epidemiol*. 2019;188(11):1878–86.

16. Horikoshi M, Beaumont RN, Day FR, et al. Genome-wide associations for birth weight and correlations with adult disease. *Nature*. 2016;538(7624):248–52.

17. Maccani JZJ, Koestler DC, Houseman EA, Marsit CJ, Kelsey KT. Placental DNA methylation alterations associated with maternal tobacco smoking at the RUNX3 gene are also associated with gestational age. *Epigenomics*. 2013;5(6):619–30.

18. Hew KM, Walker AI, Kohli A, et al. Childhood exposure to ambient polycyclic aromatic hydrocarbons is linked to epigenetic modifications and impaired systemic immunity in T cells. *Clin Exp Allergy*. 2014;45(1):238–48.

19. Breton CV, Yao J, Millstein J, et al. Prenatal air pollution exposures, DNA methyltransferase genotypes, and associations with newborn LINE1 and alu methylation and childhood blood pressure and carotid intima-media thickness in the Children's Health Study. *Environ Health Perspect*. 2016;124(12):1905–12.

20. Perera F, Tang W, Herbstman J, et al. Relation of DNA methylation of 5'-CPG island of ACSL3 to transplacental exposure to airborne polycyclic aromatic hydrocarbons and childhood asthma. *PLoS ONE*. 2009;4(2):e4488.

21. Wang IJ, Karmaus WJJ, Chen S-L, et al. Effects of phthalate exposure on asthma may be mediated through alterations in DNA methylation. *Clin Epigenet*. 2015;7(1):27.

22. Lee J, Kalia V, Perera F, et al. Prenatal airborne polycyclic aromatic hydrocarbon exposure, Line1 methylation and child development in a Chinese cohort. *Environ Int*. 2017;99:315–20.

23. Cardenas A, Rifas-Shiman SL, Agha G, et al. Persistent DNA methylation changes associated with prenatal mercury exposure and cognitive performance during childhood. *Sci Rep.* 2017;7(1):288.

24. Maccani JZJ, Koestler DC, Lester B, et al. Placental DNA methylation related to both infant toenail Mercury and adverse neurobehavioral outcomes. *Environ Health Perspect.* 2015;123(7):723–9.

25. Liu Q, Wang W, Jing W. Indoor air pollution aggravates asthma in Chinese children and induces the changes in serum level of mir-155. *Int J Environ Health Res.* 2018;29(1):22–30.

26. Kong APS, Xiao K, Choi KC, et al. Associations between microRNA (Mir-21, 126, 155 and 221), albuminuria and heavy metals in Hong Kong Chinese adolescents. *Clin Chim Acta.* 2012 Jul 11;413(13-14):1053–7.

27. Hussey MR, Burt A, Deyssenroth MA, et al. Placental lncRNA expression associated with placental cadmium concentrations and birth weight. *Environ Epigenet.* 2020;6(1):dvaa003.

28. Stephenson J, Heslehurst N, Hall J, et al. Before the beginning: nutrition and lifestyle in the preconception period and its importance for future health. *Lancet.* 2018;391(10132):1830–41.

11

Racism, Environmental Injustice, and Child Health

Danielle Laraque-Arena

Introduction

In the immediate past decade, sociopolitical context, alongside nationally publicized events of police brutality against Black men and women in the United States, have sharpened the focus of the impact of structural racism on health outcomes.[1,2] In this context, the idea of achieving health equity and environmental justice leads to a broadening of the matrix of sociopolitical determinants of health that frame the influence of the environment on children's health.[3] Presented here, beyond the social determinants of poverty and extreme poverty on health, is the layered, indolent influence of biases related to the social constructs of race and gender as well as the intersecting categorization of peoples by class, ethnicity, religion, culture, sexual orientation and a whole host of "otherings."[4]

This chapter is subdivided into six sections: the first section provides a general overview of the influence of racism and child health. The next section provides some definitions of concepts examined herein, followed by a section that examines the evolution of the bio-psycho-socio-political framework in which we now view children's environmental health. This is followed by an examination of the importance of dismantling systemic racism to advance health equity throughout the environmental research agenda. Two case studies are then presented, examining the influence of structural racism and the application of the concepts of environmental justice on firearm violence and the influence of the built environment on health. The concluding section discusses the path forward to create a system grounded in an anti-racism and justice framework.

This chapter serves as a foundation to the discussion of the branch of environmental pediatrics that underscores the unique vulnerabilities and rights of the child in a toxic world. The discussions of the chapter crosswalk to the various themes of the other chapters and urge the reader to make explicit connections between the need for comprehensive solutions to the multitude of environmental problems facing children in the 21st century.

Definitions

Environmental Justice

The environmental justice framework builds on the concepts of protected *rights* for all individuals and the need to prevent environmental degradation.[5-7] These protected rights

Danielle Laraque-Arena, *Racism, Environmental Injustice, and Child Health* In: *Textbook of Children's Environmental Health*, Second Edition. Edited by: Ruth A. Etzel and Philip J. Landrigan, Oxford University Press. © Oxford University Press 2024. DOI: 10.1093/oso/9780197662526.003.0011

include those codified by the Civil Rights Act of 1964 amended in 1988 and Voting—but not reaffirmed in its entirety as noted in the H.R. 3184-Civil Rights Modernization Act of 2021.[8] This justice framework adopts a *public health model*, stating that the elimination of harm is the optimal strategy, as opposed to solely mitigation of harm. Primary (universal) prevention dictates that we do not need to wait for harm to happen to strategize but, in an evidence-based and informed perspective, anticipate harm from the best available data and prevent exposure. One of the most salient examples comes from the history of lead poisoning. Screening of children to detect levels of blood lead is not primary prevention but secondary prevention to avert further harm and is not the model that we should adopt. The experience in Flint, Michigan, in the 21st century and countless other communities reaffirms the validity of the approach to primary prevention.[9-13] Only now is the federal government contemplating true abatement to prevent exposure, replacing the screening of already exposed children, women, and other adults.[14]

Another principle of environmental justice is articulated by Bullard, who advocates "shifting the burden of proof to polluters/dischargers who do harm, discriminate, or who do not give equal protection to racial and ethnic minorities and other 'protected' classes."[6(p. 4)] It should be noted that there are also state-specific protected classes that expand on the federal categorization. Under the current system, the burden of proof is borne by those who choose to challenge those responsible; they must prove that they have been harmed. Shifting the burden to the polluters takes a proactive approach to anticipate harm and provide legal protections in avoidance of that harm. Another concept of environmental justice is that intent is not needed as a demonstrative condition. Instead, *disparate impact and statistical weight* is sufficient.[6] *Targeting* of hotspots that bear a disproportionate burden of exposure could be prioritized to redress bias and the unfair sharing of environmentally induced health problems. This approach is inherently supportive of an anti-racism framework and challenges the stratification of people (by the social construction of race, ethnicity, status, power, gender), place (e.g., Native American lands), work (office worker vs. farm workers), etc. The distribution of risk is applied to change the current paradigm of unevenly applied enforcement of environmental protections. The key questions in an equity and justice framework would be how to prevent harm, define which groups are most affected, why they are affected, who is responsible, and, most importantly, what can be done to remedy the problem and then strategize to prevent further exposure and harm. This approach is inherently community-powered and "bottom-up," and it also looks upstream to those policies that must be changed to alleviate differential resource allocations that feed unfair treatment. Several authors point to the failure of the US Environmental Protection Agency (EPA), other regulatory bodies, industry, and academia for being structured around risk management and not focusing on discriminatory practices.[6,7,15,16] The emphasis here is on the rights of individuals and communities.

In regard to fair treatment, the concept of *equity* emerges. While equity has a number of definitions, three particular categories are most relevant to environmental justice. These include procedural equity, geographic equity, and social equity.[6] Another concept relating to the actions of organizations such as universities is that of *organizational justice*, with distinct leadership competencies needed to render racial equity, gender equity, and a future orientation that prioritizes our children and our planet.[17-19]

- *Procedural equity* refers to the uniform, fair application of rules, regulations, evaluation criteria, and enforcement in a nondiscriminatory fashion.

- *Geographic equity* refers to the location and spatial configuration of communities and the unequal distribution of the placement of, for example, noxious facilities, bus depots, landfills, incinerators, sewer plants, lead smelters, etc. The related concept of *redlining* bears mention here, in the historical segregation of Black Americans in housing and political voting blocs (gerrymandering) in an effort to influence political outcomes and the balance of power.[20]
- *Social equity* invokes the role of sociological factors such as race, ethnicity, class, culture, lifestyles, power dynamics, etc. in the decision-making process. Power differentials allow the exertion of political power to influence one community or people over another. Political determinants of health then influence health outcomes.[21,22] The argument that health is political emerges as a concept since health is known to be unevenly distributed, dependent on political action, and believed to be a dimension of human rights and citizenship in most societies.

Racism and Its Intersection with Environmental Justice

Racism is defined by some as "beliefs, attitudes, institutional arrangements, and acts that tend to denigrate individuals or groups because of phenotypic characteristics or ethnic group affiliations."[23(p. 805)] Other conceptualizations have included the relationship of those oppressed and those not oppressed. Furthermore, categorizations include attitudinal and behavioral conceptualizations of racism.[24,25] *Attitudinal racism* can refer to beliefs regarding particular groups, whereas *behavioral racism* relates to acts of discrimination and unequal treatment by individuals, institutions, and systems. Within these definitions both intergroup and intragroup racism have been described. The impact of racism can be on the individual, on systems, and on the interplay between individuals and systems.

Understanding how racism exerts its influence is important in deciding on the interventions needed to remedy the impact at the individual and/or societal levels. The intersectional impact of the various manifestations of racism also bears examination. Furthermore, the influence of various forms of racism as either moderating or mediating variables in research becomes relevant. *Moderator variables* can be defined as factors that influence the direction or magnitude of the relationship between predictor and criterion variables, whereas *mediator variables* are factors that may partially account for that relationship.[23]

Racism has not been systematically examined and often remains unmeasured in research.[23,26] This is particularly relevant given the extant literature of the physiological responses (as induced by stress) on immune, neuroendocrine, and cardiovascular functioning.[27,28] In addition, racism may affect health when it is not perceived/experienced as a stressor but is in the form of institutional racism affecting, for example, access to goods and services (e.g., food security, safe housing opportunities, environmentally safe playgrounds, etc.) that can have direct health consequences and should be factored into the analysis of the environmental impact of racism.

The topic of *implicit bias*, or *unconscious bias*, has gained recognition in the literature[29,30] as has the concept of microaggressions.[31,32] *Biases* are defined as thoughts and feelings that an individual may not be consciously aware of but that manifest in automatic responses that may influence interactions with patients and impact health outcomes; microaggressions manifest in slights and undertones of expressions of aggressions. In one systematic review, the authors concluded that "implicit attitudes were more often significantly related to patient–provider interactions and health outcomes than treatment processes."[30(p. e60)] Indirectly, their conclusions indicated the need for more research to elucidate how interventions to address these biases might influence health outcomes.

Child Rights

The discussion of environmental justice and child health must explicitly include an explanation of the rights of children as set by the United Nations Convention on the Rights of the Child (UNCRC).[33] This is particularly important given the developmental vulnerability of children to toxic exposures. The UNCRC's 54 articles, drafted in 1989, are set within a frame that recognizes the fundamental rights of children throughout the world to live in a safe and nurturing environment. The following two Articles are the most relevant to environmental justice and anti-racism.

Article 2:

1. States Parties shall respect and ensure the rights set forth in the present Convention to each child within their jurisdiction without discrimination of any kind, irrespective of the child's or his or her parent's or legal guardian's race, colour, sex, language, religion, political or other opinion, national, ethnic or social origin, property, disability, birth or other status.
2. States Parties shall take all appropriate measures to ensure that the child is protected against all forms of discrimination or punishment on the basis of the status, activities, expressed opinions, or beliefs of the child's parents, legal guardians, or family members.

Article 24, subsection c and e:

(c) To combat disease and malnutrition, including within the framework of primary health care, through, inter alia, the application of readily available technology and through the provision of adequate nutritious foods and clean drinking-water, taking into consideration the dangers and risks of environmental pollution.
(e) To ensure that all segments of society, in particular parents and children, are informed, have access to education and are supported in the use of basic knowledge of child health and nutrition, the advantages of breastfeeding, hygiene and environmental sanitation and the prevention of accidents.

It should be noted that, in 1989, the term "climate change" had not yet entered the general vernacular, although it is certainly relevant today. In an environmental case heard by the US Supreme Court, *Massachusetts v. EPA*, the court held that "the Environmental Protection Agency has the authority to regulate heat-trapping gases emitted by automobiles."[34] As noted by a number of authors, a future orientation is determinant in securing our children's futures, and brisk responses to the threats of climate change are essential components to achieving environmental justice.[17,35,36]

The Bio-Psycho-Socio-Political Framework of Environmental Health

In reviewing the bio-psycho-social-political model to explicate environmental health, an informal 2022 review using the search terms "racism," "environmental injustice," and "child health" yielded four articles in PubMed, which are summarized here.[37–40] The themes of these four articles are (1) a demonstration of case-based links between structural racism, climate change, and child health; (2) a review of the possible impact of the physical and social conditions imposed by the COVID-19 pandemic on child health related to exposures to environmental contaminants, built environment, food environment, access to educational resources, and the social environment; (3) a discourse on radical health pedagogy and its relationship to planetary medicine; and (4) a discussion of children as the targets of environmental injustices. Many more references are cited when searching on the terms "environmental justice," "structural racism," or their combination so caution is needed in not leaving out important elements and populations when examining these broad societal concepts. A search using the terms "racism," "environmental justice," and "child health" yields seven results with the themes of three articles overlapping the prior four results. The four additional articles included similar themes, as noted: (1) a review of the link between exposure to traffic-related and particulate matter air pollution and asthma and the need for clinicians to go upstream in addressing system-level determinant factors of child health[41]; (2) a review of Best Babies Zone Initiative, which implemented housing, economic, and environmental justice interventions to reduce infant mortality and that determined the need for an explicit focus on racial equity through sustained funding mechanisms[42]; (3) a review of the need for transdisciplinary efforts to address equity and mistreatment in perinatal care using a Quality Perinatal Services Hub[43]; and (4) a test of environmental management approaches using a geographic information system (GIS) that explored the concept of environmental racism. The result of this last article's multivariate analysis indicated that race was strongly associated with the number of cases of elevated blood lead levels irrespective of poverty status and that proximity to transportation corridors was the strongest indicator of lead exposure—with direct implications for environmental justice interventions.[44]

I examined the extant literature to establish our current frame of discussion for the topic of environmental justice and its intersection with racism and to synthesize how the theoretical models and various discussions yield actionable steps toward alleviating environmental and racism-related injustices, with a focus on children. I endorse the concept, illustrated below, that science needs ethical social policy and should be guided by an urgent call for actionable steps to alleviate harm; we must follow a primary preventive (universal) approach and promote the good, without academic circling to prove causality.[45–47] As articulated by Wakefield, the call for environmental justice and the research aimed at linking environmental exposures to a causal inference with health outcomes need to be separated. A scientific strategy must not lead to inaction, especially when considering complex problems; a framework for action based on sound, ethical social policies and a justice strategy must drive evidence-based interventions.[48] In discussing the health of children, a child rights framework is essential, as noted by the UNCRC. This is the essence of the precautionary stance that has been applied in other high-income countries, in contrast to the scientific approach taken in the United States. That is, we need not wait for harm to happen to use a scientific strategy to prove causality before action is taken.

A number of decades ago, in his deliberations on the ecology of child development, Bronfenbrenner proposed a converse to the notion that social policy should be based on science by stating that science needs social policy.[47] He explained that this does not mean that social policy necessarily guides organizational activities, but that it provides two necessary elements for scientific inquiry—vitality and validity. Emerging from this concept is a series of questions that merit rigorous answers but that also frame the concept of the social environment and its implications for prevention and child safety, both of which can be outgrowths of explicit ethical national policy concerning children and family life. This premise establishes the primacy of valuing the child's ecology as central to all our scientific and sociopolitical activities. The frame for scientific inquiry is grounded on the premise that the child's immediate settings of home, school, street, playground, camp, etc. are defined in three dimensions. In the first dimension are physical space and materials, people in their various roles and relationships with and toward the child, and activities with the people engaging with the child—that is the social meaning of these activities. In the next concentric layer are the institutional and social systems that affect the immediate settings in which the child lives. Hence support systems (e.g., home visitation), rules, regulations, and laws (e.g., removing toxic exposures such as lead and air pollution) operate on the premise that the child's environment must be protective of that child, their family, and their community.

Studies of child behavior undertaken in the latter part of the 20th century noted the artificial nature of many psychological studies where the child—the subject—was placed in research settings that had no relevance to the environment in which the child actually lived. These studies were made "ecologically invalid" by placing the child in an environment devoid of the physical spaces, experiences, and myriad of systems influencing their family, neighborhood, mobility, etc., and which are central to their life. Such a setting makes inferences on the potentialities of each of these factors to influence the others impossible. Ecological variables are defined as both sociological and structural, with the latter perhaps being more modifiable.

Given this background how does the concept of racism fit into the biopsychosocial model? Surprisingly, given the history of the United States, there is a paucity of emerging literature examining the effects of racism on those affected—in the context of the United States—and this tends to have its greatest impact on those identifying as African Americans (or Black). The recognition of racism as a stressor and mediating variable in the contextual model of biopsychosocial effects has been outlined by Clark.[23] In this framework, environmental stimuli (e.g., lack of opportunity, substandard housing related to racial discrimination) lead to the experience of racism and constitute stressors (acute and chronic) that are moderated by adaptive or maladaptive coping responses that may have both physiological and psychological responses that ultimately affect health outcomes. Moderator variables are also postulated to be inclusive of constitutional factors (e.g., skin tone, occupational status, personal income), sociodemographic factors (e.g., ethnicity, socioeconomic status), and psychological (e.g., perceived worth, hardiness, cynicism) and behavioral factors (e.g., anger) that come into play if an event is perceived as stressful and that influence the relationship between racism and health status. The impact of these stressors leads to an analysis of the inherently racist structural form of our institutional systems and the lack of opportunities they provide for educational and economic inclusion, which is important in building generational wealth that will influence the ability to resist environmental degradation and toxic exposures. The policies governing the environmental protection of individuals and communities are properly analyzed within the concepts of equity.[5,45,49]

Racism is not isolated to the social/cultural group of Black individuals, and although concept of racism is not new, a newer term, "structural racism," has galvanized the thoughts of many to examine the modifiable aspects of this structure and seek to ameliorate its negative impacts.

The Importance of Dismantling Systemic Racism to Advance Health Equity Throughout the Environmental Research Agenda

As noted by some investigators, there are distinct reasons for dismantling systemic racism in the research agenda as it relates to children's environmental health.[50] These authors indicate three areas that should be prioritized in response to President Biden's Executive Order on Advancing Racial Equity and Support for Underserved Communities Through Federal Funding. The three areas are (1) bridging the gap between knowledge and action to eliminate health disparities and inequities; (2) valuing all ways of knowing in research and research funding; and (3) incentivizing higher education to redefine access and advancement. In focusing on these priorities, the President could charge a White House task force to address at the federal level research focused on and undertaken by Black, Indigenous, Pacific Islander, Latine, and Asian (BIPLA) people. Notable in the published literature is the funding gap that exists in the National Institutes of Health (NIH) R01 applications submitted by Black scientists relative to White scientists.[51,52] While the NIH funding mechanism is complex, investigators have discovered several salient points for targeted interventions. In brief, based on complex analyses their results point to a need for a more diverse applicant pool, enhanced postdoctoral career support to promote faculty diversity, mentoring, and, importantly, "meritorious applications on topics that are underappreciated by review but align well with strategic priorities"[51(p. 8)] of the various institutes and centers. Black investigators, as compared to White investigators, were more likely to focus on disease prevention and intervention and human subjects (as opposed to animal research), and they tended to engage in research on health disparities and patient-focused interventions.

Closing the funding gap might begin to tilt the balance that has historically underfunded minority researchers and underfunded categories of research that may be distinct to the interests of minority populations. Environmental justice concerns that embed a participatory framework may be one of these topic areas. As noted by some,[53] participatory research with environmental justice communities may be more likely to result in structural change if there is a formal leadership role for community members, project design includes decision-makers and policy goals, and long-term partnerships are supported through multiple and robust funding mechanisms. Hiring of faculty of color at research institutions engaged in environmental justice research is needed. Funding for interventional outcomes studies that have the potential to provide structural change in the battle against environmental risks while integrating equitable recruitment strategies might also provide real-world, more immediate solutions to health problems faced by disadvantaged communities. Applying an equity lens to all research, including randomized controlled trials, has been proposed as a mechanism to improve the saliency of research to the neediest populations.[54] See Box 11.1.

Finally, prioritizing the incentives for higher education to redefine access and advancement calls for actions by the Department of Education to intentionally and explicitly remove structures that have marginalized BIPLA learners.[55] This approach

---

Box 11.1 An Equity Lens

Applying an equity lens may lead to a series of questions that challenge the current normative approach to research—especially research regarding environmental justice. These may include the following:

- Is there a health disparity noted by the social construction of race, gender, or other grouping?
- Is there a known biomedical reason—one devoid of bias—that may explain differential findings?
- Does the manifestation of an attitudinal racism approach at the practice level affect screening for specific diseases or for the selection, timing, or treatment of individuals?
- Do lower screening rates affect the recognition of specific toxic exposures for certain populations? If yes, does this lead to inaccurate data that may not display the possible causal relationship, using geo-mapping, of toxic exposures (e.g., dump sites) and elevated cancer risk in certain populations?
- Does public policy vis-à-vis a group of individuals, a neighborhood, or a community affect environmental risk for that community?
- Has the research been designed to address implicit and overt biases, thus creating an anti-racism and environmentally just framework for tackling child health opportunities and problems?
- To what extent have the communities being researched helped to construct the research question related to environmental risk?
- Are the research methods used appropriate to the study question? (e.g., use of mixed-methods reflecting all ways of knowing)?

---

challenges the admission processes for higher education (e.g., SAT, MCAT scores) as well as the "worthiness" of certain scholarly discourses.

Case Studies

Climate change poses an existential threat to our planet and its people. It will likely continue to disproportionately affect poor communities and communities of color, a blatant example of environmental injustice and its intersection with racism—and an affront to child health. The two case studies chosen for this section, however, examine the firearm-related epidemic and the epidemic of obesity/overweight. These two case studies are chosen because they are current, threaten the quality of life and health, may not always be included in the agenda regarding environmental justice, and are eminently preventable.

Firearm Violence

In the series on the War Against Children,[56] the case of Chariel Osoria, a 17-year-old who was shot in the head,[57] begins with the analysis of health equity and racism in the context of the gun violence epidemic.[20] Formica proceeds with an epidemiological view of firearm injuries and discusses the role of racial segregation as rooted in the long history of structural racism in the United States. Syracuse, New York, block groups classified by poverty rates and mean annual rate of gunshot injuries and homicides are mapped in a geospatial representation of shootings. She relates the history of redlining, a practice begun in the 1930s when the Home Owners' Loan Corporation was commissioned to create maps in 239 US cities and rate neighborhoods relative to their investment potential and risk. Minority neighborhoods were deemed the highest risk and marked in red. The use of these maps persisted to create discriminatory lending practices that reinforced concentrated areas of poverty. In 2015, Syracuse, one of the redlined areas, ranked number one in concentrated poverty out of 100 cities surveyed.[58] The corollary health outcomes of general health, asthma, myocardial infarction, maternal and child health outcomes, cancer diagnoses, and life expectancy have racially disparate outcomes.[20] Formica points to the growing body of research that associates practices of redlining and violence, specifically firearm-related injuries and deaths. This historical perspective focuses our attention on social policies that created and perpetuated environmental injustice in Syracuse and many other cities.[59] Abolishing these disparities must then go upstream to rectify the injustices codified in housing and lending practices at the national level. A complicating factor for prevention efforts is the proliferation of guns (including handguns and semi-automatic firearms), with approximately 55 million individuals in the United States owning at least one gun (the average number of guns for males and females in the United States is 5.6 and 3.6, respectively, with an estimated 270 million civilian-owned firearms).[60]

Obesity, Overweight, and Relationship to the Built Environment

A 2022 search of PubMed using the terms "built environment" and "obesity" yielded 839 results. One of these is a systematic review summarizing the findings of 30 longitudinal studies examining the built and social environments of child obesity.[61] Set within an epidemiological and ecological framework the authors describe the results of analyses for primary research articles published in 2011 through 2019 and set in a variety of countries and population areas (United States, Europe, Canada, Australia, Vietnam) in urban and rural settings, disaggregated by gender and race/ethnicity in the US studies ($N = 26$), with an age distribution of 5–15+. While the analysis had limitations, the major findings are those reported in other studies of place-based interventions,[62–65] that "greenspaces and recreational facilities appear to have a protective association with weight trajectories in children, while increases in crime appear to be a risk factor for weight gain."[61(p. 10)] Given an ecological perspective of the multiple mediators and moderators examining the impact of environmental stimuli and their association with health outcomes (inclusive of the mediating effect of racism), these findings have face validity and are not surprising. Other studies have shown the impact of social programs, such as the National School Lunch Program, that use multipronged approaches that crosswalk healthcare, public health, and educational settings to offer hope of reversing the impact of negative environmental and social policies.[66]

The framework proposed by Wakefield and Baxter affirms the connection of the roots of all injustices and a rejection of the compartmentalization of these issues to obscure the purpose of equity.[67] The precautionary stance supports a primary prevention perspective, and, while the findings noted above are encouraging and although more studies may be needed, it is important to heed the warnings of Bronfenbrenner, Gamble, and Wakefield[45,47,48] to understand the ecology of environmental health and the need for ethical social policies that lead to action and not be frozen by the need for certainty and exhaustive research requiring the demonstration of causality before acting. This does not contradict the need to apply known evidence-based interventions guided by just social policies to render environmental justice in the present.

Conclusion and the Path Forward

This chapter set a framework for understanding environmental justice and its interaction with racism and other structural inequities on child health. The chapter did not cover all areas of environmental justice but focused on the frame for analysis and a call to action based on ethical policies. An ecological approach to solutions that addresses upstream, system-level interventions is needed, one in addition to evidence- and place-based interventions, and changes at the practice level to counter embedded implicit and overt biases that feed racism. Interprofessional, interdisciplinary, and trans-departmental efforts guided by mandates offer the hope to avoid silos in the basic science, clinical care, and public health approaches to solving complex problems such as environmental health.

Acknowledgments

The chapter's author was intentional in citing the research and publications of a number of minority researchers whose voices are not sufficiently referenced.

References

1. Johnson TJ, Wright JL. Executions and police conflicts involving children, adolescents and young adults. *Pediatr Clin N Am*. 2021; 68:465–87.

2. Bailey ZD, Krieger N, Agenor M, Graves J, Linos N, Bassett MT. Structural racism and health inequities in the USA: evidence and interventions. *Lancet*. 2017 Apr 8;389:1453–62.

3. Alvarez CH. Structural racism as an environmental justice issue: a multilevel analysis of the state racism index and environmental health risk from air toxics. *J Racial Ethn Health Disparities*. https://doi.org/10.1007/s40615-01215-0.

4. Laraque-Arena D. Challenges and successes. In: Laraque-Arena D, Germain L, Young V, Laraque-Ho, A, eds., *Leadership at the Intersection of Gender and Race in Healthcare and Science: Case Studies and Tools*. Abingdon, UK: Routledge; 2022:3–15.

5. Laraque D. Global child health: reaching the tipping point for all children. *Acad Pediatr*. 2011;11:226–33.

6. Bullard RD. Environmental justice in the 21st century. *Phylon (1960)*. 2001 Autumn-Winter;49(3/4):151–71.

7. Bullard RD. Race and environmental justice in the United States. *Yale J Intl L*. 1993; HeinOnline, 319–35

8. H.R. 3184 -Civil Rights Modernization Act of 2021. Available at: https://www.congress.gov/bill/117th-congress/house-bill/3184/text.

9. Needleman HL, Gunnoe C, Leviton A, et al. Deficits in psychologic and classroom performance of children with elevated dentine lead levels. *N Engl J Med.* 1979 Mar 29;300(13):689–95.

10. Needleman HL, Schell A, Bellinger D, Leviton A, Allred EN. The long-term effect of exposure to low doses of lead in childhood: an 11-year follow-up report. *N Engl J Med.* 1990;322(2):83–8.

11. Denworth L. *Toxic Truth.* Boston: Beacon Press; 2008.

12. Laraque D, McCormick M, Norman M, Taylor A, Weller SC, Karp J. Blood lead, calcium status, and behavior in black pre-school children. *Am J Dis Child.* 1990 Feb;144:186–9.

13. Hanna-Attisha M. *What the Eyes Don't See.* New York: Random House; 2018.

14. Lead abatement. 2016. Available at: https://obamawhitehouse.archives.gov/the-press-office/2016/05/03/fact-sheet-federal-support-Flint-water-crisis-response-and-recovery.

15. Dockery DW, Pope III CA. The threat to air pollution health studies behind the Environmental Protection Agency's cloak of science transparency. *Am J Public Health.* 2020 Mar;110(3):286–7.

16. Fredickson L, Sellers C, Dillion L, et al. History of US presidential assaults on modern environmental health protection. *Am J Public Health.* 2018;108(S2):S95–103.

17. Laraque-Arena D, Etzel RA. Organizational change: helping from inside. In: Laraque-Arena D, Germain L, Young V, Laraque-Ho, A, eds., *Leadership at the Intersection of Gender and Race in Healthcare and Science: Case Studies and Tools.* Abingdon, UK: Routledge; 2023:191–220.

18. Colquitt JA. On the dimensionality of organizational justice: a construct validation of a measure. *J Appl Psychol.* 2001;86(3):386–400. doi:10.1037/0021–9010.86.3.386.

19. Colquitt JA, Rodell JB. Measuring justice and fairness. In Cropanzano RS, Ambrose ML, eds. *The Oxford Handbook of Justice in the Workplace.* Oxford, UK: Oxford University Press; 2015:187–202. Available at: https://doi.org/10.1093/oxfordhb/9780199981410.013.8.

20. Formica MK. An eye on disparities, health equity, and racism: the case of firearm injuries in urban youth in the United States and globally. *Pediatr Clin N Am.* 682021:389–99. https://doi.org/10.1016/j.pcl.2020.12.009.

21. Kickbusch I. The political determinants of health: 10 years on. *Br Med J.* 2015;350:h81 doi:10.1136/bmj.h81.

22. Montoya-Williams D, Fuentes-Afflick E. Political determinants of population health. *JAMA Network Open.* 2019;2(7):e197063. doi:10.1001/jamanetworkopen.2019.7063.

23. Clark R, Anderson NB, Clark VR, Williams DR. Racism as a stressor for African Americans. *Am Psychol.* 1999 Oct;54(10):805–16.

24. Sigelman L, Welch S. *Black American's Views of Racial Inequality: The Dream Deferred.* New York: Cambridge University Press. 1991.

25. Yetman N. Introduction: definitions and perspectives. In: Yetman N, ed. *Majority and Minority: The Dynamics of Race and Ethnicity in American Life* (6th ed.). Boston: Allyn & Bacon; 1998:1–10.

26. Krieger N, Sidney S. Racial discrimination and blood pressure: the CARDIA study of young Black and White adults. *Am J Public Health.* 1996;86:1370–8.

27. Anderson LP. Acculturative stress: a theory of relevance to Black Americans. *Clin Psychol Rev.* 1991;11:685–702.

28. Cacioppo J. Social neuroscience: autonomic, neuroendocrine, and immune responses to stress. *Psychophysiology.* 1994;31:113–28.

29. Maldonado Y, Fassiotto M, Jerome B. Implicit and overt race/gender bias: diversity of leadership in health care and health sciences. In: Laraque-Arena D, Germain L, Young V, Laraque-Ho R, eds. *Leadership at the Intersection of Gender and Race in Healthcare and Science: Case Studies and Tools.* Abingdon, UK: Routledge; 2023:67–80.

30. Hall WJ, Chapman MV, Lee KM, et al. Implicit racial/ethnic bias among health care professionals and its influence on health care outcomes: a systematic review. *Am J Public Health.* 2015 Dec;105(12):e60–76.

31. Levy SR, Stroessner SJ, Dweck CS. Stereotype formation and endorsement: the role of implicit theories. *J Pers Soc Psychol.* 1998;74(6):1421–36. https://doi.org/10.1037/0022-3514.74.6.1421.

32. Torres MB, Salles Arghavan S, Cochran A. Recognizing and reacting to microaggressions in medicine and surgery. *JAMA Surg.* 2019;154(9):868–72.

33. United Nations Convention on the Rights of the Child. 1989. https://www.ohchr.org/en/instruments-mechanisms/instruments/convention-rights-child

34. Environmental Protection Agency. *Massachusetts v Environmental Protection Agency.* 2006. Available at: https://ballotpedia.org/Massachusetts_v._Environmental_Protection_Agency#:~:text=Environmental%20Protection%20Agency%20is%20a,Environmental%20Protection%20Agency%20(EPA).

35. Jones CP. Addressing violence against children through anti-racism action. *Pediatr Clin N Am.* 2021;68:449–53.

36. Oberg C. The emergence of planetary pediatrics. *Children.* 2021;8:939. https://doi.org/10.3390/children8100939.

37. Gutchow B, Gray B, Ragavan MI, Sheffield PE, Philipsborn RP, Jee SH. The intersection of pediatrics, climate change, and structural racism: ensuring health equity through climate justice. *Curr Probl Pediatr Adolesc Health Care.* 2021 Jun;51(6):101028. doi:10.1016/j.cppeds.2021.101028. Epub 2021 Jul 5.

38. Suarez-Lopez JR, Cairns MR, Sripada K, et al. COVID-19 and children's health in the United States: consideration of physical and social environments during the pandemic. *Environ Res.* 2021;197:111160. doi:10.1016/j.envres.2021.111160.

39. McKenna B. Medical education for what? Neoliberal fascism versus social justice. *J Med Humanit.* 2021; 42(4)587–602. doi:10.1007/s10912-020-09673-z.

40. Powell DL, Stewart V. Children: the unwitting target of environmental injustices. *Pediatr Clin North Am.* 2001;48(5)1291–305. doi:10.1016/s0031-3955(05)70375-8.

41. Salas RN. Environmental racism and climate change: missed diagnoses. *N Engl J Med.* 2021;385(11):967–9. doi:10.1056/New Engl J Medp2109160.

42. Reno R, Warming E, Zaugg C, Marx K, Pies C. Lessons learned from implementing a place-based, racial justice-centered approach to health equity. *Matern Child Health J.* 2021;25:66–71. https://doi.org/10.1007/s10995-020-03076-1.

43. Vedam S, Zephyrin L, Hardtman P, et al. Transdisciplinary imagination: addressing equity and mistreatment in perinatal care. *Matern Child Health J.* 2022;26(4):674–81. doi:10.1007/s10995-022-03419-0.

44. Macey GP, Her X, Reibling ET, Ericson J. An investigation of environmental racism claims: testing environmental management approaches with a geographic information system. *Environ Manage.* 2001;27(6):893–907. doi:10.1007/s002670010197.

45. Wakefield S, Baxter J. *Linking Health Inequality and Environmental Justice: Articulating a Precautionary Framework for Research and Action*. Toronto, Canada: Geography Publications; 2010. Available at: https://ir.lib.uwo.ca/geographypub/272.

46. Bronfenbrenner U. Toward an experimental ecology of human development. *Am Psychol*. 1977 Jul;32(7):513–31.

47. Bronfenbrenner U. Developmental research, public policy and the ecology of childhood. *Child Dev*.1974;45(1):1–5.

48. Gamble VN, Stone D. US policy on health inequities: the interplay of politics and research. *J Health Polit Policy Law*. 2016 Feb;31(1):93–126.

49. Dover DC, Belon AP. The health equity measurement framework: a comprehensive model to measure social inequities in health. *Int J Equity Health*. 2019;18:36. https://doi.org/10.1186/s12939-019-0935-0.

50. Beard K, Iruka IU, Laraque-Arena D, Murry VM, Rodriguez LJ, Taylor S. Dismantling systemic racism and advancing health equity through research. Commentary, National Academy of Medicine. 2022, Jan 10. https://nam.edu/dismantling-systemic-racism-and-advancing-health-equity-throughout-research/

51. Hoppe TA, Litovitz A, Willis KA, et al. Topic choice contributes to the lower rate of NIH awards to African American/black scientists. *Sci Adv*. 2019;5:eaaw7328.

52. Ginther DK, Schaffer WT, Schnell J, et al. Race, ethnicity, and NIH research awards. *Science*. 2011 Aug 10;333:1015–9. https://www.science.org/doi/10.1126/science.1196783.

53. Davis LF, Ramirez-Andreotta MD. Participatory research for environmental justice: a critical interpretive synthesis. *Environ Health Perspect*. 2021 Feb;129(2). 026001-1-20.

54. Jull J, Whitehead M, Petticrew M, et al. When is a randomized controlled trial health equity relevant? Development and validation of a conceptual framework. *Br Med J Open*. 2017;7:e015815. doi:10.1136/bmjopen-2016–015815.

55. Garrison H. Underrepresentation by race-ethnicity across stages of US science and engineering education. *CBE Life Sci Educ*. 2013;12:357–63.

56. Laraque-Arena D, Stanton BF. A twenty-first century policy agenda: violence in the lives of children, families, and communities. *Pediatr Clin N Am*. 2021;68:xv–xviii. https://doi.org/10.1016/j.pcl.2021.01.001 0031-3955/21.

57. Dowty D. Syracuse teen shot at party graduated 6 hours before: family says he's not expected to survive. *The Post-Standard*. 2020. Available at: https://www.syracuse.com/crime/2020/06/Syracuse-teen-mortally-wounded-in-mass-party-shooting-had-graduated-high-school-6-hours-before.html.

58. Kneebone E, Nadeau C, Berube A. The re-emergence of concentrated poverty: metropolitan trends in the 2000s. *Brookings*. Available at: 2011 Nov. http://www.ewa.org/sites/main/files/file-attachments/2015-10-22-poverty-paul-jargowsky1.pdf.

59. Beard JH, Morrison CN, Jacoby SF, et al. Quantifying disparities in urban firearm violence by race and place in Philadelphia, Pennsylvania: a cartographic study. *Am J Public Health* 2017;107(3)371–3.

60. Yamane D. The sociology of US gun culture. *Sociol Compass*. 2017;11:e12497. https://doi.org/10.1111/soc4.12497.

61. Daniels KM, Schinasi LH, Auchincloss AH, Forrest CB, Diez Rouz AV. The built and social neighborhood environment and child obesity: a systematic review of longitudinal studies. *Prevent Med*. 2021;153:106790. https://doi.org/10.1016/j.ypmed.2021.106790.

62. Laraque D, Barlow B, Durkin M, Heagarty M. Injury prevention in an urban setting: challenges and successes. *J Urban Health*. 1995;72(1):16–30.

63. Durkin MS, Kuhn L, Davidson LL, Laraque D, Barlow B. Epidemiology and prevention of severe assault and gun injuries to children in an urban community. *J Trauma*. 1996;41(4):667–3.

64. Branas C, Flescher A, Formica MK, et al. Academic public health and the firearm crisis: an agenda for action. *Am J Public Health*. 2017;107(3):365–7.

65. Wolch JR, Byrne J, Newell JP. Urban green space, public health, and environmental justice: The challenge of making cities "just green enough". *Landscape and Urban Planning*. 2014;125:234–44.

66. Woo Baidal JA, Taveras EM. Protecting progress against childhood obesity: the national school lunch program. *N Engl J Med*. 2014 Nov 13;371:1862–5.

67. Laraque-Arena D. 2017. Available at: https://www.youtube.com/watch?v=5Iz6fhpaNgg

12

Economics and Children's Environmental Health

Philip J. Landrigan

Cost is an inevitable but often overlooked and insufficiently appreciated consequence of disease caused in children by toxic exposures in the environment.[1] The costs associated with disease of environmental origin include two categories—direct and indirect costs.

- The *direct or medical costs* of disease of environmental origin in children are defined as the medical expenses associated with caring for children made ill by toxic exposures. These include physician costs, hospital costs, and the costs of medications.
- The *indirect or nonmedical costs* of disease include the cost of a child's time lost from school because of illness; the cost of a parent's time lost from work while caring for a sick child; the costs of special education; the costs of rehabilitation; the costs of lifelong reduction in economic productivity in children whose brains, lungs, or other organ systems have been permanently damaged by toxic exposures; and the costs of lost productivity from premature death.

Several categories of economic analysis have been applied to the study of diseases of environmental origin in children.[2] These include

- *Cost analysis*, the systematic collection, categorization, and analysis of the direct and indirect economic costs associated with illness of environmental origin. Cost analyses are the type of economic analysis most commonly used in children's environmental health because the necessary data are the most readily available.
- *Benefit analysis*, the systematic evaluation of the direct and indirect costs of disease that are averted by a policy intervention and of the resulting societal benefits. The interventions most commonly studied in children's environmental health have been actions taken by governments to reduce children's exposures to environmental hazards such as lead and air pollution.
- *Cost-benefit analysis*, the systematic comparison of the benefits and costs of an intervention or policy. Comparison of the costs of controlling pollution versus the benefits to children's health of diminished pollution is an example.
- *Cost-effectiveness analysis*, a technique for comparing the costs and benefits of various alternative intervention strategies in order to choose the most effective.[3]

Philip J. Landrigan, *Economics and Children's Environmental Health* In: *Textbook of Children's Environmental Health*, Second Edition. Edited by: Ruth A. Etzel and Philip J. Landrigan, Oxford University Press. © Oxford University Press 2024. DOI: 10.1093/oso/9780197662526.003.0012

Cost Analyses

A methodology for estimating the costs of diseases caused by environmental contamination was developed in the United States in the early 1980s by an expert committee convened by the Institute of Medicine and chaired by Stanford University Nobel Laureate in
Economics, the late Professor Kenneth Arrow.[4] The core of this methodology is calculation of the "fractional contribution" of environmental factors to causation of a particular
disease in a particular population. This *environmentally attributable fraction* (EAF) is
defined as "the percentage of a particular disease category that would be eliminated if
environmental risk factors were reduced to their lowest feasible levels."[5]

The EAF for a particular disease, such as childhood asthma, may be converted to
an estimate of the cost of that disease in a particular population through the following
equation:

Costs = Disease rate [incidence or prevalence] × EAF × Population size × Cost per case

Costs, as noted above, are traditionally divided into direct and indirect health-related
costs, and both are included in the calculation.

The EAF model has now been used in a series of important cases to assess the costs of
environmental and occupational disease. Findings from these analyses have had powerful impacts on the health and well-being of societies around the world.

In one of its earliest applications, the EAF approach was used in 1989, in New York
state, to estimate the fractional costs of four categories of illness in adult workers attributable to occupational exposures: cancer, chronic respiratory disease, cerebrovascular and cardiovascular disease, and end-stage renal failure. Using fairly conservative
assumptions, this analysis found that occupational exposures resulting in these four diseases cost more than US$600 million per year, with the largest proportion (80%) due
to occupationally related cancer.[6] This analysis persuaded the New York state legisla-
ture to allocate funds to support a statewide network of Clinical Centers of Excellence in
Occupational Medicine that provide expert diagnostic and treatment services to workers
who had been made ill or injured while at work. These centers have provided previously
unavailable specialist care in occupational medicine to tens of thousands of workers and
contributed to striking reductions in the incidence of occupational disease and death in
New York state. These Centers of Excellence continue to provide medical care to injured
and ill workers. They played a key role in supporting the medical response to the attacks
on the World Trade Center of September 11, 2001.

The EAF approach was used by Leigh et al. to calculate the costs of occupational disease and injuries across the United States.[7] That estimate found that the total costs of occupational disease and injuries were US$171 billion (occupational injuries cost US$145
billion and illnesses US$26 billion).[7] Leigh subsequently updated these analyses and
found that, for 2007, the total estimated annual costs for occupational injuries and diseases combined had risen to approximately US$250 billion.[8]

Cost Analyses in Children's Environmental Health

The economic costs of disease in children caused by harmful exposures in the environment have been examined. A 2002 report by Landrigan et al.[9] analyzed the costs in US

children of four categories of pediatric illness for which there is published evidence of an environmental contribution: lead poisoning, asthma, cancer, and neurobehavioral disorders. In this analysis, data on incidence and rates of disease were taken from databases maintained by the US Centers for Disease Control and Prevention (CDC) and the US National Institutes of Health (NIH). Data on the size of the population at risk for each disease were obtained from the US Census.

To estimate the costs associated with the cognitive and behavioral consequences of clinical and subclinical lead poisoning, the authors relied heavily on an economic forecasting model developed by Schwartz et al.[10] and applied this model to CDC data on prevalence of lead poisoning. In this model, blood lead levels are assumed on the basis of work by Salkever[11] to produce a dose-related decrement in intelligence (IQ score).

To examine the direct medical costs of asthma, the investigators used data from multiple sources including the annual National Hospital Discharge Survey, the National Ambulatory Medical Care Survey, the annual National Health Interview Survey, the National Medical Care Utilization and Expenditure Survey, the National Medical Expenditure Survey, and a managed healthcare database of medical and pharmacy claims.[9] To calculate the indirect costs associated with childhood asthma, they utilized data on price and wage indices from the Bureau of Labor Statistics, prescription expenditure estimates from the CDC, and reports on healthcare costs from the Health Care Financing Agency.

Information on the costs of childhood cancer was difficult to obtain because more than 80% of pediatric cancer patients in the United States are participants in randomized clinical trials, and no recovery of costs occurs for these trial participants. Therefore, the investigators obtained the medical records of all patients treated for pediatric malignancies at the Mount Sinai Medical Center in New York City. Summaries of physician services, hospital charges, radiologic services, and laboratory services were abstracted from those records. Data on the costs of physician and hospital resources were taken from physician billing rates and hospital charges, adjusted by Health Care Financing Agency cost-to-charge ratios.

The most difficult aspect of this analysis was estimation of the EAFs, the proportions of cases of lead poisoning (Box 12.1), asthma, childhood cancer, and neurobehavioral disorders that are potentially attributable to toxic exposures in the environment (Box 12.2). At time of the analysis, there were no published data available estimating these EAFs. Therefore, to develop an estimate of the EAF for each of these diseases, the investigators used a formal, expert-based decision-making process, the modified Delphi technique.[12] They selected an expert panel for each disease. These panels were assembled from among prominent physicians and scientists with established national reputations and extensive records of publication in relation to the diseases under study. Each consisted of three or four persons. All panelists were asked to estimate the EAF for the disease in which they were expert on two occasions: before a meeting of the panel (by mail ballot) and again at a face-to-face meeting. Each panelist was asked to develop a best estimate, from 0% to 100%, of the EAF and also to indicate an upper and a lower bound of plausibility around that estimate. Each panel then met for 1 day and spent the day discussing the estimates of EAF that they had submitted before the meeting. The goal of the meeting was to refine the initial estimates through a consensus approach and to reduce the range of uncertainty. At the end of the day, a second vote was taken. The arithmetic means of these final estimates were used as the basis for determining EAFs.

EAFs were judged by the expert panels to be 100% for lead poisoning, 30% for asthma (range, 10–35%), 5% for cancer (range, 2–10%, and 10% for neurobehavioral disorders (range, 5–20%).[9]

The total annual costs of the fractions of these four diseases due to environmental exposures were estimated in 1997 dollars to be US$54.9 billion (range US$48.8–64.8 billion).[9] The breakdown of this total cost was as follows:

- $43.4 billion for lead poisoning,
- $2.0 billion for asthma,
- $0.3 billion for childhood cancer, and
- $9.2 billion for neurobehavioral disorders.

Box 12.1 Case Study: The Costs of Childhood Lead Poisoning

The largest component of these costs was associated with lead poisoning. These great costs reflect the lifelong reductions in earnings potential and economic productivity that result from permanent reduction in cognitive capability (IQ score) in tens of thousands of children exposed in early life to lead.[11] See Figure 12.1.

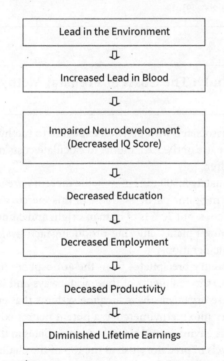

Figure 12.1. The model: ➡ Lead results in decreased ⬇ intelligence and decreased economic productivity.

The basis for the estimate that lead toxicity was responsible in 2002 for an annual cost of US$43.4 billion was a calculation by Salkever[11] that the loss of 1 IQ point is associated with an overall reduction in lifetime earnings of 2.39%. This corresponds to a loss of 1.61% of earnings potential for an IQ deficit of 0.675 points. At the time of this analysis, the present value of lifetime expected earnings was $881,027 for a 5-year-old boy and $519,631 for a 5-year-old girl.[15] Thus, the present value of economic losses attributable to lead exposure in the birth cohort of current 5-year-olds amounted to $43.4 billion per year (Table 12.1).

Table 12.1 Estimated costs of pediatric lead poisoning (United States, 1997)

EAF = 100%

Main consequence = Loss of IQ over lifetime

Mean blood lead level in 1997 among 5-year-old children = 2.7 μg/dL

A blood lead level of 1 μg/dL = Mean loss of 0.25 IQ points per child

Therefore, 2.7 μg/dL = Mean loss of 0.675 IQ points per child

Loss of 1 IQ point = Loss of lifetime earnings of 2.39%

Therefore, loss of 0.675 IQ points = Loss of 1.61% of lifetime earnings

Economic consequences

For boys: Loss of 1.61% × $881,027 (lifetime earnings) × 1,960,200 = $27.8 billion

For girls: Loss of 1.61% × $519,631 (lifetime earnings) × 1,869,800 = $15.6 billion

Total annual costs of pediatric lead poisoning = $43.4 billion

From Landrigan et al., 2002 [9].

Box 12.2 Case Study: The Costs of Prenatal Methylmercury Toxicity

A study of the costs associated with prenatal exposure to methylmercury provides a second example of the use of the EAF model to calculate costs of disease of environmental origin in children.[13]

Each year approximately 5,500 tons of metallic mercury are emitted to the Earth's atmosphere in the form of fine particles. Some of this mercury derives from natural sources such as volcanoes, but 70% is of human origin and comes from such sources as coal-fired electric power plants, chlorine-producing facilities, mercury mining and smelting, and waste incineration.

When metallic mercury precipitates from the atmosphere into rivers, lakes, and the oceans, it settles to the sediments of these waterways and is ingested by marine microorganisms. These organisms have enzyme systems that enable them to transform metallic mercury into methylmercury, a potent neurotoxin. Methylmercury is highly persistent in the environment, and it bioaccumulates in the marine food chain to reach very high levels in predatory fish at the top of the chain such as tuna, shark, king mackerel, swordfish, bluefish, and striped bass. Consumption of contaminated fish is the major route of human exposure to methylmercury. Because their brains are rapidly growing and developing, infants in the womb are the segment of the population at greatest risk of methylmercury neurotoxicity.

Acute, high-dose prenatal exposure to methylmercury from consumption of highly contaminated fish can cause devastating neurodevelopmental damage, as was documented in a tragic epidemic of congenital methylmercury poisoning that occurred in the 1950s and 1960s in Minamata, Japan[14] (see Chapters 1 and 2). More recent studies have documented that methylmercury can also cause subclinical toxicity to infants exposed prenatally to levels of methylmercury too low to produce obvious symptoms. The US National Academy of Sciences reviewed these studies and concluded that there is strong evidence for the fetal neurotoxicity of methylmercury, even at low levels of exposure.[15]

To estimate the economic costs of low-level prenatal exposure to methylmercury in US children, Trasande et al.[13] used an EAF model and focused their assessment on the costs associated with IQ loss. Using national blood mercury prevalence data from the CDC, these investigators found that between 316,588 and 637,233 US children each year have cord blood mercury levels of greater than 5.8 µg/L, a level associated with loss of IQ. The resulting loss of intelligence causes diminished economic productivity that persists over the life spans of these children. The economic cost of this lost productivity was estimated to amount to US$8.7 billion annually (range, US$2.2–43.8 billion). The model used in this analysis was similar to that described above for the lead case study.

Of the total annual costs attributed to methylmercury, Trasande et al. calculated that US$1.3 billion (range, US$0.1 billion–6.5 billion) are attributable to mercury emissions from US coal-fired power plants.

Updated Information on the Costs of Environmental Disease in Children

The cost estimates of disease in children described above by Landrigan et al.[9] were updated, expanded, and refined in 2011 by Trasande and Liu.[16] By incorporating new information on environmental causation of disease, expanding the number of disease categories considered, and updating estimates on cost of illness, these investigators found that the annual costs of lead poisoning, prenatal methylmercury exposure, childhood cancer, asthma, intellectual disability, autism, and attention deficit hyperactivity disorder in US children amount to US$76.6 billion (Table 12.2).

Additional Studies of the Costs of Environmental Disease in Children

Readers who wish to go more deeply into the literature on the costs of environmental disease in children may wish to refer to the following resources:

- Trasande et al. examined the costs of disease due to endocrine-disrupting chemicals in the European Union and estimated that exposures to these toxic chemicals are responsible for a median annual cost of €163 billion (1.28% of EU gross domestic product).[17]
- Landrigan et al. examined the economic costs of brain injury and IQ loss caused in children across Africa by early-life exposure to fine particulate ($PM_{2.5}$) air pollution. They estimated that each 1 µg/m^3 increase in annual mean $PM_{2.5}$ air pollution exposure produces an estimated loss of 0.61 IQ points—a total of nearly 2 billion IQ points lost across Africa each year. Such widespread losses in these children's cognitive function reduce the human capital and future economic productivity of entire countries.[18]
- The Lancet Commission on Pollution and Health examined the health-related costs of all forms of pollution globally and estimated that these cost amount to US$4.6

Table 12.2 Aggregate costs of environmentally mediated diseases in US children (2008)

	Base-case estimate	Low-end estimate	High-end estimate
Lead poisoning	$50.9 billion	$44.8 billion	$60.6 billion
Methylmercury toxicity	$8.4 billion	$3.2 billion	$5.1 billion
Asthma	$2.2 billion	$728.0 million	$2.5 billion
Intellectual disability	$5.4 billion	$2.7 billion	$10.9 billion
Autism	$7.9 billion	$4.0 billion	$15.8 billion
AD/HD	$5.0 billion	$4.4 billion	$7.4 billion
Childhood cancer	$95.0 million	$38.2 million	$190.8 million
Total	$76.6 billion	$59.8 billion	$105.8 billion

From Trasande and Liu, 2011 [16].

trillion per year—more than 6% of global economic output. The Commission noted that these costs are so great that they can undercut national trajectories of social and economic development.[19]

Benefits Analysis

Benefits analyses in children's environmental health quantify the economic gains that result from improvements environmental quality. These benefits include reductions in healthcare costs and also indirect benefits, most notably increases in intelligence and thus in economic productivity that result from widespread reductions in population exposure to toxic chemicals (Box 12.3 and Box 12.4).

Full-Cost Accounting

A common theme that runs through all of the analyses of the costs of environmental and occupational diseases discussed in this chapter is their focus on identifying and quantifying costs that had previously been overlooked or hidden. These analyses may, therefore, be viewed as examples of a powerful new approach to measuring the full impacts of environmental contamination, termed *full-cost accounting* or "green accounting."[23]

Full-cost accounting differs from traditional accounting which meticulously tabulates the costs of materials, labor, and energy, but fails to consider, or "externalizes," costs that fall outside conventional business accounting, such as the costs of environmental pollution or of disease caused by pollution. In essence, conventional accounting shifts the costs of environmental clean-up and the healthcare costs associated with environmental disease away from manufacturers and onto taxpayers and individual citizens. Full-cost accounting is intended to reverse that pattern and require manufacturers to take responsibility for the health and environmental damages that they cause.

The goal of full-cost accounting is to develop monetary values for previously ignored, externalized costs such as the costs of pollution, environmental degradation, and pollution-related disease. In the future, all policies—including economic

---

Box 12.3 Case Study: Benefits of Removing Lead from Gasoline

A classic example of a benefits analysis is a combined epidemiologic and economic analysis undertaken by Grosse et al. to quantify the economic benefits of removing lead from the US gasoline supply.[20] From 1922 to 1980, lead was added to automotive gasoline in the United States to improve engine performance. At peak use in 1975, approximately 100,000 tons of lead were added per year. Environmental contamination and human exposure were widespread, and the population mean blood lead level (all ages) was 17–18 μg/dL. In the face of great opposition from the lead industry, the US Environmental Protection Agency (EPA) made the decision to remove lead from gasoline beginning in 1975. Removal was phased and was largely complete by 1980.

Grosse et al. studied the health and economic impacts of the lead phase-out and found that the total decline in mean blood lead level in 1- to 5-year-old US children from 1976 to 1999 that resulted from the removal of lead from gasoline was 15.1 μg/dL. They calculated that this decline in lead levels produced a gain in the mean IQ score of American children of between 2.2 and 4.7 points. Because each IQ point is calculated to raise a child's economic productivity over a lifetime by 1.76% to 2.38%, the authors estimated that removing lead from gasoline has generated an economic benefit in each year's birth cohort born since 1980 of US$213 billion (range, US$110 to 318 billion)—an aggregate economic benefit since 1980 of more than US$8 trillion.

This striking finding has influenced countries around the world to remove lead from gasoline. Sharp declines in population blood lead levels have resulted in country after country. Algeria, the last country in the world to use lead in automotive gasoline, removed lead from their gasoline supply in 2021.[21]

---

policies—should take external costs into full account. This change will force assessments of the true costs of policy options and will send strong messages about the long-term affordability of various policies.[24–26] Full-cost accounting is growing in its application.[23–25] It has the potential to fundamentally transform business practice.

The main impediment to full-cost accounting is that the costs of pollution and pollution-related disease are not easily discerned because they are not neatly collected in one place but instead are spread out across large populations over many years, do not show up in environmental budgets, and are buried deep in healthcare costs and productivity data. New accounting practices now required in the European Union will change this paradigm and, for the first time, will force industries to systematically collect this information.[27]

Extended producer responsibility (EPR) is a legal mechanism for forcing manufacturers to take financial responsibility for previously externalized costs by requiring them to cover the full costs of the materials they produce (including the costs of disposal). EPR requires manufacturers to either take back materials that they produce at the end of the materials' useful life (e.g., deposits on returnable bottles) or to cover the costs of disposal of such materials. California has recently required manufacturers of disposable plastic items to assume financial responsibility for these products across their entire lifecycle.[28]

Box 12.4 Case Study: Benefits of the US Clean Air Act Amendments

A second example of a benefits analysis is an examination of the benefits of the cleaning of America's air that followed passage of the Clean Air Act Amendments of 1990.[22] These amendments strengthened the Clean Air Act of 1970 by increasing the stringency of some federal requirements, revamping the hazardous air pollutant regulatory program, and introducing new programs for the control of acid air pollutants.

The EPA estimated that the annual dollar value of the benefits in air quality improvements that result from these amendments is already very large and will grow over time as emissions control programs take their full effect, reaching an annual level of approximately US$2.0 trillion in 2020.[22] Most of these benefits are attributable to reductions in healthcare costs, reductions in premature mortality, and increased economic productivity associated with reductions in exposures to ozone and airborne particulates.

EPA estimated that, in 2010, reductions in fine particle and ozone pollution from the Clean Air Act Amendments prevented more than

- 160,000 cases of premature mortality
- 130,000 heart attacks
- 13 million lost work days
- 1.7 million asthma attacks.

EPA estimated that, in 2020, the Clean Air Act Amendments prevented more than

- 230,000 cases of premature mortality
- 200,000 heart attacks
- 17 million lost work days
- 2.4 million asthma attacks.

The economic gains associated with each of these health benefits far exceed the estimated costs of compliance with the Clean Air Act Amendments.[20] The ratio of benefits to costs (i.e., the return on investment) is estimated at 30:1 (95% confidence interval 4–88).[22]

Limitations of Economic Analyses

Calculations of the costs of disease of environmental origin in children that are presented in the biomedical literature and summarized in this chapter almost certainly underestimate the true economic burden of these diseases. One limitation of present-day analyses is a lack of strong surveillance systems for tracking incidence and prevalence of noncommunicable diseases in many countries. Tracking systems for noncommunicable diseases are generally much weaker than those for acute infectious disease, which have been in place for many decades. Without knowledge of the number of cases of lead poisoning, pesticide intoxication, asthma, autism spectrum disorders, childhood cancer, or other noncommunicable diseases in children that may be associated with environmental exposures, it is not possible to estimate either the burden of environmentally related disease in children or the associated costs.

Lack of knowledge of the possibly toxic effects of many chemicals to which children are exposed is a second limitation on these analyses.[29,30] It is highly likely that there are still undiscovered associations between certain commonly used chemicals and diseases in children. But until toxicological or epidemiological studies to elucidate these associations are undertaken, they will not be recognized, and thus disease and dysfunction caused by them will be undiagnosed and uncounted (see Chapters 3 and 65).

A third limitation is that none of the analyses considered here has examined the possible late complications of toxic exposures sustained in early life. For example, none of these analyses examines the possible late cardiovascular consequences of childhood lead poisoning, nor the costs of adult pulmonary disease that might be the consequence or the continuation into adult life of childhood asthma.

Finally, the estimates presented here can never capture the full burden of disease of environmental origin because none attempts to estimate the costs of the pain, deterioration in quality of life, or emotional suffering in families, friends, or affected children that are the inevitable consequences of chronic illness in a child.

Conclusion

Economic analyses can be powerful tools for disease prevention.[31] Accurate information on the previously hidden costs of diseases of environmental origin can be a persuasive adjunct to clinical, pathophysiologic, and epidemiologic data. Economic data can sway the opinions of elected officials and governmental policymakers and encourage them to prioritize the protection of children's health.

It is relatively easy for an industry to calculate the costs of installing pollution controls, and it is much more difficult for epidemiologists and economists to develop accurate information on the costs of pollution because these are hidden costs that typically are spread across populations and extend over many years.[4] But when credible data on the previously externalized costs of environmental degradation and disease are properly assembled, they can help counter one-sided arguments put forward by business interests that focus exclusively on the costs of preventing pollution.[32] They can be very important in the setting of priorities and in the allocation of scarce resources. In short, the development of information on the costs of disease caused by environmental contamination can drive major policy changes, save hundreds of billions of dollars, and improve the lives of future generations.

From the examples presented in this chapter, it is clear that diseases of toxic environmental origin make an important but underrecognized contribution to healthcare costs among children in the United States and in countries around the world. The costs of these diseases are currently estimated in the United States to amount to US$76.6 billion annually.[16]

In future years, as disease tracking systems around the world are strengthened, as more etiologic research is undertaken, and as better information becomes available on associations between environmental exposures and diseases in children, the estimates presented here will be strengthened and expanded. It may be anticipated that growing recognition of the true costs of environmental degradation and disease will force the adoption of more sustainable economic, industrial, and environmental policies in countries around the world.[26]

References

1. Goodstein ES. *Economics and the Environment* (6th ed.). New York: Wiley; 2011.

2. Trasande L. Economics of children's environmental health. *Mount Sinai J Med.* 2011;78:98–106.

3. Centers for Disease Control and Prevention. Economic evaluation of public health preparedness and response efforts. Cost effectiveness analysis. 2022. Available at: https://www.bette revaluation.org/tools-resources/cost-effectiveness-analysis-us-centers-for-disease-control-prevention

4. Institute of Medicine. *Costs of Environment-Related Health Effects: A Plan for Continuing Study.* Washington, DC: National Academy Press; 1981.

5. Smith KR, Corvalin CF, Kjellstrom T. How much global ill health is attributable to environmental factors? *Epidemiology.* 1999;10:573–84.

6. Fahs MC, Markowitz SB, Fischer E, Shapiro J, Landrigan PJ. Health costs of occupational disease in New York State. *Am J Ind Med.* 1989;16:437–49.

7. Leigh JP, Markowitz S, Fahs M, Shin C, Landrigan PJ. Costs of occupational injuries and illnesses. *Arch Intern Med.* 1997;157:1557–68.

8. Leigh JP. Economic burden of occupational injury and illness in the United States. *Milbank Q.* 2011;89:728–72.

9. Landrigan PJ, Schechter CB, Lipton JM, Fahs MC, Schwartz J. Environmental pollutants and disease in American children: estimates of morbidity, mortality, and costs for lead poisoning, asthma, cancer, and developmental disabilities. *Environ Health Perspect.* 2002;110:721–8.

10. Schwartz J, Pitcher H, Levin R, Ostro B, Nichols AL. *Costs and Benefits of Reducing Lead in Gasoline: Final Regulatory Impact Analysis.* EPA-230/05-85/006. Washington, DC: Environmental Protection Agency; 1985.

11. Salkever DS. Updated estimates of earnings benefits from reduced exposure of children to environmental lead. *Environ Res.* 1995;70:1–6.

12. Fink A, Kosecoff J, Chassin M, Brook RH. Consensus methods: characteristics and guidelines for use. *Am J Public Health.* 1984;74:979–83.

13. Trasande L, Landrigan PJ, Schechter C. Public health and economic consequences of methyl mercury toxicity to the developing brain. *Environ Health Perspect.* 2005;113:590–6.

14. Harada H. Congenital Minamata disease: intrauterine methylmercury poisoning. *Teratology.* 1978;18:285–8.

15. National Research Council. *Toxicological Effects of Methylmercury.* Washington, DC: National Academies Press, 2000.

16. Trasande L, Liu Y. Reducing the staggering costs of environmental disease in children, estimated at $76.6 billion in 2008. *Health Aff (Millwood).* 2011;30:863–70.

17. Trasande L, Zoeller RT, Hass U, et al. Burden of disease and costs of exposure to endocrine disrupting chemicals in the European Union: an updated analysis. *Andrology.* 2016 Jul;4(4):565–72.

18. Fisher S, Bellinger DC, Cropper ML, et al. Air pollution and development in Africa: impacts on health, the economy and human capital. *Lancet Planetary Health.* 2021;5:e681–8.

19. Landrigan PJ, Fuller R, Acosta NJR, et al. The Lancet commission on pollution and health. *Lancet.* 2018;391(10119):462–512.

20. Grosse SD, Matte TD, Schwartz J, Jackson RJ. Economic gains resulting from the reduction in children's exposure to lead in the United States. *Environ Health Perspect.* 2002;110:563–9.

21. UN Environment Programme (UNEP). The lead campaign. 2023. Available at: https://www. unep.org/explore-topics/transport/what-we-do/partnership-clean-fuels-and-vehicles/lead-campaign.

22. US Environmental Protection Agency. Office of air and radiation. The benefits and costs of the Clean Air Act from 1990 to 2020. 2011. Available at: https://www.epa.gov/sites/default/files/2015-07/documents/fullreport_rev_a.pdf.

23. Wikipedia. Full cost accounting. 2020. Available at: http://en.wikipedia.org/wiki/Full_cost_accounting.

24. Epstein PR, Buonocore JJ, Eckerle K, et al. Full cost accounting for the life cycle of coal. *Ann N Y Acad Sci.* 2011;1219:73–98.

25. National Research Council. *The Hidden Costs of Energy: Unpriced Consequences of Energy.* Washington, DC: National Academies Press; 2010.

26. Haines A, Alleyne G, Kickbusch I, Dora C. From the Earth Summit to Rio+20: integration of health and sustainable development. *Lancet.* 2012;379:2189–97.

27. European Union. Corporate Sustainability Reporting Directive (CSRD). 2022. Available at: https://www.consilium.europa.eu/en/press/press-releases/2022/11/28/council-gives-final-green-light-to-corporate-sustainability-reporting-directive/.

28. State of California. Plastic Pollution Prevention and Packaging Producer Responsibility Act (SB 54). 2022. Available at: https://www.brightest.io/california-plastic-law-compliance.

29. Goldman LR. Chemicals and children's environment: what we don't know about risks. *Environ Health Perspect.* 1998;106:875–80.

30. Landrigan PJ, Goldman LR. Children's vulnerability to toxic chemicals: a challenge and opportunity for health and environmental policy. *Health Aff (Millwood).* 2011;30:842–50.

31. Arrow KJ, Cropper ML, Eads GC, et al. Is there a role for benefit-cost analysis in environmental, health, and safety regulation? *Science.* 1996;272:221–2.

32. Krugman P. Springtime for toxics. *New York Times.* 2011 Dec 25. Available at: http://www.nytimes.com/2011/12/26/opinion/krugman-springtime-for-toxics.html.

13

The Global Dimension of Children's Environmental Health

Ruth A. Etzel

Although most health research and attention focuses on people in the world's wealthier countries, the greatest burden of disease falls on those in the low- and middle-income countries (LMICs).

The World Health Organization (WHO) has estimated that more than one-quarter of disease among children worldwide is due to preventable environmental risk factors.[1–3] This is likely a substantial underestimate because the WHO definition of preventable risk factors excludes many factors that involve individual choices (alcohol and tobacco consumption, drug abuse, diet), natural environments that cannot reasonably be modified (flooding rivers, rising sea levels, global warming), unemployment (provided that it is not linked to the degradation of the environment), natural biological agents (e.g., pollen), and person-to-person transmission that cannot reasonably be prevented by environmental interventions (Table 13.1). The global distribution of deaths of children under 5 is not equitable; the highest burden is being experienced in LMICs (see Figure 13.1).

The environments in which today's children live, grow, and play are dramatically different from those of their grandparents. The 21st century is sometimes called the "Urban Millennium."[4] For the first time in history, more people are living in towns and cities than in rural areas. In 1950, urban dwellers were 29.2% of the world population; in 1980, the urban population was 39.6%; and in 2018, the urban population was 58%.[5] By 2050, 68% of the world's population is projected to be urban (Figure 13.2).

Industrialization fueled the transition from rural to urban living. The production and use of man-made chemicals has skyrocketed, and more hazardous wastes began moving across national borders. With industrialization and globalization, children, an already vulnerable population, are placed at even higher risk. In LMICs, the new environmental risks associated with urbanization and industrialization are superimposed on existing environmental risk factors such as scarce food, contaminated water, polluted air, lack of sanitation and garbage disposal, inadequate housing, and social violence.[6] With industrial development, the traditional risks to children's health are compounded by newer environmentally related causes of disease such as toxic chemicals in the air children breathe, the water they drink, and food they eat.

These hazards are magnified for children living in poverty who often live and grow near polluting industries and hazardous waste sites, and they may be surrounded by violent social conditions. Throughout the world, poor communities are too often communities with degraded physical and social environments that negatively affect the health and development of infants and children.

Ruth A. Etzel, *The Global Dimension of Children's Environmental Health* In: *Textbook of Children's Environmental Health*, Second Edition. Edited by: Ruth A. Etzel and Philip J. Landrigan, Oxford University Press. © Oxford University Press 2024.
DOI: 10.1093/oso/9780197662526.003.0013

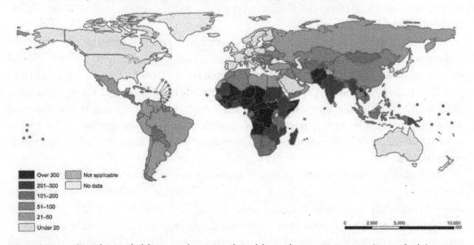

Deaths in children under five attributable to the environment (as a whole) per 100 000 people, 2012

Figure 13.1. Deaths in children under 5 attributable to the environment (as a whole) per 100,000 people, 2012.
Source: World Health Organization, 2012.

Table 13.1 WHO definitions of Modifiable Environmental Factors

Included environmental factors are the modifiable parts (or impacts) of:

Pollution of air, water or soil by chemical or biological agents

Ultraviolet (UV)[a] and ionizing radiation

Noise, electromagnetic fields

Occupational risks

Built environments, including housing; land-use patterns, roads

Major infrastructural and engineering works such as roads, dams, railways, airports

Manmade vector breeding places or breeding places catering to the specific ecological requirements of vectors, such as old tires or water containers

Agricultural methods, irrigation schemes

Man-made climate change, ecosystem change

Behavior related to environmental factors (e.g., the availability of safe water for washing hands or physical activity fostered through improved urban design)

Excluded factors are:

Alcohol and tobacco consumption, drug abuse

Diet (although it could be argued that food availability influences diet)

The natural environments of vectors that cannot reasonably be modified (e.g., rivers, lakes, wetlands)

Insecticide-impregnated mosquito nets (for this study they are considered to be nonenvironmental interventions)

Unemployment (provided that it is not related to environmental degradation, occupational disease, etc.)

Natural biological agents, such as pollen

Person-to-person transmission that cannot reasonably be prevented through environmental interventions, such as improving housing, introducing sanitary hygiene, or making improvements in the occupational environment

---

[a]Although natural UV radiation from space is not modifiable (or only in a limited way, such as by reducing substances that destroy the ozone layer), individual behavior to protect oneself against UV radiation is modifiable. UV radiation is therefore included in the assessment of the environmental disease burden.

Children in LMICs face a double burden of disease from environmental risks because they are exposed to threats related to poverty as well as to the emerging problems associated with rapid development.

Yet notable advances in health also have come with industrialization. For example, reductions in under-5 child mortality have been a major success story. Under-5 deaths decreased from 16.4 million in 1970 to 11.6 million in 1990 to 6.8 million in 2010 to 5 million in 2020[7] (Table 13.2). Globally, about 85% of child deaths in 2020 were preventable, as indicated by comparison of the worldwide child mortality rate to the observed 2020 rate in more industrialized countries alone.[8] The percentage would be higher were it conceded that a fraction of child mortality in more industrialized countries is preventable.

Many children who survive, however, grow up in environments with poor sanitation and hygiene, heavy pesticide use, and lead and mercury pollution that can lead to altered mental development, reduced functioning, and long-term disabilities. Recent global estimates reported poor environmental conditions such as air pollution and unsafe water and sanitation as among the top contributors to disability-adjusted life years (DALYs) in children under the age of 10 years.[8] An estimated 200 million children under 5 years of age in developing countries are not fulfilling their developmental potential.[9] Air, soil, and water pollution that results from rapid, uncontrolled economic development and concomitant unregulated industrialization can lead to anemia, mental

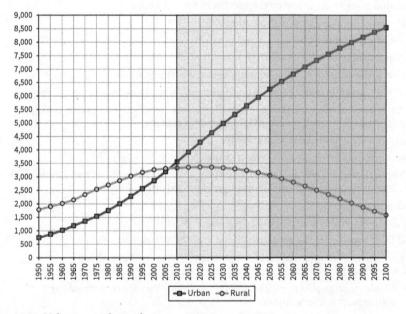

Figure 13.2. Urban—rural population estimates

Source: United Nations Department of Economic and Social Affairs Population Division. World urbanization prospects: the 2018 revision (ST/ ESA/ SER.A/ 420). New York: United Nations; 2019. Available at: https:// pop ulat ion.un.org/ wup/ Publi cati ons/ Files/ WUP2 018- KeyFa cts.pdf.

Table 13.2 Annual child mortality rate (first 5 years of life) and childhood deaths 1990–2020 (90% confidence interval [CI] in parentheses) (18)

Year	Mortality rate (deaths per 1000 live births)	Deaths (thousands in year indicated)
1990	93 (92–95)	12,526 (12,354–12,729)
2000	76 (75–77)	9,756 (9,631–9,907)
2020	37 (35–40)	5,041 (4,813–5,512)

Source: UN Inter-agency Group for Child Mortality Estimation (IGME). Under-five mortality and SDG assessment (2021). http://data.worldbank.org/data-catalog/world-development-indicators.

retardation, attention deficit hyperactivity disorder, learning disabilities, deformities, and other problems that reduce children's potential for full emotional and intellectual development.[10]

Countries with a long history of industrial development have put in place regulations that place some limits on pollutant emissions and are designed to ensure their safe storage, transportation, and use. This has not yet happened in many LMICs where there is a perceived need for fast economic growth and transport of hazardous materials across borders. The enormous burden of ill-health that falls on children in LMICs constrains the social and economic development of these countries. It stresses the healthcare system and reduces individual long-term productivity.[6]

The Burden of Environment-Related Disease

In 1990, the World Health Organization (WHO) developed a Global Burden of Disease framework that allows a wide body of scientific evidence to be used to assess in a comparable way the impact of different risks in the common currency of "lost years of healthy life" or DALYs. One DALY is equal to the loss of 1 year of healthy life. Risks that result in death reduce life expectancy; risks that result in short- or long-term morbidity mean that people stay alive, but not in full health. Healthy life expectancy is, therefore, lower than life expectancy. Measuring DALYs allows for comparison of losses occurring at different ages or from different causes of disease across geographical regions and different population groups. It provides a framework for policymakers and the public to estimate the impact of selected environmental and other risk factors on the health of the population. The initial report included estimates through 1990 and covered 107 diseases and 483 sequalae (nonfatal health consequences related to a disease) in eight regions and five age groups. In 2019, the global burden of disease estimates included 204 countries and territories, with mortality and life expectancy for 990 locations. It covered 369 diseases and injuries.[11]

In 2019, the majority of the estimated 5 million under-5 deaths were attributed to lower respiratory infections, diarrheal diseases, malaria, and meningitis, all of which have been strongly associated with poor environmental conditions.[8] Exposure to lead, secondhand smoke, and climate change also cause significant health impacts. Practically all of this burden affects children in LMICs.

Underweight is a most deadly risk factor in children; around 45% of deaths among children under 5 years of age are linked to undernutrition.[12] Underweight is closely related to the environment because it is associated with repeated diarrhea or intestinal nematode infections, many of which are caused by unsafe water, inadequate sanitation, or insufficient hygiene.

Obesity is another risk factor that is linked to the environment. Obesity results from exposure to advertising, inadequate nutrition, and lack of physical activity. Globally, at least 39 million children under the age of 5 years were overweight in 2020. Childhood obesity is associated with a higher chance of premature death and disability in adulthood.[12] The environment has a major role to play in prevention. Strategies to increase physical activity include enhanced attention to the built environment, including provision of active outdoor play facilities for children, building walking trails, and active transport to school.

The Sustainable Development Goals

In 2015, world leaders adopted the Sustainable Development Goals (SDGs)[13] (see Table 13.3). They are an urgent call to action to end poverty and other deprivations. They aim to improve health and education, reduce inequality, and spur economic growth while also tackling climate change and working to preserve the oceans and forests. The WHO has a global action plan to achieve the health-related SDGs.[14-16] Monitoring is taking place to track progress toward achieving the goals by 2030.[17,18]

Healthy environments for children are key to achieving the SDGs. Every SDG has the potential to impact the development of healthy environments for children (see Figure 13.3).

Leading Causes of Child Death

The leading causes of death for children under 5 years of age in LMICs are malaria, acute respiratory illness, diarrhea, malnutrition, and perinatal conditions. The majority of the more than 5 million children under the age of 5 who die each year throughout the world live in LMICs.[2,3,7] In children under the age of 5, as much as 26% of deaths and 25% of DALYs were attributable to the environment in 2012.[2,3] Figure 13.4 (left) shows that the environmental fraction of global burden of disease (in DALYs) is highest for most disease groups from birth (all except noncommunicable diseases which develop later in life). Figure 13.4 (right) shows the main diseases contributing to the environmental burden of disease for children under 5 years of age in 2012; respiratory infections, diarrheal diseases, and parasitic and vector-borne diseases are all easily preventable through use of clean household energy and through safe access to water, sanitation, and hygiene facilities.

The leading environmental health issues are acute respiratory illnesses, diarrhea, vector-borne diseases, unintentional injuries, and unintentional poisonings. Deaths related to the environment are largely associated with contaminated water and food, poor sanitation and hygiene, vectors, and polluted indoor and outdoor air. Figure 13.5 shows that the age-standardized DALY rate attributed to the environment in children under 5 years differs by region and disease group.

These factors often are linked to poverty and inequality. Children's mortality rates are associated with the national share of population living in extreme poverty (<US$1.90

Table 13.3 UN Sustainable Development Goals

Goal 1. End poverty in all its forms everywhere

Goal 2. End hunger, achieve food security, and improved nutrition and promote sustainable agriculture

Goal 3. Ensure healthy lives and promote well-being for all at all ages

Goal 4. Ensure inclusive and equitable quality education and promote lifelong learning opportunities for all

Goal 5. Achieve gender equality and empower all women and girls

Goal 6. Ensure availability and sustainable management of water and sanitation for all

Goal 7. Ensure access to affordable, reliable, sustainable, and modern energy for all

Goal 8. Promote sustained, inclusive, and sustainable economic growth, full and productive employment and decent work for all

Goal 9. Build resilient infrastructure, promote inclusive and sustainable industrialization, and foster innovation

Goal 10. Reduce inequality within and among countries

Goal 11. Make cities and human settlements inclusive, safe, resilient, and sustainable

Goal 12. Ensure sustainable consumption and production patterns

Goal 13. Take urgent action to combat climate change and its impacts*

Goal 14. Conserve and sustainably use the oceans, seas, and marine resources for sustainable development

Goal 15. Protect, restore, and promote sustainable use of terrestrial ecosystems; sustainably manage forests; combat desertification; halt and reverse land degradation; and halt biodiversity loss

Goal 16. Promote peaceful and inclusive societies for sustainable development, provide access to justice for all, and build effective, accountable, and inclusive institutions at all levels

Goal 17. Strengthen the means of implementation and revitalize the global partnership for sustainable development

per day; Figure 13.6). The internal distribution of national wealth is an important determinant of child health. African nations dominate the list of nations with more than 20% of population in extreme poverty, but they are not alone in experiencing child mortality rates higher than 3%.[19]

Acute Respiratory Illnesses

Respiratory illnesses kill about 1.3 million children under 5 each year.[20] These illnesses are aggravated by exposure to indoor and outdoor air pollution. Chronic exposure to air pollution increases the risk of pneumonia among children under 5 years old.[21] Household air pollution has a major impact on women and girls in developing countries who are in close proximity to open fires while preparing meals. It also poses a great risk for young children who are often carried by their mothers or older sisters during these tasks.

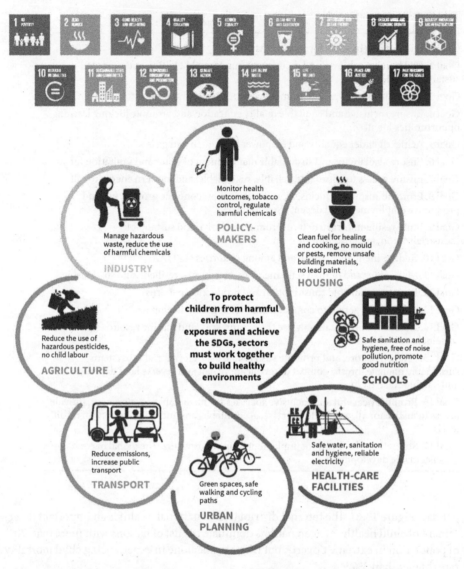

Figure 13.3. Building healthy environments for children to meet the sustainable development goals

Source: WHO (2017)

Diarrhea

Children suffer the most from diarrheal diseases, and children living in rural areas in developing countries bear a disproportionate burden of these diseases. An estimated 525,000 children under 5 die each year from diarrhea,[22] and the majority of these deaths occur in just 15 low-income countries. Worldwide, 780 million individuals lack access to improved drinking water and 2.5 billion lack improved sanitation (see Chapters 28 and 29).

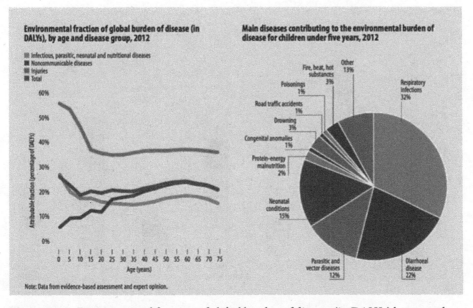

Figure 13.4. Environmental fraction of global burden of disease (in DALYs) by age and disease group (left). Main diseases contributing to the environmental burden of disease for children under 5 years, 2012 (right)

Source: WHO (2017)

Vector-Borne Diseases

The most common and most serious vector-borne diseases are transmitted by mosquitoes that breed in water close to or within the home. According to the World Malaria Report 2021, about 77% of the estimated 627,000 malaria deaths in 2020 were among children under 5, and most of these were in Africa.[23] The effects of latent malaria in children include drowsiness and difficulty focusing, which makes it difficult for infected children to be effective students. Dengue hemorrhagic fever is most common among children 15 years and younger, and it kills an estimated 10,000 children and adolescents each year. Japanese encephalitis kills an estimated 8,000 children per year (with more than 85% of deaths among children under age 15). An estimated one-third of patients who survive Japanese encephalitis remain dependent on their family's care for the rest of their lives.

Unintentional Poisoning

About 50,000 children around the world under the age of 15 years die every year as a result of unintentional poisoning. In LMICs, substances associated with poisoning include pesticides, kerosene used as household fuel, and carbon monoxide (released from faulty stoves).

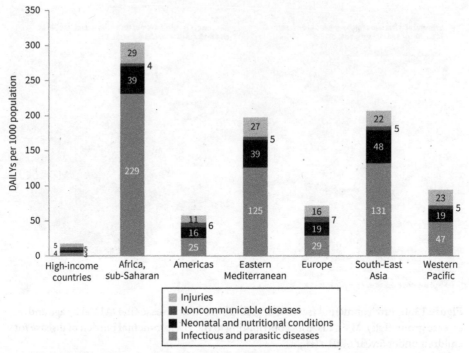

Note: High-income countries are listed separately, the remaining regions contain low- and middle-income countries only.

Figure 13.5. Total age-standardized DALY rate attributed to the environment in children under 5 years, by region and disease group, 2012

Source: WHO, Don't Pollute My Future (2017)

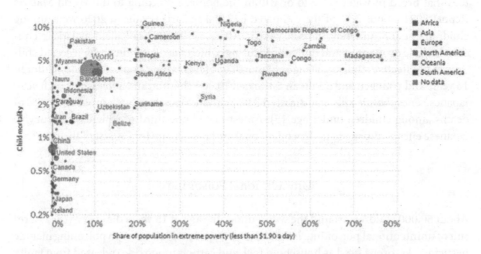

Figure 13.6. National child mortality rates as a function of fractional population living on less than US$ 1.90 per person-day. Circle sizes are proportional to population

Source: Roser M, Ritchie H, and Dadonaite B (2013) Child and infant mortality, *Our World in Data*. https://ourworldindata.org/child-mortality

Unintentional Injuries

Injury is a major cause of death for children of all age groups. It is responsible for 30% of deaths among children between 1 and 3 years, nearly 40% of deaths among 4-year-olds, and 50–60% for children 5 to 17 years. In 2015, poisonings caused 7.3 million deaths among children and adolescents.[24] Road traffic collisions, drowning, burns, falls, and poisoning account for 60% of all child injury-related deaths. More than 95% of all child deaths resulting from injuries occur in LMICs.

Health Effects of Exposure to Secondhand Smoke

More than 1 billion adults smoke worldwide. About 5 million people per year, almost 14,000 every day, are killed by tobacco—more than by any other environmental agent. By 2030, tobacco will kill 8 million people a year, and 70% of these deaths will be in LMICSs—a consequence of aggressive marketing by the tobacco industry in the world's poorest countries. Around 700 million, or almost half of the world's children, breathe air polluted by secondhand smoke.[25,26] Secondhand tobacco smoke contains more than 4,000 different chemical compounds, many of which are poisons. Exposure to high levels of secondhand smoke causes mucous membrane irritation and respiratory effects resulting in rhinitis, cough, exacerbation of asthma, headache, nausea, eye irritation, sudden infant death syndrome (SIDS), and some cancers.[26] Exposure to secondhand smoke may also increase risk of tuberculosis. There is no safe level of exposure to secondhand smoke. The WHO has urged all countries to pass laws requiring all indoor public places to be 100% smoke-free.[25] An estimated 1.4% of the burden of disease (in DALYs) in children younger than 5 years is attributable to exposure to secondhand smoke.[26]

Children at Work

An estimated 152 million children ages 5–17 were employed in the world in 2016; 73 million of them were engaged in hazardous work.[27] In LMICs, children as young as 7 or 8 years are engaged in a variety of activities such as scavenging, farm work, and commerce. Children who work in family farming may be exposed to pesticides. Children who work in construction, mining and smelting, and battery repair and reclamation may be exposed to high levels of asbestos, lead, and mercury. Children may be used for illegal forms of work, some of which are extremely dangerous. Youngsters are less experienced and less likely to be aware of risks than adults. Injuries at work occur four times more frequently among adolescents than among adult workers. Adverse health effects are more frequent and more severe because of children's and adolescents' increased sensitivity to contaminants and injuries (see Chapter 22). Street children may lack proper shelters and sufficient water and sanitation, placing them at physical and psychological risk. Working children suffer a compromised childhood, inadequate nutrition, insufficient education, long working hours, and exposure to the weather without proper clothing or shelter, all of which endanger their development.

Climate Change

Climate change has a major impact on children's health around the world (see Chapter 4). According to WHO estimates using the DALY metric, more than 88% of the existing burden of disease due to climate change occurs in children under 5 years of age.[28] In LMICs, the healthcare systems are underequipped to deal with the changes that climate change has brought.[29] Many of the main global killers, such as malaria, diarrhea, and malnutrition, are closely associated with climatic conditions.[30] Climate change affects children's health as a result of their exposure to extreme temperatures and precipitation, food insecurity, increased transmission rates of vector-borne diseases, and increases in air pollution from the burning of fossil fuels. Food insecurity, water scarcity, damp housing, water- and vector-borne diseases, mycotoxin-related illnesses, and natural disasters such as floods and hurricanes will worsen as temperatures and sea levels rise. The impacts are felt most among young children in LMICs.

Cost of Environmental Disease

In 2002, Landrigan estimated the total costs of lead poisoning, asthma, cancer, and neurobehavioral disorders related to the environment in the United States to be almost US$55 billion (representing about 2.8% of total US healthcare costs).[31] This estimate is likely low because assumptions were conservative, only four categories of illness were considered, and late complications were not quantified. The costs of pediatric environmental disease in LMICs are difficult to fully estimate due to numerous factors, including the poor recognition of the environmental burden of disease. As low-income countries undergo industrialized transition, environmental risk factors may be expected to increase as more chemicals are manufactured, used, commercialized, and stored, and as urbanization and poverty trends increase.[32] In addition to the well-known environmental risk factors, a number of emerging risks, such as those linked to exposures to persistent organic pollutants on child health and development, should be considered because they may have an impact on both industrialized and less-industrialized countries.

The Role of the World Health Organization

The Constitution of the WHO, signed in 1946, put a high priority on environmental hygiene.[33] The reason for this emphasis was that safe water and basic sanitation were deemed to constitute the foundation of all public health endeavors. One goal of the WHO, as stated in its Constitution, was "to promote, in cooperation with other specialized agencies where necessary, the improvement of nutrition, housing, sanitation, recreation, economic or working conditions and other aspects of environmental hygiene." Little did the drafters of the Constitution anticipate the need for the WHO to undertake work to protect the public from chemical contaminants in the environment. It was not until 16 years later that *Silent Spring*, the book that helped to launch the modern environmental movement, was published.[34]

International Actions

Over the past 25 years there have been acknowledgments at the highest level of the need to protect the environment in order to underpin efforts to safeguard child health. In 1989, nations pledged in the Convention on the Rights of the Child to "combat disease and malnutrition . . . taking into consideration the dangers and risks of environmental pollution."[35]

In 1997, the leaders of the G8 countries stated, "Protecting the health of our children is a shared fundamental value. Children throughout the world face significant threats to their health from an array of environmental hazards, and we recognize particular vulnerabilities of children to environmental threats."[36]

National and regional governments, international agencies, and professional coalitions responded to this challenge by establishing new initiatives on the environment and health. The WHO set up a Task Force for the Protection of Children's Environmental Health in 1999. The Task Force began developing training materials on children's health and the environment for an international audience of healthcare providers and policymakers.[37,38] An international conference in Bangkok, Thailand, produced the Bangkok Statement in 2002.[39] This statement called for raising awareness and motivating action on children's environmental health issues. A Second International Conference on Children's Health and the Environment was held in Buenos Aires, Argentina, in 2005, addressing the need for increasing knowledge for taking action.

Under the leadership of Dr. Jenny Pronczuk de Garbino (Figure 13.7), the WHO published a report *Children's Health and the Environment: A Global Perspective.*[6] A third WHO International Conference on Children's Health and the Environment, in 2009, brought together a body of health and environment researchers, scientists, practitioners, and policymakers in Busan, Republic of Korea. More than 60 countries unanimously adopted the Busan Pledge for Action on Children's Environmental Health[40] that called for increased international collaboration to translate research results, new knowledge, and international agreements into concrete political commitments and stronger practical policies for action.

Figure 13.7. Dr. Jenny Pronczuk de Garbino

International Agreements

Recognition of the special susceptibility of children to environmental hazards and global changes resulting from climate changes and new patterns of industrialization and urbanization calls for actions to ensure safe and healthy environments for children. A number of international agreements refer to the importance of protecting children from environmental threats. The G8 Siracusa High Level Environment Ministerial meeting[41] concluded that more should be done to ensure that children are born, grow, develop, and thrive in environments with clean air, clean water, safe food, and minimal exposure to harmful chemicals. Table 13.4 lists international conventions and resolutions related to children's environmental health. These international agreements are discussed in greater detail in Chapter 62.

Table 13.4 Children's environmental health conventions and resolutions

1989	United Nations Convention on the Rights of the Child https://www.ohchr.org/en/instruments-mechanisms/instruments/convention-rights-child
1990	Declaration on the Survival, Protection, and Development of Children (World Summit for Children) http://www.un-documents.net/wsc-dec.htm
1992	Agenda 21, Chapter 25 (United Nations Conference on Environment and Development) https://sustainabledevelopment.un.org/content/documents/Agenda21.pdf
1997	Declaration of the Environment Leaders of the Eight on Children's Environmental Health http://www.g7.utoronto.ca/environment/1997miami/children.html
1999	Declaration of the Third European Ministerial Conference on Environment and Health https://iris.who.int/handle/10665/108311
2001	UN Millennium Development Goals https://www.un.org/millenniumgoals/
2002	United Nations General Assembly Special Session on Children https://www.un.org/en/conferences/children/newyork2002
	The Bangkok Statement (WHO International Conference) https://www.who.int/news/item/21-03-2002-the-bangkok-statement-a-pledge-to-promote-the-protection-of-children-s-environmental-health
	World Summit on Sustainable Development: Launch of the Healthy Environments for Children Alliance and the Global Initiative of Children's Environmental Health Indicators https://iris.who.int/handle/10665/67387 https://iris.who.int/bitstream/handle/10665/42676/9241590610_eng.pdf?sequence=1
2003	Intergovernmental Forum on Chemical Safety Forum IV Recommendations on Children and Chemicals https://cdn.who.int/media/docs/default-source/chemical-safety/ifcs/forum4/exsum_en.pdf

Table 13.4 Continued

2004	Fourth Ministerial Conference on Environment and Health (Europe): Adoption of the Children's Environment and Health Action Plan for Europe https://iris.who.int/bitstream/handle/10665/107577/E83335.pdf?sequence=1
2005	International Conference on Children's Environmental Health: The Buenos Aires Commitment https://docplayer.net/56643325-Healthy-environments-healthy-children-commitment-for-action-november-16-2005-buenos-aires.html
2006	Strategic Approach to International Chemicals Management (SAICM) www.saicm.org
2007	Declaration of the commemorative high-level plenary meeting devoted to the follow-up to the outcome of the special session on children www.un.org/ga/62/plenary/children/highlevel.shtml The Faroes statement: human health effects of developmental exposure to chemicals in our environment https://doi:10.1111/j.1742-7843.2007.00114.x
2009	G8 Environmental Ministers' Meeting in Siracusa, Italy, April, 2009 http://www.g8ambiente.it/?costante_pagina=programma&id_lingua=3 Busan Pledge for Action from the 3rd WHO International Conference on Children's Health and the Environment: Busan, Republic of Korea, June, 2009 https://www.who.int/publications/m/item/busan-pledge-for-action-on-ceh-2009
2010	Parma Declaration from the Fifth Ministerial Conference on Environment and Health (Europe): Protecting Children's Health in a Changing Environment https://www.preventionweb.net/publication/protecting-childrens-health-changing-environment-report-fifth-ministerial-conference
2012	United Nations Conference on Sustainable Development, Rio+20 https://sustainabledevelopment.un.org/rio20
2013	Jerusalem Statement http://www.isde.org/Jerusalem_Statement.pdf
2015	Transforming our world: the 2030 Agenda for Sustainable Development http://www.un.org/sustainabledevelopment/sustainable-development-goals/
2016	G7 Toyama Environmental Ministers' Meeting, May 2016 http://www.env.go.jp/press/files/jp/102871.pdf
2018	Seoul Pledge of Action for Children's Health and the Environment, June 2018 https://inchesnetwork.net/conference-2018/
2019	Intergovernmental Declaration on Children, Youth and Climate Action. https://www.unicef.org/environment-and-climate-change/climate-declaration#declaration
2019	WHO Global Strategy on Health, Environment and Climate Change https://www.who.int/publications/i/item/9789240000377
2020	The Helsinki Declaration 2020 https://www.thelancet.com/journals/lanplh/article/PIIS2542-5196(20)30242-4/fulltext
2021	Pediatric societies' declaration on responding to the impact of climate change on children https://doi.org/10.1016/j.joclim.2021.100038
2023	UN affirms children's rights to live in a clean, healthy and sustainable environment https://www.ohchr.org/en/documents/general-comments-and-recommendations/general-comment-no-26-2023-childrens-rights-and

References

1. World Health Organization. *Don't Pollute My Future! The Impact of the Environment on Children's Health*. Geneva: World Health Organization; 2017.

2. World Health Organization. *Inheriting a Sustainable World? Atlas on Children's Health and the Environment*. Geneva: World Health Organization; 2017.

3. Prüss-Üstün A, Wolf J, Corvalán CF, Bos R, Neira MP. *Preventing Disease Through Healthy Environments: A Global Assessment of the Burden of Disease from Environmental Risks*. Geneva: World Health Organization; 2016. https://www.who.int/publications/i/item/9789241565196.

4. United Nations. World entering "urban millennium," secretary-general tells opening meeting of Habitat Special Session. 2001. Available at: https://press.un.org/en/2001/ga9867.doc.htm.

5. United Nations Department of Economic and Social Affairs Population Division. World urbanization prospects: the 2018 revision (ST/ESA/SER.A/420). New York: United Nations; 2019. Available at: https://population.un.org/wup/Publications/Files/WUP2018-KeyFacts.pdf.

6. World Health Organization, Pronczuk-Garbino J, ed. *Children's Health and the Environment: A Global Perspective. A Resource Manual for the Health Sector*. Geneva, Switzerland: World Health Organization; 2005.

7. UN Interagency Group for Child Mortality Estimation (IGME). Under-five mortality and SDG assessment. 2021. Available at: http://data.worldbank.org/data-catalog/world-development-indicators.

8. Murray CJL, Aravkin AY, Zheng P, et al. Global burden of 87 risk factors in 204 countries and territories, 1990–2019: a systematic analysis for the Global Burden of Disease Study 2019. *Lancet*. 2020;396(10258):1223–49. doi:10.1016/s0140-6736(20)30752-2

9. Grantham-McGregor S, Cheung YB, Cueto S, Glewwe P, Richter L, Strupp B. International Child Development Steering Group. Developmental potential in the first 5 years for children in developing countries. *Lancet*. 2007;369(9555):60–70.

10. Walker SP, Wachs TD, Grantham-McGregor S, et al. Inequality in early childhood: risk and protective factors for early child development. *Lancet*. 2011;378(9799):1325–38.

11. Institute for Health Metrics and Evaluation. Global burden of disease. 2020. Available at: https://www.healthdata.org/gbd/about.

12. World Health Organization. Obesity and overweight. WHO Factsheet. 2021. Available at: https://www.who.int/news-room/fact-sheets/detail/obesity-and-overweight.

13. United Nations. Transforming our world: the 2030 agenda for sustainable development. Seventieth United Nations General Assembly, September 25, 2015. New York: United Nations; 2015. https://sdgs.un.org/publications/transforming-our-world-2030-agenda-sustainable-development-17981.

14. World Health Organization. *Towards a Global Action Plan for Healthy Lives and Well-Being for All: Uniting to Accelerate Progress Towards the Health-Related SDGs*. Geneva: World Health Organization; 2018.

15. Mohammed AJ, Ghebreyesus TA. Healthy living, well-being, and the sustainable development goals. *Bull World Health Org*. 2018;96(9):590–590A. doi:10.2471/BLT.18.222042

16. World Health Organization. Thirteenth General Programme of Work 2019–2023. 2019. Available at: https://www.who.int/about/what-we-do/thirteenth-general-programme-of-work-2019---2023

17. Asma S, Lozano R, Chatterji S, et al. Monitoring the health-related sustainable development goals: lessons learned and recommendations for improved measurement. *Lancet.* 2020;395(10219):240–6. doi:10.1016/S0140-6736(19)32523-1

18. Racioppi F, Martuzzi M, Matić S, et al. Reaching the sustainable development goals through healthy environments: are we on track? *Eur J Public Health.* 2020;30(Suppl_1):i14–i18. doi:10.1093/eurpub/ckaa028

19. Roser M, Ritchie H, Dadonaite B. Child and infant mortality. *Our World in Data.* 2013. Available at: https://ourworldindata.org/child-mortality.

20. World Health Organization, UNICEF. Ending preventable child deaths from pneumonia and diarrhoea by 2025: the integrated global action plan for pneumonia and diarrhoea (GAPPD). 2013. https://www.who.int/publications/i/item/9789241505239

21. World Health Organization. *Air Pollution and Child Health: Prescribing Clean Air.* Summary. Geneva: World Health Organization; 2018. WHYO/CED/PHE/18.0.

22. World Health Organization. Diarrheal disease fact sheet. 2017. Available at: https://www.who.int/news-room/fact-sheets/detail/diarrhoeal-disease.

23. World Health Organization. *World Malaria Report 2021.* Geneva: World Health Organization, 2021.

24. Kassebaum N, Kyu HH, Zoeckler L, et al. Child and adolescent health from 1990 to 2015: findings from the global burden of diseases, injuries, and risk factors 2015 study. *JAMA Pediatr.* 2017;171:573–92. doi:10.1001/jamapediatrics.2017.0250.

25. World Health Organization. WHO Framework Convention on Tobacco Control. 2005. Available at: http://www.who.int/tobacco/framework/en.

26. Öberg M, Woodward A, Jaakkola MS, Peruga A, Prüss-Ustün A. *Global Estimate of the Burden of Disease from Second-Hand Smoke.* Geneva: World Health Organization; 2010.

27. International Labour Office. Global estimates of child labour: Results and trends, 2012–2016. 2017. Available at: https://www.ilo.org/wcmsp5/groups/public/---dgreports/---dcomm/documents/publication/wcms_575499.pdf.

28 Zhang Y, Bi P, Hiller JE. Climate change and disability-adjusted life years. *J Environ Health.* 2007;70:32–6.

29. UNICEF. The climate crisis is a child rights crisis: introducing the Children's Climate Risk Index. 2021. Available at: https://www.unicef.org/media/105376/file/UNICEF-climate-crisis-child-rights-crisis.pdf.

30. Colón-González FJ, Sewe MO, Tompkins AM, et al. Projecting the risk of mosquito-borne diseases in a warmer and more populated world: a multi-model, multi-scenario inter-comparison modelling study. *Lancet Planet Health.* 2021 Jul;5(7):e404–14. doi:10.1016/S2542-5196(21)00132-7.

31. Landrigan PJ, Schechter CB, Lipton JM, Fahs MC, Schwartz J. Environmental pollutants and disease in American children: estimates of morbidity, mortality, and costs for lead poisoning, asthma, cancer, and developmental disabilities. *Environ Health Perspect.* 2002;110(7):721–8.

32. Waterston T, Lenton S. Public health: sustainable development, human induced global climate change, and the health of children. *Arch Dis Child.* 2000;82(2):95–7.

33. Constitution of the World Health Organization. Geneva, Switzerland. 2006. Available at: http://www.who.int/governance/eb/who_constitution_en.pdf.

34. Carson R. *Silent Spring.* Cambridge, MA: Riverside Press, 1962.

35. Office of the United Nations High Commissioner for Human Rights. Convention on the Rights of the Child. Geneva, Switzerland. 1989. Available at: https://www.ohchr.org/en/instruments-mechanisms/instruments/convention-rights-child.

36. Environment Leaders' Summit of the Eight. 1997 Declaration of the Environment Leaders of the Eight on Children's Environmental Health. 1997 May. Available at: https://www.ohchr.org/en/instruments-mechanisms/instruments/convention-rights-child.

37. United Nations Environment Programme, United Nations Children's Fund, World Health Organization. *Children in the New Millennium: Environmental Impact on Health.* Nairobi, Kenya: United Nations Environment Programme; New York: United Nations Children's Fund; and Geneva, Switzerland: World Health Organization; 2002.

38. World Health Organization. Training package for the health sector. 2019. Available at: https://www.who.int/teams/environment-climate-change-and-health/settings-and-populations/children/capacity-building/training-modules.

39. World Health Organization. The Bangkok Statement. A pledge to promote the protection of children's environmental health. International Conference on Environmental Threats to the Health of Children: Hazards and Vulnerability, Bangkok, Thailand, March 3–7, 2002. Available at: https://www.who.int/news/item/21-03-2002-the-bangkok-statement-a-pledge-to-promote-the-protection-of-children-s-environmental-health.

40. World Health Organization. The Busan Pledge for Action on Children's Environmental Health, Busan, Republic of Korea. 2009 Jun. Available at: https://www.who.int/publications/m/item/busan-pledge-for-action-on-ceh-2009.

41. Chair's Summary: Siracusa Environment Ministerial Meeting, Italy, April 2009. Available at: http://www.g7.utoronto.ca/environment/env090424-summary.pdf.

PART II
ENVIRONMENTS

14

The Intrauterine Environment and Early Infancy

Winnie Fan, Marya G. Zlatnik, Annemarie Charlesworth, and Tracey Woodruff

Over the course of the past century, industrial chemical production has increased around 20-fold,[1] a cause and consequence of increasing industrialization and growing global demand (see Chapter 3). Approximately 700 new chemicals are introduced each year in the United States, and, as of 2021, more than 86,000 chemical substances are listed for use by the US Environmental Protection Agency (EPA) for manufacturing or processing.[2,3] For every American, over 30,000 pounds of chemicals are manufactured or imported each year.[4]

Exposure to synthetic chemicals is ubiquitous worldwide. People of every age and in every stage of development (including those in utero) are exposed daily to synthetic chemicals in air, food, water, soil, and consumer products. Globally, toxic environmental chemicals lead to millions of deaths and cost billions of US dollars each year.[5] Preconception and prenatal exposures to synthetic chemicals are especially important for the following reasons:

1. Exposure to many and varied synthetic chemicals is prevalent among pregnant people in the United States and around the world.[5]
2. Most chemical pollutants transfer from parent to child across the placenta and via breast milk.[5,6]
3. Chemical exposures are linked to adverse pregnancy outcomes and health effects, including fetal death, preterm birth, and developmental delay and/or functional deficits during the offspring's lifetime, which can potentially be transmitted to the next generation.[5,6]
4. Public health education and public policy interventions are key to mitigating and, in some cases, preventing individual exposures to synthetic chemicals.[7]
5. People from vulnerable and underserved populations, including people of color and people with low incomes, are disproportionately impacted by toxic chemical exposures during pregnancy and breastfeeding.[5]

The importance of the impacts of environmental chemicals on human health, particularly during pregnancy and the neonatal period and at other vulnerable windows of development, has generated substantial scientific literature. This chapter reviews recent data on the pregnant body burden of synthetic chemicals and addresses the importance of temporality and timing of chemical exposure during fetal development and early infancy. The varying ability of different chemicals to transfer from parents to

Winnie Fan, Marya G. Zlatnik, Annemarie Charlesworth, and Tracey Woodruff, *The Intrauterine Environment and Early Infancy*, Second Edition. In: *Textbook of Children's Environmental Health*. Edited by: Ruth A. Etzel and Philip J. Landrigan, Oxford University Press. © Oxford University Press 2024. DOI: 10.1093/oso/9780197662526.003.0014

offspring through crossing the placental barrier and/or accumulating in breast milk is also discussed. Last, factors that influence degrees of chemical exposure in vulnerable and underserved populations are considered.

Parental and Fetal Burden of Environmental Chemicals: Biomonitoring Surveys

Human exposure to environmental chemicals comes from multiple sources. These include diet (e.g., pesticides in food production and leaching from packaging chemicals), indoor environment (e.g., contaminated house dust), outdoor environment (e.g., traffic-related air pollution), and direct contact with chemical-containing products (e.g., contaminated jewelry).[7]

Biomonitoring provides an integrated measurement of environmental chemicals and their metabolites in suitable human tissues such as blood, urine, or breast milk. Biomonitoring has been widely used to monitor levels of exposure to certain pollutants, including lead, a known childhood toxicant, and persistent halogenated endocrine-disrupting chemicals (EDCs), including polychlorinated biphenyl ethers (PCBs) and dichlorodiphenyltrichloroethane (DDT). It is also used to measure exposures to nonpersistent or pseudo-persistent compounds with relatively brief residence times in pregnant people, such as phthalates, bisphenol A (BPA), and polycyclic aromatic hydrocarbons (PAHs).[8]

The US Centers for Disease Control and Prevention conduct biomonitoring surveys biennially through the National Health and Nutrition Examination Survey (NHANES) to examine synthetic chemical levels in a nationally representative sample of the US population. Analysis of NHANES surveys from 2011 to 2016 was published in the National Report on Exposure to Environmental Chemicals, with data on more than 200 chemicals.[9] The Environmental Influences on Child Health Outcomes (ECHO) research program evaluates the health effects from pre-, peri-, and postnatal exposures of these chemicals and other chemicals of concern.[10]

Biomonitoring surveys of pregnant people find widespread exposure to multiple environmental contaminants. Woodruff et al. found that among a nationally representative sample of pregnant people surveyed through NHANES 2003–2004, virtually all subjects were exposed to at least 43 different chemicals.[6] Several chemical classes, such as phthalates, were well-represented, with multiple compounds simultaneously detected in most pregnant people, while other classes were detected in few to none. PCBs, organochlorine pesticides (OCP), per- and polyfluoroalkyl substances (PFAS), phenols, polybrominated diphenyl ethers (PBDEs), phthalates, PAHs, and perchlorate were detected in 99–100% of the study population. Many of the chemical levels found in this study population are similar to those in epidemiologic and animal studies reported to be associated with adverse reproductive and developmental outcomes.

Persistent organic pollutants (POPs)—chemicals resistant to environmental degradation that remain in the environment or human body for years to decades—are prominent among the chemicals found in virtually all pregnant people today. DDTs and several PCBs were detected in nearly all pregnant people sampled in the United States, even though the manufacturing and use of these chemicals has been banned for more than 40 years.[6,11] Other frequently detected compounds include persistent and bioaccumulative chemicals that were more recently phased out, such as several PFAS and PBDEs.[6,11]

The persistent nature of POPs allows them to circulate throughout the global environment and reach remote areas of the Earth. Notably, biomonitoring studies of pregnant people in Inuit populations in the Arctic found high levels of POPs and heavy metals, indicating a global pattern of exposure to industrial pollutants.[12]

Pregnant people today are chronically exposed to multiple synthetic chemicals, a complex pattern of exposure demonstrated through measurements from motoring, modeling, and biomonitoring studies.[6,11] Chronic, low-dose exposure to multiple chemicals has more significant toxicological implication than a single chemical due to their additive, synergetic, or antagonistic effects.[11,13] Furthermore, exposure to "toxic stress" from racism or other structural determinants of health can have an additive impact to chemical exposures.[14]

Biomonitoring studies have their challenges. Identifying specific chemicals as new biomonitoring targets can be difficult. There are at least 43 chemicals measured in pregnant people that have unknown identities and sources.[15] Unlike environmental monitoring, biomonitoring does not provide direct information about exposure sources, making determining preventive measures challenging. A recent review recommended that the ECHO program conduct biomonitoring studies of new and emerging chemicals that may pose health risks, including alternative phthalates/plasticizers, organophosphate ester flame retardants (OPFR), perchlorate, and pyrethroids.[10] The review also recommended using noninvasive measurements in alternative matrices (e.g., hair, toenails, and tooth) during multiple life stages, surveying house dust, and the use of questionnaires as additional methods to understand body distribution and sources of chemicals.

Temporality and Timing of Exposure

Developmental milestones occur at highly conserved time points throughout fetal and infant development (see Chapter 2). The temporality and timing of environmental exposures can greatly influence specific health outcomes. The first trimester of pregnancy is considered most susceptible to exposures that impact early embryonic development. For example, certain antiepileptic medications, hyperglycemia, and hyperthermia during the first trimester are associated with an increased risk of neural tube defects.[16] As for environmental exposures that impact fetal organ development (e.g., brain, lungs, and kidneys) or increase risks for preterm birth, fetal growth restriction, or gestational diabetes, the risks associated with exposure may elevate throughout gestational period.[17] Current thinking about the biological importance of exposure temporality and timing during development is summarized in the developmental origins of disease theory.[18]

The importance of prenatal exposure temporality and timing is highlighted through the Dutch Hunger Winter study, which explored the role of nutrition in developmental outcomes.[19] More than 50 years after the 1944–1945 Dutch famine, researchers examined the health histories of people born to parents with significantly reduced caloric intake during early, mid, or late gestation. Children exposed to extreme famine during late gestation were born with lower birth weight and displayed lower adult obesity rates than children born before and after the famine. However, those exposed to extreme famine during early gestation later experienced elevated obesity rates, altered lipid profiles, and cardiovascular diseases as adults.

Several developmental factors increase fetal and infant susceptibility to chemicals. During the prenatal period, fetal liver and renal/biliary systems are immature and may not fully metabolize or eliminate chemicals.[20] Fetuses also have lower body fat percentage and immature blood–brain barriers, so lipid-based substances are more likely to distribute in important lipophilic structures such as the brain and other end organs. Chemicals excreted in fetal urine may be recirculated into the fetal bloodstream by fetal swallowing of amniotic fluid (amniotic recirculation). Immature dermal and gastrointestinal barriers may lead to continuous exposure of water-soluble substances.[21] In newborns, the developing microbiome in gastrointestinal tract and breastfeeding make them especially sensitive to oral route of exposure. Neonatal metabolic/excretion pathways and epidermal structures are developing rapidly, but their functions are still immature when compared to adults.[21]

Placental Transfer of Environmental Chemicals

The parental body burden of chemicals can lead to fetal exposure because many synthetic chemicals transfer across the placenta into the fetus. Direct measurement of fetal exposures during pregnancy is, however, difficult. Evaluation of fetal exposures is therefore typically achieved by the following:

- Measuring levels of chemicals in parental matrices such as serum as a surrogate for fetal exposure
- Measuring chemicals in fetal matrices such as amniotic fluid during pregnancy and umbilical cord blood, placental tissue, and neonatal meconium at birth
- Extrapolating from animal or in vitro studies[22]

The specific structure, chemical composition, and relative persistence of environmental chemicals determine the pattern of placental transfer. In a chemical analysis of maternal and cord blood samples, matched maternal and cord blood samples were found to have more similar chemical features than unmatched ones.[23] Several chemical categories, such as PAHs and phthalates, were found to preferentially partition to cord blood (Figure 14.1). In another study using machine-learning models, small compounds with high water solubility were found to more likely transfer across the placenta.[24] Additionally, more than 2,000 PFAS were expected to partition favorably to cord blood, with the potential of influencing fetus development.

Persistent chemicals (e.g., PBDEs, PFCs) are typically lipophilic or bound to proteins that enter bloodstream circulation. They can expose the fetus directly. Measurements of persistent chemicals at birth are a reasonable proxy for early and mid-gestation exposures because of their long half-lives. On the other hand, the transfer from parent to fetus of nonlipophilic compounds that are rapidly metabolized (e.g., phthalates and BPA) is more difficult to predict, as research suggests several possible metabolic determinants for fetal exposure.[25] Placental metabolism of chemicals also may alter or reduce their levels of fetal exposure; resulting metabolites may be eliminated or transferred into fetal bloodstreams.

Some synthetic chemicals bioaccumulate in the fetus to levels higher than in the parent, resulting in greater exposure to the fetus during sensitive periods of early development.[26] For example, methylmercury levels in the fetus are shown to be 1.5–3 times greater than maternal levels.[27] Perinatal health conditions also can influence the extent to which chemicals cross the placenta. Elevated blood pressure and maternal alcohol use during the second and third trimesters were associated with greater fetal blood lead

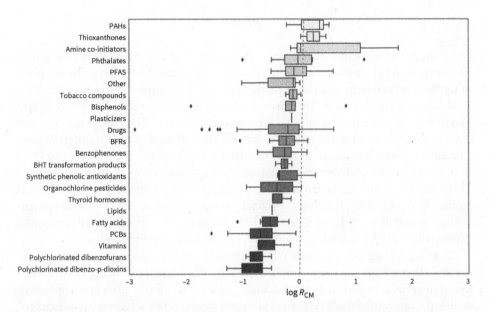

Figure 14.1. Log concentration ratio between cord and maternal blood of chemical categories. $R_{CM} = 0$ indicates that the chemical concentrations of maternal and cord blood are equal. $R_{CM} < 0$ indicates that chemicals favorably partition to maternal blood, whereas $R_{CM} > 0$ indicates that they favor cord blood (22).

concentration than maternal concentrations, while other factors reduce the degree of transfer.[28]

In the following paragraphs, current understanding of fetal exposures is reviewed for two key environmental contaminants through placental transferring during gestational development.

PBDEs

PBDEs are POPs presented in nearly all pregnant people and are commonly detected in matrices such as cord blood, placenta, and amniotic fluids.[6,11] PBDEs are a family of 209 congeners of varying degrees of bromination, with BDE 28, 47, 99, 100, 153, and 209 most commonly detected.[11] PBDEs are transferred across the placenta efficiently during pregnancy due to their lipophilicity. Because PBDEs and their metabolites are structurally similar to thyroid hormones (TH), they can interfere with placental transferring of TH and disrupt neurodevelopment dependent on TH. Additionally, PBDEs may preferentially accumulate in fetal compartments because fetal cytochrome P450 pathways are immature and cannot fully metabolize PBDEs for excretion.[29]

A previous study using ex vivo system analysis suggested that lower-brominated PBDEs are transferred more efficiently than higher-brominated congeners across placenta and thus accumulate more readily in fetal compartments.[30] However, other studies have not substantiated this finding. For example, a cross-sectional study found that all PBDE congeners were efficiently transferred across the placenta.[31]

PFAS

PFAS are chemicals with a fluorinated carbon chain and a terminal functional group. PFAS exposure in utero has been shown to increase the risk of adverse health outcomes in mother and children, including preeclampsia and fetal growth restriction. PFAS are transferred transplacental to the fetus, with several proteins produced by placental trophoblasts facilitating the transport.[32] Fetal concentrations of PFAS are generally lower than maternal concentrations, and PFAS with shorter carbon chains are transferred more efficiently across the placenta. In the Faroe Islands birth cohort, the levels of five PFAS in umbilical cord blood were measured at 29–74% those in paired maternal serum.[33] This finding was confirmed in a subcohort of the Norwegian Parent and Child Cohort Study.[34] In the 123 paired maternal and fetal plasma samples, the PFAS median concentrations in umbilical cord plasma were 30–79% lower than those in maternal serum, and shorter-chain PFAS transferred to cord blood more readily than longer ones.

There is growing evidence that PFAS target receptors in placenta, resulting in abnormal placental development and function. In a 2020 review, growing evidence was found in animal, ex vivo human placental, and in vitro human placental trophoblast studies demonstrating that PFAS target placental receptors that alter epigenetic markers, induce oxidative stress, and disrupt endocrine, lipid, and sterol signaling.[32] This indicates that many adverse outcomes associated with PFAS exposure may be due to PFAS toxicity on the placenta itself.

Exposures via Breast Milk

In early infancy, breast milk can be a significant source of environmental chemical exposure. It is well-documented globally that contaminants in maternal blood can transfer to breast milk.[35] Breast milk has a higher lipid content than serum and higher levels of certain nutrients such as calcium. Therefore, lipid-soluble pollutants (e.g., POPs) and environmental chemicals that interact with key nutrients can transfer and accumulate in breast milk.

PBDEs are lipophilic compounds that transfer to breast milk in greater proportion than maternal serum. A study of paired milk and serum samples from 34 women in the United States found that BDE-47 was the predominant congener in samples, and lower-brominated PBDEs accumulate more readily in milk than higher-brominated PBDEs.[36] PFAS are also lipophilic compounds that excrete through lactation. In a systematic review, lower serum PFAS concentrations were found in people with longer periods of breastfeeding.[37]

Lead is a chemical that transfers efficiently from parent to infant via breast milk. More than 95% of maternal lead is stored in the bone and can be mobilized into serum through bone resorption during pregnancy and lactation, during which developing fetuses and nursing infants increase demand for calcium. A study of 367 lactating mothers who were exclusively breastfeeding found that the highest lead levels in breast milk were among mothers with high lead levels in their patella bone.[38]

Breast milk is considered the optimal food for developing infants. A risk-benefit assessment on breastfeeding using the World Health Organization global surveys from 2000 to 2010 concluded that the benefits of breastfeeding, including preventing diabetes and neonatal death, outweigh the risks.[35] Last, environmental chemicals present in formulas and drinking water also should be considered in infant exposure to environmental chemicals.[39]

Factors Affecting Exposure

The levels and types of exposure vary across populations. Factors such as demographics, lifestyle, socioeconomic status (SES), and regional policy can elevate exposure or vulnerability.

Proximity to sources of pollution is an important determinant of exposure. For example, pregnant people living in the densely agricultural Salinas Valley of California had significantly higher levels of several pesticides than a representative sample living elsewhere in the United States due to high regional pesticide use.[40]

Legal standards also can influence exposure patterns. Serum levels of PBDEs in California were among the highest in the world, likely due to California's historical legal requirement for foam in furniture and children's products to meet stringent flammability standards. This policy drove high use of PBDEs as fire retardants between 1970 and the mid-2000s, resulting in the ubiquitous presence of PBDEs in California homes, particularly in house dust.[31]

Nutritional status during the prenatal and early postnatal period is another factor that influences fetal exposure to synthetic chemicals. For example, calcium deficiency and anemia are associated with increased lead absorption and retention. Hence, for pregnant and nursing people, providing dietary calcium and iron supplementation is effective in preventing elevated blood lead levels.[41]

SES and indicators of social vulnerability are consistent determinants of environmental exposures. Communities with mostly non-White racial or ethnic minority groups or people of low SES are often made vulnerable by multiple environmental stressors, including air and water pollution and a substandard built environment. Ambient air pollution from automotive traffic and industry disproportionately impacts minorities with low SES.[42] Low-income populations are also often exposed to elevated levels of indoor pollutants, including particulate matter, lead, allergens, and semi-volatile organic compounds (e.g., pesticides and PBDEs).[43] In California, non-Hispanic Black pregnant people were shown to have disproportionately higher PBDE levels than other racial/ethnic groups.[31]

Elevated environmental exposures among lower income and ethnic minority communities during the prenatal period are concerning because they can increase disparities in health outcomes such as low birth weight and preterm birth.[44] Additionally, environmental exposures may interact with other risk factors that occur disproportionately in these communities, such as chronic social stressors (e.g., poverty, food insecurity, social inequality) and poor nutrition, to adversely affect the health of children.[42]

Prevention

Pregnancy is a time when many are willing to make lifestyle and dietary changes to improve their offspring's health. Specific evidence-based preventive advice is lacking for most environmental exposures in pregnancy, but parents who take preventive measures (e.g., consume fresh/unprocessed foods, avoid tobacco products, improve indoor air quality) are likely to decrease environmental exposures during pregnancy. However, these measures will not significantly impact the body burden of persistent or bioaccumulative chemicals, making policy changes to prevent these chemicals from entering our air, water, homes, and food systems important. The International Federation of Gynecology and Obstetrics recommends that healthcare professionals actively advocate for policies to prevent toxic environmental chemicals, encourage healthy food system, incorporate environmental health into healthcare, and champion environmental justice.[45] Similarly, the American College of Obstetricians and Gynecologists calls for actions to reduce toxic agents and their consequences on reproductive and developmental health, especially on vulnerable/underserved populations.[46]

Conclusion

Environmental exposures during pregnancy and in early infancy are of great concern. In an increasingly industrialized and globalized world, pregnant people and their infants are prevalently exposed to multiple and varied chemicals. Since many chemicals can pass from parent to infant through the placenta or breast milk, the fetus and infant are at risk of exposure during vulnerable periods of early development. The extent of parental and fetal exposure depends on the source and the intrinsic properties of the chemicals in the local environment. The impact of chemical exposures on pregnancy and fetal development may vary across the population due to timing of exposure and other underlying risk factors.

There is a need for continuing research that increases the current understanding of environmental exposures during pregnancy and early infancy. At the same time, there is an urgent call for policies that reduce or prevent harmful exposures in early development, especially within vulnerable populations. Increasing practices and policies that promote a clean and healthful environment for racially, ethically, and socioeconomically diverse communities should be a priority to improve parental and child health.

References

1. American Chemistry Council. *Chemical Activity Barometer*. Washington, DC: ACC; 2014. Available at: https://www.americanchemistry.com/Jobs/CAB#

2. U.S. Environmental Protection Agency (EPA). Reviewing new chemicals under the Toxic Substances Control Act (TSCA). Office of Pollution Prevention and Toxics (OPPT). 2021 Nov. Available at: https://www.epa.gov/reviewing-new-chemicals-under-toxic-substances-control-act-tsca/statistics-new-chemicals-review#stats

3. U.S. EPA. About the TSCA chemical substance inventory. OPPT. 2021 Nov. Available at: https://www.epa.gov/tsca-inventory/about-tsca-chemical-substance-inventory

4. U.S. EPA. Chemical data reporting fact sheet: chemicals snapshot. 2014. https://www.epa.gov/sites/default/files/2014-11/documents/2nd_cdr_snapshot_5_19_14.pdf

5. Di Renzo GC, Conry JA, Blake J, et al. International Federation of Gynecology and Obstetrics opinion on reproductive health impacts of exposure to toxic environmental chemicals. *Int J Gynaecol Obstet*. 2015 Dec;131(3):219–25.

6. Woodruff TJ, Zota AR, Schwartz JM. Environmental chemicals in pregnant people in the United States: NHANES 2003–2004. *Environ Health Perspect*. 2011 Jun;119(6):878–85.

7. Sutton P, Woodruff TJ, Perron J, et al. Toxic environmental chemicals: the role of repro-ductive health professionals in preventing harmful exposures. *Am J Obstet Gynecol.* 2012 Sep;207(3):164–73.

8. Centers for Disease Control and Prevention (CDC). *Improving the Collection and Management of Human Samples Used for Measuring Environmental Chemicals and Nutrition Indicators,* Version 1.3. Atlanta, GA: CDC; 2018.

9. CDC. Fourth national report on human exposure to environmental chemicals, updated ta-bles, March 2021. CDC. 2021. Available at: https://www.cdc.gov/exposurereport/index.html.

10. Pellizzari ED, Woodruff TJ, Boyles RR, et al. Environmental influences on child health out-comes. Identifying and prioritizing chemicals with uncertain burden of exposure: oppor-tunities for biomonitoring and health-related research. *Environ Health Perspect.* 2019 Dec 18;127(12):126001.

11. Mitro SD, Johnson T, Zota AR. Cumulative chemical exposures during pregnancy and early development. *Curr Environ Health Rep.* 2015 Dec;2(4):367–78.

12. Arctic Monitoring and Assessment Programme (AMAP). *Human Health in the Arctic 2021, Summary for Policy-Makers.* Tromsø, Norway: AMAP; 2021.

13. Koman PD, Singla V, Lam J, Woodruff TJ. Population susceptibility: a vital consideration in chemical risk evaluation under the Lautenberg Toxic Substances Control Act. *PLoS Biology.* 2019 Aug 29;17(8):e3000372.

14. Williams DR, Lawrence JA, Davis BA. Racism and health: evidence and needed research. *Annu Rev Public Health.* 2019 Apr;40:105–25.

15. Wang A, Abrahamsson DP, Jiang T, et al. Suspect screening, prioritization, and confirma-tion of environmental chemicals in maternal-newborn pairs from San Francisco. *Environ Sci Technol.* 2021;55(8):5037–49.

16. Cabrera RM, Hill DS, Etheredge AJ, Finnell RH. Investigations into the etiology of neural tube defects. *Birth Defects Res C Embryo Today.* 2004 Dec;72(4):330–44.

17. Giudice LC, Llamas-Clark EF, DeNicola N, et al. Climate change, women's health, and the role of obstetricians and gynecologists in leadership. *Int J Gynecol Obstet.* 2021;155:345–56. doi:10.1002/ijgo.13958

18. Barker DJ. The developmental origins of chronic adult disease. *Acta Paediatrica.* 2004 Nov;93:26–33.

19. Roseboom T, de Rooij S, Painter R. The Dutch famine and its long-term consequences for adult health. *Early Hum Dev.* 2006 Aug;82(8):485–91.

20. Robinson JF, Hamilton EG, Lam J, et al. Differences in cytochrome p450 enzyme expression and activity in fetal and adult tissues. *Placenta.* 2020 Oct 1;100:35–44.

21. Makri A, Goveia M, Balbus J, Parkin R. Children's susceptibility to chemicals: a review by developmental stage. *J Toxicol Environ Health B Crit Rev.* 2004 Nov;7(6):417–35.

22. Codaccioni M, Bois F, Brochot C. Placental transfer of xenobiotics in pregnancy physiologically-based pharmacokinetic models: structure and data. *Comput Toxicol.* 2019 Nov;12:100111.

23. Abrahamsson DP, Wang A, Jiang T, et al. A comprehensive non-targeted analysis study of the prenatal exposome. *Environ Sci Technol.* 2021 Jul 14;55(15):10542–57.

24. Abrahamsson D, Siddharth A, Robinson JF, et al. Modeling the transplacental transfer of small molecules: a case study on per- and polyfluorinated substances (PFAS). *J Expo Sci Environ Epidemiol.* 2022;32(6):808–19.

25. Barr DB, Bishop A, Needham LL. Concentrations of xenobiotic chemicals in the maternal-fetal unit. *Reprod Toxicol.* 2007 Apr;23(3):260–6.

26. Morello-Frosch R, Cushing LJ, Jesdale BM, et al. Environmental chemicals in an urban population of pregnant women and their newborns from San Francisco. *Environ Sci Technol.* 2016 Nov;50(22):12464–72.

27. Wang A, Padula A, Sirota M, Woodruff TJ. Environmental influences on reproductive health: the importance of chemical exposures. *Fertil Steril.* 2016 Sep;106(4):905–29.

28. Harville EW, Hertz-Picciotto I, Schramm M, et al. Factors influencing the difference between maternal and cord blood lead. *Occup Environ Med.* 2005 Apr;62(4):263.

29. Vuong AM, Yolton K, Dietrich KN, et al. Exposure to polybrominated diphenyl ethers (PBDEs) and child behavior: current findings and future directions. *Horm Behav.* 2018 May 1;101:94–104.

30. Frederiksen M, Vorkamp K, Mathiesen L, et al. Placental transfer of the polybrominated diphenyl ethers BDE-47, BDE-99, and BDE-209 in a human placenta perfusion system: an experimental study. *Environ Health.* 2010;9:32.

31. Varshavsky JR, Sen S, Robinson JF, et al. Racial/ethnic and geographic differences in polybrominated diphenyl ether (PBDE) levels across maternal, placental, and fetal tissues during mid-gestation. *Sci Rep.* 2020 Jul;10(1):1–4.

32. Blake BE, Fenton SE. Early life exposure to per-and polyfluoroalkyl substances (PFAS) and latent health outcomes: a review including the placenta as a target tissue and possible driver of peri-and postnatal effects. *Toxicology.* 2020 Aug:152565.

33. Kato K, Basden BJ, Needham LL, Calafat AM. Improved selectivity for the analysis of maternal serum and cord serum for polyfluoroalkyl chemicals. *J Chromatogr A.* 2011 Apr;1218(15):2133–7.

34. Gützkow KB, Haug LS, Thomsen C, et al. Placental transfer of perfluorinated compounds is selective: a Norwegian mother and child sub-cohort study. *Int J of Hyg Environ Health.* 2012 Feb;215(2):216–9.

35. van den Berg M, Kypke K, Kotz A, et al. WHO/UNEP global surveys of PCDDs, PCDFs, PCBs, and DDTs in human milk and benefit–risk evaluation of breastfeeding. *Arch Toxicol.* 2017 Jan;91(1):83–96.

36. Marchitti SA, Fenton SE, Mendola P, et al. Polybrominated diphenyl ethers in human milk and serum from the US EPA MAMA study: modeled predictions of infant exposure and considerations for risk assessment. *Environ Health Perspect.* 2017 Apr;125(4):706–13.

37. VanNoy BN, Lam J, Zota AR. Breastfeeding as a predictor of serum concentrations of per-and polyfluorinated alkyl substances in reproductive-aged women and young children: a rapid systematic review. *Curr Environ Health Rep.* 2018 Jun;5(2):213–24.

38. Ettinger AS, Téllez-Rojo MM, Amarasiriwardena C, et al. Influence of maternal bone lead burden and calcium intake on levels of lead in breast milk over the course of lactation. *Am J Epidemiol.* 2006 Jan;163(1):48–56.

39. Carignan CC, Cottingham KL, Jackson BP, et al. Estimated exposure to arsenic in breast-fed and formula-fed infants in a United States cohort. *Environ Health Perspect.* 2015 May;123(5):500–6.

40. Castorina R, Bradman A, Fenster L, et al. Comparison of current-use pesticide and other toxicant urinary metabolite levels among pregnant women in the CHAMACOS cohort and NHANES. *Environ Health Perspect.* 2010 Jun;118(6):856–63.

41. U.S. Department of Health and Human Services. Guidelines for the identification and management of lead exposure in pregnant and lactating women. 2010 Nov. Available at: https://www.cdc.gov/nceh/lead/docs/publications/leadandpregnancy2010.pdf.

42. Morello-Frosch R, Zuk M, Jerrett M, et al. Understanding the cumulative impacts of inequalities in environmental health: implications for policy. *Health Aff.* 2011 May;30(5):879–87.

43. Adamkiewicz G, Zota AR, Fabian MP, et al. Moving environmental justice indoors: understanding structural influences on residential exposure patterns in low-income communities. *Am J.* 2011 Dec;101(S1):S238.

44. Padula AM, Huang H, Baer RJ, et al. Environmental pollution and social factors as contributors to preterm birth in Fresno County. *Environ Health.* 2018 Dec;17(1):1–21.

45. Di Renzo GC, Conry JA, Blake J, et al. International Federation of Gynecology and Obstetrics opinion on reproductive health impacts of exposure to toxic environmental chemicals. *Int J Gynaecol Obstet.* 2015 Dec 1;131(3):219–25.

46. American College of Obstetricians and Genecologists. Reducing prenatal exposure to toxic environmental agents. ACOG Committee Opinion No. 832. *Obstet Gynecol.* 2021;138:e40–54.

15

The Home Environment

Miranda Brazeal and David E. Jacobs

Overview

Housing plays a crucial and often underappreciated role in all aspects of child health. Healthy housing can and should support both physical and mental health. After all, it makes little sense to medically treat children for certain illnesses and injuries only to release them back into a home environment that initially contributed to their ailments. Renewed multidisciplinary approaches have gained traction in the United States and globally to implement evidence-based housing interventions that improve health. Three broad sectors have been challenged in recent years: healthcare finance, housing affordability, and climate. All three are linked to healthy homes.

The need for healthy housing was never clearer than during the coronavirus pandemic of 2019. Increased time at home emerged as a major strategy to minimize its spread. Yet if the home is not healthy, the increased exposure to home hazards can lead to other health problems.

Inadequate housing is a root cause of asthma exacerbation, certain cancers, lead poisoning, injuries, mold-induced illnesses, cognitive impairments, and many other physical and mental health problems.[1,2] The World Health Organization (WHO) regards healthy housing as "shelter that supports a state of complete physical, mental and social well-being."[3] The Surgeon General of the US Public Health Service defines a healthy home as one that is "sited, designed, built, and maintained to promote the health of their occupants."[4] The ten key principles of a healthy home are proper ventilation, moisture control, injury prevention, cleanliness, maintenance, pest management, avoidance of harmful chemicals and agents, accessibility, thermal comfort, and affordability.[5] Importantly, housing segregation and inequalities have been drivers of health and other disparities.[6]

Although the general connection between health and housing has been recognized for millennia,[7] the sanitation movement that implemented indoor plumbing and other improvements in the mid to late 1800s emerged during campaigns to eradicate cholera, typhoid, and tuberculosis. In short, the first housing laws in the United States were implemented primarily for health purposes.[8]

In 1999, a report to the US Congress launched the nation's modern healthy housing efforts,[9] and, in 2014, a National Healthy Housing Standard (a model housing code) was released.[10] In 2018, the WHO published the first comprehensive Housing and Health Guidelines at the international level. By 2021, the National Safe and Healthy Housing Coalition comprised more than 400 US organizations with more than 650 members, with a mission to promote healthy housing policies.[11] Through these and other

Miranda Brazeal and David E. Jacobs, *The Home Environment* In: *Textbook of Children's Environmental Health*, Second Edition. Edited by: Ruth A. Etzel and Philip J. Landrigan, Oxford University Press. © Oxford University Press 2024. DOI: 10.1093/oso/9780197662526.003.0015

processes, the health, housing, environment, and allied professional collaboration is being reestablished.

Although house calls are no longer part of routine practice for most US healthcare professionals,[12] clinicians need to understand how a child's living environment impacts health and medical management. Given the range of health problems related to housing, addressing individual residential hazards has given way to a multidisciplinary and more holistic approach that considers housing as a system. Environmental and other health professionals have played a leading role in this more enlightened approach. A national coalition and record appropriations have emerged in recent years.

This chapter reviews the various means by which the home environment influences the physical and mental health of children and other occupants, effective interventions, and the role of environmental health and allied professions.

Health Conditions Related to Housing

Inadequate housing causes and/or exacerbates a multitude of health conditions in children—some directly and others indirectly. The burden of disease associated with such housing is substantial and has been quantified by the WHO in terms of disability adjusted life years (DALYs) and premature fatalities for asthma, mold, crowding, indoor cold, radon, lead, and other housing conditions for the European region.[13] In the United States in 2009, molds in housing accounted for 13,870 DALYs and 5 deaths; leaks from interior sources accounted for 29,839 DALYs and 11 deaths; and moisture from exterior sources accounted for 36,198 DALYs and 14 deaths. The total cost of asthma for those under 15 years that may be attributed to mold and moisture in homes was almost US$2.4 billion.[14] An estimated 40% of housing (35 million housing units) in the United States had one or more health hazards in 2018.[15] Figure 15.1 summarizes how improved housing conditions support health and well-being.

Asthma

Asthma prevalence and morbidity increased from 1980 to 2010 and then reached a plateau in the United States.[16] Changes in housing likely played a role in the increase. Although environmental and genetic factors are key to the development of asthma, there is also robust evidence that certain housing conditions exacerbate asthma (see also Chapter 43). Increased amounts of time spent indoors, excessive moisture, higher mean indoor temperatures, inadequate ventilation, inadequate pest control, and widespread use of carpeting can increase exposure to indoor allergens from cockroaches, pets, rodents, dust mites, and fungi (molds). Chemicals found in the home, such as formaldehyde, nitrogen dioxide (NO_2), and fine particulate matter ($PM_{2.5}$) also are important asthma triggers that lead to allergic and other inflammatory reactions associated with asthma and allergic rhinitis.[17]

The American Housing Survey showed that 21 million homes in 2017 had water leaking from inside and outside into the home's interior.[18] Excessive moisture leads to fungal growth and supports dust mites, cockroaches, rodents, and other asthma triggers. Excessive moisture and condensation are also associated with inadequate ventilation. Mold/moisture-related problems, as well as race, income, housing type, presence

Figure 15.1. Housing conditions that impact health and well-being
Source: World Health Organization.

of smokers, pets, cockroaches, and rodents were all independent predictors of both high allergen burden in housing and increased asthma symptoms.[19] Fungi produce allergens, toxins, and irritants.[20] Up to 21% of asthma cases may be associated with dampness and molds,[21] and a systematic review found that children exposed to mold were one and a half times more likely to develop asthma or wheeze.[22] Damp, moldy, and cold indoor conditions also may be associated with anxiety and depression.[23]

Dust mites are found in bedding, pillows, mattresses, carpets, and upholstered furniture, causing and exacerbating asthma.[24,25] About 80% of homes in the United States have detectable levels of mite allergens, almost 50% have levels associated with sensitization, and over 20% have levels associated with asthma morbidity.[26,27]

Cockroach allergens come from their fecal material, saliva, secretions, and body parts. Structural deficiencies in walls, floors, and ceilings allow cockroaches to enter homes, and leaks provide them with water. Cockroach allergens are found in 13% of dwellings in the United States.[28]

Allergen levels from the fungus *Alternaria* are high enough in 56% of homes to be associated with asthma symptoms, and mouse allergen is present above the sensitization threshold in 3% of US homes.[27] Eliminating exposures to indoor allergens in homes could prevent more than 2 million cases of physician-diagnosed cases of asthma among US children, and indoor allergens have been implicated in more than half of asthma cases.[29]

Injuries

Globally, a third of injuries occur in the home.[30] In 2019, half of all unintentional injury-related deaths occurred in the home.[31] Residential injuries such as falls, fire-related inhalation injury, burns and scalds, and drowning cause thousands of deaths and millions of emergency department visits each year in the United States (see Chapter 54).

Exposure to excessive indoor temperatures leads to cardiovascular and lung disease and death notably among the poor, the elderly, the socially isolated, and persons living in homes without air conditioning. Urban heat islands and residences in poorly insulated buildings also contribute to heat-related morbidity and mortality.

Excessive noise and inadequate lighting in homes may result in sleep disturbances, hypertension, performance reduction, increased annoyance responses, and depression.[32] Homes with inadequate noise insulation that are adjacent to airports, railroad yards, and highways and in crowded neighborhoods are exposed to high noise levels.

Health Outcomes, Housing Costs, and Children who are Unhoused

When children are unhoused, they are at increased risk for infection and reduced educational opportunities.[33] One study found that 45% of children in shelters for the unhoused met criteria for special education evaluation, but only 22% received any special education testing or placement.[34] Children in families on housing waiting lists have 50% more iron deficiency anemia and six times the rate of stunted growth when compared to children in families receiving housing subsidies.[35] In colder climates, many families must choose whether to spend money on food or heat, a dilemma sometimes known as "heat or eat." Growth in young children decreased in the 3 months following the coldest parts of the year in a study of predominantly low-income children making emergency room visits.[36]

Housing that is expensive relative to income can affect health, in particular for people on low incomes. High housing costs result in cutting back on other essentials that are connected to health, including food, energy, and healthcare.[37,38] Difficulty in paying rent and mortgage costs exposes people to risks of eviction and foreclosure[39] and increases the likelihood that people must relocate frequently. Although definitions vary, affordable housing is defined as households spending less than 30% of their income on housing costs.[40] Eviction, foreclosure, and residential mobility each has been associated with adverse educational and economic effects as well as poor health outcomes. Finally, the lack of affordable housing leads to crowding, which systematic reviews have shown to be related to tuberculosis, flu-related hospitalizations and illnesses, pneumonia, acute respiratory illness, respiratory syncytial virus, diarrhea, and other infectious and mental health illness.[41]

Lead

Lead was extensively used in house paint in the United States (see Chapter 31). Although the use of residential lead paint was banned in the United States in 1978, it is still widespread in older housing. Data from the US Department of Housing and Urban Development (HUD) show that the number of homes with deteriorated lead paint increased by 4.6 million from 2005 to 2018 as the housing stock aged.[42] Children of color and those from low-income families are at much higher risk, presenting a significant environmental justice concern.[43,44] Drinking water contamination has also emerged as an area of concern. Lead in drinking water in Flint, Michigan, due to the presence of lead service lines, improper source water, and inadequate corrosion control received national attention and raised awareness for environmental justice concerns in housing.[45]

The major high-dose source for most children in the United States today is existing lead-based paint in older housing and the contaminated dust and soil it generates.[46,47,48,49] The existing limit for lead in new residential house paint set by the Consumer Product Safety Commission in the United States is 90 parts per million (ppm). Older paints already coating surfaces in housing can contain more than 500,000 ppm. These older paints can produce extraordinarily high levels of lead dust, exceeding 9,300 micrograms of lead per square foot ($\mu g/ft^2$) from only a single square foot of lead paint in an average sized room.[50] This is much higher than the dust lead standard of 10 $\mu g/ft^2$ set by the US Environmental Protection Agency (EPA).

The most recent survey of the nation's housing stock (conducted by HUD in 2019) shows that there was a statistically significant improvement in the number of US housing units with lead paint, but 29 million homes still had a lead paint hazard in the form of deteriorated lead paint, lead-contaminated house dust, or bare soil. An estimated 21.6% of Black households and 23.6% of households in poverty lived in a home with lead paint hazards. Also, 22% of homes with a child under 6 years of age had lead paint hazards.[43] Government support matters: 21% of homes with some form of subsidy had lead paint hazards compared to 25% of homes without subsidy because government standards are in place for many subsidized homes. The main problem continues to be in unassisted low-income housing where such standards are lacking. Methods of inspection, risk assessment, and remediation are described later in this chapter.

Asbestos

Asbestos is a group of naturally occurring fibrous minerals that were previously used in many building materials. Because asbestos is resistant to heat, cold, and chemicals; is easily woven; and extremely durable, it was commonly used as insulation around boilers and pipes, ceiling and floor tiles, textiles, coatings, and other products. Although asbestos is less common today in new construction in the United States and other high-income countries, old asbestos poses great danger during renovation of older homes, schools, and other buildings due to disturbance that releases microscopic asbestos fibers into the air. It also is still used in low-income countries and is not fully banned in the United States. Because of this danger, inspections and renovations involving known or suspected asbestos-containing materials must be performed only by asbestos-trained and certified professionals.[51]

Asbestos-related health problems do not occur in childhood but are manifested many years after exposure and include asbestosis, lung cancer, and mesothelioma, a rare form of cancer that affects the lining of the lungs, chest, heart, and abdomen. Mesothelioma is the most important of the asbestos-related diseases from the perspective of children's environmental health because even relatively transient, low-dose exposures to asbestos in early life have been linked to the development of mesothelioma decades later.[52]

Molds

Molds are fungi that grow on damp surfaces, including indoor building materials, and break down dead plant-based materials such as wood and drywall (see also Chapter 43). Moisture is essential for mold growth. Moisture intrusion can occur from a broken pipe, a leaky roof, or improperly drained condensate from an air conditioner and other problems. It can damage building materials such as wallboard, underlayment, and carpet. Extreme humidity also can increase moisture and mold growth. Molds can usually easily be detected by a visual exam of the household, focusing on inspection for water-damaged surfaces and a search for visible molds. There typically is a musty odor as well.

The health effects of exposure to molds depend on the type of mold, the age of the person, the duration and intensity of exposure, and the individual's preexisting allergies and sensitization. Exposure to molds can lead to many adverse health outcomes, most recently reviewed elsewhere.[53] Exposure has been associated with the development of allergies and acute pulmonary hemorrhage.[54,55] Allergic symptoms from mold exposure include those symptoms common with hay fever such as headaches, runny nose, red eyes, and breathing difficulties as well as skin irritation. Molds also produce mycotoxins, toxic substances that cling to the surface of mold spores, often in moisture-damaged buildings, and can cause receptor irritation even in nonallergic individuals. A small body of literature links molds to brain problems such as memory loss and confusion.[56]

Radon

Radon is a naturally occurring radioactive gas (see Chapter 45). Air pressure inside the home usually is lower than pressure in the soil around the home's foundation due to the "stack effect." Radon enters the home through cracks in the floor, small holes in walls and sumps, and other penetrations that may not be visible, as well as through drains. Rooms with direct contact to soil, such as basements or ground floor units in multifamily housing, are likely to have higher radon levels than the upper floors.[57]

Radon is the second leading cause of lung cancer in adults, after smoking. Children can be at higher risk because of their longer exposure.[58] Testing and remediation methods are described later.

Other Toxic Substances

Chemicals found in household cleaning products and pesticides can cause both acute or chronic poisoning and death. The number of toxic chemicals and agents is extensive and has been reviewed in detail elsewhere.[59,60] Such materials should be placed out of children's reach. Toxic cleaning chemicals such as bleach, drain cleaners, and oven clean-ers should always be stored in their original packaging and never in containers such as recycled soda or drink bottles. Ammonia and bleach cleaners should never be mixed be-cause the resulting reaction can produce dangerous gases.

Pesticides are used in many homes, and children are exposed when playing on sprayed carpets in a recently treated home or outside on newly sprayed grass (see also Chapter 35). Integrated pest management (IPM) and the use of lower-toxicity cleaners and other consumer products are described later.

Drinking Water

The health effects of drinking contaminated water vary depending on the contaminant and can be immediate or develop over a prolonged period of time. Infants have a dis-proportionately heavy consumption of drinking water relative to body weight. Wells should be tested for bacteria, inorganic chemicals, fluoride, radon, and lead. Coliforms and nitrates should be tested annually. More than 13 million US households obtain their drinking water from private wells that are not regulated by the government.[61,62]

Interventions

Evidence-based interventions exist for the housing issues identified in the previous sec-tions. Policies and the central role played by those professionals who treat children in implementing these interventions are described in the conclusions in the next section. Figure 15.2 summarizes key opportunities to improve health through better housing. This section discusses some of the main features of healthy housing interventions.

Health-based housing interventions are best conducted through four steps. First, identify housing deficiencies associated with adverse health outcomes through a system-atic structured inspection protocol using trained personnel. Several such protocols are available and have been reviewed elsewhere, such as the CDC/HUD Healthy Housing Inspection Manual, which includes resident questionnaires, visual assessment data col-lection, and environmental sampling methods.[63,64]

Second, identify methods of remediation that have been proved to be effective. Systematic reviews of the interventions have been published by the CDC[65] and WHO.[66] The specific remediation method will depend on the conditions identified in the first step as well as financing constraints. Many deficiencies have both long- and short-term solutions. The former may have high front-end costs but lower long-term costs, with the converse for the latter, analogous to capital improvements and maintenance.

Third, identify the costs and benefits of the options to finance the necessary improve-ments. With notable exceptions, health investments in housing are unlike other housing improvements because the housing market has been unable to quantify them and fi-nance their implementation. For example, housing appraisers know how to monetize the expense of upgrading an old furnace, but they have not been able to do so for healthy housing investments. This has created a classic housing market "externality" or failure, such that the normal housing financing mechanisms to improve housing do not apply. From an owner's perspective this means that it makes little financial sense to invest in healthy housing improvements because the investment cannot be recouped upon sale,

Figure 15.2. Work that is needed to improve housing conditions.
Source: World Health Organization.

unlike other housing improvements. From a societal standpoint, such healthy housing investments make enormous sense because the benefits far outweigh the costs. After all, failure to improve housing only transfers the costs of inadequate housing to the health-care, education, and other sectors. Cost and benefit calculations have been completed for lead[67] and asthma[68] and cost-benefit calculators are available for state and local jurisdictions.[69] All have shown that the economic benefits far outweigh the costs. Put another way, failure to finance remediation at scale will cost billions annually in coming years.

Fourth, implement the interventions and ensure they have been completed. For example, lead paint remediation has been evaluated and deemed successful over at least 12 years.[70,71]

Some interventions are specific. For example, IPM uses chemical pesticides only as a last resort and instead is based on meticulous cleanup of all food and water residues in a home, sealing of cracks and crevices through which insects gain entry, and regular monitoring of pest infestation levels.[72]

Testing for radon in a home often is done by using a passive sampling device that is left in the home for a specified period of time, although other methods also are available.

All ground-contact dwelling units including apartments in multifamily housing should be tested.[73] Radon levels can be reduced by specialized ventilation known as active *sub-slab depressurization* to vent radon to the outside. Merely sealing cracks and holes in the basement often is not effective.[74]

Mold remediation is accomplished by first eliminating sources of moisture and subsequent removal and has been shown to be effective.[75] Some jurisdictions require use of trained and licensed professionals when mold contamination is above certain levels.

Lead paint inspections, risk assessments, and hazard control are also required to be performed by trained and licensed professionals. Inspections identify precise locations of lead paint, risk assessments measure lead in deteriorated paint and the contaminated dust and soil it generates, and hazard controls include both long- and short-term validated methods.[76] These have been shown to reduce children's blood lead levels for at least 2 years after intervention and to reduce dust lead levels for at least 12 years.[71] In 2008, the EPA created the Lead-Based Paint Renovation, Repair, and Painting Rule, which regulates lead-based paint hazards from renovation work in homes built prior to 1978.[77]

Injury interventions significantly reduce medically attended injury rates and checklists are available to identify, and correct housing conditions related to injuries.[78,79]

Indoor air contains more pollutants than ambient air.[80] A ventilation standard for acceptable indoor air quality has been established and is gaining increasing acceptance. This is particularly important because homes require energy to heat and cool them, and they are being weatherized to help mitigate the effect of climate change.[81] Often this has an unintended consequence of decreased ventilation in homes if the weatherization is not properly designed and implemented. However, ventilation alone should be combined with eliminating sources of indoor contaminants, such as gas stoves.

The Role of Healthcare Providers and Allied Environmental and Housing Professions

Treating children with excessive exposures from unhealthy housing requires individual and systemic approaches beyond immediate medical treatment and thus requires expansion of the traditional scope of pediatric medicine. History-taking should include additional questions concerning the home (see also Chapter 56). Consultation with additional specialists such as occupational hygienists, environmental professionals, energy specialists, visiting nurses, financing experts, and the staff of pediatric environmental health specialty units (see also Chapter 56) may be helpful.

Collaborations between healthcare professionals and the legal sector, public health authorities, housing inspectors, and communities themselves can effectively mobilize actions to reduce many housing-related causes of health problems. For example, proactive housing code enforcement is being increasingly adopted, instead of using a complaint-driven system.[82] Figure 15.3 shows how a multidisciplinary approach can improve housing and health.

The legal nature of some housing problems suggests that better communications between healthcare and legal professionals can be effective in securing healthy housing for children. Although tenants have the right to report to code enforcement agencies when violations are not addressed, low-income tenants living in substandard housing are not always aware of their right to a safe and healthy home. Moreover, defending this right

Housing problems cause a ripple effect of impacts.

Here is an example of how a single housing problem – water leaks – can lead to multiple health effects and economic impacts:

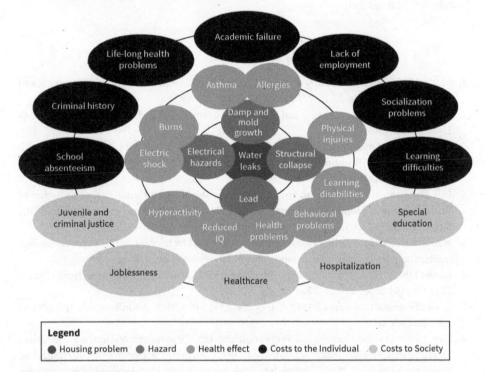

Figure 15.3. Water leaks can lead to multiple health effects and economic impacts.

often requires ample amounts of financial resources, legal information, time, and energy. Lawyers can both advise their clients to act within the guidelines of the law and, when necessary, place demands on landlord and management companies to deal with unfavorable housing conditions. The Medical Legal Partnership (http://www.medical-legal partnership.org) provides patients with free legal assistance to address environmental hazards that, if corrected, would improve the health of the child. These and other initiatives can help to overcome historic segregation and housing discrimination that results in many health and other inequities.

A healthy housing approach is far superior to merely treating individual patient illnesses or specific housing deficiencies. A comprehensive approach that integrates interventions and disciplines is far more efficient. For example, addressing moisture in homes yields many benefits simultaneously (Figure 15.3). Housing conditions directly and indirectly impact outcomes through a ripple effect.

Acknowledgments

The authors thank Johnna Murphy and Megan Sandel for their work on the chapter in the volume's first edition.

Notes

1. Matte TD, Jacobs DE. Housing and health: current issues and implications for research and programs. *J Urban Health Bull N Y Acad Med.* 2000;77.
2. Krieger J, Higgins DL. Housing and health: time again for public health action. *Am J Public Health.* 2002;92(5):758–68.
3. World Health Organization. WHO Housing and Health Guidelines 2018. World Health Organization; 2018. http://www.who.int/sustainable-development/publications/housing-health-guidelines/en/.
4. U.S. Department of Health and Human Services. The Surgeon General's Call to Action to Promote Healthy Homes. U.S. Department of Health and Human Services, Office of the Surgeon General, 2009. https://www.ncbi.nlm.nih.gov/books/NBK44192/.
5. National Center for Healthy Housing. Principles of healthy housing. 2021. https://nchh.org/information-and-evidence/learn-about-healthy-housing/healthy-homes-principles/.
6. Jacobs DE. 2011. Environmental health disparities in housing. *J Am Public Health Assoc.* 2011 Dec;101 Suppl 1:S115–22.
7. Code of Hammurabi. Circa 3000 B.C. (Section 229). https://legacy.fordham.edu/halsall/ancient/hamcode.asp.
8. Centers for Disease Control and Prevention and U.S. Department of Housing and Urban Development. *Healthy Housing Reference Manual.* Atlanta: US Department of Health and Human Services. 2006. https://www.cdc.gov/nceh/publications/books/housing/housing.htm.
9. Jacobs DE, Friedman W, Ashley PJ, McNairy M. *The Healthy Homes Initiative: A Report to Congress.* Washington, DC: U.S. Dept Housing and Urban Development, Office of Lead Hazard Control; 1999 Apr.
10. National Center for Healthy Housing and American Public Health Association. National Healthy Housing Standard. 2018 Update. https://nchh.org/tools-and-data/housing-code-tools/national-healthy-housing-standard/.
11. National Safe and Healthy Housing Coalition. 2021. https://nchh.org/build-the-movement/nshhc/.
12. Kao H, Soriano T, McCormick W. The past, present and future of house calls. *Clin Geriatr Med.* 2009;25:19–34.
13. Braubach M, Jacobs DE, Ormandy D, eds. *Environmental Burden of Disease Associated with Inadequate Housing: A Method Guide to the Quantification of Health Impacts of Selected Housing Risks in the WHO European Region.* Geneva: World Health Organization;2011 Jun.
14. Brazeal M. Health outcomes attributable to housing conditions in the United States. Doctoral Thesis. University of Illinois Chicago School of Public Health. 2021. https://doi.org/10.25417/uic.17026169.v1.
15. National Center for Healthy Housing. State of the nation's healthy housing. 2018. https://nchh.org/tools-and-data/data/state-of-healthy-housing/executive-summary/.
16. Akinbami LJ, Simon AE, Rossen LM. Changing trends in asthma prevalence among children. *Pediatrics.* 2016;137(1)e20152354. https://doi.org/10.1542/peds.2015-2354.

17. Krieger J. Home is where the triggers are: increasing asthma control by improving the home environment. *Pediatr Allergy Immunol Pulmonol.* 2010;23(2):139–45. doi:10.1089/ped.2010.0022.

18. US Bureau of the Census. American Housing Survey Data. 2017. https://www.census.gov/newsroom/press-releases/2018/ahs.html.

19. Salo PM, Arbes SJ Jr, Crockett PW, Thorne PS, Cohn RD, Zeldin DC. Exposure to multiple indoor allergens in U.S. homes and its relationship to asthma. *J Allergy Clin Immunol.* 2008;121:678–84.

20. Institute of Medicine. *Damp Indoor Spaces and Health.* Washington, DC: National Academies Press; 2004.

21. Mudarri D, Fisk WJ. Public health and economic impact of dampness and mold. *Indoor Air.* 2007;17(3):226–35.

22. Castro-Rodriguez JA, Forno E, Rodriguez-Martinez CE, Celedón JC. Risk and protective factors for childhood asthma: what is the evidence? *J Allergy Clin Immunol Pract.* 2016;4(6):1111–1122. doi:10.1016/j.jaip.2016.05.003.

23. Shenassa ED, Daskalakis, C, Liebhaber A, Braubach, Brown MJ. Dampness and mold in the home and depression: an examination of mold-related illness and perceived control of one's home as possible depression pathways. *Am J Public Health.* 2007;97:1893–99. doi:10.2105/AJPH.2006.093773.

24. Institute of Medicine. *Clearing the Air: Asthma and Indoor Air Exposures.* Washington, DC: National Academies Press. 2000.

25. Wilson JM, Platts-Mills TAE. Home environmental interventions for house dust mite. *J Allergy Clin Immunol Pract.* 2018;6(1):1–7. doi:10.1016/j.jaip.2017.10.003.

26. Arbes SJ Jr, Cohn RD, Yin M, Muilenberg ML, Burge HA, Friedman W, Zeldin DC. House dust mite allergen in US beds: results from the First National Survey of Lead and Allergens in Housing. *J Allergy Clin Immunol.* 2003 Feb;111(2):408–14. doi:10.1067/mai.2003.16. PMID: 12589364.

27. Salo PM, Arbes SJ Jr, Crockett PW, Thorne PS, Cohn RD, Zeldin DC. Exposure to multiple indoor allergens in US homes and its relationship to asthma. *J Allergy Clin Immunol.* 2008 Mar;121(3):678–84.e2.

28. Cohn RD, Arbes SJ Jr, Jaramillo R, Reid LH, Zeldin DC. National prevalence and exposure risk for cockroach allergen in U.S. households. *Environ Health Perspect.* 2006;114(4):522–26. doi:10.1289/ehp.8561.

29. Lanphear BP, Aligne C, Auinger P, Byrd R, Weitzman M. Residential exposures associated with asthma in U.S. children. *Pediatrics.* 2001;107:505–11.

30. Turner S, Arthur G, Lyons RA, et al. Modification of the home environment for the reduction of injuries. *Cochrane Database Syst Rev.* 2011;2(2).

31. National Safety Council. Injury facts. 2019. https://injuryfacts.nsc.org/all-injuries/overview/.

32. Brown MJ, Jacobs DE. Residential lighting and risk for depression and falls: results from eight European cities. *Pub Health Rep.* 2011;126(Suppl 1):131–40.

33. Wood DL, Valdez RB, Hayashi T, Shen A. Health of homeless children and housed poor children. *Pediatrics.* 1990;86:858–66.

34. Zima BT. Sheltered homeless children: their eligibility and unmet need for special education evaluations. *Am J Public Health.* 1997;87:236–40.

35. Meyers A, Rubin D, Napoleone M, Nichols K. Public housing subsidies may improve poor children's undernutrition. *Am J Public Health.* 1993;83:115.

36. Frank DA, Roos N, Meyers A, et al. Seasonal variation in weight-for-age in a pediatric emergency room. *Pub Health Rep.* 1996;111:366–71.

37. Burke T, Pinnegar S, Phibbs P, Neske C, Gabriel M, Ralston L, Ruming K. *Experiencing the Housing Affordability Problem: Blocked Aspirations, Trade-Offs and Financial Hardships.* Melbourne: Australian Housing and Urban Research Institute 2007.

38. Pollack CE, Griffin BA, Lynch J. Housing affordability and health among homeowners and renters. *Am J Prev Med.* 2010;39(6):515–21.

39. Desmond M. Evicted: Poverty and Profit in the American City. London: Penguin Books Ltd.; 2016.

40. US HUD. Glossary of terms. https://archives.hud.gov/local/nv/goodstories/2006-04-06g los.cfm.

41. World Health Organization. WHO Housing and Health Guidelines 2018. World Health Organization. 2018. Licence: CC BY-NC-SA 3.0 IGO. http://www.who.int/sustainable-deve lopment/publications/housing-health-guidelines/en/.

42. Ashley P. Presentation from HUD: Findings on lead-based paint/hazards from the American Healthy Homes Survey II. CDC Lead Exposure and Prevention Advisory Committee Meeting. 2021 May 14. https://www.cdc.gov/nceh/lead/advisory/docs/LEPAC-transcript-05-14-2021-508.pdf.

43. Dignam T, Kaufmann RB, LeStourgeon L, Brown MJ. Control of Lead Sources in the United States, 1970–2017: Public Health Progress and Current Challenges to Eliminating Lead Exposure. *J Public Health Manag Pract.* January/February 2019;25(1 Supp):S13–S22.

44. Whitehead LS, Buchanan SD. Childhood lead poisoning: a perpetual environmental justice issue? *J Public Health Manag Pract.* 2019;25(1 Supp):S115–S120.

45. Sadler RC, LaChance J, Hanna-Attisha M. Social and Built Environmental Correlates of Predicted Blood Lead Levels in the Flint Water Crisis. *Am J Public Health.* 2017;107:763–69.

46. Jacobs DE. Lead-based paint as a major source of childhood lead poisoning: A review of the evidence. In: Beard ME, Iske SDA, eds., *Lead in Paint, Soil and Dust: Health Risks, Exposure Studies, Control Measures and Quality Assurance.* Philadelphia: ASTM STP 1226, American Society for Testing and Materials 1995;175–87.

47. Wilson J, Dixon SL, Wisinski C, Speidel C, Breysse J, Jacobson M, Crisci S, Jacobs DE., Pathways and Sources of Lead Exposure: Michigan Children's Lead Determination (the Mi Child Study). Env Res. 2022;215 (114204).

48. Clark CS, Bornschein R, Succop P, Roda S, Peace B. Urban lead exposures of children in Cincinnati, Ohio. *J Chem Speciation Bioavailability.* 3(3/4):163–71.

49. President's Task Force on Environmental Health Risks and Safety Risks to Children. *Eliminating Childhood Lead Poisoning: A Federal Strategy Targeting Lead-Based Paint Hazards.* Washington DC: U.S. Department of Housing and Urban Development and U.S. Environmental Protection Agency; 2000.

50. US HUD. HUD guidelines for the evaluation and control of lead based paint hazards in housing. 1995: chapter 4. https://apps.hud.gov/offices/lead/lbp/hudguidelines/Ch04.pdf.

51. Environmental Protection Agency. Asbestos. https://www.epa.gov/asbestos. Updated March17, 2023.

52. EPA. United States Office of Chemical Safety and Pollution Prevention. Risk evaluation for asbestos. EPA Document # EPA-740-R1-8012 March 2020.

53. Mendell MJ, Mirer AG, Cheung K, Tong M, Douwes J. Respiratory and allergic health effects of dampness, mold, and dampness-related agents: a review of the epidemiologic evidence. *Environ Health Perspect.* 2011;119:748–56. doi:10.1289/ehp.1002410.

54. Dearborn DG, Smith PG, Dahms BB, Allan TM, Sorenson WG, Montana E, Etzel RA. Clinical profile of 30 infants with acute pulmonary hemorrhage in Cleveland. *Pediatrics.* 2002 Sep 1; 110(3):627–37.

55. Etzel RA, Montaña E, Sorenson WG, et al. Acute pulmonary hemorrhage in infants associated with exposure to *Stachybotrys atra* and other fungi. *Arch Pediatr Adolesc Med*. 1998 Aug;152(8):757–62.

56. National Academy of Sciences, Committee on Damp Indoor Spaces and Health. Damp indoor spaces and health. http://www.nap.edu/catalog.php?record_id=11011

57. Environmental Protection Agency. A citizen's guide to radon: the guide to protecting yourself and your family from radon. http://www.epa.gov/radon/pdfs/citizensguide.pdf).

58. National Research Council. *Health Effects of Exposure to Radon: BEIR VI*. Washington, DC: National Academies Press; 1999. https://doi.org/10.17226/5499.

59. Logue JM, McKone TE, Sherman MH, Singer BC. Hazard assessment of chemical air contaminants measured in residences. *Indoor Air*. 2011 Apr;21(2):92–109. doi:10.1111/j.1600-0668.2010.00683.x. PMID: 21392118.

60. Vardoulakis S, Giagloglou E, Steinle S, Davis A, Sleeuwenhoek A, Galea KS, Dixon K, Crawford JO. Indoor exposure to selected air pollutants in the home environment: a systematic review. *Int J Environ Res Public Health*. 2020 Dec 2;17(23):8972. doi:10.3390/ijerph17238972. PMID: 33276576; PMCID: PMC7729884.

61. U.S. Environmental Protection Agency. Private drinking water wells. 2021 Oct 12. https://www.epa.gov/privatewells.

62. United States Census Bureau. American Housing Survey (AHS). 2019. https://www.census.gov/programs-surveys/ahs/data/interactive/ahstablecreator.html?s_areas=00000&s_year=2019&s_tablename=TABLE4&s_bygroup1=1&s_bygroup2=1&s_filtergroup1=1&s_filtergroup2=1.

63. U.S. Centers for Disease Control and Prevention and U.S. Department of Housing and Urban Development. Healthy Housing Inspection Manual. 2006. https://www.cdc.gov/nceh/publications/books/inspectionmanual/default.htm.

64. U.S. Department of Housing and Urban Development. Healthy Housing Program Guidance Manual. 2012. https://www.hud.gov/program_offices/healthy_homes/HHPGM.

65. Jacobs DE, Brown MJ, Baeder A, et al. A systematic review of housing interventions and health: introduction, methods, and summary findings. *J Public Health Management Practice*. 2010 Sept;Suppl:S3–S8.

66. World Health Organization. *WHO Housing and Health Guidelines 2018*. Geneva: World Health Organization; 2018. http://www.who.int/sustainable-development/publications/housing-health-guidelines/en/.

67. Pew Charitable Trusts and Robert Wood Johnson Foundation. 10 policies to prevent and respond to childhood lead exposure. 2018. https://nchh.org/information-and-evidence/healthy-housing-policy/10-policies/.

68. Marta Gomez M, Reddy AL, Dixon SL, Wilson J, Jacobs DE. A cost-benefit analysis of a state-funded healthy homes program for residents with asthma: findings from the New York state healthy neighborhoods program. *J Public Health Manag Pract*. 2017 Mar-Apr;23(2):229–38.

69. Altarum Institute. Preventing childhood lead exposure: costs and benefits. 2019. http://valueofleadprevention.org/.

70. National Center for Healthy Housing and University of Cincinnati. Evaluation of the HUD lead hazard control grant program. 2004. https://nchh.org/research/eval-of-the-hud-lead-hazard-control-grant-program/.

71. Dixon S, Jacobs DE, Wilson J, Akoto J, Clark CS. Window replacement and residential lead paint hazard control 12 years later. *Environ Res*. 2012;113:14–20.

72. U.S. Environmental Protection Agency. Integrated pest management (IPM) principles. Available at: http://www.epa.gov/opp00001/factsheets/ipm.htm.

73. Kitto ME, Murphy C, Dixon SL, Jacobs DE, Wilson J, Malone J. Evaluating and assessing radon testing in multifamily housing. *J Public Health Manag Pract*. 2021 Jun 1. doi:10.1097/PHH.0000000000001392. Epub ahead of print. PMID: 34081671.

74. Sandel M, Baeder A, Bradman A, Hughes J, Mitchell C, Shaughnessy R, Takaro TK, Jacobs DE. Housing interventions and control of health-related chemical agents: a review of the evidence. *J Public Health Manag Pract*. 2010 Sept(Suppl):S19–S28.

75. Kercsmar CM, Dearborn DG, Schluchter M, Xue L, Kirchner HL, Sobolewski J, Greenberg SJ, Vesper SJ, Allan T. Reduction in asthma morbidity in children as a result of home remediation aimed at moisture sources. *Environ Health Perspect*. 2006 Oct;114(10):1574–80. doi:10.1289/ehp.8742. PMID: 17035145; PMCID: PMC1626393.

76. U.S. Department of Housing and Urban Development. Guidelines for the evaluation and control of lead-based paint hazards in housing, 2nd edition. 2012. https://www.hud.gov/program_offices/healthy_homes/lbp/hudguidelines.

77. Environmental Protection Agency. 40 CFR 745, Subpart E. Residential Property Renovation.

78. Phelan KJ, Khoury J, Xu Y, Liddy S, Hornung R, Lanphear BP. A randomized controlled trial of home injury hazard reduction: the HOME injury study. *Arch Pediatr Adolesc Med*. 2011 Apr;165(4):339–45. doi:10.1001/archpediatrics.2011.29. PMID: 21464382; PMCID: PMC3693223.

79. Phelan KJ, Khoury J, Xu Y, Lanphear B. Validation of a home injury survey. *Inj Prev*. 2009 Oct;15(5):300–6. doi:10.1136/ip.2008.020958. PMID: 19805597; PMCID: PMC2759088.

80. Environmental Protection Agency. The total exposure assessment methodology (TEAM) study: summary and analysis. EPA/600/6-87/002a. 1987. https://nepis.epa.gov/Exe/ZyPDF.cgi/2000UC5T.PDF?Dockey=2000UC5T.PDF.

81. American Society of Heating Refrigeration and Air Conditioning Engineers. Standard 62.2 Ventilation for Acceptable Indoor Air Quality. 2019. https://www.ashrae.org/news/esociety/2019-version-of-ashrae-standard-62-2-released.

82. Proactive Rental Inspections. https://nchh.org/resources/policy/proactive-rental-inspections/.

16

Chemicals in Food

Philip J. Landrigan

A healthy, balanced diet is essential to children's health. However, food can also be a source of children's exposure to multiple toxic chemicals. These include pesticides; toxic chemicals that enter the food supply; colorings, flavorings, and other chemicals deliberately added to foods during processing; and chemicals that migrate into foods through contact with packaging materials. These foodborne chemicals can produce a wide range of acute and chronic illnesses depending on the nature of the chemicals, the timing and route of exposure, and the age and underlying health of the exposed children.

This chapter reviews the major sources of foodborne chemical exposures for children and the health effects of these exposures. It describes the responsibilities of the US Food and Drug Administration (FDA), the US Department of Agriculture (USDA), the Centers for Disease Control and Prevention (CDC), and other federal and state agencies in protecting American children's health against toxic chemical exposures in foods. It concludes by noting the important role that alert pediatricians can play in protecting children against foodborne chemical illnesses by noting episodes of potential exposure and reporting these incidents to local and state public health agencies.

Pesticides in Food

Pesticides, including insecticides, herbicides, and fungicides are applied extensively to food crops in the United States and around the world. These synthetic chemicals are designed to kill living organisms, and acute and chronic exposure to pesticides can have multiple adverse effects on children's health including carcinogenicity, neurotoxicity, and adverse birth outcomes. Pesticides are used at all stages of food production—to protect against pests in the field, in shipping, and in storage. In the United States, an estimated 1 billion pounds of pesticides are used each year.[1] Pesticide toxicity is reviewed in Chapter 35.

Diet is the main source of American children's exposure to pesticides.[2] Children are more vulnerable to pesticides than are adults because their unique diets and age-specific behaviors result in disproportionately heavy exposures per kilogram body weight, because of their developmental vulnerability, and because they have many future years in which to develop diseases of long latency that may be triggered by toxic exposures in early life.[3] It is now recognized that exposures to pesticides and other toxic chemicals in the prenatal period and in early childhood can increase risk for multiple noncommunicable diseases at every stage of the human life span.[4]

Philip J. Landrigan, *Chemicals in Food* In: *Textbook of Children's Environmental Health*, Second Edition. Edited by: Ruth A. Etzel and Philip J. Landrigan, Oxford University Press. © Oxford University Press 2024. DOI: 10.1093/oso/9780197662526.003.0016

All pesticides must be registered with the US Environmental Protection Agency (EPA) and with state authorities before they can be used in the United States. To protect against acute and chronic pesticide toxicity, the EPA sets "tolerances," standards that define the levels of pesticides that are allowed in particular foods.[5] More than 9,000 tolerances have been established in the United States—each regulating the level of a specific pesticide allowed on a specific food. The FDA and the USDA monitor the food supply for pesticide residues and can seize foods that contain pesticides at levels that exceed tolerances. Recognizing that children are highly vulnerable to pesticides,[2] the Food Quality Protection Act (FQPA) of 1996, the US federal pesticide law, requires the EPA to take explicit cognizance of children's susceptibility and to set pesticide tolerances at levels that will protect children's health.[6]

Pesticides and Organic Food

All foods that are certified organic by the USDA are required to be grown and processed using no pesticides and no synthetic fertilizers.[7] Under the National Organic Program, food (animals and crops) labeled as "USDA-Certified Organic" must be grown and raised using no hormones, no synthetic chemicals, no antibiotics, and no irradiation. Likewise, USDA-certified organic foods are required to be genetically modified (GMO)-free, and, in organic agriculture, a farmer cannot use genetically engineered seeds and cannot feed organic animals GMO foods.[8]

Because organic foods are produced without pesticides, the levels of pesticides in organic foods are far lower than the levels in nonorganic, "conventionally grown" foods.[9] Yearly inspection of organic farms is conducted by USDA to ensure continued compliance with regulations.

The main benefit of organic food for children's health is that it substantially reduces dietary exposure to pesticides.[2] Multiple studies show that consuming a primarily organic diet reduces a child's pesticide burden by as much as 90%.[9-12] Terms such as "natural" or "homegrown" are not subject to federal regulation and thus offer no guarantee that foods are free of pesticides.

Insecticides and Foods

Insecticides are applied to multiple food crops, especially fruits and vegetables. These include organophosphates, carbamates, and neonicotinoids. Residual quantities ("residues") of these chemicals are commonly detected on and in conventionally grown foods. By federal law, these residues must be present at levels below federal tolerances.[5]

According to the Environmental Working Group (EWG) (www.ewg.org), which produces an annual "dirty dozen" list of fruits and vegetables with the highest pesticide residues, the most pesticide-heavy types of produce are strawberries, apples, peaches, celery, grapes, cherries, spinach, tomatoes, sweet bell peppers, cherry tomatoes, and cucumbers—all of which tend to be soft-skinned and thus vulnerable to insect pests. Conversely, the EWG's "clean 15," the least pesticide-burdened fruits and vegetables, are avocados, sweet corn, pineapples, cabbage, sweet peas, onions, asparagus, mangoes, papayas, kiwi, eggplant, honeydew melon, grapefruit, cantaloupe, and cauliflower.

Herbicides and GMO Foods

GMO foods are produced through genetic engineering, a process that introduces desired traits into food crop seeds by inserting novel DNA into the food crop genome.[13] Herbicides, including glyphosate, which is a "probable" human carcinogen,[14] and 2,4-D and dicamba, which are "possible" human carcinogens, are used in very large quantities in the production of GMO foods.[15] *Herbicide exposure is the principal threat to children's health associated with consumption of GMO food.*[13]

The use of genetic engineering to produce GMO food crops builds on the ancient agricultural practice of selective breeding. However, unlike selective breeding, genetic engineering vastly expands the range of genetic traits that can be moved into plants and the speed of their introduction.[13] Depending on the traits selected, genetically engineered crops could be designed to increase crop yields, incorporate essential micronutrients, tolerate drought, thrive when irrigated with salt water, or produce fruits and vegetables resistant to mold and rot.[13] Until now, the genetic traits that the seed biotechnology industries have chosen to introduce into food crops have been limited to herbicide resistance and insect resistance. This singular focus reflects the fact that the major producers of GMO seeds are the same multinational chemical corporations that manufacture some of the world's most widely used herbicides and insecticides.[16,17]

In the United States, the most commonly grown GMO food crops are corn and soybeans resistant to the herbicide glyphosate (Roundup).[18] Glyphosate resistance has also been introduced into canola, alfalfa, cotton, and sugar beets. An additional use of genetic engineering in agriculture is the introduction of genes that make crops resistant to insects. Plant geneticists have moved bacterial genes that synthesize *Bacillus thuringiensis* (Bt) toxins into corn and cotton.[19] Bt toxin accumulates in GMO crops as well as in food grain and silage derived from these crops.

Since their introduction in the 1990s, the use of corn, soybeans, and other crops with genetically engineered tolerance to the herbicide glyphosate ("Roundup-Ready" crops) has expanded steeply, and glyphosate use has increased in parallel[19] (Figure 16.1). In the United States, glyphosate use has increased more than 250-fold—from 0.4 million kilograms in 1974 to 113 million kilograms in 2014. Global use has increased more than

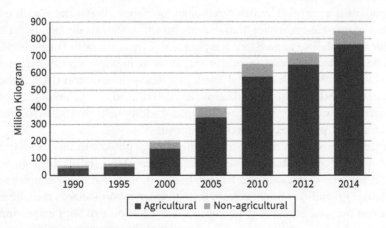

Figure 16.1. Global Glyphosate Use, 1990–2014
Source: Based on data from Benbrook 2016[18]

10-fold and continues to rise.[13] Herbicide-tolerant GMO seeds and herbicides are typically sold in tandem.[16,17] Glyphosate is now the world's most widely used herbicide.

A great advantage of herbicide-resistant GMO crops is that in the first years after their introduction, they substantially simplify weed management. Farmers are no longer required to do mechanical weeding and instead can spray herbicides before spring planting as well as at multiple times during the growing season, leaving their crops unharmed.[13] However, increasingly heavy use of herbicides on herbicide-tolerant crops, especially late in the growing season just prior to harvest, results in the accumulation of measurable quantities of glyphosate and other herbicides in GMO grains at harvest. The consequence is that processed food products derived from GMO corn and soybeans such as processed cornstarch, soybean-based oils, and high-fructose corn syrup now commonly contain measurable quantities of glyphosate and other herbicides.[20,21]

Glyphosate residues are detected with increasing frequency today in foods consumed by children, especially highly processed foods made from GMO corn and soybeans.[21,22] The Canadian Food Inspection Agency found that nearly 30% of more than 3,000 foods contained detectable levels of glyphosate.[23] Residues of glyphosate and other herbicides also have been detected in animal feeds made from herbicide-tolerant crops, thus increasing risk of glyphosate contamination of meat and dairy products.

Glyphosate contamination of foods results in human exposure. A review of 19 studies published since 2007 found glyphosate and its metabolites in urine samples obtained in the general population with levels ranging from 0.16 to 7.6 µg/L.[24] Two of these studies found increasing frequencies of glyphosate detection in more recent years. CDC reports that 80% of urine samples collected in the United States now contain detectable levels of glyphosate.[25]

A further consequence of the extensive use of glyphosate in agriculture has been the emergence of glyphosate-resistant weeds. More than 250 weed species in 70 countries are now known to be resistant to at least one herbicide.[26] In response, farmers have begun to treat crops with multiple additional herbicides such as dicamba and 2,4-D.[13] These are older, highly toxic chemicals; 2,4-D was a component of the Agent Orange defoliant used in the Vietnam War. A likely consequence of the use of multiple herbicides on GMO food crops is that residues of these multiple chemicals will be detected in GMO crops at harvest and in food products made from GMO crops, thus further increasing risk of human exposure.[13]

Glyphosate is a probable human carcinogen. Glyphosate is associated with increased incidence of non-Hodgkin lymphoma[27] and also with a statistically significant increase in incidence of diffuse large B-cell lymphoma.[28] Glyphosate also is genotoxic, and it causes dose-related increases in cancer incidence at multiple sites in experimentally exposed animals.[14] The International Agency for Research on Cancer (IARC), the cancer agency of the World Health Organization, has determined that glyphosate is "probably carcinogenic to humans."[14] IARC has determined additionally that the herbicides 2,4-D and dicamba are "possibly carcinogenic to humans."[15]

Few studies have examined the health effects of glyphosate in infants and children. A large, ongoing cohort study in Puerto Rico has reported a statistically significant association between glyphosate and its metabolites in maternal urine samples and preterm birth.[29] Preliminary findings from a multicenter cohort study in the United States suggest that prenatal exposure to glyphosate may be associated with longer anogenital distance in baby girls, but not in boys,[30] suggesting that glyphosate may be a sex-specific endocrine disruptor in humans with androgenic effects. Similar findings have been reported in rodents exposed to glyphosate.[31]

Toxic Chemicals That Enter the Food Supply

Lead

Lead can enter food crops by absorption from lead-contaminated soil.[32] Lead contamination of food can also result from use of lead arsenate pesticides, inadvertent contamination during food processing,[33,34] and leaching from improperly glazed pottery used as food storage or dining utensils.[35] In 2021, lead was found in a number of brands of commercial baby foods sold in the United States.[36] The toxicity of lead ingested via food is indistinguishable from that due to lead from other sources. Lead toxicity is discussed in Chapter 31.

Mercury

Seafood contaminated with methylmercury is the main foodborne source of children's exposure to mercury. Methylmercury is a highly neurotoxic organic compound formed in rivers, lakes, and the ocean from inorganic mercury by marine microorganisms. Methylmercury is a persistent pollutant. It bioconcentrates in the food web. Top predator species such as tuna, striped bass, and bluefish as well as marine mammals can accumulate methylmercury concentrations in their tissues that are 10 million or more times greater than those in surrounding waters.[37] Human exposure to methylmercury occurs primarily through consumption of contaminated fish and marine mammals.[37] Populations in the circumpolar region are heavily exposed to methylmercury as a consequence of their traditional consumption of a diet rich in fish and marine mammals. Methylmercury toxicity is discussed in Chapter 32.

Arsenic

Rice and seafood are the foods most commonly contaminated with arsenic. Rice grown in contaminated soil can absorb highly toxic inorganic arsenic. Rice grown in former cotton fields in the southeastern United States that were treated in decades past with lead arsenate pesticides is at particularly high risk. Rice from other sources contains little or no arsenic. Organic arsenic is the form of arsenic found in seafood and is much less toxic.

Arsenic has been found in a number of brands of commercial baby foods sold in the United States.[36] In a study that assessed arsenic content in common infant and toddler rice cereals in US supermarkets, the average total arsenic and inorganic arsenic concentrations in infant rice cereal were 174.4 and 101.4 μg/kg, respectively.[38] In response to these findings, the American Academy of Pediatrics suggested that cereals from other grains, such as oatmeal and wheat, as well as other pureed foods are equally acceptable as rice cereal to introduce to infants as first foods.[39] The toxicity of arsenic is discussed in Chapter 33.

PCBs and Dioxins

Polychlorinated biphenyls (PCBs) were manufactured for use as fire retardants and electrical insulators and were widely used in the manufacture of transformers and capacitors until their production was banned in the United States in 1977.[40] They are highly persistent and bioaccumulative. High levels of PCBS are found in fish in many lakes and rivers across the United States. Dioxins and furans are inadvertently produced during the manufacture of PCBs and other halogenated chemicals and are also generated in the incineration of chlorine-containing compounds such as PCBs and polyvinyl chloride.

Children are exposed to PCBs and dioxins through consumption of fish, crabs, and other crustaceans taken from contaminated waters and also through ingestion of contaminated meat, eggs, and dairy products. The toxicity of PCBs and dioxins is discussed in Chapter 36.

Perfluoroalkyl and Polyfluoroalkyl Substances

Perfluoroalkyl and polyfluoroalkyl substances (PFAS), produced since the 1950s, are a family of chemicals that are highly persistent in the environment. These substances are in many foods because they bioaccumulate in aquatic and terrestrial food chains. Contaminated freshwater fish can be an important source. Other sources include fruit and eggs.[41] The toxicity of PFAS is discussed in Chapter 37.

Mycotoxins

Mycotoxins are potent natural toxins produced by molds. They can accumulate to high levels in improperly stored agricultural products such as peanuts, corn, and wheat.[42] The best-known mycotoxin, aflatoxin, is produced by the fungus *Aspergillus*. Others include patulin, citrinin, zearalenone, vomitoxin, and the trichothecenes. The principal route of human exposure to aflatoxins is via the diet. Children who ingest grains contaminated with aflatoxin can develop acute aflatoxicosis,[43,44] and children who ingest vomitoxin-contaminated foods can develop vomiting with an hour or two of eating.[45] The IARC has determined that aflatoxin is a human carcinogen.[46] Clinical and epidemiological studies have found that in persons with chronic hepatitis B infection, aflatoxin is associated with high incidence of hepatocellular carcinoma.[47] Hepatitis B vaccination reduces incidence of liver cancer.[48]

Fish and Shellfish Toxins

Some species of marine algae produce powerful natural toxins.[49] Production of these toxins increases when high densities of noxious algae accumulate in an area of the ocean to form harmful algal blooms (HABs), known variously as "red tides," "green tides," or "brown tides." Algal blooms are increasing in frequency because of climate change, sea surface warming, and increasing coastal pollution.

The main route of human exposure to algal toxins is via consumption of contaminated fish or shellfish. Algal toxins can also become airborne during red tides and result in inhalation exposures in coastal regions.

The toxins produced by marine algae cause a range of diseases, some of them extremely serious. They include paralytic shellfish poisoning caused by saxitoxin, amnesic shellfish poisoning caused by domoic acid, and ciguatera fish poisoning.[50] Inhalation exposure can trigger asthma. Children are highly vulnerable to algal toxins following exposure by either ingestion or inhalation.

Routine monitoring for HAB toxins in shellfish is key to the prevention of human illness caused by these toxins. Monitoring programs are typically embedded within comprehensive shellfish safety programs.[51]

Perchlorate

Perchlorate is a small inorganic molecule found in drinking water and in some foods.[52] Perchlorate can be manmade and is used as a component of rocket fuel. It also occurs naturally. Most perchlorate probably enters foods through contact with contaminated water.

Perchlorate blocks iodine uptake in the thyroid. At higher doses it can cause thyroid dysfunction, and prenatal exposure may result in abnormalities of growth and neurologic development. If the level of perchlorate in public drinking water is greater than 15 parts per billion (ppb), bottled water or water that has perchlorate filtered out should be used to reconstitute infant formula. The EPA established a reference dose of 0.7 µg/kg body weight/day.[52]

Colorings, Flavorings, and Other Chemicals Deliberately Added to Foods During Processing

More than 10,000 chemicals are allowed to be added to food and food contact materials in the United States, either directly or indirectly, under the 1958 Food Additives Amendment to the 1938 Federal Food, Drug, and Cosmetic Act (FFDCA).[53] FFDCA allows for the use of a chemical preservative in foods if (1) it is "generally recognized as safe" or GRAS, (2) not used in a way to conceal damage or make the food appear better than it is, and (3) is properly declared on the label.[54]

Substances are determined to be safe under the conditions of their intended use by qualified experts external to the FDA. According to the 1958 Food Additives Amendment, manufacturers determine when an additive is GRAS, and substances defined as GRAS do not require premarket approval by the FDA. Manufacturers may submit a substance through a GRAS affirmation petition process, but this is not mandatory. More than 70% of GRAS petitions to the FDA are affirmed with no questions to the manufacturer. The FDA has been criticized for this program based on conflicts of interest that arise from individuals who review petitions.[55]

Some food additives may cause adverse reactions in children. Tartrazine (also known as yellow dye no. 5) is used in some foods and beverages, including cake mixes, candies, canned vegetables, cheese, chewing gum, hot dogs, ice cream, orange drinks, salad dressings, seasoning salts, soft drinks, and catsup. An estimated 0.12% of the general population is intolerant to tartrazine. Tartrazine may cause urticaria and asthma exacerbations in people who are sensitive to it.

Red dye no. 40 has been reported to be linked to neurodevelopmental disorders.[56] Petitions have been filed with FDA to ban this compound.

Monosodium glutamate (MSG) is associated with the "Chinese restaurant syndrome" of headache, nausea, diarrhea, sweating, chest tightness, and a burning sensation along the back of the neck. It seems to be linked to consuming large amounts of MSG, not only in Chinese foods, but also in other foods in which a large concentration of MSG is used as a flavor enhancer.

Sulfites are used to preserve foods and sanitize containers for fermented beverages. They may be found in soup mixes, frozen and dehydrated potatoes, dried fruits, fruit juices, canned and dehydrated vegetables, processed seafood products, jams and jellies, relishes, and some bakery products. Some beverages, such as hard cider, wine, and beer, also contain sulfites. Because sulfites can cause asthma exacerbations in sulfite-sensitive patients, the FDA ruled that packaged foods must be labeled if they contain more than 10 parts per million (ppm) of sulfites.

Toxic Chemicals that Migrate into Foods from Contact with Wrapping Materials

Indirect food additives can enter the food supply through contact with food in manufacturing, packaging, transport, and holding. More than 3,000 such substances are recognized by the FDA. These "food contact substances" include adhesives, coatings, paper and paperboard products, polymers, and an array of other materials.

The plastic chemical bisphenol A (BPA) has come under increased scrutiny as food contact item. BPA is an endocrine-disrupting chemical that acts as a weak estrogen. It is used mainly to line metal cans and in hard plastics, including hard plastic baby bottles, water bottles, and many other food containers. In a US national sample comparing data on BPA exposure from 2003 to 2004 to data from 2011 to 2012, the geometric mean urinary BPA levels significantly decreased from 2.64 ng/mL to 1.5 ng/mL; the highest intakes were in persons ages 12–19 years.[57]

BPA analogs are referred to as BPX. Bisphenol sulfate (BPS) is a widely used example. These materials have properties similar to BPA and are incorporated into products such as baby bottles and drinking cups that are marketed as "BPA-free." BPXs have not been well studied, but emerging data indicate that they bind estrogen receptors to varying degrees. Their wide-scale use as BPA replacements appears to be a classic case of "regrettable substitution," in which an alternative material is brought to market to replace a known toxicant without any independent premarket testing for safety or toxicity and is subsequently found itself to be toxic.[58] The toxicity of BPA and BPX is discussed in Chapter 40.

Phthalates, a class of plasticizers used in soft plastics, also may come into contact with food.[59] Phthalates are weakly estrogenic and anti-androgenic. They are used in inks, dyes, and adhesives in food packaging. The toxicity of phthalates is discussed in Chapter 40.

Perfluoroalkyl and polyfluoroalkyl substances (PFAS) are used in food packaging as nonstick coatings (e.g., on the inside of pizza boxes).[60] See Chapter 37 for more information about PFAS.

Responsibilities of the FDA, the USDA, the EPA, the CDC, and Other Federal and State Agencies in Protecting American Children's Health Against Toxic Chemical Exposures in Foods

Responsibility for protecting the food supply against toxic chemicals is divided in complex and sometimes confusing ways among the FDA, USDA, EPA, and CDC. State agencies have additional responsibilities. Enforcing food safety laws is important at all levels and is supported by the federal Food Safety Modernization Act.[61] Some enforcement efforts involve routine monitoring and surveillance of the food supply, such as EPA's monitoring of pesticide residue levels in fruits and vegetables under FQPA. Other efforts, often led by CDC, occur in response to reports of problems and incidents.[51]

Important Role of Alert Pediatricians

Pediatricians play a key role in recognizing new cases and outbreaks of foodborne chemical illnesses in children and reporting them to local and state public health agencies. Front-line pediatricians are often the first persons to recognize cases and outbreaks of toxic disease, and, by promptly reporting these episodes, they are in a unique position to control the spread of toxic diseases.

Robert W. Miller, MD, an early leader of the American Academy of Pediatrics Committee on Environmental Hazards (now the Council on Environmental Health and Climate Change) taught that every pediatrician must be an "alert clinician" ever open to the possibility of discovering new diseases in children caused by toxic exposures in the environment.[62] He considered diagnostic vigilance an essential component of primary care pediatrics. Dr. Miller's lesson is still valid today.

References

1. U.S. Geological Survey (USGS). Pesticides. Available at: https://www.usgs.gov/centers/ohio-kentucky-indiana-water-science-center/science/pesticides#:~:text=About%201%20billion%20pounds%20of,%2C%20insects%2C%20and%20other%20pests.

2. Forman JA, Silverman J, American Academy of Pediatrics Committee on Nutrition, Council on Environmental Health. Organic foods: health and environmental advantages and disadvantages. *Pediatrics*. 2012;130(5): e1406–15.

3. National Research Council, Committee on Pesticides in the Diets of Infants and Children. *Pesticides in the Diets of Infants and Children*. Washington, DC: National Academies Press; 1993. Available at: https://www.ncbi.nlm.nih.gov/books/NBK236275/.

4. Heindel JJ, Balbus J, Birnbaum L, et al. Developmental origins of health and disease: integrating environmental influences. *Endocrinology*. 2015;156 (10):3416–21.

5. U.S. Environmental Protection Agency (EPA). About pesticide registration. 2023. Available at: https://www.epa.gov/pesticide-registration/about-pesticide-registration.

6. EPA. Summary of the Food Quality Protection Act. 2023. Available at: https://www.epa.gov/laws-regulations/summary-food-quality-protection-act.

7. U.S. Department of Agriculture (USDA), Agricultural Marketing Service. Organic Regulations. 2023. Available at: https://www.ams.usda.gov/about-ams/programs-offices/national-organic-program.

8. USDA. Organic 101: can GMOs be used in organic products? 2013. Available at: https://www.usda.gov/media/blog/2013/05/17/organic-101-can-gmos-be-used-organic-products.

9. Curl CL, Porter J, Penwell I, Phinney R, Ospina M, Calafat AM. Effect of a 24-week randomized trial of an organic produce intervention on pyrethroid and organophosphate pesticide exposure among pregnant women. *Environ Int*. 2019;132:104957.

10. Curl CL, Fenske RA, Elgethun K. Organophosphorus pesticide exposure of urban and sub-urban preschool children with organic and conventional diets. *Environ Health Perspect.* 2003;111(3):377–82.

11. Bradman A, Quirós-Alcalá L, Castorina R, et al. Effect of organic diet intervention on pesticide exposures in young children living in low-income urban and agricultural communities. *Environ Health Perspect.* 2015;123(10):1086–93.

12. Lu C, Toepel K, Irish R, Fenske RA, Barr DB, Bravo R. Organic diets significantly lower children's dietary exposure to organophosphorus pesticides. *Environ Health Perspect.* 2006;114(2):260–3. doi:10.1289/ehp.8418

13. Landrigan PJ, Benbrook C. GMOs, herbicides, and public health. *New Engl J Med.* 2015;373:693–4.

14. Guyton KZ, Loomis D, Grosse Y, et al. Carcinogenicity of tetrachlorvinphos, parathion, malathion, diazinon, and glyphosate. *Lancet Oncol.* 2015;16:490–1.

15. Loomis D, Guyton K, Grosse Y, et al. Carcinogenicity of lindane, DDT, and 2,4-dichlorophenoxyacetic acid. *Lancet Oncol.* 2015 Aug;16(8):891–2.

16. Clapp J. Explaining growing glyphosate use: the political economy of herbicide-dependent agriculture. *Global Environmental Change* 2021;67:102239.

17. Gillam C. *Whitewash: The Story of a Weed Killer, Cancer, and the Corruption of Science.* Washington, DC: Island Press, 2017.

18. Benbrook CM. Trends in glyphosate herbicide use in the United States and globally. *Environ Sci Eur.* 2016;28(1):3.

19. U.S. Department of Agriculture (USDA), Economic Research Service. Adoption of genetically engineered crops in the U.S. Updated 2022 Sep. Available at: https://www.ers.usda.gov/data-products/adoption-of-genetically-engineered-crops-in-the-us/.

20. Center for Food Safety. About genetically engineered foods. Available at: https://www.centerforfoodsafety.org/issues/311/ge-foods/about-ge-foods.

21. Myers JP, Antoniou MN, Blumberg B, et al. Concerns over use of glyphosate-based herbicides and risks associated with exposures: a consensus statement. *Environ Health.* 2016;15(19):1–13.

22. Xu J, Smith S, Smith G, Wang W, Li Y. Glyphosate contamination in grains and goods: an overview. *Food Control.* 2019;106:106710.

23. Canadian Food Inspection Agency (CFIA). Safeguarding with science: glyphosate testing in 2015–2016. CFIA-Science Branch Survey Report, Food Safety Science Directorate. 2017. Available at: https://inspection.canada.ca/DAM/DAM-food-aliments/STAGING/text-texte/chem_testing_report_2015-2016_glyphosate_srvy_rprt_1491855525292_eng.pdf.

24. Gillezeau C, van Gerwen M, Shaffer RM, et al. The evidence of human exposure to glyphosate: a review. *Environ Health.* 2019;18:2.

25. Schütze A, Morales-Agudelo P, Vidal M, Calafat AM, Ospina M. Quantification of glyphosate and other organophosphorus compounds in human urine via ion chromatography isotope dilution tandem mass spectrometry. *Chemosphere.* 2021;274:129427.

26. Heap I. International herbicide-resistant weed database. 2020. Available at: www.weedscience.org/default.aspx.

27. Chang ET, Delzell E. Systematic review and meta-analysis of glyphosate exposure and risk of lymphohematopoietic cancers. *J Environ Sci Health B.* 2016; 51(6):402–34.

28. Zhang L, Rana I, Shaffer RM, Taioli E, Sheppard L. Exposure to glyphosate-based herbicides and risk for non-Hodgkin lymphoma: a meta-analysis and supporting evidence. *Mutat Res.* 2019;781:186–206.

29. Silver MK, Fernandez J, Tang J, et al. Prenatal exposure to glyphosate and its environmental degradate, aminomethylphosphonic acid (AMPA), and preterm birth: a nested case-control study in the PROTECT cohort (Puerto Rico). *Environ Health Perspect.* 2021;129(5):57011.

30. Lesseur C, Pirrotte P, Pathak KV, et al. Maternal urinary levels of glyphosate during pregnancy and anogenital distance in newborns in a US multicenter pregnancy cohort. *Environ Pollut.* 2021;280:117002.

31. Manservisi F, Lesseur C, Panzacchi S, et al. The Ramazzini Institute 13-week pilot study glyphosate-based herbicides administered at human-equivalent dose to Sprague Dawley rats: effects on development and endocrine system. *Environ Health.* 2019;18(1):15.

32. Kumar A, Kumar A, Cabral-Pinto MMS, et al. Lead toxicity: health hazards, influence on food chain, and sustainable remediation approaches. *Int J Environ Res Public Health.* 2020;17(7):2179.

33. Richter E, El-Sharif N, Fischbein A, et al. Re-emergence of lead poisoning from contaminated flour in a West Bank Palestinian village. *Int J Occup Environ Health.* 2000;6(3):183–6.

34. Tamayo y Ortiz M, Téllez-Rojo MM, Hu H, et al. Lead in candy consumed and blood lead levels of children living in Mexico City. *Environ Res.* 2016;147:497–502.

35. Caravanos J, Dowling R, Téllez-Rojo MM, et al. Blood lead levels in Mexico and pediatric burden of disease implications. *Ann Glob Health.* 2014;80(4):269–77.

36. U.S. House of Representatives, Subcommittee on Economic and Consumer Policy Committee on Oversight and Reform. *Baby Foods Are Tainted with Dangerous Levels of Arsenic, Lead, Cadmium, and Mercury.* Washington, DC: 2021 Feb 4. Available at: https://oversightdemocrats.house.gov/sites/democrats.oversight.house.gov/files/2021-02-04%20ECP%20Baby%20Food%20Staff%20Report.pdf.

37. World Health Organization. Mercury and health. 2017. Available at: https://www.who.int/news-room/fact-sheets/detail/mercury-and-health.

38. Juskelis R, Li W, Nelson J, Cappozzo JC. Arsenic speciation in rice cereals for infants. *J Agric Food Chem.* 2013;61(45):10670–6.

39. American Academy of Pediatrics Arsenic in Rice Expert Work Group. AAP group offers advice to reduce infants' exposure to arsenic in rice. *AAP News.* 2014; 35:13.

40. Agency for Toxic Substances and Disease Registry. *Toxicological Profile for Polychlorinated Biphenyls (PCBs).* Atlanta, GA: 2000. Available at: https://wwwn.cdc.gov/TSP/ToxProfiles/ToxProfiles.aspx?id=142&tid=26.

41. EFSA Panel on Contaminants in the Food Chain, Schrenk D, Bignami M, Bodin L, et al. Scientific opinion on the risk to human health related to the presence of perfluoroalkyl substances in food. *EFSA J.* 2020;18(9):6223. https://doi.org/10.2903/j.efsa.2020.6223.

42. Petrov V, Qureshi MK, Hille J, Gechev T. Occurrence, biochemistry, and biological effects of host selective plant mycotoxins. *Food Chem Toxicol.* 2018; 112:251–64.

43. Mwanda OW, Otieno CF, Omonge E. Acute aflatoxicosis: case report. *East Afr Med J.* 2005;82(6):320–4.

44. Lye MS, Ghazali AA, Mohan J, Alwin N, Nair RC. An outbreak of acute hepatic encephalopathy due to severe aflatoxicosis in Malaysia. *Am J Trop Med Hyg.* 1995;53(1):68–72.

45. Ruan F, Chen JG, Chen L, et al. Food poisoning caused by deoxynivalenol at a school in Zhuhai, Guangdong, China, in 2019. *Foodborne Pathog Dis.* 2020;17(7):429–33. doi:10.1089/fpd.2019.2710

46. International Agency for Research on Cancer. Aflatoxins: naturally occurring aflatoxins (group 1). Aflatoxin M1 (group 2B). *IARC Monogr.* 2002;82:171–34.

47. Wogan GN, Kensler TW, Groopman JD. Present and future directions of translational research on aflatoxin and hepatocellular carcinoma. A review. *Food Addit Contam Part A Chem Anal Control Expo Risk Assess.* 2012;29(2):249–57.

48. Chang MH. Hepatitis B virus and cancer prevention. *Recent Results Cancer Res.* 2011;188:75–84.

49. Hallegraeff G. Harmful algal blooms: a global overview. In: Hallegraeff GM, Anderson DM, Cembella AD, eds. *Manual on Harmful Marine Microalgae.* Paris: UNESCO Publishing; 2003:1–22.

50. Landrigan PJ, Stegeman JJ, Fleming LE, et al. Human health and ocean pollution. *Ann Global Health.* 2020;86(1):151.

51. Backer LC, Fleming LE. Epidemiological tools for investigating the effects of oceans on public health. In: Walsh PJ, Smith SL, Fleming LE, Solo-Gabriele HM, Gerwick WH, eds. *Oceans and Human Health: Risks and Remedies from the Seas.* Burlington, MA: Academic Press; 2008: 201.

52. US Food and Drug Administration (FDA). Perchlorate. Updated 2017 Dec. Available at: https://www.fda.gov/food/environmental-contaminants-food/perchlorate.

53. International Food Information Council Foundation and FDA. Food ingredients and colors. 2010. Available at: https://www.fda.gov/files/food/published/Food-Ingredients-and-Colors-%28PDF%29.pdf.

54. FDA. Generally recognized as safe (GRAS). Updated 2023 Jul. Available at: https://www.fda.gov/food/food-ingredients-packaging/generally-recognized-safe-gras.

55. Neltner TG, Alger HM, O'Reilly JT, Krimsky S, Bero LA, Maffini MV. Conflicts of interest in approvals to food determined to be generally recognized as safe: out of balance. *JAMA Intern Med.* 2013;173(22):2032–2036.

56. Bakthavachalu P, Kannan SM, Qoronfleh MW. Food color and autism: a meta-analysis. *Adv Neurobiol.* 2020;24:481–504.

57. Lakind JS, Naiman DQ. Temporal trends in bisphenol A exposure in the United States from 2003–2012 and factors associated with BPA exposure: spot samples and urine dilution complicate data interpretation. *Environ Res.* 2015;142:84–95.

58. Qadeer A, Kirsten KL, Ajmal Z, Jiang X, Zhao X. Alternative plasticizers as emerging global environmental and health threat: another regrettable substitution? *Environ Sci Technol.* 2022; 56(3):1482–8.

59. Carlos KS, de Jager LS, Begley TS. Investigation of the primary plasticizers present in polyvinyl chloride (PVC) products currently authorized as food contact materials. *Food Addit Contam Part A Chem Anal Control Expo Risk Assess.* 2018;35(6):1214–22.

60. FDA. Authorized uses of PFAS in food contact. 2023. Available at: https://www.fda.gov/food/process-contaminants-food/authorized-uses-pfas-food-contact-applications.

61. FDA. Food Safety Modernization Act. Updated 2023 May. Available at: https://www.fda.gov/food/guidance-regulation-food-and-dietary-supplements/food-safety-modernization-act-fsma.

62. Miller RW. How environmental hazards in childhood have been discovered: carcinogens, teratogens, neurotoxicants, and others. *Pediatrics.* 2004;113:945–51.

17

The School/Child Care Environment

Maida P. Galvez, Sarah Evans, Mana Mann, and Hester Paul

There are an estimated 48.1 million K-12 students in US public schools and 12 million preschoolers in child care facilities, spending at least 35 hours a week in learning environments.[1,2] Indoor air can be 2–5 times as polluted as outdoor air,[3] yet environments in education settings are not comprehensively addressed by child care licensing, professional development, or national accreditation. Harmful exposures identified in learning facilities include lead, environmental asthma triggers (e.g., pesticides, mold, particulate matter, and volatile organic compounds), per- and poly-fluoroalkyl substances (PFAS)-contaminated drinking water, and flame retardants. Environmental exposures can directly impact child behavior, ability to learn, academic performance, attendance, and overall health and development.[4,5] The presence of faculty, staff, and students of reproductive age further underscores the importance of optimizing learning environments to promote children's health and development across the life span.

Indoor Air Quality

Poor indoor air quality (IAQ) in school buildings and child care facilities is a pervasive problem, most notably in low-income communities and communities of color. Many school buildings are old, poorly maintained, inadequately ventilated, and at high risk for IAQ problems. Twenty-five percent of US school buildings need extensive repair, with 40% reporting at least one unsatisfactory environmental condition.[6] Thirty percent of ventilation and filtration systems were rated as fair or poor. In 2020, 50% of school districts needed to update or replace multiple systems including heating, ventilation, and plumbing, and one-third of schools needed heating, ventilation, and air conditioning (HVAC) system updates.[7]

Children spend 80–90% of their time indoors, making IAQ an important determinant of child health.[8] IAQ is impacted by outdoor air quality; underground pollution; construction; emissions from equipment and art supplies; combustion; mold and moisture; cleaning, sanitizing, disinfecting, and personal care products; pests; science laboratory materials; and new furnishings, floorings, and carpeting. Additional contributors to IAQ include temperature and humidity and ventilation rates. Art supplies can impact IAQ in learning environments. Examples include organic solvents like turpentine, kerosene, methyl alcohol, and formaldehyde used in painting, silk-screening, and shellacking. Powdered pigments, glazes, and clay may contain asbestos, silica, talc, lead, cadmium, and mercury.[9]

Maida P. Galvez, Sarah Evans, Mana Mann, and Hester Paul, *The School/Child Care Environment* In: *Textbook of Children's Environmental Health*, Second Edition. Edited by: Ruth A. Etzel and Philip J. Landrigan, Oxford University Press.
© Oxford University Press 2024. DOI: 10.1093/oso/9780197662526.003.0017

Health Effects from Poor Indoor Air Quality

Indoor air pollutants may irritate children's skin, eyes, nose, throat, and upper airways and cause headaches and affect taste. Airway irritation can cause rapid breathing, exacerbation of asthma and allergies, and flu-like symptoms. Children with asthma are particularly susceptible to air pollutants. Schools with poor ventilation are at risk for increased spread of infectious airborne pathogens such as influenza and SARS-CoV-2.[10] For example, schools that took steps to improve ventilation had a 37% lower COVID-19 incidence during the height of the pandemic.[10] Poor IAQ also affects children's academic success. A study of 100 elementary schools in two southwest US school districts found that as ventilation rates in classrooms improved, the proportion of students passing standardized tests increased.[11]

Improving Indoor Air Quality in School and Child Care Settings

The US Environmental Protection Agency (EPA) *Indoor Air Quality (IAQ) Tools for Schools* provides resources to reduce exposures to indoor environmental contaminants through effective IAQ management practices.[12] US schools with IAQ programs reported improved workplace satisfaction, fewer asthma attacks and school nurse visits, and lower absenteeism. This provides evidence that IAQ management programs improve the health and learning environment of school children.[13]

An important step toward improving IAQ is implementation of green cleaning policies and use of safer disinfectants. Many common cleaning and disinfecting products contain fragrance as well as respiratory irritants that can exacerbate asthma or chemicals associated with cancer, neurotoxicity, hormone disruption, and antibacterial resistance. Products that carry the Green Seal Certified, EcoLogo, or EPA Safer Choice Fragrance-Free label are safer, effective, and often cost the same as conventional cleaners. Schools should avoid foggers and aerosols and products that contain bleach, quaternary ammoniums,[14] and fragrance.

The Labeling of Hazardous Art Materials Act of 1988 mandates labeling of art materials that may contain lead, chromium, mercury, and solvents. Select least-toxic art materials and train students and staff in proper use. Routine use of HVAC systems and ventilation (opening doors or screened windows) improves IAQ.

Volatile Organic Compounds

Volatile organic compounds (VOCs) are carbon-based chemicals that evaporate into gaseous forms at room temperature. VOCs are found in paints, paint strippers, varnishes, lacquers, wood preservatives, glues, fuels, aerosols, cleaners, degreasers, pesticides, cigarette smoke, and dry-cleaning chemicals. Examples of common VOCs include formaldehyde, benzene, toluene, and acetone. VOCs are often higher in indoor air than outdoors. VOCs react with sunlight to form ozone or smog. The primary route of exposure to VOCs is by inhalation and less commonly via skin contact. Intentional inhalation of fumes from glues and other VOC-containing products is a source for exposure among teenagers.

Health Effects from Exposure to Volatile Organic Compounds

Health effects from VOCs include eye, nose, throat, and skin irritation; headache; nausea; dizziness; fatigue; and shortness of breath.[15] Effects vary depending on the VOCs involved and duration of exposure but usually improve once the source is eliminated. Some VOCs are known or probable human carcinogens.[16] Chronic, high-level exposure to benzene is associated with increased risk of leukemia and other blood cancers (see also Chapter 42). VOC exposure in pregnancy is associated with adverse birth outcomes including cardiac defects, low birth weight, and neurodevelopmental delay.[17]

Preventing Exposure to Volatile Organic Compounds

Good ventilation is critical to reducing VOC exposures. When using VOC-releasing products, optimize ventilation by opening doors and screened windows. Purchase low-VOC paints, furniture, and carpets when possible. School maintenance staff should ensure safe working conditions when using consumer products that contain VOCs. Store products containing VOCs securely away from public areas.

Molds

Molds (fungi) grow in moist environments such as areas that have been wet by leaks and where ventilation is inadequate. Children are exposed to molds primarily through inhalation of airborne mold spores.

Health Effects from Exposure to Molds

Mold exposure can result in allergic symptoms (e.g., rhinitis, red, itchy eyes), common cold symptoms (e.g., nasal congestion, cough), asthma attacks, and other lung problems. Children with compromised immune systems, including those with cancer or receiving chemotherapy, may be more susceptible to health problems from mold. Exposure to mold may result in a wide variety of health problems in children (see Chapter 43).

Preventing Exposure to Molds

Addressing mold in classrooms requires immediate cleanup and remediation of underlying moisture problems. Areas of mold covering less than 10 square feet can be cleaned with water and soap, with larger areas addressed by trained individuals. Mold growth can be reduced by repairing leaks, lowering air moisture content, increasing air movement, and increasing air temperature. Maintain relative humidity between 30% and 50%.

Because mold can grow in locations that aren't visible, such as under carpets and above ceiling tiles, thoroughly investigate musty odors and water damage for presence of mold. Air testing to determine the type of mold is generally not warranted because there are few specific health guidance levels for individual mold species. Remediate mold regardless of type. If mold testing is considered, hire a certified industrial hygienist. Consider hiring separate contractors to perform testing and remediation to reduce potential for conflicts of interest.

Outdoor Air Pollution

The EPA has established national standards of permissible levels of for six criteria outdoor air pollutants that impact human health (ozone, carbon monoxide, particulate matter [PM_{10} and $PM_{2.5}$], sulfur dioxide, lead, and nitrogen oxides). Outdoor air pollution sources in school environments include stationary sources such as factories, power plants, smelters, dry cleaning establishments, and hazardous waste sites, as well as mobile sources like cars, buses, and trucks. Naturally occurring sources such as windblown dust, wildfires, and volcanic eruptions further contribute to outdoor air pollution.

Children experience greater inhalation exposures to outdoor air pollutants than adults because they may spend more time outside and breathe more rapidly. Children's smaller airways are more prone to irritation by air pollutants, leading to greater airway obstruction than in adults.

Health Effects from Exposure to Outdoor Air Pollution

Acute health effects associated with exposure to outdoor air pollution include respiratory symptoms (e.g., wheezing, cough, transient decrease in lung function) and illnesses like asthma, the leading cause of school absenteeism, which results in lower academic performance. Children who live or attend school near areas of high traffic-related pollution have increased risk for wheezing, bronchitis, and asthma diagnoses and hospitalization. Diesel exhaust may worsen allergic and inflammatory responses and contribute to development of new allergies[18] and asthma.

Preventing Exposure to Outdoor Air Pollution

More than 25 million children in the US commute to school by buses, logging in about 4 billion miles each year. Some states have implemented anti-idling laws, requiring bus and other commercial heavy-duty vehicle drivers to minimize idling at public and private schools. Because diesel exhaust impacts children's health and learning, the EPA's Clean School Bus program helps communities reduce emissions by providing rebates for replacing and retrofitting older diesel school buses.[19] Some school districts are transitioning to electric school buses, which have potential to reduce air and noise pollution and lighten the load on the electric grid.[20]

During peak traffic hours, schools can close windows and doors to reduce traffic pollution. School nurses can check the daily air quality index (AQI), a scale that reports levels of air pollutants, and make recommendations for alternatives to outdoor exercise for asthmatics on poor air quality days[21] (e.g., limiting outdoor activities or playing outdoors in the early morning when pollutant levels are often lower). Ensuring proper maintenance of HVAC systems can also help to prevent outdoor air pollutants from impacting IAQ.

Artificial Turf

Artificial turf is prevalent on school playing fields and playgrounds despite concerns about their potential to expose children to harmful chemicals. The most common artificial turf infill consists of finely ground recycled automobile tires, which contain known neurotoxicants and carcinogens including styrene, butadiene, lead, black carbon, and polycyclic aromatic hydrocarbons (PAHs). Children may be exposed to chemicals from these surfaces by inhalation of chemicals that off-gas, accidental ingestion, dermal exposure, and through open wounds. Crumb rubber alternative infills exist, but several of these products have also been shown to contain chemicals of concern, and little is known about their safety.[22] Studies by the EPA are under way to determine whether harmful levels of chemicals from turf surfaces enter children's bodies under realistic playing conditions.

Health Effects of Artificial Turf

Children may be exposed to extreme heat from play on artificial turf, with temperatures over 160°F recorded on summer days.[23] Vigorous play in these conditions conveys a very real risk of burns, dehydration, heat stress, or heat stroke. Children are less able to regulate their body temperature than adults, making them particularly susceptible to conditions of extreme heat.[24] In addition, children have a higher surface area to body mass ratio, produce more body heat per unit mass, and sweat less than adults, all factors that increase susceptibility to heat injury.[25]

There is a potential for toxicants to be inhaled, absorbed through the skin, and even swallowed by children who play on recycled rubber surfaces. The major chemical components of recycled rubber are styrene and butadiene, the principal ingredients of the synthetic rubber used for tires. Styrene is neurotoxic and reasonably anticipated to be a human carcinogen.[26] Butadiene is a human carcinogen that has been shown to cause leukemia and lymphoma.[27] Shredded and crumb rubber also contain lead, cadmium, and other metals known to damage the developing nervous system. Analyses find known carcinogens and neurotoxicants including polycyclic aromatic hydrocarbons (PAHs), benzothiazole, lead, zinc, and black carbon in almost all recycled rubber alternative infill materials examined.[22] Concerns about the composition of synthetic grass blades have also been raised, with recent studies finding contamination with PFAS.

Preventing Exposures to Artificial Turf

Schools should consider grass playing fields and safer alternatives to artificial playground surfaces such as wood mulch. If a school installs artificial turf, avoid recycled rubber infill because of the potential for heat and chemical exposures. Avoid play on hot summer days, and do not use these areas for passive recreation, lounging, lunch, or snack time. Very young children prone to hand-to-mouth behaviors should not be permitted to play on artificial turf surfaces. After playing on artificial turf surfaces, children should wash their hands and shake out any infill or grass blades from hair or clothing before going indoors.

Asbestos

Asbestos, a naturally occurring fibrous mineral, was used extensively in insulation and construction materials, including in school construction in the 1950s to the 1980s. School materials that may contain asbestos include boiler wraps, ceiling tiles, drywall, floor tiles, and insulation surrounding pipes. Use of asbestos has been banned or severely restricted in most industrially developed countries because of its severe health effects.

Health Effects from Exposure to Asbestos

Schoolchildren can be exposed to asbestos when it is disturbed or fractured, leading to release of microscopic fibers that can be inhaled and lodged in small airways of the respiratory tract. If asbestos remains intact, then exposure is unlikely and asbestos does not pose a health hazard. Exposure to asbestos is not associated with acute health effects or symptoms. X-ray screening of children who may have been exposed to asbestos is therefore not justified because there are no immediate radiological signs.

Asbestos is a known human carcinogen. Chronic exposure to high levels can lead to multiple effects such as asbestosis—a fibrous disease of the lungs—and cancer of the lung, larynx, and ovary, as well as mesothelioma, a malignancy that arises in pleura and occasionally the peritoneum. Asbestosis and asbestos-related lung cancer are seen almost exclusively among industrial and construction workers with long histories of intense occupational exposure to asbestos. Risk of lung cancer is higher among asbestos workers who also smoke cigarettes in comparison to those who are nonsmokers.[28] Malignant mesothelioma, unlike other diseases caused by asbestos, may result from even brief, low-dose, nonoccupational exposures. Malignant mesothelioma usually arises 20–50 years after asbestos exposure.

Preventing Exposure to Asbestos

The Asbestos Hazard Emergency Response Act (AHERA) of 1986 requires that all US schools periodically inspect for asbestos-containing materials, create a plan to manage these materials, and regularly evaluate these locations to check for any degradation. School authorities are mandated by the federal government to make results of these inspections publicly available. In 2015, a report compiled for the Subcommittee on Superfund, Waste Management, and Regulatory Oversight concluded that the scope of asbestos in schools was likely widespread but difficult to ascertain due to incomplete reporting and lack of systematic monitoring.[29]

Use of asbestos has largely been phased out in industrialized countries but exposure may still occur, most likely during renovation or demolition projects. Asbestos that remains intact in a building typically does not represent a threat but actions that may disturb asbestos-containing material and release fibers into the air should be avoided, affected areas properly contained, and work completed by trained, certified professionals in accordance with AHERA guidelines.[30]

Pesticides

Pesticides are chemical substances used for pest control. They include insecticides, herbicides, fungicides, and disinfectants. Schools are at risk for pests because they are large, have multiple exits and entrances, prepare food for a large number of children, provide shelter for insects, and may be in poor condition. Pesticides may be sprayed routinely both indoors and outdoors for rodent and insect control, potentially exposing children to high pesticide levels.[31]

Children are exposed to pesticides by inhalation, ingestion of pesticides (from food, accidental ingestions of pesticides, or ingestions of contaminated soil in outdoor school areas), and dermal exposure (touching a surface recently applied with pesticides).

Health Effects from Exposure to Pesticides

Exposure to pesticides may cause both acute and chronic health effects and have been documented to cause illness among students and school employees. Clinical effects of acute exposures include asthma exacerbations, cough, shortness of breath, eye irritation, nausea, vomiting, and headaches.[32]

Low-dose pesticide exposure in children and in pregnant women (levels where no acute effects are seen but nonetheless organ systems are affected) has also been linked to some cancers and birth defects.[33] Prenatal and postnatal low-dose exposure to pesticides has been negatively associated with children's neurodevelopment.[33-37] Certain pesticides, such as DDT, methoxychlor, and chlordecone, have been found to disrupt endocrine function in animal studies. Details on pesticide toxicity can be found in Chapter 35.

Preventing Exposure to Pesticides

Integrated pest management (IPM) programs in learning environments can reduce pesticide exposures among children and school staff.[38,39] IPM is a cost-efficient and effective approach to control pests by preventing their access to food, water, and shelter and uses chemical pesticides only as a last resort. The EPA has a step-by-step approach for development, implementation, and evaluation of IPM. Several states have statutes requiring schools to use IPM approaches, post notices of pesticide use, and notify parents of use[40] and/or have passed legislation banning use of pesticides on fields and playgrounds at all public and private schools.[41]

Arsenic

Arsenic is a naturally occurring element and a known human carcinogen. Arsenic from geological sources contaminates drinking water in areas of the United States, Bangladesh, India, Taiwan, Chile, and Argentina and thus poses severe risks to the health of children who regularly drink contaminated water.[42] Arsenic may also be an airborne contaminant, and children who live or attend school in the vicinity of smelters are at risk of exposure to arsenic from this source. Residual arsenic is found in soils in certain agricultural areas where arsenic-containing pesticides were previously used, such as in the Wenatchee Valley, in Washington state, where lead arsenate was sprayed extensively in the cultivation of apples. Tests have found arsenic in infant food including common rice cereals. Older playground equipment made from wood treated with a preservative known as chromated copper arsenate (CCA) has been documented as a source of arsenic; this exposure is lower than for food and water sources.[43]

Health Effects from Exposure to Arsenic

Arsenic has been classified as a known human carcinogen by the International Agency for Research on Cancer (IARC). Children are particularly vulnerable to arsenic, which can affect virtually every organ system, including the gastrointestinal tract, brain, heart, kidney, liver, bone marrow, skin, and peripheral nervous system (see also Chapter 33).

Preventing Exposure to Arsenic

To reduce children's exposure to arsenic from CCA-treated playgrounds, school officials can purchase arsenic-free play equipment. In 2003, the Center for Environmental Health's Safe Playgrounds Project successfully lobbied to remove arsenic from wood intended for playgrounds. In school districts where CCA-treated wood is already in place, educators can encourage children to wash their hands after playing on such playgrounds or on the ground around them, especially before eating. Apply water-based sealant twice a year or oil-based sealant once a year to existing CCA-treated wood structures (play equipment, decks, picnic tables). Do not burn, sand, or cut CCA-treated wood as it may release arsenic into the air. Dispose of CCA-treated wood at hazardous waste sites. Routinely test the drinking water for arsenic to ensure it is below the EPA's level permitted in drinking water of 10 parts per billion. For child care facilities, if water comes from a public water system, a Consumer Confidence Report can be requested to verify compliance with state drinking water laws. Rice-free packaged snacks are preferable. Snacks made with rice flour can be high in arsenic. In child care facilities, infant rice cereal should be avoided as it is the primary source of arsenic in infant's diets.

Lead

Lead is a naturally occurring metal. Lead can be found inside a school or child care facility in paint, water, consumer products (e.g., toys and costume jewelry), and dust.[44] It can also be found in contaminated soil.[45] Although lead-based paint was banned in schools in the United States in 1977, older buildings may still contain lead-based paint. When lead-based paint deteriorates, chips and dust settle onto surfaces that children can easily reach, such as windowsills and floors.[46] Infants and toddlers are at highest risk because of their developing bodies and age-appropriate mouthing behaviors. Lead may be found in soil on school grounds that are in close proximity to busy highways or high-traffic roads, battery plants or smelters, and other industries that release lead into the air.

Lead water pipes or lead solder in the pipes[47] in older schools may leach lead into drinking water, especially if the water has been sitting overnight or over weekends or holidays.

There are three main sources of lead in water[48]: lead pipes, lead service lines (outdoor pipe connecting the water main under a street to a building's plumbing), and leaded solder (used to connect copper pipe and fittings). Congress banned these uses in 1986. Leaded fixtures—faucets and other plumbing components—may be made from brass, a metal alloy that may contain lead. In 1986, Congress limited lead in brass to 8%, and in 2014, to 0.25%.

As of January 2021, the EPA updated the Lead and Copper Rule under the Safe Drinking Water Act to require testing for lead in water at all licensed child care facilities constructed prior to 2014. The rule is effective in 2024. Lead in pipes and solder can be sharply reduced by flushing faucets that have not been used for several hours. Flushing times vary based on plumbing configuration.

Certain children's products imported to the United States and found in learning environments are known to have a higher risk of containing lead, including metal toy jewelry and toy pottery. In 1978, the United States banned lead in house products marketed to children and in dishes or cookware, but other countries still widely use lead paint. Older toys made before the ban and imported toys may contain lead, as may toys made with vinyl/PVC (e.g., teething toys, dolls, backpacks, etc.). Lead-glazed pottery or porcelain can become contaminated because lead can leach from these containers into the food or liquid. Lead may also be present in certain herbal remedies, folk medicines, and imported spices and foods. The Consumer Product Safety Commission (CPSC) has found lead in some art supplies (e.g., paints and crayons). These art materials are required to be labeled as hazardous per the Labeling of Hazardous Art Materials Act of 1988.

Health Effects from Exposure to Lead

Lead is long recognized for its negative health effects in humans. There is no lead level in the blood that is safe. Even low levels of lead exposure can result in loss of intelligence (IQ). Children exposed to lead are at elevated risk for learning difficulties, shortened attention span, dyslexia, conduct disorders, and school dropout. Children with higher levels of lead poisoning can present with constipation, anemia, and seizures, and it may cause death (see also Chapter 31).

Prevention of Lead Poisoning

Inspect learning environments for any sources of lead including deteriorating lead-based paint and ensure appropriate containment for areas that are undergoing renovation. Renovation must be undertaken by properly trained abatement workers certified by the EPA for lead safe-work practices. Flush drinking water at all taps before each use; local water utility staff can provide specific information on suggested flushing times. Routinely test water for lead, and, if found, water filtration devices certified to remove lead at the outlet can be used.[49] Handwashing is encouraged especially after coming inside and before eating. Reduce exposure to lead-contaminated soil by placing commercial walk-off mats at all entrances. Remove shoes whenever possible in infant and toddler rooms in child care facilities. Use vacuums with a high-efficiency particulate air (HEPA) filter. Clean floors daily and window frames/sills weekly using a damp mop, sponge, or paper towel with warm water and an all-purpose, fragrance-free cleaner. Test any bare soil in or around a learning environment for lead. Choose safe lead-free art supplies and toys by checking labels carefully.

Mercury

Mercury is a naturally occurring element released to the atmosphere by the combustion of coal and petroleum, both of which contain mercury. Mercury from these sources is deposited into rivers, lakes, and the sea; it is then transformed to highly toxic methyl-mercury, which accumulates in predatory fish species. The most common pathway of human exposure to mercury is through consumption of fish from contaminated water, with bigger fish such as shark, swordfish, king mackerel, and tuna containing the highest levels of mercury.

Metallic mercury was used in many places in the school environment, including lab equipment, thermometers, thermostats, batteries, and fluorescent light bulbs.[50] If handled or discarded improperly, these items can release liquid mercury into the air or ground, which then vaporizes and can then be inhaled or ingested.

Health Effects from Exposure to Mercury

There is no safe level for mercury in humans. High levels of mercury can present with acute symptoms such as hallucinations, flushing, vomiting, and vision changes. Pregnant women, fetuses, and children are particularly susceptible to the effects of low-level exposure to mercury. Exposure to mercury during pregnancy can increase the risk of miscarriage and have neurodevelopmental effects on the fetus, leading to decreased IQ and impaired memory and attention (see also Chapter 32).

Preventing Exposure to Mercury

To reduce dietary exposure to mercury, the EPA and the FDA have set standards for mercury levels in drinking water as well as in fish and bottled drinking water. Children, pregnant women, and women who may become pregnant should avoid fish with high mercury content (e.g., shark, swordfish, king mackerel, and tuna).

Schools can reduce exposures to mercury by eliminating unnecessary use of items containing mercury, replacing mercury-containing devices with safer alternatives (e.g., using LED light bulbs instead of compact fluorescent light [CFL]) bulbs), and ensuring that educators are informed about proper procedures for dealing with mercury spills. School cafeterias can work with local authorities to avoid serving species of fish known to be high in mercury (see Box 17.1).

Polychlorinated Biphenyls

Polychlorinated biphenyls (PCBs) are a family of manmade organic chemicals used widely in building materials beginning in the 1940s. PCBs were used in coolants and insulating fluid in transformers, electronics, fluorescent lighting and air conditioners, paint, and cements as plasticizers and in sealants. PCBs resist burning and persist for a long time. Because of this resistance to degradation, PCBs made more than 40 years ago are still in the environment, and humans and animals are continuously exposed. Currently, low levels of PCBs are present in most Americans, including young children.

Because of concerns of persistence in the environment as well as concerns about human health effects, a federal ban on the manufacture of PCBs was imposed in 1976 under the Toxic Substances Control Act (TSCA). In a pilot study of three New York City

---

Box 17.1 Case Study: Mercury in a Child Care Center

A child care center in Gloucester County, New Jersey, was sited in a former mercury thermometer manufacturing company[51] in operation for 20 years. The New Jersey Department of Health and Senior Services (NJDHSS) Hazardous Site Health Evaluation Program and Indoor Environments Program, with the Agency for Toxic Substances and Disease Registry/National Center for Environmental Health (ATSDR/NCEH), and the Region 2 Pediatric Environmental Health Specialty Unit investigated mercury exposures in children and staff.[52] Air and wipe samples tested positive for mercury, and the center was closed that afternoon. Two-hundred children attended the Center since its opening. Medical monitoring and chart review of a subsample of children attending the child care center did not reveal symptoms suggestive of mercury poisoning (acrodynia). No children required chelation therapy.

Legislation was passed requiring the NJDHSS to adopt procedures for the evaluation and assessment (above and beyond lead and asbestos) of buildings used for child care centers. It further prohibits construction permits for sites that do not meet the DHSS standards. Comprehensive voluntary school siting guidelines are publicly available online.[53]

(NYC) public schools, levels of PCBs were found to be above EPA guidelines. Two principal sources were deteriorating, PCB-containing fluorescent light fixtures and PCB caulking. NYC public schools subsequently removed all PCB-containing light fixtures.

PCB Exposure in Children and Fetuses

Exposure to PCBs can occur in three ways: (1) inhalation of PCBs in the air, (2) absorption of PCBs through the skin, and (3) ingestion of PCBs from contaminated dust and from fish and meats. Younger children are more likely to have exposure from ingestion of contaminated dust that settles on floors and on toys. PCBs are not easily broken down or passed from the body. PCBs can accumulate in the body over time, with half-lives ranging from months to decades.

Developing babies may be exposed to PCBs in utero if pregnant mothers breathe, eat, or touch things that contain PCBs or eat PCB-contaminated fish. PCBs can readily pass from the mother to the baby via the placenta. PCBs may also concentrate in the mother's breast milk (see also Chapter 36).

Health Effects from Exposure to Polychlorinated Biphenyls

Epidemiological studies demonstrate that babies born to mothers with elevated levels of PCBs have diminished intelligence as measured by decreased IQ scores and motor delays. While a number of other health outcomes have been associated with PCBs, this is where the strongest body of evidence exists with respect to PCB exposures. Studies have demonstrated possible links between PCBs and reproductive system effects, lowering of immune response, hypothyroidism, diabetes, and elevated blood pressure, triglycerides, and cholesterol levels.[54,55] PCBs as a group are classified as probable carcinogens, and PCB 126 was recently classified by the IARC as a known human carcinogen.

Preventing Polychlorinated Biphenyls Exposures in Schools

Although the levels of PCBs found currently in school building materials are not likely to make any child or any teacher acutely ill, it is nonetheless clear that routine assessment of PCBs-containing building materials (light fixtures, caulk, sealants) can reduce and prevent further exposure of children, teachers, and other school staff. While this situation is not a medical emergency, there is no reason to delay. There are no safe thresholds for PCB exposure during pregnancy.

Routine assessment of building materials and careful consideration for the need for containment or safe removal is warranted given the potential health concerns related to chronic exposures to PCBs in dust and air in school environments.[56,57] An overall goal is to keep environmental exposures low to minimize risk. This type of work should be done carefully and methodically to avoid a release of more PCBs into the school environment or an increase in danger to remediation workers. Prompt and thorough cleanup minimizes exposure to PCB-containing materials. An important component of this recommendation is that children, pregnant women, and women of childbearing age are not present when work is being done.

Promoting Healthy Child Care and School Environments

Learning environments offer a major but largely untapped public health opportunity to prevent childhood environmental health risks.[58] There are over 300,000 child care settings licensed in the United States. The potential impact of these facilities' environmental conditions on children's health is second only to those of the primary home. The child care workforce is predominately made up of women of color who are of childbearing age. Many providers are also low-wealth because half are part of families accessing public benefits. These factors make child care providers at higher risk for being exposed to environmental hazards, resulting in poor health outcomes.

Even when state child care licensing regulations are being met, child care facilities and schools can be located in areas or buildings where exposure to toxic chemicals is taking place. To prevent this, prior to choosing a site,[59] it is critical to perform an environmental audit.[59] An audit includes investigation of historical land and building use to assess potential for contamination with toxic or hazardous substances that may result in either indoor or outdoor environmental concerns; assessment for mold, lead, and asbestos in older buildings; determination of potential sources of noise, air pollution, and infestation; evaluation of location of the site and associated play areas from roadways, industrial emissions, and building exhaust outlets; and access to a safe drinking water supply and septic system.[60]

Architectural design and building materials can reduce toxic exposures among children and school staff. Choosing building materials that are less toxic, ensuring a well-designed ventilation system, and using cleaning products with fewer toxic chemicals can reduce exposures.[60] Optimizing natural light in classrooms as well as using energy-efficient lighting systems are well-recognized ways of promoting the well-being of students, faculty, and staff. Schools typically have centrally located stairwells to efficiently allow for movement of students from class to class. Stairs and safe walking and bicycling routes are all important features for promoting daily activity. For students requiring motor vehicle transport, many school districts have implemented "asthma-free zones" at drop-off and pickup locations to discourage idling.

Last, learning environments can foster and model sustainability by recycling, gardening, and composting. Create "Greening Committees" consisting of parents, teachers, and maintenance staff to proactively identify and address environmental problems as well as educate other community members on ways to improve the learning environment.

References

1. Children's Defense Fund. Child care basics. 2005. Available at: https://www.childrensdefe nse.org/wp-content/uploads/2018/08/child-care-basics.pdf.

2. National Center for Education Statistics. Back-to-school statistics. 2005. Available at: https://nces.ed.gov/fastfacts/display.asp?id=372#K12-enrollment.

3. Wallace LA. The total exposure assessment methodology (TEAM) study: an analysis of exposures, sources, and risks associated with four volatile organic chemicals. *J Am Coll Toxicol*. 1989;8(5):883–95.

4. Centers for Disease Control and Prevention. About CDC healthy schools 2019. Updated 2023 Jul. Available at: https://www.cdc.gov/healthyschools/about.htm.

5. Environmental Protection Agency. Executive summary of state school environmental health guidelines 2021. Updated 2022 Dec. Available at: https://www.epa.gov/schools/executive-summary-state-school-environmental-health-guidelines.

6. National Center for Education Statistics. Condition of America's public school facilities: 2012–2013. 2014. Available at: Condition of America's Public School Facilities: 2012-13 (ed.gov)

7. United States Government Accountability Office. School districts frequently identified multiple building systems needing updates or replacement. 2020. Available at: https://www.gao.gov/products/gao-20-494.

8. American Academy of Pediatrics Council on Environmental Health. Child care settings. In: Etzel RA, Balk SJ, eds. *Pediatric Environmental Health*. 4th ed. Itasca, IL: American Academy of Pediatrics; 2019:140.

9. Babin A PP, Rossol M. *Children's Art Supplies Can be Toxic*. New York: Center for Safety in the Arts; 1992.

10. Centers for Disease Control and Prevention. Ventilation in schools and childcare programs 2021. Available at: https://www.cdc.gov/coronavirus/2019-ncov/community/schools-childc are/ventilation.html.

11. Haverinen-Shaughnessy U, Moschandreas DJ, Shaughnessy RJ. Association between substandard classroom ventilation rates and students' academic achievement. *Indoor Air*. 2011;21(2):121–31.

12. Environmental Protection Agency. Indoor air quality tools for schools action kit 2021. Updated December 2022. Available at: https://www.epa.gov/iaq-schools/indoor-air-quality-tools-schools-action-kit.

13. Moglia D, Smith A, MacIntosh DL, Somers JL. Prevalence and implementation of IAQ programs in U.S. schools. *Environ Health Perspect*. 2006;114(1):141–6.

14. New York State Children's Environmental Health Centers. Green Cleaning 2021. Updated 2021 Aug. Available at: https://nyscheck.org/RXP/NYS/rx_english_GreenCleaning_NYS_WEB.pdf.

15. American Academy of Pediatrics Council on Environmental Health. Indoor air pollutants. In: Etzel RA, Balk SJ, eds. *Pediatric Environmental Health*. 4th ed. Itasca, IL: American Academy of Pediatrics; 2019:359.

16. IARC Working Group on the Evaluation of Carcinogenic Risks to Humans. *Trichloroethylene, Tetrachloroethylene, and Some Other Chlorinated Agents*. Lyon, France: IARC Publications; 2014.

17. Forand SP, Lewis-Michl EL, Gomez MI. Adverse birth outcomes and maternal exposure to trichloroethylene and tetrachloroethylene through soil vapor intrusion in New York State. *Environ Health Perspect*. 2012;120(4):616–21.

18. American Academy of Pediatrics Council on Environmental Health. Outdoor air pollutants. In: Etzel RA, Balk SJ, eds. *Pediatric Environmental Health*. 4th ed. Itasca, IL: American Academy of Pediatrics; 2019:385.

19. Environmental Protection Agency. School bus idle reduction. 2021. Available at: https://www.epa.gov/dera/school-bus-idle-reduction.

20. World Resources Institute. The state of electric school bus adoption in the US. 2021. Available at: https://www.wri.org/insights/where-electric-school-buses-us.

21. AirNow. Home of the U.S. air quality index. 2021. Available at: https://www.airnow.gov/.

22. Massey R, Pollard L, Jacobs M, Onasch J, Harari H. Artificial turf infill: a comparative assessment of chemical contents. *New Solut*. 2020;30(1):10–26.

23. Devitt D, Young M, Baghzouz M, Bird B, Devittl B. Surface temperature, heat loading and spectral reflectance of artificial turfgrass. *J Turfgrass Sports Surf Sci*. 2007;83:68–82.

24. American Academy of Pediatrics. Critical updates on COVID-19. 2023. Available at: https://www.aap.org/en-us/advocacy-and-policy/aap-health-initiatives/Children-and-Disasters/Pages/Extreme-Temperatures-Heat-and-Cold.aspx.

25. Falk B, Dotan R. Children's thermoregulation during exercise in the heat: a revisit. *Appl Physiol Nutr Metab*. 2008;33(2):420–7.

26. Agency for Toxic Substances and Disease Registry. *Toxicological Profile for Styrene*. Atlanta: Agency for Toxic Substances and Disease Registry; 2010.

27. International Agency for Research on Cancer (IARC). *Chemical Agents and Related Occupations*. Lyon: International Agency for Research on Cancer; 2008.

28. National Cancer Institute. Asbestos exposure and cancer risk. 2021. Available at: http://www.cancer.gov/cancertopics/factsheet/Risk/asbestos.

29. Markey EJ. Failing the grade: asbestos in America's schools. 2015. Available at: https://www.markey.senate.gov/imo/media/doc/2015-12-Markey-Asbestos-Report-Final.pdf.

30. US Government Publishing Office. Asbestos hazard emergency response. 2009. Available at: https://www.govinfo.gov/content/pkg/USCODE-2009-title15/html/USCODE-2009-title15-chap53-subchapII.htm.

31. American Academy of Pediatrics Council on Environmental Health. Pesticides. In: Etzel RA, Balk SJ, eds. *Pediatric Environmental Health*. 4th ed. Itasca, IL: American Academy of Pediatrics; 2019:687.

32. Alarcon WA, Calvert GM, Blondell JM, et al. Acute illnesses associated with pesticide exposure at schools. *JAMA*. 2005;294(4):455–65.

33. Grandjean P, Harari R, Barr DB, Debes F. Pesticide exposure and stunting as independent predictors of neurobehavioral deficits in Ecuadorian school children. *Pediatrics*. 2006;117(3):e546–56.

34. Bouchard MF, Bellinger DC, Wright RO, Weisskopf MG. Attention-deficit/hyperactivity disorder and urinary metabolites of organophosphate pesticides. *Pediatrics*. 2010;125(6):e1270–7.

35. Engel SM, Berkowitz GS, Barr DB, et al. Prenatal organophosphate metabolite and organochlorine levels and performance on the Brazelton Neonatal Behavioral Assessment Scale in a multiethnic pregnancy cohort. *Am J Epidemiol*. 2007;165(12):1397–404.

36. Rauh VA, Garfinkel R, Perera FP, et al. Impact of prenatal chlorpyrifos exposure on neurodevelopment in the first 3 years of life among inner-city children. *Pediatrics*. 2006;118(6):e1845–59.

37. Rohlman DS, Arcury TA, Quandt SA, et al. Neurobehavioral performance in preschool children from agricultural and non-agricultural communities in Oregon and North Carolina. *Neurotoxicology*. 2005;26(4):589–98.

38. University of California at San Francisco's Early Care and Education. Integrated pest management toolkit for facility and home-based programs (English and Spanish). 2011. Available at: https://cchp.ucsf.edu/sites/g/files/tkssra181/f/ipm_checklist.pdf

39. Environmental Protection Agency. IPM training. 2021. Available at: https://www.epa.gov/childcare/training-webinars-and-resources-child-care-providers.

40. National Institute for Occupational Safety and Health. Reducing pesticide exposure at schools. 2007. https://www.cdc.gov/niosh/docs/2007-150/default.html

41. Owens K. Schooling of state pesticide laws. 2010. Available at: https://www.beyondpesticides.org/assets/media/documents/schools/publications/Schooling2010.pdf.

42. American Academy of Pediatrics Council on Environmental Health. Arsenic. In: Etzel RA, Balk SJ, eds. *Pediatric Environmental Health*. 4th ed. Itasca, IL: American Academy of Pediatrics; 2019:891.

43. Lew K, Acker JP, Gabos S, Le XC. Biomonitoring of arsenic in urine and saliva of children playing on playgrounds constructed from chromated copper arsenate-treated wood. *Environ Sci Technol*. 2010;44(10):3986–91.

44. National Center for Healthy Housing. The Lead-Safe Toolkit for Center-Based Child Care. 2023. Available at: https://nchh.org/tools-and-data/technical-assistance/protecting-children-from-lead-exposures-in-child-care/hbcc-toolkit/.

45. Environmental Protection Agency. Soil, yards and playgrounds. 2021. Available at: https://www.epa.gov/lead/protect-your-family-sources-lead#soil.

46. Environmental Protection Agency. Renovate right: important lead hazard information for families, child care providers and schools. English. 2021. Available at: https://www.epa.gov/lead/renovate-right-important-lead-hazard-information-families-child-care-providers-and-schools.

47. LSLR Collaborative. Tool: lead service line replacement collaborative. 2021. Available at: https://www.lslr-collaborative.org.

48. Environmental Protection Agency. Drinking water. 2021.Available at: https://www.epa.gov/lead/protect-your-family-sources-lead#water.

49. Environmental Protection Agency. 3Ts for reducing lead in drinking water toolkit. 2021. Available at: https://19january2021snapshot.epa.gov/ground-water-and-drinking-water/3ts-reducing-lead-drinking-water-toolkit_.html.

50. Lanphear BP, Hornung R, Khoury J, et al. Low-level environmental lead exposure and children's intellectual function: an international pooled analysis. *Environ Health Perspect*. 2005;113(7):894–9.

51. Agency for Toxic Substances and Disease Registry. Public health implications of exposure to polychlorinated biphenyls (PCBs). 2021. https://wwwn.cdc.gov/TSP/PHS/PHS.aspx?phsid=139&toxid=26

52. Agency for Toxic Substances and Disease Registry. Health consultation mercury exposure investigation using serial urine testing and medical records review. 2007. https://www.nj.gov/health/ceohs/documents/eohap/haz_sites/gloucester/franklin_township/kiddie_kollege/kiddiekollegehc.pdf

53. Environmental Protection Agency. School siting guidelines. 2011. https://www.epa.gov/sites/default/files/2015-06/documents/school_siting_guidelines-2.pdf

54. Carpenter DO. Polychlorinated biphenyls (PCBs): routes of exposure and effects on human health. *Rev Environ Health*. 2006;21(1):1–23.

55. Johnson CL. Mercury in the environment: sources, toxicities, and prevention of exposure. *Pediatr Ann*. 2004;33(7):437–42.

56. Brown KW, Minegishi T, Cummiskey CC, Fragala MA, Hartman R, MacIntosh DL. PCB remediation in schools: a review. *Environ Sci Pollut Res Int*. 2016;23(3):1986–97.

57. Herrick RF, Stewart JH, Allen JG. Review of PCBs in US schools: a brief history, an estimate of the number of impacted schools, and an approach for evaluating indoor air samples. *Environ Sci Pollut Res Int*. 2016;23(3):1975–85.

58. Zaeh SE, Koehler K, Eakin MN, et al. Indoor air quality prior to and following school building renovation in a Mid-Atlantic school district. *Int J Environ Res Public Health*. 2021;18(22).

59. Agency for Toxic Substances and Disease Registry. Choose Safe Places for Early Care and Education (CSPECE) guidance manual. 2017. https://www.atsdr.cdc.gov/safeplacesforECE/index.html

60. American Academy of Pediatrics Council on Environmental Health. Community design. In: Etzel RA, Balk SJ, eds. *Pediatric Environmental Health*. 4th ed. Itasca, IL: American Academy of Pediatrics Council on Environmental Health; 2019:119.

18

The Shape of the Built Environment Shapes Children's Health

Richard J. Jackson

We are not only what we eat, we are also what we build. The built environment is where most humans, especially children, spend virtually all our time. It is not only buildings, but also parks, vehicles, play structures, roads—the places created or modified by people. The built environment transforms indoor and outdoor physical and social behavior, including the health and quality of life for everyone, especially children.[1] A child's life is directly affected by where she lives, and she will thrive in healthy, safe, and diverse environments that promote play and mental focus.[2] Many of the health harms that children confront have their origin, as do the prevention opportunities, in the environments we construct. Some of these threats, such as window and stairway falls, are readily apparent, but other preventable harms such as anxiety, attention-deficit/hyperactivity disorder (ADHD), and substance abuse are less directly evident and can be exacerbated by inadequate living conditions and poor urban plans.[3]

As in the workplace, the most effective way to protect individuals and change human behavior is through physical changes to the environment, for example, by good ventilation systems and prohibiting smoking and by enforcing stop light and speeding laws.[4,5] So, too, with children the most effective and cost-saving way to prevent injury and illness is to design out plausible harms in the original construction and build with a long-term and larger view of health.

The Home

Children spend their early years mostly in their homes and nearby yards. Hazards that present risks to children such as lead, mold, pesticides, radon, and volatile organic compounds like formaldehyde, are covered in other chapters. As noted in Chapter 15, inadequate housing contributes to children's poor health. Leaking pipes, peeling paint, rotting wallboard, damaged ceilings and walls, can poison children and stress the immune system.[6] These "sick buildings" increase exposures to gases, molds, moisture, lead, radon, and allergens. Housing disrepair, especially among the poor, exposes them to pests and physical harms including falls, failed plumbing, heat and cold, and fires. The building of the home, its structure, egresses, physical security, fire resistance, ventilation, heating, cooling, and the building's siting and access, all can influence a child's health. The architectural tenets that Vitruvius articulated 2,000 years ago still apply: a building should have *firmatas, utilitas,* and *venustas*.[7] *Firmatas*: buildings should be strong and

Richard J. Jackson, *The Shape of the Built Environment Shapes Children's Health* In: *Textbook of Children's Environmental Health*, Second Edition. Edited by: Ruth A. Etzel and Philip J. Landrigan, Oxford University Press. © Oxford University Press 2024.
DOI: 10.1093/oso/9780197662526.003.0018

resistant to weather and personal dangers, as well as to external natural forces such as earthquakes, hurricanes, and, in the present era, climate change. *Utilitas*: buildings must be useful; help to make life more bearable and, ideally, comfortable with effective ventilation, heating, cooling, insulation, and lighting; and be able to resist intrusion not just by harmful persons, but by pests and moisture. A good building should help us meet life's needs with adequate clean water as well as waste removal.[8] A home that lacks safe cooking facilities and refrigeration and quiet sleeping areas is not useful. *Venustas*: this often translates as beauty, which does not capture the term's richness. In real estate, the term "amenities" is often casually used; this term is inadequate. A state of good appearance and repair with adequate indoor and natural light along with near access to nature offers a sense of security and well-being: these are not "amenities"—they are comforts that every person, especially a child, deserves. Squalid places are not merely unhealthy, they are emotionally depressing not just for adults, but for children who cannot readily escape them. Crowded dwellings enable the spread of contagion, and the lack of privacy limits restfulness and life joys.[9] A badly designed neighborhood can do the same, as seen in areas where, for generations, residents were unable to acquire affordable mortgages as result of racism and "redlining."[10] It is remarkable how redlining—the limiting of mortgages and insurance three generations ago in the United States because of economic risk and racism—still leaves communities overpaved, sweltering, and economically disadvantaged.

These conditions can elevate physiological stress levels, hinder academic achievement, and increase social withdrawal and interpersonal conflict along with behavioral problems in school.[11] Loss of tree cover in neighborhoods amplifies heat stresses, elevates air pollution levels, and reduces a child's interest in being outdoors and growing their social networks and physical stamina.[12]

Indoor air pollution is a leading environmental health risk for children.[13,14] The home environment plays an important role in sensitizing children to allergens and triggering asthma attacks and other respiratory illnesses. Inadequate heating systems, as can be found in "slum" rental housing, often force residents to use space heaters or, more dangerously, a gas oven with its door open. Inadequate cooling and insulation, defective windows, and unsafe neighborhoods (where windows are nailed shut) have all worsened death rates during the heat waves inevitable with global warming. Treatment with antihypertensives, ADHD stimulants, or other medications can amplify risks in heat waves.[15]

Building structure can lead to injuries in children, such as window falls, carbon monoxide poisonings, and burns.[16] Missing stairway railings are an obvious health risks to persons of all ages, but especially to a parent trying to carry a fussing toddler. Stairway railings that are too short and do not reach beyond floor level also are risky and in violation of the Americans with Disabilities Act.[17] Injuries are a leading cause of death of children, ages 1–21 years (see Chapter 54). A portion of these deaths can be prevented when builders, developers, housing codes, and lenders offer more health-focused designs of the built environment.[7]

Environments that promote active lifestyles can help children achieve the recommended levels of physical activity.[18,19] For children, bicycling or walking to school and daily active play are essential to social and physical development.[20] As the percentage of park area within a child's neighborhood increases, so does the physical activity of young children.[21] When a neighborhood lacks parks or active play facilities, children are less physically active; this is unhealthy socially, mentally, and physically and even more hazardous for young people living in low-income neighborhoods.[22] Geographic areas with higher incomes and education levels generally have more access to youth recreational areas, including parks. Good architectural and urban plans must both preserve natural environments and create built environments to serve the mental and physical well-being of the entire population, not merely members of country clubs.

For many children, their first experience of a built environment is a hospital. The design and operation of hospitals for children merit additional consideration. Infants at birth need skin-to-skin contact with their mothers and to be kept warm. In some settings, infants are exposed to excess levels of bright light and noise, for example, in intensive care nurseries and older incubators. Ideally hospital rooms are meticulously clean and have "one-pass" ventilation systems to prevent room-to-room spread of viral infections.[23] Procedures performed on children (e.g., bone marrow aspirations) should be done in a treatment room rather than in the child's hospital room, which should be preserved as a place of more comfort. If possible, the "service wall" normally at the head of the bed should be as unobtrusive as possible and with movable covering. Art work ideally should be nature scenes, not outdated commercial cartoons. In the era of decarbonization, hospitals should strive for "net zero" energy use.[24] The furniture in a non-isolated child's room should be convertible to a sleep surface for use by a parent to convey comfort in what is often a frightening circumstance.

Transportation, Children's Health, and the Built Environment

Despite what many transportation guidelines espouse, the purpose of transportation is to bring people and goods to where they need to be: it is not merely "safely moving people and goods"—people travel to "be" somewhere.[25] Although transportation agencies traditionally have viewed their role as moving people and goods safely and efficiently, this is not enough, and increasingly transportation agencies have embraced the reality that transportation has major impacts on human health.[26]

Transportation is costly: it is one of the largest expenses for American households.[27] In much of the United States, workers must have transportation to earn an income, yet, after paying for it, the poor are forced to scrimp on food, rent, and medical care. In the United States, in 2020, households spent an average of $9,800 per year on transportation, with only $263 per capita spent on public transportation. These costs include vehicle purchases, loan interest, cost of fuel, maintenance, and insurance. Importantly, transportation costs are a greater burden to the poor, who often reside in lower cost areas at greater distance from work while contending with older, less reliable vehicles. Being "transportation disadvantaged" relates to more than money: it can exacerbate isolation, limit income opportunities, and amplify barriers related to disabilities and language differences. US households in the lowest quintile of income averaged $15,140 of after-tax income and spent $4,363 or 29% on transportation, while households in the top quintile with $176,094 after-tax income spent $16,789 or 10% of their after-tax income on transportation.

Housing people closer to community and to work and producing goods closer to where they are needed is the most efficient way to reduce the costs and pollution from the transportation system.[28] Cities developed because people, commerce, and culture function better when humans can readily access work and the services they need. Cities must have diverse incomes and workforces; it helps them to be more resilient. For example, living in a place that lacks a spread of incomes makes getting many services more diffi-cult.[29] The wealthy need workers, and the workers need income; a mix of incomes better supports services and a more robust tax base to serve community needs ranging from parks to schools to public safety.

A major US public health accomplishment of the 20th century was decreasing motor vehicle fatalities per mile, declining from 18 per million miles traveled in 1925 to about 2 per million miles traveled in 1997.[30] The improvement may be attributed to safer and better vehicles, enforcement, and new laws as well as to changes in the physical environ-ment. Regrettably, the US Department of Transportation's 2020 annual traffic crash data showed that 38,824 persons died in traffic crashes, the highest number of fatalities since 2007. The fatality rate per 100 million VMT (vehicle miles traveled) increased to 1.34, a 21% increase from 2019 and the highest since 2007.[31]

The built environment can cause or mitigate both indoor and outdoor air pollution and its effects on health. As noted in Chapter 48, asthma is the most prevalent chronic disease among children, occurring in approximately 54 of every 1,000 children. Reducing vehicular emissions improves respiratory function in children.[32] Changes in behavior regarding the built environment can lead to fewer pediatric asthma events, as evident in the study of the 1996 Olympic Games.[33] The city of Atlanta implemented a plan to re-duce the amount of automobile congestion around the Games by restricting downtown private car access and encouraging use of public transportation. The study evaluated the effects on asthma events through increased use of public transportation and found an overall 22% decrease in traffic numbers, 28% decline in daily ozone levels, and a 41% de-crease in asthma acute-care events. The Atlanta study suggests a notable positive change in healthcare utilization was associated with this change in travel patterns.

Lack of tree cover and immense "asphalt meadows" raise local temperatures and am-bient ozone levels.[34] High-density areas have approximately 50 parts per billion (ppb) ozone level difference above their low-density counterparts. Elevated levels of ambient ozone have pulmonary and other adverse health effects on children, as well as on the eld-erly, asthma sufferers, and other vulnerable populations.

Climate change is driving dramatic increases in wildfires worldwide. The physical dangers from fire are evident, but the more distant impacts of widespread wildfire smoke are increasingly documented. The health threats from indoor air pollution are also in-creasingly apparent. Most prominent among these are the high amounts of methane escaping from methane extraction, storage, piping, and industrial and domestic use, as well as its combustion byproducts. Methane ("natural gas" is a mere marketing title) is a powerful climate forcing gas that is widely released into the environment from delivery systems and domestic use. In addition, globally, many rural or undeveloped locales con-front indoor air pollution from the burning of biomass (e.g., wood or dried dung), which directly exposes those nearby, most often women and their babies.

The physical structure of a community can promote or limit access to physical activity; many suburban and more rural housing developments are designed and built for vehi-cles, not for vulnerable populations, including children, the elderly, the disabled, or the poor. Although the leading cause of death for ages 3–34 years is motor vehicle crashes, laws requiring widespread us of infant and booster seats have reduced these risks. Urban plans and suburban developments that require the use of cars increase rather than re-duce these harms.[35] Human habitation has expanded into once rural and farming areas, with lower density developments and fewer multifamily dwellings and places to gather. Sprawl is associated with increasing air contaminant levels, automobile and pedestrian injuries, auto travel,[36] and deaths. Few children in suburban and exurban areas can walk or bicycle to school due to the overall community layout, long distance to school, and lack of sidewalks or bicycle lanes. Children must depend on cars to go to events. Isolation magnifies depression and limits children's opportunities to explore, participate in physical and social challenges, and do the work of childhood by increasing autonomy and social and physical confidence.

Schools and the Built Environment

After their homes, children spend most of their time in school environments. Since World War II, the number of US schools has declined by 70%; however, the average school size has increased from 127 to 653 students.[37] There appears to be a trend toward building increasingly larger schools on increasingly larger sites. The school environment affects a child's well-being; crowding, lighting, temperature, and noise all can interfere with learning. Chemical exposures are associated with school buildings and include lead in paint, asbestos insulation, pesticides, molds (see Chapter 43), and polychlorinated biphenyls (PCBs) in lighting fixtures (see Chapter 36). Playgrounds and playing fields enable outdoor and indoor sports participation in youth; these play environments must be safe to reduce risk of injury.

Over the past two generations children are losing opportunities to get adequate levels of daily physical activity (i.e., through "active transportation" to and from school).[38] In 1969, 41% of American children walked to school, but in recent years approximately 13% do, and, in some areas, rates are as low as 5%. The rarity of children actively commuting to school reinforces parental anxiety, aggravated by sensational news reports of abductions. Parents often voice "stranger danger" about children exercising this increasing autonomy. A 2004 study found that most parents have concerns about road safety and strangers in their neighborhood.[39]

Other barriers to active transportation by children that parents report include distance to school (61%), traffic-related danger (30%), weather (19%), crime (12%), school policy (6%), and "other" barriers (15%), including weight of backpacks.[40] Traffic-related danger has become of increasing concern due to large numbers of motor vehicles. Environmental interventions to reduce pediatric pedestrian injuries include the presence of adequate sidewalks and crosswalks, bicycle paths, traffic control intersections, and traffic-calming infrastructure. These features separate children from traffic, making them more visible, and help to slow traffic near schools.[41]

School programs must offer children ways to get enough physical activity. Schools that are committed to promoting students' fitness programs see positive effects in academic performance, such as increased concentration, reduced disruptive behaviors, and greater self-control.[42,43] For example, the California Education Code mandates physical education for all students in grades 1 through 9, with one additional year in high school. Students in grades 1 through 6 are required to have 200 minutes of physical education every 10 school days, and students in grades 7 through 12 are required to have 400 minutes every 10 school days. The annual California "Fitnessgram" was designed by the California State Board of Education as the required physical fitness test administered to students in grades 5, 7, and 9.[44] It measures six major fitness areas, including aerobic capacity and body composition. In 2011, only 25% of fifth-graders, 32% of seventh-graders, and 36.8% of ninth-graders scored in the "Healthy Fitness Zone" for six out of six fitness areas. The California Department of Education State study indicates that physically fit children perform better academically.[45] Students who met minimum fitness levels in three or more physical fitness areas showed the greatest gains in academic achievement at all three grade levels.

Physical Activity and the Built Environment

In the United States, young people have a greater likelihood of being overweight or obese because of greater levels of urban sprawl. The built environment of a community directly affects its residents' levels of physical activity, especially those of children. In adults, the difference between living in highly walkable and nonwalkable communities is 6.3 pounds of weight.[46] In 2017–2021, close to 20% of children and adolescents—nearly 16 million—were obese.[47] Obesity prevalence was 26% among Hispanic children, 25% among non-Hispanic Black children, 17% among non-Hispanic White children, and 9% among non-Hispanic Asian children.

Obesity-related conditions include high blood pressure, high cholesterol, type 2 diabetes, breathing problems such as asthma and sleep apnea, and joint problems. With widely available electronic media, including television, computers, and video games, children and youth are becoming more sedentary. Children from ethnic minorities, children living in poverty, children with disabilities, and children living in neighborhoods with limited access to outdoor activities are at increased risk of having low levels of physical activity. Good built environments offer settings where people can be active daily.

Play is a critical life task for children, regardless of socioeconomic status, and essential to their growth and development. It helps children learn to cooperate, explore physical challenges, and negotiate with other children and adults; these places must be safe. Exposure to green spaces, especially at school, may promote healthy psychological development in children. Safe play cannot exist in unsafe places, and play areas that are infrequently used by families or children can quickly become magnets for litter, vandalism, and crime.

Food and the Built Environment

The food environment varies depending on location. "Food deserts" are areas with limited access to fresh food and supermarkets and usually characterized by calorie-dense, low-quality food options. Recent studies have shown unequal distribution of food resources: low-income, urban neighborhoods and rural areas have limited access to supermarkets, fresh food, and produce. Approximately 23.5 million Americans and 6.5 million children live in communities that do not have access to supermarkets. Poor-quality food environments are associated with risk factors for obesity, such as low income, absence of transport, and low education.[48] A 2009 literature review examined the relationship between food environment and obesity risk in vulnerable populations, mainly African American, Hispanic, and populations of low socioeconomic status.[49] These disadvantaged groups were found to live in environments that had little access to food stores and places to exercise, accounting for a higher risk of overweight and obesity. The greater access to fast food outlets in low-income areas and less healthy eating options may partially explain the higher obesity risk in low-income and minority populations. Foods purchased from fast food outlets and restaurants, rather than prepared at home, account for roughly 7.6% of energy intake.[50] Schools are important environments that can shape and influence good health and the eating habits of children. A New Zealand study found that less healthy food represented the majority of food sales in elementary schools and suggested that the school food environment did not provide opportunities for healthy food choices.[51]

The Need to Protect Children Through Policy

Policies to improve health in children include improved school and child care siting, "traffic calming," and enforcement. Safe Routes to School and "Complete Streets" policies promote walking and biking to school by creating safe, walkable routes to schools and promoting community involvement.[52,53] One important intervention is the "walking school bus," where students can walk side by side among fellow classmates, teachers, and parents. The US Environmental Protection Agency and affiliated government agencies have issued guidelines for the siting of school facilities to address the needs and vulnerabilities of all children.[54] The School Siting Guidelines encourage and inform consideration of environmental and public health factors in local school siting processes and many other issues, including integrated pest management strategies to control pests by removing sources of food, preventing entry into building, and minimizing use of pesticides.

The built environment is as an integral component of children's health.[55] There are increasing numbers of articles on PubMed related to children's health and the built environment since 2000 (see Figure 18.1); there were only 5 citations in 2000 but 775 in 2020.[56]

There are more than 10 joint public health and urban planning programs in the United States and a much larger number where crossover is encouraged. However, most of these are without apparent joined syllabi or co-recognized credits, which, sadly, is an effective

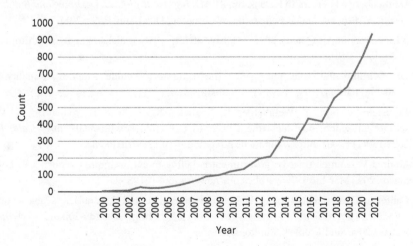

Figure 18.1. Scientific articles on the built environment and health, 2000–2021

way to discourage students from participating in shared programs. The links between urban planning, architecture, transportation design, and multiple other disciplines offer energizing and effective opportunities for those seeking to advance in their field. Every child needs to live in healthy, safe places.

References

1. Cummins SK, Jackson RJ. The built environment and children's health. *Pediatr Clin North Am.* 2001;48:1241–52.

2. Jackson RJ, Tester J. Environment shapes health, including children's mental health. *J Am Acad Child Adolesc Psychiatry.* 2008;47:129–31.

3. Jackson RJ, Sinclair S. *Designing Healthy Communities.* San Francisco: Josey Bass: 2011.

4. CDC. Motor Vehicle Prioritizing Interventions and Cost Calculator for States (MV PICCS). 2022. Available at: https://www.cdc.gov/transportationsafety/calculator/index.html.

5. NIOSH. Prevention through Design.2013. Available at: https://www.cdc.gov/niosh/topics/ptd/default.html.

6. Jacobs DE, Kelly T, Sobolewski, J. Linking public health, housing and indoor environmental policy: successes and challenges at local and federal agencies in the United States. *Environ Health Perspect.* 2007;115:976–82.

7. Vitruvius. The Ten Books on Architecture. 1960. Available at: https://www.amazon.com/Vitruvius-Ten-Books-Architecture-Bks/dp/0486206459.

8. National Center for Healthy Housing. National healthy housing standard. Available at: https://nchh.org/tools-and-data/housing-code-tools/national-healthy-housing-standard/.

9. Evans GW, Lepore SJ, Shejwal BR, Palsane MN. Chronic residential crowding and children's well-being. *Child Dev.* 1998;69:1514–23.

10. Diaz A, O'Reggio R, Norman M, Diimick J, Ibrahim A. Association of historic housing policy, modern day neighborhood deprivation, and outcomes after inpatient hospitalization. *Ann Surgery.* 2021 Dec;274(6):985–91.

11. Evans GW, Saegert S, Harris R. Residential density and psychological health among children in low-income families. *Environ Behav.* 2001;33:165–80.

12. Dannenberg AL, Frumkin H, Jackson, RJ. *Making Healthy Places: Designing and Building for Health, Well-Being, and Sustainability.* Washington, DC: Island Press; 2011.

13. NIEHS. Indoor Air Quality. 2023. Available at: https://www.niehs.nih.gov/health/topics/agents/indoor-air/index.cfm.

14. US Environmental Protection Agency. The inside story: a guide to indoor air quality. 2023. Available at: http://www.epa.gov/iaq/pubs/insidest.

15. Bernstein A, Sun, S, Weinberger K, Spangler K, Sheffield P, Wellenius, G. Warm season and emergency department visits to U.S. children's hospitals. *Env Health Persp.* 2022;130(1):17001. https://ehp.niehs.nih.gov/doi/10.1289/EHP8083.

16. Lindrea V. Grenfell tragedy: government is failing to act on inquiry report, says London mayor. 2022 Mar. https://www.bbc.com/news/uk-60816108.

17. California Department of Industrial Relations. Stair Rails and Handrails. Available at: https://www.dir.ca.gov/title8/3214.html#:~:text=Stair%20railings%20and%20handrails%20shall,inches%20beyond%20the%20bottom%20riser.

18. Howell N, Booth G. The weight of place: built environment correlates of obesity and diabetes. *Endocrine Rev.* 2022;xx:1–18.

19. Sallis JF, Floyd FF, Rodriguez DA, Saelens BE. Role of built environments in physical activity, obesity, and cardiovascular disease. *Circulation.* 2012:125:729–37.

20. Staunton CE, Hubsmith D, Kallins W. Promoting safe walking to school: the Marin County success story. *Am J of Public Health.* 2003;93:1431–4.

21. Wolch J, Jerrett M, Reynolds K, et al. Childhood obesity and proximity to urban parks and recreational resources: a longitudinal cohort study. *Health Place.* 2011 Jan;17(1):207–14.

22. Gordon-Larsen P, Nelson MC, Page P, Popkin B. Inequality in the built environment underlies key health disparities in physical activity and obesity. *Pediatrics.* 2006;117:417–24.

23. Stanford Children's Hospital. Christopher G. Dawes: How to build the best children's hospital. 2023. Available at: https://www.stanfordchildrens.org/en/about/news/thought-leaders/how-to-build-the-best-childrens-hospital.

24. Health Care Without Harm. 5 years to carbon neutral: Seattle Children's shows the way. 2020. Available at: https://noharm-uscanada.org/articles/news/us-canada/5-years-carbon-neutral-seattle-children%E2%80%99s-shows-way

25. Levine J, Grengs J, Merlin J. *From Mobility to Accessibility: Transforming Urban Transportation and Land-Use Planning.* Ithaca, NY: Cornell University Press; 2019.

26. U.S. Department of Transportation. The U.S. DoT Secretary's priorities to improve our transportation system. 2023. Available at: https://www.transportation.gov/priorities.

27. Bureau of Labor Statistics. Housing tenure and type of area: Average annual expenditures and characteristics, Consumer Expenditure Survey, 2011. https://www.bls.gov/cex/tables.htm. Consumer expenditure survey: quintiles of income before taxes: annual expenditure means, shares, standard errors, and coefficient of variation, Consumer Calculated from table 1101. Available at: http://www.bls.gov/cex/2011/Standard/tenure.pdf.

28. Litman T. *New Mobilities: Smart Planning for Emerging Transportation Technologies.* Washington, DC: Island Press; 2021. https://islandpress.org/books/new-mobilities.

29. Urban design is key to healthy environments for all. *Lancet Global Health.* Available at: https://www.thelancet.com/journals/langlo/article/PIIS2214-109X(22)00202-9/fulltext.

30. Centers for Disease Control and Prevention. Ten great public health achievements—United. States, 2001–2010. *Morb Mortal Wkly Rep.* 2011;60(19):619–23.

31. U.S. Department of Transportation. NHTSA Releases 2020 Traffic Crash Data. 2022. Available at: https://www.nhtsa.gov/press-releases/2020-traffic-crash-data-fatalities.

32. Gauderman WJ, Urman R, Avol E; et al. Association of improved air quality with lung development in children. *N Engl J Med.* 2015;372:905–13.

33. Friedman MS, Powell KE, Hutwagner L, Graham LM, Teague WG. Impact of changes in transportation and commuting behaviors during the 1996 Summer Olympic Games in Atlanta on air quality and childhood asthma. *JAMA.* 2001;285:897–905.

34. Ewing R, Pendall R, Chen D. Measuring sprawl and its impact. Available at: http://law.wustl.edu/landuselaw/Articles/measuringsprawl.pdf.

35. Jackson RJ, Tester J. Environment shapes health, including children's mental health. *J Am Acad Child Adolesc Psychiatry.* 2008;47:129–31.

36. Davoudi S, Crawford J, Mehmood A. *Planning for Climate Change: Strategies for Mitigation and Adaptation for Spatial Planners.* London: Earthscan; 2009.

37. US Environmental Protection Agency. Travel and environmental implications of school siting. Available at: http://www.epa.gov/smartgrowth/pdf/school_travel.pdf.

38. McDonald N. Active transportation to school: trends among U.S. schoolchildren, 1969–2001. *Am J Prev Med*. 2007;32:509–16.

39. Centers for Disease Control and Prevention. Barriers walking to or from school: United States. *Morb Mortal Wkly Rep*. 2004;54:949–52.

40. Centers for Disease Control and Prevention. Barriers walking to or from school: United States. *Morb Mortal Wkly Rep*. 2004;54:949–52.

41. Frumkin H, Frank L, Jackson R. *Urban Sprawl and Public Health: Designing, Planning, and Building for Healthy Communities*. Washington, DC: Island Press; 2004.

42. Shephard RJ. Curricular physical activity and academic performance. *Pediatr Exerc Sci*. 1997;9:113–26.

43. California Department of Education. Physical education model content standards for California public schools. 2005 Jan. Available at: http://www.cde.ca.gov/be/st/ss/docume nts/pestandards.pdf.

44. California Department of Education. Physical fitness test results for 2011 prompt schools: Chief Tom Torlakson and NBA All-Star Bill Walton to team up for healthy kids campaign. Available at: http://www.cde.ca.gov/nr/ne/yr11/yr11rel95.asp.

45. National Association for Sport & Physical Education. New study proves physically fit kids perform better academically. 2002 Dec. Available at: http://fairfieldschools.org/B_Level/B_RogerLudlowe/C_RogerLudlowe/05/Fitness%20Academic%20study1.htm.

46. Ewing R, Schmid T, Killingsworth R, Zlot A, Raudenbush S. Relationship between urban sprawl and physical activity, obesity, and morbidity. *Am J Health Promotion*. 2003;18:47–57.

47. CDC. Childhood obesity and overweight. 2022. Available at: https://www.cdc.gov/obesity/childhood/

48. Cummins S, Macintyre S. Food environments and obesity: neighborhood or nation? *Int J Epidemiology*. 2006;35:100–4.

49. Lovasi GS, Hutson MA, Guerra M, Neckerman KM. Built environments and obesity in disadvantaged populations. *Epidemiology Rev*. 2009;31:7–20.

50. Lake A, Townshend T. Obesogenic environments; exploring the build and food environments. *J R Soc Health*. 2006:126:262–7.

51. Carter MA, Swinburn B. Measuring the "obesogenic" food environment in New Zealand primary schools. *Health Promotion Intl*. 2004;19:15–20.

52. Staunton CE, Hubsmith D, Kallins W. Promoting safe walking and biking to school: the Marin County success story. *Am J Public Health*. 2003:93:1431–4.

53. California Department of Transportation. Complete streets program. Available at: https://dot.ca.gov/news-releases/news-release-2021-039.

54. U.S. EPA. School Siting Guidelines. 2011. Available at: https://www.epa.gov/sites/default/files/2015-06/documents/school_siting_guidelines-2.pdf

55. Bole A, Bernstein A, White MJ; American Academy of Pediatrics Council on Environmental Health and Climate Change; Section on Minority Health, Equity and Inclusion. The Built Environment and Pediatric Health. Pediatrics. 2024;153(1):e2023064773. doi: 10.1542/peds.2023-064773.

56. National Library of Medicine. Built Environment and Health. 2022. Available at: https://pubmed.ncbi.nlm.nih.gov/?term=%22built%20Environment%22%20Health&page=524&timeline=expanded.

19

The Play Environment

James R. Roberts, James T. McElligott, and Kristina K. Gustafson

Healthy children are active children. In a 2018 clinical guideline, the American Academy of Pediatrics (AAP) recommended that children meet physical activity levels as set forth by the 2018 US Department of Health and Human Services Physical Activity Guidelines for Americans (second edition). For 3- to 5-year-old children, 3 hours per day of light, moderate, and vigorous activity is recommended. For 6- to 17-year-olds, 60 minutes or more per day of moderate or vigorous aerobic activity is recommended, with at least 3 of those hours being vigorous activity. In addition, muscle- and bone-strengthening activities are recommended at least 3 days a week.[1,2] Unfortunately, adolescents, not only in the United States but around the world, fail to meet these activity levels. Among 11- to 17-year-olds, 72–94% have insufficient physical activity.[3]

A sedentary lifestyle can lead to overweight and eventually obese children and the myriad of health effects that accompany obesity, including heart disease, hypertension, asthma, and type II diabetes mellitus. Limited outdoor activity and sun exposure also is associated with vitamin D deficiency. Increasing moderate/vigorous physical activity can improve measures of metabolic function, weight, and insulin resistance and also reduce waist circumference.[4–6] Unhealthy lifestyles in childhood often translate into worsening health in adulthood.

Outdoor play in natural environments, such as parks, hiking trails, and wilderness areas, can have enormous positive influences on mental and physical health. Likewise, team sports that often occur in stadiums and ice rinks can promote a sense of belonging in the child and teenager. Children's activity levels and outdoor play have been extensively reviewed.[7–10] This chapter covers a variety of potential risks from these environments, most notably sports-related injuries, indoor and outdoor air pollution, carbon monoxide poisoning, and occasional wildlife hazards.

Parks and Green Spaces

Overview

Parks and other so-called green spaces are excellent places for children to enjoy outdoor physical activity. For the purpose of this chapter, "green spaces" is defined as any natural environment with a predominate number of trees, grass, hiking trails, and other diverse natural elements. The number of parks and green spaces purposefully established or preserved in neighborhood settings is widely variable, with an abundance of many such venues in some well-planned communities, but a minimal amount in some urban

James R. Roberts, James T. McElligott, and Kristina K. Gustafson, *The Play Environment* In: *Textbook of Children's Environmental Health*, Second Edition. Edited by: Ruth A. Etzel and Philip J. Landrigan, Oxford University Press. © Oxford University Press 2024. DOI: 10.1093/oso/9780197662526.003.0019

and suburban areas. During the development of the suburbs, tracts of land were rapidly developed, bringing more houses to areas outside the major metropolitan centers (see Chapter 18). Some cities and municipalities were better than others at building parks and saving green spaces, with newer communities often planning ahead to preserve green space and develop parks for outdoor play. A significant amount of disparity exists as well: minority and lower socioeconomic communities have fewer parks and recreational spaces.[11,12]

Scope and Nature of Problem

The most readily apparent health benefit of parks and green spaces, one that is true of almost any outdoor play environment, is basic physical activity. Active children are more likely to maintain a healthy weight and less likely to develop diseases often associated with a sedentary lifestyle. Unfortunately, there is a trend in the United States toward more indoor activity using electronic entertainment and media systems, at the expense of outdoor play activities.[10,13] Obesity, cardiovascular disease, hypertension, nonalcoholic fatty liver disease, and type 2 diabetes mellitus used to be considered diseases of adulthood. All are interrelated, are the origin of adult obesity and many other "adult disorders," and now commonly occur in children. Two other conditions, asthma and attention deficit hyperactivity disorder attention-deficit/hyperactivity disorder (ADHD), also are relevant to the discussion of play environments, although they are not necessarily directly related to obesity and its associated conditions.

Epidemiology

Rates for obesity and overweight in the United States are steadily increasing, with more than 18% of all US children and adolescents ages 2–19 years being obese[14] (see Chapter 41). Diet and physical activity both are likely to be observed behaviors that children adopt from their parents. Children are twice as likely to become obese if their parents are obese compared to children of nonobese parents.[14,15] Conversely, active kids and teenagers are more likely to be active adults, but most adolescents are not meeting minimum levels of physical activity.[3,16] Type 2 diabetes rates are rapidly rising, and the US Centers for Disease Control and Prevention (CDC) projects that unless the rate of increasing new cases is reversed, 1 in 3 children born in 2000 will develop diabetes. Children also are increasingly likely to have vitamin D deficiency. Studies vary widely and often report different measures of severity, but, in the United States, up to 40% of children have vitamin D levels that are considered deficient. Even more concerning is that approximately 70% of children and teenagers have vitamin D levels considered to be insufficient. Worldwide, the problem is even more significant.[17–19]

Approximately 5% of US children have ADHD, although in some studies and patient populations this percentage may be doubled (see Chapter 49). Self-reported stress is a major health concern of teenagers, with half acknowledging a rise in stress levels when compared to their previous school year.[7] Finally, asthma is the most common chronic condition of childhood (excluding dental caries), with more than 7 million US children being affected. Factors related to the play environment may affect asthma exacerbations. The child's play environment may positively or negatively impact the child's outcome for these conditions.

Although there are many benefits to outdoor play and physical activity, there are some potential adverse consequences as well. Approximately 200,000 US children are injured in playgrounds annually, with the majority (~70%) due to falls. Head injuries are the most likely to be fatal; however, with newer playground surfaces, the most severe injuries are less common.[6] Children also may be exposed to arsenic from older playground material made from pressure-treated wood.[9] Excessive exposure to outdoor air pollutants may exacerbate a child's asthma.[20]

Systems Affected and Benefits of Play in Parks and Natural Environments

Physical Activity

Time spent outdoors naturally correlates with an increase in physical activity. When researchers attempt to calculate physical activity, they estimate that for every hour spent outside, physical activity increased by 27 minutes per week.[9] Parks with playgrounds may facilitate this increase in activity. The proximity of residence appears to predict park use and physical activity. In one study of Los Angeles children from predominately Black or Hispanic neighborhoods with high levels of poverty, 86% of area residents visited their nearby park at least monthly and 35% didn't go to any other parks. Children who lived within 1 kilometer of a park with a playground were five times more likely to have a healthy weight than children living farther away from a park.[21] In the New York public school system, a study examined the fitness scores of adolescents and their association with area parks and other outdoor exercise environments. Medium, but not high density of recreational resources in the area surrounding a school was associated with greater annual improvements in fitness for both boys and girls.[22]

The natural environment and physical activity may have a synergistic effect. In one study, participants ran on a treadmill while being shown different photographic scenes classified as being rural pleasant, urban pleasant, rural unpleasant, or urban unpleasant. The subjects exhibited a significant reduction in blood pressure and a more positive effect on mood than exercise alone when they ran in front of the rural and urban pleasant nature pictures.[23] The largest reduction in blood pressure occurred in participants in the rural pleasant group.[7,23]

Although there are many benefits to outdoor play and physical activity, injuries at playgrounds, usually due to falls, are a potential risk at any outdoor venue. Intracranial and upper limb injuries are the most common but are becoming less frequent as communities have changed the surface from concrete or hard-packed dirt to other materials that provide more cushion and lessen the chance of a severe injury. Various components, such as asphalt, sand, wood chips, shredded rubber, and other materials have been used, producing varying degrees of protection from injuries due to falls.[24]

Mental Health

Studies among adults suggest that the outdoor natural environment has some beneficial effects on mental well-being. A cohort of physically active Canadian adolescents was assessed for self-reported mental health, as either flourishing or not flourishing. This study supported the idea that spending time outdoors will increase levels of moderate to vigorous physical activity, which will in turn improve mental well-being.[25]

The outdoor and natural environment also may improve health in general and re-covery from surgery. One study compared recovery from surgery for patients in a control group who were placed in rooms that faced a brick wall with a group who had a room with a view of the outdoor environment. Those with a view of nature had a shorter hospital stay and less pain.[26] Another study found that the use of distraction therapy (nature sights and sounds), when used as an adjunct to conscious sedation for an endoscopic procedure, significantly improved pain control and anxiety levels.[7,27]

Outdoor exercise brings more positive effects on mental well-being than does exercising indoors. In a systematic review of 11 trials among adults comparing mental well-being after a short walk or run outdoors and indoors, 9 demonstrated improved mental well-being following outdoor exercise. Those who exercised outdoors reported increased energy and vitality, and decreased tension, confusion, anger, and depression. Participants also reported a greater intent to repeat outdoor activity.[28]

While most studies of mental well-being and natural environments have been performed among adults, one study evaluated stress levels in children based on the level of natural environment at their home. Lower levels of stress in children were correlated with increased amounts of exposure to natural environments.[29]

Effects on ADHD/Inattentiveness

Anecdotal reports suggest that people's ability to concentrate is better when they are in a natural environment. Two early studies in children appear to confirm this observation. One compares children's symptoms when engaged in play in either a windowless room against various natural outdoor settings including a park, farm, or other outdoor neighborhood public space. Activities in natural settings were helpful in reducing inattentive symptoms, and increased tree cover was also associated with fewer inattentive symptoms.[30] Another study measured children's ability to recharge their mental capacity following induced mental fatigue. Following walks in a city park, an urban area, or a residential area, children were asked to complete tests of concentration and impulse control. Children demonstrated superior concentration and impulse control following a walk in a city park compared to the other environments.[31]

A systematic review found evidence to support this hypothesis of improved mental health and ADHD symptoms while noting that more rigorous studies were needed to confirm the association. Approximately half of the studies assessed had a positive effect and the other half, no effect. Studies of emotional well-being, ADHD, self-esteem, depression, and resiliency were all included. When just studies of ADHD of good quality were assessed, children had improved ADHD both when natural environments were readily accessible and when children were actively playing in the natural environment.[32]

Asthma

Asthma is linked to numerous environmental triggers including allergies to numerous plant species and outdoor air pollutants such as ozone, nitrogen dioxide (NO_2), and particulate matter as well as indoor triggers (dust mite, cockroach, mold, animal dander). It appears that children in urban areas tend to spend a large amount of time indoors, possibly increasing their exposure to indoor environmental triggers. Increased access to green spaces would serve to reduce these exposures. One study found that high tree density was correlated with a low prevalence of asthma.[33]

In general, physical activity helps children with asthma with quality of life, morbidity, and aerobic fitness, without compromising lung function.[34-36] However, if activity is taking place in an environment with very cold, dry air and/or increased particulate matter, increased bronchospasm still may occur.[37,38]

Treatment and Prevention

Healthcare professionals should consider recommending or "prescribing" outdoor play in the natural environment for their patients. The AAP recommends that children ages 6–17 have an accumulated 60 minutes of moderate to vigorous physical activity every day, including bone and muscle strengthening activities 3 days per week.[1] For children ages 3–5, there should be 3 hours a day of physical activity. Some of this activity should include a minimum of 30 minutes of unstructured/free play in an outdoor environment.[2,8] Individuals who lived in communities that have parks within a 300-meter radius of their homes were more likely to engage in walking and bicycling activities than were those without community parks.[39]

Clinicians and parents must remain aware of the potential risks of outdoor play, such as outdoor air pollution for asthma patients, and consider modifying or limiting outdoor play under poor air quality conditions or in extreme cold or hot weather.

Community associations and neighborhoods should promote safe parks and other play areas for children. Attention should be paid to the type and depth of the surfaces present in playground areas to prevent injuries. Wood chips/tanbark should be at least 8 centimeters in depth, and rubber-based surfaces of at least 7.5 centimeters yield even better protection from injury due to falls.[24] Although wood treated with arsenic was discontinued in the early 2000s, wooden structures that are older than that should have a sealant applied every 6 months to prevent the arsenic from leaching into the soil. Children should be reminded to wash their hands after playing outside and before meals.[9] Finally, although sun exposure may increase vitamin D levels, sunscreen should be applied to avoid burns and reduce the lifetime risk for skin cancer.

Wilderness

Wilderness play environments may be defined in many different ways. For the purposes of discussion, most wilderness areas consist of mountains, forests, and oceans. As such, a person is not in a designated play area but instead in a wild habitat and must accept the risks associated with that. The 2000–2002 National Survey of Recreation and the Environment revealed that an estimated 72 million Americans spent some time hiking outdoors.[40] Activities in wilderness areas carry some of the same potential physical and mental health benefits as activities in parks and other green spaces. Unique risks from wildlife encounters include bites from venomous snakes, disease-carrying ticks, and other animals; poisonous plants; and the potential for injury in an environment far from medical services.

Systems Affected

Injuries and Illness

Information regarding children's injuries and illness in a wilderness environment is limited. In one study of injuries in hikers, most participants were adults, with a small number of children. Most injuries were minor, including abrasions, blisters, sunburn, and muscle strains. Dehydration, diarrhea, and respiratory illnesses are other considerations, particularly if the hiker is generally not active and attempts a long, vigorous hike without adequate provisions.[40,41] Inexperience in a wilderness area increases the likelihood of a potentially serious injury or illness. Of those affected, most did not carry a first aid kit, have at least 1 liter of water with them, or wear appropriate footwear.[41]

Temperature Effects: Cold and Heat Injuries

Exposure to remote environments also may put children at risk for heat- and cold-related illness. Heat-related illness is discussed later in this chapter.

Wildlife Risks

A comprehensive review of wildlife risks is beyond the scope of this chapter. However, notable risks include venomous snakes and insect-related illnesses. In the United States, there are four main varieties of venomous snakes, three of which (copperhead, cottonmouth, and rattlesnake) are in the Crotalidae family (pit vipers), and the coral snake. Most bites from these snakes occur when they are accidently stepped on or touched when hiking or rock climbing. Venom from pit vipers causes significant pain, swelling, and anticoagulation at the area surrounding the bite, and venom from the coral snake is neurotoxic.

Various insects including ticks, wasps, bees, and mosquitoes are especially abundant in rural and wilderness areas. Tick bites are common among backpackers who spend several days or more on the trail.[40] Ticks may carry Lyme disease, Rocky Mountain spotted fever, and other diseases. (See chapter 21.) While most mosquitoes in the United States are more of a nuisance, some may carry viral illnesses, the most common being West Nile encephalitis. Some individuals may have a predisposition to allergic reactions to wasps or bees; otherwise, the stings generally cause local irritation.[40]

Prevention

When hiking in a wilderness setting, carry sufficient water and wear appropriate footwear. For long-term hiking and camping, bring sufficient water or have the means to properly disinfect the water in the field either by boiling or using chlorine purification tablets.[40] Hikers should be careful to watch where they step and attempt to avoid placing hands and feet into crevices where snakes might be hiding.[40] Wear long sleeves and pants for best control of insect and tick bites. Consider using an insect and tick repellent such as DEET up to 30% or a lower concentration if outdoor duration is less than an hour or two, or picaridin 10% according to product labels. Repellent should be washed off daily.[9]

Stadiums, Artificial Turf, Ice Rinks, and Skate Parks

As with other play environments, stadiums, ice rinks, and skate parks allow an area for children to be physically active and enjoy spectator sports. These venues can be either open to the outdoors, including temporary ice rinks, or completely enclosed to serve as an indoor environment that protects patrons from natural environmental factors and occurrences. Over the past 30 years, there has been an explosion of skate park development in the United States. Although there are many benefits to these play environments, injuries, heat and cold effects, and carbon monoxide poisoning are among potential hazards.

Stadiums

Scope and Nature of the Problem/Sources of Exposure

As with other play environments discussed in this chapter, outdoor air pollutants including ozone and particulate matter can adversely affect the health of children. Additional health considerations related to outdoor stadiums include noise pollution, sunlight exposure, natural weather occurrences, and sports-related injuries. Heat-related illness is a major concern, particular during the summer months at any mass gathering.[42] Injuries will be discussed in more detail throughout the chapter. The reader is also referred to Chapter 54.

Overview of Toxicity/Systems Affected

Heat-related illness in stadiums involves participants who are exerting themselves, particularly in the summer months. Children and youth at risk include athletes in competitions that have little or no break in activity, such as long-distance runners or children who are wearing heavy layers of protective gear (i.e., American football). Even marching band members who are wearing non-breathable safety equipment and uniforms and young children in crowds who may not be able to communicate that they are getting overheated may be adversely affected. The primary affected systems include the integumentary, renal, gastrointestinal, and neurologic. Physical injuries may be quite variable but primarily affect bones/joints in the extremities, head, and spine.

Clinical Effects

Heat-related illnesses include heat cramps, heat exhaustion, and heat stroke. Heat cramps are painful, involuntary large muscle spasms, particularly of the calf and hamstring muscles, caused by inadequate circulation. Heat exhaustion, the precursor to heat stroke, results from non-replenished fluid losses from perspiration that result in decreased extracellular fluid and blood volumes. Blood flow is then redirected to the muscles and skin at the expense of blood flow to the brain. Signs and symptoms include headache, nausea, vomiting, dizziness, fatigue, weakness, tachycardia, elevated core body temperature, and flushed moist skin. Heat stroke, which is life-threatening, is a result of continued severe volume depletion with failure of the thermoregulatory system. The hallmark finding is a loss of perspiration; therefore, the skin becomes pale and dry. Additional findings include a core body temperature of higher than 105°F (40.5°C), weak pulses, and further mental status alteration, potentially progressing to unconsciousness.

Players are susceptible to physical injuries, particularly those to the extremities. In particular, ankle, knee, and shoulder injuries tend to predominate with sports such as football, soccer, and baseball. Head and neck injuries, including concussion, are also of potential concern, particularly with football and ice hockey.

Diagnosis, Treatment, and Prevention

Early identification of heat-related illness is paramount to avoid the deadly progression to heat stroke. Rest breaks with fluids (water, electrolyte solutions) should be frequent, and players should be monitored by adult staff. Immediate treatment for minor injuries to the extremities should include removal from activity, ice, elevation, and compression. Referral to specialty care should be considered.

Artificial Turf

Overview

Significant improvements have been made to artificial turf since its invention in the 1960s. The latest generations of turf have significantly improved on stiffness, heat retention, and friction, thus reducing the risk of injury to users. Third-generation turf, introduced in the 1990s, has longer fibers made of polyethylene, newer versions of infill, and increased distance between fibers that leads to an overall softer surface more closely resembling natural grass. The newer turf has gained wide popularity and is replacing older versions, though first- and second-generation turf still may be in use at older sites.

From a health perspective the concerns regarding artificial turf relate to exposure, infection risk, sports injuries, and potential for heat-related illnesses. Potential health benefits include reduced airway allergies and an increased access to open space and athletic facilities.

Scope of the Exposure

The use of third-generation turf has grown significantly from a handful of fields a decade ago to more than 6,000 fields in North America by 2011.[43] In North America, the total value of installed synthetic turf systems is estimated to be US$2.7 billion in 2020. This activity translates into approximately 265 million square feet of installed turf and 436 million pounds of infill.[44] Considering their widespread use, the literature on the health impacts of these fields is limited.

Benefits of Play on Artificial Turf

The most obvious benefit of artificial turf is increased access to open space in urban environments. The field's potential for low maintenance and durability offer sustainable access to space. Additionally, environmental concerns regarding grass fields, such as pesticide use and allergen exposure, are avoided with artificial turf.

Sports Injury

Field maintenance, turf type, sport played, gender of participants, and location of injury influence the rate of injuries seen with artificial turf.[43] Athletes and trainers both should be aware of the differing injury patterns associated with different playing field types in order to inform prevention strategies.[45] Lower-extremity injuries, particularly those related to twisting and shearing injuries, may be more common with artificial turf.[43] Conversely, a reduction in concussions was seen in one study with the use of artificial turf for certain sports such American football and rugby, but not for soccer.[46] First-generation turf was consistently associated with increased injury rates, mostly due to increased abrasions and lower-extremity sprains. Second-generation turf was found to be associated with increased injury rates with the exception of one study that found a decrease in knee injuries when compared to grass surfaces.[47]

Heat Exposure

Artificial turf fields can magnify the problem of heat exposure because of efficient radiant heat absorption from sunlight and an inability to vaporize water. Artificial turf can become up to 60°F (15.5°C) hotter than grass, with surface temperatures found to reach 160°F (71.1°C). Such increased heat has led to concerns regarding the increased potential for heat exhaustion and direct heat injury. Additionally there is evidence from subjective player surveys and measures of physiologic response for an increase in the effort required to perform athletic activities while on artificial turf.[48] The turf also creates heat islands contributing to increased temperatures in urban environments that could potentially lead to secondary health impacts.[48]

Toxicant Exposure

The infill of third-generation turf is often derived from used tires, which has raised concerns regarding exposure to toxic chemicals contained in the vulcanized rubber. Compounds of concern include metals and volatile and semi-volatile organic compounds. In particular, polyaromatic hydrocarbons have been found in high levels within the rubber pellets. Older artificial turf fields (first- and second-generation) also may shed small amounts of lead as they degrade over time. A study by the US Environmental Protection Agency (EPA) demonstrated a low level of lead risk, although the number of sites tested was small. The bioavailability of these compounds has been reported as low and below the levels currently considered harmful.[43] Concern persists because there is much variability in the chemical concentrations within infill preparations, toxic levels are not known for all of the compounds, and little is known about low-level, long-term exposure for young children.[49] An EPA report on the use of crumb rubber in artificial turf, summarizing the available evidence and highlighting research gaps, is available.[50]

Infection Risk

Skin infection with methicillin-resistant *Staphylococcus aureus* has been associated with abrasions suffered on artificial turf.[51] Reducing the risk of abrasions associated with the turf, if possible, is warranted given the known associations with skin infection risk and athletics.[43]

Prevention

Advances in the design of artificial turf have increased the safety of these surfaces for athletes, although knowledge of surface-specific injury risk and regular field maintenance are important considerations to mitigate risk. Older generations of turf are more likely to increase the risk of injury, and replacement is recommended when possible.

Newer generations of turf are considered to have a low risk of toxicant exposure. As research in this arena is ongoing stakeholders should monitor reporting from national and sports association guidelines. Prudent emphasis on field maintenance and appropriate ventilation for indoor surfaces is warranted.

Ice Rinks

Epidemiology

Ice rinks are the site of competitive sports and events as well as recreational activities. Children engage in figure or recreational skating and ice hockey on both indoor and outdoor ice rinks. The AAP estimates that about 400,000 children in the United States and Canada play ice hockey on these rinks.[52]

Scope and Nature of the Problem/Sources of Exposure

To increase physical activity, temporary or seasonal outdoor ice rinks have recently increased in number and popularity. This has resulted in participants of all ages and skill levels interacting in close proximity, with potential deleterious results. The majority of ice-skating injuries at temporary rinks affected the upper limbs, often due to a fall onto the outstretched hand.[53,54] Injuries requiring medical attention most frequently involved a fracture, with the majority of these located at the distal radius.[53,54] Ice skaters sustained injury to the head 20% of the time, with the majority being a head laceration.[55] Due to the low friction of ice, ice skaters may fall differently than roller skaters, including backward or sideways, thus increasing the risk for head injury.[55,56]

The hard surface of ice rinks, with their unyielding walls or "boards" and acrylic windows or Plexiglas, coupled with the types of activities occurring in this venue lead to the potential for sports-related injury. Since the 1960s, much attention has been focused on ice hockey due to the number of activity-related injuries. Mandatory use of helmets with full face masks in ice hockey and prohibiting head-first contact decreased facial and cervical spine injuries in the 1960s and 1970s.[52] These safety interventions employing protective equipment have almost eliminated dental, eye, and facial injuries in youth hockey.[57] However, head, neck, and spinal injuries have persisted.

Several studies reported a significant percentage of head, neck, and spinal injuries occurring from body checking in youth hockey.[52] A "body check" is physical contact where the defensive player's intentional tactic is to separate the opponent from the puck. Body checking is not allowed at any level of girls' or women's ice hockey, and thus all research has focused on boys' youth ice hockey. Body checking was attributed to increased injuries in younger players and more severe injuries, such as an increased proportion of fractures and concussions.[57,58] In addition, body checking was associated with a more aggressive style of play, with this aggressive play associated with more severe injury.[57]

Another health concern arises from the production of air pollutants including carbon monoxide from the exhaust of ice resurfacing equipment, especially in indoor ice rinks. All ice surfaces must be regularly edged and resurfaced to help smooth irregularities in the ice for injury prevention. Frequency of ice resurfacing is a significant determinant of air pollutant levels.[59] Ice resurfacing is done multiple times a day while edging occurs once a day or every other day. Combustion byproducts of fossil fueled–powered edgers and resurfacing machines depend on the fuel and engine type.[60] Fossil fuel–powered ice edgers and resurfacers generate significant concentrations of carbon monoxide and NO_2.[59] Both ice resurfacers and edgers produce ultrafine particles/particulate matter, which are defined as those with an aerodynamic diameter of less than 100 nm and have adverse effects on the heart, vascular tissue, and lungs.[60] In the absence of adequate building ventilation or poor maintenance of the resurfacers, the indoor ice rink air quality may be severely compromised during operation, leading to serious health problems.[61]

Overview of Toxicology, Systems Affected, and Clinical Effects
Respiratory Disease Due to Indoor Air Pollution
Carbon monoxide and NO_2 are the two primary pollutants of severe toxicity, although ultrafine particulate matter and other exhaust components also can exacerbate asthma,

with a prevalence of asthma and exercise-induced bronchoconstriction in skating athletes documented at 20–50%.[61,62] (see Chapter 26). Inhaled carbon monoxide will bind to hemoglobin, decreasing oxygen delivery in the body. The brain, heart, and skeletal muscles are most sensitive to carbon monoxide. At low concentrations, vision problems may occur. As the carbon monoxide concentration increases, nausea, headaches, and confusion occur. Very high concentrations can lead to coma and death.[9,61]

NO_2 is an irritant to the respiratory tract mucous membrane. The effects can range from mild irritation to pulmonary edema. Noncardiogenic pulmonary edema, coined "Zamboni disease," was noted in an adolescent hockey player when the resurfacers were used prior to and halfway through a game and one of the two ventilation extractor fans wasn't working. Multiple players and a few spectators also reported mild respiratory irritation.[63] People with asthma are more like to react sooner and at a lower concentration of NO_2. NO_2 can cause decreased oxygen to the heart, resulting in ischemia and arrhythmia.[61] An immediate response to NO_2 may include coughing, fatigue, nausea, choking, headache, abdominal pain, and difficulty breathing.[64] This may be followed by a symptom-free period of 3–30 hours. Finally, onset of pulmonary edema with anxiety, mental confusion, lethargy, loss of consciousness, and possibly death. The effects of both pollutants can be confounded by the configuration of the ice rink, with a board barrier and Plexiglas causing a temperature inversion condition. Because NO_2 is heavier than other air components, it will sink toward the ice in the lower breathing zone of children on the ice, resulting in disproportionate effects.[61] Also, when exposed to the same levels of NO_2 as adults, children may receive larger doses due to their greater lung surface area to body weight ratios and increased minute volumes to weight ratios.[64]

Diagnosis and Treatment

The treatment for carbon monoxide poisoning is 100% oxygen, which will reduce the typical half-life of carbon monoxide from 4 hours to 1 hour. If hyperbaric chambers are available, they may be considered because hyperbaric oxygen lowers the half-life to 30 minutes. It is not known whether hyperbaric oxygen results in a better neurologic outcome than 100% oxygen.[9] Cardiorespiratory support should be provided to treat vaporized NO_2 exposure.[64]

Prevention

Due to the large number of upper limb injuries involving the distal radius when ice skating, protection of this area should be sought. Cadaveric biomechanical studies have demonstrated a significant reduction in forces transmitted through the forearm bones, including the radius, when wrist guards are worn.[65] The protectiveness of wrist guards has been supported by clinical studies in snowboarders and in-line skaters.[53,54] Based on the similarities in injury patterns between ice skating and skateboarding, protective gear should be worn to prevent serious injuries due to falls, and appropriate equipment should be available. In addition, mandatory helmet use laws and local ordinances to increase helmet use when ice skating, similar to those which exist for ice hockey, have been instituted in some states.[55]

Ice hockey players should continue to avoid head-first contact and should wear a properly fitting helmet with a full-face mask. Helmets and their fit and face masks should be routinely inspected prior to every game to ensure integrity of the equipment. The 2010 Ice Hockey Summit on Concussion called for a zero-tolerance policy with regard to any contact, either intentional or accidental, to the head, which was supported by the AAP in 2014.[57,66] Coach and player education have evolved to teach youth to keep their heads up, especially prior to them receiving a check, and to respect opponents by not checking from behind. This has coincided with a decrease in cervical spine injury incidence. The AAP recommends limiting body checking use and education to older ages at the elite level.[57] Other educational strategies such as ThinkFirst Canada's SMART HOCKEY initiative and USA Hockey's Heads Up Hockey Curriculum have been instituted to focus on concussion recognition and prevention.[67,68]

To decrease the exposure to carbon monoxide and NO_2, fossil fuel–powered ice resurfacing machines should be properly maintained by an experienced technician.[61] The rink doors should be opened during and following ice edging and resurfacing. Installation and continuous operation and maintenance of an effective mechanical ventilation system are imperative. Air quality standards should be established and enforced, with routine monitoring of the ambient conditions.[61] Consider a toxic chemical surveillance program and monitor for visual warning signs and symptoms of carbon monoxide and NO_2 poisoning. All rink staff, coaches, parents, and participants should receive education about the early symptoms of poisoning. Where possible, investment in electric-powered ice edgers and resurfacers should be undertaken.

Skate Parks

Epidemiology
In 1996, approximately 5.8 million US children and adolescents younger than 18 years had participated in skateboarding, with 13% doing so at least one time per week.[69] During this same time period, 17.7 million had participated in in-line skating, an increase of 24% from the previous year.[70] From 1993 to 2003, more roller skating and in-line skating–related injuries were treated in emergency departments each year as compared to ice skating–related injuries.[56] From 1990 through 2008, approximately 1.227 million children and adolescents ages 5–19 nationally were treated in emergency departments for skateboarding-related injuries.[71] According to the National Electronic Injury Surveillance System (NEISS), 27,600 pediatric scooter-related injuries presented to emergency departments in the first 11 months of 2000.[72] In the first 2 years (2015–2016) after the release of the hoverboard, there were an estimated 26,854 hoverboard injuries seen in emergency departments per the NEISS.[73]

Scope and Nature of the Problem/Sources of Exposure
Roller-skating became popular in the 1930–1950s, and again in the 1970s, and showed an uptick in popularity during the COVID pandemic. Skateboarding began in the 1950s, continually increasing in popularity, especially today, due to televised events like the "X Games." In-line skates first became part of the mainstream in the 1980s and 1990s, and have shown some popularity fluctuation due to other modalities hitting the market. As compared to roller skates, in-line skates have 3–5 low-friction wheels set in a single row, which results in greater maneuverability to allow skaters to perform more tricks and move at increased speed.

The global popularity of scooter use over the past two decades by people of all ages continues. The nonpowered scooter, introduced in 1999, is made of lightweight aluminum with a design similar to a skateboard but with handlebars and a foot brake; it is similar to in-line skates with the same small, low-friction wheels.[72] These scooters can travel at coasting speeds of 5–8 miles per hour or faster. Younger children have high centers of gravity, making the needed balancing skills more difficult. More recently, electric scooters and hoverboards, which are electric self-balancing two-wheeled boards, entered the market (in 2003 and 2015, respectively). Children's electric scooters are similar to nonpowered scooters except they have rechargeable lithium-ion batteries and can reach up to 15 miles per hour. The hoverboard is controlled by the rider's foot position and distributed body weight, with the amount of applied pressure controlling direction and speed.[74] Due to this, the hoverboard can have sudden accelerations forward, backward, or in rotation and reach speeds of 6–13 miles per hour (9.6 - 20.9 kilometers per hour), so the rider must have a good sense of balance and control.[75] Soon after its introduction, the safety of hoverboards was called into question due to fire hazard risks from the lithium-ion battery that resulted in rider injury, necessitating a US Consumer Product Safety Commission (CPSC) recall in 2016.[74]

Often children use in-line skates, skateboards, and scooters in a location of convenience: at and around their home.[71] Driveways along with neighborhood and city streets more often encompass the play environment where the majority of these children conduct this activity. This location is dangerous due to uneven road surfaces; road/driveway debris such as water, sand, gravel, or dirt; decreased visibility for those participating at night; and sharing of this play environment with motor vehicles. Younger children are at a higher risk of injury from the use of in-line skates, skateboards, and scooters because of their strength, skill level, and their ability to judge both pedestrian and vehicular traffic.[69,70] In addition to outdoor use, hoverboards are often used inside the house.[74] This indoor use can lead to potential hazards in the kitchen; falls associated with inadvertently going down stairs, over thresholds, or over different terrains/flooring; and collisions with furniture, walls, and door jambs.

Overview of Systems Affected and Clinical Effects

The most common mechanism of skate-related injury is a fall.[56] The majority of injuries among roller skaters and in-line skaters were to the upper extremity, especially the forearm, while skateboarders and children on scooters have had both upper and lower extremity injuries but have also had comparatively more head injuries.[55,56,72] These findings suggest that roller skaters and in-line skaters may be falling forward and attempting to stop the fall. Based on data from skaters seen in emergency departments, the most common types of injuries due to skating accidents are fractures/dislocations, sprains/strains, and contusions/abrasions.[56,71] Hoverboarders' fall patterns are likely a combination of all of the aforementioned directions due to the mechanism of action, with the most common body parts injured being the wrist, forearm, and head, while the ankles were the most commonly sprained body part.[73,75,76] In addition, phalangeal injuries were also seen.[75,76] Many of the more serious and fatal injuries, primarily head injuries, for skating-related activities have occurred in areas with motor vehicle traffic. Therefore, as with parks and green spaces, there may be some benefit to participating at a skate park as opposed to skating on the streets or sidewalks.[69,70]

Prevention

The AAP recommends that children younger than 10 years should not use skateboards without close supervision and those younger than 5 years should not use skateboards at all. They recommend that children younger than 8 years should not ride scooters without close adult supervision. In addition, skateboards should not be ridden in or near traffic, regardless of traffic volume, and scooters also should not be ridden in streets, in traffic, or at night. Holding on to the side or rear of a moving vehicle, known as "skitching a ride," should never be done.[52]

The AAP recommends the use of appropriately fitted safety gear for in-line skating, skateboarding, and scooter activities,[69,70] with extension also to hoverboard use.[73-76] This safety gear should include a bicycle helmet certified with the CPSC standard or a multisport helmet certified with the N-94 standard, wrist guards, elbow pads, knee pads, and possibly a mouth guard. In addition, wrist guards prevent sudden extreme hyper-extension and allow the skater to slide forward on the protective plates, which prevents palm injuries. Originally, scooters were marketed with photographs of children without a helmet or protective padding and sold without these.[72] In addition, helmet use require-ments through state legislation should be sought.[69]

Hoverboard use benefits from being limited to a controlled, outdoor environment, such as a park or rink, rather than on the street or inside the house.[74] Children on hov-erboards benefit from direct adult supervision.[76] The AAP recommends all beginners learn to skate at a skate rink or park until their skills are developed.[70] It encourages con-tinued community development and use of skate parks with controlled surface condi-tions that are separated from vehicular traffic. These parks are also preferred because they have direct supervision of participants, are more likely to monitor for safety, and have standardized guidelines for use of protective equipment.[69]

References

1. Lobelo F, Muth ND, Hanson S, Nemeth BA; Council on Sports Medicine and Fitness; Section on Obesity. Physical activity assessment and counseling in pediatric clinical settings. *Pediatrics*. 2020;145(3):e20193992. doi:10.1542/peds.2019-3992.

2. Department of Health and Human Services. Physical Activity Guidelines for Americans, 2nd Edition. 2018. Available at: https://health.gov/our-work/nutrition-physical-activity/physical-activity-guidelines/current-guidelines.

3. Guthold R, Stevens GA, Riley LM, Bull FC. Global trends in insufficient physical activity among adolescents: a pooled analysis of 298 population-based surveys with 1.6 million par-ticipants. *Lancet Child Adolesc Health*. 2020;4(1):23–35.

4. Ekelund U, Luan J, Sherar LB, Esliger DW, Griew P, Cooper A, International Children's Accelerometry Database C. Moderate to vigorous physical activity and sedentary time and cardiometabolic risk factors in children and adolescents. *JAMA*. 2012;307(7):704–12.

5. Belcher BR, Berrigan D, Papachristopoulou A, et al. Effects of interrupting children's sed-entary behaviors with activity on metabolic function: a randomized trial. *J Clin Endocrinol Metab*. 2015;100(10):3735–43.

6. Davis CL, Pollock NK, Waller JL, et al. Exercise dose and diabetes risk in overweight and obese children: a randomized controlled trial. *JAMA*. 2012;308(11):1103–12.

7. McCurdy LE, Winterbottom KE, Mehta SS, Roberts JR. Using nature and outdoor activity to improve children's health. *Curr Probl Pediatr Adolesc Health Care*. 2010;40(5):102–17.

8. American Academy of Pediatrics Council on Sports Medicine and Council on School Health. Active healthy living: prevention of childhood obesity through increased physical activity. *Pediatrics*. 2006(117):1834–42.

9. American Academy of Pediatrics Council on Environmental Health. In: Etzel RA, Balk SJ, eds. *Pediatric Environmental Health* (4th ed.). Itasca, IL: American Academy of Pediatrics; 2019.

10. Tremblay MS, Gray C, Babcock S, et al. Position statement on active outdoor play. *Int J Environ Res Public Health*. 2015;12(6):6475–505.

11. American Academy of Pediatrics Committee on Environmental Health; Tester JM. The built environment: designing communities to promote physical activity in children. *Pediatrics*. 2009;123(6):1591–8.

12. Gordon-Larsen P, Nelson MC, Page P, Popkin BM. Inequality in the built environment underlies key health disparities in physical activity and obesity. *Pediatrics*. 2006;117(2):417–24.

13. Larson L, Green G, Cordell H. Children's time outdoors: Results and implications of the national kids survey. *J Park Recreat Adm*. 2011;29(2):1–20.

14. Hales CM, Carroll MD, Fryar CD, Ogden CL. Prevalence of obesity among adults and youth: United States, 2015–2016. *NCHS Data Brief*. 2017 Oct; 288:1–8.

15. Whitaker RC, Wright JA, Pepe MS, Seidel KD, Dietz WH. Predicting obesity in young adulthood from childhood and parental obesity. *N Engl J Med*. 1997;337(13):869–73.

16. Centers for Disease Control and Prevention. Youth risk behavior surveillance—United States, 2005. *Morb Mortal Wkly Rep*. 2006;55:SS–5.

17. Kumar J, Muntner P, Kaskel FJ, Hailpern SM, Melamed ML. Prevalence and associations of 25-hydroxyvitamin D deficiency in US children: NHANES 2001–2004. *Pediatrics*. 2009;124(3):e362–70.

18. Gordon CM, DePeter KC, Feldman HA, Grace E, Emans SJ. Prevalence of vitamin D deficiency among healthy adolescents. *Arch Pediatr Adolesc Med*. 2004;158(6):531–7.

19. Taylor SN. Vitamin D in toddlers, preschool children, and adolescents. *Ann Nutr Metab*. 2020;76(Suppl 2):30–41.

20. Commodore S, Ferguson PL, Neelon B, et al. Reported neighborhood traffic and the odds of asthma/asthma-like symptoms: a cross-sectional analysis of a multi-racial cohort of children. *Int J Environ Res Public Health*. 2020;18(1).

21. Cohen DA, McKenzie TL, Sehgal A, Williamson S, Golinelli D, Lurie N. Contribution of public parks to physical activity. *Am J Public Health*. 2007;97(3):509–14.

22. Bezold CP, Stark JH, Rundle A, et al. Relationship between recreational resources in the school neighborhood and changes in fitness in New York City public school students. *J Urban Health*. 2017;94(1):20–9.

23. Pretty J, Peacock J, Sellens M, Griffin M. The mental and physical health outcomes of green exercise. *Int J Environ Health Res*. 2005;15(5):319–37.

24. Gunatilaka AH, Sherker S, Ozanne-Smith J. Comparative performance of playground surfacing materials including conditions of extreme non-compliance. *Inj Prev*. 2004;10(3):174–9.

25. Belanger M, Gallant F, Dore I, et al. Physical activity mediates the relationship between outdoor time and mental health. *Prev Med Rep*. 2019;16:101006.

26. Ulrich RS. View through a window may influence recovery from surgery. *Science*. 1984;224(4647):420–1.

27. Diette GB, Lechtzin N, Haponik E, Devrotes A, Rubin HR. Distraction therapy with nature sights and sounds reduces pain during flexible bronchoscopy: a complementary approach to routine analgesia. *Chest*. 2003;123(3):941–8.

28. Coon J. Does participating in physical activity in outdoor natural environments have a greater effect on physical and mental well-being than physical activity indoors? A systematic review. *Environ Sci Technol*. 2011(45):1761–72.

29. Evans G, Wells N. Nearby nature: a buffer of life stress among rural children. *Environ Behav.* 2003(35):311–30.

30. Taylor A, Kuo F, Sullivan W. Coping with ADD: The surprising connection to green play settings. *Environ Behav.* 2001(33):54–77.

31. Taylor AF, Kuo FE. Children with attention deficits concentrate better after walk in the park. *J Atten Disord.* 2009;12(5):402–9.

32. Tillmann S, Tobin D, Avison W, Gilliland J. Mental health benefits of interactions with nature in children and teenagers: a systematic review. *J Epidemiol Community Health.* 2018 Oct;10:958–66.

33. Lovasi GS, Quinn JW, Neckerman KM, Perzanowski MS, Rundle A. Children living in areas with more street trees have lower prevalence of asthma. *J Epidemiol Community Health.* 2008;62(7):647–9.

34. Ram FS, Robinson SM, Black PN, Picot J. Physical training for asthma. *Cochrane Database Syst Rev.* 2005(4):CD001116.

35. Flapper BC, Duiverman EJ, Gerritsen J, Postema K, van der Schans CP. Happiness to be gained in paediatric asthma care. *Eur Respir J.* 2008;32(6):1555–62.

36. Fanelli A, Cabral AL, Neder JA, Martins MA, Carvalho CR. Exercise training on disease control and quality of life in asthmatic children. *Med Sci Sports Exerc.* 2007;39(9):1474–80.

37. Stensrud T, Berntsen S, Carlsen KH. Exercise capacity and exercise-induced bronchoconstriction (EIB) in a cold environment. *Respir Med.* 2007;101(7):1529–36.

38. Anderson SD, Holzer K. Exercise-induced asthma: is it the right diagnosis in elite athletes? *J Allergy Clin Immunol.* 2000;106(3):419–28.

39. Wendel-Vos GC, Schuit AJ, de Niet R, Boshuizen HC, Saris WH, Kromhout D. Factors of the physical environment associated with walking and bicycling. *Med Sci Sports Exerc.* 2004;36(4):725–30.

40. Angert D, Schaff EA. Preventing injuries and illnesses in the wilderness. *Pediatr Clin North Am.* 2010;57(3):683–95.

41. Heggie TW, Heggie TM. Viewing lava safely: an epidemiology of hiker injury and illness in Hawaii Volcanoes National Park. *Wilderness Environ Med.* 2004;15(2):77–81.

42. Bernardo LM, Crane PA, Veenema TG. Treatment and prevention of pediatric heat-related illnesses at mass gatherings and special events. *Dimens Crit Care Nurs.* 2006;25(4):165–71.

43. Jastifer JR, McNitt AS, Mack CD, et al. Synthetic turf: history, design, maintenance, and athlete safety. *Sports Health.* 2019(1):84–90.

44. Synthetic Turf Council. Synthetic Turf Market Report: North America 2020. https://www.syntheticturfcouncil.org/page/marketreport#:~:text=%22In%20North%20America%2C%20the%20total,436%20million%20pounds%20of%20infill.%22

45. Winson DMG, Miller DLH, Winson IG. Foot injuries, playing surface and shoe design: Should we be thinking more about injury prevention. *Foot Ankle Surg.* 2020;26(6):597–600.

46. O' Leary F, Acampora N, Hand F, O' Donovan J. Association of artificial turf and concussion in competitive contact sports: a systematic review and meta-analysis. *BMJ Open Sport Exerc Med.* 2020 May;6(1):e000695.

47. Dragoo JL, Braun HJ. The effect of playing surface on injury rate: a review of the current literature. *Sports Med.* 2010;40(11):981–90.

48. Claudio L. Synthetic turf: health debate takes root. *Environ Health Perspect.* 2008;116(3):A116–22.

49. Celeiro M, Armada D, Dagnac T, de Boer J, Llompart M. Hazardous compounds in recreational and urban recycled surfaces made from crumb rubber. Compliance with current regulation and future perspectives. *Sci Total Environ.* 2021 Feb 10;755(Pt 1):142566.

50. United States Environmental Protection Agency. Federal Research Action Plan on Recycled Tire Crumb Used on Playing Fields and Playgrounds. 2016. Available at: https://www.epa.gov/chemical-research/federal-research-recycled-tire-crumb-used-playing-fields.

51. Wright JM, Webner D. Playing field issues in sports medicine. *Curr Sports Med Rep.* 2010;9(3):129–33.

52. American Academy of Pediatrics Committee on Sports Medicine and Fitness. Safety in youth ice hockey: the effects of body checking. *Pediatrics.* 2000;105(3 Pt 1):657–8.

53. Barr LV, Imam S, Crawford JR, Owen PJ. Skating on thin ice: a study of the injuries sustained at a temporary ice skating rink. *Int Orthop.* 2010;34(5):743–6.

54. Ostermann RC, Hofbauer M, Tiefenbock TM, et al. Injury severity in ice skating: an epidemiologic analysis using a standardised injury classification system. *Int Orthop.* 2015;39(1):119–24.

55. McGeehan J, Shields BJ, Smith GA. Children should wear helmets while ice-skating: a comparison of skating-related injuries. *Pediatrics.* 2004;114(1):124–8.

56. Knox CL, Comstock RD, McGeehan J, Smith GA. Differences in the risk associated with head injury for pediatric ice skaters, roller skaters, and in-line skaters. *Pediatrics.* 2006;118(2):549–54.

57. American Academy of Pediatrics Council on Sports Medicine Fitness; Brooks A, Loud KJ, Brenner JS, et al. Reducing injury risk from body checking in boys' youth ice hockey. *Pediatrics.* 2014;133(6):1151–7. doi:10.1542/peds.2014-0692.

58. Macpherson A, Rothman L, Howard A. Body-checking rules and childhood injuries in hockey. *Pediatrics.* 2006;117(2):e143–7.

59. Cox A, Sleeth D, Handy R, Alaves V. Characterization of CO and NO2 exposures of ice skating rink maintenance workers. *J Occup Environ Hyg.* 2019;16(2):101–8.

60. Kim J, Lee K. Characterization of decay and emission rates of ultrafine particles in indoor ice rink. *Indoor Air.* 2013;23(4):318–24.

61. Pelham TW, Holt LE, Moss MA. Exposure to carbon monoxide and nitrogen dioxide in enclosed ice arenas. *Occup Environ Med.* 2002;59(4):224–33.

62. Rundell KW, Smoliga JM, Bougault V. Exercise-Induced bronchoconstriction and the air we breathe. *Immunol Allergy Clin North Am.* 2018;38(2):183–204.

63. Morgan WK. "Zamboni disease." Pulmonary edema in an ice hockey player. *Arch Intern Med.* 1995;155(22):2479–80.

64. Centers for Disease Control and Prevention, Agency for Toxic Substances and Disease Registry. Medical management guidelines for nitrogen oxides. 2014. https://wwwn.cdc.gov/TSP/MMG/MMGDetails.aspx?mmgid=394&toxid=69

65. Staebler MP, Moore DC, Akelman E, Weiss AP, Fadale PD, Crisco JJ, 3rd. The effect of wrist guards on bone strain in the distal forearm. *Am J Sports Med.* 1999;27(4):500–6.

66. Smith AM, Stuart MJ, Greenwald RM, et al. Proceedings from the ice hockey summit on concussion: a call to action. *Clin J Sport Med.* 2011;21(4):281–7. Epub 2011/08/19. doi:10.1097/jsm.0b013e318225bc15. PubMed PMID: 21847809.

67. Canadian Medical Association. Position statement on bodychecking in youth ice hockey. 2013. Available at: https://policybase.cma.ca/link/policy10758.

68. Curriculum UHCE. Heads up, don't duck. 2022. Available at: https://www.usahockey.com/headsuphockey; https://cdn4.sportngin.com/attachments/document/0138/6193/HUH_Program_Guide_FINAL.pdf#_ga=2.243093568.399401467.1644266847-573333371.1644266844.

69. American Academy of Pediatrics Committee on Injury and Poison Prevention. Skateboard and scooter injuries. *Pediatrics*. 2002;109:542–3.

70. American Academy of Pediatrics Committee on Injury and Poison Prevention and Committee on Sports Medicine and Fitness. In-line skating injuries in children and adolescents. *Pediatrics*. 1998;101:720–2.

71. McKenzie LB, Fletcher E, Nelson NG, Roberts KJ, Klein EG. Epidemiology of skateboarding-related injuries sustained by children and adolescents 5-19 years of age and treated in US emergency departments: 1990 through 2008. *Inj Epidemiol*. 2016;3(1):10.

72. Levine DA, Platt SL, Foltin GL. Scooter injuries in children. *Pediatrics*. 2001;107(5):E64.

73. Bandzar S, Funsch DG, Hermansen R, Gupta S, Bandzar A. Pediatric hoverboard and skateboard injuries. *Pediatrics*. 2018;141(4):e20171253.

74. Siracuse BL, Ippolito JA, Gibson PD, Beebe KS. Hoverboards: a new cause of pediatric morbidity. *Injury*. 2017;48(6):1110–4.

75. Goldhaber NH, Goldin AN, Pennock AT, et al. Orthopedic injuries associated with hoverboard use in children: a multi-center analysis. *HSS J*. 2020;16(Suppl 2):221–5.

76. Hosseinzadeh P, Devries C, Saldana RE, et al. Hoverboard injuries in children and adolescents: results of a multicenter study. *J Pediatr Orthop B*. 2019;28(6):555–8.

20

Access to Nature and Child Health

Abby Nerlinger, Aparna Bole, and Pooja Sarin Tandon

Introduction

Exposure to nature and green environments provides many positive health and well-being benefits for children and families.[1] Contact with nature can have positive health effects throughout the life course that are sustained into adulthood. Although outdoor play has been long encouraged by the American Academy of Pediatrics,[2] there is increasing evidence that outdoor environments containing elements of nature may offer additional health benefits that come specifically from being near or engaging with the natural world. A growing body of evidence indicates that greenspaces and biophilic design can confer child health benefits, including increased physical activity, decreased obesity, and improved mental and behavioral health.[1,3] This chapter summarizes the evidence on the relationship between nature contact and child health, including the type of exposure assessed and specific health outcomes affected.

Childhood contact with nature, including greenspaces, must be considered in the context of additional environmental threats to child health and well-being. Policymakers and clinicians must be mindful of the disproportionate burden of toxic exposures borne by children. To ensure that children have access to *safe* greenspaces, policies and recommendations to increase nature contact must also aim to mitigate the potential adverse effects of exposure to environmental toxicants which may present a hazard in outdoor play.

Nature, Greenspaces, and Biophilic Design

Nature can be conceptualized broadly as "areas containing elements of living systems that include plants and nonhuman animals across a range of scales and degrees of human management."[4] The many forms of nature contact can be described based on temporal scales (varying from occasional to routine experiences), spatial scales (varying from small- to large-scale experiences), and sensory process utilized during contact (including visual vs. tactile experiences).[1] Greenspaces include parks, gardens, tree canopy, and wilderness. The term "biophilic design" is used to describe the inclusion of natural elements in the built environment that promote people's contact with nature. Biophilic design can include greenspaces such as parks, but can also include urban trees, organic elements in streetscapes, and ecological restoration.[5] Children have the potential to encounter nature in each context in which they live, learn, work, and play, including both school and residential settings. It is also important to differentiate nature contact from

Abby Nerlinger, Aparna Bole, and Pooja Sarin Tandon, *Access to Nature and Child Health* In: *Textbook of Children's Environmental Health*, Second Edition. Edited by: Ruth A. Etzel and Philip J. Landrigan, Oxford University Press. © Oxford University Press 2024. DOI: 10.1093/oso/9780197662526.003.0020

outdoor time, as stated by Fyfe-Johnson et al.[6]: "all outdoor spaces are not comparable: a parking lot is not a park, an urban playground without natural elements is not a garden."

Greenspaces and Environmental Justice

Disparities in access to greenspace are an example of environmental injustice[7] and pose barriers to achieving health equity for children. Many children facing limitations to nature contact are already at high risk of disparities in health outcomes due to a variety of social determinants of health and adverse childhood experiences (ACEs). For example, neighborhoods with primarily low-income or families of color tend to have fewer residential parks, as well as financial and transportation limitations that prevent access to more distant greenspaces.[7] As determined during primary care pediatric social determinants of health screenings, families at or below the federal poverty level show twice the odds of limited access to parks.[8] School campuses in low-income neighborhoods are less likely to include greenspaces than those in higher-income communities.[9] Using geospatial analysis, researchers have shown that persons of color had significantly fewer opportunities for greenspace access against the backdrop of the COVID-19 pandemic than did persons of White race.[10] Last, perceptions of community safety may present barriers to greenspace play and nature contact for families in certain communities, even if such greenspaces are available.[11] Communities with limited greenspace access have diminished opportunity to benefit from the positive health associations of nature contact.

Disproportionate exposure to environmental toxicants and the cumulative effects of environmental racism can all exacerbate environmental injustices for vulnerable populations.[12] Black and other minority neighborhoods bear a disproportionate burden of environmental hazards,[13] threatening safety during outdoor play. For example, households with lower income and those of nonwhite race are more likely to live near high-traffic areas and polluting industrial facilities.[13] Such balancing factors must be considered during implementation of interventions to increase greenspace contact.

Health Benefits of Greenspace Contact

Both natural and built green environments have the potential to support engagement in health-promoting behaviors and well-being through a range of proposed mechanisms.[14] A recently published systematic review focused on children's health found that there was moderate to strong evidence for a beneficial relationship between nature contact and children's physical activity, weight status, and behavioral/mental health outcomes,[6] all of which are public health priorities.

Physical Activity

A large proportion of studies examining nature contact in children from preschool through adolescence focus on physical activity and related outcomes. Most of these studies are observational and use objective measures of physical activity such as accelerometry. The majority have found positive associations.[6] Physical activity studies most commonly evaluate residential proximity to greenspaces, which often utilize land use data or geographic information systems (GIS) to identify and quantify greenspaces within a geographic boundary of a child's home. For example, one large Norwegian epidemiologic study found that having a park within 800 meters of home was associated with more leisure time physical activity in 8-year-old children.[15]

Some intervention studies, often in school settings, have found increased physical activity following a greening intervention. A natural experiment in early childhood education settings found increased physical activity in preschoolers following outdoor environment upgrades that included grassy areas.[16] A randomized controlled trial found that low-income elementary schools randomized to a school garden intervention reported children had increased moderate-to-vigorous physical activity at follow-up compared to control schools.[17] In addition to levels of physical activity, there is also evidence to support the notion that nature play has a positive impact on levels of health-related fitness and motor skill development.[18]

Weight Status and Cardiometabolic Health

There is moderately strong evidence that nature contact is associated with healthy weight status in children. The vast majority of these studies are observational, including some large epidemiological studies, such as a nationwide cross-sectional study from China. Researchers found that more school-based greenness was associated with lower body mass index (BMI) z-scores and odds of obesity in children.[19] There are a few longitudinal and randomized controlled studies; one randomized controlled trial found that increased residential park density, for those who were part of a larger behavioral weight management intervention, was associated with a decrease in BMI z-scores in overweight 8- to 14-year-old youth living in rural communities.[20] A longitudinal study that followed children for 8 years found that park access within 500 meters of home was associated with lower objectively measured BMI at age 18.[21]

There also are a small number of studies evaluating cardiovascular and metabolic outcomes, such as blood pressure, heart rate variability, and lipid levels, and the evidence is mixed. Most of these studies focus on blood pressure as an outcome. While some studies report that higher greenness is associated with lower blood pressure in children,[22] others have found null results.[23]

Asthma and Allergies

An increasing number of studies examine the relationship between greenspace and atopic conditions. Some have hypothesized that microbial exposures from the natural environment are necessary for the development of a healthy immune system.[24] In addition, trees and other greenspaces could mitigate the impact of air pollution on health; however, they may also be the source of allergens and respiratory irritants. While health-promoting positive associations are more commonly reported than negative associations, there are a notable number of studies with negative associations indicating that greenspaces may exacerbate childhood allergies or asthma in some cases.

The studies focused on asthma and allergy outcomes tend to be observational and quite heterogenous in both the nature exposures and the outcomes assessed. One study found that higher residential greenspace was associated with increased odds of wheezing, asthma, and allergic rhinitis.[25] However another large epidemiological study from New Zealand (n = 49,956) found that children who lived in greener areas were less likely to have asthma.[26] There is a need for additional research around the role of nature contact and greenspace proximity in a child's developing immune system.

Mental and Developmental Health

Access to greenspaces has been associated with improvements in physiologic markers of stress, including blood pressure and cortisol.[27] In addition, children identify greenspace as a factor promoting calmness and relaxation when surveyed about contributors to their subjective well-being.[28,29] Reduction in childhood hyperactivity and inattention, externalizing and internalizing behaviors, and depression have all been associated with access to community greenspaces including parks and playgrounds, even when accounting for other risk factors including socioeconomic status and family history.[30,31]

Exposure to nature may influence cognition, learning, and school performance. Greenspaces have been associated with improved working memory, attention, and self-regulation.[32,33] In addition, outdoor teaching has been associated with improved school motivation[34] and reading outcomes.[35] High school students with a view of greenspace from their classrooms exhibited superior stress recovery and attention compared to students without these views.[36]

Communal outdoor spaces such as parks and playgrounds can foster social cohesion, allowing children's caregivers to develop relationships and share knowledge.[37] These are also places where children can engage in physical play and develop social skills, at some distance from caregivers, which may foster development of autonomy and self-efficacy.

Greenspaces and trees can also mitigate air pollution and heat exposure.[38,39] Heat mitigation is especially important in urban environments, which may be significantly warmer than surrounding areas due to darker built surfaces that absorb heat (known as the *urban heat island effect*). Extreme heat can negatively impact school performance and learning[40] as well as mental health.[41] This may be another mechanism by which biophilic design can benefit children's mental and developmental health.

Limitations of Current Research

A number of methodological limitations to the extant literature should be noted. There is wide variability across studies in participant samples, nature exposures, and outcome measures that makes it difficult to compare and aggregate the findings. Additionally, most of the studies are observational; temporality and causality cannot be rigorously assessed using these study designs. Finally, few studies examined the impact of nature exposure on marginalized communities most at risk for inequitable access to greenspace and health disparities. This limitation is particularly important because of potential "equigenic effects": the idea that an environmental factor (e.g., lack of nature contact) could not only disrupt the existing relationship between socioeconomic disadvantage and health but also could have a greater effect on those most disadvantaged.[42]

Future Studies

Future research would inform strategies and policies that can support equitable access to nature contact for children and families. Clinical providers would benefit from research that expands on (1) strength of evidence for specific interventions, like nature prescriptions and greenspace mapping, to allow for evidence-based recommendations; (2) knowledge, attitudes, needs, and barriers of families; (3) knowledge, attitudes, needs, and barriers of providers; and (4) specific interventions that focus on communities impacted by social determinants of health and barriers to greenspace access. Policymakers would additionally benefit from an analysis of specific interventions in promoting health, highlighting which are most efficacious in terms of health outcomes, health disparities, and cost.

Policy and Clinical Implications

Role of the Clinician

Clinicians can play a role in improving child health by endorsing nature contact both during clinical encounters and through advocacy.[6] Clinicians can consider asking about access to greenspace when screening for other social and environmental determinants of health.[8] Pediatric health professionals' care for children can be enriched by knowledge about the accessibility of local parks, playgrounds, and outdoor spaces where children can safely walk, bike, play, and gather. When counseling patients about physical activity, clinicians can provide specific recommendations about accessible outdoor amenities for children and families, particularly settings that offer contact with nature.

Another strategy with potential to promote nature contact in children is the provision of guidance, resources, and even physical "prescriptions" from healthcare providers to encourage time in parks and other nature-rich environments.[43] One such resource among these is SHINE: Stay Healthy in Nature Every Day, a stress reduction and health promotion intervention available through the Center for Nature and Health at University of California San Francisco Benioff Children's Hospital Oakland.[44] The Park Rx America initiative has resources for clinicians, public health professionals, communities, and parks, including a search tool to locate parks close to various zip codes. While numerous such programs exist, the evidence evaluating these interventions is limited.[43]

It is important for clinicians to consider the barriers families may face in accessing and engaging in activities in nature and address this topic using a nonjudgmental, individualized approach. A family's barriers may include time, proximity to safe greenspaces, suboptimal outdoor gear for the weather conditions, lack of previous experiences and/or comfort with spending time in nature, and competing family priorities.

Drawing on the evidence presented here in support of the positive health effects of contact with greenspace, clinicians can play a role in advocating for the incorporation of parks, trees, walking paths, and other elements of biophilic design during community planning and development. Clinicians also can support community programs and initiatives that increase access to nature. Last, they can encourage their own workplaces to incorporate natural elements (plants, gardens, trees) into their spaces when possible and support climate-friendly practices and policies. For those involved in medical education, there also is an opportunity to engage in sharing this evidence with trainees to nurture future pediatric healthcare providers who understand and incorporate this emerging evidence into their clinical practices.

Role for Policymakers

Cross-sector approaches are necessary at local, regional, and national levels to advance policies that support equitable access to nature. Policies that support health-promoting greenspace access can include innovative approaches to utilize existing transportation infrastructure and planning and zoning practices that prioritize equitable access to parks and increase natural elements in cities and streetscapes. Public–private partnerships that promote the growth or restoration of urban tree canopy are an intervention strategy employed in a growing number of cities.[45]

However, policymakers need to be mindful that strategies that green remnant urban land or repurpose transportation infrastructure can paradoxically increase property values, making such communities cost-prohibitive to the populations they were designed to benefit (also known as *gentrification*).[7]

The majority of studies examining the relationship between nature contact and children's health utilize residential proximity to greenspace as a variable. Given potential challenges to increasing residential greenspace, including an actual physical lack of space in many urban areas, another promising strategy involves greening existing schoolyards. The research for this strategy suggests that green schoolyards are associated with increased physical activity and socioemotional health.[46]

Not only park proximity but also amenities are important in relation to public use. One study showed that public park renovations not only increased park use, but also increased perceptions of park safety.[47] Differences in park programming have been associated with lower park usage in lower-income versus higher-income neighborhoods.[11] Last, a sense of belonging and inclusion can contribute to park usage for nonwhite families.[48]

Stakeholders and communities can continue to work with local, state, and federal policymakers to adopt policies that address social and environmental determinants of health. For example, creative transportation solutions to improve access to healthcare can also be used to improve access to nature.

Greenspace Contact to Improve Health Equity

Healthcare providers, policymakers, and other stakeholders must recognize that nature contact, especially access in close proximity to children's homes and schools, is a critical environmental determinant of health with the potential to prevent disease and promote health equity in children. In parallel, adopting policies that decrease exposure to environmental toxicants, especially for nonwhite populations, can help mitigate the disproportionate burden of environmentally mediated disease that may be associated with outdoor or greenspace play for such children. For example, greening of urban environments not only provides increased opportunity for greenspace contact, but also can help mitigate air pollution[39] that exacerbates chronic lung disease related to preterm birth and asthma.

Climate change presents numerous threats to child health.[49-50] Greenspaces and trees can both mitigate climate change (by absorbing carbon pollution), and be protective against climate change effects such as extreme heat and severe precipitation events.[38,39] Increased greenspaces in urban environments could help mitigate the adverse effects of climate change that will disproportionately affect those persons most vulnerable, including children.

Nature contact also has been suggested as a potential community-based intervention for health systems to address social and environmental determinants of health.[51] The COVID-19 pandemic has highlighted the need for collaborative action to support individuals and communities in equitably accessing safe places to move, play, and socialize. Increasing access to nature for children is an opportunity for community engagement, both to identify areas of need and to address community concerns for environmental exposures.

Conclusion

Given public health priorities focused on improving physical activity, obesity, and mental health, the existing evidence for childhood contact with nature is encouraging. Although studies are variable in methods, outcomes, and rigor, nature contact offers promise both as prevention and as an intervention for health conditions starting in childhood. Interventions at both the level of the individual child and the community can enhance equitable contact with nature and improve child health outcomes.

References

1. Frumkin H, Bratman GN, Breslow SJ, et al. Nature contact and human health: a research agenda. *Environ Health Perspect*. 2017 Jul 31;125(7):075001.

2. Yogman M, Garner A, Hutchinson J, Hirsh-Pasek K, Golinkoff RM; Committee on Psychosocial Aspects of Child and Family Health; Council on Communications and Media. The power of play: a pediatric role in enhancing development in young children. *Pediatrics*. 2018 Sep;142(3):e20182058.

3. Gray C, Gibbons R, Larouche R, et al. What is the relationship between outdoor time and physical activity, sedentary behaviour, and physical fitness in children? A systematic review. *Int J Environ Res Public Health*. 2015 Jun 8;12(6):6455–74.

4. Bratman GN, Hamilton JP, Daily GC. The impacts of nature experience on human cognitive function and mental health. *Ann N Y Acad Sci*. 2012 Feb;1249:118–36.

5. Beatley T. *Biophilic Cities: Integrating Nature into Urban Design and Planning*. 2nd ed. Washington, DC: Island Press; 2011.

6. Fyfe-Johnson AL, Hazlehurst MF, Perrins SP, et al. Nature and children's health: a systematic review. *Pediatrics*. 2021 Oct;148(4):e2020049155.

7. Wolch JR, Byrne J, Newell JP. Urban greenspace, public health, and environmental justice: the challenge of making cities "just green enough." *Landsc Urban Plan*. 2014;125:234–44.

8. Razani N, Long D, Hessler D, Rutherford GW, Gottlieb L. Screening for park access during a primary care social determinants screen. *Int J Environ Res Public Health*. 2020;17(8):2777.

9. Stewart IT, Purner EK, Gusman PD. Socioeconomic disparities in the provision of school gardens in Santa Clara County, California. *Child Youth Environ*. 2013;23(2):127–53.

10. Pallathadka A, Pallathadka L, Rao S, Chang H, Van Dommelen D. Using GIS-based spatial analysis to determine urban greenspace accessibility for different racial groups in the backdrop of COVID-19: a case study of four US cities. *GeoJournal*. 2021 Nov 3:1–21.

11. Cohen DA, Han B, Park S, Williamson S, Derose KP. Park use and park-based physical activity in low-income neighborhoods. *J Aging Phys Act*. 2019 Jun 1;27(3):334–42.

12. Kaufman JD, Hajat A. Confronting environmental racism. *Environ Health Perspect*. 2021 May;129(5):51001.

13. Clark LP, Millet DB, Marshall JD. Changes in transportation-related air pollution exposures by race-ethnicity and socioeconomic status: outdoor nitrogen dioxide in the United States in 2000 and 2010. *Environ Health Perspect*. 2017 Sep 14;125(9):097012.

14. Kuo M. How might contact with nature promote human health? Promising mechanisms and a possible central pathway. *Front Psychol*. 2015 Aug 25;6:1093. doi:10.3389/fpsyg.2015.01093

15. Nordbø ECA, Raanaas RK, Nordh H, Aamodt G. Neighborhood green spaces, facilities and population density as predictors of activity participation among 8-year-olds: a cross-sectional GIS study based on the Norwegian mother and child cohort study. *BMC Public Health*. 2019 Oct 30;19(1):1426.

16. Ng M, Rosenberg M, Thornton A, et al. The effect of upgrades to childcare outdoor spaces on preschoolers' physical activity: findings from a natural experiment. *Int J Environ Res Public Health*. 2020 Jan 10;17(2):468.

17. Wells NM, Myers BM, Henderson CR Jr. School gardens and physical activity: a randomized controlled trial of low-income elementary schools. *Prev Med*. 2014 Dec;69(Suppl 1):S27–33.

18. Dankiw KA, Tsiros MD, Baldock KL, Kumar S. The impacts of unstructured nature play on health in early childhood development: a systematic review. *PLoS One*. 2020 Feb 13;15(2):e0229006.

19. Bao WW, Yang BY, Zou ZY, et al. Greenness surrounding schools and adiposity in children and adolescents: findings from a national population-based study in China. *Environ Res*. 2021 Jan;192:110289.

20. Armstrong B, Lim CS, Janicke DM. Park density impacts weight change in a behavioral intervention for overweight rural youth. *Behav Med*. 2015;41(3):123–30.

21. Wolch J, Jerrett M, Reynolds K, et al. Childhood obesity and proximity to urban parks and recreational resources: a longitudinal cohort study. *Health Place*. 2011 Jan;17(1):207–14.

22. Xiao X, Yang BY, Hu LW, et al. Greenness around schools associated with lower risk of hypertension among children: findings from the Seven Northeastern Cities Study in China. *Environ Pollut*. 2020 Jan;256:113422.

23. Abbasi B, Pourmirzaei M, Hariri S, et al. Subjective proximity to green spaces and blood pressure in children and adolescents: The CASPIAN-V Study. *J Environ Public Health*. 2020 Dec 11;2020:8886241.

24. Haahtela T. A biodiversity hypothesis. *Allergy*. 2019 Aug;74(8):1445–56.

25. Parmes E, Pesce G, Sabel CE, et al. Influence of residential land cover on childhood allergic and respiratory symptoms and diseases: evidence from 9 European cohorts. *Environ Res*. 2020 Apr;183:108953.

26. Donovan GH, Gatziolis D, Longley I, Douwes J. Vegetation diversity protects against childhood asthma: results from a large New Zealand birth cohort. *Nat Plants*. 2018 Jun;4(6):358–64.

27. Thompson CW, Roe J, Aspinall P, Mitchell R, Clow A, Miller D. More green space is linked to less stress in deprived communities: evidence from salivary cortisol patterns. *Landsc Urban Plan*. 2012;105(3):221–9.

28. Birch J, Rishbeth C, Payne SR. Nature doesn't judge you: how urban nature supports young people's mental health and wellbeing in a diverse UK city. *Health Place*. 2020 Mar;62:102296.

29. Fattore T, Mason J, Watson E. When children are asked about their well-being: towards a framework for guiding policy. *Child Indic Res*. 2009;2(1):57–77.

30. Markevych I, Tiesler CM, Fuertes E, et al. Access to urban green spaces and behavioural problems in children: results from the GINIplus and LISAplus studies. *Environ Int*. 2014 Oct;71:29–35.

31. Engemann K, Pedersen CB, Arge L, Tsirogiannis C, Mortensen PB, Svenning JC. Residential green space in childhood is associated with lower risk of psychiatric disorders from adolescence into adulthood. *Proc Natl Acad Sci U S A*. 2019 Mar 12;116(11):5188–93.

32. Dadvand P, Nieuwenhuijsen MJ, Esnaola M, et al. Green spaces and cognitive development in primary schoolchildren. *Proc Natl Acad Sci U S A*. 2015 Jun 30;112(26):7937–42.

33. Taylor AF, Butts-Wilmsmeyer C. Self-regulation gains in kindergarten related to frequency of green schoolyard use. *J Environ Psychol*. 2020;70:10144. https://doi.org/10.1016/j.jenvp.2020.101440

34. Bølling M, Otte CR, Elsborg P, Nielsen G, Bentsen P. The association between education outside the classroom and students' school motivation: results from a one-school-year quasi-experiment. *Int J Educ Res*. 2018;89(1):22–35.

35. Otte CR, Bølling M, Stevenson MP, Ejbye-Ernst N, Nielsen G, Bentsen P. Education outside the classroom increases children's reading performance: results from a one-year quasi-experimental study. *Int J Educ Res*. 2019;94(1):42–51.

36. Li D, Sullivan WC. Impact of views to school landscapes on recovery from stress and mental fatigue. *Landsc Urban Plan*. 2016;148:149–58.

37. Bedimo-Rung AL, Mowen AJ, Cohen DA. The significance of parks to physical activity and public health: a conceptual model. *Am J Prev Med*. 2005 Feb;28(2 Suppl 2):159–68.

38. United Sates Department of Agriculture, Forest Service. *Chicago's Urban Forest Ecosystem: Results of the Chicago Urban Forest Climate Project*. Gen Tech Rep. NE-186. Radnor, PA: Northeastern Forest Experiment Station; 1994.

39. Nowak DJ, Hirabayashi S, Bodine A, Greenfield E. Tree and forest effects on air quality and human health in the United States. *Environ Pollut*. 2014 Oct;193:119–29.

40. Park RJ, Behrer AP, Goodman J. Learning is inhibited by heat exposure, both internationally and within the United States. *Nat Hum Behav*. 2021 Jan;5(1):19–27.

41. Kim Y, Kim H, Gasparrini A, et al. Suicide and ambient temperature: a multi-country multi-city study. *Environ Health Perspect*. 2019 Nov;127(11):117007.

42. Mitchell RJ, Richardson EA, Shortt NK, Pearce JR. Neighborhood environments and socio-economic inequalities in mental well-being. *Am J Prev Med*. 2015 Jul;49(1):80–4.

43. Kondo MC, Oyekanmi KO, Gibson A, South EC, Bocarro J, Hipp JA. Nature prescriptions for health: a review of evidence and research opportunities. *Int J Environ Res Public Health*. 2020 Jun 12;17(12):4213.

44. Razani N, Kohn MA, Wells NM, Thompson D, Hamilton Flores H, Rutherford GW. Design and evaluation of a park prescription program for stress reduction and health promotion in low-income families: the Stay Healthy in Nature Everyday (SHINE) study protocol. *Contemp Clin Trials*. 2016 Nov;51:8–14.

45. City of Boston. The Boston urban forest plan. 2022. Available at: https://www.boston.gov/departments/parks-and-recreation/urban-forest-plan.

46. Bikomeye JC, Balza J, Beyer KM. The impact of schoolyard greening on children's physical activity and socioemotional health: a systematic review of experimental studies. *Int J Environ Res Public Health*. 2021 Jan 11;18(2):535.

47. Cohen DA, Han B, Isacoff J, et al. Impact of park renovations on park use and park-based physical activity. *J Phys Act Health*. 2015 Feb;12(2):289–95.

48. Seaman PJ, Jones R, Ellaway A. It's not just about the park, it's about integration too: why people choose to use or not use urban greenspaces. *Int J Behav Nutr Phys Act*. 2010 Oct 28;7:78.

49. Ahdoot S, Baum CR, Cataletto MB, Hogan P, Wu CB, Bernstein A; Council on Environmental Health and Climate Change; Council on Children and Disasters; Section on Pediatric Pulmonology and Sleep Medicine; Section on Minority Health, Equity, and Inclusion. Policy Statement. Climate Change and Children's Health: Building a Healthy Future for Every Child. *Pediatrics*. 2024;153(3):e2023065504.

50. Ahdoot S, Baum CR, Cataletto MB, Hogan P, Wu CB, Bernstein A; Council on Environmental Health and Climate Change; Council on Children and Disasters; Section on Pediatric Pulmonology and Sleep Medicine; Section on Minority Health, Equity, and Inclusion. Technical Report. Climate Change and Children's Health: Building a Healthy Future for Every Child. Pediatrics. 2024 1;153(3):e2023065505. e2do10.1542/peds PM.

51. South EC, Kondo MC, Razani N. Nature as a community health tool: the case for healthcare providers and systems. *Am J Prev Med*. 2020 Oct;59(4):606–10.

21

Rural and Agricultural Environments

Barbara C. Lee and Matthew C. Keifer

Rural and agricultural environments differ from urban environments in many aspects. Remote, rural settings, including farms and ranches, combine features of home life, recreational opportunities, and valuable work experiences. At the same time there are adverse environmental exposures, including extreme weather events, polluted air, animal and insect contact, and serious risk for injury and illness.

In the United States, rural communities make up tens of millions of people from diverse backgrounds, comprising 14% of the US population.[1] Within this rural population are about 2.01 million farms and ranches where nearly 1 million children live.[2,3] The rate of poverty in rural regions is 24% higher than in urban areas. Food insecurity is 22% higher than in urban areas, and the lack of healthcare insurance among rural residents is 15% greater than for their urban counterparts.[1] Children in rural and agricultural settings are affected by the absence or remoteness of health and dental care services, an unstable economy, changing and sometimes unsustainable agricultural production practices, and agricultural labor shortages. Limited communication capabilities due to poor broadband internet and cellular connectivity affect everyday life. Furthermore, options for childcare services may be absent, unaffordable, or of poor quality thus forcing parents or guardians to make difficult choices between priorities of family and work.

When compared to their urban counterparts, rural children have a higher risk of injury that is more severe and costly and with poorer outcomes, including disability or death.[4] Several factors are associated with this risk of costly injuries with their potential for disability and/or death. Rural families often lack access to reliable transportation, having to travel farther for healthcare or emergency services. This increased time on rural roadways is also associated with a higher risk of transportation crashes for rural youth who operate motorized vehicles for work, including utility quad bikes, all-terrain vehicles (ATVs), tractors, and skid steer loaders. All of these vehicles are associated with serious risk of injury or death, and estimates suggest 100 deaths and 30,000 emergency room visits per year among youth operating ATVs.[5]

Rural communities have highly diverse populations including Hispanic, Anabaptist, American Indian, Alaska Native, Black or African, Asian and White residents. Rural residents have a disproportionate level of socioeconomic vulnerability that is even greater among racial and ethnic minorities.[1] Among the vulnerabilities are financial, housing and food insecurity, poor health literacy, isolation, and transportation needs—all impacting the well-being of children.

This chapter on rural and agricultural environments cannot address the full spectrum of environmental factors affecting children's well-being. Readers are referred to chapters on outdoor air pollution, pesticides, asthma, allergy, carcinogens, and noise to expand

Barbara C. Lee and Matthew C. Keifer, *Rural and Agricultural Environments* In: *Textbook of Children's Environmental Health*, Second Edition. Edited by: Ruth A. Etzel and Philip J. Landrigan, Oxford University Press. © Oxford University Press 2024. DOI: 10.1093/oso/9780197662526.003.0021

their perspectives on environmental factors affecting children of all ages in rural and agricultural settings.

Population at Risk

Children living and working on farms have unique often dangerous exposures associated with agricultural production. In general, the US farm population has stabilized, with only a slight reduction from 2000 (2.17 million) to 2021 (2.01 million farms).[6] Food and fuel production is shifting to concentrated, commodity operations, thus the number of large farms is increasing and mid-size farms are decreasing. However, an important trend in terms of children is that the number of small, lifestyle farms is increasing.[7] Moreover, compared to productive agricultural enterprises, lifestyle farms tend to have older, more dangerous equipment being used by inexperienced operators.

A 2020 report from the US Department of Agriculture (USDA) indicated that of the nation's 2 million farms, 98% were classified as family farms.[8] National estimates of US farms with youth reveal that about 25 million youth spend time on farms, with the majority of these (95%) being frequent or occasional visitors. An estimated 892,836 youth live on farms, and another 265,604 are employed to work on farms.[9]

Fatal and Nonfatal Injuries

Since 1996, when the National Institute for Occupational Safety and Health (NIOSH) launched its National Initiative for Childhood Agricultural Injury Prevention, there has been a major effort to reduce the toll of fatal and nonfatal agriculture-related injuries affecting children.[10] Unfortunately, there is no central repository or required reporting of fatal and nonfatal childhood agricultural injuries, except for employed youth 15 years and older. Estimates are derived from various sources including an Agricultural Injury News database,[11] periodic state-based reports, and the Census of Fatal Occupational Injuries (CFOI). The US statistics most frequently referenced estimate annually about 103 deaths on farms among youth aged 0–19 years.[12] For nonworking children, the 2014 NIOSH data estimate that about 33 children are seriously injured each day on farms.[13]

Fatalities

Both working and nonworking children are victims of unintentional deaths on farms. These are referred to as "injury events" versus "accidents," because they are predictable and preventable. A NIOSH analysis of working youth fatalities from 1994 to 2013 revealed 389 agricultural work-related deaths, including 40% among children younger than 14 years. Of these, 91% were males, and 10% were reported as Hispanic. The most common sources of fatalities were transportation and machinery, contact with equipment, and exposure to harmful substances or environments.[13] When comparing youth working in agriculture with youth working in all other occupations, about 48% of deaths occurred in agriculture; in 2016, young agricultural workers were 7.8 times more likely to be killed on the job compared to all other industries combined.[14]

On small family farms and in certain populations, traumatic childhood agricultural injuries are associated with young, nonworking children being present in the hazardous worksite. An analysis of 370 pediatric farm injuries and fatalities in the United States and Canada where the child was not working revealed nearly two-thirds (63%) of cases were infants, toddlers, and preschoolers younger than 7 years and seemingly supervised by an adult.[15] An analysis of 107 pediatric (<18 years) deaths in Pennsylvania from 2000 to 2018 revealed two-thirds (66%) involved nonworking children, of whom most were males younger than 5 years, and the majority (78%) belonged to an Anabaptist (e.g., Amish or Mennonite) community.[16]

Nonfatal Injuries

Regarding nonfatal farm injuries, NIOSH conducted periodic childhood agricultural injury surveillance (CAIS) from 1998 to 2014 in cooperation with the National Agricultural Statistics Service.[3] Data revealed that, in 2014, 11,942 youth suffered agricultural injuries and, of these, 58% were incurred by males, and only one-third were work-related. The most common types of injuries were fractures or broken bones, followed by cuts, lacerations, and contusions.[9] Over the 14-year surveillance period, agricultural injuries dropped from 13.5/1,000 farms to 5.7/1,000 farms, a decline of 57.8%.[17]

Cost of Injuries

The toll of fatal and nonfatal childhood agricultural injuries can be staggering. The annual cost of fatalities was estimated at $604.8 million per year, and the cost of nonfatal farm youth injuries was estimated at $1.4 billion (converted from published 2005 dollars to 2022 dollars). Of these costs, three-fourths are associated with nonworking youth.[18,19] Similar to adult agricultural injuries, youth agricultural injuries tend to be more costly and more severe (e.g., limb amputations) than nonagricultural injuries. A related financial issue is that many children feel compelled to work alongside their parents out of economic necessity.[20]

Agricultural Safety Strategies for Youth

Given the severity and magnitude of childhood farm-related injuries, many recommendations and interventions have been developed to safeguard children. Primary prevention strategies emphasize prohibiting young children from entry to worksites altogether. However, this strategy is not culturally or practically viable in many cases. Keeping nonworking children and infants out of worksites may require off-farm childcare, passive barriers like gates, and enforced farm safety protocols. For older children, strategies address adult supervision, training in work assignments, and expectations of work versus recreation, such as in use of ATVs.[15]

In agriculture, the notion of engineering out the many environmental and physical hazards has many limitations. At the same time, there is general agreement that education alone is insufficient for safeguarding youth living and working in agricultural settings.[21] The most recognized intervention is the set of consensus-generated guidelines that match a child's developmental level with characteristics of different work tasks.

Originally released in 1999 as the North American Guidelines for Children's Agricultural Tasks, these guidelines were updated and released in 2017 as the Agricultural Youth Work Guidelines.[22,23] A controlled study found that in cases where the guidelines were applied, injuries were reduced by 50%.[24]

Respiratory Exposures and Outcomes

Children in rural and agricultural settings are exposed to respiratory irritants such as organic dusts, airborne pathogens, pesticides, cleaning agents, toxic gases, and wildfire smoke. Most studies of pediatric respiratory disease in agricultural populations have focused on asthma. However, when the general rural environment is included, other conditions like coccidiomycosis (valley fever) and even tuberculosis (potentially from *M. bovis* and *M. tuberculosis*) emerge as risks.[25] Although farmer's lung resulting from exposure to organic dusts is a serious, debilitating, somewhat uncommon experience among farming adults, it appears to be even more rare among farm children. Readers are encouraged to refer to Chapter 48 for more information about allergies and asthma.

Weather and Climate Hazards

There is no question that weather and climate conditions are changing in ways that affect everyday life. This is especially the case in rural and agricultural environments where weather and climate affect a person's daily work activities and childcare priorities. Droughts, floods, hurricanes, severe storms including tornadoes, excessive heat or cold, and wildfires have been increasing. Families living and working in rural and agricultural settings have endured changes in climate, water supply, soil conditions, and limited access to internet or text messages that provide emergency alerts. Collectively, these carry a serious economic burden that adds further stress to adults who are responsible for raising and caring for children. According to a USDA report, there were more than 90 weather-related disasters in a 30-year period, resulting in $700 billion lost revenue.[26] To exacerbate the economic losses and disruption of daily life, extreme weather events are often followed by pests and diseases associated with crops and livestock. The dire predictions about the negative impact of climate change on agricultural production is one of the reasons young people are hesitant to stay in farming.[27]

Zoonoses and Infectious Disease

For the vast majority of infections, even reportable conditions, data do not clearly distinguish rural from urban acquisition. However, the characteristics of transmission and the location of reservoir and vector species often imply the distinction. The authors used available literature to select the infectious conditions that might reasonably be ascribed more commonly to the rural and farm environment.

Zoonoses

Zoonotic agents gain access to the human body through a variety of mechanisms ranging from airborne exposures or ingestion to wound contamination. But among the most common mechanisms of zoonotic transfer is arthropod bites. While these vectors exist in the urban environment, they are more likely to contact host wildlife and farm animals in rural environments and thus transfer pathogens to human hosts. Special mention should go to mosquito-borne zoonoses, which are not emphasized herein. For most mosquito-borne zoonoses humans are dead-end hosts. However, since mosquitos can bite multiple people during their month-long lives, and urban human population density gives mosquitos ample opportunity for transmission, many of these mosquito-borne zoonoses appear more commonly in urban than rural environments (e.g., St. Louis encephalitis). This chapter focuses on the zoonoses unique to rural and agricultural environments. A full discussion of these agents and disorders is beyond the scope of this chapter but mention of many of them is made in Table 21.1. Some of the more important, less well known, and/or newly described disorders deserve mention because of their increasing prevalence or their cryptic presentation. These are briefly discussed to assure that the reader has a passing familiarity with these somewhat rare conditions.

Enteric Infections

Children who grow up on farms with frequent and early contact with domestic animals such as ruminants and swine likely develop infections and resulting immunity early in life to the common zoonotic pathogens associated with these animals. However, agritourism activities that invite contact between young, immunologically naïve human hosts and domesticated livestock, such as occurs at petting zoos and rural farm fairs, are known to be responsible for transmission of infectious agents, sometimes with serious consequences. Pathogens such as *Campylobacter*, Shiga toxin-producing *E. coli*, *Listeria monocytogenes*, Cryptosporidium, and non-typhoidal salmonella can be present in healthy farm animals. Hand-to-mouth transfer can occur after contact with animals. These pathogens can cause acute and/or chronic diarrhea and can sometimes go on to cause serious sepsis in human hosts. Careful hand hygiene is preventive, but children with frequent hand-to-mouth contact and immature immune systems are particularly susceptible to infection.[28-30] Diarrhea from *Giardia lamblia*, while associated with domesticated animal contact, is more commonly acquired from ingestion of unpurified water contaminated by wild animal feces. However, acquisition from faulty water purification in urban settings is not uncommon and even person-to-person transmission is described.[31]

Tick, Mosquito, and Vector-Borne Conditions

Many rural and farm zoonoses are transmitted by arthropod vectors such as mosquitos and ticks; these include the tick-borne disorders, such as Lyme disease, which is the most common vector-borne illness in the United States and caused by one of six species in the spirochete family Borreliaceae.[32] First described in Lyme, Connecticut, in 1975, it was discovered during the investigation of an outbreak of presumed juvenile rheumatoid arthritis. The tick vector is the black-legged or Ixodes tick. Clinical presentation is characterized by fever, chills, headache, fatigue, muscle and joint aches, and swollen lymph nodes that may occur with or without the characteristic erythema migrans (EM) rash: EM is common in Lyme disease, occurring in up to 80% of infected persons. This unique rash begins at the site of a tick bite after a delay varying from 3 to 30 days (average

Table 21.1 Rural and agricultural infectious diseases and zoonoses

Agent/Disease (organism)	Host/Carrier	Primary transmission methods/Characteristics	Symptoms	Geographic predominance
Enteric infections				
Campylobacter, Shiga toxin-producing E coli, Listeria monocytogenes, Cryptosporidium, Giardia lamblia, Nontyphoidal salmonella	Enteric organisms present in farm animals, including animals commonly found in petting zoos and wild animals such as deer	Hand-to-mouth transfer after contact with animals Ingestion of unpurified water contaminated by wild animal feces	Can cause acute and/or chronic diarrhea and can sometimes go on to cause serious sepsis in human hosts	All US states
Tick-related zoonotic diseases				
Tularemia (rabbit fever)	Rodents, rabbits, muskrats	Ingestion, inhalation, skin contact, ticks, deer fly bites	Multiple presentations from conjunctivitis to pneumonia to typhoid like illness	Pacific Northwest, south central US, and parts of Massachusetts, including Martha's Vineyard
Lyme disease (Borrelia burgdorferi)	Primary reservoir is rodents such as the white-footed mouse Humans are dead-end hosts	Infection from bite of black-legged or Ixodes tick	Clinical presentation is characterized by fever, chills, headache, fatigue, muscle and joint aches, and swollen lymph nodes which may occur in with or without the erythema migrans (EM) rash Untreated Lyme disease can result in carditis, meningitis and encephalopathy, neuropathy, and arthritis	Northeast, mid-Atlantic, and upper Midwest regions
Powassan virus	Hosted by rodents and deer	Bite of black-legged or Ixodes tick	Encephalitis	Northeastern US and Great Lakes Region
Babesiosis	Hosted by white-footed mouse	Bite of Ixodes scapularis tick	Often asymptomatic. May present with fever, aches chills, nausea. May be fatal in asplenic or immunosuppressed patients	Northeast, upper Midwest, Northwest, and Florida Few cases in various other states

Disease	Reservoir/Host	Vector/Transmission	Clinical features	Distribution
Ehrlichiosis (mostly *E chaffeensis*)	The main reservoir is the white-tail deer for *E chaffeensis*	Bite of the Lone Star and black-legged tick (*I scapularum*)	Fever and chills, malaise, myalgia, headache in most nausea, vomiting, arthralgias, and cough occur in some. Rash in fewer than half	Southeast and South-central US, from the East Coast extending westward to Texas
Anaplasmosis (mostly *A phagocytophilum*)	Hosted primarily by white-footed mouse	Bite of the black-legged tick.	Fever and chills, malaise, myalgia, headache in most Nausea, vomiting, arthralgias, and cough occur in some. Rash in fewer than half	New York, Massachusetts, Maine, Vermont, Minnesota, Wisconsin, New Hampshire, and Connecticut
Colorado tick fever, Reoviridae virus	Rodents are primary reservoirs for the virus	Bite of Rocky Mountain tick. Tick generally lives between 4,000 and 12,000 feet of elevation	Fever, chills, skin rash Aches and weakness. Biphasic fever may occur Severe illness is rare and recovery is spontaneous	Cases have been reported Colorado, Oregon, Idaho, California, Arizona, Utah, Wyoming, and Montana

Noninfectious tick-related disorders

Disease	Reservoir/Host	Vector/Transmission	Clinical features	Distribution
Alpha-gal allergy (red meat allergy)	Allergic response to tick salivary antigens Lone Star and black-legged ticks implicated in the US	Insect bite results in allergy to galactose-alpha-1,3-galactose, a sugar found in red mammal meats	May present in typical fashion with hives and airway restriction but is often atypical with recurrent abdominal pain hours after red meat consumption. Diagnosed by finding alpha-gal immunoglobulin	More common in the South, East, and Central United States
Tick paralysis	Female *Dermacentor* or *Ixodies holocyclus* tick	Tick bite results in neurotoxic effects of tick salivary constituents	Paralysis often preceded by diffuse weakness. Simple facial palsies to syndromes resembling Guillan-Barré syndrome Resolves with tick removal	Most frequent in the western US but seen periodically throughout the US

Zoonoses from mosquitos, rodents, and other vertebrate vectors

Disease	Reservoir/Host	Vector/Transmission	Clinical features	Distribution
Eastern equine encephalitis virus	Birds are the primary hosts Horses and humans are dead-end hosts	Mosquitos	The febrile version includes fever, chills, body aches, and joint pain lasting 1–2 weeks followed by complete recovery in most cases Neurologic involvement may present as fever, headache, vomiting, seizures, behavioral changes, drowsiness, and coma	Seen primarily in the Gulf Coast, Eastern seaboard, and the Great Lakes

(continued)

Table 21.1 Continued

Agent/Disease (organism)	Host/Carrier	Primary transmission methods/Characteristics	Symptoms	Geographic predominance
Rabies (*Rabies lyssavirus*)	Racoons, skunks, foxes, and bats were most of the 4,000+ rabid animals in the US in 2019	Bite or scratch	Onset is similar to the flu, but may include weakness, fever, or headache Also may include paresthesia near bite Encephalopathy and death are assured without early vaccination and antibody treatment	About half of US cases in animals were in Texas, Virginia, Pennsylvania, North Carolina, Colorado, and New York combined
Orf (Parapoxvirus)	Found in sheep and goats	Contact with an infected animal or a fomite	Localized ulcerative lesion, commonly seen on the hand which generally resolves in 8 weeks in immunocompetent individuals	Nonreportable and thus no geographic information readily available
Q fever (*Coxiella burnetii*)	Found in goats, sheep, and cattle	Direct contact with infected animals Inhalation of contaminated dust Ingestion of unpasteurized milk products Can be transmitted trans-placentally	Flu-like illness, known to cause pneumonia, hepatitis, and endocarditis in individuals with damaged cardiac valves	Most reports come from western and plains states Texas, California, and Iowa account for more than one-third of reported cases
Hantavirus	Infected rodents, importantly the deer mouse	Transmitted primarily through inhalation of rodent feces–contaminated dust or droplets	The resulting condition, known as hepatocardiopulmonary syndrome (HCPS), characterized by fever, pneumonia, and cardiovascular dysfunction	In 2019, most cases reported from the western states, especially Washington, California, Arizona, Colorado, and New Mexico but most states reported cases

Sources: Ma X, Monroe BP, Wallace RM, et al. Rabies surveillance in the United States during 2019. *J Am Vet Med Assoc.* 2021;258(11):1205–1220. doi:10.2460/javma.258.11.1205 CDC website for all the geographic distribution information.

is about 7 days). The rash expands to up to 12 inches or more over 3 to 30 days. Lyme is generally effectively treated with antibiotics. However, untreated Lyme disease can result in carditis, meningitis and encephalopathy, neuropathy, and arthritis.[33,34]

Babesiosis is caused by an intracellular parasite, *Babesia microti*, a zoonotic pathogen of the white-footed mouse. The organism behaves much like malaria by infecting the red blood cell and is diagnosed by finding the parasite in red blood cells. It is generally a benign condition in healthy individuals and often asymptomatic, clearing without intervention. However, in asplenic or immunosuppressed individuals, uncontrolled parasitemia can be fatal.[35] The primary rickettsial disorders transmitted by ticks in the United States include anaplasmosis and ehrlichiosis. Both have seen increases in occurrence over the past decade, and both are transmitted by the black-legged tick to humans. The two conditions were not distinguished until 1994, when anaplasma was identified, but the two present in similar fashion with fever, headache, chills, and, at times, a rash and nausea. Like babesiosis, these zoonoses are clinically worse and can be fatal in immunosuppressed individuals.

There exist multiple tick-borne viral illnesses, but only a few are seen with any frequency in North America. Powassan virus infects humans through tick bites and may result in encephalitis. There exist two varieties—one hosted by rodents that only rarely infects humans, and one hosted by deer and transferred by the same black-legged deer tick as Lyme disease. Cases in the United States have been seen in Minnesota, Wisconsin, New York, and Massachusetts, usually from June to September.[32,36] Colorado tick fever, a generally self-limited often biphasic fever commonly presenting with headache, myalgias, and a petechial rash, occurs in rural areas of the Rocky Mountains at altitudes between 4,000 and 10,000 feet. Its distribution is due to the limited distribution of its primary vector, the Rocky Mountain wood tick. Rodents are the natural host to the causative Reoviridae virus.

One recently identified tick-related illness results not from an infectious agent but a host immune response. Alpha-gal allergy (an allergy to galactose-α-1,3-galactose) is an allergy to the contents of tick saliva and results in an allergy to a sugar found in most red meat. The allergy may present in typical fashion with hives and airway restriction, but it often presents atypically with recurrent abdominal pain hours after the consumption of the meat. Alpha-gal may be present in dairy products, resulting in intolerance of dairy-based foods as well.[37]

Tick paralysis is a disorder resulting from a toxin in the saliva of the female *Dermacentor* or *Ixodies holocyclus* tick. Toxins from the two ticks differ in nature and mechanism, but in both paralysis is often preceded by diffuse weakness. Clinical syndromes associated with tick paralysis range from simple facial palsies to syndromes resembling Guillain-Barré syndrome. Tick removal usually results in symptom resolution within hours in *Dermacentor* ticks but more slowly with *I. holocyclus* ticks.[38]

Diseases Resulting from Direct Animal Contact

Some predominantly rural infectious disorders are the result of direct contact with animals or contaminated organic materials from domestic or wild animals that harbor pathogens. The most serious zoonosis in the United States from the perspective of lethality is rabies, which results primarily from a bite or scratch. Relatively rare in the United States today due to the widespread vaccination of dogs, this is a viral neurological

276 TEXTBOOK OF CHILDREN'S ENVIRONMENTAL HEALTH

disease that may be contracted from bites from otters, raccoons, skunks, fox, or, most commonly, from bats. Cows and other herbivores are known to harbor the virus in areas where vampire bats reside and infect them while feeding. Rabies is rare in the United States and is fully preventable with postexposure prophylaxis. It is only fatal in people who do not seek therapy. Because of the inevitably fatal outcome of a rabies infection, a bite or scratch from an otter, bat, raccoon, skunk, or fox should be addressed immediately and consideration should be given to vaccination and prophylaxis. Victims should be advised to attempt to capture the animal that bit them because examination of the animal may rule in or rule out infectivity.[39]

Orf is a zoonosis associated primarily with sheep and goats in North America. This relatively benign but very contagious skin infection caused by a parapoxvirus is contracted through contact with an infected animal or a fomite. The resulting condition is usually restricted to a localized ulcerative lesion, commonly seen on the hand, that generally resolves in 8 weeks in immunocompetent individuals. The condition can become disseminated in the immunocompromised. Therapy is supportive and focused primarily on preventing secondary infection. Wound secretions should be considered highly infective.[40]

Q fever caused by the obligate intracellular gram-negative proteobacteria *Coxiella burnetii* can be acquired through direct contact with infected animals, inhalation of contaminated dust, or ingestion of unpasteurized milk products and can be transmitted transplacentally. It is most often seen in US western states. It is commonly an occupationally acquired disease associated with the raising of cattle, goats, and sheep. The disease is most often self-limiting, manifesting as a mild flu-like illness and resolving without treatment, but it is also known to cause pneumonia, hepatitis, and endocarditis in individuals with damaged cardiac valves.[41,42] Hantavirus is another important, sometimes serious viral infection, transmitted primarily through inhalation of rodent feces–contaminated dust or droplets and most commonly seen in the western United States. Of the 28 viral variants that cause this condition, the Sin Nombre variant is the primary pathogen in the United States. The resulting condition, known as hepatocardiopulmonary syndrome (HCPS), is characterized by fever, pneumonia, and cardiovascular dysfunction, thus differing from the Old World hemorrhagic fever with renal syndrome (HFRF) that manifests as fever, renal failure, and coagulation abnormalities with bleeding.[43]

Tularemia (rabbit fever) deserves special mention for its many routes of entry and its many manifestations. It is a very infectious, uncommon (about 250 US cases per year), pleomorphic zoonotic infection caused by *Francisella tularensis*, a gram-negative coccobacillus. This organism is carried by rodents, muskrats, and rabbits. Transmissible by multiple routes including ingestion, inhalation, bites from ticks (dog, wood, and Lone Star ticks), deer flies, and contact with infected animal tissue, it displays multiple presenting syndromes. These range from ocular inflammation (occulo-glandular fever) to a typhoid-like presentation. The infection is potentially lethal but treatable with antibiotics.[44]

As is evident, many zoonotic pathogens affect young rural residents and farm visitors. Most are somewhat uncommon, but a familiarity with the conditions will facilitate the identification and prompt, appropriate treatment of affected children.

COVID-19

The COVID-19 pandemic had significant impacts on rural and agricultural populations, including unemployment, overall poorer life satisfaction, and compromised mental health.[45] Limited healthcare options and heightened reliance on telehealth posed several barriers to COVID-19 testing and access to vaccines.[45] A study of the experiences of agricultural families' during the pandemic revealed that, for more than 80% of parents, the changes in schooling and childcare arrangements resulted in having young children spend much more time on the farm. Nearly all parents reported that older children who otherwise would have been in school spent increased time doing farm chores thus adding to their risk of adverse exposures, fatigue, and isolation.[46] In addition to these hardships, many children lost a primary or secondary caregiver to a COVID-19 related death.

Mental Health

Increasing attention is being given to the mental well-being of children living in rural and agricultural environments as evidenced by publicly funded research and outreach programs. Normal adolescent development includes a tendency to have anxieties that impact one's mental state. Additionally, adolescents are strongly impacted by their parents' well-being. Research has revealed associations between financial and economic hardships internalized by youth leading to substance abuse and depression.[47] These adolescent experiences are compounded by the myriad stressors associated with rural isolation. While the agricultural work-related stresses and mental health conditions of adults have been well documented,[48] information is just beginning to surface about the mental health of adolescents in rural and farm settings.[49]

Mental well-being also is associated with risk-taking. Agricultural life is known for its culture of risk-taking because the work tasks and work environment are among the most hazardous occupational settings. This is compounded by having children of all ages living and working among the hazards. The relationship between risk-taking and the well-being of young people on farms was studied with a sample of more than 2,500 Canadian farm youth (11–16 years) and compared with other rural, non-farm youth.[49] Results revealed that male farm adolescents showed the highest levels of unsafe behavior. Risk-taking was strongly and consistently associated with indicators of poor mental health, including measures of life satisfaction, psychosomatic symptoms, and academic performance. This culture of risk-taking likely spills over into many aspects of detrimental environmental exposures.

Prevention

Safeguarding children from adverse rural and agricultural situations involves an awareness of prevailing conditions, modification of strategies for different populations, and cooperation at many levels. The socioecological model (SEM), a widely used framework for public health problems, was modified to target interventions for children living and working on farms.[21] Parents and guardians who bear legal and moral accountability for a child are at the level closest to the child's environment (Figure 21.1).

The model suggests how multilevel, repeated interventions are applied to impact behaviors and decisions made by adults responsible for children. Preventive interventions require approaches targeted to match the environmental hazards and population characteristics while accounting for barriers and motivators that prompt action or lack thereof.

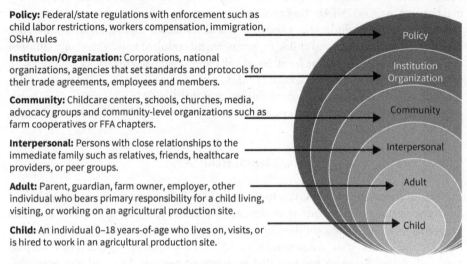

Definitions

Policy: Federal/state regulations with enforcement such as child labor restrictions, workers compensation, immigration, OSHA rules

Institution/Organization: Corporations, national organizations, agencies that set standards and protocols for their trade agreements, employees and members.

Community: Childcare centers, schools, churches, media, advocacy groups and community-level organizations such as farm cooperatives or FFA chapters.

Interpersonal: Persons with close relationships to the immediate family such as relatives, friends, healthcare providers, or peer groups.

Adult: Parent, guardian, farm owner, employer, other individual who bears primary responsibility for a child living, visiting, or working on an agricultural production site.

Child: An individual 0–18 years-of-age who lives on, visits, or is hired to work in an agricultural production site.

Figure 21.1. Socio-Ecological Model modified for children in agricultural settings. Reprinted with permission from Journal of Agromedicine: Lee BC, Bendixsen C, Liebman AK, Gallagher SS. Using the Socio-Ecological Model to Frame Agricultural Safety and Health Interventions. *J Agromedicine*. 2017;22(4):298-303. doi:10.1080/1059924X.2017.1356780

At the highest SEM level is policy, including federal and state-level laws and regulations. The Child Labor in Agriculture rules are applicable to youth employees, many of whom are migrant and seasonal youth. Children on family farms, working under the direction of a parent or guardian, are exempted from nearly all regulations. Attempts to update federal laws applied to young agricultural workers have been strongly resisted by major farm organizations for a number of reasons, including concerns over misinterpretations of laws, government intrusion into private lives, and desires to instill a strong work ethic.[50] Another public policy approach is the prosecution of adults for knowingly endangering children. For example, a parent may be charged with criminal negligence if a young child accidentally discharges a loaded firearm that harms another person. Or a parent may be charged with negligence for leaving a toddler strapped inside a hot vehicle without an adult present. Yet, within the agricultural setting, when a child is seriously injured or killed, for example being an extra rider and falling from a tractor, the adult rarely faces criminal charges.[51]

At the institution and organization level, environmental protections can be set by trade organizations and corporations. The Good Agricultural Practices (GAP) program of the USDA is an example of how a federal agency develops voluntary standards that facilitate domestic and international trade arrangements. An agricultural enterprise seeking certification must meet safety criteria and undergo periodic auditing. Many "big box" companies require their agricultural suppliers to be GAP certified. Auditing systems, including GAP certification, set requirements that exclude children from the work environment and establish age limits (e.g., 18 years) for employees. If such auditing programs applied to family farms, children would have limited access to hazardous conditions.

At the community level, environmental and safety protections are promoted by advocacy groups, schools, churches, farm cooperatives, and youth-serving organizations such as the National FFA (formerly known as Future Farmers of America). Their promotion of environmental safeguards such as pesticide safety, hearing protection, and safe operation of motorized vehicles, exert a strong influence on adult decisions affecting children's well-being in rural and agricultural settings. Interpersonal relationships, such as those that exist with relatives, friends, and healthcare providers, have a more direct impact on adult decisions regarding children's exposures. Safety practices promoted by friends and relatives typically emanate from the higher levels of the SEM and can help counteract dangerous farming practices as well as the propensity to prioritize farm production over the safety of children.

With respect to the rural and agricultural environment, an important concept to consider is the "farm-kid paradox" in which parents and guardians believe that the benefits of children living and working in these environments outweigh any risk of disease or injury.[52] Indeed, many aspects of rural and farm life, such as livestock handling, antigen exposures, physical outdoor work, and play, are beneficial in terms of gaining valuable life experiences while building stronger bodies and immune systems and developing a strong work ethic.

There are no simple or straightforward preventive measures for safeguarding children in rural and agricultural settings from unwanted environmental exposures. Educational approaches alone are weak and should be augmented with targeted policies and protocols derived from evidence-based interventions that account for the wide spectrum of environmental conditions and unique attributes of the rural and farming communities.

References

1. Bradford J, Coe E, Enomoto K, White M. COVID-19 and rural communities: protecting rural lives and health. Report by McKinsey & Company. 2021 Mar. https://www.mckinsey.com/industries/healthcare/our-insights/covid-19-and-rural-communities-protecting-rural-lives-and-health

2. U.S. Department of Agriculture, Economic Research Service. What is rural? Updated 2019 Oct. Available at: https://www.ers.usda.gov/topics/rural-economy-population/rural-classifications/what-is-rural/.

3. National Institute for Occupational Safety and Health (NIOSH). Childhood agricultural injury surveillance 1996–2015. Available at: https://www.cdc.gov/niosh/topics/childag/cais/default.html.

4. Probst JC, Barker JC, Enders A, Gardiner P. Current state of child health in rural America: how context shapes children's health. *J Rural Health*. 2016;(34) s3–s12. doi:10.1111.jrh.12222

5. Denning GM, Jennissen CA. Pediatric and adolescent injury in all-terrain vehicles. *Res Sports Med*. 2018;26(Supp1):38–56. doi:10.1080/15438627.2018.1438279

6. Statista. Agriculture > Farming. Total number of farms in the United States from 2000 to 2021. Available at: https://www.statista.com/statistics/196103/number-of-farms-in-the-us-since-2000/.

7. Barrera L. Industry consolidation Part 1. Number of US farms declines while size of farms increases. Available at: https://www.farm-equipment.com/articles/15960-number-of-us-farms-declines-while-size-of-farms-increase.

8. Whitt C, Todd JE, Donald JM. America's diverse family farms: 2020 edition. Economic Information Bulletin Number 220. Available at: https://www.ers.usda.gov/publications/pub-details/?pubid=100011.

9. Hendricks KJ, Hendricks SA, Layne LA. A national overview of youth and injury trends on U.S. farms, 2001–2014. *J Agric Saf Health*. 2021;27(3):121–34. doi:10.13031/jash.14473

10. Castillo D, Hard D, Myers J, Pizatella T, Stout N. A national childhood agricultural injury prevention initiative. *J Agric Saf Health*. Special Issue 1998;1:183–91.

11. Weichelt B, Salzwedel M, Heiberger S, Lee BC. Establishing a publicly available national database of US news articles reporting agriculture-related injuries and fatalities. *Am J Industrial Medicine*. 2018;61(8):667–74. https://doi.org/10.1002/ajim.22860

12. Rivara FP. Fatal and non-fatal farm injuries to children and adolescents in the United States, 1990–3. *Inj Prev*. 1997;3(3):190–94. doi:10.1136/ip.3.3.190

13. Perritt KR, Hendricks KJH, Goldcamp EM. Young worker injury deaths: a historical summary of surveillance and investigative findings. Morgantown, WV: National Institute of Occupational Safety and Health. Publication No. 2017-168. 2017. https://www.cdc.gov/niosh/docs/2017-168/pdfs/2017-168.pdf?id=10.26616/NIOSHPUB2017168

14. National Children's Center for Rural and Agricultural Health and Safety. 2022 Fact sheet: Childhood agricultural injuries. Marshfield, WI: Marshfield Clinic Health System; 2022. doi.org/10.21636/nfmc.nccrahs.injuryfactsheet.r.2022.

15. Pickett W, Brison RJ, Berg RL, Zentner J, Linneman J, Marlenga B. Pediatric farm injuries involving non-working children injured by a farm work hazard: five priorities for primary prevention. *Inj Prev*. 2005;11(1):6–11. doi:10.1136/ip.2004.005769

16. Pate ML, Görücü S. Agricultural work-related fatalities to non-working youth: implications for intervention development. *J Agric Saf Health*. 2020;26(1):31–43. doi:10.13031/jash.13691

17. Centers for Disease Control and Prevention (CDC). Ten great public health achievements—United States, 2001–2010. *MMWR Morb Mortal Wkly Rep*. 2011;60(19):619–23.

18. Zaloshnja E, Miller TR, Lee BC. Incidence and cost of nonfatal farm youth injury, United States, 2001–2006. *J Agromedicine*. 2011;16(1):6–18. doi:10.1080/1059924X.2011.534714

19. Zaloshnja E, Miller TR, Lawrence B. Incidence and cost of injury among youth in agricultural settings, United States, 2001–2006. *Pediatrics*. 2012;129(4):728–34. doi:10.1542/peds.2011-2512

20. National Center for Farmworker Health. Child labor in agriculture fact sheet. NCFH, Inc; 2018. https://www.ncfh.org/child-labor-fact-sheet.html

21. Lee BC, Bendixsen C, Liebman AK, Gallagher SS. Using the socio-ecological model to frame agricultural safety and health interventions. *J Agromedicine*. 2017;22(4):298–303. doi:10.1080/1059924X.2017.1356780

22. Lee B, Marlenga B, eds. *Professional Resource Manual: North American Guidelines for Children's Agricultural Tasks*. Marshfield, WI: Marshfield Clinic; 1999.

23. Swenson AV, Salzwedel M, Peltier C, Lee BC. Safety guidelines for youth agricultural work in the United States: A description of the development and updating process. Front Public Health. 2023;11:1048718. doi: 10.3389/fpubh.2023.1048718.

24. Gadomski A, Ackerman S, Burdick P, Jenkins P. Efficacy of the North American guidelines for children's agricultural tasks in reducing childhood agricultural injuries. *Am J Public Health*. 2006;96(4):722–27. doi:10.2105/AJPH.2003.035428

25. Medel-Herrero A, Martínez-López B, Silva-Del-Rio N, Pires AF, Edmondson A, Schenker M. Tuberculosis prevalence among us crop-workers, 2000 to 2012: trends and contributing factors. *J Occup Environ Med*. 2018;60(7):603–11. doi:10.1097/JOM.0000000000001257

26. Motha RP. The impact of extreme weather events on agriculture in the United States. In: Attri SD, Rathore LS, Sivakumar MVK, Dash SK, eds. *Challenges and Opportunities in Agrometeorology*. Heidelberg: Springer; 2011:397–408.

27. Bangkok RR. "Our children may not want to be farmers": living on the frontline of global heating. *The Guardian*. Available at: https://www.theguardian.com/environment/2021/nov/11/children-may-not-want-to-be-farmers-living-on-the-frontline-of-global-heating.

28. Conrad CC, Stanford K, Narvaez-Bravo C, Callaway T, McAllister T. Farm fairs and petting zoos: a review of animal contact as a source of zoonotic enteric disease. *Foodborne Pathog Dis*. 2017;14(2):59–73. doi:10.1089/fpd.2016.2185

29. Oliver SP, Jayarao BM, Almeida RA. Foodborne pathogens in milk and the dairy farm environment: food safety and public health implications. *Foodborne Pathog Dis*. 2005;2(2):115–29. doi:10.1089/fpd.2005.2.115

30. Bender JB, Shulman SA; Animals in Public Contact subcommittee; National Association of State Public Health Veterinarians. Reports of zoonotic disease outbreaks associated with animal exhibits and availability of recommendations for preventing zoonotic disease transmission from animals to people in such settings. *J Am Vet Med Assoc*. 2004;224(7):1105–9. doi:10.2460/javma.2004.224.1105

31. Conners EE, Miller AD, Balachandran N, Robinson BM, Benedict KM. Giardiasis outbreaks—United States, 2012–2017. *MMWR Morb Mortal Wkly Rep*. 2021;70(9):304–7. doi:10.15585/mmwr.mm7009a2

32. Stanek G, Wormser GP, Gray J, Strle F. Lyme borreliosis. Lancet. 2012 Feb 4;379(9814):461–73. doi:10.1016/S0140-6736(11)60103-7.

33. Long AD, Baldwin-Brown J, Tao Y, et al. The genome of Peromyscus leucopus, natural host for Lyme disease and other emerging infections. *Sci Adv*. 2019;5(7):eaaw6441. doi:10.1126/sciadv.aaw6441

34. Centers for Disease Control and Prevention (CDC). Tick paralysis—Washington, 1995. *MMWR Morb Mortal Wkly Rep*. 1996;45(16):325–6.

35. CDC, Centers for Disease Control and Prevention. Babesiosis. Available at: https://www.cdc.gov/parasites/babesiosis/ Surveillance for Babesiosis.

36. Fatmi SS, Zehra R, Carpenter DO. Powassan virus-a new reemerging tick-borne disease. *Front Public Health*. 2017;5:342. doi:10.3389/fpubh.2017.00342

37. Wilson JM, Schuyler AJ, Workman L, et al. Investigation into the α-Gal syndrome: characteristics of 261 children and adults reporting red meat allergy. *J Allergy Clin Immunol Pract*. 2019;7(7):2348–58.e4. doi:10.1016/j.jaip.2019.03.031

38. Haas E. 1996. Tick paralysis – Washington, 1995. *MMWR Weekly*. 1996 Apr 26;45(16):325–6. https://www.cdc.gov/mmwr/preview/mmwrhtml/00040975.htm

39. Biggs HM, Behravesh CB, Bradley KK, et al. Diagnosis and management of tickborne rickettsial diseases: rocky mountain spotted fever and other spotted fever group rickettsioses, ehrlichioses, and anaplasmosis—United States. *MMWR Recomm Rep*. 2016;65(2):1–44. doi:10.15585/mmwr.rr6502a1

40. Bergqvist C, Kurban M, Abbas O. Orf virus infection. *Rev Med Virol*. 2017;27(4). doi:10.1002/rmv.1932

41. Morroy G, van der Hoek W, Albers J, Coutinho RA, Bleeker-Rovers CP, Schneeberger PM. Population screening for chronic Q-fever seven years after a major outbreak. *PLoS One*. 2015;10(7):e0131777. doi:10.1371/journal.pone.0131777

42. Maurin M, Raoult D. Q fever. *Clin Microbiol Rev.* 1999 Oct;12(4):518–53. doi:10.1128/CMR.12.4.518.

43. Hjelle B, Glass GE. Outbreak of hantavirus infection in the Four Corners region of the United States in the wake of the 1997–1998 El Nino-southern oscillation. *J Infect Dis.* 2000;181(5):1569–73. doi:10.1086/315467

44. Choi E. Tularemia and Q fever. *Med Clin North Am.* 2002;86(2):393–416. doi:10.1016/s0025-7125(03)00094-4.

45. Mueller JT, McConnell K, Burow PB, Pofahl K, Merdjanoff AA, Farrell J. Impacts of the COVID-19 pandemic on rural America. *Proc Natl Acad Sci U S A.* 2021;118(1):2019378118. doi:10.1073/pnas.2019378118

46. Becot FA. Children, work, and safety on the farm during COVID-19: a harder juggling act *J Agromedicine.* 2022;27(3):315–28. doi:10.1080/1059924X.2022.2068716

47. Conger RD, Ge X, Elder GH Jr, Lorenz FO, Simons RL. Economic stress, coercive family process, and developmental problems of adolescents. *Child Dev.* 1994;65(2 Spec No.):541–61.

48. Hagen BNM, Albright A, Sargeant J, et al. Research trends in farmers' mental health: A scoping review of mental health outcomes and interventions among farming populations worldwide. *PLoS ONE.* 2019;14(12): e0225661.

49. Pickett W, Berg RL, Marlenga B. Health and well-being among young people from Canadian farms: associations with a culture of risk-taking. *J Rural Health.* 2017;34(3):275–82. Https://doi.org/10.111/jrh.12281

50. Reid-Musson E, Strauss K, Mechler M. "A virtuous industry": the agrarian work-family ethic in US rulemaking on child agricultural labour. *Globalizations.* 2022. https://doi.org/10.1080/14747731.2022.2031795

51. Benny CP, Beyer D, Krolczyk M, Lee BC. Legal responses to child endangerment on farms: Research methods. *Front Public Health.* 2022;10:1015600. doi: 10.3389/fpubh.2022.1015600.

52. Bendixsen C. A farm kid paradox. *J Anthropol N Am Fall* 2019;22(2):139–142. doi:10.1002/nad.12118.

22

The Work Environment and Children's Health, Safety, and Well-Being

Kimberly Rauscher and Jennifer Fuller

Introduction

Work is something that few can avoid. It is how we provide for our basic needs and wants and is not optional for most people. But when should work begin, and what should that work look like? Is work a wholly positive thing that should be encouraged, or should it be delayed as long as possible to allow young people to grow outside the often negative influences of the world of work? In most countries around the world, children labor, some legally and some illegally, some with protections and some without. In most high-income countries, work looks very different for young people than it does for their peers in low- and middle-income nations. Young workers in the United States and similar nations are protected by laws that help ensure their safety and keep them in school as long as possible. Although the working conditions in industrialized countries are superior to those of children working around the world in the brick-making industry, sewing soccer balls or rugs by hand, or scavenging for scrap metal, for example, they can present very real risks to the physical and mental health of young workers. This chapter describes the working conditions of youth in the United States and the risks posed to their health and well-being. It ends with several recommendations for how these conditions and experiences can be improved.

Employment Patterns of Young People

Labor Force Participation and Employment

The number of young people who are employed in formal jobs has been steadily increasing over the past decade (Figure 22.1) and the youth labor force participation rate in 2019 (61.9%) was the highest it has been in 9 years.[1] The 19.3 million working 16- to 24-year-olds made up 12% of the total workforce in that year, or just over 1 in every 10 workers.[2] In the United States, people younger than 16 are legally allowed to work under limited conditions,[3] but it is challenging to estimate exactly how many. The US Bureau of Labor Statistics (BLS) does not consider these youth part of the labor force, and they are generally excluded from published BLS employment statistics.[4] (The BLS defines the labor force as including *only* non-institutionalized people ages 16 and older who are working or looking for work and who are not students or retirees.[4]) This limitation leaves us lacking a full picture of the employment patterns of all working youth.

Kimberly Rauscher and Jennifer Fuller, *The Work Environment and Children's Health, Safety, and Well-Being*
In: *Textbook of Children's Environmental Health*, Second Edition. Edited by: Ruth A. Etzel and Philip J. Landrigan,
Oxford University Press. © Oxford University Press 2024. DOI: 10.1093/oso/9780197662526.003.0022

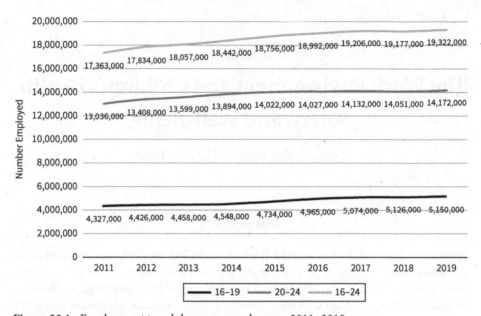

Figure 22.1. Employment trends by young worker age, 2011–2019

Source: U.S. Bureau of Labor Statistics. Household data annual averages, Table 3. Employment status of the civilian noninstitutionalized population by age, sex and race for the years 2011–2019

Young people in the United States are employed in a handful of industries and occupations. Table 22.1 shows that a majority of 16- to 19-year-olds worked in only two industries in 2019: "leisure and hospitality" and "wholesale and retail trade."[5] The majority of 20- to 24-year-olds also were found in these same two industries, with the addition of

Table 22.1 Top industries and occupations of young workers ages 16–24, by age group (2019)

	16–19 years old (N = 5,150,000)		20–24 years old (N = 14,172,000)	
	Number	Percent	Number	Percent
Major industry categories				
Education and health services	595,000	11.6	2,803,000	19.8
Leisure and hospitality	2,124,000	41.2	2,666,000	18.8
Wholesale and retail trade	1,181,000	22.9	2,671,000	18.8
Professional and business services	219,000	4.3	1,372,000	9.7
Major occupation categories				
Sales and office occupations	1,717,000	33.3	3,836,000	27.1
Service occupations	2,126,000	41.3	3,815,000	26.9
Management, professional, and related Occupations	428,000	8.3	3,402,000	24.0
Production, transportation, and material moving occupations	54,300	10.5	180,800	12.8

Source: US Bureau of Labor Statistics (BLS), Employment Data from the Current Population Survey.

"education and health services"[5] rounding out the top three industries. In terms of occupations, both age groups worked largely in "sales and office occupations" as sales clerks, cashiers, and office workers, as well as in "service occupations" working mainly as food servers and preparation workers (e.g., prep-chefs, cooks).[5]

Demographic Differences

After the 1980s, as more women moved into the workplace, the gender gap in youth employment also began to close and today young males and females are just as likely to be employed before the age of 25.[2,6] In 2019, the leisure and hospitality industry employed the largest proportions of both females (44.6%) and males (39.2%) ages 16–19 years.[7] Among the 20- to 24-year-olds, females worked mainly in education and health services (30.4%) and leisure and hospitality (21.0%) while males were dispersed across a wider range of industries, with the largest proportion being in wholesale and retail trade (21.0%).[7] This shows that the gendered work we see among adults also is evident even among young workers.[8] Even with improvements over time in employment rates between youth of differing races, racial differences have been persistent, with White youth still being more likely than all other racial groups to be employed.[2,6]

Legal Protections for Young Workers

Child Labor Regulations

In the United States, minors under the age of 18 are protected by child labor regulations born from federal legislation passed in 1938: the Fair Labor Standards Act (FLSA). These regulations were put in place to protect the health, morals, and safety of children.[9] These protections saved children from work in places like coal mines, canning factories, and other jobs that posed serious risks to their physical health. The passage of child labor protections is considered among the most significant public health victories of the time, and these laws still safeguard young workers today.[9]

Occupation/Task Prohibitions
Last updated in 2010, the child labor regulations limit the types of *jobs* young people are allowed to perform and the *hours* they are allowed to work. There are separate rules for children working in agricultural and nonagricultural industries, and the rules differ by age. The regulations that restrict jobs for all minors are called Hazardous Occupations Orders (HO) and include 17 nonagricultural jobs/tasks too dangerous for anyone under the age of 18 to perform and 11 agricultural jobs/tasks too dangerous for anyone under 16.[3,10] There are also 19 additional job/task restrictions that apply to 14- and 15-year-olds working only in nonagricultural industries.[3] Table 22.2 shows a sampling of these restrictions.

Hour Restrictions
The hour restrictions of the FLSA apply only to 14- and 15-year-olds working in nonagricultural industries and differ depending on the time of year, with greater protections in place during the school year compared to summer.[3] These laws also limit the times of

Table 22.2 Select occupation restrictions for minor employees in nonagricultural industries

---

Hazardous occupations orders applicable to employees under age 18

Occupations of motor-vehicle driver and outside helper (570.52-HO2)

Occupations involved in the operation of power-driven woodworking machines (570.55-HO5)

Occupations in the operation of power-driven meat-processing machines and occupations involving slaughtering, meat packing or processing, or rendering (570.61-HO10).

Occupations involved in the operation of bakery machines (570.62-HO11).

Occupations involved in the operation of paper-products machines, scrap paper balers, and paper box compactors (570.63-HO12)

Occupations involved in the operations of circular saws, band saws, and guillotine shears (570.65-HO14)

Occupations in roofing operations and on or about a roof (570.67-HO16)

Occupational standards applicable to employees ages 14 and 15

Occupations in connection with transportation of persons or property by rail, highway, air, water, pipeline, or other means [570.33 (f1)]

Occupations in connection with communications and public utilities [570.33 (f3)]

Occupations in connection with construction (including demolition and repair) [570.33 (f4)]

Outside window washing that involves working from window sills, and all work requiring the use of ladders, scaffolds, or their substitutes [570.34 (b4)]

Baking and cooking are prohibited with limited exceptions [570.34 (b5)]

Occupations which involve operating, setting up, adjusting, cleaning, oiling, or repairing power-driven food slicers and grinders [570.34 (b6)]

Loading and unloading goods to and from trucks, railroad cars, or conveyors [570.34 (b8)]

All occupations in warehouses except office and clerical work [570.34 (b9)]

For the full list, see original source: US Child Labor Bulletin 101.Child Labor Provisions for Non-agricultural Occupations Under the Fair Labor Standards Act. 2010.

day when 14- and 15-year-olds can work, to prohibit what is referred to as "night work," or working after the sun has gone down and dangers increase. Hour limits are shown in Table 22.3.

Hour restrictions for 16- and 17-year-olds do not exist at the federal level, yet these workers may be protected by state level regulations. Children who work in their family's business and all youth employed in agriculture (even those not working on their family's farm)[3,10] are not subject to hour restrictions. Parents who employ their children in non-agricultural businesses, however, must comply with the applicable job/task restrictions.

Health and Safety Standards

Although they are not youth-specific, young workers also are protected by the safety regulations established under the 1970 Occupational Safety and Health Act. This Act was established to "assure safe and healthful working conditions for men and women" and led to the creation of the Occupational Safety and Health Administration (OSHA).[11] In addition to hazard-specific regulations, the OSH Act includes the General Duty

Table 22.3 Child labor regulations on work hours for employees ages 14 and 15 in nonagricultural industries, by time of year

---

Maximum weekly hours of work

40 when school is not in session
18 when school is in session
8 per day when school is not in session
3 per day when school is in session

Permissible hours of work

7 AM to 7 PM when school is in session
7 AM to 9 PM when school is not in session

Source: US Child Labor Bulletin 101.Child Labor Provisions for Non-agricultural Occupations Under the Fair Labor Standards Act. 2010.

Clause (5(a)(1)), which puts forth that employers must provide a workplace free from recognized hazards.[11] While these regulations affect all workers regardless of age, some standards are particularly relevant to young workers. These include those pertaining to training and hazard communication.

Hazard Communication

Also referred to as the "right-to-know" standard, the Hazard Communication Standard (1910.1200 (b)(1)) requires employers to inform employees about the health risks associated with the chemicals they are using, what protective measures must be taken when using the chemicals, and the first aid procedures to follow should an exposure occur. In general terms, this is mainly accomplished through employee training and the labeling of chemicals.

Safety Training

When young people are starting out in the workforce, they tend to try out a variety of jobs. This means every new job is a new experience in which their surroundings are unfamiliar, making training in safety procedures critical for young workers. OSHA regulations require workers to be trained when first hired.[11] This training must be provided in a language and method the learner will understand, which speaks directly to the need for age-appropriate training for young people (discussed later in this chapter).

The protections provided by both the FLSA and the OSH Act are enforced by the US Department of Labor (DOL) and, together, provide the foundation for young worker safety. Once a young worker reaches the age of 18, the child labor protections no longer apply and their only protections are the OSHA regulations.

The Working Conditions of Young People

As mentioned in the Introduction, the work experiences of young people in the United States are superior to those of many others around the world, yet not without some significant drawbacks. While many view work as a good thing for youth, teaching them independence, skills, and maturity, the empirical evidence suggests that the extent to which youth benefit from work has much more to do with the quality of the work experience than simply the act of working itself. While many do reap positive psychosocial benefits from their work experience, not all youth are as lucky. This section provides an overview of research on the working conditions young people experience that put them at risk for a host of negative outcomes, both mental and physical.

Injury Risks

Prevalence of Work-Related Injury

Despite the safety and child labor protections in place, hundreds of thousands of young people are injured on the job every year, some fatally. Between 2012 and 2018, 3.2 million work-related injuries among young people were treated in emergency departments[12]; that is approximately 500,000 injuries per year (not injured workers).[12] Across this period, the injury rates among 15- to 24-year-olds were 1.2 to 2.3 times higher than the rate for 25- to 44-year-olds who are at the prime working age.[12] Fatalities, while the most serious of cases, are much more rare. Between 2001 and 2012, there were 406 fatalities among workers under the age of 18 across the United States.[13] The authors noted that these fatalities translated into "24,790 years of potential life lost; 12,241 of which were in agriculture alone."[13]

The literature on the epidemiology of adolescent work-related injury is small in comparison to that of adult workers but spans more than 30 years. Age, gender, race, and socioeconomic status all are associated with young worker injury risk,[6,14–16] and multiple work-based factors lead to greater injury risk as well as negative impacts on mental health and psychosocial development of young workers. These are discussed below.

Injury Risk Factors

Enumerated in the NIOSH publication, Proceedings of a US and Canadian Series of Symposia, *Health and Safety of Young Workers: Proceedings of a US and Canadian Series of Symposia*[17] work-based risk factors for young worker injury include being new to the job, working without supervision or without safety training, working long hours or at a fast pace, and using equipment or tools, handling hot liquids and grease, using cutting tools, lifting or moving heavy objects, operating tractors and other farm vehicles (e.g., all-terrain vehicles), and working late at night and/or alone.

Working at night is a risk factor for crime-related assaults and fatalities, as are cash handling and customer interaction.[18,19] This is particularly troubling because these risks are common in the retail and service sectors[20,21] where the large majority of young people work. Data from the National Crime Victimization Survey show that the highest rates of workplace violence among young people were in protective service occupations, transportation, and retail sales.[22] Last, when youth are employed in violation of the child labor regulations, they are at increased risk for both injury and death.[23–26]

Health Risks

Although far less common than safety hazards, young workers may also be exposed to health hazards.[27–29] In one of the most recent comprehensive studies on young worker exposures, Runyan et. al found that 55% of young workers were exposed to thermal hazards and 54% to chemical hazards[29] (mainly cleaning products). Biological hazards are common in healthcare settings where young people mainly work as dietary aides, orderlies, or janitors. These are jobs in which they may come in contact with bloodborne pathogens, bacteria, and viruses from contaminated surfaces, needles, and other medical instruments.

Risks to Positive Psychosocial Development

Poor Work Quality

Young workers are still developing in all ways—socially, emotionally, cognitively, and physically,[6,30]—and their work experiences should be ones that promote positive development. However, many work experiences have the opposite effect. Two of the major working conditions that negatively impact youth development are work stress and long working hours.

Stressful Work Environments

When young workers experience stress at work they are more likely to experience a variety of mental health problems.[31] Work stress can come from many sources including understaffing and fast-paced work as well as conflict with coworkers and supervisors or angry customers.[32] Customers are among the most common perpetrators of workplace violence (e.g., verbal abuse, threats of violence) against young workers.[33,34] This is particularly problematic for youth, given that they largely work in public-facing occupations where they have a lot of contact with the public.[5] It may not be surprising then to learn that between 33% and 74% of young people report having experienced sexual harassment at work, mainly from customers and coworkers.[35,36]

Sexual harassment, interpersonal conflict, and bullying at work all have significant emotional consequences for youth[37] including depression and decreased self-esteem,[4,38] as well as nightmares and thoughts of suicide.[38] Young women who are sexually harassed at work face similar effects and also are at increased risk of anxiety and eating disorders.[39,40]

Working Long Hours

Multiple studies show that working more than 20 hours per week during high school has negative contemporaneous and future impacts.[41,42] High work intensity is associated with dropping out of school, truancy, delinquency, substance use, smoking, and early problem behaviors.[41]

Improving the Work Experiences of Young People

Below are some key strategies that can improve work experiences for youth.

Eliminate or Reduce Hazards in the Workplace

Studies characterizing the youth workplace show that young workers are using equipment they feel is dangerous, including power equipment/tools, motor vehicles, and heavy machinery.[27] They also use ladders or scaffolding, forklifts, electric food slicers, and hot oil fryers.[27] Adding to the risk is the fact that many young workers often feel rushed at work and feel pressure from supervisors, coworkers, and customers to work quickly.[32]

To keep workers safe, these and other hazards must be eliminated or, at a minimum, reduced. Using engineering solutions and technological advances in the workplace that do not rely on individual effort or action are the best options. These approaches are prioritized in the hierarchy of controls used to evaluate various methods to mitigate risk. At the bottom of the hierarchy are administrative controls, such as training and personal protective equipment (PPE), which take hazards for granted. Many of the interventions targeted at young workers occur at the lower levels of the hierarchy. Relying on the individual behaviors of young workers to follow administrative procedures and use PPE is the least effective way to protect them.

Expand the Child Labor and OSHA Regulations

Although these regulations have protected countless lives, they nevertheless have significant limitations. Children who work on their family's farm are not subject to a single child labor regulation.[10] Children under the age of 17 who work on farms not owned by their parents are protected by the agricultural child labor regulations,[10] but if they are any older, they have no restrictions placed on their work.

The laws on both hours and tasks need to be brought into alignment with the level of protection given every other child who has *chosen* to work. Adding requirements for supervision and safety training for minors, particularly when working under known hazardous conditions, would help encourage employers to take the need for these supports more seriously. Together with increased penalties for child labor and safety violations, these measures would improve safety for young people.

Improve Enforcement

Violations of the federal child labor regulations are far more frequent than one might imagine in the United States. Every year, hundreds of thousands of youth are given tasks and asked to work hours that are in violation of the child labor regulations.[43] In one survey of working youth, 37% of those in retail and service occupations reported performing a task that was prohibited by the child labor regulations.[43] In a population study of worker fatalities, child labor violations were responsible for 43% of the deaths reported from 2001 to 2012.[26]

Regulatory protections are only as effective as the efforts put forth to enforce them. The lack of adequate enforcement activities by the DOL's Workers Health Division is a perennial problem and tends to fluctuate with the goals of each presidential administration. The focus on enforcement gives way to "compliance assistance" and then back to enforcement again. Even when resources are plentiful, enforcement activities generally operate on a complaint-driven basis. It is problematic if workplace safety enforcement relies on the reporting of problems by the young people who experience them. The DOL must recommit its efforts to the mission of enforcing employer compliance with the child labor and safety regulations.

Provide Safety Training and Adequate Supervision

When working in the same jobs, coworkers of different ages may have the same exposures to these hazards, yet the younger workers are less equipped to respond in a manner that protects their health, safety, and well-being. As a group, young workers do not have the life experiences, education, and knowledge of their more advanced coworkers. As such, they need training and supervision to develop the skills necessary to identify hazards, engage controls, and respond appropriately to risks.[44]

Young people, particularly those in early and middle adolescence, are in a developmental stage in which they are striving to gain competence and become more self-reliant. In the work setting, this often leads youth to try to prove that they are reliable, dependable workers who do not complain or cause trouble.[6,45,46] Given these completely normal developmental goals, it is not surprising that youth often do not speak up when they have safety concerns,[46,47] blame themselves,[48] and do not report their injuries to their supervisors.[34,48,49] It is therefore important to ensure that young workers are given high-quality supervision and safety training.

Unfortunately, many young people do not receive either. In some studies, as many as half of young workers reported they had not received any health and safety training.[23,50-53] Additionally, there is a lack of adequate supervision in the youth workplace.[27,54,55] The most recent comprehensive study of working teens showed that more than one-quarter of youth worked with no adult supervisor present at least 1 day a week and that only 25% had supervisors who checked on them once a day to see if they were doing their jobs correctly.[27]

All employers have both a moral and a legal right to keep young employees safe, and part of that responsibility involves providing safety training on any dangerous tasks they should expect to encounter on the job and monitoring their work to ensure they are working safely and that supervisors are available should questions or concerns arise.

When supervisors are supportive and model good safety behaviors, young workers are more likely to adopt those behaviors and feel more confident to speak up[56,57]; this in turn moderates injury risk.[58] The most successful supervisors teach young workers through hands-on experiences and demonstrations of safety procedures, and they model safety practices as well as inculcate within each worker the shared expectation of safety.

Improve the Overall Quality of the Work

Good work quality can have positive effects on the mental health and psychosocial development of young workers. By far the most influential job quality is skill use. The positive impacts of skill use have been summarized.[32] When young people learn and use new skills on the job, their feeling of competence increases their hope for the future, and life satisfaction also increases. When youth view the skills they are learning as relevant to their career goals, their self-esteem increases, their depression lessens, and their alcohol use is lower.

Helping young people find more fulfilling jobs that provide true growth opportunities should be prioritized, particularly for those who are not college-bound. It is likely, however, that the quality of work available to young people may continue to deteriorate. As employers demand increasing "flexibility," more young workers may find themselves part of the precarious "gig economy."[32,59]

Work experiences that are positive still should be limited in their intensity because too much work can be harmful to youth. Based on the large body of evidence on the effects of work intensity on young people, the National Academy of Sciences has recommended limiting youth work to 20 hours or less per week.[6,41]

References

1. US Bureau of Labor Statistics, US Department of Labor. Youth labor force participation rate at 61.8 percent in July 2019, a 9-year high. *The Economics Daily*. 2019. Available at: https://www.bls.gov/opub/ted/2019/youth-labor-force-participation-rate-at-61-point-8-percent-in-july-2019-a-9-year-high.htm.

2. US Bureau of Labor Statistics. Labor Force Statistics from the Current Population Survey. Employment status of the civilian noninstitutional population 16 to 24 years of age by sex, race, and Hispanic or Latino ethnicity, July 2018–2021. Updated 2023. https://www.bls.gov/cps/cpsaat03.htm

3. United States Department of Labor. *Child Labor Bulletin 101: Child Labor Provisions for Nonagriculture Occupations Under the Fair Labor Standards Act*. Washington, DC: US Department of Labor, Wage and Hour Division; 2010 July. Report No.: WH-1330.

4. US Department of Labor. Design and methodology: current population survey. Technical Paper 66. 2006. https://www.census.gov/housing/hvs/files/tp-66.pdf

5. US Bureau of Labor Statistics. Household data annual averages. 11b. Employed persons by detailed occupation and age. 2019. Available at: https://bls.gov/cps/aa2019/cpsaat11b.htm.

6. National Research Council, Institute of Medicine, Committee on the Health and Safety Implications of Child Labor. *Protecting Youth at Work: Health, Safety, and Development of Working Children and Adolescents in the United States*. Washington, DC: National Academies Press; 1998.

7. US Bureau of Labor Statistics. Household data annual averages. Employed persons by industry, sex, and age. 2021. Available at: https://www.bls.gov/cps/cpsa2021.pdf.

8. Sheppard LC, Raby R, Lehmann W, Easterbrook R. Grill guys and drive-thru girls: discourses of gender in young people's part-time work. *Journal of Childhood Studies*. 2019;44(3):56–69.

9. Trattner WI. *Crusade for the Children: A History of the National Child Labor Committee and Child Labor Reform in America*. Chicago: Quadrange Books; 1970.

10. United States Department of Labor, Wage and Hour Division. *Child Labor Bulletin 102: Child Labor Requirements in Agricultural Occupations Under the Fair Labor Standards Act*. Washington, DC; US Department of Labor, Wage and Hour Division; 2007. Report No.: WH-1295.

11. The Occupational Safety and Health Act. Public Law 91-596 84 STAT. 1590. 91st Congress, S.2193. December 29, 1970, as amended through January 1, 2004. 1970. Available at: https://www.osha.gov/laws-regs/oshact/completeoshact.

12. Guerin RJ, Reichard AA, Derk S, Hendricks KJ, Menger-Ogle LM, Okun AH. Nonfatal occupational injuries to younger workers—United States, 2012–2018. *MMWR Morb Mortal Wkly Rep*. 2020;69(35):1204–9.

13. Rauscher KJ, Myers DJ. Occupational fatalities among young workers in the United States: 2001–2012. *Am J Ind Med*. 2016;59(6):445–52.

14. Apostolico AA, Shendell DG. Injury surveillance and associations with socioeconomic status indicators among youth/young workers in New Jersey secondary schools. *Environ Health*. 2016;15:1–9.

15. Rauscher KJ, Myers DJ. Socioeconomic disparities in the prevalence of work-related injuries among adolescents in the United States. *J Adolesc Health*. 2008;42(1):50–7.

16. Zierold KM, Anderson HA. Racial and ethnic disparities in work-related injuries among teenagers. *J Adolesc Health*. 2006;39(3):422–6.

17. Rauscher K, Runyan C. The prevalence of working conditions associated with adolescent occupational injury in the US: a review of the literature. In: Runyan C, Lewko J, Rauscher K, Castillo D, Brandspigel S, eds. *Health and Safety of Young Workers: Proceedings of a US and Canadian Series of Symposia* DHHS(NIOSH) (pp. 126–136). Publication No. 2013-144. Washington, DC: National Institute for Occupational Safety and Health; 2013.

18. National Institute for Occupational Safety and Health. *NIOSH Alert: Preventing Deaths, Injuries and Illnesses of Young Workers*. Washington, DC: US Department of Health and Human Services, Centers for Disease Control and Prevention; 2003. Report No.: DHHS (NIOSH) 2003-128.

19. Richardson S, Windau J. Fatal and nonfatal assaults in the workplace, 1996 to 2000. *Clin Occup Environ Med*. 2003;3(4):673–89.

20. Moracco KE, Runyan CW, Loomis D, Wolf SH, Napp D, Butts JD. Killed on the clock: a population-based study of workplace homicide, 1977–1991. *Am J Ind Med*. 2000;37(6):629–36.

21. Peek-Asa C, Runyan CW, Zwerling C. The role of surveillance and evaluation research in the reduction of violence against workers. *Am J Prevent Med*. 2001;20(2):141–8.

22. Toussaint M, Ramirez MR, Peek-Asa C, Saftlas A, Casteel C. Workplace violence victimization in young workers: an analysis of the US National Crime Victimization Survey, 2008 to 2012. *Am J Ind Med*. 2019;62(8):691–700.

23. Zierold K, Anderson H. The relationship between work permits, injury, and safety training among working teenagers. *Am J Ind Med*. 2006;49(5):360–6.

24. Suruda A, Philips P, Lillquist D, Sesek R. Fatal injuries to teenage construction workers in the US. *Am J Ind Med*. 2003;44(5):510–4.

25. Rauscher K, Runyan C. Adolescent occupational fatalities in North Carolina (1990–2008): an investigation of child labor and OSHA violations and enforcement. *New Solut*. 2012;22(4):473–88.

26. Rauscher KJ, Myers DJ, Miller ME. Work-related deaths among youth: understanding the contribution of US child labor violations. *Am J Ind Med*. 2016;59(11):959–68.

27. Runyan C, Schulman M, Dal Santo J, Bowling J, Agans R, Ta M. Work-related hazards and workplace safety of US adolescents employed in the retail and service sectors. *Pediatrics*. 2007;119(3):526–34.

28. Woolf A. Adolescents' descriptions of hazards in the workplace. *Pediatrics*. 2007;120(3):685; author reply, 6.

29. Runyan CW, Vladutiu CJ, Rauscher KJ, Schulman M. Teen workers' exposures to occupational hazards and use of personal protective equipment. *Am J Ind Med*. 2008;51(10):735–40.

30. Sudhinaraset M, Blum R. The unique developmental considerations of youth-related work injuries. *Int J Occ Environ Health*. 2010;16(2):216–22.

31. Law PCF, Too LS, Butterworth P, Witt K, Reavley N, Milner AJ. A systematic review on the effect of work-related stressors on mental health of young workers. *Int Arch Occup Environ Health*. 2020;93(5):611–22.

32. Rauscher KJ, Wegman DH, Wooding J, Davis L, Junkin R. Adolescent work quality: a view from today's youth. *J Adolesc Res*. 2013;28(5):557–90.

33. Rauscher KJ. Workplace violence against adolescent workers in the US. *Am J Ind Med*. 2008;51(7):539–44.

34. Brown B, Myers D, Casteel C, Rauscher K. Exploring differences in the workplace violence experiences of young workers in middle and late adolescence in the United States. *J Safety Res*. 2020;74:263–9.

35. Fineran S, Gruber JE. Youth at work: adolescent employment and sexual harassment. *Child Abuse Negl*. 2009;33(8):550–9.

36. Leslie CA. Extent and nature of sexual harassment in the fashion retail workplace: 10 years later. *Fam Consum Sci Res J*. 2005;34(1):8–34.

37. Shields M, Dimov S, Kavanagh A, Milner A, Spittal MJ, King TL. How do employment conditions and psychosocial workplace exposures impact the mental health of young workers? A systematic review. *Soc Psychiatry Psychiatr Epidemiol*. 2021;56(7):1147–60.

38. Kenway J, Fitzclarence L. Toxic shock: understanding violence against young males in the workplace. *J Men Stud*. 2000;8(2):131.

39. Clark JP. "The more lady you are, the more they treat you like a lady": sexual harassment and health risk for young women in a male-dominated work setting. *Can Womens Stud*. 1998;18(1):82.

40. Tang TL-P, McCollum SL. Sexual harassment in the workplace. *Public Pers Manage*. 1996;25(1):53–8.

41. Staff J, Yetter AM, Cundiff K, Ramirez N, Vuolo M, Mortimer JT. Is adolescent employment still a risk factor for high school dropout? *J Res Adolesc*. 2020;30(2):406–22.

42. Leadbeater B, Ames ME, Contreras A. Male-dominated occupations and substance use disorders in young adulthood. *Am J Mens Health*. 2020;14(2).

43. Rauscher KJ, Runyan CW, Schulman MD, Bowling JM. US child labor violations in the retail and service industries: findings from a national survey of working adolescents. *Am J Public Health*. 2008;98(9):1693–9.

44. Laberge M, Ledoux E. Occupational health and safety issues affecting young workers: a literature review. *Work*. 2011;39(3):215–32.

45. Curtis Breslin F, Polzer J, MacEachen E, Morrongiello B, Shannon H. Workplace injury or "part of the job"?: towards a gendered understanding of injuries and complaints among young workers. *Soc Sci Med*. 2007;64(4):782–93.

46. Runyan CW, Schulman M, Ta M. *Adolescent Employment: Relationships to Injury and Violence*. Liller K, ed. Washington, DC: American Public Health Association; 2005.

47. Zakocs RC, Runyan CW, Schulman MD, Dunn KA, Evensen CT. Improving safety for teens working in the retail trade sector: opportunities and obstacles. *Am J Ind Med*. 1998;34(4):342–50.

48. Zierold KM, McGeeney TJ. Communication breakdown: how working teens' perceptions of their supervisors impact safety and injury. *Work*. 2016;54(1):3–9.

49. Tucker S, Diekrager D, Turner N, Kelloway EK. Work-related injury underreporting among young workers: prevalence, gender differences, and explanations for underreporting. *J Safety Res*. 2014;50(0):67–73.

50. Massachusetts Department of Public Health Occupational Health Surveillance Program. *Teens at Work. Work-Related Injuries to Teens in Massachusetts, 2000–2004.* Boston: Massachusetts Department of Public Health; 2007 Dec.

51. Knight EB, Castillo DN, Layne LA. A detailed analysis of work-related injury among youth treated in emergency departments. *Am J Ind Med*. 1995;27(6):793–805.

52. Zierold KM. Safety training for working youth: methods used versus methods wanted. *Work*. 2016;54(1):149–57.

53. Zierold KM, Anderson HA. Severe injury and the need for improved safety training among working teens. *Am J Health Behav*. 2006;30(5):525–32.

54. Zierold KM. Perceptions of supervision among injured and non-injured teens working in the retail or service industry. *Workplace Health Saf*. 2016;64(4):152–62.

55. Runyan CW, Zakocs RC. Epidemiology and prevention of injuries among adolescent workers in the United States. *Annu Rev Public Health*. 2000;21:247–69.

56. Rohlman DS, TePoel M, Campo S. Evaluation of an online training for supervisors of young agricultural workers. *Int J Environ Res Public Health*. 2021;18(19).

57. Turner N, Tucker S, Deng C. Revisiting vulnerability: comparing young and adult workers' safety voice intentions under different supervisory conditions. *Accid Anal Prev*. 2020;135.

58. Tucker S, Turner N. Sometimes it hurts when supervisors don't listen: the antecedents and consequences of safety voice among young workers. *J Occup Health Psychol*. 2015;20(1):72–81.

59. Santilli S, Di Maggio I, Ginevra MC, Nota L, Soresi S. Stories of courage among emerging adults in precarious work. *Career Dev Q*. 2021;69(3):248–62.

23

Intrauterine Nutrition and Children's Health

Alicia Cousins, Nicholas Rickman, and Cynthia F. Bearer

Adequate intrauterine nutrition is essential for fetal growth, development, and survival. Infant birth weight is the primary clinical marker of excess or deficient intrauterine nutrition. Low birth weight (LBW) describes infants weighing less than 2,500 grams and occurs at rates of approximately 8% in the United States, 7% in all developed countries, and in excess of 15% for developing countries. LBW may reflect normal fetal growth in constitutionally small infants, fetal growth restriction (FGR, fetal growth <10th percentile for gestational age), or premature birth (<37 weeks gestation). Premature birth accounts for approximately two-thirds of all LBW deliveries in developed countries. Factors such as gender, birth order, number of fetuses, maternal diet, maternal age, and maternal socioeconomic status all contribute to birth weight and future health. In the presence of FGR (also referred to as intrauterine growth restriction [IUGR] and small for gestational age [SGA]) and/or preterm birth, LBW is associated with increased infant morbidities and death.[1]

FGR results from decreased nutrient delivery to the fetus due to maternal disease, smoking, impaired intrauterine blood flow, or altered maternal dietary intake. Deficiencies in maternal intake can lead to a global reduction in energy, protein, fat, vitamins, and minerals to the fetus. FGR can also be seen in infants of mothers with an identified single micronutrient deficiency such as choline, iron, selenium, zinc, folate, or vitamin D deficiency.[2]

Alterations in the aforementioned nutrients during critical periods of fetal growth and development affect not only birth weight, but an infant's short- and long-term health. In the immediate post-birth timeframe, infants born FGR are more susceptible to infection, hypothermia, and hypoglycemia. Those who are high birth weight (>4,000 g) have a greater incidence of delivery-related complications and injury.[1]

The long-term health repercussions are thought to be secondary to permanent alterations to tissue structure and function as a result of intrauterine insults. The term "programming," coined in 1991, describes the process by which specific environmental stimuli or insults trigger adaptations that result in permanent changes to the physiology of the organism, particularly during critical phases of development.[3] These changes are likely the consequence of nutritionally programmed efforts to improve the infant's chances of survival and reproduction.

Epidemiologic Evidence of Intrauterine Nutrition's Effects

Epidemiological studies have identified a series of associations between birth weight and disease in children through adults. For example, maternal undernutrition during

Alicia Cousins, Nicholas Rickman, and Cynthia F. Bearer, *Intrauterine Nutrition and Children's Health* In: *Textbook of Children's Environmental Health*, Second Edition. Edited by: Ruth A. Etzel and Philip J. Landrigan, Oxford University Press.

pregnancy is associated with dyslipidemia, insulin resistance, hypertension, cardiovascular disease, vascular dysfunction, osteoporosis, and neurobehavioral disorders (Table 23.1)[1]

A historical tragedy led to one of the first epidemiological studies on maternal nutrition. The Dutch Famine of World War II, the result of a food embargo by Nazi forces, abruptly reduced the daily calorie intake of the population to 500–600 kcal/day. Following the 6-month famine, caloric intake returned to normal. Studies of the effects of the Dutch Famine on survivors and their descendants has allowed epidemiologists to measure the result of famine on human health and confirm an epidemiological link between poor maternal nutrition in pregnancy and susceptibility of offspring to disease.[3] Research on this topic is plagued with the unavoidable issue of being primarily based on retrospective cohort studies because there is no role for randomized control studies in this area. For this reason, confirmation of these theories has relied on experimental animal models, where a diverse range of species from rodents to sheep and nonhuman primates have shown relationships between maternal nutritional status and blood pressure, feeding behavior, adiposity, and glucose homeostasis in offspring.[3]

Epidemiological studies of maternal overnutrition also suggest a link to an increased risk for offspring to develop morbidities such as obesity, insulin resistance, and cardiovascular disease. Italian women, with a presumed high-fat dietary intake marked by hypercholesterolemia during pregnancy, had offspring with fatty streaks in their arteries, suggesting increased risk of cardiovascular disease.[4] Pregnant Pima Indian women with diabetes mellitus had offspring who were more likely to also develop diabetes mellitus than the offspring of nondiabetic mothers.[5]

Table 23.1 Effects of fetal or extrauterine growth restriction on infant, child, and adult phenotype

	Infant	Child	Adult
Body morphometry			
Weight	↓	↓↑	↑↓
Length/height	↓	↓	↓
Body fat	↓	↑	↑
Metabolism			
Dyslipidemia		↑	↑
Glucose intolerance	↑	↑	↑
Insulin resistance			↑
Type 2 diabetes			↑
Cardiovascular			
Hypertension		↑	↑
Coronary artery disease			↑
Neurodevelopment			
Cerebral vascular accident			↑
Cognitive deficits		↑	
Neurobehavioral disorders		↑	
Psychiatric disorders			↑

Epidemiological Evidence and Timing of Malnutrition

Epidemiological evidence suggests that both the timing and duration of intrauterine malnutrition play an important role in outcomes. The Dutch Famine provided insight into how the timing of maternal dietary restriction during pregnancy influenced the health consequences of the offspring. Adults who experienced famine in early gestation were more likely to experience dyslipidemia, clotting abnormalities, coronary artery disease, and adult obesity. Impaired renal function and microalbuminemia were more common in offspring who experienced mid-gestation famine. Famine experienced during late gestation was associated with glucose intolerance, type 2 diabetes, and, similar to early gestation famine, higher rates of adult obesity. The proposed theory is that organ development is affected during critical windows, causing lasting damage. For example, nephron numbers in the kidney multiply during the second trimester and thus kidneys affected by malnutrition during this time could have declining function earlier than those receiving adequate nutrition in that window.[3]

Epidemiologic Evidence of Maternal Undernutrition and Adult Disease

Infant recovery from maternal nutrient deprivation in pregnancy appears to play a significant role in the development of adult disease. A prospective study of 11,593 individuals, ages 15–41 years, residing in five low- to middle-income countries (Brazil, Guatemala, India, the Philippines, and South Africa) found associations between maternal and child undernutrition and adult body size and metabolic indicators of disease. Maternal and child nutrition was assessed by maternal height, birth weight, FGR, and child weight, height, and body mass index (BMI, body weight kg/length cm^2) at 2 years.[6]

Poor fetal growth and also stunting in the first 2 years of life results in irreversible damage, including decreased adult height and lower offspring birth weight. Greater birth size is positively associated with childhood and adult body mass index (BMI), less associated with blood pressure, and unrelated to blood glucose. Lower birth weight and child undernutrition during the first 2 years of life was linked to increased blood glucose, blood pressure, and dyslipidemia. Most notably, rapid postnatal weight gain, especially after infancy, had the strongest association with obesity, hyperglycemia, and hypertension. There was no evidence in these studies that rapid weight gain during the first 2 years of life increases risk of chronic disease. Rather, the greatest risk of chronic disease appears to be to children who are undernourished through the first 2 years of life and then put on weight rapidly during childhood and adolescence.[6]

Data from the Dutch Famine studies also suggested the potential for poor intrauterine nutrition to result in psychiatric impairment. Those experiencing famine in utero have been shown to have increased rates of schizophrenia and schizophrenia spectrum disorders as well as antisocial personality disorders.[7] Some of these psychiatric outcomes may be linked to specific micronutrient deficiencies, such as choline. As many as 90% of pregnant women are not meeting the daily adequate intake for choline, defined by the National Institutes of Health (NIH), which is not only dangerous for themselves, but also can be detrimental to the development of their baby. Choline deficiency during pregnancy has been linked to a wide range of early neurodevelopmental defects and diseases, such as neural tube defects, schizophrenia, autism spectrum disorder (ASD), attention deficit/hyperactivity disorder (ADHD), and Down syndrome. This is likely secondary to the critical role choline plays in cell division and growth, particularly in brain cells.[8]

The wide variety and the severity of neurodevelopmental diseases that have been associated with choline intake, combined with the widespread estimated deficiency especially in pregnant women, is strong evidence supporting the hypothesis that undetected choline deficiency is a major silent contributor to neurodevelopmental disease and even to less severe developmental impairment.

Epidemiologic Evidence of Undernutrition and Disease in Preterm Children

The majority of LBW infants delivered in industrialized countries are born preterm. Postnatal growth in these infants after premature birth does not mirror in utero growth. Despite an increased focus on optimal nutrition for the preterm neonate, it remains exceedingly difficult to achieve weight, length, and head growth in LBW infants that match standard growth parameters of the fetus. This extrauterine growth restriction further challenges growth and development of the premature LBW infant. In one study, an estimated 20% of very LBW (VLBW, <1,500 g) infants were less than 2 standard deviations from average for gestational age at time of birth, but when they corrected to term gestation, 53% were less than 2 standard deviations from the norm.[9]

Long-term follow-up of preterm infants shows a gender-specific difference in growth. Although preterm females have similar weight, height, and BMI to population-based peers by age 20, preterm males are significantly shorter and lighter, with lower BMI at both 8 and 20 years of age compared to term controls with normal birth weight.[9]

Perhaps some of the most worrisome differences seen in the preterm LBW infant are the complications that persist into adulthood and that predispose to adult-onset diseases. Persistent effects on body composition and metabolism have, for example, been noted in preterm infants when they become young adults. Fat tissue provides an example. Preterm infants have less fat tissue at birth than term infants because fat tissue is predominantly deposited in the fetus during the third trimester of pregnancy. But despite their decreased total body fat, LBW infants have altered deposition of fat tissue stores compared to term-born infants. In particular, they have increased intraabdominal visceral fat tissue depots, as demonstrated by magnetic resonance imaging (MRI). This is key because increased intraabdominal adiposity correlates with markers of disease severity and plays a role in insulin resistance and dyslipidemia.[10] Metabolic derangements such as impaired glucose tolerance and insulin resistance have been seen in preterm infants when they become adults. Young adults born preterm have higher fasting insulin as well as higher insulin and higher glucose concentrations in response to oral glucose tests. Furthermore, lower insulin sensitivity has been confirmed in this population by hyperglycemic euglycemic clamp studies.[11]

Premature delivery and LBW are consistently associated with adverse neurodevelopmental outcomes. The severity of neurodevelopmental impairment is inversely proportional to gestational age and birth weight. While poor postnatal weight gain negatively impacts neurodevelopment, poor linear growth is also associated with lower cognitive function.[12] This association remains significant even in the absence of neurologic abnormalities. Conversely, head circumference catch-up growth is shown to be positively correlated with neurodevelopmental outcomes in childhood.[13] Using MRIs obtained throughout childhood, it has been shown that brain volume is also impacted by preterm birth and LBW. Preterm infants have significantly reduced total brain volume, white matter volume, and gray matter volume compared to term-born children, all by greater than 0.5 standard deviations, compared to term-born children. Additionally, they have significantly decreased cerebellar volume, hippocampus volume, and size of the corpus callosum. The functional correlates of reduced brain volumes range from lower IQ to decreased language, memory, motor skills, and executive functioning.[14]

Respiratory distress is a frequent complication of premature birth. Normal lung development is interrupted with preterm delivery, and the structure and function are equally immature. The premature lung is highly susceptible to additional damage due to mechanical ventilation often required for survival in the youngest, most premature infants. Both respiratory distress syndrome (RDS) and chronic lung disease (CLD) increase the metabolic requirements for an infant, making achievement of optimal postnatal nutritional support even more difficult.

Epidemiologic Evidence of Transgenerational Effects of FGR or Prematurity

The transgenerational effects of intrauterine nutrition have been explored. Women exposed to famine in utero have offspring with lower birth weight, greater adiposity, and poorer health. However, there is no association between offspring birth weight or the incidence of cardiovascular or metabolic disease and grandmaternal exposure to famine during gestation.[7] Mothers born small in both length and weight themselves have smaller infants.[6]

These epidemiological studies provide indispensable insight into the role of intrauterine nutrition on long-term health. Their findings imply that the detrimental effects of poor nutrition during gestation on later health do indeed pass down to subsequent generations. Additionally, these landmark studies provide impetus to investigate possible mechanisms or markers associated with the changed risk of disease in offspring.

Fetal Growth Restriction and Lung Disease

The development of postnatal lung disease is increased in infants with FGR. One well-studied example is the development of CLD of infancy, or bronchopulmonary dysplasia (BPD).[15] More severe FGR increases the risk of BPD.[15] FGR infants developing BPD tend to be preterm and subject to postnatal lung insults such as mechanical ventilation and oxygen exposure. Male gender also contributes to the development of lung disease following FGR.[16] While the mechanisms linking FGR to increased susceptibility for BPD are still being elucidated, altered development of the lung is implicated. It can be theorized that the same process that limits fetal growth also limits fetal lung growth and maturation.[15]

In human lungs, alveolar formation begins at approximately 28 weeks of gestation and continues into postnatal life. FGR, preterm birth, and subsequent postnatal lung insults often occur during the time frame of alveolar formation. Indeed, BPD is characterized histologically by alterations in alveolar formation, specifically fewer, larger airspaces with a decreased surface area for gas exchange. Other histological characteristics of BPD include thicker airspace walls indicating an altered balance of cellular proliferation and apoptosis.[15]

Animal models of FGR and BPD share these histological findings. Reduced alveolar number, increased airspace wall thickness, and reduced lung apoptosis have been demonstrated in rat and sheep models of FGR.[17,18] Characteristics such as lung structure, apoptosis, and proliferation rely on precisely timed expression of genes that govern these functions. Animal models of FGR have demonstrated programmed changes in lung gene expression that persist into the postnatal period.[18]

In humans, a study in Bangladesh of more than a thousand children showed that SGA status at time of birth, defined as a z-score of 2 or greater, was significantly related to development of childhood asthma.[19] This implies that nutrition as well as prematurity is responsible for inappropriate lung development, thus supporting the theory that the same processes leading to FGR limit fetal lung growth. The mechanisms by which prenatal growth and nutrition predispose to the development of postnatal lung disease and BPD are incompletely understood. However, programmed changes in gene expression accompanying lung development clearly contribute. Changes in programmed gene expression are, in part, produced by altered epigenetic regulation of gene expression.

Epigenetics and Developmental Plasticity

The role of epigenetics in the regulation of gene expression has been well described, particularly during embryonic development. Newer to the field of epigenetics, however, is understanding of the role of epigenetic modulation in providing developmental plasticity, or the ability of a fixed genotype to respond to changing environmental conditions during development.[16] Developmental plasticity is imparted by subtle changes in gene expression profiles initiated by changes in the epigenetic profile of genes. These epigenetic profiles consist primarily of patterns of DNA methylation and covalent histone modifications that occur across a gene. Other post-transcriptional epigenetic mechanisms, such as microRNA (miRNA), are also important for developmental plasticity. In combination, DNA methylation, histone modifications, and miRNA direct the transcription machinery and associated factors to appropriate locations within the gene to regulate gene expression resulting in plasticity.

DNA methylation occurs on the cytosine (C) of a C-guanine(G) dinucleotide. This dinucleotide is referred to as a *CpG*, where the C and G are on the same DNA strand. CpGs are often clustered in CpG "islands" consisting of a greater than 200 base pair region with a CG content of at least 50%. Methylation of CpG islands is often associated with gene silencing, although methylation outside of promoter regions is also important in the regulation of transcription. Frequently, CpGs found elsewhere within a gene (intragenic) and in other intergenic regions are methylated. As well as enhancing transcription, these inter- and intragenic CpGs appear to be involved in the regulation of alternative promoter usage.

Covalent modifications of histone proteins also contribute to the regulation of gene transcription. These covalent modifications include acetylation, methylation, ubiquitination, and phosphorylation and occur largely on the N-terminal tails of the histone proteins. The effects of a particular histone modification on gene transcription depend on the location within a gene, context, and surrounding histone modifications.[16]

Gene expression is also affected by miRNAs, small noncoding RNAs that function post-transcriptionally by promoting mRNA degradation or preventing translation. Interestingly, cell stress has a significant impact on miRNA levels and processing, suggesting that they may contribute to resolution of cell stress and thus developmental plasticity.

Alterations in epigenetic regulation of gene expression that occur during early development may alter the final structure and function of an organ by altering expression of apoptotic or proliferative genes, thus leading to plasticity. Furthermore, when the epigenetics of a gene are altered during development, the "new" epigenetic code becomes the platform for future epigenetic modifications that would normally accompany development. The discovery of long-term changes in epigenetics is a critically important finding because such changes have the potential to alter the expression of susceptible genes in the face of subsequent stressors long after the initial insult.

Epigenetic Responses to Fetal Growth Restriction in the Lung

Alterations in fetal nutrition have been shown to alter expression profiles of genes important for alveolar formation in the lungs. One such susceptible gene is the transcriptional regulator peroxisome proliferator activated receptor gamma (PPARγ).[20] PPARγ is important in the development of the lung. Targeted lung epithelial cell knockout of PPARγ alters postnatal lung development in mice.[21] Importantly, changes in expression of PPARγ in lungs affected by FGR (1) persist in the postnatal period and (2) are accompanied by epigenetic changes to the PPARγ gene.

In the lungs of male and female rats, FGR decreases mRNA transcript levels of PPARγ variants. These changes in PPARγ mRNA are accompanied by changes in histone (H)3 and H4 methylation along the PPARγ gene. Interestingly, these changes are frequently sex-specific. For example, H3 lysine 9 trimethylation (H3K9Me3) along the PPARγ gene is decreased by FGR in male neonatal rats. Conversely, in female neonatal rats, FGR increases H3K9Me3 along the PPARγ gene. Given that, in control rats, H3K9Me3 is not different between males and females, this demonstrates that both male and female rats have similar basal epigenetic regulation of PPARγ but respond differently to FGR. This is an important point as well as a common response to FGR.[20] The molecular basis for this sex-specific epigenetic response in the face of a stressor such as FGR is currently unknown and represents an important area of ongoing research.

In summary, prenatal insults such as FGR, impart sex-specific, lasting changes to the epigenetic profile of genes important for alveolar formation during a critical period developmental period.

Epigenetics Responses to Fetal Growth Restriction in the Brain

Alterations in intrauterine nutrition also have critical impacts on brain development. Being FGR predisposes infants to neurodevelopmental impairment with cognitive dysfunction, memory impairment, and structural changes in the brain. Animal models of FGR demonstrate molecular and phenotypic changes in the brains, again with sex-specific differences. Male FGR infants have more severe neurodevelopmental deficits than female infants. Molecular changes include persistent alterations in gene expression and sex-specific epigenetic modifications.

In the adolescent rat, female brains are characterized by increased site-specific acetylation, whereas male brains become characterized by decreased acetylation at K9 and K14 of H3. Gene-specific epigenetic changes are also seen in FGR rat brains, and, again, the changes are sex-specific. For example, in FGR rat hippocampus, the glucocorticoid receptor (GR) gene has sex-specific differences in the variant mRNA transcript levels as well as sex-specific changes in histone modifications along the GR gene.[22]

Conclusion

Alterations in intrauterine nutrition can have profound and lifelong impacts on human health, development of the brain and lungs, and risk of asthma, obesity, diabetes, and heart disease. Many of these effects appear to be the consequence of nutritionally programmed, epigenetically mediated changes in gene expression in response to changing nutritional circumstances. These changes can result in impaired health in childhood and increased risk of disease across the entire human life span.

Because so many different factors influence epigenetic programming during intrauterine development, the result is a high-fidelity, individualized epigenetic code that is a precise record of a person's nutritional and environmental history. This code, although still poorly understood, has the potential to be predictive of future disease risk.

The apparent relationships between perturbations of intrauterine nutrition, epigenetics, and childhood disease present an opportunity to pursue "personalized medicine." This opportunity arises from the observation that while FGR predisposes to multiple postnatal morbidities, including CLD, neurodevelopmental impairment, and insulin resistance, it does not inevitably dictate those consequences. Understanding the factors that influence individual variations in risk will advance the "personalized" care of children. It will enable identification of infants at high risk who require the most intense and invasive interventions. But it will also enable identification of those infants who are likely not in need of intensive or invasive interventions.

References

1. Wardlaw T, Blanc A, Zupan J, Ahman E. Low birthweight: country, regional and global estimates. UNICEF; 2004. Available at: https://data.unicef.org/resources/low-birthweight-country-regional-and-global-estimates/.

2. Abu-Saad K, Fraser D. Maternal nutrition and birth outcomes. *Epidemiologic Reviews.* 2010; 5(1):5--25.

3. Langley-Evans SC. Nutrition in early life and the programming of adult disease: a review. *J Hum Nutr Diet.* 2015;28 (Suppl. 1):1–14 doi:10.1111/jhn.12212

4. Napoli C, Infante T, Casamassimi A. Maternal-foetal epigenetic interactions in the beginning of cardiovascular damage. *Cardiovasc Res.* 2011;92(3):367–74.

5. Pettitt DJ, Javanovic L. The vicious cycle of diabetes and pregnancy. *Curr Diab Rep.* 2007;7(4):295–7.

6. Victora OG, Adair L, Fall C, et al. Maternal and child undernutrition: consequences for adult health and human capital. *Lancet.* 2008;371(9609):340–57.

7. Roseboom TJ, Painter RC, van Abeelen AFM, Veenendaal MVE, de Rooij SR. Hungry in the womb: what are the consequences? Lessons from the Dutch famine. *Maturitas* 2011;70(2):141–5.

8. Ross RG, Hunter SK, McCarthy L, et al. Perinatal choline effects on neonatal pathophysiology related to later schizophrenia risk [published correction appears in Am J Psychiatry. 2013 May 1;170(5):566]. *Am J Psychiatry.* 2013;170(3):290–8.

9. Hack M, Schluchter M, Cartar L, Rahman M, Cuttler L, Borawski E. Growth of very low birth weight infants to age 20 years. *Pediatrics* 2003 Jul;112(1 Pt 1):e30–8.

10. Uthaya S, Thomas EL, Hamilton G, Doré CJ, Bell J, Modi N. Altered adiposity after extremely preterm birth. *Pediatr Res.* 2005 Feb;57(2):211–5.

11. Rotteveel J, van Weissenbruch MM, Twisk JW, Delemarre-Van de Waal HA. Infant and childhood growth patterns, insulin sensitivity, and blood pressure in prematurely born young adults. *Pediatrics.* 2008 Aug;122(2):313–21.

12. Ramel SE, Demerath EW, Gray HL, Younge N, Boys C, Georgieff MK. The relationship of poor linear growth velocity with neonatal illness and two-year neurodevelopment in preterm infants. *Neonatology.* 2012 Mar 22;102(1):19–24. [Epub ahead of print]

13. Ghods E, Kreissl A, Brandstetter S, Fuiko R, Widhalm K. Head circumference catch-up growth among preterm very low birth weight infants: effect on neurodevelopmental outcome. *J Perinat Med.* 2011 Sep;39(5):579–86.

14. de Kieviet JF, Zoetebier L, van Elburg RM, Vermeulen RJ, Oosterlaan J. Brain development of very preterm and very low-birthweight children in childhood and adolescence: a meta-analysis. *Dev Med Child Neurol.* 2012 Apr;54(4):313–23. doi:10.1111/j.1469-8749.2011.04216.x. Epub 2012 Jan 28

15. Bose C, Van Marter LJ, Laughon M, et al. Fetal growth restriction and chronic lung disease among infants born before the 28th week of gestation. *Pediatrics.* 2009;124(3):e450–8.

16. Joss-Moore LA, Albertine KH, and Lane RH. Epigenetics and the developmental origins of lung disease. *Mol Genet Metab.* 2011;104(1-2):61–6.

17. O'Brien EA, Barnes V, Ahao L, et al. Uteroplacental insufficiency decreases p53 serine-15 phosphorylation in term IUGR rat lungs. *Am J Physiol Regul Integr Comp Physiol.* 2007; 293(1):R314–22.

18. Rozance PJ, Seedorf GJ, Brown A, et al. Intrauterine growth restriction decreases pulmonary alveolar and vessel growth and causes pulmonary artery endothelial cell dysfunction in vitro in fetal sheep. *Am J Physiol Lung Cell Mol Physiol.* 2011; 301(6): 860–71.

19. Nozawa Y, Hawlader MDH, Ferdous F, et al. Effects of intrauterine growth restriction and postnatal nutrition on pediatric asthma in Bangladesh. *J Dev Orig Health Dis.* 2019;10:627–35. doi:10.1017/S2040174419000096

20. Joss-Moore LA, Metcalfe DB, Albertine KH, McKnight RA, Lane RH. Epigenetics and fetal adaptation to perinatal events: diversity through fidelity. *J Anim Sci.* 2010; 88(13 Suppl):216–22.

21. Simon DM, Arikan MC, Srisuma S, et al. Epithelial cell PPARγ contributes to normal lung maturation. *FASEB J.* 2006; 20(9):1507–9.

22. Ke X, Schober ME, McKnight RA, et al. Intrauterine growth retardation affects expression and epigenetic characteristics of the rat hippocampal glucocorticoid receptor gene. *Physiol Genomics.* 2010; 42(2):177–89.

24

Biodiversity and Children's Health

Keith Martin

Introduction

What is biodiversity, and why is it important to children's health? The answer to that question tragically played out in 2014. A 2-year-old boy living in the small village of Meliandou, Guinea, was playing in a hollow tree housing a colony of insectivorous bats (*Mops condylurus*), which are known to be a vector for the Ebola virus. The child encountered the bats, contracted the virus, and died. But before passing away he had infected others and thus began the 2014–2016 Ebola outbreak that spread to several countries in West Africa, Europe, and the United States. By the time the last person recovered, 28,616 people had been infected, 11,325 of them had died, with the majority in West Africa. This was the largest recorded Ebola virus outbreak in history.

That virus is part of Earth's dizzying array of biological diversity. It is estimated that there are approximately 8.7 million species of plants and animals. Only 1.2 million have been identified, the majority of which are insects.[1] Some species are deeply beneficial to our lives, others are pathogens. This web of life is a pillar of our natural world and provides benefits known as *ecosystem services* that are crucial to our health and well-being.[2] These services fall into four categories.

Regulating services: Modifying the climate, cleaning the air, purifying freshwater, mitigating floods, controlling erosion, detoxifying soils, controlling diseases and pests
Provisioning services: Providing products we use, such as food, water, medicines, fuel, oxygen, wood, and fiber
Supporting services: Nutrient cycling, pollination
Cultural services: Nonmaterial benefits such as a sense of place, intellectual stimulation, aesthetics

Biodiversity contributes to many of these services including food, livelihoods, freshwater, clean air, climate regulation, pest and disease control, medicinal products, and a platform for medical research, and it contributes to our physical, spiritual, and psychosocial well-being. It also includes the human microbiota, which are the symbiotic microbial communities that live on our skin and in our gut, respiratory, and genitourinary tracts. They make important contributions to our nutrition, immune system, and the prevention of infections.

Despite the vital, life-supporting functions biodiversity provides, human activity is destroying it at a nearly unprecedented rate. In fact, we are in the midst of a sixth extinction crisis, the largest and most rapid loss of life since a meteor crashed into the Yucatán Peninsula 60 million years ago precipitating the extinction of the dinosaurs. It is estimated

Keith Martin, *Biodiversity and Children's Health* In: *Textbook of Children's Environmental Health*, Second Edition.
Edited by: Ruth A. Etzel and Philip J. Landrigan, Oxford University Press. © Oxford University Press 2024.
DOI: 10.1093/oso/9780197662526.003.0024

that 1 million of Earth's 8.7 million species are at risk of disappearing, a rate of loss 1,000 times greater than what would normally be expected.[3] We are not only losing species, but we are also seeing a massive reduction in the number of individuals within a species and their size. Overall, wildlife populations (excluding humans) have plummeted by 68% over the past 50 years.[4] This is a massive, global loss of nonhuman biomass.

Plummeting biodiversity, along with climate change, represent two existential threats to the child and, more broadly, human health and security. Unlike climate change though, extinction is forever. Once a species disappears it cannot come back. Because we barely understand the interrelationships between species, the loss of a single species can have profound impacts on the web of life, creating downstream effects across entire ecosystems. Since biodiversity also regulates the climate, freshwater circulation, and weather patterns, species losses affect many aspects of child health.

Our food security is also deeply dependent on the interrelationships of species. Seventy-five percent of the crops produced depend, at least in part, on pollinators like bees, birds, and bats for their reproduction.[5] The diversity and density of these pollinators affect crop yields and thus access to nutrition. Pollinator-dependent food products contribute to healthy diets for children. They are also an important source of economic security for children's families.

The root causes of the extinction crisis, like climate change, are anthropogenic so it is up to us to stop it. Let's look more closely at how biodiversity impacts child health.

Food Security and Nutrition

Nutrition has been called the single greatest environmental influence in childhood development, particularly from the prenatal stage through the first years of life. Since global biodiversity is a foundation for our food systems, reductions in the number and types of species that make up the food chain threaten children's food security. This can lead to protein-energy malnutrition of which there are three types: wasting (acute malnutrition), stunting (chronic malnutrition), and undernutrition (combination of both wasting and stunting).[6]

Malnutrition has a significant, negative impact on a child's cognitive and physical development. Shortages of nutrients such as iron and iodine impair cognitive and motor development, and these effects are often irreversible. In addition, there is growing evidence to show that docosahexaenoic acid (DHA), an essential fatty acid, is a key component in the production of synapses during the first years of life, a critical period for learning and development.[7,8]

Children who are severely malnourished can suffer from stunting, which is having a height that falls 2 standard deviations below the median height for that age. This negatively affects cognitive and physical development and is considered to be irreversible beyond the first thousand days of life.[9]

The World Health Organization (WHO) estimates that more than 2 billion people suffer from micronutrient deficiencies worldwide. Essential micronutrients include but are not limited to iron, zinc, calcium, iodine, vitamin A, B vitamins, vitamin C, and vitamin E. Children are particularly vulnerable to micronutrient deficiencies because of their high physiological requirements. Globally, approximately 42% of children have anemia, and rates can be as high as 60% in sub-Saharan Africa. Vitamin A deficiency is a leading cause of preventable blindness in children.[10]

Nutrition deficits during pregnancy and in the early years of a child's life increase the risk of that child becoming obese later in life by causing metabolic changes in how energy is used and stored. This increases the risk for that child acquiring noncommunicable diseases like type 2 diabetes and cardiovascular disease as an adult. In addition, the nutritional status of a woman impacts the health of her children.[11,12]

Micronutrient deficits in pregnant women can have a significant impact on the fetus. Iron-deficiency anemia will contribute to higher rates of spontaneous abortions, and a lack of folic acid is a risk factor for a child being born with spina bifida. Worldwide, approximately 40% of pregnant women are anemic, with the highest prevalence found in South Asia and sub-Saharan Africa where rates can be as high as 60%. It is estimated that 20% of maternal deaths are attributed to anemia. Mortality rates among mothers affect mortality rates for their children. The infants of women who died during or after childbirth are at a significantly higher risk of death in the first 2 years of life when compared to those whose mothers survived.[13,14]

Why does biodiversity matter to food security and children's health? Bees, butterflies, other insects, birds, and bats are critically important pollinators of the foods we consume. According to the UN Environment Program, of the 100 crop varieties that provide 90% of the world's food, 71 are pollinated by bees. Pollinators are declining worldwide at alarming rates. Approximately 40% of global insect populations are in decline, and one-third are in danger of extinction. Since the 1970s, North America alone has lost 29% of its total bird population. In 2010, of the 11,000 species of birds worldwide, 40% of them are in decline. Colony collapse disorder has resulted in large declines in bee populations. Intensive farming, urbanization, pesticide use, alien species, climate change, eutrophication, fertilizers, and pollution also contribute to this loss of pollinators.[15–17]

Biological diversity in the oceans also affects child health. Ten percent of the world relies on fisheries for their livelihoods, and 3.3 billion people rely on fish for 20% of their average per capita intake of animal proteins.[18] However, nearly all commercial fish species are maximally exploited or overexploited. Fish are declining in number and size, which means the total biomass in the oceans is in sharp decline. This is due to overfishing, pollution, habitat destruction, and climate change. The latter is causing ocean acidification and warming seas, which is damaging marine ecosystems such as corals that are vital for fish reproduction. The loss of marine biodiversity includes the foundation of the food chain in the oceans, phytoplankton. Globally, phytoplankton has fallen by about 40% since 1950 due to climate change warming the oceans. Since it also engages in photosynthesis, absorbing carbon dioxide and releasing 50% of the world's oxygen output, this loss of phytoplankton will exacerbate climate change and ocean warming and contribute to food insecurity for children who depend on fish as a key source for their nutrition. It is predicted that an additional 10% of the world's population will experience deficiencies in important micronutrients including zinc, iron, vitamin A, vitamin B_{12}, and fatty acids if global fish catches continue to decline. This will be particularly important for populations close to the Equator, where nutrition is highly dependent on wild seafood and where fish stocks are at greatest risk from overfishing, pollution, and climate change. This has critical implications for the children of families that are economically dependent on fishing.[19,20]

Since the 1900s, we have seen a reduction in the genetic diversity in the plants we consume: 75% of the world's food comes from only 12 plant and 5 animal species. Of the 4% of the 300,000 known edible plant species, only 200 are used by humans. Rice, maize, and wheat contribute nearly 60% of the calories and proteins obtained by humans from plants.[21] Reducing the diversity of the foods produced is an effort to maximize production. However, this poses a risk. Maintaining genetic diversity in the food we eat lessens the chance that a single disease can have a devastating impact on our food security. Also, commercial food production involves the increased use of pesticides and other chemical inputs to protect crops and maximize output. However, pesticides have significant consequences for child health and wildlife. Twenty-five million people per year suffer from acute pesticide poisoning in developing countries. The impact of pesticides on children is particularly dangerous due to their immature and developing nervous systems. This is also a risk factor for developing cancer, asthma, and learning disabilities. In the United States, 50% of the 2 million poisoning incidents each year involve children younger than 6 years of age.[22]

Across Africa and South Asia animal carcasses are deliberately seeded with poisons to kill carnivores and reduce predation on livestock. This is not only killing apex predators like lions, leopards, and tigers but this is also wiping out vulture populations, which are nature's most efficient scavengers. They consume and remove disease-ridden animals and are essential to healthy ecosystems. In Africa, over the past 50 years, populations of 7 out of 11 vulture species have declined by 80–97%. In Southeast Asia, poisoning by a veterinary drug has caused vulture populations to decline by 99%. Child health can be affected directly by exposure to these poisons or indirectly due to the spread of diseases from animals that are not cleared away by nature's carcass cleaners, vultures, which are on the cusp of extinction.[23,24]

Infectious Diseases

Different species can act as pathogens, as vectors for diseases, or can provide protection by keeping harmful pathogens in check. Intact ecosystems have diverse species which naturally check and balance each other. Declining biodiversity results in increased infectious disease transmission. In degraded ecosystems with fewer species, the internal checks that species exert on each other are reduced, which increases the risks for disease outbreaks.[25]

Zoonotic diseases are spread from animals to humans and are caused by a variety of bacteria, viruses, fungi, protozoa, parasites, and other pathogens. Sixty percent of human pathogens are zoonotic. It is estimated that 70% of emerging infectious diseases will come from nonhuman animals. Existing zoonoses include HIV, severe acute respiratory syndrome (SARS), rabies, Lyme disease, dengue fever, and Zika virus. SARS-COV-2 (source of the COVID-19 pandemic) likely originated in a bat. Pathogens can also be an enormous problem in agriculture. Diseases such as mad cow disease, H1N1, and avian flu can have devastating impacts on livestock, which threatens our food supply. Who can forget the swine flu outbreaks in Europe that resulted in the culling of millions of pigs?[26]

In North America, approximately 30 significant diseases can be contracted from pets. These include salmonellosis, toxoplasmosis, roundworm, and hookworm. Any child can be affected but children with compromised immune systems are at high risk for these infections.

Unregulated "wet markets" pose a serious risk factor for disease spillover. Many of these markets are in Africa and South and East Asia. In these markets, adults and children can live side-by-side with multiple species of animals, many of which have come from the wild. This proximity of wild animals to children lends itself to the easy transmission of potential pathogens. Epidemics have started in these markets. Some suspect that the SARS-CoV-2 pandemic began in a wet market in Wuhan, China.[27]

Wildlife trafficking, a driver of biodiversity losses and often connected to organized criminal gangs, also poses a risk for zoonotic spillover. The commercial chain across which wildlife is moved and sold brings people, including children, close to the animals being trafficked.[28]

Freshwater and Biodiversity

Access to potable water is essential for life. In Africa, more than two-thirds of the population must leave their home to fetch water for drinking and domestic use. Time spent walking to access water is a determinant of under-5 child health. A 15-minute decrease in one-way walk time to a water source is associated with a 41% relative reduction in diarrhea prevalence, improved childhood nutrition status, and an 11% reduction in under-5 child mortality. Increasing access to potable water can decrease the time and distance it takes for people to acquire the water they need, thereby improving child health.[29]

Plant biodiversity plays an important role in the freshwater and nutrient cycles, preventing soil erosion and purifying freshwater. Intact ecosystems with their complement of indigenous biodiversity reduces erosion, prevents floods, and thus helps to deliver cleaner water in a reliable manner. Intact floodplains contribute to the regulation of seasonal water flows, thus securing water availability for consumption, agriculture, and industry. It also reduces the runoff of nutrients and sediments, which improves water quality.

In urban centers, trees reduce temperatures and provide shade, which helps to decrease heat-related injuries in children. Greenspaces in urban environments are becoming increasingly important in reducing ambient temperatures as the planet warms because of climate change.[30,31]

A particular bio-disaster currently occurring that impacts global water (and food) security is the rapid deforestation of the Amazon and Congo Basins. The Amazon Basin alone has lost 30% of its forest cover over the past 20 years. This has local, regional, and global implications for child health. Deforestation disrupts the natural water cycle. As trees are removed the evaporation-transpiration of water from those trees into the atmosphere is reduced. This leads to less rainfall and drier and warmer conditions.[32]

Deforestation in the Congo Basin threatens food and water supplies not only locally but throughout the African continent. This vast region of biodiversity encompasses the second largest tropical primary rainforest on Earth. It is also home to 80 million people. Many of them rely on local biodiversity for their essential needs. The Basin regulates rainfall across the continent. It is estimated that at current deforestation rates, driven primarily by small-scale slash-and-burn farming, the Democratic Republic of Congo's primary rainforest will be completely gone by 2100. Since the Basin is a major source of rainfall in the Sahel region, this will have a major negative impact on food and water security in a region that is chronically food insecure. Nations in this region are some of the world's poorest, where children suffer from some of the world's worst health outcomes. In the Sahel, millions of children are malnourished, a third of them are also stunted. This affects their immune system, making them susceptible to an array of diseases including malaria, pneumonia, and diarrheal diseases. Food insecurity and drought can also lead to migration and conflict, as we have seen in this belt across Africa.

Noncommunicable Diseases and Biodiversity

Noncommunicable diseases (NCDs: diabetes, cardiovascular disease, pulmonary disorders, cancer, and mental health disorders challenges) are responsible for 70% of mortality worldwide and are the leading causes of disability. They affect populations at a younger age and are more severe in low-income settings. Healthy ecosystems can have important impact on NCDs.

Mental health disorders are the world's leading cause of disability. Among adolescents, mental health disorders such as depression, anxiety, and suicidal ideation are increasing. Spending time in nature reduces depression and anxiety. Studies have also shown that it improves cognitive function, brain activity, blood pressure, and sleep. Nature provides children with opportunities for discovery, creativity, risk-taking, mastery, and control which positively influence different aspects of their brain development. Being physically active in nature will also improve fitness and reduce obesity.[33] According to the WHO, in 2020, 39 million children under age 5 and more than 340 million children between 5 and 19 years of age were considered obese.[34]

Interesting associations have been identified between the acquisition of autoimmune diseases such as type I diabetes, multiple sclerosis, and inflammatory bowel disease and depleted microbial diversity in the human gut microbiome. The gut microbiome affects our ability to digest foods. Its functionality depends on our ability to be exposed to and acquire organisms with the relevant genes necessary to produce the enzymes we need. Our ability to be exposed to these genes is threatened by losses of biodiversity in the gene reservoir of environmental microbes. Urbanization reduces access to greenspaces and exposure to this genetic biodiversity. This can negatively impact the immune system and lead to autoimmune diseases. Maintaining exposure to a diverse microbial environment is important in maintaining a healthy immune system. Time spent in nature, exposed to appropriate microbial diversity, can be important in preventing autoimmune diseases and maintaining healthy nutrition.[35]

Pollution and Biodiversity

A biodiverse ecosystem cleans the water, purifies the air, and maintains the health of our soils. The quality of the air we breathe is affected in three important ways.

1. Plants remove air pollution through the absorption or intake of gases through leaves and directly absorb particulate matter.
2. The burning of vegetation, including forests, can exacerbate respiratory problems in children, including asthma, and predispose them to pneumonia. In areas where forest fires are prevalent, such as in Southeast Asia, millions of children's health is at risk. Small children are particularly vulnerable to air pollution because they breathe more rapidly and their immune system is not fully developed, which makes it easier for them to contract pneumonia and asthma.[36,37]
3. Microorganisms in the soil are also able to break down and clean up certain types of pollutants and decompose some organic pollutants, turning them into nontoxic substances.

Soil Biodiversity and Health

Soils are rich ecosystems of living organisms. Healthy soils house vertebrates and invertebrates, viruses, bacteria, fungi, lichens, and plants and are home to more than 25% of the planet's biodiversity. It is estimated that 95% of our food is directly or indirectly produced in soil. The level of biodiversity in the soil directly affects the nutrient content of the foods we consume.

Soil biodiversity is under threat from anthropogenic activities including farming practices, the misuse of agrochemicals, climate change, tillage, and soil pollution. Sustainable soil management practices must improve to increase soil biodiversity and thereby increase nutrients in the soil. This in turn will improve the nutrient content of the food children consume. Practices that will improve soil biodiversity include planting diverse crops, crop rotation, avoiding monocultures, composting, and using natural shelters such as hedges to prevent erosion.[38]

New Medicines and Biodiversity

It is estimated that 50,000–70,000 plant species are utilized to produce medicines today. Eighty percent of registered medicines come from plants, and 50% of modern medical therapies come from natural products. As biodiversity declines, so, too, do potential sources of new medicines.[39]

Traditional medicine plays an important role in healthcare, particularly in primary healthcare. Many antibiotics, analgesics, cancer chemotherapeutics, and other drug classes are derived from animal or plant sources. However, the unsustainable use of wild medicinal plants is contributing to biodiversity losses. The ongoing loss of global biodiversity is depriving the world of undiscovered therapeutics that could be applied to an array of pediatric and adult illnesses.

Climate Change and Biodiversity

Climate change impacts biodiversity, and biodiversity impacts climate change. Both affect child health. The drivers of climate change and biodiversity losses overlap. Addressing these common drivers will achieve the triple benefit of arresting climate change and biodiversity losses while improving child health.

Global warming is having a significant impact on climate variability. Biodiversity plays a key role in reducing greenhouse gases and in climate adaptation. It is estimated that more than 50% of anthropogenic carbon dioxide emissions are absorbed through photosynthesis and stored in organic material. Ecosystem degradation is damaging the ability of natural carbon sinks containing plants to sequester carbon. As global temperatures increase, permafrost in the Arctic will melt, releasing stored methane which is 24 times more powerful than carbon dioxide as a greenhouse gas. This will act as a dangerous feedback loop to worsen global warming.[40]

Protecting Biodiversity Improves Child Health

In this chapter, we have seen how Earth's extraordinary biodiversity is foundational to the health and well-being of children. Food and water security, protection against disease outbreaks, a source of medicines, prevention of noncommunicable diseases, cleaning our environment, and providing adaptation and mitigation to address the existential threat of climate change are all provided by Earth's web of life. Despite the extraordinary benefits the planet's species provide, we are witnessing a massive extinction crisis. This is, like climate change, an existential threat to child health. The more than 1 million species that are threatened with extinction will have untold downstream impacts on the ecosystem benefits that biodiversity provides. Addressing the drivers of extinction will also have a significant impact on climate change while improving child and human health: a triple benefit.

References

1. International Union for the Conservation of Nature (IUCN). The IUCN red list of threatened species. Version 2021–3. Gland, Switzerland; 2021. Available at: www.iucnredlist.org.

2. Kalam T. What are ecosystem services. 2022 Jan 22. Available at: https://www.scienceabc.com/nature/what-are-ecosystem-services.html.

3. Intergovernmental Policy-Platform on Biodiversity and Ecosystem Services (IPBES). Global Assessment Report on Biodiversity and Ecosystem Services. Brondizio E, Settele J, et al. Bonn, Germany: IPBES; 2019. Available at: https://www.ipbes.net/global-assessment.

4. World Wildlife Federation (WWF). Living Planet Report 2020: Bending the Curve of Biodiversity Loss. Almond R, Grooten M, Petersen T, eds. Gland, Switzerland: WWF; 2020. Available at: www.livingplanet.panda.org.

5. Ritchie H. How much of the world's food production is dependent on pollinators? Our World in Data. 2021 Aug 2. Available at: www.ourworldindata/pollinator-dependence.org.

6. Types of malnutrition. The use of epidemiologic tools in conflict affected populations. LSHTM. 2009. www.conflict.lshtm.ac.uk

7. Rosales FJ, Reznik JS, Zeisel SH. Understanding the role of nutrition on the brain and behavioral development in toddlers and preschool children. *Nutr Neurosci.* 2009 Oct;12(5):190–202.

8. Rivera JA, Hotz C, Gonzales-Cossio T et al. Effect of micronutrient deficiencies on child growth: a review of results from community-based supplementation trials *J Nutr.* 2003 Nov;133(11 Suppl 2):4010S–20S.

9. World Health Organization (WHO). *Malnutrition.* Geneva: WHO; 2021 Jun. Available at: www.who.int.

10. Bhan MK, Sommerfelt AH, Strand T. Micronutrient deficiencies in children. *Br J Nutr.* 2001 May;85 Suppl 2:S199–203.

11. Grey K, Gonzales G, Abera M, et al. Severe malnutrition or famine exposure in childhood and cardio metabolic noncommunicable disease later in life: a systematic review. *BMJ Global Health.* 2021 Mar;6(3):e003161.

12. Black RE, Victora CG, Walker SP, et al. Maternal, child undernutrition and overweight in low-income and middle-income countries. *Lancet* 2013;382:427–51.

13. Kanasaki K, Kumagai A. Impact of micronutrient deficiency on pregnancy complications and development origins of health and disease. *J Obstet Gynecol Res.* Mar 2021;47(6):1965–72.

14. Farias PM, Marcelino G, Santana LF, et al. Minerals in pregnancy and their impact on child growth and development. *Molecules.* 2020 Dec;25(23):5630.

15. Kopec K, Burd L. Pollinators in peril: a systematic review of North American and Hawaiian native bees. 2017 Feb. Available at: https://www.biologicaldiversity.org/campaigns/nat ive_pollinators/pdfs/Pollinators_in_Peril.pdf.

16. IPBES. Assessment Report on Pollinators, Pollination and Food Production: Summary for Policymakers. 2016 Feb. Bonn, Germany. Available at: www.ipbes.net.

17. Rosenberg KV, Dokter AM, Blancher PJ, et al. Decline of North American avifauna. *Science.* 2019 Sep 19;366(6461):120–24.

18. FAO. 2020. The State of World Fisheries and Aquaculture 2020. Sustainability in action. Rome. https://doi.org/10.4060/ca9229en

19. Golden C, Allison E, Cheung W, et al. Fall in fish catch threatens human health. *Nature.* 2016 Jun 16;534:317–20.

20. Morello L. Phytoplankton population dropped 40% since 1950. 2010 Jul. Available at: www. scientificamerican.com.

21. Lachat C, Raneri JE, Smith KW, et al. Dietary species richness as a measure of food biodiversity and nutritional quality of diets. *PNAS.* 2017 Dec 18;115(1):127–32. Available at: www. pnas.org.

22. Boedeker W, Watts M, Clausing P, Marquez E. The global distribution of acute unintentional pesticide poisoning: estimations based on a systematic review. *BMC Public Health.* 2020;20:1875.

23. Botha A. A clear and present danger: impacts of poisoning a vulture population and the effects on poison response activities. *Oryx.* 2018;52(3):552–8.

24. Keesing F, Belden LK, Daszak P, et al. Impacts of biodiversity on the emergence and transmission of infectious diseases. *Nature.* 2017;468:647–52.

25. Centers for Disease Control and Prevention. Zoonotic diseases. 2021 Jul. Available at: https://www.cdc.gov/onehealth/basics/zoonotic-diseases.html.

26. UNEP. Preventing the next pandemic: zoonotic diseases and how to break the chain of transmission. 2020. Available at: www.unep.org.

27. Ruiz-Aravena M, McKee C. Ecology and evolution and spillover of coronavirus from bats. *Nat Rev Microbiol.* 2022;20:299–314.

28. WWF. Reducing zoonotic disease risk from the wildlife trade. Summer 2022. Available at: www.worldwildlife.org.

29. Pickering AJ, Davis J. Fresh water availability and water fetching distance affect child health in sub-Saharan Africa. *Environ Sci Technol.* 2020 Jul 21;54(14):9143.

30. Convention on Biological Diversity. Global biodiversity outlook 5. Montreal. 2020. Available at: www.cbd.int/GBO5.

31. WHO. Urban green spaces and health interventions—a review of impacts and effectiveness. Copenhagen. 2017. Available at: https://www.who.int/andorra/publications/m/item/urban-green-space-interventions-and-health--a-review-of-impacts-and-effectiveness.-full-report.

32. Thompson E. Amazon deforestation and fires are a hazard to public health. *EOS* 102. 2021 Aug. Available at: https://eos.org/research-spotlights/amazon-deforestation-and-fires-are-a-hazard-to-public-health

33. Jimenez M, Deville NV, Elliott EG, et al. Association between nature exposure and health: a review of the evidence. *Int J Environ Res Public Health*. 2021;18(9):479.

34. WHO. Obesity and Overweight. Geneva. 2021 Jun 9. Available at: www.who.int.

35. Andersen L, Corazon S. Nature exposure and effects on immune system functioning: a systematic review. *Int J Environ Res Public Health*. 2021 Feb;18(4):1416.

36. Fuller R, Landrigan P, et al. Pollution and health: a progress update. *Lancet Planet Health*. 2022 Jun 1;6(6):E535–47.

37. Holm S, Miller M. Health effects of wildlife smoke in children and public health tools: a narrative review. *J Expo Sci Environ Epidemiol*. 2021;31:1–20.

38. FAO, ITPS, GSBI, SCBD and EC. 2020. State of knowledge of soil biodiversity – Status, challenges and potentialities, Summary for policymakers. Rome, FAO. https://doi.org/10.4060/cb1929en

39. Neergheen-Bhujun V, Awan AT, Baran Y, et al. Biodiversity, drug discovery, and the future of global health: Introducing the biodiversity to biomedicine consortium, a call to action. *J Glob Health*. 2017;7(2):020304.

40. A climate scientist explains what the melting Arctic means for the world. *World Economic Forum*. Jan 23, 2020. https://www.weforum.org/agenda/2020/01/arctic-ice-melting-climate-change-global-impact/

PART III
ENVIRONMENTAL HAZARDS

25
Outdoor Air Pollution

Frederica Perera

Introduction

The horrific air pollution episodes of the mid-20th century in London, England, and Donora, Pennsylvania, highlighted the acute dangers of air pollution, due largely to coal-burning emissions. Combustion of coal, oil, and gas continues to be the predominant source of air pollution as well as the major contributor to climate change. Although a detailed review is outside the scope of this chapter, interactions between these two threats can magnify the harm to children's health.

Understanding of the range of acute and chronic health risks of air pollution has led to substantial improvements in air quality over the past few decades in the United States, Europe, Japan, and other high-income countries. Yet, despite this improvement, there is evidence that the health effects of air pollution extend down to the lowest concentrations of air pollution that can be measured in developed countries today. And in many low- and middle-income countries (LMICs), air pollution—from fossil fuel combustion and from burning biomass fuels in poorly vented cookstoves—remains a major health hazard of global concern.

The effects of air pollution are not distributed equally across populations. Along with the elderly and persons with compromised defense mechanisms, the developing fetus and child are especially vulnerable to the acute and chronic effects of air pollution episodes. According to the World Health Organization (WHO), more than 90% of the world's children breathe air that puts their health and development at serious risk, and more than 500,000 children die from acute lower respiratory infections every year as a result of polluted air.[1]

Communities of color and low-income communities tend to have higher exposure to air pollution.[2,3] The disproportionate exposure of children in those communities is a contributor to the observed disparities in health, along with the lack of adequate nutrition, healthcare, education, and social support. For example, in the United States, the preterm birth rate is more than 50% higher for Black compared to White women; for childhood asthma, the rate is more than double.[4,5] The rate of pediatric emergency room visits for asthma in Chicago was highest in Black children.[6]

This chapter discusses the exposures of children to air pollution, why they are susceptible, and the multiple health effects observed in them.

Air Pollution

Air pollution is regulated based on its specific chemical constituents. For example, the US Environmental Protection Agency (EPA) has issued air quality standards that set

Frederica Perera, *Outdoor Air Pollution* In: *Textbook of Children's Environmental Health*, Second Edition. Edited by: Ruth A. Etzel and Philip J. Landrigan, Oxford University Press. © Oxford University Press 2024. DOI: 10.1093/oso/9780197662526.003.0025

maximum limits for six so-called criteria air pollutants—sulfur dioxide (SO_2), nitrogen dioxide (NO_2), carbon monoxide (CO), ozone (O_3), particulate matter less than 10 μm in aerodynamic diameter (PM_{10}), and a separate standard for fine particles less than 2.5 μm in aerodynamic diameter ($PM_{2.5}$). The WHO has published guidelines for these same pollutants.[7] Table 25.1 summarizes the US EPA standards. Historically these criteria pollutants have been treated independently because they were viewed as having different sources and independent health effects. However, in practice, air pollution is a mixture of these criteria air pollutants plus other toxic and nontoxic components, and the health effects observed are a response to this mixture. Understanding the health effects of air pollution in children therefore requires a more nuanced understanding of air pollution exposures than the traditional pollutant-specific approach. Nevertheless, control of air pollution is based on control of specific sources, so understanding linkages of exposure

Table 25.1 National Ambient Air Quality Standards

Pollutant	Primary/ secondary	Averaging time	Level	Form
Carbon monoxide (CO)	Primary	8 hours	9 ppm	Not to be exceeded more than once per year
		1 hour	35 ppm	
Lead (Pb)	Primary and secondary	Rolling 3 month average	0.15 μg/m^3	Not to be exceeded
Nitrogen dioxide (NO_2)	Primary	1 hour	100 ppb	98th percentile of 1-hour daily maximum concentrations, averaged over 3 years
	Primary and secondary	1 year	53 ppb	Annual Mean
Ozone (O_3)	Primary and secondary	8 hours	0.070 ppm	Annual fourth-highest daily maximum 8-hour concentration, averaged over 3 years
Particle pollution (PM) $PM_{2.5}$	Primary	1 year	9.0 μg/m^3	Annual mean, averaged over 3 years
	Secondary	1 year	15.0 μg/m^3	Annual mean, averaged over 3 years
	Primary and secondary	24 hours	35 μg/m^3	98th percentile, averaged over 3 years
PM_{10}	Primary and secondary	24 hours	150 μg/m^3	Not to be exceeded more than once per year on average over 3 years
Sulfur dioxide (SO_2)	Primary	1 hour	75 ppb	99th percentile of 1-hour daily maximum concentrations, averaged over 3 years
	Secondary	3 hours	0.5 ppm	Not to be exceeded more than once per year

Source: Environmental Protection Agency. (2024). NAAQS table. https://www.epa.gov/criteria-air-pollutants/naaqs-table

to sources remains necessary. Current research is focused on understanding the role of mixtures in defining exposures and responses to air pollution.

Meteorology

On the broad scale, air pollution exposure is defined by time and place. At a given location, for example outside a child's home, air pollution—outdoor or "ambient" concentration—will vary with time over multiple scales—hour by hour, day by day, season by season, and year by year. The ambient concentration is the result of emissions coupled with dilution, transport, and atmospheric transformation, all driven by meteorology. Wind transports air pollution from its sources to the child. Thus, only children downwind of a source are affected. Sources can be in the immediate neighborhood (tens of meters to kilometers) or far upwind (hundreds to thousands of kilometers). Wind also dilutes the concentration of air pollution. The stronger the wind, the greater the dilution. Concentrations of ambient air pollution are generally proportional to the inverse of the wind speed. High wind speeds can lift large particles and wind-blown dust into the air by resuspension; these, however, are less harmful than the smaller respirable particles.

Vertical mixing is another factor affecting dilution. The atmosphere near the earth's surface, where children live, is stratified by temperature. The air is transparent to most solar energy. The air column is heated from below; temperature therefore decreases with increasing height (altitude) above the ground. Because warm air is less dense than cold air, the warm air at the surface rises and upper cold air sinks. This produces vertical mixing and dilution of surface emissions of air pollutants. The depth of this mixing—the mixing height—can be an important determinant of concentrations of ambient air pollution. There are situations in which the vertical temperature structure of the atmosphere does not decrease, but rather increases with height. In this case, warm air (less dense) is above cold air (more dense) and there is no vertical mixing; that is, there is a stable atmosphere limiting dilution of air pollution. Because this vertical temperature structure is the opposite of normal, it is called a *temperature inversion*. An inversion can be caused by a warm air mass moving over a cold surface. It is also observed in the morning or on clear nights when radiational cooling of the ground cools the surface air compared to air above.

In another scenario, cold air draining into valleys can set up temperature inversion conditions. These conditions are often accompanied by ground-level fog. The combination of low wind speed and low mixing heights typically seen in mountain valleys in the early morning before the sun starts effectively heating the surface can lead to low dilution and some of the highest ambient air pollution concentrations of the day. Valley inversions can create particularly serious situations for community health when there are large local sources (industry) or many small air pollution sources (domestic wood/ biomass burning for heating and/or cooking).

On a larger scale, meteorologic conditions that produce regional, multiday inversion conditions can lead to the build-up of high levels of air pollutants over many days. These regional inversions were important contributors to historical air pollution disasters (London 1952, Donora 1948) as well as to modern air pollution episodes.

Weather affects the transformation of pollutants in the air. As pollutants are mixed in the air they interact and react with one other to form new chemical species. Air pollutants are classified as primary (those that come directly out of smokestacks and tail-pipes) and secondary (those that are formed in the atmosphere after leaving the source).

CO is an example of a primary pollutant. CO is produced by the incomplete combustion of hydrocarbons. Carbon dioxide is also a primary air pollutant.[8] While it is the major anthropogenic cause of global warming, it has not been implicated in causing health effects at concentrations found in the atmosphere and is therefore not considered a direct risk to children; however, indirectly its effects on climate are a concern for children's health[9] (see Chapter 4). Sulfur dioxide (SO_2), which is produced by oxidation of sulfur contamination of fossil fuels, is a primary pollutant. In contrast, ozone (O_3) is not directly emitted from industrial, domestic, or transportation sources but instead is produced as a secondary pollutant by photochemical reactions of other pollutants in the atmosphere. In this complex chemistry, nitrogen oxides (NO_x) and volatile organic compounds (VOCs) from motor vehicle and industrial sources react in the presence of ultraviolet (UV) sunlight to produce hydroxyl radicals (OH^-). These radicals then react with normal molecular oxygen (O_2) to form tropospheric ozone (O_3) (Box 25.1).

Because ground-level (tropospheric) ozone formation is driven by sunlight, there is a strong diurnal pattern to ozone concentrations in urban regions. Ozone concentrations typically are highest at midday. The other diurnal factor affecting this pattern is the source of the primary O_3 precursor pollutants. The morning and afternoon rush hours add excess NO_x and VOCs to the air. Nitrogen oxide (NO) reacts strongly with O_3 to form NO_2 and molecular oxygen (O_2). This scavenging of O_3 by NO_x peaks during the morning and evening rush hours and produces deficits of O_3, thus accentuating the diurnal pattern. These photochemical reactions are promoted by warm temperatures. The combination of sunlight and temperature means these O_3 events are more common in summer.

While this diurnal O_3 pattern is seen in urban areas, different patterns can be seen outside of cities. Urban air masses that become laden with precursor pollutants from the morning rush hour are transported downwind to suburban or rural communities during the day, reacting along the way. The highest O_3 concentrations are therefore often found in communities downwind of major urban centers. Indeed, because of the scavenging of O_3 by primary NO_x emissions in the source regions, average inner-city O_3 concentrations

---

Box 25.1 Stratospheric (High-Altitude) Versus Tropospheric (Low-Altitude) Ozone

---

UV light reacts strongly with molecular oxygen in the upper atmosphere to produce high concentrations of O_3 in the stratosphere. The heat generated by this reaction produces a warm layer of air high in the atmosphere (10–20 km) that creates a strong temperature inversion called the *tropopause*. This inversion limits mixing of air between the upper stratosphere and the lower troposphere. Thus, the high O_3 concentrations in the stratosphere do not mix down to the lower atmosphere, except during extreme convective storms. The high stratospheric O_3 concentrations form a protective layer absorbing most of the cancer-causing UV radiation from the Sun before it reaches the Earth's surface.

Stratospheric (high-altitude) O_3 is protective of health while tropospheric (low-altitude) O_3 is a hazard to health.

tend to be lower than those in downwind suburban or rural locations. Locally, O_3 ambient concentrations will be lower close to major roadways, and NO_2 concentrations higher compared to upwind and downwind locations.

Another consequence of this transport is that peak exposures at downwind sites may not follow the typical diurnal pattern. While O_3 is highly reactive when it contacts surfaces, including respiratory tissue, suspended in the air with little surface contact it can be transported intact over long distances. Thus, peak O_3 concentrations downwind of source regions can occur late in the day, in the evening, or even late at night.

Particulate matter and specifically fine particulate matter ($PM_{2.5}$) is a combination of primary and secondary pollutants. Fine particulates come directly out of smokestacks and tailpipes as unburned hydrocarbons (inorganic and organic particles) or as condensates of volatilized materials or other liquid or solid aerosols; that is, as particles or droplets suspended in the air. Fine particulates are also formed by secondary reactions in the atmosphere of primary gaseous emissions. For example, the primary pollutant SO_2 is oxidized in the atmosphere to form sulfur trioxide (SO_3), which reacts with water and cations (positively charged ions) in the atmosphere to form sulfate (SO_4) compounds. Thus, the primary gaseous emissions are converted over many hours or days to fine particles far downwind of the source. The hydroxyl radicals (OH^-) that are involved in the formation of ozone are also important in the oxidation of SO_2 and formation of secondary sulfate fine particles. Thus, air masses containing high summer O_3 concentrations also often have high sulfate fine particle concentrations. Similarly, the primary gaseous nitrogen oxides (NO_x) are oxidized by these same processes to form secondary nitrate aerosols. These fine particles are seen as the visibility-reducing white haze ("smog") characteristic of summer air pollution episodes (see Figure 25.1). Note that although the secondary particles are photochemically produced, they are not scavenged at night and do not show the same diurnal pattern. Rather, they continue to accumulate in a stagnant air mass over multiple days.

Meteorology also can affect emissions of primary pollutants. Traditionally there were higher emissions during the colder winter months due to domestic heating. This is still the case for communities where coal, wood, or other biomass fuels are used for heating. These communities continue to have frequent episodes of high air pollution during cold periods.

But those traditional seasonal patterns of pollution are changing. Today, in high-income countries and increasingly in LMICs, cleaner burning fuels are replacing dirtier, inefficient fuels. At the same time, increased power demand for air conditioning during the warmer months means that in high-income countries air pollutant emissions from power plants now peak in the summer rather than in winter, leading to summer peak formation of secondary pollutants. Modern pollution controls are another factor that have substantially reduced seasonal variation in pollution emissions. These controls have increased the efficiency of modern motor vehicles and reduced emissions, particularly during the cold start-up phase.

Biomass emissions for heating and cooking are a major source of air pollution exposures for children in LMICs. The indoor exposure of children, particularly neonates, to highly inefficient biomass fires has been identified as one of the major risk factors for morbidity and mortality among children worldwide. The indoor concentrations of air pollution (CO and $PM_{2.5}$) attained from incomplete combustion in poorly vented cookstoves can be extremely high. Moreover, widespread use of such stoves leads to very high ambient air pollution levels in communities (Box 25.2). Although the introduction of

Figure 25.1. Air pollution in New York City
Source: © iStock/terrababy

cleaner cookstoves and chimneys has reduced indoor air pollution, these control strategies do not necessarily lead to improved community air quality (see Chapter 26).

Climate change has affected air pollution levels in multiple ways. For example, the formation of O_3 from its precursors is accelerated at higher temperatures, severe forest fires have become more common, releasing vast amounts of particulate matter and other air pollutants, and changes in air stagnation and precipitation frequency due to climate change can lead to higher loadings of air pollutants.

Box 25.2 Indoor Versus Outdoor Air Pollution

Although the focus of this chapter is on ambient, outdoor air pollution, in fact, children spend the majority of their time indoors (see Chapters 26 and 27). Thus, the air pollution exposure of a child is determined by how much outdoor air pollution penetrates indoors, how effectively this pollution is filtered or removed, and what air pollutants are generated indoors.

Ventilation, measured as air changes per hour, is a key factor governing indoor air quality. Ventilation determines how much outdoor air pollution enters indoors and also how effectively pollutants that are generated indoors are diluted and removed. High ventilation, such as occurs when doors and windows are open, dilutes and removes pollutants generated indoors. In situations of high ventilation, indoor air pollution levels therefore mirror those outside. By contrast, low ventilation rates, when buildings are sealed for heating or cooling, result in trapping and accumulation of indoor air pollutants, and the linkage with outdoor pollution is reduced.

Even in low-ventilation systems, outdoor air is the source of the *make-up air*—the air from outdoors that enters a building—and, on average, the indoor concentrations of outdoor air pollutants will correlate with outdoor concentrations.

Air pollutants of indoor origin can be major exposures for children. For example, secondhand smoke can be a major indoor environmental exposure for children and is considered specifically in Chapter 27. Airborne exposures to pesticides and other household chemicals can be important; these are discussed in Chapters 35 and 39). Airborne biological materials (e.g., fungal spores; see Chapter 43) can be important indoor respiratory risks. Indoor radiation exposure through inhaled radon can cause a significantly elevated lifetime risk for cancer (Chapter 45).

The criteria air pollutants can also be generated indoors. Faulty combustion systems and chimneys lead to elevated indoor levels of CO and accidental deaths. Unvented combustion sources (gas stoves and unvented space heaters) produce elevated indoor levels of nitrogen dioxide.

The most extreme and most frequent indoor air pollution exposures for children globally arise from the uncontrolled burning of solid biomass fuels (wood, crop residues, dung, charcoal, coal, etc.) for cooking and/or heating. Primitive indoor fires have been shown to produce indoor concentrations of particles (PM_{10}) of 300–3000 $\mu g/m^3$ and CO concentrations of 16–500 parts per million.[25] These indoor concentrations are orders of magnitude above international guidelines. They are comparable to outdoor concentrations observed in the great historical air pollution events. The hazard associated with such exposures is extreme. Death from bronchitis and pneumonia is the frequent consequence. There is no question that action should be taken to prevent such exposures to young children.

Susceptibility of the Fetus, Infant, and Child to the Multiple Impacts of Air Pollution

The fetus, infant, and child are uniquely vulnerable to air pollution due to a number of biological and behavioral factors.[9,10] The developing lung, brain, and other organs are vulnerable to disruption by toxic pollutants due to their rapid growth and the complex developmental programming required during these stages. Furthermore, in the infant and child, the biological defense mechanisms to detoxify chemicals, repair DNA damage, and provide immune protection are immature, heightening their vulnerability to air pollution.

Children have a larger lung surface area and higher ventilation rates for their body weight than adults. Thus, more air pollution is delivered to their lungs per unit of body weight than to adults. Moreover, children spend more time outdoors participating in active play, often during daytime periods when air pollutants, specifically photochemical pollutants, tend to be at their highest levels. Even indoors, due to their greater physical activity, children may be exposed to higher air pollution levels than adults. Finally, children have a long remaining lifetime during which an illness such as asthma or a neurodevelopmental disorder may persist, affecting health and functioning throughout the life course.

The lungs provide an example of developmental vulnerability. These organs are not fully developed at birth. Eighty percent of the gas exchange alveoli are formed between birth and approximately age 6 years. Functional development of the lungs continues through adolescence. In addition, the child's immune system is immature at birth and continues to develop during early childhood. Because of the immaturity of their lungs and immune systems, children have much higher rates of respiratory infections than adults, and chronic respiratory conditions such as asthma are much more prevalent in children. These conditions make them particularly susceptible to further effects of air pollution. Small changes during vulnerable periods in early development may lead to higher health risks later in life.[5,6] For example, damage to the developing lung may lead to reduced maximal functional capacity and decreased lifelong ability to cope with respiratory challenges.

Similarly, the developing brain is highly vulnerable to air pollution due to many of the same susceptibility factors that are operating in the lung. The first is the rapid and complex development of the brain, making it vulnerable to disruption by toxic exposures such as air pollution. Already by the time of birth, the basic structures of the brain are in place and most of the 100 billion neurons, or nerve cells, that will be present in the adult brain have been formed. To accomplish this, the brain has grown at a phenomenal rate of about 250,000 nerve cells per minute, on average, throughout pregnancy.[11] At the end of 9 months, the brain is a functioning organ; however, the processes of myelination of axons, the formation of synapses, and further changes to maximize performance continue all the way through childhood and adolescence. Thus, there is increased vulnerability throughout these stages.

Health Effects of Air Pollution

As noted earlier, this section focuses on respiratory health effects because they are the earliest and most studied of the impacts of air pollution on children's health. Effects on birth outcomes and the developing brain are also described.

Respiratory Health Effects

The respiratory system is a clear target for the health effects of air pollution in children. The development of the lung from a very immature organ at birth that does not reach full functionality until post-adolescence, the immature status of the immune system, and frequent respiratory infections mean that children are set up to be particularly responsive to acute and chronic exposures to air pollution. Repeated exposures can lead to long-term cumulative lung damage.

Lung Function

Some of the clearest evidence of the health effects of air pollution comes from studies of lung volume and air flows in children. An early panel study in Utah had third- and fourth-grade school children measure and record their own pulmonary peak flow every day for 4 months using a hand-held peak flow meter before going to bed.[12] The difference between each child's daily peak flow and that child's overall mean was calculated, and these deviations were averaged over all children for each day. A strong negative correlation was found between air pollution measured as PM_{10} and the mean deviations in self-measured peak flow. This finding suggests that children's lung function drops acutely following air pollution exposures. Similar approaches have been applied in numerous other studies around the world. In quantitative reviews of these studies, lung function of children in the community is consistently found to decrease acutely following episodes of particulate and ozone pollution.[13]

Children living in communities with higher ambient levels of air pollution or closer to major roadways have lower average lung function.[14] In follow-up studies of the growth of children, children living in communities with chronically elevated levels of ambient air pollution have slower rates of growth of lung function.[15] These deficits are small, only a few percent. However, even small deficits in lung growth over childhood can accumulate into substantial deficits in maximum attained lung function at maturity in the 20s, thus reducing pulmonary reserve capacity and placing individuals at elevated risk of clinical respiratory impairment later in adult life.

Reduction in air pollution exposure can benefit children's lung function. A follow-up study of 10-year-old children in Southern California who relocated to areas with differing levels of particulate matter were tested for lung function 5 years later: those who had moved to areas with lower particulate levels had improved lung function growth rates.[16]

Acute Respiratory Infections

Acute respiratory infections are a major cause of morbidity and mortality among children globally. The burden of disease for respiratory infections is particularly high in LMICs, but it is also a major contributor to morbidity in high-income countries.[17]

While respiratory infections are common in children, specific clinical diagnosis is not. Thus, most of the epidemiologic evidence for associations with air pollution is based on symptom reports. Respiratory symptoms are divided into two main classes: upper respiratory symptoms (cough, runny nose, sore throat, ear discharge) are usually viral and are usually managed with supportive care at home. Lower respiratory symptoms (cough, dyspnea, wheeze, bronchitis) can be more serious. Treatment can range from supportive home care to hospital admission. Community-based observational studies around the world have examined associations of ambient air pollution with respiratory symptom reports. While these population studies lack clinical confirmation, they provide insights into the burden of respiratory infections attributable to air pollution.

Numerous studies have reported that both upper and lower respiratory symptoms increase on days with higher air pollution.[18,19] School absences increase modestly during and following high air pollution days.[20] Similarly, respiratory hospital admissions of children increase on days with elevated air pollution. This is true for both primary pollutants such as SO_2, NO_2, and a fraction of fine particulates and for secondary pollutants such as O_3 and a fraction of fine particulates.[21]

In communities with higher ambient concentrations of air pollution, children are reported to have elevated rates of both upper and lower respiratory symptoms.[22] Among children living in homes using biomass fuels for cooking and/or heating, a meta-analysis of 25 studies found a 3.5-fold increased risk of acute respiratory infections.[23] There is evidence that community air pollution has a greater effect among children with chronic respiratory disease.

Improvements in air quality have been shown to lead to reduced respiratory symptoms in children. For example, reductions in outdoor SO_2 and particulate pollution in the former German Democratic Republic following reunification were accompanied by parallel reductions in the reported rate of infectious respiratory illness.[24] In the Utah Valley, the closure of a steel mill by a 14-month labor strike led to dramatic reductions in both winter particulate air pollution concentrations and respiratory hospital admissions of children.[25] These and other observational studies provide compelling evidence of the beneficial effects of reducing air pollution on children's respiratory symptoms and infections.

Asthma

Children with asthma would be expected to be the sentinel population for the adverse respiratory effects of air pollution. Indeed, children with asthma exhibit the same acute responses to episodes of air pollution noted above: that is, decreased lung function, increased respiratory symptoms, and increased hospital emergency department visits and admissions. In addition, they show increased use of asthma medications such as rescue inhalers. However, in some observational studies of children with asthma in the community, their responses are comparable to or even blunted compared to those in the general population of children. This is not an indication that children with asthma are less sensitive. Rather, it reflects the fact that a child with properly managed asthma is often able to minimize the effect of environmental challenges through prophylactic or rescue medications. Indeed, children with asthma are observed to use their rescue medications more frequently in response to air pollution episodes. Thus, air pollution episodes are clearly a risk factor for triggering an attack among children with asthma, which can be serious if not appropriately managed.

More recently, studies have found that air pollution is a risk factor for the development of asthma in children.[26] An estimated 1.85 million new pediatric asthma cases were attributable to combustion-related NO_2 globally in 2019, with developing countries and urban communities disproportionately impacted.[27]

Effects on Birth Outcomes

Globally, an estimated 2.7 million premature births per year (18% of preterm births) have been linked to exposure to $PM_{2.5}$; most of these preterm births, but not all, are in developing countries.[28] A review of data from studies in the United States concluded that exposure to $PM_{2.5}$ is a risk factor for both preterm birth and low birth weight.[29] A large epidemiological study found evidence of an interaction between climate change and air pollution: that is, combined prenatal exposure to heat and air pollution had a multiplicative effect on risk of preterm birth.[30] Preterm babies are at higher risk for subsequent health problems, including lower respiratory infections, other infectious diseases, and asthma in childhood, as well as for long-term intellectual disabilities.

Effects on Infant/Child Mortality

The risk of air pollution for the health of children was first shown during the London Fog episode of 1952. During the second week of December 1952, cold temperature, low wind speed, and a temperature inversion over the city of London led to extremely high levels of black smoke and SO_2. The principal source was coal fires used for heating. There were at least 4,000 excess deaths during the week of the fog; more than two-thirds of these deaths were among persons 65 years of age or older. However, there also was a 1.8-fold increase in deaths for children under 4 weeks of age and a 2.2-fold increase for infants 4 weeks to 1 year during the fog.[31]

Infant mortality and sudden infant death syndrome (SIDS) have been shown to be highly associated with exposure to air pollution during the first months of life.[32] Today, air pollution contributes to an estimated 500,000 deaths globally among children; pneumonia and other respiratory infections are among the principal causes, along with preterm birth, which is also linked to air pollution (see below).[33,34]

There is evidence that elevated exposures to community air pollution during pregnancy may trigger stillbirth (intrauterine mortality). For example, a study in Ohio found that exposure to high levels of fine particulate air pollution in the third trimester of pregnancy was associated with increased stillbirth risk.[35] For women exposed to indoor air pollution from uncontrolled biomass fuel burning for cooking, there is about a 50% increased risk of stillbirth.[36]

Neurodevelopmental and Mental Health Effects

Prenatal and postnatal exposure to various air pollutants has been associated with effects on the developing brain including reduced cognition, attention problems, attention-deficit/hyperactivity disorder (ADHD), and autistic traits in childhood.[37,38] Studies using magnetic resonance imaging (MRI) have found structural and functional changes in the brains of children who were more highly exposed to air pollution before birth; these changes were linked to worse neurodevelopmental outcomes.[39,40]

Recent epidemiologic research indicates that air pollution is a risk factor for mental health conditions in children and adolescents. For example, lifetime exposure to traffic-related air pollution in Cincinnati was associated with self-reported depression and anxiety symptoms in 12-year-old children.[41] In studies in New York City and Krakow,

Poland, higher prenatal exposure to polycyclic aromatic hydrocarbons (PAHs), a component of fine particulate matter, was linked to more anxiety and depression symptoms in childhood.[37]

Pediatric Cancer

Cancer is a leading cause of mortality in children. The etiology of pediatric cancer is largely unknown. An early review of the literature found that studies of cancer and air pollution were generally of low power and had weak exposure assessment and other limitations.[42] Most of these studies focused on traffic volume or proximity to roadways, rather than on direct measures of air pollution. This review concluded that the observational studies published to date did not show evidence of a causal association between air pollution and pediatric cancer. The evidence continues to be mixed and inconclusive. However, prenatal exposure to $PM_{2.5}$ has been linked in some studies to childhood leukemia.[43] Additionally, a nationwide case-control study in Denmark found that residential exposure to $PM_{2.5}$ increases the risk of childhood non-Hodgkin lymphoma.[44]

Solutions and Benefits of Action for Children's Health

Solutions to the serious problem of air pollution include a transition from the reliance on fossil fuel to renewable sources of energy (e.g., wind, solar, geothermal) and programs to shift from indoor biomass and coal burning for cooking and heating to cleaner alternatives. Government policies that have directly targeted emissions from fossil fuel burning and improve air quality can both benefit children's health and be cost-saving. Examples include a congestion charging program to reduce transportation emissions in London and the Regional Greenhouse Gas Initiative (RGGI), a program in the northeastern United States aimed at reducing emissions from the electric power sector.[45,46] For example, between 2009 and 2014, the RGGI prevented an estimated 26,000 asthma attacks and respiratory illnesses, mainly in children, and prevented more than 540 new asthma cases and 800 preterm or low-birth-weight births—in addition to hundreds of premature adult deaths.[46,47] Estimated economic savings from these and other health benefits of RGGI amounted to almost $US6 billion.

Conclusion

Research, such as that reviewed in this chapter, has clearly demonstrated that children are highly vulnerable to air pollution and that the burden of disease is not shared equitably among them. Although all children are at potential risk, those in developing countries and low-income communities and communities of color in industrialized countries bear the brunt. Perhaps no other environmental exposure is associated with the diversity of adverse health effects as air pollution. The health and economic toll is very large, especially as early health damage can have lifelong impact. Fortunately, solutions are available to reduce exposures, with major health and economic benefits.

References

1. World Health Organization (WHO). More than 90% of the world's children breathe toxic air every day. 29 October 2018. Available at: https://www.who.int/news/item/29-10-2018-more-than-90-of-the-worlds-children-breathe-toxic-air-every-day.

2. Chambliss, SE, Pinon CPR, Messier KP, et al. Local- and regional-scale racial and ethnic disparities in air pollution determined by long-term mobile monitoring. *PNAS*. 2021;118(37):e2109249118.

3. Tessum CW, Paolella DA, Chambliss SE, Apte JS, Hill JD, Marshall JD. PM2.5 polluters disproportionately and systemically affect people of color in the United States. *Sci Adv*. 2021;7(18):eabf4491.

4. Centers for Disease Control and Prevention (CDC). Preterm birth. 2020. Available at: https://www.cdc.gov/reproductivehealth/maternalinfanthealth/pretermbirth.htm.

5. CDC. Most recent national asthma data. 2019. Available at: https://www.cdc.gov/asthma/most_recent_national_asthma_data.htm.

6. Respiratory Health Association. Persisting racial disparities among Chicago children with asthma. 2018. Available at: https://resphealth.org/wp-content/uploads/2018/05/Asthma-Report-Final.pdf.

7. World Health Organization. WHO global air quality guidelines: particulate matter (PM2.5 and PM10) , ozone, nitrogen dioxide, sulfur dioxide, and carbon monoxide. 2021. Available at: https://apps.who.int/iris/handle/10665/345329.

8. Bayer P, Aklin M. The European Union emissions trading system reduced CO2 emissions despite low prices. *PNAS*. 2020;117(16):8804.

9. Perera FP. Multiple threats to child health from fossil fuel combustion: impacts of air pollution and climate change. *Environ Health Perspect*. 2017;125(2):141–8.

10. Goldman R, Woolf A, Shannon M. Principles of pediatric environmental health. 2014. Available at: https://www.atsdr.cdc.gov/csem/ped_env_health/docs/ped_env_health.pdf.

11. Grandjean P. *Only One Chance: How Environmental Pollution Impairs Brain Development—and How to Protect the Brains of the Next Generation*. New York: Oxford University Press; 2013.

12. Pope CA 3rd, Dockery DW, Spengler JD, Raizenne ME. *Respiratory health and PM10 pollution. A daily time series analysis. Am Rev Respir Dis*. 1991;144(3 Pt 1):668–74.

13. Garcia E, Rice MB, Gold DR. *Air pollution and lung function in children. J Aller Clin Immunol*. 2021;148(1):1–14.

14. Rice MB, Rifas-Shiman SL, Litonjua AA, et al. Lifetime exposure to ambient pollution and lung function in children. *Am J Resp Crit Care Med*. 2016;193(8):881–8.

15. Gauderman WJ, Urman R, Avol E, et al. Association of improved air quality with lung development in children. *N Engl J Med*. 2015;372(10):905–13.

16. Avol E, Gauderman WJ, Tan SM, London SJ, Peters JM. Respiratory effects of relocating to areas of differing air pollution levels. *Am J Resp Crit Care Med*. 2002;164:2067–72.

17. Smith KR, Mehta S, Maeusezahl-Feuz M. Indoor air pollution from household use of solid fuels. In: Ezzati M, Lopez AD, Rodgers A, Murray C, eds., *Comparative Quantification of Health Risks: Global and Regional Burden of Disease Attributable to Selected Major Risk Factors*. 2004:1435–93.

18. Ratajczak A, Badyda A, Czechowski PO, Czarnecki A, Dubrawski M, Feleszko W. Air pollution increases the incidence of upper respiratory tract symptoms among Polish children. *J Clin Med*. 2021;10(10):2150.

19. Horne BD, Joy EA, Hofmann MG, et al. Short-term elevation of fine particulate matter air pollution and acute lower respiratory infection. *Am J Resp Crit Care Med*. 2018;198(6):759–66.

20. Mendoza D, Pirozzi CS, Crosman ET, et al. Impact of low-level fine particulate matter and ozone exposure on absences in K-12 students and economic consequences. *Environ Res Lett*. 2020;15:114052.

21. Zhao Y, Kong D, Fu J, et al. Increased risk of hospital admission for asthma in children from short-term exposure to air pollution: case-crossover evidence from Northern China. *Front Public Health*. 2021;9:798746.

22. Chen Z, Salam MT, Eckel SP, Breton CV, Gilliland FD. Chronic effects of air pollution on respiratory health in Southern California children: findings from the Southern California Children's Health Study. *J Thorac Dis*. 2015;7(1):46–58.

23. Po JY, FitzGerald JM, Carlsten C. Respiratory disease associated with solid biomass fuel exposure in rural women and children: systematic review and meta-analysis. *Thorax*. 2011;66(3):232–9.

24. Krämer U, Behrendt H, Dolgner R, et al. Airway diseases and allergies in East and West German children during the first 5 years after reunification: time trends and the impact of sulphur dioxide and total suspended particles. *Int J Epidemiol*. 1999;28(5):865–73.

25. Pope CA, 3rd. Respiratory hospital admissions associated with PM10 pollution in Utah, Salt Lake, and Cache Valleys. *Arch Environ Health*. 1991;46(2):90–7.

26. Achakulwisut P, Brauer M, Hystad P, Anenberg SC. Global, national, and urban burdens of paediatric asthma incidence attributable to ambient NO2 pollution: estimates from global datasets. *Lancet*. April 10 2019;3(4):E166–78.

27. Anenberg SC, Mohegh A, Goldberg DL, et al. Long-term trends in urban NO$_2$ concentrations and associated paediatric asthma incidence: estimates from global datasets. *Lancet Planet Health*. 2022;6(1):e49–58.

28. Malley CS, Kuylenstierna JC, Vallack HW, Henze DK, Blencowe H, Ashmore MR. Preterm birth associated with maternal fine particulate matter exposure: A global, regional and national assessment. *Environ Int*. April 2017;101:173–82.

29. Bekkar B, Pacheco S, Basu R, DeNicola N. Association of air pollution and heat exposure with preterm birth, low birth weight, and stillbirth in the US: a systematic review. *JAMA Netw Open*. 2020;3(6):e208243.

30. Wang Q, Li B, Benmarhnia T, et al. Independent and combined effects of heatwaves and pm2.5 on preterm birth in Guangzhou, China: a survival analysis. *Environ Health Perspect*. 2020;128(1):17006.

31. Logan WP. Mortality in the London fog incident, 1952. *Lancet*. 1953;1(6755):336–8.

32. Litchfield IJ, Ayres JG, Jaakkola JJK, Mohammed NI. Is ambient air pollution associated with onset of sudden infant death syndrome: a case-crossover study in the UK. *BMJ Open*. 2018;8(4):e018341.

33. United Nations Children's Fund (UNICEF). Clear the air for children. 2016 Oct. Available at: https://www.unicef.org/media/60106/file.

34. State of Global Air. Impacts on newborns. 2020. Available at: https://www.stateofglobalair.org/health/newborns.

35. DeFranco E, Hall E, Hossain M, et al. Air pollution and stillbirth risk: exposure to airborne particulate matter during pregnancy is associated with fetal death. *PLoS One.* 2015;10(3):e0120594.

36. Stieb DM, Chen L, Eshoul M, Judek S. Ambient air pollution, birth weight and preterm birth: a systematic review and meta-analysis. *Environ Res.* 2012;117:100–11.

37. Perera FP. Pollution from fossil-fuel combustion is the leading environmental threat to global pediatric health and equity: solutions exist. *Int J Environ Res Public Health.* 2018;15(1):16.

38. Volk HE, Lurmann F, Penfold B, Hertz-Picciotto I, McConnell R. Traffic-related air pollution, particulate matter, and autism. *JAMA Psychiatry.* 2013;70(1):71–7.

39. Peterson BS, Rauh VA, Bansal R, et al. Effects of prenatal exposure to air pollutants (polycyclic aromatic hydrocarbons) on the development of brain white matter, cognition, and behavior in later childhood. *JAMA Psychiatry.* 2015;72(6):531–40.

40. Guxens M, Lubczyńska MJ, Muetzel RL, et al. Air pollution exposure during fetal life, brain morphology, and cognitive function in school-age children. *Biol Psychiatry.* 2018;84(4):295–303.

41. Yolton K, Khoury JC, Burkle J, LeMasters G, Cecil K, Ryan P. Lifetime exposure to traffic-related air pollution and symptoms of depression and anxiety at age 12 years. *Environ Res.* 2019;173:199–206.

42. Raaschou-Nielsen O, Reynolds P. Air pollution and childhood cancer: a review of the epidemiological literature. *Int J Cancer.* 2006;118(12):2920–9.

43. Schraufnagel DE, Balmes JR, Cowl CT, et al. Air pollution and noncommunicable diseases: a review by the forum of international respiratory societies' environmental committee, Part 2: air pollution and organ systems. *Chest.* 2019;155(2):417–26.

44. Hvidtfeldt UA, Erdmann F, Urhoj SK, et al. Residential exposure to PM(2.5) components and risk of childhood non-Hodgkin lymphoma in Denmark: a nationwide register-based case-control study. *Int J Environ Res Public Health.* 2020;17(23):8949.

45. Mayor of London. World's first ultra low emission zone to save NHS billions by 2050. 2020. Available at: https://www.london.gov.uk/press-releases/mayoral/ulez-to-save-billions-for-nhs.

46. Perera FP, Cooley D, Berberian A, Mills D, Kinney P. Co-benefits to children's health of the U.S. Regional Greenhouse Gas Initiative. *Environ Health Perspect.* 2020;128(7):77006.

47. Manion M, Zarakas C, Wnuck S, et al. *Analysis of the Public Health Impacts of the Regional Greenhouse Gas Initiative.* Cambridge, MA: Abt Associates; 2017.

26

Household Air Pollution

Roshan Wathore and Archana Patel

Scope of the Problem

Household air pollution (HAP) is a major health problem. There is a relationship between the extent of solid fuel use, economic development, resulting HAP, and health impacts; this is evident especially in low- and middle-income countries (LMICs) in Southern Asia, sub-Saharan Africa, and Central America (Figure 26.1).

Despite the fact that millions of people in LMICs are moving out of poverty and into the middle class, continued population growth results in more people being exposed to HAP today than at any previous period in human history. Therefore, this chapter focuses on sources of exposure to HAP, toxicology, systems affected, clinical effects, diagnosis, prevention, and proposed public health interventions.

Epidemiology of Household Air Pollution

HAP is a significant public health hazard predominantly affecting poor rural and urban communities in LMICs. While the majority of people at risk of exposure live in rural areas of the world's poorest countries, HAP is increasingly becoming a problem of poor urban dwellers, a trend likely to increase in the future with rapidly increasing urbanization in countries around the world. Energy from household fuel combustion is needed for cooking, space heating, lighting, small-scale income generation, various household tasks, and entertainment. Solid fuels are burned in simple three-stone fireplaces or traditional stoves that are highly inefficient and usually unvented. This leads to high levels of emissions of a wide range of health-damaging pollutants. Since these stoves are used mainly indoors or in semi-enclosed spaces, this results in very high exposures of household members—particularly young children and women.[1] Whereas cities in industrialized countries infrequently exceed the EPA's 24-hour standard for PM_{10}, in rural homes in LMICs the standard may be exceeded on a regular basis by a factor of 10, 20, and sometimes up to 50. Typical 24-hour mean levels of fine particulate matter (PM_{10}) in homes using biomass fuels may range from 300 to greater than 3,000 mg/m^3 depending on the type of fuel, stove, and housing. Measured concentration levels depend on where and when monitoring takes place, given that significant temporal and spatial variations (within a house, including from room to room) may occur.[2] Concentrations of 50,000 μg/m^3 or more can be recorded in the immediate vicinity of a fire, with levels falling significantly with increasing distance.[3] Levels of carbon monoxide (CO) and other health-damaging pollutants also often exceed international guidelines. Even when used outdoors, these stoves may produce surprisingly high levels of exposure for women and children.

Roshan Wathore and Archana Patel, *Household Air Pollution* In: *Textbook of Children's Environmental Health*,
Second Edition. Edited by: Ruth A. Etzel and Philip J. Landrigan, Oxford University Press. © Oxford University Press 2024.
DOI: 10.1093/oso/9780197662526.003.0026

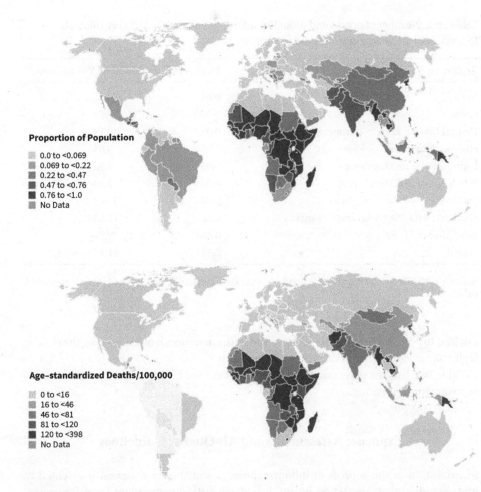

Figure 26.1. Global maps of the proportion of the population (top) and age standardized death rates (bottom) attributable to household air pollution in 2019

A global map of age standardized death rates attributable to household air pollution in 2019 is depicted in Figure 26.1. It is evident that dependence on solid fuels, especially in the regions of South Asia, Southeast Asia, East Asia, Oceania, and sub-Saharan Africa were associated with a larger number of deaths and disability-adjusted life years (DALYs) in 2019 (Table 26.1). Nearly 42% of all global deaths and 32% of DALYs are from India and China alone, although this is partly due to their large contribution to the total world population.

Short-term exposures, ranging from a few hours to a few days can lead to irritation in the throat, eyes, ears, and nose; these effects subside with reductions in exposure. Lower respiratory and chronic conditions such as allergies, bronchitis, chronic obstructive pulmonary disease (COPD), and asthma also may be trigged in vulnerable people. Short-term exposure of more vulnerable people to air pollution may result in arrhythmias, heart attacks, and even deaths.[5] Long-term exposure to airborne pollutants induces some of the most significant effects of air pollution in terms of public health. According to growing scientific evidence, a mother's exposure to air pollution during pregnancy

Table 26.1 Number of deaths and disability-adjusted life years (DALYs) attributable to household air pollution from solid fuels in 2019

Region	Deaths (millions)	DALYs (millions)
India	0.61	20.90
China	0.36	8.74
Central Europe, Eastern Europe, and Central Asia	0.03	0.79
High-income	0.00	0.05
Latin America and Caribbean	0.06	1.92
North Africa and Middle East	0.05	2.34
South Asia	0.84	30.48
Southeast Asia, East Asia, and Oceania	0.65	17.84
Sub-Saharan Africa	0.68	38.05
Global	2.31	91.47

Source: Health Effects Institute. State of Global Air 2020. Data source: Global Burden of Disease Study 2019. IHME, 2020.[4]

can lead to poor birth outcomes for her child in the first month of life, putting the child at high risk for various major disorders and mortality.[5]

Table 26.2 lists the percentage of deaths from selected health conditions attributed to exposure to air pollution.

Exposure Assessments and Air Quality Guidelines

Identification of the sources of indoor exposures and their assessment is essential to understanding their impact on health. It is also a critical component for risk management, which involves identifying strategies to reduce people's exposure through the use of cleaner fuels and technologies. The World Health Organization (WHO) has synthesized evidence on health aspects of air quality and developed Air Quality Guidelines (AQG) for indoor air quality to protect public health from the health risks presented

Table 26.2 Percentage of global deaths from selected health conditions attributed to exposure to air pollution

	Health condition	Percentage of deaths due to air pollution
1	Chronic obstructive pulmonary disease (COPD)	40%
2	Diabetes	20%
3	Ischemic heart disease	20%
4	Lung cancer	19%
5	Stroke	26%
6	Lower respiratory infections	30%
7	Neonatal	20%

Source: Health Effects Institute. Factsheet - Health Effects of Air Pollution. 2020.[5]

by a number of chemicals commonly found in indoor air. A multidisciplinary group of experts studying the toxic properties and health effects of these pollutants developed these guidelines based on a comprehensive review and evaluation of toxicological and epidemiological data of exposure levels causing health concerns. These guidelines also guide specialists and authorities involved in the design and use of buildings, indoor materials, and products. The AQG provide a scientific basis for legally enforceable standards to protect public health.

The WHO guidelines were revised in 2021, based on overwhelming evidence over the past decade on associations between exposure to air pollution and human health. The pollutants include $PM_{2.5}$, PM_{10}, ozone (O_3), nitrogen dioxide (NO_2), sulfur dioxide (SO_2) and CO. Additionally, best practices also cover specific components of PM, which include black carbon (BC; also known as elemental carbon [EC]), ultrafine particles, and sand and dust storms.[6]

Individual exposure to HAP can be determined by microenvironmental (household-level determinants) or macroenvironmental (socioeconomic, demographic, and geographic) determinants. The interactions within and between these categories of determinants can be complex. Microenvironmental determinants usually influence exposure directly in households and often vary over space and time. These include house location (near traffic or industrial area), household income, availability and accessibility of free biomass, fuel choice, cookstove choice and location(indoor or outdoor), stove stacking (using multiple stove and fuel cooking combinations within the same household), kitchen layout, ventilation, season (e.g., in winters, the stove also may be used for space heating and be kept indoors), lighting choice, tobacco smoke, incense, mosquito coils, type of food cooked (e.g., frying vs. boiling), and cultural preferences. Macroenvironmental determinants usually influence exposures in indirect ways. Additional uncontrollable factors include local meteorology (temperature, relative humidity, wind speed, etc.) that influence the choice of fuel as well as affect aerosol dispersion or deposition. Patterns of vegetation (e.g., tropical rainforests vs. scrub) could contribute to household decisions on seeking alternative energy sources.

Technologies for Measuring Household Air Pollution Exposure Concentration

PM exposures have most often been measured *gravimetrically*, where samples are collected on Teflon filters via a pump with an impactor cyclone to remove larger undesired particles. The weight difference of the filter before and after sampling divided by the volume of air sampled from the pump provides the mass concentration of PM. This is a standard reference method but is not well-suited for assessing child and infant exposures because the equipment used is bulky, cumbersome and resource-intensive due to pre- and post-conditioning, transport, handling, and analysis of the filters. This method also provides a time-averaged concentration.

Nephelometers are miniature and easier to deploy than gravimetric samplers and offer time-resolved exposure concentrations. However, because nephelometers are sensitive to differences in environmental conditions (such as temperature and relative humidity) as well as the optical properties of different aerosols, they must be calibrated against a reference method in the target aerosol if they are to produce accurate concentration estimates.

To account for these differences, recent personal samplers for children and infants, such as the Enhanced Children's MicroPEM (RTI, Research Triangle Park, North Carolina, US) incorporate both gravimetric and nephelometric measurements in tandem, along with environmental parameters to provide accurate estimates of PM exposures.[7]

Personal exposure measurements of CO have largely been conducted using technologies based on electrochemical cells or diffusion tubes. Diffusion tubes, which change color as CO interacts with the chemical in the tube, have been used for some of the largest exposure studies. They are passive sampling systems which require no power. The contaminant (CO) molecule diffuses into the tube and reacts with the reagent inside the tube, resulting in discoloration of the tube. The CO exposure is determined by the markers provided across the length of discoloration in the tube. One challenge associated with diffusion tubes is their inability to provide real-time information of exposure: a tube is used only once and has limited precision (the color change demarcation is not distinct). Electrochemical cell-based monitors are generally small, can log exposure data for long durations, and record semicontinuous concentrations, making them ideal for exposure assessments of children. They are more accurate than diffusion-based measurements. However, electrochemical cells undergo loss in sensitivity to the target pollutant over time and need to be frequently calibrated. Measurement of personal CO exposure is more feasible than for PM, especially for small children or infants, and thus has been used as a proxy for PM concentrations given the strong correlations among various pollutant levels. The key consideration for these models is that relationships between CO and PM change as a function of the source and even across the different combustion phases during cooking, so care must be taken to ensure that prediction of PM exposure takes into account relevant HAP sources and behavioral patterns. BC, colloquially known as *soot*, is measured via *aethalometers*, where the sample is collected in quartz filters. The light attenuation of the collected sample, measured as absorption, empirically correlates to the BC concentration. Alternatively, analysis can also be done on polytetrafluoroethylene (PTFE) filters and analyzed in the Sootscan (Magee Scientific, US), which provides a nondestructive optical method of analysis. The PTFE filters can subsequently be used for other analyses.[8] Exposures to other pollutants such as PAHs are less common because their measurement is mostly done by chromatography techniques that are more complex and expensive.

Sources, Toxicology (Including Biomarkers), and Health Effects of Main Household Air Pollutants

Particulate Matter

$PM_{2.5}$ (particles with diameters <2.5 microns) is a component of PM_{10} (particles with diameters <10 microns) that can travel into and deposit on the surface of the deeper parts of the lung. PM_{10} can deposit on the surfaces of the larger airways of the upper region of the lung. Combustion sources are the major contributors of $PM_{2.5}$, whereas dust comprises a significant portion of PM_{10}. $PM_{2.5}$ also is known as fine particulate matter and strongly correlates with adverse health effects in children.[8]

The WHO released its Global AQG in 2005 for PM and other pollutants, with three interim targets (IT-1, IT-2, and IT-3) to facilitate stepwise improvement in air quality and significant health benefits for the population. In 2019, outdoor and household air pollution together accounted for approximately 12% of all deaths. In view of overwhelming evidence demonstrating the health effects of air pollution over the past two decades, the AQG were revised in 2021, wherein an additional interim target (IT-4) (same as the previous AQG of 2005) was introduced, making the updated AQG is more stringent. It is estimated that the proposed updated guidelines can avoid nearly 80% of the deaths related to $PM_{2.5}$ exposure. The WHO AQG 2021 guidelines are summarized in Table 26.3.[6]

Carbon Monoxide

CO is a toxic, colorless, and odorless gas primarily arising from incomplete indoor combustion of fuels for cooking or heating or due to infiltration from outdoor air. It accumulates due to poorly maintained combustion devices, leaky chimneys and furnaces, and back-drafting from furnaces, woodstoves and fireplaces. CO poisoning results in symptoms that range from headache to unconsciousness to even death depending on the dose: exposure at low concentrations causes fatigue in healthy people and chest pain in people with heart disease; exposure to moderate concentrations may result in angina, impaired vision, and reduced brain function; exposure to higher concentrations results in impaired vision and coordination, headaches, dizziness, confusion, nausea, and flu-like symptoms; exposure to very high concentration can be fatal.[9] CO is transported across the lungs via inhalation, diffuses into the bloodstream, and binds preferentially to hemoglobin (Hb) in the blood, forming carboxyhemoglobin (COHb). The most important variables governing the formation of COHb are the concentration and duration of CO in inhaled air and the rate of alveolar ventilation. The affinity of Hb for CO is around 240 times greater than that for oxygen. CO competes with oxygen for Hb binding sites, but, unlike oxygen, which is quickly and easily dissociated from its Hb bond, CO remains bound (leftward shift in the oxyhemoglobin dissociation curve) for a much longer

Table 26.3 Revised World Health Organization Air Quality Guidelines 2021 and Interim Targets for particulate matter

Interim targets	Averaging period	PM_{10} $(\mu g/m^3)$	$PM_{2.5}$ $(\mu g/m^3)$
Interim target-1	Annual	70	35
	24 Hour	150	75
Interim target-2	Annual	50	25
	24 Hour	100	50
Interim target-3	Annual	30	15
	24 Hour	75	37.5
Interim target -4 (Previously WHO AQG 2005)	Annual	20	10
	24 Hour	50	25
AQG (WHO,2021)	Annual	15	5
	24 Hour	45	15

Source: World Health Organization. WHO global air quality guidelines: particulate matter ($PM_{2.5}$ and PM_{10}), ozone, nitrogen dioxide, sulfur dioxide and carbon monoxide. World Health Organization; 2021. xxi, 273 p. Available at: https://apps.who.int/iris/handle/10665/345329.[6]

time. COHb levels continue to increase with continued exposure, leaving progressively less Hb available for carrying oxygen to the tissues, resulting in arterial hypoxemia of the tissues. CO also reduces the diffusion of oxygen into tissue via myoglobin by formation of carboxymyoglobin.[10,11] CO is produced endogenously in humans as well, and the body has a compensatory mechanism that increases red cell volume and blood flow to the brain when COHb levels are less than 20%. This compensatory activity also occurs in neonates and fetuses.[12,13] The recommended WHO indoor AQG for CO exposures relevant to typical indoor exposures is 100 mg/m^3 for 15 minutes and 35 mg/m^3 for 1 hour (assuming light exercise and that such exposure levels do not occur more often than one per day); 10 mg/m^3 for 8 hours (arithmetic mean concentration, light to moderate exercise); and 7 mg/m^3 for 24 hours (arithmetic mean concentration, assuming that the exposure occurs when the people are awake and alert but not exercising).[13]

Polycyclic Aromatic Hydrocarbons

PAHs are hydrocarbons consisting of multiple fused aromatic rings. Human exposures occur mainly through dermal contact, ingestion of contaminated food/water, and respiration of contaminated air due to incomplete combustion of fossil fuels and biofuels. Concentrations of PAHs and benzo[a]pyrene (B[a]P) in indoor air increases in the order of fuel used for household cooking (i.e., LPG < kerosene < coal < wood < dung cake/wood mixture < dung cake).[13,14] PAHs bound to ultrafine particles (i.e., PM$_{0.1}$, particles with diameters ≤0.1 μm) can travel further in the lungs and enter the bloodstream and reach other organs.[15] There are more than 200 PAHs, 16 of which have been identified as high priority by the US Environmental Protection Agency (EPA). The 16 priority PAHs are further divided into low-molecular-weight (2–3 rings; exist in gaseous phase), medium-molecular-weight, and high-molecular-weight (5–6 rings), the latter two of which exist majorly in the particulate phase (bound to PM). PAHs, especially particulate-bound PAHs, pose a fairly high health risk to humans due to their deposition and accumulation in the body. The level of cancer risk through inhalation of particulate PAHs is almost four orders of magnitude more severe than that of exposure by ingestion and dermal contact. Their toxic, immunogenic, embryotoxic, carcinogenic, genotoxic, and mutagenic effects are manifested through formation of DNA adducts.[16] B[a]P, a high-molecular-weight PAH found in ambient and indoor air, is used as an indicator of exposure to carcinogenic PAHs in epidemiological studies. Maternal exposure to PAHs affects fetal development, hormone balance, placental physiology, neonate head circumference, birth length and weight, neurodevelopmental delays, behavioral problems, and childhood asthma.[15] PAH exposure is assessed usually by measurement of urine 1-hydroxypyrene levels or by measuring aromatic bulky DNA adducts in peripheral lymphocytes. A good correlation has been observed between the PAH concentration in air and urine 1-hydroxypyrene in several occupational environments.[17]

Benzene

Benzene (C_6H_6) is the simplest aromatic compound; it is colorless and volatile and remains in the vapor phase in the atmosphere, where it can last for a few hours to days depending on environmental conditions and the presence of other pollutants. Indoor air contamination occurs from outdoor air and from indoor sources such as building materials and furniture (furnishing materials; polymeric materials such as vinyl, PVC, and rubber floorings; and nylon carpets, styrene butadiene resin-latex-backed carpets, particleboard furniture, plywood, fiberglass, flooring adhesives, paints, wood paneling, caulking, and paint remover), attached garages (potential source of gasoline vapor owing to evaporation and exhaust emissions), smoking tobacco, and heating and cooking using either solid fuels, coal, kerosene, or liquid petroleum gas (LPG), especially with use of stoves of low efficiency and inadequate ventilation. Exposure to benzene is associated with hematotoxicity, bone marrow depression, genotoxic effects, carcinogenic effects, myeloproliferative disorders, immunological effects, leukemia, and other lymphohematopoietic malignancies: there is no safe level of exposure of benzene. The annual standard for benzene is 5 $\mu g/m^3$ in India, whereas in Vietnam, the 1-hour guideline is 22 $\mu g/m^3$.[18] Kitchen concentrations during cooking times in rural homes using cow dung and wood for cooking ranged from 31.2 to 75.3 $\mu g/m^3$ depending on the fuel and kitchen layout.[19]

Naphthalene

Naphthalene ($C_{10}H_8$) is a highly volatile PAH consisting of two fused benzene rings. Common indoor sources of naphthalene are wood smoke, fuel oil, gasoline, and tobacco smoke. Other major sources are mothballs, various solvents, herbicides, lubricants, charcoal lighters, unvented kerosene heaters, etc. Naphthalene is a possible human carcinogen.

Acute exposure to naphthalene can occur by inhalation, ingestion, and dermal contact. Naphthalene sniffing during pregnancy can cross the placenta and cause adverse effects on the fetus. Persons with glucose 6-phosphate dehydrogenase deficiency can develop hemolytic anemia. Even small amounts can cause intravascular hemolysis in neonates and resulting severe hyperbilirubinemia and liver injury.[20] Frequent inhalation can lead to nasal lesions, and prolonged high exposures can cause severe inflammation and tumors at this site. Acute exposure to naphthalene by inhalation, ingestion, and dermal contact is associated with damage to the liver, and, in infants, neurological damage. Symptoms include headache, nausea, vomiting, diarrhea, malaise, confusion, anemia, jaundice, convulsions, and coma. Cataracts have been reported in humans acutely exposed to naphthalene by inhalation and ingestion.[21] The indoor air guideline of 0.01 mg/m^3 has been established to minimize the carcinogenic risk to the respiratory tract. Urinary and breast milk 1- and 2-naphthol are well-established human biological exposure indices to evaluate naphthalene exposure in pregnant women and mothers.[22]

Formaldehyde

Formaldehyde (CH_2OH) is found ubiquitously in the environment. It arises from combustion processes such as smoking, heating, cooking, candle or incense burning, and also from numerous building materials and consumer products. The possible routes of exposure are inhalation, ingestion, and dermal absorption. Ventilation can reduce indoor exposure to formaldehyde. Formaldehyde is water soluble, so it is rapidly absorbed in the respiratory and gastrointestinal tracts and metabolized to formate. When it is mixed with particles, more of it is retained by the respiratory tract than when it is inhaled alone. When it exceeds 0.1 mg/m^3 (recommended indoor air guideline levels) individuals may experience burning sensations in the eyes, nose, throat, coughing, wheezing, nausea, and skin irritation. Exposure to high levels of formaldehyde causes myeloid leukemia and rare cancers of the paranasal sinuses, nasal cavity, and nasopharynx.[23] A study in Korea looked into the urinary excretion of formaldehyde and its major metabolite (thiazolidine-4-carboxylic acid [TZCA]) of 620 schoolchildren (10–12 years) and concluded that formaldehyde exposure was significantly associated with rhinitis, peripheral olfactory dysfunction, increased pharmacologically reversible nasal obstruction, small airway impairment, asthma, and elevated FeNO.[24]

Nitrogen Dioxide and Sulfur Dioxide

NO_2 is one of a group of highly reactive gases known as oxides of nitrogen or nitrogen oxides (NO_x). Primary indoor sources of NO_2 are combustion sources such as fireplaces, water heaters, gas, stoves, and other poorly maintained appliances. NO_2, being a free radical, has the potential to deplete tissue antioxidant defenses and cause inflammation and injury. Indoor levels can be influenced by high outdoor levels arising from local traffic or other combustion sources. Exposures over short periods can aggravate respiratory diseases, particularly asthma, leading to respiratory symptoms, hospital admissions, and visits to emergency rooms. Longer exposures to elevated concentrations may contribute to the development of asthma and potentially increase susceptibility to respiratory infections. One of the most comprehensive assessments was conducted as part of the Six Cities study. The association of respiratory symptoms with indoor NO_2 level was examined in more than 1,500 children[25] who were followed for 1 year. At the end of follow-up, the annual cumulative incidence of any lower respiratory symptom (shortness of breath, chronic wheeze, chronic cough, chronic phlegm or bronchitis) was higher in those children living in homes with a source (gas stove or kerosene heater) and was higher with increasing annual average indoor NO_2 (1.40; 95% confidence interval [CI]: 1.14, 1.72 per 28-µg/m^3 increase). Hasselblad et al. published a meta-analysis in the early 1990s of the association of indoor NO_2 levels with respiratory illness in children.[26] In children under the age of 12 years, a 30 µg/m^3 increase was equivalent to a 20% increased risk of symptoms. This analysis provided the basis for setting outdoor AQGs by the WHO in 1997[27] of 40 µg/m^3 (0.023 ppm). In a meta-analysis of 66 studies, pooled hazard ratios (HR) for long-term exposure to NO_2 were significantly associated with mortality from all/natural (1.047), cardiovascular disease (1.091), lung cancer (1.083), respiratory disease (1.062), and ischemic heart disease (1.111) per 10 ppb increment.[28]

Similarly, SO_2 is a major component of oxides of sulfide (SO_x). Primary sources of anthropogenic SO_2 are coal-based powerplants, refineries, and other industrial processes utilizing fossil fuels. Short-term exposures to SO_2 can harm the human respiratory system and make breathing difficult. Zheng et al. performed a systematic review and meta-analysis of short-term exposures to SO_2. The pooled relative risk (RR) per 10 µg/m^3 increase of ambient concentrations was 1.008 for maximum 8-hour daily or average 24-hour O_3, 1.014 for average 24-hour NO_2, and 1.010 for 24-hour SO_2, with children and elderly people being more susceptible to the adverse effects of air pollution.[27] The AQG has revised the 24-hour guideline to 20 µg/m^3 as a prudent precautionary approach to protect health by simultaneous reduction in exposure to a causal and correlated substance. Studies have shown that respiratory effects are experienced at a 10-minute rise of SO_2, so the AQG provided a short-term health effects guideline of 500 $\mu g/m^3$.[13]

Health Effects of Combined Household Air Pollution Exposures in Children

Individual pollutants have their own effects, as mentioned previously. However, in the real world, these pollutants are co-emitted. These also include various trace metals (Zn, Ar, Cr, Pb, Cd, Ni, Cu, V), which are associated with a myriad of health risks including birth defects, cardiovascular diseases, neurological health problems, cancer, damage to organs, and many other disorders; this has been comprehensively reviewed elsewhere.[29] Once emitted, these pollutants subsequently undergo secondary reactions, for instance, nitric oxide (NO) oxidizes to NO_2 after release, which can further react with O_3. Various sulphates and nitrates are also formed during the combustion process. Combustion processes are continuous and dynamic in nature, and the emissions characteristics depend on factors such as the environmental/meteorological conditions, the type of food cooked, fuel characteristics, stove and vessel designs, ventilation, and many other factors. Fine and ultrafine PM provide adequate surface areas for reactions and depositions. Hence the characteristics of PM emitted during the combustion process play an important role in influencing surface chemistry and eventual health impacts. The health effects of HAP in children may originate in the womb. Prenatal exposures of mothers and continued exposure to HAP throughout childhood increases their risk to acute lower respiratory infections, tuberculosis, asthma in childhood, and, subsequently, to adult health diseases such as lung cancer, cancer of the upper aero-digestive tract, cardiovascular disease, cataracts, and cervical cancer in women. Methylation maintains genomic stability and regulation of gene functions; this can be disrupted by environmental exposures and result in the etiology of complex diseases. In a large-scale epigenome-wide meta-analysis, DNA methylation in newborns was found to have significant associations for PM_{10} and $PM_{2.5}$ exposure during pregnancy with methylation differences in several genes of relevance for respiratory health.[30] The periconception period exposures provide the highest risk for congenital malformations compared to exposures during any other period.[31] Although the risk is modest in populous countries, this has a substantial impact on burden of anomalies. Exposure to different air pollutants during pregnancy also increases the risk of low birth weight (LBW).[32] A meta-analysis of 54 cohort studies examining $PM_{2.5}$, PM_{10}, NO_2, CO, SO_2, and O_3 exposure on LBW were 1.081 (95% CI: 1.043, 1.120), 1.053 (95% CI: 1.030, 1.076), 1.030 (95% CI: 1.008, 1.053), 1.007 (95% CI: 1.001, 1.014), 1.125 (95% CI: 1.017, 1.244), and 1.045 (95% CI: 1.005, 1.086), respectively. Consistent with these findings, another systematic review of 45 epidemiologic studies from 29 countries published up to August 2020 examined the association between ambient and household particulate exposure and small for gestational age (SGA) at birth that results in childhood stunting. They found significant positive associations between SGA and a 10 $\mu g/m^3$ increase in $PM_{2.5}$ exposure over the entire pregnancy [odds ratio [OR] = 1.08; 95% CI: 1.03, 1.13]. A 19% increased risk of postnatal stunting (95% CI: 1.10, 1.29) was also associated with postnatal exposure to HAP.[33] Children born LBW are subsequently more susceptible to poor health outcomes and neurodevelopmental disorders.[34] HAP exposure increases the risk of all respiratory health problems, particularly pneumonia and asthma in children. The increased risk of pneumonia accounts for about a million deaths globally. A systematic review investigated the individual effect of CO, BC, and $PM_{2.5}$ on under-5 pneumonia in LMICs. Although exposure to solid fuels showed significant associations, the individual pollutants failed to show association, suggesting the need to improve study designs and the measurement of additional pollutants.[35] A study in Indonesia evaluated associations between NO_2, SO_2 and LBW; infant death; neonatal death; and acute respiratory infection (ARI) in 4,931 children (0–3 years). An interquartile range increase in mean NO_2 exposure increased the risk for ARI by 18%. NO_2 exposure increased the risk for respiratory infections in early childhood. Interestingly, SO_2 was not associated with the examined health outcomes, suggesting that there may have been exposure misclassification because the area of the study was dominated by NO_2 levels.[36]

The Human Early Life Exposome (HELIX) study measures multiple environmental exposures during early life (pregnancy and childhood) in a prospective cohort and associates these exposures with molecular omics signatures and child health outcomes across six longitudinal population-based birth cohorts in six European countries (France, Greece, Lithuania, Norway, Spain, and the United Kingdom).[37] The biological mechanisms of exposures to both short- and long-term air pollution on the health of 1,170 children ages 6–11 years was examined. Hepatocyte growth factor (HGF) and interleukin-8 (IL-8) levels were positively associated with a 1-week home exposure to some of the pollutants (NO_2, PM_{10}, or $PM_{2.5}$). NO_2 1-week home exposure was also related to higher systolic blood pressure (SBP).[38] These are important findings because recent evidence found that children with higher BP are more likely to develop cardiovascular diseases during adulthood.[39]

Policy and Intervention Measures That Could Improve Children's Health

The fact that HAP impacts a wide range of interrelated issues that include children's health, women's lives, the environment, and socioeconomic development underscores the urgent need for a collaborative approach by local and national governments to implement effective and sustainable interventions. A wide range of interventions include choosing improved thermal and fuel-efficient cooking devices, use of low-polluting fuel, improved ventilation, pollution reduction measures through improved kitchen designs and placement of stoves, and user behavior change to reduce exposure. These actions could help reduce the burden of illness due to HAP. Unfortunately, this collaborative action has by and large been lacking. The Global Alliance for Clean Cookstoves (the Alliance) is a public–private partnership hosted by the United Nations Foundation to save lives, improve livelihoods, empower women, and combat climate change by creating a thriving global market for clean and efficient household cooking solutions. The goal of the Alliance is to work with public, private, and nonprofit partners to build collaborations and help overcome the market barriers that currently impede the production, deployment, and use of clean cookstoves in the developing world. Evidence on health effects and on cost-effectiveness still needs further strengthening. Concerted global action on this major preventable public health hazard impacting predominantly on the poor is long overdue.

The largest cooking intervention initiative was undertaken by Prime Minister Ujjwala Yojana in India to promote subsidized liquid petroleum gas (LPG) for clean cooking in poor communities. However, complete adoption of LPG has been a challenge. Primary barriers for completely shifting to LPG are affordability, accessibility, taste, and several other social, economic, and cultural factors that lead stove-stacking. Improved cookstoves have the potential to offer benefits, both in terms of health and climate, but the real-world performance of these cookstoves is significantly poorer than in the lab (e.g., $PM_{2.5}$ and CO emission factors in the field can be 2–5 times higher than in the laboratory for the same improved cookstove). Fuel processing is another barrier to complete adoption, especially for forced-draft gasifier stoves, which are considered the cleanest solid fuel–burning stoves.[40] Even LPG stoves exhibit significantly higher emissions in the field.[41] However, a study in Rwanda introduced an ecosystem for upstream feedstock sourcing, fuel processing, and marketing/distribution of pellets that can be used with forced-draft cookstoves. This study demonstrated that pellet-fed forced draft stoves approach the performance equivalent of LPG stoves in terms of both health and climate impacts.[42] The ventilation and presence of chimneys in rural kitchens also play a significant role in reducing exposures by venting out toxic emissions.[43] Clean cooking interventions also present an attractive opportunity to incorporate carbon finance. The availability of appropriate population exposure monitoring is critical to determine health effects. The increase in the number of continuous ambient air quality monitoring stations over the past decade has enabled estimations of exposures. However, there is still a lack of data from remote and rural areas. Recent approaches incorporating satellite-based data and other modeling-based tools have been gaining widespread attention due to their ability to support epidemiological studies.

To summarize, although there have been significant endeavors toward clean cooking, they have usually been one-dimensional in nature, that is, focusing on only one (or some) aspect(s) of a spectrum of interventions. A complete clean cooking solution involves a cleaner cooking device; readily available, clean, processed, sustainably sourced fuel; an efficient ventilation system; proper information, education and communication programs; and, most importantly, a solution that is well accepted by the communities of a specific geographic location as meeting their needs.

References

1. World Health Organization. Air pollution and child health: prescribing clean air. Summary. Geneva: World Health Organization; 2018 (WHO/CED/PHE/18.01). https://www.who.int/publications/i/item/WHO-CED-PHE-18-01

2. Bilsback KR, Dahlke J, Fedak KM, et al. A laboratory assessment of 120 air pollutant emissions from biomass and fossil fuel cookstoves. *Environ Sci Technol.* 2019 Jun 18;53(12):7114–25.

3. Ezzati M, Mbinda BM, Kammen DM. Comparison of emissions and residential exposure from traditional and improved cookstoves in Kenya. *Environ Sci Technol.* 2000 Feb 1;34(4):578–83.

4. Health Effects Institute. *State of Global Air 2020.* Data source: Global Burden of Disease Study 2019. Boston: HEI; 2020. https://www.healthdata.org/research-analysis/library/state-global-air-2020

5. Health Effects Institute. Factsheet—Health Effects of Air Pollution. 2020.

6. World Health Organization. WHO global air quality guidelines: particulate matter (PM2.5 and PM10), ozone, nitrogen dioxide, sulfur dioxide and carbon monoxide. World Health Organization; 2021. xxi. Available at: https://apps.who.int/iris/handle/10665/345329.

7. Chartier R, Newsome R, Rodes C. Development and evaluation of an enhanced children's micropem (ECM) to support indoor air pollution studies. ISEE Conference Abstracts. 2014 Oct 20. Available at: https://ehp.niehs.nih.gov/doi/abs/10.1289/isee.2014.P3-802.

8. Grange SK, Lötscher H, Fischer A, Emmenegger L, Hueglin C. Evaluation of equivalent black carbon source apportionment using observations from Switzerland between 2008 and 2018. *Atmos Meas Tech.* 2020 Apr 14;13(4):1867–85.

9. US Environmental Protection Agency (EPA). Carbon monoxide's impact on indoor air quality. 2014. Available at: https://www.epa.gov/indoor-air-quality-iaq/carbon-monoxides-impact-indoor-air-quality.

10. Kaufman DP, Kandle PF, Murray I, Dhamoon AS. Physiology, oxyhemoglobin dissociation curve. Treasure Island, FL: StatPearls Publishing; 2021. Available at: http://www.ncbi.nlm.nih.gov/books/NBK499818/.

11. Pan KT, Leonardi GS, Croxford B. Factors contributing to CO uptake and elimination in the body: a critical review. *Int J Environ Res Public Health.* 2020 Jan;17(2):528.

12. Jones MD, Traystman RJ. Cerebral oxygenation of the fetus, newborn, and adult. *Semin Perinatol.* 1984 Jul;8(3):205–16.

13. WHO. *WHO Guidelines for Indoor Air Quality: Selected Pollutants.* Copenhagen: WHO; 2010.

14. Zhang L, Yang L, Zhou Q, et al. Size distribution of particulate polycyclic aromatic hydrocarbons in fresh combustion smoke and ambient air: a review. *J Environ Sci.* 2020 Feb 1;88:370–84.

15. Huang X, Xu X, Dai Y, Cheng Z, Zheng X, Huo X. Association of prenatal exposure to PAHs with anti-Müllerian hormone (AMH) levels and birth outcomes of newborns. *Sci Total Environ.* 2020 Jun 25;723:138009.

16. US Environmental Protection Agency. *Polycyclic Aromatic Hydrocarbons (PAHs).* Washington, DC: US Environmental Protection Agency; 2008. https://archive.epa.gov/epawaste/hazard/wastemin/web/pdf/pahs.pdf

17. Gromadzińska J, Wąsowicz W. Health risk in road transport workers. Part I. Occupational exposure to chemicals, biomarkers of effect. *Int J Occup Med Environ Health.* 2019 Apr 2. Available at: http://www.journalssystem.com/ijomeh/Health-risk-in-road-transport-workers-Part-I-Occupational-exposure-to-chemicals-biomarkers,99520,0,2.html.

18. Sekar A, Varghese GK, Varma MR. Analysis of benzene air quality standards, monitoring methods and concentrations in indoor and outdoor environment. *Heliyon.* 2019;5(11):e02918.

19. Sinha SN, Kulkarni PK, Shah SH, et al. Environmental monitoring of benzene and toluene produced in indoor air due to combustion of solid biomass fuels. *Sci Total Environ.* 2006 Mar 15;357(1):280–7.

20. Sahni M, Vibert Y, Bhandari V, Menkiti O. Newborn infant with mothball toxicity due to maternal ingestion. *Pediatrics.* 2019 Jun;143(6):e20183619.

21. Yost EE, Galizia A, Kapraun DF, et al. Health effects of naphthalene exposure: a systematic evidence map and analysis of potential considerations for dose–response evaluation. *Environ Health Perspect.* 2021 Jul;129(7):076002.

22. Wheeler AJ, Dobbin NA, Héroux ME, et al. Urinary and breast milk biomarkers to assess exposure to naphthalene in pregnant women: an investigation of personal and indoor air sources. *Environ Health.* 2014 Apr 27;13(1):30.

23. National Cancer Institute. Formaldehyde. 2019. Available at: https://www.cancer.gov/about-cancer/causes-prevention/risk/substances/formaldehyde.

24. Yon DK, Hwang S, Lee SW, et al. Indoor exposure and sensitization to formaldehyde among inner-city children with increased risk for asthma and rhinitis. *Am J Respir Crit Care Med.* 2019 Aug;200(3):388–93.

25. Huang S, Li H, Wang M, et al. Long-term exposure to nitrogen dioxide and mortality: a systematic review and meta-analysis. *Sci Total Environ.* 2021 Jul 1;776:145968.

26. Hasselblad V, Eddy DM, Kotchmar DJ. Synthesis of environmental evidence: nitrogen dioxide epidemiology studies. *J Air Waste Manage Assoc.* 1992 May;42(5):662–71.

27. Zheng X-Y, Orellano P, Lin H-L, Jiang M, Guan W-J. Short-term exposure to ozone, nitrogen dioxide, and sulfur dioxide and emergency department visits and hospital admissions due to asthma: a systematic review and meta-analysis. *Environ Int.* 2021 May 1;150:106435.

28. Stieb DM, Berjawi R, Emode M, et al. Systematic review and meta-analysis of cohort studies of long term outdoor nitrogen dioxide exposure and mortality. *PLoS One.* 2021 Feb 4;16(2):e0246451.

29. Ali MU, Yu Y, Yousaf B, et al. Health impacts of indoor air pollution from household solid fuel on children and women. *J Hazard Mater.* 2021 Aug;416:126127.

30. Gruzieva O, Xu CJ, Yousefi P, et al. Prenatal particulate air pollution and DNA methylation in newborns: an epigenome-wide meta-analysis. *Environ Health Perspect.* 2019 May;127(5):057012.

31. Ren S, Haynes E, Hall E, et al. Periconception exposure to air pollution and risk of congenital malformations. *The Journal of Pediatrics.* 2018 Feb;193:76–84.e6.

32. Li C, Yang M, Zhu Z, et al. Maternal exposure to air pollution and the risk of low birth weight: a meta-analysis of cohort studies. *Environ Res.* 2020 Nov;190:109970.

33. Pun VC, Dowling R, Mehta S. Ambient and household air pollution on early-life determinants of stunting—a systematic review and meta-analysis. *Environ Sci Pollut Res.* 2021 Jun;28(21):26404–12.

34. Chun H, Leung C, Wen SW, McDonald J, Shin HH. Maternal exposure to air pollution and risk of autism in children: a systematic review and meta-analysis. *Environ Pollut.* 2020 Jan 1;256:113307.

35. Adaji EE, Ekezie W, Clifford M, Phalkey R. Understanding the effect of indoor air pollution on pneumonia in children under 5 in low- and middle-income countries: a systematic review of evidence. *Environ Sci Pollut Res.* 2019 Feb 1;26(4):3208–25.

36. Suryadhi MAH, Abudureyimu K, Kashima S, Yorifuji T. Nitrogen dioxide and acute respiratory tract infections in children in Indonesia. *Arch Environ Occup Health.* 2020 Jul 3;75(5):274–80.

37. Maitre L, de Bont J, Casas M, et al. Human Early Life Exposome (HELIX) study: a European population-based exposome cohort. *BMJ Open.* 2018 Sep;8(9):e021311.

38. de Prado-Bert P, Warembourg C, Dedele A, et al. Short- and medium-term air pollution exposure, plasmatic protein levels and blood pressure in children. *Environ Res.* 2022 Aug 1;211:113109.

39. Yang L, Magnussen CG, Yang L, Bovet P, Xi B. Elevated blood pressure in childhood or adolescence and cardiovascular outcomes in adulthood: a systematic review. *Hypertension.* 2020 Apr;75(4):948–55.

40. Wathore R, Mortimer K, Grieshop AP. In-use emissions and estimated impacts of traditional, natural- and forced-draft cookstoves in rural Malawi. *Environ Sci Technol.* 2017 Feb 7;51(3):1929–38.

41. Islam MM, Wathore R, Zerriffi H, Marshall JD, Bailis R, Grieshop AP. In-use emissions from biomass and LPG stoves measured during a large, multi-year cookstove intervention study in rural India. *Sci Total Environ.* 2020 Nov;758:143698.

42. Champion WM, Grieshop AP. Pellet-fed gasifier stoves approach gas-stove like performance during in-home use in Rwanda. *Environ Sci Technol.* 2019 Jun 4;53(11):6570–9.

43. Islam MM, Wathore R, Zerriffi H, Marshall JD, Bailis R, Grieshop AP. Assessing the effects of stove use patterns and kitchen chimneys on indoor air quality during a multiyear cookstove randomized control trial in rural India. *Environ Sci Technol.* 2022 May 13;acs.est.1c07571.

27

Tobacco Smoke

Active and Passive Smoking

Meghan Buran, Kelsey Phinney, and Jonathan M. Samet

Tobacco smoking has long been the leading cause of avoidable morbidity and mortality worldwide, even though some of its many adverse effects on health were documented decades ago. Worldwide, there are more than 1 billion tobacco smokers, including smokers of cigarettes and bidis (tobacco rolled loosely in a leaf; smoked widely in India and Bangladesh).[1] More than 80% of adult smokers live in low- and middle-income countries (LMICs). Survey data from 14 high-burden countries collected around 2010, in the initial wave of the Global Adult Tobacco Survey (GATS), showed that about 41% of men and 5% of women smoked cigarettes.[2] While tobacco use is declining in many high-income Western countries, it has risen in some LMICs as the multinational tobacco companies seek new markets and new smokers.

Survey data show the enormity of the smoking problem for children. Almost half of children younger than age 15 are exposed to smoking in their homes[3]; globally, an estimated 24 million youth ages 13–15 use combustible tobacco products and 13 million use smokeless tobacco products.[3] While smoking has declined among adolescents in many countries, the rate of decline has slowed and there is concern that the tobacco industry's targeting of young adults with novel products, particularly various types of electronic cigarettes (e-cigarettes), could increase future rates of smoking. In the United States, 9.3% (2.55 million) of middle and high school students participating in the 2021 National Youth Tobacco Survey reported current use of tobacco products.[4] E-cigarettes have overtaken cigarettes as the tobacco product of choice among US youth and were the most commonly used tobacco product overall among high school students who are current users of tobacco products (11.3%) in 2021. In the 2021 survey, dual use was common, and 1.9% of high school students were current smokers of cigarettes only.

The significance of the worldwide epidemic of tobacco smoking for child health is well recognized. Tobacco smoking has adverse effects that extend across the full life span, beginning before conception and ending with old age and death. We have learned that exposures in early life, such as to tobacco smoke, may have long-term implications for health across the life span. These observations have given rise to the "developmental origins of adult disease" hypothesis or "Barker" hypothesis, in reference to David Barker, who documented associations between early-life nutrition and subsequent risk for hypertension and cardiovascular disease.[5-7] These clinical and epidemiological findings are corroborated by experimental research.[8] The proposed underlying mechanisms emphasize genetic and epigenetic changes that could have lasting implications across the life span.[9]

Meghan Buran, Kelsey Phinney, and Jonathan M. Samet, *Tobacco Smoke* In: *Textbook of Children's Environmental Health*, Second Edition. Edited by: Ruth A. Etzel and Philip J. Landrigan, Oxford University Press. © Oxford University Press 2024.
DOI: 10.1093/oso/9780197662526.003.0027

Childhood also is a critical period for experimenting with tobacco use and becoming addicted to nicotine. The tobacco industry has long targeted youths with aggressive marketing strategies to assure that its customer base is maintained.[10] The entry of e-cigarettes into the tobacco market, along with advertising reaching to youth,[11] has resulted in rising e-cigarette use in many countries—overtaking cigarette use as the most commonly used tobacco product among middle and high school students.[4,12]

Scope and Nature of the Problem

This chapter provides an overview of the adverse effects of tobacco smoking on child health, taking a life course perspective (Figure 27.1). The chapter begins by providing an introduction to the mechanisms by which tobacco smoke causes disease. The chapter first covers passive and then active smoking. It ends with a discussion of strategies for protecting children from tobacco smoke.

Although the primary focus of this chapter is on adolescence and young adulthood, it also is important to consider early-life exposures to tobacco smoke. In childhood and adolescence, the sperm and oocytes of young people who smoke and who are destined to become future parents are exposed to the DNA-damaging constituents of tobacco smoke years before they conceive a pregnancy.[13] The fetus also is exposed to these materials and often will have a reduced birth weight if the mother smokes or is exposed to secondhand smoke (SHS) during pregnancy.[13,14] One study demonstrated epigenetic changes in children with in utero exposure to maternal smoking,[15] a finding consistent with one proposed mechanism for the later consequences of early-life exposures. Because exposure to smoke can occur across all phases of human development, exposures during pregnancy may affect the expression of later risk of disease through mechanisms figuring in the developmental origins of disease as well as through the more direct pathways that have been extensively studied.

For many of the chronic diseases caused by smoking, risks increase with increasing duration and cumulative amount of smoking, and thus the age of smoking initiation has consequences for the age at which the risks of smoking become manifest. In the United States, the age of smoking initiation regularly became increasingly younger in the late 20th century,[16] first for males and then for females; more recently, this age has been stable and even increased.[17] By the early 1990s, the mean age of first trying a cigarette was about 16 years for those who ever smoked; in many other countries, the mean age of uptake is similarly low. More recently in the United States, the age of smoking initiation

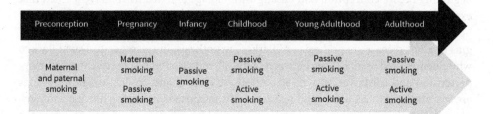

Figure 27.1. Smoking across the life course

has shifted upward into young adulthood (ages 18–24), perhaps reflecting the impact of more restrictive policies on youth access.

Through adolescence, growth is not yet complete, and susceptibility to the ill effects of tobacco smoke may be enhanced. For the major chronic diseases caused by smoking, the epidemiologic evidence indicates that risk rises progressively with increasing duration of smoking; indeed, for lung cancer, the risk rises more steeply with duration of smoking than with number of cigarettes smoked per day.[13,18,19] For chronic obstructive pulmonary disease (COPD), risk varies directly with the total number of cigarettes consumed over a lifetime,[13] which would suggest greater risk for longer duration or higher intensity. There is little direct evidence, however, on whether the age of smoking initiation, by itself, modifies the risk of smoking-related disease later; that is, whether starting to smoke during adolescence versus young adulthood increases the subsequent risk for such disease given the same cumulative amount smoked.[20]

Sources of Exposure

Tobacco smoke is a rich chemical mixture containing thousands of chemical compounds, many with well-documented toxicity such as benzene (a leukemogen), formaldehyde (an irritant and carcinogen), benzo-a-pyrene (a carcinogen), carbon monoxide and cyanide (asphyxiants), acrolein (an irritant), and polonium (a radioactive carcinogen). The smoke from manufactured cigarettes is generated by the burning of tobacco, a complex organic material, along with the various additives and paper comprising the wrapping, all at a very high temperature that can reach about 1,000°C at the burning core of the cigarette.[21] Myriad toxic components can cause injury through inflammation and irritation, asphyxiation, carcinogenesis, and other mechanisms.

Active smokers inhale mainstream smoke (MS), the smoke which is drawn directly through the end of the cigarette. This fresh smoke is inhaled without dilution and consequently the smoker receives high doses of many compounds that are known to be toxic. Passive smokers (i.e., nonsmokers exposed to smoke from smoking by others) inhale smoke that is often referred to as "secondhand" smoke, comprising a mixture of mostly sidestream smoke (SS) given off by the smoldering cigarette and some exhaled MS. SHS is diluted as the SS and exhaled MS mix with air, and, consequently, concentrations of tobacco smoke components in SHS are well below the levels in the MS inhaled by the active smoker. Nonetheless, qualitative similarities between MS and SHS support generalizing the findings on the health risks of active smoking to those of passive smoking.[22] The term "thirdhand smoke" has been used to refer to surface-deposited tobacco smoke components that may return to the air because of their volatility. The components of thirdhand smoke can persist in indoor environments and undergo chemical transformations.[23] Exposure to thirdhand smoke may prove to be particularly significant for toddlers and young children who are in contact with contaminated surfaces and who are also likely to put objects into their mouths.

Tobacco smoke, both MS and SHS, is a mixture of gases and small particles in a size range that penetrates to the deepest portion of the lungs, reaching the bronchioles (small airways) and the gas-exchanging alveoli (air sacs). Most inhaled particles are deposited in these regions, where they cause inflammation and release bound components, including nicotine, which cross the lining of the lung and enter the circulation. The particles also contain toxic components, such as carcinogens, that move into the cells lining the lung's airways. Reactive gases in tobacco smoke, such as formaldehyde, are removed in the upper airways, while insoluble and unreactive gases, such as carbon monoxide, reach the alveoli, where they are absorbed into the bloodstream. Some components of smoke undergo metabolic transformation into their active forms, and there is evidence that these transformations are affected by various genes.[20] The genitourinary system is exposed to toxicants in tobacco smoke through the excretion of tobacco smoke compounds and metabolites in the urine, including carcinogens. The gastrointestinal tract is exposed through direct deposition of smoke in the upper airway and the clearance of smoke-containing mucus from the trachea through the glottis into the esophagus.

There is substantial scientific literature on the mechanisms by which tobacco smoking causes disease.[24] Broad classes of mechanisms include carcinogenesis, inflammation, irritation, asphyxiation, and altered blood coagulability. There are also effects on the functioning of the immune system. Nicotine in tobacco smoke is a potent cause of addiction. Nicotine is an alkaloid that binds to nicotinic cholinergic receptors in the brain, leading to the release of dopamine.[24] Sustained exposure to nicotine leads to tolerance, driven by an increase in the number of nicotinic cholinergic receptors and other changes within the brain.[25] As a result, when addicted smokers stop smoking, there is a reduced production of dopamine and other neurotransmitters in response to stimuli. Withdrawal symptoms result. The most prominent of these symptoms is a craving to smoke, a symptom that may persist for years. Nicotine is highly addictive, and relatively limited contact can lead to addiction.

Epidemiology and Systems Affected

Passive Smoking and Child Health

Overview
From a policy perspective, the identification of passive smoking as a cause of lung cancer in adult never-smokers provided the public health foundation to push for smoking bans in public places and workplaces. However, the first studies to find adverse effects of SHS exposure were directed at the consequences of smoking within a household for family members, including children. These early investigations, dating to the late 1960s, focused on household smoking and respiratory symptoms in household members and parental smoking and lower respiratory illnesses in infants; studies of lung function and respiratory symptoms in children soon followed.[22,26] At about the same time, the first measurements of tobacco smoke components in indoor environments were made, and, citing these data, the 1972 report of the US Surgeon General called attention to passive smoking as a public health threat.[27]

Exposure of Children to Secondhand Smoke
Burning tobacco products are strong sources of indoor pollution, contaminating the air with particles and gases; some smoke components, like nicotine, are specific to tobacco smoke and their concentrations are a useful indicator of the contribution of smoking to indoor air contamination. Studies using exposure monitoring for nicotine, particles, and other tobacco smoke components have documented the role of various environments where smoking takes place in contributing to the exposures of nonsmokers, particularly children, to SHS. For children and for nonsmoking women, the home is a dominant locus of smoking because of the substantial time spent there. Since smoking by men predominates in homes, women and children comprise the majority of passive smokers

worldwide. Smoking bans and changing social norms have greatly reduced the contributions of public places and workplaces to SHS exposure.

Biomarkers also have been used to document exposures of children to SHS. The most widely used is cotinine, a metabolite of nicotine, which can be measured in blood, saliva, and urine. Concentrations of biomarkers of exposure to tobacco smoke, including cotinine, tend to be significantly greater in SHS-exposed nonsmokers compared to unexposed nonsmokers and to increase with an increasing number of smokers within the household.[14] For children, smoking by the mother is associated with the greatest increment in cotinine concentration, perhaps reflecting the amount of time that children spend in proximity to their mothers.

Sudden Infant Death Syndrome

Exposure to SHS increases the risk of sudden infant death syndrome (SIDS), the unexplained death of an infant 12 months of age or younger. This increased risk has been shown in epidemiologic studies conducted throughout the world. All of these studies examined the relationship between exposure from maternal smoking and SIDS and found that infants whose mothers smoked were more likely to die of SIDS. Smoking in the house by the father and other smokers was also shown to increase risk. The evidence that SHS exposure causes SIDS is both "consistent" and "strong."[14] The increased risk of SIDS associated with SHS exposure may be caused by exposure to nicotine and other components of SHS that are neurotoxic. Such components may interfere with brain development and breathing regulation, which in turn may increase an infant's risk of SIDS. In addition, SHS exposure also makes an infant more susceptible to respiratory infections and lung irritation, which may impair breathing and contribute to SIDS.

Smoking by the mother during pregnancy increases the risk of low birth weight (<2,500 g or 5.5 lb). Carbon monoxide and nicotine limit the flow of oxygen to the fetus and also decrease the flow of blood through the uterus into the umbilical cord and thereby slow the development of the growing fetus. Newborns born to mothers exposed to SHS are approximately 20% more likely to have low birth weight than infants of unexposed mothers. Numerous epidemiologic studies show that even if the mother herself does not smoke during pregnancy, being around other people who smoke may have a similar, although smaller, effect on birth weight. Finally, infants born to women exposed to SHS are on average 30 grams lighter than infants born to mothers who were not exposed.[14]

Ear Problems

Ear infections, although common in childhood, are more frequent among children exposed to SHS. Ear infections typically affect the middle ear and can lead to temporary or permanent hearing loss if the infections are severe enough. SHS increases the risk of recurring middle ear infections, especially among children with a history of ear infection. On average, children whose mothers smoke have an almost 40% higher risk of buildup of fluid behind the eardrum and ear infections than children whose mothers do not smoke. Of six studies conducted in six different countries, all but one showed greater risk of middle ear effusion when at least one parent smoked.[14]

Lung Growth and Development

Substantial evidence from cross-sectional and cohort studies demonstrates the harmful effects of SHS exposure on the structural and functional development of children's lungs. For more than two decades there has been sufficient evidence to conclude that

SHS exposure before and after birth is associated with reduced lung function in children. The 1984 US Surgeon General's report concluded that children of smoking parents have reduced lung function compared to children of nonsmoking parents,[28] and the 1986 US Surgeon General's report concluded that SHS exposure reduces the rate of lung function growth during childhood.[22] A pooled analysis of 26 studies, published from 1979 to 2001, found that the lung function of children exposed to SHS at home is significantly reduced for three out of four key indicators compared to children not exposed.[29] These findings were updated and reaffirmed in the 2006 Surgeon General's report.[14]

Respiratory Infections

Infants and young children exposed to SHS are at increased risk for respiratory infections and are more likely than children who are unexposed to be hospitalized for a serious respiratory infection. Parental smoking is consistently associated with an increased risk of lower respiratory illnesses, such as bronchitis and pneumonia, particularly in children aged 2 years or younger.[14] Of 34 studies reported in the 2006 Surgeon General's report that were based in several different countries and using a variety of study designs, all but one found an elevated risk of lower respiratory illness in young children whose parents smoked. On average, smoking by the mother was associated with a 60% increase in risk of a lower respiratory illness, and smoking by the father was associated with a 30% increase. Seventeen of 22 studies found that each additional smoker in a household increased the risk of illness, as did increasing smoking intensity (the number of cigarettes smoked in the household). In addition, young children who are exposed to SHS were more likely to be hospitalized for a serious respiratory illness.[14]

Asthma

Of 41 studies in the 2006 Surgeon General's report examining the risk of asthma in relation to SHS exposure in school-aged children (ages 5–16), all but three found increased asthma risk among children exposed to SHS. In pooled analyses of all studies, the risk of asthma was 23% higher among exposed than unexposed children.[14]

Among 58 studies in the 2006 Surgeon General's report that measured wheeze using different definitions, all but one showed increased risk associated with exposure to SHS. Risk of wheeze was 25% higher among school-aged children exposed to SHS in studies designed to control for other characteristics that might affect risk estimates (e.g., age, gender, socioeconomic status). Out of 44 studies that examined chronic cough and SHS exposure, the risk of chronic cough was 27% higher in school-aged children exposed to SHS among the studies designed to control for other risk factors.[14]

The risks of asthma, wheeze, and cough are higher when both parents smoke than when only one parent smokes. The Surgeon General of the US Public Health Service has determined that SHS exposure causes asthma, wheeze, and chronic cough during childhood.[14]

Summary on Passive Smoking and Child Health

The evidence on child health and passive smoking was comprehensively reviewed in 1986 by the US Surgeon General and the US National Research Council.[22,30] At that time, the evidence documented increased risk for lower respiratory illnesses, increased respiratory symptoms, and reduced lung function in children of smokers compared with nonsmokers.

Mounting evidence since then has affirmed and expanded these earlier findings. SHS exposure has now been causally associated with SIDS, exacerbation of asthma in children, various ear problems, and slowing of lung growth (Table 27.1).[14] The evidence was

Table 27.1 Adverse health effects in children from exposure to tobacco smoke

Health effect	SGR	SGR	EPA	CAL/EPA	UK	WHO	IARC	CAL/EPA[a]	SGR
	1984	1986	1992	1997	1998	1999	2004	2005	2006
Increased prevalence of chronic respiratory symptoms	Yes/a	Yes/a	Yes/c	Yes/c	Yes/c	Yes/c	Yes/c		Yes/c
Decrement in pulmonary function	Yes/a	Yes/a	Yes/a	Yes/a		Yes/c	Yes/a		Yes/c
Increased occurrence of acute respiratory illnesses	Yes/a	Yes/a	Yes/a	Yes/c		Yes/c	Yes/c		Yes/c
Increased occurrence of middle ear disease		Yes/a	Yes/c	Yes/c	Yes/c	Yes/c	Yes/c		Yes/c
Increased severity of asthma episodes and symptoms			Yes/c	Yes/c		Yes/c	Yes/c		Yes/c
Risk factor for new asthma			Yes/a	Yes/c		Yes/c	Yes/c		Yes/c
Risk factor for sudden infant death syndrome				Yes/c	Yes/a	Yes/c	Yes/c		Yes/c

Yes/a = association; Yes/c = cause.

SGR 1984: US Department of Health and Human Services (1984); SGR 1986: US Department of Health and Human Services (1986); EPA 1992: US Environmental Protection Agency (1992); Cal/EPA 1997: California Environmental Protection Agency and Office of Environmental Health Hazard Assessment (1997); UK 1998: Scientific Committee on Tobacco and Health and HSMO (1998); WHO 1999: World Health Organization (1999); IARC 2004: International Agency for Research on Cancer (2004); Cal/EPA 2005: California Environmental Protection Agency and Air Resources Board (2005); SGR 2006: US Department of Health and Human Services (2006).

[a] Only effects causally associated with secondhand smoke exposure are included.

Source: The 2006 US Surgeon General's report.[14]

once again comprehensively reviewed in the 2006 Surgeon General's report with affirmation of the conclusions of prior reports (Table 27.1). Since then, there has not been another authoritative report on the topic.

We emphasize that maternal smoking during pregnancy represents passive smoking for the fetus. The fetus of a woman who smokes during pregnancy is exposed to numerous components of tobacco smoke that are absorbed through the mother's lungs into her bloodstream and then reach the placental circulation. While this route of exposure is not traditionally considered a form of exposure to SHS, the reality is that the fetus of a mother who smokes during pregnancy is exposed to nicotine and carbon monoxide, affecting the developing brain and oxygen delivery, respectively. Smoking during pregnancy reduces birth weight by an average of 200 grams, and the degree of reduction is related to the amount smoked. Smoking also increases rates of placenta previa and perinatal mortality. Smoking during pregnancy, along with SHS exposure, is a cause of SIDS. There is some evidence suggesting that smoking by the mother may increase cancer risk for some types of childhood cancer and risk for congenital defects, specifically cleft lip and palate.

The 2014 report of the Surgeon General offered several strong conclusions on nicotine exposure to the fetus: "The evidence is sufficient to infer that nicotine exposure during fetal development, a critical window for brain development, has lasting adverse consequences for brain development" and "The evidence is sufficient to infer that nicotine adversely affects maternal and fetal health during pregnancy, contributing to multiple adverse outcomes such as preterm delivery and stillbirth."[31]

Active Smoking and Child Health

Overview

The injurious processes that lead to the diseases caused by smoking begin with the first cigarettes smoked and continue over the life course for the regular smoker. Over approximately the past half-century, the age smoking initiation experimentally and then regularly moved steadily downward toward younger ages in the United States until the more recent increase in the age of initiation over the past decade.[31] The age of starting to smoke has been moving downward in LMICs. This shift is significant as risks for most smoking-caused diseases vary directly and positively with the duration of smoking. Smokers who start younger tend to smoke for a longer duration across their lives.

Epidemiological studies show convincingly that adverse effects of smoking are manifest in regularly smoking adolescents and young adults. Consequences have been documented for both the respiratory and cardiovascular systems. Nicotine addiction also begins in adolescence for young people who smoke, and longitudinal studies have documented the range of trajectories of smoking across adolescence and their association with onset of nicotine addiction.

The problem of smoking among youth and young adults has been greatly complicated by the arrival of generations of noncombustible e-cigarettes, which have evolved greatly since reaching the United States in 2009.[32,33] At the time that this chapter was written, devices on the market were effective in delivering nicotine to their users with doses that matched the expectations of cigarette smokers and that also posed a risk of addiction for non–cigarette smokers. Some products used chemical formulations that reduced

irritation and enhanced nicotine delivery and many used fruit, menthol, and other flavorings to attract young users and mask the irritating properties of nicotine.

The availability of these products had substantial impact on use of tobacco products,
with an overall rise in their use even though the prevalence of smoking combustible
cigarettes continued to decline slowly. Dual use (i.e., use of e-cigarettes and smoking
combustible cigarettes) became the most frequent pattern. So far, we have only a brief
window on the consequences of the introduction of electronic cigarettes into the marketplace. The products have been dynamic, as has been their impact on use of tobacco
products by youth and young adults. Of particular concern is whether their use for initiation will enhance risk for nicotine addiction and lifelong use of tobacco products.[34]

Respiratory Health

Active smoking that starts during adolescence slows lung growth, leading to a lower
level of lung function by the end of adolescence. At any age, including adolescence, tests
of lung function show reduced lung function in smokers compared with nonsmokers. Sustained smoking leads to an immediate start of the decline of lung function that
accompanies aging; in never-smokers, this decline does not begin until early middle age
(Figure 27.2). This general pattern has been documented in multiple cohort studies.[10]
As a result, young people who smoke enter adulthood with lesser pulmonary reserve,
they begin immediately to lose lung function, and they are therefore at increased risk for
early development of COPD. The 2012 Report of the Surgeon General concluded: "The
evidence is sufficient to conclude that there is a causal relationship between active smoking and both reduced lung function and impaired lung growth during childhood and
adolescence."[10]

Young smokers have more respiratory symptoms than nonsmokers. Increased rates
for the major respiratory symptoms—cough, phlegm, wheezing, and shortness of

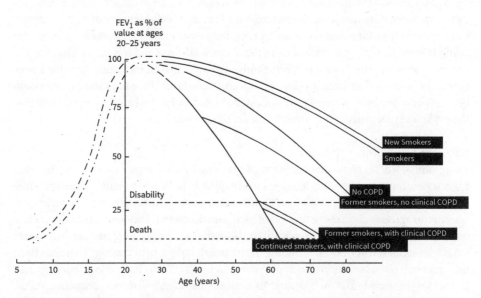

Figure 27.2. Decline of lung function as measured by levels of forced respiratory volume in
1 second (FEV$_1$) across the life course. COPD, chronic obstructive pulmonary disease.
Source: The 2004 US Surgeon General's report [10].

breath—have been documented in numerous cross-sectional and cohort studies. These associations of active smoking with increased respiratory symptoms have been considered causal in the reviews of the 1994 and 2004 Surgeon Generals' reports. Studies have addressed the role of active smoking by adolescents and young adults in causing asthma and not just increasing frequency of wheezing. Accumulating evidence from cohort studies shows that active smoking contributes to incident asthma in susceptible children, adolescents, and young adults by increasing the already greater risk of recurrent, persistent, or new-onset persistent wheeze in children with underlying airway hyperreactivity and atopy. The 2012 report of the Surgeon General found that "The evidence is sufficient to conclude that there is a causal relationship between active smoking and wheezing severe enough to be diagnosed as asthma in susceptible child and adolescent populations."

The evidence does not suggest, however, that smoking increases atopy or allergic sensitization. There is some indication that the additional airway inflammation caused by smoking in atopic adolescents and young adults may increase resistance to therapy for asthma.

Cardiovascular Health

Atherosclerosis, an inflammatory process involving the walls of arteries that results in buildup of plaques and narrowing, underlies the occurrence of coronary heart disease, cerebrovascular disease, and abdominal aortic aneurysm in adults. All of these common diseases of adults are causally linked to cigarette smoking, as is atherosclerosis. Autopsy studies of young veterans of the Korean War showed the unexpected presence of early atherosclerosis. Since then, studies of young adults have confirmed that atherosclerosis begins in young adults and that active smoking is associated with early atherosclerosis.

The Pathobiological Determinants of Atherosclerosis in Youth (PDAY) study has provided particularly critical evidence that smoking accelerates atherosclerosis. In this study, specimens of coronary arteries and the abdominal aorta were obtained from almost 3,000 15- to 34-year-olds (Whites and Blacks) who had died of external causes (accidents, homicides, suicides).[30] Standardized measurements of the extent and severity of atherosclerosis were made in the abdominal aorta and the coronary arteries. Smoking was associated particularly strongly with atherosclerosis in the abdominal aorta and to a lesser extent in the coronary arteries. The Bogalusa Heart Study, a prospective cohort study, provided confirmatory findings. The Surgeon General has concluded that there is a causal relationship between smoking by adolescents and young adults and abdominal atherosclerosis and that the evidence is suggestive for atherosclerosis of the coronary arteries.[31]

Nicotine Addiction

Longitudinal studies show that adoption of smoking by adolescents may follow differing trajectories. Some young experimenters move quickly to become regular smokers while others may become regular smokers only after some years, continue to experiment, or quit. There is no agreement on the definition of nicotine addiction for adolescents as the criteria used for adults are of uncertain applicability. Nonetheless, some young smokers who cease smoking manifest withdrawal, including the hallmark symptom of craving. There has been little research on cessation methods for young smokers. We do have information on risk factors for smoking initiation by youth; in most populations, lower socioeconomic status, less education, and having family members who smoke are associated with increased risk.

However, as noted above, the changes in the marketplace for tobacco products with the addition of many different e-cigarettes and other devices have greatly increased the prevalence of exposure to nicotine among youth and young adults. Many anecdotes and survey findings suggest that some of these devices have been remarkably effective in addicting youth to nicotine. The 2014 report of the Surgeon General highlighted risks of nicotine exposure of adolescents, concluding that "The evidence is suggestive that nicotine exposure during adolescence, a critical window for brain development, may have lasting adverse consequences for brain development."[31]

Clinical Implications

The effects of active and passive smoking on the health of children are of great relevance to both public health and clinical medicine. Attention to tobacco smoke exposure of children should begin with pregnancy; at the first prenatal visit, the active smoking status of the mother and her exposure to SHS should be assessed. Appropriate education and counseling should be provided, directed specifically, as needed, at the smoking of the mother and father. Birth provides a further point for education and intervention with an opportunity for providing information about the severity and range of adverse effects of SHS for infants and young children.

Some especially important "teachable moments" in life should always be points for intervention against smoking: diagnosis of a lower respiratory illness, particularly with wheezing; middle ear infections and chronic serous otitis; and onset of asthma, as well as routine management of asthma and of exacerbations specifically. Guidance is available to healthcare providers.[35,36]

For adolescents, an assessment of smoking behavior should be a routine component of all visits to healthcare providers. Guidance and interventions should be provided immediately to those youngsters who have become occasional or regular smokers. There has been little research, however, to identify effective strategies for cessation among adolescents.[10]

Prevention

After decades of research and intervention, a powerful array of evidence-based approaches is available to protect children and young adults from passive and active smoking. Since its creation by the Master Settlement Agreement in 1998, the Truth Initiative (originally the American Legacy Foundation) has conducted nationwide campaigns and supported more local activities directed at preventing use of tobacco products by youth.[37] In the United States, the 2009 Family Smoking Prevention and Tobacco Control Act provides a set of approaches to tobacco control for the Food and Drug Administration (FDA) that are based in regulation and also in education of the public. For example, the Tips from Former Smokers (Tips) campaign, the first national education campaign on tobacco to receive federal funding, was initiated by the Centers for Disease Control and Prevention (CDC) in 2012.[38] To date, the FDA's use of its regulatory authority has been limited, although it has eliminated flavors in some e-cigarettes and is considering other measures. Of course, the tobacco industry, which now includes those marketing and selling electronic cigarettes and other noncombustible products, remains adversarial and adapts its tactics in a dynamic fashion to counter tobacco control initiatives. Research has documented the powerful effects of its advertising and promotion on youth as well as the influence of smoking in the movies on the initiation of smoking by adolescents.

The foundation for tobacco control globally has been summarized by the World Health Organization in its MPOWER package: M for monitoring or surveillance, P for protection of nonsmokers from exposure to secondhand smoke, O for offering effective cessation, W for warn through appropriate pack warnings, E for enforcement of tobacco control policies and regulations, and R for raise taxes to an appropriate level.[39] These general elements of tobacco control are relevant to prevention of active and passive smoking among children in countries around the world. Surveillance is needed to guide tobacco control programs directed at youth and to identify industry strategies that may reach them. Protection of nonsmokers, while possible through regulation in workplaces and public places, needs to be achieved also in homes through education of families, particularly male household members, and by establishing a social norm that makes smoking around pregnant women and children unacceptable. Prohibitions on sales of tobacco to minors need to be enforced if they are to be effective. Graphic pack warnings reduce youth smoking. Young people are particularly sensitive to the pricing of cigarettes, and raising prices through tax increases reduces initiation of smoking by young people.

A global tobacco surveillance system is currently in place, the Global Tobacco Surveillance System (GTSS), which includes the Global Youth Tobacco Survey (GYTS). The GYTS is a school-based survey of 13- to 15-year-olds that has now been conducted in 140 countries and 11 territories worldwide for up to two rounds per country. It has proved valuable, providing data for tracking patterns of experimentation and smoking initiation worldwide.[40] The GTSS also includes the GATS, which reaches down to age 18, thus providing coverage of the critical window of initiation in young adulthood, a group now targeted by the tobacco industry. In the United States, cigarette smoking and use of other nicotine-containing products is tracked closely through several surveys.[10] The prevalence of tobacco use continues to decline among adolescents in the United States, but the pace of decline has slowed. Eliminating experimentation and initiation by youth is critical if we are to achieve a tobacco-free world.

The healthcare sector has a critical role in reducing exposure to SHS and also in assessing use of tobacco products by youth and guiding parents to resources on use of tobacco products by their children. With regard to SHS exposure, assessment and intervention should begin with pregnancy and continue once the child is born. For children at risk for asthma or with asthma, avoidance of SHS exposure is critical and beneficial. Clinicians providing care for infants and children with lower respiratory illnesses and ear infections should inquire about SHS exposure, a reversible risk factor. Inquiries concerning active smoking and electronic cigarette use should be part of routine pediatric care. Healthcare providers should be equipped to provide education on the consequences of tobacco products for use and offer resources. Resources are available for healthcare providers. For example, through its Julius B. Richmond Center of Excellence, the American Academy of Pediatrics provides "education, training, and tools" to protect children from active and passive smoking.[36]

References

1. Reitsma MB, Kendrick PJ, Ababneh E, et al. Spatial, temporal, and demographic patterns in prevalence of smoking tobacco use and attributable disease burden in 204 countries and

territories, 1990–2019: a systematic analysis from the Global Burden of Disease Study 2019. *Lancet*. 2021 Jun 19;397(10292):2337–60.

2. Giovino GA, Mirza SA, Samet JM, et al. Tobacco use in 16 countries with 4 billion population: the GATS Collaborative Group. *Lancet*. 2012;380(9842):668–79.

3. World Health Organization. Tobacco fact sheet. 2019 Jul 26. Available at: https://www.who.int/news-room/fact-sheets/detail/tobacco.

4. Gentzke AS, Wang TW, Cornelius M, et al. Tobacco product use and associated factors among middle and high school students: National Youth Tobacco Survey, United States, 2021. *MMWR Surveill Summ*. 2022;71(No. SS-5):1–29. doi:http://dx.doi.org/10.15585/mmwr.ss7105a1external icon.

5. Barker DJ. Developmental origins of adult health and disease. *J Epidemiol Community Health*. 2004;58:114–5.

6. Huxley RR, Shiell AW, Law CM. The role of size at birth and postnatal catch-up growth in determining systolic blood pressure: a systematic review of the literature. *J Hypertens*. 2000;18:815–31.

7. Barker DJ, Osmond C, Forsen TJ, Kajantie E, Eriksson JG. Trajectories of growth among children who have coronary events as adults. *N Engl J Med*. 2005;353:1802–9.

8. Nuyt AM. Mechanisms underlying developmental programming of elevated blood pressure and vascular dysfunction: evidence from human studies and experimental animal models. *Clin Sci (Lond)*. 2008;114:1–17.

9. Gicquel C, El-Osta A, Le Bouc Y. Epigenetic regulation and fetal programming. *Best Pract Res Clin Endocrinol Metab*. 2008;22:1–16.

10. US Department of Health and Human Services. *Preventing Tobacco Use Among Youth and Young Adults: A Report of the Surgeon General*. Atlanta, GA: US Department of Health and Human Services, Centers for Disease Control and Prevention, National Center for Chronic Disease Prevention and Health Promotion, Office on Smoking and Health; 2012.

11. Czaplicki L, Kostygina G, Kim Y, et al. Characterising JUUL-related posts on Instagram *Tobacco Control*. 2020;29:612–617.

12. World Health Organization. Summary results of the Global Youth Tobacco Survey in selected countries of the WHO European Region. Copenhagen: WHO Regional Office for Europe; 2020. License: CC BY-NC-SA 3.0 IGO.

13. US Department of Health and Human Services. *The Health Consequences of Smoking. A Report of the Surgeon General*. Atlanta, GA: US Department of Health and Human Services, Centers for Disease Control and Prevention, National Center for Chronic Disease Prevention and Health Promotion, Office on Smoking and Health; 2004.

14. US Department of Health and Human Services. *The Health Consequences of Involuntary Exposure to Tobacco Smoke. A Report of the Surgeon General*. Atlanta, GA: US Department of Health and Human Services, Centers for Disease Control and Prevention, Coordinating Center for Health Promotion, National Center for Chronic Disease Prevention and Health Promotion, Office on Smoking and Health; 2006.

15. Breton CV, Byun HM, Wenten M, Pan F, Yang A, Gilliland FD. Prenatal tobacco smoke exposure affects global and gene-specific DNA methylation. *Am J Respir Crit Care Med*. 2009;180:462–7.

16. Thun MJ, Day-Lally C, Myers DG, et al. Trends in tobacco smoking and mortality from cigarette use in Cancer Prevention Studies I (1959–1965) and II (1982–1988). In Burns DM, Garfinkel L, Samet JM, eds., *Changes in Cigarette Related Disease Risks and Their Implication*

for Prevention and Control: Monograph 8. Bethesda, MD: US Government Printing Office; 1997: 305–82.

17. Barrington-Trimis JL, Braymiller JL, Unger JB, et al. Trends in the age of cigarette smoking initiation among young adults in the US from 2002 to 2018. *JAMA Netw Open*. 2020;3(10):e2019022.

18. Doll R, Peto R. Cigarette smoking and bronchial carcinoma: dose and time relationships among regular smokers and lifelong non-smokers. *J Epidemiol Community Health*. 1978;32:303–13.

19. Peto R. Influence of dose and duration of smoking on lung cancer rates. In Zaridze D, Peto R, eds., *IARC Scientific Publication 74*. Lyon, France: World Health Organization, International Agency for Research on Cancer; 1986: 23–33.

20. International Agency for Research on Cancer. *Tobacco Smoke and Involuntary Smoking*. Lyon, France: International Agency for Research on Cancer; 2004. IARC Monograph 83.

21. US Department of Health Education and Welfare. *Smoking and Health. Report of the Advisory Committee to the Surgeon General*. Washington, DC: US Government Printing Office; 1964. DHEW Publication No. [PHS] 1103.

22. US Department of Health and Human Services. *The Health Consequences of Involuntary Smoking. A Report of the Surgeon General*. Washington, DC: US Department of Health and Human Services, Public Health Service, Office on Smoking and Health; 1986. DHHS Publication No. (CDC) 87-8398.

23. Matt GE, Quintana PJ, Destaillats H, et al. Thirdhand tobacco smoke: emerging evidence and arguments for a multidisciplinary research agenda. *Environ Health Perspect*. 2011;119:1218–26.

24. US Department of Health and Human Services. *How Tobacco Smoke Causes Disease: The Biology and Behavioral Basis for Smoking-Attributable Disease. A Report of the Surgeon General*. Atlanta, GA: US Department of Health and Human Services, Centers for Disease Control and Prevention, National Center for Chronic Disease Prevention and Health Promotion, Office on Smoking and Health; 2010.

25. Benowitz NL. Nicotine addiction. *N Engl J Med*. 2010;362:2295–303.

26. Samet JM, Neta GI, Wang SS. Secondhand smoke. In Lippmann M, ed. *Environmental Toxicants: Human Exposures and Their Health Effects*. 3rd ed. Hoboken, NJ: Wiley; 2009: 709–61.

27. US Department of Health Education and Welfare. *The Health Consequences of Smoking. A Report of the Surgeon General*. Atlanta, GA: US Government Printing Office; 1972.

28. US Department of Health and Human Services. *The Health Consequences of Smoking: Chronic Obstructive Lung Disease. A report of the Surgeon General*. Washington, DC: US Department of Health and Human Services, Public Health Service, Office on Smoking and Health; 1984.

29. Cook DG, Strachan DP, Carey IM. Health effects of passive smoking. 9. Parental smoking and spirometric indices in children. *Thorax*. 1998;53(10):884–893. doi:10.1136/thx.53.10.884.

30. McGill HC Jr, McMahan CA, Gidding SS. Preventing heart disease in the 21st century: implications of the Pathobiological Determinants of Atherosclerosis in Youth (PDAY) study. *Circulation*. 2008;117:1216–27.

31. US Department of Health and Human Services. *The Health Consequences of Smoking: 50 Years of Progress. A Report of the Surgeon General*. Atlanta, GA: U.S. Department of Health

and Human Services, Centers for Disease Control and Prevention, National Center for Chronic Disease Prevention and Health Promotion, Office on Smoking and Health; 2014.

32. National Academies of Sciences, Engineering, and Medicine. 2018. *Public Health Consequences of E-Cigarettes*. Washington, DC: The National Academies Press. https://doi.org/10.17226/24952.

33. U.S. Department of Health and Human Services. *E-Cigarette Use Among Youth and Young Adults: A Report of the Surgeon General*. Atlanta, GA: U.S. Department of Health and Human Services, Centers for Disease Control and Prevention, National Center for Chronic Disease Prevention and Health Promotion, Office on Smoking and Health; 2016.

34. Samet JM, Barrington-Trimis J. E-cigarettes and harm reduction: an artificial controversy instead of evidence and a well-framed decision context. *Am J Public Health*. 2021;111(9):1572–1574. doi:10.2105/AJPH.2021.306457.

35. Hawthorne MA, Hannan LM, Thun MJ, Samet JM. *Protecting Our Children from Second-Hand Smoke*. Geneva, Switzerland: International Union Against Cancer (UICC); 2008.

36. American Academy of Pediatrics Julius B. Richmond Center of Excellence. 2013. Available at: http://www2.aap.org/richmondcenter/.

37. Truth Initiative. Our history. 2022 May 11. https://truthinitiative.org/who-we-are/our-history.

38. Centers for Disease Control and Prevention. Tips from former smokers: about the campaign. 2022. Available at: https://www.cdc.gov/tobacco/campaign/tips/about/index.html.

39. World Health Organization. *WHO Report on the Global Tobacco Epidemic 2021: Addressing New and Emerging Products*. Geneva: World Health Organization; 2021. License: CC BY-NC-SA 3.0 IGO.

40. Warren CW, Jones NR, Peruga A, et al. Global youth tobacco surveillance, 2000–2007. *MMWR Surveill Summ*. 2008 Jan 25;57(1):1–28.

28

Water Pollution and Children's Health

Katherine Arnold

Polluted water is among the greatest threats to child health, particularly in the developing world. Many factors contribute. The fractional contribution of water to body weight is relatively high in children, their immune systems are not fully developed, play habits provide opportunities for fecal–oral exposures, (see Figure 28.1) and organs are less developed and capable of regulating water content and internal temperature. Finally, young children lack basic knowledge in areas of hygiene. It is not surprising that (1) children are disproportionately affected by waterborne and water-related diseases, and (2) the burden of such diseases is greatest in underdeveloped areas (Figure 28.2).

Water is required for all forms of life and comprises about 60% of the human body by volume. It has been described as "the universal solvent," a feature that helps account for the presence of many biological and chemical contaminants that contribute to the burden of disease in humans. Primary waterborne and water-related threats to child health arise from pathogenic diarrheal diseases, pathogenic nondiarrheal diseases, vector-borne diseases, contact with contaminated soil, and malnutrition. Chemical pollution of water and excessive concentrations of some naturally occurring chemicals are also growing threats. Impacts on children's health range from impaired growth and cognitive development to severe illness and death.

Figure 28.1. Child playing in polluted water, India

Katherine Arnold, *Water Pollution and Children's Health* In: *Textbook of Children's Environmental Health*,
Second Edition. Edited by: Ruth A. Etzel and Philip J. Landrigan, Oxford University Press.
© Oxford University Press 2024. DOI: 10.1093/oso/9780197662526.003.0028

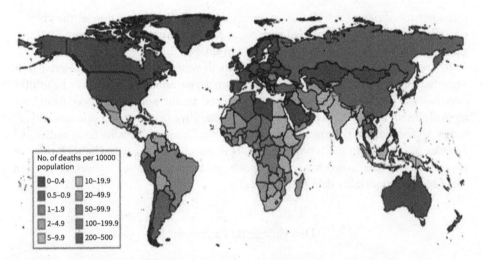

No. of deaths per 10000 population
- 0–0.4
- 0.5–0.9
- 1–1.9
- 2–4.9
- 5–9.9
- 10–19.9
- 20–49.9
- 50–99.9
- 100–199.9
- 200–500

Figure 28.2. Geographic distribution of rotavirus-associated mortality rates among children younger than 5 years in 2016 (15)

Diarrheal diseases far outnumber those of all other water-related diseases. However, the gastrointestinal system is not the only system affected by contaminated water, and essentially no part of the human body is immune to the effects of water-borne or water-related threats.[1]

Nondiarrheal diseases are caused by organisms which can cause serious gastrointestinal symptoms, pneumonia, polio, hepatitis, encephalitis, and other illnesses. Diarrhea also can be a symptom of these diseases but typically is not the primary one.

Vector-borne diseases are considered "water-related" since ingestion of water is not the route of infection. Instead, vector organisms breed in water and transmit a parasite (bacterial, protozoan, or helminth [worm]) to a human host. Transmittal can be direct, through bites (as with malaria or onchocerciasis), or indirect, where a vector organism sheds larvae into water or soils and subsequent contact allows entrance through human skin.[2]

Malnutrition can result from repeated episodes of water-related diarrhea as well as lack of water for food production. Finally, lack of drinking water contributes to malnutrition, leading to recognition that water itself is a critical component of diet.[3]

Unlike microbial contamination, chemical contamination leads to health problems primarily through chronic exposure. Significant, possibly irreversible health problems can develop before contamination is detected.[2] Arsenic, lead, and nitrate are examples of such contaminants.

Effective water management reduces the health-related burden of waterborne and water-related diseases and of chemical pollution. It provides additional benefits by protecting food crop production, thereby mitigating childhood malnutrition.

Health Threats from Polluted Water

This section describes primary diseases and health issues associated with waterborne and water-related pathogens, modes of transmission, and, where available, their pathogen-specific burdens of disease among children. A later section of the chapter addresses health threats associated with chemical contaminants. Interventions to mitigate these water-related threats are described in Chapter 29.

To create an illusion of order, contaminants in water are classified as microbial or parasitic; microbes as bacteria, viruses, or protozoa; parasites as protozoa or helminths (worms); and diseases as waterborne, water-related, vector-borne or sourced from contaminated soil. Table 28.1 lists many of the primary water-associated diseases and illnesses with causative pathogens. Where available, childhood deaths from individual diseases, as reported in the World Health Organization (WHO) 2016 Global Burden of Disease study (GBD) and other recent reports, are included.[4] In some cases, as noted in Table 28.1, only mortality data are available.

Diarrheagenic Pathogens

Most waterborne diarrhea is a direct consequence of drinking water contaminated with bacteria, viruses, or protozoa associated with human and animal wastes.

Diarrhea is an intestinal disorder clinically defined by the WHO as the passage of three or more loose or liquid stools per day.[5] There are three levels associated with a diagnosis of diarrhea:

- Acute watery diarrhea: Lasts several hours or days
- Acute bloody diarrhea: Also called dysentery
- Persistent diarrhea: Lasts 14 days or longer.

The worldwide number of childhood diarrhea-related deaths in 2017 has been estimated at more than 500,000, mostly in low-income countries.[4] Survivors of repeated episodes of diarrhea can experience stunted growth, weakened immune function, and impaired intellectual development. Uncontrolled diarrhea can cause death by depleting body fluids, causing profound dehydration.

Gastroenteritis is a related condition resulting from a viral or bacterial infection that causes inflammation of both the gastric and intestinal systems. Effects can include stomach pain, diarrhea, and vomiting. Viral gastroenteritis generally produces watery diarrhea without blood or mucus, while bacterial gastroenteritis often includes diarrhea with blood and/or mucus.

Waterborne Diarrheagenic Bacteria

Waterborne diarrheagenic bacteria include those responsible for cholera, salmonellosis, shigella, and numerous other diseases. Since the 1880s, cholera has been a source of multiple pandemics and is currently thought to cause 21,000–143,000 deaths worldwide annually.[6] The primary symptoms of cholera include profuse diarrhea that can quickly spread the causative bacterium (*Vibrio cholera*), particularly when housing is cramped and sanitation is inadequate. Diarrhea can rapidly lead to dehydration and kidney failure. Those with untreated, symptomatic infections suffer about a 50% fatality rate. However, asymptomatic infections comprise about 80% of all cases. Those without overt symptoms remain capable of excreting *V. cholera* in their feces for days or weeks, adding to disease spread. The spread of cholera and other waterborne diseases is greatly enhanced by infrastructure failures due to natural events (e.g., severe weather, earthquakes) and/or

Table 28.1 Summary of major waterborne, water-related, and soil-related diseases

Disease	Infective agent	Childhood deaths
Diarrheagenic diseases/pathogens: bacteria (B), virus (V), protozoa (P)		
Cholera	*Vibrio cholera* (B)	52,232
Salmonellosis	*Salmonella spp.* (B)	37,410
Shigellosis	*Shigella* spp. (B)	63,713
Campylobacteriosis	*Campylobacter* spp. (B)	40,854
Gastroenteritis	*Aeromonas spp.* (B); *Leptospira* spp.(B)	6,332
Dysentery	Enterotoxigenic *Escherichia coli (ETEC)*	51,186
Gastroenteritis	Rotaviruses (V)	128,515
	Noroviruses (V)	10,629
	Adenovirus (V)	52,613
Dysentery	*Giardia lamblia* (P)	48,301
	Cryptosporidium parvum (P)	
Amoebiasis	*Entamoeba histolytica* (P)	4,567
Microsporidiosis	*Microsporidia* spp.(P)	
Nondiarrheagenic diseases/pathogens: bacteria (B), virus (V), protozoa (P)		
Typhoid fever	*Salmonella typhi (B)*	128 - 161000 (all ages)
Legionnaire's disease	*Legionella pneumonophia* (B)	
Melioidosis	*Burkholderia pseudomallei* (B)	89,000 (all ages)
Environmental enteropathy	Various fecal bacteria (B)	
Hepatitis	Hepatitis A virus and Hepatitis E virus (V)	
Poliomyelitis	Poliovirus (V)	
Encephalitis	*Naegleria fowleri; Acanthamoeba* spp.(P)	
Vector-related diseases/pathogens: bacteria (B), protozoa (P), helminths (H)		
Malaria	*Plasmodium* spp. (P) (vector: Anopheles mosquitos)	670000 (all ages, >475,000 children)
Onchocerciasis	*Onchocerca volvulus* (H) (filarial worm) (vector: *Simulium* spp.; black flies)	
Trachoma	*Chlamydia trachomatis (B) (vector: black flies)*	
Schistosomiasis	*Schistosoma* spp. (H) (blood fluke) (vector: snails)	24,000 – 200,000 (all ages)
Dracunculiasis	*Dracunculus mediensis (H) (vector: cyclops flea)*	
Lymphatic filariasis	*Wuchereria bancrofti; Brugia malayi (H)* (nematode worms) (vector: mosquitos)	
Dengue	Dengue virus (vector: mosquitos)	4,000 (younger ages)
Soil-related diseases/pathogens: helminths (H)		
Hookworm	*Nicator americanus; Ancylostoma duodenale (H))*	
Ascariasis	*Ascaris lumbricoides (H)*	

anthropogenic causes (e.g., conflict, mass migration) leading to breakdown in basic hygiene and sanitation. In such situations, transmission of cholera and other diarrheal diseases through food and person-to-person contact is accelerated. That was the sequence of events following the post-earthquake 2010 cholera outbreak in Haiti that resulted in more than 820,000 cases and nearly 10,000 deaths.[6]

Other conspicuous diarrheagenic bacteria include several *Salmonella* serotypes that cause salmonellosis (diarrhea, fever, and stomach cramps). Infection occurs from contaminated foods or water and most commonly infects children. Most cases are mild, but some can be life-threatening. Salmonellosis is one of four key global causes of diarrhea worldwide.[7] *Shigella* spp., are the causative agents for shigellosis (diarrhea, fever, and stomach cramps). *Shigella disenteriae* type 1, in particular, can be deadly in low-resource areas.[8] *Campylobacter* spp. in contaminated food or water are the sources of campylobacteriosis (diarrhea, fever, stomach cramps). The disease is rarely fatal, but is estimated to cause 5–14% of all diarrheas worldwide, more commonly in low-resource countries.[2,9] *Aeromonas* spp., which survive well even in treated potable water, cause gastroenteritis in infected individuals. More severe symptoms can occur, including death, primarily in younger patients and those with underlying disease. In recent decades, improved data collection regarding prevalence of infection and outbreaks have identified *Aeromonas* spp. as increasingly important food- and waterborne pathogens.[10] *Leptospira* spp. are the causative agents of leptospirosis (fever, headache, diarrhea, jaundice), which can be transmitted through the urine of infected animals and hence spread by flooding events. Infection may result in minor symptoms. More significant infections, if left untreated, can lead to kidney damage, meningitis, liver failure, and death.[11] Several virulent strains of the normally benign *Escherichia coli* cause diarrhea or dysentery. They are collectively termed enterotoxigenic *E. coli* (ETEC). Infection with ETEC causes "traveler's diarrhea" and is a major cause of childhood diarrhea in low-income countries.[12] The GBD study attributed more than 50,000 deaths in children under 5 years to ETEC diarrhea in 2016.[4]

Secondary effects linked to some diarrhea-causing bacteria include such conditions as reactive arthritis (*Salmonella, Campylobacter, Shigella*)[13] and Guillain-Barré syndrome (*Campylobacter* spp.), in which the immune system attacks the nerves and leads to varying degrees of paralysis.[9] A specific strain of ETEC, *E. coli* 0157:H7 produces a powerful Shiga toxin that can damage the small intestine and cause severe stomach cramps, bloody diarrhea, and vomiting. Some affected patients, particularly young children or older adults develop hemolytic-uremic syndrome, a complication that can result in renal failure in addition to severe dehydration.[14]

Waterborne Diarrheagenic Viruses

Waterborne diarrheagenic viruses include rotavirus, norovirus, and adenovirus and are collectively responsible for a large fraction of worldwide diarrheal disease. Rotavirus infection is the most common source of diarrhea worldwide, particularly among children. Rotavirus infections are characteristically accompanied by fever, watery diarrhea, vomiting, and abdominal pain. Like other diarrheal diseases, the principal cause of death is extreme dehydration. Successes with rotavirus vaccines greatly reduced the overall number of infections and deaths over the past decade, but they remain very common.[15] The 2016 GBD Study placed the number of deaths among children under 5 due to rotavirus infections at 128,000, primarily in low- and middle-income countries (LMICs).[4]

Norovirus is a highly contagious virus that most people contract in their lifetime. Transmission occurs from contaminated food or water and from person-to-person contact. Norovirus causes acute gastroenteritis, with serious symptoms primarily in children and the elderly. Accurate data on occurrence and deaths due to norovirus in developing countries have been challenging to obtain; estimates put the overall annual death rate for children under 5 at 50,000.[16,17]

Adenovirus infections cause acute gastroenteritis that is often difficult to distinguish from other causes of gastroenteritis, particularly in LMICs where tools and funding for detection and record keeping may be limited. Until recently, adenovirus was not considered a major contributor to the overall burden of pediatric diarrhea. However, recent case-control studies along with reanalysis of previous studies of diarrheal etiology showed that the number of adenovirus infections in children has been substantially underestimated.[18] In those studies, adenovirus was found to be the second most common cause of moderate to severe diarrhea. The Global Burden of Disease (modelling) study estimated that, in 2016, enteric adenovirus infections caused 75 million episodes of diarrhea globally among children under 5 years of age, with an associated mortality of 11.78% (95% confidence interval [CI]: 8.19, 16.13), behind only rotavirus and *Shigella*.[4]

Waterborne Diarrheagenic Protozoa

Giardia lamblia (synonyms *Giardia intestinalis*, *Giardia duodenalis*) is the agent of giardiasis and the most frequently isolated protozoan pathogen associated with waterborne diarrheal diseases in LMICs as well as in high-income countries. Symptoms include watery diarrhea, dehydration, stomach pain, fever, and nausea that can persist for 2 or more weeks. Complications include diarrheal-associated dehydration, while chronic cases can be associated with failure to thrive.[19] *Giardia* occurs in high-income countries with an overall occurrence of 2–5%. In LMICs, *Giardia* infects infants and young children and is a major cause of childhood diarrhea. Occurrence in children is estimated at 15–20%.[20]

Cryptosporidium parvum is a protozoan parasite that causes cryptosporidiosis, an infection of the intestinal system. In healthy people, it most often causes watery diarrhea. In children and/or immunocompromised adults, however, the infection can lead to prolonged diarrhea and death. The 2016 GBD Study listed cryptosporidium as causing more than 48,000 deaths in children under 5 (95% uncertainty interval: 24,600, 81,900). In children, cryptosporidium infections can also lead to lagging growth and cognitive function.[21]

In the tropics, amoebiasis (*Entamoeba histolytica*) is among the diarrheagenic maladies contracted through water when sanitation and hygiene measures are lacking. Amoebiasis infection is often asymptomatic but can cause serious gastrointestinal illness (amebic colitis) with associated diarrhea. The protozoan parasite can also cause amebic liver abscess. Amebic colitis is a leading cause of diarrhea in the first 2 years of life in children in the LMICs[22] Due to limited diagnostic and surveillance capabilities in these countries, the exact burden of amebiasis is difficult to determine. However, the 2016 GBD study lists the number of childhood deaths from amebiasis (in 2016) as 4,567.[4]

Microsporidiosis is a collection of diseases caused by Microsporidia, a class of unicellular intracellular parasites, closely related to fungi. These are opportunistic pathogens that can be fatal to the immune-compromised. There are hundreds of species of microsporidia, with at least 15 identified as human pathogens. Most cases of microsporidiosis are associated with *Enterocytozoon bieneusi* and present as diarrhea. However, other species can infect different areas of the body such as the eyes or muscles or cause a more systemic infection. These organisms are increasingly recognized as opportunistic infectious agents worldwide. However, their overall impact is difficult to determine.[23]

Nondiarrheagenic Pathogens in Water

Bacterial, viral, and protozoan diseases having primary symptoms other than diarrhea include Legionnaire's disease, typhoid fever, polio, hepatitis, malaria, and schistosomiasis.

Waterborne Bacteria

The pathogenic bacteria in this group cause pneumonia, malabsorption, high fever, and intestinal symptoms.

Salmonella typhi and *Salmonella paratyphi* are two species of *Salmonella* that cause typhoid fever (sustained fever, weakness, diarrhea, intestinal bleeding or perforation). Infection is primarily from contaminated food or water. Antibiotics and, to a lesser extent, vaccines are effective in controlling typhoid fever, but in LMICs these tools are less available, so cases and deaths from typhoid remain significant. Typhoid-related deaths for 2017 were estimated at 128,000–161,000 (all ages).[24]

Legionella pneumophila is the causative agent of Legionnaire's disease and the milder Pontiac fever. It is highly infectious, transmitted via aerosol inhalation, and thrives in hot water storage tanks, where its thermotolerance enables it to out-compete other bacteria. However, large outbreaks are not common. It can be a source of hospital-acquired pneumonias.[2]

Burkholderia pseudomallei is the causative agent for melioidosis, a disease of concern primarily in the tropics. Melioidosis is transmitted through contact with contaminated soil or water, placing children at particular risk. There are several types of melioidosis, with differing symptoms. Local, pulmonary, bloodstream, and disseminated infections produce symptoms of fever, respiratory distress, headache and muscle pain, and seizures. The diverse symptoms make melioidosis difficult to diagnose and track. A 2016 modeling study estimated that there are 165,000 total cases with 89,000 fatalities per year (all age groups). However, underreporting is likely.[25]

Environmental (tropical) enteropathy is not associated with a single organism but rather is a syndrome occurring among children under conditions of poor sanitation and hygiene who habitually ingest water containing fecal bacteria. Infected children develop a chronically inflamed gut with associated malabsorption and malnutrition. Impaired development, stunting, and increased susceptibility to disease can result.[26]

Waterborne Viruses

Waterborne viruses that produce nondiarrheal symptoms include those responsible for hepatitis A, hepatitis E, and poliomyelitis.

Hepatitis is a viral disease that causes liver inflammation. Two forms of the disease, hepatitis A and E, are caused by ingestion of fecally contaminated drinking water. Symptoms include yellowing of the skin and eyes, dark urine, fatigue, and nausea. Childhood symptoms are typically mild, and most patients recover completely.[2]

Poliomyelitis, a highly infectious viral disease, mainly affects children under 5. The virus is transmitted through contaminated food and water and multiplies in the intestine. Many of those infected present no symptoms but continue to excrete the virus in their feces, thus transmitting it to others in areas where there is inadequate sanitation.

Initial symptoms of polio include fever, fatigue, headache, vomiting, neck stiffness, and pain in the extremities. One in 200 infections leads to irreversible paralysis (usually in the legs). Thanks to a global eradication campaign orchestrated by the World Health Organization, polio cases have decreased by more than 99% worldwide since 1988, from an estimated 350,000 cases per year in more than 125 countries to 33 reported cases in 2018.[2,27]

Waterborne Protozoan Infections

Naegleria fowleri is an amoeba that causes primary amebic meningoencephalitis, a brain infection that destroys brain tissue. Infection occurs when water containing the amoeba enters the body through the nose and travels to the brain. This rare occurrence happens when people are swimming or diving in warm freshwater lakes and rivers where *Nagleria fowleri* is living. The disease is rare: only 33 infections were reported in the United States between 2011 and 2020, but mortality is greater than 97%.[28]

Toxoplasmosis, a disease caused by *Toxoplasma gondii*, one of the world's most common parasites. It can be transmitted through contact with infected soil or water, ingestion of contaminated food, or transplacentally (mother to fetus). It is estimated that those with healthy immune systems will usually not develop symptoms. When disease occurs prenatally, it can lead to visual and mental disabilities in the newborn.[29]

Vector-Borne Diseases Related to Water

Diseases transmitted via vectors can be more difficult to control than waterborne diseases, primarily due to vector mobility. Vector organisms include mosquitos, flies, snails, and water fleas while the infective organisms can be bacterial, viral, protozoan, or helminth.

Malaria is caused by infection with *Plasmodium* spp., parasitic protozoa that are transmitted by a bite from *Anopheles* mosquitos. The mosquitos reproduce in small bodies of stagnant water. Malaria is characterized by fever, chills, and flu-like symptoms that can progress in severity and cause death. Worldwide malaria-related deaths in 2020 were estimated at 627,000, the majority of which occurred in children in sub-Saharan Africa. South East Asia, especially India, also continues to have significant malarial disease.[30]

Onchocerciasis, or river blindness, is an infection by the parasitic worm *Onchocerca volvulus* and is transmitted by blackflies (*Somilium* spp.) that breed in moving water. The main burden occurs in sub-Saharan Africa, but the disease is also found in parts of South America and in Yemen. The GBD study estimated that there were about 21 million cases of onchocerciasis in 2017, with more than 1 million cases of vision loss.[31]

Trachoma is a chronic infection of the eyelids caused by the bacterium *Chlamydia trachomatis*. It is transmitted by flies and by person-to-person contact, mostly among children via contaminated fingers and cloths. Untreated, the infection can lead to blindness. Globally, trachoma is a public health problem in 42 countries, with 2.5 million people affected by trachomatous trichiasis in 2019. Antibiotics can treat trachoma, but improved sanitation and hygiene contribute significantly to its prevention.[32]

Schistosomiasis is a chronic disease in tropical regions that is caused by infection with *Schistosoma* spp. (flatworm or blood flukes). It is endemic in 74 countries, with the highest burden in sub-Saharan Africa. Certain freshwater snails act as the initial host for *Schistosoma* parasites. The parasite is shed in larval form by infected snails, and humans become infected through contact with the larvae. In humans, the larvae mature into adult worms and live in the blood vessels of the body, where the females produce eggs. The eggs travel throughout the body causing inflammation and scarring. Some are passed through the urine or stool and can recontaminate water. Hygiene and play habits make children particularly vulnerable to infection. In children, schistosomiasis can cause anemia, stunting, and a reduced ability to learn. Repeated infections damage the liver, intestine, bladder, and lungs. Schistosomiasis disables more than it kills. The number of deaths attributable to schistosomiasis is difficult to determine, in part because of overlapping disease pathologies. Current estimates are broad and range from 24,000 to 200,000. In areas with adequate health facilities, schistosomiasis can be treated.[33,34]

Also known as "guinea-worm disease," dracunculiasis is caused by the parasitic worm *Dracunculus medinensis*. Guinea worm larvae are ingested by the *Cyclops* water flea, and people become infected by drinking water containing the flea. The larvae migrate through the victim's subcutaneous tissues and grow to adult worms up to 600–800 millimeters in length, causing severe pain, especially in joints. The worm eventually emerges (from the feet in most cases) causing an intensely painful edema, a blister, and an ulcer accompanied by fever, nausea, and vomiting. Intense eradication efforts have reduced guinea worm infections. In 1989, almost 900,000 cases were reported from 15 of the 20 endemic countries. By 2011, only four countries continued to experience local transmission of dracunculiasis, and slightly more than a thousand cases were reported. In 2020, only 27 cases were reported.[35]

Lymphatic filariasis, commonly known as elephantiasis, is transmitted to humans through mosquitos. The infectious agents are nematodes of the family Filariodidea, most commonly *Wuchereria bancrofti*. These are thread-like worms that enter through the skin as larvae and migrate to the lymphatic vessels, where they mature into adult worms. Most infections are externally asymptomatic, although damage to the lymphatic system, kidneys, and immune system can occur. Symptomatic filariasis leads to lymphedema (tissue swelling) or elephantiasis (skin and tissue thickening). In 2000, WHO began the Global Programme to Eliminate Lymphatic Filariasis using preventive chemotherapy. Fifty-one million people were infected in 2018, a 74% decline from year 2000 case numbers. WHO reported that, in 2019, at least 36 million people were afflicted with chronic lymphatic filariasis.[36]

Dengue is a viral infection transmitted through the bite of an infected mosquito. The virus is from the Flaviviridae family, and its serotypes are collectively called dengue virus (DENV). More than 80% of infections are mild and asymptomatic, but occasionally dengue can become an acute or lethal flu-like illness. In areas with proper medical care, fatality can be lower than 1%. However, many low-income areas of the world don't have such access. Before 1970, only nine countries had reported severe dengue outbreaks, but the disease is now endemic in more than 100 countries, with Asia providing approximately 70% of the cases. The number of reported dengue cases has increased over the past two decades, from 500,000 in 2000 to 2.4 million in 2010 and 5.2 million in 2019. Reported deaths, primarily in younger age groups, grew from 960 in 2000 to just over 4,000 in 2015.[37]

Diseases Acquired via Direct Contact or Contact with Feces-Contaminated Soils

These diseases are also termed "water-washed diseases" because they arise from inadequate use of water for domestic and personal hygiene. More specifically they arise from lack of hand washing after contact with infective parasites. Trachoma, described earlier, is considered a water-washed disease, as are the soil-transmitted helminths described below.[2]

People become infected with soil-transmitted helminths through contact with their eggs in fecally contaminated soils. The main species that infect people are *Necator americanus* and *Ancylostoma duodenale* (hookworms), *Ascaris lumbricoides* (roundworm), and *Trichuris trichiura* (whipworm). The WHO estimates that 1.5 billion people worldwide are infected with soil-transmitted helminths. Eggs are ingested through work or play in soil, with subsequent contact between hand and mouth, or via ingestion of contaminated water or crops watered with contaminated water but not washed. Adult worms live in the intestine and continue to produce eggs that are passed in stools to continue the infective cycle. The worms feed on host tissues, including blood, leading to loss of iron and protein. They cause malabsorption of nutrients, and hookworms can cause chronic intestinal blood loss resulting in anemia. Chronic helminth infection impairs physical and intellectual development in children and adversely affects adult productivity. Heavier infections can cause intestinal blockages.[38]

Health Threats from Chemical Pollution

During the past few decades, advanced instrumentation revealed that a vast array of anthropogenic and natural chemicals are present at trace concentrations in waters impacted by industrial and/or agricultural activity. Those chemicals are sometimes broadly classified as petroleum and coal hydrocarbons, synthetic organics, metals and metalloids, radionuclides, disinfection by-products, endocrine disruptors, pharmaceuticals, pesticides, and other pollutant categories capable of producing health-related effects in exposed animals, including humans. Because conventional water treatment processes were designed primarily to prevent human contact with microbial pathogens, trace organics of health concern frequently survive such treatments, although generally at very low (ng/L) concentrations. The following chemical pollutants deserve attention in the present context.

Pharmaceuticals, consequent to drug manufacturing and use, agricultural activity, and hospital waste discharges, are routinely detected in treated municipal wastewaters that may subsequently impact drinking water quality. In most cases, the concentrations detected are too low to affect human health. However, organisms living within these waters can be at higher risk.

Endocrine-disrupting compounds (EDCs) include chemicals that stimulate or diminish normal hormone response in exposed organisms. EDCs include pesticides; plasiticizers such as phthalates and bisphenol A, which can leach into infant formula from plastic containers; and natural or synthetic hormones. Because endocrine system components respond to hormones at very low concentrations, there are potential consequences for human and animal development.

Endocrine system disruption has been widely observed in fish populations chronically exposed to low nanogram-per-liter levels of 17α-ethinyl estradiol, an active ingredient in birth control medications. Fish and other exposed organisms develop recognizable intersex characteristics in response to the presence of chemical estrogens and androgens at ng/L concentrations in water. Effects on human health among those chronically exposed remain largely speculative.

Nutrients (nitrogen and phosphorus). Agricultural runoff and municipal wastewater effluent can contribute nutrients such as ammonia, nitrate, and phosphate species to receiving waters, stimulating algal growth and eutrophication. Toxins released by certain algae in recreational water bodies and sources of drinking water can reach dangerous concentrations.

In aerobic waters, ammonia nitrogen is readily oxidized to nitrate ion (NO_3^-) by ammonia-oxidizing bacteria. Unfortunately, nitrate can be reduced to nitrite ion (NO_2^-) by nitrate-reducing bacteria in the upper gastrointestinal tracts of infants. Nitrite is then absorbed into the infant's bloodstream, where it reacts spontaneously with hemoglobin to form methemoglobin, which is unable to transport molecular oxygen. This is the basis of methemoglobinemia or "blue baby syndrome," which has been observed when nitrate fertilizers are leached into groundwater sources of domestic water supply. After about age 4 months, pH decreases in the upper gastrointestinal tracts of infants to levels that inhibit nitrate-reducing bacteria. WHO guideline values for nitrate and nitrite are 50 and 3 mg/L, respectively, to protect against methemoglobinemia in bottle-fed infants.[2]

Arsenic. Arsenic is naturally present in some geological strata and can dissolve over time into associated groundwaters, leading to human exposure in drinking water, See Figure 28.3. Consistent exposure to elevated levels of arsenic in drinking water can cause arsenic poisoning (arsenicosis). Initial symptoms include skin lesions and discoloration, while longer-term exposure causes cancers of the skin and/or internal organs.[2] Prolonged exposure during childhood increases cancer risk, particularly when accompanied by malnutrition. There is evidence that early exposure to arsenic also impairs intellectual development. The WHO concentration limit for total arsenic in groundwater is 10 µg/L (as As). Technical difficulties in reaching this concentration in many waters led to its identification as a provisional limit. WHO estimates that millions of people around the world are exposed to arsenic at concentrations of 100 µg/L or higher.[39]

Fluoride. Along with arsenic, fluoride is a serious contaminant found naturally in drinking water. Unlike arsenic, low levels of fluoride are beneficial (to dental and bone health). This beneficial effect is evident up to concentrations of 2 mg/L.[4] Surface waters generally contain less than 0.3 mg/L, but groundwater can contain up to 10 mg/L. or more. High concentrations of fluoride cause dental fluorosis, characterized by pitting and staining of dental enamel in young children. At higher concentrations, skeletal fluorosis causes stiffness, pain, and decreased mobility. In severe cases, ligaments can calcify and bone structure may change, causing crippling skeletal fluorosis. As with arsenic, fluorosis can be a product of bore-well access to groundwater, a source that is frequently favored over available surface waters.[2]

Metals. Metals of concern in drinking water include cadmium, chromium, lead, mercury, and selenium. These metals are generally not at elevated levels in drinking water unless there is significant nearby industrial activity. The exception is lead, which merits individual attention due to its former widespread use in household plumbing materials. Despite widespread reduction in the use of lead for this purpose, legacy problems persist in locations where the removals of lead-containing plumbing are incomplete.

Progress in Reducing Waterborne and Water-Related Disease

Notable reductions in diarrheal disease deaths occurred following the introduction and spread of oral rehydration therapy (ORT). Vaccine development and dissemination for such diseases as polio and rotavirus infections have been highly successful. Medications

Figure 28.3. Children pumping water from arsenic-contaminated tube well, Bangladesh

can treat once-rampant diseases such as lymphatic filariasis (elephantiasis), and eradication of disease vectors has progressed.

Despite progress in understanding, quantifying, and treating water pollution and associated diseases, significant uncertainties and difficulties remain. Areas of effort going forward will include (1) continued improvement in the quality of disease surveillance and reporting, (2) attention to the impacts of multiple and sometimes interacting disease exposures, and (3) acknowledgment of climate change effects on the spread and persistence of disease.

Disease Surveillance and Reporting

Waterborne disease is chronically underreported in LMICs, as are mild cases in industrialized nations. It has been suggested that the burden of diarrheal and other water-related diseases may be underestimated by an order of magnitude or more for the following reasons:

1. Deaths due to water-related diseases likely go undiagnosed, particularly in the poorest communities of LMICs.
2. In disability-adjusted life year (DALY) calculations, diarrheal disease is considered a short-term infection. However, childhood diarrhea can lead to long-term physiological impairments including malnutrition, stunted growth, immunodeficiency, and impaired intellectual development. These consequences are not captured in the DALY methodology.
3. Specific bacterial and viral infections that cause waterborne diarrhea are also linked to chronic diseases. These consequences are also not captured in the DALY methodology.
4. Although chemical contaminants in drinking water can cause symptoms of diarrhea, they are responsible for additional health outcomes, both acute and chronic. Of particular concern are fluoride, arsenic, and nitrates. Other chemicals whose

health impacts remain incompletely understood include disinfection by-products, endocrine disrupters, and a vast array of pharmaceutical products. Routine monitoring for most of these chemicals is difficult to impossible, and their burden to human health is unknown.

Multiple Disease Exposures and Interacting Pathways

The reality in many rural, low-income communities is that multiple disease threats occur simultaneously. Exposure to contaminated water and soils can be coupled with under-nutrition and crowded living conditions. The Ganges River in India, for example, is the recipient of municipal sewage from the 29 cities it passes, but is also used for drinking, bathing, crop irrigation, recreation, and washing of clothes and dishes (Figure 28.4) In the same places, malaria and dengue are additional threats. As noted at the beginning of this chapter, developmental and behavioral characteristics make children particularly susceptible to disease from these types of exposures.

Impacts of Climate Change

Reports pursuant to the 2014 Fifth Assessment Report of the Intergovernmental Panel on Climate Change (IPCC) provide an assessment of climate change impacts on vector- and waterborne diseases.[40] Climate change–related extremes in weather include excessive heat, drought, and heavy rain. High temperatures can alter pathogen survival,

Figure 28.4. Multiple uses of the ganges: commercial laundries on the ghats in Varanasi, India
Source: Photo courtesy of Steve Hamner.

replication, and virulence; heavy rainfall events can mobilize pathogens and compromise water and sanitation infrastructure; and drought can concentrate pathogens in vital water supplies.[41] The IPPC report states with "very high confidence" that increased risks of vector- and waterborne diseases, especially diarrheal diseases, can be expected if climate change continues as projected across the representative concentration pathway (RCP) scenarios until mid-century.[40]

Further analysis and discussion of successes and uncertainties in mitigating and quantifying the global burden of waterborne and water-related disease, particularly among children in LMICs, is found in Chapter 29.

References

1. Centers for Disease Control and Prevention. Diarrhea: common illness, global killer. n.d. Available at: https://www.cdc.gov/healthywater/pdf/global/programs/globaldiarrhea508c.pdf

2. United Nations Children's Fund (UNICEF). *UNICEF Handbook on Water Quality*. New York, UNICEF; 2008.

3. Kleiner, SM. Water: an essential but overlooked nutrient. *J Am Dietetic Assoc*. 1999;99:200–6.

4. GBD 2016 Diarrhoeal Disease Collaborators. Estimates of the global, regional, and national morbidity, mortality and aetiologies of diarrhoea in 195 countries: a systematic analysis for the Global Burden of Disease Study 2016. *Lancet Infect Dis*. 2018 Nov 1;18(11):1211–28.

5. National Institute of Diabetes and Digestive and Kidney Diseases. Definition and facts for diarrhea. 2016 Nov. Available at: https://www.niddk.nih.gov/health-information/digestive-diseases/diarrhea/definition-facts.

6. Centers for Disease Control and Prevention. Cholera in Haiti. 2020 Oct 30. Available at: https://www.cdc.gov/cholera/haiti/index.html.

7. World Health Organization. Salmonella (non-typhoidal). 2018 Feb 20. Available at: https://www.who.int/news-room/fact-sheets/detail/salmonella-(non-typhoidal).

8. Centers for Disease Control and Prevention. Shigella, shigellosis. 2021 Jun 29. Available at: https://www.cdc.gov/shigella/index.html.

9. Centers for Disease Control and Prevention. Campylobacter. 2019 Dec 23. Available at: https://www.cdc.gov/campylobacter/faq.html#serious.

10. Igbinosa IH, Igumbur EU, Aghdasi F. Emerging aeromonas species infections and their significance in public health. *Sci World J*. 2012 Jun 4.

11. Centers for Disease Control and Prevention. Leptospirosis. 2019. Available at: https://www.cdc.gov/leptospirosis/.

12. Kaper JB, Nataro JP, Mobley HLT. Pathogenic *Escherichia coli*. *Nat Rev Microbiol*. 2004;2:123–40.

13. Mayo Clinic. Reactive arthritis. 2011. Available at: https://www.mayoclinic.org/diseases-conditions/reactive-arthritis/symptoms-causes/syc-20354838.

14. Ameer MA, Wasey A, Salen P. Escherichia coli (E coli O157 H7). 2021. Available at: https://www.ncbi.nlm.nih.gov/books/NBK507845/#_article-20825_s4_.

15. Troeger C, Khalil IA, Rao PC, et al. Rotavirus vaccination and the global burden of rotavirus diarrhea among children younger than 5 years. *JAMA Pediatr*. 2018;172(10):958–65.

16. Centers for Disease Control and Prevention. Norovirus. 2021. Available at: https://www.cdc.gov/norovirus/trends-outbreaks/worldwide.html.

17. Lopman B, Steele D, Kirkwood C, Parashar U. The vast and varied global burden of norvirus: prospects for prevention and control. 2016 Apr 26. Available at: http://dx.plos.org/10.1371/journal.pmed.1001999.

18. Lee B, Damon CF, Platts-Mills JA, et al. Pediatric acute gastroenteritis due to adenovirus 40/41 in low- and middle-income countries. *Curr Opin Infect Dis*. 2020 Oct;33(5):398–403.

19. Mayo Clinic. Giardia infection. 2022. Available at: https://www.mayoclinic.org/diseases-conditions/giardia-infection/symptoms-causes/syc-20372786.

20. Medscape. Global incidence of giardiasis. 2018. Available at: https://www.medscape.com/answers/176718-120626/what-is-the-global-incidence-of-giardiasis.

21. Khalil IA, Troeger C, Roa PC, et al. Morbidity, mortality and long-term consequences associated with diarrhoea from Cryptosporidium infection in children younger than 5 years: a meta-analyses study. *Lancet Global Health*. 2018 Jul;6(7):E758–68. https://www.thelancet.com/journals/langlo/article/PIIS2214-109X(18)30283-3/fulltext

22. Shirley DT, Farr L, Watanabe K, Moonah S. A review of the global burden, new diagnostics, and current therapeutics for amebiasis. *Open Forum Infect Dis*. 2018 Jul 5;5(7):ofy161. doi:10.1093/ofid/ofy161

23. Centers for Disease Control and Prevention. DPDx: Laboratory identification of parasites of public health concern. Microsporidiosis. 2019 May 29. Available at: https://www.cdc.gov/dpdx/microsporidiosis/index.html.

24. World Health Organization. Health topics, fact sheets, detail, typhoid. 2023. Available at: https://www.who.int/news-room/fact-sheets/detail/typhoid.

25. Wiersinga WJ, Virk HS, Torres AG, et al. Melioidosis. *Nat Rev Dis Prim*. 2018;4:17107. https://doi.org/10.1038/nrdp.2017.107.

26. Petri WA, Naylor C, Haque R. Environmental enteropathy and malnutrition: do we know enough to intervene? *BMC Med*. 2014;12:187. https://doi.org/10.1186/s12916-014-0187-1.

27. World Health Organization. Fact sheets, detail, poliomyelitis. 2019 Jul 22. Available at: https://www.who.int/news-room/fact-sheets/detail/poliomyelitis.

28. Centers for Disease Control and Prevention. 2023. Parasites – Naegeria fowleri – primary amebic meningoencephalitis (PAM) – Amebic encephalitis. Available at: https://www.cdc.gov/parasites/naegleria/index.html.

29. World Health Organization. Toxoplasmosis: Greater Awareness Needed. 2016. Available at: https://www.euro.who.int/en/health-topics/disease-prevention/food-safety/news/news/2016/11/toxoplasmosis-greater-awareness-needed.

30. World Health Organization. Fact sheets, malaria. 2021. Available at: https://www.who.int/news-room/fact-sheets/detail/malaria.

31. Centers for Disease Control and Prevention. Parasites: Onchocerciasis (also known as river blindness). 2019. Available at: https://www.cdc.gov/parasites/onchocerciasis/epi.html.

32. Pan American Health Organization/World health Organization. Trachoma. 2022. Available at: https://www.paho.org/en/topics/trachoma.html.

33. World Health Organization. Fact sheets, schistosomiasis. 2022. Available at: https://www.who.int/en/news-room/fact-sheets/detail/schistosomiasis.

34. Centers for Disease Control and Prevention. Parasites: Schistosomiasis. 2020. Available at: cdc.gov/parasites/schistosomiasis/

35. World Health Organization. Fact sheets, Dracunculiasis (guinea-worm disease). 2020. Available at: https://www.who.int/news-room/fact-sheets/detail/dracunculiasis-(guinea-worm-disease).

36. World Health Organization. Fact sheets, lymphatic filariasis. 2021. Available at: https://www.who.int/news-room/fact-sheets/detail/lymphatic-filariasis

37. World Health Organization. Fact sheets, dengue and severe dengue. 2022. Available at: https://www.who.int/news-room/fact-sheets/detail/dengue-and-severe-dengue

38. World Health Organization. Fact sheets, soil-transmitted helminth infections. 2022. Available at: https://www.who.int/news-room/fact-sheets/detail/soil-transmitted-helminth-infections

39. World Health Organization. Fact sheets, arsenic. 2018. Available at: https://www.who.int/news-room/fact-sheets/detail/arsenic

40. Smith KR, Woodward A, Campbell-Lendrum D, et al. 2014: Human health: impacts, adaptation, and co-benefits. In: Field CB, Barros VR, Dokken DJ, et al. eds., *Climate Change 2014: Impacts, Adaptation, and Vulnerability. Part A: Global and Sectoral Aspects. Contribution of Working Group II to the Fifth Assessment Report of the Intergovernmental Panel on Climate Change.* Cambridge: Cambridge University Press; 2014:709–54.

41. Levy K, Smith SM, Carlton EJ. Climate change impacts on waterborne diseases: moving toward designing interventions. *Curr Envir Health Rep.*2018;5:272–82. https://doi.org/10.1007/s40572-018-0199-7

29

Interventions to Mitigate Burdens of Waterborne and Water-Related Disease

Robert G. Arnold

Overview

In 2020, the World Health Organization (WHO) estimated that more than 800 million people lacked access to basic or safely managed water services in their homes. About 1.7 billion lacked access to safely managed or basic sanitation services, and more than 2.3 billion could not effectively wash their hands at home.[1] Relative to 1990, when activities on behalf of the United Nations Millennium Development Goals (MDGs) were initiated, 2020 estimates represent enormous achievements. Nevertheless, regional differences in the availability of safe water, sanitation, and hygiene (WASH) facilities remain discouraging. At the beginning of this decade, 74% of sub-Saharan Africans, for example, lacked basic handwashing facilities in their homes.[1]

Inadequate water quality, improper sanitation, and poor hygiene increase human exposure to pathogenic bacteria, viruses, and helminths in excreta, polluted water, and contaminated soils. These exposures damage human health, particularly children's health (see Chapter 28). Pathogen transmission occurs through contaminated water and soil, but also via contaminated fingers, food, fomites, field crops, and various fluids. Mosquitoes and flies are vectors for specific water-related diseases.[2] Figure 29.1 illustrates both routes of exposure and route-specific interventions to mitigate the burden of waterborne and water-related diseases. The figure provides context for discussion of interventions on behalf of disease mitigations, the primary topic of this chapter.

The burden of waterborne diseases, particularly diarrheal diseases among children, remains high wherever a substantial proportion of the population lacks access to WASH services.

Since 1990, the UN has engaged in formal programs to broaden the global availability of WASH components, using indicators and related metrics to assess goal-specific progress. Improvements are reviewed here, as are the trajectories of child health indicators, the costs and benefits of WASH interventions, and durable impediments to universal eradication of preventable disease.

Water Availability, Human Needs

Chronic dehydration elevates risks of kidney disease, heart disease, hypertension, embolism, cerebral infarction, and dental diseases.[4-6] At 2.8% dehydration, there can be a

Robert G. Arnold, *Interventions to Mitigate Burdens of Waterborne and Water-Related Disease* In: *Textbook of Children's Environmental Health*, Second Edition. Edited by: Ruth A. Etzel and Philip J. Landrigan, Oxford University Press.
© Oxford University Press 2024. DOI: 10.1093/oso/9780197662526.003.0029

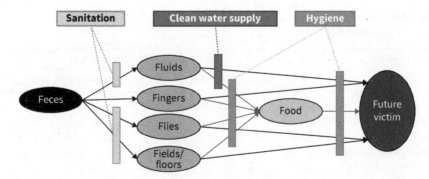

Figure 29.1. Routes of fecal disease transmission; points for interventions via water, sanitation, and hygiene (WASH) program measures (3)

reduction in short-term memory, visual perception, and hand–eye coordination.[7-9] Loss of 10% of body weight through dehydration can be fatal.[10,11]

Water is consumed both directly and in foods/other beverages. Factors that affect health-related recommendations for daily water intake (all sources) include stage of life, sex, climate, and level of exertion.[12] Temporarily ignoring other factors, daily water requirements per unit body weight in children are higher than those of adults due to a greater contribution of water to total body mass (50–60% in adults vs. 75% in infants[12]) and a higher rate of daily water turnover (4% vs. 15% per day).[13] Recommendations for total water intake appropriate to temperate climates in the industrialized world are summarized in Table 29.1.[12]

Water intake recommendations considering both ambient temperature and intensity of activity were developed on behalf of the United States Army (Table 29.2; modified from Kolka et al.[14]). Manual agricultural activity and hauling 20 liters of water were equated to moderate work, as defined in Table 29.2. The minimum daily quantity of water suggested by the WHO for drinking is 5.3 L/person, enough to satisfy water requirements of lactating women engaged in moderate activity at moderately high temperatures.[12] Based on the relationship between subtropical latitudes and low national

Table 29.1 Recommended total daily water intake as a function of life stage and gender

Gender/stage of life	Recommended total water intake (L/day)
Infants 6–12 months; both sexes	0.8–1.0
Children 2–3 years; both sexes	1.1–1.3
Children 4–8 years; both sexes	1.6
Children 9–13; boys	2.1
Children 9–13; girls	1.9
Adult females	2.0
Adult males	2.5
Pregnant women	Add 0.3
Lactating women	Add 0.6–0.7

Source: Howard et al., 2020.[12]

Table 29.2 Hourly water intake recommendations as a function of ambient temperature and level of activity

Activity intensity	Water intake (liters/hour)			
	25.6–27.7°C	27.8–29.4°C	29.5–32.2°C	>32.2°C
Easy work (walking, hard surface at 4 km/hr; carrying <13.6 kg)	0.47	0.47	0.71	0.94
Moderate work (walking, sand at 4 km/hr or hard surface at 5.5 km/hr; carrying <18.1 kg)	0.71	0.71	0.71	0.94
Hard work (walking, sand at 4 km/hr or hard surface at 5.5 km/hr; carrying ≥18.1 kg)	0.71	0.94	0.94	0.94

Source: Howard et al., 2020.[12]

income, water intake requirements for avoidance of dehydration effects are potentially greater in disadvantaged parts of the world.

The volume of water available for basic hygiene also is critically important to human health. The WHO suggested that major shifts in the volume of water used at home for drinking, cooking, and hygiene occur at definable levels of water access.[12] The WHO then defined access levels and consequent volumes of use (Table 29.3), acknowledging that per capita use depends on the time and effort required to obtain water, the continuity of piped supply, and the unit cost of water. The authors also indicated that, based on expert consensus, 20 L/person-day is a sufficient volume to support household ingestion, cooking, food hygiene, handwashing, and face washing, but not other hygiene needs. The Sphere Project[15] recommended a minimum of 15 L/person-day for drinking, cooking, and basic hygiene. Gleick[16] indicated that 50 L/person-day is appropriate for drinking, cooking, bathing, and sanitation.

Table 29.3 Summary of World Health Organization water access definitions—related water use expectations and level of health concern related to volume of use

Access level descriptor/estimate of per capita water use at each access level	Accessibility of water supply (definitions)	Level of concern to health (based on volume used)
Inadequate (≤5.3 L/person-day)	>1,000 m or 30 min total collection time	Very high
Basic (≤20 L/person-day)	100–1000 m or 5–30 min total collection time	High
Intermediate (~50 L/person-day)	Delivery via one tap; on-plot or ≤ 100 m and 5 min total collection time	Medium
Optimal (≥100 L/person-day)	Delivery via multiple taps w/ continuous availability	Low

Source: Howard et al., 2020.[12]

The Global Burden of Disease in Children: Relationship to Water, Sanitation, and Hygiene

Unsafe water, inadequate sanitation, and poor hygiene practices contribute heavily to childhood disease. Statistics relevant to the global trajectory of children's health are provided in Chapter 13.

Among the primary contributors to the worldwide burden of childhood disease, diarrheal diseases are linked most closely to environmental (WASH) conditions.[17] Chronic diarrhea and intestinal nematode infections due to inadequate sanitation and/or hygiene are also linked to underweight, a condition that, in 2004, contributed to the deaths of an estimated 3.8 million children worldwide.[18]

Clean water, sanitation, and hygiene are essential to child health.

Water, Sanitation and Hygiene Interventions on Behalf of Child Health: Early Evidence

United States Experiences

In 1900, the US child mortality rate was almost one-in-four.[19] "Water closets" were broadly introduced in the 1870s, and improved sanitation systems were widespread by the middle of the 20th century.[20] In 1972, a minimum of secondary treatment was mandated for all publicly owned (sewage) treatment works discharging treated wastewater to navigable US waters.[21] Twentieth-century progress in terms of improved health was substantial. In 1990, the mortality rate for US children under 5 was 12 per 1,000 live births—just 5% of the 1900 rate. By 2019, the child mortality rate had further declined to 7 per 1,000 live births.[19]

Nevertheless, problems persist. Handwashing, for example, remains a challenge in areas lacking indoor plumbing, the importance of which was amplified by the COVID-19 pandemic.[22] In the United States, about 180 million cases of acute gastroenteritis are identified each year. In 2009–2010, the suspected or confirmed cause for 89% of 1,419 US outbreaks of gastroenteritis with a reported etiology was norovirus infection.[23] Careful handwashing reduces the spread of norovirus.

Regions of Alaska with lower proportions of in-home water service have experienced significantly higher hospitalization rates for pneumonia and influenza, skin or soft tissue infection, and respiratory syncytial virus among children under 5.[24] The most likely explanation is that piped water increases the ease and frequency of handwashing.[25]

Sanitation Interventions (Worldwide)

An abundance of evidence indicates that management of human waste interrupts transmission pathways of bacteria, viruses, and helminths. Thirty early investigations in Bangladesh, Brazil, Chile, Guatemala, Kenya, Malaysia, and Panama focused on sanitation interventions to reduce diarrhea.[26] In a subset of 11 studies, the median reduction in diarrhea morbidity was 22%. The more rigorous studies (5) yielded a median reduction in diarrhea morbidity of 36%. A subsequent meta-analysis, using results from 2 interventions[27,28] placed reduction of diarrheal morbidity at 32%.[29] A case-control study found that children under 5 suffering from diarrhea were less likely to live in homes with improved latrines than were healthy children in the comparison group (odds ratio 0.57; 95% confidence interval [CI]: 0.42, 0.77).[30]

A Cochrane review evaluated the effectiveness of improved disposal of human excreta in preventing diarrheal disease.[31] Thirteen studies from six low- or middle-income countries involving 33,400 children and adults in rural, urban, and school settings met inclusion criteria.[32-44] Eleven of the thirteen interventions produced statistically significant protective effects. Due to heterogeneity among studies and the unavailability of reliable confidence intervals, a pooled effect was not calculated.

Hygiene-Related Interventions and Diarrheal Disease

Structural measures unaccompanied by behavioral changes among the populations served are of limited consequence to health outcomes. In six studies, handwashing reduced diarrhea morbidity by 14–48% with a median reduction of 33%.[26]

A systematic review and meta-analysis of health impacts due to handwashing with soap showed a 47% reduction in diarrheal morbidity.[45] Another pooled analysis of contemporary interventions included data from 17 studies for children under 5. Handwashing produced a 37% reduction in diarrheal morbidity.[46] The importance of handwashing, even as a standalone measure, to reduction of diarrheal disease was made clear in several additional, contemporary studies.[47-50]

The impact of multiple, simultaneous WASH interventions is significantly less than the summed effects of individual measures.[26,47,51] This is particularly relevant when measures on behalf of human health are financially constrained.

A 2005 summary of available data representing effects of standalone and multiple WASH interventions on diarrheal disease in less industrialized countries speaks plainly on behalf of WASH measures (Table 29.4).[51,52,53,54] Absent such evidence, common sense nevertheless commands the introduction or strengthening of WASH interventions in areas where the burden of enteric disease is greatest.

Breastfeeding is particularly helpful in homes without adequate sanitation and hygiene. In Malaysia, for example, breastfeeding in homes that lacked piped water and sanitation halved the risk of child death during post-perinatal infancy.[55] Breastfeeding remains an effective way to protect infants from diarrheal diseases, particularly when other measures are not immediately available.[55-58]

The Millennium Development Goals

Acknowledging both (1) the dependence of human health on clean water, sanitation, and hygienic practices and (2) the immense inequality between disease outcomes in developed and developing countries, the UN/WHO pursued a global, decades-long series of WASH-related improvements—the Millennium Development Goals (MDGs) during 1990 to 2015 and the Sustainable Development Goals (SDGs) thereafter.

MDGs consisted of 8 goals, 21 related targets, and 60 measurable indicators linked to the quality of human life. Indicators tracked national, regional, and global progress toward MDG objectives.[59] Several MDG goals, targets, and indicators (Table 29.5) were particularly relevant to clean water, sanitation, hygiene, and child health.

Table 29.4 Diarrheal disease reductions due to water, sanitation and hygiene improvements

Intervention	Number of studies reviewed	Risk reduction (-)	95% confidence interval (-)	References
Improved drinking water	6	0.25	(0.09–0.38)	51
	5	0.20	-	26
	8	0.02	(−0.06–0.11)	46
Improved sanitation	2	0.32	(0.13–0.47)	51
	5	0.36	-	26
	6	0.37	(0.07–0.57)	46
Improved hygiene	8	0.45	(0.25–0.60)	51
	6	0.33	-	26
	17	0.31	(0.23–0.39)	46
Handwashing (only)	4	0.32	(0.10–0.48)	47
	24	0.31	(0.19–0.42)	48
	10	0.47	(0.24–0.63)	49
	22	0.28	(0.17–0.38)	50
In-home water treatment[a]	8	0.39	(0.19–0.54)	51
	4	0.15	-	26
	31	0.42	(0.33–0.50)	46
	33	0.35	(0.29–0.41)	52
	10	0.29	(0.13–0.42)	53
	4	−0.09	(−0.22–0.02)	54
Multiple interventions[b]	5	0.33	(0.24–0.41)	51
Water + sanitation (only)	7	0.38	(0.17–0.54)	46
	2	0.30	-	26

[a] A.k.a., water quality or point-of-use interventions.
[b] Unless otherwise indicated, multiple interventions consisted of water, sanitation, and hygiene measures in combination.

Source: Ejemot et al. 2009.[51]

Indicators of progress related to the MDGs during 1990–2015 and, subsequently, to the SDGs (see below) were tracked by the WHO/UNICEF Joint Monitoring Programme for Water Supply and Sanitation (JMP). Those directly related to child mortality were the time-dependent national and global child and infant mortality rates.

Table 29.5 MDGs, targets, and indicators for (1) global reduction in child mortality and (2) introduction of interventions on behalf of safe water and basic sanitation

Goal	Target	Indicator
MDG 4—reduce child mortality	4A—reduce by two-thirds, between 1990 and 2015, the (1) under-5 and (2) infant mortality rates	Time-dependent child and infant mortality rates
MDG 7—ensure environmental sustainability	7C—halve, by 2015, the proportion of people without sustainable access to (1) safe drinking water and (2) basic sanitation	Time-dependent, fractional access to "safe" drinking water and "basic" sanitation. (Note: see definitions in text)

Source: World Health Organization. Millenium Development Goals (MDGs).[59]

Table 29.6 Characteristics of improved and unimproved drinking water sources and sanitation facilities. "Improved" systems were accepted as sources of safe drinking water or basic sanitation for measurement of progress toward MDG targets

Water supply characteristics		Sanitation system characteristics	
Improved	Unimproved	Improved	Unimproved
Piped water in dwelling	Vendor-provided water	Flush toilet	Public or shared latrine
Piped water into yard	Cart w/tank or drum	Connection to a piped sewer system	Flush/pour flush toilet not connected to pit, septic tank, or sewer
Public tap/standpipe	Tanker truck	Flush/pour-flush toilet connected to a pit latrine	Pit latrine w/out slab
Tubewell/boreholes	Surface water	Pit latrine w/ slab	Bucket latrines
Protected dug wells	Unprotected dug wells	Ventilated improved pit latrine	Hanging toilet/latrine
Protected springs	Unprotected springs	Composting toilet	Open defecation
Rainwater collection			
Bottled water			

Source: World Health Organization. Millenium Development Goals (MDGs).[59]

To measure progress in areas of water supply and sanitation, definitions related to the characteristics of *safe drinking water* sources and *basic sanitation* were required. Sanitation was described as *improved* if human excreta was hygienically separated from subsequent human contact. The opposite was *unimproved sanitation*. For the sanitation indicator, the adjectives "improved" and "basic" were synonymous.

An *improved water source*, by the nature of its construction or through active intervention, was likely to be protected from outside contamination—particularly fecal contamination—and hence *safe*. Facilities that qualified as *improved* (water sources and sanitation) are summarized in Table 29.6.[60]

JMP data were obtained via household surveys and censuses. By 2012, the safe drinking water portion of MDG 7C (Table 29.7) had already been met, but the sanitation goal had not. At that point, only 63% of the world's population had access to basic sanitation, and the projection for 2015 was 67%. The 2015 MDG target was 77%. Projections indicated that the child mortality target for 2015 (MDG 4A) would also not be reached. In fact, it was not. Eliminating open defecation remained an important unmet priority.

The lowest levels of WASH service coverage are in the "least developed countries" (LDCs), a UN designation based on composite, objective criteria (Table 29.7). In the 46 nations included among the LDCs, households are (1) more than five times more likely to lack access to an improved water supply and (2) twice as likely to lack basic sanitation as the global population 2020 estimates. In LDCs, the fractions of population practicing

Table 29.7 Water, sanitation and hygiene indicators for fractions of population served in 1990, 2000, 2015, and 2019–2020—global and least developed countries

Target or indicator	1990	2000	2015	2019–2020
(a) Global estimates (all countries)				
Fraction with improved water supply	0.76	0.81	0.88	0.94
Fraction with access to basic sanitation	0.54	0.59	0.68	0.68
Fraction practicing open defecation	0.24	0.22	0.10	0.06
Fraction with adequate household hygiene	ND	ND	0.67	0.71
(b) Least developed countries (only)				
Fraction with improved water supply	ND	ND	0.62	0.67
Fraction with access to basic sanitation	ND	ND	0.34	0.37
Fraction practicing open defecation	ND	ND	0.20	0.16
Fraction with adequate household hygiene	ND	ND	0.32	0.37

The numerical score leading to identification of Least Developed Countries involves three indices that are themselves composites: gross national income per capita; the human asset index, based on six measurable indices; and economic and environmental vulnerability, composed of eight measurable indices. The designation is fluid, based on calculated composite values. The current LDC list consists of 46 nations. ND indicates that data are not available as composite statistics and/or not considered reliable.

Source: Ritchie et al., 2018[60]; Roser et al., 2019.[61]

open defecation and households without adequate hygiene are more than twice those of the global community.

The Sustainable Development Goals

Despite significant progress toward MDGs from 1990 to 2015, or possibly because of it, the UN recalibrated its objectives within the SDGs, to be pursued from 2015 to 2030. These are considerably broader than the MDGs, although those related to child health and WASH practices are similar in nature (Table 29.8).[62] and [63]

For purposes of the SDGs, types of water services were expanded to include those that were "safely managed" and "basic" in character. *Basic* water sources were off-site, improved sources from which water could be obtained with 30 minutes or less of (round trip) effort.[64] A second newly defined level, "limited" water service, was used to describe off-site, improved water sources when travel times were more than 30 minutes. *Limited* water services did not qualify as basic services. Service levels were thus defined from lowest to highest as surface water, unimproved, limited, basic, and safely managed.

Sanitation levels also were reclassified. A new term, *basic sanitation service*, was defined to include improved sanitation facilities not shared with other households.[64] *Limited sanitation service* consists of improved sanitation facilities that are shared with other households. *Safely managed sanitation services* remain those in which excreta are safely disposed of in situ or transported and treated offsite.[64] Rungs on the sanitation

Table 29.8 Sustainable Development Goals (SDGs) related to child mortality and availability of water, sanitation and hygiene facilities

Goal	Target	Indicator
SDG 1—Universal access to basic services.	1.4—By 2030, ensure that all have equal access to basic services for water, sanitation, and hygiene.	1.4.1. Populations with basic drinking water, sanitation, and hygiene services.
SDG 3—Ensure healthy lives and promote well-being for all ages.	3.2—By 2030, end preventable deaths of newborns and children under five. 3.2 (cont.)—all countries aiming to reduce . . . under-5 mortality to ≤ 25/1000 live births.	Infant and child mortality statistics—global and national.
SDG 6—Ensure availability and sustainable management of water and sanitation for all.	6.1—By 2030, achieve universal and equitable access to safe, affordable drinking water for all. 6.2—By 2030, achieve access to adequate and equitable sanitation and hygiene for all. 6.2—By 2030, end open defecation.	6.1.1. Population lacking safely managed drinking water services. 6.2.1. Population without safely managed or basic sanitation services. 6.2.1. Population without basic handwashing facility with soap. 6.2.1. Population practicing open defecation.

Source: United Nations General Assembly, 2015[62]; Sachs, 2012.[63]

ladder, from lowest to highest, are open defecation, unimproved, limited, basic, and safely managed services.

Hygiene

Basic handwashing facilities include a device to contain, transport, or regulate the flow of water and facilitate handwashing with soap and water. *Limited* household hygiene lacks either soap or water. Thus, the hygiene ladder progresses from "no service," through "limited service," to "basic service."[64]

Progress Toward the Sustainable Development Goals

Progress toward the SDGs can be defined relative to the Table 29.8 targets. The 15-year term for SDG pursuit/activity is beyond its mid-point. Trajectories of WASH indicators beyond year 2015 show that progress has continued under the SDGs (Table 29.9). Importantly, however, geographic and economic disparities in the provision of WASH measures persist, decreasing the likelihood that the SDG target related to reduction of child mortality rates (≤25 deaths among children under 5 per 1,000 live births; *each nation*) will be achieved by 2030 (Table 29.7; Figures 29.2 and 29.3).

Table 29.9. Progress on the "handwashing ladder" by geographic region during 2015-2020; global coverages at bottom. Values are percentages of home handwashing coverages of the types indicated (at top) that were present in years 2015 and 2020.

Region Represented	Basic (improved)		Limited		No household facility	
	2015	2020	2015	2020	2015	2020
Sub-Saharan Africa	25	26	40	40	35	34
Oceania	36	36	29	28	35	36
Central & Southern Asia	65	69	31	28	4	3
Northern Africa and Western Asia	84	90	10	7	6	3
Least Developed Nations	32	37	39	36	29	27
Global	67	71	23	21	10	8

Notes: "Basic"—homes in which handwashing facilities are present and used with soap and water. "Limited"—homes in which either soap or water was lacking on site.

"No household facility"—self-explanatory

Source: WHO, UNICEF, JMP. https//data unicef.org/resources/jmp-report-2023.[64]

WASH improvements since 1990 have greatly advanced the health of children— and likely all age groups. Nevertheless, satisfaction of SDGs related to child health and WASH coverages remains unlikely. Relatively high rates of infant and child mortality persist, particularly in the LDCs (Chapter 13). Worldwide elimination of open defecation remains challenging. Despite improvement in access to basic handwashing facilities, disparities in access based on economic conditions remain apparent Table 29.9.

The Benefits and Costs of Interventions on Behalf of Human Health

The benefits and costs of providing universal improved water and/or sanitation coverage have been examined and compared.[65-69] Results are summarized in Table 29.10 and Figure 29.2. Contributions to total benefits are attributed to differential (1) health-care costs, (2) productivity, (3) mortality, and (4) time saved due to improved access. Year 2010 was the baseline for measurement of improvements in sanitation and/or clean water coverages and estimation of related costs and benefits.

More than three-fourths of the benefits attributed to universal extension of improved sanitation and water supply resulted from time saved (travel and waiting times averted). Annualized capital costs were far greater than the present values of recurring maintenance expenditures. A total of 136 countries for which sufficient data were available were included in the calculations. Worldwide estimates of benefits and costs were based on those of individual nations, weighted for populations.

The estimated benefits of universal improved water and sanitation *in combination* and resultant cost-benefit ratios are incomplete in Table 29.10 because risk reductions due to multiple simultaneous interventions (and therefore benefits) are significantly lower than

Table 29.10 Estimated annualized costs and benefits of universal extension of improved sanitation and water supply (2011–2015)

Benefit, Cost Category/Source	Universal sanitation	Universal improved water	Water plus sanitation
Benefits/yr (2010 US$-millions)	194,857	34,479	
Healthcare (%)	8	12	
Productivity (%)	5	6	
Mortality (%)	6	10	
Time savings (%)	81	71	
Cost/yr (2010 US$-millions)	35,277	17,476	52,753
Capital cost (%)	62	66	63
Cost for rural improvements (%)	39	28	35
Benefit-cost ratio (-)	5.5	2.0	

Costs and benefits are in millions of 2010 dollars per year, annualized over a 5-year period (2011–2015) Noncapital costs are for system operation and maintenance.

Source: World Health Organization, 2012[66]; Hutton, 2013.[67]

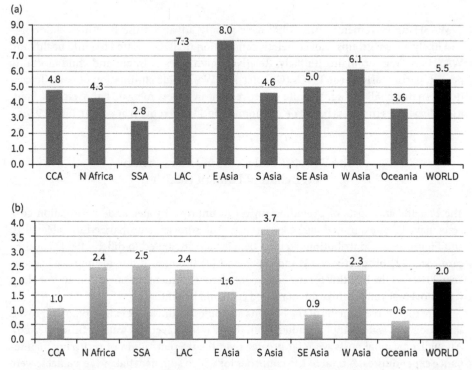

Figure 29.2. Benefit-cost ratios for universal access to improved (a) sanitation, and (b) water supply, by region. Reported 2010 coverages were used as a baseline.[66] Key: CCA = Caucasus and Central Asia, N Africa = Northern Africa, SSA = Sub- Saharan Africa, LAC = Latin America and the Caribbean, E Asia = Eastern Asia, S Asia = Southern Asia, SE Asia = Southeast Asia, W Asia = Western Asia, Oceania = Oceania

Notes:

1. "Basic" hygiene implies handwashing with soap and water in the home; "limited" hygiene implies soap and/or water is lacking; "no service" is self-explanatory.

2. Source of "Least Developed Countries" designation is explained in Table 29.7.

Table 29.11 Determinants of cost-benefit (C/B): bases for sensitivity analyses in Table 29.12

Determinants of C/B	High value (each determinant)	Baseline value (each determinant)	Low value (each determinant)
Value of premature deaths avoided	Based on the value of statistical life	Based on the human capital approach	Based on half the human capital approach
Value of time saved due to reduced morbidity	Full value of per capita GDP; 50% for children	30% of per capita GDP; 15% for children	15% of per capita GDP; zero for children
Time savings due to improved water and sanitation (access time)	Double baseline value	30% of hourly GDP for adult hours saved, 15% for children	Half baseline value
Unit costs for water or sanitation service (tech. cost)	High values from literature	Average values from literature	Low values from literature
Discount rate (yr^{-1})	0.12	0.08	0.03

Source: Hutton, 2013.[67]

the additive risk reductions of individual interventions, as acknowledged in discussions of relative risk.[66,67]

Regional cost-benefit ratios attributed to universal provision of improved sanitation varied from 2.8 in sub-Saharan Africa to 8.0 in East Asia. The global economic return on sanitation spending was US$5.5 per dollar invested. Benefit-cost ratios for universal access to improved water supply varied regionally from 0.9 in Southeast Asia to 3.7 in Southern Asia. The estimated global cost-benefit ratio was 2.0.

To add perspective, the total economic benefit from global provision of improved water supply and sanitation was estimated at about US$ 230 billion annually, or 1.54% of the total gross domestic product of the 136 countries included in the analysis.[67]

The sensitivity of results to the primary drivers of benefits and costs was determined from variation in cost-benefit ratios arising from alternate assumptions regarding the primary determinants of costs or benefits (Table 29.11)—that is, by setting value parameters at high, intermediate (baseline), or low estimated values in calculations of costs or benefits.

Ranges of cost-benefit ratios for universal sanitation or water service (Table 29.12) arise from variation in individual determinants of cost or benefit, while other determinants remain at their respective baseline values (one-way sensitivity). Results indicate that cost-benefit estimates are highly sensitive to assumptions regarding: (1) access time to safe water and sanitation services and (2) the unit value of time saved through improved access and avoided morbidity.

The Durability of Preventable Health Problems

The focus here is on financial, institutional, and social impediments to effective, sustainable reductions in childhood disease.

Table 29.12 One-way sensitivity analysis. Global cost-benefit ratios for improved sanitation or water as a function of high, baseline, and low values for individual determinants of costs and benefits. See Table 29.11 for sources of C/B determinants.

Cost and benefit determinants	Sanitation cost/benefit			Safe water cost/benefit		
	High	Base	Low	High	Base	Low
Premature deaths avoided	6.6	5.5	5.4	2.7	2.0	1.9
Opportunity cost of time lost due to illness/access	16.6	5.5	3.1	5.5	2.0	1.2
Access time saved	10.1	5.5	3.3	3.9	2.0	1.2
Unit cost of water or sanitation service	4.8	5.5	10.9	1.6	2.0	4.1
Discount rate	5.4	5.5	6.3	1.9	2.0	2.4

Source: Hutton, 2013.[67]

Global Wealth Distribution

The relationship between national mortality rates attributable to environment (preventable diseases) and gross national income per capita is striking (Figure 29.3).

For nations with per capita incomes greater than about US$25,000, the mortality rate attributable to the environment is essentially fixed at about 50 per 100,000 per year. At lower income levels, mortality rates linked to the environment increase rapidly as per capita income decreases, to a high of 450–500 per 100,000 each year among those with per capita annual incomes below about US$2,000.[68, 69]

Issues related to the dependence of health outcomes (e.g., child mortality) on persistent, extreme poverty are discussed in Chapter 13.

Social Sciences and Water, Sanitation and Hygiene Interventions

All the ideas presented in this section arise from a single source[70] although they represent intuition derived from decades of experience acquired by dozens of investigators. It is held that when designing and executing WASH interventions, the participation of engineers, health experts, and social scientists (anthropologists) enhances benefits among the populations served. That is, meaningful attention to anthropological principles throughout the conception, implementation, and use of WASH-related interventions enhances their probable value, shaping them to serve within human or cultural contexts.

These contentions arise from differences in the professional perspectives of engineers, health scientists, and anthropologists. The focus of engineers is on creation or application of technology and infrastructure that efficiently address recognizable problems in health science. The focus of anthropologists is on defining the problem and problem solutions from the perspectives of the population served. As such, engagement of social scientists in the identification and resolution of WASH deficiencies should be organic, as should that of the public. Interdisciplinary approaches are held to increase probabilities that (1) projects will be fully embraced by the populations served, (2) targeted improvements in human (particularly children's) health outcomes will be fully realized, and (3) the durability of preventable, water-related diseases will be minimized.

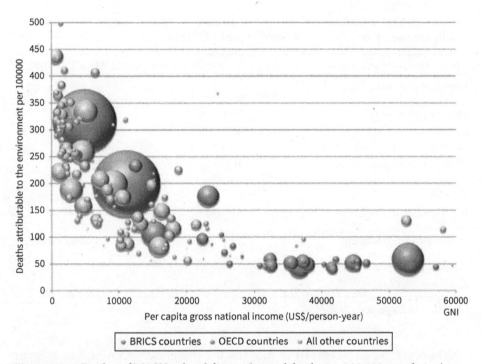

Figure 29.3. Burden of WASH-related disease (annual deaths per 100,000 population) as a function of per capita gross national income. Bubbles represent individual nations—bubble size is proportional to population.[68] BRICS: Brazil, Russia, India, China, South Africa; OECD: Organization for Economic Co-operation and Development

Conclusion

The distribution of wealth among nations impedes the universal achievement of goals on behalf of WASH intervention—despite their apparent cost effectiveness. The trajectories of progress suggest that SDGs will not be achieved in the nations where they would be of the most benefit to human health—particularly child health. Necessary investment to achieve universal coverage may be within human reach, but the appetite for such investment by wealthy nations on behalf of the poorest countries appears to be missing in a conflicted world that is afflicted by pandemic in addition to other preventable diseases. If the whole apple is in fact unattainable, at least in a single bite, what course toward mitigation of child morbidity and mortality remains most attractive?

Short of concession of any kind, it may be useful to concentrate initial effort in areas that are likely to provide the greatest return on investment—both to benefit those directly served and to highlight the potential benefits of worldwide investment in human health. Statements put forward here are potentially important. First, alternative interventions on behalf of human health are not equally cost effective. Access to improved water supply, basic sanitation, or basic household hygiene each provides a marked health improvement. Simultaneous provision of multiple WASH interventions, however, diminishes the individual benefits of improved water supply, sanitation, and hygiene. Decades of study reviewed here are clear on that point.

Table 29.13 Cost-effectiveness of interventions on behalf of public health in low- and middle-income countries (in DALYs averted per US$1,000 invested)

Disease control intervention	Cost-effectiveness ratio (DALYs/US$1,000)
Diarrheal diseases	
Hygiene promotion	200
Sanitation promotion	90
Water regulation and advocacy	12
Cholera/retrovirus immunization	0.5
Malaria	
Insecticide-treated bed nets	80–140
HIV avoidance/treatment	
Condom promotion/distribution	10–12
Antiretroviral therapy	1–3

Source: Laxminarayan et al., 2006.[73]

Cost-benefit ratios developed to compare the efficiencies of investment in water supply versus sanitation on behalf of human health clearly recommend investment in sanitation. Unfortunately, measures related to household hygiene are missing from many of the studies reviewed. However, multiple studies indicate that household hand-washing alone lowers the relative risk of diarrheal diseases by a factor of 0.28–0.47,[51–54] reductions that can be improved by educating the public regarding the importance of in-home hygiene practices.

The cost of providing universal household hand hygiene services throughout the 46 countries listed as least developed nations was recently estimated at US$12.2–15.3 billion over a 10-year period.[71] The cost of hand hygiene promotion for behavioral change would comprise about 40% of the total program cost or US$0.47 per year for each person served. A 2002 study in Burkina Faso[72] estimated the cost of each disability adjusted life year (DALY) saved through handwashing at US$88–225, placing handwashing at a level equivalent to oral rehydration therapy and most childhood vaccinations in terms of cost effectiveness.

Finally, a 2006 comparison of intervention cost-effectiveness for disease avoidance[73] provided the Table 29.13 ratios (DALYs averted per US$1,000 expended). Again, the relevance of judgment in health-related program selection is obvious. It should be noted that costs related to avoidance of diarrheal diseases were developed ahead of the COVID-19 pandemic.

Material presented here builds on chapter 28's descriptions of water-related diseases that impair childhood health and on chapter 13's child morbidity and mortality data. The natures of interventions designed to mitigate the burden of water-related disease, particularly among children in low-income countries, were described in areas of water supply, sanitation, and hygiene—the WASH elements. Large differences in the burdens of water-related diseases in low- and high-income countries reinforce the idea that these forms of disease are preventable.

From 1990 to 2015, the UN and the WHO pursued goals and quantifiable targets (the MDGs) established to improve health outcomes arising from water-related disease. The JMP was co-created to track health improvements arising from the MDGs. Despite successes, the MDGs were not fully achieved, and, in 2015, they were supplemented by the SDGs, a diverse and challenging set of objectives for the period 2015–2030. Data from 2020 suggest that, although additional progress is evident has been made, WASH-related SDGs will not be satisfied by the end of 2030 without increased investment.

Estimates of economic benefits attributable to human health improvements arising from WASH interventions greatly exceed the estimated costs of those interventions. Costs (capital and recurring for global WASH improvements) remain, however, greater than the ability or willingness of the global community to pay and out of reach for the developing world on its own. Inadequate funding for WASH improvements contributes to the resilience of water-related disease, including much of the childhood burden of disease, as do poverty, data inadequacies in low-income countries, and a lack of interdisciplinary participation in the design and execution of interventions on behalf of human health.

References

1. World Health Organization. Progress on household drinking water, sanitation and hygiene 2000–2020. Five years into the SDGs. Geneva: World Health Organization (WHO), United Nations Children's Fund (UNICEF), 2021. License: CC BY-NC-SA 3.0 IGO.

2. World Health Organization. Health through safe drinking water and basic sanitation. Available at: http://www.who.int/water_sanitation_health/mdg1/en/index.html.

3. World Bank. http://siteresources.worldbank.org/INTTOPSANHYG/Images/The_F_Diagram.jpg

4. Armstrong LE. Challenges of linking chronic dehydration and fluid consumption to health outcomes. *Nutr Rev.* 2012;70(Suppl 2):121–7.

5. Chan J, Knutsen SF, Blixx GG, Lee JW, Fraser GE. Water, other fluids, and fatal coronary heart disease: the Adventist health study. *Am J Epidemiol.* 2002;155(9):827–33.

6. Johnson RL, Wesseling C, Newman LS. Chronic kidney disease of unknown cause in agricultural communities. *N Engl J Med.* 2019;380:1843–52.

7. Cian C, Koulmann N, Barraud PA, Raphel C, Jimenez C, Melin B. Influence of variations in body hydration on cognitive function. *J Psychophysiol.* 2000;14(1):29–36.

8. Cian C, Barraud PA, Melin B, Raphel C. Effects of fluid ingestión on cognitive function after heat stress or exercise-induced dehydration. *Int J Psychophysiol.* 2001;42(3):243–51.

9. Popkin BM, Rosenberg IH. Water, hydration and health. *Nutr Rev.* 2010;68(8):439–58.

10. World Health Organization. *Nutrients in Drinking Water.* Geneva: WHO; 2005.

11. European Food Safety Authority (EFSA). Scientific opinion on dietary reference values for water. *EFSA J.* 2010;8(3):48.

12. Howard G, Bartram J, Williams A, Overbo A, Fuente D, Greene J. *Domestic Water Quantity, Service Level and Health.* 2nd ed. Geneva: WHO; 2020.

13. Kleiner SM. Water: an essential but overlooked nutrient. *J Am Diet Assoc.* 1999;99:200–6.

14. Kolka MA, Latzka WA, Montain SJ, Sawka MN. Current US military fluid replacement guidelines. Paper presented at the RTO HFM Specialists' meeting on "Maintaining Hydration: Issues, Guidelines, and Delivery." Boston, 10–11 December 2003 (RTO-MP-_HFM-086).

15. Sphere Project. *The Sphere Handbook: Humanitarian Charter and Minimum Standards in Humanitarian Response*, vol. 1. United Kingdom: Practical Action Publishing; 2018.

16. Gleick P. Basic water requirements. *Water Int.* 1996;21(2):83–92.

17. Prüss-Üstün, Corvalan C. & World Health Organization Preventing disease through healthy environments: towards an estimate of the environmental burden of disease: executive summary. 2006. Available at: https://apps.who.int/iris/handle/10665/43457.

18. World Health Organization. *Global Health Risks: Mortality and Burden of Diseases Attributable to Selected Major Risks.* Geneva: WHO; 2009. https://apps.who.int/iris/handle/10665/44203.

19. O'Neill A. Child mortality in the United States 1800–2020. 2021. Available at: https://www.statista.com/statistics/1041693/united-states-all-time-child-mortality-rate/.

20. Cutler D, Miller G. The role of public health improvements in health advances: the 20th century United States. *Demography.* 2005;42:1–22.

21. United States Environmental Protection Agency. Clean Water Act. 1972; last updated June 22, 2023. Available at: https://www.epa.gov/laws-regulations/summary-clean-water-act.

22. Schmidt CW. Lack of handwashing access: a widespread deficiency in the age of COVID-19. *Environmental Health Perspectives.* Science Selection. 2020. Available at: https://doi.org/10.1289/EHP7493.

23. Centers for Disease Control. Outbreaks of acute gastroenteritis transmitted by person-to-person contact—United States, 2009–2010. *Morb Mortal Wkly Rep.* 2012;61:1–12.

24. Hennessy TW, Ritter T, Holman RC, et al. The relationship between in-home water service and the risk of respiratory tract, skin, and gastrointestinal tract infections among rural Alaska Natives. *Am J Public Health.* 2008;98:2072–8.

25. Gessner BD. Lack of piped water and sewage services is associated with pediatric lower respiratory tract infections in Alaska. *J Pediatr.* 2008;152:666–70.

26. Esrey SA, Potash JB, Roberts L, Schiff C. Effects of improved water supply and sanitation on ascariasis, diarrhoea, dracunculiasis, hookworm infection, schistosomiasis, and trachoma. *Bull World Health Organ.* 1991;69:609–21.

27. Azurin JC, Alvero M. Field evaluation of environmental sanitation measures against cholera. *Bull World Health Organ.* 1974;51:19–26.

28. Daniels DL, Cousens SN, Makoae LN, Feachem RG. A case-control study of the impact of improved sanitation on diarrhoea morbidity in Lesotho. *Bull World Health Organ.* 1990;68: 455–63.

29. Fewtrell L, Colford JM Jr. Water, sanitation, and hygiene in developing countries: interventions and diarrhoea: a review. *Water Sci Technol.* 2005;52:133–42.

30. Meddings DR, Ronald LA, Marion S, Pinera JF, Oppliger A. Cost effectiveness of a latrine revision programme in Kabul, Afghanistan. *Bull World Health Organ.* 2004;82:281–9.

31. Clasen TF, Bostoen K, Schmidt WP, et al. Interventions to improve disposal of human excreta for preventing diarrhoea. *Cochrane Database of Syst Rev.* 2010;6:1–32.

32. Aziz KMA, Hoque BA, Hasan KZ, et al. Reduction in diarrhoeal diseases in children in rural Bangladesh by environmental and behavioural modifications. *Trans R Soc Trop Med Hyg.* 1990;84:433–8.

33. Garrett V, Ogutu P, Mabonga P, et al. Diarrhoea prevention in a high-risk rural Kenyan population through point-of-use chlorination, safe water storage, sanitation, and rainwater harvesting. *Epidemiol Infect.* 2008;136:1463–71.

34. Hu X-R, Liu G, Liu S-P, et al. [Field evaluation of the effect of diarrhoea control of methanogenesis treatment of human and animal faeces and rubbish of cellulose nature in the rural areas of Xiang Cheng]. *Henan Yu Fang Yi Xue Za Zhi [Henan Journal of Preventative Medicine].* 1988:11–13.

35. Huttly SR, Blum D, Kirkwood BR, et al. The Imo State (Nigeria) drinking water supply and sanitation project, 2: impact on dracunculiasis, diarrhoea and nutritional status. *Trans R Soc Trop Med Hyg.* 1990;84:316–21.

36. McCabe LJ, Haines TW. Diarrheal disease control by improved human excreta disposal. *Public Health Rep.* 1957;72:921–8.

37. Messou E, Sangare SV, Josseran R, Le Corre C, Guelain J. Effect of hygiene and water sanitation and oral rehydration on diarrhea and mortality of children less than five years old in the south of Ivory Coast. *Bull Soc Pathol Exot.* 1997;90:44–7.

38. Rubenstein A, Boyle J, Odoroff CL, Kunitz SJ. Effect of improved sanitary facilities on infant diarrhea in a Hopi village. *Public Health Rep.* 1969;84:1093–7.

39. Wei X-N, Shao B-C, Fang Z-H, Gao M, Zhu X-L. [Evaluation of the effect of prevention of diarrhoea and ascariasis among students of health intervention measures in schools in rural areas. *Zhongguo Gong Gong Wei Sheng [China Journal of Public Health].* 1998;14:228.

40. Xu J-Z. [Observation on the efficacy of three squares septic tank lavatory for disease prevention]. *Huan Jing Yu Jian Kang Za Zhi [Journal of Environment and Health].* 1990:250–2.

41. Xu G-X, Zhu X-L. [The assessment of the effects for prevention of diseases by non-hazardous treatment of night soil at experimental spots in rural areas]. *Wei Sheng Yan Jiu [Journal of Hygiene Research].* 1994;23:23–7.

42. Yan Z-S, Wang G-F, Cui C, et al. [An observation of the effect on reducing the fly density and diarrhea of the use of double urn funnel lavatory in faeces management]. *Henan Yu Fang Yi Xue Za Zhi [Henan Journal of Preventive Medicine].* 1986:73–6.

43. Zhang W-P, Liu M-X, Yin W-H, et al. [Evaluation of a long-term effect on improving drinking water and lavatories in rural areas for prevention of diseases]. *Ji Bing Kong Ji Za Zhi [Chinese Journal of Disease Control & Prevention].* 2000;4:76–8.

44. Zhu X-L, Xia Q-Y, Meng Z-Y, et al. [Assessment of effects of disease prevention by intervention measures of school environmental hygiene in rural areas]. *Wei Sheng Yan Jiu [Journal of Hygiene Research].* 1997;26:378–80.

45. Curtis V, Cairncross S. Effect of washing hands with soap on diarrhea risk in the community: a systematic review. *Lancet Infect Dis.* 2003;3:275–81

46. Waddington H, Snilstveit B. Effectiveness and sustainability of water, sanitation, and hygiene interventions in combating diarrhea. *J Dev Effect.* 2009;1(3):295–335.

47. Fewtrell L, Kauffmann RB, Kay D, Enanoria W, Haller L, Colford Jr JM. Water, sanitation, and hygiene interventions to reduce diarrhea in less developed countries: a systematic review and meta-analysis. *Lancet Infect Dis.* 2005;5:42–52.

48. Clasen T, Schmidt WP, Rabie T, Roberts I, Cairncross S. Interventions to improve water quality for preventing diarrhoea: systematic review and meta-analysis. *Br Med J.* 2007;334(7597):782–5.

49. Arnold BF, Colford JM. Treating water with chlorine at point-of-use to improve water quality and reduce child diarrhea in developing countries: a systematic review and meta-analysis. *Am J Trop Med Hyg.* 2007;76(2):354–64.

50. Schmidt WP, Cairncross S. Household water treatment in poor populations: is there enough evidence for scaling up now? *Environ Sci Technol.* 2009;43(4):986–92.

51. Ejemot RI, Ehiri JE, Meremikwu MM, Critchley JA. Cochrane review: Hand washing for preventing diarrhoea. *Evid Based Child Health.* 2009;4(2):893–939.

52. Aiello AE, Coulborn RM, Perez V, Larson EL. Effect of hand hygiene on infectious disease risk in the community setting: a meta-analysis. *Am J Public Health*. 2008;98(8):1372–81.

53. Cairncross S, Hunt C, Boisson S, Bostoen K, Curtis V, Fung ICH, et al. Water, sanitation, and hygiene for the prevention of diarrhoea. *Int J Epidemiol*. 2010;39(Suppl 1):i193–205.

54. Ejemot-Nwadiaro RI, Ehiri JE, Arikpo D, Meremikwu MM, Critchley JA. Hand washing promotion for preventing diarrhoea. *Cochrane Database Syst Rev*. 2015;(9):CD004265.

55. Habicht J-P, DeVanzo J, Butz WP. Mother's milk and sewage: their interactive effects on infant mortality. *Pediatr* 1988;81:456–61.

56. Dai D, Walker W. Protective nutrients and bacterial colonization of the immature human gut. *Advances in Pediatrics* 1999;46:353–82.

57. Quigley MA, Kelly YJ, Sacker A. Breastfeeding and hospitalization for diarrheal and respiratory infection in the United Kingdom Millennium Cohort Study. *Pediatrics*. 2007;119(4):e837–42.

58. Lamberti LM, Walker CLF, Noiman A., Victoria C., Black RE (2011) Breastfeeding and the risk for diarrhea morbidity and mortality *BMC Public Health*. 2011;11(3):1.

59. World Health Organization. Millenium Development Goals (MDGs). Available at: https://www.who.int/news-room/fact-sheets/detail/millenium-development-goals-mdgs.

60. Ritchie, Roser, Mispy, Oritiz-Ospina. Measuring progress toward Sustainable Development Goals. *SDG-Tracker.org. 2018.* https://ourworldindata.org/sdgs

61. Roser M, Ritchie H, Dadonaite B. Child and infant mortality. *OurWorldInData.org.* 2019. Available at: https://ourworldindata.org/child-mortality.

62. United Nations General Assembly. Transforming our world: The 2030 Agenda for Sustainable Development, Resolution 70/1. 2015. Available at https://sdgs.un.org/2030agenda

63. Sachs J. From Millennium Development Goals to Sustainable Development Goals, *Lancet*. 2012;379: 2206–11.

64. WHO, UNICEF, JMP. https://data.unicef.org/resources/jmp-report-2023.

65. Brauer M, Zhao J, Bennitt F, Stanaway J. Global access to handwashing: Implications for COVID-19 control in low-income countries. *EHP.* 2020;128:5. https://doi.org/10.1289/EHP7200.

66. World Health Organization. *Global Costs and Benefits of Drinking-Water Supply and Sanitation Interventions to Reach the MDG Target and Universal Coverage*. Geneva: WHO/HSE/WSH/12.01; 2012.

67. Hutton G. Global costs and benefits of reaching universal coverage of sanitation and drinking-water supply. *J. Water Health*. 2013;11(1):1–12.

68. Prüss-Ustün A, Wolf J, Corvalán C, Neville T, Bos R, Neira M. Diseases due to unhealthy environments: an updated estimate of the global burden of disease attributable to environmental determinants of health. *J Public Health (Oxf)*. 2017 Sep 1;39(3):464–475. doi:10.1093/pubmed/fdw085. PMID: 27621336; PMCID: PMC5939845.

69. Prüss-Üstün A, Bos R, Gore F, Bartram J. *Safer Water, Better Health: Costs, Benefits, and Sustainability of Interventions to Protect and Promote Health*. World Health Organization, Geneva; 2008.

70. Workman CL, Cairnes MR, de los Reyes III FL, Verbyla ME. Global water, sanitation, and hygiene approaches: anthropological contributions and future directions for engineering. *Environ Eng Sci*. 2021;38(5):402–17.

71. Ross I, Esteves Mills J, Slaymaker T, et al. Costs of hand hygiene for all I household settings: estimating the price tag for the 46 least developed countries. *BMJ Global Health*. 2021;6(12):e007361. doi:10.1136/bmjgh-2021-007361.

72. Black R. et al. *Disease Control Priorities*. 3rd ed. Vol. 2. Reproductive, Maternal, Newborn, and Child Health. Washington DC: World Bank; 2016:319–34.

73. Laxminarayan R, Chow J, Shahid-Salles SA. Intervention cost-effectiveness: overview of main messages. In: Jamison DT, Breman JG, Measham AR, et al., eds., *Disease Control Priorities in Developing Countries*. 2nd ed. Washington DC: World Bank; 2006:35–86.

30

Hazardous Waste and Toxic Hotspots

Richard Fuller

Concern about the potential effects on human health of exposures to hazardous waste sites first arose in the 1970s, in the United States. The initial catalysts were well-publicized episodes of community exposure to discarded toxic chemicals at the Love Canal in Niagara Falls, New York, and in Times Beach, Missouri. The chemicals involved in these sentinel episodes included polychlorinated biphenyls (PCBs) and dioxins.

Further understanding of the adverse effects of abandoned toxic chemicals coupled with the recognition that waste sites were widespread across the United States led to passage in 1980 of the Comprehensive Environmental Response, Compensation, and Liability Act (CERCLA), commonly referred to as the Superfund Act.[1] Individual US states often developed analogous legislation to address smaller sites, such as leaks at dry cleaning shops and gasoline stations. The US Department of Defense undertook the clean up of its properties, such as the massively contaminated Rocky Mountain Arsenal in Colorado. Other industrialized countries, notably the European nations, have also developed waste site remediation programs. As a result, tens of thousands of contaminated sites in the United States, Western Europe, and Japan have been assessed and remediated over the past three decades (Figure 30.1). Waste site evaluation in the industrialized countries remains an ongoing and continuously improving process, with discoveries of newly emerging contaminants, development of new analytical methods, and a focus on ever smaller sites.[2]

By contrast, in many low- and middle-income countries (LMICs), the prevalence, geographic distribution, and impacts on human health of hazardous waste sites are only beginning to be documented.[3,4]

Quantification of the disease burden associated with hazardous waste sites is essential for public health planning and for prioritization of remediation efforts. Burden of disease estimates are typically expressed in disability-adjusted life years (DALYs). A DALY is a measure of the severity and duration of a given disease as well as of any lost years of life resulting from premature death.[5] DALY-based estimates can be used to rank hazardous waste sites and compare the health impacts of hazardous waste sites with those due to other hazards.

Prior calculations of the burden of disease resulting from toxic exposures have not been able to include estimates of the burden of disease associated with hazardous waste sites due to the absence of data on their health impacts. Thus, in 2001, Fewtrell et al. estimated that lead causes nearly 1% of the global burden of disease (GBD).[6] In 2011, Prüss-Üstün et al. calculated that exposure to a variety of chemicals, including lead, secondhand smoke, asbestos, and arsenic, accounts for 5.7% of total global DALYs and 8.3% of the total number of global deaths.[7] Because of insufficient data, neither of these

Richard Fuller, *Hazardous Waste and Toxic Hotspots* In: *Textbook of Children's Environmental Health*,
Second Edition. Edited by: Ruth A. Etzel and Philip J. Landrigan, Oxford University Press.
© Oxford University Press 2024. DOI: 10.1093/oso/9780197662526.003.0030

Figure 30.1. Barrels of toxic waste, hazardous waste site, New Jersey, USA

important studies was able to include data on death and disease resulting from exposures to hazardous waste sites.

Sources of Exposure

Toxic hotspots and hazardous waste sources in the LMICs fall into categories that are common from country to country and similar across similar industries. Dumpsites leaching chemicals into groundwater can be found in most large urban areas in LMICs. Hazardous waste dumped illegally on riverbanks or in other open areas contaminate local neighborhoods. Acute lead poisoning resulting from informal car battery recycling has been shown to kill quickly at high levels, and, at lower levels, lead has debilitating effects on neurological development, resulting in lives made even harder in some of the poorest countries in world.[8] Factories making chemicals for tanneries and other industries can release large amounts of hexavalent chromium waste. Artisanal gold mining often results in mercury-contaminated waste sites. Metal smelters and heavy industrial complexes can generate toxic hotspots if improperly managed. An emerging source of toxic exposures in LMICs has been electronic waste, or e-waste, which can release a variety of neurotoxicants, including lead, mercury, and PCBs.[4] Children, even small children, often are employed in e-waste sites scavenging precious metals from other electronic components (Figure 30.2).

Lead, mercury, cadmium, and chromium are among the most pervasive and persistent toxicants affecting children in LMICs. Sources of exposure to these toxicants and

Figure 30.2. Informal recycling in an e-waste site

to other hazardous wastes that are particularly hazardous to children include abandoned factories, mine tailings, artisanal mining activities, dumpsites, smelters, and lead-acid battery recycling activities.

Other chemicals of concern include arsenic (found naturally in bore well water in certain parts of the world), asbestos, hazardous pesticides, per- and polyfluoroalkyl substances (PFAS, a group of chemicals that do not break down in the environment), brominated flame retardants (BFRs,) and PCBs.

Urban Dump Sites

Across the world, each person on average produces between 0.2 and 1 kilogram of solid waste (garbage) per day. Industrial and construction activities produce additional, even larger quantities of waste. In developing countries LMICs, much of this solid waste ends up in large open piles, which fester under rain and sun. Major cities in LMICs such as Lagos, Manila, and Dhaka, all with populations above 10 million, produce 5,000–7,000 tons per day of solid waste. Between 30% and 60% of this waste is ultimately deposited in a recognized "dumping place." It is reliably estimated that a city of 1 million people—of which there are about 500 worldwide—produces approximately 500 tons of solid waste per day, every day, not including industrial waste.[8] As communities grow larger, denser, and economically better off, more goods are consumed and more material is thrown away.

In every city, there are groups of people who will take waste from households, either for a small fee or for the value of what can be scavenged. However, much of the waste is dumped in drainage ditches or on any patch of wasteland. Eventually, by a process of natural selection, large dumps appear and become semi-formalized.

These dumps can be several hectares in size and 20 meters or more high. They typically burn and smoke (due to the methane and other gases released as the material rots). They seep large quantities of corrosive liquid ("leachate"), which pollutes both ground and surface water. They are generally noxious and pose health hazards to local residents, contaminate the atmosphere, support vermin, and pollute local waterways.

The most immediate impacts of these urban dump sites are on the "ragpickers" who live on or beside the dumps.[9] These are often women or children who have migrated into cities from impoverished rural areas. People living on a large dump may number in the hundreds, while people involved with the "informal recycling" industry in a large city, including those who collect from homes or pick up along streets, may number in the tens of thousands. These people are directly exposed to the general filth of urban dumps, including medical waste and human excreta, as well as sharp edges and sometimes-toxic materials.

Most dumps are located in or close to densely populated areas, often slums, since immigrants also seek the uncontrolled wastelands as a place to camp and eventually build shanties. These people, who can number in the many thousands, are exposed daily to the dust, fumes, and smoke from the dumps. Fumes and smoke, notably from burning plastics and tires, present significant pulmonary risks. In addition, the dumps often present direct toxic chemical exposure risks to the ragpickers and area residents. It is common for materials such as paints, cleaning solvents, acids, spoiled pharmaceuticals, and medical waste to be dumped—basically any chemical used by small shops and factories that is no longer usable. These materials also leach into groundwater or drain to nearby streams and ponds, with the result that chemical concentrations in well or surface waters used for potable or washing purposes is highly contaminated. In the worst cases, dumps also pose a direct physical threat to nearby residents. Pockets of gas can explode and injure people, and the slow but relentless expansion of dumps can physically displace people. In an infamous case in Manila, a landslide of a 15-meter-high dump is estimated to have buried and killed about 200 people in the adjoining shanty town.[10]

Health problems reported commonly among ragpickers include a high prevalence of respiratory illness in children, impaired lung function in adults, headache, diarrhea, skin diseases, back pain, cuts and punctures from sharp materials, poor nutritional status, and infection by intestinal protozoa and helminths, which are particularly harmful in young children. The health problems encountered in people living around urban dumps include respiratory problems from smoke inhalation and chronic diarrhea from drinking and growing crops in contaminated water.

E-Waste Recycling Sites

E-waste refers to electronic and electrical goods that are past their useable life span. The rate at which new technology is released and older items become obsolete is increasing, and therefore so is the amount of e-waste produced.

E-waste comprises used computers (circuit boards, power supplies, plastic casings); cathode ray tubes (CRTs), which have large quantities of leaded glass; flat screens (which may contain mercury); used phones (trace amounts of various metals); scrap wire (with PVC or other plastic coatings); and various other components. Although a great deal of electronic equipment is recycled in large and safer facilities in the United States and

Figure 30.3. Burning plastic wires in Agbogbloshie, Accra, Ghana

Europe, some of this waste stream, especially older and less valuable equipment, ends up in Asia and Africa in poorly managed often informal recycling facilities.

Activities at recycling sites pose threats of contamination to the soil and air, which can leach into waterways and food sources. This can have adverse effects on the health of e-waste workers (including children) and the surrounding communities.

E-waste products contain intricate blends of plastics and chemicals that, when not properly handled, can be harmful to people and the environment. In Ghana, e-waste is dismantled and recycled by hand, and harmful pollutants are introduced into the environment via water, air, and soil. Workers dismantling and burning e-waste to retrieve valuable metals and other materials are exposed to heavy metals, PAHs, and inorganic acids, which have the potential for long-term and serious health risks. Similar situations can be found in China, India, and throughout much of Africa.

Sources of contamination occur throughout the cycle of recovery. In the most basic recycling communities, wires are burned in the open, producing clouds of hazardous smoke. CRTs are broken in open spaces, and the lead and plastics within the glass contaminate surrounding soils. Efforts to leach valuable metals from circuit boards also contaminate local soils with a mix of acids and heavy metals (Figure 30.3).

These industries generate income for local communities, and simply banning these practices is ineffective. The most effective strategies involve moving the most contaminating segments of the industry to nonresidential areas and formalizing the sector, assisting local recyclers to improve their methods to cleaner technologies. Much work is needed to improve this situation.

Illegal Hazardous Dumps

Illegal dumpsites of industrial waste have also become a major problem in LMICs, perhaps even more common today than informal municipal waste dumps, although the two may often overlap. Illegal hazardous waste dumps are found in many industrial neighborhoods in LMICs. They are the result of poorly managed and regulated local industry and mining operations. Infrequently, but often visibly, hazardous waste imported from developed countries will be dumped illegally in developing countries in contravention of the Basel Convention, an international treaty designed to protect human health and the environment against the adverse effects of international transport of hazardous wastes.[11]

LMICs are often ill-equipped to handle hazardous waste. Some countries, including Mexico, India, and China, have put in place regulations and established facilities of varying degrees of complexity and capability. However, many countries do not yet have such systems. Often there is no approved facility in an entire country to accept hazard wastes, and, even if there is, due to cost or other factors, illegal dumping of hazardous waste is common.[12] Even where disposal facilities are available, contractors who collect hazardous waste from a factory or facility may actually dump in an open area and pocket the disposal fees. Tracing the waste to the owner of the materials can be difficult for environmental agencies, many of which lack resources and technical expertise in this area. Developing countries do not have the powers or capacity for enforcement throughout the ownership chain of hazardous waste.

Used Lead-Acid Battery Processing Sites

Lead-acid batteries are made of lead plates suspended in a sulfuric acid solution and contained in a plastic casing. The batteries can be recharged many times but the lead plates eventually deteriorate, causing the battery to lose its ability to hold stored energy for any period of time. At that point the battery is likely to be discarded. Such batteries are known as used lead-acid batteries (ULABs). They need to be treated as hazardous waste.

About three-quarters of the 6 million tons of lead used annually worldwide goes into the production of lead-acid batteries. These batteries are used in automobiles, in industry, and in a wide range of other applications. Their global production is increasing. Much of the demand for lead is met through the recycling of secondary materials[9] and in particular through recovery of lead from ULABs. This high level of recycling is very effective in reducing the volumes of lead dumped in the environment and in minimizing the need for mining more ore. But the health impacts on persons involved in recycling can be severe.

Recycled lead is a valuable commodity. Therefore, for many people in LMICs, the recovery of car and similar batteries (ULABs) can be a viable and profitable business. The market for reclaiming secondary lead has been growing, especially in LMICs. Many LMICs have entered the business of buying ULABs in bulk to recycle them for lead recovery. These ULABs are often shipped over long distances for recycling, sometimes from industrialized countries. But the most common source of ULABs in any country is from domestic consumption.[13] Many ULAB recycling facilities adhere to strong environmental performance standards, ensuring that releases of lead to the environment are minimized and workers protected. However, more than 40% of ULABs are not recycled safely, and instead are recycled in informal and highly toxic circumstances.

Figure 30.4. Used lead-acid battery recovery operation

Informal ULAB recycling occurs in almost every city in LMICs, and even in some countries in transition. These small-scale ULAB recycling and smelting operations are often located in densely populated urban areas with few (if any) pollution controls. In many cases, the operations result in contaminated waste being released into the local environment and ecosystems in significant quantities (Figure 30.4). The releases include dust, contaminated waste and soil, and lead-bearing acid discharges. This results in many exposure paths, including for children when playing on the waste and handling rocks or dirt containing lead, as well as by bringing objects covered with lead dust back into the home.[14]

Acute lead poisoning can occur when people are directly exposed to large amounts of lead through inhaling dust or fumes dispersed in air. However, chronic poisoning from absorbing low amounts of lead over long periods of time is a much more common and pervasive problem. Health risks include impaired physical growth, kidney damage, mental retardation, and, in extreme cases, convulsions, coma, and death.[15] Children are more susceptible to lead poisoning than adults and may suffer permanent neurological damage (see Chapter 31). Exposure of women during pregnancy can result in irreversible fetal brain damage.

Lead is not only an issue at contaminated sites. Newer research finds lead exposures to be very high from leaded pottery (the glaze is sometimes made with lead), lead in aluminum cookware, lead in paint, and lead added to foodstuffs to make them more colorful. In many countries, lead is added to turmeric to improve its yellow appearance. These exposure routes mean than large percentages of a country can be exposed to lead at very high levels, not only those living around poorly run battery recycling locations.[15,16]

Tannery Sites

The leather industry uses trivalent chromium compounds in the tanning of animal hides and skins to preserve the leather and produce a tough, supple texture that is resistant to biodegradation and ready to be dyed. The methods employed by tanneries often produce large amounts of residues that can be harmful to the environment, including chromium wastes, hair, salts, and fleshing residues. In small-scale or uncontrolled tanneries, these residues and the wastewater from the processes are sometimes dumped into the surrounding area. In the environment, the chromium can convert into the more toxic hexavalent form. Tanneries are often located in clusters, and the cumulative impact on the environment of these tannery clusters can be very serious.

Artisanal Mining Sites

"Artisanal" mining and smelting is defined as small-scale mining and smelting operations that use rudimentary methods to extract and process minerals and metals. Artisanal mining and smelting sites are often located in rural areas in LMICs and even in family homes. Artisanal miners frequently use toxic materials, notably mercury, in their efforts to recover metals and gems. Such miners work in difficult and often very hazardous conditions, and the crude methods they use can result in release of toxic materials into the environment, posing great health risks to the miners, their families, and surrounding communities.

Gold mining operations are particularly dangerous as they often use mercury amalgam to extract gold from ore.[17] A key step in gold ore processing is to grind the ore to a powder, then wash it in water through sluices. In the sluices, the heavier particles of gold settle out, and the lighter sand and other particles are washed away into surrounding waters. Mercury is often added to the ground gold ore during sluicing to amalgamate it. Much of this mercury is washed out of the factories into streams and rivers in finely particulate form to cause widespread contamination. In rivers and streams, microorganisms can convert the metallic mercury to methylmercury, which is environmentally persistent, bioaccumulative, and highly neurotoxic. Exposure of pregnant women to methylmercury through consumption of contaminated, locally caught fish can lead to fetal brain injury (see Chapter 32).

Artisanal gold mining is estimated to produce at least a quarter of the world's total gold supply[18] and is one of the most significant sources of mercury release into the global environment. Either before or after sluicing, artisanal gold miners combine mercury with gold-carrying silt to capture the gold and form it into a hardened amalgam. This hardened amalgam is later heated with blowtorches or over an open flame to evaporate the mercury, leaving small pieces of gold. The gaseous mercury given off during the burning is often inhaled by the miners and their immediate families, including their children, or it is released into the environment. Because mercury condenses easily in the atmosphere, it often settles in nearby areas, thus resulting in additional contamination of waterways. Mercury amalgamation results in more than 30% of the world's anthropogenic releases of mercury, affecting an estimated 15 million or more miners, including 4.5 million women and 600,000 children. Children exposed to mercury are at particularly high risk for developmental problems. Exposure to mercury can cause kidney problems, arthritis, memory loss, miscarriages, psychotic reactions, respiratory failure, neurological damage, and even death.

Abandoned Factories

Abandoned factories are now recognized as a major source of toxic pollution throughout the globe. As industrial plants become inefficient, newer factories are built or companies close their doors. Many older factories never properly dispose of abandoned hazardous materials. Chemicals left behind can include unusable solvents, chemical sludges, pesticides, PCBs, asbestos, acids, and even munitions. These materials pose problems not only for disposal, but they also leach into soil and groundwater and evaporate into the air. Fire is another risk at abandoned sites that store significant amounts of flammable chemicals. Fires may release large volumes of hazardous smoke to the atmosphere and then into nearby communities. Clean-up of these sites poses a particularly difficult task because no party will take responsibility for the damage. The polluters may be unknown, bankrupt, or defunct. Sometimes, the sites become ensnared in legal battles, and bureaucratic debates rage over who is responsible while all the while the toxic pollution continues.

Smelters

Metal processing plants and smelters refine ores that contain metals such as copper, nickel, lead, zinc, silver, cobalt, gold, and cadmium. Their functions are to separate metal from ore, usually by heating the ore in large furnaces at high temperatures, and to create refined, highly pure versions of metals for use in manufacturing. The smelting process involves heating the ore with a reducing agent such as coke or charcoal to reduce oxides and metal compounds to pure metals and adding fluxes to separate impurities. Sulfuric acid and other acids are often created.

Metallurgical and smelting processes can be highly polluting. They often emit large quantities of airborne pollutants containing lead, cadmium, zinc, and arsenic as well as hydrogen fluoride, sulfur dioxide, oxides of nitrogen, and large quantities of airborne particulates.[19] Unless well controlled with scrubbers or fabric filter "baghouses," smelters can produce high levels of air pollution and contribute to the formation of acid rain.

Although smelting takes place around the world, approximately 60% of the world's smelters are located today in LMICs. These countries often have less stringent pollution control requirements than industrialized nations and poorer enforcement. Too frequently, the pollution control equipment is either nonexistent or poorly maintained. The consequence is extensive environmental contamination.[20] Many companies in LMICs, particularly local or national companies (as opposed to well financed multinationals) do not have the resources to upgrade old smelters or to replace them with cleaner technologies. Highly populated communities develop near smelters because they offer employment opportunities. The consequence is that entire communities, including young children and pregnant women, who live near smelters are exposed to high levels of toxic air pollution. These communities can be heavily impacted by the metal dust and the acidity. Also the land can be contaminated with heavy metals, which in some circumstances reach human food supplies. Similar situations prevailed around smelters in the United States until only a generation ago.[21]

The Way Forward

Industrial wastes, airborne emissions, and legacy pollution from hazardous waste sites affect millions of people around the world. Tens of thousands are poisoned and killed each year. Many more suffer reduced intelligence, disrupted behavior, damaged immune systems, and long-term health problems. Women and children are especially at risk.

Much of the damage to human health and the environment that is caused by toxic pollution from hazardous waste sites can be affordably and effectively averted. There exist culturally and economically responsible interventions that have been documented to save lives. Many of these interventions have been developed at the local level with input from technical experts. Others adapt more complex technologies to be more appropriate for developing country environments.

Despite the efficacy and cost-effectiveness of pollution controls and environmental remediation, the international response to toxic pollution from hazardous waste sites is still in its infancy. There is very little understanding of the extent of the contamination or of the severity of the effects on human health, especially on the health of children. The Global Alliance on Health and Pollution (GAHP) and the United Nations Environment Programme (UNEP) have some involvement in the remediation of toxic pollution, as does the World Bank. To date only a tiny fraction of international aid, however, is allocated to remediation of critical sites or toxic pollution in general despite the significant threat posed by pollution and the known efficacy of interventions.

Pure Earth (formerly known as Blacksmith Institute), a global nonprofit with expertise in the remediation of contaminated waste sites in LMICs, has developed a methodology and database for rapid identification and assessment of toxic hotspots around the world. The Pure Earth database assesses and tabulates toxic sites by their principal pollutants, determines credible pathways of human exposure, and estimates the size of the population at risk at each site.[22] Pure Earth's database of global toxic hotspots represents the first investigation of the global scope of hazardous waste pollution (www.contaminatedsites.org). Pure Earth's work indicates that well over 100 million people worldwide are currently affected by toxic pollution, and this number is probably conservative. There is much work to be done to correct this situation.

References

1. Johnson BL, DeRosa CT. The toxicologic hazard of Superfund hazardous waste sites. *Rev Environ Health*. 1997;12:235–51.

2. Ela WP, Sedlak DL, Barlaz MA, et al. Toward identifying the next generation of superfund and hazardous waste site contaminants. *Environ Health Perspect*. 2011;119:6–10.

3. Yáñez L, Ortiz D, Calderón J, et al. Overview of human health and chemical mixtures: problems facing developing countries. *Environ Health Perspect*. 2002;110:901–9.

4. Chen A, Dietrich KN, Huo X, Ho SM. Developmental neurotoxicants in e-waste: an emerging health concern. *Environ Health Perspect*. 2011;119:431–8.

5. World Health Organization. Disability-adjusted life years. 2023. Available at: https://www.who.int/data/gho/indicator-metadata-registry/imr-details/158.

6. Fewtrell LJ, Prüss-Ustün A, Landrigan P, Ayuso-Mateos JL. Estimating the global burden of disease of mild mental retardation and cardiovascular diseases from environmental lead exposure. *Environ Res.* 2004;94:120–33.

7. Prüss-Ustün A, Vickers C, Haefliger P, Bertollini R. Knowns and unknowns on burden of disease due to chemicals: a systematic review. *Environ Health.* 2011;10:9.

8. Kong V, Ericson B, Hanrahan D, et al. The world's worst pollution problems: the top ten of the toxic twenty. 2016. Available at: http://www.worstpolluted.org/.

9. da Silva, Fassa A, Siqueira C, Kriebel D. World at work: Brazilian ragpickers. *Occup Environ Med.* 2005;62:736–40.

10. Mydans S. Before Manila's garbage hill collapsed: living off scavenging. *New York Times.* 2000 Jul 18. Available at: http://query.nytimes.com/gst/fullpage.html?res=9803E2DA133 BF93BA25754C0A9669C8B63&sec=&spon=&pagewanted=all.

11. The Basel Convention. New York: United Nations. 2011. Available at: http://www.basel.int/.

12. Clapp J. The toxic waste trade with less-industrialised countries: economic linkages and political alliances. *Third World Q.* 1994;15:505–18.

13. Commercial Diplomacy. The Basel Ban and batteries, a teaching case. n.d. Available at: http://www.commercialdiplomacy.org/case_study/case_batteries.htm.

14. Initiative for Responsible Battery Recycling. Pure Earth, n.d. Available at: https://www.pureearth.org/lead-poisoning-and-car-batteries-initiative/

15. United Nations. New Basel guidelines to improve recycling of old batteries. United Nations Environment Programme. 2002. Available at: http://www.unep.org/Documents.Multiling ual/Default.asp?DocumentID=248&ArticleID=3069&l=en.

16. UNICEF. The toxic truth. 2020. Available at: https://www.unicef.org/reports/toxic-truth-childrens-exposure-to-lead-pollution-2020.

17. Hilson G, Hilson CJ, Pardie S. Improving awareness of mercury pollution in small-scale gold mining communities: challenges and ways forward in rural Ghana. *Environ Res.* 2007;103:275–87.

18. Veiga MM. Pilot project for the reduction of mercury contamination resulting from artisanal gold mining fields in the Manica district of Mozambique. 2005. https://archive.iwle arn.net/globalmercuryproject.org/countries/mozambique/docs/Moz_Final_Report_Aug_2005.pdf

19. Blacksmith Institute. Top 10 worst pollution problems. 2008. Available at: http://www.worstpolluted.org/projects_reports/display/61.

20. Yanez L, Ortiz D, Calderon J, et al. Overview of human health and chemical mixtures: problems facing developing countries. *Environ Health Perspect.* 2002;110:901–9.

21. Landrigan PJ, Gehlbach SH, Rosenblum BF, et al. Epidemic lead absorption near an ore smelter: the role of particulate lead. *N Engl J Med.* 1975;292:123–9.

22. Ericson B, Caravanos J, Chatham-Stephens K, Landrigan P. *Approaches to Systematic Assessment of Environmental Exposures Posed at Hazardous Waste Sites in the Developing World: The Global Inventory Project.* New York: Blacksmith Institute; 2012.

31

Lead

Nicholas C. Newman and Bruce P. Lanphear

Lead was used in a large variety of industrial applications over the past century. Lockhart Gibson, an ophthalmologist in Queensland, Australia, described the first cases of childhood lead poisoning more than a century ago in 1904.[1] A group of children presented to hospital with a constellation of unexplained symptoms consistent with lead poisoning: abdominal cramps (lead colic), persistent vomiting, headache, and visual disturbance without a fever. Gibson showed that lead-based paint was the primary source of their exposure by documenting high levels of lead in wipe samples collected from the children's homes. Lead-based paint had been introduced to consumer markets in Australia only a few years earlier. Efforts to prevent lead poisoning by discouraging oral behaviors alone were not sufficient to prevent lead poisoning. Approximately 4 years later Gibson's colleague, Dr. A. J. Turner, concluded, "Prevention is easy. Paint containing lead should never be employed . . . where children, especially young children, are accustomed to play."[2] Turner argued that "legislative interference" was necessary to protect children against lead-contaminated paint.

Over the next three decades, some countries banned the use of lead-based paints. But others, the United States among them, were swayed by the marketing strategies of the lead industry and failed to ban lead paint. As result, the use of lead paint spread worldwide. Lead was also added to gasoline and other consumer products. Annual consumption of lead in gasoline for automobiles in the United States peaked at 270,000 tons in the 1970s and approximately 375,000 tons in other countries around the world.[3] A global pandemic of childhood lead poisoning followed.

In 1979, Herbert Needleman and his colleagues found in a landmark study that children with higher dentine lead concentrations in deciduous teeth had significantly lower IQ scores and were more likely to be rated unfavorably by their teachers on the dimensions of distractibility, persistence in work, organizational ability, dependence, and impulsivity.[4] The level of lead in dentine in deciduous teeth reflects a child's cumulative absorption of lead over the first 5–7 years of life because lead is a bone-seeking element with long residence time in skeletal tissues such as dentine.

In an 11-year follow-up study of these same children, Needleman and others reported that higher dentine lead levels were associated with lower reading scores, lower class rank, and increased absenteeism in adolescence. Children in the highest dentine lead group were five times more likely to have a reading disability and seven times less likely to graduate from high school than children in the lowest dentine lead group.[5] Needleman, whose work was instrumental in shifting the focus of research, regulation, and public health practice from overt toxicity to subclinical toxicity, is one of the seminal leaders in pediatric environmental health (Figure 31.1). His work was central to understanding

Nicholas C. Newman and Bruce P. Lanphear, *Lead* In: *Textbook of Children's Environmental Health*, Second Edition. Edited by: Ruth A. Etzel and Philip J. Landrigan, Oxford University Press.
© Oxford University Press 2024. DOI: 10.1093/oso/9780197662526.003.0031

Figure 31.1. Herbert Needleman, MD, a pediatrician and child psychiatrist, conducted seminal research over four decades that shifted the focus of research and public health activities from overt lead poisoning and treatment to the prevention of subclinical lead toxicity
Source: Heinz Foundation

that lead can cause chronic, low-level, "subclinical" toxicity at blood levels too low to trigger the overt symptoms of acute lead poisoning.

Beginning in the 1980s, regulations to reduce lead exposure from gasoline, paints, canned foods, and other consumer products were gradually enacted in countries around the world, and, over the subsequent decades, blood lead levels plummeted (see Chapter 1). Concurrently, the problem of lead toxicity in children shifted from overt lead poisoning to low-level, subclinical, or "silent" toxicity.

Extensive evidence now indicates that lead-associated cognitive deficits and psychopathology occur at blood lead levels well below 5 µg/dL. In 2012, the US National Toxicology Program concluded that blood lead concentrations of less than 5 µg/dL are associated with intellectual deficits (lower IQ score), academic disabilities, attention-related behaviors, and problem behaviors[6] (Table 31.1).

An additional striking finding was that the decrements in IQ associated with lead were proportionately greater at the lowest blood lead levels. The IQ decrement associated with an increase in blood lead concentration from less than 1 µg/dL to 30 µg/dL was 9.2 IQ points, but the decrement associated with an increase in blood lead concentration from less than 1 µg/dL to 10 µg/dL was 6 IQ points.[7] No apparent threshold exists for lead-associated decrements in intellectual abilities and academic abilities. Both the World Health Organization (WHO) and the US Centers for Disease Control and Prevention

Table 31.1 Evidence linking various health effects to low-level blood lead
concentrations in children, national toxicology program

<5 µg/dL	Sufficient	Decreased academic achievement Deficits in intellectual abilities (IQ scores) Attention-related behavior problems Antisocial behaviors
	Limited	Delayed puberty Decreased kidney function in children >12 years
<10 µg/dL	Sufficient	Delayed puberty Reduced postnatal growth Decreased hearing
	Limited	Hypersensitivity by skin prick test
	Inadequate	Asthma and eczema Cardiovascular effects Kidney function <12 years

Source: Adapted from the National Toxicology Monograph on Health Effects of Low-level
Lead, National Institute of Environmental Health Sciences, Department of Health and Human
Services, 2012.

(CDC) affirmed that there is no safe level of lead in children.[8] The CDC adopted a lead
"reference level" of the 97.5th percentile for blood lead concentration in the National
Health and Nutrition Examination Survey (NHANES), 5 µg/dL in 2012, and, in 2021,
the reference level in the United States was reduced to 3.5 µg/dL.[9,10]

Epidemiology of Lead Poisoning

Widespread lead exposure was the result of its use in both industrial and nonindustrial
applications (Table 31.2). In the United States, the American Academy of Pediatrics
reports that the major sources of childhood lead exposure are lead-containing dust
(40%) and drinking water (20%), and between 10% and 20% each from soil ingestion,
renovation, and soil lead.[11] Lead-based paints, which are still used in many countries,
and legacy contamination from leaded gasoline are both important sources. A systematic review of studies examining sources of lead exposure in low- and middle-income
countries found that the major sources of lead were informal lead acid battery recycling and manufacture, metal mining and processing, electronic waste, and use of lead
as a food adulterant, especially spices.[12] A limitation to this systematic review was that,
unlike the United States where lead testing of children and case investigation is routine, population-level lead exposure data are not available in many countries.

Over the past three decades, blood lead concentrations among children in high-
income countries have declined dramatically.[13-15] From 1976 to 1980, blood lead concentrations among US children declined much more sharply than anticipated following
the phaseout of leaded gasoline.[16] Prior to 1977, the allowable level of lead in gasoline
was 0.78 g/L in the United States; in 1977, it was lowered to 0.026 g/L. Similar reductions
in blood lead levels were reported following the phase-out of leaded gasoline in other
countries.[13,15,17,18] In 2000, about 100 countries were still using leaded gasoline. By 2011,
only six countries in the world continued to use leaded gasoline (Figure 31.2). In 2021,
Algeria became the last country to phase out leaded gasoline for automobiles.[19]

Table 31.2 Current and historical uses of lead

Storage batteries (automobile)	Paint (house, bridge, industrial)	Traditional Medications
Electronics (components and solder)	Rubber materials	Pesticides
Cosmetics	Munitions	Printing presses
Art paints and supplies (including crayons)	Pottery glaze	Plumbing (solder, fixtures, service lines)
Gasoline additive	Stabilizer for plastics	Toys painted with lead paint

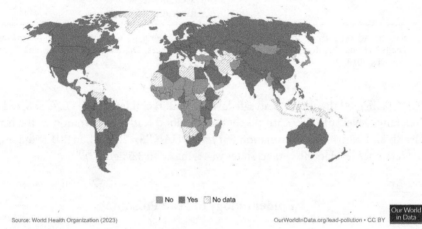

Lead paint is **unregulated** in over 71 countries

■ No ■ Yes ▨ No data

Source: World Health Organization (2023)   OurWorldInData.org/lead-pollution · CC BY   Our World in Data

Figure 31.2. More than 70 countries in the world do not regulate lead content in paint. Many LMICs make up this list. This lack of regulation of lead paint may push these countries unnecessarily through the same "lead pandemic" that the U.S. and other countries faced in the 20th century.

Leaded gasoline is still used in aviation gasoline (avgas). Over half of current lead emissions in the United States are from leaded aviation gasoline.[20] In a cross-sectional study of proximity to airports and childhood blood lead levels, Zahran and colleagues found that, after adjusting for aircraft volumes, wind direction, season, industrial sources of lead, lead-contaminated housing, and population density (measure of previous leaded automotive gasoline use), children living less than 1 kilometer and 1–2 kilometers from an airport are 45% and 25%, respectively, more likely to exceed a blood lead level of 10 µg/dL than those living more than 4 kilometers away.[21] They also estimated that the cost to society per gallon of avgas was $10.

Other regulations contributed to the decline in blood lead concentrations among children in industrialized countries. In 1976, the US Consumer Product Safety Commission (CPSC) restricted the allowable level of lead in residential paint to 0.06% (600 parts per million (ppm)); in 2008, this was lowered to 90 ppm. Efforts to reduce lead contamination in water systems were widely implemented over the past three decades, and the use of lead solder in canned foods was banned. Finally, the number of older housing units

containing lead-based paint has decreased. It is difficult to quantify the blood lead decline attributable to specific sources, but the combined effect of these changes clearly led
to dramatic declines in children's blood lead concentrations (Figure 31.3).

Despite these reductions in children's blood lead concentration, the prevalence of
subclinical lead toxicity remains high in some populations, with the most severe persistent problems occurring among children from low-income households.[13]

In a national survey of housing conducted in 1998, Jacobs and his coworkers estimated
that more than 25 million (27%) of US housing units contain one or more lead hazards;
the prevalence was higher in older housing.[22] For housing built from 1978 to 1998, 1%
contained lead hazards, whereas the prevalence of residential hazards increased to 10%
in housing built between 1960 and 1978, 51% of housing built from 1940 to 1959, and
67% of housing units built before 1940. More than 30% of housing units with lead-based
paint contained one or more lead hazards even when the paint was in good condition.
Notably, these estimates of lead hazards were defined by levels of lead in paint, dust, and
soil that were expected to pose a risk for children having a blood lead concentration of
greater than 10 µg/dL; they are obviously too high to protect children from having blood
lead concentrations at the lower levels now known to be toxic.

Globally, the WHO reports that 41% of countries have legally binding controls on lead
paint (https://www.who.int/news-room/fact-sheets/detail/lead-poisoning-and-health).
The lack of surveillance for lead poisoning in many parts of the world impairs our ability
to determine the true extent of the problem and impedes efforts to determine and mitigate sources of lead exposure.

Pathophysiology of Lead

The adage that "children are not just little adults" is appropriate when thinking about
routes of exposure, absorption, and distribution of lead. Children younger than 6 years
have a higher internal dose of lead because they consume 2–3 times the amount of food
and water per kilogram bodyweight and absorb 3 times the amount of lead as compared
with adults.[23] The exact mechanisms for lead absorption are not known, but at least three
mechanisms have been proposed: paracellular mechanisms, diffusion, and active transport. One active transport mechanism, divalent metal transporter 1 (DMT1), is an H+
-coupled transport protein that transports non-heme iron, and a polymorphism in this
transporter is associated with increased blood lead levels.[24,25]

Neurobehavioral Toxicity

Low-level lead toxicity does not exhibit a specific "behavioral signature," but a consistent
pattern of lead-induced disorders has emerged. These disorders result from a variety of
environmental and hereditary factors, but lead toxicity clearly plays a role in their genesis.[26–30] Antisocial behaviors—such as conduct disorder, delinquency, and criminal
behaviors—are important elements of this pattern. Needleman and coworkers found that
adolescents with higher bone lead concentrations had higher scores for delinquency and
aggression.[26] Higher childhood blood lead levels or tooth lead levels have been associated
with higher rates of self-reported delinquent behaviors and arrests in two prospective,
longitudinal studies.[28–30] In the NHANES, Braun and others found a striking increase in

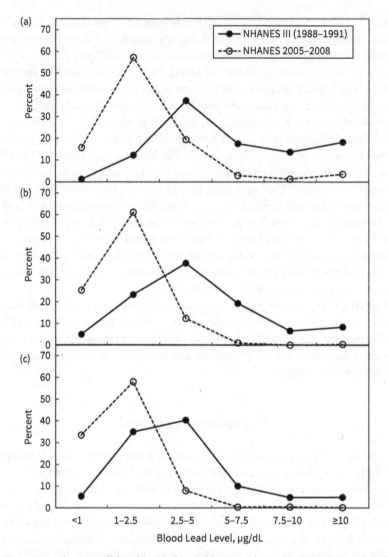

Figure 31.3. Distribution of blood levels for children aged 1 to 5 in NHANES III and NHANES 2005–2008, living in high-risk, medium-risk, and low-risk homes.

the risk of conduct disorder even at lower levels of lead exposure.[30] Finally, the decline in homicides and violent behaviors correlated with the decline in blood lead levels.[31]

Lead exposure is a risk factor for attention-related behavioral problems and for attention-deficit/hyperactivity disorder (ADHD), including poorer measures of attention and hyperactivity.[32,33] Moreover, the joint effects of exposures to lead and tobacco were synergistic. Prenatal tobacco exposure and childhood lead exposure have both been associated with a two-fold increased risk for ADHD, but Froehlich and her colleagues found that children who had higher childhood blood lead concentrations and were exposed to tobacco in utero were eight times more likely than their peers to meet criteria for ADHD[32] (see Chapter 49).

Perhaps most telling of the combined neurodevelopmental impacts of lead on both the brain in childhood and through adulthood is explained through a combination of neuroimaging and longitudinal cohort studies. Cecil and others found that the brain volume was diminished in more heavily in young adults in the Cincinnati Longitudinal Study.[34] Marshall and colleagues carried out a cross-sectional study of 9,712 US children ages 9–10 years examining the combined effect of socioeconomic factors and lead exposure (as measured by census tract-level lead risk) on cognitive test scores and brain structure. Low-income children with higher lead exposure had lower mean cortical surface area and scored more poorly on cognitive tests compared with higher-income children also at high risk for lead exposure.[35] In a longitudinal study from New Zealand (Dunedin Multidisciplinary Health & Development Study) of 565 participants, investigators studied the association between blood lead levels at age 11 years and IQ scores and socioeconomic status at age 38 years. They found a significant association between blood lead levels at age 11 and decreased IQ and decreased socioeconomic mobility at age 38.[36]

Renal Toxicity

Lead was first recognized as a cause of kidney failure in children more than 50 years ago among survivors of an epidemic of overt lead poisoning in Queensland, Australia. Subsequently, lead was found to be a risk factor for reduced kidney function and chronic kidney failure in adults. Low-level lead exposure has now been implicated in diminished kidney function in adolescents. Using the NHANES, Fadrowski and colleagues found, that among 769 adolescents with a median blood lead concentration of 1.5 μg/dL, a doubling of blood lead was associated with a significant reduction in the glomerular filtration rate.[37] Collectively, these and other studies indicate that chronic, low-level lead exposure is an important risk factor for chronic renal failure (see Chapter 53).

Reproductive Toxicity

Lead is a reproductive toxicant. It has been associated with spontaneous abortion and low birth weight. In a case-control study in Mexico City, the increase in the odds for spontaneous abortion rose by 1.8 for every 5 μg/dL increase in maternal blood lead concentrations.[38]

Early longitudinal studies that examined the association of prenatal lead exposure and low birth weight or preterm birth, measured using either maternal or cord blood lead concentrations, were inconsistent. In a study involving more than 34,000 live births, investigators found that blood lead concentrations of 5 μg/dL were associated with a 61 gram decrement in birth weight.[39] Pregnant women who had a blood lead level of greater than 5 μg/dL were almost twice as likely to give birth preterm.[40] Lead has also been associated with delayed onset of puberty in both girls and boys.[41,42]

Diagnosis and Treatment of Lead Poisoning in Children

A description of the diagnosis and treatment of lead poisoning is beyond the scope of this chapter, but guidelines are available from the American Academy of Pediatrics[43] and the CDC.

Sources of Lead Exposure

Heterogenous Sources in Different Countries and Regions

The elimination of lead-based paint and leaded gasoline has resulted in dramatic decreases in childhood blood concentrations in many countries.[13-15] The story for LMICs is quite different. Although there has been a phase-out of leaded gasoline, rapid industrialization may mitigate this effect somewhat.[18] The lack of population-level surveillance data has limited our ability to identify and mitigate sources of childhood lead exposure. A systematic review estimated that the average blood lead level in children in LMICs is much higher than in children in high-income countries. An estimated 632 million children in LMICs have blood lead levels of greater than 5 μg/dL; the major sources of lead were informal lead-acid battery recycling and manufacturing, metal mining and processing, electronic waste, and use of lead as an adulterant in spices.[12] In China, industrial emissions, automobile exhaust, contaminated soil/play areas, e-waste recycling, paint (half of new enamel paints have lead levels of >600 ppm), and traditional medicines are common sources. Children in India were exposed through paint (16.67% of samples had >5,000 ppm), floor dust, cosmetics, industrial sources (particularly battery manufacturing), and traditional medicines. In Nigeria, leaded gasoline, battery recycling, e-waste, paint (96% of samples >600 ppm, especially yellow), dust, well water, and traditional remedies are commonly cited sources of lead. In Mexico, lead-containing ceramic glazes are a common source of lead exposure, in addition to industrial sources.[44] See Table 31.3 for lead content limits in various products.

Lead-Based Paint

In the United States, paint is the most common, highly concentrated source of lead exposure for children who live in older housing. Paint that was used on both the interiors and exteriors of houses throughout the 1950s, and, to a lesser extent, through the 1970s, often contained high concentrations of lead.[45] Paint containing lead is still used in many parts of the world and should be banned immediately.

Lead-Contaminated House Dust

Lead-based paint is the major source of lead intake for most children, but lead-contaminated house dust and residential soil are often the major pathways.[46] In 1974, Sayre, Charney, and others showed that normal mouthing behaviors led to ingestion of lead-contaminated house dust.[47] The primary source of lead for most children is house dust contaminated by lead-based paint that is damaged or in disrepair. Scraping,

Table 31.3 Limits for lead in different items

Item	Limit	Authority
Infant formula	0.02 mg/kg	World Health Organization Codex Standard 193–1995
Salt	2 mg/kg	World Health Organization Codex Standard 193–1995
Natural source food additives	10 mg/kg	US FDA
Candy	0.1 mg/kg	US FDA
Other foods	0.5 mg/kg	US FDA
Shellfish	1.5 mg/kg	UK Foods Standards Agency https://www.food.gov.uk/business-guidance/chemical-contaminant-monitoring

sanding, or construction during painting, renovation, and abatement also increases lead contamination of house dust.

Lead-Contaminated Soil

Lead-contaminated soil is an important source of lead intake for children.[48–50] Children who live in former or active mining, milling, and smelter communities are at particular risk for ingestion of lead-contaminated soil. In a pooled analysis of 12 studies, children's blood lead concentration increased by an estimated 3.8 µg/dL for every 1,000 ppm increase in soil lead concentration.[50] Variations in the reported relationship of lead-contaminated soil and blood lead concentrations are due to several factors, including the age of children studied and adjustment for the contribution of lead intake from other sources.

Drinking Water

Water is an important source of lead intake for young children in many communities. The maximum contaminant level (MCL) set by the US Environmental Protection Agency (EPA) for lead in water is 15 µg/L (15 ppb).[51] The MCL is a level used as an administrative tool to evaluate community-level exposure; the EPA did not intend it to be used as a health-based standard. The maximum contaminant level goal, the value the EPA deems acceptable for health, is zero. The WHO has recommended a guideline value of 10 µg/L (10 ppb), but it is outdated.[52] Health Canada has set a guidance level of 5 ppb. A health action level for water should be established at less than 5 ppb of lead in the short term and less than 1 ppb in the longer-term.

Water is an important source of lead exposure for young children. Among infants who lived in high-risk neighborhoods of Washington, DC, the incidence of elevated blood lead concentration increased 2.4 times after the change in water disinfectant from chlorine to chloramine.[53] During an incident in Flint, Michigan, in which there was both a change in the municipal water supply combined with a disruption in water treatment practices, increases in blood lead levels were observed in children within Flint.[54,55] In a prospective study of 248 children followed from 6 to 24 months, children who were exposed to water lead of greater than 5 ppb had blood lead concentrations 1 μg/dL higher than children who had water lead levels of less than 5 ppb.[45] Replacing lead service lines and reducing the allowable amount of lead in brass fixtures is necessary to eliminate lead intake from water.

Other Sources of Lead Exposure

Other sources of lead intake are important for children in countries around the world, such as folk medicines, ceramics, and cosmetics (Table 31.4).[56] Lead brought into the home from a worksite by a parent can also be a major source of exposure.[57] Products such as crayons and window blinds that are contaminated with lead receive a great deal of attention. While these products constitute only a small source of lead intake for most children, they are nonetheless entirely unnecessary and should be eliminated. Furthermore, because lead exposure is cumulative and there is no apparent threshold for the adverse effects of lead exposure, all sources of lead exposure must be eliminated. Finally, by reducing the allowable levels of lead in consumer products, we will protect workers who produce lead-contaminated products, as well as their children.

Table 31.4 Common sources of lead exposure in children and mitigation strategies

Source	Primary prevention recommendations
Dust	Identify and control sources Maintain floor dust loadings <5 μg/ft^2
Water	Point of use filtration and/or replacement of lead-containing plumbing components Maintain water lead concentration <5 ppb
Soil	Identify and mitigate contaminated soil Maintain soil lead concentration <100 ppm
Paint/Renovation	Avoid use of lead-based paint; Identify and properly mitigate lead paint hazards Maintain floor dust loadings <5 μg/ft^2
Paraoccupational exposure	Remove work clothing and shower before leaving work
Folk remedies	Avoid items known to contain lead such as: Greta, Azarcon, many Ayurveda, Ghasard, Ba-baw-san, Saoott, Cebagin, Bint al dahab, and Santrinj
Cosmetics and ceremonial powders	Avoid items that commonly contain lead such as: Sindoor, Tiro, Kohl, and Surma
Spices and seasonings	Contamination of multiple items have been reported ranging from dried herbs to chili powders and turmeric. Avoid any items that are known to contain lead

Ingestion and Absorption of Lead

Ingestion and absorption of lead varies throughout the first 2 years of life. Children's blood lead levels rise rapidly between 6 and 12 months of age, peak between 18 months to 36 months, and then gradually decline. The peak in children's blood lead levels is primarily due to the confluence of normal mouthing behaviors and increasing mobility.[46] Lead-contaminated water and floor dust are sources of lead intake throughout early childhood, but lead-contaminated dust on windowsills is not a major source of intake until the second year of life, when children stand upright.[46] Soil ingestion, as reported by parents, peaks between 12 and 18 months and diminishes thereafter. Younger children absorb lead more efficiently than older children and adults. Other factors, like iron status and a genetic predisposition to absorb calcium and other metals, also can impact children's lead absorption.

Residential Standards for Lead in Paint and Dust

Under section 403 of Title X, the US Congress mandated the EPA to promulgate residential health-based lead standards.[58] Standards are necessary to identify lead hazards before a child is unduly exposed and for managing children who have excessive lead exposure.[59] Attempts to reduce lead exposure—such as abatement or renovation—can result in increased contamination and elevation in a child's blood lead concentration.[60,61] Clearance dust tests—wipe samples to quantify the dust lead loadings in the housing unit to make sure a lead hazard doesn't exist—should therefore be conducted after remodeling, renovation, or abatement to protect children from lead hazards. Finally, standards serve as a benchmark to compare the effectiveness and duration of various lead hazard controls.

Most existing lead standards are inadequate to protect children from lead toxicity. In 1976, the Consumer Product Safety Commission set the residential paint lead concentration at .06% (600 ppm) because there was evidence that paint could be manufactured with this lower level of contamination, not because it was shown to be protective.[62] Similarly, the MCL of 15 ppb of lead in water, which is used to regulate water systems in the United States, will not adequately protect children or pregnant women.[46,51] In 1988, the US Department of Housing and Urban Development established a post abatement floor standard of 200 $\mu g/ft^2$ because there was evidence that it was feasible to attain, not because it was shown to be safe or protective. In 2020, the EPA promulgated residential lead standards of 10 $\mu g/ft^2$ for floors and 100 $\mu g/ft^2$ for window sills.[58] Unfortunately, these standards—which are insufficient to protect children from having a blood lead concentration of less than 5 $\mu g/dL$—dictate the levels of lead contamination considered "normal" or "low," and they provide an illusion of safety (https://www.epa.gov/lead/hazard-standards-and-clearance-levels-lead-paint-dust-and-soil-tsca-sections-402-and-403).

Control of Lead Exposure

Several trials have evaluated the effect of dust control on children's blood lead levels. In a meta-analysis of various dust control trials, investigators did not find a significant reduction in children's mean blood lead concentration.[63] Importantly, none of these studies attempted to achieve dust lead levels now known to be associated with excess lead exposure. Data from a national survey showed that 5% of children have a blood lead level of greater than 5 μg/dL at a median floor dust lead level of 1.5 μg/ft^2.[64] At a floor standard of 40 μg/ft^2—the current EPA standard for floors—50% of children were estimated to have blood lead levels of greater than 5 μg/dL.[64] This study reinforces earlier reports showing that the floor lead standard set by the EPA is inadequate to protect children from lead toxicity.

Lead hazard controls can result in sizable reductions in dust lead loading. Dust lead levels immediately following abatement were 8.5 μg/ft^2, 8.0 μg/ft^2, and 21 μg/ft^2 for floors, interior windowsills, and window troughs, respectively—representing reductions of more than 80% compared with pre-abatement levels.[65] In a study of more than 2,600 housing units, post abatement dust lead levels were 12 μg/ft^2, 31 μg/ft^2, and 32 μg/ft^2 for floors, windowsills, and window troughs, respectively.[66] Because these levels were achieved with dust clearance testing set at 100 μg/ft^2 or higher, it is likely that floor dust lead levels below 10 μg/ft^2 can routinely be achieved. Floor lead levels below 5 μg/ft^2 can routinely be met.[67] Children will not be protected if we do not achieve these lower dust lead standards.[68]

Hazards of Lead Hazard Controls

Lead poisoning is a preventable disease, but measures intended to control or reduce residential lead hazards can increase the risk of lead poisoning if not done carefully. Studies conducted since the 1988 post abatement dust lead standards promulgated by the HUD show that paint abatement can cause children's blood lead levels to rise, possibly because the abatement process can generate substantial quantities of lead-contaminated dust. In a controlled study of children with baseline blood lead concentrations below 22 μg/dL, Aschengrau et al. reported a 6.5 μg/dL increase in blood lead for children whose homes had undergone paint abatement.[49] Clark and others reported that 6-month-old infants were 11 times more likely to experience a greater than 5 μg/dL increase in blood lead concentrations after abatement than were older children.[61] The rise in blood lead levels after abatement or renovation is due to lead contamination from removal or scraping of leaded paint. Collectively, these studies indicate that the levels of lead-contaminated dust generated by lead hazard controls are sufficient to result in excessive lead exposure and absorption for children.

Conclusion: Elimination of Lead Poisoning

Despite dramatic reductions in children's blood lead levels, childhood lead poisoning remains a major public health problem. To end childhood lead poisoning globally, societal efforts to prevent lead exposure among children must emphasize "primary prevention"—the elimination of lead hazards *before* children are exposed. This approach contrasts with current practices and policies in too many countries that continue to rely on early detection of children after these children already have been exposed to lead and have developed lead poisoning—"secondary prevention."

There are many reasons why primary prevention of childhood lead exposure should remain a high public health priority.

- Despite the dramatic decline in children's blood lead levels, a substantial number of children around the world continue to be exposed to excessive amounts of lead.[69]
- Lead is a systemic toxicant, and therefore preventing exposure will have many health benefits in addition to preventing brain injury.
- No effective treatments exist for ameliorating the developmental effects of lead toxicity. Investigators who conducted a randomized controlled trial of succimer (DMSA) did not find any neurobehavioral benefit of chelation for children who had blood lead levels between 20 μg/dL and 44 μg/dL.[43,70]
- The economic benefits of preventing childhood lead toxicity from residential hazards are substantial.

Primary prevention is cost-effective. Yet, for too long, we have simply passed out brochures or instructed mothers to "clean their houses better" to reduce their children's risk of lead poisoning. As foretold by Turner in 1909, education alone is not sufficient; "legislative interference" is necessary to prevent lead toxicity in children.[71]

The following actions are needed to achieve the goal of primary prevention of lead poisoning in children worldwide:

1. Identify the major sources of lead exposure in each country and community.
2. Pinpoint the most relevant sources of lead exposure. Because lead is ubiquitous, it can always be found in house dust, water, and residential soil. It is therefore necessary to identify unacceptable or hazardous levels of lead in relevant sources that children encounter.
3. Conduct blood lead surveys in children to establish baseline patterns of lead exposure and detect geographical areas and subpopulations at increased risk of exposure.
4. Conduct targeted screening in countries and communities with a lead paint problem to identify housing and products, such as toys, that may contain lead paint hazards. Older housing units should be screened before a child is unduly exposed; before occupancy, after lead hazard controls, or renovation. Plans for the remediation of lead-contaminated housing, including the gradual elimination of lead hazards during renovation or demolition of older housing.
5. Conduct special screening programs for children in highly contaminated communities such as smelter communities or cities with a high prevalence of lead exposure to identify those who have undue lead exposure and establish programs to prevent future exposure in these "hotspots."
6. Promulgate and enforce strict legal standards based on empirical data that regulate allowable levels of lead in air, water, soil, house dust, and consumer products.
7. Enforce regulations to identify lead hazards and require their remediation. These standards should address all major sources of lead exposure, including industrial emissions, lead paint in older housing, lead-contaminated soil, lead service lines, and other consumer products.

It is clear that a focus on behavioral change to prevent childhood lead exposure will not eliminate this problem. Until we recommit our efforts to the primary prevention of childhood lead exposure, we will inadvertently but knowingly continue to use children as biologic indicators of contaminated environments. Now is the time for global elimination of childhood lead poisoning through primary prevention.

References

1. Gibson JL. A plea for painted railings and painted rooms as the source of lead poisoning amongst Queensland children. *Australasian Med Gaz*. 1904;23:149–53.
2. Turner AJ. On lead poisoning in childhood. *Br Med J*. 1909;1(2519):895–7. https://pubmed. ncbi.nlm.nih.gov/20764402   https://www.ncbi.nlm.nih.gov/pmc/articles/PMC2318612/. doi:10.1136/bmj.1.2519.895.
3. Nriagu JO. The rise and fall of leaded gasoline. *Sci Total Environ*. 1990;92(C):13–28.
4. Needleman HL, Gunnoe C, Leviton A, et al. Deficits in psychologic and classroom performance of children with elevated dentine lead levels. *N Engl J Med*. 1979;300(13):689–95.
5. Needleman HL, Schell A, Bellinger D, Leviton A, Allred EN. The long-term effects of exposure to low doses of lead in childhood: an 11-year follow-up report. *N Engl J Med*. 1990;322(2):83–8. doi:10.1056/NEJM199001113220203.
6. Program NT. Monograph on health effects of low-level lead. 2012. Available at: http:// ntp. niehs. nih. gov/ntp/ohat/lead/final/monographhealtheffectslowlevellead_newissn_ 508. pdf.
7. Lanphear BP, Hornung R, Khoury J, et al. Low-level environmental lead exposure and children's intellectual function: an international pooled analysis. *Environ Health Perspect*. 2005;113(7):894–9.
8. WHO. Lead poisoning. Updated 2021. Available at: https://www.who.int/news-room/fact-sheets/detail/lead-poisoning-and-health.
9. CDC. CDC response to advisory committee on childhood lead poisoning prevention recommendations in "Low level lead exposure harms children: a renewed call of primary prevention." Updated 2012. Available at: http://www.cdc.gov/nceh/lead/ACCLPP/CDC_Resp onse_Lead_Exposure_Recs.pdf.
10. Ruckart PZ, Jones RL, Courtney JG, et al. Update of the blood lead reference value: United States, 2021. *MMWR Morb Mortal Wkly Rep*. 2021;70(43):1509–12. https://www.cdc.gov/ mmwr/volumes/70/wr/mm7043a4.htm. doi:10.15585/mmwr.mm7043a4.
11. American Academy of Pediatrics Council on Environmental Health, Lanphear BP, Lowry JA, et al. Prevention of childhood lead toxicity. *Pediatrics*. 2016;138(1):e20161493. https:// doi.org/10.1542/peds.2016-1493. doi:10.1542/peds.2016-1493.
12. Ericson B, Hu H, Nash E, Ferraro G, Sinitsky J, Taylor MP. Blood lead levels in low-income and middle-income countries: a systematic review. *Lancet Planetary Health*. 2021;5(3):e145–53.
13. Bierkens J, Smolders R, Van Holderbeke M, Cornelis C. Predicting blood lead levels from current and past environmental data in Europe. *Sci Total Environ*. 2011;409(23):5101–10.
14. Egan KB, Cornwell CR, Courtney JG, Ettinger AS. Blood lead levels in U.S. children ages 1–11 years, 1976–2016. *Environ Health Perspect*. 2021;129:37003.
15. Gottlieb S. Sustained fall in UK blood lead levels reported. *BMJ*. 1998;317(7151):99.

16. Annest JL, Pirkle JL, Makuc D, Neese JW, Bayse DD, Kovar MG. Chronological trend in blood lead levels between 1976 and 1980. *N Engl J Med*. 1983;308(23):1373-7.

17. Ahamed M, Verma S, Kumar A, Siddiqui MKJ. Blood lead levels in children of Lucknow, India. *Environ Toxicol*. 2009;13; 25(4; 1):48-54.

18. Han Z, Guo X, Zhang B, Liao J, Nie L. Blood lead levels of children in urban and suburban areas in China (1997-2015): temporal and spatial variations and influencing factors. *Sci Total Environ*. 2018;625:1659-66.

19. Vasquez-Peddie A. Last country on earth to use leaded gasoline in cars bans its sale. *CNN Wire*. 2021. Available at: https://go.exlibris.link/NpDVmqRx.

20. Rebecca K. Sunset for leaded aviation gasoline? *Environ Health Perspect*. 2013;121(2):a54-7. https://doi.org/10.1289/ehp.121-a54. doi:10.1289/ehp.121-a54.

21. Zahran S, Iverson T, McElmurry SP, Weiler S. The effect of leaded aviation gasoline on blood lead in children. *J Assoc Environ Resource Econ*. 2017;4(2):575-610.

22. Jacobs DE, Clickner RP, Zhou JY, et al. The prevalence of lead-based paint hazards in US housing. *Environ Health Perspect*. 2002;110(10):A599-A606.

23. Guzelian PS, Henry CJ, Olin S. *Similarities and Differences Between Children and Adults: Implications for Risk Assessment*. Washington, DC: ILSI Press/International Life Sciences Institute; 1992: 285.

24. Kayaalti Z, Akyuzlu DK, Soylemezoglu T. Evaluation of the effect of divalent metal transporter 1 gene polymorphism on blood iron, lead and cadmium levels. *Environ Res*. 2015;137:8-13.

25. Bressler JP, Olivi L, Cheong JH, Kim Y, Bannona D. Divalent metal transporter 1 in lead and cadmium transport. *Ann N Y Acad Sci*. 2004;1012(1):142-52.

26. Needleman HL, Riess JA, Tobin MJ, Biesecker GE, Greenhouse JB. Bone lead levels and delinquent behavior. *JAMA*. 1996;275(5):363-9.

27. Dietrich KN, Douglas RM, Succop PA, Berger OG, Bornschein RL. Early exposure to lead and juvenile delinquency. *Neurotoxicol Teratol*. 2001;23(6):511-8.

28. Wright JP, Dietrich KN, Ris MD, et al. Association of prenatal and childhood blood lead concentrations with criminal arrests in early adulthood. *PLoS Med*. 2008;5(5):e101.

29. Fergusson DM, Boden JM, Horwood LJ. Dentine lead levels in childhood and criminal behaviour in late adolescence and early adulthood. *J Epidemiol Community Health*. 2008;62(12):1045-50.

30. Braun JM, Froehlich TE, Daniels JL, et al. Association of environmental toxicants and conduct disorder in US children: NHANES 2001-2004. *Environ Health Perspect*. 2008;116(7):956-62.

31. Reyes JW. Environmental policy as social policy? the impact of childhood lead exposure on crime. *BE J Econ Analysis Policy*. 2007;7(1):1-70.

32. Froehlich TE, Lanphear BP, Auinger P, et al. Association of tobacco and lead exposures with attention-deficit/hyperactivity disorder. *Pediatrics*. 2009;124(6):e1054-63.

33. Nigg JT, Knottnerus GM, Martel MM, et al. Low blood lead levels associated with clinically diagnosed attention-deficit/hyperactivity disorder and mediated by weak cognitive control. *Biol Psychiatry*. 2008;63(3):325-31.

34. Cecil KM, Brubaker CJ, Adler CM, et al. Decreased brain volume in adults with childhood lead exposure. *PLoS Med*. 2008;5(5):e112. doi:10.1371/journal.pmed.0050112.

35. Marshall AT, Betts S, Kan EC, McConnell R, Lanphear BP, Sowell ER. Association of lead-exposure risk and family income with childhood brain outcomes. *Nat Med*. 2020;26(1):91–7.

36. Reuben A, Caspi A, Belsky DW, et al. Association of childhood blood lead levels with cognitive function and socioeconomic status at age 38 years and with IQ change and socioeconomic mobility between childhood and adulthood. *JAMA*. 2017;317(12):1244–51.

37. Fadrowski JJ, Navas-Acien A, Tellez-Plaza M, Guallar E, Weaver VM, Furth SL. Blood lead level and kidney function in US adolescents: the third national health and nutrition examination survey. *Arch Intern Med*. 2010;170(1):75–82.

38. Borja-Aburto V, Hertz-Picciotto I, Lopez MR, Farias P, Rios C, Blanco J. Blood lead levels measured prospectively and risk of spontaneous abortion. *Am J Epidemiol*. 1999;150(6):590–7.

39. Zhu M, Fitzgerald EF, Gelberg KH, Lin S, Druschel CM. Maternal low-level lead exposure and fetal growth. *Environ Health Perspect*. 2010;118(10):1471–5.

40. Taylor CM, Golding J, Emond AM. Adverse effects of maternal lead levels on birth outcomes in the ALSPAC study: a prospective birth cohort study. *BJOG: Int J Obstet Gynecol*. 2015;122(3):322–8. https://doi.org/10.1111/1471–0528.12756. doi:https://doi.org/10.1111/1471–0528.12756.

41. Selevan SG, Rice DC, Hogan KA, Euling SY, Pfahles-Hutchens A, Bethel J. Blood lead concentration and delayed puberty in girls. *N Engl J Med*. 2003;348(16):1527–36.

42. Hauser R, Sergeyev O, Korrick S, et al. Association of blood lead levels with onset of puberty in Russian boys. *Environ Health Perspect*. 2008;116(7):976–80.

43. American Academy of Pediatrics Committee on Environmental Health. Lead exposure in children: prevention, detection, and management. *Pediatrics*. 2005;116(4):1036–46.

44. Obeng-Gyasi E. Sources of lead exposure in various countries. *Rev Environ Health*. 2019;34(1):25–34.

45. Clark CS, Bornschein RL, Succop P, Hee SQ, Hammond PB, Peace B. Condition and type of housing as an indicator of potential environmental lead exposure and pediatric blood lead levels. *Environ Res*. 1985;38(1):46–53.

46. Lanphear BP, Hornung R, Ho M, Howard CR, Eberly S, Knauf K. Environmental lead exposure during early childhood. *J Pediatr*. 2002;140(1):40–7.

47. Sayre JW, Charney E, Vostal J, Pless IB. House and hand dust as a potential source of childhood lead exposure. *Am J Dis Child*. 1974;127(2):167–70.

48. Duggan MJ, Inskip MJ. Childhood exposure to lead in surface dust and soil: a community health problem. *Public Health Rev*. 1985;13(1-2):1–54.

49. Aschengrau A, Beiser A, Bellinger D, Copenhafer D, Weitzman M. Residential lead-based-paint hazard remediation and soil lead abatement: their impact among children with mildly elevated blood lead levels. *Am J Public Health*. 1997;87(10):1698–1702.

50. Lanphear BP, Matte TD, Rogers J, et al. The contribution of lead-contaminated house dust and residential soil to children's blood lead levels: a pooled analysis of 12 epidemiologic studies. *Environ Res*. 1998;79(1):51–68.

51. USEPA. Drinking water regulations; maximum contaminant level goals and national primary drinking water regulations for lead and copper. *Final Rule. 40 CFR, Parts 141, 142. Fed. Reg*. 1991;56(110):26505.

52. World Health Organization, ed. *Guidelines for Drinking-Water Quality: Fourth Edition Incorporating the First Addendum*. 4th ed. Geneva, Switzerland: World Health Organization; 2017.

53. Edwards M, Triantafyllidou S, Best D. Elevated blood lead in young children due to lead-contaminated drinking water: Washington, DC, 2001–2004. *Environ Sci Technol*. 2009;43(5):1618–23.

54. Kennedy C, Yard E, Dignam T, et al. Blood lead levels among children aged <6 years: Flint, Michigan, 2013–2016. *Morb Mortal Weekly Rep*. 2016;65(25):650–4.

55. Hanna-Attisha M, LaChance J, Sadler RC, Champney Schnepp A. Elevated blood lead levels in children associated with the flint drinking water crisis: a spatial analysis of risk and public health response. *Am J Public Health*. 2016;106(2):283–90.

56. Levin R, Brown MJ, Kashtock ME, et al. Lead exposures in US children, 2008: implications for prevention. *Environ Health Perspect*. 2008;116(10):1285–93.

57. Roscoe RJ, Gittleman JL, Deddens JA, Petersen MR, Halperin WE. Blood lead levels among children of lead-exposed workers: a meta-analysis. *Am J Ind Med*. 1999;36(4):475–81.

58. US Environmental PA. 40 CFR part 745. Lead: identification of dangerous levels of lead. 2001:1206–40. https://www.federalregister.gov/documents/2001/01/05/01-84/lead-identification-of-dangerous-levels-of-lead

59. Lanphear BP. The paradox of lead poisoning prevention. *Science*. 1998;281(5383):1617–8.

60. Amitai Y, Brown MJ, Graef JW, Cosgrove E. Residential deleading: effects on the blood lead levels of lead-poisoned children. *Pediatrics*. 1991;88(5):893–7.

61. Clark S, Grote J, Wilson J, et al. Occurrence and determinants of increases in blood lead levels in children shortly after lead hazard control activities. *Environ Res*. 2004;96(2):196–205.

62. National Research Council Committee on Toxicology, and US Consumer Product Safety Commission. Recommendations for the prevention of lead poisoning in children. 1976. https://nap.nationalacademies.org/catalog/18520/recommendations-for-the-prevention-of-lead-poisoning-in-children

63. Yeoh B, Woolfenden S, Lanphear B, Ridley GF, Livingstone N, Jorgensen E. Household interventions for preventing domestic lead exposure in children. *Cochrane Database Syst Rev*. 2014(12):CD006047. doi:10.1002/14651858.CD006047.pub4. Update in: Cochrane Database Syst Rev. 2016 Oct 16;10:CD006047.

64. Dixon SL, Gaitens JM, Jacobs DE, et al. Exposure of US children to residential dust lead, 1999–2004: II. the contribution of lead-contaminated dust to children's blood lead levels. *Environ Health Perspect*. 2009;117(3):468–74.

65. Farfel MR, Rohde C, Lees P, Rooney B, Bannon DL, Derbyshire W. *Lead-Based Paint Abatement and Repair and Maintenance Study in Baltimore: Findings Based on Two Years of Follow-Up*. Washington, DC: US Environmental Protection Agency; 1998.

66. Gaulke W, Clark S, Wilson J, et al. Evaluation of the HUD lead hazard control grant program: early overall findings. *Environ Res*. 2001;86(2):149–56.

67. Braun JM, Hornung R, Chen A, et al. Effect of residential lead-hazard interventions on childhood blood lead concentrations and neurobehavioral outcomes: a randomized clinical trial. *JAMA Pediatr*. 2018;172(10):934–42. https://doi.org/10.1001/jamapediatrics.2018.2382. doi:10.1001/jamapediatrics.2018.2382.

68. Braun JM, Yolton K, Newman N, Jacobs DE, Taylor M, Lanphear BP. Residential dust lead levels and the risk of childhood lead poisoning in United States children. *Pediatr Res.* 2021;90(4):896–902. https://doi.org/10.1038/s41390-020-1091-3. doi:10.1038/s41390-020-1091-3.

69. Jones RL, Homa DM, Meyer PA, et al. Trends in blood lead levels and blood lead testing among US children aged 1 to 5 years, 1988–2004. *Pediatrics.* 2009;123(3):e376-85.

70. Dietrich KN, Ware JH, Salganik M, et al. Effect of chelation therapy on the neuropsychological and behavioral development of lead-exposed children after school entry. *Pediatrics.* 2004;114(1):19–26.

71. Taylor MP, Schniering CA, Lanphear BP, Jones AL. Lessons learned on lead poisoning in children: one hundred years on from turner's declaration. *J Paediatr Child Health.* 2011;47(12):849–56.

32

Mercury

Philippe Grandjean

Two forms of mercury are most likely to contribute to human exposure during early development. One is the elemental or metallic mercury, a liquid at room temperature with a vapor pressure that allows buildup of potentially fatal concentrations of mercury vapor. The other is methylmercury, an organometallic compound that bioaccumulates in aquatic and marine food chains. Mercury vapor has resulted in numerous cases of poisoning in children. Exposure to methylmercury interferes with brain development and occurs from contaminated food, often via the mother's diet during pregnancy. Methylmercury easily passes the placenta and the neurotoxicity results in deficits in cognitive development. This chapter provides an overview of mercury toxicity and refers to key references. The interested reader is referred to more detailed accounts, such as the US Environmental Protection Agency (EPA) webpage,[1] the most recent Global Mercury Assessment from the United Nations Environment Programme,[2] the still relevant review by the US National Research Council,[3] and recent review articles.[4,5]

Scope and Nature of the Problem

Methylmercury is the most important form of mercury from the perspective of children's environmental health. It is widely distributed in the environment and results in human exposure worldwide. Methylmercury forms when, for example, elemental mercury from the atmosphere deposits in rivers, lakes, and the ocean and is converted by microorganisms in the water or sediment to organic methylmercury. Methylmercury bioaccumulates in aquatic and marine food chains (see Figure 32.1), reaching the highest concentrations in large, predatory species at the top of the food chain such as shark, swordfish, and tuna. By analyzing historical tissue samples from centuries past, it is now known that methylmercury levels in the environment seem to have increased by 10-fold above the preindustrial background.[6] In addition, mercury may contaminate flooded rice paddies, where the anaerobic microorganisms can convert the mercury to methylmercury, which accumulates in the rice grain and thereby results in widespread exposure.[7]

The developing brain of the human fetus is highly vulnerable to methylmercury. Methylmercury readily crosses the placenta from mother to child, and a pregnant woman therefore shares her exposure with the fetus. Even at exposures that are completely safe to the mother, adverse effects on fetal brain development may ensue. Additional exposure, mainly from inorganic mercury, occurs via breastfeeding.[8] Because complete integrity of the brain is necessary for optimal functioning throughout life, even a minor degree of methylmercury neurotoxicity can have long-term consequences. A child has only one

Philippe Grandjean, *Mercury* In: *Textbook of Children's Environmental Health*, Second Edition. Edited by: Ruth A. Etzel and Philip J. Landrigan, Oxford University Press. © Oxford University Press 2024. DOI: 10.1093/oso/9780197662526.003.0032

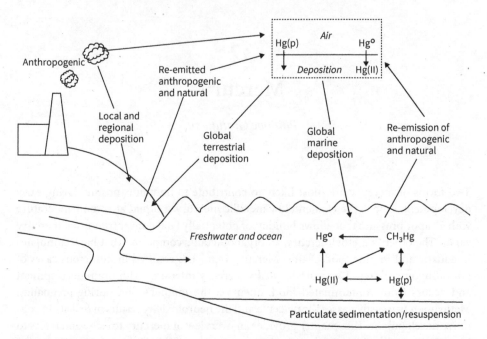

Figure 32.1. Mercury is mined from mercury-rich ores. Its uses and mercury-containing waste generate mercury releases to the environment as mercury vapor (Hg^0) or as particulate mercury [Hg(p)]. Airborne mercury particles and oxidized mercury [Hg(II)] later settle in sediments. The mercury released by anthropogenic activities is supplemented by natural sources, such as volcanic eruptions, weathering of mercury-containing rocks, and resuspension of previously deposited mercury. environmental dissemination of mercury occurs on a global scale via movements of air and water. The cycle is repeated when mercury is chemically converted, deposited, revolatized, and transported again. In the water environment, mercury can be metabolized by microorganisms to form methylmercury (CH_3Hg), which accumulates in aquatic and marine food chains. The presence of methylmercury in fish and seafood is increased worldwide near pollution sources, especially in lakes, rivers, and coastal waters.

chance to develop a healthy brain, and toxic brain injury that occurs during early development can interfere with that chance and is likely to cause permanent brain damage. Even relatively low-level exposures to methylmercury previously thought to be safe are now known to be associated with cognitive deficits in children, and these appear to be permanent and irreversible. Given that these exposures usually are due to consumption of contaminated fish and seafood, pregnant women and children should completely avoid large, predatory fish species that accumulate the highest levels of methylmercury.

Elemental (metallic) mercury is the second form of mercury that is important for children's environmental health. Exposure to elemental mercury can result from amalgam dental fillings, skin-lightening creams, and release from broken light bulbs. Adverse effects occasionally occur from exposures to mercury from these sources, and these sources contribute to environmental mercury pollution.

Sources of Exposure

Small amounts of elemental mercury occur naturally in the Earth's crust. Mercury from the Earth is constantly being added to the biosphere through volcanic emissions and through releases from other natural sources such as deposits of mercuric sulfide (cinnabar) ore. Industrial emissions of mercury are superimposed on and far exceed these natural releases. Coal combustion, the use of amalgamation for extraction and purification of gold and other noble metals from ores, losses incurred during use of mercury and its compounds in industry, and emissions from waste disposal and incineration are the main industrial sources of environmental mercury. On a global scale, coal-fired power plants are a major contributor, and China is the source of the greatest mercury emissions. In addition, the artisanal and small-scale gold mining (ASGM) that uses mercury to amalgamate gold involve large numbers of children as workers directly exposed to mercury vapors.

Direct human exposure to inorganic mercury can also occur from its application for cosmetic and therapeutic purposes. Some mercury compounds are used for topical treatment in skin lighteners.[9] Mercury is also used in dental amalgams, and amalgam fillings remain a continuing source of internal exposure to elemental mercury. Because of its low price, dental amalgams continue to be widely implemented in the world.

In addition, in some ethnic groups, metallic mercury is used for magical purposes that may cause substantial exposure to mercury vapor.[10] Hazardous releases of elemental mercury at home or at school can result from broken thermometers or other instruments. Energy-saving lights (compact fluorescent lamps) use mercury, although the amount is often small and bound as amalgam. However, broken bulbs should be meticulously cleaned up using adhesive tape (never by vacuum cleaner).

Mercury Vapor

In its elemental form, mercury (often denoted Hg^0) is a liquid, dense, and shiny metal, with a vapor pressure at room temperature of 0.17 Pa (0.0013 mm Hg). A saturated atmosphere therefore contains a mercury concentration of 14 mg Hg/m^3, a concentration that is more than 100 times the occupational exposure limit. Uptake through the lungs of inhaled mercury vapor is almost complete. Hg^0 may penetrate the blood–brain barrier, although most will be oxidized to mercuric mercury in the blood and will eventually be eliminated through urine and feces with a half-life of 30–40 days. The kidneys and the brain are the main retention sites after inhalation of Hg^0. Acute poisoning with mercury vapor may cause airway irritation, chemical pneumonitis, and even pulmonary edema. Accidental ingestion of liquid mercury is much less of a hazard because Hg^0 is poorly absorbed from the gastrointestinal tract and typically passes through.

Information on the adverse effects of Hg^0 has mainly been obtained through studies of exposed adult workers. The classical form of mercury vapor poisoning is called *erethism* and consists of emotional instability, irritability, and extreme shyness along with tremor and gingivitis. The central nervous system is critically affected; deficits in concentration and memory are well documented following chronic exposures.

Adverse effects on pregnancy outcomes have been reported in women occupationally exposed to Hg^0, but insufficient details are available to evaluate dose–response relationships. However, because mercury vapor can pass the placenta and because the brain is considered a main target organ, there can be no doubt that Hg^0 exposure during pregnancy is a risk to brain development in the fetus. When exposure occurs during early brain development, the effects will likely be permanent and similar to those associated with methylmercury exposure.

In children exposed to inorganic mercury, a dramatic clinical syndrome termed "pink disease" or *acrodynia* occasionally may occur.[11] A generalized skin eruption develops, and the hands and feet show a characteristic scaly, reddish appearance. Exposed children become irritable, sleep badly, fail to thrive, sweat profusely, and show photophobia. Hypertension develops and may be the reason for hospitalization. Acrodynia was first discovered in children treated with (now obsolete) "teething powders" that contained calomel (mercury chloride), but it has also been seen in children after inhalation of mercury vapor (e.g., from broken thermometers or mercury released from indoor paint that contained a mercury fungicide).[12] The condition was rather common until 30 years ago, when the calomel etiology was finally recognized and the teething powders were discontinued.

A common source of Hg^0 exposure is mercury dental amalgams. Although technically bound to other metals, small amounts of mercury are released from dental amalgams during chewing. However, a clinical trial found no detectable adverse effects on neurodevelopment in children with amalgam fillings as compared to children who received mercury-free fillings.[13] Although definite evidence of adverse effects has not emerged,[14] the use of amalgam in dentistry is decreasing due to other concerns about mercury applications in society and its dissemination to the environment.

Methylmercury

Methylmercury is somewhat lipophilic and also soluble in water. Absorption from the gut of ingested methylmercury is virtually complete. An L-cysteine complex of methylmercury is transported across membranes by the neutral amino acid carrier system, and passage across the placenta and the blood–brain barrier is thereby facilitated. Inorganic mercury compounds do not pass these cell layers as easily as methylmercury.

The neurotoxicity of methylmercury was first demonstrated in occupationally exposed adults, while the vulnerability of the developing brain was discovered only much later. The first recognition of the developmental toxicity of methylmercury was made in the 1950s, when a severe epidemic of congenital methylmercury poisoning occurred in Minamata, Japan.[15] In this setting, methylmercury pollution of the Minamata Bay was mainly due to leakage of the compound from a nearby chemical factory. The methylmercury accumulated in fish and, when eaten by pregnant women, resulted in a congenital poisoning termed "Minamata disease."[16] Affected children suffered severe sensory impairments (blindness and deafness), general paralysis, hyperactive reflexes, cerebral palsy, and impaired mental development. The mothers remained healthy or suffered no more than minor symptoms.

Methylmercury and related compounds had been used as fungicides for seed dressing, but this application was abandoned after a disastrous episode of human methylmercury poisoning in Iraq, where seed grain treated with methylmercury had been erroneously used to bake bread for local consumption.

To examine the developmental neurotoxic effects of chronic low-level in utero exposure to methylmercury, more than a dozen prospective studies have been carried out in several countries, including China,[7] the Faroe Islands,[17] and the United States.[18] This extensive database shows that developmental delays occur in children born to mothers whose hair mercury concentrations during pregnancy were at levels of about 1 μg/g and above (or maternal blood concentrations >4 μg/L), levels that are commonly exceeded in populations that rely on seafood or freshwater fish. Adverse effects occur in neurological, cognitive, and motor functions. In older children, deficits occur in specific neuropsychological functions, such as motor speed, reaction time, attention, language, and visuospatial and memory functions. Neurophysiological testing (e.g., measurement of evoked potentials) supports these associations. Postnatal exposure is usually at lower levels than prenatal but has nonetheless been linked to a reduction of motor speed and increased latencies on evoked potentials. These effects may be masked in part by benefits from essential nutrients in the seafood, such as fatty acids.[19] Conversely, the benefits from a healthy diet may not be fully apparent when the diet contains contaminated fish, thus resulting in negative confounding.[20] Genetic predisposition may worsen the adverse impact of the methylmercury exposure.[21] Elevated blood pressure may be an additional effect of the neurotoxicity[22] that may have implications for development of cardiovascular disease in adulthood.[23] Also, accelerated aging of nervous system functions has been demonstrated in subjects who were developmentally exposed to methylmercury.[24]

In summary, numerous individual and community studies have contributed clear evidence that maternal consumption of methylmercury during pregnancy can have serious and irreversible effects on the physical and mental development of children, even if the mother herself exhibited no outward symptoms. While heavy exposures constitute a clear hazard with adverse neurological signs and test deficits, the effects of more common medium- and low-level exposures from ocean and freshwater fish, or rice are less pronounced and more difficult to document and quantify. Nonetheless, to protect optimal brain development, pregnant women should be advised to abstain from contaminated seafood or other foods and to seek essential nutrients from safe food sources.

Other Mercury Sources

Some multidose vaccines still contain thimerosal, a mercury compound used as a preservative.[25] Thimerosal breaks down to ethylmercury, which is metabolized and eliminated from the body much more rapidly than methylmercury. This mercury compound is still used for certain multidose vaccines where refrigeration cannot be guaranteed, but elimination of this source of exposure has been successful in most parts of the world.

Skin lightening creams containing mercury have been popular in some parts of the world and have caused numerous cases of poisoning.[9]

Diagnosis and Treatment

Even in cases of serious methylmercury poisoning, most affected children were not immediately diagnosed in the past because the spastic paresis-like syndrome seen in these children was far less distinctive of mercury poisoning than the clinical picture of poisoned adults, in whom tunnel vision was highly characteristic. Early signs in an infant with congenital poisoning included mental retardation, movement problems, seizures, primitive reflexes, and speech difficulty and could easily be mistaken for some other pediatric disease; mild stages are even more easily overlooked. Thus, diagnosis was usually made only later in infancy when children fail to achieve developmental milestones.

Neuropathologic examinations support these clinical impressions. Adult poisoning with methylmercury is associated with localized lesions in certain brain areas (such as the calcarine, postcentral and precentral cortex, and the cerebellum) which correlate well with the clinical presentation. In contrast, methylmercury poisoning in children results in more widely distributed damage to the brain, while congenital poisoning leads to a completely diffuse pattern of damage with disruption of normal structures.[16] Accordingly, diagnosis in children must be accompanied by proper documentation of exposure, and this conclusion also applies to cases of mercury vapor exposure. Only acrodynia could be considered reasonably specific to indicate mercury as the cause.

For methylmercury, the pattern of tissue distribution in the body is fairly uniform, except that the red blood cells contain mercury concentrations that are 10–20 times greater than those in serum or plasma. Because the blood–brain and placental barriers are readily crossed by methylmercury and by mercury vapor, these forms of mercury will accumulate in the brain; the kidneys also retain high tissue concentrations of mercury. The elimination half-life for methylmercury in blood is about 45–60 days and for inorganic mercury about 1 month. Thus, when exposures are long-term or occurred mainly in the past, current blood tests may not reflect mercury retention in the target organs.

Detection of recent exposures to mercury can be reliably based on an analysis of whole blood (or red blood cells) or scalp hair.[26,27] If exposure to mercury vapor or other inorganic mercury compounds is suspected, the concentration in serum should be determined. Mercury in hair, when measured segmentally along the length of a hair strand, can be used as an indicator of past methylmercury exposures. On average, hair grows about 1 centimeter per month, and mercury concentrations in long hair strands can therefore be used to generate a calendar of past exposure. Methylmercury accumulates in hair in the process of hair strand formation, with average concentrations being roughly 250-fold higher than in blood. However, hair mercury concentrations may be augmented by the binding of exogenous mercury to the surface. Prenatal exposure to methylmercury is best determined as the mercury concentration in cord blood or cord tissue. The cord blood concentration at birth is about 30% higher than in maternal blood and mainly reflects the exposure during the third trimester, which also seems to be the most vulnerable period in regard to neurotoxic effects.

Dietary information can support the assessment of prenatal exposure. Methylmercury exposure depends on frequency of fish intake, the size of each meal, and the particular species consumed. The highest concentrations of methylmercury occur in predatory fish and marine mammals because of the biomagnification of methylmercury through marine food chain. However, the contamination varies much within individual species, due to size, age, and location, thus rendering dietary questionnaires somewhat imprecise for assessment of methylmercury exposure.

The highest methylmercury exposures are found in Arctic populations who eat marine mammals,[28] but increased levels also frequently occur in populations who eat other species high in the food chain, such as halibut, sea bream, and sea bass. Average exposures are lower in countries such as the United States, with less frequent fish intakes. From a public health point of view, fish and seafood contain essential nutrients and are recommended as an important part of a varied diet. Thus, dietary advisories need to encourage consumption of the types of fish that are low in mercury while advising against species high in mercury (see, e.g., https://www.nrdc.org/stories/smart-seafood-buy ing-guide).

Treatment options for mercury poisoning are limited. In situations of acute exposure to elemental mercury, chelation therapy with succimer (dimercaptosuccinic acid) has been shown to be successful in speeding elimination of mercury from the body via the kidneys, but removal of mercury from deep tissue compartments, such as the brain, would seem unrealistic. Thus, prevention of exposure is critically important.

Prevention

Mercury should be regarded as an extremely toxic element. All unnecessary use should be eliminated, and necessary uses should be enclosed to eliminate the risk of human exposure and releases to the environment. The Minamata Convention, a global UN treaty to protect human health and the environment from the adverse effects of mercury, entered into force in 2017.[2] The objective is "to protect human health and environment from anthropogenic emissions and releases of mercury and mercury compounds." New mercury mines are banned and existing ones should be phased-out, and international trade in mercury must be controlled. Likewise, mercury uses in products and processes must be phased-out or phased-down. Finally, the Convention requires controls on artisanal and small-scale gold mining. This international agreement is a major victory for public health. Coal burning is an additional and major source of mercury pollution, and the growth of this means of energy production has now decreased.

To control dietary exposure to mercury, the EPA has established a Reference Dose (RfD, i.e., the daily permissible intake over a lifetime) for methylmercury of 0.1 µg/kg body weight per day. Methylmercury intake at this level results in a hair mercury concentration of about 1 µg/g and a blood concentration of about 4 µg/L. However, more recent data on neurodevelopmental deficits emerging from recent studies indicate that toxicity may occur at still lower levels of exposure.

A very worrisome finding in the adverse effects regarding neurobehavioral development is that exposures are usually determined with some degree of imprecision, thereby resulting in an underestimation of the dose–effect relationship. In the presence of confounders, even further underestimation may occur. Although this is a general problem in population research, the implications have been studied in particular detail in regard to methylmercury. One of the conclusions is that hair mercury concentrations vary substantially and involve a relative imprecision of as much as 50%. The cord blood mercury concentration is a better indicator of prenatal methylmercury exposure, but it still involves an imprecision of about 25%. Taking into account these uncertainties in exposure assessment, it is clear that the RfD for methylmercury is too high to fully protect brain development in children. It should be decreased by at least 50%.[29] Furthermore, recent results indicate that neurodevelopment is also affected by postnatal methylmercury exposure,[30] which would also require attention.

When considering prevention from a historical perspective, it is clear that recognition of the developmental toxicity of methylmercury has taken many years. Obvious cases of congenital methylmercury poisoning occurred in the 1950s, and, since then, abundant evidence has accumulated to document the adverse effects of exposures from contaminated seafood even at low levels of exposure. Yet experts for the World Health Organization remained unconvinced that developmental neurotoxicity needed special consideration when establishing exposure limits. Thus, until 2003, the recommended exposure limit for methylmercury was based solely on adult toxicity.[31] Future research may well reveal toxicity at low exposure levels that are currently considered "safe," and any exposure to this toxic element therefore ought to be minimized.

An unavoidable problem that arises in considering fish and seafood as important components of a balanced diet, in particular during pregnancy and early childhood, is that the omega-3 fatty acids that occur in many types of fish benefit fetal brain development. A dietary intake of two fish dinners (400 g or 14 oz) per week is therefore considered optimal. However, the revised RfD for mercury will be exceeded by such a diet if the average mercury concentration in fish is higher than 0.1 µg/g. Nonetheless, regulatory agencies in many countries use allowable limits for mercury concentrations in fish that are far too high—up to 1 µg/g for species such as tuna and swordfish (i.e., 10-fold higher than the average that would allow two dinners per week without excess exposure to methylmercury). In some countries, dietary advice to pregnant women recommends avoidance of species known to be high in methylmercury, and they also advise the general population to limit the frequency of such meals. Such advisories must be carefully balanced so they do not scare women away from healthy fish and sea-food.[5] Fish advisories are posted at large numbers of freshwater lakes and waterways, and mercury is the most common reason for those fish advisories in the United States that warn against consumption of locally caught fish. Unfortunately, different state agencies use different approaches, and the advisories in the United States therefore differ substantially.[32] Of course, such advisories are of little help when, for example, mercury releases from a factory have contaminated a whole riverine environment that represents the livelihood of a population, as in the case of a First Nation community in Canada.[33]

The United States and the European Union have already banned export of mercury, and the use of mercury in thermometers and scientific instruments is being phased out. However, due to centuries of mercury releases to the environment, it is going to take a very long time to bring down mercury concentrations in the biosphere. For a foresee-able future, we therefore have to limit exposures by secondary prevention means, such as choosing seafood items wisely.

References

1. Resources for Mercury Science and Research U.S. EPA. 2022. Available at: https://www.epa.gov/mercury/resources-mercury-science-and-research#profile.

2. United Nations Environment Programme. *Global Mercury Assessment 2018*. Geneva: UN Environment Programme, Chemicals and Health Branch; 2019.

3. National Research Council. *Toxicological Effects of Methylmercury*. Washington, DC: National Academy Press; 2000.

4. Karagas MR, Choi AL, Oken E, et al. Evidence on the human health effects of low-level methylmercury exposure. *Environ Health Perspect*. 2012;120:799–806.

5. Groth E, 3rd. Scientific foundations of fish-consumption advice for pregnant women: epidemiological evidence, benefit-risk modeling, and an integrated approach. *Environ Res*. 2017;152:386–406.

6. Riget F, Braune B, Bignert A, et al. Temporal trends of Hg in Arctic biota, an update. *Sci Total Environ*. 2011;409:3520–6.

7. Rothenberg SE, Yu X, Liu J, et al. Maternal methylmercury exposure through rice ingestion and offspring neurodevelopment: a prospective cohort study. *Int J Hyg Environ Health*. 2016;219:832–42.

8. Cherkani-Hassani A, Ghanname I, Mouane N. Total, organic, and inorganic mercury in human breast milk: levels and maternal factors of exposure, systematic literature review, 1976–2017. *Crit Rev Toxicol*. 2019;49:110–21.

9. Chan TY. Inorganic mercury poisoning associated with skin-lightening cosmetic products. *Clin Toxicol (Phila)*. 2011;49:886–91.

10. Rogers HS, Jeffery N, Kieszak S, et al. Mercury exposure in young children living in New York City. *J Urban Health*. 2008;85:39–51.

11. Lai O, Parsi KK, Wu D, et al. Mercury toxicity presenting as acrodynia and a papulovesicular eruption in a 5-year-old girl. *Dermatol Online J*. 2016;22.

12. Johnson-Arbor K, Tefera E, Farrell J, Jr. Characteristics and treatment of elemental mercury intoxication: a case series. *Health Sci Rep*. 2021;4:e293.

13. Bellinger DC, Trachtenberg F, Barregard L, et al. Neuropsychological and renal effects of dental amalgam in children: a randomized clinical trial. *JAMA*. 2006;295:1775–83.

14. Scientific Committee on Emerging and Newly Identified Health Risks (SCENIHR). *Opinion on the Safety of Dental Amalgam and Alternative Dental Restoration Materials for Patients and Users*. Bruxelles: European Commission; 2015.

15. Harada M. Minamata disease: methylmercury poisoning in Japan caused by environmental pollution. *Crit Rev Toxicol*. 1995;25:1–24.

16. Takeuchi T, Eto K. *The Pathology of Minamata Disease. A Tragic Story of Water Pollution*. Fukuoka: Kyushu University Press; 1999.

17. Grandjean P, Weihe P, White RF, et al. Cognitive deficit in 7-year-old children with prenatal exposure to methylmercury. *Neurotoxicol Teratol*. 1997;19:417–28.

18. Oken E, Radesky JS, Wright RO, et al. Maternal fish intake during pregnancy, blood mercury levels, and child cognition at age 3 years in a US cohort. *Am J Epidemiol*. 2008;167:1171–81.

19. Strain JJ, Bonham MP, Duffy EM, et al. Nutrition and neurodevelopment: the search for candidate nutrients in the Seychelles Child Development Nutrition Study. *Neurotoxicology*. 2020;81:300–6.

20. Choi AL, Cordier S, Weihe P, Grandjean P. Negative confounding in the evaluation of toxicity: the case of methylmercury in fish and seafood. *Crit Rev Toxicol*. 2008;38:877–93.

21. Julvez J, Smith GD, Golding J, et al. Prenatal methylmercury exposure and genetic predisposition to cognitive deficit at age 8 years. *Epidemiology*. 2013;24:643–50.

22. Farzan SF, Howe CG, Chen Y, et al. Prenatal and postnatal mercury exposure and blood pressure in childhood. *Environ Int*. 2021;146:106201.

23. Hu XF, Singh K, Chan HM. Mercury exposure, blood pressure, and hypertension: a systematic review and dose-response meta-analysis. *Environ Health Perspect*. 2018;126:076002.

24. Yorifuji T, Takaoka S, Grandjean P. Accelerated functional losses in ageing congenital Minamata disease patients. *Neurotoxicol Teratol*. 2018;69:49–53.

25. Department of Health & Human Services. *Understanding Thimerosal, Mercury, and Vaccine Safety*. Washington, DC: DHEW; 2013.

26. Gaskin J, Rennie C, Coyle D. Reducing periconceptional methylmercury exposure: cost-utility analysis for a proposed screening program for women planning a pregnancy in Ontario, Canada. *Environ Health Perspect*. 2015;123:1337–44.

27. Kirk LE, Jorgensen JS, Nielsen F, Grandjean P. Public health benefits of hair-mercury analysis and dietary advice in lowering methylmercury exposure in pregnant women. *Scand J Public Health*. 2017;45:444–51.

28. Dietz R, Sonne C, Basu N, et al. What are the toxicological effects of mercury in Arctic biota? *Sci Total Environ*. 2013;443:775–90.

29. Grandjean P, Budtz-Jørgensen E. An ignored risk factor in toxicology: the total imprecision of exposure assessment. *Pure Appl Chem*. 2010;82:383–91.

30. Lozano M, Murcia M, Soler-Blasco R, et al. Exposure to mercury among 9-year-old children and neurobehavioural function. *Environ Int.* 2021;146:106173.

31. Joint Expert Committee on Food Additives. Summary and conclusions. Sixty-first meeting of the Joint FAO/WHO Expert Committee on Food Additives held in Rome, 10–19 June 2003.

32. Cleary BM, Romano ME, Chen CY, Heiger-Bernays W, Crawford KA. Comparison of recreational fish consumption advisories across the USA. *Curr Environ Health Rep.* 2021;8:71–88.

33. Takaoka S, Fujino T, Hotta N, et al. Signs and symptoms of methylmercury contamination in a First Nations community in Northwestern Ontario, Canada. *Sci Total Environ.* 2014;468-469:950–7.

33

Arsenic Exposure in Children

Antonio J. Signes-Pastor, Fen Wu, Shohreh F. Farzan, Yu Chen, and
Margaret R. Karagas

Arsenic Exposure in the Environment

Arsenic is a ubiquitous metalloid element. It is found in various chemical forms in soil, groundwater and foods. Inorganic arsenic has been classified as a group 1 carcinogen by the International Agency for Research on Cancer (IARC), initially based on evidence of its effects on skin cancer. Arsenic exposure has also been related to increased risk of several internal cancers, including lung and bladder, as well as to systemic health effects, especially effects on the respiratory, cardiometabolic, and nervous systems.[1]

Ingestion and inhalation are the two primary routes of entry of arsenic into the body. Children are most likely to be exposed to arsenic via ingestion of contaminated drinking water and foods, juice, or infant formula made with arsenic-contaminated water. Ingestion of contaminated soil is also a source of arsenic exposure for children. Children often play in the soil and put their hands in their mouths and sometimes intentionally eat soil.

Arsenic in Drinking Water

Contaminated groundwater is a widespread source of human exposure to inorganic arsenic. Elevated concentrations of arsenic in groundwater have been reported from Bangladesh, India, Nepal, Cambodia, Myanmar, Taiwan, Mongolia, Vietnam, China, Afghanistan, Pakistan, Argentina, Mexico, Chile, Argentina, and areas of the United States. This contamination results most commonly from natural sources, such as erosion and leaching from geological formations. Sometimes, arsenic contamination of groundwater arises from anthropogenic sources, such as uses of arsenic for industrial purposes, mining activities and metal processing, and application of pesticides and fertilizers containing arsenic.

An epidemic of arsenic poisoning—referred to as the largest mass poisoning in human history—is currently taking place in South Asia, where nongovernmental organizations promoted a switch of drinking sources in the late 1960s and early 1970s from surface water to groundwater obtained from hand-pumped tube wells. This action was taken to reduce the burden of mortality from cholera and other diarrheal diseases related to microbially contaminated surface water. Concentrations of arsenic in drinking water in Bangladesh and West Bengal often measured 2–3 orders of magnitude above 10 μg/L, the World Health Organization (WHO) drinking water guideline.

Antonio J. Signes-Pastor, Fen Wu, Shohreh F. Farzan, Yu Chen, and Margaret R. Karagas, *Arsenic Exposure in Children* In: *Textbook of Children's Environmental Health*, Second Edition. Edited by: Ruth A. Etzel and Philip J. Landrigan, Oxford University Press.

Arsenic in Treated Wood

Wood treated with chromated copper arsenate (CCA) to prevent rotting due to growth of microorganisms is commonly used in marine applications, patio decks, and recreational structures for children's playgrounds. It can be a source of children's exposure to arsenic. Arsenic has been detected in soil under CCA-treated wood play structures and on the hands of children after playing on such wood structures. Children who have contact in their play with CCA-treated wood have been reported to have increased urine levels of arsenic, reflecting increased exposure.[2] In December 2003, chromated arsenicals manufacturers voluntarily discontinued manufacturing chromated arsenicals-treated wood products for homeowner uses to protect human health; however, the US Environmental Protection Agency (EPA) does not require the removal of existing structures made with wood treated with chromated arsenicals or the surrounding soil.[3]

Sources of Accidental Ingestion

Arsenic-containing rodenticides have been reported to be a source of accidental ingestion of arsenic by children. Accidental ingestion has also been reported to occur when an arsenic-containing herbicide had been transferred to a bottle that was previously used for drinking water or cola and was mistakenly ingested. Arsenic has been shown to leach from some crayons and some art paints when exposed to acidic solutions that mimic saliva.

Dietary Exposure

Consumption of food is an important source of arsenic exposure for children. In comparison with drinking water, where arsenic occurs almost exclusively in inorganic forms, the species profile for food is complex, comprising inorganic arsenic and more than 100 organic arsenic species that vary within and across foods for children. Inorganic arsenic is considered more toxic than the organic forms of arsenic, such as arsenobetaine that relates to fish/seafood consumption and is excreted in the urine unchanged. To accurately assess dietary exposure to inorganic arsenic, it is crucial to perform arsenic speciation in food and for biomarkers analysis.[4]

Among populations with access to water that complies with the WHO drinking water guideline of less than 10 µg/L, food intake is the major source of exposure.[5] Rice accumulates higher levels of arsenic from soil and water than do other crops, and this is of concern because rice is the staple food for half of the world's population and is often a first food for infants and fed to young children. The arsenic burden in rice has a complex interaction between plant physiology, genetics, and environmental factors. Rice grain accumulates most of the arsenic in the outer layers of the pericarp and aleurone, thus brown rice may contain higher concentrations of arsenic than polished/white rice.[6] Children's consumption of rice, including during the weaning period, is associated with increased urinary arsenic concentrations.[7,8] A maximum level of 100 µg/kg inorganic arsenic in rice has been advised or regulated to reduce infant's and young children's exposure.[9,10] Root tubers such as carrots, radish, and potatoes may also accumulate high arsenic contents.[11] Chicken feed might be supplemented with organoarsenical compounds (e.g., roxarsone) for control of coccidial parasites, promoting rapid weight gain, and enhancing an attractive color of the chicken's flesh; however, chickens can metabolize organoarsenic compounds to inorganic arsenic that may be transferred to the food chain. In 2011, the US Food and Drug Administration (FDA) found that the livers of roxarsone-treated chicken had elevated levels of inorganic arsenic; roxarsone's manufacturer voluntarily pulled the drug from the US market, although it is still sold overseas.[12,13] In 2013, the FDA withdrew marketing approvals for roxarsone and other arsenic-based feed additives.

Arsenic Exposure in Utero

Both inorganic arsenic and its methylated metabolites, methylarsonic acid (MMA) and dimethylarsinic acid (DMA), can cross the placenta.[14] Arsenic has been detected in cord blood, and arsenic concentrations in cord blood correlated with arsenic concentrations in maternal blood. Indeed, in a study from Bangladesh, concentrations of inorganic arsenic, MMA, and DMA in newborn blood closely approximated levels in maternal blood.[15]

Epidemiology of Health Effects of Arsenic Exposure in Children

Neurotoxicity

Several epidemiologic studies among highly exposed populations in which arsenic exposure was measured at the individual level have reported lower IQ and cognitive functioning in school children exposed to arsenic from drinking water. A study from Shanxi province in China conducted among children age 8–12 years observed a 10-point reduction in IQ in children consuming high levels of arsenic in drinking water as compared to children exposed to lower levels (190 µg/L vs. 2 µg/L).[16] In Bangladesh, a dose-response relationship between arsenic exposure and reduced intellectual function was observed among children ages 10 and 6 years in two separate cross-sectional studies.[17,18] Children 10 years of age exposed to water arsenic concentrations of greater than 50 µg/L had significantly poorer performance and full-scale IQ scores compared to children with water arsenic levels of less than 5.5 µg/L. Among children 6 years of age, adverse effects of arsenic exposure on verbal IQ and full-scale IQ were observed in girls, but not boys.[19] Among 8- to 11-year-olds, arsenic exposure was inversely associated with motor function scores.[20]

Low to moderate concentrations of arsenic in drinking water have also been associated with neurocognitive deficits in children.[1] In the United States, in a cross-sectional study of school children in grades 3–5, home tap water with arsenic levels of 5 µg/L or higher was associated with reductions in full-scale IQ and all index scores (i.e., working memory, perceptual reasoning, and verbal comprehension).[21] In Spain, detectable placenta arsenic concentrations were associated with impaired neurodevelopment,[22] and urinary arsenic with a median of 4.85 µg/L was related to a decrease in motor function in children of 4–5 years of age.[23]

Arsenic exposure from industrial activities is also of concern. In an early study in Mexico, an inverse association was observed between elevated urinary arsenic levels and decreased verbal intelligence in a small group of children living near a lead smelter. Higher concentrations of urinary arsenic were also associated with diminished long-term memory and decreased linguistic abstraction in these children.[24]

In summary, the literature is consistent in reporting adverse effects of arsenic exposure on cognitive development and IQ in children of different ages. These adverse effects are manifest in overall intelligence and are especially striking in verbal intelligence. Evidence for effects of arsenic on other aspects of neurological function, such as motor function, warrants further investigation. The decrements in neurological function in children that are associated with exposure to arsenic are related to impaired school performance and disruptions of daily activities at home and school. Therefore, there is the potential for critical impacts on children's lifelong achievement even at relatively low exposure levels.

Respiratory Effects

Several epidemiologic studies in adults have reported that arsenic exposure at high levels influences risk of lung disease, including lung cancer; adversely affects lung function; and increases risk of respiratory infections. Fewer studies have investigated the pulmonary effects of arsenic in children. In an arsenic-contaminated area of Antofagasta, Chile, the arsenic concentration in drinking water averaged 598 µg/L between 1958 and 1971, and the prevalence of cough and/or dysphonia was 37.9%. After arsenic filtration systems were installed in 1971, the prevalence of these symptoms fell to less than 7.5%.[25] From 1989 to 2000, significantly increased rates of mortality from lung cancer and bronchiectasis were observed for young adults exposed to arsenic in utero and in early childhood in comparison to a nonexposed region of Chile. Similarly, among adults exposed to arsenic in utero or in early childhood, lung function tests in adulthood revealed lower forced vital capacity in 1 second (FEV_1), lower forced vital capacity (FVC), and increased breathlessness, with effects similar in size to smoking in adulthood.[26] A prospective US study also observed a decreased FVC and FEV_1 in children (about 7 years of age) related to prenatal arsenic exposure.[27]

The mechanisms by which arsenic induces lung disease are not clear, but recent investigations have shown that arsenic may increase frequency of respiratory infection among very young children, particularly those exposed in utero. A study from Matlab, Bangladesh, observed that elevated levels of maternal urinary arsenic in pregnancy were associated with longer duration of acute respiratory infection episodes among infant boys born to these mothers.[28] In a larger group of children, higher maternal urinary arsenic level in pregnancy (262–977 µg/L) was associated with an elevated risk of lower respiratory tract infection (LRTI) and of more severe LRTI.[29] Among offspring of pregnant women who used a private well in New Hampshire, gestational urinary arsenic concentrations were associated with infant's risk of respiratory symptoms and infections at 4 months of age and increased risk of respiratory symptoms over the first year of life.[30] In this cohort, infant urinary arsenic concentrations were also associated with their gut microbiome, which plays a critical bidirectional role in the developing immune system.[31] These studies provide some evidence that warrants additional research on immunological and respiratory effects of arsenic exposure in children.

Cardiovascular Effects

Growing evidence from epidemiological studies of adults supports a role for arsenic exposure in cardiovascular disease (CVD) mortality and morbidity, as well as CVD risk factors, such as carotid atherosclerosis and hypertension. Arsenic is thought to promote CVD pathogenesis through key etiologic mechanisms, including greater inflammation, endothelial damage, and vascular lesion development.[32] Interestingly, some experimental studies have observed a greater degree of atherosclerosis and adverse effects on vascular smooth muscle and endothelial cells at doses much lower than are required to induce cancer, suggesting that the cardiovascular system may be particularly sensitive to arsenic exposure.[33]

While few children develop overt CVD, subclinical indicators such as blood pressure, lipid levels, and carotid intima media thickness can be detected in young children and may inform cardiovascular health risk in adulthood.[34] Among the first work to indicate a possible link between in utero arsenic exposure and CVD was a set of autopsy reports showing vascular lesions and death from acute myocardial infarction among infants who had lived in Antofogasta, Chile, during the 1958–1970 period of public water supply contamination with arsenic.[35] A later ecological study in this region observed that the rate of acute mortality from myocardial infarction among young adult men born during Antofogasta's period of highest arsenic contamination, and likely exposed in utero, was triple that of the general Chilean population.[36]

More recently, studies of children exposed to relatively high levels of arsenic in utero and early in life have provided additional evidence linking arsenic exposure with subclinical cardiovascular health indicators. In a cohort of 1,887 children from Bangladesh, in utero exposure to arsenic was associated with elevated blood pressure at 4.5 years of age,[37] and similar observations were reported among Bangladeshi adolescents, where both current arsenic exposure and early childhood arsenic exposure were associated with higher blood pressure and with stronger effects observed among those with higher body mass index.[38] In a pair of studies of children in Mexico with relatively stable residence, urinary arsenic was associated with increases in carotid intima-media thickness, asymmetric dimethylarginine levels, blood pressure, left ventricular mass, and ejection fraction.[39,40] In New Hampshire, where average arsenic exposure levels from water tend to be somewhat lower than those in Bangladesh and Mexico, positive associations between maternal prenatal urinary arsenic and infant plasma markers of intercellular adhesion molecules and vascular adhesion molecules were observed.[41]

Overall, these studies begin to suggest a role for early-life arsenic exposure on subclinical cardiovascular health outcomes in childhood, but more work is needed to understand whether these effects persist into adulthood and whether effective interventions to reduce arsenic during critical windows could reduce later-life CVD risk.

Childhood Cancer

Although the effect of arsenic exposure on cancer risk in adults is well documented, only a few studies have examined the relationship between arsenic exposure and risk of childhood cancer. In a population-based case-control study of childhood acute lymphoblastic leukemia in Québec, Canada, arsenic exposure, measured using a tap water survey, was

unrelated to disease risk.[42] In an ecological study of childhood cancer undertaken between 1979 and 1999 in all 17 Nevada counties, leukemia risks were not increased at
the concentrations of arsenic in water encountered (10–90 μg/L).[43] The standardized
incidence ratio for childhood cancer was, however, elevated in the most highly exposed
group in this study, thus raising the possibility of increased risk in populations with
higher levels of arsenic exposure.

In an ecological study from Antofagasta, Chile, when arsenic levels in drinking water
were highly elevated (870 μg/L), mortality rates from the most common childhood cancers, leukemia, and brain cancer were not elevated.[44] However, mortality from liver cancer among children occurred at higher rates than expected.

In a study conducted in Okayama Prefecture, Japan, investigators examined cancer
risk later in life among children who had been exposed in 1955 as infants to arsenic in
contaminated milk powder. This was a very severe episode of mass poisoning in which
roughly 2,000 infants were poisoned, and many died. The data showed that both the
total numbers of cancers and the number of liver cancers were elevated among children
exposed before 1 year of age. In addition, pancreatic and hematopoietic cancers were elevated among children exposed before 5 years of age.[45] In 2006, hundreds of surviving
victims, then in their 50s, reportedly suffered from mental retardation, neurological diseases, and other disabilities.[46] An attempt to follow-up causes of death in the survivors by
abstracting cancer mortality data for the Okayama Prefecture suggests increased mortality ratios for leukemia and pancreatic cancer in young adults.[45]

In summary, although the literature is limited, there is some evidence that high-level
arsenic exposure in early life increases risk of childhood liver cancer. The evidence is
scant and inconsistent on whether arsenic exposure in early life is associated with other
common childhood cancers or to cancers that have been related to arsenic exposure in
adults, such as bladder and lung cancer. But given the well-established carcinogenicity of
arsenic exposure in adult life, further long-term follow-up study of populations exposed
to arsenic in early life is clearly warranted.

Reproductive Health Outcomes

Epidemiologic studies of arsenic exposure in relation to fetal loss include investigations
of spontaneous abortions (loss up to 28 weeks of pregnancy) and stillbirths (loss after 28
weeks).

In a large study from Matlab, Bangladesh, consumption during pregnancy of well
water containing more than 50 μg/L significantly increased the risks of fetal loss and infant death.[47]

Reductions in birth weight have been reported in three studies from Chile, Taiwan,
and Bangladesh. In Chile, the estimated reduction in average birth weight associated
with prenatal exposure to arsenic was 57 g, in Taiwan 30 g, and, in Bangladesh, it was
estimated that each increase of 1 μg/g in maternal hair arsenic level during pregnancy
was associated with a 193 g reduction in birth weight.[48]

In a US study of lower arsenic levels, maternal urinary arsenic concentration was
associated with increased birth length and decreased Ponderal Index for infants of
overweight/obese, but not normal weight, mothers.[49] Maternal toenail arsenic concentrations were related to reduced newborn's head circumference, especially among
males, and an increase in birth length and weight among females.[50] A doubling in hair

arsenic was associated with a 72.2 g lower birth weight in Latina mothers in urban Los Angeles.[51]

More research on the reproductive toxicity of arsenic is needed with better-defined outcomes and more comprehensive data on timing and level of the exposure before and during pregnancy as well as other major prenatal risk factors.

Toxicology of Arsenic

The specific mechanisms by which arsenic causes toxicity and carcinogenicity are not well understood. Experiments in animals and in vitro indicate that arsenic acts at the cellular level at low doses of exposure.[52] The valence state and form of arsenic are of great importance for the toxicity. The trivalent forms of arsenic are the principal toxic forms that react with thiol-containing enzymes or protein sulfhydryls and could lead to the inhibition of essential biochemical reactions, alteration of cellular redox status, and eventual cytotoxicity.[53]

Several possible mechanisms of action of the carcinogenic effect of arsenic have been proposed: chromosomal abnormalities, oxidative stress, altered DNA repair, altered DNA methylation, altered growth factors, enhanced cell proliferation, altered cell signaling, promotion/progression, gene amplification, p53 gene suppression, induction of co-carcinogenicity, and perturbations of histone modifications and miRNAs expression.[54] Previous studies indicated that inorganic arsenic does not act through classic genotoxic or mutagenic mechanisms but rather may be a tumor promoter that modifies signal transduction pathways involved in cell growth and proliferation.[55] Arsenite has been shown to modulate expression and/or DNA binding activities of several key transcription factors, including nuclear factor kappa B, p53, and activating protein-1 (AP-1).[52] Mechanisms of AP-1 activation by trivalent inorganic arsenic include stimulation of the mitogen-activated protein kinase cascade with a consequent increase in the expression and/or phosphorylation of the two major AP-1 constituents, c-Jun and c-Fos.[55]

A direct toxic effect of arsenic on developing brain cells is possible because arsenic is a pro-oxidant and may cause oxidative stress in the sensitive developing brain. In situations of prolonged arsenic exposure, the neurotoxic effect is associated with changes in the function of brain cell membranes caused by the generation of reactive oxygen species and nitrogen oxide. It has also been demonstrated that arsenite at low concentrations and in a dose-dependent manner induces reactive oxygen species and activates early transcription factors, such as NF-kappa B and AP-1 in a dopamine-producing cell line.[55] Accumulating evidence now indicates that arsenic also alters the epigenome in the brain cortex and hippocampus in mice. The latter is notable in that the hippocampus plays roles in spatial memory and in the consolidation of short-term working memory into long-term memory.[56]

For respiratory effects, although the mechanism is not known, several human and animal studies have shown that arsenic is deposited and stored in the lung, especially in the epithelium.[57] It is possible that the arsenic deposited in the lung acts like some other metals to enhance tissue inflammation or increase pulmonary fibrosis, leading to impaired respiratory function. Some in vitro evidence indicates inorganic arsenic exposure may alter the expression of an important protease in lung function, matrix metalloproteinase (MMP-9), and adenosine triphosphate (ATP)-dependent Ca^{2+} signaling.[58] Both could affect the signaling pathways for cell migration and alter the airway epithelial barrier by restricting proper wound repair and compromising innate airway defense mechanisms. Higher permeability of the bronchial epithelium and the resultant loss of Clara cell protein (CC16) and other lung proteins as well as oxidative stress may also be involved in arsenic-induced lung toxicity.[59]

Arsenic Metabolism and Toxicity

It is becoming increasingly evident that the toxicity and carcinogenicity of arsenic is likely to be closely associated with metabolic processes. The major metabolic pathway for inorganic arsenic in humans is methylation. Methylation of arsenic (III) first generates monomethylarsonic acid [MMA (V)]. After the reduction of MMA (V) to monomethylarsonous acid [MMA (III)], a second methylation can occur to generate dimethylarsinic acid [DMA (V)]. It is unclear to what extent DMA (V) is reduced to dimethylarsinous acid [DMA (III)] in vivo because DMA (III) is an unstable intermediate. Methylation of inorganic arsenic has long been considered a detoxification process because the methylated species most commonly found in human urine, MMA (V) to DMA (V), are more water soluble, more readily excreted, and often less acutely toxic than inorganic arsenic based on evidence from in vitro studies.

There is considerable experimental evidence suggesting that arsenic methylation—specifically the formation of MMA (III)—leads to increased arsenic toxicity. MMA (III) has been found to be more toxic than inorganic arsenic or any of the pentavalent metabolites to human hepatocytes, epidermal keratinocytes, and bronchial epithelial cells. Both MMA (III) and DMA (III) have been shown to cause enzyme inhibition, cell toxicity, genotoxicity, and clastogenecity.[52] In addition, DMA (III) induces the activation of oncogenes and produces bladder cancer in rats, suggesting that it may also be carcinogenic in humans.

Emerging experimental data indicate that gut microbiome may contribute to the pre-systemic biotransformation of ingested arsenic (systemic metabolism being defined as all metabolic reactions carried out by human cells).[60] For instance, Rowland and Davies[61] reported the reduction of arsenate to arsenite and the formation of MMA and DMA by rat intestinal and cecal bacteria. Also in this study, when the gut contents were autoclaved or antibiotics were added prior to incubation, no arsenate metabolism took place, indicating that the bacterial flora of the gut was critical. Several studies also reported that in vitro anaerobic microbiota of mouse cecum can catalyze production of MMA (V) and DMA (V) from inorganic arsenic.[62] Van de Wiele et al. found a high level of inorganic arsenic methylation by in vitro cultured human colon microbiota, including the formation of MMA (V) and the highly toxic MMA (III).[63] Recently, higher MMA levels and MMA-to-DMA ratios were also observed in mice with disrupted gut microbiome.[64] These data suggest that methylation of arsenic may actually take place before arsenic crosses into the internal environment of our bodies. In addition, the gut microbiome can also indirectly be involved in inorganic arsenic metabolism through the host, such as altering inorganic arsenic absorption, cofactors, and genes related to inorganic arsenic metabolism.[60]

Susceptibility to Arsenic Toxicity

The epidemiologic evidence is overwhelming that individuals who are relatively ineffective in converting MMA to DMA, as evidenced by a relatively high MMA% in urine or a high MMA-to-DMA ratio, are at increased risk for a host of adverse health outcomes from arsenic, including skin lesions, skin cancer, lung and bladder cancer, and CVD.[65] Conversely, individuals who can efficiently methylate MMA to DMA are at lower risk for disease, presumably because DMA has a short half-life and is preferentially eliminated by the kidney. A genome-wide association study of 1,313 arsenic-exposed Bangladeshi individuals found five single nucleotide polymorphisms near the gene for arsenic methyltransferase that were associated with poor conversion of MMA to DMA, one of which was also associated with an increased risk for skin lesions.[66] Another recent study of 1,660 arsenic-exposed individuals that measured protein-coding variants across the human exome identified an additional genetic variation in formiminotransferase cyclodeaminase, which was associated with increased urinary inorganic arsenic%, MMA%, and decreased DMA%, as well as increased skin lesion risk.[67] Moreover, genetic polymorphisms in genes involved in potential pathways by which inorganic arsenic leads to various diseases, particularly CVD, may modify the effects of inorganic arsenic exposure on disease risk.[32]

Other studies in the Bangladeshi population have reported that dietary intake of nutrients influences the toxicity of arsenic by modulating its metabolism.[68] For instance, administration of folic acid to folate-deficient persons has been shown to facilitate the methylation of MMA to DMA and thereby lower blood arsenic levels, presumably by facilitating the synthesis of S-adenosylmethione, the universal methyl donor for all biochemical methylation reactions.[69] In addition to methylation capacity, human susceptibility to arsenic toxicity is significantly modulated by many other intrinsic and extrinsic factors that may increase its health risks.

Although a substantial number of studies have been conducted on susceptibility to the health effects of arsenic exposure in adults, little is known about the influence of these factors on the health effects of arsenic exposure in children. A recent study suggests that children's metabolism of inorganic arsenic differs from that in adults, which might explain the lack of data on arsenic metabolism as a susceptibility factor for arsenic toxicity in children. The main influencing factor for the urinary arsenic metabolites in children was selenium status, which unexpectedly was associated with increasing inorganic arsenic% and MMA% but decreasing DMA%. Plasma folate was associated with lower inorganic arsenic% and higher DMA%. Other nutritional markers (zinc, copper, and manganese, hemoglobin, vitamin B_{12}, and child anthropometry) had no apparent impact, nor did children's age or gender. Studies on susceptibility in children are needed because such investigation may help identify high-risk subpopulations and future intervention.

Prevention

Arsenic exposure is related to significant adverse health effects in children through several potential mechanisms. Earlier studies mainly assessed effects of exposure at high levels, often using ecological measures of arsenic exposure. A growing number of studies have employed modern epidemiologic methods and technological advances, with the inclusion of biomarkers for exposure such as arsenic levels in urine, toenail, or blood samples and biologic response, including changes in the epigenome and microbiome. Use of more sophisticated approaches have enabled the examination of toxic effects of arsenic at lower levels of exposure. The available epidemiology studies indicate positive associations between arsenic exposure and neurological effects in children, especially diminished verbal IQ scores, increased risks of respiratory diseases, and adverse reproductive outcomes. Arsenic is associated with cardiometabolic outcomes and cancer among adults, and emerging evidence suggests effects on early indicators of CVD and potential risk of childhood liver cancer among the highly exposed. Research questions on susceptibility, in utero and early-life exposures, and mixture effects need to be addressed.

Food

Rice cultivar selection and agronomic practices dominate as driving factors of grain arsenic concentrations. The influence of genetic variation on grain arsenic indicates that breeding low-arsenic rice cultivars is a viable approach and is currently under way. Manipulating flooding cycles in rice production, tied to sustainable agricultural practices for reducing water use, can also reduce grain arsenic. Fertilization strategies, particularly using silicon, have potential as effective strategies to reduce arsenic and prevent certain plant diseases (e.g., straighthead disease in rice).[70]

Rice polishing and washing reduces the arsenic burden in the rice grain, and greater volumes of cooking water facilitates the removal of inorganic arsenic from rice during cooking. Root tubers accumulate most of the arsenic in the outer skin by absorption, thus washing and peeling before consumption reduces exposure.[71] Several countries have already phased out the use of organosenical compounds in chicken feed to reduce the inorganic arsenic burden in chicken flesh.[12]

Water

Prevention of exposure to arsenic can be achieved via pretreatment of water systems. Treatments for arsenic removal include preoxidation, ion exchange, pH adjustment, and reverse osmosis. In rural areas or countries like Bangladesh, where treatment of public water supply is costly and not immediately feasible, common arsenic remediation approaches include comprehensive well testing, promotion of well switching through health education, the provision of filtration devices, and the utilization of deeper aquifers that are low in arsenic.

Several studies have shown effectiveness of arsenic mitigation programs in reducing arsenic exposure, measured as reduction in urinary arsenic over time. However, studies that assess long-term compliance with well-switching strategies are needed.

In a prospective study from the United States, informing participants of their water test results reduced the use of tap water for drinking and cooking and for mixing formula among those with arsenic concentrations of greater than 10 μg/L compared to those with concentrations of less than 5 μg/L.[72] A practice-based study targeting pediatric practices was designed to encourage people to test their household private water. Using a randomized design, the intervention arm with reminders and a structured follow-up versus discussion yielded higher testing completion rates. In both arms, kits distributed by clinicians (vs. nursing staff) were the most effective.[73]

Clearly, successful prevention of arsenic-related health effects in children relies on mitigation strategies that are feasible and sustainable. With accumulating evidence of adverse health effects of arsenic due to exposure in utero and/or early in life, such programs should target women of childbearing age and young children.

References

1. National Research Council. *Critical Aspects of EPA's IRIS Assessment of Inorganic Arsenic: Interim Report.* Washington, DC: National Academies Press; 2013.

2. Shalat SL, Solo-Gabriele HM, Fleming LE, et al. A pilot study of children's exposure to CCA-treated wood from playground equipment. *Sci Total Environ.* 2006;367(1):80–8.

3. Morais S, Fonseca HMAC, Oliveira SMR, et al. Environmental and health hazards of chromated copper arsenate-treated wood: a review. *Int J Environ Res Public Health.* 2021;18(11):5518; 1–12.

4. European Food Safety Authority (EFSA). Scientific opinion on arsenic in food. EFSA panel on contaminants in food chain (CONTAM). *EFSA J.* 2009;7(10):1351; 1–199.

5. Kurzius-Spencer M, Burgess JL, et al. Contribution of diet to aggregate arsenic exposures-an analysis across populations. *J Exposure Sci Environ Epidemiol.* 2014;24(2):156–62.

6. Meharg AA, Lombi E, Williams PN, et al. Speciation and localization of arsenic in white and brown rice grains. *Environ sci technol.* 2008;42(4):1051–7.

7. Davis MA, Signes-Pastor AJ, Argos M, et al. Assessment of human dietary exposure to arsenic through rice. *Sci Total Environ.* 2017;586:1237–44.

8. Signes-Pastor AJ, Cottingham KL, Carey M, et al. Infants' dietary arsenic exposure during transition to solid food. *Sci Rep.* 2018;8:2–9.

9. European Communities. Commission Regulation 2015/1006 of 25 June 2015 amending Regulation (EC) No. 1881/2006 as regards maximum levels of inorganic arsenic in foodstuffs. *Official J Eur Comm.* 2015;161:14–16.

10. FDA. FDA proposes limit for inorganic arsenic in infant rice cereal. 2016. https://www.fda.gov/news-events/press-announcements/fda-proposes-limit-inorganic-arsenic-infant-rice-cereal. Accessed on 09/05/2023.

11. Signes-Pastor AJ, Mitra K, Sarkhel S, et al. Arsenic speciation in food and estimation of the dietary intake of inorganic arsenic in a rural village of West Bengal, India. *J Agricult Food Chem.* 2008;56(20):9469–74.

12. Mondal NK. Prevalence of arsenic in chicken feed and its contamination pattern in different parts of chicken flesh: a market basket study. *Environ Monitor Assess.* 2020;192(9):590.

13. Schmidt CW. Arsenical association: inorganic arsenic may accumulate in the meat of treated chickens. *Environ Health Perspect.* 2013;121(7):A226.

14. Concha G, Vogler G, Lezcano D, Nermell B, Vahter M. Exposure to inorganic arsenic metabolites during early human development. 1998;44:185–90.

15. Hall M, Gamble M, Slavkovich V, et al. Determinants of arsenic metabolism: blood arsenic metabolites, plasma folate, cobalamin, and homocysteine concentrations in maternal-newborn pairs. *Environ Health Perspect.* 2007;115(10):1503–9.

16. Wang SX, Wang ZH, Cheng XT, et al. Arsenic and fluoride expose in drinking water: children's IQ and growth in Shanyin Country, Shanxi Province, China. *Environ Health Perspect.* 2007;115(4):643–7.

17. Wasserman GA, Liu X, Parvez F, et al. Water arsenic exposure and children's intellectual function in Araihazar, Bangladesh. *Environ Health Perspect.* 2004;112(13):1329–33.

18. Wasserman GA, Liu X, Parvez F, et al. Water arsenic exposure and intellectual function in 6-year-old children in Araihazar, Bangladesh. *Environ Health Perspect*. 2007;115(2):285–9.

19. Hamadani JD, Tofail F, Nermell B, et al. Critical windows of exposure for arsenic-associated impairment of cognitive function in pre-school girls and boys: a population-based cohort study. *Int J Epidemiol*. 2011;40(6):1593–604.

20. Parvez F, Wasserman GAGA, Factor-Litvak P, et al. Arsenic exposure and motor function among children in Bangladesh. *Environ Health Perspect*. 2011;119(11):1665–70.

21. Wasserman GA, Liu X, LoIacono NJ, et al. A cross-sectional study of well water arsenic and child IQ in Maine schoolchildren. *Environ Health*. 2014;13(1):1–0.

22. Freire C, Amaya E, Gil F, et al. Prenatal co-exposure to neurotoxic metals and neurodevelopment in preschool children: the Environment and Childhood (INMA) Project. *Sci Total Environ*. 2018;621:340–51.

23. Signes-Pastor AJ, Vioque J, Navarrete-Muñoz EM, et al. Inorganic arsenic exposure and neuropsychological development of children of 4–5 years of age living in Spain. *Environ Res*. 2019;174(Oct 2018):135–42.

24. Calderón J, Navarro ME, Jimenez-Capdeville ME, et al. Exposure to arsenic and lead and neuropsychological development in Mexican children. *Environ Res*. 2001;85(2):69–76.

25. Zaldivar R, Gullier A. Environmental and clinical investigations on endemic chronic arsenic poisoning in infants and children. *Zentralblatt fur bakteriologie mikrobiologie und hygiene serie b-umwelthygiene krankenhaushygiene arbeitshygiene praventive medizin*. 1977;165(2):226–34.

26. Dauphiné DC, Ferreccio C, Guntur S, et al. Lung function in adults following in utero and childhood exposure to arsenic in drinking water: preliminary findings. *Int Arch Occ Environ Health*. 2011;84(6):591–600.

27. Signes-Pastor AJ, Martinez-Camblor P, Baker E, Madan J, Guill MF, Karagas MR. Prenatal exposure to arsenic and lung function in children from the New Hampshire Birth Cohort Study. *Environ Int*. 2021;155:106673.

28. Rahman A, Vahter M, Ekström E-C, Persson L-Å. Arsenic exposure in pregnancy increases the risk of lower respiratory tract infection and diarrhea during infancy in Bangladesh. *Environ Health Perspect*. 2011 Dec;119(5):719–24.

29. Raqib R, Ahmed S, Sultana R, et al. Effects of in utero arsenic exposure on child immunity and morbidity in rural Bangladesh. *Toxicol Lett*. 2009;185(3):197–202.

30. Farzan SF, Li Z, Korrick SA, et al. Infant infections and respiratory symptoms in relation to arsenic exposure in a U.S. Cohort. *Environ Health Perspect*. 2016;124(6):840–7.

31. Hoen AG, Madan JC, Li Z, et al. Sex-specific associations of infants' gut microbiome with arsenic exposure in a US population. *Sci Rep*. 2018;8(1):12627.

32. Wu F, Molinaro P, Chen Y. Arsenic exposure and subclinical endpoints of cardiovascular diseases. *Curr Environ Health Rep*. 2014;1(2):148–62.

33. Lemaire M, Lemarié CA, Molina MF, Schiffrin EL, Lehoux S, Mann KK. Exposure to moderate arsenic concentrations increases atherosclerosis in ApoE-/- mouse model. *Toxicol Sci*. 2011;122(1):211–21.

34. Chen X, Wang Y. Tracking of blood pressure from childhood to adulthood: a systematic review and meta-regression analysis. *Circulation*. 2008;117(25):3171–80.

35. Rosenberg HG. Systemic arterial disease and chronic arsenicism in infants. *Arch Pathol*. 1974;97(6):360–5.

36. Yuan Y, Marshall G, Ferreccio C, Steinmaus C, Selvin S, Liaw J, Bates MN, Smith AH. Original Contribution Acute Myocardial Infarction Mortality in Comparison with Lung and Bladder Cancer Mortality in Arsenic-exposed Region II of Chile from 1950 to 2000. *Am J Epidemiol.* 2007;166(12):1381–91.

37. Hawkesworth S, Wagatsuma Y, Kippler M, et al. Early exposure to toxic metals has a limited effect on blood pressure or kidney function in later childhood, rural Bangladesh. *Int J Epidemiol.* 2013;42(1):176–85.

38. Chen Y, Wu F, Liu X, et al. Early life and adolescent arsenic exposure from drinking water and blood pressure in adolescence. *Environ Res.* 2019;178:108681.

39. Osorio-Yáñez C, Ayllon-Vergara JC, Aguilar-Madrid G, et al. Carotid intima-media thickness and plasma asymmetric dimethylarginine in Mexican children exposed to inorganic arsenic. *Environ Health Perspect.* 2013;121(9):1090–6.

40. Osorio-Yáñez C, Ayllon-Vergara JC, Arreola-Mendoza L, et al. Blood pressure, left ventricular geometry, and systolic function in children exposed to inorganic arsenic. *Environ Health Perspect.* 2015;123(6):629–35.

41. Farzan SF, Brickley EB, Li Z, et al. Maternal and infant inflammatory markers in relation to prenatal arsenic exposure in a U.S. pregnancy cohort. *Environ res.* 2017;156:426–33.

42. Infante-Rivard C, Olson E, Jacques L, Ayotte P. Drinking water contaminants and childhood leukemia. *Epidemiology.* 2001;12(1):13–19.

43. Moore LE, Lu M, Smith AH. Childhood cancer incidence and arsenic exposure in drinking water in Nevada. *Arch Environ Health.* 2002;57(3):201–6.

44. Liaw J, Marshall G, Yuan Y, Ferreccio C, Steinmaus C, Smith AH. Increased childhood liver cancer mortality and arsenic in drinking water in northern Chile. *Cancer Epidemiol Biomarkers Prevent.* 2008;17(8):1982–7.

45. Yorifuji T, Tsuda T, Grandjean P. Unusual cancer excess after neonatal arsenic exposure from contaminated milk powder. *J Nat Cancer Inst.* 2010;102(5):360–1.

46. Dakeishi M, Murata K, Grandjean P. Long-term consequences of arsenic poisoning during infancy due to contaminated milk powder. *Environ Health.* 2006;5(1):31.

47. Rahman A, Vahter M, Ekström E-C, et al. Association of arsenic exposure during pregnancy with fetal loss and infant death: a cohort study in Bangladesh. *Am J Epidemiol.* 2007;165(12):1389–96.

48. Smith AH, Steinmaus CM. Health effects of arsenic and chromium in drinking water: recent human findings. *Ann Rev Public Health.* 2009;30:107–22.

49. Gilbert-Diamond D, Emond JA, Baker ER, Korrick SA, Karagas MR. Relation between in utero arsenic exposure and birth outcomes in a cohort of mothers and their newborns from New Hampshire. *Environ Health Perspect.* 2016;124(8):1299–1307.

50. Signes-Pastor AJ, Doherty BT, Romano ME, et al. Prenatal exposure to metal mixture and sex-specific birth outcomes in the New Hampshire Birth Cohort Study. *Environ Epidemiol.* 2019;3(5):e068.

51. Howe CG, Farzan SF, Garcia E, et al. Arsenic and birth outcomes in a predominately lower income Hispanic pregnancy cohort in Los Angeles. *Environ Res.* 2020;184:109294.

52. Tchounwou PB, Patlolla AK, Centeno JA. Carcinogenic and systemic health effects associated with arsenic exposure–a critical review. *Toxicol Pathol.* 2003;31(6):575–88.

53. Hughes MF. Arsenic toxicity and potential mechanisms of action. *Toxicol Lett.* 2002;133(1):1–6.

54. Tchounwou PB, Yedjou CG, Udensi UK, et al. State of the science review of the health effects of inorganic arsenic: perspectives for future research. *Environ Toxicol*. 2019;34(2):188–202.

55. Simeonova PP, Luster MI. Mechanisms of arsenic carcinogenicity: genetic or epigenetic mechanisms? *J Environ Pathol Toxicol Oncol*. 2000;19(3):281–6.

56. Tyler CR, Allan AM. The effects of arsenic exposure on neurological and cognitive dysfunction in human and rodent studies: a review. *Curr Environ Health Rep*. 2014;1(2):132–47.

57. Saady J, Blanke R, Poklis A. Estimation of the body burden of arsenic in a child fatally poisoned by arsenite weedkiller. *J Analytical Toxicol*. 1989;13(5):310–2.

58. Sherwood CL, Lantz RC, Burgess JL, Boitano S. Arsenic alters ATP-dependent Ca²+ signaling in human airway epithelial cell wound response. *Toxicol Sci*. 2011;121(1):191–206.

59. Parvez F, Chen Y, Brandt-Rauf PW, et al. Nonmalignant respiratory effects of chronic arsenic exposure from drinking water among never-smokers in Bangladesh. *Environ Health Perspect*. 2008;116(2):190–5.

60. Yang Y, Chi L, Lai Y, Hsiao Y-C, Ru H, Lu K. The gut microbiome and arsenic-induced disease-iAs metabolism in mice. *Curr Environ Health Rep*. 2021;8(2):89–97.

61. Rowland IR, Davies MJ. In vitro metabolism of inorganic arsenic by the gastrointestinal microflora of the rat. *J Appl Toxicol*. 1981;1(5):278–83.

62. Pinyayev TS, Kohan MJ, Herbin-Davis K, Creed JT, Thomas DJ. Preabsorptive metabolism of sodium arsenate by anaerobic microbiota of mouse cecum forms a variety of methylated and thiolated arsenicals. *Chem Res Toxicol*. 2011;24(4):475–7.

63. Van de Wiele T, Gallawa CM, Kubachka KM, et al. Arsenic metabolism by human gut microbiota upon in vitro digestion of contaminated soils. *Environ Health Perspect*. 2010;118(7):1004–9.

64. Chi L, Xue J, Tu P, Lai Y, Ru H, Lu K. Gut microbiome disruption altered the biotransformation and liver toxicity of arsenic in mice. *Arch Toxicol*. 2019;93(1):25–35.

65. Kuo C-CC, Moon KA, Wang S-LL, Silbergeld EK, Navas-Acien A. The association of arsenic metabolism with cancer, cardiovascular disease, and diabetes: a systematic review of the epidemiological evidence. *Environ Health Perspect*. 2017;125(8):87001.

66. Pierce BL, Kibriya MG, Tong L, et al. Genome-wide association study identifies chromosome 10q24.32 variants associated with arsenic metabolism and toxicity phenotypes in Bangladesh. *PLoS Genet*. 2012;8(2):e1002522.

67. Pierce BL, Tong L, Dean S, et al. A missense variant in FTCD is associated with arsenic metabolism and toxicity phenotypes in Bangladesh. *PLoS Genet*. 2019;15(3):e1007984.

68. Kile ML, Ronnenberg AG. Can folate intake reduce arsenic toxicity? *Nutrit Rev*. 2008;66(6):349–53.

69. Gamble MV, Liu X, Slavkovich V, et al. Folic acid supplementation lowers blood arsenic. *Am J Clin Nutrit*. 2007;86(4):1202–9.

70. Norton GJ, Islam MR, Deacon CM, et al. Identification of low inorganic and total grain arsenic rice cultivars from Bangladesh. *Environ Sci Technol*. 2009;43(15):6070–5.

71. Carbonell-Barrachina AA, Signes-Pastor AJ, Vázquez-Araújo L, Burló F, Sengupta B. Presence of arsenic in agricultural products from arsenic-endemic areas and strategies to reduce arsenic intake in rural villages. *Mol Nutrit Food Res*. 2009;53(5):531–41.

72. He X, Karagas MR, Murray C. Impact of receipt of private well arsenic test results on maternal use of contaminated drinking water in a U.S. population. *Science Total Environ*. 2018;643:1005–12.

73. Murray CJ, Olson AL, Palmer EL, et al. Private well water testing promotion in pediatric preventive care: a randomized intervention study. *Prevent Med Rep*. 2020;20:101209.

34

Fluoride and Manganese

Yenny Fariñas Diaz, Somaiyeh Azmoun, and Roberto G. Lucchini

Fluorides and manganese are widely diffused in the environment and can be harmful for newborn infants and young children. These metallic elements are well-known occupational hazards and have been recognized for decades to pose health risks to exposed workers. More recently, the spread of these metals beyond the workplace as the result of both natural and anthropogenic activities has been documented. Once they have been released into the general environment, metals can reach children via multiple routes. Emissions of metals from industrial and agricultural operations can lead to contamination of air, soil, water, and the food supply. Drinking water and dietary products in contaminated localities can contain high levels of fluorides and manganese. Airborne particles and soil can contain high concentration especially near "hotspots" such as mines and industrial facilities. Children can absorb metals from the environment via multiple routes including inhalation and through inadvertent ingestion of contaminated dust and soil as a consequence of their normal, age-related hand-to-mouth behavior. Metals can cause toxic effects in multiple organ systems in children. The nervous system is particularly sensitive to metal toxicity during early development. Manganese and fluoride target motor and cognitive functions in the developing nervous system. A specific effect of excessive exposure to fluorides in early life is fluorosis of the teeth and bones.

Fluoride

Sources of Fluoride Exposure

Fluoride [F^-] is abundant in the earth's crust, and fluorides are naturally occurring components of rocks, soil, plants, and animals. In industrial settings, fluorine and its compounds are used in producing uranium, plastics, ceramics, pesticides, and pharmaceuticals. Fluorochlorohydrocarbons are used in refrigeration and aerosol propellant applications. Fluorides are present in drinking water through natural releases from these sources (e.g., weathering of rocks and runoff from soil). Other sources of fluoride include toothpaste, milk, tea, table salt, seafood, and medicinal supplements. Fluoride compounds can also be found because of industrial processes that use the mineral apatite.[1] Anthropogenic sources of fluoride emissions include the combustion of fluorine-containing materials, which releases hydrogen fluoride and particulate fluorides into the air, thus becoming a potential inhalation contaminant. Coal contains small amounts of fluorine, and coal-fired powerplants constitute the largest source of anthropogenic hydrogen fluoride emissions. According to the US Environmental Protection Agency's

Yenny Fariñas Diaz, Somaiyeh Azmoun, and Roberto G. Lucchini, *Fluoride and Manganese* In: *Textbook of Children's Environmental Health*, Second Edition. Edited by: Ruth A. Etzel and Philip J. Landrigan, Oxford University Press. © Oxford University Press 2024.
DOI: 10.1093/oso/9780197662526.003.0034

Toxic Chemical Release Inventory,[2] electrical utilities are the single largest industrial source of fluoride emissions. Additional major industrial sources of fluoride emissions are aluminum production and phosphate fertilizer plants; both emit hydrogen fluoride and particulate fluorides. Other sources of hydrogen fluoride release are chemical production, steel making, magnesium production, and brick and structural clay products manufacture. Hydrogen fluoride also is released by municipal incinerators as a consequence of the presence of fluoride-containing material in the waste stream. Fluorides also are released into surface water from municipal wastewater as a result of water fluoridation (usually obtained with the addition of hexafluorosilicic acid) to prevent dental caries.[3,4] Therefore, fluoride absorption into the human body can take place through both inhalation and ingestion. Occupational exposure to these industrial processes, if not controlled, can result in exposure to potentially harmful levels of fluoride, thus environmental and safety controls are key in preventing such exposures.

Children are exposed to fluoride through consumption of drinking water, foods, dental products such as fluoride-containing toothpaste, and medical drugs, and through exposure to airborne contaminants. The fluorides used in toothpastes and dental products are sodium fluoride, sodium mono-fluorophosphate, and stannous fluoride. Fluoridated dentifrices and mouth rinses are important sources of fluoride, particularly for small children who do not have complete control of their swallowing reflex. Children living in areas with natural and/or anthropogenic high fluoride levels in water[5] and soil may be exposed to high levels of fluoride in drinking water. This is especially true if drinking water is derived from private wells, water sources where water quality is unregulated, or from water contaminated sources in industrial areas. Tap or bottled water that does not conform to regulatory water quality standards for different biological, chemical, and physical parameters can be a source of toxicity to humans. Some bottled water supplies still contain high fluoride levels, as found in a study conducted in the United Arab Emirates (UAE), where bottled water is the major source for drinking. The research focused on measuring fluoride levels compared to international and local standards for different bottled water brands of bottled water sold in UAE to determine if fluoride was a health problem. Seven out of the 23 brands evaluated contained fluoride, with one containing fluoride levels higher than all standards.[6]

Fluoride can be absorbed by the fetus through passive diffusion from the placenta[7] and cross the blood-brain barrier.[8] It is not excreted through human breast milk, thus babies that are exclusively breastfed will receive virtually no fluoride exposure.[9] For most children, water is the predominant source of fluoride intake. This is due in large part to children receiving fluoridated water mixed with infant formula concentrate. Children who frequently drink tea may also be exposed to high levels of fluoride in their diets. Children living near industrial sources of fluoride may be exposed to elevated levels of fluorides in the air they breathe. Vegetables and fruits grown near such sources may contain higher levels of fluorides as a consequence of fluoride-containing dust settling on food plants.[10] Fluoride retention varies in children, infants, and adults. Children and infants retain higher proportions of absorbed fluoride (about 80–90%) compared to adults (50–60%).[2,11]

Health Effects of Fluorides: Fluorosis

Dental fluorosis, which is characterized by an increased porosity or hypomineralization of the tooth enamel, results from excess exposure to fluoride during tooth development (ages 1–8 years). The development and severity of dental fluorosis depend on the amount of fluoride ingested, the duration of exposure, and the stage of enamel development at the time of exposure. In the more severe cases, the tooth enamel is discolored, pitted, and prone to fracture and wear. Chronic exposure to high levels of fluoride results in muscle impairment, mineralization of tendons and muscle attachments, and bone thickness in the central skeleton.[12] Fluoride can stimulate bone cell proliferation and help optimize bone mineral density, which is essential in sustaining healthy bones throughout life.[12] Fluoride is known to increase osteoblast activity and bone density, especially in the lumbar spine. However, excess fluoride also causes skeletal fluorosis, which results from long-term ingestion of fluoride during the bone modeling (growth) and/or remodeling stages. Most bone growth takes place between 12 and 26 weeks of age, while bone remodeling actively continues until about 40 years of age. During this process, the ingestion and excessive retention of fluoride in the body can lead to skeletal fluorosis.

Fluoride is commonly found in water and air, but the major dietary source of fluoride is drinking water, thus water fluoridation was a flagship method of adding fluoride to the diet in the 20th century. There is no recommended dietary fluoride allowance, thus fluoride intake is expressed as an adequate intake. The adequate intake for fluoride for toddlers is 0.7 mg/day, rising to 3 mg/day for adult women and 4 mg/day for adult men. Fluoride dietary intake is absorbed rapidly in the stomach and small intestine. Calcified tissues absorb one-quarter to one-third of the intake of fluoride and the rest is lost in urine. Bones and teeth contain 99% of total body fluoride.[1]

Per the National Institutes of Health (NIH) Fluoride Fact Sheet for Health Professionals, typical daily fluoride intakes in the United States from foods and beverages (including fluoridated drinking water) are 1.2–1.6 mg for infants and toddlers younger than 4 years, 2.0–2.2 mg for children ages 4–11 years, 2.4 mg for those ages 11–14 years, and 2.9 mg for adults.[13] Optimal [F^-] levels to protect from dental decay are 0.7–1.2 mg/L. According to the US Environmental Protection Agency (EPA), the current non-enforceable Maximum Contaminant Level Goal in drinking water for fluoride is 4.0 mg/L. The EPA established also a Secondary Maximum Contaminant Level of 2 mg/L to protect from dental fluorosis.

A study in 2007 suggested that drinking water with fluoride levels of greater than 2.0 mg/L also can cause damage to liver and kidney functions in children, without dental fluorosis.[14]

Fluoride in the environment and in drinking water are the major contributors to fluorosis. A recent study of skeletal fluorosis and the role of oxidative stress in its development, highlighted the ways in which fluoride has become a serious global public health concern; it also discussed recent fluoride mitigation strategies.[15]

A review of case studies in 2020 revealed that consumption of fluoride at concentrations of 1.5 ppm (1.5 mg/L) is largely responsible for skeletal fluorosis. Skeletal fluorosis is prevalent in various regions of Africa and Asia, where it has been accounted to affect approximately 100 million people with little or no access to the limited treatment options available to reverse fluoride toxicity. In the absence of accessible and effective specific treatments to treat skeletal fluorosis, recent strategies to address this health issue focus on prevention and fluoride removal from water as some of the safest and best approaches to fight fluorosis.[15] In a 2011 longitudinal intervention study conducted in three villages in India with high occurrences of various dental, skeletal, and nonskeletal manifestations of fluorosis, it was determined that after 5 months of consuming safe drinking water from supplied domestic filters and a community filter, with fluoride concentrations below the permissible limit in India (1.5 mg/L per the Bureau of Indian Standard [BIS]), there was a decrease in manifestations of dental fluorosis and in various manifestations of skeletal and nonskeletal fluorosis. The use of fluoride-containing toothpaste and consumption of black lemon tea and tobacco, which are sources of fluoride ingestion causing fluorosis, was also considered among the participants during the study.[16]

Despite the scientific benefits of fluoride in bone health and cavity prevention, major concerns about excessive fluoride intake and related toxicity continue to be raised worldwide. This has led to a ban on fluoridation in various countries. Much controversy still exists regarding whether the benefits outweigh the risks associated with fluoride exposure and intake. The EPA's drinking water fluoride permissible limits enforceable standards have been challenged on grounds that they might not provide enough protection to potentially susceptible populations. Potentially susceptible populations may include those with underlying medical conditions; for example, those with chronic kidney disease [CKD] causing renal impairment who may retain more fluoride than healthy people and thus be at a higher risk of skeletal fluorosis, bone fractures, and severe enamel fluorosis.[17]

Cognitive Neurotoxicity

More recently, concerns have been raised on [F⁻] neurotoxicity. In 2019, the National Toxicology Program (NTP) indicated that "fluoride is presumed to be a cognitive neurodevelopmental hazard to humans." NTP is proceeding with a further systematic review on fluoride's neurodevelopmental impacts based on human epidemiology, experimental, and mechanistic studies. Data suggest that exposure to fluoride levels above 1.5 mg/L in drinking water may negatively impact the development of intelligence in children.[18] A 2016 comprehensive review of various published animal studies conducted by the NTP reviewed potential neurobehavioral effects in rats and mice that had been exposed to fluoride in water and diet at levels higher than 0.7 ppm (the recommended level for water fluoridation in the United States) during development and adulthood. The review indicated that there was a low to moderate level of evidence supporting adverse effects on learning and memory, being strongest among adult animals and weaker in animals during development.[19]

The NTP 2019 draft monograph "Systematic Review of Fluoride Exposure and Neurodevelopmental and Cognitive Health Effects" was questioned because of gaps in supporting evidence regarding the risks and benefits associated with fluoride exposure. As a result, the NTP engaged the National Academies of Sciences, Engineering, and Medicine (NASEM) to review the draft monograph. In 2021, the NASEM published its review of the Revised NTP monograph that raised concerns regarding the analysis of various aspects of some studies and the analysis, summary, and presentation of the data in the NTP monograph.[20]

Studies document an association between IQ impact, behavioral disorders, and [F⁻] levels below 1.5 mg/L in drinking water. Elevated fluoride uptake is suspected of causing neurological toxicity and adverse effects during early child development. Research findings have shown evidence of fluoride crossing the placenta, reaching the fetus and the amniotic fluid, and passing through the blood-brain barrier.[21,22] Evidence of fluoride passing the blood-brain barrier has been seen in samples where fluoride concentrations in human cerebrospinal fluid approach those in serum.[23] Adult brains seem to be somewhat protected from toxic agents by the blood-brain barrier; however, fetuses and small children with an incompletely formed barrier are less likely to be protected.[24]

A cross-sectional and prospective study of pregnant women in Spain has provided evidence that exposure to fluoride in drinking water during the early stages of pregnancy has a neurotoxic effect on children, negatively affecting their cognitive development during childhood. This neurotoxic effect is said to occur because, as fluoride crosses the placenta and penetrates the fetal blood-brain barrier, it accumulates in cerebral tissue. The study followed mothers throughout their pregnancy, and urinary samples were collected during their first and third trimesters of pregnancy. Their children's cognitive domains and intelligence indexes were evaluated using Bayley Scales at age 1 year and at age 4 years. There was a positive association between maternal urinary fluoride (MUF) and General Cognitive Index (GCI) scores and other cognitive functions at 4 years of age. There was a direct positive effect for males and no significant effect for females. These results warrant further research to rule out an effect at lower levels of fluoride in water.[25]

Other techniques, such has fetal blood sampling and maternal fluoride sampling, are also used to document fluoride concentrations in fetuses as indicators or biomarkers of fetal or early-life fluoride exposures.[26] Accordingly, assessment of fluoride in maternal samples during pregnancy may be used as an indicator of fetal exposure.

A meta-analysis on fluoride exposure and intellectual disability in children was supportive of previous findings that demonstrated that children exposed to elevated levels of fluoride in groundwater had cognitive deficits. The study consisted of 14 cross-sectional studies from endemic areas with naturally high fluoride concentrations in groundwater. The meta-analysis was published in 2012. It assessed 27 research reports, all but two from China, and supported previous findings that children with elevated fluoride exposures were found to have cognitive deficits. IQ performance at school age was also assessed, and all but one study suggested that consumption of residential drinking water with a higher fluoride content was associated with poorer IQ.[2] Similarly, in Mexico and Canada, prospective studies of children's cognitive performance demonstrated that children's neurotoxicity was dose-dependent, supporting that elevated fluoride intake in water during early development can result in considerable IQ deficit.[11]

The association between early-life fluoride exposure and IQ in children also was seen in 14 cross-sectional studies, where all but one of the 14 studies reported apparent associations between elevated fluoride exposure and reduced intelligence. Several new Chinese-language studies also showed similar associations between fluoride exposure and reduced IQ.[27]

A study in Mexico investigated the negative association between prenatal fluoride exposure and longitudinal IQ. The research indicated that prenatal fluoride exposure had more impact on visual-spatial and perceptual reasoning abilities than on verbal abilities.[28]

A study that administered neuropsychological tests to 49 adults suffering from skeletal fluorosis showed cognitive problems and neurological symptoms that included deficits in language fluency, recognition, similarities, associative learning, and working memory when compared to control groups.[29] Neurological symptoms such as headaches, insomnia, and lethargy have also been reported in both children and adults in a study conducted in areas with waterborne fluorosis, though evaluation of past exposures of the study population was not assessed.[30]

Despite the benefits of water fluoridation, developmental neurotoxicity is a serious risk associated with fluoride exposure, especially when the exposure occurs during early development. Exposure monitoring and control is of significant importance. Given the increasing evidence of harm, water fluoridation may be abandoned in favor of more effective oral health prevention.

Manganese

Manganese (Mn) is an essential element required for several biological processes, including brain and skeletal development, neuronal function, immune integrity, blood clotting, and lipid or carbohydrate metabolism in the human body.[31,32] It is the third most abundant element in the human body, and it plays a key function as a cofactor, gene modulator, or catalyst in a variety of critical physiological processes.[33,34] As with all essential metals, both deficiency and excess of manganese cause toxicity. Manganese deficiency has been associated with skeletal defects, dermatitis, hypocholesterolemia, and abnormal lipid or glucose metabolism.[33,35] With growing environmental and occupational exposure, toxicity related to manganese overexposure is of far more concern than rare deficiency conditions.[36] Diet is the primary physiological source of manganese, and an intake range of 2–9 mg/day is recommended. This can be obtained from nutrition including rich dietary sources of manganese such as vegetables, grains, fruits (i.e., avocados, blueberries), nuts, tea or plant-derived beverage, egg yolks, and legumes.[34,36] Infant formula has been reported to contain a high level of manganese, up to 195 times as much as breast milk with a range of 2–6 µg/L.[37]

Manganese toxicity has been related to ingestion of high-manganese tap water used to reconstitute infant formula.[38] In fact, drinking water may contain high manganese levels from natural or anthropogenic sources, as shown in Inner Mongolia,[39] China,[5] Nepal,[40] Bangladesh,[41] and Quebec.[42] Other products consumed by young children, especially soy-based, rice-based, goat milk–based, chocolate-flavored, and nutritional beverages, contain different levels of manganese, and multiple studies have reported an excessive level of manganese in these products. Parenteral nutrition, which is routinely supplemented with manganese, also represents a potential exposure source for neonates. The transiently diminished biliary excretion and a higher rate of manganese absorption in neonates—16–37% compared to roughly 3% in adults—make them more susceptible to manganese toxicity.[37] Recommended manganese intake for infants is 600 µg or less, the median intake for the children 1–3 years old is 1.22 mg/day, for ages 4–13 years it is 1.48 mg/day, and for ages 14–18 years it is 1.6 mg/day.[43]

Industrial emissions of manganese are generated by mining, steel and ferroalloy industries, and dry battery production. Occupational exposure occurs during welding and during the production of potassium permanganate ($KMnO_4$), electrical coils, glass, and fireworks.[34] High airborne concentrations of manganese can be found near these manganese-emitting industrial sources and lead to child environmental exposure. Moreover, gasoline with the manganese-containing fuel additive methylcyclopentadienyl manganese tricarbonyl (MMT)[44] and manganese-containing fungicides such as Maneb and Mancozeb can cause manganese exposure.[34,45]

Health Effects of Manganese

The developing brain is much more sensitive to manganese than the adult brain due to enhanced absorption of the metal in early life, relatively lower biliary excretion, and the continuing development of synapses throughout childhood and into adolescence, particularly in the prefrontal cortex. All these factors increase the potential for neurological injury from excessive manganese exposure in early life. During development, manganese readily crosses the placenta and can accumulate in the brain. Manganese readily crosses also the blood-brain barrier in the developing fetus, neonate, and the mature neonate making the central nervous system more sensitive to manganese toxicity despite the low distribution of manganese in the brain.[46] Overexposure to manganese during pregnancy can affect the fetus's development. Manganese superoxide dismutase (MnSOD) is a mitochondrial enzyme that scavenges reactive oxygen species (ROS). An elevated level of MnSOD mRNA due to antioxidant or inflammatory responses has been reported in fetal membranes of women with spontaneous preterm labor compared to in-term cases.[47] Neonates receiving total parenteral nutrition (TPN) can accumulate more than 200% more manganese in the brain compared to children not receiving TPN. Although likely governed by content in commercially available products, the dose of manganese may regularly exceed the recommendations of clinical guidelines and should be limited to 55 µg/day.[48,49]

Neurodevelopmental Impacts of Manganese from Drinking Water

Neurodevelopmental effects associated with elevated manganese exposure in children include impairment of cognitive function, with decrements in memory, verbal learning, and intelligence. Studies documenting these effects have reported that neurotoxicity occurs in children with drinking water manganese levels considered very low[42,50] compared to existing guidelines. The U.S Environmental Protection Agency (EPA) has developed a health advisory level of 0.3 mg/L for manganese in drinking water. The level of 50 µg/L was established as the advisory secondary maximum contaminant level (SMCL) by the EPA in 2004 to avoid aesthetic, cosmetic, and technical problems, but it was not based on health effects.[51] Health Canada has proposed a health-based guideline of 100 µg/L for manganese in drinking water.[52]

A potential association has been reported between ingestion of elevated levels of manganese in drinking water and learning problems, based on the higher manganese levels found in the hair of learning-disabled and hyperactive children compared to normally functioning children.[53] Another study in Quebec examined the relationship between exposure to chronic levels of manganese in drinking water (610 µg/L) and hyperactive behavior in children.[54] Bouchard's study has shown a sex-specific relationship in manganese exposure from drinking water in children. Higher toenail manganese was associated with poorer performance IQ in girls, whereas higher manganese in water was associated with better IQ in boys.[53] Hair manganese concentration was correlated with hyperactive and oppositional behaviors, and the high exposure group had significantly higher levels of hair manganese and a stronger association with hyperactive behaviors. Adverse cognitive, neurodevelopment, and behavioral effects of manganese exposure from drinking water on school-aged children have been confirmed by a systematic review of multiple studies between 2006 and 2017.[55] Studies conducted in Bangladesh showed that manganese in drinking water affects intellectual function, resulting in lower IQ.[38] In Bangladesh, Wasserman's cross-sectional study of 142 10-year-old children whose well water supply was contaminated with a mean manganese concentration of 793 µg/L has reported a significant association between manganese exposure and reduced full-scale, performance and verbal raw scores on the Wechsler Intelligence Scale for Children in a dose-response fashion. An ecological study in North Carolina reported a positive association between manganese concentration in the groundwater and county-level infant mortality but not birth weight.[56] Collectively, these data suggest that increased levels of manganese in drinking water directly affect neurobehavioral development and function in children.

Neurodevelopmental Impacts of Manganese Through Inhalation

Neurodevelopmental effects have been reported from inhaled manganese.[31] Elevated inhaled manganese concentrations were negatively associated with intellectual function in school-aged children living near a manganese mining and processing facility in Mexico.[57] Haynes's study of 404 children ages 7–9 years in Ohio who were exposed to manganese from ferroalloy emissions demonstrated a negative association between both low and high manganese concentrations (in blood and hair) and child IQ scores.[58] A high level of manganese in toenail samples after chronic exposure to manganese has been associated with behavioral disorders such as aggressiveness and rule-breaking in school-aged children (7–12 years old) living near a ferro-manganese alloy plant in Brazil.[59] Environmental exposure to manganese from ferroalloy emission also has been associated with tremor, impaired motor skills, and diminished ability to identify odors in children in northern Italy.[60] Another study conducted by Lucchini et al. in southern Italy documented an inverse association between hair manganese and cognitive functioning in schoolchildren ages 6–12 years. This correlation was analyzed based on the proximity of these children's residences to a large steel plant operation.[61] In a cross-sectional study in Italian adolescents, significant associations also were found between ambient manganese and adverse respiratory outcomes such as asthma.[62]

The combined effect of manganese with other neurotoxic agents has been addressed in multiple studies. At low levels of copper, co-exposure of manganese, lead, and chromium were associated with lower IQ scores, especially at low copper levels in an Italian adolescent cohort exposed to a mixture of metals.[63] Another study of the neurotoxicant effect of metal co-exposure to manganese, lead, chromium, and copper among 188 Italian adolescents (ages 10–14 years) showed a sex-specific impact on visuospatial learning.

Early adolescence is a sensitive developmental window, and metal mixtures resulted in a different magnitude and direction of effect between sexes. This study showed that in girls, slower visuospatial learning was more driven by manganese and copper, whereas in boys, faster visuospatial learning was mostly driven by chromium within the mixture.[64] A sex-specific exposure window has been documented in the association of manganese exposure during prenatal and early life with neuromotor function in a group of adolescents.[65] Another study shows that not only neuromotor function but also olfactory functions are influenced in adolescents living in manganese-contaminated regions, resulting in reduced the limbic system responses. In this condition, manganese exposure may lead to emotional dysregulation due to the interruption of brain networks.[66]

Animal studies reported that mice exposed to manganese during juvenile development experience greater neuroinflammatory injury and behavioral dysfunction upon subsequent adult exposure to manganese compared to mice without early exposure.[67]

Therefore, it is plausible to hypothesize that manganese exposure during critical years in early development could predispose individuals to more severe effects of neurotoxicity as they age.

Mechanisms of Neurodevelopmental Impacts

The precise mechanisms of manganese toxicity are not fully understood, but manganese intoxication could be associated with interactions with other metals, especially iron, due to certain similarities between their absorption and transport processes.[68] Altered blood and cerebrospinal fluid iron concentrations have been demonstrated in cases of chronic manganese exposure. This interaction could be due to the role of both manganese and iron in the function of some proteins, such as metalloproteins, that catalyze critical reactions in the body. Both in vivo and in vitro studies have documented that dietary iron deficiency increases manganese levels in various tissues.[69,70] Manganese accumulates mainly in those parts of the brain with high concentrations of non-heme iron, including subthalamic nuclei, caudate-putamen, substantia nigra, and globus pallidus.[71] Manganese deposition in these central nervous system parts could alter multiple neurotransmitter systems such as glutamate and dopamine, causing multiple brain disorders. Altogether, it has been shown that disturbances in iron hemostasis impact manganese toxicity, and iron deficiency increases manganese toxicity. Accordingly, the level of iron and genetic variation that impact serum ferritin (iron storage protein) could affect manganese bioavailability in children.[72]

Manganese transporter genetic alterations, especially mutations in the solute carrier (SLC) family that transports manganese across cell membranes, may contribute to manganese toxicity. Multiple studies have demonstrated that SLC30A10 and SLC39A14 mutations lead to manganese overload, while mutations in SLC39A8 cause manganese deficiency.[73,74] Different studies have reported that SLC30A10 mutations are associated with severe hypermagnesemia, and neurological disorders such as dystonia, gait and speech disturbances, and liver diseases.[75–77] An investigation in a population of 645 Italian children with a wide range of environmental manganese exposures has demonstrated that polymorphism in manganese transporter genes contributed to different susceptibilities to manganese exposure. They also reported a sex-dependent association between soil manganese levels and neurobehavioral outcome, suggesting that girls who are genetically less efficient at regulating manganese may be a particularly vulnerable group in this study.[78] Another study of 195 Italian children with variable environmental manganese exposure showed that common single nucleotide polymorphisms in manganese transporters were associated with manganese concentrations in primary teeth, suggesting an important effect of manganese on early development. Their result also reported various impacts of these transports on manganese hemostasis based on developmental stage (fetal, early postnatal, and early childhood) and gender.[77]

Children's exposure to manganese is a public health concern. one that needs to be addressed through research into the complexities of manganese exposure at the different developmental windows and considering sex differences and genetic influence. This issue requires detailed investigation into the complexities of manganese exposure at the different developmental windows, taking into account potential differences based on sex and genetic factors.[79]

Summary of Toxic and Clinical Effects

- Excessive exposure to fluorides may cause dental and bone fluorosis.
- High levels of fluorides in drinking water reduce cognitive abilities in children.
- Manganese intoxication is mainly an occupational disease that principally affects the nervous system. It is characterized by an extrapyramidal syndrome and parkinsonism. Motor, cognitive, and behavioral deficits may be caused by children's exposure to manganese.

Prevention of Toxic Effects

A variety of preventive strategies are required to protect children against metal toxicity because different metals have different environmental sources and reach children by a multitude of different routes. Several overall principles guide the protection of children against metal toxicity.

Primary prevention that reduces unnecessary uses of metals and prevents environmental contamination at source is the most effective strategy against all forms of metal toxicity.

Home is an important source of metal toxicity. Young children are especially sensitive to environmental exposures at home—a reflection of the large amount of time they spend in their homes, their patterns of hand-to-moth behavior, and their developmental immaturity.

Groundwater, especially when it is taken from unregulated private wells, can be an important source of exposure to toxic metals, especially for young children because of the large volume of water they consume relative to their body mass. Consuming groundwater from private wells can cause potential hazards related to high levels of manganese or fluorides. Well water can be tested for a wide range of potential contaminants. Public water must not contain levels of manganese and fluoride that may impact neurodevelopment.

Residence in the vicinity of mines or industrial facilities may lead to excessive exposure to these metals and should be avoided. In addition to inhalation, ingestion and dermal absorption of contaminated dust and soil are relevant routes of children's exposures to toxic metals in the vicinity of active or abandoned industrial "hotspots." Take-home exposure from contaminated workplaces is another potential source of children's exposure to toxic metals. The most common vehicle is contaminated work clothing brought home for cleaning.[21]

Manganese toxicity of dietary origin in neonates and young children can be prevented by avoiding formulas containing high manganese levels and by reducing the manganese content of materials used in TPN. Prevention of fluoride toxicity requires striking a balance between the prevention of dental caries and avoidance of excessive exposure. Accordingly, the US Public Health Service has made the following recommendations: "(1) the fluoride content of foods and beverages, particularly infant formulas and water used in their reconstitution, should continue to be monitored closely in an effort to limit excessive fluoride intake; (2) ingestion of fluoride from dentifrice by young children should be controlled, and the use of only small quantities of dentifrice by young children should be emphasized; and (3) dietary fluoride supplements should be considered a targeted preventive regimen only for those children at higher risk for dental caries and with low levels of ingested fluoride from other sources."[22]

References

1. Aoun A, Darwiche F, Al Hayek S, Doumit J. The fluoride debate: the pros and cons of fluoridation. *Prevent Nutrit Food Sci*. 2018;23(3):171–80.

2. Choi AL SG, Zhang Y, Grandjean P. Developmental fluoride neurotoxicity: a systematic review and meta-analysis. *Environ Health Perspect*. 2012;120:1362–8.

3. McDonagh MS, Whiting PF, Wilson PM, et al. Systematic review of water fluoridation. *BMJ*. 2000;321(7265):855–9.

4. Peckham S, Awofeso N. Water fluoridation: a critical review of the physiological effects of ingested fluoride as a public health intervention. *Sci World J*. 2014;2014:293019.

5. Tang J, Zhu Y, Xiang B, et al. Multiple pollutants in groundwater near an abandoned Chinese fluorine chemical park: concentrations, correlations and health risk assessments. *Sci Rep*. 2022;12(1):3370.

6. Y. AM. Evaluation of fluoride levels in bottled water and their contribution to health and teeth problems in the United Arab Emirates. *Saudi Dental J*. 2016;28(4):194–202.

7. Gupta S, Seth AK, Gupta A, Gavane AG. Transplacental passage of fluorides. *J Pediatrics*. 1993;123(1):139–41.

8. McPherson CA, Zhang G, Gilliam R, et al. An evaluation of neurotoxicity following fluoride exposure from gestational through adult ages in Long-Evans hooded rats. *Neurotox Res*. 2018;34(4):781–98.

9. Org FA. Infant exposure. 2022. Available at: https://fluoridealert.org/issues/infant-exposure/protect/.

10. Agency for Toxic Substances and Disease Registry. Toxicological profile for fluorides, hydrogen fluoride, and fluorine. 2003. Available at: http://www.atsdr.cdc.gov/toxprofiles/tp.asp?id=212&tid=38#bookmark10.

11. Grandjean P. Developmental fluoride neurotoxicity: an updated review. *Environ health*. 2019;18(1):110.

12. Zohoori F, Duckworth RM. Fluoride intake and metabolism, therapeutic and toxicological consequences. In *Molecular, Genetic, and Nutritional Aspects of Major and Trace Minerals*. Elsevier; 2016:539–550. https://doi.org/10.1016/B978-0-12-802168-2.00044-0

13. National Institutes of Health (NIH). Supplements. Fluoride. Fact sheet for health professionals. 2020. Available at: https://ods.od.nih.gov/factsheets/Fluoride-HealthProfessional/#en10.

14. Xiong XZ LJ, He WH, et al. Dose-effect relationship between drinking water fluoride levels and damage to liver and kidney functions in children. *Environ Res*. 2007;103:112–6.

15. Srivastava S FS. Fluoride in drinking water and skeletal fluorosis: a review of the global impact. *Curr Environ Health Rep*. 2020;7(2):140–6.

16. Majumdar KK. Health impact of supplying safe drinking water containing fluoride below permissible level on fluorosis patients in a fluoride-endemic rural area of West Bengal. *Indian J Public Health*. 2011;55(4):303–8.

17. National Research Council. *Fluoride in Drinking Water: A Scientific Review of EPA's Standards*. Washington, DC: National Academies Press; 2006.

18. Saxena S, Sahay A, Goel P. Effect of fluoride exposure on the intelligence of school children in Madhya Pradesh, India. *J Neurosci Rural Pract.* 2012;3:144–9.

19. National Toxicology Program. NTP Research Report on Systematic Literature Review on the Effects of Fluoride on Learning and Memory in Animal Studies: Research Report 1 [Internet]. Research Triangle Park (NC): National Toxicology Program; 2016.

20. National Academies of Sciences, Engineering, and Medicine; Division on Earth and Life Studies; Board on Environmental Studies and Toxicology; Exposure and Neurodevelopmental and Cognitive Health Effects; Committee to Review the NTP Monograph on the Systematic Review of Fluoride. Review of the Draft NTP Monograph: Systematic Review of Fluoride Exposure and Neurodevelopmental and Cognitive Health Effects. Washington (DC): National Academies Press (US); 2020.

21. Fawell J, Bailey K, Chilton J, Dahi E, Fewtrell L, Magara Y. Fluoride in drinking-water. Geneva, World Health Organization, 2006.

22. O'Mullane DMBR, Jones S, Lennon MA, Petersen PE, Rugg-Gunn AJ, Whelton H, Whitford GM. Fluoride and oral health. *Community Dent Health.* 2016;33(2):69–99.

23. Hu YH, Wu SS. Fluoride in cerebrospinal fluid of patients with fluorosis. *J Neurol Neurosurg Psychiatry.* 1988;51(12):1591–3.

24. Adinolfi M. The development of the human blood-CSF-brain barrier. *Dev Med Child Neurol.* 1985;27(4):532–7.

25. Ibarluzea J, Gallastegi M, Santa-Marina L, Zabala AJ, et al. Prenatal exposure to fluoride and neuropsychological development in early childhood: 1- to 4-year-old children. *Environ Res.* 2022;207:112181.

26. Forestier F DF, Said R, Brunet CM, Guillaume PN. The passage of fluoride across the placenta: an intra-uterine study. *J Gynecol Obstet Biol Reprod (Paris).* 1990;19(2):171–5.

27. Wei NLY, Deng J, Xu S, Guan Z. The effects of comprehensive control measures on intelligence of school-age children in coal-burning-borne endemic fluorosis areas. *Chinese J Epidemiol.* 2014;33(3):320–4.

28. Goodman CV, Bashash M, Green R, et al. Domain-specific effects of prenatal fluoride exposure on child IQ at 4, 5, and 6–12 years in the ELEMENT cohort. *Environ Res.* 2022;211:112993.

29. Shao QL WY, Li L, Li J. Initial study of cognitive function impairment as caused by chronic fluorosis. *Chinese J Epidemiol.* 2003;22(4):336–8.

30. Sharma JD SD, Jain P. Prevalence of neurological manifestations in a human population exposed to fluoride in drinking water. *Fluoride.* 2009;42(2):127–32.

31. Lucchini R, Placidi D, Cagna G, et al. Manganese and developmental neurotoxicity. *Adv Neurobiol.* 2017;18:13–34.

32. Li L, Yang X. The essential element manganese, oxidative stress, and metabolic diseases: links and interactions. *Oxid Med Cell Longev.* 2018;2018:7580707.

33. Martins AC, Jr, Morcillo P, Ijomone OM, et al. New insights on the role of manganese in Alzheimer's disease and Parkinson's disease. *Int J Environ Res Public Health.* 2019;16(19):3546.

34. Klaassen CD, Casarett LJ, Doull J. *Casarett and Doull's Toxicology: The Basic Science of Poisons.* 8th ed. New York: McGraw-Hill Education; 2013: xiii.

35. Finley JW, Davis CD. Manganese deficiency and toxicity: are high or low dietary amounts of manganese cause for concern? *Biofactors.* 1999;10(1):15–24.

36. Bjorklund G, Chartrand MS, Aaseth J. Manganese exposure and neurotoxic effects in children. *Environ Res*. 2017;155:380–4.

37. Frisbie SH, Mitchell EJ, Roudeau S, Domart F, Carmona A, Ortega R. Manganese levels in infant formula and young child nutritional beverages in the United States and France: comparison to breast milk and regulations. *PLoS One*. 2019;14(11):e0223636.

38. Wasserman GA, Liu X, Parvez F, et al. Arsenic and manganese exposure and children's intellectual function. *Neurotoxicology*. 2011;32(4):450–7.

39. Zhao C, Zhang X, Fang X, et al. Characterization of drinking groundwater quality in rural areas of Inner Mongolia and assessment of human health risks. *Ecotoxicol Environ Saf*. 2022;234:113360.

40. Sarkar B, Mitchell E, Frisbie S, Grigg L, Adhikari S, Maskey Byanju R. Drinking water quality and public health in the Kathmandu Valley, Nepal: coliform bacteria, chemical contaminants, and health status of consumers. *J Environ Public Health*. 2022;2022:3895859.

41. Saxena R, Gamble M, Wasserman GA, et al. Mixed metals exposure and cognitive function in Bangladeshi adolescents. *Ecotoxicol Environ Saf*. 2022;232:113229.

42. Kullar SS, Shao K, Surette C, et al. A benchmark concentration analysis for manganese in drinking water and IQ deficits in children. *Environ Int*. 2019;130:104889.

43. Trumbo P, Yates AA, Schlicker S, Poos M. Dietary reference intakes: vitamin A, vitamin K, arsenic, boron, chromium, copper, iodine, iron, manganese, molybdenum, nickel, silicon, vanadium, and zinc. *J Am Diet Assoc*. 2001;101(3):294–301.

44. Smith D, Woodall GM, Jarabek AM, Boyes WK. Manganese testing under a clean air act test rule and the application of resultant data in risk assessments. *Neurotoxicology*. 2018.

45. Mora AM, Córdoba L, Cano JC, et al. Prenatal mancozeb exposure, excess manganese, and neurodevelopment at 1 year of age in the Infants' Environmental Health (ISA) study. *Environ Health Perspect*. 2018;126(5):057007.

46. Aschner M, Aschner JL. Manganese transport across the blood-brain barrier: relationship to iron homeostasis. *Brain Res Bull*. 1990;24(6):857–60.

47. Than NG, Romero R, Tarca AL, et al. Mitochondrial manganese superoxide dismutase mRNA expression in human chorioamniotic membranes and its association with labor, inflammation, and infection. *J Matern Fetal Neonatal Med*. 2009;22(11):1000–13.

48. Fell JM, Reynolds AP, Meadows N, et al. Manganese toxicity in children receiving long-term parenteral nutrition. *Lancet*. 1996;347(9010):1218–21.

49. Reinert JP, Forbes LD. Manganese toxicity associated with total parenteral nutrition: a review. *J Pharm Technol*. 2021;37(5):260–6.

50. Khan K, Factor-Litvak P, Wasserman GA, et al. Manganese exposure from drinking water and children's classroom behavior in Bangladesh. *Environ Health Perspect*. 2011;119(10):1501–6.

51. EPA. Drinking water health advisory for manganese. 2004.Available at: https://www.epa.gov/sites/default/files/2014-09/documents/support_cc1_magnese_dwreport_0.pdf.

52. Health Canada. Manganese in drinking water. Document for Public Consultation Ottawa, Canada. 2016. Available at: https://www.canada.ca/en/health-canada/programs/consultation-manganese-drinking-water/manganese-drinking-water.html.

53. Bouchard MF, Surette C, Cormier P, Foucher D. Low level exposure to manganese from drinking water and cognition in school-age children. *Neurotoxicology*. 2018;64:110–7.

54. Bouchard M, Laforest F, Vandelac L, Bellinger D, Mergler D. Hair manganese and hyperactive behaviors: pilot study of school-age children exposed through tap water. *Environ Health Perspect*. 2007;115(1):122–7.

55. Iyare PU. The effects of manganese exposure from drinking water on school-age children: a systematic review. *Neurotoxicology*. 2019;73:1–7.

56. Spangler AH, Spangler JG. Groundwater manganese and infant mortality rate by county in North Carolina: an ecological analysis. *Ecohealth*. 2009;6(4):596–600.

57. Riojas-Rodriguez H, Solis-Vivanco R, Schilmann A, et al. Intellectual function in Mexican children living in a mining area and environmentally exposed to manganese. *Environ Health Perspect*. 2010;118(10):1465–70.

58. Haynes EN, Sucharew H, Kuhnell P, et al. Manganese exposure and neurocognitive outcomes in rural school-age children: the Communities Actively Researching Exposure Study (Ohio, USA). *Environ Health Perspect*. 2015;123(10):1066–71.

59. Rodrigues JLG, Araujo CFS, Dos Santos NR, et al. Airborne manganese exposure and neurobehavior in school-aged children living near a ferro-manganese alloy plant. *Environ Res*. 2018;167:66–77.

60. Lucchini RG, Guazzetti S, Zoni S, et al. Tremor, olfactory and motor changes in Italian adolescents exposed to historical ferro-manganese emission. *Neurotoxicology*. 2012;33(4):687–96.

61. Lucchini RG, Guazzetti S, Renzetti S, et al. Neurocognitive impact of metal exposure and social stressors among schoolchildren in Taranto, Italy. *Environ Health*. 2019;18(1):67.

62. Rosa MJ, Benedetti C, Peli M, et al. Association between personal exposure to ambient metals and respiratory disease in Italian adolescents: a cross-sectional study. *BMC Pulm Med*. 2016;16:6.

63. Bauer JA, Devick KL, Bobb JF, et al. Associations of a metal mixture measured in multiple biomarkers with IQ: evidence from Italian adolescents living near ferroalloy industry. *Environ Health Perspect*. 2020;128(9):97002.

64. Rechtman E, Curtin P, Papazaharias DM, et al. Sex-specific associations between co-exposure to multiple metals and visuospatial learning in early adolescence. *Transl Psychiatry*. 2020;10(1):358.

65. Chiu YM, Claus Henn B, Hsu HL, et al. Sex differences in sensitivity to prenatal and early childhood manganese exposure on neuromotor function in adolescents. *Environ Res*. 2017;159:458–65.

66. Iannilli E, Gasparotti R, Hummel T, et al. Effects of manganese exposure on olfactory functions in teenagers: a pilot study. *PLoS One*. 2016;11(1):e0144783.

67. Moreno JA, Yeomans EC, Streifel KM, Brattin BL, Taylor RJ, Tjalkens RB. Age-dependent susceptibility to manganese-induced neurological dysfunction. *Toxicol Sci*. 2009;112(2):394–404.

68. Farina M, Avila DS, da Rocha JB, Aschner M. Metals, oxidative stress and neurodegeneration: a focus on iron, manganese and mercury. *Neurochem Int*. 2013;62(5):575–94.

69. Bjorklund G, Dadar M, Peana M, Rahaman MS, Aaseth J. Interactions between iron and manganese in neurotoxicity. *Arch Toxicol*. 2020;94(3):725–34.

70. Ye Q, Park JE, Gugnani K, Betharia S, Pino-Figueroa A, Kim J. Influence of iron metabolism on manganese transport and toxicity. *Metallomics*. 2017;9(8):1028–46.

71. Aschner M, Guilarte TR, Schneider JS, Zheng W. Manganese: recent advances in understanding its transport and neurotoxicity. *Toxicol Appl Pharmacol*. 2007;221(2):131–47.

72. Meltzer HM, Brantsaeter AL, Borch-Iohnsen B, et al. Low iron stores are related to higher blood concentrations of manganese, cobalt and cadmium in non-smoking, Norwegian women in the HUNT 2 study. *Environ Res.* 2010;110(5):497–504.

73. Mercadante CJ, Prajapati M, Conboy HL, et al. Manganese transporter Slc30a10 controls physiological manganese excretion and toxicity. *J Clin Invest.* 2019;129(12):5442–61.

74. Anagianni S, Tuschl K. Genetic disorders of manganese metabolism. *Curr Neurol Neurosci Rep.* 2019;19(6):33.

75. Quadri M, Kamate M, Sharma S, et al. Manganese transport disorder: novel SLC30A10 mutations and early phenotypes. *Mov Disord.* 2015;30(7):996–1001.

76. Tuschl K, Clayton PT, Gospe SM Jr, et al. Syndrome of hepatic cirrhosis, dystonia, polycythemia, and hypermanganesemia caused by mutations in SLC30A10, a manganese transporter in man. *Am J Hum Genet.* 2012;90(3):457–66.

77. Wahlberg K, Arora M, Curtin A, et al. Polymorphisms in manganese transporters show developmental stage and sex specific associations with manganese concentrations in primary teeth. *Neurotoxicology.* 2018;64:103–9.

78. Broberg K, Taj T, Guazzetti S, et al. Manganese transporter genetics and sex modify the association between environmental manganese exposure and neurobehavioral outcomes in children. *Environ Int.* 2019;130:104908.

79. Coetzee DJ, McGovern PM, Rao R, Harnack LJ, Georgieff MK, Stepanov I. Measuring the impact of manganese exposure on children's neurodevelopment: advances and research gaps in biomarker-based approaches. *Environ Health.* 2016;15(1):91.

35

Pesticides

Catherine J. Karr and Virginia A. Rauh

"Pesticide" is a broad term and is generally defined as any substance or mixture of substances intended for preventing, destroying, repelling, or mitigating any pest. The benefits of pesticides for human health include positive impacts on crop yields and prevention of human and animal disease. Yet the potential for harm is substantial because of the large numbers of chemicals involved and sheer volume of use. More than 600 unique chemical products and more than 20,000 commercial pesticide products are currently on the market. More than 1 billion pounds of pesticides are applied annually in the United States, and many billions of additional pounds are used in countries around the world, making oversight, regulation, and public health policy a daunting task.

Short-term, high-dose exposures to pesticides may pose hazards for acute toxicity, while relatively low-level chronic exposures may pose hazards for chronic toxicity. Vulnerability to the toxic effects of pesticides is greatest during the 9 months of the prenatal period and in the first months and years of postnatal life. The acute toxicity of pesticide active ingredients and formulations is highly variable but generally well described. By contrast, data on chronic toxicity, particularly developmental toxicity, are less robust.

Despite the complexity of pesticide toxicity, the accumulated evidence focuses attention on particular groups of chemicals and on particular pesticide formulations (e.g., broadcast sprays and foggers), which appear to present the greatest danger to children because of either their widespread use or multiple sources of exposure (Figure 35.1). Some groups and compounds, such as organophosphates (OPs) and organochlorines, have a strong evidence base for concern. Others have received relatively less attention in child health research despite widespread exposure and accumulating evidence for adverse impacts on child health (e.g., pyrethroids, neonicatinoids, glyphosate).

Pesticides often are classified by intended pest target or use, for example, insecticide, herbicide, fungicide, rodenticide, fumigant, or repellant. Each product use category may be further subdivided according to similarities in chemical structure, typically reflecting common mechanisms of action, with implications for human toxicity. For example, the OPs, carbamates, and pyrethroids (chemical classes of insecticides), because of their widespread use and/or toxicity, have raised important child health concerns.

Regulated labeling practices in the United States require that the pesticide active ingredient be included on all product labels. Specific wording also is required that indicates the potential for acute toxicity. Although the label may be an important source of information in acute exposure events, it lacks much crucial information. For example, pesticide labels in the United States do not specify pesticide class, they do not specify safety levels for chronic toxicity, and they do not list the nonpesticide ingredients in commercial pesticide products, which are misleadingly termed "other" or "inert" ingredients

Catherine J. Karr and Virginia A. Rauh, *Pesticides* In: *Textbook of Children's Environmental Health,*
Second Edition. Edited by: Ruth A. Etzel and Philip J. Landrigan, Oxford University Press.
© Oxford University Press 2024. DOI: 10.1093/oso/9780197662526.003.0035

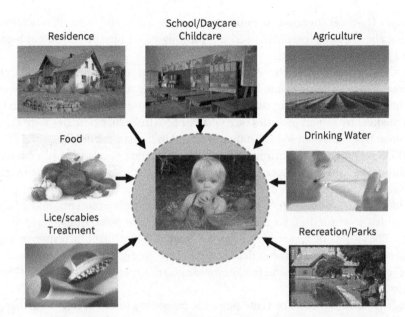

Residence

School/Daycare
Childcare

Agriculture

Food

Drinking Water

Lice/scabies
Treatment

Recreation/Parks

Figure 35.1. Multiple sources and pathways through which children are exposed to pesticides

because they are not designed to kill pests. Some "inert" ingredients have significant toxicity and can account for up to 99% of the total weight or volume of many commercial pesticide products.

Acute Pesticide Poisoning

Children typically experience acute poisonings through unintentional ingestion of improperly stored pesticide products (e.g., pesticide stored in a soft drink bottle). Acute pesticide poisoning can also result from inhalation and skin contact with airborne pesticide drift following agricultural spraying or from spills. Immediate symptoms of intoxication range from mild and subtle to severe (e.g., nausea, headaches, skin rashes, eye irritation, seizures, coma, and death) depending on the inherent toxicity of the pesticide product and the dose received.

Acute pesticide poisoning is a worldwide problem (see Chapter 13). A major impediment to control of acute pesticide poisoning is lack of comprehensive national or international surveillance data. This lack of data limits accurate characterization of the scope and tracking of the problem. Based on unintentional pesticide poisoning data derived from systematic review and national population information, it is estimated that approximately 385 million cases of pesticide poisoning occur annually (all ages), including 11,000 fatalities.[1] Although much of this burden of disease is occupationally related, there also are substantial effects on children who often accompany parents to workplaces, live in farm regions, come in contact with post-application residue, and/ or participate themselves in agricultural production, including pesticide application. These estimates exclude information from studies specifically focused on intentional ingestions; pesticides remain a significant agent of suicide in low- and middle-income countries (LMICs). Because of poor regulatory systems and limited education and training about pesticides, LMICs account for only 25% of pesticide use but represent 99% of pesticide-related deaths.[2]

In the United States, the American Association of Poison Control Centers' National Poison Data System (NPDS) provides annual summaries of self-reports of acute pesticide poisoning cases from patients and/or family members and calls from medical treatment facilities (see Chapter 55). In 2020, pesticides were the eleventh most frequently named substance involved in human exposure reports (3.0% of all NPDS reports), the eighth most common substance encountered in poisonings of children 5 years or younger (3.4% of pediatric NPDS reports), and the fourth for pregnancy exposures (4.8%).[3] Regulatory oversight (including removal of particularly hazardous products from home use), packaging improvements, and education have all led to reductions in the frequency of pesticide poisonings reported to the NPDS. Phase-outs of residential usage for many OP products in response to child health concerns was reflected in declines in poison control catchment area estimated "rates" of moderate, major, and fatal pesticide poisonings by approximately 42% from 1995 to 2004.[4]

Inadequate recognition of acute pesticide poisoning in children is a further challenge to surveillance and prevention. Misdiagnosis and delayed diagnosis have been clearly documented in children even when signs of acute pesticide toxicity closely followed high-dose pesticide exposures. A fundamental problem is that clinical training on topics in pediatric environmental health remains minimal even despite recent significant growth in the research emphasis on children's environmental health (see Chapter 56). Many clinicians have limited knowledge and experience in handling pesticide exposures and their effects on children's health. To remedy this gap in training, excellent comprehensive and up-to-date reviews and resources regarding pesticide acute toxicity, including recognition and management of high-dose, short-term exposures, are available.[5]

Chronic Exposure and Toxicity

A problem that is much more common among children today than acute pesticide poisonings is daily, cumulative, chronic lower-level exposures to pesticides. These chronic, lower-level exposures may be associated with subclinical health effects that become evident only with special testing (see Chapter 49) or with delayed health consequences that do not appear until weeks, months, or even years later. In high-income countries, concern regarding these chronic exposures and their health implications remains an important focus of research and public health activity.

Children encounter pesticides daily in air, food, dust, and soil and on surfaces from residential and public lawn or garden applications, household insecticide use, pet applications, and agricultural product residues. Pesticides, most often from the classes of pesticides known as pyrethroids, are purposefully applied directly to children's skin to treat lice or scabies. Glyphosate has become the most widely used pesticide worldwide, with escalating use over the past two decades.[6] In addition to widespread agricultural uses, it is applied commonly to lawns and gardens. Available biomonitoring studies find that levels detected in children exceed those measured in adults.[7] For most children, the majority of their exposure to pesticides comes from two sources: pesticide residues in the food supply and home pesticide use. For subgroups of young children and youth, particularly in rural areas, proximity to agricultural production activities where pesticides are heavily used or workplace-related exposures may be particularly important (see Chapter 22). Agriculture accounts for a high proportion (80%) of all pesticide usage, with herbicides representing the largest pesticide use category (43%).

Food and Water

Measurement of pesticides and pesticide metabolites in human urine or blood samples, performed biannually in the United States through the nationally representative biomonitoring program conducted by the Centers for Disease Control and Prevention (CDC), provides useful data on exposure of subpopulations and temporal trends. For example, in the CDC's Fourth National Report on Human Exposure to Environmental Chemicals, urinary concentrations of common OP and pyrethroid insecticide metabolites were observed to be higher in the youngest age group sampled (ages 6–11 years) as compared to older children and adults.[8] Differences in doses received via diet are suspected to explain much of this variation. Apples, grapes, and carrots, all of which are heavily consumed by children, are among the foods with frequent detection of OP residues. In addition to fruits and vegetables, pyrethroids are commonly found in rice, pasta, and breakfast cereals. The relative importance of diet as a source of exposure to these pesticides is clearly illustrated by intervention trials that compare organic diets versus conventional diet phases in children. These studies demonstrate that organic diets lead to significant reductions in several insecticides (OP, pyrethroid, neonicatinoid) and the herbicides 2,4-D and glyphosate.[9,10]

In contrast, US national biomonitoring data for serum levels of organochlorine insecticides and their metabolites, such as p,p'-dichlorodiphenyl-trichloroethane (DDT), dieldrin, and chlordane, many of which were banned in the United States in the 1970s and 1980s, reveal lower concentrations in the youngest age group monitored (12–19 years)[8] and higher levels in older persons. The long persistence of these legacy pollutants in the bodies of older persons reflects lifelong storage in adipose tissue following early-life exposure.

To protect children against exposures to pesticides, the 1996 Food Quality Protection Act, the US federal law on pesticides (see Chapter 1), fostered a uniquely responsive approach to pesticide control that specifically recognized the importance of the dietary contribution and children's greater toxicological vulnerability. Under the Act, the US Environmental Protection Agency (EPA) must presume a 10-fold margin of safety in risk assessments in support of acceptable food residues (tolerances) to account for potential pre- and postnatal developmental toxicity and data completeness on pediatric exposure and toxicity. In addition, cumulative impacts of pesticides with common toxicity mechanisms must be addressed.

Pesticide contamination of drinking water presents another potentially important source of exposure for children.[11] National sampling efforts in the United States find herbicides most frequently in drinking water in agricultural areas, in particular herbicides of the triazine class. In urban areas, both herbicides and insecticides are commonly detected. Detected concentrations rarely exceed water quality health reference levels, although these reference levels are based on single pesticides while sampling often reveals the co-occurrence of multiple pesticides.

Residential and Community Exposure

Residential use of pesticides presents another major source of pediatric exposure and is particularly important for indoor use of pyrethroid insecticides and for lawn and garden use of rodenticides, herbicides, and fumigants. Approximately 4.4 billion pesticide applications are made each year to American homes, gardens, and yards. The EPA notes that about three-quarters of US households use pesticides and that 80% of exposure occurs indoors.[12] A review of this topic for pediatric providers has been published.[13]

Broadcast applications of pesticides, including sprays, "flea bombs," and foggers, in indoor environments can leave lingering residues in air, carpet, toys, and house dust that persist for weeks or months. Children's typical oral exploratory behavior and frequent hand-to-mouth activity, crawling across and playing on the floor, increases their dermal, inhalation, and oral exposure to pesticide residues on surfaces and to pesticide vapor layers in air that are suspended close to the floor. Parents who work with pesticides may introduce pesticide residues to home surfaces or to the child directly through "take-home" exposures. Herbicides applied on the lawn or garden can be tracked into the home by residents and pets, and residues may accumulate over time. This source of exposure can be greatly reduced by removing contaminated shoes and clothing before entering the home. Local patterns of pest problems and cultural practices in pest control will vary geographically and influence the relative magnitude and types of pesticide exposure a child encounters.

Children may be exposed to pesticides in other settings outside the home where they spend time, including school, child care, a relative's home, recreational areas, and so on, depending on indoor and outdoor pesticide use patterns and proximity to pesticide use sites.

The Agricultural Setting

Children who reside in settings near livestock or crop production face unique risks (Figure 35.2). Regular pesticide application, with periodic airborne drift, may occur

Figure 35.2. Proximity of farm families to pesticide application

proximal to their homes, schools, and play areas (see Chapter 21). The opportunity for "take-home" exposure from occupationally exposed parents or a child's own participation in farm work activities may be particularly salient.

Studies utilizing biological markers of pesticide exposure such as urinary metabolite monitoring have clearly demonstrated uptake in farm children and pregnant women that exceeds levels observed in reference populations.[14]

Chronic Health Effects

Pesticide chemicals include neurotoxicants, mutagens, carcinogens, immunotoxicants, and endocrine disruptors. This wide range of toxicity provides biological plausibility for a rapidly expanding base of epidemiological evidence that supports links between pesticide exposure in early life and a number of the most prevalent and severe chronic health conditions in children.[11] Epidemiological studies of children most frequently have addressed associations of pesticides with pediatric cancer (brain and leukemia) and neurodevelopmental/neurobehavioral effects. Comprehensive reviews of these topics are available.[11] Adverse birth outcomes (premature birth, low birth weight, reduced head circumference, and birth defects) also have been explored in epidemiological studies. The most compelling evidence for associations between pesticide exposure and adverse effects comes from studies of prenatal maternal exposures to OP and organochlorine pesticides. In addition, hypothesis-generating studies propose a role for pesticides in asthma and type 2 diabetes mellitus.

Many insecticides have well-described acute neurotoxicant properties that are important causes of human poisoning events. However, information on the potential neurodevelopmental toxicity arising from chronic, low-level exposure in gestational or postnatal life is inadequate or lacking for most pesticides in current use. There is sizable and growing evidence supporting neurodevelopmental toxicity from the organochlorines (specifically DDT and its metabolite p,p' dichlorodiphenyldichloroethylene (delete space between o and r) [DDE]) and, most recently, OPs. Several recent reviews describe this evidence base. The OPs provide an illustrative case study.

Early Evidence for Organophosphate Toxicity

Studies of human populations first reported adverse effects of OP exposures in agricultural settings, where exposures were both occupational and community-wide. Developmental problems were first seen among 4- to 5-year-old children in a Mexican agricultural community using high OP and organochlorine pesticides, including disturbances in stamina, hand-eye coordination, drawing ability, and short-term recall.[15] More recent studies of agricultural OP exposures, using biological measures of exposure, have also documented developmental deficits in exposed children. In one of the first human studies to show persistent effects, Grandjean et al. reported cognitive deficits among Ecuadorian children exposed to multiple pesticides as a result of maternal floriculture employment.[16]

Mechanisms of Organophosphate Toxicity

All OPs produce systemic toxicity by inhibiting the enzyme acetylcholinesterase (AChE), which rapidly leads to overt symptoms of cholinergic hyperstimulation. Because these effects were assumed to be the common mechanism producing neurodevelopmental deficits, the EPA set human exposure limits using guidelines for detection of AChE inhibition. However, subsequent animal studies showed that the OPs are developmental neurotoxicants at much lower exposure levels—levels that are below the threshold for systemic toxicity due to AChE inhibition and that cause no significant AChE inhibition.

A Case Study of Chronic Toxicity and the Vulnerability of Children: Organophosphate Insecticides

The organophosphate insecticides have received much attention as pediatric environmental health hazards because of their inherent acute toxicity, historical importance in acute intoxication events, and widespread use in residential pest control and food production. The maturation of an increasingly robust science base regarding child health and OP insecticides provides important examples of the principles of pediatric pesticide public health science and policy.

Evidence now suggests that many mechanisms are involved in the pathogenesis of the neurodevelopmental disabilities associated with OP pesticides. These mechanisms involve the direct targeting of developmental processes in the brain—targets unrelated to AChE inhibition and targets disrupted at levels of exposure below those that inhibit AChE.[17] Thus, via multiple mechanisms of disrupted brain development, OPs likely produce a wider variety of pathogenic effects, each operating at different stages of development and in different brain regions. The discovery of these additional mechanisms of toxicity suggests strongly that the current use of AChE enzymatic activity as a biomarker of chlorpyrifos neural toxicity is insufficient to assess the risk of neural toxicity in children.

Epidemiologic Evidence for Organophosphate Toxicity

In 2011, three epidemiologic studies reported independent and corroborating evidence that prenatal OP pesticide exposure, including exposure to chlorpyrifos, has persistent deleterious effects on cognitive function through the age of 7 years. These studies, based on different populations, located in distinct geographical regions of the United States, with different routes of exposure, and using different biomarkers, had strongly convergent results. One study using urinary metabolites as biomarkers of exposure reported OP effects on cognition among the children of agricultural workers in the Salinas Valley in California.[18] The second study found similar effects in a New York City Hispanic population with residential exposures.[19] The third study, in New York City among African American and Dominican children, used a biomarker of exposure in umbilical cord blood, specific to chlorpyrifos, and found similar magnitude of effects on cognitive domains.[20] These findings indicate that OP effects from a variety of exposure sources persist into the school years, when they may have implications for learning and academic success.

This evidence was followed by a study using magnetic resonance imaging (MRI) showing that even low to moderate levels of prenatal exposure to chlorpyrifos may lead to long-term, potentially irreversible changes in the structure of the developing brain.[21] This study was the first to use MRI to identify the structural evidence for these cognitive deficits in humans, confirming earlier findings in animals. This study also reported evidence that chlorpyrifos may eliminate or reverse the male–female differences that are ordinarily present in the brain. Notably, all brain abnormalities were apparent at exposure levels below the EPA threshold for toxicity, which was based on exposures high enough to inhibit AChE.

Subsequent epidemiologic research in the United States and internationally showed additional links between OP exposure and neurodevelopmental disorders including autism spectrum disorder.[22-25] Specifically, results from the CHARGE Study[22] reported that proximity to agricultural application of OPs at some point during gestation was associated with a 60% increased risk for autism spectrum disorder, higher for third-trimester exposures (odds ratio [OR] = 2.0; 95% confidence interval [CI]: 1.1, 3.6), and second-trimester chlorpyrifos applications (OR = 3.3; 95% CI: 1.5, 7.4). The von Ehrenstein study showed that risk of autism spectrum disorder was associated with prenatal exposure to several different pesticides: glyphosate (OR = 1.16; 95% CI: 1.06, 1.27), chlorpyrifos (OR = 1.13; CI: 1.05, 1.23), diazinon (OR = 1.11; CI: 1.01, 1.21), malathion (OR = 1.11; CI: 1.01, 1.22), avermectin (OR = 1.12; CI: 1.04, 1.22), and permethrin (OR = 1.10; CI: 1.01, 1.20). It is relevant to note that a review of raw data on chlorpyrifos[26] concluded that there may have been some bias in the reporting of industry-sponsored toxicity studies supporting the safety of the pesticide, further complicating the regulatory challenges.

In conclusion, recent studies suggest that a wide range of exposures to OP pesticides, at levels commonly found in homes and agricultural settings, are harmful to the developing brain and nervous system. These findings underscore the great vulnerability of the human brain to toxic environmental exposures during the prenatal period and in early childhood when the brain is still undergoing developmental changes.

Regulatory Issues

In 2000, the EPA phased out residential use of OPs, including chlorpyrifos, but continued to permit agricultural and commercial uses of these compounds.[27] In 2007, the Pesticide Action Network North America and the Natural Resources Defense Council (NRDC) filed a petition requesting that the EPA revoke all chlorpyrifos tolerances (the maximum allowed residue levels in food) because those tolerances were not safe due to the potential for neurodevelopmental harm. In 2011, the EPA convened a meeting of the Federal Insecticide, Fungicide, and Rodenticide Act (FIFRA) Scientific Advisory Panel (SAP) to review the physiologically based pharmacokinetic pharmacodynamic model for chlorpyrifos. In 2012, a reconvening of the FIFRA SAP reviewed epidemiologic data from the three major US children's health cohort studies that demonstrated converging evidence (as described in the "Epidemiological Evidence" section above). Although the EPA did revise its human health risk assessment in 2016, the Agency eventually denied the petition to revoke all pesticide tolerances for chlorpyrifos and cancel all chlorpyrifos registrations, stating that the epidemiologic evidence addressing neurodevelopmental effects remained equivocal. This denial was challenged in the Ninth Circuit Court of Appeals in 2019 by a coalition of farmworker, health, environmental, and other groups. In April 2021, the Court found that "EPA had abdicated its statutory duty under the Federal Food, Drug and Cosmetic Act" to "conclude, to the statutorily required standard of reasonable certainty, that the present tolerances caused no harm," and ordered the

EPA to issue a final rule in which the agency either modifies the chlorpyrifos tolerances with a supporting safety determination or revokes the tolerances, and modify or cancel food-use registrations of chlorpyrifos. It was not until August 2021 that the EPA revoked all tolerances for chlorpyrifos and announced it will stop the use of chlorpyrifos on all food to better protect human health, particularly that of children and farmworkers[28]

A number of other countries, including the European Union and Canada, and some states including California, Hawaii, New York, Maryland, and Oregon have taken similar action to restrict the use of this pesticide on food. These state-level restrictions had already discouraged the use of chlorpyrifos by farmers prior to the recent EPA ruling, and some alternatives have been registered in recent years. As of 2021, the EPA is committed to reviewing replacements and alternatives to chlorpyrifos and is proceeding with registration review for the remaining non-food uses (www.regulations.gov).

In conclusion, the legacy of acute poisonings and recent evidence for the adverse health consequences of pesticide exposure among children highlight the importance of reductions in pesticide use and exposure. A number of approaches have been taken to achieve exposure reduction. These include mandatory premarket testing and risk assessment that incorporates assessments of developmental toxicity and child-specific vulnerability, integrated pest management techniques of low- or no-toxicity pest control, community right-to-know policies, and use-restriction areas. Improved medical education of healthcare professionals also plays a role through enhanced prevention guidance to families and children at risk (see Chapter 56).

Last, the heavy burden of pesticide exposure and toxicity in LMICs cannot be overlooked. Prevention approaches similar to those developed in the United States are needed in these settings, along with enhanced capacity-building in regulatory protections and bans on products that are banned or restricted in the United States or other nations based on health risks (see Chapter 62).

References

1. Boedeker W, Watts M, Clausing P, et al. The global distribution of acute unintentional pesticide poisoning: estimations based on a systematic review. *BMC Public Health*. 2020;20:1875. https://doi.org/10.1186/s12889-020-09939-0.

2. World Health Organization. *Children's Health and the Environment WHO Training Package for the Health Sector. Pesticide Module*. Geneva, Switzerland: World Health Organization; 2010.

3. Gummin, DD, Mowry JB, Beuhler MC, et al. 2020 Annual Report of the American Association of Poison Control Centers' National Poison Data System (NPDS): 38th Annual Report. *Clin Toxicol*. 2021;59(12):1282–501, doi:10.1080/15563650.2021.1989785.

4. Blondell JM. Decline in pesticide poisonings in the United States from 1995 to 2004. *Clin Toxicol*. 2007;45(5):589–592.

5. American Academy of Pediatrics Council on Environmental Health. Pesticides. In: Etzel RA, Balk SJ, eds., *Pediatric Environmental Health*. 4th ed. Elk Grove Village, IL: American Academy of Pediatrics; 2018:687–728.

6. Benbrook CM. Trends in glyphosate herbicide use in the United States and globally. *Environ Sci Eur*. 2016;28:3. https://doi.org/10.1186/s12302-016-0070-0.

7. Gillezeau C, Lieberman-Cribbin W, Taioli E. Update on human exposure to glyphosate, with a complete review of exposure in children. *Environ Health*. 2020;19:115. https://doi.org/10.1186/s12940-020-00673-z.

8. Centers for Disease Control and Prevention. National report human exposure to environmental chemicals. 2023. Available at: https://www.cdc.gov/exposurereport/index.html.

9. Hyland C, Laribi O. Review of take-home pesticide exposure pathway in children living in agricultural areas. *Environ Res*. 2017 Jul;156:559–70. doi:10.1016/j.envres.2017.04.017.

10. Fagan J, Bohlen L, Patton S, Klein K. Organic diet intervention significantly reduces urinary glyphosate levels in U.S. children and adults. *Environ Res*. 2020 Oct;189:109898. doi:10.1016/j.envres.2020.109898

11. Roberts JR, Karr CJ, American Academy of Pediatrics Council on Environmental Health. Pesticide exposure in children. Technical report. *Pediatrics*. 2012;130:e1765–88.

12. Centers for Disease Control. Healthy housing reference manual. Chapter 5. Indoor air pollutants and toxic materials. Available at: https://www.cdc.gov/nceh/publications/books/housing/cha05.htm.

13. Karr C, Solomon GM, Brock-Utne A. Health effects of common home, lawn and garden pesticides. *Pediatr Clin North Am*. 2007;54:63–80.

14. Dereumeaux C, Fillol C, Quenel P, Denys S. Pesticide exposures for residents living close to agricultural lands: a review. *Environ Int*. 2020;134:105210. doi:10.1016/j.envint.2019.105210. Epub 2019 Nov 16. PMID: 31739132.

15. Guillette EA, Meza MM, Aquilar MG, Soto AD, Garcia IE. An anthropological approach to the evaluation of preschool children exposed to pesticides in Mexico. *Environ Health Perspect*. 1998;106:347–53.

16. Grandjean P, Harari R, Barr DB, Debes F. Pesticide exposure and stunting as independent predictors of neurobehavioral deficits in Ecuadorian school children. *Pediatrics*. 2006;117:e546–56.

17. Slotkin TA. Cholinergic systems in brain development and disruption by neurotoxicants: nicotine, environmental tobacco smoke, organophosphates. *Toxicol Appl Pharmacol*. 2004;198:132–51.

18. Bouchard MF, Chevrier J, Harley KG, et al. Prenatal exposure to organophosphate pesticides and IQ in 7-year-old children. *Environ Health Perspect*. 2011;119:1189–95.

19. Engel SM, Wetmur J, Chen J, et al. Prenatal exposure to organophosphates, paraoxonase 1, and cognitive development in childhood. *Environ Health Perspect*. 2011;119:1182–8.

20. Rauh VA, Arunajadai S, Horton M, et al. Seven-year neurodevelopmental scores and prenatal exposure to chlorpyrifos, a common agricultural pesticide. *Environ Health Perspect*. 2011;119:1196–201.

21. Rauh VA, Perera FP, Horton MK, et al. Brain anomalies in children exposed prenatally to a common organophosphate pesticide. *Proc Natl Acad Sci USA*. 2012;109:7871–6.

22. Shelton JF, Geraghty EM, Tancredi DJ, et al. Neurodevelopmental disorders and prenatal residential proximity to agricultural pesticides: the CHARGE study. Environ Health Perspect. 2014;122(10):1103–9. doi:10.1289/ehp.1307044.

23. von Ehrenstein OS, Ling C, Cui X, et al. Prenatal and infant exposure to ambient pesticides and autism spectrum disorder in children: population based case-control study. *BMJ*. 2019 Mar 20;364:l962.

24. Muñoz-Quezada MT, Lucero BA, Iglesias VP, et al. Chronic exposure to organophosphate (OP) pesticides and neuropsychological functioning in farm workers: a review. *Int J Occup Environ Health*. 2016;22(1):68–79.

25. He X, Tu Y, Song Y, Yang G, You M. The relationship between pesticide exposure during critical neurodevelopment and autism spectrum disorder: A narrative review. *Environ Res.* 2022;203:111902.

26. Mie A, Rudén C, Grandjean P. Safety evaluation of pesticides: developmental neurotoxicity of chlorpyrifos and chlorpyrifos-methyl. *Environ Health.* 2018;17:77. https://doi.org/10.1186/s12940-018-0421-.

27. U.S. Environmental Protection Agency. *Chlorpyrifos Revised Risk Assessment and Agreement with Registrants.* Washington, DC: US Environmental Protection Agency; 2000.

28. Docket identification number EPA-HQ-OPP-2021-0523; Office of Pesticide Programs Regulatory Public Docket in the Environmental Protection Agency Docket Center (EPA/DC), West William Jefferson Clinton Bldg., Rm. 3334, 1301 Constitution Ave. NW, Washington, DC 20460-0001.

36

PCBs, Dioxins, Furans, DDT, Polybrominated Compounds, Per- and Polyfluoroalkyl Substances, and Other Halogenated Hydrocarbons

Philip J. Landrigan

The halogenated hydrocarbons are a large and diverse family of synthetic chemicals, all based on a carbon-halogen bond. They include polychlorinated biphenyls (PCBs), dioxins, furans, the pesticide DDT, polybrominated biphenyls (PBBs), polybrominated diphenyl ethers (PBDEs), and the per- and polyfluoroalkyl substances (PFAS).

Halogenated hydrocarbon compounds have become widely disseminated in the environment.[1] They are found today in biological and environmental specimens from diverse locations around the globe and in the bodies of humans in all countries surveyed. The insolubility of most of these compounds in water, their high boiling point, thermal stability, and resistance to chemical degradation—the same properties that contributed to their utility—have made the halogenated hydrocarbon compounds extremely persistent environmental pollutants.

Most halogenated hydrocarbons are fat-soluble and bioaccumulative.[1] They can reach especially high concentrations in predator species at the top of the food chain, such as game fish, marine mammals, and humans.

Polychlorinated Biphenyls

The PCBs are a class of 209 structurally similar halogenated hydrocarbon compounds (congeners) each consisting of two linked phenyl rings with between 1 and 10 attached chlorine atoms (see Figure 36.1). The commercial PCBs that are the source of most of today's pollution were produced as oily liquids consisting of mixtures of 50–90 congeners with various levels of chlorination.

Industrial use of PCBs began in 1929.[2] Because of their extreme stability, nonflammability, and resistance to heat and chemicals, PCBs were used widely in electricity-generating power plants and other industries as insulating (dielectric) fluids in transformers and capacitors. PCBs were also used in consumer products such as hydraulic fluids, immersion oil for microscopes, carbonless copy paper, electrical appliances, caulking compound, and ballasts for fluorescent lighting. In the United States, because of concerns about their toxicity and environmental persistence, Congress

Philip J. Landrigan, *PCBs, Dioxins, Furans, DDT, Polybrominated Compounds, Per- and Polyfluoroalkyl Substances, and Other Halogenated Hydrocarbons* In: *Textbook of Children's Environmental Health*, Second Edition. Edited by: Ruth A. Etzel and Philip J. Landrigan, Oxford University Press. © Oxford University Press 2024. DOI: 10.1093/oso/9780197662526.003.0036

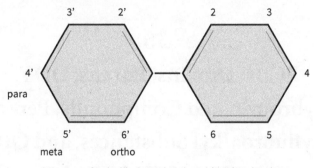

Structure of Polychlorinated Biphenyl (PCB) Molecule

Figure 36.1. Structure of polychlorinated biphenyl (PCB) molecule

banned the manufacture and use of PCBs under the Toxic Substances Control Act of 1977. PCB production was banned internationally in 2001, under the Stockholm Convention on Persistent Organic Pollutants (POPs). See chapter 62 for more on details on the Stockholm Convention.

Although PCBs have not been manufactured in the United States for more than 40 years, they remain widespread in the environment as a consequence of their extensive past use, improper containment, and environmental stability.[1] The US Environmental Protection Agency (EPA) has estimated that 150 million pounds of PCBs were dispersed in air and water in the United States and an additional 290 million pounds discarded into landfills.[3]

Once released into the environment, PCBs degrade only very slowly and therefore may persist for decades. The most heavily chlorinated PCBs are generally the most highly persistent. They can cycle between air, water, and soil, and PCB vapors can be transported long distances in the atmosphere.[1] Very high concentrations are found in the circumpolar regions as a consequence of long-range atmospheric transport and subsequent precipitation.

Many waterways in the United States and in other countries are contaminated today with PCBs. The sources of the PCBs in lakes, rivers, and oceans include industrial releases, influx from streams and sewers, and atmospheric fallout. The Hudson River in New York State is especially heavily contaminated as the result of leakage of PCBs from General Electric manufacturing plants above Albany, New York, and PCBs have been detected in sediment, water, fish, and other aquatic species along a 200-mile (322-kilometer) stretch of the river from above Albany to the southern tip of Manhattan. Studies monitoring changes in levels PCB contamination in Hudson River sediments over time have shown that PCBs degrade only minimally in nature.[4]

PCBs in water and sediment are taken up by marine microorganisms and enter the food chain, where they concentrate (bioaccumulate) as they move upward. PCB levels in fish can exceed by 100,000 times the levels in the water in which they live. For example, in a PCB-contaminated region of the Hudson River north of Troy, New York, water concentrations of PCBs averaged less than 1 part per billion (ppb), while fish had levels greater than 100 parts per million (ppm).[4] Statistical surveys of Lake Michigan fish have shown that most specimens contained detectable PCBs and that levels were greater in larger, older fish and highest in predator (sport fish) species.

Human Exposure to Polychlorinated Biphenyls

Human exposure to PCBs is widespread. Population-based surveys such as the US National Health and Nutrition Examination Survey (NHANES) show detectable levels of PCBs in the serum of virtually all persons examined.[1] Because PCBs are remarkably stable in the human body, older adults tend to have the highest concentrations.

Consumption of contaminated food items, most notably fish, crabs, and shellfish from contaminated waters, is the main route of human exposure.[1] PCB concentrations tend to be higher in sport fish than in market fish. Red meat and chicken may also be contaminated, and, although these food items are much less highly contaminated than fish, they are eaten in greater quantities and thus contribute significantly to overall population exposure. Inhalation and transdermal absorption are not important routes of exposure to PCBs, except for working populations with occupational exposures.

A survey of Great Lakes fish eaters undertaken from 1993 to 1995 found that all had detectable levels of PCBs as well as of dioxins and furans. PCB levels were higher in sport fish consumers than in a referent population. Levels increased with increasing levels of sport fish consumption.[6]

Because they are fat-soluble, PCBs concentrate in human milk, and, in population surveys, PCB concentrations in whole milk have ranged as high as 960 ppb, while mean levels have varied from 3.0 to 100 ppb.[5] Breast milk is therefore an important source of exposure for nursing infants.

Developmental Neurotoxicity of Polychlorinated Biphenyls

Prenatal exposure to PCBs and other halogenated hydrocarbons during vulnerable periods in early development can cause brain injury resulting in neurodevelopmental deficits.[7,8] From the perspective of children's environmental health, these are the most important effects of the halogenated hydrocarbons.

The first recognition of the developmental neurotoxicity of PCBs arose in studies of children exposed to high levels of PCBs in poisoning incidents in Asia. One of these incidents occurred in Japan in the late 1960s and involved approximately 1,800 persons.[9] The syndrome was termed *Yusho,* the Japanese term for "rice oil disease." It resulted from PCB contamination of rice oil used in cooking. The PCBs entered the rice oil in the factory where the oil was produced as the result of a leak in a heat exchange system containing liquid PCB. Follow-up investigation demonstrated that the PCBs were cross-contaminated with highly toxic polychlorinated dibenzofurans (PCDFs) that had formed during the heating of the PCBs. In a second, similar incident in Taiwan, termed *Yu-cheng* (oil disease), rice oil was again contaminated by a mix of PCBs and PCDFs. Infants exposed prenatally to high levels of PCBs and PCDFs in utero in the Yusho and Yu-cheng episodes exhibited a variety of symptoms including low birth weight, abnormal skin pigmentation, delayed developmental milestones, and lower IQs than unexposed siblings. These children also had abnormal tooth and nail development (see Figure 36.2).

Neurodevelopmental toxicity was quantified in the Yu-cheng episode. The difference in mean, full-scale IQ between exposed and unexposed children ranged from 9 to 19 points during 6 years of annual testing. Exposed boys, but not girls showed deficits in spatial reasoning.[8]

To investigate the possibility that early-life exposure to PCBs might cause subclinical neurodevelopmental toxicity in children, the National Institutes of Health (NIH) launched a prospective epidemiologic study in North Carolina in the 1970s.[8] This study was triggered by a 1976 report from the EPA that found that PCB levels in breast milk of US women approached the levels that had been associated with developmental impairment in rhesus monkeys.

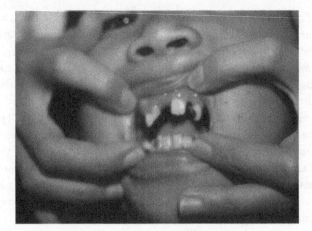

Figure 36.2. Abnormal tooth development in child with Yu-cheng disease

The NIH North Carolina study found an association between prenatal exposures of children to PCBs and delayed neurobehavioral development. Exposed children had lower scores on the Bayley Scales of Psychomotor Development at ages 18 and 24 months than did unexposed children. Scores on the Psychomotor Development Index at age 24 months were reduced by an average of 8 points in the most heavily exposed children.[8]

To further examine the effects of prenatal exposure to PCBs in a second population—infants born to mothers who had consumed Great Lakes fish during pregnancy—the EPA initiated a parallel study in Michigan.[7] Prenatal PCB exposure was assessed by measurement of PCB levels in umbilical cord blood at birth. This study found decreased visual recognition memory at 7 months, decreased McCarthy scale memory at 4 years, and decreased IQ at 11 years in children who had been exposed prenatally to PCBs. These effects were strongest at highest levels of exposure, but also extended down to quite low levels.

In both the North Carolina and Michigan studies, the association between diminished neurobehavioral development and PCB exposure was found principally for transplacental prenatal exposure and was much less evident among children exposed only postnatally via breast milk.

More recent epidemiologic studies of infants exposed prenatally to PCBs with longer follow-up periods extending later into childhood have found evidence for motor problems, cognitive deficits, problems with visual recognition, deficits in executive functioning, attention-deficit/hyperactivity disorder (ADHD), and autism spectrum disorder (ASD). The most serious of these deficits have generally occurred among the most heavily exposed children.[10]

Studies in animals of the neurodevelopmental toxicity of PCBs are generally consistent with the human data. They show that prenatal exposure is more dangerous gram for gram than postnatal exposure via breast milk, and they document damage to similar functional domains. For example, learning and visual recognition memory in rats showed an effect of prenatal, but not of postnatal PCB exposure. Spatial learning deficits in adult rats were associated with prenatal PCB exposure, though the effects were limited to females. Perinatal exposure of rats to PCBs from gestational day 6 through postnatal day 21 produced altered brainstem auditory evoked responses, suggesting a sensorineural hearing loss.

Polychlorinated Biphenyls and Endocrine Disruption

Recent investigations have examined the possibility that PCBs may act as endocrine disruptors. For instance, PCB exposure can lead to disruption in thyroid function,[11] and it has been suggested that the neurobehavioral toxicity of prenatal exposure to PCBs may be mediated, at least in part, by prenatal disruption of thyroid function.[12] This relationship is, however, complex, and epidemiologic studies have found that dietary iodine, cigarette smoking, and other factors can modify the relationship between PCBs and thyroid function.[13]

Polychlorinated Biphenyls and Cardiovascular Health

Emerging evidence indicates that PCBs, dioxins, and other halogenated hydrocarbons increase the risk for cardiovascular disease. A key initial finding was the observation in animal toxicology studies that PCB-126 (a dioxin-like PCB congener) can produce cardiovascular lesions.[14] This congener has also been shown to elevate serum cholesterol levels and raise blood pressure.[15] Further evidence for the cardiovascular toxicity of halogenated hydrocarbons comes from animal studies showing that ApoE-/- mice exposed to dioxin (TCDD) develop earlier and more severe atherosclerosis than unexposed mice[16] and from studies showing that certain PCB congeners such as PCB-77 stimulate the secretion of proinflammatory adipokines, which promote obesity.[17]

Large-scale epidemiologic studies have shown a dose-response association between serum PCB levels and prevalence of type 2 diabetes among US adults at current levels of environmental exposure.[18] A systematic review of studies of adults occupationally exposed to dioxins and dioxin-like PCBs concluded that dioxin exposure is associated with increased mortality from ischemic heart disease as well as from cardiovascular disease.[19]

Polychlorinated Biphenyls and Immune Function

Animal data indicate that exposure in early life to dioxin-like PCBs and to tetrachlorodibenzo-p-dioxin depresses immune function. To explore the potential immunotoxicity of PCBs in developing infants, studies of immune response to standard childhood immunizations were conducted in a birth cohort in the Faroe Islands, where PCB exposures can be very heavy because traditional diets include PCB-contaminated whale blubber. Prenatal PCB exposure was determined by measuring maternal levels of PCBs in serum during pregnancy and in milk postnatally. The investigators found that the antibody response to diphtheria toxoid was depressed at age 18 months by 24.4% for each doubling of the cumulative PCB exposure.[20]

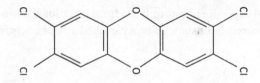

Figure 36.3. Chemical formula of dioxin

By contrast, a study undertaken in Slovakia, where average PCB levels are 10 times higher than in the United States, found that maternal, cord blood, or 6-month infant PCB concentrations were not associated with total serum immunoglobulin levels at 6 months, regardless of the timing of PCB exposure, PCB congener, or specific immunoglobulin.[21]

Dioxins and Furans

Dioxins and furans are highly toxic halogenated hydrocarbon compounds.[22] Although dioxins and furans can form naturally as the consequence of volcanic emissions and forest fires, most dioxin is formed as a by-product of industrial processes including smelting, chlorine bleaching of paper pulp, and the manufacturing of certain herbicides, most notably the trichlorophenols. Combustion of PCBs and of vinyl chloride plastics can produce dioxins. The chemical formula of dioxin is shown in Figure 36.3.

Dioxins, like PCBs, have multiple congeners with varying numbers of chlorine atoms and differing levels of toxicity. The most highly toxic and best-known dioxin is 2,3,7,8-tetrachloro-dibenzo-dioxin (2,3,7,8-TCDD). Like PCBs, dioxins are fat-soluble, highly persistent in the environment, and bioaccumulative.

Sources of Exposure to Dioxins

More than 90% of the daily intake of dioxins and furans for the general population, including children, comes from food, primarily meat and dairy products, and, to a lesser extent, from fish. Dioxins have been measured in human milk, cow's milk, and infant formula, and therefore infants are at risk of exposure. To protect against dietary exposure to dioxins and furans, countries around the world have established food monitoring programs, and, from time to time, these programs detect excess levels of dioxins in certain foodstuffs leading to product recalls.[23]

Major episodes of acute human exposure to dioxins have occurred over the past five decades. The first of these was the exposure of Vietnamese civilians and US service members during the Vietnam War to 2,3,7,8-TCDD, which was a contaminant in the trichlorophenol herbicide known as Agent Orange used as a defoliant. A second episode was the release of large amounts of dioxin and other chemicals in an explosion at a chemical factory in Seveso, Italy, in 1976, that exposed an estimated 37,000 people of all ages, including children.[23] Industrial workers in herbicide manufacturing plants and other industries are occupationally exposed to dioxins on an ongoing basis.

Health Effects of Dioxins

Studies of persons exposed to dioxins in the above episodes, studies of workers exposed occupationally, and experimental studies in animals have produced much information on the toxicology and health effects of dioxins. Many of the toxic effects of dioxins and furans parallel those described above for PCBs but tend to be more severe and to occur at much lower levels of exposure.

Short-term exposure to high levels of dioxins can result in skin lesions, such as chloracne and patchy darkening of the skin, as well as altered liver function.[23]

Prenatal exposures to dioxins and furans are linked to impairment of the immune system, the developing nervous system, the endocrine system, and reproductive functions. These effects were seen in the infants exposed prenatally to a mixture of PCBs and furans in the Yusho and Yu-cheng episodes.[8]

Dioxins have been linked to delayed growth and decreased body mass in a population of Russian boys highly exposed to dioxin.[24] This finding parallels the observation in animal studies that high-dose exposures to dioxins can cause a wasting disease marked by significant loss of body weight.

The International Agency for Research on Cancer (IARC) has classified dioxin as a "known human carcinogen."[23]

DDT (Dichlorodiphenyltrichloroethane)

DDT (dichlorodiphenyltrichloroethane) is a chlorinated hydrocarbon pesticide. It is the best known of this class of pesticides, which were previously heavily used and include such highly toxic, but now banned products as aldrin, dieldrin, endrin, and hexachlorobenzene (HCB). DDT is a synthetic chemical and does not occur in nature. It is highly persistent in the environment, and it is bioaccumulative.[25] DDT was used extensively in the US from the 1940s until its banning in 1972 to control insects in agriculture and insect vectors of disease. In World War II, DDT contributed importantly to the control of malaria.

DDT was banned in the United States following the discovery by Rachel Carson that DDT is an endocrine disruptor and reproductive toxicant. Carson described this discovery and its consequences for the environment and potentially for human health her iconic book *Silent Spring*.[26] Carson observed that DDT had accumulated to reach high levels in predatory bird species at the top of the food chain, most notably ospreys and bald eagles. In those species, DDT disrupted estrogen function, which in turn interfered with egg formation and led to the production of thin-walled, nonviable eggs that cracked prematurely leading to fetal death. The result was the near extinction of ospreys and bald eagles in many areas of the United States. These species have rebounded since the banning of DDT (Figure 36.4). The ban on DDT marked the beginning of the modern environmental movement and was one of the first official acts of the then newly formed EPA.

DDT continues to be manufactured in India, China, and North Korea. The World Health Organization recommends DDT for indoor malarial vector control under certain circumstances.[25] Except for this restricted use, DDT is banned internationally under the Stockholm Convention on Persistent Organic Pollutants.

Figure 36.4. Rachel Carson, the mother of the environmental movement

In the environment, DDT breaks down slowly into DDE and DDD as a consequence of the action of soil microorganisms. DDT, DDE, and DDD have long environmental half-lives and can potentially persist in soil for hundreds of years. DDT and its breakdown products are found at many hazardous waste sites.[25]

Sources of Exposure to DDT

Diet is the major route of exposure to DDT for most people in the United States today. Meat, poultry, dairy products, and fish, especially sport fish, are the principal food sources. Residual levels of DDT persist in soil in many areas of the United States as a legacy of past use and can enter food crops. DDT can leach into waterways to contaminate fish. The Agency for Toxic Substances and Disease Registry (ATSDR) reported that from 1986 to 1991 the average US adult consumed 0.8 micrograms of DDT per day.[25] Food imported from countries that still use DDT is an additional source of human exposure.

People who live in countries that continue use DDT in agriculture or vector control can be directly exposed through inhalation, skin contact, and ingestion of recently spayed foods and contaminated drinking water.

Children's Exposure to DDT

DDT freely crosses the placenta from the mother to the fetus and can therefore result in prenatal exposure. DDT, like PCBs and other halogenated hydrocarbons, is fat-soluble and concentrates to reach high levels in lipid-rich human milk. Breast milk is therefore an important source of exposure for nursing infants. Cow's milk and infant formula can also contain measurable levels of DDT.

Children are exposed to DDT, DDE, and DDD by eating food contaminated with these chemicals. Because children eat more food per pound of body weight than adults they are proportionately more heavily exposed. The ATSDR reported that between 1985 and 1991 the average 8½-month-old infant in the United States consumed 4 times more DDT per pound of body weight than the average adult.[25]

Health Effects of DDT

Acute high-dose exposure to DDT can cause neurotoxicity with irritability and, in extreme cases, convulsions. Such exposure is not seen today except in locations where DDT is still directly applied as a pesticide.

DDT is an endocrine disruptor and developmental toxicant, as first demonstrated in the work of Rachel Carson.[26] Experimental studies in animals have shown that exposure to DDT during pregnancy can slow the growth of the fetus. Male rats exposed to the DDT breakdown product, p,p'-DDE as fetuses or while nursing showed changes in development of their reproductive systems and delayed onset of puberty. Neurobehavioral problems in later life have been reported in mice exposed to DDT during the first weeks of life.[25]

DDT exposure has been linked to increased risk of obesity. Of five published prospective studies examining associations between maternal levels of DDE and either body mass index (BMI) or body weight in children, four found a positive association between DDE exposure and measures of obesity. However, the association was statistically significant in only two of these studies. Prenatal DDE exposure had a positive dose-dependent relationship with both BMI and body weight among adult daughters in the Michigan fish-eater cohort.[27] Ongoing experimental studies are exploring the possible obesogenicity of DDT in rodent species. This information is examined in detail in Chapter 41.

DDT and Breast Cancer

The possibility that DDT exposure increases risk of breast cancer has been the subject of intense study. Previous cross-sectional epidemiological studies such as the Long Island Breast Cancer Study that examined this question by measuring serum DDT levels in adult women with and without breast cancer found little evidence for an association. However, those studies were limited by their inability to measure exposure in young women during previous periods of heavy DDT use, and thus they generally found very low levels of DDT and its metabolites in serum.

A more recent study from California investigated this question using blood samples that had been obtained in the 1960s, before DDT was banned and use of DDT was very high.[28] In an analysis based on these stored samples, investigators found that high serum levels of DDT in childhood were associated with a five-fold increased risk of breast cancer among women born after 1931. These women were under 14 years of age in 1945, when DDT came into widespread use, and mostly under 20 years of age at the time when DDT use peaked. Among women who were not exposed to DDT before age 14 years no association was observed between DDT levels and breast cancer.

Brominated Compounds

Polybrominated Biphenyls

The PBBs, like the PCBs, are comprised of 209 congeners each based on two linked phenyl rings. In the PBBs, bromine atoms rather than chlorine atoms are attached to the rings. The physical and chemical properties of the PBBs as well as their potential for persistence in the environment and bioaccumulation resemble those of the PCBs.

PBBs were employed previously as flame retardants in the plastics industry. Their commercial production in the United States began in the 1970s but was discontinued in 1976 after the Michigan episode described below. Consequently, PBBs have not been widely dispersed in the environment, and exposure of most human populations is minimal.[29]

In 1973, several hundred pounds of PBBs were introduced into cattle feed in Michigan as the result of a labeling accident in a chemical factory. A devastating and often fatal syndrome developed in exposed cattle—anorexia, weight loss, skin changes, decreased milk production, and increased fetal wastage. Widespread human exposure resulted from contact with contaminated feed and from consumption of contaminated meat, eggs, and dairy products. The most intense exposure was noted in persons who had lived or worked on contaminated farms and in workers at the chemical plant that produced PBBs. Some people who ate the contaminated food as well as some exposed workers developed acne and hair loss, similar to the chloracne seen in the Yusho episode.

To evaluate the chronic health effects of this episode of PBB exposure, the Michigan Department of Public Health and the US Public Health Service established a cohort of 4,545 persons who are being followed prospectively.[30] Because PBBs are highly persistent in human adipose tissue and have been found in chronic rat feeding studies to induce cancerous liver nodules, the public health agencies have followed the Michigan population for many decades. IARC considers PBBs to be possible human carcinogens.[29]

Polybrominated Diphenyl Ethers

PBDEs have a two-ringed chemical structure similar to that of the PCBs and PBBs. Like PCBs, there are 209 congeners with varying numbers of attached bromine atoms and differing levels of toxicity.

The PBDEs are used extensively as flame retardants in plastics and textiles.[31] They can be found today in computers, television sets, mobile phones, electronics and electrical items, automotive equipment, construction materials, polyurethane foam mattresses, cushions, upholstered furniture, carpets, and draperies. PBDEs are not chemically bound to the plastics and other products in which they are used, making it relatively easy for them to leach out of these products and into the environment. Commercial production of PBDEs began in the 1970s and continues to the present. Global production of PBDEs is reported to be 67,390 metric tons. The United States phased out the manufacturing of penta and octa BDEs in 2004.

PBDEs, like PCBs, are disseminated extensively in the environment. They are an important component of electronic waste (see Chapter 30). Like PCBs and other halogenated hydrocarbons, PBDEs can persist in the environment for decades, travel far, and bioaccumulate. As a result, PBDEs have been detected in sediments, as well as in fish and other marine and terrestrial species.

Sources of Exposure to Polybrominated Diphenyl Ethers

Routes of human exposure to PBDEs are not fully defined.[29] Inhalation and ingestion of indoor dust contaminated by the shedding of PBDEs from computers, carpets, mattresses, and other household furnishings appears to be a major source. Food intake is another route of exposure.

Human body burdens of PBDEs have increased markedly over the past several decades.[32] Levels in the United States are generally 10- to 100-fold higher than levels observed in Europe, Asia, or New Zealand. Substantial increases in PBDE levels in human breast milk have also been documented. Breast milk levels are much higher in North American than in Swedish women (see Figure 36.5). As is the case for PCBs, PBDEs bioaccumulate with age and have been documented to be present in higher concentration in older adults.

Infants are exposed to PBDEs transferred during pregnancy from the mother to the fetus. Nursing infants are exposed via breast milk.

Health Effects of Polybrominated Diphenyl Ethers

Neurodevelopmental disorders[33] and endocrine disruption[34] are two adverse outcomes of PBDE exposure reported to date. Animal studies have shown that PBDEs affect thyroid function and can impair the developing central nervous system and brain. There is evidence that PBDEs may be more toxic when combined with PCBs.

Neurodevelopmental Toxicity of Polybrominated Diphenyl Ethers

An ongoing prospective birth cohort study of children in New York City found an association between prenatal exposures to PBDEs during pregnancy and subsequent deficits

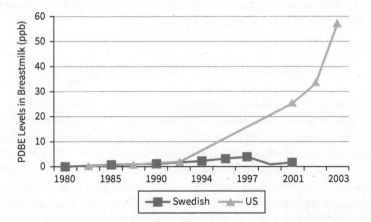

Figure 36.5. PBDE levels in breast milk, United States versus Sweden. Triangles indicate US data and squares indicate Swedish data.

in neurobehavioral development, especially in memory development.[35] Children with higher concentrations of PBDEs scored lower on tests of mental and physical development at 12–48 and 72 months. Associations were significant for the 12-month Psychomotor Development Index, the 24-month Mental Development Index (MDI), the 36-month MDI, the 48-month full-scale and verbal and performance IQ, and the 72-month performance IQ. These findings parallel the effects on neurobehavioral development seen in infants exposed prenatally to PCBs.[7,8]

Per- and Polyfluoroalkyl Substances

The PFAS are fluorine-based halogenated hydrocarbons with powerful surfactant and water-repelling properties. These chemicals have found multiple applications in industrial and consumer products including nonstick cooking pans, water-resistant upholstery, textiles, carpets, food packaging, lubricants, and electronics.[36]

PFAS are environmentally persistent. They are found in surface water, groundwater, soil, and sediment, especially near facilities that have made or used these substances. Like PCBs, they have been found at remote locations such as the Arctic.

Food and drinking water are the major sources of PFAS exposure for adults. Measurable levels of PFAS are present in the serum of a high proportion of US residents studied through the National Health and Nutrition Examination Survey (NHANES). PFAS have been shown to be able to cross the human placenta from mother to child. Measurable levels have been documented in the umbilical cord blood of a very high proportion of newborn infants in the United States.[36]

In experimental studies of the developmental toxicity of perfluorooctane sulfonic acid (PFOS), birth defects were observed in both rats and mice including cleft palate, ventricular septal defect, and enlargement of the right atrium, primarily in the groups that received the higher dosages—10 and 20 mg/kg.[37] In several studies, statistically significant developmental delays were observed. Exposures earlier in gestation appeared to produce stronger responses.[38]

Studies of pregnant women exposed to perfluorooctanoic acid (PFOA) suggest that higher levels of maternal PFOA are associated with lower birth weight and possibly fetal growth restriction.[39] Studies have examined the possibility of an association between early-life exposure to PFAS and ADHD, but have found little to no evidence for an association.[40,41] Epidemiologic studies indicate that PFAS depresses immune responsiveness in children, resulting in lower levels of antibody titers in children at age 5 years.[42–44] See Chapter 37 for more detailed information about PFAS.

Prevention of Exposure to Halogenated Hydrocarbons

National governments can reduce exposures to halogenated hydrocarbons by curtailing and banning their manufacture and use (as has occurred in the United States for PCBs and DDT and internationally for POPs), controlling industrial sources of their environmental release, controlling their release from landfills and hazardous waste sites, requiring incineration of PCB-contaminated waste materials at temperatures of higher than 850° C in specially constructed incinerators in order to prevent formation of dioxins, and establishing and enforcing strict programs for monitoring of these chemicals in the food supply.[45]

State and local governments can reduce exposures by controlling commercial sources of release and securing landfills. Local governments can control the spread of PCBs in school environments by requiring timely removal of all PCB-containing fluorescent light ballasts. Parents can accelerate this process through citizen action.

Families can reduce their exposures to halogenated hydrocarbons by consuming low-fat dairy products, limiting consumption of contaminated meat and fish, trimming fat from meat before cooking (halogenated hydrocarbons concentrate in fat), and avoiding products that contain brominated flame retardants. A balanced diet that includes adequate amounts of fruits, vegetables, and cereals will help to avoid excessive exposure from a single source. This is a long-term strategy to reduce body burdens of persistent pollutants and is especially important for girls and young women to reduce current and future exposures to fetuses and breastfeeding infants.

To guide families in the selection of fish species low in PCBs and other persistent pollutants such as methylmercury, various groups—the Natural Resources Defense Council (NRDC), the Environmental Working Group (EWG) and the Monterrey Bay Aquarium—have published authoritative tables listing safer and less safe fish.

Halogenated Hydrocarbons and Breastfeeding

There has been much discussion about whether it is safe for women to breastfeed their infants given the documented high concentrations of halogenated hydrocarbons and other toxic environmental chemicals in human breast milk. Authoritative bodies that have studied this question including the American Academy of Pediatrics have concluded that the benefits of breastfeeding significantly outweigh any risks from exposures to halogenated hydrocarbons or toxic chemicals in mother's milk.[46] Nevertheless, women with unusually high amounts of DDT, PCBs, or other halogenated hydrocarbons in their bodies should be informed of the potential exposure of the fetus if they become pregnant and the potential risks of breastfeeding and should consult with their physician.

References

1. Agency for Toxic Substances and Disease Registry. Toxicological profile for polychlorinated biphenyls (PCBs). 2000. Available at: https://wwwn.cdc.gov/TSP/ToxProfiles/ToxProfiles.aspx?id=142&tid=26.

2. World Health Organization. Environmental health criteria. 2. *Polychlorinated biphenyls and terphenyls*. 1977. https://www.who.int/publications/i/item/9241540621.

3. US Environmental Protection Agency. EPA bans PCB manufacture; phases out use. 1979. Available at: https://www.nrc.gov/docs/ML0832/ML083260462.pdf.

4. US Environmental Protection Agency. Hudson River PCBs superfund site. 2023. Available at: https://www.epa.gov/hudsonriverpcbs.

5. Turyk M, Anderson HA, Hanrahan LP, et al. Relationship of serum levels of individual PCB, dioxin, and furan congeners and DDE with Great Lakes sport-caught fish consumption. *Environ Res.* 2006;100:173–83.

6. Landrigan PJ. General population exposure to environmental concentrations of halogenated biphenyls. In: Kimbrough R, ed., *Halogenated biphenyls, terphenyls. naphthalenes, dibencodioxins and related products*. Amsterdam: Elsevier/North-Holland Biomedical Press, 1980.

7. Jacobsen JL, Jacobson SW. Intellectual impairment in children exposed to polychlorinated biphenyls in utero. *N Engl J Med*. 1996; 335:783–9.

8. Longnecker MP, Rogan WJ, Lucier G. The human health effects of DDT (dichlorodiphenyl-tricholorethane) and PCBs (polychlorinated biphenyls) and an overview of organochlorines in public health. *Ann Rev Pub Health*. 1997; 18:211–44.

9. Kuratsune M, Yoshimura T, Matsuzaka, Yamaguchi A. Epidemiologic study on Yusho, a poisoning caused by ingestion of rice oil contaminated with a commercial brand of polychlorinated biphenyls. *Environ Health Perspect*. 1972;1:119–28.

10. Ribas-Fito N, Sala M, Kogevinas M, Sunyer J. Polychlorinated biphenyls (PCBs) and neurological development in children: a systematic review. *J Epidemiol Community Health*. 2001;55:537–46.

11. Little CC, Barlow J, Alsen M, van Gerwen M. Association between polychlorinated biphenyl exposure and thyroid hormones: a systematic review and meta-analysis. *J Environ Sci Health C Toxicol Carcinog*. 2022 Dec 14:1–20.

12. Winneke G, Walkowiak J, Lilenthal H. PCB-induced neurodevelopmental toxicity in human infants and its potential mediation by endocrine dysfunction. *Toxicology*. 2002;181:161–5.

13. Zoeller RT. Endocrine disrupting chemicals and thyroid hormone action. *Adv Pharmacol*. 2021;92:401–417. doi:10.1016/bs.apha.2021.05.002.

14. Jokinen MP, Walker NJ, Brix AE, Sells DM, Haseman JK, Nyska A. Increase cardiovascular pathology in female Sprague-Dawley rats following chronic treatment with 2,3,4,8-tetrachlorodibenzo-*p*-dioxin and 3,3',4,4',5-penatachlorobiphenyl. *Cardiovas Toxicol*. 2003;3:299–310.

15. Lind PM, Orberg J, Edlund UB, Sjoblom L, Lind L. The dioxin-like pollutant PCB 126 (3,3',4,4',5-pentachlorobiphenyl) affects risk for cardiovascular disease in female rats. *Toxicology Lett*. 2004; 150:293–9.

16. Dalton TP, Kerzee JK, Wang B, et al. Dioxin exposure is an environmental risk factor for heart disease. *Cardiovas Toxicol*. 2001;1:285–98.

17. Arsenescu V, King V, Swanson H, Cassis LA. Polychlorinated biphenyl-77 induces adipocyte differentiation and proinflammatory adipokines and promotes obesity and atherosclerosis. *Environ Health Perspect*. 2008;116:761–8.

18. Lee DH, Lee IK, Porta M, Steffes M, Jacobs DR. Relationship between serum concentrations of persistent organic pollutants and the prevalence of metabolic syndrome among non-diabetic adults: results from the National Health and Nutrition Examination Survey 1999–2002. *Diabetologia*. 2007;50:1841–51.

19. Humblet O, Birbaum L, Rimm E, Mittleman MA, Hauser R. Dioxins and cardiovascular disease mortality. *Environ Health Perspect*. 2008; 116:1143–8.

20. Heilmann C, Grandjean P, Weihe P, Nielsen F, Budtz-Jørgensen E. Reduced antibody responses to vaccinations in children exposed to polychlorinated biphenyls. *PLoS Med*. 2006;3:e311.

21. Jusko TA, De Roos AJ, Schwartz SM, et al. Maternal and early postnatal polychlorinated biphenyl exposure in relation to total serum immunoglobulin concentrations in 6-month-old infants. *J Immunotoxicol*. 2011;1:95–100.

22. Agency for Toxic Substances and Disease Registry. Toxicological profile for chlorinated dibenzo-*p*-dioxins (CDDs). 2012. Available at: https://wwwn.cdc.gov/TSP/ToxProfiles/ToxProfiles.aspx?id=366&tid=63.

23. World Health Organization. Dioxins and their effects on human health. Fact sheet. 2016. Available at: https://www.who.int/news-room/fact-sheets/detail/dioxins-and-their-effects-on-human-health.

24. Burns JS, Williams PL, Sergeyev O, et al. Serum dioxins and polychlorinated biphenyls are associated with growth among Russian boys. *Pediatrics* 2011;127:e59–68.

25. Agency for Toxic Substances and Disease Registry. Public health statement for DDT, DDE, and DDD. 2002. Available at: https://wwwn.cdc.gov/TSP/ToxProfiles/ToxProfiles.aspx?id=81&tid=20.

26. Carson R. *Silent Spring*. Boston: Houghton Mifflin: 1962.

27. Karmaus W, Osuch JR, Eneli I, et al. Maternal levels of dichlorodiphenyl-dichloroethylene (DDE) may increase weight and body mass index in adult female offspring. *Occup Environ Med.* 2009;66:143–9.

28. Cohn BA, Cirillo PM, Terry MB. DDT and breast cancer: prospective study of induction time and susceptibility windows. *J Natl Cancer Inst.* 2019 Aug 1;111(8):803–810.

29. Agency for Toxic Substances and Disease Registry. Toxicological profile for polybrominated biphenyls (PBBs). 2004. Available at: https://wwwn.cdc.gov/TSP/ToxProfiles/ToxProfiles.aspx?id=529&tid=94.

30. Humphrey HEB, Hayner NS. Polybrominated biphenyls: an agricultural incident and its consequences, an epidemiological investigation of human exposure. Presented at the Ninth Annual Conference of Trace Substances in Environmental Health, Columbia, Missouri, June 1975.

31. Agency for Toxic Substances and Disease Registry. Toxicological profile for polybrominated diphenyl ethers (PBDEs). 2017. Available at: https://wwwn.cdc.gov/TSP/ToxProfiles/ToxProfiles.aspx?id=901&tid=183.

32. Betts KS. Rapidly rising PBDE levels in North America. *Environ Sci Technol.* 2002;36:50A–52A.

33. Branchi I, Capone F, Alleva E, et al. Polybrominated diphenyl ethers: neurobehavioral effects following developmental exposure. *Neurotoxicology.* 2003;24:449–62.

34. Turyk ME, Persky VW, Imm P, Knobeloch L, Chatterton R, Anderson HA. Hormone disruption by PBDEs in adult male sport fish consumers. *Environ Health Perspect.* 2008;116:1635–41.

35. Herbstman JB, Sjödin A, Kurzon M, et al. Prenatal exposure to PBDEs and neurodevelopment. *Environ Health Perspect.* 2010 May;118:712–9. doi:10.1289/ehp.0901340.

36. National Academies of Sciences, Engineering, and Medicine. 2022. *Guidance on PFAS Exposure, Testing, and Clinical Follow-Up*. Washington, DC: National Academies Press. https://doi.org/10.17226/26156.

37. Thibodeaux JR, Hanson RG, Rogers JM, et al. Exposure to perfluorooctane sulfonate during pregnancy in rat and mouse. I: maternal and prenatal evaluations. *Toxicol Sci.* 2003;74:369–81.

38. Leubker DJ, Case MT, York RG, Moore JA, Hansen KJ, Butenhoff JL. Two-generation reproduction and cross-foster studies of perfluorooctanesulfonate (PFOS) in rats. *Toxicology.* 2005;215:126–48.

39. Savitz DA, Stein CR, Elston B, et al. Relationship of perfluorooctanoic acid exposure to pregnancy outcome based on birth records in the mid-Ohio valley. *Environ Health Perspect.* 2012;120:1201–7.

40. Hoffman K, Webster TF, Weisskopf MG, Weinberg J, Vieira VM. Exposure to polyfluoro-alkyl chemicals and attention deficit/hyperactivity disorder in U.S. children 12–15 years of age. *Environ Health Perspect.* 2010;118:1762–7.

41. Dalsager L, Jensen TK, Nielsen F, Grandjean P, Bilenberg N, Andersen HR. No association between maternal and child PFAS concentrations and repeated measures of ADHD symptoms at age 2½ and 5 years in children from the Odense Child Cohort. *Neurotoxicol Teratol.* 2021 Nov-Dec;88:107031.

42. Forns J, Verner MA, Iszatt N, et al. Early life exposure to perfluoroalkyl substances (PFAS) and ADHD: a meta-analysis of nine European population-based studies. *Environ Health Perspect.* 2020 May;128(5):57002.

43. Timmermann CA, Budtz-Jørgensen E, Jensen TK, et al. Association between perfluoroalkyl substance exposure and asthma and allergic disease in children as modified by MMR vaccination. *J Immunotoxicol.* 2017 Dec;14(1):39–49.

44. Grandjean P, Heilmann C, Weihe P, Nielsen F, Mogensen UB, Timmermann A, Budtz-Jørgensen E. Estimated exposures to perfluorinated compounds in infancy predict attenuated vaccine antibody concentrations at age 5-years. *J Immunotoxicol.* 2017 Dec;14(1):188–195

45. U.S. Environmental Protection Agency. EPA actions to address PFAS. 2023. Available at: https://www.epa.gov/pfas/epa-actions-address-pfas.

46. Meek JY, Noble L; American Academy of Pediatrics Section on Breastfeeding. Policy statement: breastfeeding and the use of human milk. *Pediatrics.* 2022 Jul 1;150:e2022057988.

37

Perfluoroalkyl and Polyfluoroalkyl Substances

Philippe Grandjean

Overview

Per- and polyfluoroalkyl substances (PFAS) do not occur naturally but have been globally disseminated from countless sources. Following early production of PFAS chemicals from about 1950 onward, the substances have been applied in the production of multiple consumer, commercial, and industrial products including stain repellants, fire extinguishing products, and many others. Shortly after 2000, major PFAS producers in the United States announced their intention of phasing out the production of common long-chain PFAS. Because of the strong chemical bond between carbon and fluorine (see Figure 37.1), many PFAS are highly stable; they are also somewhat water soluble and can potentially leach through soil to reach the groundwater, and they can accumulate in aquatic and marine food chains. The long-chain PFAS are known to be persistent in the human body and therefore stay in the blood and organs for a long time.

A main source of information on adverse health effects from PFAS exposures is the C8 Health Project that collected data from approximately 70,000 Ohio and West Virginia residents with at least 1 year of exposure to drinking water severely contaminated with perfluorooctanoic acid (PFOA) from a local production facility.[1] Data were collected on serum PFOA concentrations and their relationships with concentrations in drinking water and a variety of diagnoses and biological changes. Exposures to perfluorooctane sulfonate (PFOS) also were elevated, though much less so. Since then, multiple population studies have been carried out; many with child populations, many of them prospectively followed since birth, and mostly in regard to background exposures to a variety of PFAS.

Regulatory agencies have carried out evaluations and risk assessments in recent years; these include the US Environmental Protection Agency (EPA),[2] the Agency for Toxic Substances and Disease Registry (ATSDR),[3] and the European Food Safety Association (EFSA).[4] In addition, other important reviews have emphasized particular aspects, such as immunotoxicity[5] and cancer risks.[6] Due to the fact that more than 1,000 scientific articles on PFAS are now published every year, the documentation is expanding and changing, and adverse effects have repeatedly been found at exposure levels previously thought to be safe. This has especially been true for developmental PFAS exposures in children. The PFAS may be useful, but they endanger children's health.

Philippe Grandjean, *Perfluoroalkyl and Polyfluoroalkyl Substances* In: *Textbook of Children's Environmental Health*, Second Edition. Edited by: Ruth A. Etzel and Philip J. Landrigan, Oxford University Press. © Oxford University Press 2024.
DOI: 10.1093/oso/9780197662526.003.0037

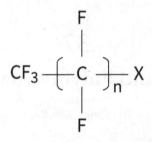

Figure 37.1. Chemical structure of perfluoroalkyl substances (PFAS), in which the strong bond between carbon and fluorine results in strong resistance to break-down and therefore persistence in the environment and in the human body

Scope and Nature of the Problem

The PFAS have been dubbed "forever-chemicals" because they tend to be highly stable due to the strong chemical bond between carbon and fluorine. Thus, the major PFAS do not break down, thus potentially causing continued dissemination to surface water bodies or groundwater, from, for example, former production sites, fire-training areas, or waste dumps. Municipal incinerators may not reach sufficiently high temperatures to decompose PFAS, thereby potentially adding to the dispersal of these compounds. At sewage treatment plants, much of the PFAS may end up in the sludge, which is often used for soil treatment purposes in agriculture. Once released into the environment, PFAS accumulate in food chains and in wildlife.

While early research focused on worker safety, environmental dissemination and population health attracted less attention. Industry claims inspired the belief that the PFAS were almost innocuous. However, as early toxicology studies were finally released in 2000, information emerged on transmission from mother to progeny and a variety of adverse health effects, thereby inspiring independent academic research. New evidence showed that PFAS exposure at lower and lower levels could represent a health hazard. At the same time, PFAS pollution was found to occur globally, disseminated through industrial products, air pollution, marine currents, and other means. The persistence of these compounds, in both the environment and the human body, added to the concern that these compounds might represent a new environmental hazard.

The usefulness of the PFAS for a variety of purposes due to their surfactant properties (e.g., in rain gear, food-wrapping materials, and fire-fighting foams) means that these chemicals are considered highly useful or perhaps even indispensable. While the most commonly used PFAS in the past, PFOA and PFOS, have now been phased out in much of the world, their substitutes are usually similar members of the PFAS group of chemicals and are often found, when properly tested, to be far from innocuous.

Because these compounds are considered particularly worrisome in regard to early-life exposures, the PFAS constitute a prime example where the current environmental pollution with these persistent compounds may be of some concern to present-day adults, but the major risk will be carried by the next generation and perhaps also by children in the future.

Sources of Exposure

Due to their surfactant properties, the PFAS have been used for numerous applications in surface coating and protectant formulations (e.g., on paper and cardboard packaging

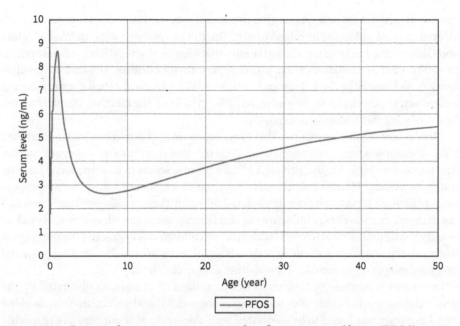

Figure 37.2. Estimated serum concentrations of perfluorooctane sulfonate (PFOS) in a female exposed prenatally and then via breastfeeding for 12 months, followed by background levels from food intake through adulthood (reproduced from the European Food Safety Authority (4), with permission)

Source: https://efsa.onlinelibrary.wiley.com/doi/epdf/10.2903/j.efsa.2020.6223 page 144

products, carpets, clothing, and other textiles). They have also been used in fire-fighting foams and in the manufacture of nonstick coatings on cooking utensils. Environmental dissemination and human exposures have resulted from the primary production or application of these materials, leaching from their use (e.g., in food-packaging materials and from waste deposition or incineration). The PFAS therefore occur in drinking water, house dust, and many foods on a global scale.

Due to the (slight) water solubility of most PFAS, they can migrate into groundwater, and runoff water may reach local water bodies that, together with airborne emissions, can contaminate coastal waters and thereby impact seafood. In addition to drinking water, dietary sources of PFAS exposure include fish and other seafood as the most common source, but also to some degree in eggs, meats, and fruit, while vegetable products appear to be less commonly affected.

PFAS pass the placenta and therefore cause prenatal exposures, and an additional concern is that these substances are excreted in human milk.[7] The main source of exposure during infancy is therefore breastfeeding that may result in the infant's serum PFAS concentration exceeding the mother's level by a factor of 10 (see Figure 37.2). If the drinking water is contaminated, bottle-fed infants may also become highly exposed. For these reasons, and due to the vulnerability of early-life development, the PFAS are of particular concern for children's health.

Adverse Health Effects

Elevated exposure to PFAS is unlikely to cause clinical signs of poisoning. Rather, the PFAS contribute to a range of adverse influences on body functions, thereby causing developmental deficits or increasing the risk of noncommunicable disease generally considered to be of unknown or mixed origin. The PFOA contamination in West Virginia and Ohio resulted in large population studies that identified several likely adverse effects in adults, such as ulcerative colitis, preeclampsia, thyroid disease, elevated serum cholesterol, and cancer of the kidney and testicles.[1] While elevated blood pressure during pregnancy, or preeclampsia, may adversely affect the fetus, the focus was initially on the health of adults with long-term exposures.

More recent studies have shown that PFAS exposure in children can lead to increased risks of multiple adverse outcomes, none of them being specific to the exposure. Studies first focused on birth weight, although PFAS exposure seems to be responsible for only minor decreases.[8] However, studies also have found a tendency toward larger proportions of preterm birth or birth weight below 2,500 g.[9,10] Further follow-up indicated that the early-life exposures may influence the child's subsequent growth and possible risk of obesity.[11] Additional research on health risks to children has begun to take into regard that peak exposures often occur in early infancy, thereby potentially affecting postnatal programming of organ functions that mature during childhood.

One focus is immune system responses to antigen challenges, as illustrated by the generation of specific antibodies following routine childhood vaccinations. A doubled PFAS exposure was found to be associated with a decrease of about 50% in the specific antibody concentration against the tetanus and diphtheria toxoids.[12] The importance of this finding is illustrated by the fact that a substantial number of exposed children later had antibody concentrations below the threshold for long-term protection despite a full series of vaccinations. Because similar results have been obtained in several countries and with other vaccines as well, immunotoxicity is now regarded as the most sensitive outcome of PFAS toxicity.[4] As a consequence of these adverse effects, elevated PFAS exposure also is associated with more frequent infections in the child[13] as well as more frequent hospitalization for pneumonia and other airway infections.[14]

Changes in endocrine functions are reflected by reports on delayed puberty development.[15,16] The thyroid gland is a possible target for several environmental chemicals, and the ATSDR concludes that PFAS exposure is linked to the occurrence of thyroid disease. However, it is doubtful whether such ailments affect children, and the major health impact in early life is more likely thyroid dysfunctions that occur during gestation and influence early brain development.[17] Interestingly, PFAS-associated endocrine disruption in the mother may impede her capability of breastfeeding,[18] the ironic consequence of which is a potential protection of the infant against exposure from the maternal PFAS excretion through the milk.

Furthermore, while type 2 diabetes and obesity are likely outcomes in adults with elevated PFAS exposures, preclinical changes may also occur in exposed children (e.g., in the form of increased body mass index [BMI] and deficient glucose metabolism).[11,19] Thus, the propensity of developing PFAS-associated metabolic changes, including elevated serum lipids,[20] may begin in early life. Another outcome that may be affected by endocrine disruption is osteoporosis. Given the need to achieve the necessary bone mineral mass during childhood growth, PFAS-associated reductions in bone mass in children[21] may imply risks of disease in later life.

PFAS-associated increases in serum lipids and other parameters may indicate involvement of aberrant liver functions. Although the evidence is not yet clear in regard to children's health, a multinational European study focused on early signs of nonalcoholic fatty liver disease (NAFLD) in more than 1,000 children and found metabonomic signs of PFAS-mediated hepatotoxicity.[22] Given the public health importance of NAFLD, the adverse impact of hepatotoxic agents like PFAS should not be underestimated. Adverse effects on kidney function are suspected, but data on childhood kidney disease are too few to interpret.

Current evidence supports the notion that at least some common PFAS are possible carcinogens.[23] The best evidence is on kidney cancer,[24] but other likely sites may include testicles, prostate, breast, and bladder. In Minnesota, exposure to PFAS-contaminated drinking water was associated with an increased incidence of childhood cancers, but limited data are available to evaluate the importance of this possible risk.

Prevention

National regulatory agencies vary somewhat in their approaches to risk assessment. Thus, the ATSDR has often considered human data too uncertain[3] because they are difficult to relate to exposure levels, and the ATSDR therefore has relied primarily on animal toxicology results. However, species differences occur, particularly in regard to PPAR receptor expression and renal excretion, thus complicating the translation of rodent data to the human situation. When regulatory agencies rely on animal toxicology, the advantage is that the results provide a clear association to intake levels, and comparisons based on serum concentrations can potentially link with epidemiological studies.

Over time, the tolerable (or permissible) intake limits for PFOA and PFOS have been lowered substantially. For example, the lowest current limits represent a decrease by more than 1,000-fold, as compared with more optimistic limits from 2008 or earlier. As first proposed by scientists at the Minnesota Department of Health,[25] proper protection of the next generation requires that population exposures be controlled so that women in fertile age groups do not accumulate and transfer hazardous amounts of PFAS to the next generation. In agreement with this approach, the EFSA's updated risk assessment aimed at protecting the next generation against developmental immunotoxicity from the four major PFAS and established a tolerable weekly intake (TWI) at 4.4 ng/kg body weight per week.[4] This TWI will also protect against other potential adverse effects, and it corresponds to a drinking water limit of 2 ng/L. Being the lowest internationally, and thus the most protective e.g., when compared to the guideline issued by the EPA in 2016 at 70 ng/L for the sum of PFOS and PFOA. Many municipal water supply systems in the United States exceed the EPA guideline value, and elevated exposures are thought to affect a large proportion of Americans. Given that PFAS are persistent, sources of current drinking water contamination often include former production sites or previous fire training areas, where kilogram amounts of PFAS have accumulated and begun to leach. In some communities, the water is now being filtered, and untold households choose to use activated carbon filters on the kitchen faucets to decrease the PFAS concentrations.

In 2023 the EPA proposed to regulate PFAS under the Safe Drinking Water Act (SDWA) through issuance of a proposed National Primary Drinking Water Regulation (NPDWR) for six PFAS compounds. The NPDWR would establish nationwide, legally enforceable drinking water Maximum Contaminant Levels (MCLs) for six PFAS compounds. The proposed NPDWR would also set nonenforceable, health-based maximum contaminant level goals (MCLGs) for the same six PFAS compounds. Once finalized, public water systems would have 3 years to implement required monitoring and reporting. The six specific PFAS compounds covered include PFOA; PFOS; perfluorononanoic acid (PFNA) and the GenX chemicals—hexafluoropropylene oxide dimer acid (HFPO-DA), perfluorohexane sulfonic acid (PFHxS), and perfluorobutane sulfonic acid (PFBS). The EPA proposed to set the health-based value, the MCLG, for PFOA and PFOS at zero. Considering feasibility, including currently available analytical methods to measure and treat these chemicals in drinking water, the EPA proposed individual MCLs of 4.0 ng/L or parts per trillion (ppt) for PFOA and PFOS.[26]

In regard to foods, particularly high PFAS concentrations (usually PFOS is the most prevalent) occur in some fish, especially fish liver.[27] Although formal advice has not been issued, the consumer should be advised not to eat species like seabass and crustaceans from contaminated waters. Farmed salmon, mackerel, cod, and saithe contain much less PFAS. Eel and freshwater fish like perch and carp can hold quite high concentrations that would argue against any regular intake of these species. Many countries now post fishing bans at contaminated fresh waters or warnings against eating locally caught fish. Because of the persistence of the PFAS, such dietary advice may have to remain for a foreseeable future.

Little can be done if an individual has accumulated PFAS in the body. Women usually have lower blood PFAS concentrations than men, most likely due to monthly blood losses and to a history of pregnancy and breastfeeding. Also, blood donors tend to show lower PFAS burdens, although these observations do not suggest any realistic approach to removal of PFAS from the body. The hope is that some pharmaceutical that prevents gastrointestinal uptake of lipids, such as cholesterol, may at some point be shown to be both efficient and safe in lowering PFAS burdens.

In 2021, five EU Member States officially proposed restrictions to the use of PFAS, with likely exceptions of PFAS needed for so-called essential uses, because "the consequences of [PFAS] persistence include that the presence of these substances in the environment is practically irreversible, and pose an unacceptable risk to the environment and humans." At present, most documentation relates to PFOS, PFOA, and a few other PFAS, but the similarity of many other PFAS would suggest that restriction of future uses of a much wider range of these substances would be wise.[28] In fact, Belgium, in 2021, demanded that a polluting PFAS-producing facility be closed down, despite the fact that traditional PFAS were no longer manufactured there. While restriction of PFAS production and use will of course be advantageous in the long run, the persistence of these compounds promises that PFAS will remain an environmental hazard for years to come, particularly in regard to children's health.

References

1. C8 Science Panel. The Science Panel Website/ 2013. http://www.c8sciencepanel.org/

2. U.S. Environmental Protection Agency. *Drinking Water Health Advisories for PFOA and PFOS. Raleigh, NC: EPA;* 2016.

3. Agency for Toxic Substances and Disease Registry. *Toxicological Profile for Perfluoroalkyls.* Atlanta, GA: ATSDR; 2021.

4. EFSA CONTAM Panel. Risk to human health related to the presence of perfluoroalkyl substances in food. *EFSA J.* 2020/10/01 ed2020;18:e06223.

5. National Toxicology Program. *Systematic Review of Immunotoxicity Associated with Exposure to Perfluorooctanoic Acid (PFOA) or Perfluorooctane Sulfonate (PFOS).* Research Triangle Park, NC: NTP, National Institute of Environmental Health Sciences; 2016.

6. Zahm S, Bonde JP, Chiu WA, Hoppin J, Kanno J, Abdallah M, et al. Carcinogenicity of perfluorooctanoic acid and perfluorooctanesulfonic acid. *Lancet Oncol* 2024;25:16-17.

7. Fiedler H, Sadia M. Regional occurrence of perfluoroalkane substances in human milk for the global monitoring plan under the Stockholm Convention on Persistent Organic Pollutants during 2016–2019. *Chemosphere*. 2021;277:130287.

8. Wikstrom S, Lin PI, Lindh CH, Shu H, Bornehag CG. Maternal serum levels of perfluoroalkyl substances in early pregnancy and offspring birth weight. *Pediatr Res*. 2020;87:1093–9.

9. Waterfield G, Rogers M, Grandjean P, Auffhammer M, Sunding D. Reducing exposure to high levels of perfluorinated compounds in drinking water improves reproductive outcomes: evidence from an intervention in Minnesota. *Environ Health*. 2020;19:42.

10. Cao T, Qu A, Li Z, Wang W, Liu R, Wang X, et al. The relationship between maternal perfluoroalkylated substances exposure and low birth weight of offspring: a systematic review and meta-analysis. *Environ Sci Pollution Res Int*. 2021;28:67053–65.

11. Bloom MS, Commodore S, Ferguson PL, et al. Association between gestational PFAS exposure and children's adiposity in a diverse population. *Environ Res*. 2022;203:111820.

12. Grandjean P, Andersen EW, Budtz-Jørgensen E, et al. Serum vaccine antibody concentrations in children exposed to perfluorinated compounds. *JAMA*. 2012;307:391–7.

13. Granum B, Haug LS, Namork E, et al. Prenatal exposure to perfluoroalkyl substances may be associated with altered vaccine antibody levels and immune-related health outcomes in early childhood. *J Immunotoxicol*. 2013;10:373–9.

14. Dalsager L, Christensen N, Halekoh U, et al. Exposure to perfluoroalkyl substances during fetal life and hospitalization for infectious disease in childhood: a study among 1,503 children from the Odense Child Cohort. *Environ Int*. 2021;149:106395.

15. Lopez-Espinosa MJ, Fletcher T, et al. Association of perfluorooctanoic acid (PFOA) and perfluorooctane sulfonate (PFOS) with age of puberty among children living near a chemical plant. *Environ Sci Technol*. 2011;45:8160–6.

16. Carwile JL, Seshasayee SM, Aris IM, et al. Prospective associations of mid-childhood plasma per- and polyfluoroalkyl substances and pubertal timing. *Environ Int*. 2021;156:106729.

17. Coperchini F, Croce L, Ricci G, et al. Thyroid disrupting effects of old and new generation PFAS. *Front Endocrinol (Lausanne)*. 2020;11:612320.

18. Romano ME, Xu Y, Calafat AM, et al. Maternal serum perfluoroalkyl substances during pregnancy and duration of breastfeeding. *Environ Res*. 2016;149:239–46.

19. Valvi D, Hojlund K, Coull BA, Nielsen F, Weihe P, Grandjean P. Life-course exposure to perfluoroalkyl substances in relation to markers of glucose homeostasis in early adulthood. *J Clin Endocrinol Metab*. 2021;106:2495–504.

20. Blomberg AJ, Shih YH, Messerlian C, Jorgensen LH, Weihe P, Grandjean P. Early-life associations between per- and polyfluoroalkyl substances and serum lipids in a longitudinal birth cohort. *Environ Res*. 2021;200:111400.

21. Buckley JP, Kuiper JR, Lanphear BP, et al. Associations of maternal serum perfluoroalkyl substances concentrations with early adolescent bone mineral content and density: the Health Outcomes and Measures of the Environment (HOME) Study. *Environ Health Perspect*. 2021;129:97011.

22. Stratakis N, D VC, Jin R, Margetaki K, et al. Prenatal exposure to perfluoroalkyl substances associated with increased susceptibility to liver injury in children. *Hepatology*. 2020;72:1758–70.

23. Steenland K, Winquist A. PFAS and cancer, a scoping review of the epidemiologic evidence. *Environ Res*. 2021;194:110690.

24. Shearer JJ, Callahan CL, Calafat AM, et al. Serum concentrations of per- and polyfluoroalkyl substances and risk of renal cell carcinoma. *J Natl Cancer Inst*. 2021;113:580–7.

25. Goeden HM, Greene CW, Jacobus JA. A transgenerational toxicokinetic model and its use in derivation of Minnesota PFOA water guidance. *J Expo Sci Environ Epidemiol*. 2019;29:183–95.

26. U.S. Environmental Protection Agency. PFAS national primary drinking water regulation rulemaking. *Federal Register*. 2023 Mar 29. Available at: https://www.govinfo.gov/content/pkg/FR-2023-03-29/pdf/2023-05471.pdf

27. Pasecnaja E, Bartkevics V, Zacs D. Occurrence of selected per- and polyfluorinated alkyl substances (PFAS) in food available on the European market: a review on levels and human exposure assessment. *Chemosphere*. 2022;287(Pt 4):132378.

28. Cousins IT, DeWitt JC, Gluge J, et al. The high persistence of PFAS is sufficient for their management as a chemical class. *Environ Sci Process Impacts*. 2020;22:2307–12.

38

Nanomaterials and Child Health

Kam Sripada

Introduction

Nanomaterials are all around us. COVID tests and advanced medicines depend on nanomaterials. Nanomaterials are used in everything from paint to batteries, car tires, sunscreen, and water filters, and generated unintentionally through combustion, friction, volcanic eruptions, and countless other processes in nature. Nanomaterials represent a growing category of potentially toxic substances for children.[1]

Some nanoscale particles may be more toxic than larger forms, and some nanomaterials are smaller than barriers of the human body. It is therefore important to understand the potential for early exposures. Even if just a small fraction crosses the body's protective barriers, this can still be a large number of particles.

While research on and development of nanomaterials race ahead, our understanding of the potential effects of nanomaterials on health gives reason for concern. This chapter introduces key terms and uses for the wide variety of nanomaterials and summarizes the emerging evidence of health effects during pregnancy and early life.

What Are Nanomaterials?

Materials with any external dimension in the nanoscale of 1 nanometer (nm, one billionth of a meter) to 100 nm are called nanomaterials (Figure 38.1). Compared to larger

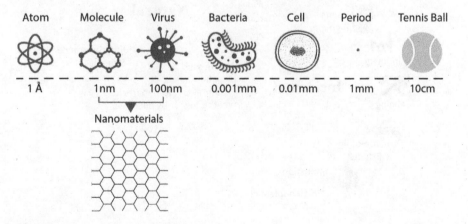

Figure 38.1. Size comparison of nanomaterials
Source: Copyright 2016 © European Chemicals Agency.

Kam Sripada, *Nanomaterials and Child Health* In: *Textbook of Children's Environmental Health*, Second Edition.
Edited by: Ruth A. Etzel and Philip J. Landrigan, Oxford University Press. © Oxford University Press 2024.
DOI: 10.1093/oso/9780197662526.003.0038

materials, nanomaterials have unique features such as shape, stability, transparency, color, strength, conductivity, melting point, and magnetic properties. They have a high ratio of surface area to volume. Nanomaterials can be engineered (manufactured), incidental, or naturally occurring. Figure 38.2 provides examples of nanomaterials from each of these three categories, which are described further below. This chapter deals with nano-objects, rather than larger materials with nanoscale features.

Engineered nanomaterials. Nanomaterials could—in theory—be produced from nearly any material; in practice, most nanoparticles manufactured today are made from transition metals, metal oxides, silicon, and carbon. As of 2019, more than 3,000 firms globally were active in nanotechnology,[2] including many of the world's largest food and medical corporations. More than 100,000 types of engineered nanomaterials exist today. Although these manmade nanomaterials have been produced in much smaller amounts than the other two categories, their potential impacts on human and planetary health—both beneficial and hazardous—are of great scientific interest. Around the world each year, up to 1 teragram (around 100 times the mass of the Eiffel Tower) of engineered nanomaterials are released into the environment.[3]

Incidental nanomaterials. Not all nanomaterials are produced intentionally. Abundant and largely unregulated are incidental nanomaterials, created by combustion, friction, erosion, and degradation. This is a very diverse category, ranging from ultrafine air

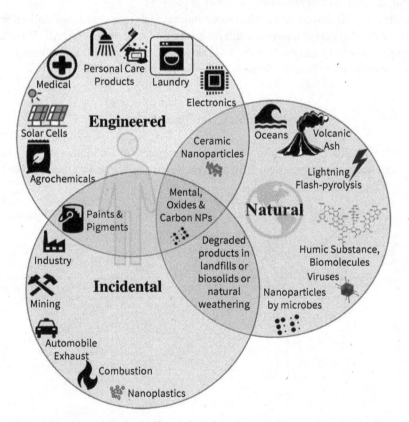

Figure 38.2. Examples of engineered, incidental, and naturally occurring nanomaterials.

Source: Adapted from Malakar et al.[2], Science of The Total Environment (2021), and reprinted with permission

pollution (e.g., solid fuel smoke, car diesel exhaust, factory emissions, e-waste burning) to welding fume particles, nanoplastics, and diverse other degradation products. Each year, an estimated 1–10 teragrams of incidental nanomaterials are released globally.[3]

Naturally occurring nanomaterials. Nanomaterials made by nature through biological, geological, and mechanical processes are found around the world. This is by far the largest category of nanomaterials, with thousands of teragrams present globally each year.[3] For example, nano-silica is formed in rice husks, and sea-spray and volcanic ash contain nanomaterials. This chapter focuses on engineered and incidental nanomaterials because they are rapidly proliferating.

See the Glossary at the end of the chapter for key terms and definitions.

Sources of Exposure

Children may be exposed to engineered nanomaterials via foods, packaging, cosmetics including sunscreen, antibacterial textiles, and plastic coatings. Nanoplastics degraded from larger plastics, ultrafine air pollution, and even carbon nanoparticles from burnt toast are common incidental nanomaterials. Estimating children's exposure to nanomaterials is challenging, and most exposure estimates today are based on generic assumptions rather than empirical measurements.[4]

Table 38.1 provides examples of some common nanomaterials with their sources or (for engineered nanomaterials) their uses.

Airborne Nanomaterials

Inhalation is considered a critical route for exposure to nanomaterials. Inhalation is likely a more direct exposure route for dispersed nanoparticles compared to oral or dermal exposure, and nanomaterials may remain deposited in the lungs for extended periods.[5] Fuel combustion produces gases, liquid-like carbon nanomaterials, fullerenes (carbon mesh nanomaterials with low solubility, such as buckyballs), polycyclic aromatic hydrocarbons, soot, and other substances[6] and contributes large quantities of incidental nanomaterials into the environment.

Children are routinely exposed to airborne carbon nanotubes and other nanomaterials that reflect the types of particles found in dust and vehicle exhaust near where they live.[7,8] The placenta,[9] lungs,[7] and developing brain[8] appear to be targets in the human body for metal-bearing nanoparticles from inhaled combustion- and friction-derived air pollution.

Nanomaterials in Food, Breastmilk, and Baby Bottles

Nanomaterials are used in foods to enhance taste, color, or texture. One common food additive is titanium dioxide (TiO_2) nanoparticles (referred to as E171 in the European Union), which is frequently added as a whitener to chocolate, candy, chewing gum, and snack foods. The French government banned nano TiO_2 in foods in 2019.[10] In response, the French subsidiary of Mars Inc. stated that it would spend nearly $100 million to phase nano TiO_2 out of its products, which include M&Ms and Wrigley chewing gum.[10]

Table 38.1 Examples of some common nanomaterials

Category	Name	Source/use
Inorganic nanomaterials	Titanium dioxide	Food additive, whitener, sunscreen, paint additive, self-cleaning glass and paint, wastewater purification, cosmetics, medicines
	Gold	Catalytic applications
	Silver	Antimicrobial properties, wound dressings, consumer products
	Copper	Treatment for wood products
	Quantum dots	Fluorescence in medical applications, flat screen displays
Carbon nanomaterials	Fullerenes	Solar cells, electronics, cosmetics
	Carbon nanotubes	Single-wall: additive for polymer composites, water filtration; multi-wall: flame-retardant coatings, electronics, fuel cells, batteries, water filtration
	Carbon black	Reinforcement component in rubber (primarily automobile tires), plastics, inks, coatings, and paints
Other manmade nanomaterials	Nanoclays	Food packaging, agricultural and environmental applications
	Nanoplastics	Degradation of larger plastics, including infant feeding bottles
	Ultrafine air pollution	Emissions from industry, transport; degradation products
Naturally-occurring nanomaterials	Mercury nanoparticles	Soil and water resources
	Carbon nanotubes and fullerenes	Atmosphere, soot
	Nickel, zinc, cadmium, silver, tin, selenium, lead, bismuth	Volcanic eruptions, atmosphere

In 2021, the European Food Safety Authority concluded that TiO_2 is no longer considered safe as a food additive, estimating that children 3–9 years old ingest 1.9–11.5 mg/kg body weight per day via food.[11] The agency determined that an acceptable daily intake cannot be established for E171.

Both powder and liquid infant formula have been found to contain nanoparticles of TiO_2 as well as SiO_2 (texture additive) and needle-shaped nanomaterials made of hydroxyapatite (a calcium-rich mineral),[12] which the European Commission's Scientific Committee on Consumer Safety warned against using in cosmetics due to potential toxicity.[13]

Labeling of nanoscale ingredients in foods varies. In the European Union, engineered nanomaterials in food require a safety assessment and must be labelled on the package. The European Food Safety Authority's Scientific Committee has developed guidance on risk assessment of nanomaterials in foods.[14] In 2020, the Indian government released guidelines that aimed to balance promotion of nanotechnology in agriculture and food with needs for risk assessment.[15] Overall, monitoring of food nanomaterials and requirements for consumer labeling are in their infancy.

Food packaging and food contact materials contain both incidental nanomaterials—like nanoplastics—and engineered nanomaterials—such as nanomaterials (such as nanoclays and nanoparticles) to prevent spoiling. Plastic infant feeding bottles also are a source of nanomaterials. Trillions of nanoplastics per liter have been found in plastic baby bottles after shaking with warm water.[16] Because nanoplastics tend to aggregate in water, accurately quantifying them is a challenge.

Metal and inorganic nanoparticles, such as Ag[17,18] and TiO_2,[19] have been found to pass through mammary glands in rodents. Determining whether and which nanomaterials are expressed in human milk requires additional research.

Children's Products

Manufactured nanomaterials are added to children's products to achieve a range of desired properties. For example, children's products advertised as antimicrobial are commonly treated with silver nanoparticles, which are known for their antifungal, antibacterial, antiviral, and antimicrobial properties. Silver nanoparticles have been detected in baby blankets, wet wipes, disinfectant sprays, plush toys, sippy cups, and toothbrush bristles.[20,21] A toy teddy bear advertised to contain silver nanoparticles in its foam stuffing was found to contain 77 mg of silver.[22]

As long as the engineered nanomaterials are embedded in the polymers or other matrices in the item, most products are not expected to cause exposure.[23] However, nanomaterials have been found to leach from children's products. Quadros et al.[24] studied common children's items in the United States and ranked them from most to least likely to be a source of bioavailable silver: plush toy and fabric products (e.g., teddy bear, clothing, blanket), cleaning products (e.g., disinfectant spray, wipes), sippy cups, humidifiers, breastmilk storage bags, and kitchen scrubbers. These items leached silver after contact with tap water, saliva, sweat, urine, and drinks. In the study, the highest amount of silver leached was 18.5 mg silver per kilogram of product. Silicone pacifiers have also been shown to leach silver.[25] Although silver is added in nanomaterial form to these products, it is expected that the silver leaches primarily in ionic form (Ag^+), with a smaller fraction released as nanoparticles or aggregates. This may influence its potential toxicity.

Toxicity

Because of their unique nanoscale features, nanomaterials have been studied for their ability to cause oxidative stress, inflammation, genotoxicity,[26] and cancer.[27,28] Exposure to nanomaterials can produce a pro-oxidant and pro-inflammatory environment in the tissue.[23] As with other contaminants discussed in this textbook, multiple routes of exposure are relevant for nanomaterials (Figure 38.3). Because of children's special vulnerability (see Chapter 2) and immature defense mechanisms, it is expected that the developing child has a reduced ability to detoxify any toxic effects that may arise from exposure to nanoparticles.[29]

Inhalation and oral exposure are emphasized in this field. Nanomaterials deposited in the alveoli may be removed relatively slowly and constitute a continuous source of

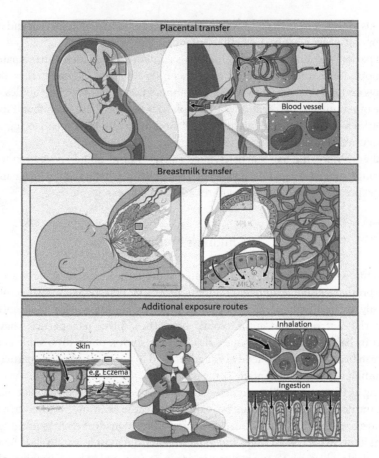

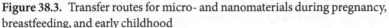

Figure 38.3. Transfer routes for micro- and nanomaterials during pregnancy, breastfeeding, and early childhood

Source: Figure originally published in Sripada et al. (2022)[4] and reproduced with permission. © 2020 Dorothy Fatunmbi

exposure, whereas passage is much faster in the gastrointestinal tract.[5] After escaping clearance, some nonbiodegradable nanomaterials have been found in blood circulation and can accumulate inside organs and cells.[30] Although substantial dermal uptake of nanoparticles is not expected for healthy skin, children born preterm or those with eczema or other skin conditions may be more sensitive.

It is generally accepted that nanotoxicity cannot be deduced from toxicological profiles of chemicals or larger particles—even microparticles.[23] When bulk (macro) materials are made into nanoparticles, they tend to become chemically more reactive.[1] Even chemically inert materials like gold or platinum can—in nano-powder form—catalyze chemical reactions. Bulk materials and nanomaterials have different properties, like size, shape, or surface characteristics, even if they are made of the same substance. Inorganic nanoparticles (e.g., gold, silver, titanium, manganese) demonstrate dominant quantum effects at nanoscale which do not exist in larger dimensions.

Particle toxicity is influenced by particle composition, potential for dissolution, surface area, shape, chemistry, crystal type, and presence of coatings, as well as reactivity and transport within the body. The shape of the nanomaterials (e.g., spheres, shards, fibers, tubes) and their solubility (i.e., tendency to ionize) in the human body also influences their absorption, distribution, metabolism, and elimination. Some particles are more toxic in smaller sizes, and others are not. If, for example, inhaled carbon nanotubes are too large for a macrophage to enclose, the macrophage will be unable to remove the fiber and will become pro-inflammatory.[31] Nanoparticles, in contrast to microparticles, are more likely to cluster together and form aggregates.[32]

Nanomaterials are potential carriers for other chemicals.[4] The toxicants polycyclic aromatic hydrocarbons and polychlorinated biphenyls, for example, have been shown to sorb to various nanomaterials such as nanoplastics, multiwalled carbon nanotubes, and C_{60} fullerene.[33] Indeed, the ability of inorganic nanoparticles to absorb and carry probes and proteins makes them relevant for advanced drug delivery systems. Some quantum dots are produced with cadmium.[5] Carbon black, by contrast, contains low levels of associated chemicals.[23]

Because there are limited data on human nanotoxicity, the following sections discuss studies from human research as well as animal models. Note that some findings of reproductive or developmental toxicity may be better explained by chemical exposures rather than exposures to the nanoform of the chemical. For example, cadmium, cobalt, nickel, and aluminum have demonstrated toxicity in their nanoforms and are also known toxicants in their chemical forms. Disentangling chemical from particle toxicity is challenging; it is therefore prudent to consider the chemical and particle characteristics of a substance (and mixtures) together.

Nanomaterials Crossing the Placenta

Can nanoparticles freely cross the placenta? Although nanoparticles encounter less resistance than microparticles,[30] their translocation across membranes is still limited by their composition, surface chemistry, and particle size. Almost all types of nanoparticles studied seem able to reach and/or cross the placenta, though sometimes only in trace amounts.[34] Translocation of the placenta has been reported for nanomaterials of silver (Ag), aluminum trioxide (AlO_3), gold (Au), cerium dioxide (CeO_2), silicon dioxide (SiO_2), TiO_2, zinc oxide (ZnO), zirconium oxide (ZrO), iron trioxide (FeO_3), polystyrene, and poly(glycidyl methacrylate), as well as C_{60} fullerenes, quantum dots, and nanoscale diesel exhaust particles.[34] Timing of exposure during pregnancy likely plays a role in distribution via the placenta into fetal tissue. Even if nanomaterials have not themselves crossed the placenta or entered the maternal bloodstream, they potentially could impact development in utero indirectly if the mother experiences oxidative stress, inflammation, impaired placental growth and vascularization, and hormonal disruption following her own exposure to nanomaterials.[35,36] The mechanisms of transplacental transfer of nanomaterials and their effects on the placenta and fetus are a growing field of research.[37-39]

During pregnancy, nanomaterials accumulate in the fetal brain, liver, lung, kidney, gastrointestinal tract, and blood.[34] It is possible that nanoparticles that are resistant to excretion might accumulate in the mother's body over time and be slowly released during pregnancy, even if exposure during pregnancy is avoided.[23]

Inhaled Nanofibers

Could toxicology of asbestos—which has been extensively studied—teach us something about newly emerging high aspect ratio (i.e., fiber-like) nanomaterials like carbon nanotubes and metallic nanowires? Nanoparticles deposit onto tissue differently than larger particles, and long, thin, rigid fibers can penetrate deep into the respiratory system. The "fiber pathogenicity paradigm" has been used to understand the links between lung disease and inhalation of asbestos and other fibers. Under this paradigm, width, length, biopersistence, and mechanical bending stiffness are determinants of pathogenicity and carcinogenicity.[40] Toxicologists indicate that any fiber or nanofiber longer than 5 μm, less than 1 μm in diameter, and biopersistent in the lungs (i.e., not dissolving or breaking into shorter fibers) poses a risk of causing asbestos-like pathogenicity.[31,41] These thresholds may be different for children.

Inhaled nanoparticles may deposit in the maternal alveoli and generate reactive oxygen species, resulting in oxidative stress and inflammation; inflamed lung tissue may then release inflammatory mediators such as cytokines to the maternal bloodstream for circulation.[42]

Estimates of the amount of particle deposition in children's lungs are mainly based on scaling adult models. But, based on our knowledge of children's special physiology and behavior, infants and children are risk groups for inhaling nanomaterials.[43] Children's smaller airway geometry, higher flow rates through their entire respiratory system, short stature, and tendency to breathe through the mouth lead to more particles depositing in children's lungs, compared to adults.[29] Normal lung development in infancy and childhood requires tightly synchronized development together with nerve connections, making children's lungs more susceptible to injury than adults'. Repairing injured airways following damaging exposure to inhaled nanomaterials in early life may provide the basis for hyperactivity resulting in decreased airway function, asthma, and potentially other long-term respiratory dysfunction.[29]

Cancer

The International Agency for Research on Cancer (IARC) has assessed evidence for carcinogenicity in humans of various nanomaterials, including for childhood cancers. Outdoor air pollution (which includes nanoparticles) has been determined by IARC to be carcinogenic to humans.[28] TiO_2, carbon black,[27] and the multiwalled carbon nanotube MWCNT-7 are listed as possibly carcinogenic to humans. Evaluation of other carbon nanotubes is ongoing.

Reproductive Nanotoxicity

Maternal exposure to nanoparticles via the airways (Cu, Fe) or orally (ZnO) has been associated with decreased survival and/or growth and viability of the offspring in rodent pups.[5] Adverse effects on cardiac tissue and functioning have been observed in offspring from pregnant female rats exposed by inhalation to TiO_2 nanoparticles.[44] Several studies have indicated that nanoparticles may affect male reproductive development, but results are mixed.[23] Following subcutaneous maternal exposure, male rodents in one study were found to have aggregates of TiO_2 particles located in testicular tissue 6 weeks after birth.[45] There is currently a lack of research on nanomaterials and female fertility.

Because the placenta contains many cytokines and other immune mediators, there is a large potential for nanomaterials to influence fetal immune development in complex and unpredictable ways.[29] Immune response may be attenuated or amplified by nanoparticles, depending on their size and physical–chemical properties.[29]

Neurotoxicity

Nanomaterials may reach the central nervous system directly via the nose-to-brain olfactory system, bypassing the protective blood–brain barrier.[46] Moreover, the developing brain may be more vulnerable to oxidating insults than the adult brain. Nanoparticles of Ag, carbon black, and Al have been linked to effects in the developing brain or signs of neurotoxicity after exposure.[5]

Highly reactive metals are often bound to nanoparticles derived from traffic-related combustion and friction (e.g., vehicle brakes). Therefore, especially in urban areas, exposure to pervasive metal-rich airborne nanoparticles is seen as a risk for brain development. Based on human postmortem pathology, Calderón-Garcidueñas et al. (2020)[8] located Fe-, Al- and Ti-rich nanomaterials in the brainstems of children and adolescents. The presence of nanomaterials was associated with substantia nigra pathology, including mitochondrial damage indicating early neurodegeneration. The study suggests that such reactive, cytotoxic, and magnetic nanomaterials may act as catalysts for reactive oxygen species formation, altered cell signaling, and protein misfolding in the brain.

The notion that early nanomaterial exposure may be linked to premature brain aging is supported by evidence from animal research. Maternal inhalation of a type of carbon black (Printex 90) at a level corresponding with Danish occupational exposure limits was linked to increased expression levels of glial fibrillary acidic protein in astrocytes in the cerebral cortex and hippocampus in 6-week-old mouse offspring, which is generally a response after damage to the nervous system.[47] Enlarged lysosomal granules were also seen in brain perivascular macrophages, and behavioral alterations were observed in an open field test. Umezawa et al.[47] compared the observed changes to premature brain aging. Emerging evidence in mice shows an association between TiO_2 nanoparticles inhaled[48] or injected during pregnancy[49] with subtle changes in offspring behavior.

Toxicity Testing for Particle Pollutants

As nanotechnology took off, the chemical risk assessment framework was selected as a starting point for testing developmental toxicity. But standardized guidelines such as the Organisation for Economic Cooperation and Development (OECD) Guidelines for the Testing of Chemicals,[50] which were developed for conventional soluble chemicals, are not suited to studying the behavior of engineered nanomaterials. Therefore, in the absence of methods adapted to testing developmental toxicity of particle pollutants, only a handful of relevant studies have followed such test guidelines.[5] Instead, nanomaterials have been studied with a wide range of study methods, examining nanomaterial properties (type, size, dose, modifications: e.g., coatings), timing and routes of exposure, and health outcomes. This has resulted in a diverse but fragmented evidence base for assessing child health risks. The vast majority of nanomaterials have never been tested for impacts on child development.

Prevention

Nanomaterials can have both beneficial and hazardous effects on children's health and well-being. For example, water filtration of contaminants and next-generation energy solutions are using nanotechnology. The key is to encourage development of safe nanotechnologies that will benefit society while preventing exposure to nanomaterials that may be unsafe during the sensitive early years.

Still, only a few types of engineered nanomaterials, such as TiO_2 and carbon nanotubes, have undergone extensive toxicological testing by the US National Institute for Occupational Safety and Health (NIOSH), the European Food Safety Authority, or other agencies.[11,27,51] As of 2020, companies selling products in the EU are required by REACH legislation to register nanomaterials (nanoforms), and the EU has developed consumer safety reports on numerous nanomaterials.[13,52]

In terms of manufactured nanomaterials, the health and safety of pregnant workers in nanotechnology should be prioritized based on a precautionary principle approach and out of concern for development in utero. Although occupational exposures tend to be low, there are instances of higher release of nanomaterials (close to milligram per cubic meter) during handling of nanosized powders and during mixing into paints.[23] In general, dry powder forms of nanomaterials typically present the highest potential for exposure in occupational settings.

Continuous, widespread exposures to incidental nanomaterials and ultrafine air pollution worldwide by children and adults alike are a major public health hazard. Scientists are encouraged to work across disciplines to find ways to locate and monitor nanomaterials in children's air, water, food, and products.[4] Ultimately, the responsibility for surveillance and protection of children's health belongs to governments and is a key aspect of fulfilling each child's rights. The fast-emerging field of nanotechnology is no exception.

References

1. OECD, Allianz. Small sizes that matter: opportunities and risks of nanotechnologies. 2005. Available at: https://www.oecd.org/science/nanosafety/44108334.pdf

2. OECD. Key nanotechnology indicator 1: number of firms active in nanotechnology, 2008–2019. 2021.Available at: https://www.oecd.org/sti/nanotechnology-indicators.htm

3. Hochella Jr MF, Mogk DW, Ranville J, et al. Natural, incidental, and engineered nanomaterials and their impacts on the Earth system. *Science*. 2019.Available at: https://www.science.org/doi/abs/10.1126/science.aau8299

4. Sripada K, Wierzbicka A, Abass K, et al. A children's health perspective on nano- and microplastics. *Environ Health Perspect*. 2022;130(1):015001.

5. Larsen PB, Mørch TA, Andersen DN, Hougaard KS. A critical review of studies on the reproductive and developmental toxicity of nanomaterials. European Chemicals Agency (ECHA). 2020. Available at: https://data.europa.eu/doi/10.2823/421061

6. Murr LE, Guerrero PA. Carbon nanotubes in wood soot. *Atmospheric Sci Lett*. 2006;7(4):93–5.

7. Kolosnjaj-Tabi J, Just J, Hartman KB, et al. Anthropogenic carbon nanotubes found in the airways of Parisian children. *EBioMedicine*. 2015;2(11):1697–704.

8. Calderón-Garcidueñas L, González-Maciel A, Reynoso-Robles R, et al. Quadruple abnormal protein aggregates in brainstem pathology and exogenous metal-rich magnetic nanoparticles. *Environ Res.* 2020;191:110139.

9. Liu NM, Miyashita L, Maher BA, et al. Evidence for the presence of air pollution nanoparticles in placental tissue cells. *Sci Total Environ.* 2021;751:142235.

10. USDA. France bans titanium dioxide in food products by January 2020. 2019.Available at: https://apps.fas.usda.gov/newgainapi/api/report/downloadreportbyfilename?filename= France%20bans%20Titanium%20Dioxide%20in%20food%20products%20by%20Janu ary%202020_Paris_France_5-3-2019.pdf

11. EFSA, Younes M, Aquilina G, et al. Safety assessment of titanium dioxide (E171) as a food additive. *EFSA J.* 2021;19(5):e06585.

12. Schoepf JJ, Bi Y, Kidd J, Herckes P, Hristovski K, Westerhoff P. Detection and dissolution of needle-like hydroxyapatite nanomaterials in infant formula. *NanoImpact.* 2017;5:22–8.

13. Scientific Committee on Consumer Safety (SCCS). Opinion on hydroxyapatite (nano). 2021. Available at: https://ec.europa.eu/health/system/files/2021-08/sccs_o_246_0.pdf

14. EFSA, More S, Bampidis V, et al. Guidance on risk assessment of nanomaterials to be applied in the food and feed chain: human and animal health. *EFSA J.* 2021;19(8):e06768.

15. DBT. Guidelines for Evaluation of nano-based agri-input and food products in India. *Department of Biotechnology, Government of India.* 2020. Available at: https://static.pib.gov. in/WriteReadData/userfiles/NanoAgri_15.6.2020.pdf

16. Li D, Shi Y, Yang L, et al. Microplastic release from the degradation of polypropylene feeding bottles during infant formula preparation. *Nature Food.* 2020;1(11):746–54.

17. Morishita Y, Yoshioka Y, Takimura Y, et al. Distribution of silver nanoparticles to breast milk and their biological effects on breast-fed offspring mice. *ACS Nano.* 2016;10(9):8180–91.

18. Melnik EA, Buzulukov Yu P, Demin VF, et al. Transfer of silver nanoparticles through the placenta and breast milk during in vivo experiments on rats. *Acta Naturae.* 2013;5(3):107–15.

19. Cai J, Zang X, Wu Z, Liu J, Wang D. Translocation of transition metal oxide nanoparticles to breast milk and offspring: the necessity of bridging mother-offspring-integration toxicological assessments. *Environ Int.* 2019;133:105153.

20. Tulve NS, Stefaniak AB, Vance ME, et al. Characterization of silver nanoparticles in selected consumer products and its relevance for predicting children's potential exposures. *Int J Hygiene Environ Health.* 2015;218(3):345–57.

21. U.S. EPA. Nanomaterial case study: nanoscale silver in disinfectant spray (final report). U.S. Environmental Protection Agency. 2012. Available at: https://hero.epa.gov/hero/index.cfm/ reference/details/reference_id/6773979.

22. Cleveland D, Long SE, Pennington PL, et al. Pilot estuarine mesocosm study on the environmental fate of silver nanomaterials leached from consumer products. *Sci Total Environ.* 2012;421–422:267–72.

23. Campagnolo L, Møller P, Jacobsen NR, Hougaard KS. Developmental toxicity of engineered nanomaterials. In: Gupta R, ed., *Reproductive and Developmental Toxicology* (3rd ed.). Academic Press; 2022:258–305. Available at: https://doi.org/10.1016/ B978-0-323-89773-0.00016-3.

24. Quadros ME, Pierson R, Tulve NS, et al. Release of silver from nanotechnology-based consumer products for children. *Environ Sci Technol.* 2013;47(15):8894–901.

25. Choi JI, Chae SJ, Kim JM, et al. Potential silver nanoparticles migration from commercially available polymeric baby products into food simulants. *Food Additives Contaminants: Part A*. 2018;35(5):996–1005.

26. NIOSH. Current Intelligence Bulletin 70: Health effects of occupational exposure to silver nanomaterials. U.S. DHHS National Institute for Occupational Safety and Health (NIOSH). 2021. Available at: https://www.cdc.gov/niosh/docs/2021-112/.

27. IARC. Carbon black, titanium dioxide, and talc. 2010. Available at: https://publications.iarc.fr/Book-And-Report-Series/Iarc-Monographs-On-The-Identification-Of-Carcinogenic-Hazards-To-Humans/Carbon-Black-Titanium-Dioxide-And-Talc-2010.

28. IARC. Outdoor air pollution. 2013. Available at: https://publications.iarc.fr/Book-And-Report-Series/Iarc-Monographs-On-The-Identification-Of-Carcinogenic-Hazards-To-Humans/Outdoor-Air-Pollution-2015.

29. Schüepp K. The potential harmful and beneficial effects of nanoparticles in children. In: Marijnissen JC, Gradon L, eds., *Nanoparticles in Medicine and Environment*. Dordrecht: Springer Netherlands; 2010: 211–26. Available at: http://link.springer.com/10.1007/978-90-481-2632-3.

30. Geiser M, Kreyling WG. Deposition and biokinetics of inhaled nanoparticles. *Particle Fibre Toxicol*. 2010;7(1):2.

31. Donaldson K, Poland CA, Murphy FA, MacFarlane M, Chernova T, Schinwald A. Pulmonary toxicity of carbon nanotubes and asbestos: similarities and differences. *Adv Drug Deliv Rev*. 2013;65(15):2078–86.

32. Prüst M, Meijer J, Westerink RHS. The plastic brain: neurotoxicity of micro- and nanoplastics. *Particle Fibre Toxicol*. 2020;17(1):24.

33. Velzeboer I, Kwadijk CJAF, Koelmans AA. Strong sorption of PCBs to nanoplastics, microplastics, carbon nanotubes, and fullerenes. *Environ Sci Technol*. 2014;48(9):4869–76.

34. Bongaerts E, Nawrot TS, Van Pee T, Ameloot M, Bove H. Translocation of (ultra)fine particles and nanoparticles across the placenta; a systematic review on the evidence of in vitro, ex vivo, and in vivo studies. *Particle Fibre Toxicol*. 2020;17(1):56.

35. Dugershaw BB, Aengenheister L, Hansen SSK, Hougaard KS, Buerki-Thurnherr T. Recent insights on indirect mechanisms in developmental toxicity of nanomaterials. *Particle Fibre Toxicol*. 2020;17:31.

36. Dugershaw BB, Nowak-Sliwinska P, Hornung R, Sturla S, Buerki-Thurnherr T. Exploring indirect embryo-fetal risks of nanoparticles: impact on human placental function, the release of placental signaling factors and subsequent alterations on angiogenic and neurodevelopmental processes. *Placenta*. 2021;112:e64–5.

37. Buerki-Thurnherr T, Mandach U von, Wick P. Knocking at the door of the unborn child: engineered nanoparticles at the human placental barrier. *Swiss Medical Weekly*. 2012. Available at: https://smw.ch/article/doi/smw.2012.13559

38. Grafmueller S, Manser P, Diener L, et al. Bidirectional transfer study of polystyrene nanoparticles across the placental barrier in an ex vivo human placental perfusion model. *Environ Health Perspect*. 2015b;123(12):1280–6.

39. Poulsen MS, Mose T, Maroun LL, Mathiesen L, Knudsen LE, Rytting E. Kinetics of silica nanoparticles in the human placenta. *Nanotoxicology*. 2015;9(0 1):79–86.

40. Kane AB, Hurt RH, Gao H. The asbestos-carbon nanotube analogy: an update. *Toxicol Appl Pharmacol*. 2018;361:68–80.

41. Poland CA, Duffin R, Kinloch I, et al. Carbon nanotubes introduced into the abdominal cavity of mice show asbestos-like pathogenicity in a pilot study. *Nature Nanotech.* 2008;3(7):423–8.

42. Hougaard KS, Campagnolo L, Chavatte-Palmer P, et al. A perspective on the developmental toxicity of inhaled nanoparticles. *Reprod Toxicol.* 2015;56:118–40.

43. Semmler-Behnke M, Kreyling WG, Schulz H, et al. Nanoparticle delivery in infant lungs. *Proc Nat Acad. Sci.* 2012;109(13):5092–7.

44. Stapleton PA, Minarchick MsVC, YI J, Engels MrK, McBride MrCR, Nurkiewicz TR. Maternal engineered nanomaterial exposure and fetal microvascular function: does the Barker hypothesis apply? *Am J Obstet Gynecol.* 2013;209(3):227.e1e11.

45. Takeda K, Suzuki K, Ishihara A, et al. Nanoparticles transferred from pregnant mice to their offspring can damage the genital and cranial nerve systems. *J Health Sci.* 2009;55(1):95–102.

46. Lucchini RG, Dorman DC, Elder A, Veronesi B. Neurological impacts from inhalation of pollutants and the nose–brain connection. *NeuroToxicology.* 2012;33(4):838–41.

47. Umezawa M, Onoda A, Korshunova I, et al. Maternal inhalation of carbon black nanoparticles induces neurodevelopmental changes in mouse offspring. *Particle Fibre Toxicol.* 2018;15(1):36.

48. Hougaard KS, Jackson P, Jensen KA, et al. Effects of prenatal exposure to surface-coated nanosized titanium dioxide (UV-Titan): a study in mice. *Particle Fibre Toxicol.* 2010;7(1):16.

49. Notter T, Aengenheister L, Weber-Stadlbauer U, et al. Prenatal exposure to TiO2 nanoparticles in mice causes behavioral deficits with relevance to autism spectrum disorder and beyond. *Transl Psychiatry.* 2018;8(1):1–10.

50. OECD. Revised Guidance Document 150 on Standardised Test Guidelines for Evaluating Chemicals for Endocrine Disruption. OECD. 2018. Available at: https://www.oecd-ilibrary.org/docserver/9789264304741-en.pdf?

51. NIOSH. Protecting workers during the handling of nanomaterials. U.S. DHHS, National Institute for Occupational Safety and Health (NIOSH). 2018. Available at: https://www.cdc.gov/niosh/docs/2018-121/pdfs/2018-121.pdf?id=10.26616/NIOSHPUB2018121

52. Scientific Committee on Consumer Safety (SCCS). Guidance on the safety assessment of nanomaterials in cosmetics. 2019.Available at: https://data.europa.eu/doi/10.2875/40446

Glossary

Aggregate. Particle comprising strongly bonded or fused particles.

Carbon nanotubes. Hollow nano-objects with two similar external dimensions in the nanoscale and the third dimension significantly larger (i.e., high aspect ratio), composed of carbon.

Manufactured nanomaterial (i.e., engineered nanomaterial). Nanomaterial intentionally produced to have selected properties or composition.

Nanomaterial. Material with any external dimension in the nanoscale (between 1 and 100 nanometers [nm]) or having an internal structure or surface structure in the nanoscale. Some substances have multiple "nanoforms" based on their different characteristics.

Nanoscale. Descriptor principally referring to the size range 1–100 nm. "Ultrafine" is also used to describe airborne particles in this size range.

Nano-object. Discrete piece of material with one, two, or three external dimensions in the nano-scale, typically classified based on their size and shape:

- In one dimension (nanoplates, nanofilms);
- In two dimensions (nanofibers, nanotubes, nanowires); or
- In all three dimensions (nanoparticles, quantum dots).

Quantum dots. Semiconducting nanoscale crystals that emit light under certain conditions and can be used to create thin displays that are vivid and energy efficient.

39

Volatile Organic Chemicals

Dana Boyd Barr, Parinya Panuwet, and P. Barry Ryan

Volatile organic compounds (VOCs) are a diverse group of hydrocarbon compounds that all have low boiling points, usually less than 250°C. They therefore evaporate readily at ambient temperatures and contaminate both indoor and outdoor air.[1]

VOCs are pervasive environmental contaminants because they arise from a wide variety of industrial and residential sources.[1] Some of the most common VOCs include traffic fumes from automobile and truck exhaust containing benzene, toluene, ethyl benzene, and xylene, known collectively as BTEX.[1,2] Propane, benzene, and other components of gasoline are all VOCs.[1,2] Other important VOCs include the formaldehyde and volatile solvents that result from off-gassing of manufactured materials such as carpets and pressed wood furniture and from use in household products such as paints, wood preservatives, aerosol sprays, and craft supplies.[1,2] These include chlorofluorocarbons, formaldehyde, tetrachloroethylene, and *para*-dichlorobenzene.[1,2]

VOCs have been associated with a variety of health outcomes in children including upper respiratory illness and exacerbations of asthma.[3-6] Benzene is classified as a known human carcinogen because of its role in the development of leukemia and other lymphohematopoietic cancers.[7,8] Other VOCs such as carbon tetrachloride, formaldehyde, and 1,3-butadiene also have been linked to cancers.[7,9-11]

A variety of studies have reported effects associated with selected VOC exposures in children.[3,12-16] Children's activities tend to increase their exposure to VOCs, and their immature metabolism increases their susceptibility to toxic effects. Children participate in a wide variety of outdoor activities and sports that increase their respiratory rates for several hours each day and thus increase their inhalation of VOC-contaminated air. Children's bodies are still developing and their ability to detoxify VOCs enzymatically is still not fully mature.

Table 39.1 shows a list of common VOCs. This chapter focuses on the VOCs that have been most closely studied, for which the most data exist, and to which children are highly exposed. Table 39.2 shows selected VOCs and their sources and outcomes.

Scope and Nature of the Problem

Because VOCs are ubiquitous in the environment, exposure to them is inevitable. Many are related to nonspecific symptoms such as irritation of the eyes, nose, and throat; headaches; dizziness; fatigue; and allergic reactions on the skin.[1,17-20]

Benzene is among the top 20 industrial chemicals produced in the United States. Benzene is a precursor in many manufacturing processes and chemical synthesis, and it is a fundamental building block in the manufacture of plastics, resins, and synthetic

Dana Boyd Barr, Parinya Panuwet, and P. Barry Ryan, *Volatile Organic Chemicals* In: *Textbook of Children's Environmental Health*, Second Edition. Edited by: Ruth A. Etzel and Philip J. Landrigan, Oxford University Press. © Oxford University Press 2024. DOI: 10.1093/oso/9780197662526.003.0039

Table 39.1 Common volatile organic compounds

1,1,1-Trichloroethane	2-Butanone	Chloroform	n-Hexane
1,1,2,2-Tetrachloroethane	2-Hexanone	Chloromethane	Nitrobenzene
1,1,2-Trichloroethane	Acetone	Dichlorobenzenes	Stoddard solvent
1,1-Dichloroethane	Acrolein	Dichloromethane (methylene chloride)	Styrene
1,1-Dichloroethene	Benzene	Dichloropropenes	Tetrachloroethylene (aka perchloroethylene or PERC)
1,2,3 Trichloropropane	Bromodichloromethane	Ethylbenzene	Toluene
1,2-Dibromo-3-Chloropropane	Bromoform and dibromochloromethane	Formaldehyde	Trichloroethylene (TCE)
1,2-Dibromoethane	Bromomethane	Gasoline, automotive	Vinyl chloride
1,2-Dichloroethane	Carbon disulfide	Hexachlorobutadiene	Xylenes
1,2-Dichloroethene	Carbon tetrachloride	Hexachloroethane	
1,2-Dichloropropane	Chlorobenzene	Hydrazines	
1,3-Butadiene	Chloroethane	Methyl mercaptan	

fibers. Benzene is also used to make certain rubbers, lubricants, dyes, detergents, drugs, and pesticides. Benzene is a component of crude oil and gasoline. It can be emitted from natural combustion processes such as volcanoes and forest fires and also from human-made combustion processes such as cigarette smoke and automobile exhaust.[19]

Inhalational exposure to high levels of benzene (i.e., >10,000 µg/m^3) for only a few minutes can result in death.[19] Exposures to lower concentrations (1–2 orders of magnitude lower) can result in drowsiness, dizziness, rapid heart rate, tremors, headaches, confusion, and unconsciousness.[19] Although the inhalational route of exposure predominates, people also can be exposed dermally or through drinking water. Chronic inhalational exposure to benzene can reduce the production of red blood cells, resulting in aplastic anemia and ultimately leukemia and other hematologic tumors. Effects from exposure by ingestion are less well understood. Children can experience the same effects as adults, but these effects are often more severe in children because of their small body size and immature systems. Prenatal benzene exposure can occur because benzene readily crosses the placenta.[19]

Dichlorobenzenes are mixtures of the *ortho-* (i.e., 1,2-dichloro), *meta-* (i.e., 1,3-dichloro), and *para-* (i.e., 1,4-dichloro) isomers of dichlorobenzene. Typically, *para*-dichlorobenzene is the primary isomer of concern because of its widespread use in mothballs and deodorizers.[20] The other isomers are used as precursors in pesticide manufacture or occur as contaminants in technical *para*-dichlorobenzene. *Para*-dichlorobenzene can sublime at room temperature and so is effective as a deodorizer or fumigant. *Para*-dichlorobenzene is a possible human carcinogen.[20]

Table 39.2 Target volatile organic compounds for this chapter, their sources, and related disease

Volatile organic compound	Sources	Systems affected	Outcomes
Benzene	Secondhand smoke (SHS), automobile exhaust, industrial emissions, gasoline	Hematological, immunological, neurological, respiratory	Leukemia (especially acute myelogenic leukemia); asthma exacerbation; known human carcinogen
Dichlorobenzenes	Air fresheners, mothballs, toilet deodorizer blocks	Dermal, developmental, hepatic, ocular, renal	Reasonably anticipated human carcinogen; hematological problems, respiratory irritant
Ethylbenzene	SHS, gasoline, automobile exhaust, solvents, printing inks, paints, varnishes, contaminated water	Developmental, neurological, respiratory	Eye and throat irritation, dizziness, hepatotoxicity, eye and dermal irritation, decreased growth; possible human carcinogen
Formaldehyde	Smog, SHS, gas cookers, open fireplaces, industrial solvent, off-gassing from treated wood, carpets, permanent press fabrics	Dermal, gastrointestinal, immunological, respiratory	Irritation of eyes, nose, throat, and skin; asthma exacerbation, pain, vomiting, coma, death, nasal cancer; reasonably anticipated human carcinogen
Tetrachloroethylene (aka perchloroethylene or PERC)	Dry-cleaning solvents, contaminated water	Developmental, neurological, respiratory	Dizziness, headaches, lethargy, confusion, nausea, death; reasonably anticipated human carcinogen
Toluene	Automobile exhaust, gasoline, kerosene, heating oil, paints, lacquers, contaminated water	Cardiovascular, neurological, respiratory	Tiredness, confusion, weakness, memory loss, dizziness, nausea, loss of appetite, kidney toxicity, hearing and vision loss
Xylenes	Gasoline, paint, varnishes, shellac, rust preventatives, SHS	Developmental, hepatic, neurological, renal, respiratory	Headaches; loss of muscle coordination; breathing difficulty; memory problems; skin, eye, nose, and throat irritant; delayed reaction time; liver and kidney toxicity; unconsciousness; death

Children have widespread exposure to *para*-dichlorobenzene.[21,22] In 2015–2016, more than half of the children and adolescents tested in the United States had measurable levels of *para*-dichlorobenzene in their blood.[21] Additionally, the vast majority of the general US population has measurable levels in blood of 2,5-dichlorophenol, the primary metabolite of *para*-dichlorobenzene, suggesting widespread exposure.[21,23]

Ethylbenzene is a common component of traffic-related air pollution and has the characteristic odor of gasoline.[1,17] It is found in natural products such as coal tar and petroleum and is also found in manufactured products such as inks, insecticides, and paints. Ethylbenzene is used in the synthesis of styrene for polymeric uses.[17]

Ethylbenzene is most often found in traffic-dense areas. It is a possible human carcinogen.[17] Most adults in the US population have measurable levels of ethylbenzene in their blood[24]; however, Sexton et al. found that children in an economically disadvantaged area of Minneapolis had levels up to two times lower than adults.[25] This may be due to the increased commute patterns that may expose adults to more traffic-related VOCs.[26] Regardless, childhood exposure to ethylbenzene is still of concern.[1,27–29]

Formaldehyde is a reactive chemical gas used extensively in the production of fertilizer, paper, plywood, and urea-formaldehyde resins. It is also used as a preservative in some foods and in many products used around the house, such as antiseptics, medicines, and cosmetics.[30] Formaldehyde is a known human carcinogen.[11,30] Small amounts of formaldehyde are produced in the human body in normal metabolism.[30]

Formaldehyde gained notoriety after displaced Hurricane Katrina victims were relocated by the Federal Emergency Management Agency into mobile homes with high indoor levels of formaldehyde.[31,32] It remains a public health concern for government-provided temporary housing as a part of its disaster response.[32]

Tetrachloroethylene (also known as perchloroethylene or PERC) is a manufactured chemical that is widely used for dry cleaning of fabrics and for degreasing of metal.[33] It is reasonably anticipated to be a human carcinogen, and the International Agency for Research on Cancer (IARC) concluded that "there is evidence for consistently positive associations between exposure to tetrachloroethylene and the risks for esophageal and cervical cancer and non-Hodgkin's lymphoma."[33] There have also been weaker links to kidney and bladder cancers.[33]

Exposure to tetrachloroethylene comes primarily from contact with recently dry-cleaned fabrics or from living in close proximity to dry-cleaning facilities. Dry-cleaned fabrics stored in homes can out-gas tetrachloroethylene, and the levels in indoor air may persist for some time if the home is not well ventilated.[33] Fewer than 10% of the adolescents tested in the National Health and Nutrition Examination Survey (NHANES) during 2017–2018 had measurable levels of tetrachloroethylene in their blood, and the concentrations in adolescents were about 30% lower than those observed in adults.[24] This represents a marked decline over the past several decades.[24]

Toluene is a high production volume chemical that is widely used as an industrial solvent and as a precursor in the synthesis of many other chemicals.[18] Toluene occurs naturally in crude oil and thus is isolated in the production of gasoline.[18] It is also produced in the manufacture of coke from coal. Toluene is used in making paints, paint thinners, fingernail polish, lacquers, adhesives, and rubber and in some printing and leather tanning processes. Toluene is a common traffic-related air pollutant, but indoor concentrations can sometimes exceed outdoor concentrations because of poor ventilation after use of paints and polishes[18] It is not believed to be carcinogenic.[18]

Xylenes have three predominant isomers in which the methyl groups vary in their position on the benzene ring: *meta*-xylene, *ortho*-xylene, and *para*-xylene. Xylenes occur naturally in petroleum and coal tar. Chemical industries produce xylene from petroleum. It is one of the top 30 chemicals produced in the United States in terms of volume.[34] Xylenes are used as solvents, cleaning agents, and thinners for paints and varnishes. They are also used as solvents in the printing, rubber, and leather industries. Xylenes can be found in small amounts in airplane fuel and gasoline. Xylenes are not believed to be carcinogenic. Most adolescents studied in the NHANES during 2005–2006 had measurable levels of all xylene isomers, but the levels were about half of those found in adults.[24]

Sources of Exposure to Volatile Organic Compounds

The sources of exposure to VOCs are numerous, and even a single VOC can arise from multiple sources.[1] For example, benzene exposure can occur from cigarette smoking or traffic-related pollution.[1,19] Exposure to VOCs occurs in both the outdoor and indoor environments.[1]

Outdoor sources of VOCs include automobile combustion, emissions from industrial plants, coal burning and other combustion processes, industrial solvents, and improperly disposed refrigerants. VOC concentrations are usually highest near cities or metropolitan areas where traffic is intense or in close proximity to industrial emissions.[1,2]

Indoor sources of VOCs include secondhand smoke, emissions from fireplaces or coal-burning stoves, use of aerosols and paints, off-gassing from new carpets and upholstery, and automobile emissions within attached garages. Because indoor ventilation is often poor, the concentrations of VOCs can be two- to five-fold higher in indoor air than in outdoor air. Further, if intensive use of industrial solvents is occurring indoors, for example, use of acetone in nail salons or paint stripping, the concentrations can be several orders of magnitude higher.[1]

Health Effects of Volatile Organic Compounds

Benzene is one of the best studied VOCs.[1,8] Epidemiologic studies have focused especially on carcinogenic effects from chronic exposures to benzene in occupational settings.[8] The human carcinogenicity of benzene is well documented, and it is classified as a group 1 carcinogen (a known human carcinogen). A causal relationship has been demonstrated between chronic exposures to benzene and acute myelogenous leukemia (AML).[8] Benzene also has been linked to a wide range of other leukemias as well as to a series of lymphomas. These effects have been confirmed in a large collaborative study of Chinese workers with cumulative benzene exposure levels of 40–99 ppm-years.[35,36] This study found an increased risk for all leukemias, acute nonlymphocytic leukemia (ANLL), and combined ANLL and precursor myelodysplastic syndromes.[8]

Benzene can affect multiple organ systems, including the liver, brain, and bone marrow. It is also genotoxic. Hematological effects of chronic exposure in addition to leukemia include anemia, leukopenia, and thrombocytopenia. Exposure can also cause aplastic anemia leading to myelofibrosis, which may be an early indicator of leukemia.[8] Benzene exposure also causes neurological effects.[8,19,27] Lower-dose chronic exposure can cause distal neuropathy and memory loss, and high-level acute exposure can be fatal.

Vascular congestion in the brain and/or ventricular fibrillation have been suggested as the cause of death in settings of acute exposure.[19]

Additional effects of benzene exposure include irritation of the eyes and skin with dermal exposure. Myalgia was reported in steel plant workers exposed to benzene vapors.[19]

Prenatal exposure to benzene has not been associated with increased risk of acute lymphoblastic leukemia, but the risk of AML following prenatal exposure has not been evaluated.[19] The likelihood of childhood leukemia may be increased in children born to parents employed in industries such as painting and printing in which benzene exposure is prevalent.[19]

Benzene has been well studied as an air toxicant from traffic-relation pollution and in relation to respiratory illness.[4,6,26,37] These studies suggest that exposure to benzene and other traffic-related air toxicants is related to inflammation in the nasal passageways and to exacerbations of asthma. These studies recognize that benzene may merely be a marker of inflammatory processes caused by other components of traffic pollution.

Exposure to *para*-dichlorobenzene has been linked to pulmonary effects[20,38,39] and liver toxicity[38] in the general population, while exposed workers had elevated white blood cell counts and serum alanine aminotransferase (ALT) levels.[39] Prenatal exposures to *para*-dichlorobenzene have been linked with lower birth weight among male children.[22]

. Several cases of mothball ingestion and inhalation with subsequent intoxication have been reported.[40] A small boy who ingested *para*-chlorobenzene mothballs developed hemolytic anemia with subsequent hemolysis and mild methemoglobinemia[41] while twin adolescent girls developed neurocutaneous symptoms after exposure.[42] An adult female developed encephalopathy with cognitive features after experiencing withdrawal from mothball intoxication.[43]

Ethylbenzene effects have typically been studied in conjunction with other traffic-related air toxicants (e.g., BTEX toxicants). Acute exposures to ethylbenzene at levels of greater than 1,000 parts per million (ppm) via inhalation resulted in ocular irritation, a burning sensation, and extensive eye watering.[17] In Japanese homes containing dampness and mold, airborne ethylbenzene concentrations were associated with self-reported eye irritations.[17]

Epidemiologic studies of formaldehyde have focused largely on occupational exposures. Occupational exposures have been associated with increased rates of cancer mortality and with spontaneous abortions in some studies, but not others.[10,44] Women who were exposed to levels of formaldehyde at concentrations greater than 47 ppb had higher odds of having atopic eczema.[45] A recent meta-analysis found that 10 μg/m³ increase in formaldehyde exposure was associated with increased childhood asthma diagnosis.[46]

Tetrachloroethylene has been associated with liver toxicity in the general US population.[24,33]

Toluene has been associated with exacerbation of symptoms among children with asthma. In addition, in Japanese homes with dampness and mold, airborne toluene measurements were associated with family report of skin irritation symptoms.[45] Other outcomes of toluene exposure have been evaluated among occupationally exposed persons or habitual abusers of toluene inhalation ("glue sniffers"). These effects include reversible neurological symptoms such as fatigue, headache, and decreased manual dexterity. Narcosis occurs at higher exposure levels. Deficits in neurological function and neuromuscular performance, including hearing, and color discrimination have been reported in chronically exposed workers.[18] Several case reports of birth defects and developmental delays among children of mothers who abused toluene during pregnancy suggest that exposure to high levels of toluene may be toxic to the developing fetus.[18]

Xylenes have been linked with self-reported eye and throat irritations in Japanese homes with mold and dampness.[34]

Overview of the Toxicology of Volatile Organic Compounds

Most studies of the toxicology of VOCs have focused on exposure via inhalation.[1] Inhalation is the primary route of exposure for VOCs and is also the primary route of excretion.[1] As much as 97% of VOCs in the human body are expelled in exhaled breath.[1]

Acute exposures to high concentrations of all VOCs are associated with respiratory irritation. Most VOCs also have some neurotoxic properties and can cause a range of neurobehavioral effects, including death from acute, high-dose exposure. VOCs also can cause well-described hematoxic, hepatoxic, and nephrotoxic effects that create significant risks to human health. Many VOCs are also carcinogenic or are suspected carcinogens. Additional less common forms of toxicity include the ototoxic effects of ethyl benzene.

The common features in the toxicology of the many different VOCs may in part reflect the commonality of their metabolism. Most VOCs are thought to be metabolized by P450 enzymes, specifically CYP2E1. The fact that many VOCs share common metabolic pathways suggests that exposure to mixtures of VOCs can cause competitive metabolic inhibition.[47]

While VOCs share many toxicological properties, their differences are too great to create a single toxicological profile. The toxicological mechanisms of individual compounds are therefore described next.

Benzene exposures in rats and mice can result in multiple tumors, including Zymbal gland carcinomas, squamous cell papillomas, carcinomas of the oral cavity and skin, and lung alveolar/bronchiolar adenomas and carcinomas, Harderian gland adenomas, preputial gland squamous cell carcinomas, and mammary gland carcinomas.[19] Benzene is also fetotoxic to animals at inhaled concentrations above 47 ppm.[19]

Reactive hepatic metabolites of benzene, including phenolic metabolites such as hydroquinone, catechol, 1,2,4-benzenetriol, and 1,2- and 1,4-benzoquinone, are thought to be transported to the bone marrow, where bone marrow peroxidases can cause further metabolism to highly reactive oxidative species. These reactive oxygen species can damage tubulin, histone proteins, topoisomerase II, other DNA associated proteins, and DNA itself.[19]

The central role of the enzyme CYP2E1 in benzene metabolism was demonstrated in animal studies in which CYP 2E1 was either inhibited or knocked out, reducing the benzene toxicity.[19]

While no single metabolite of benzene is thought to be completely responsible for its carcinogenic and other toxic effects, the phenol metabolites and trans,trans-muconaldehyde are thought to play a role.[48] Unconjugated metabolites of benzene (except phenol and 1,2,4-benzenetriol) decrease erythropoiesis.[48] Hemoglobin and albumin adducts of benzene oxide have been detected in the blood of workers exposed to benzene,[19,48] while trans,trans-muconic acid (oxidative product of trans,

trans-muconaldehyde) is often found in the urine of exposed individuals, but the alde-hyde has not been detected in other compartments such as blood.

Para-dichlorobenzene is oxidized to 2,5 dichlorophenol via a CYP 2E1 mechanism and then further metabolized to phase II products: mercapturate, glucuronide, and sulfate conjugates. Metabolism is thought to primarily occur in the liver but may also occur (to a much lesser extent) in the kidney and lung. Although limited data are available, the liver, nervous system, and hematopoietic system are systemic targets in humans.[20]

Hepatotoxicity and nephrotoxicity have been observed in studies of animals exposed to *para*-dichlorobenzene, and these effects are thought to be caused by reactive inter-mediates during oxidation to the phase I metabolite (2,5-dichlorophenol). The intermediate metabolites are not yet identified but are thought to occur intracellularly and to be detoxified by glutathione (GSH).[20]

Para-dichlorobenzene has been reported to have hematological affects in both animals and humans. These effects are limited to white and red blood cell abnormalities in rats and mice. In an unusual case of acute exposure, a 21-year-old pregnant woman consumed 1–2 toilet air freshener blocks containing *para*-dichlorobenzene. She developed severe microcytic, hypochromic anemia with excessive polychromasia and marginal nuclear hypersegmentation of the neutrophils. Heinz bodies were seen in a small number of the red blood cells. After she discontinued eating these blocks at about 38 weeks of gestation, her hemoglobin levels began to rise steadily.[20] No birth outcomes of consequence were reported. In high concentrations, *para*-dichlorobenzene is a respiratory irritant. The rare reported deaths from *para*-dichlorobenzene exposure were attributed to hepatic necrosis.[20]

Ethylbenzene undergoes phase I and phase II metabolism and is excreted in the urine as multiple glucuronide and sulfate metabolites. Phase I metabolism is catalyzed by the enzymes CYP2E1 and CYP2B6. The major metabolites are mandelic acid (64–71%) and phenylglyoxylic acid (19–25%).[17] Minor metabolites include 1-phenylethanol (4%), *p*-hydroxyacetophenone (2.6%), *m*-hydroxyacetophenone (1.6%), and trace amounts of 1-phenyl-1,2-ethanediol, acetophenone, 2-hydroxyacetophenone, and 4-ethylphenol.[17]

The systemic effects of ethylbenzene observed in humans are respiratory tract and ocular irritation, possible ototoxicity (hearing loss), and hematological alterations.[17]

Ethylbenzene is carcinogenic in laboratory animals, based on observations of renal tubule neoplasms in rats, alveolar/bronchiolar neoplasms in mice, and hepatocellular neoplasms in female mice. Carcinogenic activity may be attributed to the parent compound and/or reactive oxidative metabolites in the 4-ethylphenol pathway, but not the 1-phenylethanol pathway.[17,28]

Ethylbenzene itself rather than its major metabolites is thought to be the primary ototoxicant.[17]

In vitro studies have demonstrated that ethylbenzene can cause changes in cell membranes, particularly in astrocytes. Changes in the integrity of the cell membrane are caused by portioning of the parent into the lipid bilayer and may affect the integrity of the membrane barrier, energy transduction, and the formation of a matrix for proteins and enzymes.

Acute exposure to ethyl benzene causes eye, skin, and mucous membrane irritation, with tearing of the eyes, irritation of the nose and upper respiratory tract, and skin irritation.[17] Symptoms of narcosis include fatigue, drowsiness, staggering gait, and incoordination. Chronic exposure causes fatigue, headache, and eye and upper respiratory tract irritation.[17]

Formaldehyde is an animal carcinogen. Inhalational dosing studies in laboratory animals induced squamous cell carcinomas and polyploid adenomas in the upper respiratory tract. Evidence suggests that exposure concentration is more important than duration in determining the extent of formaldehyde-induced nasal epithelial damage. However, repeated exposures for protracted durations are required to induce nasal cancer in rats.[30] Formaldehyde can form cross-links between protein and DNA in vivo by electrophilic addition to the amine group.[30]

Formaldehyde is metabolized to formate by formaldehyde dehydrogenase via GSH intermediates. Metabolism takes place in all tissues, and formate is removed rapidly by the blood supply. Metabolic incorporation of formaldehyde and its metabolites into macromolecules (DNA, RNA, and proteins) was demonstrated in multiple animal inhalation studies in which DNA cross-links were discovered post exposure. These cross-links were most abundant in the middle turbinate tissues of the nose and lowest in the nasopharyngeal tissues. Some were observed in the larynx, trachea, and carina and major intrapulmonary airway tissues.[30]

Formaldehyde dehydrogenase activity is not inducible in response to formaldehyde exposure, and toxicity occurs when intracellular levels saturate formaldehyde dehydrogenase activity, allowing the unmetabolized intact molecule to act locally. Formate is subject to excretion as a salt in the urine and not as reactive as the parent compound. Formaldehyde forms amino acid derivatives, which may be related to its germicidal properties. High concentrations will precipitate protein in biological fluids.[30]

Toxic effects of formaldehyde are generally found to be restricted to portal-of-entry tissue in animal studies. Target organs of airborne formaldehyde are thus the nose and the eyes, with the lungs being a secondary target at high concentrations. Distant site effects were occasionally observed and may occur only when the capacity for local metabolism of formaldehyde is exceeded. Irritation at the point of contact is seen with exposures by the inhalation, oral, and dermal routes. Irritation is believed to be caused by reaction of the electrophilic site on formaldehyde with nucleophile sites (amine groups) on cell membranes. High doses of formaldehyde are cytotoxic, causing degeneration of mucosal and epithelial cell layers. Toxic effects are believed to be mediated by formaldehyde itself and not by metabolites causing DNA-protein cross-links.[30]

Nasal and eye irritation, neurological effects, and increased risk of asthma and/or allergy have been observed among humans breathing formaldehyde at concentrations of 0.1–0.5 ppm. Eczema and changes in lung function have been observed at air concentrations of 0.6–1.9 ppm.[30]

Tetrachloroethylene causes hepatocellular carcinomas and adenomas in experimental animals following oral administration. Liver cancers are far less likely in humans because humans are relatively insensitive to peroxisome proliferators such as tetrachloroethylene and therefore produce very little trichloroacetic acid (~2%), the principal metabolite of tetrachloroethylene.[33]

At high concentrations, tetrachloroethylene is an anesthetic agent and a cardiac epinephrine sensitizer, and it can potentially cause sudden death. At high sublethal air levels (232–385 ppm) it causes respiratory irritation.[33]

The liver is a major target organ at lower sustained exposure levels. Hepatocellular damage has been documented in one case study with liver damage having been diagnosed by the presence of hepatomegaly, icterus, and elevations of serum glutamic oxaloacetic transaminase, bilirubin, and urinary urobilinogen in other studies.[33]

The nervous system is also a target organ in acute exposure to tetrachloroethylene. Neurologic symptoms of acute exposure include loss of consciousness, dizziness, headache, sleepiness, confusion, nausea, difficulty in speaking and walking, narcosis, and death.[33] Behavioral studies among workers demonstrated short-term memory deficits in a high-exposure group (40.8 ppm) relative to the low-exposure group (11.2 ppm) and workers with a total weighted average concentration of 12 or 54 ppm tetrachloroethylene had lost perceptual function, attention, and intellectual function. Rodent studies demonstrated tetrachloroethylene alters the fatty acid pattern of brain phospholipids and amino acids, but tetrachloroethylene is also incorporated into brain membranes, potentially altering neural transmission velocity. Both processes may explain its neurotoxic effects.[33]

Respiratory (nasal) irritation may result from repeated or extended inhalation, and skin irritation may occur with extended contact. Ambient concentrations in air and water are not thought to create a health risk.[33]

Tetrachloroethylene is thought to be metabolized by P450 enzymes to form an epoxide intermediate to trichloroacetic acid. However, this pathway can be saturated by high-dose exposure, and metabolism is inhibited by ethanol consumption. The CYP isozymes can create free radical metabolic intermediates by cleavage of the carbon-chlorine bond. These free radicals may react with unsaturated lipids and proteins in the endoplasmic reticulum of hepatocytes and cause cellular dysfunction.[33]

Blood concentration screening values for both oral and inhalation exposure range from 0.0005 to 100 ng/L. The primary route of excretion up to 98% is through exhalation of the parent tetrachloroethylene.[33]

Toluene toxicity primarily affects the central nervous system, and high doses cause narcosis. The lipophilic nature of toluene, which enables it to interact with lipids and proteins in the nervous system, is thought to be the mechanism primarily responsible for its generally reversible neurological effects. The acute anesthetic actions of toluene may involve intercalation of toluene into the lipid bilayer of nerve membranes. More rapid metabolic clearance creates a shortened recovery time from toluene-induced narcosis.[18]

Magnetic resonance imaging studies show diffuse central nervous system demyelination and white matter abnormalities in some solvent abusers chronically exposed to toluene. These effects appear to be related to loss of neurological function. Repeated interaction of toluene with membrane proteins and/or phospholipids changes enzyme activities involved in the synthesis and/or degradation of neurotransmitters and thus affects levels of neurotransmitter in particular regions of the brain. Another hypothesized mechanism of neurotoxicity is that toluene changes the binding of neurotransmitters to membrane receptors.[18]

Neurological impairment (speech abnormalities) and brain atrophy have been observed in chronic toluene-inhalation abuse due to preferential atrophy in lipid-rich regions of the brain. Other neurologic effects include hearing loss, changes in brainstem auditory-evoked potentials, color vision impairment, and changes in brainstem visual-evoked potentials. Deaths from exposure have been attributed to cardiac arrhythmias, central nervous system depression, asphyxia, and hepatic and renal failure but are generally associated with multiple solvents from solvent abusers.[18]

High levels of toluene exposure may be nephrotoxic and also irritate the respiratory tract. Low to moderate levels cause classic solvent intoxication symptoms, including tiredness, confusion, weakness, drunken-type actions, memory loss, nausea, loss of appetite, and hearing and color vision loss.[18] Many of these are present the day after exposure.

Toluene is believed to be primarily metabolized in the liver. Metabolism to multiple products includes oxidation to benzyl alcohol, then benzoic acid and eventually hippuric acid, epoxide formation by CYPs to cresols and conjugates, GSH conjugation to benzyl-glutathione intermediate and eventually a benzylmercapturic acid, and others. The final metabolites include in order of prevalence of the major metabolite, hippuric acid and minor metabolites, the glucuronyl conjugate of benzoic acid, sulfate, and glucuronide conjugates of *ortho-* and *para*-cresol.[18]

Acute exposure to high levels of xylene may cause irritation of the skin, eyes, nose, and throat; difficulty in breathing; impaired function of the lungs; delayed response to a visual stimulus; impaired memory; stomach discomfort; and possible changes in the liver and kidneys.[34] Long-term exposure can cause headaches, lack of muscle coordination, dizziness, confusion, and changes in the sense of balance. In some instances, very high levels of xylene exposure have caused acute narcosis and death. No health effects have been noted at background levels. Animal studies indicate that high doses of xylene can be hepatotoxic and produce harmful effects on the kidneys, lungs, heart, and nervous system.[34]

Dissolution of lipophilic membranes within cells is thought to be the mechanism primarily responsible for both the irritant and the neurotoxic effects of xylenes. In neurons, these changes in membrane properties can affect transmission of nerve impulses either by disruption of the lipid environment or by direct interaction with the hydrophobic/hydrophilic conformation of proteins in the neuronal membrane. *Meta*-xylene may cause inhibition of gamma-aminobutyric acid (GABA)ergic neurotransmission in the cerebellum to produce effects on motor coordination. Oxidation of xylene by microsomal enzyme systems within brain cells to arene oxides or methylbenzaldehyde also may contribute to its neurotoxicity.[34]

Both inhalational and oral exposures to high concentrations of *para*-xylene result in death of cochlear hair cells and hearing loss in rats. Xylene exposure also appears to target hippocampal neurons in the brain. The combination of toxicity to vestibular structures and to the hippocampus results in impaired motor coordination and problems with spatial navigation.[34]

Xylene inhibits pulmonary microsomal enzymes. The mechanism of this inhibition is unknown, but it has been attributed to the formation of a toxic reactive metabolite. Reduction in cytochromes CYP 2B1, 2E1, and 4B1 in the lung and 2B1 and 2E1 in the nasal cavity has been observed following xylene exposure.[34]

Nephrotoxicity may be related to formation of reactive metabolites by CYP2E1 and subsequent irritation or direct membrane fluidization and subsequent necrosis. It may also involve induction of apoptosis through the activation of mitochondrial caspase-9 and caspase-3.[34]

The developmental toxicity of *meta*-and *para*-xylenes in animal bioassays is largely related to maternal toxicity and possibly isomer-specific.

Xylene metabolism employs oxidation of the side chain methyl groups by mixed function oxidases in the liver to form methylbenzoic acids, which are conjugated to form methylhippuric acids (glycine conjugate). This metabolite accounts for roughly 90% of the metabolism/excretion (regardless of dose) with the remaining 10% excreted unchanged either in exhaled breath or urine.[34]

Organ Systems Affected by Volatile Organic Compound Exposure

Because the primary route of exposure to VOCs is via inhalation, many parts of the respiratory system are affected. The liver, as the primary organ of detoxification, is involved in VOC detoxification. The systems affected for the target VOCs discussed are given in Table 39.2.

Prevention of Volatile Organic Compound Toxicity

Prevention of VOC-related disease is best achieved by limiting exposure to VOCs. Control of VOCs in outdoor air is achieved through enforcement of regulatory standards that either limit VOC emissions from automotive or industrial sources or regulate the permissible levels of VOCs in ambient air. The US Environmental Protection Agency (EPA) requires industry to monitor emissions of certain hazardous air pollutants, including many VOCs. The following guidance will reduce VOC concentrations in indoor air:

- Cigarette smoking should be eliminated.
- Use of aerosolized products should be minimized.
- Products such as artificial carpets and pressed wood furniture containing VOCs should be avoided whenever possible and replaced with natural products.
- Other consumer products containing VOCs should be bought in small quantities when possible and disposed of immediately after use.
- Low-VOC paints and varnishes should be used whenever possible. Dry cleaning that is fresh from the cleaning shop should be aired for a few hours before it is put in a closet.
- Homes should be ventilated well with open windows and fans, especially when large quantities of VOCs are being generated, as when painting.
- The ignition in automobiles should be turned off immediately after entering a garage and before closing the garage door, and it should not be started until the garage door is open.
- Using carbon filters and changing them per manufacturers guideline may reduce indoor VOC concentrations.

The EPA has made a series of recommendations to guide consumers who want to reduce indoor air exposure to VOCs.[49]

References

1. Montero-Montoya R, López-Vargas R, Arellano-Aguilar O. Volatile organic compounds in air: sources, distribution, exposure and associated illnesses in children. *Ann Global Health.* 2018;84:225–38.

2. Gokhale S, Kohajda T, Schlink U. Source apportionment of human personal exposure to volatile organic compounds in homes, offices and outdoors by chemical mass balance and genetic algorithm receptor models. *Sci Total Environ.* 2008;407:122–38.

3. van Vliet D, Smolinska A, Jöbsis Q, et al. Can exhaled volatile organic compounds predict asthma exacerbations in children? *J Breath Res* 2017;11:016016.

4. Zora JE, Sarnat SE, Raysoni AU, et al. Associations between urban air pollution and pediatric asthma control in El Paso, Texas. *Sci Total Environ.* 2013;448:56–65.

5. Nurmatov U, Tagieva N, Semple S, Devereux G, Sheikh A. Volatile organic compounds and risk of asthma and allergy: a systematic review and meta-analysis of observational and interventional studies. *Primary Care Respir J.* 2013;22:Ps9–15.

6. Delfino RJ, Gong H, Linn WS, Hu Y, Pellizzari ED. Respiratory symptoms and peak expiratory flow in children with asthma in relation to volatile organic compounds in exhaled breath and ambient air. *J Expo Anal Environ Epidemiol.* 2003;13:348–63.

7. Xiong Y, Zhou J, Xing Z, Du K. Cancer risk assessment for exposure to hazardous volatile organic compounds in Calgary, Canada. *Chemosphere.* 2021;272:129650.

8. Galbraith D, Gross SA, Paustenbach D. Benzene and human health: a historical review and appraisal of associations with various diseases. *Crit Rev Toxicol.* 2010;40 Suppl 2:1–46.

9. Tsai WT. An overview of health hazards of volatile organic compounds regulated as indoor air pollutants. *Rev Environ Health.* 2019;34:81–9.

10. Goldstein BD. Hematological and toxicological evaluation of formaldehyde as a potential cause of human leukemia. *Hum Exp Toxicol.* 2011;30:725–35.

11. Collins JJ, Esmen NA, Hall TA. A review and meta-analysis of formaldehyde exposure and pancreatic cancer. *Am J Ind Med.* 2001;39:336–45.

12. Bayati M, Vu DC, Vo PH, et al. Health risk assessment of volatile organic compounds at daycare facilities. *Indoor Air.* 2021;31:977–88.

13. Nakaoka H, Hisada A, Matsuzawa D, et al. Associations between prenatal exposure to volatile organic compounds and neurodevelopment in 12-month-old children: The Japan Environment and Children's Study (JECS). *Sci Total Environ.* 2021;794:148643.

14. Garcia E, Rice MB, Gold DR. Air pollution and lung function in children. *J Allergy Clin Immunol.* 2021;148:1–14.

15. Le Cann P, Bonvallot N, Glorennec P, Deguen S, Goeury C, Le Bot B. Indoor environment and children's health: recent developments in chemical, biological, physical and social aspects. *Int J Hygiene Environ Health.* 2011;215:1–18.

16. Khalequzzaman M, Kamijima M, Sakai K, Chowdhury NA, Hamajima N, Nakajima T. Indoor air pollution and its impact on children under five years old in Bangladesh. *Indoor Air.* 2007;17:297–304.

17. ATSDR. *Ethylbenzene.* Atlanta, GA: Agency for Toxic Substances and Disease Registry; 2010. https://www.atsdr.cdc.gov/toxprofiles/tp110.pdf

18. ATSDR. *Toluene.* Atlanta, GA: Agency for Toxic Substances and Disease Registry; 2017. https://wwwn.cdc.gov/TSP/ToxProfiles/ToxProfiles.aspx?id=161&tid=29

19. ATSDR. *Benzene.* Atlanta, GA: Agency for Toxic Substances and Disease Registry; 2007. https://www.atsdr.cdc.gov/toxprofiles/tp3.pdf

20. ATSDR. *Dichlorobenzenes.* Atlanta, GA: Agency for Toxic Substances and Disease Registry; 2006. https://www.atsdr.cdc.gov/toxprofiles/tp10.pdf

21. Ye X, Wong LY, Zhou X, Calafat AM. Urinary concentrations of 2,4-dichlorophenol and 2,5-dichlorophenol in the U.S. population (National Health and Nutrition Examination Survey, 2003-2010): trends and predictors. *Environ Health Perspect.* 2014;122:351–5.

22. Wolff MS, Engel SM, Berkowitz GS, et al. Prenatal phenol and phthalate exposures and birth outcomes. *Environ Health Perspect.* 2008;116:1092–7.

23. Hill RH, Jr., Ashley DL, Head SL, Needham LL, Pirkle JL. p-Dichlorobenzene exposure among 1,000 adults in the United States. *Arch Environ Health*. 1995;50:277–80.

24. National Report on Human Exposure to Environmental Chemicals. 2022. www.cdc.gov/exposurereport.)

25. Sexton K, Adgate JL, Church TR, et al. Children's exposure to volatile organic compounds as determined by longitudinal measurements in blood. *Environ Health Perspect*. 2005;113:342–9.

26. Miri M, Rostami Aghdam Shendi M, Ghaffari HR, et al. Investigation of outdoor BTEX: Concentration, variations, sources, spatial distribution, and risk assessment. *Chemosphere*. 2016;163:601–9.

27. Webb E, Moon J, Dyrszka L, et al. Neurodevelopmental and neurological effects of chemicals associated with unconventional oil and natural gas operations and their potential effects on infants and children. *Rev Environ Health*. 2018;33:3–29.

28. Sweeney LM, Kester JE, Kirman CR, et al. Risk assessments for chronic exposure of children and prospective parents to ethylbenzene (CAS No. 100-41-4). *Crit Rev Toxicol*. 2015;45:662–726.

29. Rumchev K, Spickett J, Bulsara M, Phillips M, Stick S. Association of domestic exposure to volatile organic compounds with asthma in young children. *Thorax*. 2004;59:746–51.

30. ATSDR. *Formaldehyde*. Atlanta, GA: Agency for Toxic Substances and Disease Registry; 1999. https://www.atsdr.cdc.gov/toxprofiles/tp111.pdf

31. Murphy MW, Lando JF, Kieszak SM, et al. Formaldehyde levels in FEMA-supplied travel trailers, park models, and mobile homes in Louisiana and Mississippi. *Indoor Air*. 2013;23:134–41.

32. Parthasarathy S, Maddalena RL, Russell ML, Apte MG. Effect of temperature and humidity on formaldehyde emissions in temporary housing units. *J Air Waste Manag Assoc*. 2011;61:689–95.

33. ATSDR. *Tetrachloroethylene*. Atlanta, GA: Agency for Toxic Substances and Disease Registry; 2019. https://www.atsdr.cdc.gov/ToxProfiles/tp18.pdf

34. ATSDR. *Xylenes*. Atlanta, GA: Agency for Toxic Substances and Disease Registry; 2007. https://www.atsdr.cdc.gov/toxprofiles/tp71.pdf

35. Qu Q, Shore R, Li G, et al. Hematological changes among Chinese workers with a broad range of benzene exposures. *Am J Ind Med*. 2002;42:275–85.

36. Rappaport SM, Waidyanatha S, Qu Q, et al. Albumin adducts of benzene oxide and 1,4-benzoquinone as measures of human benzene metabolism. *Cancer Res*. 2002;62:1330–7.

37. Vardoulakis S, Giagloglou E, Steinle S, et al. Indoor exposure to selected air pollutants in the home environment: a systematic review. *Int J Environ Res Public Health*. 2020;17.

38. Liu J, Drane W, Liu X, Wu T. Examination of the relationships between environmental exposures to volatile organic compounds and biochemical liver tests: application of canonical correlation analysis. *Environ Res*. 2009;109:193–9.

39. Hsiao PK, Lin YC, Shih TS, Chiung YM. Effects of occupational exposure to 1,4-dichlorobenzene on hematologic, kidney, and liver functions. *Int Arch Occup Environ Health*. 2009;82:1077–85.

40. Shafer G, Arunachalam A, Lohmann P. Newborn with perinatal naphthalene toxicity after maternal ingestion of mothballs during pregnancy. *Neonatology*. 2020;117:127–30.

41. Sillery JJ, Lichenstein R, Barrueto F, Jr., Teshome G. Hemolytic anemia induced by ingestion of paradichlorobenzene mothballs. *Pediatr Emerg Care*. 2009;25:252–4.

42. Feuillet L, Mallet S, Spadari M. Twin girls with neurocutaneous symptoms caused by moth-ball intoxication. *N Engl J Med*. 2006;355:423–4.

43. Cheong R, Wilson RK, Cortese IC, Newman-Toker DE. Mothball withdrawal enceph-alopathy: case report and review of paradichlorobenzene neurotoxicity. *Subst Abuse*. 2006;27:63–7.

44. Conolly RB, Kimbell JS, Janszen D, et al. Human respiratory tract cancer risks of inhaled formaldehyde: dose-response predictions derived from biologically-motivated computa-tional modeling of a combined rodent and human dataset. *Toxicol Sci*. 2004;82:279–96.

45. Saijo Y, Kishi R, Sata F, et al. Symptoms in relation to chemicals and dampness in newly built dwellings. *Int Arch Occup Environ Health*. 2004;77:461–70.

46. Lam J, Koustas E, Sutton P, et al. Exposure to formaldehyde and asthma outcomes: a system-atic review, meta-analysis, and economic assessment. *PLoS One*. 2021;16:e0248258.

47. Pohl HR, Scinicariello F. The impact of CYP2E1 genetic variability on risk assessment of VOC mixtures. *Regul Toxicol Pharmacol*. 2011;59:364–74.

48. Snyder R, Hedli CC. An overview of benzene metabolism. *Environ Health Perspect*. 1996;104 Suppl 6:1165–71.

49. Technical overview of volatile organic chemicals. EPA. 2023. Available at: https://www.epa.gov/indoor-air-quality-iaq/technical-overview-volatile-organic-compounds.

40

Endocrine Disruptors

Annemarie Stroustrup and Shanna H. Swan

"Endocrine disruptor" is a term used to describe a phenomenon that has been recognized for many years.[1] Endocrine disruptors are synthetic chemicals that act on the body's natural system of hormone-responsive tissues. In many cases, an endocrine-disrupting chemical (EDC) has a chemical structure similar to that of naturally occurring hormones, such as the sex steroid hormones estrogen and testosterone, and thus can mimic, block, or alter normal hormonal signaling. Although "modern" EDCs, including bisphenol A (BPA), phthalates, and per- and polyfluoroalkyl substances (PFAS), which are discussed later in this chapter, have become an important focus of recent research, recognition that synthetic chemicals can possess hormonal activity is much older.[2]

The risks associated with the industrial EDCs such as polychlorinated biphenyls (PCBs) and DDT were first recognized more than 50 years ago by Rachel Carson.[3] Many of the originally identified EDCs also are classified today as persistent organic pollutants (POPs) because they are still contaminating the environment decades after having been eliminated from common use in much of the world. Consequences of human and animal exposure to POPs and other EDCs range from reproductive abnormalities, to neurocognitive deficits, to cancer. Some EDCs can also affect adiposity and weight gain (see Chapter 41).

It is illustrative to consider the human health effects of EDCs in the perspective provided by experience with the synthetic estrogen diethylstilbestrol (DES), which was administered to women during pregnancy. The DES story provides a framework for understanding the multiple impacts of hormonally active chemicals on human health, particularly human reproductive health. The multiple health effects of DES exposure that affected a high percent of exposed offspring underscore the dangers of exposure to hormonally active chemicals during developmentally vulnerable periods in early life.

The Example of Diethylstilbestrol

DES is a synthetic estrogen first synthesized in 1933.[2] Between 1947 and 1971, DES was prescribed to pregnant women in the United States and other countries in the erroneous belief that it prevented spontaneous abortion. In high-use areas, such as Boston where the first human studies of DES were conducted, physicians frequently recommended prophylactic use of DES for all women at the time of confirmation of pregnancy and as "pre-conceptional therapy" for women with a history of multiple miscarriages.[4] Elsewhere, DES was prescribed for use in the late first trimester or early second trimester to "protect the pregnancy," particularly for women with a history of miscarriage.

Annemarie Stroustrup and Shanna H. Swan, *Endocrine Disruptors* In: *Textbook of Children's Environmental Health*, Second Edition. Edited by: Ruth A. Etzel and Philip J. Landrigan, Oxford University Press. © Oxford University Press 2024.
DOI: 10.1093/oso/9780197662526.003.0040

In 1971, Dr. Arthur Herbst and colleagues reported a series of eight women, ages 15–22 years, who had been diagnosed with vaginal clear-cell adenocarcinoma (CCAC) during a 3-year period at a single hospital in New England.[5] This finding was noteworthy because CCAC of the vagina is an extremely rare cancer that had previously never been seen in this age group. These authors published the startling finding that seven of the eight cases—but none of 32 matched controls—had been exposed prenatally to DES. Seven months after this publication, the US Food and Drug Administration (FDA) withdrew approval for use of DES by pregnant women. In the decade following publication of this small case series, the causal relationship between prenatal exposure to DES in the first trimester of pregnancy and CCAC of the vagina and cervix was firmly established.[6]

Experimental and clinical studies of DES conducted since the early 1970s have revealed that in utero exposure during critical periods of organogenesis can profoundly disrupt the differentiation of a variety of estrogen-sensitive tissues.[7] Because numerous tissue types are sensitive to estrogen in early development, these studies have documented that prenatal exposure to EDCs can produce a wide range of adverse outcomes.

In the 75 years since DES was first used by pregnant women, research has shown that this exposure has had adverse effects on three generations. Risk to the DES mothers (F1) themselves was limited to a small increase in the risk of developing[8] and dying[9] from breast cancer. Here we discuss adverse effects on the children (F2) and grandchildren (F3) of these exposed women.

The overwhelming proportion of DES research in humans has been conducted in female offspring. It is now known that the relative risk of CCAC among DES daughters is very high, while the absolute risk of this cancer is low, with incidence estimated to be 1 case in 1,000 DES exposures[10] (see Table 40.1). In the process of screening DES daughters for CCAC, physicians unexpectedly found a range of other histological and structural genital tract anomalies including vaginal adenosis, cervical ectropion, and transverse cervical and vaginal ridges as well as abnormalities of the uterus, ovaries, and

Table 40.1 Hazard ratios for adverse health outcomes following in utero diethylstilbesterol (DES) exposure

Adverse outcome	Hazard ratio (95% confidence interval)
Infertility	2.37 (2.05–2.75)
Spontaneous abortion	1.64 (1.42–1.88)
Ectopic pregnancy	3.72 (2.58–5.38)
Loss of second-trimester pregnancy	3.77 (2.56–5.54)
Preterm delivery	4.68 (3.74–5.86)
Preeclampsia	1.42 (1.07–1.89)
Stillbirth	2.45 (1.33–4.54)
Neonatal death	8.12 (3.53–18.65)
Early menopause	2.35 (1.67–3.31)
Cervical intraepithelial neoplasia	2.28 (1.59–3.27)
Breast cancer	1.82 (1.04–3.18)
Clear-cell adenocarcinoma	∞ (0.37–∞)

Source: Hoover RN, Hyer M, Pfeiffer RM. et al. Adverse health outcomes in women exposed in utero to diethylstilbesterol. *N Engl J Med* 2011;365:1304–14.

fallopian tubes.[7] In utero exposure to DES therefore has both carcinogenic and terato-genic effects on the developing female fetus. These anatomical abnormalities of the female reproductive tract led to high rates of infertility, miscarriage, ectopic pregnancy, and preterm delivery among DES daughters. Based on follow-up of the DES cohort and unexposed controls, it is estimated that, among DES daughters, 41% of pregnancies ended in an adverse pregnancy outcome, compared with 15% of pregnancies in non-DES exposed women.[7]

In addition to reproductive problems, DES daughters experienced an increase in breast cancer risk, which was greatest in those exposed to high doses of DES.[11] More recent follow-up of DES daughters also has identified a somewhat elevated increased risk of cardiovascular disease.[12] Although the focus of most studies in DES daughters has been on the reproductive organs, effects on other tissues, such as brain and bone, also have been noted in the F2 generation[13] as well as a nearly doubled risk of pancreatic cancer.[14]

Follow-up of DES sons, although more limited than that of DES daughters, has shown that prenatal exposure to DES increased the risk of male urogenital abnormalities, an association that is strongest when exposure began early in gestation, increasing the risk of cryptorchidism, epididymal cyst, and inflammation/infection of the testes.[15] A recent meta-analysis of six studies that investigated testicular cancer in DES sons identified a three-fold increased risk of this cancer.[16]

The National Cancer Institute's Third Generation Study is a cohort of DES-exposed and -unexposed granddaughters (F3). Compared to the unexposed, DES-exposed third-generation women had an increased risk of irregular menses and amenorrhea and increased risk of preterm delivery, particularly in women whose prenatally DES-exposed mothers were affected by vaginal epithelial changes.[17] An overall increase of birth defects in the DES exposed grandsons (F3) was seen on follow-up, mostly due to an increased risk of genitourinary conditions, with some European cohorts reporting an elevated frequency of hypospadias in these men.[18]

The DES story provides an unfortunate but powerfully informative example of the devastating consequences for human health that can result from exposures to manmade chemicals with endocrine activity. This story also underscores the importance of the developmental timing of exposure. In fact, children of women who took DES only after the mid-second trimester show few or none of the adverse effects seen in children exposed during the first half of pregnancy. Additionally, far fewer adverse effects were seen in the mothers who took the drug than their offspring, thus documenting that the embryo and fetus are much more sensitive to EDCs than is the adult. The DES story also demonstrates delayed effects of prenatal exposure because the effects of exposure to EDCs in early life did not become apparent until a later developmental stage—no adverse health effects were seen in the DES offspring until they had passed through puberty. Finally, recent follow-up of DES grandchildren has demonstrated that prenatal exposure can influence development in the third generation, a "transgenerational" effect. Whether this will persist in subsequent generations is not known.

Endocrine-Disrupting Chemicals of Current Interest

We now focus on two EDCs of great current interest: BPA and phthalates. These chemicals are ubiquitous in modern society (Figure 40.1). Exposures in utero and in early

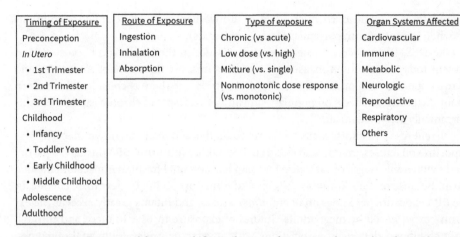

Timing of Exposure	Route of Exposure	Type of exposure	Organ Systems Affected
Preconception	Ingestion	Chronic (vs acute)	Cardiovascular
In Utero	Inhalation	Low dose (vs. high)	Immune
• 1st Trimester	Absorption	Mixture (vs. single)	Metabolic
• 2nd Trimester		Nonmonotonic dose response (vs. monotonic)	Neurologic
• 3rd Trimester			Reproductive
Childhood			Respiratory
• Infancy			Others
• Toddler Years			
• Early Childhood			
• Middle Childhood			
Adolescence			
Adulthood			

Figure 40.1. Conceptual framework for endocrine-disrupting chemical analysis. Human exposure to endocrine-disrupting chemicals is typically chronic at multiple time points, low-dose, and as mixtures rather than single entity exposures. Unlike many other toxic exposures, dose-response is often non-linear, and epigenetic mechanisms as well as prenatal or early life exposure have potential for altering developmental pathways

childhood to these chemicals have been associated with a range of adverse health effects at birth and later in life.

Conceptual Framework for Endocrine-disrupting Chemical Analysis

Human exposure to EDCs is typically chronic, occurring at multiple time points, low-dose, and to mixtures rather than single-entity exposures. Unlike many other toxic exposures, dose-response is often non-monotonic, and epigenetic mechanisms as well as prenatal or early-life exposure to these have the potential to alter developmental pathways.

Bisphenol A

Like DES, BPA was originally designed to be a synthetic estrogen. Sir Charles Dodds, credited with first synthesizing DES, also demonstrated the estrogenic properties of BPA. Because BPA is a much weaker estrogen than DES, which was developed at about the same time, it was relatively ineffective clinically and little used in hormone replacement.

But BPA possesses chemical properties that make it very useful as an industrial chemical. By the mid-1950s, BPA had come into use as an adjuvant to epoxy resins and polycarbonate plastics. Currently, BPA is a ubiquitous component of plastic and nonplastic materials in the industrialized environment, including the inner lining of canned food, aluminum cans, the shiny coating of thermal merchandise receipts, plastic tableware, plastic toys, food storage containers, water supply pipes, and even medical equipment. As such, BPA is present in the human environment worldwide. Because BPA is an additive rather than an integral component of these materials, it can leach out of the matrices in which it is embedded to enter foods and liquids, particularly when stressed by high temperatures (as in the oven, microwave, or dishwasher) or by acidic or basic conditions (as when in contact with particular foods, such as fats).

Like DES, BPA acts on estrogen-sensitive tissues in the body. And because BPA is present today in small but measurable amounts in more than 90% of Americans and in the populations of other industrialized countries,[19,20] there is considerable concern about the potential effects on population health of widespread chronic exposure to this hormonally active chemical.

Current research efforts focus on the reproductive and neurocognitive effects of BPA exposures on fetuses, infants, and children. Like DES, exposure to BPA during the particularly vulnerable stages of early development has potential for high impact and lifelong harm. Because of their low body weights and immature ability to clear toxic chemicals, the BPA exposure per kilogram of embryos, fetuses, and infants greatly exceeds the exposure experienced by most adults. Studies of exposure to BPA in utero and through early childhood document associations with adverse neurodevelopmental outcomes, including increased aggression and hyperactivity[21]; obesity (see Chapter 41); increased risk of cardiovascular disease and diabetes; and reproductive abnormalities, including decreased levels of endogenous sex hormones, sexual dysfunction, and precocious puberty.[22]

A 2008 review stated that "evidence from animal models is accumulating that perinatal exposure to . . . low doses of . . . BPA alters breast development and increases breast cancer risk."[23] Another 2008 review concluded that "animal experiments and epidemiological data strengthen the hypothesis that fetal exposure to xenoestrogens may be an underlying cause of the increased incidence of breast cancer observed over the last 50 years."[24]

BPA is a short-lived, water-soluble compound. In humans, the half-life of BPA can be as short as 5 hours. Levels can be easily measured in both blood and urine but accurately reflect only very recent exposure. The rapidity of BPA clearance, the ubiquitous nature of human BPA exposure, and the significant potential for contamination during testing complicate studies of BPA exposure and effects.

Phthalates

The diesters of phthalic acid, commonly referred to as phthalates, have become ubiquitous in modern society. Phthalate esters are a complex class of chemicals and include both high- (e.g., diethylhexyl phthalate) and low- (e.g., diethyl phthalate) molecular weight phthalates. These two groups of phthalates find very different uses in industrial and commercial products. Their routes of human exposure are therefore quite different from one another, as are their health effects.

When added to otherwise hard and brittle plastics, phthalates confer softness and flexibility and are therefore referred to as "plasticizers." Phthalates are present in a wide variety of soft and flexible plastic products and notably in virtually all polyvinyl chloride (PVC). Additionally, phthalates help plastics and other materials to retain scents and colors, so they are present in perfumes, cosmetics, shampoos, creams, and other health and beauty products. Like BPA, phthalates are not chemically bound to the plastics they impregnate. They are therefore readily liberated into adjacent food, liquids, or air by heat, humidity, or prolonged exposure. Like BPA, phthalates are present in measurable amounts in industrialized populations worldwide.

Also like BPA, phthalates are short-lived in the human body. Because they are water-soluble, they are readily metabolized and excreted via the urine. Phthalates can be measured in the urine, with levels reflecting exposure in the preceding 24 hours. Peak urine concentrations of the primary phthalate metabolites are seen approximately 4 hours after exposure via ingestion, inhalation, or skin absorption and are mostly cleared after about 24 hours. Secondary and tertiary metabolites take somewhat longer to be excreted. Since exposure is largely the consequence of the regular use of home products, most people are continuously exposed. "Spot" measurements of phthalates in urine are therefore felt to accurately represent average exposure for most people.[25]

The action of phthalates on steroid-responsive mammalian tissues varies. Some phthalates, such as diethyl phthalate (DEP), have not been shown to be hormonally responsive. Others, such as diethyl hexyl phthalate (DEHP) and dibutyl phthalate (DBP), were originally considered to be estrogenic but subsequently have been shown to be anti-androgenic. Early studies noted the pronounced effect of high-dose exposure of DHEP and DBP on reproductive development and function in male rodents, which was termed the "phthalate syndrome."[26]

Studies in humans document a strong association between in utero and early-life phthalate exposure and disruption of male reproductive structure and function. Decreased anogenital distance (the distance between the base of the genitalia and the anus) is a well-documented consequence of prenatal exposure to phthalate in humans. Anogenital distance is a measure of the length of the perineum. Prenatal growth of the perineum is stimulated by androgen, and anogenital distance is normally 50–100% longer in males than in females of most mammalian species, including humans.[27] Decreased anogenital distance therefore represents "incomplete masculinization" of the normal male reproductive tract, reflecting a reduction in prenatal androgen induced by phthalates during the critical window of male programming.[28]

Shortened anogenital distance has been associated with a series of adverse effects on reproduction in human males. These include decreased sperm count in adult males[29] as well as infertility.[30]

Additional studies document associations between in utero phthalate exposure and neurodevelopmental anomalies in later childhood, including feminized play in boys[31] and difficulties with attention and with aggressive and defiant behavior in both boys and girls.[32] Most of these associations are stronger in boys than in girls, consistent with the hypothesis that phthalates act as anti-androgens and exert powerful effects on hormonally sensitive neurologic tissue when exposure occurs early in development.

Common Pathways

Although many endocrine disruptors react directly with sex steroid receptors in sensitive tissues, the effects of BPA and phthalates on the developing reproductive and nervous systems appear to be multifaceted and likely are mediated through several different mechanisms. DEHP, for example, disrupts sexual differentiation in the male rat by reducing testosterone levels in blood to levels similar to those typically found in females rather than by inhibiting the action of androgen by binding to the androgen receptor.[33]

Monoaminergic neurons in the brain, including the dopaminergic system, are sensitive to endogenous sex steroids and have been shown in animal models to be altered by synthetic chemicals, including BPA. Neurotransmitter levels, including dopamine, serotonin, acetylcholine, and others, also are affected by exposure to BPA and other endocrinologically active compounds in animal models.[34]

Endocrine disruptors also may act by impairing thyroid function to cause abnormal neurodevelopment at the cellular level. Thyroid hormones are critically important to the developing central nervous system. If left untreated, congenital hypothyroidism leads to cretinism, a syndrome of severe cognitive and growth impairment. BPA has been shown to cause functional alterations in the developing thyroid gland in animal models and to alter thyroid hormone levels in humans.[35] Additionally, due to structural similarity to thyroid hormone, EDCs may affect thyroid hormone function by competing for receptor binding sites and/or interfering with normal thyroid hormone–regulating systems.[35]

Epigenetic Mechanisms

In addition to affecting health through direct effects on sensitive tissues, EDCs also may cause heritable changes in protein expression through epigenetic mechanisms that modify gene expression in tissues. Epigenetic marks regulate gene expression through the selective addition of methyl, acetyl, and phosphoryl groups to DNA, as well as via the specific actions of DNA-regulating proteins such as ubiquitin and histones, and noncoding microribonucleic acids (miRNAs). Epigenetic marks can alter protein levels and function for the exposed individual and can be transmitted in a heritable fashion to offspring for several generations.

Exposure to environmental toxicants, including EDCs, change epigenetic marks in animal and in vitro models of early (in utero) human development and in human placental cells. For example, specific relationships have been demonstrated between exposure to BPA or certain phthalates and methylation levels of a number of genes involved in male reproductive development.[36] In some cases where in which in utero exposure to an EDC is associated with hypomethylation at the target site of embryonic DNA, treatment of the mother with folate, a pro-methylation medication, can counteract the EDC effect. This finding, in particular, demonstrates the critical interplay of various environmental exposures on in utero epigenetic effects.[30] The multigenerational effects of DES described above illustrate these points.

Mixtures

Today there are more than 85,000 synthetic chemicals in commerce, and at least 3,000 are produced in quantities greater than 1,000,000 pounds per year. It should therefore not be surprising that large numbers of substances occur together in ecosystems and even in human tissues. Such multiple exposures may result from the intended use of chemicals in personal care products, pesticides, and pharmaceuticals or be the consequence of exposures to unintentionally contaminated media, for example, residues in food and feed or pollutants in groundwater. Research demonstrates that exposures to such mixtures may have greater effects than isolated exposures to their individual chemical components.[37] For example, a 2007 study showed that exposure to a mixture of nine anti-androgens, at doses that individually appeared to exert only small effects, significantly altered the anogenital distance of male rats.[38] In an analysis of cord blood from 10 babies, a total of 287 contaminants were identified, with an average of 200 in the blood of each infant.[39] The effects of chemical mixtures on the developing fetus are largely unknown but are likely to be underestimated by current studies that examine only one chemical at a time.

Exposure Reduction

Although BPA and phthalates are present in an overwhelming number of common materials worldwide, reduction of exposure to these chemicals, especially by particularly vulnerable pregnant women and young children, can be achieved and has been well documented. Reduction in consumption of canned foods, reduction in use of plastic food and beverage storage containers, and reduction in use of fragranced creams and beauty products can measurably reduce exposures in adults and children. Because of the short half-life of these nonpersistent chemicals, reductions in exposure result almost immediately in reduction of body burden. Unfortunately, such rapid reduction in body burden is not seen following reductions in exposure to other EDCs with longer half-lives, such as the brominated flame retardants (PBDEs).

Due to increasing public pressure, many manufacturers have electively chosen to reduce exposures to EDCs by removing BPA from many baby care products, including baby bottles. Efforts are under way by some major canned food production companies to implement alternatives to BPA as can liners. Although governmental regulation lags behind elective manufacturing efforts and individual consumer awareness, resulting in spotty adoption and great confusion for the general consumer, hope exists that gains in BPA and phthalate exposure reduction may be possible in the near term.

Conclusion

Much remains to be learned about the specific biologic activity and clinical consequences of exposures in early life to current ubiquitous EDCs such as BPA and phthalates. Early indications of the potentially toxic effects of these chemicals, especially on the developing fetus and infant, indicate that further epidemiologic, pharmacologic, clinical, and public health research is needed. As was the case with DES, clinical effects of exposure to modern EDCs may be delayed, subtle, and likely multifactorial.[40]

Although specific public health measures to reduce exposures to EDCs, such as the elective removal of BPA from baby bottles and of phthalates from certain types of medical equipment, represent prudent steps, the ubiquitous nature of EDCs in the industrialized environment makes complete avoidance of these chemicals impossible at the present time. Future public health investigation will demonstrate whether such avoidance is necessary.

The characteristics of EDCs, summarized in Figure 40.1, including multiplicity of exposure routes, typically low-level long-term exposure to individual chemicals, nonmonotonic dose response, low-dose effects, and the common occurrence of EDC mixtures, make the paradigms under which carcinogens and reproductive toxicants are currently regulated inadequate for EDCs.[41] New regulatory strategies and new paradigms for toxicity testing required to limit exposures to these widespread, health-impacting toxicants are being developed, as can be seen in the EU Chemical Strategy for Sustainability.[42]

References

1. Colborn T, vom Saal FS, Soto AM. Developmental effects of endocrine-disrupting chemicals in wildlife and humans. *Environ Health Perspect.* 1993;101:378–84.

2. Cook JW, Dodds EC. Sex hormones and cancer producing compounds. *Nature.* 1933;131:205–6.

3. Carson R. *Silent Spring.* Boston, MA: Houghton Mifflin Company; 1962.

4. Randall CL, Baetz RW, Hall DW, Birtch PK. Pregnancies observed in the likely-to-abort patient with or without hormone therapy before or after conception. *Am J Obstet Gynecol.* 1955;69:643–56.

5. Herbst AL, Ulfelder H, Poskanzer DC. Adenocarcinoma of the vagina: association of maternal stilbestrol therapy with tumor appearance in young women. *N Engl J Med.* 1971;284:878–81.

6. IARC. Diethylstilbestrol and diethylstilbestrol dipropionate. *IARC Monogr Eval Carcinog Risk Chem Hum.* 1979;21:173–231.

7. Swan SH. Intrauterine exposure to diethylstilbestrol: long-term effects in humans. *APMIS.* 2000;108:793–804.

8. Titus-Ernstoff L, Hatch EE, Hoover RN, et al. Long-term cancer risk in women given diethylstilbestrol (DES) during pregnancy. *Br J Cancer.* 2001;84(1):126–33. doi:10.1054/bjoc.2000.1521.

9. Titus-Ernstoff L, Troisi R, Hatch EE, et al. Mortality in women given diethylstilbestrol during pregnancy. *Br J Cancer.* 2006;95(1):107–11. doi:10.1038/sj.bjc.6603221.

10. Melnick S, Cole P, Anderson D, Herbst A. Rates and risks of diethylstilbestrol-related clear-cell adenocarcinoma of the vagina and cervix. An update. *N Engl J Med.* 1987;316:514–16.

11. Tournaire M, Devouche E, Espié M, et al. Cancer risk in women exposed to diethylstilbestrol in utero. *Therapie.* 2015 Sep-Oct;70(5):433–41.

12. Troisi R, Titus L, Hatch EE, et al. A prospective cohort study of prenatal diethylstilbestrol exposure and cardiovascular disease risk. *J Clin Endocrinol Metab.* 2018;103(1):206–12. doi:10.1210/jc.2017-01940.

13. Migliaccio S, Newbiold RR, Bullock BC, McLachlan JA, Korach KS. Developmental exposure to estrogens induces persistent changes in skeletal tissue. *Endocrinology.* 1992;130:1756–8.

14. Troisi R, Hyer M, Titus L, et al. Prenatal diethylstilbestrol exposure and risk of diabetes, gallbladder disease, and pancreatic disorders and malignancies. *J Dev Orig Health Dis.* 2021 Aug;12(4):619–26. doi:10.1017/S2040174420000872.

15. Palmer JR, Herbst AL, Noller KL, et al. Urogenital abnormalities in men exposed to diethylstilbestrol in utero: a cohort study. *Environ Health.* 2009 Aug 18;8:37. doi:10.1186/1476-069X-8-37.

16. Hom M, Sriprasert I, Ihenacho U, et al. Systematic review and meta-analysis of testicular germ cell tumors following in utero exposure to diethylstilbestrol. *JNCI Cancer Spect.* 2019;3(3):pkz045.

17. Titus L, Hatch EE, Drake KM, et al. Reproductive and hormone-related outcomes in women whose mothers were exposed in utero to diethylstilbestrol (DES): a report from the US National Cancer Institute DES Third Generation Study. *Reprod Toxicol.* 2019 Mar;84:32–8. doi:10.1016/j.reprotox.2018.12.008.

18. Titus L. Evidence of intergenerational transmission of diethylstilbestrol health effects: hindsight and insight. *Biol Reprod.* 2021;105(3):681–6. doi:10.1093/biolre/ioab153.

19. Calafat AM, Ye X, Wong LY, et al. Exposure of the U. S. population to bisphenol A and 4-tertiary-octylphenol: 2003–2004. *Environ Health Perspect.* 2008;116:39–44.

20. Vandenberg LN, Chauhoud I, Heindel JJ, et al. Urinary, circulating and tissue biomonitoring studies indicate widespread exposure to bisphenol A. *Environ Health Perspect.* 2010;118:1055–70.

21. Braun JM, Yolton K, Dietrich KN, et al. Prenatal bisphenol A exposure and early childhood behavior. *Environ Health Perspec.t* 2009;117:1945–52.

22. Braun JM, Hauser R. Bisphenol A and children's health. *Curr Opin Pediatr.* 2011;23:233–9.

23. Brisken C. Endocrine disruptors and breast cancer. *CHIMIA Int J Chemi.* 2008;62:406–9.

24. Soto Am VL, Vandenberg LN, Maffini MV, Sonnenschein C. Does breast cancer start in the womb? *Basic Clin Pharm Toxicol* 2008;102:125–33.

25. Swan SH, Weiss B. Phthalates: what they are and why they raise concerns about human health. In: Friis RH, ed., *Praeger Handbook of Environmental Health.* Santa Barbara, CA: ABC-CLIO; 2012: 453–73.

26. Foster PMD. Disruption of reproductive development in male rat offspring following in utero exposure to phthalate esters. *Int J Andrology.* 2006;29:140–7.

27. Swan SH, Main KM, Liu F, et al. Decrease in anogenital distance among male infants with prenatal phthalate exposure. *Environ Health Perspect.* 2005;113:1056–61.

28. MacLeod DJ, Sharpe RM, Welsh M, et al. Androgen action in the masculinization programming window and development of male reproductive organs. *Int J Andrology.* 2010;33:279–87.

29. Mendiola J, Stahlhut RW, Jørgensen N, Liu F, Swan SH. Shorter anogenital distance predicts poorer semen quality in young men in Rochester, New York. *Environ Health Perspect.* 2011;119:958–63.

30. Eisenberg ML, Hsieh MH, Walters RC, Krasnow R, Lipshultz LI. The relationship between anogenital distance, fatherhood, and fertility in adult men. *PLoS One.* 2011;6:e18973.

31. Swan SH, Liu F, Hines M, et al. Prenatal phthalate exposure and reduced masculine play in boys. *Int J Androl.* 2010;33:259–69.

32. Miodovnik A, Engel SM, Zhu C, et al. Endocrine disruptors and childhood social impairment. *Neurotoxicology.* 2011;32:261–7.

33. Parks LG, Ostby JS, Lambright CR, et al. The plasticizer diethylhexyl phthalate induces malformations by decreasing fetal testosterone synthesis during sexual differentiation in the male rat. *Toxicol Sci.* 2000;58:339–49.

34. Masuo Y, Ishido M. Neurotoxicity of endocrine disruptors: possible involvement in brain development and neurodegeneration. *J Toxicol Environ Health B.* 2011;14:346–69.

35. Chevrier J, Gunier RB, Bradman A, et al. Maternal urinary bisphenol A during pregnancy and maternal and neonatal thyroid function in the CHAMACOS Study. *Environ Health Perspect.* 2013;121:138–44.

36. Singh S, Shoei-Lung Li. Epigenetic effects of environmental chemicals bisphenol A and phthalates. *Int J Mol Sci.* 2012;13:10143–53.

37. Evans RM, Scholze M, Kortenkamp A. Additive mixture effects of estrogenic chemicals in human cell-based assays can be influenced by inclusion of chemicals with differing effect profiles. *PLoS One.* 2012;7:e43606.

38. Hass U, Scholze M, Christiansen S, et al. Combined exposure to anti-androgens exacerbates disruption of sexual differentiation in the rat. *Environ Health Perspect.* 2007;115:122–8.

39. Houlihan J, Kropp T, Wiles R, Gray S, Campbell C. Body burden: the pollution in newborns. 2005. Available at: http://www.ewg.org/research/body-burden-pollution-newborns.

40. Bergman A, Heindel JJ, Jobling S, Kidd KA, Zoeller RT. eds. *State of the Science of Endocrine Disrupting Chemicals—2012*. Geneva, Switzerland: World Health Organization and United Nations Environment Programme; 2013.

41. Myers JP, Zoeller RT, vom Saal FS. A clash of old and new scientific concepts in toxicity, with important implications for public health. *Environ Health Perspect*. 2009;117:1652–5.

42. European Commission. Chemicals Strategy. 2020. Available at: https://ec.europa.eu/environment/pdf/chemicals/2020/10/Strategy.pdf.

41

Chemical Obesogens and Obesity

Michele La Merrill and Leda Chatzi

The possibility that certain toxic chemicals in the environment may contribute to childhood obesity is a concept that, in the past decade, has gained substantial interest in the research community as well as among policymakers. In the United States, both the White House and the National Institutes of Health have called for further research to examine the role of environmental exposures in the causation of obesity.[1]

To facilitate development of a research strategy on chemical obesogens, the US National Toxicology Program (NTP) organized a Workshop on the "Role of Environmental Chemicals in the Development of Diabetes and Obesity" in 2011. This Workshop updated and critically evaluated the published evidence for associations between environmental chemicals and obesity and/or diabetes and developed a summary of current knowledge of environmental chemical exposures—environmental "obesogens"—in early development that may increase risk of obesity.[1]

While obesity is closely associated with metabolic syndrome and diabetes, these topics are outside the scope of this chapter.

Scope and Nature of the Problem

The global prevalence of children, adolescents, and adults with obesity has been increasing for more than half a century.[2] The increasing global prevalence of people with obesity is cause for grave concern. Risks of diabetes, cardiovascular disease, and cancer are all increased in people with obesity, and these risks appear to be extended to children and adolescence with overweight or obesity.[3]

Fundamentally, obesity results from an imbalance in energy—an imbalance between the number of calories consumed and the number of calories expended. Thus, the increasing global prevalence of obesity is usually attributed to a combination of excess caloric consumption and deficiency in physical activity coupled with underlying genetic susceptibility. The possibility that certain toxic chemicals may also contribute to the risk of obesity by altering energy balance in the body is a newer concept and is the focus of this chapter.

Thousands of new synthetic chemicals have been invented and disseminated into the environment in the past half century. The majority of these new chemicals have not been tested for toxicity (see Chapter 3). Moreover, there are no standardized testing protocols currently available to evaluate whether certain chemicals can contribute to increased adiposity. In the absence of standardized protocols that can be utilized by regulatory agencies as a basis for rule-making, investigation into the potential role of environmental

Michele La Merrill and Leda Chatzi, *Chemical Obesogens and Obesity* In: *Textbook of Children's Environmental Health*,
Second Edition. Edited by: Ruth A. Etzel and Philip J. Landrigan, Oxford University Press. © Oxford University Press 2024.
DOI: 10.1093/oso/9780197662526.003.0041

chemicals in the genesis of obesity has largely been undertaken by academic laboratories and nonregulatory research groups within government.

A number of common chemical contaminants of people, including persistent organic pollutants such as organochlorine pesticides (OCPs), toxic metals, tobacco smoke, and additives used in plastics and cosmetics such as phthalates and phenols, appear to perturb the formation of adipocytes (adipogenesis) as well as energy balance, especially when exposure occurs during in utero or early life.[1,4,5] The body of evidence indicating that environmental chemicals can act as obesogens has increased notably over the past decade, especially for chemical exposures during prenatal development. Fetal exposure is a key feature of the chemical obesogen hypothesis because maternal body burdens of synthetic chemicals pass through the placenta to the developing fetus when tissues essential to the healthy maintenance of energy balance are forming.[5,6] In this chapter, we highlight a subset of synthetic chemical obesogens relevant to children with obesity.

Sources of Exposure to Chemical Obesogens

Maternal Smoking During Pregnancy

In the United States, it is estimated that 7–17% of women smoke during pregnancy, and passive smoking during pregnancy has been detected biochemically in up to 50% of women.[7-9] Maternal tobacco smoking during pregnancy in the United States is more common in relatively young and less educated women.[8] The prevalence of smoking during pregnancy is heterogenous around the world, with it being about eight times higher in the European region compared to the African region.[7] Furthermore, a study found that 50% of their sample of had cotinine levels indicating exposure to second-hand smoke (SHS) when although only 12% reported active smoking during pregnancy.[10] Sources of SHS include fathers who smoked and other smokers who lived with the mother.[11,12]

Cigarette smoke is a complex mixture containing 599 known additives in addition to the chemicals that naturally occur in tobacco.[13] Maternal smoking and SHS exposure during pregnancy and in the first months after childbirth expose children to these chemicals trans-placentally and via maternal breast milk. For further reading on the health hazards of smoking and SHS exposure, please see Chapter 27.

Bisphenol A and Analogs

Bisphenol A (BPA) is a synthetic chemical produced in large quantities primarily for use in the synthesis of epoxy resins and polycarbonate plastics. Polycarbonate plastics have extensive applications, including use in food and beverage packaging such as water bottles and infant bottles, linings of metal food cans, impact-resistant safety equipment, compact discs, and medical devices. Epoxy resins are used as lacquer coating in metal products such as bottle tops, food cans, and pipes that supply drinking water as well as in dental sealants and composites. BPA can leach into food from the epoxy resins used to line food cans and from polycarbonate consumer products such as drinking water bottles. The primary source of most daily exposure to bisphenol A is believed to be orally through the diet.

BPA is also used in some thermal papers such as cash register and credit card receipts.[14] BPA from receipt paper appears to transfer readily from receipts to skin and to be absorbed transdermally.[15] Analogs of BPA are increasingly being used in place of BPA in consumer products, with resulting oral and transdermal exposure. Additional information on BPA can be found in Chapter 40.

Organochlorine Pesticides

OCPs have not been in commercial use in the United States for decades, with dichlorodiphenyltrichloroethane (DDT) and hexachlorobenzene (HCB) use ending in the 1970s. Most adults born in the United States prior to the ban of OCPs were highly exposed to OCPs during the developmental window that programs lifetime metabolic function—prenatal to adolescence—and are now of age for heightened risk of associated chronic diseases such as obesity.[16] OCPs or their metabolites are found in most people around the world today because they are resistant to breakdown, are transported long distances in the environment, and are commonly found throughout the global food supply.[17] Their concentrations are particularly high in animal fats because OCPs are lipophilic, although they are detected in human foods of all types.[18]

Globally, few OCPs are still being used and manufactured. For example, under certain circumstances the World Health Organization (WHO) recommends the use of DDT for indoor malarial vector control. Migration globalizes these ongoing DDT exposures among people from areas using or manufacturing DDT.[19] The ban of DDT in most countries has been a public health success in that DDT is infrequently detected in people who have been born and lived in places where DDT is banned. However, DDT is metabolized in the body to dichlorodiphenyldichloroethylene (DDE). DDE is even more persistent than DDT and so DDE can be found in people even when DDT is no longer in use or otherwise detected. This is due to DDE contamination of the food supply, particularly foods rich in animal fat. There are race and ethnic disparities in exposure to both DDT and DDE. For example, White people have lower levels of DDE exposure than do Black, non-White Hispanic or Asian people.[19,20]

Epidemiological Evidence Associating Chemical Exposures with Obesity

Maternal Smoking During Pregnancy

Maternal smoking during pregnancy increases risk of low birth weight and of small-for-gestational-age births. Paradoxically, smoking during pregnancy is also a consistent risk factor for childhood overweight and obesity.[4] In a meta-analysis, maternal smoking during pregnancy was associated with a 1.64 (pooled odds ratio [OR]; 95% confidence interval [CI]: 1.42, 1.90) increased odds of obesity (body mass index [BMI] >95th percentile) relative to normal weight among toddlers, children, adolescents, and adults.[21] The modest evidence for publication bias was insufficient to negate the overall increased risk of obesity and overweight associated with maternal smoking during pregnancy. Furthermore, the association between maternal smoking during pregnancy and offspring obesity is unlikely to be due to measurement bias or the limitations of BMI as an estimate of body fat. When adiposity is measured directly by magnetic resonance imaging, adolescents exposed to maternal smoking during pregnancy had 33% and 26% higher intra-abdominal and subcutaneous fat, respectively, than adolescents without exposure to maternal smoking during pregnancy.[22] Even SHS exposure during the prenatal period is associated with increased childhood obesity in a meta-analysis of eight studies (pooled OR = 1.90; 95% CI: 1.23, 2.94).[12]

In the first "exposome" evaluation of potential obesogens, where the potential impact of more than 100 environmental chemical and nonchemical stressors on childhood obesity was assessed, the only two prenatal stressors associated with childhood obesity were maternal smoking and maternal urinary cotinine concentration.[23] Maternal active smoking was associated with an increase in the child's z BMI score of 0.28 (95% CI: 0.09, 0.48), and maternal secondhand smoking with an increase of 0.16 (95% CI: -0.002, 0.32), compared to nonsmoking.[23] Identification of active smoking and SHS exposure through maternal urinary cotinine levels gave similar associations.

Fewer studies have evaluated the impact of SHS on childhood obesity. A recent systematic review and meta-analysis showed a positive association between prenatal exposure to SHS and childhood obesity (OR = 1.905; 95% CI: 1.23, 2.94) and no association between SHS exposure and overweight (OR = 1.51; 95% CI: 0.49, 4.59).[12] Studies have also showed an association between child exposure to SHS and childhood obesity,[24–26] but have found it difficult to disentangle effects of child SHS exposure from correlated maternal smoking during pregnancy. In the study by Vrijheid et al, adjustment of the childhood associations for maternal pregnancy smoking did not change results. The study showed also that the association between child cotinine levels and BMI was observed exclusively in the low/medium maternal educational classes where detectable child cotinine levels were far more prevalent (30% vs. 6% in high maternal educational class).

Bisphenol A and Analogs

The association between exposure to BPA and measures of adiposity has been examined in multiple meta-analyses of epidemiological studies and each found a positive association between BPA exposure and various measures of obesity.[27–29] For example, a meta-analysis of 13 published epidemiology studies concluded that children with elevated exposures to BPA had a significantly greater risk of childhood obesity compared to children with lower BPA exposures (pooled OR = 1.57; 95% CI: 1.10, 2.23).[27] Another meta-analysis (n = 10 epidemiology studies) estimated that, for each increase of 1 ng BPA per mL urine, the risk of obesity increased by 11%.[28]

Emerging evidence in human studies indicates that BPA replacements, such as bisphenol F (BPF) are also associated with obesity in children and adolescents.[30]

Organochlorine Pesticides

A recent systematic review and meta-analysis of the associations between prenatal exposure to OCPs and childhood obesity showed a positive association of prenatal exposure to DDE with BMI-z (beta = 0.12; 95% CI: 0.03, 0.21) (Figure 41.1A), with no

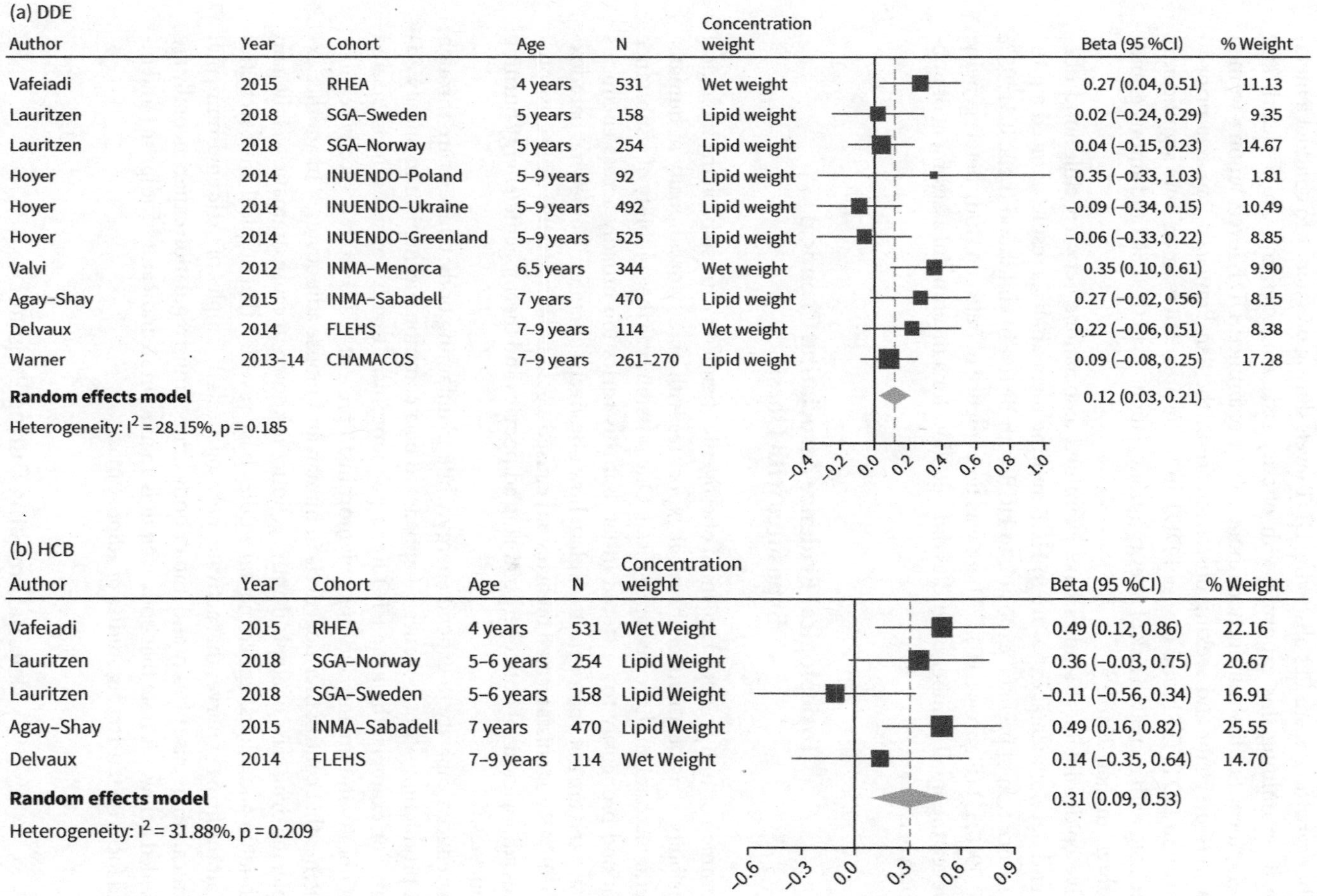

(a) DDE

Author	Year	Cohort	Age	N	Concentration weight	Beta (95 %CI)	% Weight
Vafeiadi	2015	RHEA	4 years	531	Wet weight	0.27 (0.04, 0.51)	11.13
Lauritzen	2018	SGA–Sweden	5 years	158	Lipid weight	0.02 (−0.24, 0.29)	9.35
Lauritzen	2018	SGA–Norway	5 years	254	Lipid weight	0.04 (−0.15, 0.23)	14.67
Hoyer	2014	INUENDO–Poland	5–9 years	92	Lipid weight	0.35 (−0.33, 1.03)	1.81
Hoyer	2014	INUENDO–Ukraine	5–9 years	492	Lipid weight	−0.09 (−0.34, 0.15)	10.49
Hoyer	2014	INUENDO–Greenland	5–9 years	525	Lipid weight	−0.06 (−0.33, 0.22)	8.85
Valvi	2012	INMA–Menorca	6.5 years	344	Wet weight	0.35 (0.10, 0.61)	9.90
Agay–Shay	2015	INMA–Sabadell	7 years	470	Lipid weight	0.27 (−0.02, 0.56)	8.15
Delvaux	2014	FLEHS	7–9 years	114	Wet weight	0.22 (−0.06, 0.51)	8.38
Warner	2013–14	CHAMACOS	7–9 years	261–270	Lipid weight	0.09 (−0.08, 0.25)	17.28

Random effects model   0.12 (0.03, 0.21)

Heterogeneity: $I^2 = 28.15\%$, p = 0.185

(b) HCB

Author	Year	Cohort	Age	N	Concentration weight	Beta (95 %CI)	% Weight
Vafeiadi	2015	RHEA	4 years	531	Wet Weight	0.49 (0.12, 0.86)	22.16
Lauritzen	2018	SGA–Norway	5–6 years	254	Lipid Weight	0.36 (−0.03, 0.75)	20.67
Lauritzen	2018	SGA–Sweden	5–6 years	158	Lipid Weight	−0.11 (−0.56, 0.34)	16.91
Agay–Shay	2015	INMA–Sabadell	7 years	470	Lipid Weight	0.49 (0.16, 0.82)	25.55
Delvaux	2014	FLEHS	7–9 years	114	Wet Weight	0.14 (−0.35, 0.64)	14.70

Random effects model   0.31 (0.09, 0.53)

Heterogeneity: $I^2 = 31.88\%$, p = 0.209

Figure 41.1. Meta-analysis results of epidemiology studies showing a positive association of prenatal exposure to DDE and HCB with BMI-z in childhood.[28]

evidence of important between-study heterogeneity.[31] The same meta-analysis showed no significant association between prenatal exposure to DDE and BMI-z during infancy (0–2 years), suggesting that the effects of prenatal exposure to DDE are more pronounced in late childhood.

There are too few studies observing DDT levels during pregnancy to conduct a meta-analysis examining this exposure with obesity risk. However, the Child Health and Development Studies multigenerational cohort examined DDT levels among women who were pregnant and seeking obstetric care in the San Francisco Bay area prior to the ban on DDT. The blood levels of DDT in those pregnant women were significantly associated with increased risk of obesity among daughters in their 50s and among granddaughters in their 20s.[16,32]

Five epidemiological studies have examined associations between childhood obesity and developmental exposure to HCB and the meta-analysis results showed a positive association of prenatal exposure to HCB with BMI-z in childhood (beta: 0.31; 95% CI: 0.09, 0.53) (Figure 41.1B).[31] There was no evidence of between study heterogeneity; however, the small number of studies did not allow for a meaningful assessment of publication bias.

Toxicological Evidence Associating Chemical Exposures with Obesity

A common or unifying mechanism of obesogenic chemicals can establish the biological plausibility of associations of chemical exposures with obesity or adiposity in humans. Several mechanisms have been proposed. One possibility is the solubility of a substance in fat and propensity to be stored in fat, but this varies too broadly across chemicals that are and are not obesogens to explain how obesogens operate.[33] Somewhat paradoxically, PPARγ agonists treated diabetes but caused a gain in fat mass; this drug class promotes adipogenesis of both energy storing adipocytes and thermogenic/energy burning adipocytes.[34]

A reduced capacity to expend energy while maintaining body temperature through heat production—thermogenesis—appears to be a common mechanism in obesity etiology. For example, the gene FTO has a polymorphism that is consistently associated with obesity in humans, and this polymorphism prevents FTO from correctly regulating genes that control thermogenesis in human and mouse adipocytes.[35] In another example, the drug class of atypical antipsychotics is known to cause dramatic weight gain in children and adults. Experimental studies have revealed that atypical antipsychotics impair the firing of nerves that activate thermogenesis. Examples of toxic environmental chemicals that cause obesity and impair body temperature regulation experimentally are provided below. We note that body weight is an inaccurate indicator of adiposity and instead focus on illustrating studies of adipose mass.

Maternal Smoking During Pregnancy

The toxicological literature strongly supports the epidemiological findings of increased adiposity in offspring exposed to maternal cigarette smoke during pregnancy.[4] All of the developmental studies of the impact of cigarette smoking on obesity that have been conducted in rats have focused on nicotine exposure, and most of these studies found that perinatal exposure to nicotine increased fat mass.[4] Although body weight was also generally increased in rodents with perinatal exposure to nicotine or cigarette smoke, the effect size was more modest than that seen for adiposity. This may reflect body weight's inferior status as an indicator of rodent adiposity. Body composition effects were frequently observed by the time of weaning and persisted into adulthood. Not all of the studies reporting increased body weight in adulthood included an assessment of food intake, but food intake was unaffected in studies that did examine this endpoint.

Two studies examined whether a diet high in fat would modify the effects of cigarette smoke or nicotine.[36,37] In both these studies, a high-fat diet exacerbated the effects of smoking and nicotine on the body mass of male rodents. Furthermore, male rats exposed to both prenatal nicotine and adult high-fat diet had increased fat mass, decreased physical activity, and decreased body temperature.[36]

Bisphenol A and Analogs

A meta-analysis of extensive experimental studies revealed that developmental exposure to BPA increases adiposity in both mice and rats.[38] An explanation of how BPA acts as an obesogen is that BPA reduces body heat production (thermogenesis).[39] Furthermore, chemicals that have been used as BPA replacements, such as bisphenol S (BPS), also cause obesity in rodents.[40] Indeed, BPA, BPS, and BPF all downregulate levels of micro-RNA 26, a regulator of thermogenesis.[41] Experiments with cultured preadipocytes and pluripotent progenitor cells from rodents and humans further demonstrate that BPA, BPS, and BPF act as obesogens by causing the accumulation of lipids in these differentiating cells.[42]

Organochlorine Pesticides

A handful of studies have exposed pregnant rodents to DDT and observed obesity in subsequent generations of both rats and mice.[43] Because DDT is metabolized to DDE, doses of DDT result in exposure to both DDT and DDE; this merits bearing in mind given that many human studies report on DDE exposure while DDT is not detected. Female mice dosed with DDT or DDE during the perinatal period have decreased body temperature and increased adipose tissue as adults.[44,45] This is, at least in part, caused by decreased sympathetic innervation of thermogenic adipose tissue and decreased synaptic connections in the sympathetic ganglia that innervates thermogenic adipose tissue.[45] The expression of many genes that instigate thermogenesis was substantially reduced in the adipose tissue of adult female mice that were dosed with DDT during the perinatal period.[44] These changes in thermogenic gene expression of mouse adipose tissue likely result both from decreased sympathetic tone as well as adipocyte cell autonomous toxicities. The later is supported by a handful of studies that show that DDT increases the accumulation of fat in differentiating adipocytes.[43]

We are unaware of any direct experimental investigation of a relationship between developmental exposure to HCB OCPs and obesity. In rats, HCB reduced deiodinase type II activity in thermogenic adipose tissue, suggestive of reduced triiodothyronine (T3) in the adipose.[46] Because T3 is known to activate thermogenesis in adipose tissue, such a change is expected to reduce metabolic rate and increase risk of obesity. While this has not yet been confirmed, a human study of a mixture of organochlorines, including DDT, DDE, and HCB, found that elevated organochlorines attenuated the beneficial effect of weight loss on the resting metabolic rate.[47]

Systems Affected

Obesity is the result of energy imbalance. Energy balance is dynamically regulated by multiple organs and tissues. Constant coordination between adipose tissue, liver, muscle, thyroid, adrenals, pancreas, hypothalamus, pituitary, and the peripheral nervous system is involved in the homeostasis of lipids, glucose, and energy. All of these tissues and organs may be targets of metabolic perturbation by environmental influences, including toxic chemicals.

Vulnerability to environments that may influence risk of obesity is especially great in early development. Exposures to toxic chemicals during the formation and development of tissues such as adipose and brain can increase risk of obesity. It is noteworthy that adipose tissue, skeletal muscle, pancreas, and brain continue to develop postnatally, extending the window of susceptibility of these tissues well into childhood.

The metabolic programming of infants by their environment during vulnerable windows of pre- and perinatal development can exert powerful influence over energy balance and risk of obesity that lasts throughout life. The impact of the early nutritional environments is dramatically illustrated by study of the consequences on children's health of severe maternal caloric restriction in pregnancy during the Dutch Famine of World War II. Maternal undernutrition during the Dutch Famine was associated with a greater incidence of obesity in the adult children of the mothers who were malnourished, and this risk persisted throughout life.[48] Another example of metabolic programming in early development is that babies born small for gestational age have an increased lifelong risk of obesity. A third example is that higher maternal pre-pregnancy BMI is associated with increased birth weight, increased fat mass at birth, and increased BMI in childhood and in adult life. A fourth example is the association between prenatal DDT exposures and increased obesity in women in their 50s.[16]

Obesity increases risk of many diseases and abnormalities: hyperlipidemia, hypertension, fatty liver, insulin resistance, diabetes, metabolic syndrome, asthma, sleep apnea, cancers, and cardiovascular diseases. Hence many of the body's systems can be affected as a consequence of obesity. Obesity creates a chronic low-grade inflammatory state characterized by inflammatory cell infiltrates and elevated cytokine signaling.[49] This chronic inflammatory state may contribute to the observed increased incidence of asthma in obese children. Furthermore, adipose tissue is an active endocrine organ, and the production of hormones such as estrogen and leptin by adipocytes is more pronounced in obese people. This excess estrogen may also explain in part why childhood obesity is also associated with earlier onset of puberty.

Clinical Effects

Alert pediatricians have the opportunity through their clinical practice to identify new chemical causes of childhood obesity. Pediatricians may observe, for example, that excessive weight gain or change in growth trajectory is associated with use of certain prescribed medications, including off-label uses. Observation of such sentinel events by astute clinicians has often provided the first recognition of new chemical hazards in the modern environment (see Chapter 3).

Childhood obesity is most often assessed in clinical practice by measurement of BMI. It is important to bear in mind, however, that BMI is not a direct measure of adiposity because it does not distinguish between fat and lean mass. Human adiposity can be assessed indirectly by measuring waist circumference, skin fold thickness, and fluid displacement and directly assessed by magnetic resonance imaging (MRI) and dual emission x-ray absorbance (DXA).

The majority of the epidemiological studies reviewed in this chapter have relied on indirect assessments of obesity. Therefore it is perhaps not surprising that associations between these indirect measures of adiposity (i.e., body fat, abdominal fat, skin fold thickness) and early environmental exposures at times appear inconsistent. Given the limitations of indirect measures of adiposity, direct measures of adiposity/fat mass are important to clearly identify the contribution of environmental chemicals to childhood obesity.

Prevention

Obesity is a complex disorder of multifactorial origin. Known risk factors include excessive caloric consumption, insufficient physical activity, and poorly designed built environments. Multiple strategies have been developed and evaluated for addressing those risk factors, and those strategies are discussed elsewhere in this volume.

The possibility that exposures to certain toxic chemicals, especially exposures to such chemicals in early development, may also contribute to obesity is still a new concept. It is not well understood and requires further research. But it is a concept that is supported by a growing body of epidemiological and toxicological literature. This evidence suggests the need for prudent action to reduce exposures to chemical obesogens, especially during windows of vulnerability in early life.

For preventing or minimizing exposure of children and pregnant women to known and suspected chemical obesogens, three prudent strategies are suggested:.

Dietary modification in early childhood as well as before and during pregnancy is one approach to reducing exposures to chemical obesogens. Fatty animal-based foods contain high levels of many POPs, some of which may be chemical obesogens. As consumption of fatty animal-based foods increases, so do the levels of many POPs in human serum and milk. Children who consume large quantities of fatty-animal based foods may be at increased risk of obesity not only because of the caloric density of their diets, but also because of the "density" of their exposures to toxic chemicals in these foodstuffs. Prudent avoidance of a high-fat diet during pregnancy and childhood has multiple health benefits, and reduction in exposure to potential chemical obesogens may be a further advantage.

Reducing exposure to BPA during pregnancy and childhood is another strategy for prudent avoidance of suspect chemical obesogens. Although developmental exposure to BPA has not been sufficiently studied to directly assess its effect on risk of human obesity, numerous rodent studies strongly indicate that developmental exposure to BPA increases adiposity. Even though this evidence is incomplete, it would be prudent at this time for children, adolescents, and pregnant women to exercise the following precautions to reduce BPA exposure:

1. *Minimize use of polycarbonate plastics containing BPA.* When such plastics are used, avoid consuming food or beverages from scratched or heated polycarbonate plastics because heating increases the leakage of BPA from the plastic into the food.
2. *Minimize consumption of acidic foods, such as tomatoes and citrus, from metal cans* because acidity has been shown to increase the release of BPA from the lining of these cans. When possible buy fresh food.
3. *Minimize handling thermal paper used in cash register and credit card receipts,* and promptly wash hands after touching these receipts. Many of these receipts are printed on paper that contains BPA.

Smoking cessation. The NTP Workgroup on Chemical Obesogens concluded that the "epidemiological data strongly support a positive association between maternal smoking and increased risk of obesity or overweight in offspring and that this association is likely causal."[1] The increased adiposity consistently seen in the children, adolescent, and adult offspring of women who smoke during pregnancy is supported by experimental studies in rodents.

Given that maternal smoking in pregnancy is associated with premature birth, small for gestational age, and obesity in offspring, as well as numerous health risks to the mother herself, no woman should smoke while pregnant.

Additional published studies suggest that maternal exposure during pregnancy to secondhand cigarette smoke from a smoking parent, partner, or other person in the home increases risk of childhood obesity. These sources of exposure to cigarette smoke also should be avoided by pregnant women.

References

1. Thayer KA, Heindel JJ, Bucher JR, Gallo MA. Role of environmental chemicals in diabetes and obesity: a National Toxicology Program workshop review. *Environ Health Perspect.* 2012;120:779–89.
2. Collaboration NCDRF. Worldwide trends in body-mass index, underweight, overweight, and obesity from 1975 to 2016: a pooled analysis of 2416 population-based measurement studies in 128.9 million children, adolescents, and adults. *Lancet.* 2017;390:2627–42.
3. Twig G, Yaniv G, Levine H, et al. Body-mass index in 2.3 million adolescents and cardiovascular death in adulthood. *N Engl J Med.* 2016;374:2430–40.
4. Behl M, Rao D, Aagaard K, et al. Evaluation of the association between maternal smoking, childhood obesity, and metabolic disorders: a national toxicology program workshop review. *Environ Health Perspect.* 2013;121:170–80.
5. La Merrill M, Birnbaum LS. Childhood obesity and environmental chemicals. *Mt Sinai J Med.* 2011;78:22–48.
6. Needham LL, Grandjean P, Heinzow B, et al. Partition of environmental chemicals between maternal and fetal blood and tissues. *Environ Sci Technol.* 2011;45:1121–6.

7. Lange S, Probst C, Rehm J, Popova S. National, regional, and global prevalence of smoking during pregnancy in the general population: a systematic review and meta-analysis. *Lancet Glob Health*. 2018;6:e769–76.

8. Kondracki AJ. Prevalence and patterns of cigarette smoking before and during early and late pregnancy according to maternal characteristics: the first national data based on the 2003 birth certificate revision, United States, 2016. *Reprod Health*. 2019;16:142.

9. Pineles BL, Park E, Samet JM. Systematic review and meta-analysis of miscarriage and maternal exposure to tobacco smoke during pregnancy. *Am J Epidemiol*. 2014;179:807–23.

10. Braun JM, Daniels JL, Poole C, et al. Prenatal environmental tobacco smoke exposure and early childhood body mass index. *Paediatr Perinat Epidemiol*. 2010;24:524–34.

11. Harris HR, Willett WC, Michels KB. Parental smoking during pregnancy and risk of overweight and obesity in the daughter. *Int J Obesity*. 2013;37:1356–63.

12. Qureshi R, Jadotte Y, Zha P, et al. The association between prenatal exposure to environmental tobacco smoke and childhood obesity: a systematic review. *JBI Database System Rev Implement Rep*. 2018;16:1643–62.

13. Rogers JM. Tobacco and pregnancy. *Reprod Toxicol*. 2009;28:152–60.

14. Biedermann S, Tschudin P, Grob K. Transfer of bisphenol A from thermal printer paper to the skin. *Anal Bioanal Chem*. 2010;398:571–6.

15. Zalko D, Jacques C, Duplan H, Bruel S, Perdu E. Viable skin efficiently absorbs and metabolizes bisphenol A. *Chemosphere*. 2011;82:424–30.

16. La Merrill MA, Krigbaum NY, Cirillo PM, Cohn BA. Association between maternal exposure to the pesticide dichlorodiphenyltrichloroethane (DDT) and risk of obesity in middle age. *Int J Obesity*. 2020;44:1723–32.

17. Guo W, Pan B, Sakkiah S, et al. Persistent organic pollutants in food: contamination sources, health effects and detection methods. *Int J Environ Res Public Health*. 2019;16.

18. La Merrill M, Emond C, Kim MJ, et al. Toxicological function of adipose tissue: focus on persistent organic pollutants. *Environ Health Perspect*. 2013;121:162–9.

19. La Merrill MA, Johnson CL, Smith MT, et al. Exposure to persistent organic pollutants (POPs) and their relationship to hepatic fat and insulin insensitivity among Asian Indian immigrants in the United States. *Environ Sci Technol*. 2019;53:13906–18.

20. Nguyen VK, Kahana A, Heidt J, et al. A comprehensive analysis of racial disparities in chemical biomarker concentrations in United States women, 1999–2014. *Environ Int*. 2020;137:105496.

21. Ino T. Maternal smoking during pregnancy and offspring obesity: meta-analysis. *Pediatr Int*. 2010;52:94–9.

22. Syme C, Abrahamowicz M, Mahboubi A, et al. Prenatal exposure to maternal cigarette smoking and accumulation of intra-abdominal fat during adolescence. *Obesity (Silver Spring)*. 2010;18:1021–5.

23. Vrijheid M, Fossati S, Maitre L, et al. Early-life environmental exposures and childhood obesity: an exposome-wide approach. *Environ Health Perspect*. 2020;128:67009.

24. McConnell R, Shen E, Gilliland FD, et al. A longitudinal cohort study of body mass index and childhood exposure to secondhand tobacco smoke and air pollution: the Southern California Children's Health Study. *Environ Health Perspect*. 2015;123(4):360–6. doi:10.1289/ehp.1307031.

25. Raum E, Küpper-Nybelen J, Lamerz A, et al. Tobacco smoke exposure before, during, and after pregnancy and risk of overweight at age 6. *Obesity (Silver Spring)*. 2011;19(12):2411–7. doi:10.1038/oby.2011.129.

26. Robinson O, Martínez D, Aurrekoetxea JJ, et al. The association between passive and active tobacco smoke exposure and child weight status among Spanish children. *Obesity (Silver Spring)*. 2016;24(8):1767–77. doi:10.1002/oby.21558.

27. Kim KY, Lee E, Kim Y. The association between bisphenol A exposure and obesity in children: a systematic review with meta-analysis. *Int J Environ Res Public Health*. 2019;16 (14):2521.

28. Wu W, Li M, Liu A, et al. Bisphenol A and the risk of obesity a systematic review with meta-analysis of the epidemiological evidence. *Dose Response*. 2020;18:1559325820916949.

29. Ribeiro CM, Beserra BTS, Silva NG, et al. Exposure to endocrine-disrupting chemicals and anthropometric measures of obesity: a systematic review and meta-analysis. *BMJ Open*. 2020;10:e033509.

30. Liu B, Lehmler HJ, Sun Y, et al. Association of bisphenol A and its substitutes, bisphenol F and bisphenol S, with obesity in United States children and adolescents. *Diabetes Metab J*. 2019;43:59–75.

31. Stratakis N, Rock S, La Merrill MA, et al. Prenatal exposure to persistent organic pollutants and childhood obesity: a systematic review and meta-analysis of human studies. *Obes Rev*. 2021:e13383.

32. Cirillo PM, La Merrill MA, Krigbaum NY, Cohn BA. Grandmaternal perinatal serum DDT in relation to granddaughter early menarche and adult obesity: three generations in the child health and development studies cohort. *Cancer Epidemiol Biomarkers Prevent*. 2021;30(8):1480–88.

33. Griffin MD, Pereira SR, DeBari MK, Abbott RD. Mechanisms of action, chemical characteristics, and model systems of obesogens. *BMC Biomed Eng*. 2020;2:6.

34. Tontonoz P, Spiegelman BM. Fat and beyond: the diverse biology of PPARgamma. *Annu Rev Biochem*. 2008;77:289–312.

35. Claussnitzer M, Dankel SN, Kim KH, et al. FTO obesity variant circuitry and adipocyte browning in humans. *N Engl J Med*. 2015;373:895–907.

36. Somm E, Schwitzgebel VM, Vauthay DM, et al. Prenatal nicotine exposure alters early pancreatic islet and adipose tissue development with consequences on the control of body weight and glucose metabolism later in life. *Endocrinology*. 2008;149:6289–99.

37. Ng SP, Conklin DJ, Bhatnagar A, Bolanowski DD, Lyon J, Zelikoff JT. Prenatal exposure to cigarette smoke induces diet- and sex-dependent dyslipidemia and weight gain in adult murine offspring. *Environ Health Perspect*. 2009;117:1042–8.

38. Wassenaar PNH, Trasande L, Legler J. Systematic review and meta-analysis of early-life exposure to bisphenol A and obesity-related outcomes in rodents. *Environ Health Perspect*. 2017;125:106001.

39. Batista TM, Alonso-Magdalena P, Vieira E, et al. Short-term treatment with bisphenol-A leads to metabolic abnormalities in adult male mice. *PloS One*. 2012;7:e33814.

40. Brulport A, Vaiman D, Chagnon MC, Le Corre L. Obesogen effect of bisphenol S alters mRNA expression and DNA methylation profiling in male mouse liver. *Chemosphere*. 2020;241:125092.

41. Verbanck M, Canouil M, Leloire A, et al. Low-dose exposure to bisphenols A, F and S of human primary adipocyte impacts coding and non-coding RNA profiles. *PloS One.* 2017;12:e0179583.

42. Mohajer N, Du CY, Checkcinco C, Blumberg B. Obesogens: how they are identified and molecular mechanisms underlying their action. *Front Endocrinol (Lausanne).* 2021;12:780888.

43. Cano-Sancho G, Salmon AG, La Merrill MA. Association between exposure to p,p'-DDT and its metabolite p,p'-DDE with obesity: integrated systematic review and meta-analysis. *Environ Health Perspect.* 2017;125:096002.

44. La Merrill M, Karey E, Moshier E, et al. Perinatal exposure of mice to the pesticide DDT impairs energy expenditure and metabolism in adult female offspring. *PloS One.* 2014;9:e103337.

45. vonderEmbse AN, Elmore SE, Jackson KB, et al. Developmental exposure to DDT or DDE alters sympathetic innervation of brown adipose in adult female mice. *Environ Health.* 2021;20:37.

46. Alvarez L, Hernandez S, Martinez-de-Mena R, Kolliker-Frers R, Obregon MJ, Kleiman de Pisarev DL. The role of type I and type II 5' deiodinases on hexachlorobenzene-induced alteration of the hormonal thyroid status. *Toxicology.* 2005;207:349–62.

47. Tremblay A, Pelletier C, Doucet E, Imbeault P. Thermogenesis and weight loss in obese individuals: a primary association with organochlorine pollution. *Int J Obes Relat Metab Disord.* 2004;28:936–9.

48. Roseboom T, de Rooij S, Painter R. The Dutch famine and its long-term consequences for adult health. *Early Hum Dev.* 2006;82:485–91.

49. Nyambuya TM, Dludla PV, Mxinwa V, Nkambule BB. Obesity-related asthma in children is characterized by T-helper 1 rather than T-helper 2 immune response: a meta-analysis. *Ann Allergy Asthma Immunol.* 2020;125:425–32 e4.

42

Environmental Carcinogens and Childhood Cancer

Kurt Straif

Introduction and Overview

Cancer is the second leading cause of death globally and in many, particularly high-income countries ranks first in the age group of 30–69 years.[1] Childhood cancers are distinctively different from adult cancers in many aspects: certain histologies occur only in children, while the vast majority of others occur among adults; for histologies diagnosed in children and in adults, their clinical behavior, descriptive epidemiology, and risk factor profiles—genetic and environmental—may be quite different; and even among children, noticeable differences are observed across age groups. This chapter provides an overview on the descriptive epidemiology of childhood cancers and established and emerging environmental risk factors. Environmental factors are broadly defined and include any factor that is not genetic and could therefore, at least in principle, be mitigated for the purpose of primary prevention of childhood cancer. For an overview on genetic risk factors, clinical diagnosis, treatment, and survival of childhood cancers the reader is referred to other overviews.[2,3]

Classification and Descriptive Epidemiology

A consensus classification of tumors is a necessary prerequisite for diagnosis and treatment, as well as for descriptive and etiologic epidemiology. The World Health Organization (WHO) Classification of Tumors provides the scientific foundation; international expert groups periodically review and evaluate the published literature and develop and update consensus classifications that are published as editions of the WHO Classification of Tumors. Typically, these co-called WHO Blue Books (currently in their 5th edition) are organized by organ system with emphasis on primary site for adults; childhood cancers have been integrated by organ system. In turn, the WHO Blue Books inform the evolution and future editions of the International Classification of Diseases (ICD; currently in its 11th revision).

However, the classification of childhood cancer is traditionally based on tumor morphology and primary site with an emphasis on morphology, a system that is thought to be better adapted to classify tumors occurring in adolescence than the ICD coding system used in Cancer in 5 Continents (CI5). Therefore the classification of childhood tumors has typically relied on the International Childhood Cancer Classification (ICCC), which

Kurt Straif, *Environmental Carcinogens and Childhood Cancer* In: *Textbook of Children's Environmental Health*,
Second Edition. Edited by: Ruth A. Etzel and Philip J. Landrigan, Oxford University Press. © Oxford University Press 2024.
DOI: 10.1093/oso/9780197662526.003.0042

is derived from the International Classification of Oncology (ICD-O), with its latest revision ICCC-3 being based on ICD-O3.

With the ongoing development of the 5th edition, the Blue Books systematically incorporate, as far as supported by scientific evidence, morphology, immunohistochemistry, and molecular characteristics for all tumors, and—as a new addition—a volume dedicated to childhood cancer will become part of the series.[4]

Cancer Registries and Descriptive Cancer Epidemiology

Population-based cancer registries utilize international cancer classification systems and record incident cases of cancer for a defined population. Historically, first (unsuccessful) efforts to undertake a cancer survey date back to the first half of the 18th century in London, with more attempts and refinements at city, regional, or national level in the first half of the 20th century. In 1946, a conference on cancer registration in Copenhagen recommended the worldwide establishment of cancer registries to the Interim Commission for the WHO with adherence to the following principles: collection of data about cancer patients from as many different countries as possible, such data should be recorded on an agreed plan to be comparable, each nation should have a central registry for the recording and collection, and an international body should be established to correlate the data and statistics.

Soon after the establishment of the International Agency for Research on Cancer (IARC) as a specialized cancer research center of the WHO in 1965, and the foundation of the International Association of Cancer Registries (IACR) in 1966, the first volume of Cancer Incidence in Five Continents was published, since then updated about every 5 years. The series provides high-quality and comparable data on cancer incidence for countries for which data have been made available by population-based cancer registries and passed detailed quality checks. For instance Volume 11[5] contains information from 343 cancer registries in 65 countries for cancers diagnosed from 2008 to 2012. Despite the proportion of the world's total population covered by the registries, Volume 10[6] covers only 14%, and coverage varies significantly between 95% for North America and 2% for Africa. Data include statistics on cancer occurrence by anatomic site, stage at diagnosis, and survival.

Similarly informative data from international childhood cancer registries were only published more than two decades after the first volume of CI5.[7] In the book's preface Dr. Tomatis, the late Director of IARC, noted that this volume was the most extensive collection of data on the incidence of cancer in childhood that has ever been published and highlighted the importance of cancer registries for the study of geographical and ethnic variation in the risk of childhood cancer and its great value in formulating hypotheses concerning the relative contribution of genetic susceptibility and environmental exposure to etiology. A second volume was published 10 years later and covered the age range 0–14 years, classified tumors primarily by histology rather than by the site of the tumor, and contained data from specialized children's tumor registries that are not included in Cancer Incidence in Five Continents. In summary, the overall incidence of childhood tumors was only about 1% of that in adults, and the most common adult cancers hardly ever occurred in children.[8]

The third and most recent International Incidence of Childhood Cancer[9] includes data from more than 300 registries from five continents that met the inclusion criteria and provided high-quality comparable datasets. The targeted age range was extended to 19 years, accounting for the observation that some tumor types that are common among children peak in the age group of 15–19 years. Some experts argue that the upper age limit should be even further increased to the age of 24 years because the distribution of malignancies occurring in 15- to 24-year-olds more closely resembles that seen in younger children than in older adults.[2] The target period started in 1990 and the length of the reporting period was up to 20 years, and tumors were classified according to the ICCC-3. Overall, compared to cancer in adults, cancers in children were marked by lower incidence rates, a relatively small variety of typical histologies, and often favorable survival, particularly in resource-rich countries, which results in long-term survivorship issues with special medical, ethical, psychological, and societal concerns. Despite these encouraging data, childhood cancers are the most common cause of disease-related death in children.[10]

As noted above, cancers occurring in childhood and adolescence differ markedly from the cancer patterns in adults. Globally, the most common cancers in children are leukemia and lymphoma, while major cancers that are typical among adults, such as cancers of the lung, breast or colon, are extremely rare in children. Among the younger children (ages 0–14 years) leukemia and tumors of the central nervous system (CNS) are most frequent and lymphoma ranks third. Among adolescents (ages 15–19 years), incidence rates of lymphoma surpass those of leukemia and tumors of the CNS; furthermore, epithelial tumors and melanoma emerge in this older age group. Similarly, as in adults, important variations are seen across regions and ethnicities. By region, the highest age-standardized rates for any group of childhood tumors are for epithelial tumors and melanoma in Oceania (>70/1,000,000 person-years) compared to fewer than 15/1,000,000 person-years for the same group of tumors in India. Note that some of the low rates recorded by population-based cancer registries in some low-income countries, such as leukemia and CNS tumors in sub-Saharan Africa, may at least partially result from underdiagnosis (Figure 42.1). Childhood cancer rates by ethnicity in the United States indicate highest leukemia rates in White Hispanics among both age groups, while the highest rates for CNS cancer in the younger and the highest lymphoma rates in the older age group are reported for White Non-Hispanics ("not shown").[1]

Trends of incidence rates for childhood tumor over time have been reported for the IICC-3 subset of European countries for calendar years 1991–2010.[11] In children (ages 0–14 years) and in adolescents (age 15–19 years), the overall incidence of tumors increased by 0.54% (0.44–0.65) per year, and by 0.96% (0.73–1.19) per year, respectively. Specifically, the incidence of leukemia increased by 0.66% in children and 0.93% in adolescents; the incidence of malignant CNS tumors increased by 0.49% in children; and the incidence of lymphoma increased by 1.04% in adolescents. Enhanced accuracy of the classification and registration may partially explain these increases. However, the results also call for extended research into environmental causes of childhood cancer.

Environmental Causes of Childhood Tumors

The IARC Monographs and the IARC Handbooks of Cancer Prevention

Several international and national programs have several decades-long experience and expertise in the evaluation of evidence on environmental carcinogens.[12,13] Specifically,

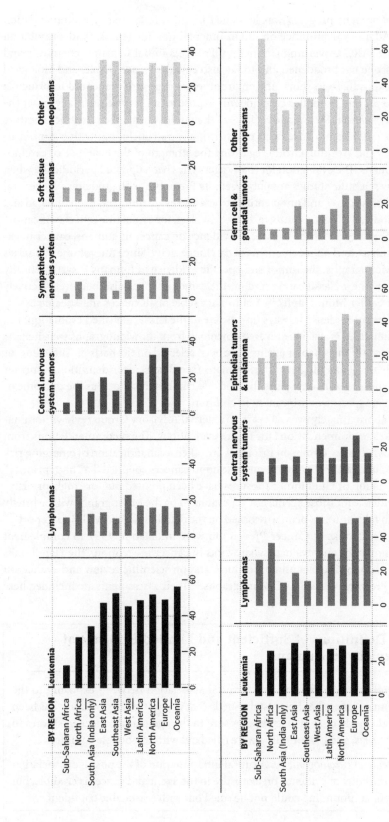

Figure 42.1. Age-standardized cancer incidence rates per million population, 2001–2010 among children and adolescents of ages 0–14 years (upper chart) and ages 15–19 years (lower chart)

the IARC Monographs program was launched in 1971, following a recommendation by the IARC Advisory Committee on Environmental Carcinogenesis (and a resolution adopted by the IARC Governing Council).[14] From its initial focus on chemicals and mixtures, its scope has broadened and today also covers personal habits and biological and physical exposures—in short, any environmental exposure that could in principle be modified to prevent or reduce carcinogenic exposures and consequentially lower the global burden of cancer. Systematic reviews of the publicly available scientific literature on more than 1,000 agents have been published, always concluding with a consensus classification by the Working Group regarding the strength of the evidence of carcinogenicity in humans. The evaluation for many agents has been updated periodically when important new scientific studies are published. Its Preamble, with guidelines on general principles and procedures and on scientific review and evaluation, has been updated and amended periodically, most recently in 2018.[15] In brief, for evaluation and classification, the Monographs integrate three streams of evidence on cancer in humans, cancer in experimental animals, and on mechanisms of carcinogenicity. Since the volume 100 series of the IARC Monographs, the tumor site-specific evidence in humans is systematically evaluated for any new classification and, at the same time, has also been retroactively completed for earlier Monographs.[16] A summary table of all cancer site-specific evaluations of the human evidence (always up-to-date to the latest published evaluation; i.e., currently Volume 132) is available for free download from the Monographs website.[12]

For the purpose of this chapter agents have been selected on the basis of "sufficient" or "limited" evidence of carcinogenicity in humans (see Box 42.1 for definitions), further restricted to exposure–outcome pairs that include childhood cancers as the outcome or exposures that are specific or prominent in children.

Furthermore, approximately every 5 years an external Advisory Group reviews nominations and makes recommendations for future evaluations. The recommendations from the most recent Advisory Group on Priorities are taken as an indication of emerging risk factors and included here if evidence on childhood cancers supported a "high priority" for future evaluation or childhood exposures have been pointed out for "high-priority" agents. The evidence for those agents not yet evaluated by the Monographs will be briefly summarized in tabular form, primarily based on the detailed Advisory Group Report.[17]

The IARC Handbooks of Cancer Prevention were initiated in 1997 to complement the IARC Monographs and evaluate what works in cancer prevention. The Handbooks follow guidelines on principles and procedures and on scientific review and evaluation similar to the Preamble of the IARC Monographs.[18] Their assessments are included here

---

Box 42.1 Definitions of Sufficient and Limited Evidence of Carcinogenicity in Humans

Sufficient evidence of carcinogenicity: A causal association between exposure to the agent and human cancer has been established. That is, a positive association has been observed in the body of evidence on exposure to the agent and cancer in studies in which chance, bias, and confounding were ruled out with reasonable confidence.

Limited evidence of carcinogenicity: A causal interpretation of the positive association observed in the body of evidence on exposure to the agent and cancer is credible, but chance, bias, or confounding could not be ruled out with reasonable confidence.

if pertinent to the topic of environmental factors for childhood cancer, following the same selection criteria as indicated above for the IARC Monographs.

The summaries on environmental exposure–childhood cancer pairs are organized by type of environmental exposure (physical, chemical, and biological) and by highest level of evidence ("sufficient" or "limited") for any cancer site. Critical periods of exposure range from preconception to childhood and adolescence, sometimes with cancer occurring only later in adult life. Cumulative exposure is often thought as the biologically most plausible exposure metric for carcinogenesis, and, together with level of exposure, duration of exposure is the critical component. Naturally, compared to cancers in adults, the maximum duration of exposure and also its latency (from first exposure to clinical diagnosis) are considerably shorter for childhood cancers. Conversely, children are thought to be more susceptible to carcinogens because of rapidly dividing cell populations and developing organ systems well into puberty and adolescence. Beyond cumulative exposure, the timing of exposure may be decisive: for example, for a suspected mammary carcinogen the critical window of exposure may be during the development of the female during puberty. Children are not little adults; from the same level of exposure they may experience relatively higher exposure per body weight than adults (also see Chapter 2 and Chapter 45). Furthermore, children may be exposed in a different manner (e.g., oral exposure pathways during infancy and early childhood), which may result in relatively higher exposure.

Improved survival of childhood cancers also comes with higher opportunity of secondary cancers via shared genetic or environmental risk factors or due to adverse effects of the curative cancer treatment from exposure to ionizing radiation and chemotherapy. Evidence on effective preventive measures other than exposure reduction (e.g., vaccination) will be included in the context of the agent-specific summaries.

Radiation

All types of ionizing radiation, alpha- and beta-particle emitters, x-rays and gamma-rays, and neutron radiation, have been classified as carcinogenic to humans (Group 1). Some of the first case reports on cancer were communicated soon after the discovery of x-rays.[19] Some 50 years later, in the 1950s, exposure to diagnostic x-rays of pregnant women was the first in utero exposure linked to an increased cancer risk in children, specifically acute leukemia.[20] Later, studies of the Japanese A-bomb survivors reported statistically significant dose-related increases in incidence rates of various solid cancers in children and adults (at ages 12–55 years) among survivors exposed in utero and during early childhood (age <6 years at exposure).[21] The increased risk of cancer following in utero exposure to radiation starts in childhood and persists long into adulthood.

Concerning mixed beta-particle emitters, epidemiological studies in the aftermath of the Chernobyl disaster have linked the iodine fallout to an increased risk of thyroid cancer.[22] Doses appear to be mainly due to and were highest in those who were youngest at the time of the accident. Epidemiological studies on in utero exposures and thyroid cancer in children are based on small numbers but suggest a strong increase of risk in a dose-related manner and are consistent with estimates linking postnatal exposure with thyroid cancer. On the population level, a massive increase of thyroid cancer became evident starting soon after the accident in the early 1990s, first in Belarus, then in Ukraine and the Russian Federation.[23] For instance, in Belarus, the incidence of childhood thyroid cancer increased from 0.03–0.05 cases per 100,000 per year before the accident to 4 cases per 100,000 per year in 1995. While informative studies on early-life exposure and thyroid cancer have been facilitated due to the massive exposures after the Chernobyl accident, it is important to note other sources of exposure to radioiodines, and particularly ^{131}I, including from atmospheric nuclear weapons tests, accidental or routine emissions from nuclear power plants, and nuclear weapons production facilities.

For non-ionizing radiation, specifically electromagnetic fields from extremely-low-frequency and radiofrequency sources, see Chapter 46.

Diethylstilbestrol

Diethylstilbestrol (DES) is one of the first chemicals where in utero exposure has been linked with cancer in childhood or adolescence.[24] Commercial production of DES started in the early 1940s. The drug was widely prescribed, particularly in the United States, to treat various medical conditions (such as amenorrhea or postmenopausal osteoporosis), but also specifically to pregnant women who were at higher risk of miscarriage. In 1971, Herbst et al. described seven patients ages 15–22 years with adenocarcinoma of the vagina or cervix, extremely rare tumors, especially in young women.[25] These patients were born between 1946 and 1951, a period when DES was introduced for the management of high-risk pregnancies, and had been exposed prenatally to DES medication for prior pregnancy loss or for threatened abortion during the relevant pregnancy. Another study reported on seven cases of these rare cancers in girls ages 7–19 years; four mothers were successfully contacted and recalled DES use during pregnancy or taking a hormone for vaginal bleeding.[26] Several additional studies supported a causal link.

Since this first assessment of the carcinogenicity of DES and up to the most recent IARC Monograph on DES[27] epidemiological studies have reported new associations between DES prescribed for various medical conditions and cancer risk. Of special interest here is the association between in utero exposure and testicular cancer in men. A case-control study from California reported a strong association between using DES during pregnancy and testicular cancer in the sons, albeit based on only two exposed case mothers.[28] In the National Cancer Institute (NCI) Combined Cohort Study including 1,365 exposed and 1,394 unexposed sons,[29] a total of seven cases were observed in men exposed prenatally compared to 3.4 expected on the basis of an external reference population. Overall results from the epidemiological studies were mixed and based on small numbers, and the Working Group concluded that there was "limited evidence."

First studies have started to investigate cancer risk in the third generation (i.e., among children of women who were exposed to DES in utero). In the NCI Combined Cohort, two cases of ovarian cancer were diagnosed at the young ages of 7 and 20 years, respectively, in the daughters of prenatally exposed women, compared to 0.38 cases expected and no cases among the unexposed third-generation daughters.[30]

Parental Smoking

Tobacco smoking was causally linked with lung cancer in the early 1960s. Since then, the evidence on the carcinogenicity of tobacco smoking has incrementally strengthened and substantially broadened to other cancer sites and other exposure scenarios. While the odds of lung cancer among long-term and heavy smokers may be a hundred times those in never-smokers,[31] the relative risks for cancer sites more recently linked with tobacco smoking are smaller (e.g., for colorectal cancer around 1.3) and therefore it has been more difficult to firmly establish causality.[32,33] Similarly, for involuntary smoking and lung cancer, the relative risk for never-smoking spouses exposed to secondhand tobacco smoke from their partner is about 1.3 compared to nonexposed spouses.[32] Following this line of reasoning, it took even longer for parental smoking to be associated with childhood cancer. In Volume 100E of the Personal Habits and Indoor Combustions report,[34] increased risk of hepatoblastoma, a rare embryonal tumor, in children born to parents who smoked preconception, during pregnancy, and/or perinatally was observed in four epidemiological studies. For instance, the United Kingdom Childhood Cancer study found relative risks of hepatoblastoma of 1.86 (0.5–7.6), 2.01 (0.4–10.2), and 4.74 (1.68–13.4) if only the father, only the mother, or both parents smoked, respectively.

Tobacco smoking has also been linked with leukemia, particularly acute myeloid leukemia (AML), with relative risks in the order of 2, and further supported by patterns of specific cytogenetic damage (e.g., translocations involving chromosomes 8 and 21) in a subset of adult AML patients who were smokers, similar to findings among workers heavily exposed to benzene.[32] Several studies have investigated the association between parental smoking and childhood leukemia, in particular acute lymphoblastic leukemia (ALL), which is one of the most frequent cancers in children. A meta-analysis of 11 studies reported a meta-RR for paternal smoking of 1.12 (1.04–1.21). Mechanistic data of increased frequencies of HPRT mutations, chromosomal translocations, and DNA strand break in newborns of smoking mothers and increased frequencies of aneuploidy, DNA adducts, strand breaks, and oxidative damage in sperm of smokers corroborate the findings of the epidemiologic studies.

Smokeless Tobacco, Betel Quid and Areca Nut

Another tobacco habit is smokeless tobacco that is chewed or snuffed. Particularly in India and some neighboring countries in Southeast Asia smokeless tobacco is widely consumed in the form of betel quid formed by wrapping areca nut, catechu, slaked lime, and often tobacco in betel leaf. Depending on local, ethnic, and cultural traditions betel quids may also be chewed without tobacco, and sometimes areca nuts are chewed without betel leaves, but in different preparations. Each of these habits, consumption of betel quid with or without added tobacco, and of areca nut, has been classified as carcinogenic to humans (Group 1). Its use may start among children and adolescents, and prevalence of use in Southeast Asia may vary between less than 10% and almost 50%; use is typically higher in boys than girls and depends on age group, ethnicity, and region.[35] Its use is also prevalent in Asian migrant communities, including among children and adolescents. Because of traditional beliefs regarding the beneficial effects of areca nuts, these may sometimes be given to young children to soothe the discomfort of teething or aid digestion. Case reports of children who developed oral submucous fibrosis (a precancerous lesion with a high risk to progress to oral cancer) following areca nut use have been published (cases include children as young as 4 years in Asian migrants to Canada and Great Britain).[36,37]

Benzene

Benzene was among the first established environmental carcinogens, and, as for tobacco, the evidence has strengthened and broadened over time. In its first evaluation by the IARC Monographs,[38] a causal link with AML among adults occupationally exposed to benzene was strongly suggested, but only the latest evaluation[39] of results from several recent studies showed positive associations between AML in children and environmental exposure to benzene. Benzene is a ubiquitous air pollutant; a component of gasoline, vehicle exhaust, industrial emissions, and tobacco smoke; and an occupational exposure in diverse industries. In terms of childhood cancers, studies on childhood exposure to benzene in outdoor air were most informative. Seven studies, mainly case-control, investigated leukemia, albeit with inconsistent groupings of the outcome. Four of these studies reported results for both ALL and AML separately.[40–43] These studies consistently showed associations between benzene and AML, but no or weaker associations with ALL.

In the context of the evaluation of the carcinogenicity of occupational exposures as a painter,[44,45] a subset of case-control studies on parental occupation as a painter or parental exposure to paints and childhood leukemia was reviewed. Five of seven studies reported significant increases in childhood leukemia, particularly myeloid leukemia, primarily associated with maternal exposure to paint before or during pregnancy or both; two of the studies showed some evidence of an exposure-response relationship.[45]

Epstein-Barr Virus and Malaria

Epstein-Barr virus (EBV), a ubiquitous oncogenic herpes virus, is a necessary risk factor for the most frequent childhood cancer in sub-Saharan Africa, now called endemic Burkitt lymphoma. Two case-control studies among children in holoendemic areas reported significantly increased risks for endemic Burkitt lymphoma with high titers of total IgG antibodies specific to whole schizont extracts jointly with high EBV antibody titers. African children are infected by EBV early in life (<3 years of age), and the timing of EBV and Plasmodium coinfection and the intensity of malarial infections at an individual level seem to influence EBV dysregulation, which may lead to endemic Burkitt lymphoma. EBV was classified as carcinogenic to humans in 1997,[46] and infection with *Plasmodium falciparum* in holoendemic areas was classified as "probably carcinogenic to humans" in 2012 (Group 2A).[47] Since this evaluation, new data relevant to the link between malaria infection and animal carcinogenicity, mechanistic data on other cancer types in humans (particularly Kaposi sarcoma together with Kaposi sarcoma-associated herpesvirus) have been published. These new results supported a recommendation by the IARC Advisory Group for a high priority reevaluation.[17]

Pharmaceuticals

Several pharmaceuticals have been classified as carcinogenic to humans (Group 1), as probably carcinogenic (Group 2A), or possibly carcinogenic to humans (Group 2B) (Table 42.1). Several of these drugs may be a component of highly effective cancer

Table 42.1 Pharmaceuticals classified into Group 1, 2A, or 2B

Pharmaceuticals	Classification	Mechanistic upgrade	Mechanism	Prescribed for (in children)	Cancer site(s) with sufficient evidence	Cancer site(s) with limited evidence	Latest volume of evaluation[a]
Cyclophosphamide	1		Activated to alkylating agent	Various childhood cancers	AML, urinary bladder		100A
Chlorambucil	1		Alkylating agent	Various childhood cancers	AML		100A
Etoposide	1	X	Translocations MLL gene	Various childhood cancers	AML (including in children) for combined regimen		100A
MOPP	1		Except for prednisone, drugs of this treatment regime are genotoxic	E.g. Hodgkin disease	AML, lung		100A
Procarbazin component of MOPP regime	2A		Genotoxic	E.g. Hodgkin disease		AML	100A
Bischloroethyl nitrosourea	2A		Alkylating agent	Childhood brain tumors, high-grade glioma		AML	Supplement 7
Teniposide	2A	X	Distinctive cytogenetic lesions in leukemic cells	Various childhood cancers		AML	76
Mitoxantrone	2B	X		Childhood ALL relapse		AML	76
Chloramphenicol	2A	X	Induces aplastic anaemia	Serious infections with restricted antimicrobial options	Childhood leukemia		50

[a] Respective IARC Monographs are available at https://publications.iarc.fr/Book-And-Report-Series/Iarc-Monographs-On-The-Identification-Of-Carcinogenic-Hazards-To-Humans?sort_by=year_desc&limit=208&page=1

treatments, often in combination with other anticancer drugs and other treatment modalities, including radiotherapy. Some of these drugs have been instrumental in improving survival from childhood cancer. Often stimulated by case reports, epidemiologic studies among cancer survivors investigated the association of secondary cancers with anticancer drugs or components of multidrug treatment regimes. Due to combined treatment modalities it has been challenging to identify the carcinogenic drug in such a regimen, but knowledge about the other components and mechanistic evidence (e.g., on characteristic genetic damage) helped to pinpoint certain drugs.

Cyclophosphamide, chlorambucil, etoposide, and MOPP (combined chemotherapy with chlormethine, vincristine, procarbazine, and prednisone) are used for treatment of cancer in children and adolescents and have been classified as carcinogenic to humans (Group 1), typically with sufficient evidence for AML. For etoposide, a case-control study on secondary leukemia in children contributed to the conclusion of limited evidence for AML; data on balanced translocations of the MLL gene on chromosome 11 supported a mechanistic upgrade of etoposide to Group 1.

Additional anticancer treatment drugs used in children and adolescents have been classified as probably carcinogenic (procarbazine, bischloroethyl nitrosourea, and teniposide) or possibly carcinogenic to humans (mitoxantrone), each based on positive associations with AML. Finally, chloramphenicol, a drug used for serious infections with restricted antimicrobial options, induces aplastic anemia, and case reports and epidemiological studies reported associations with childhood leukemia. It is important to note that the improved survival from cancer or serious infections outweighs the risk of typically rare (secondary) cancers. Nevertheless, evidence about the carcinogenic effects may help to stimulate the search for similarly effective drugs with less or no carcinogenic potential.

Exposure in Childhood and Cancer in Adults

For all of the pharmaceuticals reviewed above, use in children has been documented, but only for some of these is the carcinogenicity classification based on an increased risk of childhood tumors. This group of agents therefore marks a transition from the environmental carcinogens, where early-life exposures have been linked with childhood tumors, to the agents summarized in Table 42.2 that are characterized by well-documented and widespread exposures during childhood but classification of the carcinogen typically is based on epidemiological studies among adults. For some of the carcinogenic infectious agents, vaccination (e.g., against hepatitis B virus) or antibiotic treatment to prevent chronic infection (e.g., with *Heliobacter pylori* or *Schistosoma haematobium*) are also effective to prevent cancer later in life. Furthermore, public health policies have been shown to be effective in some settings (e.g., taxation of sugar-sweetened beverages has been reported to reduce the prevalence of obesity, particularly among girls).[48]

All the exposure–outcome pairs discussed so far have been selected on the basis of at least "limited evidence" in humans for a childhood tumor or as a Group 1 or 2A carcinogen with documented widespread exposure, including in children. However, of the more than 1,000 agents evaluated by the IARC Monographs program only some 20% have been classified in Group 1 or 2A; yet another 30% have been classified in Group 2B, "possibly carcinogenic to humans." These have been typically classified based on sufficient

Table 42.2 Exposures during childhood and cancer among adults

Agent	Specifics about early life exposure	Cancer sites with sufficient evidence	Other notable characteristics
Indoor air pollution from combustion of solid fuels	Significant exposures particularly in low-medium-income countries; young children and women receive the highest exposures	Lung	Among the leading environmental risk factors in terms of global burden of disease (GBD)
Outdoor air pollution	Ubiquitous exposure, in most cities WHO air quality guidelines are exceeded	Lung	Among the leading environmental risk factors in terms of global burden of disease (GBD)
Diesel engine exhaust	Exposure in children primarily from proximity to roads	Lung	
Arsenic	Exposure in children primarily via drinking water, particularly in areas with high arsenic concentrations	Lung, urinary bladder, non-melanoma skin	
Aflatoxins	Dietary intake, mainly through consumption of contaminated maize and peanuts, particularly in some tropical countries.	Hepatocellular carcinoma	Mycotoxin control in low- and middle-income countries. IARC Working Group Report Volume 9. Wild CP, Miller JD, Groopman JD (eds), 2015
Asbestos	Environmental exposures in vicinity of asbestos mines and factories;	Mesothelioma	The evidence for environmental asbestos exposure is strongest for mesothelioma; Case report about a son who brought his father hot lunch to the factory (Schneider et al.);
Erionite	Environmental exposure in certain villages in Cappadocia/Turkey	Mesothelioma	Incidence similarly increased in men and women, some diagnosed as young as 26 years.
Fluoro-edenite	Biancavilla, Italy; village and street built of stones from nearby quarry	Mesothelioma	No indication of occupational or environmental asbestos exposure; increased incidence of mesothelioma at relatively young age, and similar for men and women, pointed toward an environmental exposure

(continued)

Table 42.2 Continued

Agent	Specifics about early life exposure	Cancer sites with sufficient evidence	Other notable characteristics
Solar radiation	Exposure to solar ultraviolet radiation varies by time of day (highest around mid-day), geographic latitude (higher at lower degrees latitude) and season, altitude (higher at higher altitude), cloud coverage (usually clouds decrease exposure), and surface reflections (particularly from snow and water surfaces).	Melanoma and non-melanoma skin	Sunburns, particularly during childhood and adolescence strongly associated with skin cancer, particularly melanoma
Chronic infection with Hepatitis B virus	Predominant modes of transmission of HBV infection include perinatal.	Liver	The risk of chronic carriage is related to the age at infection; the risk is highest among people infected as infants, of whom about 90% become chronic carriers; vaccination also as primary prevention of cancer.
Helicobacter pylori	Initial infection typically during childhood, associated with low socioeconomic status, once acquired infection is typically lifelong (unless treated)	Stomach, non-Hodgkin lymphoma (MALT gastric lymphoma)	Antibiotic treatment to eradicate chronic *H. pylori* reduces incidence of gastric cancer
Schistosoma haematobium	Primarily in sub-Saharan Africa and along the Nile River, contact with contaminated freshwater as major risk factor of infection, which may begin in early childhood and peaks between the ages of 5 and 15 years.	Urinary bladder	Infection can be treated with prescription medicine, due to frequent reinfections in endemic areas, repeated treatment is essential to prevent cancer as an outcome of chronic infection
Obesity	Worldwide, an estimated 110 million children and adolescents in 2013 were obese. Among children the estimated age-standardized prevalence of obesity in 2014 was 5.0%.	Esophagus (adenocarcinoma) gastric cardia colon and rectum liver, gallbladder pancreas breast (postmenopausal), Corpus uteri, ovary kidney (renal-cell) Meningioma Thyroid Multiple myeloma	Some epidemiological studies investigated the link between BMI or body shape in childhood or adolescence with cancer in adult life. For several cancers the associations were similar to those found in studies on adult BMI.

Box 42.2   Melamine

Melamine is a chemical that has been widely used in the production of specific plastic materials (e.g., kitchenware, coatings, and adhesives). Exposure has been primarily occupational by inhalation of melamine dust. Despite sufficient evidence in experimental animals based on increased incidences of tumors of the urinary bladder and the ureter in mice and of urinary bladder transitional cell carcinoma in rats, melamine was classified in Group 3 (v73,[a] 1999), instead of Group 2B. The rationale for a mechanistic downgrade was that the tumors in experimental mice occurred as a consequence of chronic irritation from bladder calculi and only at very high doses, a mechanism that would not operate in humans. However, in 2008, in China, melamine had been used illegally to adulterate foods, including infant milk formula, and heavily exposed children developed urinary tract inflammation and stones that contained melamine, similar to those in the experimental animals, where the incidence of urinary stones correlated with the tumor incidence. Therefore, a recent IARC Monographs Working Group concluded that melamine is possibly carcinogenic to humans (Group 2B) (v119).[b]

[a] IARC. *Some Chemicals That Cause Tumors of the Kidney or Urinary Bladder in Rodents and Some Other Substances.* Vol. 73. 1999. Available at: https://publications.iarc.fr/Book-And-Report-Series/Iarc-Mon ographs-On-The-Identification-Of-Carcinogenic-Hazards-To-Humans/Some-Chemicals-That-Cause-Tumors-Of-The-Kidney-Or-Urinary-Bladder-In-Rodents-And-Some-Other-Substances-1999
[b] IARC. *Some Chemicals That Cause Tumors of the Urinary Tract in Rodents.* Vol. 119. 2019. Available at: https://publications.iarc.fr/Book-And-Report-Series/Iarc-Monographs-On-The-Identification-Of-Carcinogenic-Hazards-To-Humans/Some-Chemicals-That-Cause-Tumors-Of-The-Urinary-Tract-In-Rodents-2019

---

evidence of carcinogenicity in experimental animals. Therefore, it is difficult to predict which tumors (if any) may be increased in humans and if this would include childhood tumors. For the purpose of this chapter, this group of carcinogens has not been systematically reviewed for potential widespread exposure in children. However, for illustration, one such agent with well-documented exposure in children is summarized in Box 42.2.

Primarily based on the evaluations of the IARC Monographs program, this chapter summarizes the evidence on environmental exposures and cancer as pertinent to children, either directly in terms of childhood cancers or in terms of prevalent exposures during childhood linked with cancer later in life. New scientific publications may indicate new and emerging environmental exposures or additional cancer sites for known carcinogens that may warrant an updated evaluation by the Monographs program. To this end, about every 5 years, the secretariat calls for nominations of agents for future evaluation. Nominated agents are reviewed by an external Advisory Group, which makes recommendations for future priority evaluations. The most recent Advisory Group was convened in 2019. Table 42.3 summarizes recommendations with high priority for evaluation that mention childhood cancer as a reported outcome of new epidemiological studies or pertinent exposure during childhood.

Table 42.3 Selected recommendations by the Advisory Group on priorities for future evaluation with high priority by the IARC Monographs

---

Combustion of Biomass (Group 2A, IARC 2006), for heating and cooking, particularly in low- and middle-income countries, a meta-analysis including 14 case-control studies of biomass cooking or heating reported increased odds ratios for lung cancer.

Holo-endemic malaria (Group 2A, IARC 2013), evidence suggests that malaria enhances lytic replication of KSHV, which is thought to be a major risk factor, both for transmission to other people and for pathogenesis to Kaposi sarcoma. KSHV seroprevalence was higher in adults and children with malaria parasitaemia than in children without malaria parasitaemia.

Some anthracyclines (Group 2A and 2B, IARC, 1987), anticancer drugs, a cohort study of childhood cancer survivors reported dose-dependent doxorubicin-related increased risks of all solid cancers and breast cancer.

Acetaldehyde (Group 2B, IARC 1999), air pollutant, positive associations of prenatal or early-life exposure to acetaldehyde (as well as other correlated pollutants) in ambient air and childhood central nervous system primitive neuroectodermal tumor (PNET), Wilms tumor, and retinoblastoma were observed in case-control studies conducted in California, USA.

Automotive gasoline (Group 2B, IARC 1989), acute childhood leukemia in residents living near gasoline stations or repair garages in a case-control study in France; acute myeloid leukemia in young adults that increased with car density in Sweden; an elevated risk of childhood leukemia with proximity to main roads and gasoline stations in the UK.

Permethrin (Group 3, IARC 1991), worldwide as an insecticide, a case-control study of infant and childhood leukemia conducted in Brazil found that prenatal exposure was associated with acute lymphoblastic leukemia and acute myeloid leukemia for children ages 0–11 months.

Neonatal Phototherapy (not previously evaluated), visible light used to treat severe jaundice in the neonatal period. Treatment with phototherapy is implemented to prevent the neurotoxic effects of high serum levels of unconjugated bilirubin. A retrospective cohort study in Canada found that infants who received neonatal phototherapy had more than 2 times the risk of any solid tumor between age 4 years and 11 years compared with non- exposed children. In a cohort of infants born in hospitals in California, USA, infants with diagnosis codes for phototherapy had a 60% increased risk of overall cancer, myeloid leukemia and kidney cancer, compared with children without such codes. In a retrospective cohort study of children born in Kaiser Permanente Northern California hospitals, cancer incidence in children exposed to phototherapy was increased by 40% compared with that in non-exposed children, and specifically any leukemia, non-lymphocytic leukemia, and liver cancer.

Conclusion

Childhood cancer shows a pattern that differs from cancer in adults, and, altogether, cancer rates in children are lower than in adults and survival is better. But every single diagnosis of a childhood cancer is traumatizing to the child, the whole family, and beyond. Also, the benefits of increased survival entail new risks, ranging from impairment of cognitive development to secondary cancers, at least in part caused by the initial curative cancer treatment. Also, because childhood cancers occur earlier in life, the potential loss of life years and disability-adjusted life years may be far larger from a cancer occurring during childhood compared to one occurring in an adult.

More importantly, the current evidence about environmental risk factors for childhood cancer is probably only the tip of the iceberg. Although genetic risk factors proportionally probably play a more important role for childhood cancers than for cancers in old age, there are many particular challenges for the identification of environmental risk factors of childhood cancer, including an imbalance of funding favoring clinical and genetic research over research into environmental and social determinants: with the relative rarity of specific childhood cancers, very large multinational cohort studies are needed to achieve reasonable precision, while case-control studies of childhood cancers may be particularly prone to validity threats, such as exposure misclassification and recall bias.

The above summary considerations apply primarily to childhood cancers. Childhood exposures linked to cancer later in life may dominate the overall burden, yet are even more difficult to investigate due to the long latency from exposure to cancer, particularly for solid tumors. For what is known or at least very plausible as a hazard regarding exposure in childhood and cancer later in life, quantification of the attribution of childhood exposures to adult cancer is currently not well characterized for many exposure–cancer pairs.

Nevertheless, there is sufficient evidence to prioritize more effective public health interventions to prevent childhood cancer and carcinogen exposures during childhood. Organized population-wide vaccination against HBV, particularly in countries with high rates of infection in early childhood, is a low-hanging fruit but requires a functioning public health infrastructure and resources for vaccination programs, often not sufficiently available in concerned low- and middle-income countries (LMICs). Indoor and outdoor air pollution are among the top-ranking environmental risk factors for the global burden of disease, and disproportionately affect children in LMICs and those of low socioeconomic status. And, despite some progress over the past two decades, indoor air pollution still disproportionately affects women and children. Together, the exposure to carcinogens contributes to social inequalities. Finally, universal adherence to and enforced compliance with the latest WHO Air Quality Guidelines would also help mitigate what is probably the biggest health crisis of the generation of today's children, the emerging public health crisis due to climate change.

References

1. Straif K. The global cancer pandemic: trends and disparities. In: Landrigan PJ, Vicini A, eds., *The Rising Global Cancer Pandemic*. Online: Pickwick Publications; 2022:13–21.

2. Roman E, Lightfoot T, Picton S, Kinsey S. Childhood cancers. In: Thun M, et al., eds., *Cancer Epidemiology and Prevention*. New York: Oxford University Press; 2017:1119–54. doi:10.1093/oso/9780190238667.003.0059.

3. Caron HN, Biondi A, Boterberg T, Doz F. *Oxford Textbook of Cancer in Children*. 7th ed. New York: Oxford University Press; 2020.

4. Pfister SM, Reyes-Múgica M, Chan JKC, et al. A summary of the inaugural WHO classification of pediatric tumors: transitioning from the optical into the molecular era. *Cancer Discov.* 2022;12(2):331–55. doi:10.1158/2159-8290.CD-21-1094.

5. Bray F, Colombet M, Mery L, et al. *Cancer Incidence in Five Continents.* Vol 11. Lyon: IARC Scientific Publications; 2021.

6. Forman D, Bray F, Brewster DH, et al. *Cancer Incidence in Five Continents.* Vol 10. Lyon: IARC Scientific Publications; 2014.

7. Parkin D, Stiller C, Draper G, Bieber C, Terracini B, Young J. *International Incidence of Childhood Cancer*. Vol. 1. Lyon: IARC Scientific Publications, ScP 87; 1988.

8. Parkin D, Kramárová E, Draper G, et al. *International Incidence of Childhood Cancer*. Vol 2. Lyon: IARC Scientific Publications, ScP 144; 1998.

9. Steliarova-Foucher E, Colombet M, Ries LAG, et al. International incidence of childhood cancer, 2001–10: a population-based registry study. *Lancet Oncol*. 2017;18(6):719–31. doi:10.1016/S1470-2045(17)30186-9.

10. Pui CH, Gajjar AJ, Kane JR, Qaddoumi IA, Pappo AS. Challenging issues in pediatric oncology. *Nat Rev Clin Oncol*. 2011;8(9):540–9. doi:10.1038/nrclinonc.2011.95.

11. Steliarova-Foucher E, Fidler MM, Colombet M, et al. Changing geographical patterns and trends in cancer incidence in children and adolescents in Europe, 1991–2010 (Automated Childhood Cancer Information System): a population-based study. *Lancet Oncol*. 2018;19(9):1159–69. doi:10.1016/S1470-2045(18)30423-6.

12. International Agency for Research on Cancer. Agents Classified by the IARC Monographs, Vols. 1–132. IARC Monographs on the Identification of Carcinogenic Hazards to Humans. Available at: https://monographs.iarc.who.int/agents-classified-by-the-iarc/.

13. National Toxicology Program, U.S. Department of Health and Human Services. Report on Carcinogens Process & Listing Criteria. Available at: https://ntp.niehs.nih.gov/whatwestudy/assessments/cancer/criteria/index.html.

14. Baan R, Straif K. The Monographs Programme of the International Agency for Research on Cancer. *ALTEX*. 2021. doi:10.14573/altex.2004081.

15. Samet JM, Chiu WA, Cogliano V, et al. The IARC Monographs: updated procedures for modern and transparent evidence synthesis in cancer hazard identification. *JNCI J Natl Cancer Inst*. 2020;112(1):30–7. doi:10.1093/jnci/djz169.

16. Cogliano VJ, Baan R, Straif K, et al. Preventable exposures associated with human cancers. *JNCI J Natl Cancer Inst*. 2011;103(24):1827–39. doi:10.1093/jnci/djr483.

17. International Agency for Research on Cancer. Report of the Advisory Group to Recommend Priorities for the IARC Monographs during 2020–2024. World Health Organization. Available at: https://www.iarc.who.int/news-events/report-of-the-advisory-group-to-recommend-priorities-for-the-iarc-monographs-during-2020-2024/.

18. International Agency for Research on Cancer. Preamble for primary prevention. World Health Organization. October 2019. Available at: https://handbooks.iarc.fr/documents-handbooks/hb-preamble-primary-prevention.pdf.

19. Doll R. Hazards of ionising radiation: 100 years of observations on man. *Br J Cancer*. 1995;72(6):1339–49. doi:10.1038/bjc.1995.513.

20. Stewart A, Webb J, Hewitt D. A survey of childhood malignancies. *BMJ*. 1958;1(5086):1495–1508. doi:10.1136/bmj.1.5086.1495.

21. Preston DL, Cullings H, Suyama A, et al. Solid cancer incidence in atomic bomb survivors exposed in utero or as young children. *JNCI J Natl Cancer Inst*. 2008;100(6):428–36. doi:10.1093/jnci/djn045.

22. Cardis E, Howe G, Ron E, et al. Cancer consequences of the Chernobyl accident: 20 years on. *J Radiol Prot*. 2006;26(2):127–40. doi:10.1088/0952-4746/26/2/001.

23. United Nations Scientific Committee on the Effects of Atomic Radiation, United Nations. *Sources and Effects of Ionizing Radiation: Volume 2: Effects*. New York: United Nations; 2000.

24. International Agency for Research on Cancer. Sex hormones. In: *IARC Monographs on the Evaluation of Carcinogenic Risk of Chemicals to Man Volume 6*. Vol 6. Geneva: World Health Organization; 1974:1–238.

25. Herbst AL, Ulfelder H, Poskanzer DC. Adenocarcinoma of the vagina: association of maternal stilbestrol therapy with tumor appearance in young women. *N Engl J Med.* 1971;284(16):878–81. doi:10.1056/NEJM197104222841604.

26. Noller KL, Decker DG, Lanier AP, Kurland LT. Clear-cell adenocarcinoma of the cervix after maternal treatment with synthetic estrogens. *Mayo Clin Proc.* 1972;47(9):629–30.

27. IARC. Pharmaceuticals. Vol. 100A. 2012. Available at: https://publications.iarc.fr/Book-And-Report-Series/Iarc-Monographs-On-The-Identification-Of-Carcinogenic-Hazards-To-Humans/Pharmaceuticals-2012.

28. Depue RH, Pike MC, Henderson BE. Estrogen exposure during gestation and risk of testicular cancer. *J Natl Cancer Inst.* 1983;71(6):1151–5.

29. Strohsnitter WC, Noller KL, Hoover RN, et al. Cancer risk in men exposed in utero to diethylstilbestrol. *JNCI J Natl Cancer Inst.* 2001;93(7):545–51. doi:10.1093/jnci/93.7.545.

30. Titus-Ernstoff L, Troisi R, Hatch EE, et al. Offspring of women exposed in utero to diethylstilbestrol (DES): a preliminary report of benign and malignant pathology in the third generation. *Epidemiol Camb Mass.* 2008;19(2):251–7. doi:10.1097/EDE.0b013e318163152a.

31. Pesch B, Kendzia B, Gustavsson P, et al. Cigarette smoking and lung cancer-relative risk estimates for the major histological types from a pooled analysis of case-control studies. *Int J Cancer.* 2012;131(5):1210–9. doi:10.1002/ijc.27339.

32. IARC. Tobacco smoke and involuntary smoking. Vol. 83. 2004. Available at: https://publications.iarc.fr/Book-And-Report-Series/Iarc-Monographs-On-The-Identification-Of-Carcinogenic-Hazards-To-Humans/Tobacco-Smoke-And-Involuntary-Smoking-2004.

33. Vineis P, Alavanja M, Buffler P, et al. Tobacco and cancer: recent epidemiological evidence. *J Natl Cancer Inst.* 2004;96(2):99–106. doi:10.1093/jnci/djh014.

34. IARC. Personal habits and indoor combustions. Vol. 100E. 2012. Available at: https://publications.iarc.fr/Book-And-Report-Series/Iarc-Monographs-On-The-Identification-Of-Carcinogenic-Hazards-To-Humans/Personal-Habits-And-Indoor-Combustions-2012.

35. IARC. Betel-quid and areca-nut chewing and some areca-nut-derived nitrosamines. Vol. 85. Available at: https://publications.iarc.fr/Book-And-Report-Series/Iarc-Monographs-On-The-Identification-Of-Carcinogenic-Hazards-To-Humans/Betel-quid-And-Areca-nut-Chewing-And-Some-Areca-nut-derived-Nitrosamines-2004.

36. Hayes PA. Oral submucous fibrosis in a 4-year-old girl. *Oral Surg Oral Med Oral Pathol.* 1985;59(5):475–8. doi:10.1016/0030-4220(85)90087-8.

37. Shah B, Lewis MAO, Bedi R. Oral submucous fibrosis in a 11-year-old Bangladeshi girl living in the United Kingdom. *Br Dent J.* 2001;191(3):130–2. doi:10.1038/sj.bdj.4801117.

38. IARC. Some anti-thyroid and related substances, nitrofurans and industrial chemicals. Vol 7. 1974. Available at: https://publications.iarc.fr/Book-And-Report-Series/Iarc-Monographs-On-The-Identification-Of-Carcinogenic-Hazards-To-Humans/Some-Anti-Thyroid-And-Related-Substances-Nitrofurans-And-Industrial-Chemicals-1974.

39. IARC. Benzene. Vol, 120. 2018. Available at: https://publications.iarc.fr/Book-And-Report-Series/Iarc-Monographs-On-The-Identification-Of-Carcinogenic-Hazards-To-Humans/Benzene-2018.

40. Vinceti M, Rothman KJ, Crespi CM, et al. Leukemia risk in children exposed to benzene and PM10 from vehicular traffic: a case-control study in an Italian population. *Eur J Epidemiol.* 2012;27(10):781–90. doi:10.1007/s10654-012-9727-1.

41. Heck JE, Park AS, Qiu J, Cockburn M, Ritz B. Risk of leukemia in relation to exposure to ambient air toxics in pregnancy and early childhood. *Int J Hyg Environ Health*. 2014;217(6):662–8. doi:10.1016/j.ijheh.2013.12.003.

42. Houot J, Marquant F, Goujon S, et al. Residential proximity to heavy-traffic roads, benzene exposure, and childhood leukemia: the GEOCAP study, 2002–2007. *Am J Epidemiol*. 2015;182(8):685–93. doi:10.1093/aje/kwv111.

43. Janitz AE, Campbell JE, Magzamen S, Pate A, Stoner JA, Peck JD. Benzene and childhood acute leukemia in Oklahoma. *Environ Res*. 2017;158:167–73. doi:10.1016/j.envres.2017.06.015.

44. IARC. Painting, firefighting, and shiftwork. Vol. 98. 2010. Available at: https://publications.iarc.fr/Book-And-Report-Series/Iarc-Monographs-On-The-Identification-Of-Carcinogenic-Hazards-To-Humans/Painting-Firefighting-And-Shiftwork-2010.

45. IARC. Chemical agents and related occupations. Vol. 100F. 2012. Available at: https://publications.iarc.fr/Book-And-Report-Series/Iarc-Monographs-On-The-Identification-Of-Carcinogenic-Hazards-To-Humans/Chemical-Agents-And-Related-Occupations-2012.

46. IARC. Epstein-Barr virus and Kaposi's sarcoma herpesvirus/human herpesvirus 8. Vol. 70. 1997. Available at: https://publications.iarc.fr/Book-And-Report-Series/Iarc-Monographs-On-The-Identification-Of-Carcinogenic-Hazards-To-Humans/Epstein-Barr-Virus-And-Kaposi%E2%80%99s-Sarcoma-Herpesvirus-Human-Herpesvirus-8-1997.

47. IARC. Malaria and some polyomaviruses (SV40, BK, JC, and Merkel cell viruses). Vol. 104. 2013. Available at: https://publications.iarc.fr/Book-And-Report-Series/Iarc-Monographs-On-The-Identification-Of-Carcinogenic-Hazards-To-Humans/Malaria-And-Some-Polyomaviruses-SV40-BK-JC-And-Merkel-Cell-Viruses--2013.

48. Rogers NT, Cummins S, Forde H, et al. Associations between trajectories of obesity prevalence in English primary school children and the UK soft drinks industry levy: an interrupted time series analysis of surveillance data. Popkin BM, ed. *PLOS Med*. 2023;20(1):e1004160. doi:10.1371/journal.pmed.1004160.

43

Mold and Population Health

J. David Miller

The current best estimate of the number of species of fungi is between 2.2 and 3.8 million.[1] It would be fair to say that although most people are aware of particular aspects of the role of fungi in the global ecosystem, most are surprised to know that they comprise about 25% of the biomass of the planet. The fungal kingdom currently comprises 19 phyla (groups) of fungi.[2] Of these, 4–5 are important for public health, including those that produce toxins in foods, allergens, and fungal diseases. Among the most recognized fungi are those within the Ascomycota and Basidiomycota phyla. The Ascomycota includes fungi that are commonly called "molds," and mushrooms are within the Basidiomycota. "Mold" is a colloquial term for fungi that are most commonly found growing on food or wet materials. This includes the *Penicillium* species that produces penicillin; most of the fungi that spoil our bread, fruit, and cheese; and, most importantly, the fungi that recycle dead organic matter including wood from fallen trees and leaves. This group of organisms contains the species that produce agricultural mycotoxins, species that dominate the outdoor air mycoflora, and many species that grow on damp building materials. Depending on where one lives, this represents three important aspects of population health.

Fungi in Outdoor Air

Except when there is snow cover on the ground or in deserts, we are all exposed to the very high concentrations of the fungi that are found in outdoor air. These are almost exclusively those that grow on the surfaces of leaves (phylloplane fungi), crop diseases, mushrooms, and related species. Soil fungi such as *Aspergillus* and *Penicillium* represent only a few percent of the total above grade except during the fall.[3,4] More than 10% of the population is allergic to the fungi in outdoor air, mainly phylloplane species and mushroom spores.[5,6] Fungi in outdoor air account for an appreciable fraction of emergency room admissions for asthma.[7] This represents an important burden for population health and a considerable amount is spend on treatment and over the counter drugs. Unless removed by vacuums equipped with high-efficiency particulate air (HEPA) filters, the allergens from the dominant fungi in outdoor air accumulate indoors and result in the corresponding allergic response.[8,9] Climate change is changing the pattern of aeroallergens associated with warming.[10]

Dampness and Mold in the Built Environment

Problems associated with building dampness have been around since antiquity. However, health issues associated with mold and rot in homes were rare prior to the

J. David Miller, *Mold and Population Health* In: *Textbook of Children's Environmental Health*, Second Edition. Edited by: Ruth A. Etzel and Philip J. Landrigan, Oxford University Press. © Oxford University Press 2024. DOI: 10.1093/oso/9780197662526.003.0043

mid 1970s. Studies on fungi in indoor air were published by European researchers from the late 1940s. By the late 1970s, European allergy researchers made recommendations to reduce mold exposures in residential housing but this remained contested in North America.[11,12]

Dampness and mold problems in buildings increased in the United States and Canada in the 1980s for two important reasons: making them tighter to save energy and changing the nature of the building materials. A family of four adds 15–20 kg of water vapor per day to a house and a new house rather more in the first year or two after construction.[13] Adding insulation to the air leaky homes built after World War II and to new construction particularly, after the first energy crisis in the 1970s, reduced natural ventilation rates. By the mid-1990s, approximately a quarter of residential homes were inadequately ventilated and prone to condensation failure.[14,15] In addition, the prevalence of air conditioning in homes increased from 13% in 1960 to 80% in 1980.[16] This increased the risk of condensation failure in hot, humid areas. Perhaps the most dramatic change was that from using durable, mold-resistant building materials to those that were easily biodegradable, most notably the shift from traditional plaster to paper-faced gypsum wallboard and less wood. Paper-faced wallboard is susceptible to mold growth whereas wood and traditional plaster are highly resistant.[12] Apart from nutrient specializations, different fungi can tolerate different amounts of available moisture, which is called *water activity* or Aw. For example, high water activity fungi like *Stachybotrys* or *Chaetomium* require an Aw of nearly 1, whereas a fungus that is surprisingly common growing on dust and debris in carpet, *Wallemia* grows well down to 0.65. The change in materials had a major impact on the amount of water that might result in mold growth. For example, to get wood to an Aw of 0.80—that which supports most common damp building fungi—requires 17% by weight, and wood is resistant to mold growth. Traditional plaster made from lime is inherently antifungal when wet. In contrast, to get paper-faced gypsum wallboard to Aw of 0.80 requires 0.5% by weight.[17] Since the 1960s, building materials have transitioned from highly resistant to mold and rot to susceptible with even a modest amount of water. Another large factor affecting the prevalence of mold and dampness is the failure to design, build, and properly maintain buildings to prevent water infiltration.[18,19]

From the mid-1980s, the impact of these changes leading to increased mold damage in residential housing and public buildings was being felt first in Europe and then in the United States and Canada.[11,12,18–22] In the intervening years, much time and effort has been spent to understand the disease burden associated with damp buildings and fungi. This work has demonstrated the attributable risk for asthma associated with mold and dampness in houses is approximately 20% and associated with a considerable burden on the healthcare system.[23,24] Comprehensive remediation measures to reduce mold exposure as well as the underlying moisture conditions reduce respiratory symptoms and asthma morbidity.[25–28]

In North America, the epidemiological studies linking mold and dampness and respiratory health in children began in communities that had been extensively studied for the impact of outdoor air pollution and health. In the United States, the long-running Harvard Six Cities studies collected exhaustive data on the health effects of air pollution in six northeastern cities.[29,30] A decision was made to examine the role that the emerging problem of dampness and mold played in the respiratory health of the children within those well-studied communities. Based on both on a questionnaire and objective measures, Brunekreef et al.[31] found a strong and consistent association between measures of home dampness and both respiratory symptoms and other non-chest illness in 4,600 8- to 12-year-old children after adjustment for maternal smoking, age, gender, city of residence, and parental education. Importantly, there was a high prevalence of homes reporting mold problems (20–30% depending on community).

Studies on outdoor air pollution and child respiratory health were under way in Canada during the same period.[32,33] In 1987, a large study of dampness, mold, and child respiratory health was initiated using a similar instrument to the Harvard US studies just noted. This involved 30 Canadian communities from sea to sea, comprising approximately 13,000 5- to 8-year-old children. The Canadian data showed that dampness and mold were associated with respiratory and upper respiratory disease in children, and there was evidence of a dose-response effect. The impact on respiratory health did not stratify according to atopic status. Some 37% of the homes reported mold and/or dampness.[34]

The results from both population health studies were surprising, and there was uncertainty about their value, in part because there were no obvious mechanisms for some of the findings and because of the potential limitations of questionnaire-based data. Regardless, because European studies were reporting similar outcomes,[12] as well as the fact that the studies involved such a large number of children with an apparent dose-response effect, a major effort was made to pursue the findings. This initially led to two large studies involving children in communities representative of the original thirty Canadian communities. The first involved 700 children 5 to 8 years old in a rural town in southern Ontario. When the same survey instrument was administered, the data were found to be highly reproducible.[35] A significant investment was made to collect air and dust samples from 400 homes as well as conduct engineering inspections of a subset of the homes. The presence of endotoxins and house dust mite allergens in the dust did not affect the impact of mold and dampness on health.[36] More than 270 species of fungi were isolated from the homes.[37] The measure of colony-forming units fungi per gram of dust was only weakly related to objective measures, and occupant reports of water damage and mold? were subject to bias.[38] A detailed analysis of 59 of the homes found that moisture sources in the houses rather than relative humidity were associated with mold levels and dust mite antigen. Much of the visible mold present was hidden, mostly in the wall cavity around window frames.[39]

A second large study involved 400 women, following the health of their newborn children for several years. Detailed engineering studies were made to quantitatively assess areas of moisture and mold damage in their houses, and settled dust samples were assessed for fungi and dust mite allergens in the children's bedrooms and in living rooms. Air samples were collected over 5–7 days in the bedrooms and analyzed for mold glucan (see below) and endotoxins. Endotoxin in air was associated with hospitalizations of the infants.[40] In buildings with mold and dampness problems, mycelial and spore fragments were dominant. The fungal biochemicals measured (ergosterol and mold glucan) were related to the area of visible mold and dampness.[41] The extent of mold damage in these homes was below the threshold for a population health effect.[42] This is important because all homes have some mold, and exposures depend on area, cleaning, and ventilation rate.[43,44]

As discussed in greater detail in the reviews cited above, broadly similar relationships between mold and dampness and respiratory disease have been seen in many studies. The causes of the mold damage vary according to the climate of the community.[17,18] Another major issue is that homes in underprivileged areas (e.g., Indigenous communities) are more likely to have mold and moisture problems.[45–47]

Mold and dampness in residential housing has been clearly associated with increased respiratory symptoms regardless of atopic status, and not just mold but other allergens as well.[31,34,48] There is a good understanding of the common species that grow on damp building materials in the United States and Canada.[17] As with other fungi, the common damp building fungi make human antigens and allergens, including, for example, *Aspergillus herbariorum* (formerly *Eurotium herbariorum*),' *A. versicolor*, *Chaetomium globosum*, *Penicillium rubens*, *Stachybotrys chartarum*, and *Wallemia sebi*.[49–58] However, inhalation exposures to damp building fungi affect lung biology differently than do the fungi common in outdoor air, not just because of the respective allergens and other proteins but because the form of beta glucan differs.[59] It is important to note that it never happens that an exposure is to one fungus: it is always a mixture depending on the nature of the water event (e.g., flood vs. condensation failure) and the nature of the building material.[18,60]

In moldy buildings, the primary exposure is to mycelial and spore fragments.[41,61–64] As much fungal biomass is present in the small particle size fractions as that containing intact spores.[44] The majority of the small particles penetrate deeply into the tracheobronchial and pulmonary regions.[65] Mold fungi produce substances that contain damage-associated molecular patterns. These bind to pattern recognition receptors, stimulating innate immune responses in humans.[66–69] Spore and mycelial fragments in damp buildings mainly comprise cell wall constituents, particularly beta 1,3 D glucan and fragments of fungal chitin; these are fungal proteins that are similar to human proteins and secrete fungal enzymes, many of which are allergens.[69,70] Different fungi and plants produce different glucans, but the form in molds—beta 1,3 D glucan—is by far the most potent effector of lung biology of all the chemicals found in the spores and mycelial fragments of damp building molds.[59,69,71,72] Aside from impact on respiratory health, there is good evidence that high inhalation exposure to *S. chartarum* glucan/cell walls affects neural and cognitive functions in the murine brain, not the low-molecular-weight compounds.[73]

As with the allergens, quite a lot is now known about the low-molecular-weight compounds that are produced by damp building fungi and their effects on lung biology in relevant animal models.[59,74–76] Importantly, none of the low molecular compounds important in agriculture (i.e., mycotoxins) is produced by fungi that grow on building materials. Outside occupational and extreme building exposure, the impact of low-molecular-weight compounds is believed to be modest compared to glucans, chitin fragments, allergens, and other proteins.[66–69,77] A detailed discussion of the available information on fungal compounds associated with damp building fungi and health is found in Miller.[59]

There are clear examples where exposures in adults working with water-soaked building contents with extensive growth of the cellulolytic fungus *S. chartarum* without personal protection suffered serious lung damage plausibly associated with one class of its low-molecular-weight compounds, macrocyclic trichothecenes. Similar lung damage has been seen in workers handling *Stachybotrys*-contaminated hay and straw without personal protection.[78] Other toxigenic fungi that only occur in agricultural workplaces have, in the past, been associated with serious disease associated with the toxins until occupational hygiene measures were improved.[59,79] Similarly, extreme exposures to high Aw fungi including *S. chartarum* growing in houses because of persistent and largely hidden water damage resulted in lung damage in infants.[78,80–81] In a murine model, the impact of spores and mycelia fragments of *S. chartarum* on respiratory function was ascribed to cell wall components, including glucan, as opposed to trichothecenes.[82]

Public health efforts to alert the public to mold and dampness by the US Environmental Protection Agency (EPA), the Centers for Disease Control and Prevention (CDC), the National Institute for Occupational Safety and Health (NIOSH), Health Canada, the American Industrial Health Association (AIHA), and the American Society of Heating, Refrigerating, and Air Conditioning Engineers (ASHRAE) have created awareness and provided strategies to help consumers and professionals address mold and moisture problems.[83-86]

Outside research studies, there are no measurements of fungi that provide useful information on the risk associated with mold and dampness in a building.[86-89] Measurements can be useful during the investigation process but have no value for health assessments.[17,84,89] Commercial tests that claim to measure low-molecular-weight compounds from damp building fungi in human sera or urine are not reliable.[77,90,91] In 2008, the AIHA published comprehensive protocols for assessing and documenting the extent of visible mold and dampness; these are now in their second edition.[17] Another source of useful information on specific information on mold and dampness tailored to physicians reading home inspection reports emerged from the American Academy of Allergy, Asthma, and Immunology (AAAAI) panel that worked on clinical practice parameters for allergens in homes.[86,92] A useful trigger to prompt an investigation includes patient reports of mold odor.[87] Chew et al.[92] outline a two-part questionnaire to evaluate whether a clinician should recommend a home inspection for mold and dampness.

Apart and aside from mold and dampness, it is important to consider other environmental allergens that are important in buildings. AAAAI Clinical Practice Parameters were developed and include furry animals,[93] rodents,[94] cockroaches,[95] and house dust mites.[96] In addition, burdens of bacterial endotoxin originating from pets, outdoor air, and firewood stored inside affect child respiratory health,[40,46] and higher versus lower endotoxin exposure indoors affects allergic response.[97,98] Unless effectively cleaned to reduce particle burdens and appropriately ventilated and maintained, houses always have a mixture of agents known to affect respiratory health.[8,44,99] Climate change is increasing the risks for mold and dampness in buildings due to increased risks of storms and because buildings are built to a code that no longer reflects the local environment.[100]

Mycotoxins in Food

From the first use of the term in the literature (ca. 1960), "mycotoxin" refers to secondary metabolites produced by microfungi that are known to cause disease and death in humans and other animals. In the United States and Canada, more cereal-based foods, nuts, and grain samples are analyzed for mycotoxins than the sum of all other contaminants combined. The analysis of cereals alone costs more than $US 200 million per year, more in bad years, costs mostly paid by farmers and the private sector.[101] In high-income countries, few people outside of national regulators such as the US Food and Drug Administration, Health Canada, and the agrifood value chain have heard of mycotoxins. In bad years, costs to the agrifood system in the United States and Canada range from hundreds of millions to billions of dollars for individual mycotoxins, with total annual costs in the United States of $US 1–2 billion.[102-107] Climate change is greatly increasing the risk of mycotoxin contamination in crops globally.[108]

Five important mycotoxins pose challenges to the agrifood system globally: aflatoxin, deoxynivalenol, fumonisin, ochratoxin, and zearalenone. In the United States and

Canada, aflatoxin in corn is produced by *Aspergillus flavus* and in peanuts by *A. parasiticus*. In the United States and Canada, deoxynivalenol and zearalenone are produced by *Fusarium graminearum*, and, in stored grain, ochratoxin is produced by *Penicillium verrucosum*.[109] Aflatoxin has the unfortunate distinction of being the only chemical having a tolerable daily intake (TDI) from the World Health Organization (WHO)/Food and Agriculture Organization (FAO) committee based on population data on cancer in humans.[110] The TDI of fumonisin is based on a study of liver toxicity in male transgenic mice and renal toxicity in a 90-day rat study.[111] Because deoxynivalenol is a high human exposure toxin in Europe, China, and North America which that cannot be eliminated, there is a comprehensive understanding of its effects. Its TDI is based on weight reduction in mice associated with its impact on the brain.[112,113]

Which toxins are most challenging varies considerably by the crop and where it is grown. In the United States, aflatoxin, fumonisin, and deoxynivalenol are the most important. Aflatoxin is normally confined to areas that are experiencing hot drought conditions in corn and nuts, especially peanuts. Fumonisin occurs in corn under similar conditions. Whether fumonisin or aflatoxin will occur in corn in the United States depends on the timing of the weather events and the presence of insect damage. In Canada, neither toxin occurs in corn because it is not warm enough in the country's major corn-producing areas. Deoxynivalenol is a serious problem in some years, mainly around the Great Lakes, in wheat and/or corn depending on environmental conditions. In contrast to Europe, ochratoxin is less of a problem in the major grain areas of the United States and Canada because it is too cold in the winter for the producing fungi to grow in storage.[109,113] Over the past decades, the pattern of toxins in crops in the United States and Canada has changed. Several decades ago, it was more common to find material concentrations of zearalenone and T-2/HT-2 toxin in crops in the United States and Canada than it is now because it is warmer.[114] The prevalence of aflatoxin in the US corn crop is anticipated to continue to increase because of further warming.[115]

There is a sophisticated understanding of the impact of mitigation steps and the effect of processing on the important mycotoxins.[113] Monitoring of foods and commodities ensures that mycotoxin concentrations are well below tolerable limits.[116-119] There are rare examples where the US Food and Drug Administration or the Canadian Food Inspection Agency will seize cereal products found to be noncompliant. These are often imported products or products damaged during shipping and storage.

Exposure can be assessed by validated methods of serum and/or urine samples. No commercial laboratories can analyze urinary or serum biomarkers for the agricultural mycotoxins to legal standards: only the US CDC, US Department of Agriculture, and Agriculture AgriFood Canada have this capacity.[120-122]

Low- and Middle-income Countries

The immense effort to keep mycotoxins out of the North American food system means that, in bad years, the only impact is that food prices go up the following season. For example, in the last very serious year for aflatoxin in corn in the United States (2012), global prices for milled corn and inverted sugar rose by 10% worldwide. In stark contrast, 500–750 million people, mainly in Africa and parts of Latin America and Asia are exposed to the above mycotoxins, often at orders of magnitude above tolerable limits. Part of the reason for this is that, over the past 50 years, the food system in in sub-Saharan

Africa has shifted from several indigenous crops to corn and peanuts, both of which are highly susceptible to aflatoxin contamination. In many parts of Africa, corn can comprise 80% of calories.[123] Aside from this is the fact that crop storage is extremely poor in Africa, with high losses, and aflatoxin increases under such circumstances.[124,125]

Aflatoxin is the most potent human carcinogen by far, accounting for 30% of liver cancer on the planet, much of it in Africa.[126] Aside from liver cancer, aflatoxin exposure has a dramatic impact on children. In bad years, African children die from acute and painful liver toxicosis.[110] At sublethal exposures, aflatoxin contributes to the impacts of the severe malnutrition syndromes kwashiorkor and marasmus.[127] Aside from the increased relative risk for liver cancer, aflatoxin contributes to child stunting, an extremely serious issue in food-insecure countries.[110,123] Corn is essentially the only important source of fumonisin in the diet. In all but South Africa, subsistence farmers are typically exposed to both fumonisin and aflatoxin well above tolerable limits.[123] This combination is the only example of two compounds that are truly synergistic in relevant animal models. The carcinogenicity and toxicity of aflatoxin is greater with concurrent fumonisin exposure.[128]

Another challenge unique to most of Africa is that concurrent exposures occur of not just fumonisin and aflatoxin but also of dexoynivalenol and zearalenone. A recent hospital-based study of 139 pregnant women in Rwanda illustrates this problem. Most of the women (81%) had serum aflatoxin adduct concentrations indicating exposure 1–2 orders of magnitude above current guidance. Zearalenone and its metabolic products were detected in the urine of 61% of the women, most of whom had exposures 2–5 times tolerable limits, with one-third having levels more than an order of magnitude above limits. Urinary deoxynivalenol and metabolic products were found in 77% of the participants. Of these, 28% were twice the tolerable limits and 12% were more than ten-fold higher. Urinary fumonisin was detected in 30% of the women. Approximately one-fifth of the women were above tolerable limits for all four toxins.[120] Aside from the interaction of fumonisin and aflatoxin on liver cancer noted above for adults and children, the impact on the future cancer burden increases as result of in utero exposure. In relevant animal models, in utero exposure to fumonisin can cause neural tube birth defects. The mechanism is relevant to humans.[111] The WHO TDI for zearalenone is based on its reproductive impact in young female pigs and is known to affect young women when present in the diet above the TDI.[113,129] Of the women in the study, 35% had exposures more than an order of magnitude above the tolerable limit, close to the lowest observed adverse effect level. At these levels, exposures would likely affect the development of the fetus.[120]

Three Important Health Issues

From a population health perspective, allergies to fungi in outdoor air, allergic and respiratory disease resulting from mold and dampness indoors, and mycotoxins in low- and middle-income countries represent three of the health effects associated with fungi. A number of fungi are included in the realm of molds that cause invasive disease in immune-compromised individuals but the population health burden in the United States and Canada is low.[130] All three of these issues are being made worse due to climate change, more costly to manage in high-income countries, and, in the case of mycotoxins in low- and middle-income countries, increasing an already serious impact. The only proven methods to alleviate this situation are to increase food diversity, which would require a decade of effort with massive investment, and to improve crop storage methods.[123]

References

1. Hawksworth DL, Lücking R. Fungal diversity revisited: 2.2 to 3.8 million species. *Microbiol Spectr.* 2017;5(4):5–4.

2. Wijayawardene NN, Hyde KD, Al-Ani LKT, et al. Outline of fungi and fungus-like taxa. *J Fungal Biol.* 2020;11(1):1060–456.

3. Khattab A, Levetin E. Effect of sampling height on the concentration of airborne fungal spores. *Ann Allergy Asthma Immunol.* 2008;101(5):529–34.

4. Mullins J, Flannigan B. Microorganisms in outdoor air. In: Flannigan B, Samson RA, Miller JD, eds., *Microorganisms in Home and Indoor Work Environments: Diversity, Health Impacts, Investigation and Control.* 2nd ed. New York: Taylor & Francis; 2011: 1–24.

5. Barnes C. Fungi and atopy. *Clin Rev Allergy Immunol.* 2019;57(3):439–48.

6. Horner WE, Helbling A, Salvaggio JE, Lehrer SB. Fungal allergens. *Clin Microbiol Rev.* 1995;8(2):161–79.

7. Dales RE, Cakmak S, Burnett RT, Judek S, Coates F, Brook JR. Influence of ambient fungal spores on emergency visits for asthma to a regional children's hospital. *Am J Respiratory Crit Care Med.* 2000;162(6):2087–90.

8. Salo PM, Arbes Jr SJ, Sever M, Jaramillo R, Cohn RD, London SJ, Zeldin DC. Exposure to *Alternaria alternata* in US homes is associated with asthma symptoms. *J Allergy Clin Immunol.* 2006 118(4):892–8.

9. Salo PM, Arbes Jr SJ, Crockett PW, Thorne PS, Cohn RD, Zeldin DC. Exposure to multiple indoor allergens in US homes and its relationship to asthma. *J Allergy Clin Immunol.* 2008;121(3):678–84.

10. Beggs PJ. Climate change, aeroallergens, and the aeroexposome. *Environ Res Let.* 2021; 16(3):035006.

11. Flannigan B, Miller JD. Health implications of fungi in indoor environments: an overview. In: Samson R, Flannigan B, Flannigan ME, Graveson S, eds., *Health Implications of Fungi in Indoor Environments.* Amsterdam: Elsevier; 1994: 3–28.

12. Miller JD. Health effects from mold and dampness in housing in western societies: early epidemiology studies and barriers to further progress. In: Adan O, Samson RA, eds., *Molds, Water, and the Built Environment.* Wageningen, The Netherlands: Wageningen Academic Press; 2011: 211–43.

13. Christian JE. A search for moisture sources. In: *Bugs Mold & Rot II Proceedings.* Building Environment and Thermal Envelope Council; 1993: 71–82.

14. Johnson L, Miller JD. Consequences of large-scale production of marijuana in residential buildings. *Indoor Built Environ.* 2012;21(4):595–600.

15. Murray DM, Burmaster DE. Residential air exchange rates in the United States: empirical and estimated parametric distributions by season and climatic region. *Risk Analysis.* 1995;15(4):459–65.

16. Biddle J. Explaining the spread of residential air conditioning, 1955–1980. *Explor Econ Hist.* 2008;45(4):402–23.

17. Hung LL, Caufield S, Miller JD, eds. *Recognition, Evaluation and Control of Indoor Mold.* 2nd ed. Fairfax, VA: American Industrial Hygiene Association; 2020.

18. Miller JD. Mycological investigations of indoor environments. In: Flannigan B, Samson RA, Miller JD, eds., *Microorganisms in Home and Indoor Work Environments: Diversity, Health Impacts, Investigation and Control.* 2nd ed. New York: Taylor & Francis; 2011: 229–45.

19. National Academies of Science. *Damp Indoor Spaces and Health Institute of Medicine Committee on Damp Indoor Spaces and Health.* Washington, DC: National Academies Press; 2004: chapter 2.

20. Gravesen S, Nielsen PA, Iversen R, Nielsen KF. Microfungal contamination of damp buildings-examples of risk constructions and risk materials. *Environ Health Perspect.* 1999;107(S3):505–8.

21. Miller JD. Fungi as contaminants of indoor air. *Atmosph Environ.* 1992;26(12): 2163–72.

22. Miller JD. Fungi and the building engineer. In: Geshwiler M, ed., *ASHRAE IAQ '92, Environments for People.* Atlanta, GA: American Society Health Refrigeration Air-Conditioning Engineers; 1993: 147–62.

23. Dekker C, Dales R, Bartlett S, Brunekreef B, Zwanenburg H. Childhood asthma and the indoor environment. *Chest.* 1991;100(4):922–6.

24. Mudarri D, Fisk WJ. Public health and economic impact of dampness and mold. *Indoor Air.* 2007;17(3):226–35.

25. Krieger J, Jacobs DE, Ashley PJ, et al. Housing interventions and control of asthma-related indoor biologic agents: a review of the evidence. *J Public Health Man Practice.* 2010;16(5):S11–S20.

26. Mendell MJ, Mirer AG, Cheung K, Tong M, Douwes J. Respiratory and allergic health effects of dampness, mold, and dampness-related agents: a review of the epidemiologic evidence. *Environ Health Perspect.* 2011;119(6):748–56.

27. Quansah R, Jaakkola MS, Hugg TT, Heikkinen SA, Jaakkola JJ. Residential dampness and molds and the risk of developing asthma: a systematic review and meta-analysis. *PLoS One.* 2012;7(11): e47526.

28. Sauni R, Uitti J, Jauhiainen M, Kreiss K, Sigsgaard T, Verbeek JH. Remediating Buildings damaged by dampness and mould for preventing or reducing respiratory tract symptoms, infections and asthma. *Cochrane Database Syst Rev.* 2012;9:CD007897.

29. Dockery DW, Pope CA, Xu X, Spengler JD, Ware JH, Fay ME, Speizer FE. An association between air pollution and mortality in six US cities. *N Engl J Med.* 1993;329(24):1753–59.

30. Laden F, Schwartz J, Speizer FE, Dockery DW. Reduction in fine particulate air pollution and mortality: extended follow-up of the Harvard Six Cities study. *Am J Respir Crit Care Med.* 2006;173(6):667–72.

31. Brunekreef B, Dockery DW, Speizer FE, Ware JH, Spengler JD, Ferris BG. Home dampness and respiratory morbidity in children. *Am Rev Respir Dis.* 1989;140(5):1363–7.

32. Franklin CA, Burnett RT, Paolini RJ, Raizenne ME. Health risks from acid rain: a Canadian perspective. *Environ Health Perspec.* 1985;63:155–68.

33. Stern BR, Raizenne ME, Burnett RT, Jones L, Kearney J, Franklin CA. Air pollution and childhood respiratory health: exposure to sulfate and ozone in 10 Canadian rural communities. *Environ Res.* 1994;66:125–42.

34. Dales RE, Zwanenburg H, Burnett R, Franklin CA. Respiratory health effects of home dampness and molds among children. *Am J Epidemiol.* 1991;134(2):196–203.

35. Dales RE, Schweitzer I, Bartlett S, Raizenne M, Burnett R. Indoor air quality and health: reproducibility of respiratory symptoms and reported home dampness and molds using a self-administered questionnaire. *Indoor Air.* 1994;4(1):2–7.

36. Dales RE, Miller JD. Residential fungal contamination and health: microbial cohabitants as covariates. *Environ Health Perspect.* 1999;107(s.3):481–3.

37. Miller JD. Indoor mold exposure: epidemiology, consequences and immunotherapy. *Can J Allergy Clin Immunol.* 1997;2:25–32.

38. Dales RE, Miller JD, McMullan E. Indoor air quality and health: validity and determinants of reported home dampness and moulds. *Int J Epidemiol.* 1997;26(1):120–25.

39. Lawton MD, Dales RE, White J. The influence of house characteristics in a Canadian community on microbiological contamination. *Indoor Air.* 1998;8(1):2–11.

40. Dales RE, Miller JD, Ruest K, Guay M, Judek S. Airborne endotoxin is associated with respiratory illness in the first two years of life. *Environ Health Perspect.* 2006;14(4):610–14.

41. Foto M, Vrijmoed LLP, Miller JD, Ruest K, Lawton M, Dales RE. Comparison of airborne ergosterol, glucan and Air-O-Cell data in relation to physical assessments of mold damage and some other parameters. *Indoor Air.* 2005;15(4):256–66.

42. Dales RE, Ruest K, Guay M, Marro K, Miller JD. Residential fungal growth and incidence of respiratory illness during the first two years of life. *Environ Res.* 2010;110(7):692–8.

43. Miller JD, Dales RE, White J. Exposure measures for studies of mold and dampness and respiratory health. In: Johanning E, ed., *Bioaerosols, Fungi and Mycotoxins: Health Effects, Assessment, Prevention and Control.* Albany, NY: Eastern New York Occupational and Environmental Health Center; 1999: 298–305.

44. Salares VR, Hinde CA, Miller JD. Analysis of settled dust in homes and fungal glucan in air particulate collected during HEPA vacuuming. *Indoor Built Environ.* 2009;18(6):485–91.

45. Jacobs DE. Environmental health disparities in housing. *Am J Public Health.* 2011;101(S1):115–22.

46. Kovesi T, Mallach G, Schreiber Y, et al. Housing conditions and respiratory morbidity in Indigenous children in remote communities in Northwestern Ontario, Canada. *Can Med Assoc J.* 2022;194(3):E80–8.

47. Patino EDL, Siegel JA. Indoor environmental quality in social housing: a literature review. *Building Environ.* 2018;131:231–41.

48. Cox-Ganser JM, White SK, Jones R, et al. Respiratory morbidity in office workers in a water-damaged building. *Environ Health Perspect.* 2005;113(4):485–90.

49. Desroches T, McMullin DR, Miller JD. Extrolites of *Wallemia sebi*, a very common fungus in the built environment. *Indoor Air.* 2014;24(5):533–42.

50. Levac SA. Isolation, purification and characterization of proteins from indoor strains of *Eurotium amstelodami, Eurotium rubrum* and *Eurotium herbariorum* that are antigenic to humans. MSc Thesis. Carleton University, Ottawa, Ontario; 2011.

51. Liang Y, Zhao W, Xu J, Miller JD. Characterization of two related exoantigens from the biodeteriogenic fungus *Aspergillus versicolor. Int Biodegrad Biodeterior.* 2011;65(1):217–26.

52. Luo W, Wilson AW, Miller JD. Characterization of a 52 kDa exoantigen of *Penicillium chrysogenum* and monoclonal antibodies suitable for its detection. *Mycopathologia*. 2010;169(1):15–26.

53. Provost NB, Shi C, She, Y-M, Cyr TD, Miller JD. Characterization of an antigenic chitosanase from the cellulolytic fungus *Chaetomium globosum*. *Med Mycol*. 2013;51(3):290–9.

54. Shi C, Smith ML, Miller JD. Characterization of human antigenic proteins SchS21 and SchS34 from *Stachybotrys chartarum*. *Int Arch Allergy Immunol*. 2011;155(1): 74–85.

55. Shi C, Miller JD. Characterization of the 41 kDa allergen Asp v 13, a subtilisin-like serine protease from *Aspergillus versicolor*. *Mol Immunol*. 2011;48(15–16):1827–34.

56. Shi C, Miller JD. Sta c 3 epitopes and their application as biomarkers to detect specific IgE. *Mol Immunol*. 2012;50(4):271–7.

57. Wilson, AW, Luo W, Miller JD. Using human sera to identify a 52 kDa exoantigen of *Penicllium chrysogenum* and implications of polyphasic taxonomy of anamorphic ascomyce- tes in the study of allergens. *Mycopathologia*. 2009;168(5):213–26.

58. Xu J, Jensen JT, Liang Y, Belisle D, Miller JD. The biology and immunogenicity of a 34 kDa antigen of *Stachybotrys chartarum sensu lato*. *Int Biodegrad Biodeterior*. 2007;60(5):308–18.

59. Miller JD. Fungal metabolites. In: Marcham C, Springston JP, eds., *Bioaerosols: Assessment and Control*. 2nd ed. Cincinnati, OH: American Conference of Government Industrial Hygienists; 2022: in press.

60. Flannigan B, Miller JD. Microbial growth in indoor environments. In: Flannigan B, Samson RA, Miller JD, eds., *Microorganisms in Home And Indoor Work Environments: Diversity, Health Impacts, Investigation and Control*, 2nd ed. New York: Taylor & Francis; 2011: 7–107.

61. Adhikari A, Reponen T, Rylander R. Airborne fungal cell fragments in homes in relation to total fungal biomass. *Indoor Air*. 2013;23(2):142–47.

62. Green BJ, Schmechel D, Summerbell RC. Aerosolized fungal fragments. In: *Fundamentals of Mold Growth in Indoor Environments and Strategies for Healthy Living*. Wageningen, The Netherlands: Wageningen Academic Publishers; 2011: 211–43.

63. Lu R, Pørneki AD, Lindgreen JN, Li Y, Madsen AM. Species of fungi and pollen in the PM1 and the inhalable fraction of indoor air in homes. *Atmosphere*. 2021;12(3):404.

64. Reponen T, Seo SC, Grimsley F, Lee T, Crawford C, Grinshpun SA. Fungal fragments in moldy houses: a field study in homes in New Orleans and Southern Ohio. *Atmosph Environ*. 2007;41(37):8140–9.

65. Phalen RF, Oldham MJ, Wolff RK. The relevance of animal models for aerosol studies. *J Aerosol Med Pulmon Drug Delivery*. 2008;21(1):113–24.

66. Lambrecht MD, Hammad H. Allergens and the airway epithelium response: gateway to al- lergic sensitization. *J Allergy Clin Immunol*. 2014;134(3):499–507.

67. Portnoy JM, Williams P, Barnes CS. Innate immune responses to fungal allergens. *Curr Allergy Asthma Rep*. 2016;16(9):1–6.

68. Rudert A, Portnoy J. Mold allergy: is it real and what do we do about it? *Exp Rev Clin Immunol*. 2017;13(8):823–35.

69. Williams PB, Barnes CS, Portnoy JM. Innate and adaptive immune response to fungal prod- ucts and allergens. *J. Allergy Clin Immunol Pract*. 2016;4(3):386–95.

70. Lee CM. Chitin, chitinases and chitinase-like proteins in allergic inflammation and tissue remodelling. *Yonsei Med J*. 2009;50(1):22–30.

71. Rand TG, Sun M, Gilyan A, Downey J, Miller JD. Dectin-1 and inflammation-associated gene transcription and expression in mouse lungs by a toxic (1,3)-beta-D: glucan. *Arch Toxicol.* 2010;84(3):205–20.

72. Rand TG, Robbins C, Rajaraman D, Sun M, Miller JD. Induction of Dectin-1 and asthma-associated signal transduction pathways in RAW 264.7 cells by a triple helical (1, 3)-β- D glucan, curdlan. *Arch Toxicol.* 2013;87(10):1841–50.

73. Harding CF, Pytte CL, Page KG, Ryberg KJ, Normand E, Remigio GJ, Abreu N. Mold inhalation causes innate immune activation, neural, cognitive and emotional dysfunction. *Brain Behav Immun.* 2020;87:218–28.

74. Miller JD, McMullin DW. Fungal secondary metabolites as harmful indoor air contaminants: 10 years on. *Appl Microb Biotechnol.* 2014;98(24):9953–66.

75. McMullin DR, Renaud JB, Barasubiye T, Sumarah MW, Miller JD. Small molecules and peptaibols of *Trichoderma* species from damp building materials. *Can J Microbiol.* 2017;63(7):621–32.

76. Rand TG, Chang CT, McMullin DR, Miller JD. Inflammation-associated gene expression in RAW 264.7 macrophages induced by toxins from fungi common on damp building materials. *Toxicol Vitro.* 2017;43:16–24.

77. Larenas-Linnemann MD, Baxi S, Phipatanakul W, Portnoy JM. Clinical evaluation and management of patients with suspected fungus sensitivity. *J Allergy Clin Immunol Pract.* 2016;4(3):405–14.

78. Miller JD, Rand T, Jarvis BB. *Stachybotrys chartarum*: cause of human disease or Media darling? *Med Mycol.* 2003;41(4):271–91.

79. Miller JD. Mycotoxins. In: Rylander R, Jacobs RR, eds., *Organic Dusts Exposure, Effects, and Prevention.* Boca Raton, FL: CRC; 1994: 87–92.

80. Etzel RA, Montaña E, Sorenson WG, et al. Acute pulmonary hemorrhage in infants associated with exposure to *Stachybotrys atra* and other fungi. *Arch Pediatr Adolesc Med.* 1998;152(8):757–62.

81. Dearborn DG, Smith PG, Dahms BB, Allan TM, Sorenson WG, Montaña E, Etzel RA. Clinical profile of 30 infants with acute pulmonary hemorrhage in Cleveland. *Pediatrics.* 2002;110(3):627–37.

82. Croston TL, Lemons AR, Barnes MA, et al. Inhalation of *Stachybotrys chartarum* fragments induces pulmonary arterial remodeling. *Am J Respir Cell Mol Biol.* 2020;62(5):563–76.

83. American Industrial Hygiene Association. Mold resource center. 2021. Available at: https://www.aiha.org/public-resources/consumer-resources/disaster-response-resource-center/mold-resource-centerp.

84. Health Canada. Health Canada Residential Indoor Air Quality Guidelines: Moulds. 2007. Available at: http://www.hc-sc.gc.ca/ewh-semt/alt_formats/hecs-sesc/pdf/pubs/air/mould-moisissures-eng.pdf.

85. National Institute for Occupational Safety and Health. *Preventing Occupational Respiratory Disease from Exposures Caused by Dampness in Office Buildings, Schools, and Other Nonindustrial Buildings.* Publication 2013–102. Cincinnati, OH: National Institute for Occupational Safety and Health, 2012.

86. Barnes CS, Horner WE, Kennedy K, Grimes C, Miller JD. Home assessment and remediation. *J Allergy Clin Immunol Pract.* 2016;4(3):423–31.

87. Mendell MJ, Kumagai K. Observation-based metrics for residential dampness and mold with dose–response relationships to health: a review. *Indoor Air.* 2017;27(3):506–17.

88. Mendell MJ, Adams RI. Does evidence support measuring spore counts to identify dampness or mold in buildings? A literature review. *J Exposure Sci Environ Epidemiol.* 2021;32:177–87.

89. Hung LL, Miller JD, Dillon HK, eds. *Field guide for the Determination of Biological Contaminants in Environmental Samples,* 2nd ed. Fairfax, VA: American Industrial Hygiene Association; 2005.

90. Kawamoto M, Page E. Use of unvalidated urine mycotoxin tests for the clinical diagnosis of illness – United States, 2014. *MMWR.* 2015;64(6):157–8.

91. Nielsen KF, Frisvad JD. Mycotoxins on building materials. In: Adan O, Samson RA, eds., *Molds, Water, and the Built Environment.* Wageningen, The Netherlands: Wageningen Academic Press; 2011: 245–75.

92. Chew GL, Horner WE, Kennedy K, Grimes C, Portnoy J, Barnes C, Miller JD. Procedures to assist healthcare providers to determine when home assessments for potential mold exposure are warranted. *J Allergy Clin Immunol Pract.* 2016;4(3):417–22.

93. Portnoy JM, Kennedy K, Sublett JL, et al. Environmental assessment and exposure control: a practice parameter: furry animals. *Ann Allergy Asthma Immunol.* 2012;108(4):223–38.

94. Phipatanakul W, Matsui E, Portnoy J, et al. Environmental assessment and exposure reduction of rodents: a practice parameter. *Ann Allergy Asthma Immunol.* 2012;109(6):375–87.

95. Portnoy J, Sublett J, Kennedy K, et al. Environmental assessment and exposure reduction of cockroaches: a practice parameter. *J Allergy Clin Immunol.* 2013;132(4):802–32.

96. Portnoy J, Miller J, Williams B, et al. Environmental assessment and exposure control of dust mites: a practice parameter. *Ann Allergy Asthma Immunol* 2013;111(6):465–507.

97. Michel O, Ginanni R, Duchateau J, Vertongen F, Le Bon B, Sergysels R. Domestic endotoxin exposure and clinical severity of asthma. *Clin Exper Allergy.* 1991;21(4):441–8.

98. Michel O, Kips J, Duchateau J, Vertongen F, Robert L, Collet H. Severity of asthma is related to endotoxin in house dust. *Am J Respir Crit Care Med.* 1996;154(6):1641–6

99. Sun L, Miller JD, Van Ryswyk K, Wheeler AJ, Héroux ME, Goldberg MS, Mallach G. Household determinants of biocontaminant exposures in Canadian homes. *Indoor Air.* 2021;32, e12933.

100. National Academies of Science. *Climate Change, the Indoor Environment, and Health.* Washington, DC: National Academies Press; 2011.

101. Renaud JB, Miller JD, Sumarah MW. The mycotoxin testing paradigm: challenges and opportunities for the future. *J AOAC Int.* 2019;102(6):1681–8.

102. Mitchell NJ, Bowers E, Hurburgh C, Wu F. Potential economic losses to the US corn industry from aflatoxin contamination. *Food Add Contam: Part A.* 2016;33(3):540–50.

103. Sassi A, Vardon PJ, Flannery B. Economic impact of mycotoxin contamination in U.S. food and feed production. 2017. *Annual Meeting, Agricultural and Applied Economics Association* #259955. Available at: https://EconPapers.repec.org/RePEc:ags:aaea17:259955.

104. Vardon PJ, McLaughlin C, Nardinelli C. Economic impact of mycotoxin contamination in U.S. food and feed production. *Council for Agricultural Science and Technology (CAST) Task Force Report.* 2003;139:136–42.

105. Wilson W, Dahl B, Nganje W. Economic costs of *Fusarium* head blight, scab and deoxynivalenol. *World Mycotox J.* 2018;11(2):291–302.

106. Xia R, Schaafsma AW, Wu F, Hooker DC. The change in winter wheat response to deoxynivalenol and *Fusarium* head blight through technological and agronomic progress. *Plant Dis.* 2021;105(4):840–50.

107. Yu J, Hennessy DA, Wu F. The impact of Bt corn on aflatoxin-related insurance claims in the United States. *Sci Rep.* 2020;10(1):1.

108. Food and Agriculture Organization. Climate change: unpacking the burden on food safety. Food safety and quality series No. 8. Food & Agriculture Organization *of the United Nations, Rome.* 2020. Available at: https://www.fao.org/documents/card/en?details=ca8185en.

109. Miller JD. Mycotoxins in food and feed: a challenge for the 21st century. In: Li D-W, ed., *Biology of Microfungi.* New York: Springer International; 2016:469–93.

110. Doerge DR, Shephard GS, Adegoke GO, et al. Aflatoxins. Safety evaluation of certain contaminants in food: prepared by the eighty-third meeting of the Joint FAO/WHO Expert Committee on Food Additives (JECFA). WHO Food Additives Series, No. 74. 2016; 3–279.

111. Riley RT, Edwards SG, Aidoo K, et al. Safety evaluation of certain contaminants in food: prepared by the eighty-third meeting of the Joint FAO/WHO Expert Committee on Food Additives (JECFA). WHO Food Additives Series, No. 74. 2018; 415–571.

112. Joint FAO/WHO Expert Committee on Food Additives. Deoxynivalenol. 72nd Joint FAO/WHO Expert Committee on Food Additives and Contaminants. *WHO Food Additives Series 63.* Geneva: World Health Organization; 2011.

113. Miller JD. Mycotoxins: still with us after all these years. In: Popping B, Knowles M, Boobis A, eds., Present Knowledge in Food Safety: A Risk Based Approach. Amsterdam: Elsevier; 2022: in press.

114. Miller JD. Changing patterns of fungal toxins in crops: challenges for analysts. *J AOAC Int.* 2016;99(4):837–41.

115. Yu J, Hennessy D, Tack J, Wu F. Climate change will increase aflatoxin risk in US corn. *Environ Res Lett.* 2022;17(5):054017.

116. Bianchini A, Horsley R, Jack MM, et al. DON occurrence in grains: a North American perspective. *Cereal Foods World.* 2015;60(1):32–56.

117. Kolakowski B, O'Rourke SM, Bietlot HP, Kurz K, Aweryn B. Ochratoxin A concentrations in a variety of grain-based and non–grain-based foods on the Canadian retail market from 2009 to 2014. *J Food Protection.* 2016;79(12):2143–59.

118. Lee HJ, Ryu D. Worldwide occurrence of mycotoxins in cereals and cereal-derived food products: public health perspectives of their co-occurrence. *J Agri Food Chem.* 2017;65(33):7034–51.

119. Zhang K, Flannery BM, Oles CJ, Adeuya A. Mycotoxins in infant/toddler foods and breakfast cereals in the US retail market. *Food Add Contam: Part B.* 2018;11(3):183–90.

120. Collins SL, Walsh JP, Renaud JB, et al. Improved methods for biomarker analysis of the big five mycotoxins enables reliable exposure characterization in a population of childbearing age women in Rwanda. *Food Chem Toxicol.* 2021;147:111854.

121. Riley RT, Torres O, Showker JL, et al. The kinetics of urinary fumonisin B1 excretion in humans consuming maize-based diets. *Mol Nutr Food Res.* 56(9):1445–55.

122. Schleicher RL, McCoy LF, Powers CD, Sternberg MR, Pfeiffer CM. Serum concentrations of an aflatoxin-albumin adduct in the National Health and Nutrition Examination Survey (NHANES) 1999–2000. *Clinica Chimica Acta.* 2013;423:46–50.

123. Wild C, Miller JD, Groopman JD. *Mycotoxin Control in Low and Middle Income Countries.* IARC Working Group Report #9. Lyon: International Agency for Research on Cancer; 2015.

124. Anonymous. Mycotoxins and postharvest losses in Sub-Saharan Africa. European Commission, Brussels. 2021. Available at: https://knowledge4policy.ec.europa.eu/publicat

ion/mycotoxins-postharvest-losses-sub-saharan-\\ africa_en" https://knowledge4policy.ec.europa.eu/publication/mycotoxins-postharvest-losses-sub-saharan-\ africa_en.

125. Pitt JI, Wild CP, Baan RA, Gelderblom WCA, Miller JD, Riley RT, Wu F. *Improving Public Health Through Mycotoxin Control.* No. 158. Lyon: International Agency for Research on Cancer Scientific Publications Series; 2012.

126. Liu Y, Wu F. Global burden of aflatoxin-induced hepatocellular carcinoma: a risk assessment. *Environ Health Perspect.* 2010;118(6):818–24.

127. McMillan A, Renaud JD, Burgess KMN, et al. Aflatoxin exposure in Nigerian children with severe acute malnutrition. *Food Chem Toxicol.* 2018;111:356–62.

128. Riley RT, Hambridge T, Alexander J, et al. Co-exposure of fumonisins with aflatoxins. Safety evaluation of certain contaminants in food: prepared by the eighty-third meeting of the Joint FAO/WHO Expert Committee on Food Additives (JECFA). WHO Food Additives Series, No. 74. 2018: 879–948.

129. Eriksen GS, Pennington J, Schlatter J, Alexander J, Thuvander A. Zearalenone. Safety evaluation of certain food additives and contaminants. Prepared by the fifty-third meeting of the Joint FAO/WHO Expert Committee on Food Additives (JECFA), *WHO Food Additives Series,*44, 2000; 393–482.

130. Miller JD. The basis for health concerns. In: Marcham C, Springston JP, eds., *Bioaerosols: Assessment and Control.* 2nd edition. Cincinnati, OH: American Conference of Government Industrial Hygienists; 2022: in press.

44

Physical Exposures

Sophie J. Balk

The Earth has evolved to support plant and animal life. It is warmed and lit by the Sun's rays. People and other living beings are surrounded by sound, both naturally occurring and manmade. Climates vary by season and geographic location. We are all exposed to energy from sunlight, sound, and temperature.

Exposure to ultraviolet radiation (UVR) has the beneficial effects of increasing levels of vitamin D circulating in the body, regulating biorhythms, and improving mood. Excessive or prolonged exposure to sunlight or to artificial sources of UVR can, however, result in skin cancer, considered an epidemic, and other adverse effects. There is evidence that these adverse consequences are increasing due, in part, to the effects of modern life (such as thinning of the ozone layer and use of artificial tanning devices) and because people are living longer, thus allowing for the expression of cumulative, prolonged, or intense exposures. UVR exposure in childhood and adolescence generally is thought to increase the risk of skin cancer.

Perception of sound is often needed for survival, communication, and other reasons, but exposure to excessive sound can result in hearing loss. Hearing loss early in life can adversely affect learning, potentially altering a child's life trajectory. Very young children rely on adults to remove them from excessively noisy situations. Because of developmental immaturity, older children and adolescents may not understand the consequences of loud sounds, including listening to loud music delivered through headphones and other personal listening devices. Environmental noise (mainly from transportation sources) has physiologic and psychological effects and also has effects on children's learning.

Temperature extremes (including increasingly common heat extremes as the result of global climate change) can kill; infants and young children are at higher risk compared to older children and adults and thus depend on adults to recognize the hazards and take steps to prevent them.

Ultraviolet Radiation, Sun Protection, and Skin Cancer

People are exposed to UVR via natural sunlight and artificial sources. UVR, a part of the Sun's electromagnetic spectrum, ranges from short-wave, high-energy ultraviolet C radiation (UVC, < 290 nanometers [nm] in length) to longer wave, lower energy UVB radiation (290–320 nm), to UVA radiation (320–400 nm). UVC emanating from sunlight is completely absorbed by stratospheric ozone and has no biologic significance. UVB and UVA are transmitted to the Earth's surface. The amount of UVB reaching the Earth has increased as a result of depletion of the stratospheric ozone layer;

Sophie J. Balk, *Physical Exposures* In: *Textbook of Children's Environmental Health*, Second Edition. Edited by: Ruth A. Etzel and Philip J. Landrigan, Oxford University Press. © Oxford University Press 2024. DOI: 10.1093/oso/9780197662526.003.0044

chlorofluorocarbons used as aerosol propellants and in refrigeration and air condition-ing can destroy ozone.[1] The main source of artificial UVR exposure is through tanning booths and tanning lamps.

Health Effects of Exposure to Ultraviolet Radiation

Sunlight is beneficial to humans because exposure helps to regulate biorhythms and pro-mote feelings of well-being. Exposure to UVR also is needed for vitamin D production from precursors found in skin. Excessive UVR exposure, however, has deleterious effects on the skin, eyes (such as cataract development), and immune system (immune suppres-sion). Adverse effects on the skin include sunburning, tanning, phototoxic and photoal-lergic reactions, skin aging (also known as photoaging), non-melanoma skin cancer, and melanoma skin cancer.[1]

Skin cancers are the most common cancers in the Unites States and far exceed the combined total of other human cancers.[2] The majority of skin cancers comprise basal cell carcinoma (BCC), squamous cell carcinoma (SCC), and melanoma. BCC and SCC (grouped together as keratinocyte carcinoma [KC]]) are the major types of non-melanoma skin cancer (NMSC). Many epidemiologists are of the opinion that skin can-cer has reached epidemic proportions. In the United States, about 5.4 million BCCs and SCCs are diagnosed each year in more than 3.3 million Americans.[3] BCC and SCC are highly curable if detected early and removed. Nonetheless, the American Cancer Society (ACS) estimates that, in the United States, about 2,000 – 8000 people die of BCC and SCC each year, with most of these deaths from SCC.[3] Unlike most other cancers, BCC and SCC are not reported to or tracked by cancer registries; thus, the exact number of people developing or dying from these cancers is not precisely known.[3] Treatment results in substantial morbidity and cost. BCCs and SCCs usually occur in maximally sun-exposed areas (most commonly on the head and neck) of fair-skinned people.[4] These cancers are uncommon in people with larger amounts of natural pigmentation. Most cases occur in people older than age 50, with the incidence in these people increasing.[2] These can-cers also are increasing in young adults. Between 1976 and 2003, the incidence of BCC increased significantly among young women (<40 years of age), and the incidence of SCC increased significantly among both men and women.[5] Cumulative exposure to sun-light over long periods is considered important in the pathogenesis of SCC.[6]

Melanoma usually occurs in skin but can be found in other locations such as the eyes. Approximately 132,000 melanoma skin cancers occur globally each year.[7] After increas-ing for decades, the incidence rates of invasive melanoma stabilized in women under 50 and declined by about 1% per year in men under 50 since the early 2000s; incidence rates in women ages 50 and older continued to increase by about 1% per year from 2015 to 2019, but stabilized in men.[8]

Melanoma is most likely to occur in males and at older ages but is not uncommon among people under age 30.[9] People at highest risk have light skin and eyes and sunburn easily. Approximately 25% of reported sun exposure occurs during childhood and ado-lescence.[10] Sunlight exposure during these vulnerable periods, especially in the form of blistering sunburns, is generally[11,12] considered to result in increased risk of melanoma. Although cancer is not common among teenagers and young adults ages 15–39 years, melanoma was the 4th most common cancer in this age group between 2016 and 2020.[13] Melanoma occurs in children, although it is very rare. The incidence in children is decreasing.[14]

Melanoma is highly curable when detected and treated in its early stages. Although accounting for only 1% of skin cancer cases, melanoma is responsible for a majority of skin cancer deaths.[9] In contrast to past years when there were few good treatments for metastatic melanoma, mortality rates dropped steeply from 2015 to 2019 by approximately 4% yearly due to advances in treatment including targeted and immune therapies.[8,15]

Abundant evidence links UVR exposure to an increased risk of skin cancer. In 1992, the International Agency for Research on Cancer (IARC, a part of the World Health Organization) concluded that "solar radiation causes cutaneous malignant melanoma and non-melanocytic skin cancer."[16] Since that time, numerous epidemiologic, biologic, and cellular studies have strengthened the link between sun exposure and skin cancer.[1]

Artificial Tanning

Skin cancer risk also is raised when people use tanning beds and tanning lamps to expose themselves to artificial UVR in tanning salons and other venues such as health clubs and apartment buildings. Figure 44.1 shows a tanning bed.

In 2006, the IARC issued a report based on a meta-analysis of 19 studies that examined the associations between tanning bed use and skin cancer. This analysis revealed a 15% increase in risk of melanoma for individuals who had ever used an artificial tanning device ("sunbed") compared with those who never did, although there was no consistent evidence of a dose-response relationship. The relative risk of melanoma was 75% greater with first use of the tanning bed before age 35.[17] In 2009, the IARC declared that UVR from salons was a Group 1 carcinogen—known to cause cancer in humans.[18] Three large studies added strength to the association between tanning bed exposure and increased skin cancer risk.[19–21] It is estimated that 25% of melanomas observed in young women might be attributable to tanning beds.[22]

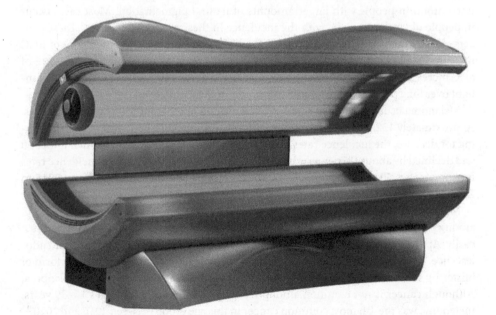

Figure 44.1. Tanning bed

The prevalence of indoor tanning has declined significantly in adults and adolescents since the publication of the 2009 IARC statement. A meta-analysis of indoor tanning prevalence worldwide during 1986–2012 found a past-year prevalence of 18% in adolescents and 14% in adults, with higher prevalence during 2007–2012. During 2013–2018, indoor tanning prevalence dropped to 6.7% in adolescents and 11.9% in adults. This decrease most likely reflects an increase in regulations designed to limit this exposure, especially in adolescents.[23] Nonetheless, these data indicate that a significant percentage of teens and adults continue to expose themselves to carcinogenic artificial UVR.

Tanning Addiction

Many people continue to tan despite risks that are increasingly well known. Sunbathing and artificial tanning persist widely, in part because of a belief that a tanned appearance is attractive and healthy. In one study, 68% of youth strongly agreed or agreed with the statement "I look better when I have a tan," and 55% strongly agreed or agreed with the statement "I feel healthy when I have a nice tan."[24] Motivations in some tanners include relaxation and a better mood. Indoor tanning is associated high levels of physical activity, playing on sports teams, and attempts at weight control by healthy or unhealthy methods. This indicates a possible contradiction whereby generally beneficial behaviors such as increased physical activity or participation in sports are coupled with a behavior that is carcinogenic.[25] In view of potentially severe consequences, it has been suggested that the continued and frequent use of tanning beds by some individuals may be motivated, at least in part, by a type of UV light substance-related disorder,[26] and the reinforcing properties of artificial tanning have been conceptualized within an addiction framework.[27]

Prevention

Since most skin cancers are caused by UVR exposures, most are preventable. "The Surgeon General's Call to Action to Prevent Skin Cancer," issued in 2014, identified skin cancer as a major public health issue and set forth prevention goals, including for the healthcare sector.[28] In March 2018, the United States Preventive Services Task Force (USPSTF) recommended that clinicians counsel fair-skinned individuals ages 6 months to 24 years about ways to decrease skin cancer risk.[29] This recommendation received a USPSTF "B" rating, indicating "that here is high certainty that the net benefit is moderate or there is moderate certainty that the net benefit is moderate to substantial."[30] Most public health and professional organizations—including the American Academy of Pediatrics—generally recommend that children and families avoid deliberate sunburning and sun tanning, wear protective clothing and hats with brims, seek shade, time activities outside the hours of peak sun exposure (10 AM–4 PM), and apply sunscreen. Wearing of sunglasses is recommended. Consumers should look for a label stating that the sunglasses block at least 97% of UVA and UVB rays. Adolescents and young adults should avoid exposure to artificial sources of UVR at tanning salons, gyms, or other indoor venues.[1]

Sunscreen prevents sunburn. One randomized controlled trial demonstrated that properly applied sunscreen significantly decreased the incidence of SCC and melanoma.[31,32] No studies have conclusively demonstrated that sunscreen use decreases the incidence of BCC.

It is recommended that sunscreen be applied whenever a person might sunburn. The US Food and Drug Administration (FDA) regulates sunscreens as nonprescription drugs and governs sunscreen safety and effectiveness in the United States. Seventeen chemicals have been approved by the FDA for sunscreen. In 2011, the FDA set an updated rule governing labeling of sunscreen. The rule established a standard "broad-spectrum" test procedure to measure UVA protection in relation to UVB protection. Manufacturers whose products pass the broad-spectrum test are allowed to use "Broad Spectrum" on their label. Manufacturers of products with a sun protection factor (SPF, a measure of protection the chemical offers against the burning effects of UVB) of 15 or greater are allowed to state that the product protects the user from sunburn and skin cancer when used with other sun protection measures. For sunscreens labeled with an SPF but not "Broad Spectrum," manufacturers are allowed to state that their product protects against sunburn only. The rule also prohibits manufacturers from labeling sunscreens as "waterproof" or "sweatproof" or from using "sunblock" because these terms overstate effectiveness. Manufacturers also must provide information about water resistance.[6] In 2019[33] (with another update in September 2021),[34] the FDA proposed updates to sunscreen regulations. Just two ingredients—zinc oxide and titanium dioxide—were classified as safe and effective. Concern has been raised about several other sunscreen chemicals—including oxybenzone, a commonly used ingredient—because they are absorbed through the skin[35,36]; several may have hormonal properties, raising concern particularly about potential effects on infants and the fetus.[37,38] The American Academy of Pediatrics recommends avoiding oxybenzone if other formulations without this ingredient are available.[1] Sunscreen use is, however, strongly recommended as part of the total program of sun protection. There is no benefit to sunburning, and sunburning raises skin cancer risk.

Certain sunscreens have been implicated as sources of harm to marine ecosystems, including coral reefs.[39] As of January 2021, the state of Hawaii prohibited selling and distributing sunscreens containing the chemicals oxybenzone and octinoxate in the Hawaiian Islands.[40]

Because of the link of artificial tanning to skin cancer risk, the World Health Organization, the American Academy of Pediatrics, the American Academy of Dermatology, and many other authoritative agencies in countries around the world strongly recommend that legislation be passed to prohibit minors from tanning in salons. At least 44 states and the District of Columbia regulate minors' indoor tanning,[41] as do some cities and counties. Several nations have banned teen tanning. Brazil and Australia[42] banned tanning for all age groups.

Deliberate exposure to natural and/or artificial sources with the goal of increasing levels of vitamin D is discouraged because the risk of skin cancer is increased with UVR exposure. People are encouraged to obtain vitamin D through the diet and through supplements if needed.[1,6]

Noise

Noise has often been defined as loud, unpleasant, unexpected, or undesired sound. A new definition—noise is "unwanted and/or harmful sound"—was adopted by the International Commission on Biological Effects of Noise (ICBEN) in June 2023.[43] Since harmful noise levels are sometimes sought during leisure activities, this definition is more accurate. "Sound" usually refers to the form of energy that produces hearing.

Sound results from vibration, usually in air. Hearing and localizing sounds are of survival value to humans, other mammals, and other species. Sound is measured in terms
of frequency (pitch), amplitude (loudness), and time pattern. The *frequency* of sound
is measured in cycles per second, as designated by the term Hertz (Hz): 1 Hz equals 60
cycles per second. Humans respond to a range of 20–20,000 Hz and are most sensitive
at 500–3000 Hz, the range of frequencies that includes human speech. Sound volume or
amplitude is measured in pascals (pa) or decibels (dB). The dB is a unit that indicates the
ratio of one sound level to another (the relative measure of sound intensity). The most
commonly used unit is dB SPL, indicating that the ratio of sound pressure levels (SPL) is
being used.[6] The dB scale is logarithmic, not linear, to give a manageable measure of the
wide range of sound intensities from the threshold of human hearing to the threshold
of pain at many thousand times that intensity.[44] With regard to *time pattern*, sound can
be continuous or intermittent. Detailed information about sound characteristics and
hearing is available.[44] Table 44.1 lists the decibel ranges and effects of common sounds.

Table 44.1 Decibel Ranges and Effects of Common Sounds

Example	Sound Pressure (dBA)	Effect from Exposure
Breathing	0–10	Threshold of hearing
Whisper, rustling leaves	20	Very quiet
Quiet rural area at night	30	
Library, soft background music	40	
Quiet suburb (daytime), conversation in living room	50	Quiet
Conversation in restaurant or average office, background music, chirping bird	60	Intrusive
Freeway traffic at 15 m, vacuum cleaner, noisy office or party, TV audio	70	Annoying
Garbage disposal, clothes washer, average factory, freight train at 15 m, food blender, dishwasher, arcade games	80	Possible hearing damage
Busy urban street, diesel truck	90	Hearing damage
Power lawn mower, iPod or other MP3 player, motorcycle at 8 m. outboard motor, farm tractor, printing plant, jack hammer, garbage truck, jet takeoff (305 m away), subway	100	
Automobile horn at 1 m, boom box stereo held close to ear, steel mill, riveting	110	
Front row at live rock music concert, siren, chain saw, stereo in cars thunderclap, textile loom, jet takeoff (161 m away)	120	Human pain threshold
Earphones at maximum level, armored personnel carrier, jet takeoff (100 m away)	130	
Aircraft carrier deck	140	
Toy cap pistol, firecracker, jet takeoff (25 m)	150	Eardrum rupture

dBA indicates decibels weighted by the A scale; m, meters.
Source: *Pediatric Environmental Health*, 4th Edition (6).

Health Effects of Exposure to Noise

Hearing sound is important for human communication and often for survival. Sounds perceived as pleasant can have beneficial health effets, including in healthcare settings. For example, a Cochrane review of the effects of music therapy on hospitalized cancer patients showed positive impacts on anxiety, fatigue, pain, and quality of life. Only a few studies in the review concerned pediatric patients so the results should be interpreted cautiously.[45]

Excessive noise exposure can diminish hearing and has other adverse health effects. *Occupational noise* is experienced in workplaces. The most information about the effects of noise on hearing comes from studies in occupational settings. *Environmental noise* ("noise pollution") includes outdoor sources such as road traffic, airport and airplane noise, railways, and wind turbines, and indoor sources including appliances. Environmental noise adversely affects the lives of millions of people; environmental noise exposure has been characterized as "one of the most important environmental stressors affecting public health throughout the world."[46]

Problems related to environmental noise include stress-related illnesses, high blood pressure, speech interference, sleep disruption, and lost productivity.[6] Sources of *recreational noise* ("leisure noise") include personal listening devices and activities including celebratory event and concerts.

Until recently, noise-related trauma to the ears was considered a condition affecting only adults engaged in certain occupations or on battlefields. With the increasing popularity of personal listening devices, there is concern about hearing loss in children and adolescents from acute and chronic noise exposures. Even young children now use devices to listen to music and for in-person and remote learning. Children and teenagers often do not understand the dangers of excessive noise through their devices. They may not remove themselves from noisy situations including events such as concerts, dances, and sporting events, or playing with fireworks, noisy toys, and video games. Infants are unable to remove themselves from noisy environments without help from adults.

Noise-induced hearing loss (NIHL) occurs when excessive noise causes damage to the cochlear hair cells, the sound receptors of the inner ear. NIHL is usually most severe at 4,000 Hz and extends downward to speech frequencies with prolonged exposure. NIHL affects millions of children and adolescents in the United States.[47] *Tinnitus* is a ringing or buzzing in the ears that often results from exposure to loud noise and is often associated with hearing loss. *Hyperacusis*, a heightened sensitivity to a range of sounds that usually do not bother most people, also can result after inner ear damage from loud noise. Exposure to loud noise also may result in a temporary loss in hearing sensitivity known as temporary *noise-induced threshold shift* (NITS). A threshold shift is a decrease in the softest sound a person can hear. NITS may become permanent depending on the duration and severity of noise exposure.

A study using data from the National Health and Nutrition Examination Survey III (NHANES, a periodic survey of the health and nutritional status of the US civilian population—NHANES III was conducted from 1988 to 1994) was one of the first to estimate the prevalence of NITS among US children. Twelve percent to 15% of school-aged children had hearing deficits attributable to noise exposure.[48] Other researchers investigated trends in NITS and hearing loss from 1988–1994 to 2005–2006, to test the hypothesis that more recent cohorts had increased hearing impairment. Audiometric testing was done in a total of 4,310 adolescents 12–19 years of age as part of the 1988–1994 and 2005–2006 NHANES surveys. The overall prevalence of reported exposure to loud noise or listening to music through headphones in the previous 24 hours increased from approximately 20% to 35%. There were no significant overall increases in NITS, high-frequency hearing loss, or low-frequency hearing loss between survey periods. A significant increase in the prevalence of NITS occurred among female teens, erasing previous gender-related differences in the estimated NITS prevalence. In 2005–2006, females had a similar prevalence of exposure to recreational noise and a lower prevalence of using hearing protection compared with males. Although no increase in hearing loss was demonstrated in this study, hearing damage from childhood noise exposure may not manifest fully by the age of 19 years. The effects of noise exposure are thought to be cumulative. Although effects may not be immediately evident in hearing tests, cumulative effects might become evident in the mid-20s.[49] A more recent compilation of NHANES data showed that approximately 1 in 6 middle and high school students had evidence of hearing loss although there was no consistent association with loud music exposure.[50]

Noise exposure can harm physical health. Noise causes a stress response at levels as low as 65 dBA. Noise that is sudden and unexpected can be startling, resulting in a physiologic stress response. At levels of 40–45 dBA, noise contributes to sleep deprivation by increasing awakenings or arousals. Noise has effects on the cardiovascular system; levels greater than 70 dBA cause increases in constriction of blood vessels, heart rate, and blood pressure. Psychological effects of moderate levels of noise include annoyance and feelings of bother, interference with activities, and symptoms such as headache and fatigue.[6] Even if not loud enough to cause hearing loss, repeated exposure to noise during critical periods of development may affect a child's acquisition of speech, language, and language-related skills, such as reading and listening. The inability to concentrate in a noisy environment can affect a child's capacity to learn.[51] Guidelines for noise levels in classrooms, based on the child's age and any vulnerable conditions, have been proposed.[6]

Prevention

Federal and state regulations control noise levels in the workplace. Hearing protection is required for certain levels of noise for specified lengths of time. It is often difficult to control environmental noise. In 1972, Congress passed the Noise Control Act, resulting in the establishment of the Office of Noise Abatement and Control (ONAC) charged with carrying out investigations and studies on noise and its effect on public health and welfare. The US Environmental Protection Agency (EPA) coordinated all federal noise control activities through ONAC. In 1981, however, the Administration concluded that noise pollution matters were best handled at the state and local levels. The ONAC was closed and primary responsibility of addressing noise issues was transferred to state and local governments.[6]

It is recommended that pediatricians and other child health clinicians discuss noise exposure and noise reduction during health supervision visits.[52,53] The most common excessive exposure is likely to be from personal listening devices such as headphones and earbuds. Because teens often feel that they are not vulnerable to harm, they may not understand that excessive noise exposure may have the long-term effects of hearing loss and tinnitus, as well as other health effects. They can be advised to reduce the volume on their devices. If the music at events such as concerts or dances is loud enough to pro-duce discomfort or pain, using hearing protection (ear plugs or ear muffs) is advised, or leaving the event. For younger children, parents can utilize volume control and noise-cancelling features available on headphones and other devices. They can avoid buying toys that make loud noises, including cap pistols. Firecrackers should be avoided to pre-vent injuries to hearing and other injuries.

Heat and Cold

Extreme heat and cold are both environmental health hazards. Children often are more susceptible to these hazards compared to adults. The body normally maintains its tem-perature ("thermoregulates") at approximately 98.6°F (37°C). When people are exposed to heat, thermoregulatory mechanisms enable the body to become cooler through sweating and dilating of blood vessels in the skin. With excessive cold, body temperature is raised through shivering.

Heat Extremes

Extremes of heat can occur as a result of weather events and in manmade environments such as overheated automobiles. A *heat wave* is defined as at least 3 consecutive days with temperatures greater than 90°F (32.2°C). Heat waves are becoming increasingly common due to global climate change (see Chapter 4).[54-55] During heat waves, heat-related hospitalizations and deaths are most common in people working or exercising outdoors in the heat; older adults; and people with mental illnesses, chronic medical conditions, and those with disabilities. Infants and young children, however, also are affected. Social determinants of health play a part in determining risk[56]; this includes access to air conditioning. About 700 heat-related deaths occur in the United States every year.[57]

Compared to adults, children (especially young children) are less well able to regulate body temperatures to normal levels when they are exposed to heat extremes. Infants and young children cannot remove themselves from overheated environments and must rely on adults to keep them cool and hydrated. Older children generally spend more time outdoors compared to adults during play and sports activities. Children sweat less than do adults, thus limiting their capacity to cool the body. Adolescents may be employed in agricultural, landscaping, or other outdoor environments, leading to prolonged expo-sure to hot weather environments.

Heat extremes may result in dehydration. Heat-related illnesses include sunburn, heat cramps, heat rash, heat exhaustion, and heatstroke. Heat exhaustion and heatstroke can result in death when the body is not able to sweat to cool down enough and body temper-ature rises rapidly. Symptoms of heat exhaustion include heavy sweating, muscle cramps, fatigue, weakness, paleness, cold or clammy skin, dizziness, headache, nausea or vomit-ing, and fainting. Untreated heat exhaustion can progress to heatstroke with symptoms of stupor and coma, rapid heart rate, and high or low blood pressure. Even with prompt medical care, 15% of heatstroke cases are fatal.[58]

Injury and death from heat extremes are preventable. Public health messages delivered through the media can make the public aware of the signs and symptoms of heat-related illness and can prevent heat-related illness, injury, and death.[59] Parents and other care-givers should be educated about the heat sensitivity of young children. They should provide adequate hydration during summer months and refrain from overdressing. Parents should never leave children in hot cars even briefly because of the risk of heat illness and death. Children and adolescents involved in play and athletic activities during hot weather should be allowed to have periods of rest and liberal access to fluids.

Cold Extremes

Temperature extremes may occur during prolonged periods of cold weather. Power failures during winter storms can result in cold interior temperatures. Climate change contributes to conditions, such as more severe hurricanes, that may increase potential for exposure to cold. Cold-related hospitalizations and deaths are related to social factors including poverty and being unhoused.[60]

Cold temperatures disproportionately affect children. Contact with cold water can quickly result in hypothermia. Compared to adults, children are more likely to become hypothermic (defined as a core body temperature of <95°F (35°C) when exposed to cold extremes. Children have a greater ratio of body surface area to mass, also predisposing them to hypothermia. Newborn infants have a small amount of subcutaneous fat and have a decreased ability to shiver.

The physiologic effects of cold are categorized as mild, moderate, and severe. Symptoms range from shivering, a slowed or rapid pulse, and confusion seen with mild hypothermia to a severely slowed pulse, heart rhythm abnormalities, and coma seen with severe hypothermia. Cold injuries to body extremities range from frostnip, the mildest form of cold injury, to frostbite, a severe form that results in permanent injury. Treatment of cold injury consists of rewarming.[6]

Cold injuries may be prevented by wearing proper clothing such as hats, mittens, and other clothing that creates a static layer of warm air, providing a barrier against the wind and keeping the body dry,[61] avoiding severe cold situations, and finding alternative housing if the residence has lost its heat. Families should be counseled about the hazards of using potentially dangerous heating sources during cold weather or power outages. Gas stoves, fireplaces, and space heaters are sometimes used to provide supplemental heat, possibly resulting in increased indoor air pollution as well as potentially fatal fire hazards[62] and carbon monoxide poisoning. It has been recommended that local health departments implement strategies tailored to address the needs of vulnerable populations when there is extreme cold. Families trying to save money on fuel may deliberately keep their homes cold; public service announcements advising persons to maintain thermostats at warmer than 60°F (15.6°C) may prevent cases of indoor hypothermia.[60]

References

1. Balk SJ, American Academy of Pediatrics Council on Environmental Health and Section on Dermatology. Technical report: UVR: A hazard to children and adolescents. *Pediatrics.* 2011;127:e791–817.

2. Rogers HW, Weinstock MA, Feldman SR, Coldiron BM. Incidence estimate of nonmelanoma skin cancer (keratinocyte carcinomas) in the US Population, 2012. *JAMA Dermatol.* 2015;151(10):1081–6. doi:10.1001/jamadermatol.2015.1187.

3. Key statistics for basal and squamous cell skin cancers. Available at: https://www.cancer.org/cancer/basal-and-squamous-cell-skin-cancer/about/key-statistics.html. Last Revised: October 31 2023.

4. American Cancer Society. What are basal and squamous cell skin cancers? Available at: https://www.cancer.org/cancer/basal-and-squamous-cell-skin-cancer/about/what-is-basal-and-squamous-cell.html. Last Revised: October 31, 2023.

5. Christenson LJ, Borrowman TA, Vachon CM, et al. Incidence of basal cell and squamous cell carcinomas in a population younger than 40 years. *JAMA*. 2005;294(6): 681–90.

6. American Academy of Pediatrics Council on Environmental Health and Climate Change. In: Etzel RA, Balk SJ, eds., *Pediatric Environmental Health*. 4th ed. Itasca, IL: American Academy of Pediatrics; 2019: Chapter 26: Cold and Heat, pp. 463–73; Chapter 35: Noise, pp. 611–26; Chapter 44: Ultraviolet Radiation, pp. 769–98.

7. World Health Organization. UVR and the INTERSUN Programme. 16 October 2017. Available at: http://www.who.int/uv/faq/skincancer/en/index1.html.

8. American Cancer Society. Cancer facts and figures. 2023. Available at: https://www.cancer.org/content/dam/cancer-org/research/cancer-facts-and-statistics/annual-cancer-facts-and-figures/2023/2023-cancer-facts-and-figures.pdf.

9. American Cancer Society. Key statistics for melanoma skin cancer. Available at: https://www.cancer.org/cancer/melanoma-skin-cancer/about/key-statistics.html. Last Revised: January 17, 2024.

10. Godar DE, Wengraitis SP, Shreffler J, Sliney DH. UV doses of Americans. *Photochem Photobiol*. 2001;73(6):621–9.

11. Whiteman DC, Whiteman CA, Green AC. Childhood sun exposure as a risk factor for melanoma: a systematic review of epidemiologic studies. *Cancer Causes Control*. 2001;12(1):69–82.

12. Pfahlberg A, Kolmel KF, Gefeller O. Timing of excessive UVR and melanoma: epidemiology does not support the existence of a critical period of high susceptibility to solar UVR-induced melanoma. *Br J Dermatol*. 2001;144(3):471–5.

13. National Cancer Institute. Surveillance, Epidemiology and End Results Program. Cancer Stat Facts: Cancer among adolescents and young adults (AYAs) (ages 15–39). Available at: https://seer.cancer.gov/statfacts/html/aya.html.

14. Campbell LB, Kreicher KL, Gittleman HR, Strodtbeck K, Barnholtz-Sloan J, Bordeaux JS. Melanoma incidence in children and adolescents: decreasing trends in the United States. *J Pediatr*. 2015;166(6):1505–13. doi:10.1016/j.jpeds.2015.02.050.

15. Islami F, Ward EM, Sung H, et al. Annual report to the nation on the status of cancer, part 1: national cancer statistics. *J Natl Cancer Inst*. 2021;113(12):1648–69. doi:10.1093/jnci/djab131.

16. International Agency for Research on Cancer. IARC monographs on the evaluation of carcinogenic risks to humans. Vol. 55. Solar and UVR. Summary of data reported and evaluation. Lyon: World Health Organization; 1997. Available at. http://monographs.iarc.fr/ENG/Monographs/vol55/mono55.pdf.

17. International Agency for Research on Cancer Working Group on Artificial Ultraviolet (UV) Light and Skin Cancer. The association of use of sunbeds with cutaneous malignant melanoma and other skin cancers: a systematic review. *Int J Cancer*. 2007;120:1116–22.

18. International Agency for Research on Cancer. Sunbeds and UV radiation. 2009. Available at: http://www.iarc.fr/en/media-centre/iarcnews/2009/sunbeds_uvradiation.php.

19. Lazovich D, Vogel RI, Berwick M, Weinstock M, Anderson KE, Warshaw EM. Indoor tanning and risk of melanoma: a case-control study in a highly exposed population. *Cancer Epidemiol Biomarkers Prev.* 2010;19:1557–68.

20. Cust AE, Armstrong BK, Goumas C, et al. Sunbed use during adolescence and early adulthood is associated with increased risk of early-onset melanoma. *Int J Cancer.* 2011;128:2425–35.

21. Veierød MB, Adami H, Lund E, Armstrong BK, Weiderpass W. Sun and solarium exposure and melanoma risk: effects of age, pigmentary characteristics, and nevi. *Cancer Epidemiol Biomarkers Prev.* 2010; 19;111–20.

22. Diffey B. Sunbeds, beauty and melanoma. *Br J Dermatol.* 2007; 157: 215–6.

23. Rodriguez-Acevedo AJ, Green AC, Sinclair C, van Deventer E, Gordon LG. Indoor tanning prevalence after the International Agency for Research on Cancer statement on carcinogenicity of artificial tanning devices: systematic review and meta-analysis. *Br J Dermatol.* 2020;182(4):849–59. doi:10.1111/bjd.18412.

24. Cokkinides V, Weinstock M, Glanz K, Albano J, Ward E, Thun M. Trends in sunburns, sun protection practices, and attitudes toward sun exposure protection and tanning among US adolescents, 1998–2004. *Pediatrics.* 2006;118(3):853–6.

25. Heckman CJ, Manning M. The relationship between indoor tanning and body mass index, physical activity, or dietary practices: a systematic review. *J Behav Med.* 2019 Apr;42(2):188–203. doi:10.1007/s10865-018-9991-y.

26. Harrington CR, Beswick TC, Leitenberger J, Minhajuddin A, Jacobe HT, Adinoff B. Addictive-like behaviours to ultraviolet light among frequent indoor tanners. *Clin Exp Dermatol.* 2010;36:33–8.

27. Nolan BV, Taylor SL, Liguori A, Feldman SR. Tanning as an addictive behavior: a literature review. *Photodermatol Photoimmunol Photomed.* 2009;25(1):12–9.

28. U.S. Department of Health and Human Services. *US Surgeon General's Call to Action to Prevent Skin Cancer.* Washington, DC: U.S. Department of Health and Human Services; 2014.

29. United States Preventive Services Task Force. Skin cancer prevention: behavioral counseling. March 20, 2018. Available at: https://www.uspreventiveservicestaskforce.org/uspstf/rec ommendation/skin-cancer-counseling.

30. US Preventive Services Task Force. October 2018. Grade definitions. Available at: https://www.uspreventiveservicestaskforce.org/Page/Name/grade-definitions#brec.

31. Green A, Williams G, Neale R, et al. Daily sunscreen application and betacarotene supplementation in prevention of basal cell and squamous-cell carcinomas of the skin: a randomised controlled trial. *Lancet.* 1999;354(9180):723–9.

32. Green AC, Williams GM, Logan V, Strutton GM. Reduced melanoma after regular sunscreen use: randomized trial follow-up. *J Clin Oncol.* 2010; 29:257–63.

33. US Federal Register. Sunscreen drug products. 2019. Available at: https://www.federalregis ter.gov/documents/2019/02/26/2019-03019/sunscreen-drug-products-for-over-the-coun ter-human-use.

34. US Food and Drug Administration. An update on sunscreen requirements: The deemed final order and the proposed order. Content current as of: 12/16/2022. Available at: https://www.fda.gov/drugs/news-events-human-drugs/update-sunscreen-requirements-deemed-final-order-and-proposed-order.

35. Matta MK, Florian J, Zusterzeel R, et al. Effect of sunscreen application on plasma concentration of sunscreen active ingredients: a randomized clinical trial. *JAMA.* 2020;323(3):256–67. doi:10.1001/jama.2019.20747.

36. Matta MK, Zusterzeel R, Pilli NR, et al. Effect of sunscreen application under maximal use conditions on plasma concentration of sunscreen active ingredients: a randomized clinical trial. *JAMA*. 2019;321(21):2082–91. doi:10.1001/jama.2019.5586.

37. Schlumpf M, Cotton B, Conscience M, Haller V, Steinmann B, Lichtensteiger W. In vitro and in vivo estrogenicity of UV screens. *Environ Health Perspect*. 2001;109(3):239–44.

38. Schlumpf M, Kypkec K, Vöktd CC, et al. Endocrine active UV filters: developmental toxicity and exposure through breast milk. *Chimia (Aarau)*. 2008;62:345–51.

39. Watkins YSD, Sallach JB. Investigating the exposure and impact of chemical UV filters on coral reef ecosystems: review and research gap prioritization. *Integr Environ Assess Manag*. 2021;17(5):967–81. doi:10.1002/ieam.4411.

40. Moulite M. Hawaii bans sunscreens that harm coral reefs. July 3, 2018. Available at: https://www.cnn.com/2018/07/03/health/hawaii-sunscreen-ban/index.html

41. AIM at Melanoma. Indoor tanning legislation 2023. Available at: https://www.aimatmelanoma.org/legislation-policy-advocacy/indoor-tanning/

42. Skin Cancer Foundation. Sun and Skin News. Indoor Tanning Legislation: Here's Where We Stand. May 30, 2023. Available at: https://www.skincancer.org/blog/indoor-tanning-legislation-heres-stand/

43. Fink D. A new definition of noise: noise is unwanted and/or harmful sound. Noise is the new 'secondhand smoke'. *Proc Mtgs Acoust*. 2019;39:050002. https://doi.org/10.1121/2.0001186

44. Nave CR. Hyperphysics: Sound and hearing. Available at: http://hyperphysics.phy-astr.gsu.edu/hbase/HFrame.html.

45. Bradt J, Dileo C, Magill L, Teague A. Music interventions for improving psychological and physical outcomes in cancer patients. *Cochrane Database Syst Rev*. 2016;15(8):CD006911.

46. Murphy E, King EO. *Environmental Noise Pollution: Noise Mapping, Public Health and Policy*. Amsterdam: Elsevier; 2014.

47. Centers for Disease Control and Prevention. Preventing noise-induced hearing loss. Available at: https://www.cdc.gov/ncbddd/hearingloss/noise.html

48. Niskar AS, Kieszak SM, Holmes AE, Esteban E, Rubin C, Brody DJ. Estimated prevalence of noise-induced hearing threshold shifts among children 6 to 19 years of age: the Third National Health and Nutrition Examination Survey, 1988–1994, United States. *Pediatrics*. 2001;108:40–3.

49. Henderson E, Testa MA, Hartnick C. Prevalence of noise-induced hearing-threshold shifts and hearing loss among US youths. *Pediatrics*. 2011;127:e39–e46.

50. Su BM, Chan DK. Prevalence of hearing loss in US children and adolescents: Findings from NHANES 1988–2010. *JAMA Otolaryngol Head Neck Surg*. 2017;143(9):920–7. doi:10.1001/jamaoto.2017.0953.

51. Stansfeld S, Clark C. Health effects of noise exposure in children. *Curr Envir Health Rpt*. 2015;2:171–8. https://doi.org/10.1007/s40572-015-0044-1.

52. Balk SJ, Bochner RE, Ramdhanie MA, Reilly BK; American Academy of Pediatrics Council on Environmental Health and Climate Change; Section on Otolaryngology - Head and Neck Surgery. Preventing excessive noise exposure in infants, children, and adolescents. [Policy statement]. *Pediatrics*. 2023;152(5):e2023063752. doi: 10.1542/peds.2023-063752.

53. Balk SJ, Bochner RE, Ramdhanie MA, Reilly BK; American Academy of Pediatrics Council on Environmental Health and Climate Change; Section on Otolaryngology - Head and Neck Surgery. Preventing excessive noise exposure in infants, children, and adolescents. [Technical report]. *Pediatrics*. 2023;152(5):e2023063753. doi: 10.1542/peds.2023-063753.

54. Ahdoot S, Baum CR, Cataletto MB, Hogan P, Wu CB, Bernstein A; Council on Environmental Health and Climate Change; Council on Children and Disasters; Section on Pediatric Pulmonology and Sleep Medicine; Section on Minority Health, Equity, and Inclusion. [Policy statement]. Climate Change and Children's Health: Building a Healthy Future for Every Child. *Pediatrics.* **2024**;153(3):e2023065504.

55. Ahdoot S, Baum CR, Cataletto MB, Hogan P, Wu CB, Bernstein A; Council on Environmental Health and Climate Change; Council on Children and Disasters; Section on Pediatric Pulmonology and Sleep Medicine; Section on Minority Health, Equity, and Inclusion. [Technical report]. Climate Change and Children's Health: Building a Healthy Future for Every Child. *Pediatrics.* **2024**;153(3):e2023065505.

56. US Centers for Disease Control and Prevention. Heat-related deaths – United States, 2004–2018. Available at: https://www.cdc.gov/mmwr/volumes/69/wr/mm6924a1.htm.

57. US Centers for Disease Control and Prevention. Heat-related emergency department visits during the northwestern heat wave – United States. 2021 Jun. Available at: https://www.cdc.gov/mmwr/volumes/70/wr/mm7029e1.htm.

58. US Centers for Disease Control and Prevention. Heat-related deaths – Four states, July–August 2001, and United States, 1979–1999. *MMWR Morb Mortal Wkly Rep.* July 5, 2002. Available at: http://www.cdc.gov/mmwr/preview/mmwrhtml/mm5126a2.htm.

59. US Centers for Disease Control and Prevention. Infographic: beat the heat. Last Reviewed: July 26, 2021. Available at: https://www.cdc.gov/cpr/infographics/beattheheat.htm.

60. Lane, K, Ito, K, Johnson, S, Gibson, EA, Tang, A, Matte, T. Burden and risk factors for cold-related illness and death in New York City. *Int J Environ Res Public Health.* **2018**;15:632. https://doi.org/10.3390/ijerph15040632.

61. US Centers for Disease Control and Prevention. Hypothermia-related deaths – United States, 1999–2002 and 2005. *MMWR Morb Mortal Wkly Rep.* Date Last Reviewed: 3/16/2006. Available at: http://www.cdc.gov/mmwr/preview/mmwrhtml/mm5510a5.htm#box.

62. CNN. Space heater sparked fire in the Bronx that killed 17 people, including 8 children. 2022 Jan 10. Available at: https://www.cnn.com/2022/01/10/us/nyc-bronx-apartment-fire-monday/index.html.

45

Ionizing Radiation

Eric J. Grant and David G. Hoel

X-rays and gamma rays (high-energy photons), alpha particles, beta particles, protons, and neutrons are all forms of ionizing radiation. These different types of radiation are considered ionizing because they all have sufficient energy to displace electrons from molecules, a process termed "ionization." Electrons freed by ionization are highly energized, and, when they are released from molecules within cells in the body, these highly energized electrons can damage other molecules and result in intracellular injury. In particular, they can disrupt DNA to cause genetic mutations. The health effects of radiation are mainly the result of these events.

Cancer is the best studied health effect of ionizing radiation. Ionizing radiation is a well-known cause of cancer, and cancer induced by radiation has been examined epidemiologically as well in the laboratory. Epidemiological studies of radiation include follow-up investigations of the survivors of the atomic bombings in Hiroshima and Nagasaki, studies of nuclear workers with occupational exposures, studies of persons exposed to background environmental radiation, and follow-up studies of persons exposed to medical radiation. Other health effects beyond cancer that appear to be associated with exposure to lower doses of ionizing radiation include cardiovascular effects, cataracts, and disruption of the immune system.

Due to increasing diagnostic examinations using x-ray machines and computed tomography (CT) scanners, the amount of radiation received since the 2000s is increasing. Currently 50% of ionizing radiation exposures are from medical exposures, 48% from environmental exposures (of which two-thirds is from radon), and 2% from consumer products. The average yearly total effective exposure to individuals in the United States is 6.2 millisieverts (mSv). An extensive and detailed analysis of current patterns of exposure to radiation is available.[1]

The major issues in radiation health research today derive from the need to better quantify the risks at lower levels of exposure and to elucidate the effects of acute versus chronic exposures. There is also need for better radiobiological understanding of potential health effects in order to improve estimates of low-dose effects and of risks to sensitive subgroups, especially for radiation types and energy levels that have not been adequately examined in epidemiological studies. New research findings on genomic instability (higher rates of alterations occurring during cell division), adaptive response (where an initial radiation exposure may prime the cell to better withstand a future exposure), bystander effects (changes in cells that were not directly exposed to radiation but react to the cellular milieu that has been affected by exposed cells), and varying gene expression at different dose levels have complicated understanding of low-dose radiation effects. Research also is needed on the noncarcinogenic effects of radiation on health.

Eric J. Grant and David G. Hoel, *Ionizing Radiation* In: *Textbook of Children's Environmental Health*, Second Edition. Edited by: Ruth A. Etzel and Philip J. Landrigan, Oxford University Press.

The National Research Council report on Biological Effects of Ionizing Radiation (BEIR VII)[2] reviews these concepts as they relate to radiation dose and cancer risk.

To assess risks of radiation exposure, a linear no-threshold (LNT) model is most often used. This model has been applied in estimates of environmental exposures to radiation and for assessing risk of cancer in relation to medical diagnostic tests. The rationale for using the LNT model is based on two fundamental concepts: first, that the deposition of physical energy in tissues by ionizing radiation increases cancer risk linearly with increasing dose and, second, that the carcinogenic effectiveness of radiation is constant independent of dose. Although the LNT assumption has been criticized, the BEIR VII Committee concluded "The Committee judged that the linear no-threshold model (LNT) provided the most reasonable description of the relation between low-dose exposure to ionizing radiation and the incidence of solid cancers that are induced by ionizing radiation."

Due to complexities that include numerous different types of ionizing radiation (which can be released from a variety of nature and manmade sources), inconsistent technical terms used to quantify exposure, and the stochastic nature of the effects that take years to manifest, communicating the risks of radiation exposure to the public is often difficult. Communicating those risks can be especially fraught when they are coupled with an adverse event, such as a release of radioactive materials from a nuclear power plant.

Epidemiology of Radiation Cancer

Epidemiologic studies of cancer induced by ionizing radiation have been extensive and provide the basis for much of our current understanding of radiation carcinogenesis. These studies provide information on radiation effects across a wide range of exposures. They include populations with high-dose exposures to natural background radiation, workers exposed occupationally, populations exposed to nuclear testing and to reactor accidents, survivors of the atomic bombings, and patients who have received high-dose medical radiation therapies.

Radiation cancer studies are very similar to other types of occupational and environmental cancer studies because radiation-induced cancers are not distinguishable pathologically from cancers caused by other exposures. Radiation cancer studies must deal with methodological issues common to all epidemiological investigations, such as exposure confounding, long latencies (e.g., 10–20 years for solid tumors), accurate measurement of levels of exposure, and study size (see Chapter 6). The most informative radiation cancer studies have been those that involve large numbers of individuals of both sexes, exposed to large radiation doses at different ages at exposure, and followed over several decades.

Atomic Bomb Survivor Studies

The Radiation Effects Research Foundation (RERF) in Hiroshima and Nagasaki, Japan has published a series of cancer mortality and incidence studies covering the years

1950–2009 on a stable, well characterized population—the Life Span Study cohort—consisting of more than 100,000 atomic bomb survivors. The RERF analyses of incidence for solid (nonleukemia) cancers cover the years 1958–1998 and provided the basis for the BEIR VII cancer risk models,[2,3] while the most recent solid cancer study follows the survivors through 2009.[4] RERF has published a series of cancer and non-cancer mortality reports. Report 14 is the most recent[4] with a total of 10,929 deaths due to solid cancers with an estimated excess relative risk (ERR; see the Glossary for an explanation of "ERR") of 0.47 (95% confidence interval [CI]: 0.38, 0.56) per Gray (Gy).[5] RERF data show that the radiation risks differ by organs. Certain organs such as the thyroid, breast, urinary bladder, and lung have relatively high risks while other organs (e.g., the pancreas and rectum) are relatively lower. Details of the RERF study cohorts, publications, and downloadable data can be accessed at the RERF website (http://www.rerf.or.jp).

The Life Span Study population includes many persons who were exposed to the atomic bombings as children and adolescents—24,000 persons who were exposed under the age of 10 years, and another 25,000 who were exposed between the ages of 10 and 19 years. It also includes 1,000 persons who were exposed in utero; the findings on persons exposed in utero to the atomic bombings are described in the next section. RERF also publishes analyses of bombing survivors who are enrolled in the Adult Health Study, a subcohort of the Life Span Study consisting of 22,400 persons who are invited on a biannual basis to visit RERF's clinics to monitor non-cancer morbidities.

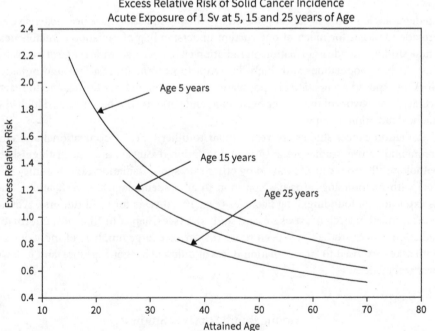

Figure 45.1. Excess relative risk of solid cancer incidence after acute exposure of 1 Sv at 5, 15 and 25 years of age

The RERF cancer incidence studies of persons exposed to the atomic bombings as children and adolescents show clearly that the ERR for solid cancers is greatest for persons who were exposed as children. For example, it is estimated that a child exposed at age 5 will have an ERR of cancer at future ages that is about twice as great as that of an adult exposed at age 40–45 years (see Figure 45.1). These studies also show that the ERR for solid cancers decreases with advancing age as the background risks of cancer increase exponentially with advancing age.

In addition to these general findings regarding age at exposure for all solid cancers in aggregate, recent results have demonstrated that radiation risks in particular organs may be particularly sensitive to age at exposure as well as other life events. In a paper on the risks of breast cancer, RERF scientists showed that girls who experienced earlier menarche were at higher risks of radiation-induced breast cancer than were their peers who experienced menarche at an older age. Further, it was demonstrated that girls who experienced exposure to radiation coinciding with menarche had even higher risks.[6]

Studies of Radiation Exposure in Utero and in Early Childhood

Dr. Alice Stewart, an English physician and pioneering radiation epidemiologist, reported in 1956 that diagnostic x-rays of the developing fetus may be carcinogenic. This finding was based on a case-control study of children under the age of 10 who had died of cancer in England and Wales during the mid-1950s. Dr. Stewart observed nearly a doubling in risk for childhood leukemia among children who had received radiation in utero—a statistically significant relative risk of 1.9 based on 42 deaths. She also reported that risk of solid (non-leukemia) cancer was more than doubled in this population—a relative risk of 2.3 based on 43 deaths.[7] Six years later, Professor Brian MacMahon found a similar effect among children born in New England. His study used medical records, an approach that eliminated the question of possible recall bias and increased the accuracy of information on radiation dose. The most striking finding in the MacMahon study is that it reported an increased risk of childhood cancer at a radiation dose of only 10 milliGray (mGy)—the lowest dose of radiation that has ever been reported to show a cancer effect in humans. The estimated absolute leukemia risk based on these studies is about 6% per Gy. MacMahon found that the most vulnerable period in pregnancy is the third trimester, but there is also some evidence of increased risk from exposures earlier in gestation. A detailed review and history of these studies is available.[8]

An initial study of persons exposed in utero to ionizing radiation from the atomic bombings reported no increased deaths from childhood leukemia or from solid cancers (neurological effects of in utero exposure are discussed later in the chapter). More recently[9] the incidence of solid cancers has been reanalyzed among a population of 2,452 atomic bombing survivors who were exposed in utero. This reanalysis found that there is a significant dose-response relationship between radiation in utero and risk of cancer, with an estimated ERR of cancer per Sv of radiation dose of 1.0 (95% CI: 0.2, 2.3) at age 50. Cancer risk in survivors exposed in utero was then compared with risk in 15,338 children exposed to the atomic bombings at the ages of 0–5 years.

The estimated ERR of cancer in persons exposed as young children was somewhat greater than that among survivors exposed in utero—ERR of 1.7 (95% CI: 1.1, 2.5) — and showed a significantly positive dose-response. The excess absolute risk of cancer increased with age among the exposed children but not among the survivors exposed in utero. These findings led the study's authors to conclude that exposures to ionizing radiation in utero and in early childhood increase the risk of solid cancers in adult life. They concluded further and most importantly that lifetime risk of solid cancer is considerably lower for persons exposed to ionizing radiation in utero compared to persons exposed as young children.

More frequent CT imaging coupled with higher long-term risks among those exposed at younger ages is a growing public health issue. Recent studies have tried to quantify possible risks to the population. These studies are particularly difficult because diagnostic radiation doses are low, resulting in the need for very large studies to overcome the poor statistical power inherent when studying small effect sizes. In a recent review article, a meta-analysis of childhood exposures indicated an ERR/100 mGy of 1.78 (95% CI: 0.01, 3.53) for leukemia/myelodysplastic syndrome and an ERR/100 mGy of 0.80 (95% CI: 0.48, 1.12) for brain tumors.[10]

Studies of Hereditary Risks of Ionizing Radiation Exposure

In non-human systems, there is ample evidence of germline mutations being introduced in the progeny of exposed species via ionizing radiation exposure. A major concern after the atomic bombings (and subsequently, from other radioactive releases) was whether children born to exposed parents would suffer deleterious consequences. A reanalysis of an early study of children born to atomic bomb survivors was recently published.[11] Among children born in Hiroshima and Nagasaki ($N = 71,603$) between 1948 and 1954, 90% were examined for major congenital malformations and perinatal deaths. The study showed some evidence of increased risk of perinatal deaths within 14 days of birth, but many questions remain whether the risks are truly due to radiation exposure or are confounded with poor living conditions in the aftermath of the bombings. Long-term follow-up of a cohort of children born to atomic bomb survivors continues. The most recent study of mortality among 75,327 singleton children showed no increased risks of either cancer or non-cancer death related to parental exposure.[12] However, the mean age at follow-up (through 2009) was only 53 years, with a total of 5,183 disease-related deaths. A whole-genome sequencing study is being planned using blood collected from trios (mother-father-child) to study whether increased mutation rates due to ionizing radiation exposure exist.

Studies of Radiation Exposure near Nuclear Power Plants

Ecological studies of cancer have been carried out in the vicinity of nuclear power plants in many countries. An unexplained increase in childhood leukemia often has been reported in these investigations.[13] A case-control study of childhood cancer has been conducted in the areas in the vicinity of 16 German nuclear power plants.[14,15] During the period 1980–2003, 1,592 cancer cases were observed, including 593 cases of leukemia among children under the age of 5. Children residing within 5 kilometers of a nuclear facility had a significant increase reported for leukemia (based on 37 leukemia cases), but not for other cancers. The authors found that radiation exposures near these German nuclear power plants are below natural background radiation by a factor of 1,000–100,000.

In France, a study of childhood leukemia among children living in the vicinity of 29 nuclear sites reported 670 childhood cancer cases with 729 cases expected from national rates.[16] The authors did not observe a trend in leukemia rates as a function of distance from the plants. For those children living within 5 kilometers of a nuclear facility, there were 65 leukemia cases with 75 expected. For children less than 5 years of age, there were 39 leukemia cases, with 40 cases expected. In a follow-up study of this population, researchers estimated the doses to the bone marrow based on the nuclear plants' radioactive discharges.[17] The radionuclide discharge data were combined with the local climate data to model environmental exposure levels. Estimated doses to the red marrow for children in the vicinity of the nuclear sites ranged from 0.06 to 1.33 μSv per year. There was no trend of increased leukemia incidence based on yearly exposure rates. As in the German studies, the estimated doses of radiation associated with the nuclear plants were approximately 1,000–100,000 times below the average dose due to natural radiation.

In an effort to explain these reported increases in leukemia among young children living near nuclear plants, Kinlen[18,19] invoked the hypothesis of population mixing, the notion that nuclear workers who relocate into the typically rural areas where nuclear plants are often situated will import foreign infectious agents, which in turn will affect local children and produce leukemia. More details and discussions of these issues is available in the 2011 COMARE report focused on leukemia in children residing near nuclear power plants.[20]

Radionuclides

Radionuclides, also termed "radioactive isotopes" or "radioisotopes," are radioactive atoms with unstable nuclei. Radionuclides occur naturally and can also be produced artificially in nuclear reactors. They include radium, uranium, strontium, radioactive iodine, and many more. Radionuclides undergo radioactive decay, which results in the release of gamma rays or radioactive particles such as alpha and beta particles.

Radon

Radon is a naturally occurring radioactive gas that results from the radioactive decay of uranium and radium in rocks in the Earth's crust. It is a Class A human carcinogen.

The first epidemiologic studies of the carcinogenicity of radon examined risk of lung cancer mortality among heavily exposed underground uranium miners. These studies

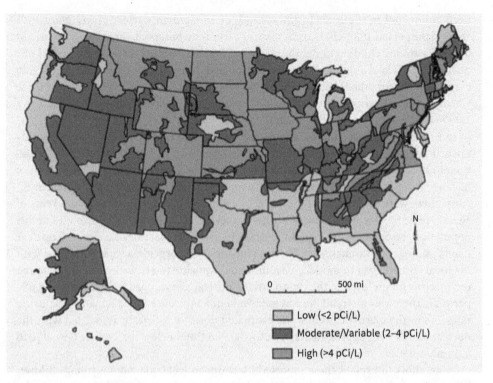

N

	Low (<2 pCi/L)
	Moderate/Variable (2–4 pCi/L)
	High (>4 pCi/L)

Figure 45.2. Generalized geologic radon potential of the United States (US Geological Survey). Units are picocuries per liter (pCi/L).

established that high-dose exposures to radon and its daughters are a clear risk for lung cancer. There are also some suggestions that leukemia risk may be elevated in persons exposed to radon (see the BEIR VI[21] and IARC[22] reports).

Recently many case-control studies have been conducted in the United States and Europe of lung cancer risk in persons exposed to lower doses of radon in the home environment. Joint analyses of these multiple studies have been undertaken by Darby (13 European studies[23]) and by Krewski (7 North American studies[24]). Both of these joint analyses found a positive dose-response relationship between home radon exposure and risk of lung cancer. The estimated ERR of lung cancer ranged from 0.08 to 0.10 per 100 Bq/m³. These estimated lung cancer risks were consistent with the risks projected from the uranium miners exposed to radon at higher doses (i.e., 0.12 per 100 Bq/m³). It has been estimated that, in the United States, 10–14% of all lung cancer deaths can be attributed to radon exposures. Smoking greatly increases the risk of lung cancer at a given level of radon exposure.

BEIR IV provides tables giving the excess lifetime relative risk of lung cancer for various ages at first exposure to radon, exposure level, and period of exposure. These tables have since been updated using the BEIR VI models[21] and published by Chen.[25] From these tables, the calculated excess lifetime relative risk of lung cancer for a male nonsmoker after a radon exposure of 100 Bq/m^3 for 10 years steadily increases from 8.3% for those beginning exposure at age 0 to 12.2% for those beginning exposure at age 40. The corresponding risks are somewhat less for female nonsmokers—7.5% and 10.8%, respectively. These findings indicate that exposure to radon that begins in early childhood carries risk of future lung cancer. Such exposure needs to be reduced.

The average home in the United States has about 50 Bq/m^3 of radiation from radon. The US Environmental Protection Agency (EPA) action level for radon in the home environment of 4 pCi/L is equivalent to a radiation dose of about 150 Bq/m^3. Figure 45.2 is a map of radon levels in the United States. Radon levels tend to be highest in the Appalachian and Rocky Mountain states due to the geology of the area, but there are wide fluctuations, and no area is entirely free of radon.

Radium

Isotopes of radium have been used for many decades as luminescent paint on watches and instruments (radium-226 and -228) and in medical applications (radium-223 and -224). These uses have resulted in occupational exposures to workers. Radium also is found in drinking water in various parts of the world, and this has resulted in environmental exposures of persons of all ages, including young children and pregnant women.

Epidemiological studies of cancer risk among persons exposed to radium have been carried out among workers and also among populations with environmental exposures. The best known of these studies were investigations that began in the 1920s of young women who worked painting the dials of watches with paint containing radium-226 and -228. Many of these women "pointed" the tips of their paint brushes by placing the brushes with radium paint in their mouths to moisten the tips. This practice resulted in the ingestion of relatively large amounts of radium. A cohort of 1,476 dial painters has been established. Radium exposure doses in this population have been carefully reconstructed.[26] The two main findings are increases in bone cancer (because radium is a bone-seeker) and increases in cancers of the paranasal sinuses. Although about one-half of the persons in the cohort had cumulative radiation exposures of less than 1 Gy, there were no bone cancers at doses below 10 Gy.

Some of the young women in this population began work at ages as young as 14 years. The risk of a future bone cancer was twice as great for women who were first exposed at ages 14–16 years compared to women who were first exposed at ages 16–19 years and considerably greater than for those first exposed in their mid-20s.[27] This clearly shows that risk of radium cancer is much greater for children with growing bones who are exposed to a bone-seeking radionuclide such as radium. This finding of greater risk for radium cancer among children was confirmed in a study by Nekolla[28] of young persons treated for bone tuberculosis with injections of the short-lived isotope Ra^{224}. The lifetime attributable risk for a bone cancer per Gy was estimated in this population to be 0.67% for treatment at the age of 5 years compared to 0.40%, 0.18%, and 0.08% for treatments at ages 15, 30, and 45 years, respectively.

Radioiodine

Radioactive iodine, iodine-131, is avidly absorbed by the thyroid gland, especially among people who have an iodine-deficient diet. It is an emitter of beta radiation with a half-life of only 8 days, meaning that while the exposure to the thyroid gland may be short-lived, the doses can be very high. Iodine-131 is a known cause of thyroid cancer, and because iodine is taken up exclusively by the thyroid gland, its cancer risk is limited to the thyroid.[29]

Exposures to iodine-131 and also to cesium-134 and -137 occurred as the result of atmospheric releases of these radionuclides in the Chernobyl and Fukushima nuclear reactor accidents. Persons of all ages were exposed, including young children and pregnant women. Uptake of iodine-131 was especially intense around Chernobyl because the diet there was typically low in iodine and due to poor intervention strategies to reduce consumption of food products, particularly milk, resulting in large doses. Fukushima doses of iodine-131 were low compared to Chernobyl due to a smaller initial release and immediate mitigation procedures implemented by the Japanese government to halt food production/ingestion. Population exposure to iodine-131 also occurred in the 1950s and 1960s as the result of exposure to fallout from atmospheric nuclear testing at the Nevada, South Pacific, and former Soviet test sites.

Cardis conducted a case-control study of thyroid cancer among European children less than 15 years of age who were exposed to iodine-131 following the Chernobyl accident.[30] Radiation doses were estimated for each child based on their location and their diets at the time of the accident. A positive dose-response relationship was observed between iodine-131 exposure and risk of thyroid cancer, with an estimated odds ratio (OR) of 5.5–8.4 at 1 Gy depending on the risk model used. The risk of radiation-related thyroid cancer was three times greater in iodine-deficient areas. Cardis has predicted that, by 2065, about 16,000 cases of thyroid cancer may be expected within Europe due to radiation from the Chernobyl accident.[31] A review of the Chernobyl thyroid studies is available.[32]

Children and adolescents are also at high risk of thyroid cancer caused by external radiation. This risk was documented in studies of external radiation exposures in the atomic bomb survivor studies and also in studies of children who were treated by radiation for tinea capitis (ringworm of the scalp). A pooled analysis of studies of thyroid cancer associated with childhood exposures to external radiation estimated that there is an ERR of more than 7-fold per Gy of radiation (relative risk = 7.7; 95% CI 2.1, 28.7).[29]

Non-Carcinogenic Effects of Ionizing Radiation

Cardiovascular Disease

Cardiovascular disease (CVD) is much more common than cancer among populations in the industrialized countries. For this reason, even a relatively small increased relative risk for CVD caused by ionizing radiation—a relative risk far below that observed for radiation-induced cancer—can have important effects on CVD prevalence and result in many total deaths.[33]

The most recent RERF report on mortality among survivors of the atomic bombings found that mortality due to circulatory disease had a small excess in estimated relative risk (ERR = 0.11; 95% CI: 0.05, 017) but found also there were a large number of circulatory deaths—19,054. These data show a linear dose-response relationship between radiation exposure and mortality from heart disease. These same data show a curvilinear dose-response relationship between radiation exposure and mortality from stroke.[34] It is not clear whether there are CVD effects at acute exposures below 0.5 Gy.

Although they were not statistically significant, the greatest ERRs for both heart disease and stroke were observed among persons under 10 years of age exposed to the atomic bombings. A review of scientific understanding of the relationship between CVD and radiation[35] and a meta-analysis of the epidemiological studies examining this question are available.[36]

Atherosclerosis is a possible mechanism for the increases in CVD seen among persons exposed to ionizing radiation. Atherosclerosis is an inflammatory disease of the arteries that can lead to occlusion or complete obstruction of blood vessels and thus to ischemia of the heart or to ischemic stroke. In a study of inflammatory biomarkers (tumor necrosis factor [TNF]-α, interleukin [IL]-10, immunoglobulin [IgM, IgA], and interferon [IFN]-γ) as well as of erythrocyte sedimentations rates (a marker of chronic inflammation) among atomic bombing survivors, it was shown that these markers were associated with radiation exposure. It was estimated that a radiation exposure of 1 Sv is equivalent to an increase in immunological age of 9–10 years (see Hoel[37] for a discussion).

Respiratory Disease

Among survivors of the atomic bombings, death due to respiratory disease had a significantly increased ERR of 0.21 (95% CI: 0.10, 0.33) and a large number of deaths 5,119.[3]

Cataracts

Cataracts—opacities in the lenses of the eyes—were one of the earliest radiation-associated effects to be discovered. The connection between radiation exposure and cataracts was first recognized shortly after the discovery of x-rays.

Previously, it was believed that cataracts could result only from high doses of radiation to the lens. More recent studies of the atomic bomb survivors have shown, however, that radiation exposures to the lens at doses well under 1 Sv also result in an increased risk of opacities. Thus, in ophthalmological screening of atomic bomb survivors exposed as adults, statistically significant dose-response relationships were observed for posterior subcapsular opacities (OR = 1.4) and for cortical opacities (OR = 1.3). The risk for posterior subcapsular opacities was found to be greater for persons exposed at a younger age. In a subsequent study of atomic bomb survivors who had lens removal surgery, the radiation risk was estimated to be OR = 1.4 at 1 Gy with an estimated threshold of 0.1 Gy. It is not known whether the risk for cataracts is increased in persons with continuous exposures to radiation.[38]

Neurodevelopmental Effects

The effects on neurobehavioral development of exposure to ionizing radiation in early life have been examined among the atomic bomb survivors exposed in utero and in early childhood. In these analyses, the most significant effects on the developing brain were observed for infants exposed in utero at 8–15 weeks of gestation.[39] This appears to be a period of great sensitivity. It is a time in gestation when there is a rapid increase in the number of neurons that subsequently migrate to the cerebral cortex and become nondividing perennial cells.

Among the infants exposed to radiation at 8–15 weeks of gestation, there was an increased frequency of severe mental retardation, a diminution in IQ scores, a reduction in school performance, and an increase in the occurrence of seizures. Among 30 exposed children with severe retardation (<64 IQ points), 60% had small head sizes. The dose-response relationship between radiation exposure and mental retardation appears linear above an estimated threshold of about 0.2 Gy.[40] The data suggest that at a radiation dose of 1 Gy there is a loss of about 25–30 IQ points. No effect was observed at doses below 0.1 Gy (see COMARE 2004[41] for a further discussion).

Conclusion

Ionizing radiation is a known cause of human cancer, and persons exposed at younger ages are generally at greater risk due to more rapid cellular turnover at the time of exposure. It is important to recognize, however, that the carcinogenic effects of radiation exposure take years to be expressed. The same may also be true for non-cancer effects outside the scope of deterministic medical effects due to high-dose, acute exposures.

With the increased medical use of radiation in diagnostic procedures there is concern that CT scans, which involve much higher exposures than traditional radiographs, may be a growing public health risk, especially for children.

Studies of low-dose exposure are inherently difficult due to the small number of additional cases above background rates that are highly variable due to many environmental and lifestyle factors. Statistical inference from such studies requires very large sample sizes, careful dosimetry, and long follow-up periods.

For an in-depth reading of the potential adverse health effects of ionizing radiation, the reader is referred to textbooks by E. Hall and A. Giacca *Radiobiology for the Radiologist* (2005) and by F. Mettler and A. Upton *Medical Effects of Radiation* (2008).

References

1. National Council on Radiation Protection. *NCRP Report No. 160: Ionizing Radiation Exposure of the Population of the United States*. Bethesda, MD: National Council on Radiation Protection and Measurements; 2009.

2. BEIR VII. *Health Risks from Exposure to Low Levels of Ionizing Radiation*. National Research Council Advisory Committee on the Biological Effects of Ionizing Radiation. Washington, DC: National Academies Press, 2006.

3. Preston DL, Ron E, Tokuoka S, et al. Solid cancer incidence in atomic bomb survivors: 1958–1998. *Radiat Res.* 2007;168:1–64.

4. Grant EJ, Brenner A, Sugiyama H, Sakata R, Sadakane A, Utada M, et al. Solid Cancer Incidence among the Life Span Study of Atomic Bomb Survivors: 1958–2009. *Radiat Res.* 2017;187(5):513–37.

5. Ozasa K, Shimizu Y, Suyama A, Kasagi F, Soda M, Grant EJ, et al. Studies of the mortality of atomic bomb survivors, Report 14, 1950-2003: an overview of cancer and noncancer diseases. *Radiat Res.* 2012;177(3):229–43.

6. Brenner AV, Preston DL, Sakata R, Sugiyama H, de Gonzalez AB, French B, et al. Incidence of Breast Cancer in the Life Span Study of Atomic Bomb Survivors: 1958-2009. *Radiat Res.* 2018;190(4):433–44.

7. Stewart A, Webb J, Giles D, Hewitt D. Malignant disease in childhood and diagnostic irradiation in utero. *Lancet.* 1956;268(6940):447.

8. Doll R, Wakeford R. Risk of childhood cancer from fetal irradiation. *Brit J Radiol.* 1997;70:130–9.

9. Preston DL, Cullings H, Suyama A. Solid cancer incidence in atomic bomb survivors exposed in utero or as young children. *J Nat Cancer Inst.* 2008;100:428–36.

10. Berrington de Gonzalez A, Pasqual E, Veiga L. Epidemiological studies of CT scans and cancer risk: the state of the science. *Br J Radiol.* 2021;94(1126):20210471.

11. Yamada M, Furukawa K, Tatsukawa Y, Marumo K, Funamoto S, Sakata R, et al. Congenital Malformations and Perinatal Deaths among the Children of Atomic Bomb Survivors: A Reappraisal. *Am J Epidemiol.* 2021.

12. Grant EJ, Furukawa K, Sakata R, Sugiyama H, Sadakane A, Takahashi I, et al. Risk of death among children of atomic bomb survivors after 62 years of follow-up: a cohort study. *The lancet oncology.* 2015;16(13):1316–23.

13. Baker PJ, Hoel DG. Meta-analysis of standardized incidence and mortality rates of childhood leukaemia in proximity to nuclear facilities. *Eur J Cancer Care.* 2007;16:355–63.

14. Spix C, Schmiedel S, Kaatsch P, Schulze-Rath R, Blettner M. Case-control study on childhood cancer in the vicinity of nuclear power plants in Germany 1980–2003. *Eur J Cancer.* 2008;44:275–84.

15. Kaatsch P, Spix C, Schulze-Rath R, et al. Leukaemia in young children living in the vicinity of German nuclear power plants. *Int J Cancer.* 2008;122:721–6.

16. White-Koning ML, Hemon D, Laurier D, et al. Incidence of childhood leukaemia in the vicinity of nuclear sites in France, 1990–1998. *Br J Cancer.* 2004;91:916–22.

17. Evrard AS, Hemon D, Morin A, et al. Childhood leukaemia incidence around French nuclear installations using geographic zoning based on gaseous discharge dose estimates. *Br J Cancer.* 2006;94:1342–7.

18. Kinlen, L. Evidence for an infective cause of childhood leukaemia: comparison of a Scottish new town with nuclear reprocessing sites in Britain. *Lancet.* 1988;2(8624):1323–7.

19. Kinlen, L. Childhood leukemia, nuclear sites, and population mixing. *Br J Cancer.* 2011;104:12–18.

20. Committee on Medical Aspects of Radiation in the Environment (COMARE), Fourteenth Report. *Further Consideration of the Incidence of Childhood Leukemia Around Nuclear Power Plants in Great Britain.* London: Health Protection Agency for the Committee on Medical Aspects of Radiation in the Environment; 2011.

21. BEIR VI. *Health Effects of Exposure to Radon.* National Research Council Advisory Committee on the Biological Effects of Ionizing Radiation. Washington, DC: National Academies Press; 1999.

22. International Agency for Research on Cancer. A review of human carcinogens: part D radiation. *IARC Monogr Eval Carcinog Risks Hum.* 2009;100D:1–362.

23. Darby S, Hill D, Deo H, et al. Residential radon and lung cancer: detailed results of a collaborative analysis of individual data on 7148 subjects with lung cancer and 14208 subjects without lung cancer from 13 epidemiologic studies in Europe. *Scand J Work Environ Health.* 2006;32(Suppl1):1–83.

24. Krewski D, Lubin JH, Zielinski JM, et al. A combined analysis of North American case-control studies of residential radon and lung cancer. *J Toxicol Environ Health A.* 2006; 69:533–97.

25. Chen J. Estimated risks of radon-induced lung cancer for different exposure profiles based on the new EPA model. *Health Phys.* 2005;88:323–33.

26. Rowland RE, Sterhney AF, HF Lucas. Dose-response relationships for female radium dial workers. *Radiat Res.* 1978;76:368–83.

27. Hoel DG, Carnes B. Cancer dose-response analysis of the radium dial workers. Proceedings from HEIR 2004 9th International Conference on Health Effects Incorporated Radionuclides, 2004:169–73.

28. Nekolla EA, Kreisheimer M, Kellerer AM, Kuse-Isinggschulte M, Grossner W, Spiess H. Induction of malignant bone tumors in radium-224 patients: risk estimates based on the improved dosimetry. *Rad Radiat.* 2000;153:93–103.

29. Ron E, Lubin JH, Shore RE, et al. Thyroid cancer after exposure to external radiation: a pooled analysis of seven studies. *Radiat Res.* 1995;141:259–77.

30. Cardis E, Kesminiene A, Ivanov V, et al. Risk of thyroid cancer after exposure to [131] I in childhood. *J Nat Cancer Inst.* 2005;97:724–32.

31. Cardis E, Howe G, Ron E, et al. Cancer consequences of the Chernobyl accident: 20 years on. *J Radio Prot.* 2006;26:127–40.

32. Ron E. Thyroid cancer incidence among people living in areas contaminated by radiation from the Chernobyl accident. *Health Phys.* 2007;93(5):502–11.

33. Shimizu Y, Kodama K, Nishi N, et al. Radiation exposure and circulatory disease risk: Hiroshima and Nagasaki atomic bomb survivor data, 1950–2003. *BMJ.* 2010; 340:193.

34. Takahashi I, Shimizu Y, Grant EJ, Cologne J, Ozasa K, Kodama K. Heart Disease Mortality in the Life Span Study, 1950–2008. *Radiat Res.* 2017;187(3):319–32.

35. Darby SC, Cutter DJ, Boerma M, et al. Radiation-related heart disease: current knowledge and future prospects. *Int J Radiat Oncol Biol Phys.* 2010;76:656–65.

36. Little MP, Azizova TV, Bazyka D, et al. Systematic review and meta-analysis of circulatory disease from exposure to low-level ionizing radiation and estimates of potential population mortality risks. *Environ Health Perspect.* 2012;120:1503–11.

37. Hoel DG. Ionizing radiation and cardiovascular disease. *Ann of NY Acad of Sci* 2006; 1076:309–17.

38. Neriishi K, Nakashima E, Akahoshi M, et al. Radiation dose and cataract surgery incidence in atomic bomb survivors 1986–2005. *Radiol.* 2012;65:167–74.

39. Otake M, Schull WJ, Yoshimaru T. Brain damage among the prenatally exposed. *J Radiat Res.* 1991;32(sup):249–64.

40. Ishihara K, Kato N, Misumi M, Kitamura H, Hida A, Yamada M. Radiation Effects on Late-life Neurocognitive Function in Childhood Atomic Bomb Survivors: A Radiation Effects Research Foundation Adult Health Study. *Radiat Res.* 2022.

41. Committee on Medical Aspects of Radiation in Environment (COMARE). *Eighth Report: Review of Pregnancy Outcomes Following Preconceptional Exposure to Radiation.* Chilton: National Radiological Protection Board; 2004.

Glossary

Ionizing radiation. Ionizing radiation has sufficient energy to displace electrons from atoms and molecules, a process termed "ionization." Electrons freed by ionization are highly energized, and, when they are released from molecules within cells, these highly energized electrons can damage other molecules and result in intracellular injury.

Non-ionizing radiation. Non-ionizing radiation is without enough energy to remove electrons from atoms and molecules. Examples are microwaves and visible light.

Linear no-threshold model (LNT). Increased rates of disease are often modeled as a proportional increase over the background (i.e., non-exposed) rate. The manner in which this proportion increases, especially at low doses, is often debated. LNT model assumes that the increased proportional risk increases linearly with radiation dose. This implies that any radiation dose increases the risk of disease (usually cancer when discussing radiation exposure). Alternatives to the LNT model may include a threshold model (i.e., no increased risk of disease until the dose exceeds a threshold level for which the body is not able to repair), sublinear model, supra-linear model, linear-quadratic or quadratic models, etc.

Excess relative risk (ERR): Models describing radiation risks of exposure (particularly for cancer) often model the rate of cancer occurrence in the exposed population as a proportional increase over the background rate that grows in a linear (i.e., not exponential) fashion. This ERR model is written as:

$$\text{Rate}_{\text{exposed}} = \text{Rate}_{\text{background}} * (1 + \text{ERR})$$

where the ERR is often:

$$\text{ERR} = \beta * \text{dose}.$$

In this parameterization, β is the increased risk per unit dose. Note that if the dose is zero, the observed rate is the background rate. As the dose increases, the rate of disease in the exposed group is proportionally higher than the background risk. The ERR term could be written as a threshold or other type of term (quadratic, etc.), but the increased risk is proportionally higher and not exponentially higher as dose increases. Most epidemiological or statistical models assume exponential increases in rates; this is generally not the case in radiation risk models.

Types of Ionizing Radiation

Gamma rays: Electromagnetic waves or photons emitted from the nucleus of an atom.
X-rays: Electromagnetic waves or photons released by energy changes in electrons.

Alpha radiation: Consists of particles emitted from the nucleus of an atom. Alpha particles are heavy and thus can cause much tissue damage. Each alpha particle contains two protons and two neutrons.

Beta radiation: Consists of small, high-speed particles emitted from the nucleus of an atom. Each beta particle is identical to an electron.

Neutrons: Particles that are normally contained in the nuclei of all atoms. Neutrons may be removed from atomic nuclei by high-energy processes such as nuclear fission.

Radiation Units

Curie (Ci): A measure of radioactivity. One curie of a radioactive material such as radium will produce 3.7×10^{10} radioactive disintegrations per second. Radioactivity is often expressed in smaller units such as thousandths of a curie (mCi), millionths (uCi), or even ten-billionths (pCi). There are 3.7×10^{10} becquerels (Bq) in one curie.

Becquerel (Bq): This is the new unit of radioactivity, which replaces the curie in many applications. It is defined as the amount of radioactive material that will produce one disintegration per second.

Rad (radiation absorbed dose): A measure of the amount of radiation energy, sometimes termed the "radiation dose," absorbed by some material such as human tissue. One rad of absorbed dose of ionizing radiation is equal to 100 ergs of energy per gram of tissue.

Gray (Gy): The new measure of radiation dose which replaces the rad in many applications. It is defined as 1 joule of energy absorbed per kilogram of tissue. One gray is equal to 100 rads.

Rem (radiation equivalent man): A unit developed to account for the fact that some forms of ionizing radiation are more harmful to human tissue than others. The rem measures a quantity called *radiation equivalent dose*. One rem is equal to the absorbed dose in rads multiplied by the radiation quality factor Q of a particular form of radiation.

Quality Factor (Q): A measure of the degree of tissue damage produced by a particular form of radiation. It is a measure of the fact that some forms of ionizing radiation are more harmful to human tissue than others. For typical x-rays and gamma rays, Q = 1. For alpha particles Q = 20, and for protons Q = 5. Q factors for neutrons range from 5 to 20 depending on their energy level.

Sievert (Sv): The new unit of equivalent dose, which replaces the rem in many applications. One sievert is equal to the absorbed dose in Grays multiplied by a radiation quality factor Q that depends on the type of radiation. Thus, 1 Sv = 100 rem. For x-ray and gamma exposures, Q = 1, therefore, 1 gray = 1 sievert.

Ultraviolet radiation (UVR): Electromagnetic radiation with a wavelength of 100–400 nm. The range of wavelengths is further subdivided into UVA (315–400 nm), UVB (280–315 nm), and UVC (100–280 nm). Ultraviolet radiation's photons can damage molecules in cells by altering chemical bonds without causing ionization of atoms.

46

Electromagnetic Fields

Denis L. Henshaw, Fiorella Belpoggi, Daniele Mandrioli, and Alasdair Philips

Introduction

When the Earth was formed 4.5 billion years ago, *electric fields* (EFs) and *magnetic fields* (MFs) were already present. Life on Earth has evolved in the presence of these fields and all forms of life are known to detect MFs and many also detect EFs.

Children living in modern society are exposed ubiquitously to both electric *and* magnetic fields (EMFs) associated with our electricity supply. Whether such exposure may result in adverse health effects has been the subject of much research. Residential exposure to the *magnetic* component of such extremely low-frequency fields (ELF-MFs) has been associated with an increased risk of childhood leukemia in many early epidemiological studies and remains present in a series of pooled analyses of studies (Table 46.1).

Children and adolescents around the world are today exposed extensively to *electromagnetic fields* (EMF), namely in the radio frequency range (RF-EMF) and to *electromagnetic radiation* (EMR) through their use of mobile phones, Wi-Fi, and related sources.

Although the potential health effects in these age groups have yet to be comprehensively addressed in epidemiological studies, exposure assessments demonstrate that the average radiofrequency energy-deposition is two times higher in the brain and up to 10 times higher in the bone marrow of the skull of children compared with that from mobile phone use by adults. At the same time, repeated studies have shown that exposure to mobile phone EMF/EMR adversely affects sperm quality in men.

Recent years have seen significant advances in elucidating the mechanisms by which MFs interact with biological systems. Particular advances concern magnetoreception in animals and its use in navigation and migration. The important role of cryptochrome protein molecules and the demonstration that human cryptochromes are magneto-sensitive is particularly relevant to understanding adverse health effects associated with EMF exposures.

Definitions and Sources of Exposure

In physics, the terms "electrical field" and "magnetic field" are used to indicate areas where one experiences an electric or magnetic force, in much the same way as we experience the gravitational force in a gravitational field. Electric charges and magnetic moments are basic properties of the fundamental particles that make up our universe.

Denis L. Henshaw, Fiorella Belpoggi, Daniele Mandrioli, and Alasdair Philips, *Electromagnetic Fields* In: *Textbook of Children's Environmental Health*, Second Edition. Edited by: Ruth A. Etzel and Philip J. Landrigan, Oxford University Press.

Table 46.1 Pooled analyses of childhood leukaemia studies

Published analysis	Included studies	ORs at magnetic flux levels				
		Ref	<0.1	0.2–0.29	0.3–0.39	≥0.4uT
Ahlbom et al. 2000[13]	9	a	1.0	1.0	1.0	2.0
Greenland et al. 2000[14]	15	b	1.0	1.1	1.7	1.6
Greenland et al. 2000[14]	(ALL only)	c	1.0	-	-	2.4
Schuz et al. 2007[15]	9	d	1.0	1.1	1.2	2.0
Kheifets et al. 2010a,b[16,17]	7 since 2000	e	1.0	1.1	1.2	1.4
Seomun et al. 2021[19]	33	g	1.0	1.3	1.2	1.7
Zhao et al. 2014[18]	9	f	1.0	1.0	1.4	-

EFs and MFs can be static, as in the natural electric field in the atmosphere and the Earth's magnetic field. However, as soon as these fields vary with time, then the one field can start to generate the other.

However, at power frequencies associated with our electricity supply—50 or 60 Hertz (Hz) and up to at least 10 kHz—collectively known as *extremely low frequency* (ELF), EFs and MFs are not directly related and are not *electromagnetic radiation* (EMR).

At higher frequencies including *radio frequencies* (RF), there is a *near-field* region related to frequency where EFs and MFs act reactively and must be considered separately, which transitions into the *far-field* where the EFs and MFs intimately interact and result in EMR—for mobile phones, Wi-Fi, and Bluetooth this occurs from about two wavelengths away from the source.

Even at microwave frequencies such as 2.5 GHz, used for mobile phones and WiFi, the purely reactive near-field still extends 20 mm from the source and the transmitted power does not become true EMR until more than 2 meters away. Both ELF and RF EMFs/EMR are part of the non-ionizing radiation spectrum that includes visible light (Figure 46.1).[1]

EFs, expressed in volts per meter (V m⁻¹), are generated by differences in voltage. The Earth has a natural EF of around 150 V m⁻¹ but can increase to a few kV m⁻¹ during thunderstorms. MFs are expressed in units of Tesla (T) or Gauss (G) and are associated with the flow of electric currents. The Earth has a static MF whose intensity varies across the Earth's surface from around 25 to 65 microtesla (μT).

Exposure

Exposure to Extremely Low-Frequency Electromagnetic Fields

Geometric-mean ELF MFs in homes range between 0.025 and 0.07 μT in Europe and from 0.055 to 0.11 μT in the US.[2–9] Sources of elevated exposure include proximity to powerlines, especially high-voltage transmission lines where levels can reach approximately 20 μT. [7–8] Hair dryers and shavers can exceed 20 μT close to, but exposures are

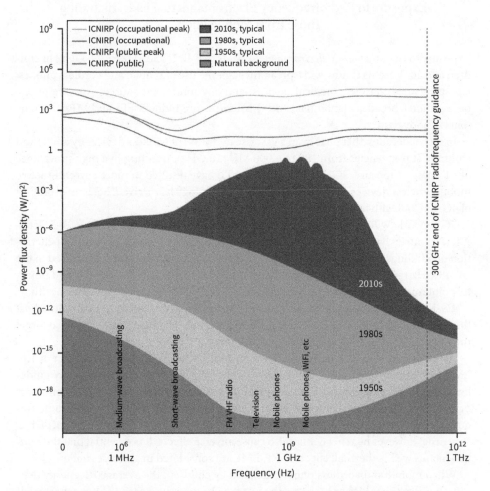

Figure 46.1. Typical maximum daily exposure to radiofrequency electromagnetic radiation from man-made and natural power flux densities in comparison with International Commission on Non-Ionizing Radiation Protection safety guidelines.[90]

for short periods. Few children have time-averaged exposures to residential 50- or 60-Hz MFs in excess of the levels associated with an increased incidence of childhood leukemia: approximately 1–4% have mean exposures above 0.3 μT and only 1–2% have median exposures in excess of 0.4 μT.[5,9]

The mean values of residential EFs are in the range of several tens of V m^{-1} and can reach up to several thousand V m^{-1}, notably near powerlines.

EFs and MFs are complex and vary in time and space. Many different parameters are necessary to characterize them completely. It is not known which parameters or their combination, if any, are relevant for inducing health effects. EFs vary in intensity both around the body, which is partially conducting, and around conducting objects in general. For this reason, epidemiological studies have concentrated on the MF exposure. Most studies use the time-weighted average (TWA), defined as a weighted average of exposure measurements taken over a period of time, with the weighting factor equal to the time interval between measurements.[2]

Exposure to Radiofrequency Electromagnetic Fields, Including Those from Mobile Phones

Exposure to *radiofrequency electromagnetic fields* (RF-EMF: 30 kHz to 300 GHz) in children and adolescents is now widespread through the use of mobile and cordless phones, laptops, games consoles, medical devices, and environmental sources (mobile-phone base stations, broadcast antennas).[6] RF-EMF exposure induces EFs and MFs and currents in body tissues.

In recent decades, there has been a vast rise in the use of wireless devices by adults and children, at frequencies from 180 to 6,000 MHz (6 GHz), resulting in a high prevalence of RF energy exposure. 2G, 3G, 4G, and 4G-LTE have resulted in more currently active mobile wireless devices than people. Now 5G technology has arrived. 5G is an evolution of 4G-LTE but with faster and more complex modulation.

5G networks will use different frequency bands. Most 5G RF is "mid-band" (3–5 GHz). A few countries have authorized low band (600–2,700 MHz), which gives better penetration of buildings, and short-range "millimeter-wave" bands are starting to be used in the 24–48 GHz range. These latter bands have traditionally been used for radar and point-to-point microwave links and few have been studied for their impact on public health.

4G-LTE and 5G use complex modulation and timing schemes that dynamically adjust the power on an individual basis many times a second. The base station can also target the user's handset with beams of microwave energy.

As explained later, the harmful effects of nonthermal biological interaction of RF-EMF with human and animal tissues have not been included in official exposure guidelines, despite the huge number of scientific publications demonstrating the harmfulness or potential harmfulness of those effects. Athermal bioresponses exist, and some frequencies are being used for therapeutic purposes in a number of branches of medicine. Any drug, however beneficial, may also cause adverse effects. It is essential that the nonthermal, as well as thermal, effects of RF-EMF are considered in risk assessment.

When mobile and cordless phones are used by children, the average RF energy deposition is two times higher in the brain and up to 10 times higher in the bone marrow of the skull, compared with adults.[10] Use of hands-free kits lowers exposure to the brain to below 10% of the exposure from use at the ear, but might increase exposure to other parts of the body.[11]

At present most RF exposure assessment has been based on the heating power of the energy (Specific Absorption Rate or SAR). It is now clear from recent research that far more detail is needed for meaningful risk analysis (frequency, modulation, timing pulses, etc.). As the technology has changed over the past 30 years, research into health effects has decreased and virtually no research has used 5G signals.

Epidemiological Studies

Extremely Low-frequency Electromagnetic Fields (ELF-EMF)

Residential Exposure and Childhood Cancer

Childhood cancer is mercifully rare. In the general population, few cases are exposed to elevated levels of ELF-EMF. As a result, even national studies involving cases over many years often lack the resolving power to detect a potential association with ELF-EMF exposure. A further difficulty lies in the assessment of average exposures, especially in historic cases.

Wertheimer and Leeper were the first to suggest that residential ELF-EMF exposure may be associated with cancer in children.[12] Since then, numerous epidemiological studies have examined the association between childhood cancer (particularly leukemia and brain cancer) and exposure to the magnetic component of ELF-EMF, measured by a variety of metrics, as well as through reviews, meta-analyses, and pooled analyses.[13–17]

The largest body of evidence pertains to leukemia, which comprises around 30% of the childhood cancers. The majority of individual studies have demonstrated an increased risk among small numbers of exposed cases in the highest percentile of exposure to magnetic fields. Individual studies have been pooled to overcome the limitations of small sample size. All seven pooled analyses conducted to date have been consistent in demonstrating an approximate doubling of risk of childhood leukemia associated with exposure to higher than 0.3–0.4 μT[13–19] and, overall, around a 20% increased risk with exposure higher than 0.2 μT (Table 46.1).

No consistent relationship has been observed between residential exposure to ELF-EMF and cancers of the brain or other sites in children. The studies have generally experienced methodological limitations, including limited sample size in these much rarer cancers.[17]

Radiofrequency Electromagnetic Fields (RF-EMF)

Exposure to Radio and Television Broadcast Transmitters and Childhood Leukemia

There have been several ecological studies in which cancer incidence or mortality rates in young people have been compared between defined populations living near exposure sources such as TV or radio broadcast stations or other RF fixed-site transmitters or transmission towers. These studies consistently showed a pattern of increased risk for childhood leukemia with greater proximity to the exposure sources when using distance as a proxy for exposure[20–24] whereas there was no increase in risk among studies using analytical designs with better exposure assessment.[25–26] No conclusions can be drawn from these limited and conflicting data.[6]

Mobile Phone Use and Childhood Brain Tumors

There are few studies describing the health effects in children associated with exposure to radiofrequency EMF through the use of personal devices such as mobile phones.

CEFALO is a multicenter case-control study conducted in Denmark, Sweden, Norway, and Switzerland that includes all children and adolescents ages 7–19 years who were diagnosed with a brain tumor between 2004 and 2008.[27] Control subjects were randomly selected from population registries and matched by age, sex, and geographical region. Children who used mobile phones had a nonstatistically significant increased risk of brain tumors, but confidence intervals were wide when assessing exposure-response. When operator-recorded data were available, brain tumor risk was related to the time elapsed since the mobile phone subscription was started, but the pattern of risk was inconsistent.

MOBI-Kids, a 14-country (Australia, Austria, Canada, France, Germany, Greece, India, Israel, Italy, Japan, Korea, the Netherlands, New Zealand, Spain) case-control study, was conducted to evaluate whether wireless phone use (and particularly resulting from exposure to ELF-EMF and RF-EMF) increases risk of brain tumors in young people.[28] Between 2010 and 2015, the study recruited 899 people with brain tumors ages 10–24 years and 1,910 controls. A critical review was published in May 2022,[29] reporting that

> [o]verall no increased risk was found although for brain tumors in the temporal region an increased risk was found in the age groups 10–14 and 20–24 years. Most odds ratios (ORs) in MOBI-Kids were <1.0, some statistically significant, suggestive of a preventive effect from RF radiation; however, this is in contrast to current knowledge about radiofrequency (RF) carcinogenesis. The MOBI-Kids results are not biologically plausible and indicate that the study was flawed due to methodological problems. For example, not all brain tumor cases were included since central localization was excluded. Instead, all brain tumor cases should have been included regardless of histopathology and anatomical localization. Only surgical controls with appendicitis were used instead of population-based controls from the same geographical area as for the cases. Additional case-case analysis should have been performed. The data from this study should be reanalyzed using unconditional regression analysis adjusted for potential confounding factors to increase statistical power. Then all responding cases and controls could be included in the analyses. The results, as reported, in this paper seem uninterpretable and should be dismissed.

Mobile Phones and Non-Cancer Health Effects

A systematic review and updated meta-analysis concluded that exposure to mobile phones is associated with reduced sperm motility, viability, and concentration.[30] Accumulated data from in vivo studies show that mobile phone usage is harmful to sperm quality.

Animal Studies

Extremely Low-frequency Electromagnetic Fields (ELF-EMF)

Up to now, experimental studies on rodents have failed to provide conclusive confirmation of the carcinogenicity of ELF magnetic fields (ELF-EMF, 50 Hz). Two recent studies performed in the laboratory of the Italian Ramazzini Institute demonstrated that concurrent life span exposure to ELF-EMF in combination with two known carcinogens at low doses, formaldehyde or γ-radiation, might enhance their carcinogenic effects in Sprague-Dawley rats.[31-32] In the same laboratory, life span exposures to continuous and intermittent sinusoidal-50 Hz ELF-EMFs, when administered alone, did not represent a significant risk factor for neoplastic development in the used experimental rat model.[33] In light of the previous results on the carcinogenic effects of ELF-EMF in combination with formaldehyde and γ-radiation, further experiments are necessary to elucidate the possible role of ELF-EMF as a cancer enhancer in the presence of other chemical and physical carcinogens.

Radiofrequency Electromagnetic Fields (RF-EMF)

New data in experimental animals for exposure to RF-EMF have been published since the IARC Monographs evaluation in 2011,[6] when they were classified as possible carcinogens.

The large study by the US National Toxicology Program (NTP) found an increased risk of malignant schwannomas of the heart in male rats with high exposure to radiofrequency radiation at frequencies used by 2G and 3G cell phones, as well as possible increased risks of gliomas in the brain and tumors of the adrenal glands; an equivocal increased risk of tumors was found in mice and in female rats.[34,35] An expert peer-review panel concluded that the NTP studies were well designed and that the results demonstrated that both global system modulation (GSM)- and code division multiple access (CDMA)-modulated RFR were carcinogenic to the heart (schwannomas) and brain (gliomas) of male rats (final evaluation: clear evidence of carcinogenicity).

The Ramazzini Institute in Italy performed a life span carcinogenicity study on Sprague-Dawley rats to evaluate the carcinogenic effects of RF-EMF in the far field situation, reproducing the environmental exposure to RF-EMF generated by 1.8 GHz GSM antennae at radio base stations for mobile phones. This is the largest long-term study ever performed in rats on the health effects of RF-EMF, including 2,448 animals. The authors reported the final results regarding brain and heart tumors, confirming and strengthening the same observation as NTP on rats: a statistically significant increase in schwannomas of the heart in males and an increase in glial malignant tumors in females.[36]

Lerchl et al., in a promotion study, found that incidences of tumors of the lung and liver in exposed animals were significantly higher than in sham-exposed controls. In addition, lymphomas were also found to be significantly elevated by exposure, suggesting a promotional effect of RF-EMF.[37]

The recent NTP and Ramazzini RF-EMF studies presented similar findings in heart schwannomas and brain gliomas, strengthening the reciprocal results. Schwannomas are tumors arising from the Schwann cells, which are peripheral glial cells that cover and protect the surface of all nerves diffused throughout the body, so vestibular (acoustic nerve) and heart schwannomas have the same tissue of origin. The findings of the two laboratories could not be interpreted as occurring "by chance"; they are the same kind of tumors and show increased incidence in epidemiological studies.

In conclusion, for RF-EMF-exposed experimental animals, positive associations with sufficient evidence have been observed between exposure and the incidence of gliomas and schwannomas.

Concerning the evidence from animal studies on non-cancer health effects, RFR exposure might negatively affect male fertility. Seminiferous tubules, spermatozoa, and Leydig cells are the main targets of this damage, and sperm count, motility, and morphology represent the more frequently affected parameters.[38]

Mechanisms

The EMFs from power frequency ELF to mobile phone radio frequencies fall in the non-ionizing part of the EMF spectrum (Figure 46.1) and have their own mode of interaction with biological systems. It has been demonstrated that power frequency MFs engender *genomic instability* in biological cells, testifying to their carcinogenic potential, with the delayed de novo appearance of genetic alterations multiple cell generations after exposure, a characteristic of most cancer cells.[39–43] These actions of EMFs and EMR leading to biological changes (and hence any positive or negative health effects) should be evaluated similarly to agents such as carcinogenic chemicals, cancer viruses, and asbestos particles, which are also non-ionizing.

The Radical Pair Mechanism and Cryptochromes

The past 20 years have seen considerable advances in understanding how the MF component of EMFs interacts with biological systems in a manner that may explain a host of experimental observations, including adverse health effects.

Foremost is the development in understanding of the so-called radical pair mechanism (RPM), whereby the lifetime of free radical pairs may be increased by altering their spin states from the short-lived (~nanoseconds) singlet state to the long-lived (~microseconds) triplet state[44–45] and consequently their reaction products.

The radical pair mechanism is, to date, the only well-established mechanism by which the rates and yields of chemical reactions are known to be influenced by MFs. It has been particularly successful in explaining magnetoreception in animals and how avian and other species detect tiny changes in the Earth's MF for the purposes of navigation and migration.

A seminal paper[46] proposed that the radical pair mechanism acting on blue light–activated cryptochrome protein molecules[47–49] in the eye could form the basis of a magnetic compass in birds. The authors argued that cryptochromes contain the necessary structures to produce radical pairs. They suggested that radiofrequency MFs could disrupt the intrinsic singlet-triplet spin interconversion of these radical pairs, which otherwise result in changes to reaction products and constitute a magnetoreceptor response.

Other authors[50] then reported that oscillating MFs in the RF range—broadband (0.1–10 MHz) or single-frequency (7-MHz)—disrupt the magnetic orientation behavior of migratory robins using the Earth's MF. The findings are consistent with a resonance effect on singlet-triplet transitions and a magnetic compass based on a radical pair mechanism in cryptochromes.[51]

These findings have since been replicated in birds and other species. What is remarkable is the very low levels of RF fields required to induce orientation disruption.[52,53] Together, the findings suggest an MF detection threshold of approximately 10 nT.

The purported action of the RPM in cryptochromes has several implications. Human cryptochromes (specifically CRY2) have been shown to be magneto-sensitive.[54] Fruit flies, when engineered without their corresponding *CRY* genes, lose their magneto-sensitivity. However, the sensitivity is restored with the introduction of human *CRY2* genes. Human CRY2 is present in most tissues,[55] and while blue light activation of radical pairs is widely discussed in the context of the animal compass, there is evidence that relevant radical pair formation can take place in the dark.[56–59] Mammalian-type cryptochromes appear to function independently of light and would be expected to retain the characteristics to respond to MFs.[60] Thus, the ability of MFs to increase the lifetime of free radicals may allow those radicals to be more available to cause biological damage in general.

Electromagnetic Fields and the Release of Free Radicals

Reactive oxygen species (ROS) such as superoxide ($O_2^{\bullet-}$) and hydrogen peroxide (H_2O_2) are a class of free radicals produced as potentially toxic cellular metabolites with multiple roles in DNA damage and cell signaling, stress response, and cellular aging.

A seminal paper[60] reported that RF (1.4 MHz) MFs alter relative yields of cellular $O_2^{\bullet-}$ and H_2O_2 ROS products in primary human umbilical vein endothelial cells (HUVECs), indicating coherent singlet-triplet spin-state mixing at the point of ROS formation. The findings are an example of the RPM operating in live cell cultures that bridge atomic and cellular levels by connecting ROS partitioning to cellular bioenergetics.

Other authors[61] reported that 2 mT, 10 Hz pulsed MFs induce human cryptochromes to modulate intracellular ROS. This profound observation makes it conceivable that carcinogenesis associated with power lines, pulsed MF-induced ROS generation, and animal magnetoreception share a common mechanistic basis.[62]

The findings are of immediate relevance to a range of observations spanning recent decades. Of relevance to childhood leukemia, 50 Hz MFs have the cell activating capacity to release reactive oxygen intermediates in human umbilical cord blood-derived monocytes and in mono mac 6 cells.[63] If monocytes can respond to MFs exposure by producing radicals, we have a reasonable carcinogenic mechanism. These cells would be recruited to sites of inflammation and proliferation where their ROS could generate genetic aberrations.

More generally, a plausible biological mechanism to account for MF-induced carcinogenesis is through intracellular formation of free radicals. These reactive species kill cells by damaging macromolecules, such as DNA, protein, and membrane structures. Furthermore, free radicals play an essential role in the activation of certain signaling pathways. Several reports have indicated that MFs enhance free radical activity in cells particularly via the Fenton reaction, a catalytic process during which iron converts hydrogen peroxide, a product of oxidative respiration in the mitochondria, into the hydroxyl radical, a potent and toxic reactive oxygen species. MF-induced free radical release has been reported across a wide range of cell types, and increased free radical production has long been considered a plausible biological mechanism for carcinogenesis induced by prolonged low-intensity MF exposures.[64–67]

In summary, the radical pair mechanism has seen considerable success in accounting for magnetoreception in animals, for which the presence of cryptochrome is a requirement. Pulsed MFs have been shown to release ROS in human cryptochrome 2, consistent with the radical pair mechanism. MFs have also been shown to engender genomic instability, a characteristic of most cancer cells. Together, these findings underpin the long-standing plausible hypothesis that MF-induced carcinogenesis, notably childhood leukemia, can act through intracellular formation of free radicals.

Cryptochromes, Melatonin, and Circadian Rhythm Disruption

Cryptochromes are best known as a key component in the control of circadian rhythms,[68] an aspect of which is the nocturnal production in the pineal gland of the powerful natural antioxidant and anti-cancer agent, melatonin. Nocturnal pineal melatonin is fully suppressed once exposure to light at night (LAN) exceeds 200 lux (about 300 mW m^{-2}; e.g., in a low-lighted hallway).[69] Cryptochromes are extremely sensitive to blue light and even a single blue LED light visible from the bed will affect nocturnal melatonin synthesis (and hence sleep quality) in a young child.

A seminal paper[70] proposed that exposure to LAN and ELF EFs and/or MFs increased the risk of breast cancer. There is now considerable evidence to support this hypothesis in relation to LAN, notably in shift workers but also in aircraft cabin crew. The International Agency for Research on Cancer (IARC) has classed nightshift work as a 2A probable carcinogen.[71]

Human circadian rhythmicity begins forming in early gestation but is not robust at birth and takes several more months to become consolidated.[72] This raises the question of whether LAN stimulates cancer development in children.[73]

Electromagnetic Fields Disruption of Melatonin and Circadian Rhythms

EFs have a demonstrable effect on circadian rhythms. In a long series of experiments, human volunteers were exposed for several weeks to 10 Hz square wave EFs of only 2.5 V/m. Volunteers were immediately entrained to the external signal, and the 24-hour circadian rhythm was shortened by about 1 hour. The effect lasted for a few days, indicating that EFs act as *zeitgebers*.[74] Zeitgebers (literally, time givers) are environmental variables that are capable of acting as circadian time cues.

In contrast to LAN, MFs appear less effective in melatonin suppression, with maximum suppression in the range 20–30%. However, people exposed to elevated fields living under high-voltage powerlines, for example, may be chronically exposed, so the overall effect may be greater than LAN.

Nocturnal melatonin levels are typically assayed by analysis of the metabolite 6-sulfatoxymelatonin in morning urine. Demonstration of MF suppression must overcome the wide natural person-to-person variation in nocturnal melatonin production.[75] This suggests that studies with only a few tens of subjects will lack the power to detect an effect. In a well-conducted study involving 203 women and a dose-response design, melatonin suppression was noted for nocturnal exposure as low as 0.2 µT, with an overall 14% reduction.[76]

Thus, reviews of studies suggest mixed findings,[77,78] but grouping by the number of subjects reveal that those reporting no effect employed on average 28 subjects (range 7–60), whereas those reporting an effect had on average 150 subjects (range 44–416).[79] Overall, studies of MF disruption of melatonin and circadian rhythms are inconsistent with no effect, suggesting that effects do occur.

The hypothesis that MFs may at least in part increase childhood leukemia risk by melatonin and circadian rhythm disruption has been progressively developed to include action via the radical pair mechanism on cryptochromes, DNA damage, and genomic instability,[77–81] adding to the pathway of ROS release discussed above.

Magnetic Particles in the Body

Biogenic magnetic particles are found widely in living systems. They occur in the human brain in some abundance, where sizes vary up to 600 nm, with many clumped in groups of up to 100 particles.[82] While most appear to be biogenic in origin, in modern conditions evidence suggests a component from pollution sources.[83] The larger particles could respond to a 50 Hz MF at 0.4 µT, putting mechanical stress on neighboring cells.

MFs in the millitesla range produced around intracellular superparamagnetic nanoparticles could enhance free radical concentration in hematopoietic stem cells via an RPM and act as a mechanism of increased childhood leukemia risk.[84] A proposed action via the RPM is found in the observation that 1 mm magnetite particles encapsulated in polystyrene dramatically decreased the time for 50% hemolysis of UV-irradiated human erythrocytes.[85]

Magnetic particles in the brain or elsewhere in the body can also transduce EMFs in the GHz region and dissipate energy in their vicinity.[86,87]

While the presence of magnetic particles in the body is clearly able to transduce both MFs and EMFs across a wide range of frequencies, their potential role in explaining adverse health effects remains unclear.

Applicability of the Radical Pair Mechanism to Radiofrequency Electromagnetic Fields/Electromagnetic Radiation

The radical pair mechanism has applicability to MFs ranging from ELF up to the MHz region. While the radical pair mechanism is inapplicable to GHz carrier wave frequencies, in particle applications, such frequencies are modulated with much lower frequencies.

Summary

In respect of the effects of MF exposure, the radical pair mechanism acting on cryptochrome protein molecules accounts for the release of free radicals widely reported in the research literature. The radical pair mechanism also accounts for circadian rhythm disruption, especially melatonin disruption in MF-exposed subjects. The ubiquitous presence of magnetic particles in the body would be expected to enhance these effects, although the precise mechanism by which they may do so remains unclear.

Overall, there are now clear mechanistic pathways by which MF and EMF exposures may result in the adverse health effects reported in the epidemiological literature.

Position of Governmental Agencies

The IARC of the World Health Organization (WHO), an international authority on classification of carcinogens, has classified ELF MF as well as RF-EMFs as possibly carcinogenic to humans (Group 2B).[6] The IARC has also declared that static electric and magnetic fields and ELF electric fields are not classifiable as to their carcinogenicity to humans (Group 3).

Almost 10 years later many new studies have been published and an update is necessary. An Advisory Group of 29 scientists from 18 countries met at the IARC in March 2019 to recommend priorities for the IARC Monographs program during 2020–2024, among them RF-EMF.[88]

The WHO is undertaking a health risk assessment of RF-EMF, to be published as a monograph in the Environmental Health Criteria Series.[89]

The International Commission on Non-Ionizing Radiation Protection (ICNIRP) in March 2020 published new guidelines covering several new technologies, including 5G.[90] The guidelines refer only to thermal effects caused by 6 minutes and 30 minutes of exposure to RF-EMF, so the guidelines concern only short-term exposure. Other authoritative governmental agencies have conducted reviews to evaluate the health risks associated with electromagnetic radiation.[2,6,91,92] Essentially, they have concluded that although the overall evidence for health effects is weak, exposure to EMF cannot be regarded as entirely safe and that the status of both ELF-EMF and RF-EMF should be updated and reevaluated.

Prevention

Simple measures can be taken to reduce exposure to sources of ELF-EMF and RF-EMF. These measures include limiting unnecessary exposure to cell phones, text-messaging instead of calling, reducing duration of calls, and using earphones while keeping the phone away from the body.

As regards ELF-EMF, overhead power lines may be placed underground to eliminate electric fields and reduce the strength of the magnetic fields and the extent of human exposure at the ground surface.

Recently, European policies have promoted the sustainability of a new economic and social development model that uses new technologies to constantly monitor the planet's state of health, including climate change, the energy transition, agro-ecology, and the preservation of biodiversity. Using the lowest frequencies of 5G and adopting precautionary exposure limits such as those used in Italy, Switzerland, China, and Russia, among others—and that are significantly lower than those recommended by ICNIRP—could help achieve these European sustainability objectives.[93] The same should be done in the United States and worldwide.

Much of the remarkable performance of new wireless 5G technology can also be achieved by using optical fiber cables and by adopting engineering and technical measures to reduce exposures from 2–4G systems. This would minimize exposure wherever connections are needed at fixed sites. For example, we could use optical fiber cables to connect schools, libraries, workplaces, houses, public buildings, all new buildings, etc.

Unfortunately, there is a lack of information on the potential harms of RF-EMF. The information gap creates scope for deniers as well as alarmists, giving rise to social and political tension in many countries.[94] Campaigns to inform citizens should be therefore a priority. Electric power, radio transmission, mobile phones, and other sources of exposure to ELF-EMF and RF-EMF offer obvious social and economic benefits and will continue to be part of our daily lives. Advances in technology have brought considerable health benefits. The costs and benefits of preventive measures must therefore be carefully considered.

Acknowledgments

We are grateful to Kurt Straif and the other authors of this chapter in the previous edition which provided a good basis for our work.

References

1. Bandara P, Carpenter DO. Planetary electromagnetic pollution: it is time to assess its impact. *Lancet Planet Health.* 2018 Dec;2(12):e512–14. PMID: 30526934. doi:10.1016/S2542-5196(18)30221-3. *Image by permission Philips & Lamburn.*

2. IARC. Non-ionizing radiation, Part I: static and extremely low-frequency (ELF) electric and magnetic fields. *IARC Monogr Eval Carcinog Risks Hum.* 2002;80:1–395.

3. Miller AB, Green LM. Electric and magnetic fields at power frequencies. *Chronic Dis Can.* 2010;29(Suppl 1):69–83.

4. WHO. *Extremely Low Frequency Fields.* World Health Organization, editor. Environmental Health Criteria Monograph No.238. 2007. WHO: Geneva.

5. WHO. World Health Organisation. Electromagnetic fields and public health. Exposure to extremely low frequency fields. Fact sheet No. 322. 2007. https://www.who.int/teams/environment-climate-change-and-health/radiation-and-health/non-ionizing/exposure-to-extremely-low-frequency-field

6. IARC. Non-Ionizing radiation, Part II: Radiofrequency Electromagnetic Fields [includes mobile telephones]. In prep. *IARC Monogr Eval Carcinog Risks Hum* 2013;102.

7. Schuz J, Grigat JP, Stormer B, Rippin G, Brinkmann K, Michaelis J. Extremely low frequency magnetic fields in residences in Germany: distribution of measurements comparison of two methods for assessing exposure and predictors for the occurrence of magnetic fields above background level. *Radiat Environ Biophys.* 2000;39:233–40.

8. Maslanyj MP, Mee TJ, Renew DC, et al. Investigation of the sources of residential power frequency magnetic field exposure in the UK Childhood Cancer Study. *J Radiol Prot.* 2007;27(1):41–58.

9. Greenland S, Kheifets L. Leukemia attributable to residential magnetic fields: results from analyses allowing for study biases. *Risk Anal.* 2006;26(2):471–82.

10. Christ A, Gosselin MC, Christopoulou M, Kuhn S, Kuster N. Age-dependent tissue-specific exposure of cell phone users. *Phys Med Biol.* 2010;55(7):1767–83.

11. Kühn S, Jennings W, Christ A, Kuster N. Assessment of induced radio-frequency electromagnetic fields in various anatomical human body models. *Phys Med Biol.* 2009;54(4):875–90.

12. Wertheimer N, Leeper E. Electrical wiring configurations and childhood cancer. *Am J Epidemiol.* 1979;109(3):273–84.

13. Ahlbom A, Day N, Feychting M, et al. A pooled analysis of magnetic fields and childhood leukaemia. *Br J Cancer.* 2000;83:692–8.

14. Greenland S, Sheppard AR, Kaune WT, Poole C, Kelsh MA. A pooled analysis of magnetic fields wire codes and childhood leukemia. *Epidemiology.* 2000;11:624–34.

15. Schüz J, Svendsen AL, Linet MS, et al. Nighttime exposure to electromagnetic fields and childhood leukemia: an extended pooled analysis. *Am J Epidemiol.* 2007;166(3):263–9.

16. Kheifets L, Ahlbom A, Crespi CM, et al. Pooled analysis of recent studies on magnetic fields and childhood leukaemia. *Br J Cancer.* 2010a;103(7):1128–35.

17. Kheifets L, Ahlbom A, Crespi CM, et al. A pooled analysis of extremely low-frequency magnetic fields and childhood brain tumors. *Am J Epidemiol.* 2010b;172(7):752–61.

18. Zhao L, Liu X, Wang C, et al. Magnetic fields exposure and childhood leukemia risk: a meta-analysis based on 11,699 cases and 13,194 controls. *Leukemia Res.* 2014;38;269–74.

19. Seomun G, Lee J, Park J. Exposure to extremely low-frequency magnetic fields and childhood cancer: a systematic review and meta-analysis. *PLoS One.* 2021;16(5):e0251628. https://doi. org/10.1371/journal.pone.0251628.

20. Hocking B, Gordon IR, Grain HL, Hatfield GE. Cancer incidence and mortality and proximity to TV towers. *Med J Aust.* 1996;165(11–12):601–5.

21. Cooper D, Hemming K, Saunders P. Re: Cancer incidence near radio and television transmitters in Great Britain. I. Sutton Coldfield transmitter; II. All high power transmitters. *Am J Epidemiol.* 2001;153(2):202–4.

22. Michelozzi P, Capon A, Kirchmayer U, et al. Adult and childhood leukemia near a high-power radio station in Rome, Italy. *Am J Epidemiol.* 2002;155(12):1096–103.

23. Park SK, Ha M, Im HJ. Ecological study on residences in the vicinity of AM radio broadcasting towers and cancer death: preliminary observations in Korea. *Int Arch Occup Environ Health.* 2004;77(6):387–94.

24. Ha M, Im H, Lee M, et al. Radio-frequency radiation exposure from AM radio transmitters and childhood leukemia and brain cancer. *Am J Epidemiol.* 2007;166(3):270–9.

25. Merzenich H, Schmiedel S, Bennack S, et al. Childhood leukemia in relation to radio frequency electromagnetic fields in the vicinity of TV and radio broadcast transmitters. *Am J Epidemiol.* 2008;168(10):1169–78.

26. Elliott P, Toledano MB, Bennett J, et al. Mobile phone base stations and early childhood cancers: case-control study. *BMJ.* 2010;340:c3077.

27. Aydin D, Feychting M, Schuz J, et al. Mobile phone use and brain tumors in children and adolescents: a multicenter case-control study. *J Natl Cancer Inst.* 2011;103(16):1264–76.

28. Castaño-Vinyals G, Sadetzki S, Vermeulen R, et al. Wireless phone use in childhood and adolescence and neuroepithelial brain tumours: results from the international MOBI-Kids study. *Environ Int.* 2022 Feb;160:107069.

29. Hardell L, Moskowitz J. A critical analysis of the MOBI-Kids study of wireless phone use in childhood and adolescence and brain tumor risk, *Rev Environ Health*. 2022 May. https://doi.org/10.1515/reveh-2022-0040.

30. Kim S, Han D, Ryu J, Kim K, Kim YH. Effects of mobile phone usage on sperm quality - no time-dependent relationship on usage: a systematic review and updated meta-analysis. *Environ Res*. 2021 Nov;202:111784.

31. Soffritti M, Tibaldi E, Padovani M, et al. Synergism between sinusoidal-50 Hz magnetic field and formaldehyde in triggering carcinogenic effects in male Sprague-Dawley rats. *Am J Ind Med*. 2016 Jul;59(7):509–21.

32. Soffritti M, Tibaldi E, Padovani M, et al. Life-span exposure to sinusoidal-50 Hz magnetic field and acute low-dose γ radiation induce carcinogenic effects in Sprague-Dawley rats. *Int J Radiat Biol*. 2016;92(4):202–14.

33. Bua L, Tibaldi E, Falcioni L, et al. Results of lifespan exposure to continuous and intermittent extremely low frequency electromagnetic fields (ELFEMF) administered alone to Sprague Dawley rats. *Environ Res*. 2018 Jul;164:271–9.

34. National Toxicology Program. Toxicology and carcinogenesis studies in Hsd:Sprague Dawley SD rats exposed to whole body radio frequency radiation at a frequency (900 MHz) and modulations (GSM and CDMA) used by cell phones. 2018. Research Triangle Park, NC: National Toxicology Program; NTP TR-595.

35. National Toxicology Program. Toxicology and carcinogenesis studies in B6C3F1/N mice exposed to whole-body radio frequency radiation at a frequency (1900 MHz) and modulations (GSM and CDMA) used by cell phones. 2018. Research Triangle Park, NC: National Toxicology Program; NTP TR-596.

36. Falcioni L, Bua L, Tibaldi E, et al. Report of final results regarding brain and heart tumors in Sprague-Dawley rats exposed from prenatal life until natural death to mobile phone radiofrequency field representative of a 1.8 GHz GSM base station environmental emission. *Environ Res*. 2018 Aug;165:496–503.

37. Lerchl A, Klose M, Grote K, et al. Tumor promotion by exposure to radiofrequency electromagnetic fields below exposure limits for humans. *Biochem Biophys Res Commun*. 2015 Apr 17;459(4):585–90.

38. Vornoli A, Falcioni L, Mandrioli D, Bua L, Belpoggi F. The contribution of in vivo mammalian studies to the knowledge of adverse effects of radiofrequency radiation on human health. *Int J Environ Res Public Health*. 2019 Sep 12;16(18):3379.

39. Cho YH, Jeon HK, Chung HW. Effects of extremely low-frequency electromagnetic fields on delayed chromosomal instability induced by bleomycin in normal human fibroblast cells. *J Toxicol Environ Health A*. 2007 Aug;70(15-16):1252–8.

40. Mairs RJ, Hughes K, Fitzsimmons S, et al. Microsatellite analysis for determination of the mutagenicity of extremely low-frequency electromagnetic fields and ionising radiation in vitro. *Mutat Res*. 2007 Jan 10;626(1–2):34–41.

41. Luukkonen J, Liimatainen A, Juutilainen J, Naarala J. Induction of genomic instability, oxidative processes, and mitochondrial activity by 50Hz magnetic fields in human SH-SY5Y neuroblastoma cells. *Mutat Res*. 2014 Feb;760:33–41. doi:10.1016/j.mrfmmm.2013.12.002.

42. Kesari KK, Luukkonen J, Juutilainen J, Naarala J. Genomic instability induced by 50Hz magnetic fields is a dynamically evolving process not blocked by antioxidant treatment. *Mutat Res Genet Toxicol Environ Mutagen*. 2015 Dec;794:46–51. doi:10.1016/j.mrgentox.2015.10.004.

43. Naarala J, Kolehmainen M, Juutilainen J. Electromagnetic fields, genomic instability and cancer: a systems biological view. *Genes (Basel).* 2019 Jun 25;10(6):479. doi:10.3390/genes10060479.

44. Brocklehurst B, McLauchlan KA. Free radical mechanism for the effects of environmental electromagnetic fields on biological systems. *Int J Radiat Biol.* 1996 Jan;69(1):3–24.

45. Rodgers CT. Magnetic field effects in chemical systems. *Pure Appl Chem.* 2009;81(1):19–43.

46. Ritz T, Adem S, Schulten K. A model for photoreceptor-based magnetoreception in birds. *Biophys J.* 2000Feb;78(2):707–18.

47. Ahmad M, Cashmore AR. HY4 gene of A. thaliana encodes a protein with characteristics of a blue-light photoreceptor. *Nature.* 1993 Nov 11;366(6451):162–6.

48. Cashmore AR, Jarillo JA, Wu YJ, Liu D. Cryptochromes: blue light receptors for plants and animals. *Science.* 1999 Apr 30;284(5415):760–5.

49. Brudler R, Hitomi K, Daiyasu H, et al. Identification of a new cryptochrome class: structure, function, and evolution. *Mol Cell.* 2003 Jan;11(1):59–67.

50. Ritz T, Thalau P, Phillips JB, Wiltschko R, Wiltschko W. Resonance effects indicate a radical-pair mechanism for avian magnetic compass. *Nature.* 2004 May 13;429(6988):177–80.

51. Maeda K, Robinson AJ, Henbest KB, et al. Magnetically sensitive light-induced reactions in cryptochrome are consistent with its proposed role as a magnetoreceptor. *Proc Natl Acad Sci U S A.* 2012 Mar 27;109(13):4774–9.

52. Engels S, Schneider NL, Lefeldt N, et al. Anthropogenic electromagnetic noise disrupts magnetic compass orientation in a migratory bird. *Nature.* 2014 May 15;509(7500):353–6.

53. Pakhomov A, Bojarinova J, Cherbunin R, et al. Very weak oscillating magnetic field disrupts the magnetic compass of songbird migrants. *J R Soc Interface.* 2017 Aug;14(133):20170364.

54. Foley LE, Gegear RJ, Reppert SM. Human cryptochrome exhibits light-dependent magneto-sensitivity. *Nat Commun.* 2011 Jun 21;2:356.

55. Protein Atlas. Human CRY2 Atlas, version 23 June 19 2023. Available at: https://www.proteinatlas.org/ENSG00000121671-CRY2/tissue

56. Vieira J, Jones AR, Danon A, et al. Human cryptochrome-1 confers light independent biological activity in transgenic Drosophila correlated with flavin radical stability. *PLoS One.* 2012;7(3):e31867.

57. Wiltschko R, Ahmad M, Nießner C, Gehring D, Wiltschko W. Light-dependent magnetoreception in birds: the crucial step occurs in the dark. *J R Soc Interface.* 2016 May;13(118):20151010.

58. Höytö A, Herrala M, Luukkonen J, Juutilainen J, Naarala J. Cellular detection of 50 Hz magnetic fields and weak blue light: effects on superoxide levels and genotoxicity. *Int J Radiat Biol.* 2017 Jun;93(6):646–52.

59. Pooam M, Arthaut LD, Burdick D, Link J, Martino CF, Ahmad M. Magnetic sensitivity mediated by the Arabidopsis blue-light receptor cryptochrome occurs during flavin reoxidation in the dark. *Planta.* 2019 Feb;249(2):319–32.

60. Usselman RJ, Chavarriaga C, Castello PR, et al. The quantum biology of reactive oxygen species partitioning impacts cellular bioenergetics. *Sci Rep.* 2016 Dec 20;6:38543.

61. Sherrard RM, Morellini N, Jourdan N, et al. Low-intensity electromagnetic fields induce human cryptochrome to modulate intracellular reactive oxygen species. *PLoS Biol.* 2018 Oct 2;16(10):e2006229.

62. Landler L, Keays DA. Cryptochrome: the magnetosensor with a sinister side? *PLoS Biol.* 2018 Oct 2;16(10):e3000018

63. Lupke M, Rollwitz J, Simkó M. Cell activating capacity of 50 Hz magnetic fields to release reactive oxygen intermediates in human umbilical cord blood-derived monocytes and in Mono Mac 6 cells. *Free Radic Res.* 2004 Sep;38(9):985–93.

64. Lai H, Singh NP. Acute exposure to a 60 Hz magnetic field increases DNA strand breaks in rat brain cells. *Bioelectromagnetics.* 1997;18(2):156–65.

65. Simkó M, Mattsson MO. Extremely low frequency electromagnetic fields as effectors of cellular responses in vitro: possible immune cell activation. *J Cell Biochem.* 2004 Sep 1;93(1):83–92.

66. Wolf FI, Torsello A, Tedesco B, Fasanella S, et al. 50-Hz extremely low frequency electro-magnetic fields enhance cell proliferation and DNA damage: possible involvement of a redox mechanism. *Biochim Biophys Acta.* 2005 Mar 22;1743(1–2):120–9.

67. Mattsson MO, Simkó M. Grouping of experimental conditions as an approach to evaluate effects of extremely low-frequency magnetic fields on oxidative response in in vitro studies. *Front Public Health.* 2014 Sep 2;2:132.

68. Sancar A. Regulation of the mammalian circadian clock by cryptochrome. *J Biol Chem.* 2004 Aug 13;279(33):34079–82.

69. Zeitzer JM, Dijk DJ, Kronauer R, Brown E, Czeisler C. Sensitivity of the human circadian pacemaker to nocturnal light: melatonin phase resetting and suppression. *J Physiol.* 2000 Aug 1;526 Pt 3(Pt 3):695–702.

70. Stevens RG. Electric power use and breast cancer: a hypothesis. *Am J Epidemiol.* 1987 Apr;125(4):556–61.

71. International Agency for Research on Cancer. IARC Monographs of the Evaluation of Carcinogenic Risks to Humans, Vol. 98. Painting, Firefighting and Shiftwork. Lyon: International Agency for Research on Cancer; 2010.

72. Rivkees SA. Developing circadian rhythmicity in infants. *Pediatrics.* 2003 Aug;112(2):373–81.

73. Stevens RG. Does electric light stimulate cancer development in children? *Cancer Epidemiol Biomarkers Prev.* 2012 May;21(5):701–4.

74. Wever R. *The Circadian System of Man: Results of Experiments Under Temporal Isolation.* New York: Springer-Verlag; 1979.

75. Sack RL, Lewy AJ, Erb DL, Vollmer WM, Singer CM. Human melatonin production decreases with age. *J Pineal Res.* 1986;3(4):379–88.

76. Davis S, Kaune WT, Mirick DK. Residential magnetic fields, light-at-night, and nocturnal urinary 6-sulfatoxymelatonin concentration in women. *Am J Epidemiol* 2001;154:591–600.

77. Henshaw DL, Reiter RJ. Do magnetic fields cause increased risk of childhood leukemia via melatonin disruption? *Bioelectromagnetics.* 2005;Suppl 7:S86–97.

78. Touitou Y, Selmaoui B. The effects of extremely low-frequency magnetic fields on melatonin and cortisol, two marker rhythms of the circadian system. *Dialogues Clin Neurosci.* 2012 Dec;14(4):381–99.

79. Jones J. Magnetic fields and childhood leukaemia: candidate mechanistic pathways. Report of a Think Tank sponsored by Children with Cancer UK held at the Holiday Inn Regent's Park, London, 22–23 September 2014. Available at: https://www.childrenwithcancer.org.uk/childhood-cancer-info/we-fund-research/for-researchers/conferences-workshops/

80. Vanderstraeten J, Burda H, Verschaeve L, De Brouwer C. Could magnetic fields affect the circadian clock function of cryptochromes? Testing the basic premise of the cryptochrome hypothesis (ELF magnetic fields). *Health Phys.* 2015 Jul;109(1):84–9.

81. Juutilainen J, Herrala M, Luukkonen J, Naarala J, Hore PJ. Magnetocarcinogenesis: is there a mechanism for carcinogenic effects of weak magnetic fields? *Proc Biol Sci.* 2018 May 30;285(1879):20180590.

82. Kirschvink JL, Kobayashi-Kirschvink A, Woodford BJ. Magnetite biomineralization in the human brain. *Proc Natl Acad Sci U S A.* 1992 Aug 15;89(16):7683–7.

83. Maher BA, Ahmed IA, Karloukovski V, et al. Magnetite pollution nanoparticles in the human brain. *Proc Natl Acad Sci U S A.* 2016 Sep 27;113(39):10797–801.

84. Binhi V. Do naturally occurring magnetic nanoparticles in the human body mediate increased risk of childhood leukaemia with EMF exposure? *Int J Radiat Biol.* 2008 Jul;84(7):569–79.

85. Chignell CF, Sik RH. Effect of magnetite particles on photoinduced and nonphotoinduced free radical processes in human erythrocytes. *Photochem Photobiol.* 1998 Oct;68(4):598–601.

86. Kirschvink JL, Kobayashi-Kirschvink A, Diaz-Ricci JC, Kirschvink SJ. Magnetite in human tissues: a mechanism for the biological effects of weak ELF magnetic fields. *Bioelectromagnetics.* 1992;Suppl 1:101–13.

87. Kirschvink JL. Microwave absorption by magnetite: a possible mechanism for coupling non-thermal levels of radiation to biological systems. *Bioelectromagnetics.* 1996;17(3):187–94.

88. IARC Monographs Priorities Group. Advisory Group recommendations on priorities for the IARC Monographs. *Lancet Oncol.* 2019 Jun;20(6):763–4.

89. Verbeek J, Oftedal G, Feychting M, et al. Prioritizing health outcomes when assessing the effects of exposure to radiofrequency electromagnetic fields: a survey among experts. *Environ Int.* 2021 Jan;146:106300.

90. International Commission on Non-Ionizing Radiation Protection (ICNIRP). Guidelines for limiting exposure to electromagnetic fields (100 kHz to 300 GHz). *Health Phys.* 2020 May;118(5):483–524.

91. NRPB. Electromagnetic fields and the risk of cancer: report of an advisory group on non-ionizing radiation. 3. 1992. Chilton, Didcot, UK.

92. NIEHS. *Assessment of Health Effects from Exposure to Power-Line Frequency Electric and Magnetic Fields.* Portier C, Wolf MS, eds. NIH Publication No. 98-3981. Research Triangle Park, NC: National Institute of Environmental Health Sciences, National Institutes of Health, Department of Health and Human Services, Public Health Service; 2008.

93. EU-STOA. *Health Impact of 5G.* Brussels: European Union; 2021. ISBN: 978-92-846-8030-6

94. OECD. *Trust and Public Policy: How Better Governance Can Help Rebuild Public Trust.* OECD Public Governance Reviews, Paris: OECD Publishing; 2017.

PART IV

THE ENVIRONMENT AND DISEASE IN CHILDREN

47

Prematurity and Low Birth Weight Associated with Environmental Exposures

Margaret Kuper-Sassé, Cansu Tokat, Hilal Yildiz Atar, and Cynthia F. Bearer

The quality of the intrauterine environment is a critically important determinant of an infant's short- and long-term health and well-being. In utero exposure to chemical, biological, nutritional, and social stressors can lead to intrauterine growth retardation (IUGR) and premature delivery, two common complications in pregnancy.[1,2] These conditions may occur together and both result in low birth weight (LBW) (see Box 47.1).

More than 20 million LBW babies are born worldwide each year, representing 14.6% of all deliveries. Ninety-five percent of LBW babies are born in developing countries. The percent of babies who are LBW in developing countries is 16.5% as compared to 7% in industrially developed countries.[3] Malnutrition, infectious diseases, and exposures to very high levels of indoor air pollution are the major contributors responsible for LBW babies in the developing world.

Premature and LBW infants are at increased risk of morbidity and mortality in the newborn period. They require a high amount of healthcare resources. A high prevalence of LBW within a population is associated with increased rates of infant mortality.

Box 47.1   Definitions

Low birth weight (LBW) is defined as a birth weight less than 2,500 g. LBW can be the result of premature birth (birth before 37 complete weeks of gestation) and impaired in utero growth.

These two conditions may occur together, and babies born of such double-jeopardy pregnancies have LBW because they did not grow well in utero and were born prematurely. Some studies differentiate between these two types of LBW, whereas others look only at birth weight but not at gestational age.

Infants weighing less than the 10th percentile for gestational age are classified as *small for gestational age* (SGA). Those weighing more than the 90th percentile are considered *large for gestational age* (LGA).

Intrauterine growth restriction (also referred to as intrauterine growth retardation) (IUGR) refers to a fetus that is less than 10% of predicted fetal weight for gestational age. Aberrant intrauterine growth may be due to maternal, placental, and fetal factors, any of which can be affected by environmental exposures.

Margaret Kuper-Sassé, Cansu Tokat, Hilal Yildiz Atar, and Cynthia F. Bearer, *Prematurity and Low Birth Weight Associated with Environmental Exposures* In: *Textbook of Children's Environmental Health*, Second Edition. Edited by: Ruth A. Etzel and Philip J. Landrigan, Oxford University Press. © Oxford University Press 2024. DOI: 10.1093/oso/9780197662526.003.0047

Prematurity and LBW also have long-term impacts on health. Impaired intrauterine growth leads to maladaptive developmental patterns that may increase future risk for hypertension, coronary heart disease, obesity, and type 2 diabetes. The association between intrauterine undernutrition and subsequent increased risk for chronic disease was first described by Barker et al.[4] in classic epidemiological studies conducted in England. This is now referred to as the "developmental origins of health and disease hypothesis" (see Chapters 1, 2, and 23). The association between intrauterine malnutrition and increased risk of adult disease was further confirmed in long-term epidemiologic studies conducted among a population in Holland who were severely malnourished in utero during the Dutch Winter Famine of World War II.[5]

This chapter describes the toxic environmental exposures associated with prematurity and LBW. The long-term health consequences of LBW caused by toxic environmental exposures are still the subject of active investigation.

Assessment of Growth

Researchers have used several measures to monitor infants' physical growth: weight, length, head circumference, weight for length, and ponderal index (weight divided by length cubed).

Intrauterine growth is assessed during pregnancy by regular physical examination (fundal height) and fetal ultrasound. Early detection and management of abnormal fetal growth is essential to improve outcomes, though incorrect diagnosis occurs in up to 30% of cases.[6]

Growth charts are used to track the physical growth of fetuses, infants, and children. Growth charts follow an infant's weight, length, head circumference, and weight for length. In the United States, it is recommended that the World Health Organization (WHO) growth charts be used to assess postnatal growth in children younger than 2 years of age, the Fenton Growth Charts for Premature Infants be used for premature infants, and the Centers for Disease Control and Prevention (CDC) charts be used for children 2–19 years of age.[7] The WHO charts are based on international data from 1,743 mother–infant pairs. All of the infants whose data are included in the WHO charts were breastfeeding, and their growth trajectory is therefore considered to represent optimal normal growth. Because breastfeeding babies have slower growth before 18 months of age than babies on other diets, the use of the WHO charts may enable earlier detection of childhood obesity.

Assessment of Gestational Age

Gestational age (GA) of the fetus prior to delivery is calculated from the date of the mother's last menstrual period (LMP). This assessment is often facilitated by the use of a "wheel" (Figure 47.1). Term delivery is defined as 37 to 42 weeks post LMP. Premature delivery is defined as before 37 weeks post LMP.

Ultrasound also is used to estimate fetal GA. Both crown–rump length (prior to 12 weeks) and biparietal diameter (13–30 weeks) can be used to estimate GA. GA is estimated post-delivery by the Dubowitz/Ballard physical examination, which scores physical and neuromuscular maturity. Electroencephalogram (EEG) is also used, though rarely, to assess GA.

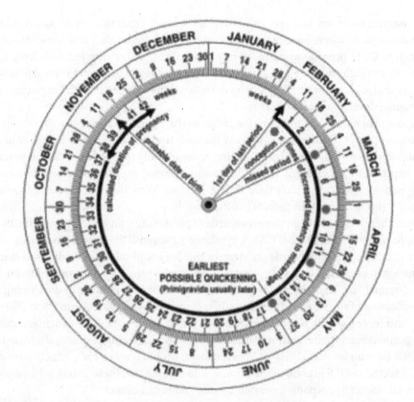

Figure 47.1. The classic obstetrics wheel, or pregnancy calculator.
Source: Created with biorender.com by Hilal Yildiz Atar.

Toxic Environmental Exposures, Prematurity, and Low Birth Weight

Chemical exposures are widespread in the modern environment (Chapter 3). In a small study of chemicals in umbilical cord blood samples, more than 200 chemicals—most of the synthetic chemicals invented since 1950—were detected in the cord blood of each of 10 newborn infants.[8] The Fourth National Report on Human Exposure to Environment Chemicals issued by the CDC (with updated tables placed online in September, 2023) presents data on levels of more than 400 chemicals in the urine and blood of a nationally representative sample of US women of childbearing age.[9] Paired maternal and cord blood samples collected from urban dyads in 2010–2011 showed that 80% of 59 harmful chemicals found in the mother also were found in the newborn, some of which were found in higher concentrations, indicating that some toxic chemicals are concentrated in the intrauterine environment.[10] Some of the chemicals widely detected in the bodies of American women are recognized to be harmful and have previously been banned (see Chapter 14).

Certain toxic chemicals in the environment are known to cause intrauterine growth disorders, preterm birth, and both functional and structural abnormalities in the fetus (see Figure 47.2). Knowledge of the underlying mechanisms by which chemicals bring about these effects is still incomplete. Establishing the associations between environmental exposures in early life and adverse developmental outcomes is challenging because it requires long-term epidemiologic follow-up studies, and those studies are often very complex to design and very expensive to conduct. Evaluation of an infant at birth will miss most functional anomalies (such as learning disabilities) and some structural anomalies. Only prenatal assessment of chemical exposures coupled with long-term follow-up through childhood and even into adult life that incorporates careful neurodevelopmental assessment can fully establish associations between prenatal exposures and subsequent developmental problems.

Exposure of either the mother or the father to toxic chemicals in the environment before conception or during pregnancy may increase the risk of disease in a child either by causing mutations in DNA or by producing epigenetically mediated changes in parental gene expression. An example is preconception paternal exposure to Agent Orange (a chlorphenoxy herbicide widely used in the Vietnam War) that was associated with an increased risk of neural tube defects in offspring.[11]

Toxic chemicals have been shown capable of producing transgenerational effects. An example is diethylstilbestrol (DES), a synthetic estrogen. DES is well known to cause direct toxic effects in the female children of mothers who took the medication during pregnancy to prevent miscarriage. Young women who were directly exposed to DES in utero (termed "DES daughters") have an increased risk after puberty of developing adenocarcinoma of the vagina. They also suffer from infertility, genitourinary malformations, and increased risk of both cervical and breast carcinoma. DES daughters, due to their genitourinary malformations, are at high risk for delivering prematurely. Male DES children have an increased risk of breast cancer. More recently, toxic effects have also been detected in DES grandchildren. Some data suggest that these grandchildren are at increased risk for hypospadias, ovarian cancer, and breast cancer.[12]

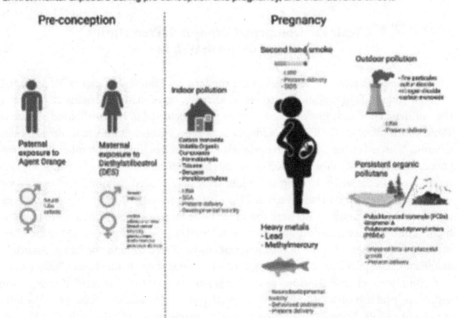

Figure 47.2. Environmental exposures and their effects on pregnancy and the unborn fetus. Pre-conception effects on offspring listed based on the gender difference.
Source: Created with biorender.com by Hilal Yildiz Atar.

Many factors influence how toxic chemicals contribute to harmful health effects in the mother and fetus, including dose, genetically determined individual susceptibility, nutrition, socioeconomic status, and concurrent exposures to other drugs and chemicals. Timing of exposure is important during pregnancy because the fetal brain develops through a series of critical stages, each of which is much more sensitive to toxic chemicals in the environment than is the adult brain (see Chapter 2).

Meconium analysis is a new and promising technology for the time-integrated assessment of fetal exposures to environmental toxicants during the third trimester.

The following sections describe toxic environmental exposures that have been linked in clinical and epidemiological studies to prematurity, IUGR, and abnormal postnatal growth and development.

Indoor Air Pollution

Indoor air pollution arises from many sources. In low- and middle-income country settings, poorly ventilated cook stoves that burn biomass are the major source of indoor air pollution (see Chapter 26). Sources of indoor air pollution in industrialized countries include oil and gas stoves, building materials and furnishings, cigarette smoking, and household products used in cleaning. Indoor air quality is an important public health issue. Poor indoor air quality is associated with poor health and long-term health problems. Specific toxic chemicals that can be present in indoor air are discussed next.

Carbon Monoxide

Carbon monoxide (CO) is a colorless, odorless, and highly toxic gas produced by cigarette smoking, automotive combustion, improperly maintained furnaces, charcoal burners, and wood and kerosene stoves. Acute or chronic CO exposure can cause severe and even fatal poisoning by binding to hemoglobin with a higher affinity than oxygen and displacing oxygen. This leads rapidly to tissue hypoxia and brain death. CO readily crosses the placenta from the maternal to the fetal circulation, and fetal hemoglobin has a higher affinity for CO than does adult hemoglobin. Elimination of CO from the fetus is therefore slow, leading to CO accumulation in the fetus and tissue damage.

Acute CO poisoning in a pregnant woman may cause severe adverse effects on her fetus, including fetal death. In a study done in Los Angeles County, CO exposure from air pollution was associated with preterm birth.[13] In a mouse model, chronic exposure to a CO concentration of 400 ppm parts per million (ppm) was shown to have adverse effects on fetal growth and development.[14] Acute CO poisoning harms both mother and fetus, but chronic exposure to CO may immensely harm the fetus during its development.[15]

Volatile Organic Compounds

Volatile organic compounds (VOCs) are a diverse group of chemicals that all have a low boiling point and therefore readily evaporate at room temperature and are respirable (see also Chapter 39). They include formaldehyde (used in glues and pressed wood furniture), toluene (used in aerosols), benzene (a contaminant of gasoline), and perchloroethylene (dry cleaning solution). Exposures to VOCs have been associated with LBW, SGA, prematurity, and developmental toxicity in both human epidemiologic and animal mechanistic studies.[16-18]

Secondhand Smoke

Tobacco smoke contains at least 60 chemicals known or suspected to be carcinogens and teratogens (see also Chapter 27). Some of these toxic chemicals, such as nicotine and benzo(a)pyrene, readily cross the placenta.

A recent study in Canada showed that nonsmoking pregnant women exposed to SHS had babies with lower mean birth weight (−53.7 g: 95% confidence interval [CI]: −98.4, −8.9 g) and a dose-related trend toward increased prematurity (odds ratio [OR] 1.87; 95% CI: 1.00, 3.53; P = 0.05).[19] These same associations were found in a meta-analysis of 76 published studies.[20]

A recent birth cohort study from China observed that continued exposure to second-hand smoke (SHS) during pregnancy was associated with medically indicated preterm delivery compared to spontaneous preterm delivery, which suggests that secondhand smoking increases risk of placental malperfusion, IUGR, preeclampsia, or other maternal health problems that result in the decision to deliver prematurely.[21]

Exposure to SHS during pregnancy is associated with sudden infant death syndrome (SIDS).[22] Certain maternal genetic polymorphisms in cytochrome P450 enzymes and in glutathione transferase may increase the risk for SIDS and for other adverse outcomes in fetuses exposed to SHS.[23,24] Fetal injury and death caused by maternal smoking and exposure to SHS are highly preventable.

Outdoor Air Pollution

Numerous studies have investigated the adverse effects of ambient air pollution on pregnant women and their children (see also Chapter 25). There is suggestive evidence that exposures to outdoor air pollution during pregnancy are associated with an increased risk of LBW and stillbirth.[25] The association between exposure to air pollution and LBW was confirmed in a recent study in the New England states that linked LBW with maternal exposure in pregnancy to exposures to airborne particulate matter, sulfur dioxide, nitrogen dioxide, and carbon monoxide.[26] A large population study from Africa that included 15 countries shows exposure to fine particulate matter also was associated with LBW and preterm delivery in low-income countries.[27] IUGR has been associated in multiple studies with exposures to air pollution.[24]

Due to the combustion of fossil fuels, air pollution exposure has increased and is significantly associated with adverse outcomes in pregnancy, including preterm delivery, LBW, and stillbirth.[28]

Metals

Metals have been known since ancient times to have adverse effects on pregnancy. In ancient Rome, lead was used to induce abortions, and in China, mercury was used for the same purpose.

Lead exposure has decreased sharply in the United States in the past 25 years because of the coordinated efforts of national, state, and local agencies. The mean blood lead concentration began to decrease in the United States after 1976, when the US Environmental Protection Agency (EPA) promulgated standards reducing the lead content in gasoline. In 1977, lead-based paint was banned from use in all residential and public properties. Of note, small aircraft, off-road vehicles, such as farm equipment, and racing cars in the United States, may still use leaded gasoline. Recently, an ayurvedic remedy was found to have contributed to elevated blood lead concentrations (29–64 μg/dL) in a number of pregnant women.[30] Maternal blood lead concentrations of even less than 10 μg/dL are associated with an increased risk of preterm delivery.[31] Lead exposure is known to cause developmental learning and behavioral problems in children at the lowest blood lead levels that can be measured. There is no known safe level for lead in blood (see Chapter 31).

Maternal consumption of contaminated fish during pregnancy increases the risk of exposure to methylmercury and hence of neurodevelopmental toxicity (see Chapter 32). Methylmercury also appears to increase the risk of preterm delivery. In one recent study, women recruited from 52 prenatal clinics in five Michigan communities had hair measurements performed for mercury analysis. Consumption of canned fish, bought fish, and sport-caught fish all were associated with elevated hair mercury concentrations. Women in this population who delivered before 35 weeks gestation were significantly more likely to have hair mercury levels at or above the 90th percentile than were women delivering at term.[32] Furthermore, a study from Japan showed that mercury concentration was doubled in cord blood as compared to maternal blood, and there was an inverse relationship between mercury concentration and birth weight.[33]

The predominant source of arsenic exposure around the world is drinking water (see Chapter 33). In a study in Bangladesh, where arsenic contamination of drinking water is widespread, women exposed to arsenic during pregnancy were significantly more likely to deliver preterm than were unexposed women of the same socioeconomic status, education, and age at marriage.[34]

Persistent Organic Pollutants

Persistent organic pollutants are a group of halogenated hydrocarbon compounds that are highly persistent in the environment and the human body. They are lipophilic and tend to accumulate in the food chain to reach high levels in top predator species. They include polychlorinated biphenyls (PCBs), the pesticide DDT, dioxin, and brominated flame retardants.

Polychlorinated biphenyls (PCBs) persist in the environment today despite having been banned in the United States in 1976 (see also Chapter 36). When incorporated into the food chain, PCBs bioaccumulate to reach very high levels in predatory species. The food supply is the major source of human exposure to PCBs, and the food principally responsible for human PCB exposure is predatory fish. The Danish national birth cohort study showed impaired fetal and placental growth in infants exposed to PCBs in utero[35] (see Box 47.2).

Polybrominated diphenyl ethers (PBDEs), commonly known as brominated flame retardants, are another group of persistent organic pollutants. PBDE levels were measured in pregnant women at 20 weeks of gestation in the CHAMACOS study in California. A negative association was seen between birth weight and levels of several PBDE congeners. However, the statistical significance of this association

---

Box 47.2 Fish Consumption in Pregnancy

Women are strongly advised to eat fish during pregnancy because the omega-3 fatty acids contained in high quantities in many fish are important for infant brain development. But pregnant women should avoid those fish species that tend to be contaminated by PCBs or by methylmercury. A reliable source of information on which species of fish are safe and unsafe to eat during pregnancy is a list prepared by the US Food and Drug Administration, is found at https://www.fda.gov/food/consumers/advice-about-eating-fish

---

was lost when maternal weight gain was added to the models. PBDEs also were measured in pregnant women and the cord blood of their infants in San Francisco, California, and found to be transferred across the placenta and concentrated in newborn blood.[10]

Bisphenol A is a common constituent of polycarbonate plastics such as plastic drinking water bottles.[36] A study done in Mexico demonstrated that women who delivered prior to 37 weeks of pregnancy had higher urinary concentrations of bisphenol A than women who delivered at term.

Pesticides

Pesticides are a diverse group of chemicals that include herbicides, insecticides, bactericides, and fungicides. The EPA registers and regulates pesticides in the United States. In their most recent national report from 2018, the EPA reported that the United States uses more than 1.1 billion pounds of pesticide chemicals per year. Many pesticides are persistent in the environment.

Chlorpyrifos, an organophosphate insecticide, has drawn much attention in the past decade. Accumulating data show a relationship between prenatal exposure to chlorpyrifos and adverse neurodevelopmental outcomes (see Chapter 35). Three large cohort studies, two in New York and one in the Salinas Valley of California, showed an association between chlorpyrifos exposure and diminished head circumference at birth, a finding suggesting an adverse effect of chlorpyrifos on brain development in utero.[37]

Mechanisms of Toxicity

Developmental Origins of Adult Disease

Multiple population studies have demonstrated significant correlations between LBW and increased risks in later life of the heart and cardiovascular disease, obesity, and type 2 diabetes mellitus, and a relationship now referred to as the *developmental origins of adult disease* hypothesis.

LBW is an adaptive phenomenon. It is the consequence of the fetus' attempts to survive in an adverse intrauterine environment that is not conducive to normal growth. IUGR fetuses show decreased insulin sensitivity to protein, decreased pancreatic development, and increased hepatic glucose production. All of these changes are adaptations designed to maintain basal metabolic function in the fetus at the expense of body growth in the face of insufficient intrauterine nutrition.

Many of these adaptive changes appear to be brought about by epigenetic changes that alter gene expression to favor survival over somatic growth. For example, studies in a rat model of IUGR suggest that fetal growth restriction induces epigenetic changes in key genes regulating pancreatic β-cell development. A decrease in histone acetylation plays a major role in these changes. If not reversed, these changes can ultimately lead to type 2 diabetes. However, it has been shown that this decreased histone acetylation can be reversed with proper intervention and normal expression of promoter of *Pdx1*.[38] Thus, it seems that this process of epigenetic regulation is reversible at an early neonatal stage. This finding suggests that there may be a novel therapeutic window for the prevention of common diseases such as type 2 diabetes.

It is currently unclear whether environmental pollutants that result in LBW have the same long-term effects on adult blood pressure, diabetes, and risks of other chronic illnesses as intrauterine undernutrition.

Epigenetics and Fetal Development

The approximately 9 months between conception and birth encompass a period of extraordinarily rapid growth, proliferation, and differentiation of cells. The single cell of the fertilized ovum must in these 9 months differentiate into an enormous variety of different cell types with very diverse structures and functions—neurons, cardiomyocytes, lung cells, hepatocytes, and skin cells. The great rapidity and complexity of these developmental processes create great vulnerability, and any critical insult during this vulnerable period may lead to altered cellular programming which, in turn, causes changes in gene expression and, hence, bodily composition. Some of these changes can result in health effects that become evident in childhood, but others may not become manifest until adult life.

Epigenetic alterations in gene expression that vary from cell to cell and organ to organ are now thought to be a critical mechanism that guides cells in different tissues—all of which have the same DNA—to develop in different directions. Epigenetic mechanisms have been studied in both humans and animal models. Mechanisms of epigenetic alteration of gene expression include *DNA methylation* and *histone modification*. Methylation of DNA produces a modification of cytosine by the enzyme DNA methyltransferase and is tightly coupled to modification of the overlying histone. This is a normal phenomenon, and approximately 70–80% of the cytosines found next to guanines (CpG) are constitutively methylated in the human genome. Most of the unmethylated CpG islands are located near the coding sequences of genes. These CpG islands are either methylated or unmethylated based on their developmental programming.

Environmental exposures may cause either hypo- or hypermethylation of these islands, along with changes in modifications in the overlying histones, thus altering gene expression without changing the DNA base sequence.

Variations in maternal gene expression can alter the fetal environment by changing the characteristics of the placenta.

Gene–Environment Interactions

Complex interactions take place between the environment and an individual's genetic makeup at every stage of development. Genetic polymorphisms may lead to great differences in susceptibly to toxic environmental exposures and very different health outcomes in the face of similar exposures. Genetic polymorphisms have been shown to increase the risk for various cancers and other chronic diseases.

Epidemiologists are trying to define relationships between environmental exposures, genetic variations, and developmental outcomes. For example, a study examining the relationship between newborn exposure to organochlorine pesticides, genetically determined polymorphisms in the enzyme, glutathione S-transferase, and IUGR found that infants with IUGR had higher levels of pesticide metabolites in their cord blood than control infants and that infants with IUGR also had a higher frequency than control infants of the null phenotype for GSTM1–/GSTT1–.[39] A significant association was found between the null genotype and pesticide metabolite level, which was linked to a 213 gram reduction in birth weight.

Climate Change

Although climate change has been widely accepted as the largest public health threat in the modern world, attention to its effects on the fetus and neonate has only recently increased. Climate change has caused increasing frequency and severity of heat waves, wildfires, and extreme weather events, causing flooding, drought, pollution, power outages, and physical displacement. The direct and indirect effects of these exposures are associated with adverse effects on fetal growth and birthweight, preterm birth, spontaneous abortion, and neonatal death.[40]

Zika virus is an arbovirus transmitted to humans by various species of *Aedes* mosquitoes; it is known to cause fetal microcephaly when a pregnant mother is infected. The Zika virus pandemic in the Americas in 2015–2017 drew attention to the high cost a virus that causes birth defects can inflict. Zika virus currently is limited geographically by where weather conditions are suitable for the mosquitoes through which it is transmitted. Using thermal transmission curves it is predicted that, as climate change causes an increase in global temperature, the areas that can support *Aedes* mosquitoes will expand; by 2050, up to 1.3 billion new people will be exposed.[41]

Conclusion

Many environmental pollutants are associated with reduced fetal growth and shortened gestation. These effects may in turn lead to health problems in adult life. Pregnant people should be educated about those adverse pregnancy outcomes and should be encouraged to avoid exposure to environmental pollutants whenever possible. More research is necessary to ascertain mechanisms of toxicity during pregnancy and guide evidence-based strategies for treatment and prevention.

References

1. Gu H, Wang L, Liu L, et al. A gradient relationship between low birth weight and IQ: A meta-analysis. *Sci Rep*. 2017;7(1):18035.

2. Jornayvaz FR, Vollenweider P, Bochud M, Mooser V, Waeber G, Marques-Vidal P. Low birth weight leads to obesity, diabetes and increased leptin levels in adults: the CoLaus study. *Cardiovasc Diabetol*. 2016;15:73.

3. WHO/UNICEF. Birthweight: country, regional and global estimates. UNICEF. 2015. Available at: https://data.unicef.org/topic/nutrition/low-birthweight/.

4. Barker DJ, Osmond C, Golding J, Kuh D, Wadsworth ME. Growth in utero, blood pressure in childhood and adult life, and mortality from cardiovascular disease. *BMJ* 1989 Mar 4; 298(6673):564–7.

5. Friedman P, Guo XM, Stiller RJ, Laifer SA. Carbon monoxide exposure during pregnancy. *Obstet Gynecol Surv*. 2015 Nov;70(11):705–12.

6. Debost-Legrand A, Laurichesse-Delmas H, Francannet C, Perthus I, Lémery D, Gallot D, Vendittelli F. False positive morphologic diagnoses at the anomaly scan: marginal or real problem, a population-based cohort study. *BMC Pregnancy Childbirth*. 2014 Mar 24;14:112.

7. Centers for Disease Control and Prevention. Grummer-Strawn LM, Reinold C, Krebs NF, eds. Use of World Health Organization and CDC growth charts for children aged 0–59 months in the United States. MMWR. 2010. Available at: http://www.cdc.gov/mmwr/preview/mmwrhtml/rr5909a1.htm.

8. Houlihan J, Kropp T, Wiles R, Gray S, Campbell C. Body burden: The pollution in newborns. Environmental Working Group. 2005 Jul 14. Available at: http://www.ewg.org/reports/body burden2/execsumm.php.

9. Centers for Disease Control and Prevention. National report on human exposure to environmental chemicals. Centers for Disease Control and Prevention. 2022 Mar 24. Available at: https://www.cdc.gov/exposurereport/.

10. Morello-Frosch R, Cushing L, Jesdale B, et al. Environmental chemicals in an urban population of pregnant women and their newborns from San Francisco. *Environ Sci Technol*. 2016;50(22):12464–72.

11. Ngo AD, Taylor R, Roberts CL. Paternal exposure to Agent Orange and spina bifida: a meta-analysis. *Eur J Epidemiol*. 2010;25(1):37–44.

12. Reed CE, Fenton SE. Exposure to diethylstilbestrol during sensitive life stages: a legacy of heritable health effects. *Birth Defects Res C Embryo Today*. 2013 Jun;99(2):134–46.

13. Ritz B, Wilhelm M, Hoggatt KJ, Ghosh JK. Ambient air pollution and preterm birth in the environment and pregnancy outcomes study at the University of California, Los Angeles. *Am J Epidemiol*. 2007;166(9):1045–52.

14. Venditti CC, Casselman R, Smith GN. Effects of chronic carbon monoxide exposure on fetal growth and development in mice. *BMC Pregnancy Childbirth*. 2011;11:101.

15. Friedman P, Guo XM, Stiller RJ, Laifer SA. Carbon monoxide exposure during pregnancy. *Obstet Gynecol Surv*. 2015 Nov;70(11):705–12.

16. Forand SP, Lewis-Michl EL, Gomez MI. Adverse birth outcomes and maternal exposure to trichloroethylene and tetrachloroethylene through soil vapor intrusion in New York State. *Environ Health Perspect.* 2012;120(4):616–21.

17. Carney EW, Thorsrud BA, Dugard PH, Zablotny CL. Developmental toxicity studies in Crl:CD (SD) rats following inhalation exposure to trichloroethylene and perchloroethylene. *Birth Defects Res B Dev Reprod Toxicol.* 2006;77(5):405–12.

18. Bergstra AD, Brunekreef B, Burdorf A. The influence of industry-related air pollution on birth outcomes in an industrialized area. *Environ Pollut.* 2021 Jan 15;269:115741.

19. Crane JM, Keough M, Murphy P, Burrage L, Hutchens D. Effects of environmental tobacco smoke on perinatal outcomes: a retrospective cohort study. *BJOG.* 2011;118(7):865–71.

20. Salmasi G, Grady R, Jones J, McDonald SD, Knowledge Synthesis Group. Environmental tobacco smoke exposure and perinatal outcomes: a systematic review and meta-analyses. *Acta Obstet Gynecol Scand.* 2010;89(4):423–41.

21. Chen X, Huang L, Zhong C, et al. Association between environmental tobacco smoke before and during pregnancy and the risk of adverse birth outcomes: a birth cohort study in Wuhan, China. *Environ Sci Pollut Res Int.* 2021 Jun;28(21):27230–7.

22. Liebrechts-Akkerman G, Lao O, Liu F, et al. Postnatal parental smoking: an important risk factor for SIDS. *Eur J Pediatr.* 2011;170(10):1281–91.

23. Wang X, Zuckerman B, Pearson C, et al. Maternal cigarette smoking, metabolic gene polymorphism, and infant birth weight. *JAMA.* 2002;287(2):195–202.

24. Tsai HJ, Liu X, Mestan K, et al. Maternal cigarette smoking, metabolic gene polymorphisms, and preterm delivery: new insights on GxE interactions and pathogenic pathways. *Hum Genet.* 2008;123(4):359–69.

25. Lee KK, Bing R, Kiang J, et al. Adverse health effects associate with household air pollution: a systematic review, meta-analysis, and burden estimation study. *Lancet Glob Health.* 2020;8(11):e1427–34.

26. Sram RJ, Binkova B, Dejmek J, Bobak M. Ambient air pollution and pregnancy outcomes: a review of the literature. *Environ Health Perspect.* 2005;113(4):375–82.

27. Bachwenkizi J, Liu C, Meng X, et al. Maternal exposure to fine particulate matter and preterm birth and low birth weight in Africa. *Environ Int.* 2022 Feb;160:107053.

28. Bekkar B, Pacheco S, Basu R, DeNicola N. Association of air pollution and heat exposure with preterm birth, low birth weight, and stillbirth in the US: a systematic review. *JAMA Netw Open.* 2020 Jun 1;3(6):e208243.

29. Bell ML, Ebisu K, Belanger K. Ambient air pollution and low birth weight in Connecticut and Massachusetts. *Environ Health Perspect.* 2007;115(7):1118–24.

30. Centers for Disease Control and Prevention (CDC). Lead poisoning in pregnant women who used ayurvedic medications from India—New York City, 2011–2012. *MMWR Morb Mortal Wkly Rep.* 2012 Aug 24;61:641–6.

31. Vigeh M, Yokoyama K, Seyedaghamiri Z, et al. Blood lead at currently acceptable levels may cause preterm labour. *Occup Environ Med.* 2011;68(3):231–4.

32. Xue F, Holzman C, Rahbar MH, Trosko K, Fischer L. Maternal fish consumption, mercury levels, and risk of preterm delivery. *Environ Health Perspect.* 2007;115(1):42–7.

33. Vigeh M, Nishioka E, Ohtano K, et al. Prenatal mercury exposure and birth weight. *Reprod Toxicol.* 2018;76:78–83.

34. Ahmad SA, Sayed MH, Barua S, et al. Arsenic in drinking water and pregnancy outcomes. *Environ Health Perspect.* 2001;109(6):629–31.

35. Halldorsson TI, Thorsdottir I, Meltzer HM, Nielsen F, Olsen SF. Linking exposure to polychlorinated biphenyls with fatty fish consumption and reduced fetal growth among Danish pregnant women: a cause for concern? *Am J Epidemiol.* 2008;168(8):958–65.

36. Cantonwine D, Meeker JD, Hu H, et al. Bisphenol A exposure in Mexico City and risk of prematurity: a pilot nested case control study. *Environ Health.* 2010;9:62.

37. Rauh VA, Perera FP, Horton MK, et al. Brain anomalies in children exposed prenatally to a common oraganophosphate pesticide. *Proc Natl Acad Sci USA.* 2012;15109(20):7871–6.

38. Park JH, Stoffers DA, Nicholls RD, Simmons RA. Development of type 2 diabetes following intrauterine growth retardation in rats is associated with progressive epigenetic silencing of Pdx1. *J Clin Invest.* 2008 Jun;118(6):2316–24.

39. Sharma E, Mustafa M, Pathak R, et al. A case control study of gene environmental interaction in fetal growth restriction with special reference to organochlorine pesticides. *Eur J Obstet Gynecol Reprod Biol.* 2012;161(2):163–9.

40. Sandie HA. The changing climate and pregnancy health. *Curr Environ Health Rep.* 2022 Feb 22. doi:10.1007/s40572-022-00345-9.

41. Ryan SJ, Carlson CJ, Tesla B, et al. Warming temperatures could expose more than 1.3 billion new people to Zika virus risk by 2050. *Glob Change Biol.* 2021;27:84–93.

48

Asthma, Allergy, and the Environment

Dwan Vilcins and Peter D. Sly

Asthma can be thought of as a developmental disease in which the normal development of the respiratory and immune systems is altered by the impacts of environmental exposures acting on underlying genetic predispositions.[1-4]

The respiratory and immune systems are both immature at birth and both undergo prolonged periods of postnatal maturation. This long development creates prolonged vulnerability to environmental exposures. Thus, pre- as well as postnatal exposures to environmental factors may disturb lung growth, delay maturation of the immune system, and increase susceptibility to wheeze in early life. In addition, delayed immune maturation increases the risk of primary sensitization to aeroallergens.

Epidemiological data show that major early-life risk factors for persistent asthma are a family history of asthma and allergies, reflecting a genetic predisposition; an increased frequency and severity of respiratory viral infections; early allergic sensitization to ubiquitous aeroallergens; and low lung function. Adverse environmental exposures contribute to at least three of these risk factors.

Environmental exposures contribute significantly to exacerbations of clinical asthma and impairment of lung function. Most acute asthma exacerbations result from respiratory viral infections. Exposures to traffic-related pollution from cars, trucks, and buses; secondhand cigarette smoke; and other contaminants in indoor air also increase asthma symptoms and result in deterioration of asthma control. Aeroallergens in the form of bioaerosols contribute to pulmonary inflammation and worsen the clinical expression of asthma.

Environmental control, as a means of preventing or controlling asthma, is controversial, with most studies demonstrating that allergen avoidance does not prevent asthma. There is some evidence that avoiding air pollutants may result in improved asthma control.

Scope and Nature of the Problem

Asthma is a huge global problem. It affects an estimated 339 million people worldwide.[5] Asthma is the most common chronic disease of children. It also affects millions of adults. Asthma prevalence varies markedly around the world, with the highest rates seen in high-income countries. Prevalence has been increasing rapidly in low- and middle-income countries (LMICs), however, especially as they undergo industrialization. These recent widespread increases in the prevalence of asthma are far too rapid to be of genetic

Dwan Vilcins and Peter D. Sly, *Asthma, Allergy, and the Environment* In: *Textbook of Children's Environmental Health,* Second Edition. Edited by: Ruth A. Etzel and Philip J. Landrigan, Oxford University Press. © Oxford University Press 2024. DOI: 10.1093/oso/9780197662526.003.0048

origin. They support the hypothesis that environmental triggers must be important causes of asthma. This recognition that there is a powerful environmental contribution to allergy and asthma provides opportunity for preventive intervention. If the responsible environmental factors can be identified and understood, they may be able to be modified and the global burden of asthma reduced.

Asthma is not a single unified condition but rather is a collection of related phenotypes. In high-income countries, the type of asthma that most commonly persists from childhood into adult life is associated with allergic sensitization to perennial aeroallergens, especially house dust mite, cat dander, cockroach, and molds. This is commonly known as *atopic asthma*. In LMICs, by contrast, *nonatopic asthma*, that is, asthma not associated with allergic sensitization, is more common. This global variation in phenotype complicates the understanding of asthma, especially as most asthma research is conducted in high-income countries.

Sources of Exposure: Exposure Pathways

While it might be natural to assume that the most important environmental exposures involved in the causation or exacerbation of asthma are airborne, the situation is far more complex. Multiple exposure pathways are relevant to the development of asthma and at different developmental stages, different exposures predominate (reviewed in Sly and Flack[6]). They include the following:

- *Transplacental exposures.* We now recognize that many environmental toxicants either directly cross the placenta or have indirect impacts on the developing fetus. Maternal smoking and maternal exposure to secondhand smoke are among the prenatal exposures associated with asthma. Inhaled particulate matter can travel to the placenta and is measurable on the fetal side.[7,8]
- *Breast milk.* Many lipophilic compounds can be found in breast milk. Some, including polychlorinated biphenyls, are likely to modify maturation of the immune system.
- *Non-nutritive ingestion.* The "hand-to-mouth" behavior of infants and toddlers results in their exposure to a wide range of environmental toxicants that settle in dust or on surfaces, especially in the home environment. These include pesticides, plasticizers, flame retardants, and personal care products. The mechanisms by which these toxicants contribute to asthma are not fully understood, although evidence is emerging on their ability to modify immune functioning.[9]
- *Dermal exposure.* Young children have a higher surface area to body mass ratio than adults. This factor, together with the fact that young children spend a great deal of their time on the floor in close contact with potentially contaminated floor coverings, make the skin a potentially important exposure pathway.
- *Inhalational exposure.* Young children have a higher ventilation requirement than older children or adults. This means that they breathe more air per unit body weight and potentially receive a proportionately higher dose of airborne contaminants.

Exposures relevant to the development of asthma and allergies are those that induce inflammation in the airways or that interfere with normal lung growth and immune system maturation. These include the following:

- *In utero exposures*: Active maternal smoking and secondhand smoke exposure; ambient air pollution; household chemicals and personal care products, especially aerosol sprays; maternal consumption of alcohol and some drugs; maternal microbiome and microbial exposure; and poor maternal nutrition.
- *Indoor air exposures*: Secondhand smoke and other combustion sources (e.g., candles, mosquito coils, incense); air pollutants such as nitrogen dioxide (NO_2), carbon monoxide (CO), polyaromatic hydrocarbons, ozone, volatile organic compounds, formaldehyde; particulate matter; and bioaerosols.
- *Ambient air exposures*: Traffic-related pollution; industrial pollution; and living in the vicinity of a municipal incinerator or toxic waste site. While epidemiological associations have been reported for many individual pollutants, the evidence is stronger for exposures to diesel exhaust particles, ozone, NO_2, and "acid vapor."

Epidemiology

Much of the information linking exposures in early life to risk factors for asthma has come from longitudinal birth cohort studies, either community-based studies or studies specifically targeting children at high risk for developing asthma and allergies. In addition, most chronic adult diseases have their origin in childhood; this is especially true for asthma.

Major early-life risk factors for persistent asthma that have been identified and examined through epidemiological studies are recurrent severe lower respiratory infections, early allergic sensitization, and a family history of asthma and allergies, probably reflecting a genetic predisposition. In addition, asthma is associated with low lung function, with a deficit in lung function being seen as the first measurement in most cohort studies.[10] This observation raises the possibility that low lung function at birth is an independent risk factor for the development of asthma.

Lung function "tracks" along percentiles during the period of rapid lung growth in childhood. Thus, the lung function with which an infant is born is a major determinant of lung function throughout life. Factors that limit lung function at birth are likely therefore to have lifelong consequences. In addition, the lungs are immature at birth and have a prolonged period of postnatal maturation. Thus, factors that limit lung growth, especially during the rapid growth period in early childhood, may reduce an individual's peak lung function.

There appear to be two pathways by which prenatal factors can result in low lung function at birth: decreased somatic growth and decreased airway growth (Figure 48.1). Development of the airway tree is completed by 16–18 weeks gestation, so exposures occurring before this period may have an adverse impact on airway growth and lung function at birth. Such exposures include maternal smoking during pregnancy[11]; exposure to ambient air pollution[12]; and potentially exposure to secondhand smoke, household chemicals,[13] or maternal alcohol consumption.[14] Prenatal exposures to the mother that result in decreased somatic growth of the fetus are likely to result in low lung function at birth. Having a low birth weight for gestational age is associated with a reduced lung function, independent from any effects on somatic growth.[15] Furthermore, birth weight is a predictor of adult lung function.[16] The best-understood examples of exposures that act through this mechanism are cigarette smoking, alcohol consumption, and maternal consumption during pregnancy of prescribed and illicit drugs.[17] Inadequate nutrition during pregnancy can also influence fetal lung growth, with growth restriction being associated with abnormal lung function. The impact of maternal respiratory

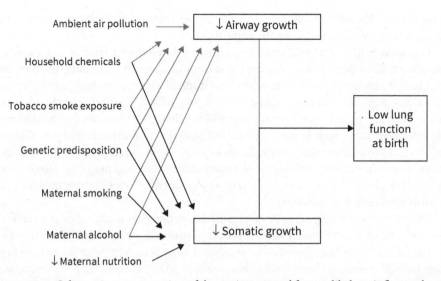

Figure 48.1. Schematic representation of the environmental factors likely to influence lung function at birth

infections on the infant's lung function has recently been reported. Infants born to mothers with a history of respiratory infections during pregnancy had a reduced respiratory compliance measured with passive mechanics.[18] There are increasing data to support a link between maternal exposure to ambient air pollution and decreased fetal growth.[19] Newer evidence has found that a lack of green vegetation, with a subsequent rise in dry/dead vegetation or bare earth, is associated with lower birth weight.[20]

There is compelling evidence for the role of the maternal microbiome in immune development and risk of infections and allergic disease. Immune ontogeny is heavily developmentally regulated during fetal development and early postnatal life and is influenced by a variety of factors. Among these, the microbiota plays a fundamental role in the induction, education, and function of the host immune system.[21] Immune development is heavily influenced by fetal immune priming from the maternal gut microbiome, which in turn is influenced by maternal environmental exposures.[22] Research on immune programming has examined the impact on allergic and immune outcomes in the offspring. Disturbances of the maternal gut microbiome during pregnancy worsen infectious outcomes in offspring. A key example is that of maternal antibiotic use during pregnancy, especially in mid to late gestation, which is associated with an increased risk of infection-related hospitalization,[23] increased risk of otolaryngologic surgery,[24] and increased risk of childhood asthma.[25] The diversity of maternal microbial exposure is important in studies of the "farm effect." The farm effect refers to a lower risk of asthma and atopy seen for children growing up on farms. This effect is strongest when exposure occurs in early life, including the prenatal period, and the protective mechanism is believed to be through sustained maternal exposure to diverse microbial stimuli.[26,27]

Our increasing knowledge about chemical exposures indicates a role for altered immune function in the child as a result of maternal exposure. Exposure to perfluorobutane sulfonic acid has been associated with an increase in the number of respiratory tract infections by age 5 years, as well as a decrease in IgG concentration.[28] Children exposed to higher levels of dioxins during the prenatal period have a higher odds of wheezing later in life.[29] These chemicals are persistent and widespread in our environments, indicating a high risk of exposure for the population.

Our knowledge about postnatal lung growth is limited by a relative lack of data in healthy children. Most knowledge has come from measurements of lung function, supplemented with limited anatomical data. Few studies have systematically measured lung function during the period of rapid growth during infancy and early childhood. A true assessment of lung growth is virtually impossible in living children, and we need to make do with substitutes, such as measurements of lung function using techniques that only indirectly reflect airway size; lung volumes using plethysmographic or gas dilution techniques; or radiographic assessments of airway dimensions and lung size. However, the limited longitudinal data available show that lung function "tracks" along trajectories, at least as represented by spirometry.[30]

The extent to which genetically determined lung function is altered by postnatal environmental exposures is uncertain. A variety of studies have investigated the impact of postnatal insults on lung growth. Factors shown to be associated with lower lung function include prematurity and low birth weight for gestation,[15] exposure to secondhand smoke,[11] ambient air pollution,[31,32] and indoor air pollution associated with burning biomass fuel.[33] In addition, the evidence for early-life viral respiratory infections in limiting lung growth and/or increasing the risk of asthma is increasing. The argument as to whether early-life infections damage the lungs (viral-induced effect) or whether such infections unmask vulnerable individuals (susceptible host) has not been settled. Severe viral infections requiring hospitalization, especially with adenoviruses and respiratory syncytial virus (RSV), can damage the developing lung and lead to recurrent respiratory problems, including recurrent wheeze in childhood. Longitudinal birth cohort studies have shown that wheezing in early life associated with human rhinovirus infection is both a risk factor for subsequent asthma and associated with lower lung function in childhood.[34] However, none of these studies provides definitive evidence to determine whether these postnatal exposures increase the risk for asthma by limiting lung growth. Again, there is a role for the microbiome in postnatal immunity. At birth, transmission of maternally derived bacteria likely leverages in utero immune programming to accelerate postnatal immune maturation. In turn, postnatal environmental exposures influence the rate and nature of innate immune training.[35] Beneficial exposures, including breastfeeding, result in well-regulated responses that favor health, but adverse exposures result in dysregulated responses that favor inflammation and disease.[36] Last, there is a mechanism of interaction between environmental exposures and infections. Air pollutants induce oxidation at the lung barrier and damage the epithelium—either directly or from oxidative damage—and this damage can increase the risk of respiratory infections.[37]

The oxidative effect of air pollutants and other environmental hazards is emerging as a key mechanism that may be the mediator of several adverse respiratory outcomes. It also may be the missing link to understanding epigenetic risk because impaired function in genes encoding for antioxidant defense has been linked to a higher risk of asthma in children exposed to air pollution.[38] Environmentally persistent free radicals (EPFRs), recently discovered by-products of combustion, are formed from the binding of chemicals and metals to the surface of particulate matter, causing the creation of an oxidant compound that can remain stable in the environment for prolonged periods of time.[39] Preliminary work has found that the presence of high EPFRs can predict wheeze outcomes in children,[40] while animal models have linked EPFRs to airway hyperresponsiveness.[41]

Changes in our environment and climate are increasing the risk from environmental hazards. It is predicted that bushfires will be longer, more frequent, and more severe. Ground-level ozone will increase due to temperature rise, which is predicted to increase

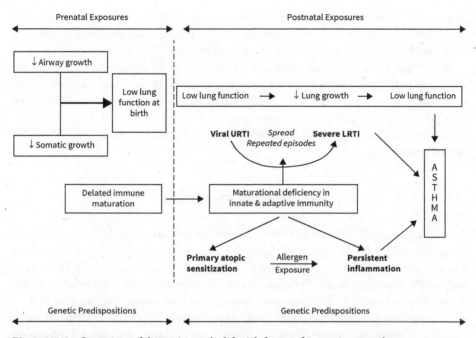

Figure 48.2. Overview of the major early-life risk factors for persistent asthma. Environmental exposures and genetic predispositions contribute to asthma risk via several pathways, including limiting lung growth, delaying immune system maturation, increasing the risk of allergic sensitization, and increasing the number and severity of lower respiratory infections . URTI = upper respiratory tract infection. LRTI = lower respiratory tract infection

the prevalence of asthma.[42] Higher levels of carbon dioxide in the atmosphere can trigger increased growth of allergenic plant species, leading to an increase in aeroallergens.[43] Last, the risk of severe weather events also may play a role in increased risk of epidemic "thunderstorm asthma." Thunderstorm asthma occurs when aeroallergens rupture due to severe meteorological conditions. Children with a history of allergic rhinitis are at risk of experiencing thunderstorm asthma, even in the absence of a history of asthma.[44]

A model demonstrating how a variety of environmental exposures interacting with genetic predispositions may result in persistent asthma is shown in Figure 48.2. This schema provides a series of research areas that will hopefully be pursued in the near future and result in a more complete understanding of how asthma develops.

Clinical Effects

Asthma is a clinical syndrome in which a variety of signs and symptoms may be encountered. There is not simply one type of asthma and no simple diagnostic test that diagnoses all types of asthma (also see next section). Clinical features that are commonly seen in children with asthma include cough and wheeze, most commonly occurring at times of respiratory viral infections; night cough; wheeze and shortness of breath occurring with exercise; and exercise limitation. For the vast majority of children with asthma, these symptoms are intermittent and children generally spend long periods of time without any symptoms. Asthma symptoms may be accompanied by low lung function, increased airway responsiveness to a variety of chemical or physical stimuli, and evidence of atopy or airway inflammation on tests of blood or exhaled air samples. Although the signs and symptoms differ from person to person, the variable nature of the condition is relatively stable and is included in most attempts to define asthma.

In addition to more chronic or persistent, albeit variable symptoms, asthmatics are also prone to acute exacerbations, especially at times of acute viral respiratory infections or increased air pollution. Not all children with asthma have acute exacerbations but an "exacerbation-prone" phenotype is more commonly being recognized. Acute exacerbations of asthma not only impose an economic burden but also are a major risk factor for disease progression.

Diagnosis and Treatment

Asthma, especially in children, is a clinical diagnosis with no single test or group of tests being able to "rule in" or "rule out" all cases of asthma. The diagnosis is made by taking a careful clinical history from the child and the family. Wheezing in early life is quite common, and not all children who wheeze in the first years of life will go on to have asthma. A positive family history of asthma and a history of personal atopy (e.g., eczema or food allergies) increases the likelihood that a child with wheezing will have persistent asthma.

Much of the management of asthma focuses on obtaining and maintaining good asthma control. Asthma control is characterized by two domains: current impairment and future risk. Management of current impairment seeks to reduce the frequency and intensity of symptoms and roll back the functional limits that asthma imposes on the individual child. Control of future risk, by contrast, is focused on managing factors that increase the likelihood of future health care utilization.

Medications to treat asthma generally fall into one of three classes:

- Drugs aimed at acute relief of symptoms, such as short-acting bronchodilators
- Drugs aimed at controlling symptoms, such as long-acting bronchodilators
- Drugs aimed at preventing symptoms from occurring, such as inhaled corticosteroids

While these drugs can be effective in controlling asthma and allowing children with asthma to live normal lives, none has true disease-modifying or "curing" actions. Symptoms generally recur if medication is stopped.

Prevention

The long-term solution to the burden of disease imposed by asthma is to prevent asthma from progressing to the persistent form. Current treatment options, however, provide only symptomatic relief, require continuous use, and do not have any truly disease-modifying effects.

As discussed earlier, the major risk factors for persistent asthma are a family history of asthma and allergies (genetic predisposition), allergic sensitization to aeroallergens in early life, and recurrent severe lower respiratory infections during the first 1–3 years of life. Furthermore, synergistic interactions between allergic sensitization and viral infections substantially magnify the risk. Recognition of these risk factors provides a series of testable hypotheses for primary prevention of asthma by preventing either allergic sensitization or severe lower respiratory infections in high-risk children. Allergen avoidance is not the correct strategy to prevent allergic sensitization; in fact, data from a number of studies in various parts of the world have demonstrated that low-dose allergen exposure in early life increases the risk of allergic sensitization, whereas high-dose exposure is more likely to induce immunological tolerance.

Prevention of severe lower respiratory infections would likely reduce the risk of developing persistent asthma. However, there are no current vaccines against the common respiratory viruses, especially human rhinovirus, which is associated with the majority of severe lower respiratory infections in many studies. An alternative approach has been used in Europe, where immuno-stimulatory products derived from bacteria are used to reduce wheezing in children with recurrent respiratory infections.[45] These products appear to be safe and are inexpensive. They have never been used, however, in a study designed to prevent the development of asthma. The maternal diet offers an opportunity to improve the maternal gut microbiome and potentially the offspring's immune function and health.[46] The gut microbiota is a key component of human homeostasis and helps maintain the intestinal barrier to prevent pathogenic bacteria and products crossing into systemic circulation.[47] Microbe-derived metabolites, such as bile acid derivatives, short-chain fatty acids, amino acid derivatives, and lipopolysaccharides, are key signaling molecules that link gut–host responses.[22] Bacterial diversity in the gut influences the range of metabolites, including short-chain fatty acids, produced. Maternal intake of certain fibers and carbohydrates, collectively known as *microbiota-accessible carbohydrates*, are metabolized to short chain fatty acids, which influence immune function and microbiome health.

Environmental control may not be able to prevent asthma, but there are some well-established asthma risk factors that are amenable to environmental control. The most important of these is tobacco smoke. Maternal smoking during pregnancy and the exposure of pregnant women, infants, and young children to secondhand smoke are completely avoidable, and eliminating such exposures reduces the risk of developing asthma. Similarly, avoiding use of or exposure to household chemicals, personal care aerosol products, and volatile organic compounds during pregnancy is likely to reduce the risk to the developing fetus. Renovating the room the new baby is to occupy appears to be a common practice but is one that may result in exposure of mother, fetus, and subsequently the infant to air pollutants if appropriate care is not taken. Exposure to bushfire smoke can be reduced through simple interventions such as staying indoors, closing windows and reducing drafts, and wearing an N95/P2 mask if needing to leave the house. Staying indoors during and immediately prior to storms can reduce the risk of thunderstorm asthma.

Environmental design may play a role in ameliorating harmful environmental exposures. There is evidence that children living in homes exposed to high volumes of traffic are less likely to be diagnosed with asthma if they live in an area with more than 40% green space.[48] However, the type of green space is important, with grasslands and areas with uncontrolled weed populations likely to increase the number of aeroallergens and therefore allergic sensitization.[49] Trees are more likely to trap pollutants and are associated with lower asthma hospitalizations than are smaller shrubs found in most private gardens.[50]

Environmental determinants of asthma and allergic disease are complex, multifactorial, and historically have been poorly understood. Current research is beginning to probe at the potential causal mechanisms, thus opening up new avenues for prevention.

References

1. Guerra S, Martinez FD. Asthma genetics: from linear to multifactorial approaches. *Annu Rev Med*. 2008;59:327–41. doi:10.1146/annurev.med.59.060406.213232.

2. London SJ, Romieu I. Gene by environment interaction in asthma. *Annu Rev Public Health* 2009;30:55–80. doi:10.1146/annurev.publhealth.031308.100151.

3. Martinez FD. The connection between early life wheezing and subsequent asthma: The viral march. *Allergol Immunopathol (Madr)*. 2009;37(5):249–51. doi:10.1016/j.aller.2009.06.008.

4. Sly PD, Boner AL, Björksten B, et al. Early identification of atopy in the prediction of persistent asthma in children. *Lancet*. 2008;372(9643):1100–6. doi:10.1016/s0140-6736(08)61451-8.

5. Global Asthma Network. *The Global Asthma Report 2018*. Auckland, New Zealand: Global Asthma Network; 2018.

6. Sly PD, Flack F. Susceptibility of children to environmental pollutants. *Ann N Y Acad Sci.* 2008;1140:163–83. doi:10.1196/annals.1454.017.

7. Bové H, Bongaerts E, Slenders E, et al. Ambient black carbon particles reach the fetal side of human placenta. *Nature Commun*. 2019;10(1):3866. doi:10.1038/s41467-019-11654-3.

8. Liu NM, Miyashita L, Maher BA, et al. Evidence for the presence of air pollution nanoparticles in placental tissue cells. *Sci Total Environ*. 2021;751:142235. doi:https://doi.org/10.1016/j.scitotenv.2020.142235.

9. Robinson L, Miller R. The impact of bisphenol A and phthalates on allergy, asthma, and immune function: a review of latest findings. *Curr Environ Health Rep*. 2015;2(4):379–87. doi:10.1007/s40572-015-0066-8.

10. Robertson CF. Long-term outcome of childhood asthma. *Med J Aust*. 2002;177(S6):S42–4. doi:10.5694/j.1326-5377.2002.tb04813.x.

11. Le Souëf PN. Pediatric origins of adult lung diseases. 4. Tobacco related lung diseases begin in childhood. *Thorax*. 2000;55(12):1063–7. doi:10.1136/thorax.55.12.1063.

12. Latzin P, Röösli M, Huss A, Kuehni CE, Frey U. Air pollution during pregnancy and lung function in newborns: a birth cohort study. *Eur Respir J*. 2009;33(3):594–603. doi:10.1183/09031936.00084008.

13. Henderson J, Sherriff A, Farrow A, Ayres JG. Household chemicals, persistent wheezing and lung function: effect modification by atopy? *Eur Respir J*. 2008;31(3):547–54. doi:10.1183/09031936.00086807.

14. Lazic T, Sow FB, Van Geelen A, Meyerholz DK, Gallup JM, Ackermann MR. Exposure to ethanol during the last trimester of pregnancy alters the maturation and immunity of the fetal lung. *Alcohol*. 2011;45(7):673–80. doi:10.1016/j.alcohol.2010.11.001.

15. Hoo AF, Stocks J, Lum S, et al. Development of lung function in early life: influence of birth weight in infants of nonsmokers. *Am J Respir Crit Care Med*. 2004;170(5):527–33. doi:10.1164/rccm.200311-1552OC.

16. Tennant PW, Gibson GJ, Pearce MS. Lifecourse predictors of adult respiratory function: results from the Newcastle Thousand Families Study. *Thorax*. 2008;63(9):823–30. doi:10.1136/thx.2008.096388.

17. Delemarre-van de Waal HA. Environmental factors influencing growth and pubertal development. *Environ Health Perspect*. 1993;101(Suppl 2):39–44. doi:10.1289/ehp.93101s239.

18. Van Putte-Katier N, Uiterwaal CS, De Jong BM, Kimpen JL, Verheij TJ, Van Der Ent CK. The influence of maternal respiratory infections during pregnancy on infant lung function. *Pediatr Pulmonol*. 2007;42(10):945–51. doi:10.1002/ppul.20688.

19. Hansen CA, Barnett AG, Pritchard G. The effect of ambient air pollution during early pregnancy on fetal ultrasonic measurements during mid-pregnancy. *Environ Health Perspect*. 2008;116(3):362–9. doi:10.1289/ehp.10720.

20. Vilcins D, Scarth P, Sly PD, Jagals P, Knibbs LD, Baker P. The association of fractional cover, foliage projective cover and biodiversity with birthweight. *Sci Total Environ*. 2021;763:143051. doi:https://doi.org/10.1016/j.scitotenv.2020.143051.

21. Belkaid Y, Harrison OJ. Homeostatic Immunity and the Microbiota. *Immunity.* 2017;46(4):562–76. doi:https://doi.org/10.1016/j.immuni.2017.04.008.

22. Feng J, Cavallero S, Hsiai T, Li R. Impact of air pollution on intestinal redox lipidome and microbiome. *Free Radical Biol Med.* 2020;151:99–110. doi:https://doi.org/10.1016/j.freera dbiomed.2019.12.044.

23. Miller JE, Wu C, Pedersen LH, de Klerk N, Olsen J, Burgner DP. Maternal antibiotic exposure during pregnancy and hospitalization with infection in offspring: a population-based cohort study. *Int J Epidemiol.* 2018;47(2):561–71. doi:10.1093/ije/dyx272.

24. Lovern C, Todd I, Haberg S, Magnus M, Burgner D, Miller J. Association of prenatal antibiotics and mode of delivery with otolaryngology surgery in offspring: a national data linkage study. *J Pediatr Infect Dis.* 2022;41(5):368–74.

25. Uldbjerg CS, Miller JE, Burgner D, Pedersen LH, Bech BH. Antibiotic exposure during pregnancy and childhood asthma: a national birth cohort study investigating timing of exposure and mode of delivery. *Arch Dis Child.* 2021;106(9):888–94. doi:10.1136/archdischild-2020-319659.

26. Holt PG, Strickland DH, Custovic A. Targeting maternal immune function during pregnancy for asthma prevention in offspring: harnessing the "farm effect"? *J Allergy Clin Immunol.* 2020;146(2):270–2. doi:10.1016/j.jaci.2020.04.008.

27. Wlasiuk G, Vercelli D. The farm effect, or: when, what and how a farming environment protects from asthma and allergic disease. *Curr Opin Allergy Clin Immunol.* 2012;12(5):461–6. https://journals.lww.com/co-allergy/Fulltext/2012/10000/The_farm_effect,_or__when,_what_and_how_a_farming.5.aspx.

28. Huang H, Yu K, Zeng X, et al. Association between prenatal exposure to perfluoroalkyl substances and respiratory tract infections in preschool children. *Environ Res.* 2020;191:110156. doi:https://doi.org/10.1016/j.envres.2020.110156.

29. Miyashita C, Bamai YA, Araki A, et al. Prenatal exposure to dioxin-like compounds is associated with decreased cord blood IgE and increased risk of wheezing in children aged up to 7 years: the Hokkaido study. *Sci Total Environ.* 2018;610–11:191–9. doi:https://doi.org/10.1016/j.scitotenv.2017.07.248.

30. Hopper JL, Hibbert ME, Macaskill GT, Phelan PD, Landau LI. Longitudinal analysis of lung function growth in healthy children and adolescents. *J Appl Physiol (1985).* 1991;70(2):770–7. doi:10.1152/jappl.1991.70.2.770.

31. Calderón-Garcidueñas L, Mora-Tiscareño A, Fordham LA, et al. Lung radiology and pulmonary function of children chronically exposed to air pollution. *Environ Health Perspect.* 2006;114(9):1432–7. doi:10.1289/ehp.8377.

32. Wichmann FA, Müller A, Busi LE, et al. Increased asthma and respiratory symptoms in children exposed to petrochemical pollution. *J Allergy Clin Immunol.* 2009;123(3):632–8. doi:10.1016/j.jaci.2008.09.052.

33. Woodcock A, Lowe LA, Murray CS, et al. Early life environmental control: effect on symptoms, sensitization, and lung function at age 3 years. *Am J Respir Crit Care Med.* 2004;170(4):433–9. doi:10.1164/rccm.200401-083OC.

34. Guilbert TW, Singh AM, Danov Z, et al. Decreased lung function after preschool wheezing rhinovirus illnesses in children at risk to develop asthma. *J Allergy Clin Immunol.* 2011;128(3):532–8.e10. doi:10.1016/j.jaci.2011.06.037.

35. Troy N, Strickland D, Serralha M, et al. Protection against severe infant lower respiratory infections by immune training: mechanism-of-action studies. *J Allergy Clin Immunol.* 2022;150(1):93–103.

36. Netea MG, Domínguez-Andrés J, Barreiro LB, et al. Defining trained immunity and its role in health and disease. *Nature Rev Immunol.* 2020;20(6):375–88. doi:10.1038/s41577-020-0285-6.

37. Woodby B, Arnold MM, Valacchi G. SARS-CoV-2 infection, COVID-19 pathogenesis, and exposure to air pollution: what is the connection? *Ann N Y Acad Sci.* 2021;1486(1):15–38. doi:10.1111/nyas.14512.

38. MacIntyre EA, Brauer M, Melén E, et al. GSTP1 and TNF Gene variants and associations between air pollution and incident childhood asthma: the traffic, asthma and genetics (TAG) study. *Environ Health Perspect.* 2014;122(4):418–24. doi:10.1289/ehp.1307459.

39. Saravia J, Lee GI, Lomnicki S, Dellinger B, Cormier SA. Particulate matter containing environmentally persistent free radicals and adverse infant respiratory health effects: a review. *J Biochem Mol Toxicol.* 2013;27(1):56–68. doi:https://doi.org/10.1002/jbt.21465.

40. Sly PD, Cormier SA, Lomnicki S, Harding JN, Grimwood K. Environmentally persistent free radicals: linking air pollution and poor respiratory health? *Am J Respir Crit Care Med.* 2019;200(8):1062–3. doi:10.1164/rccm.201903-0675LE.

41. Balakrishna S, Saravia J, Thevenot P, et al. Environmentally persistent free radicals induce airway hyperresponsiveness in neonatal rat lungs. *Particle Fibre Toxicol.* 2011;8(1):11. doi:10.1186/1743-8977-8-11.

42. Sheffield PE, Knowlton K, Carr JL, Kinney PL. Modeling of regional climate change effects on ground-level ozone and childhood asthma. *Am J Prev Med.* 2011;41(3):251–A3. doi:10.1016/j.amepre.2011.04.017.

43. D'Amato G, Chong-Neto HJ, Monge Ortega OP, et al. The effects of climate change on respiratory allergy and asthma induced by pollen and mold allergens. *Allergy.* 2020;75(9):2219–28. doi:10.1111/all.14476.

44. Xu Y-Y, Xue T, Li H-R, Guan K. Retrospective analysis of epidemic thunderstorm asthma in children in Yulin, northwest China. *Pediatr Res.* 2021;89(4):958–61. doi:10.1038/s41390-020-0980-9.

45. Del-Rio-Navarro BE, Espinosa Rosales F, Flenady V, Sienra-Monge JJ. Immunostimulants for preventing respiratory tract infection in children. *Cochrane Database Syst Rev.* 2006(4):Cd004974. doi:10.1002/14651858.CD004974.pub2.

46. Gao Y, Nanan R, Macia L, et al. The maternal gut microbiome during pregnancy and offspring allergy and asthma. *J Allergy Clin Immunol.* 2021;148(3):669–78. doi:https://doi.org/10.1016/j.jaci.2021.07.011.

47. Nicholson JK, Holmes E, Kinross J, et al. Host-gut microbiota metabolic interactions. *Science.* 2012;336(6086):1262–7. doi:10.1126/science.1223813.

48. Feng X, Astell-Burt T. Is neighborhood green space protective against associations between child asthma, neighborhood traffic volume and perceived lack of area safety? multilevel analysis of 4447 Australian children. *Int J Environ Res Public Health.* 2017;14(5):543. (https://www.mdpi.com/1660-4601/14/5/543).

49. Aerts R, Dujardin S, Nemery B, et al. Residential green space and medication sales for childhood asthma: a longitudinal ecological study in Belgium. *Environ Res.* 2020;189:109914. doi:https://doi.org/10.1016/j.envres.2020.109914.

50. Alcock I, White M, Cherrie M, et al. Land cover and air pollution are associated with asthma hospitalisations: a cross-sectional study. *Environ Int.* 2017;109:29–41. doi:https://doi.org/10.1016/j.envint.2017.08.009.

49

Environmental Chemicals and Neurodevelopmental Disorders in Children

David C. Bellinger

Neurodevelopmental disorders (NDDs) are among the most common chronic health conditions among children around the world today. Based on the 2015–2018 National Health Interview Survey, NDDs affect at least one of every six 3- to 17-year-old children in the United States (17.8%),[1] and rates are similar in other industrially developed countries. The most common of these disorders are attention-deficit/hyperactivity disorder (ADHD) (9.6%), learning disabilities (7.7%), and autism spectrum disorders (ASD) (2.5%). Other NDDs of lower prevalence included sensory (visual, auditory) deficits, intellectual disability, seizures, speech disorders, and cerebral palsy. The treatment of NDDs is difficult, and the disabilities they cause can be permanent. They are extremely costly to families and society.

Some NDDs are reportedly increasing in prevalence. Between the 2009–2011 and 2015–2017 cycles of the National Health Interview Survey, significant increases were found for certain disorders, such as ADHD, ASD, and intellectual disability.[2] The increase in ASD prevalence was particularly striking. Early epidemiological studies estimated the prevalence to be approximately 5/10,000 children.[3] Using data collected at the Autism and Developmental Disabilities Monitoring Network sites, which has employed consistent and rigorous diagnostic procedures based on record review or individual examination rather than on caretaker report, the prevalence of ASD in 8-year-old children increased from 6.7/1,000 (1 in 150) in the period of 2000–2002, to 18.5/1,000 (1 in 54) in 2016, to 23/1,000 (1 in 44) in 2018.[4] In interpreting any trends in prevalence, however, the potential influences of increased awareness, diagnostic substitution, changes in diagnostic in criteria, and better access to health care must be considered.

Evidence has been accumulating over several decades that toxic chemicals in the environment can cause NDDs and that subclinical stages of these conditions may be common. The suspicion of a link between chemicals and widespread neurobehavioral damage was first raised by research demonstrating that lead is toxic to the developing brain across a wide range of exposures. The number of chemicals purportedly linked to the risk of an NDD has increased dramatically in recent decades. An early report from the US National Research Council concluded that 3% of developmental disabilities are the direct consequence of neurotoxic environmental exposures and that another 25% arise through interplay between environmental factors (chemical and other) and individual genetic susceptibility.[5] These estimates were based on the limited neurotoxicity information available at the time and are therefore likely to underestimate the true prevalence of chemically induced abnormalities.

David C. Bellinger, *Environmental Chemicals and Neurodevelopmental Disorders in Children* In: *Textbook of Children's Environmental Health*, Second Edition. Edited by: Ruth A. Etzel and Philip J. Landrigan, Oxford University Press.
© Oxford University Press 2024. DOI: 10.1093/oso/9780197662526.003.0049

Neurobehavioral damage caused by exposure to toxic chemicals is, in principle, preventable. An essential prerequisite to prevention is recognition of a chemical's ability to harm the developing brain. Knowing that a chemical is neurotoxic can trigger efforts to restrict its use and control exposures. Previous evidence-based programs of exposure prevention, such as those directed against children's exposure to lead, have ultimately been highly successful, although they were initiated only after substantial delay.

Most research to discover the environmental causes of neurodevelopmental disabilities in children has focused on continuously distributed indicators of "subclinical" dysfunction, such as IQ loss or impaired social skills, rather than on clinical diagnoses. Characterization of exposures to chemicals has posed substantial challenges for several reasons. Critical windows of vulnerability for chemical exposures that might be of etiological significance with respect to NDDs are not known. Ideally, children should be followed prospectively and biomarkers of exposures of interest collected over the entire interval from conception to the time of diagnosis. A mismatch between the time interval captured by an exposure biomarker and the interval in which the developing brain is most vulnerable to the chemical will tend to bias estimates of the association toward the null, increasing the likelihood of false-negative inferences. As an alternative to measuring exposure biomarkers at the level of the individual child, some studies of air pollution, for example, have relied on proxy indices of exposure, such as distance of residence from roadways or land use regression models, or on administrative databases such as the US Environmental Protection Agency (EPA)'s Toxics Release Inventory (TRI), which identifies the amounts of chemicals released by facilities located in a defined area. The obvious disadvantage of relying on such data is that they pertain solely to external rather than internal exposure and often do not take into account time-varying activity patterns that might influence exposure.

In this chapter, the evidence regarding the potential contribution of chemical exposures to clinical disorders such as ASD, ADHD, learning disabilities, and other disorders is reviewed. The methodological challenges that attend the conduct of research that focuses on such endpoints are also identified.

Autism Spectrum Disorders

The diagnosis of ASDs currently includes disorders that, until the publication in 2013 of the 5th edition of the *Diagnostic and Statistical Manual of Mental Disorders* (DSM), were considered to be separate diagnoses: autistic disorder, Asperger syndrome, and pervasive developmental disorder not otherwise specified. An explanation for the apparent increase in recent decades in the prevalence of ASDs has proved elusive. As suggested above, at least some of it can likely be attributed to artifacts such as diagnostic substitution and accretion, changes in diagnostic practices (e.g., lower age at diagnosis, broadening of criteria), and increased awareness and surveillance. The proportion of individuals with an ASD who also have an intellectual disability (i.e., IQ ≤70) has declined from approximately one-half in 2000 to approximately one-third. In addition, racial and ethnic disparities in ASD prevalence have also declined in recent years.[6] Both of these findings suggest that changes in diagnostic practices might have played a role in the apparent increase in prevalence. However, the possibility of a true increase due to other factors cannot be ruled out.

Although genetics clearly contribute to a child's risk of an ASD,[7] an increase over a period as brief as one or two decades is unlikely to reflect a classical genetic effect. Indeed, to date, efforts to identify the genetics of ASDs have generally been rather disappointing, with candidate genes and abnormalities (duplications, deletions, and copy number variations) usually accounting for only a small percentage of cases. Moreover, attempts at replication have often been disappointing. The fact that exposure to thalidomide, valproic acid, or the rubella virus in the first trimester is associated with an increased risk of an ASD or autism in animal models supports the concept that exposure to a chemical at a critical time in development can result in an ASD. If exposure to environmental chemicals has played a role in the recent increase in ASDs, the mechanism might be a direct neurotoxic process involving an exposure that has increased dramatically over this period, an epigenetic process involving chemical-induced alterations in gene expression, or gene–environmental interactions that have become more important due to changes in exposure patterns. The high heritability of ASDs among twins[8] is sometimes cited to bolster the argument for the primacy of genetics in the etiology, but twins share the same prenatal environment. Some portion of the concordance could reflect shared prenatal chemical exposures.

Efforts to evaluate the associations between environmental chemicals and ASDs must address challenges in both defining the outcome and characterizing exposures, as described in the following sections.

Diagnosis

The diagnoses of conditions included among the ASDs are based solely on the behavioral phenotype. To date, no laboratory findings can confirm the diagnoses, and there are no "bright lines" separating typically developing children from children who display the three critical features of autism: social impairment, language impairment, and repetitive stereotyped behaviors. Children assigned an ASD diagnosis simply display more (or less) of a behavior that is usually present to some degree in children. So-called autistic traits show a smooth rather than bimodal distribution, providing little evidence for the existence of distinct "clinical" and "nonclinical" groups.[9] For disorders such as ASDs, therefore, the key question is not "does one have it" but, "how much of it does one have?" Indeed, parents in families that include two or more children with autism show more features of what is termed the "broader autism phenotype" than do parents in families with no or one child with autism.[10] Studies that rely on rating scales as the basis for outcome classification thus assess only the hypothesis that a chemical exposure is associated with the underlying traits that form the broader autism phenotype.

Identifying children who meet diagnostic criteria for an ASD is highly resource-intensive. To minimize outcome misclassification, all children in a study sample should be evaluated using gold-standard, but lengthy, standardized instruments (e.g., the Autism Diagnostic Observation Scale, the Autism Diagnostic Interview-Revised). It would be inefficient to apply such measures in a cohort study involving several hundred children recruited from the general population due to the low yield of cases that would be expected. Despite the apparent recent increase in prevalence, ASDs are still relatively rare so that the number of cases identified would be small unless the sample were enriched with children at increased risk (e.g., siblings of cases). As a result, many studies of chemical exposures have relied on administrative databases, such as state or regional listings of children with an ASD diagnosis made for purposes of surveillance or service allocation rather than research. A disadvantage of this strategy is a lack of certainty that diagnostic criteria have been applied in a standard manner to all children.

As an alternative to using a formal diagnosis as the outcome, some studies have relied on parent reports of behaviors that are relevant to ASDs, such as the Social Responsiveness Scale and the Child Behavior Checklist (CBCL). For example, the CBCL for children 1.5–5 years of age includes a DSM-oriented Pervasive Developmental Problems scale. An elevated score on this scale is not equivalent to a diagnosis, however. There is considerable overlap in the behaviors associated with different disorders, and children who are developmentally delayed, but who do not have an ASD, also tend to have elevated scores on several CBCL/1.5–5 scales and on the Social Responsiveness Scale.[11,12] Diagnostic misclassification as a result of using a screening instrument with limited positive predictive value or sensitivity would most likely bias toward the null an estimate of the risk associated with a chemical exposure.

In the following sections, selected recent evidence suggesting links between different chemical exposures and ASDs is summarized.

Air Pollution

Numerous recent studies have investigated prenatal and early life exposure to air pollution, most notably particulate matter, as a risk factor for ASDs. Several potential mechanisms have been suggested for this association, including altered amyloid precursor protein processing, oxidative stress leading to neuroinflammation, impairments in mitochondrial respiration, and changes in maternal immune function.[13–16]

Multiple meta-analyses have yielded largely consistent results of modest associations for at least some components of this complex exposure. Lam et al.[17] found an odds ratio (OR) of 2.3 (95% confidence interval [CI]: 2.15, 2.51) for each 10 μg/m^3 increase in prenatal exposure to $PM_{2.5}$ and an OR of 1.07 (1.06, 1.08) for a similar increase in prenatal PM_{10} exposure. Overall, they characterized this as "limited evidence of toxicity" because possible roles for chance, bias, and confounding could not be ruled out due to limitations in the data. They also noted heterogeneity in study findings that could not be explained by differences in study design. In a more recent meta-analysis, which included studies published prior to 2019, Chun et al.[18] concluded that there was "some" evidence for an association between ASD and prenatal $PM_{2.5}$ exposure and "little" evidence for PM_{10} exposure. The OR that Chun et al. estimated for $PM_{2.5}$, 1.06 (1.01, 1.11) was considerably smaller than the OR estimated by Lam et al., however, which might be due to the meta-analytic methods used or to differences in the specific studies included in the meta-analysis. Trimester-specific analyses suggested stronger associations with third-trimester exposures, but studies were inconsistent on this.

Since these meta-analyses were conducted, additional studies have been published, with some providing additional evidence supporting prenatal $PM_{2.5}$ as an ASD risk factor,[19,20] and others not.[21] It has become apparent that some of the heterogeneity in study findings might be due to effect modification. Jo et al.[22] found stronger associations between prenatal air pollution exposure and ASD in boys than in girls, while McGuinn et al.[23] reported evidence suggesting that the association might be stronger for children who experience greater neighborhood deprivation during the first year of life in association with ASD. Volk et al.[24] found that the association between $PM_{2.5}$ was greater among children with a mutation in the MET receptor tyrosine kinase gene. Nutritional factors might also be important, as greater periconceptional folic acid intake may reduce ASD risk in children with higher prenatal exposure to air pollution.[25] Recent studies suggest that exposure to ultrafine particles (<100 nanometers) should also receive attention as a potential risk factor for ASD.[26]

Pesticides

The evidence regarding pesticide exposure as a risk factor for ASD is mixed. Studies have varied in terms of the source of the study cohort (e.g., population-based or high-risk of an ASD) and in terms of how pesticide exposure was measured (e.g., residential proximity to agricultural applications, urinary biomarkers of pesticide metabolites, reported use of household pesticides). Von Ehrenstein et al.[27] estimated prenatal and infant exposure to numerous pesticides using a statewide reporting system used in California. Using a case-control design, they found significantly elevated ORs (between 1.1 and 1.6) for prenatal exposure to glyphosate, malathion, avermectin, and permethrin. ORs were increased by approximately 30% with regard to children with ASD who also had intellectual disability. A Finnish national birth cohort study[28] found that significantly increased odds of ASD associated with prenatal exposure to the DDT metabolite p,p'-DDE, but, as in the von Ehrenstein et al. study, the odds were approximately two-fold greater for children with ASD and comorbid intellectual disability.

While some studies have reported significant associations between prenatal organophosphate (OP) metabolites (dialyl phosphates) and autistic traits,[29-31] others have not, including the large population-based Generation R study.[32] In the high-risk MARBLES cohort, Philippat et al.[33] did not find significant associations between seven OP metabolites during pregnancy and ASD risk. A recent meta-analysis concluded that the odds of ASD reported in studies that estimated maternal exposure by residential proximity were greater than those reported in studies that relied on measurement of exposure biomarkers, perhaps reflecting the challenge of characterizing exposure to chemicals with short biological half-lives.

Just as higher periconceptional intake of folic acid was reported to reduce the risk of ASD associated with air pollution, it appears to reduce the risk associated with pesticide exposure as well.[34]

Phthalates

The evidence regarding a potential role of phthalate exposure in ASD is also mixed, with some studies reporting increased risk for autism trait behaviors[35,36] but not for a formal diagnosis of ASD.[37] Larson et al.[38] found that parents of children in houses with PVC flooring, which is one source of phthalate exposure, were more likely than parents of children in homes with other types of floorings to report that the child had "autism, Asperger or Tourette's syndrome," although the diagnosis was not confirmed clinically. A study in which phthalate concentrations were measured in house dust and in which a child's ASD diagnosis was confirmed clinically did not find significant associations.[39]

Per- and Polyfluoroalkyl Substances

Whether increased exposure to per- and polyfluoroalkyl substances (PFAS) increases the risk of an ASD is uncertain, with some studies reporting positive results and others negative results.[40-43] Studies tend to differ, however, in which of the thousands of chemicals classified as PFAS were measured and at which developmental time points, so additional research is needed in order to draw any conclusions.

Other Chemicals

Although the hypothesis that use of ethyl mercury (thimerosal) as a vaccine preservative increases a child's risk of an ASD has been hotly debated, the evidence does not support it,[44] nor do studies implicate mercury exposure from consumption of contaminated seafood as a risk factor.[45] Neither prenatal[46] nor postnatal exposure[47] to polybrominated diphenyl ethers increases risk of an ASD. Risk has also not been found to be increased among children exposed prenatally to alcohol[48] or maternal smoking during pregnancy.[49]

Attention-Deficit/Hyperactivity Disorder

Diagnosis

Like ASDs, ADHD is a symptom complex, the diagnosis of which is based solely on a behavioral phenotype that is heterogeneous in its presentation and represents the extreme of a continuously distributed trait.[50] Three subtypes are recognized in the American Psychiatric Association's DSM-5: predominantly inattentive, predominantly hyperactive/impulsive, and combined. Many parent- and teacher-completed questionnaires and rating scales (e.g., CBCL, Connors' Rating Scales, Behavior Assessment System for Children-2) include scales that assess behaviors relevant to ADHD, and all have been used in studies of chemical exposures. As with ASDs, a high score on the scales of these questionnaires that pertain to attention and activity is not sufficient to establish a diagnosis of ADHD. Additional evaluation is required because the DSM-5 criteria include several features that are not assessed by the questionnaires. Some signs must have been present for at least 6 months and were present prior to age 12, some impairment must be present in at least two settings (e.g., home, school), the impairments must not be the result of a comorbid disorder, and there must be clear evidence that the impairments interfere with developmentally appropriate social or academic functioning.

In the following sections, selected evidence linking different chemical exposures to ADHD or its associated behaviors is summarized.

Metals

Lead is the chemical for which the evidence of a link to ADHD is strongest. Since the 1970s, many studies have reported that higher childhood lead exposure is associated with behaviors associated with ADHD, such as distractibility and impulsivity. In its most recent (2013) Integrated Science Assessment for Lead, the EPA[51] concluded that the relationship between elevated exposure to lead and reduced attention, impulsivity, and hyperactivity can be considered causal. Several studies have also suggested that greater childhood exposure is a risk factor for the clinical diagnosis of ADHD. Braun et al.[52] reported, in analyses of 6- to 15-year-olds in the National Health and Nutrition Examination Survey (NHANES) 1999–2002, an adjusted odds ratio of 4 for parent-reported ADHD among children with a blood lead level greater than 2 µg/dL. Cases consisted of children who were reported to be taking stimulant medication and whose parents had been told by a medical professional that they met diagnostic criteria for ADHD. In a case-control study that included children in whom a diagnosis of ADHD was clinically confirmed,[53] higher concurrent blood lead level was significantly associated with the risk of ADHD combined subtype and, to lesser extent, the hyperactive-impulsive subtype. A recent systematic review and meta-analysis[54] found that in 16 of 18 studies, a significant association was found between blood lead concentration (10 µg/dL) and at least one subtype of ADHD (combined, inattentive, hyperactive-impulsive).

Per- and Polyfluoroalkyl Compounds

Early studies tended to report that early-life exposure to PFAS was associated with risk of an ADHD diagnosis or increases in ADHD-related behaviors. Using NHANES 1999-2000 and 2003-2004 data, Hoffman et al.[55] found that 12- to 15-year-old children with higher serum levels of three polyfluoroalkyl compounds—perfluorooctane sulfonic acid (PFOS), perfluorooctanoic acid (PFOA), and perfluorohexane sulfonic acid (PFHxS)—were at increased risk of having been diagnosed with ADHD. In a cross-sectional study, Stein and Savitz[56] reported a higher rate of parent-reported ADHD in children with greater PFHxS, but not PFOA, concentrations in serum. However, more recent studies have generally been negative. A meta-analysis of nine European population-based studies published in 2020[57] found no associations between PFOS or PFOA concentrations in maternal serum during pregnancy or in breast milk and ADHD in children. In the Odense Child Cohort, five PFAS compounds measured during pregnancy or in childhood were not associated with ADHD symptoms at 2 ½ or 5 years of age.[58] In a Japanese cohort (the Hokkaido Study), Itoh et al.[59] measured 11 PFAS compounds in maternal serum and found only inverse associations with ADHD symptoms in children at age 8 years (i.e., higher PFAS, lower risk). One study[60] did report consistent associations between some PFAS measured during in pregnancy or at delivery (PFOS, PFNA, and PFHxS, but not PFOA) and ADHD-related symptoms, as measured by the Behavioral Assessment System for Children-2 at 5 and 8 years or the Diagnostic Interview Schedule for Children-Young Child at age 5 years. The available studies differ in many ways, including which PFAS compounds are measured, at what developmental stages, the assessment tool used to classify children's outcomes, and, for symptom checklists, the cutoffs used.

Fluoride

Fluoride has received increasing attention as a potential neurotoxicant. A cross-sectional study of children of 6–17 years in Canada[61] reported that a 1 mg/L increase in tap water fluoride concentration led to a 6.1-fold increase in the odds of a child's ADHD diagnosis. However, the confidence interval was wide (1.60–22.8), and no association was found between a biomarker of exposure, urinary fluoride concentration, and risk of ADHD. In a Mexican study, higher maternal urinary fluoride concentrations during pregnancy were associated with significantly more ADHD-related behaviors in children at 6–12 years of age, although analyses of ADHD diagnosis was not reported.[62]

Chlorinated Compounds

A pooled analysis of seven European birth cohort studies[63] evaluated the associations between three persistent organic pollutants, PCB-153, p-p'-DDE, and hexachlorobenzene, and ADHD in children. Exposures between pregnancy and 24 months of age were considered. The criteria for ADHD differed among studies (e.g., medical diagnosis vs. parent or teacher ratings). No associations were found between any of the exposures, whether measured prenatally or postnatally, and ADHD-related outcomes up to 10 years of age.

Pesticides

Among pesticide classes, the strongest evidence of a link with ADHD outcomes in children is for exposure to pyrethroids. In the Odense Child Cohort, Dalsager et al.[64] reported modest associations between the concentration of the metabolite 3-phenoxybenzoic acid (3-PBA) in maternal urine and ADHD traits at 2–4 years of age. Each doubling of maternal 3-PBA was associated with a 3% increase in ADHD score and a 13% higher risk of having an ADHD diagnosis. Wagner-Schuman et al.[65] found that among 8-to 15-year-olds in the NHANES 2001–2002 survey, having a concurrent 3-PBA concentration above the limit of detection was associated with an OR of 2.4 (95% CI: 1.1, 5.6) for an ADHD diagnosis. In another cross-sectional study, Lee et al.[66] similarly found that greater urinary 3-PBA concentration was associated with greater ADHD symptoms on a rating scale in preschool children. Some studies have reported associations between prenatal exposures to organophosphates (or metabolites) and ADHD or its associated behaviors.[30,67,68] In the large Dutch Generation R study ($N = 781$), however, maternal concentrations of 6 dialkyl phosphates (OP metabolites) were not associated with ADHD-traits, as measured using the CBCL at ages 3, 6, and 10 years.[32]

Flame Retardants

In a systematic review, Lam et al.[69] noted that studies tend to report a positive association between biomarkers of exposure to polybrominated diphenyl ethers and ADHD or attention-related behaviors, but that the number of studies was insufficient to support a meta-analysis and that the confidence intervals of the estimates tended to include the null, thus precluding any definitive conclusion. In a case-control study within a Norwegian birth cohort, Choi et al.[70] found that odds of ADHD were increased among children with greater prenatal exposure to organophosphate esters, another class of flame retardant compounds.

Alcohol

The risk of ADHD among children with fetal alcohol spectrum disorders has been found in several studies[71] although some uncertainty exists about the possibility of residual confounding by prenatal smoking and social disadvantage.[72]

Secondhand Tobacco Smoke

Secondhand tobacco smoke (SHS) has long been regarded as a risk factor for ADHD[73] but recent studies using a variety of designs (cohort, case-control, related and unrelated mother–offspring pairs) have cast doubt on the validity of this association.[74–78] These studies suggest that adjustments for environmental or inherited factors, such as parental psychopathological vulnerability, have not been adequate.

Phthalates

Higher concurrent urinary phthalate metabolite concentrations[79] or higher maternal urinary metabolite concentrations prenatally[80] have been linked to more parent- or teacher-reported ADHD-related behaviors. More recent studies have reported increased risk of clinically defined ADHD in children whose mothers had greater urinary phthalate metabolite concentrations.[81,82]

Air Pollutants

Greater exposures to polycyclic aromatic hydrocarbons (PAHs),[83] PM_{10},[84] and nitrogen dioxide (from household gas appliances)[85] have been linked either to increased risk of ADHD or to the behaviors on which the diagnosis is based. In a study linking Danish national registers to modeled concentrations of nitrogen dioxide (NO_2) and $PM_{2.5}$ during the first 5 years of life, Thygesen et al.[86] found that both exposures were associated with an increased risk of clinically diagnosed ADHD. However, in an analysis of eight European birth cohort studies, pregnancy NO_2 and PM exposures, estimated using land-use regression models, were not associated with teacher- or parent-reported ADHD symptoms in 3- to 10-year-old children.[87] In 2019, a systematic review concluded that, overall, there is "limited evidence" that air pollution exposures are associated with risk of ADHD.[88] As observed for other potential exposures, risk might be modified by co-exposures. Perera et al.[89] reported that the risk of ADHD-related behavior problems associated with PAH exposure was increased among children from families who experienced greater material hardship.

Specific Learning Disabilities

Various criteria are used to identify an SLD, with the most common pertaining to difficulty in a skill (e.g., thinking, speaking, reading, writing, spelling, calculations) which is not due to intellectual deficiency, emotional disorder, or socioeconomic disadvantage.

Sometimes a discrepancy between aptitude (i.e., intelligence) and achievement is considered to indicate an SLD. However, the fact that exposure to many chemicals causes intellectual deficits complicates the task of determining whether such exposures also cause a specific deficit above and beyond this. Surkan et al.[90] found that children with a blood lead level of 5–10 μg/dL achieved significantly lower reading and mathematics scores than did children with a level of 1–2 μg/dL, even when adjustment was made for Full-Scale IQ. Most studies reporting an association between higher blood lead levels and poorer academic achievement have not adjusted for IQ, however. Miranda et al.[91] found that the percentage of North Carolina children classified as "exceptional due to learning or behavioral" difficulties increased monotonically between blood lead concentrations of 1 and 10+ μg/dL. Studies using geostatistical methods have shown that learning disability rates are higher in areas with historically higher levels of lead exposure.[92] Using NHANES survey data, Lee et al.[93] showed that higher serum levels of certain persistent organic pollutants (e.g., polychlorinated dibenzo-p-dioxins and polychlorinated dibenzofurans) were associated with a greater risk of parent-reported learning disability. Ciesielski et al.[94] found that children with higher urinary levels of cadmium were at increased risk of both parent-reported learning disability and receipt of special education services, and Anderko et al.[95] reported that prenatal exposure to ETS was a risk factor for parent-reported learning disability.

Psychiatric Disorders

Opler et al.[96] reported that young adults diagnosed with schizophrenia were more likely than controls to have mothers who had elevated levels of δ-aminolevulinic acid, a biomarker of lead exposure, in archived second-trimester serum samples. Guilarte et al.[97] suggested lead's inhibition of the N-methyl-D-aspartate (NMDA) receptor as a basis for this association. Brown[98] marshaled a variety of indirect evidence to support the hypothesis that endocrine disruption resulting from exposure to chemicals such as bisphenol A is a pathway for the pathogenesis of schizophrenia. Prenatal and early-life exposure to tetrachloroethylene-contaminated drinking water was associated with increased risk of several psychiatric disorders in adulthood, including bipolar disorder, posttraumatic stress disorder, and schizophrenia, but not depression.[99] The confidence intervals of all risk ratios included 1, however. These investigators also found that early exposure increased the risk of later illicit drug use.[100] A meta-analysis focused on particulate matter and mental health outcomes suggested associations between $PM_{2.5}$ and depression and anxiety.[101] Childhood lead exposure has been associated with an increased risk of aggression,[102] delinquent behavior,[103] and likelihood of being arrested in young adulthood.[104] Prenatal alcohol exposure has been linked to increased risk of childhood depression, oppositional defiant disorder, conduct disorder, and delinquent behavior.[105]

Neurodegeneration

Studies in rodents and nonhuman primates suggest that developmental exposure to lead is a risk factor for neurodegenerative disease in adulthood. Animals exposed only in early life showed elevations of β-amyloid protein precursor (APP) mRNA, APP, and its amyloidogenic product, Abeta, in old age, particularly in the frontal cortex.[106] DNA methylation was decreased and oxidative damage to DNA increased, suggesting a role for epigenetic influence in these latent effects. Studies in humans have just begun.[107]

As noted earlier, greater exposure to air pollution has been linked to increased levels of biomarkers of neuroinflammation and neurodegenerative disorders, such as α-synuclein and β-amyloid.[14] Adverse effects on the brain appear to begin early in life, as a recent neuroimaging study of 9- and 10-year-old children showed $PM_{2.5}$- related alterations in white matter microarchitecture.[108]

Conclusion

At present, the evidence linking children's chemical exposures to clinically defined NDDs such as ASDs or ADHD is suggestive but limited and, for many chemicals, consists of only a few studies. However, a wide variety of chemicals have been linked, in at least some studies, to increased risk. For ADHD, for example, the list of suspect chemicals includes lead, PFAS, phthalates, fluoride, pyrethroids, air pollutants, and prenatal alcohol, among others. Given that the modes of action differ among these chemicals, it seems unlikely that all of these associations are causal, raising the question of the extent to which at least some of them are artifactual, resulting from chance, residual confounding, publication bias, or other factors. For some exposures, the accumulation of additional studies seems to have muddied rather than clarified the waters. This might reflect the fact that, in general, even for non-null associations that are true, the effect sizes reported in early studies tend to be inflated compared to the true effect size.[109]

Many challenges confront the investigator seeking to evaluate associations between NDDs and chemical exposures. One set of challenges pertains to case definition and identification. The use of screening methods rather than full diagnostic methods increases the likelihood of outcome misclassification and thus the risk of bias in the effect estimate. Another important issue is that the behaviors on which many NDDs are based are not simply "present" or "absent" in a child. Rather they are behaviors that typically developing children display to some extent. Children who meet diagnostic criteria just display much more extreme forms or frequencies of these behaviors. At community exposure levels to chemicals, it seems more likely that children will show subdiagnostic increases in these behaviors than to exhibit the extremes required to meet diagnostic criteria. This might explain why more studies show exposure-related increases in ASD- or ADHD-related traits than exposure-related increases in the prevalence of children with these diagnoses. Most NDDs are also based on behavioral phenotypes that consist of behaviors that are final common pathways for the expression of many factors other than chemical exposures. In addition to requiring that these factors be measured and considered as potential confounders, which most studies do to some extent, it is also important that they be considered as potential effect modifiers. The latter is less commonly explored in studies, but, as noted, there are reports that the strength of the associations between certain chemical exposures and NDDs differ depending on factors such as sex, socioeconomic status, nutritional status, genotype, and others. More thorough consideration of effect modification might help to resolve some of the apparent inconsistencies in the evidence currently available.

A second set of challenges pertains to exposure classification. When little is known about the windows of vulnerability, the risk of failing to detect a true association mandates the measurement of exposure over the entire period preceding diagnosis of the NDD. Because NDDs tend to be diagnosed only when a child reaches several years of age (or even decades later in the case of schizophrenia or neurodegeneration), this poses a formidable logistical hurdle that often cannot be cleared. For chemicals with a short biological residence time (e.g., OP pesticides), and for which early-life, perhaps even prenatal exposure is most important in predicting risk, it might be very difficult to reconstruct a child's exposure history over potentially important intervals. In instances in which exposure is based on ecologic measurements that pertain to external rather than internal exposure (e.g., land-use regression models) and which therefore assume equivalent exposures for individuals who share a characteristic such as census tract residence, the risk of exposure misclassification is high. Finally, most studies have considered only one chemical or one chemical class at a time. Children are not exposed only to single chemicals, however, but to complex mixtures, which could alter the dose-response relationships with regard to neurodevelopmental risk.

In conducting systematic reviews of different chemicals and their associations with NDDs, many authors commented that it was not feasible to conduct formal pooled or meta-analyses due to the heterogeneity across studies in terms of so many critical aspects of study design, including the populations studied (e.g., national birth cohorts, cohorts known to be at enhanced risk of a particular NDD), the way in which exposure was quantified (e.g., internal dose using biomarkers, external dose using ecologic models), the developmental stage(s) at which exposure was measured, the methods used to define child outcomes (e.g., parental or teacher report, diagnostic interviews and assessments), the ages at which outcomes were assessed, and so on.

Little success has been achieved in identifying neurodevelopmental "signatures" for chemicals, that is, a unique constellation of findings that identify a particular chemical as the responsible agent. Indeed, this goal might forever remain beyond reach, given that the behavioral phenotypes that underlie most NDDs consist of characteristics that occur, to a lesser degree, in the general population of children. Some clarity on this might be gained from the results of studies using neuroimaging modalities to identify the anatomical and functional expressions of chemical exposures.[110–112] These studies have reported novel exposure-related differences in regional volumes, levels of metabolic activity, white matter organization, and connectivity.

The global burden of disease in children attributable to environmental chemical exposures is likely to be substantial. The Institute for Health Metrics and Evaluation estimated that, in 2019, lead exposure accounted for 62.5% of the global burden of idiopathic developmental intellectual disability and that its overall adverse effects on human health accounted for 21.7 million years of healthy life lost (i.e., disability-adjusted life years) (https://www.healthdata.org/results/gbd_summaries/2019/lead-exposure-level-3-risk). Given the high prevalence of NDDs in children, suggestions that some disorders might be increasing in prevalence and the fact that children's exposures to environmental chemicals are, in principle, preventable, it behooves us, as a society, to determine the role that such exposures play in the etiology of NDDs.

References

1. Zablonsky B, Black LI. Prevalence of children aged 3–17 years with developmental disabilities, by urbanicity: United States: 2015–2018. *Nat Health Stat Rep.* 2020 Feb 19;139: 1–7.

2. Zablonsky B, Black LI, Maenner MJ, et al. Prevalence and trends of developmental disabilities among children in the United States: 2009–2017. *Pediatrics.* 2019; 144(4): 1–11.

3. Irwin JK, MacSween J, Kerns KA. History and evolution of the autism spectrum disorders. In J.L Matson, P Sturmey, eds., *International Handbook of Autism and Pervasive Developmental Disorders.* New York: Springer Science + Business Media; 2011: 3–16.

4. Maenner MJ, Shaw KA, Bakian AV, et al. Prevalence and characteristics of autism spectrum disorder among children aged 8 years: Autism and Developmental Disabilities Monitoring Network, 11 sites, United States, 2018. *MMWR.* 2021;70(11):1–16.

5. Maenner MJ, Shaw KA, Bakian AV, et al. Prevalence and characteristics of autism spectrum disorder among children aged 8 years: Autism and Developmental Disabilities Monitoring Network, 11 Sites, United States, 2018. *MMWR Surveill Summ.* 2021;70(11):1–16.

6. National Research Council. *Scientific Frontiers in Developmental Toxicology and Risk Assessment.* Washington, DC: National Academies Press, 2000.

7. Bai D, Yip BHK, Windham GC, et al. Association of genetic and environmental factors with autism in a 5-country cohort. *JAMA Psychiatry.* 2019;76(10):1035–43.

8. Sandin S, Lichtenstein P, KujaHaloka R, et al. The heritability of autism spectrum disorder. *JAMA.* 2017;318(12):1182–84.

9. Happe F, Ronald A, Plomin R. Time to give up on a single explanation for autism. *Nature Neurosci.* 2006;9:1218–20.

10. Bernier R, Gerdts J, Munson J, et al. Evidence for broader autism phenotype characteristics in parents from multiple-incidence autism families. *Autism Res.* 2012;5(1):13–20.

11. Hus V, Bishop S, Gotham K, et al. Factors influencing scores on the social responsiveness scale. *J Child Psychol Psychiatry.* 2013;54(2):216–24.

12. Ha EH, Kim SY, Song DH, et al. Discriminant validity of the CBCL 1.5-5 in diagnosis of developmental delayed infants. *J Korean Acad Child Adolesc Psychiatry.* 2011;22:120–7.

13. Shabani S. A mechanistic view on the neurotoxic effects of air pollution on central nervous system: risk for autism and neurodegenerative diseases *Environ Sci Pollut Res Int.* 2021;28(6):6349–73.

14. Costa LG, Cole TB, Dao K, et al. Effects of air pollution on the nervous system and its possible role in neurodevelopmental and neurodegenerative disorders. *Pharmacol Ther.* 2020;210:107523.

15. Frye RE, Cakir J, Rose S, et al. Prenatal air pollution influences neurodevelopment and behavior in autism spectrum disorder by modulating mitochondrial physiology. *Mol Psychiatry.* 2021;26(5):1561–77.

16. Volk HE, Park B, Hollingue C, et al. Maternal immune response and air pollution exposure during pregnancy: insights from the Early Markers for Autism (EMA) study. *J Neurodev Disord.* 2020;12(1):42.

17. Lam J, Sutton P, Kalkbrenner A, et al. A systematic review and meta-analysis of multiple airborne pollutants and autism spectrum disorder. *PLoS One.* 2016;11(9):e0161851.

18. Chun H, Leung C, Wen SW, et al. Maternal exposure to air pollution and risk of autism in children: a systematic review and meta-analysis. *Environ Pollut.* 2020; 256:113307.

19. Carter SA, Rahman MM, Lin JC, et al. In utero exposure to near-roadway air pollution and autism spectrum disorder in children. *Environ Int.* 2022;158:106898.

20. Wang SY, Cheng YY, Guo HR, et al. Air pollution during pregnancy and childhood autism spectrum disorder in Taiwan. *Int J Environ Res Public Health.* 2021;18(18):9784.

21. Pagalan L, Bickford C, Weikum W, et al. Association of prenatal exposure to air pollution with autism spectrum disorder. *JAMA Pediatr.* 2019;173(1):86–92.

22. Jo H, Eckel SP, Wang X, et al. Sex-specific associations of autism spectrum disorder with residential air pollution exposure in a large Southern California pregnancy cohort. *Environ Pollut.* 2019;254(Pt A):113010.

23. McGuinn LA, Windham GC, Messer LC, et al. Air pollution, neighborhood deprivation, and autism spectrum disorder in the Study to Explore Early Development. *Environ Epidemiol.* 2019;3(5):e067.

24. Volk HE, Kerin T, Lurmann F, et al. Autism spectrum disorder: interaction of air pollution with the MET receptor tyrosine kinase gene. *Epidemiology.* 2014;25(1):44–7.

25. Goodrich AJ, Volk HE, Tancredi DJ, et al. Joint effects of prenatal air pollutant exposure and maternal folic acid supplementation on risk of autism spectrum disorder. *Autism Res.* 2018;11(1):69–80.

26. Allen JL, Oberdorster G, Morris-Schaffer K, et al. Developmental neurotoxicity of inhaled ambient ultrafine particle air pollution: parallels with neuropathological and behavioral features of autism and other neurodevelopmental disorders. *Neurotoxicology.* 2017;59:140–54.

27. Von Ehrenstein OS, Ling C, Cui X, et al. Prenatal and infant exposure to ambient pesticides and autism spectrum disorder in children: population based case-control study *BMJ.* 2019;364:l962.

28. Brown AS, Cheslack-Postava K, Rantakokko P, et al. Association of maternal insecticide levels with autism in offspring from a national birth cohort. *Am J Psychiatry.* 2018;175(11):1094–1101.

29. Eskenazi B, Huen K, Marks A, et al. PON1 and neurodevelopment in children from the CHAMACOS study exposed to organophosphate pesticides in utero. *Environ Health Perspect.* 2010;118:1775–81.

30. Rauh V, Garfinkel R, Perera FP, et al. Impact of prenatal chlorpyrifos exposure on neurodevelopment in the first 3 years of life among inner-city children. *Pediatrics.* 2006;118:e1845–e1859.

31. Sagiv SK, Harris MH, Gunier RB, et al. Prenatal organophosphate pesticide exposure and traits related to autism spectrum disorders in a population living in proximity to agriculture. *Environ Health Perspect.* 2018;126(4):047012.

32. van den Dries MA, Guxens M, Pronk A, et al. Organophosphate pesticide metabolite concentrations in urine during pregnancy and offspring attention-deficit hyperactivity disorder and autistic traits. *Environ Int.* 2019;131:105002.

33. Philippat C, Barkoski J, Tancredi DJ, et al. Prenatal exposure to organophosphate pesticides and risk of autism spectrum disorders and other non-typical development at 3 years in a high-risk cohort. *Int J Hyg Environ Health.* 2018;221(3):548–55.

34. Schmidt RJ, Kogan V, Shelton JF, et al. Combined prenatal pesticide exposure and folic acid intake in relation to autism spectrum disorder. *Environ Health Perspect.* 2017;125(9):097007.

35. Miodovnik A, Engel SM, Zhu C, et al. Endocrine disruptors and childhood social impairment. *Neurotoxicology.* 2011;32:261–7.

36. Oulhote Y, Lanphear B, Braun JM, et al. Gestational exposures to phthalates and folic acid, and autistic traits in Canadian children. *Environ Health Perspect.* 2020;128(2):27004.

37. Shin HM, Bennett DH, Calafat AM, et al. Modeled prenatal exposure to per- and polyfluoroalkyl substances in association with child autism spectrum disorder: a case-control study. *Environ Res.* 2020;186:109514.

38. Larsson M, Weiss B, Janson S, et al. Associations between indoor environmental factors and parental-reported autistic spectrum disorders in children 6-8 years of age. *Neurotoxicology.* 2009;30:822–31.

39. Philippat C, Bennett DH, Krakowiak P, et al. Phthalate concentrations in house dust in relation to autism spectrum disorder and developmental delay in the CHildhood Autism Risks from Genetics and the Environment (CHARGE) study. *Environ Health.* 2015;14:56.

40. Skogheim TS, Weyde KVF, Aase H, et al. Prenatal exposure to per- and polyfluoroalkyl substances (PFAS) and associations with attention-deficit/hyperactivity disorder and autism spectrum disorder in children. *Environ Res.* 2021;202:111692.

41. Lyall K, Yau VM, Hansen R, et al. Prenatal maternal serum concentrations of per- and polyfluoroalkyl substances in association with autism spectrum disorder and intellectual disability. *Environ Health Perspect.* 2018;126(1):017001.

42. Oh J, Bennett DH, Calafat AM, et al. Prenatal exposure to per- and polyfluoroalkyl substances in association with autism spectrum disorder in the MARBLES study. *Environ Int.* 2021;147:106328.

43. Oh J, Shin HM, Kannan K, et al. Childhood exposure to per- and polyfluoroalkyl substances and neurodevelopment in the CHARGE case-control study. Environmental Research 2022;215(Pt 2):114322.

44. Landrigan PJ. What causes autism? Exploring the environmental contribution. *Curr Opin Pediatrics* 2010;22:219–25.

45. van Wijngaarden E, Davidson PW, Smith TH, et al. Autism spectrum disorder phenotypes and prenatal exposure to methylmercury. *Epidemiology* 2013;24(5):651–9.

46. Lyall K, Croen LA, Weiss LA, et al. Prenatal serum concentrations of brominated flame retardants and autism spectrum disorder and intellectual disability in the Early Markers of Autism Study: a population-based case-control study in California. *Environ Health Perspect.* 2017;125(8):087023.

47. Hertz-Picciotto I, Bergman A, Fängström B, et al. Polybrominated diphenyl ethers in relation to autism and developmental delay: a case-control study. *Environ Health.* 2011;10(1):1.

48. Eliasen M, Tolstrup JS, Nybo Andersen AM, et al. Prenatal alcohol exposure and autistic spectrum disorders: a population-based prospective study of 80,552 children and their mothers. *Int J Epidemiol.* 2010;39:1074–81.

49. Lee BK, Gardner RM, Dal H, et al. Maternal smoking during pregnancy and autism spectrum disorders. *J Autism Dev Disord.* 2012;42(9):2000–5.

50. Larsson M, Anckarsater H, Rastam M, et al. Childhood attention-deficit hyperactivity disorder as an extreme of a continuous trait: a quantitative genetic study of 8,500 twin pairs. *J Child Psychol Psychiatry.* 2012;53:73–80.

51. United States Environmental Protection Agency. Integrated Science Assessment for Lead. National Center for Environmental Assessment RTP Division, Office of Research and Development, EPA/600/R-10/075F, June, 2013.

52. Braun JM, Kahn RS, Froehlich T, et al. Exposures to environmental toxicants and attention deficit hyperactivity disorder in U.S. children. *Environ Health Perspect.* 2006;114:1904–9.

53. Nigg JT, Nikolas M, Knottnerus M, et al. Confirmation and extension of association of blood lead with attention-deficit/hyperactivity disorder (ADHD) and ADHD symptom domains at population-typical exposure levels. *J Child Psychol Psychiatry*. 2010;51:58–65.

54. Daneshparvar M, Mostafavi S-A, Jeddi MZ, et al. The role of lead exposure on attention-deficit/ hyperactivity disorder in children: a systematic review. *Iran J Psychiatry*. 2016; 11(1): 1–14.

55. Hoffman K, Webster TF, Weisskopf MG, et al. Exposure to polyfluoroalkyl chemicals and attention deficit/hyperactivity disorder in U.S. children 12–15 years of age. *Environ Health Perspect*. 2010;118:1762–7.

56. Stein CD, Savitz DA. Serum perfluorinated compound concentration and attention deficit/hyperactivity disorder in children 5–18 years of age. *Environ Health Perspect*. 2011;119:1466–71.

57. Forns J, Verner M-A, Iszatt N, et al. Early life exposure to perfluoroalkyl substances (PFAS) and ADHD: a meta-analysis of nine European population-based studies. *Environ Health Perspect*. 2020;128(5):57002.

58. Dalsager L, Jensen TK, Nielsen F, et al. No association between maternal and child PFAS concentrations and repeated measures of ADHD symptoms at age 2½ and 5 years in children from the Odense Child Cohort. *Neurotoxicol Teratol*. 2021;88:107031.

59. Itoh S, Yamazaki K, Suyama S, et al. The association between prenatal perfluoroalkyl substance exposure and symptoms of attention-deficit/hyperactivity disorder in 8-year-old children and the mediating role of thyroid hormones in the Hokkaido study. *Environ Int*. 2021;159:107026.

60. Vuong AM, Webster GM, Yolton K, et al. Prenatal exposure to per- and polyfluoroalkyl substances (PFAS) and neurobehavior in US children through 8 years of age: the HOME study. *Environ Res*. 2021;195:110825.

61. Riddell JK, Malin AJ, Flora D, et al. Association of water fluoride and urinary fluoride concentrations with attention deficit hyperactivity disorder in Canadian youth. *Environ Int*. 2019;133(Pt B):105190.

62. Bashash M, Marchand M, Hu H, et al. Prenatal fluoride exposure and attention deficit hyperactivity disorder (ADHD) symptoms in children at 6–12 years of age in Mexico City. *Environ Int*. 2018;121(Pt 1):658–66.

63. Forns J, Stigum H, Høyer BB, et al. Prenatal and postnatal exposure to persistent organic pollutants and attention-deficit and hyperactivity disorder: a pooled analysis of seven European birth cohort studies. *Int J Epidemiol*. 2018;47(4):1082–97.

64. Dalsager L, Fage-Larsen B, Bilenberg N, et al. Maternal urinary concentrations of pyrethroid and chlorpyrifos metabolites and attention deficit hyperactivity disorder (ADHD) symptoms in 2–4-year-old children from the Odense Child Cohort. *Environ Res*. 2019;176:108533.

65. Wagner-Schuman M, Richardson JR, Auinger P, et al. Association of pyrethroid pesticide exposure with attention-deficit/hyperactivity disorder in a nationally representative sample of U.S. children. *Environ Health*. 2015;14:44.

66. Lee W-S, Lim Y-H, Kim B-Y, et al. Residential pyrethroid insecticide use, urinary 3-phenoxybenzoic acid levels, and attention-deficit/hyperactivity disorder-like symptoms in preschool-age children: the Environment and Development of Children study. *Environ Res*. 2020;188:109739.

67. Marks AR, Harley K, Bradman A, et al. Organophosphate pesticide exposure and attention in young Mexican-American children: the CHAMACOS Study. *Environ Health Perspect.* 2010;118:1768–74.

68. Bouchard MF, Bellinger DC, Wright RO, et al. Attention-deficit/hyperactivity disorder and urinary metabolites of organophosphate pesticides. *Pediatrics.* 2010;125:e1270–7.

69. Lam J, Lanphear BP, Bellinger D, et al. Developmental PBDE exposure and IQ/ADHD in childhood: a systematic review and meta-analysis. *Environ Health Perspect.* 2017;125(8):086001.

70. Choi G, Keil AP, Richardson DB, et al. Pregnancy exposure to organophosphate esters and the risk of attention-deficit hyperactivity disorder in the Norwegian mother, father and child cohort study. *Environ Int.* 2021;154:106549.

71. Mattson SN, Crocker N, Nguyen TT. Fetal alcohol spectrum disorders: neuropsychological and behavioral features. *Neuropsychol Review.* 2011;21:81–101.

72. Rodriguez A, Olsen J, Kotimaa AJ, et al. Is prenatal alcohol exposure related to inattention and hyperactivity symptoms in children? Disentangling the effects of social adversity. *J Child Psychol Psychiatry.* 2009;50:1073–83.

73. Cornelius MD, Day NL. Developmental consequences of prenatal tobacco exposure. *Curr Opinion Neurol.* 2009;22:121–5.

74. Ball SW, Gilman SE, Mick E, et al. Revisiting the association between maternal smoking during pregnancy and ADHD. *J Psychiat Res.* 2010; 44:1058–62.

75. Yoshimasu K, Kiyohara C, Minami T, et al. Maternal smoking during pregnancy and offspring attention-deficit hyperactivity disorder: a case-control study in Japan. *Atten Defic Hyperact Disord.* 2009;1:223–31.

76. Knopik VS. Maternal smoking during pregnancy and child outcomes: real or spurious effect? *Dev Neuropsychol.* 2009;34:1–36.

77. Thapar A, Rice F, Hay D, et al. Prenatal smoking might not cause attention-deficit/hyperactivity disorder: evidence from a novel design. *Biol Psychiatry.* 2009;66:722–7.

78. Gustavson K, Ystrom E, Stoltenberg C, et al. Smoking in pregnancy and child ADHD. *Pediatrics.* 2017;139(2):e20162509.

79. Kim B-N, Cho S-C, Kim Y, et al. Phthalates exposure and attention deficit/hyperactivity disorder in school-age children. *Biol Psychiatry.* 2009;66:958–63.

80. Engel SM, Miodovnik A, Canfield RL, et al. Prenatal phthalate exposure is associated with childhood behavior and executive functioning. *Environ Health Perspect.* 2010;118:565–71.

81. Engel SM, Villanger GD, Nethery RC, et al. Prenatal phthalates, maternal thyroid function, and risk of attention-deficit hyperactivity disorder in the Norwegian Mother and Child Cohort. *Environ Health Perspect.* 2018;126(5):057004.

82. Kamai EM, Villanger GD, Nethery RC, et al. Gestational phthalate exposure and preschool attention deficit hyperactivity disorder in Norway. *Environ Epidemiol.* 2021;5(4):e161.

83. Perera FP, Wang S, Vishnevetsky J, et al. Polycyclic aromatic hydrocarbons-aromatic DNA adducts in cord blood and behavior scores in New York City children. *Environ Health Perspect.* 2011;119:1176–81.

84. Siddique S, Banerjee M, Ray MR, et al. Attention-deficit hyperactivity disorder in children chronically exposed to high level of vehicular pollution. *Eur J Pediatr.* 2011;170:923–9.

85. Morales E, Julvez J, Torrent M, et al. Association of early life exposure to household gas appliances and indoor nitrogen dioxide with cognition and attention behavior in preschoolers. *Am J Epidemiol.* 2009;169:1327–36.

86. Thygesen M, Holst GJ, Hansen B, et al. Exposure to air pollution in early childhood and the association with attention- deficit hyperactivity disorder. *Environ Res.* 2020;183:108930.

87. Forns J, Sunyer J, Garcia-Esteban R, et al. Air pollution exposure during pregnancy and symptoms of attention deficit and hyperactivity disorder in children in Europe. *Epidemiology.* 2018;29(5):618–26.

88. Aghaei M, Janjani H, Yousefian F, et al. Association between ambient gaseous and particulate air pollutants and attention deficit hyperactivity disorder (ADHD) in children; a systematic review. *Environ Res.* 2019;173:135–56.

89. Perera FP, Wheelock K, Wang Y, et al. Combined effects of prenatal exposure to polycyclic aromatic hydrocarbons and material hardship on child ADHD behavior problems. *Environ Res.* 2018;160:506–513.

90. Surkan PJ, Zhang A, Trachtenberg F, et al. Neuropsychological function in children with blood lead levels <10 microg/dL. *Neurotoxicology.* 2007;28:1170–7.

91. Miranda ML, Maxson P, Kim D. Early childhood lead exposure and exceptionality designations for students. *Int J Child Health Hum Dev.* 2010;3:77–84.

92. Margai F, Henry NA. A community-based assessment of learning disabilities using environmental and contextual risk factors. *Soc Sci Med.* 2003;56:1073–85.

93. Lee D-H, Jacobs DR, Porta M. Association of serum concentrations of persistent organic pollutants with the prevalence of learning disability and attention deficit disorder. *J Epidemiol Comm Health.* 2007;61:591–6.

94. Ciesielski T, Weuve J, Bellinger DC, et al. Cadmium exposure and neurodevelopmental outcomes in U.S. children. *Environ Health Perspect.* 2012;120:758–63.

95. Anderko L, Braun J, Auinger P. Contribution of tobacco smoke exposure to learning disabilities. *J Obstet Gynecol Neonatal Nurs.* 2010;39:111–7.

96. Opler MG, Buka SL, Groeger J, et al. Prenatal exposure to lead, delta-aminolevulinic acid, and schizophrenia: further evidence. *Environ Health Perspect.* 2008;116(11):1586–90.

97. Guilarte TR, Opler M, Pletnikov M. Is lead exposure in early life an environmental risk factor for schizophrenia? Neurobiological connections and testable hypotheses. *Neurotoxicology.* 2012;33(3):560–74.

98. Brown JS. Effects of bisphenol-A and other endocrine disruptors compared with abnormalities of schizophrenia: an endocrine-disruption theory of schizophrenia. *Schizophrenia Bull.* 2009;35:256–78.

99. Aschengrau A, Weinberg JM, Janulewicz PA, et al. Occurrence of mental illness following prenatal and early childhood exposure to tetrachloroethylene (PCE)-contaminated drinking water: a retrospective cohort study. *Environ Health.* 2012;11:2.

100. Aschengrau A, Weinberg JM, Janulewicz PA, et al. Affinity for risky behaviors following prenatal and early childhood exposure to tetrachloroethylene (PCE)-contaminated drinking water: a retrospective cohort study. *Environ Health.* 2011,10:102.

101. Braithwaite I, Zhang S, Kirkbride JB, et al. Air pollution (particulate matter) exposure and associations with depression, anxiety, bipolar, psychosis and suicide risk: a systematic review and meta-analysis. *Environ Health Perspect.* 2019;127(12):126002.

102. Nkomo P, Naicker N, Mathee A, et al. The association between environmental lead exposure with aggressive behavior, and dimensionality of direct and indirect aggression during mid-adolescence: Birth to Twenty Plus cohort. *Sci Total Environ.* 2018;612:472–9.

103. Needleman HL, McFarland C, Ness RB, et al. Bone lead levels in adjudicated delinquents. A case control study. *Neurotoxicol Teratol.* 2002;24:711–7.

104. Wright JP, Dietrich KN, Ris MD, et al. Association of prenatal and childhood blood lead concentrations with criminal arrests in early adulthood. *PLoS Med.* 2008;5:e101.

105. Lange S, Rehm J, Anagnostou E, Popova S. Prevalence of externalizing disorders and autism spectrum disorders among children with fetal alcohol spectrum disorder: systematic review and meta-analysis. Biochemistry and Cell Biology 2018;96(2):241–251.

106. Wu J, Basha MR, Brock B, et al. Alzheimer's disease (AD)-like pathology in aged monkey after infantile exposure to environmental lead (Pb): evidence for a developmental origin and environmental link for AD. *J Neuroscience.* 2008;2:3–9.

107. Mazumdar M, Xia W, Hofmann O, et al. Prenatal lead levels, plasma amyloid β levels and gene expression in young adulthood. *Environ Health Perspect.* 2012;120:702–7.

108. Burnor E, Cserbik D, Cotter DL, et al. Association of outdoor ambient fine particulate matter with intracellular white matter microstructural properties among children. *JAMA Netw Open.* 2021;4(12):e2138300.

109. Ioannidis JP. Why most discovered true associations are inflated. *Epidemiology.* 2008;19(5):640–8.

110. Sagiv SK, Bruno JL, Baker JM, et al. Prenatal exposure to organophosphate pesticides and functional neuroimaging in adolescents living in proximity to pesticide application. *Proc Natl Acad Sci U S A.* 2019;116(37):18347–56.

111. Thomason ME, Hect JL, Rauh VA, et al. Prenatal lead exposure impacts cross-hemispheric and long-range connectivity in the human fetal brain. *Neuroimage.* 2019;191:186–92.

112. Peterson BS, Rauh VA, Bansal R, et al. Effects of prenatal exposure to air pollutants (polycyclic aromatic hydrocarbons) on the development of brain white matter, cognition, and behavior in later childhood. *JAMA Psychiatry.* 2015;72(6):531–40.

50

Prenatal Environmental Exposures and Birth Defects

Stephanie Ford and Cynthia F. Bearer

Up to 6% of pregnancies worldwide result in a child with a major birth defect.[1] It is estimated that about 10% of birth defects are attributable to environmental exposures, that approximately 30% have a known genetic etiology, and that the remaining 60% are due to unknown causes. However, there is a complicated interplay between environmental factors, genetics, and epigenetics, and the relationships among these factors are not fully understood.

Many environmental exposures, such as exposures to air pollutants, are ubiquitous in pregnancy. Others are also surprisingly common. One recent study showed that more than 73% of women took a medication (excluding vaccines, supplements, and vitamins) in pregnancy, with 55% of women taking one in the first trimester.[2] Clearly, not all environmental exposures are teratogenic. The challenge is to determine which of these exposures, and at what dose and timing in gestation, has the potential to cause birth defects.

Characteristics of Teratogens

A theoretical framework for evaluating environmental exposures for their teratogenic potential was outlined in the early 1970s by James Wilson.[3] These principles of teratology are the following:

1. *Susceptibility to teratogenesis depends on the genotype of the conceptus (the embryo or fetus) and the manner in which the conceptus interacts with environmental factors.* Genetic (or epigenetic) characteristics of the mother and/or fetus may alter the way a drug or chemical is metabolized or may alter the susceptibility of a developmental process to being disturbed by a chemical exposure.
2. *Susceptibility to teratogenic agents varies with the developmental stage at the time of exposure.* The principle of gestational timing—of critical *windows of vulnerability* in early development—requires that a teratogenic exposure occur during the stage in development when the targeted developmental process is most susceptible (see Chapter 2). For example, the critical window for interference with closure of the neural tube in the human embryo is approximately 21–28 days post conception. An exposure that occurs more than 28 days after conception would not have the potential to cause a defect of neural tube closure.

Stephanie Ford and Cynthia F. Bearer, *Prenatal Environmental Exposures and Birth Defects* In: *Textbook of Children's Environmental Health*, Second Edition. Edited by: Ruth A. Etzel and Philip J. Landrigan, Oxford University Press. © Oxford University Press 2024. DOI: 10.1093/oso/9780197662526.003.0050

Furthermore, risks for specific birth defects may differ based on the specific gestational timing of the exposure. For example, maternal treatment with warfarin
as an anticoagulant is associated with a pattern of nasal hypoplasia and skeletal
abnormalities when prenatal exposure occurs during the latter portion of the first
trimester, whereas exposure later in gestation is associated with abnormalities in
the central nervous system (CNS).[4]

3. *Teratogenic agents act in specific ways (through specific mechanisms) on developing
   cells and tissues to initiate abnormal embryogenesis.* There is no known teratogenic
   exposure that increases the risk of all birth defects. Teratogenic exposures typically
   act on specific targets to produce characteristic patterns of effects. This principle
   forms the basis for the diagnostic strategy that has been used to discover many
   human teratogens, namely the recognition of a particular pattern of abnormalities
   in a group of infants that is then linked to a particular teratogenic exposure during
   pregnancy. A classic historical example is the epidemic of phocomelia in the 1960s
   that was traced to thalidomide exposure.

4. *The final manifestations of abnormal development are death, malformation, growth
   retardation, and functional disorder.* Teratogenic exposures may induce birth
   defects but also can be associated with increased risks for spontaneous abortion or
   stillbirth, prenatal and/or postnatal growth deficiency, preterm delivery, and functional deficits or learning disabilities. For example, moderate to heavy maternal
   alcohol consumption in pregnancy increases the risks for spontaneous abortion,
   stillbirth, a characteristic pattern of craniofacial abnormalities, certain types of
   heart defects and oral clefts, prenatal and postnatal growth deficiency, deficits in
   global IQ, and specific behavioral and learning abnormalities.[5]

5. *The access of adverse environmental influences to developing tissues depends on the
   nature of the influence (the agent).* The potential of an agent to increase the risks for
   birth defects depends on the dose that is available to the embryo or fetus, and this
   may be influenced by the route of administration. For example, a topical exposure
   versus a systemic exposure to the same agent might result in quite different doses
   to the embryo or fetus.

6. *Manifestations of deviant development increase in degree as dosage increases from
   the no-effect to the totally lethal level.* This principle suggests that if an environmental exposure is teratogenic at a given threshold dose, higher doses would presumably produce more severe or more frequent effects.

Identification of New Human Teratogens

For most exposures, the available data are comprised exclusively of experimental animal
studies. Interpretation of this information for human pregnancy is not always straightforward, and thus it is ultimately desirable to have human data to evaluate risk. While
it is almost never possible to evaluate an exposure for teratogenicity using a randomized clinical trial design, a number of sources and types of human data may exist, each
with strengths and weaknesses in terms of determining risk to the developing embryo
or fetus.

1. *Case reports.* Clinical case reports published in the literature or adverse event
   reporting systems such as state-based Birth Defects Monitoring Programs may

describe birth defects occurring in pregnancies following specific prenatal exposures. These reports have the potential to generate hypotheses regarding teratogenicity. However, there often is no way to determine if the birth defect occurred independently of the exposure (i.e., is part of the baseline risk in every pregnancy). However, in the uncommon circumstance in which a series of case reports links a very rare or unusual clustering of birth defects with a specific exposure, case reports can be highly suggestive of a causal relation.

2. *Cohort studies*. Epidemiologic studies that compare pregnancy outcomes in women who have had a specific exposure (at a specific time in pregnancy and specific dose) to pregnancy outcomes in women of similar background who have not had the exposure of interest can test hypotheses related to teratogenicity. An advantage of cohort studies is that they can address many of the principles of teratology described above, including evaluating an exposure for a range of outcomes. One disadvantage of this approach is that cohort studies can be costly, and the usual sample sizes are typically too small to rule out any but the most dramatically increased prevalence of specific major birth defects.

3. *Case-control studies*. Epidemiologic studies that compare the prevalence of exposure to an agent (at a specific time and dose) in women who have had an infant with a specific birth defect to pregnancy outcomes in women who are similar but have had a healthy child also can test hypotheses related to teratogenicity. A major strength of case-control studies is that, with proper numbers of cases and controls, they can provide sufficient power to detect increased risks for rare outcomes, such as specific birth defects.

No single study approach is usually sufficient to confirm or rule out the teratogenic potential of an environmental exposure. The establishment of causation is further complicated for many environmental exposures by the difficulties in actually confirming that an exposure occurred and at what dose. From a public health perspective, a combination of complementary human studies is desirable, ideally including studies that can directly measure or validate maternal exposure.

Epidemiology and Clinical Effects of Selected Human Teratogens

Medications

Vitamin K Antagonists

A specific pattern of birth defects, or embryopathy, has been identified in an estimated 6% of infants of mothers who have used vitamin K antagonist medications such as warfarin as anticoagulants during pregnancy.[6] This pattern includes an underdeveloped nose, unusual stippling of the bones, and prenatal growth deficiency. CNS and eye abnormalities also can occur. The critical period for the nasal and skeletal effects caused by warfarin seems to be between 6 and 9 weeks gestation. In addition, increased risks for spontaneous abortion, stillbirth, and preterm delivery have been noted.[7]

The risk for adverse outcome was found to be greater in one small study with warfarin doses of greater than 5 mg/day.[7] In addition, in one study that evaluated child IQ, low scores (IQ <80) occurred more frequently in children who were prenatally exposed to the drug in only the last two trimesters of pregnancy.[8]

Valproic Acid

Based on cohort and case-control studies, there is an estimated 1–2% dose-related risk of neural tube defects in infants whose mothers are treated with valproic acid for seizure disorders or for psychiatric conditions during the critical window for neural tube closure—the first month after conception. By comparison, the risk of neural tube defects in the general population is 1/1,000. In addition, some studies have noted increased risks for cardiovascular, limb, and genital anomalies.[9] Furthermore, valproic acid is linked to deficits in intellectual performance in children exposed prenatally.[10]

Angiotensin I Converting Enzyme Inhibitors and Angiotensin II Receptor Blockers

Angiotensin I converting enzyme (ACE) inhibitor exposure in the first trimester is associated with an increased risk of cardiac and CNS malformations.[11] Cohort and case-control studies have identified a specific pattern of birth defects and other adverse outcomes associated with second- and third-trimester exposure to ACE inhibitors and angiotensin II receptor blockers (ARBs) classes. Perhaps the most serious consequence is fetal and neonatal renal failure and hypotension, which is distinctive for being refractory to both volume expansion and vasopressors. The fetal oliguria or anuria results in reduced amniotic fluid, leading to fetal growth restriction, joint contractures, underdeveloped skull bones, underdeveloped lungs, and often stillbirth or neonatal death.[12]

Isotretinoin

Prenatal exposure to isotretinoin used for the treatment of severe cystic acne is associated with a specific pattern of birth defects that includes CNS, cardiac, and craniofacial defects, as well as abnormalities of the thymus.[13,14] With current worldwide guidelines emphasizing contraception and counselling for the use of isotretinoin in reproductive age women, there are far fewer children born with isotretinoin-related congenital defects than when the drug was introduced and therefore few recent studies with large numbers of women using isotretinoin in pregnancy. Previous publications indicate that there is an estimated 28% risk for structural defects, a 47% risk for mental deficiency, and about a 22% risk for spontaneous abortion even with treatment for only a few days in the first trimester.[15]

Physical Agents and Chemicals

Ionizing Radiation

Based on data from a small sample of pregnancies exposed to high-dose radiation from atomic bombs in Nagasaki and Hiroshima, there is an estimated increased risk of small head circumference, intellectual disability, and growth deficiency associated with exposure between 8 and 15 weeks gestation at doses of 50 rad or greater. Higher doses are required to produce similar defects with exposures between 15 and 25 weeks gestation. Conservative estimates suggest that there is no measurable increased risk if the cumulative dose to the uterus is less than 5–10 rads, and there are no currently used routine scans with radiation exposure that exceeds this level.[16,17] Recent studies have not shown a significant correlation between current fetal x-ray exposures and cancers, though international recommendations continue to recommend limiting ionizing radiation exposure to the fetus whenever possible.[18]

Methylmercury

Neurodevelopmental effects including a cerebral palsy-like disorder and deficits in intellectual performance have been linked to prenatal exposure to methylmercury following environmental contaminations that occurred in the past century in Minamata, Japan, and in Iraq (see Chapter 32).[19,20]

At present, the major source of fetal exposure to organic mercury in most parts of the world is through maternal consumption of contaminated fish. Some studies have suggested that prenatal exposure to methylmercury at levels obtained through routine fish consumption results in deficits in intellectual performance.[21] For these reasons, authorities at the US Food and Drug Administration as well as the US Environmental Protection Agency have advised pregnant women to limit fish consumption in general and to avoid eating certain types of fish such as shark, swordfish, tuna, king mackerel, and tilefish, in which high levels of mercury are more likely. However, as evidence accumulates that there are measurable benefits of fish consumption during pregnancy for both the mother and the developing fetus, it would appear wise for pregnant women to consume fish, but to avoid those species that are high in mercury.[22,23]

Lead

Although prenatal exposure to lead has not been definitively linked to an increased risk for structural birth defects, there is evidence of an increased risk for pregnancy loss, premature delivery, and intellectual deficits in pregnancies/infants with prenatal exposure to more than 30 μg/dL in maternal blood. Recent studies indicate that even the lowest measurable blood lead levels are associated with neurocognitive impairment, behavioral difficulties, and poorer executive function in infants and children. [24,25] Lead can be stored in the maternal skeleton for many years after exposure, and fetal lead exposure may therefore result from maternal bone stores that are mobilized during pregnancy as requirements for calcium increase.[26]

Chemicals, Pesticides, Ambient Air Pollution

Emerging evidence is beginning to link exposures to a variety of environmental chemicals to risks for birth defects and other adverse developmental outcomes. Stimulated in part by the known teratogenic effects of alcohol on fetal development, occupational exposure to organic solvents has been a focus of some studies, with mixed results. A large number of studies also have focused on prenatal exposure, through either occupation or residence, to suspected endocrine-disrupting chemicals. These include organochlorine pesticides, polychlorinated biphenyls, polybrominated biphenyls, bisphenol A, phthalates, and dioxins. Case-control investigations of exposures to these chemicals have reported increased risks of male genital malformations, such as hypospadias, cryptorchidism, and micropenis. The strongest and most consistent associations seen in most but not all studies are for hypospadias.[27–29] Although most studies have ascertained exposure through parental interview and estimated the "dose" from sources such as occupational titles, other studies have obtained direct measurements of chemical exposure from cord blood. Obtaining individual-level exposure data continues to be a challenge, and persistence in the environment and/or the body of some of these chemicals can lead to exposures during pregnancy that are difficult to assess without individual biomonitoring. Prospective birth cohort studies that enroll women during pregnancy and measure exposures prenatally may in the future help to resolve some of these difficulties in exposure assessment.

Exposures to ambient air pollution have been linked to certain congenital heart defects in some studies,[30-32] though not in others. Common approaches to evaluating the association involve case-control methods, using residential addresses of the mothers of cases and controls assigned to pollutant levels from the closest stationary air monitor.[33,34] Air pollution exposures also have been linked to sudden infant death syndrome (SIDS), childhood asthma, and abnormal neurodevelopment.[35-37] Newer studies with observations from China during periods of quarantine and transportation/work restriction during the COVID-19 pandemic may provide more insight.[38]

Polycyclic aromatic hydrocarbons (PAHs) are air pollutants formed during the incomplete burning of organic compounds, especially in oil and gas production, coal power plants, and restaurants.[39] Exposure to PAH pollution in pregnancy is related to low birth weight, premature delivery, and delayed development.[40,41] Because PAHs are lipophilic, they penetrate cellular membranes (including the placenta) and form covalent bonds with DNA, which may result in teratogenic mutations.[42] Intrauterine exposure to PAHs has been linked to gastroschisis, cleft lip and cleft palate, and neural tube defects.[43-45] High exposure is associated with low IQ at age 3, increased behavioral problems ages 6–8, and childhood asthma.[46-48]

Recreational Drugs

Alcohol

Fetal alcohol spectrum disorder (FASD) encompasses a characteristic pattern of adverse effects associated with prenatal exposure to alcohol. Features of FASD include prenatal and/or postnatal growth deficits, neurologic abnormalities, cognitive and behavioral deficits, and a range of birth defects.[49] The more severe end of the spectrum is thought to occur in the US and Western European populations at a rate of about 1 to 7 per 1,000 live births.[50,51] However, as many as 5 in 100 children worldwide are thought to have some effects that fall within the FASD spectrum, though the rates are 10–40 times higher in at-risk populations.[452,53] In addition, craniofacial, eye, heart, gastrointestinal, and kidney malformations occur more commonly following prenatal alcohol exposure, and there is an increased risk for spontaneous abortion, stillbirth, SIDS, and preterm delivery in alcohol-exposed pregnancies.[54,55]

The women at highest risk of delivering babies with FASD appear to be those who have already had an affected child and who continue to consume alcohol in subsequent pregnancies, particularly in a heavy episodic manner. Lower levels of alcohol exposure have been associated with less severe neurobehavioral outcomes and growth effects, though lower levels of exposure may play a larger role during critical periods of fetal development. These effects may be modified by maternal age, nutrition, genetic factors, and length of exposure in gestation.[56]

Because there is no known threshold dose, the Surgeon General of the US Public Health Service recommends that the safest course is to avoid alcohol entirely during pregnancy, and any reduction in exposure is thought to be beneficial.

Tobacco

Specific structural birth defects, including oral clefts, congenital heart disease, gastroschisis, musculoskeletal defects, and craniosynostosis have been linked to maternal tobacco use in a variety of studies. In addition, there is strong evidence that maternal tobacco use is associated with increased risks for spontaneous abortion, placental complications that may lead to stillbirth, preterm delivery, reduced birth weight, and SIDS.[57,58] Some studies have demonstrated gene–environment interaction for these adverse outcomes, and, for some outcomes, there is a clear dose gradient, also reflected in exposures due to secondhand smoke.[59] Conversely, evidence indicates that any reduction in the number of cigarettes smoked may reduce risk for low birth weight, preterm birth, and placental complications.[60]

Global Climate Change

Global climate changes lead to increased pollution, severe storms, flooding, wildfires, extreme heat, and drought, all of which lead to political instability. These effects in turn lead to secondary effects, such as resource scarcity, disruption of health services, failed crops, and malnutrition. The food chain is particularly vulnerable, which results in a lack of biodiversity, increased toxic metals and chemicals in food sources, increased pesticide use, increased dangerous microbes, and increased microplastic contamination. These secondary effects are thought to increase birth defects through increased maternal exposure to these contaminants in food and water and by increasing geographical distribution of disease vectors. One of the most obvious effects of climate change is the increase in average temperatures across the world. Fevers in the first trimester have long been known to be associated with congenital heart disease.[61] New data suggest that increased environmental temperatures, especially in the spring and summer, are associated with congenital heart defects as well.[62] Increased rates of hypospadias, neural tube defects, and gastroschisis also have been associated with higher environmental temperatures.[63] Related to increased temperature are wildfires, which are known to be associated with toxic air pollution and a more than two times greater risk of gastroschisis in the 30 days before pregnancy and a 28% higher risk with exposure in the first trimester.[64]

Environmental changes also have expanded areas and/or time frames in which mosquito-borne diseases, such as dengue fever and malaria, are spread.[65] Infection with either during pregnancy is associated with prematurity, being small for gestational age, and low birth weight.[66,67] One notorious example is the 2015 El Niño weather pattern, which "caused exceptional climatic conditions in north-eastern South America during winter and spring in the Southern Hemisphere." The southern El Niño weather oscillation is made more pronounced by warmer water temperatures, which in 2015–2016 led to the expansive Zika virus outbreak in Latin America.[68] Zika virus has since been linked to microcephaly, neurodevelopmental abnormalities, eye abnormalities, and hearing loss.[69] While this is only one specific example, modeling indicates that, as global temperatures increase, vector-borne illnesses are likely to spread.[70]

Prevention of Birth Defects

Several opportunities and strategies have been developed and used by primary care practitioners and public health agencies for prevention of environmentally induced birth defects and other developmental problems in children.

Clear communication to women in advance of pregnancy about the importance of avoiding recreational drugs and therapeutic agents that are known teratogens is a critically important strategy for preventing birth defects. Clear communication is also critical for informing women about the importance of eating a healthy, balanced diet and of taking vitamins, especially folic acid, that will reduce risk of birth defects. This type of prevention effort relies to a great extent on improving the rate of planned pregnancy from the current rate of about 50% and on improvements in the delivery of preconception counseling.

Better government policy decisions about acceptable levels of certain chemicals and about the conditions under which pregnant women should avoid or reduce exposure, especially regarding chemical exposures in the workplace and in the general environment, represent a second opportunity for prevention of birth defects. However, advances in prevention policy rely on advances in knowledge about which exposures are harmful and at what levels of exposure. Current epidemiologic studies can add incrementally to this knowledge base, but a quantum increase in knowledge about the impacts of chemicals in the modern environment on the risk of birth defects and other developmental problems in children will require large-scale efforts with individual level sampling, large sample sizes, careful phenotyping, and long-term follow-up.

References

1. Lobo I, Zhaurova K. Birth defects: causes and statistics. *Nature Educ.* 2008;1(1):18.

2. Haas DM, Marsh DJ, Dang DT, et al. Prescription and other medication use in pregnancy. *Obstet Gynecol.* 2018 May;131(5):789–98.

3. Wilson JG, Fraser FC. *Handbook of Teratology. Vol. 1: General Principles and Etiology.* New York, Plenum Press; 1977: 49–62.

4. Schaefer C, Hanneman D, Meister R, et al. Vitamin K antagonists and pregnancy outcome: a multi-centre prospective study. *Thromb Haemost.* 2006;95:949–57.

5. Floyd RL, O'Connor MJ, Sokol RJ, et al. Recognition and prevention of fetal alcohol syndrome. *Obstet Gynecol.* 2005;106:1059–64.

6. Beyer-Westendorf J, Tittl L, Bistervels I, et al. Safety of direct oral anticoagulant exposure during pregnancy: a retrospective cohort study. *Lancet Haematol.* 2020;7(12):e884–91.

7. van Driel D, Wesseling J, Sauer PJ, et al. Teratogen update: fetal effects after in utero exposure to coumarins overview of cases, follow-up findings, and pathogenesis. *Teratology.* 2002;66:127–40.

8. van Driel D, Wesseling J, Sauer PJ, et al. In utero exposure to coumarins and cognition at 8–14 years. *Pediatrics.* 2001;107:123–9.

9. Jentink J, Loane MA, Dolk H, et al. Valproic acid monotherapy in pregnancy and major congenital malformations. *N Engl J Med.* 2010:362:2185–93.

10. Adab N, Jacoby A, Smith D, et al. Additional educational needs in children born to mother with epilepsy. *J Neurol Neurosurg Psychiatry.* 2001;70:15–21.

11. Bateman BT, Patorno E, Desai R, et al. Angiotensin-converting enzyme inhibitors and the risk of congenital malformations. *Obstet Gynecol.* 2017;Jan 129(1):174–84.

12. Quan A. Fetopathy associated with exposure to angiotensin converting enzyme inhibitors and angiotensin receptor antagonists. *Early Hum Develop.* 2006;82(1):23–8.

13. Lammer EJ, Chen DT, Hoar RM, et al. Retinoic acid embryopathy. *N Engl J Med.* 1985; 313:837–41.

14. Suuberg A. Psychiatric and developmental effects of isotretinoin (retinoid) treatment for acne vulgaris. *Curr Ther Res Clin Exp.* 2019;90:27–31.

15. Adams J, Lammer EJ. Neurobehavioral teratology of isotretinoin. *Reprod Toxicol.* 1993;7:175–7.

16. Brent RJ. The effect of embryonic and fetal exposure to x-ray, microwaves, and ultrasound: counseling the pregnant and nonpregnant patient about these risks. *Semin Oncol.* 1989;16:347–68.

17. Brent RL. Utilization of developmental basic science principles in the evaluation of reproductive risks from pre- and postconception environmental radiation exposures. *Teratology.* 1999;59(4):182–204.

18. Wit F, Vroonland CJJ, Bijwaard H. Prenatal X-ray exposure and the risk of developing pediatric cancer: a systematic review of risk markers and a comparison of international guidelines. *Health Phys.* 2021;121(3):225–33.

19. Sakamoto M, Tatsuta N, Izumo K, et al. Health Impacts and biomarkers of prenatal exposure to methylmercury: lessons from Minamata, Japan. *Toxics.* 2018;6(3):45.

20. Matsumoto H, Koya G, Takeuchi T. Fetal Minamata disease: a neuropathological study of two cases of intrauterine intoxication by a methylmercury compound. *J Neuropath Exp Neurol.* 1965;24:563–74.

21. Karagas MR, Choi AL, Oken E, et al. Evidence of the human effects of low-level methylmercury exposure. *Environ Health Perspect.* 2012;120(6):799–806.

22. Klebanoff MA, Harper M, Lai Y, et al. Fish consumption, erythrocyte fatty acids, and preterm birth. *Obstet Gynecol.* 2011;117:1071–7.

23. US EPA. *Advice About Eating Fish: For Those Who Might Become or Are Pregnant or Breastfeeding and Children Ages 1–11 Years.* Washington, DC: U.S. Environmental Protection Agency and U.S. Food and Drug Administration. 2021. https://www.fda.gov/food/consumers/advice-about-eating-fish.

24. Fruh V, Rifas-Shiman SL, Amarasiriwardena C, et al. Prenatal lead exposure and childhood executive function and behavioral difficulties in project viva. *Neurotoxicology.* 2019;75:1056–115.

25. WHO. Lead poisoning. World Health Organization. 2021. Available at: https://www.who.int/news-room/fact-sheets/detail/lead-poisoning-and-health.

26. Hertz-Picciotto I, Schramm M, Watt-Morse M, et al. Patterns and determinants of blood lead during pregnancy. *Am J Epidemiol.* 2000;152:829–37.

27. Rignell-Hydbom A, Lindh CH, Dillner J, et al. A nested case-control study of intrauterine exposure to persistent organochlorine pollutants and the risk of hypospadias. *PLoS One.* 2012;7:e44767.

28. Rocheleau CM, Romitti PA, Sanderson WT, et al. Maternal occupational pesticide exposure and risk of hypospadias in the National Birth Defects Prevention Study. *Birth Defects Res A Clin Mol Teratol.* 2011;91:927–36.

29. Bonde JP, Flachs EM, Rimborg S, et al. The epidemiologic evidence linking prenatal and postnatal exposure to endocrine disrupting chemicals with male reproductive disorders: a systematic review and meta-analysis. *Hum Reprod Update.* 2016;23(1):104–25.

30. Yuan X, Liang F, Zhu J, et al. Maternal Exposure to $PM_{2.5}$ and the Risk of Congenital Heart Defects in 1.4 Million Births: A Nationwide Surveillance-Based Study. *Circulation.* 2023;147(7):565–74. doi:10.1161/CIRCULATIONAHA.122.061245.

31. Wan X, Wei S, Wang Y, et al. The association between maternal air pollution exposure and the incidence of congenital heart diseases in children: a systematic review and meta-analysis. *Sci Total Environ.* 2023;892:164431. doi:10.1016/j.scitotenv.2023.164431.

32. Buteau S, Veira P, Bilodeau-Bertrand M, Auger N. Association between First Trimester Exposure to Ambient $PM_{2.5}$ and NO_2 and Congenital Heart Defects: a Population-Based Cohort Study of 1,342,198 Live Births in Canada. *Environ Health Perspect.* 2023;131(6):67009. doi:10.1289/EHP11120.

33. Strickland MJ, Klein M, Correa A, et al. Ambient air pollution and cardiovascular malformations in Atlanta, Georgia, 1986–2003. *Am J Epidemiol.* 2009;169:1004–14.

34. Vrijheid M, Martinez D, Manzanares S, et al. Ambient air pollution and risk of congenital anomalies: a systematic review and meta-analysis. *Environ Health Perspect.* 2011;119:598–606.

35. Woodruff TJ, Darrow LA, Parker JD. Air pollution and postneonatal infant mortality in the United States, 1999–2002. *Environ Health Perspect.* 2008;116(1):110–5.

36. Veras MM, de Oliveira Alves N, Fajersztajn L, Saldiva P. Before the first breath: prenatal exposures to air pollution and lung development. *Cell Tissue Res.* 2017;367(3):445–55.

37. Frye RE, Cakir J, Rose S, et al. Prenatal air pollution influences neurodevelopment and behavior in autism spectrum disorder by modulating mitochondrial physiology. *Mol Psychiatry.* 2021;26(5):1561–77.

38. Peden DB. Prenatal exposure to particulate matter air pollution: a preventable risk for childhood asthma. *J Allergy Clin Immunol.* 2021;148(3):716–18.

39. Sjaastad AK, Svendsen K. Exposure to polycyclic hydrocarbons (PAHs), mutagentic aldehydes, and particulate matter in Norwegian al la carte restaurants. *Ann Occup Hyg.* 2009;53(7):723–9.

40. Choi H, Rauh V, Garfinkel R, et al. Prenatal exposure to airborne polycyclic aromatic hydrocarbons and risk of intrauterine growth restriction. *Environ Health Perspect.* 2008;116(5):658–65.

41. Perera FP, Rauh V, Whyatt RM, et al. Effect of prenatal exposure to airborne polycyclic aromatic hydrocarbons on neurodevelopment in the first 3 years of life among inner-city children. *Environ Health Perspect.* 2006;114(8):1287–92.

42. Blaszczyk E, Mielzynska-Svach D. Polycyclic aromatic hydrocarbons and PAH-related DNA adducts. *J Appl Genet.* 2017;58(3):321–30.

43. Lupo PJ, Langlois PH, Reefhuis J, et al. Maternal occupational exposure to polycyclic aromatic hydrocarbons: effects on gastroschisis among offspring in the national birth defects prevention study. *Environ Health Perspect.* 2012;120(6):910–5.

44. Langlois PH, Hoyt AT, Lupo PJ, et al. Maternal occupational exposure to aromatic hydrocarbons and risk of oral cleft-affected pregnancies. *Cleft Palate Craniofac J.* 2013;50(3):337–46.

45. Yuan Y, Jin L, Wang L, et al. Levels of PAH-DNA adducts in placental tissue and the risk of fetal neural tube defects in a Chinese population. *Reprod Toxicol.* 2013:37;70–5.

46. Edwards SC, Jedrychowski W, Butscher M, et al. Prenatal exposure to airborne polycyclic aromatic hydrocarbons and children's intelligence at 5 years of age in a prospective cohort study in Poland. *Environ Health Perspect.* 2010;118(9):1326–31.

47. Perera FP, Tang D, Wang S, et al. Prenatal polycyclic aromatic hydrocarbon (PAH) esposure and child behavior at age 6–7 years. *Environ Health Perspect.* 2012:120(6):921–6.

48. Perera F, Herbstman J. Prenatal environmental exposures, epigenetics, and disease. *Reprod Toxicol.* 2011:31(3):363–73.

49. Wozniak JR, Riley EP, Charness ME. Diagnosis, epidemiology, assessment, pathophysiology, and management of fetal alcohol spectrum disorders. *Lancet Neurol.* 2019;18(8):760–70.

50. Demiguel V, Laporal S, Quatremere G, et al. The frequency of severe fetal alcohol spectrum disorders in the neonatal period using data from the French hospital discharge database between 2006 and 2014. *Drug Alcohol Depend.* 2021;225:108748.

51. May PA, Fiorentino D, Coriale G, et al. Prevalence of children with severe fetal alcohol spectrum disorders in communities near Rome, Italy: new estimated rates are higher than previous estimates. *Int J Environ Res Public Health.* 2011;8(6):2331–51.

52. Lange S, Probst C, Gmel G, et al. Global prevalence of fetal alcohol spectrum disorder among children and youth: a systematic review and meta-analysis. *JAMA Pediatr.* 2017;171(10):948–56.

53. Popova S, Lange A, Shield KK, et al. Prevalence of fetal alcohol spectrum disorder among special subpopulations: a systematic review and meta-analysis. *Addiction.* 2019;114(7):1150–72.

54. Caputo C, Wood E, Jabbour L. Impact of fetal alcohol exposure on body systems: a systematic review. *Birth Defects Res C Embryo Today.* 2016;108(2):174–80.

55. Sampson PD, Streissguth AP, Bookstein FL, et al. Incidence of fetal alcohol syndrome and prevalence of alcohol-related neurodevelopmental disorder. *Teratology.* 1997;56:317–26.

56. Jacobson SW, Jacobson JL, Sokol RJ, et al. Maternal age, alcohol abuse history, and quality of parenting as moderators of the effects of prenatal alcohol exposure on 7.5-year intellectual function. *Alcohol Clin Exp Res.* 2004;28:1732–45.

57. Mei-Dan, E, Walfisch A, Weisz B, et al. The unborn smoker: association between smoking during pregnancy and adverse perinatal outcomes. *J Perinat Med.* 2015;43(5):553–8.

58. Hackshaw A, Rodeck C, Boniface S. Maternal smoking in pregnancy and birth defects: a systematic review based on 173 687 malformed cases and 11.7 million controls. *Hum Reprod Update.* 2011;17(5):589–604.

59. Shi M, Christensen K, Weinberg CR, et al. Orofacial cleft risk is increased with maternal smoking and specific detoxification-gene variants. *Am J Hum Genet.* 2007;80:76–90.

60. Wang X, Zuckerman B, Pearson C, et al. Maternal cigarette smoking, metabolic gene polymorphism, and infant birth weight. *JAMA.* 2002;287:195–202.

61. Shi QY, Zhang JB, Mi YQ, et al. Congenital heart defects and maternal fever: systematic review and meta-analysis. *J Perinatol.* 2014;34:677–82.

62. Auger N, Fraser WD, Sauve R, et al. Risk of congenital heart defects after ambient heat exposure early in pregnancy. *Environ Health Perspect.* 2017;125(1):8–14

63. Mosalman Haghighi M, Wright CY, Ayer J, et al. Impacts of high environmental temperatures on congenital anomalies: a systemic review. *Int J Environ Res Public Health.* 2021;18(9):4910.

64. Park BY, Boles I, Monavvari S, et al. The association between wildfire exposure in pregnancy and foetal gastroschisis: a population-based cohort study. *Paediatr Perinat Epidemiol.* 2022;36(1):45–53.

65. Watts N, Amann M, Arnell N, et al. The 2019 report of The Lancet Countdown on health and climate change: ensuring that the health of a child born today is not defines by a changing climate. *Lancet.* 2019;394(10211):16–22.

66. Rijken MJ, De Livera AM, Lee SJ, et al. Quantifying low birth weight, preterm birth, and small-for-gestational-age effects of malaria in pregnancy: a population cohort study. *PLoS One.* 2014;9(7):e1000247.

67. Paixao ES, Campbell OM, Teixeira MG, et al. Dengue during pregnancy and live birth outcomes: a cohort of linked data from Brazil. *BMJ Open.* 2019;9(7):e023529.

68. Paz S, Semenza JC. El Niño and climate change: contributing factors in the dispersal of Zika virus in the Americas? *Lancet.* 2016;387(10020):745.

69. Benavides-Lara A, de la Paz Barboza-Arguello M, Gonzales-Elizondo M, et al. Zika virus-associated birth defects, Costa Rica, 2016–2018. *Emerg Infect Dis.* 2021;27(2)360–71.

70. Caminade C, McIntyre KM, Jones AE. Impact of recent and future climate change on vector-borne disease. *Ann N Y Acad Sci.* 2019;1436(1):157–73.

51

The Environment and Cardiovascular Disease in Children

Philip J. Landrigan

Cardiovascular diseases (CVD) are the world's leading cause of morbidity and mortality. CVD were responsible in 2019 for an estimated 18.6 million deaths globally and for 957,000 deaths in the United States.[1,2]

Environmental exposures are important but often overlooked causes of CVD.[3] Diseases of the heart and cardiovascular system that are linked to hazardous exposures in the environment include acute myocardial infarction (MI), stroke, arrhythmias, peripheral vascular injury, cardiomyopathy, and sudden death. Three CVD risk factors—hypertension, diabetes, and dyslipidemia—also are linked to toxic environmental exposures. Limited evidence suggests that some birth defects of the heart and great vessels may be linked to environmental exposures.

Most CVD occurs in adults. However, toxic environmental exposures also can cause CVD in infants and children. Additionally, it is now understood that environmental exposures in early life can initiate multistage pathophysiologic cascades that increase CVD risk across the life span and result ultimately in disease and premature death.[4] For this reason, early intervention strategies that begin in childhood and aim to reduce risk of obesity, type 2 diabetes, dyslipidemia, and hypertension are of great importance.

Cardiotoxic environmental exposures reviewed in this chapter include air pollution[3,5]; carbon monoxide; toxic metals[6]; halogenated hydrocarbons, including dioxins, polychlorinated biphenyls (PCBs) and per- and polyfluoroalkylated substances (PFAS); organophosphate insecticides; and the plastics chemicals, phthalates, and bisphenol A.[7,8] This chapter also examines the cardiovascular impacts of noise and climate change.

Likelihood is high that, beyond these known hazards, there are other exposures in the modern environment whose toxicity to the heart and cardiovascular system has not yet been recognized.[9] Those still undiscovered cardiotoxicants will be found hidden in plain sight among the tens of thousands of new synthetic chemicals that have been invented in the past half century.[10] These chemicals are used widely in consumer products and result in extensive human exposure, but, because of failure of stewardship by the chemical industry as well as by governments, few have ever been tested for safety or toxicity and thus their potential dangers to health are not yet known.[9]

Air Pollution

A strong and rapidly expanding body of research, especially large, prospective clinical and epidemiologic studies that combine state-of-the-art assessments of pollution

Philip J. Landrigan, *The Environment and Cardiovascular Disease in Children* In: *Textbook of Children's Environmental Health,* Second Edition. Edited by: Ruth A. Etzel and Philip J. Landrigan, Oxford University Press. © Oxford University Press 2024. DOI: 10.1093/oso/9780197662526.003.0051

exposure with sophisticated markers of CVD, has substantially advanced knowledge of the effects of air pollution on health and disease.[5,11]

These studies consistently find that fine particulate matter ($PM_{2.5}$) air pollution is closely linked to increased CVD prevalence and mortality.[12-16] The exposure-response relationship between $PM_{2.5}$ pollution and CVD extends down to $PM_{2.5}$ concentrations well below the current US National Ambient Air Quality Standard of 12 $\mu g/m^3$.[17,18]

Short-term peaks in $PM_{2.5}$ levels (hours to days) also are associated with increased risk for myocardial infarction, stroke, and CVD death[5,19] as well as with increased risk for atrial fibrillation and ventricular arrhythmias.[5,20]

In addition to its association with CVD, $PM_{2.5}$ air pollution increases the incidence and prevalence of multiple CVD risk factors, most notably hypertension and diabetes.[5] Both short- and long-term $PM_{2.5}$ exposures are associated with increased incidence of new-onset hypertension[21] as well as with increased carotid intima media thickness, coronary artery calcification, abdominal aortic calcification, increased atherosclerotic plaque vulnerability, and left ventricular hypertrophy.[22] Evidence from clinical, epidemiological, and experimental studies supports links between $PM_{2.5}$ exposure, insulin resistance, and type 2 diabetes. These associations extend down to $PM_{2.5}$ concentrations below 5 $\mu g/m^3$.[1,5,23]

The mechanisms through which $PM_{2.5}$ air pollution produces CVD are complex, and they proceed through multiple stages over many years, even beginning in childhood.[3,4] The most common first step appears to be the induction by toxic particulates of oxidative stress and inflammation. Chronic oxidative stress then triggers a series of mechanisms that include the activation of rapid neural pathways and the release of biologically active intermediates such as inflammatory cytokines, oxidized lipids, immune cells, microparticles, and microRNA. Over time, the cumulative impacts of these developments result in increased vascular tone, hypertension, insulin resistance, and type 2 diabetes, all of which increase risk for endothelial damage, plaque formation, and eventually acute or chronic CVD. See Box 51.1.

Box 51.1 Air Pollution and Congenital Heart Disease

Prenatal exposures to ambient air pollution have been linked to congenital heart defects in some studies, but the association has been inconsistently observed and the exposure-response relationship does not appear to be linear.[24-27] (See Chapter 50) Possible explanations for these inconsistencies include a biologically weak association; inadequate measurement of air pollution exposures during pregnancy since exposures have generally not been measured at the individual level; and the possibility that cardiac abnormalities caused by higher levels of air pollution are not consistent with life and thus end in miscarriage.[28] Prospective epidemiologic studies that enroll women during pregnancy and use state-of-the art technologies to measure individual-level air pollution exposures during pregnancy may help to resolve some of these difficulties.[29]

Carbon Monoxide

Carbon monoxide (CO) is a colorless, odorless, highly toxic, gaseous air pollutant produced by the incomplete combustion of hydrocarbons. Automotive exhaust is the major environmental source. Other sources include improperly adjusted gas heaters, old appliances, wall ovens, stoves, tobacco smoking, and charcoal grilles. CO exerts its toxicity by binding strongly to hemoglobin to block oxygen transport to tissues. The heart and brain are especially vulnerable. At high doses, CO is acutely toxic, and inhalation exposure can be rapidly fatal.

Initial recognition of the relationship between chronic CO exposure and heart disease came from studies of cigarette smokers who had stopped smoking. Although the incidence of nonmalignant lung disease and lung cancer took years to return to baseline following smoking cessation, the incidence of acute cardiac events decreased within 24 hours, a time course consistent with acute reduction in CO exposure.[30] The relationship between CO exposure and heart disease is further substantiated by studies showing that daily variation in ambient CO pollution levels is associated with concomitant variation in hospital admissions for ischemic heart disease and heart failure.[31]

Metals

Lead, cadmium, mercury, arsenic, cobalt, and thallium have all been documented to cause disease and dysfunction of the heart and cardiovascular system.[32]

Lead

Although lead is best known as a neurotoxicant, it also has a long history of causing CVD. Increased prevalence of hypertension and stroke was reported more than a century ago among workers with poorly controlled, high-dose occupational exposures to lead.[33] Recent large-scale epidemiologic studies in the general US population have confirmed a relationship between lead and hypertension even at very low blood lead levels and have concluded that the association is causal.[34,35] The hypertensive effects of lead have been confirmed experimentally.[31] Additional positive associations have been identified between lead and coronary heart disease, stroke, alterations in cardiac rhythm, and peripheral arterial disease.[36]

Until recently, lead was thought to have little cardiotoxicity in adults at blood levels below 40 µg/dL. However, data from a long-term follow-up study of 14,289 American adults in the Third National Health and Nutrition Examination Survey (NHANES3) with a mean blood lead concentration of 2.71 µg/dL suggest that a relationship between lead and CVD mortality extends down to blood lead concentrations below 3 µg/dL.[37] Based on these findings, the number of CVD deaths in the United States could be as much as 10 times greater than current estimates.[1,2] Lead is discussed in detail in Chapter 31. See also Box 51.2.

Box 51.2 Lead and Congenital Heart Disease

Prenatal exposure to lead has been associated with cardiac birth defects in a limited number of studies. A study in an area of Italy polluted with lead from ceramic factories documented an elevated incidence of cardiovascular defects.[38] An investigation of paternal employment in a US microelectronics/business machine manufacturing facility found that lead exposure was associated with increased risk of ventricular septal defect, with an odds ratio (OR) of 2.7 (95% confidence interval [CI]: 1.09, 6.67).[39]

Cadmium

Tobacco smoke, occupational exposures, and certain green leafy vegetables including lettuce, spinach, cabbage, carrots, beets, eggplant, and peas are the major sources of cadmium exposure. A positive dose-response association between cadmium levels and increased blood pressure has been found in the general US population.[40] Increased risk of coronary artery disease, peripheral arterial disease, and stroke, as well as increased CVD mortality, have been noted in epidemiologic studies of exposed populations.[32] Cadmium has been found to be an independent risk factor for atherosclerosis. A recent systematic review found a dose-dependent increased risk for all CVD endpoints except stroke, even down to very low exposures, defined as urine cadmium levels below 0.5 µg/g creatinine.[40]

Mercury

Methylmercury is the form of mercury most highly toxic to humans. Methylmercury is created when microscopic airborne particles of metallic mercury released to the atmosphere by industrial sources such as coal-burning power plants precipitate into lakes, rivers, and oceans and are transformed by marine microorganisms. Methylmercury is lipophilic and highly persistent in the environment. It bioaccumulates to reach particularly high levels in predatory fish at the top of the aquatic food chain. Consumption of contaminated fish is the major route of human exposure to methylmercury.

Methylmercury is a potent developmental neurotoxicant[41] and also a cardiovascular toxicant.[32] It is associated with disturbances in cardiac rhythm, specifically decreased heart rate variability, hypertension, increased carotid arterial intima-media thickness, accelerated progression of carotid atherosclerosis, increased risk of myocardial infarction, and increased risk of coronary and cardiovascular death.[42,43]

Epidemiologic studies of populations chronically exposed to methylmercury through heavy consumption of fish and marine mammals have had to disentangle the adverse effects of methylmercury from the potentially beneficial effects of omega-3 fatty acids.[44] Experimental studies corroborate the cardiovascular toxicity of methylmercury and suggest that methylmercury may contribute to progression of CVD by causing oxidative stress, enhancing coagulation, and activating inflammatory mediators.[45,46] Mercury is discussed in detail in Chapter 32.

Arsenic

Drinking water is the principal source of nonoccupational human exposure to arsenic, and major problems of arsenic in drinking water are found in Southeast Asia, Taiwan, Chile, Argentina, northern New England, and the American Southwest. High concentrations of arsenic also are released into the environment by polluting industries, and inhalation can be an important exposure route for people living near such industries, as well as for workers exposed occupationally.

Arsenic exposure is strongly associated with increased risk of CVD.[32,47] Positive dose-response relationships have been documented between chronic arsenic exposure and carotid atherosclerosis, hypertension, and ischemic heart disease. Arsenic exposure also has been linked to increased risk of diabetes.[48]

Chronic arsenic exposure is strongly associated with peripheral vascular disease. The severity appears related to cumulative dose and is greatest when exposure begins in utero or in early childhood. The most severe cases can progress to endarteritis obliterans with frank gangrene of the extremities—"black foot disease".[49] Arsenic is discussed in detail in Chapter 33; see also Box 51.3.

Cobalt

Cobalt is an essential trace element necessary for the formation of vitamin B_{12}. Excessive exposure to cobalt has been linked to cardiac disease. In 1966, a syndrome labeled "beer drinker's cardiomyopathy" was recognized among heavy beer drinkers in Quebec City, Canada, and was characterized by pericardial effusion, elevated hemoglobin concentrations, and congestive heart failure. The appearance of the syndrome coincided temporally with addition of cobalt to beer.[52] A similar cardiomyopathy has been reported in other groups chronically exposed to cobalt, among them beer-drinking populations and workers producing "hard metal", an alloy containing cobalt.[53]

Thallium

Thallium is a highly toxic metallic element. It has been used as a rodenticide, although this use has not been permitted in the United States since 1972. It can be absorbed orally

---

Box 51.3 Arsenic and Congenital Heart Disease

Prenatal exposure to arsenic via maternal consumption of contaminated drinking water has been associated with congenital heart disease in several studies.[28] An early study reported an association of arsenic exposure with coarctation of the aorta, with OR 3.4 (95% CI: 1.3, 8.9)[50] A more recent, large, case-control study found an association between an arsenic level in drinking water of >10 μg/L and total congenital heart disease incidence with an OR of 1.41 (95% CI: 1.28, 1.56) and atrial septal defects with an OR of 1.79 (95% CI: 1.59, 2.01)[51].

and transdermally. It has neither an odor nor a taste and has been used in poisonings and assassinations.[55]

Symptoms of acute intoxication include gastrointestinal symptoms, polyneuropathy, and dermatologic changes. Alopecia is a hallmark of thallium poisoning and typically develops 3–4 weeks after exposure.[55] Cardiovascular manifestations of acute thallium poisoning are hypotension and bradycardia, apparently secondary to direct toxic effects of thallium on the sinus node and myocardium. Diagnosis is made by toxicological screen.

Halogenated Hydrocarbons

The halogenated hydrocarbons are a large and diverse family of synthetic chemicals that have as their common feature a chemical bond between a carbon atom and one or more halogen atoms (chlorine, bromine, or fluorine). The carbon-halogen bond is very strong. Consequently, many halogenated hydrocarbons are extremely persistent in the human body and in the environment. Halogenated hydrocarbons are discussed in detail in Chapter 36.

Halogenated hydrocarbons have been used in a wide range of products from insecticides (DDT), electrical insulation (PCBs) to flame retardants (polybrominated diphenyl ethers [PBDEs]) and water repellents (PFAS).[56] Some, such as dioxins and furans, are produced as inadvertent by-products in chemical manufacture or released to the environment through combustion of materials containing chlorine. Many of the halogenated hydrocarbons are legacy chemicals such as DDT and PCBs released into the environment decades ago, while others such as PFAS are gaining in significance. Production of many halogenated hydrocarbons has been banned under the Stockholm Convention on Persistent Organic Pollutants.[57] Consumption of contaminated meat and fish is the main route of human exposure.

Because of their lipophilic nature, halogenated hydrocarbons readily cross the placenta and the blood–brain barrier. They concentrate and may persist for years in adipose tissues. Depending on their chemical composition and structure, halogenated hydrocarbons have multiple and varied toxicities, among them neurotoxicity, carcinogenicity, and ability to disrupt endocrine function. A number of halogenated hydrocarbons are toxic to the heart and cardiovascular system.

Box 51.4 Halogenated Solvents and Congenital Heart Disease

Because maternal consumption of ethyl alcohol during pregnancy is associated with a range of birth defects, studies have investigated the possible association of solvent exposure with birth defects. The National Birth Defects Prevention Study, which assessed the impact of solvent exposure from 1 month before conception to the end of the first trimester, found a significant association between exposure to any chlorinated solvent in the first trimester and ventricular septal defect, with an OR of 1.7 (95% CI: 1.0, 2.8). In the same study, an alternative approach to assessing solvent exposure found strong associations between solvent exposure and aortic stenosis, transposition of the great arteries, right ventricular outflow tract obstruction, and pulmonary valve stenosis, with some ORs exceeding 2.0.[59]

Halogenated Solvents

High-level, brief exposures to halogenated solvents (e.g., trichloroethylene 1,1,1-trichloroethane) have been recognized for many decades to be associated with acute death, caused apparently by sudden cardiac arrhythmias. High-dose exposures to these compounds may occur in occupational settings and also among persons who "sniff" solvents as a means of intoxication. The suspected mechanism of toxicity appears to involve disruption of potassium, calcium, and sodium channels (see Box 51.4).[58]

Methylene Chloride

The halogenated solvent dichloromethane (methylene chloride) is uniquely toxic to the heart and cardiovascular system because its metabolism produces significant amounts of CO.[60] Epidemiologic studies of chemical workers occupationally exposed to dichloromethane have observed excess mortality from heart disease.[61]

Dioxins

Dioxins are a family of chlorinated hydrocarbon compounds with a common cyclic chemical structure.[62] Dioxins are highly persistent in humans and the environment. They are lipophilic and bioaccumulative. Exposures to dioxin, and specifically to the highly toxic dioxin congener 2,3,7,8-tetracloro-p-dibenzodioxin (TCDD), have occurred among workers in the chemical industry, particularly workers producing herbicides,[63] and in military personnel in the Vietnam War who applied dioxin-contaminated herbicide (Agent Orange). Exposure in the general population results most commonly from consumption of foods, especially meats, in which TCDD has accumulated. Dioxins are discussed in detail in Chapter 36.

TCDD has been linked in multiple studies to increased prevalence of CVD risk factors, specifically increased insulin resistance and type 2 diabetes.[64,65] TCDD was associated with increased mortality from diabetes among community residents in Seveso, Italy, exposed after an industrial explosion.[66] TCDD and other persistent organic pollutants are associated with hypertension in the general population[67] and with increased risk of obesity and the metabolic syndrome.[68]

A systematic review of 12 epidemiologic studies (10 of them in occupationally exposed populations) found a consistently positive, dose-related, statistically significant association between high-level TCDD exposure and mortality from ischemic heart disease as well as a more modest association with mortality from all forms of CVD.[69]

Polychlorinated Biphenyls

The most highly cardiotoxic PCBs are congeners closely resembling TCDD (e.g., PCB-126). Dioxin-like PCBs have been linked to cardiovascular risk factors, specifically to components of the metabolic syndrome, high blood pressure, elevated triglycerides, and glucose intolerance[67] as well as to increased risk for obesity and diabetes.[68] PCBs are discussed in detail in Chapter 36.

Epidemiologic studies of industrial workers exposed occupationally to PCBs find excess mortality from CVD that increases with the duration of employment.[70] Studies of Native American populations heavily exposed to PCBs through a traditional diet high in PCB-contaminated fish find associations of serum PCB levels with elevated serum lipids and CVD risk.[71]

Polybrominated Diphenyl Ethers

The PBDEs are a class of bromine-based synthetics used as flame retardants in polymers and textiles.[72] They are incorporated into numerous consumer products, including computers, television sets, mobile phones, electronics and electrical items, polyurethane foam mattresses, carpets, upholstered furniture, and draperies. Their use has increased substantially in the past decade, and PBDEs have become major environmental pollutants. PBDEs are lipophilic, accumulate in adipose tissue, and are biologically persistent. PBDEs are discussed in detail in Chapter 36.

PBDEs are not firmly bound to the materials into which they are incorporated and consequently are released from consumer products to contaminate the home environment. Exposures to PBDEs in house dust are estimated to account for more than 80% of current US intake.[30] Because of their normal, age-appropriate, hand-to-mouth behavior children are at particularly high risk of PBDE exposure from house dust.[74]

PBDEs are endocrine disruptors, and they disrupt thyroid function.[75] PBDE levels are associated with the metabolic syndrome and with diabetes at exposure levels currently seen in the general US population.[76] A recent report suggests an association between PBDE exposure and children's cardiovascular responses to stress.[77]

Per- and Polyfluoroalkyl Substances

The PFAS—perfluorooctanyl sulfonate (PFOS), perfluorooctanoic acid (PFOA), and their congeners—have been used extensively in the production of fluoropolymers. Members of this class of synthetic chemicals have powerful surfactant and water-repelling properties and have been used in textiles, carpets, upholstery, food packaging, lubricants, electronics, and frying pans.[78] PFAS are discussed in detail in Chapter 37.

PFAS act as endocrine disruptors and appear to increase CVD risk by increasing rates of obesity, dyslipidemia, and type 2 diabetes.[79] A cross-sectional study undertaken in Denmark found that increased levels of perfluorinated compounds in overweight 8- to 10-year-old children were associated with elevated insulin and triglyceride concentrations.[80] A cohort mortality study of US workers occupationally exposed to PFAS found borderline significant increases in mortality from diabetes and cerebrovascular disease (standardized mortality ratio, 1.8).[81]

Plastic-Associated Chemicals

Plastic-associated chemicals include bisphenol A (BPA) and phthalates. These materials are used in the manufacture of an enormous array of consumer plastic products. They also are abundant in personal care products, food preservatives, pharmaceuticals, and paper products. Both are endocrine disruptors. They increase CVD risk by increasing risks of diabetes and obesity.[79] A meta-analysis estimated the pooled relative risk (RR) for type 2 diabetes to be 1.45 (95% CI: 1.13, 1.87) in relation to BPA exposure and 1.48 (95% CI: 0.98, 2.25) in relation to phthalate exposure.[82] Analysis of the dose-response relationship between BPA and diabetes showed an increase in RR of 1.09 per 1 ng/mL increase (95% CI: 1.03, 1.15) in urinary BPA level.[82]

Box 51.5 Organophosphate Insecticides and Congenital Heart Disease

Multiple studies have examined potential associations between parental exposure to pesticides and risk of congenital malformations.[28] These studies have not focused specifically on organophosphates but instead have examined all pesticides as a class and thus lack specificity. Another shortcoming is that few have utilized individual-level biomarkers of exposure but instead have relied on questionnaires, self-reported occupational exposures, or proximity of residence to sprayed fields to assess exposure.

A case-control study of more than 300,000 infants used maternal agricultural pesticide exposure based on application to crops within 500 meters of maternal residence at birth as a surrogate for exposure. This investigation found a significant association of pesticide exposure with total birth defects with an OR of 1.98 (95% CI: 0.69, 5.66), and found specific associations for atrial septal defects with an OR of 1.70 (95% CI: 1.34, 2.14) and patent ductus arteriosus with an OR of 1.50 (95% CI: 1.22, 1.85) at the highest level of exposure.[84]

In the National Birth Defects Prevention study, higher maternal occupational insecticide exposure from 1 month prior to conception to the end of the first trimester was found to be associated with atrial septal defects, while higher exposure to both insecticides and herbicides was associated with hypoplastic left heart syndrome, pulmonary valve stenosis, and tetralogy of Fallot.[85]

None of these studies specifically examined organophosphate exposure.

Organophosphate Insecticides

The organophosphates are a large family of synthetic pesticides that include malathion, chlorpyrifos, and diazinon. The organophosphates were deliberately designed to be neurotoxic. The organophosphates are discussed in detail in Chapter 35 (see also Box 51.5).

In addition to their neurotoxicity, the organophosphates are cardiotoxic. Key cardiac manifestations of acute organophosphate poisoning are bradycardia, ST-segment elevation, and atrioventricular conduction disturbances. Sinus tachycardia is seen in some cases.[83] Longer-lasting cardiac changes include QT prolongation and polymorphic tachycardia (torsade de pointes).

Complex ventricular arrhythmias are a frequently overlooked and potentially lethal aspect of acute organophosphate poisoning that can cause sudden cardiac death. Cardiac monitoring of acutely intoxicated patients for relatively long periods of time and early aggressive treatment of arrhythmias are critically important to patient survival.[83]

Noise

A growing body of evidence indicates that chronic exposure to noise can increase risk for ischemic heart disease. A recent meta-analysis of 10 studies examining the cardiovascular effects of chronic road and aircraft noise exposure found that relative risk for ischemic heart disease increased by 6% with each 10-dB increase in average noise exposure. The exposure-response relationship was linear and appeared to begin at a noise level of 50 dB.[86] Concomitant exposure to air pollution does not appear to account for the association.[87] Noise is discussed in detail in Chapter 44.

Noise appears to exert its adverse cardiovascular effects by increasing risk for hypertension, which in turn increases risk for myocardial infarction and stroke. Chronic nighttime noise appears to be especially hazardous and is associated with disruptions of sleep and increases in stress hormone levels and oxidative stress, which in turn may result in endothelial dysfunction and arterial hypertension.[88] These epidemiological findings are corroborated by experimental studies.[89]

Climate Change

The principal driver of global climate change is an increase in atmospheric levels of carbon dioxide, methane, and other greenhouse gases that trap heat above the surface of the earth. Fossil fuel combustion is the major underlying cause of the increase in greenhouse gases and thus the major driver of climate change.[87] Climate change is discussed in detail in Chapter 4.

Climate change has multiple negative impacts on cardiovascular health. Many of these cardiovascular consequences are the consequence of increased surface temperatures and heat waves. It is projected that, by 2050, major US. cities such as New York and Chicago may experience as many as three times their current average number of days hotter than 32°C (90°F). High temperatures are strongly associated with elevated levels of air pollution.[91] The combination of extreme heat and increased air pollution will lead to increased incidence and mortality of heart disease and stroke.[11]

Wildfires, which are increasing in frequency and severity in areas of the world that are becoming hotter and drier, such as the western United States, will further exacerbate the cardiac impacts of climate change. Levels of $PM_{2.5}$ pollution associated with wildfire smoke can be extremely high and are associated with elevated risk of CVD.[92]

The posttraumatic stress disorder and depression that are associated with climate-related natural disasters such as extreme storms, coastal flooding, and forced migration also will exacerbate the impacts of heat and pollution and may be expected to further increase risk of CVD, especially in highly vulnerable populations.[93]

References

1. Global burden of 87 risk factors in 204 countries and territories, 1990–2019: a systematic analysis for the Global Burden of Disease Study 2019. *Lancet*. 2020;396:1223–49.

2. Roth GA, Mensah GA, Johnson CO, et al. Global burden of cardiovascular diseases and risk factors, 1990–2019: Update from the GBD 2019 Study. *J Am Coll Cardiol*. 2020;76(25):2982–3021. doi:10.1016/j.jacc.2020.11.010. Erratum in: *J Am Coll Cardiol*. 2021 Apr 20;77(15):1958–9.

3. Rajagopalan S, Landrigan PJ. Pollution and the heart. *N Engl J Med*. 2021;385(20):1881–92. doi:10.1056/NEJMra2030281

4. Kim JB, Prunicki M, Haddad F, et al. Cumulative lifetime burden of cardiovascular disease from early exposure to air pollution. *J Am Heart Assoc*. 2020;9(6):e014944.

5. Al-Kindi SG, Brook RD, Biswal S, Rajagopalan S. Environmental determinants of cardiovascular disease: lessons learned from air pollution. *Nat Rev Cardiol*. 2020;17:656–72.

6. Solenkova NV, Newman JD, Berger JS, Thurston G, Hochman JS, Lamas GA. Metal pollutants and cardiovascular disease: mechanisms and consequences of exposure. *Am Heart J*. 2014;168:812–22.

7. Rao X, Montresor-Lopez J, Puett R, Rajagopalan S, Brook RD. Ambient air pollution: an emerging risk factor for diabetes mellitus. *Curr Diab Rep*. 2015;15(6):603.

8. Magliano DJ, Loh VH, Harding JL, Botton J, Shaw JE. Persistent organic pollutants and diabetes: a review of the epidemiological evidence. *Diabetes Metab*. 2014;40(1):1–14.

9. Landrigan PJ, Goldman L. Children's vulnerability to toxic chemicals: a challenge and opportunity to strengthen health and environmental policy. *Health Aff*. 2011; 30:842–50.

10. Wang Z, Walker GW, Muir DCG, Nagatani-Yoshida K. Toward a global understanding of chemical pollution: a first comprehensive analysis of national and regional chemical inventories. *Environ Sci Technol*. 2020 Mar 3;54(5):2575–84.

11. Brook RD, Franklin B, Cascio W, et al.; Expert Panel on Population and Prevention Science of the American Heart Association. Air pollution and cardiovascular disease: a statement for healthcare professionals from the Expert Panel on Population and Prevention Science of the American Heart Association. *Circulation*. 2004;109(21):2655–71.

12. Krewski D, Jerrett M, Burnett RT, et al. *Extended Follow-Up and Spatial Analysis of the American Cancer Society Study Linking Particulate Air Pollution and Mortality*. Cambridge, MA: Health Effects Institute; 2009.

13. Stafoggia M, Cesaroni G, Peters A, et al. Long-term exposure to ambient air pollution and incidence of cerebrovascular events: results from 11 European cohorts within the ESCAPE project. *Environ Health Perspect*. 2014;122(9):919–25.

14. Kaufman JD, Adar SD, Allen RW, et al. Prospective study of particulate air pollution exposures, subclinical atherosclerosis, and clinical cardiovascular disease: the Multi-Ethnic Study of Atherosclerosis and Air Pollution (MESA Air). *Am J Epidemiol*. 2012;176(9):825–37.

15. Liang F, Liu F, Huang K, et al. Long-term exposure to fine particulate matter and cardiovascular disease in China. *JACC*. 2020;75:707–17.

16. Alexeeff SE, Liao NS, Liu X, Van Den Eeden SK, Sidney S. Long-term PM2.5 exposure and risks of ischemic heart disease and stroke events: review and meta-analysis. *JAHA*. 2021;10:e016890.

17. US Environmental Protection Agency (EPA). Integrated Science Assessment (ISA) for particulate matter. 2019. Available at: https://cfpub.epa.gov/ncea/isa/recordisplay.cfm?deid=347534.

18. Christidis T, Erickson AC, Pappin AJ, et al. Low concentrations of fine particle air pollution and mortality in the Canadian Community Health Survey cohort. *Environmental Health*. 2019;18:84.

19. Liu C, Chen R, Sera F, et al. Ambient particulate air pollution and daily mortality in 652 cities. *N Engl J Med*. 2019;381:705–15.

20. Peralta AA, Link MS, Schwartz J, et al. Exposure to air pollution and particle radioactivity with the risk of ventricular arrhythmias. *Circulation*. 2020;142:858–67.

21. Cai Y, Zhang B, Ke W, et al. Associations of short-term and long-term exposure to ambient air pollutants with hypertension: a systematic review and meta-analysis. *Hypertension*. 2016;68:62–70.

22. Bevan GH, Al-Kindi S, Brook RD, Munzel T, Rajagopalan S. Ambient air pollution and atherosclerosis: insights into dose, time, and mechanisms. *Arterioscler Thromb Vasc Biol*. 2021 Feb;41(2):628–37.

23. Rajagopalan S, Park B, Palanivel R, et al. Metabolic effects of air pollution exposure and reversibility. *JCI*. 2020;130:6034–40.

24. Vrijheid M, Martinez D, Manzanares S, et al. Ambient air pollution and risk of congenital anomalies: a systematic review and meta-analysis. *Environ Health Perspect*. 2011;119:598–606.

25. Yuan X, Liang F, Zhu J, et al. Maternal Exposure to PM 2.5 and the Risk of Congenital Heart Defects in 1.4 Million Births: a Nationwide Surveillance-Based Study. *Circulation*. 2023;147(7):565–74. doi:10.1161/CIRCULATIONAHA.122.061245.

26. Wan X, Wei S, Wang Y, et al. The association between maternal air pollution exposure and the incidence of congenital heart diseases in children: a systematic review and meta-analysis. *Sci Total Environ*. 2023;892:164431. doi:10.1016/j.scitotenv.2023.164431.

27. Buteau S, Veira P, Bilodeau-Bertrand M, Auger N. Association between First Trimester Exposure to Ambient PM2.5 and NO2 and Congenital Heart Defects: a Population-Based Cohort Study of 1,342,198 Live Births in Canada. *Environ Health Perspect*. 2023;131(6):67009. doi:10.1289/EHP11120.

28. Nicoll R. Environmental contaminants and congenital heart defects: a re-evaluation of the evidence. *Int J Environ Res Public Health*. 2018 Sep 25;15(10):2096.

29. Stingone JA, Luben TJ, Daniels JL, et al.; National Birth Defects Prevention Study. Maternal exposure to criteria air pollutants and congenital heart defects in offspring: results from the national birth defects prevention study. *Environ Health Perspect*. 2014 Aug;122(8):863–72.

30. Thomas D. Cardiovascular benefits of smoking cessation. *Presse Med*. 2009;38(6):946–52.

31. Burnett RT, Cakmak S, Raizenne ME, et al. The association between ambient carbon monoxide levels and daily mortality in Toronto, Canada. *J Air Waste Manag Assoc*. 1998;48(8):689–700.

32. Chowdhury R, Ramond A, O'Keeffe LM, et al. Environmental toxic metal contaminants and risk of cardiovascular disease: systematic review and meta-analysis. *BMJ*. 2018;362:k3310.

33. Lancereaux E. Nephrite et arthrite saturnine: coincidences de ces affections;parallele avec la nephrite etl'arthrite gouttesses. *Transact Int Med Congr*. 1881;2:193–202.

34. Schwartz J. Lead, blood pressure, and cardiovascular disease in men. *Arch Environ Health*. 1995;50:31–7.

35. Navas-Acien A, Guallar E, Silbergeld EK, et al. Lead exposure and cardiovascular disease: a systematic review. *Environ Health Perspect*. 2007;115:472–82.

36. Schober SE, Mirel LB, Graubard BI, et al. Blood lead levels and death from all causes, cardiovascular disease and cancer: results from NHANES III Mortality Study. *Environ Health Perspect*. 2006;114:1538–41.

37. Lanphear BP, Rauch S, Auinger P, Allen RW, Hornung RW. Low-level lead exposure and mortality in US adults: a population-based cohort study. *Lancet Public Health*. 2018;3:e177–84.

38. Vinceti M, Rovesti S, Bergomi M, et al. Risk of birth defects in a population exposed to environmental lead pollution. *Sci Total Environ*. 2001 Oct 20;278(1–3):23–30.

39. Silver SR, Pinkerton LE, Rocheleau CM, Deddens JA, Michalski AM, Van Zutphen AR. Birth defects in infants born to employees of a microelectronics and business machine manufacturing facility. *Birth Defects Res A Clin Mol Teratol*. 2016;106:696–707.

40. Tellez-Plaza M, Jones MR, Dominguez-Lucas A, et al. Cadmium exposure and clinical cardiovascular disease: a systematic review. *Curr Atheroscler Rep.* 2013;15(10):356.

41. National Academy of Sciences. *Toxicological Effects of Methyl Mercury.* Washington, DC: National Academies Press; 2000.

42. Choi AL, Weihe P, Budtz-Jergensen E, et al. Methylmercury exposure and adverse cardiovascular effects in Faroese whaling men. *Environ Health Perspect.* 2009;117(3):367–72.

43. Hu XF, Lowe M, Chan HM. Mercury exposure, cardiovascular disease, and mortality: a systematic review and dose-response meta-analysis. *Environ Res.* 2021;193:110538.

44. Choi AL, Cordier S, Weihe P, et al. Negative confounding in the evaluation of toxicity: the case of methylmercury in fish and seafood. *Crit Rev Toxicol.* 2008;38(10):877–93.

45. Lim KM, Kim S, Noh JY, et al. Low-level mercury can enhance procoagulant activity of erythrocytes: a new contributing factor for mercury-related thrombotic disease. *Environ Health Perspect.* 2010;118(7):928–35.

46. Kempuraj D, Asadi S, Zhang B, et al. Mercury induces inflammatory mediator release from human mast cells. *J Neuroinflammation.* 2010;7:20.

47. Stea F, Bianchi F, Cori L, Sicari R. Cardiovascular effects of arsenic: clinical and epidemiological findings. *Environ Sci Pollut Res Int.* 2014;21(1):244–51.

48. Moon K, Guallar E and Navas-Acien A. Arsenic exposure and cardiovascular disease: an updated systematic review. *Curr Atheroscler Rep.* 2012;14:542–55.

49. Tseng WP. Black foot disease in Taiwan: a 30-year follow-up study. *Angiology.* 1989;40:547–58.

50. Zierler S, Theodore M, Cohen A, Rothman KJ. Chemical quality of maternal drinking water and congenital heart disease. *Int J Epidemiol.* 1988;17:589–94.

51. Rudnai T, Sándor J, Kádár M et al. Arsenic in drinking water and congenital heart anomalies in Hungary. *Int J Hyg Environ Health.* 2014;217:813–18.

52. Morin Y, Daniel P. Quebec beer-drinkers' cardiomyopathy: etiological considerations. *Can Med Assoc J.* 1967;97(15):926–8.

53. Perez AL, Tang WH. Contribution of environmental toxins in the pathogenesis of idiopathic cardiomyopathies. *Curr Treat Options Cardiovasc Med.* 2015;17(5):381.

54. Zhao G, Ding M, Zhang B, et al. Clinical manifestations and management of acute thallium poisoning. *Eur Neurol.* 2008;60 (6):292–7.

55. Agency for Toxic Substances and Disease Registry (ATSDR). Toxicological profile for thallium. ATSDR. 1992. Available at: https://www.atsdr.cdc.gov/ToxProfiles/tp54.pdf

56. Agency for Toxic Substances and Disease Registry (ATSDR). Toxicological profile for polychlorinated biphenyls (PCBs).ATSDR, 1998. Available at: https://wwwn.cdc.gov/TSP/ToxP rofiles/ToxProfiles.aspx?id=142&tid=26.

57. UN Environment. The Stockholm Convention on Persistent Organic Pollutants. 2001 May 22. Available at: http://chm.pops.int/TheConvention/Overview/TextoftheConvention/tabid/2232/Default.aspx.

58. Himmel HM. Mechanisms involved in cardiac sensitization by volatile anesthetics: general applicability to halogenated hydrocarbons? *Crit Rev Toxicol.* 2008;38(9):773–803.

59. Gilboa SM, Desrosiers TA, Lawson C, et al; National Birth Defects Prevention Study. Association between maternal occupational exposure to organic solvents and congenital heart defects, National Birth Defects Prevention Study, 1997–2002. *Occup Environ Med.* 2012;69(9):628–35. doi:10.1136/oemed-2011-100536.

60. Agency for Toxic Substances and Disease Registry (ATSDR). *Toxicological Profile for Methylene Chloride*. Atlanta, GA: Centers for Disease Control and Prevention. http://www.atsdr.cdc.gov/substances/toxsubstance.asp?toxid=42.

61. Centers for Disease Control and Prevention (CDC). Fatal exposure to methylene chloride among bathtub refinishers: United States, 2000–2011. *MMWR Morb Mortal Wkly Rep*. 2012;61(7):119–22.

62. Agency for Toxic Substances and Disease Registry (ATSDR). *Toxicological Profile for Chlorinated Dibenzo-p-Dioxins (CDDs)*. Atlanta, GA: Centers for Disease Control and Prevention. 2012. http://www.atsdr.cdc.gov/toxprofiles/tp.asp?id=366&tid=63.

63. Fingerhut MA, Sweeney MH, Halperin WE, et al. The epidemiology of populations exposed to dioxin. *IARC Sci Publ*. 1991;108:31–50.

64. Consonni D, Pesatori AC, Zocchetti C, et al. Mortality in a population exposed to dioxin after the Seveso, Italy, accident in 1976: 25 years of follow-up. *Am J Epidemiol*. 2008;167(7):847–58.

65. Lee DH, Lee IK, Song K, et al. A strong dose-response relation between serum concentrations of persistent organic pollutants and diabetes: results from the National Health and Nutrition Examination Survey 1999–2002. *Diabetes Care*. 2006;29(7):1638–44.

66. Pesatori AC, Consonni D, Bachetti S, et al. Short- and long-term morbidity and mortality in the population exposed to dioxin after the "Seveso accident." *Ind Health*. 2003;41(3):127–38.

67. Lee DH, Lee IK, Porta M, et al. Relationship between serum concentrations of persistent organic pollutants and the prevalence of metabolic syndrome among non-diabetic adults: results from the National Health and Nutrition Examination Survey 1999–2002. *Diabetologia*. 2007;50:1841–51.

68. Vafeiadi M, Georgiou V, Chalkiadaki G, et al. Association of prenatal exposure to persistent organic pollutants with obesity and cardiometabolic traits in early childhood: the Rhea Mother-Child Cohort (Crete, Greece). *Environ Health Perspect*. 2015;123(10):1015–21.

69. Humblet O, Birnbaum L, Rimm E, et al. Dioxins and cardiovascular disease mortality. *Environ Health Perspect*. 2008;116(11):1443–8.

70. Gustavsson P, Hogstedt C. A cohort study of Swedish capacitor manufacturing workers exposed to polychlorinated biphenyls (PCBs). *Am J Ind Med*. 1997;32 (3):234–9.

71. Goncharov A, Haase RF, Santiago-Rivera A, et al. High serum PCBs are associated with elevation of serum lipids and cardiovascular disease in a Native American population. *Environ Res*. 2008;106 (2):226–39.

72. Agency for Toxic Substances and Disease Registry (ATSDR). *Toxicological Profile for Polybrominated Diphenyl Ethers (PBDEs)*. Atlanta, GA: Centers for Disease Control and Prevention. 2017. https://www.atsdr.cdc.gov/ToxProfiles/tp207.

73. Lorber M. Exposure of Americans to polybrominated diphenyl ethers. *J Expo Sci Environ Epidemiol*. 2008;18(1):2–19.

74. Costa LG, Giordano G, Tagliaferri S, et al. Polybrominated diphenyl ether (PBDE) flame retardants: environmental contamination, human body burden and potential adverse health effects. *Acta Biomed*. 2008;79 (3):172–83.

75. Gore AC, Chappell VA, Fenton SE, et al. Executive summary to EDC-2: the Endocrine Society's second scientific statement on endocrine-disrupting chemicals. *Endocr Rev*. 2015;36(6):593–602.

76. Lim JS, Lee DH, Jacobs DR Jr. Association of brominated flame retardants with diabetes and metabolic syndrome in the U.S. population, 2003–2004. *Diabetes Care*. 2008;31(9):1802–7.

77. Gump BB, Yun S, Kannan K. Polybrominated diphenyl ether (PBDE) exposure in children: possible associations with cardiovascular and psychological functions. *Environ Res.* 2014;132:244–50.

78. Agency for Toxic Substances and Disease Registry (ATSDR). *Draft Toxicological Profile for Perfluoroalkyls*. Atlanta, GA: Centers for Disease Control and Prevention, 2015 Aug. http://www.atsdr.cdc.gov/toxprofiles/tp200.pdf.

79. Kahn LG, Philippat C, Nakayama SF, Slama R, Trasande L. Endocrine-disrupting chemicals: implications for human health. *Lancet Diabetes Endocrinol.* 2020;8:703–18.

80. Timmermann CA, Rossing LI, Grøntved A, et al. Adiposity and glycemic control in children exposed to perfluorinated compounds. *J Clin Endocrinol Metab.* 2014;99(4):E608–14.

81. Lundin JI, Alexander BH, Olsen GW, et al. Ammonium perfluorooctanoate production and occupational mortality. *Epidemiology.* 2009;20(6):921–8.

82. Song Y, Chou EL, Baecker A, et al. Endocrine-disrupting chemicals, risk of type 2 diabetes, and diabetes-related metabolic traits: a systematic review and meta-analysis. *J Diabetes.* 2016;8(4):516–32. doi:10.1111/1753-0407.12325.

83. Bar-Meir E, Schein O, Eisenkraft A, et al. Guidelines for treating cardiac manifestations of organophosphates poisoning with special emphasis on long QT and torsades de pointes. *Crit Rev Toxicol.* 2007;37(3):279–85.

84. Carmichael SL, Yang W, Roberts E, et al. Residential agricultural pesticide exposures and risk of selected congenital heart defects among offspring in the San Joaquin Valley of California. *Environ Res.* 2014 Nov;135:133–8.

85. Rocheleau CM, Bertke SJ, Lawson CC, et al.; National Birth Defects Prevention Study. Maternal occupational pesticide exposure and risk of congenital heart defects in the National Birth Defects Prevention Study. *Birth Defects Res A Clin Mol Teratol.* 2015 Oct;103(10):823–33.

86. Vienneau D, Schindler C, Perez L, et al. The relationship between transportation noise exposure and ischemic heart disease: a meta-analysis. *Environ Res.* 2015;138:372–80.

87. Tétreault LF, Perron S, Smargiassi A. Cardiovascular health, traffic-related air pollution and noise: are associations mutually confounded? A systematic review. *Int J Public Health.* 2013;58(5):649–66.

88. Münzel T, Gori T, Babisch W, Basner M. Cardiovascular effects of environmental noise exposure. *Eur Heart J.* 2014;35(13):829–36.

89. Basner M, Babisch W, Davis A, et al. Auditory and non-auditory effects of noise on health. *Lancet.* 2014;383(9925):1325–32.

90. Patz JA, Frumkin H, Holloway T, et al. Climate change: challenges and opportunities for global health. *JAMA.* 2014;312(15):1565–80.

91. Kinney P. Climate change, air quality, and human health. *Am J Prev Med* 2008;35:459–67.

92. Chang AY, Tan AX, Nadeau KC, Odden MC. Aging hearts in a hotter, more turbulent world: the impacts of climate change on the cardiovascular health of older adults. *Curr Cardiol Rep.* 2022 Jun;24(6):749–60.

93. McMichael AJ, Campbell-Lendrum D, Kovats S, et al., eds. *Comparative Quantification of Health Risks, Global and Regional Burden of Disease Attributable to Selected Major Risk Factors*. Geneva, Switzerland: World Health Organization; 2004.

52

The Environment and Liver Disease
in Children

Frederick J. Suchy

The liver is highly susceptible to structural and functional injury caused by toxic chemicals in the environment because it is the organ principally responsible for the metabolism of drugs and other chemicals absorbed into the body by any route.

Exposure to a toxic chemical in the home, community, or school or transported home from a parent's workplace should be considered in every child with unexplained dysfunction of the liver. A careful history taken by an alert clinician is frequently required to discern the connection between exposure to a toxic chemical and liver disease in a child.

Toxic injury to the liver can produce a wide spectrum of illness. This range extends from asymptomatic abnormalities in the serum biochemical tests used in clinical practice to assess liver function to fulminant hepatic failure, which can result in death. Liver injury may be the only clinical feature of exposure to a toxic chemical in the environment, or it may be accompanied by systemic manifestations and damage to other organs such as the brain or kidney.

Certain toxic chemicals have been recognized for many decades to cause acute and chronic disease of the liver. Carbon tetrachloride, an organic solvent, is one of these "classical" hepatotoxicants. Inhalation or ingestion of carbon tetrachloride can cause significant liver injury. White phosphorus—used in the manufacture of explosives and fireworks—is another potent hepatotoxicant. It has caused liver injury or even liver failure in children through ingestion or inhalation of vapors. Because the toxic potential of these chemicals is well known, liver injury caused by them is now relatively rare, although it may still occur accidentally.

Other more recently recognized hepatotoxicants may have more insidious consequences on children's health. Table 52.1 presents data on a large and growing number of chemicals that are recognized today to be toxic to the liver. Among them are a number of chemicals that act as endocrine disruptors and chemical obesogens (see also Chapters 40 and 41). Chemical obesogens increase risk of obesity and nonalcoholic fatty liver disease (NAFLD), now the most common chronic liver disease affecting children and adults.

The role of environmental toxicants in the pathogenesis of pediatric liver diseases such as biliary atresia, autoimmune hepatitis, and primary sclerosing cholangitis is unknown. There have, however, been instances of geographic or seasonal clustering of such cases, a pattern that suggests an environmental contribution. Some of these diseases likely have an autoimmune etiology. Certain environmental exposures may cause the loss of self-tolerance that is a cardinal feature of autoimmune disorders through molecular mimicry or self-antigen modification that leads, in turn, to cross-reactivity and tissue injury.[1]

Frederick J. Suchy, *The Environment and Liver Disease in Children* In: *Textbook of Children's Environmental Health*,
Second Edition. Edited by: Ruth A. Etzel and Philip J. Landrigan, Oxford University Press. © Oxford University Press 2024.
DOI: 10.1093/oso/9780197662526.003.0052

Table 52.1 Examples of environmental hepatotoxicants and hepatotoxins

Agent	Sources/Uses	Type of liver injury
Inorganic arsenic salts	Pesticides, herbicides, dyes, ceramics	Acute necrosis, fibrosis/cirrhosis, cancer
Volatile industrial solvents Carbon tetrachloride Trichloroethene 1,1,1-trichloroethane Tetrachloroethene Toluene and xylene	Solvents, degreasing, other	Fatty change, necrosis, cirrhosis, and cancer in some cases
Vinyl chloride	Solvents and resins	Insulin resistance and fatty liver, fibrosis, cirrhosis, and cancer
Dioxane	Solvent, degreasing, adhesives, paints, polishes	Variable hepatocellular injury, probable endocrine disruptors
Polychlorinated biphenyls	Electrical insulation, herbicides, resins, paper textiles	Acute hepatocellular injury, cancer
Phosphorus	Explosives, fireworks, rodenticides	Acute necrosis, steatosis
Plant-derived toxins	Herbal teas, contaminated grain and cereal, meat	Veno-occlusive disease
Pyrrolizidine alkaloids, phytoestrogens		
Aflatoxins	Contaminated food	Acute hepatocellular injury, hepatocellular carcinoma (possible synergy with hepatitis B)
Microcystins	Blue-green algae blooms	Acute necrosis, hepatocellular carcinoma
Phthalate esters	Plasticizing agents	Possible fatty change, endocrine disruptors leading to nonalcoholic fatty liver disease (NAFLD)
Bisphenol A	Plasticizing agent	Acute hepatocellular injury possible, endocrine disruptors contributing to obesity and NAFLD
Secondhand cigarette smoke		Hepatoblastoma, systemic inflammation, fatty liver disease
Air pollution		Systemic inflammation, fatty liver disease
Organochlorides, organophosphates, carbamates	Pesticides	Acute hepatocellular injury possible but emerging as endocrine disruptors
Organotin compounds	Antifouling agents in marine paints and industrial water systems; fungicides on crops	Endocrine disruptors promoting obesity and NAFLD

Liver Development and Environmental Hepatotoxicity

Infants and children may paradoxically be both more and at the same time less vulnerable to toxic liver injury than adults. The degree of toxic injury depends on the particular chemical to which a child is exposed. Immaturity of pathways for biotransformation may prevent efficient degradation and elimination of certain toxic compounds, thus increasing risk of liver damage. But, by contrast, this same immaturity may limit the formation of certain reactive metabolites that can damage the liver, thus reducing toxicity.[2]

Many developmental variables influence the capacity of the liver to metabolize drugs and toxic chemicals, including liver size, blood flow, plasma protein binding, and intrinsic clearance (a product of the enzymatic and transport capacity of the liver).[3] The liver must undergo significant maturation both pre- and postnatally before it attains full adult capacity for such functions as detoxification, energy metabolism, and excretion of wastes.

There is increasing recognition that genetic factors may influence vulnerability of the liver to toxic chemicals in the environment. Even in the "mature" liver, the functional capacity of the organ or the ability of the liver enzymes responsible for metabolism and biotransformation of drugs and toxic chemicals is governed by genetic factors. Many phase I and phase II enzymes, as well as phase III transporters, that are critically important for liver biotransformation are polymorphically expressed, developmentally regulated, and subject to considerable inducibility. Because of this variability, therapeutic drug use can potentially influence susceptibility to the effects of certain environmental toxicants.[2] The xenobiotic nuclear receptors (NRs), the pregnane X receptor (PXR, also known as the steroid and xenobiotic receptor, or SXR), the constitutive androstane receptor (CAR), and the aryl hydrocarbon receptor (AhR) are all developmentally regulated and coordinately induce genes involved in the three phases of xenobiotic metabolism. Many environmental toxicants, including certain endocrine disruptors, are ligands for NRs, CAR, and PXR that transcriptionally activate the promoters of many genes involved in xenobiotic metabolism and may contribute to some forms of environmentally induced liver disease.

Because of the complexity of these mechanisms and the difficulty in obtaining normal liver tissue from human fetuses and children, there are many gaps in our understanding of the pathways for biotransformation during early life.

Epigenetics and Liver Disease in Children

Epigenetic changes are defined as heritable changes in phenotype or gene expression that are caused by mechanisms other than mutations in DNA (see Chapter 10).[4] Toxic chemicals and other environmental factors such as stress, diet, behavior, and disease can cause epigenetic changes by activating chemical switches such as DNA methylations and histone tail modifications that regulate gene expression. Environmental toxicants, such as bisphenol A, that are ligands for nuclear receptors also have been shown to alter epigenetic profiles during development. Altered expression of certain key genes in the liver is likely to predispose to cancer and to the metabolic dysfunction that leads to obesity.

Prototypical
Environmental Hepatotoxicants and Hepatotoxins

Carbon Tetrachloride

Carbon tetrachloride is an example of a predictable hepatotoxicant that will produce liver damage in all persons if given in a sufficient dose acutely or chronically.[5] Carbon tetrachloride has been used mainly in the production of chlorofluorocarbon refrigerants, foam-blowing agents, and solvents (including cleaning agents). At one point early in the past century carbon tetrachloride was used effectively to treat hookworm infestation in children. Carbon tetrachloride is now considered so toxic that the US Food and Drug Administration banned the chemical and any mixture containing it from domestic use in the United States in 1970.

Today, in the United States, exposures to carbon tetrachloride at levels above "background" are likely to occur only at industrial locations where the chemical is still used or near chemical waste sites where emissions into air, water, or soil have not been controlled. Exposures at such sites could occur by inhalation of carbon tetrachloride vapor in the air, by drinking water contaminated with carbon tetrachloride, or by ingesting contaminated soil. Small children who live near sites contaminated by carbon tetrachloride could accidentally ingest some of the chemical from putting soiled hands in their mouths, but the level of carbon tetrachloride in soil is generally too low to be harmful.[6]

Carbon tetrachloride produces cellular injury and even necrosis in the liver following absorption into the body by any route—inhalation, dermal absorption, or oral ingestion. In many respects, it is representative of a large class of chemically related chlorinated hydrocarbon solvents (see Table 52.1). The biotransformation of carbon tetrachloride to its toxic form is initially catalyzed by cytochrome P450 enzymes (mainly CYP2E1), leading to the formation of a highly reactive trichloromethyl radical.[5] This intermediate is oxidized further to the even more reactive trichloromethylperoxyl free radical, which can react further to form phosgene. These reactive metabolic intermediates of carbon tetrachloride, particularly the trichloromethylperoxyl radical, cause tissue damage by binding covalently to macromolecules and causing lipid peroxidation.

Within 24 hours after significant exposure to carbon tetrachloride, the patient often presents with neurologic and gastrointestinal symptoms. Tender hepatomegaly may be present. Liver blood test results may be abnormal, and clinically apparent jaundice can occur within 48–96 hours. Liver failure, reflecting massive hepatocellular necrosis, may occur in severely poisoned patients. Liver histology shows diffuse microvesicular fatty change and variable zone 3 hepatocellular necrosis.

Aflatoxins

Aflatoxins are a family of mycotoxins produced by molds (see Chapter 43); they are widespread contaminants of nuts, soy beans, grain, and other foodstuffs, particularly in warm, humid tropical and subtropical regions.[7] Acute aflatoxin poisoning is associated with vomiting, abdominal pain, lung injury, and fatty degeneration and necrosis of the liver. Although aflatoxin can be detected in products such as peanut butter in high-income countries, levels are generally well below concentrations expected to produce acute hepatocellular injury. Of greater concern, however, is the role of aflatoxins in the causation of hepatocellular carcinoma, perhaps the commonest solid tumor worldwide.

Aflatoxin is activated to its toxic and carcinogenic form predominantly by the cytochrome P450 (CYP) monooxygenase system. A highly reactive epoxide intermediate is catalyzed by CYP1A2 and CYP3A4, and it reacts further with DNA to yield a mutagenic aflatoxin-N-guanine adduct.[7] Malnourished children seem to be particularly susceptible to hepatotoxicity and carcinogenesis from aflatoxin exposure. In the setting of undernutrition, a child's capacity to detoxify aflatoxin epoxide metabolites through conjugation with glutathione is likely to be compromised. Chronic infection with hepatitis B virus, also prevalent in low- and middle-income countries, further increases the risk of developing liver cancer in children. Immunization against hepatitis B reduces risk of liver cancer, even in populations heavily exposed to aflatoxin.

A range of interventions can be implemented to reduce aflatoxin exposure and dosage. These strategies include planting mold-resistant crops, lowering mold growth in harvested crops, and improving methods for storage of harvested grain, nuts, and soybeans. In low- and middle-income countries, where foodstuffs are often scarce and high-technology storage facilities in short supply, these preventive strategies may be difficult to implement.

Arsenic

Despite major efforts at mitigation, arsenic has been found at high levels in drinking water in many parts of the world, including Bangladesh, India, Argentina, and Chile (see also Chapter 33). Persistent exposure to arsenic in drinking water in rural Bangladesh is associated with high urinary excretion, particularly during the first 5 years of life.[8] Arsenic trioxide is the most prevalent inorganic arsenical found in air, while a variety of inorganic arsenates occur in water, soil, or food.[9]

Arsenic was first identified as a epatotoxicant in the 18th century. Arsenic induces the formation of oxidized lipids in the liver and other organs, which in turn generate reactive oxygen species, aldehydes, and other toxic intermediates. Acute toxicity has been associated with biochemical evidence of liver damage and, in some cases, veno-occlusive disease. Prolonged exposure can produce cirrhosis and noncirrhotic portal hypertension. A recent study in Chile found marked increases in liver cancer mortality in children 10–19 years of age who had exposure to high concentrations of arsenic in drinking water starting soon after birth.[10]

Arsenic acts as an endocrine disruptor at very low, noncytotoxic, but still environmentally relevant concentrations through interaction with six nuclear receptors: the receptors for glucocorticoid (GR), androgen (AR), progesterone (PR), mineralocorticoid (MR), estrogen (ER), and thyroid (TR) hormones. It is not clear whether these adverse effects extend to the liver.[11]

Air Pollution and Secondhand Smoke

Exposure to particulate air pollution and secondhand smoke have been associated with low-grade systemic inflammation, oxidative stress, and insulin resistance (see also

Chapters 25, 26, and 27).[12] All of these factors have been implicated in the development of obesity and NAFLD.

In a recent European longitudinal population-based birth cohort study of 1,102 children, prenatal and childhood exposures to air pollution and traffic were not associated with child liver injury biomarkers.[13] However, there was strong association between prenatal exposure to ambient particulate matter of measuring less than 10 μm in aerodynamic diameter (PM_{10}) and alanine transaminase (ALT) among children who were overweight or obese compared with children who were not overweight.

Diesel exhaust fumes, which are major constituents of atmospheric particulate matter in urban areas, generate reactive oxygen species that are catalyzed enzymatically by the cytochrome P450 system in the liver. Lipid peroxides and inflammatory cytokines such as interleukin (IL)-6 can be generated in the lung and may transit to the liver where they may exacerbate steatohepatitis.[12] Airborne particulate matter also selectively activates the endoplasmic reticulum stress response in rodent liver, leading to hepatocyte apoptosis.[14] Exposure to air pollution may be a risk factor for obesity and NAFLD.[12] In a recent study of 374 children, a measure of air pollution including exposure to fine particulate matter was significantly associated with markers of systemic inflammation, oxidative stress, and insulin resistance after adjustment for age, gender, body mass index, waist circumference, healthy eating index, and physical activity level.[15]

Experimental studies have shown that exposure to secondhand smoke, which contains thousands of toxic compounds, can stimulate lipid accumulation in the liver by stimulating activation of sterol response element binding protein 1 (SREBP-1) and inactivating the activity 5′ AMP-activated protein kinase (AMPK). These are both critical mediators controlling triglyceride synthesis and accumulation in the liver.[16]. In a 31-year population-based cohort study, both child and adult passive smoking were associated with an increased risk of fatty liver in adulthood.[17] Parental cigarette smoking is also a risk factor for developing hepatoblastoma, the most common malignant liver tumor in infants.[18] Thirdhand smoke from accumulation and aging of secondhand smoke toxicants on surfaces where smoking has occurred is an emerging environmental health hazard, but data on liver injury are largely limited to animal studies.[19]

Mercury

Mercury (Hg) has been designated as a high-priority pollutant by the US Environmental Protection Agency. Hg makes its way into the ecosystem through a number of human activities including waste incineration, mining, coal-fired power plants, and disposal of a number of consumer products such as electronic devices. Fish and shellfish concentrate mercury from sea water, often in the form of highly toxic methylmercury. The consumption of fish is the most significant source of ingestion-related mercury exposure in humans and is a particular health risk for women who are or may become pregnant, nursing mothers, and young children.[20] The European Human Early-Life Exposome study has recently demonstrated in a multicenter mother–child cohort that maternal blood mercury concentrations during pregnancy were associated with a phenotype in children characterized by elevated serum levels of ALT and circulating inflammatory cytokines.[21] A previous study using data from the National Health and Nutrition Examination Survey (NHANES) 1999–2014 also demonstrated a positive association between blood Hg concentrations and elevated

ALT levels in US adolescents.[22] These results suggest that exposure to Hg during pre- and postnatal life can contribute to chronic inflammation, liver dysfunction, and the risk for NAFLD. However, it remains unclear how Hg leads to liver injury. Direct cytotoxic effects and immune mechanisms are likely to be operative.

Cadmium

Cadmium (Cd) is a toxic metal with a long biological life. There are a number of natural environmental sources of the metal, such as volcanic activity and forest fires. Human exposure occurs primarily from fossil fuel combustion, cigarette smoke, plastic stabilizers, pesticides, fertilizers, and discarded batteries.[23] Cd is readily absorbed in the gastrointestinal tract from contaminated water and food. The liver is one of the first organs to be exposed to the metal, and it may be damaged by the generation of reactive species and oxidative stress.[24] Cd-induced liver injury may be a risk factor for hepatocellular carcinoma.

Organophosphate Pesticides

Organophosphate pesticides are widespread in the modern environment and include agents such as malathion, parathion, diazinon, fenthion, dichlorvos, chlorpyrifos, and ethion (see also Chapter 35).[25] These compounds have been shown to cause acute and chronic neurotoxicity, and acute liver injury may occur through lipid peroxidation and generation of reactive oxygen species.

Malathion is one of the most commonly used organophosphates.[26] Mechanisms of malathion toxicity include inhibition of acetylcholinesterase, change of oxidants/antioxidants balance, DNA damage, mitochondrial dysfunction, autophagy, activation of the innate immune system, and facilitation of apoptotic cell damage. Histopathology of malathion injury includes degenerative changes in the form of necrosis, congestion of the central vein, hemorrhage, and steatohepatitis.

Exposures to organophosphate pesticides in early life may cause long-lasting metabolic disruption at doses that do not produce acute signs of toxicity. Substantial new experimental evidence in animals has shown that exposures to organophosphates during fetal and neonatal life dramatically alter hepatic adenyl cyclase/cyclic adenosine monophosphate (AMP) signaling in the liver.[27] It is proposed that these effects persist into adult life and predispose to type 2 diabetes, abnormal lipid metabolism, and obesity.

Environmental Obesogens/Endocrine Disruptors

Although excess caloric consumption and a sedentary lifestyle are key risk factors for obesity, evidence is mounting that exposures to certain toxic chemicals that have become widespread in the modern environment—"chemical obesogens"—are likely also contributing to the global epidemic of obesity and to its complications, such as diabetes and NAFLD (see also Chapter 41).[28,29] Exposures to these compounds may occur in utero and have effects that last throughout life. The most serious variant of NAFLD, non-alcoholic steatohepatitis (NASH), currently affects 5–10% of all children in the United States and has the potential to evolve to cirrhosis and liver failure.[30]

Some chemical obesogens appear to act directly on adipocytes to increase their number and/or promote their storage of fat, resulting in weight gain. Chemical obesogens may also act indirectly to increase adiposity by altering mechanisms through which the body regulates appetite, the basal metabolic rate, and energy balance to favor the storage of calories. The most convincing evidence for the role of chemical obesogens is derived from animal and cell culture studies, and direct proof of their action in humans is still lacking.

Chemical obesogens of current concern include the organotin compounds and the plasticizer chemicals, phthalates and bisphenol A. These chemicals are lipophilic and may have toxic effects even at extremely low concentrations. They are structurally similar to many hormones and other natural ligands for nuclear receptors.[31] In fact, all of these compounds interact with nuclear receptors in liver cells whose target genes likely contribute to endocrine disruption, hepato-carcinogenesis, and the metabolic syndrome. Less well understood are effects on other pathways including transcriptional coactivators, histone modifying enzymes, and enzymatic pathways involved in metabolism.[32] Aberrant epigenetic programming, such as changes in DNA methylation in the fetus may also be operative.[4]

Chemical obesogens regulate many target genes involved in adipocyte differentiation, adipokine release, insulin sensitivity, and energy balance by acting as potent ligands for the retinoid (RXR) X receptor-α and peroxisome proliferator-activated receptor-γ (PPARγ).[32] The ligand-binding pocket of PPARγ is larger than required for natural endogenous ligands such as free fatty acids and eicosanoids, and it can be considered promiscuous in its accommodation of chemicals of dissimilar size and structure. RXRα functions as the common heterodimeric partner to many other nuclear receptors in hormonal signaling pathways. Aberrant activation of RXRα and PPARγ may lead to wide-ranging disturbances in the body's homeostatic hormonal controls, contributing to obesity and associated NAFLD. It is of note that mutations and naturally occurring polymorphisms in PPARγ are associated with obesity and insulin resistance.

Per- and polyfluoroalkyl substances (PFAS) are a diverse group of manmade chemicals used in a wide range of consumer and industrial products. PFAS are highly resistant to degradation and bioaccumulate in food chains and drinking water. Prenatal exposure to PFAS has been associated with increases in child adiposity.[33] PFAS have been associated with liver dysfunction in children and adults. In a well-characterized multicenter cohort of European mothers and their children, maternal exposure to a PFAS mixture during pregnancy was associated with increased liver injury risk in childhood.[33] A recent study found that higher plasma PFAS concentrations were associated with increased risk of NASH in children diagnosed with NAFLD.[34] More advanced stages of fibrosis and liver inflammation were associated with higher plasma concentrations of PFAS in these children. Moreover, plasma PFAS chemical analysis and untargeted metabolomics analyses found that PFAS exposure was associated with changes in key amino acids and lipids pathways underlying NAFLD pathophysiology.

Organotin compounds are persistent and ubiquitous contaminants found in fish, shellfish, and some crops. These synthetic chemicals are used as antifouling agents in marine paints and industrial water systems and also as antifungal agents in agriculture

Organotins stimulate adipogenesis in vitro and by targeting key transcription factors in the adipogenic pathway.[35] Of particular concern, in utero exposure to tributyltin, a potent, dual agonist for RXRα and PPARγ, is associated with markedly elevated lipid accumulation in adipose tissue and livers of neonate mice.[31] Tributyltin also disrupts cholesterol homeostasis by activating the LXR/RXR heterodimer. Agonist binding to NRs leads to conformational changes at the promoters of target genes that in turn produce modifications to chromatin, dissociations of co-repressors, and recruitment of co-activators.

Phthalates are used as plasticizers in a broad array of industrial applications including plasticized polyvinyl chloride, cosmetics, residential construction, toys, automotive components, and medical devices (see also Chapter 40). Since phthalates are not covalently bound into these products, they can leach from plastics, leading to significant environmental contamination of food, indoor air, soil, and sediments. As is the case with other environmental contaminants, children appear to be more heavily exposed to phthalates than adults through mouthing of toys and other phthalate-containing items, through inhalation of dust, and through an increased food intake per body weight. In a German study, exposure of nursery school children to di-*n*-butylphthalate and butyl-benzylphthalate as assessed by urinary excretion was approximately two to four times higher than their parents and teachers.[36] High levels of exposure may also occur in children and neonates exposed to blood storage bags, intravenous tubing and containers, and hemodialysis tubing.[37]

There is experimental evidence for acute liver injury caused by phthalates through mitochondrial depolarization and generation of reactive oxygen species, leading to apoptosis.[38]

The role of phthalates as endocrine disruptors has raised great concern. Phthalates modulate expression of genes that are targets of PPARs, leading to adipogenesis, obesity, and possibly NAFLD. CAR binds to phthalates leading to dysregulation of cytochrome P450 enzymes. This likely contributes to phthalates' deleterious carcinogenic and metabolic effects.

Bisphenol A (BPA) is another plastics chemical, also produced in high volume (see also Chapter 40). It is a key component of polycarbonate plastics and is commonly found in food and beverage containers including infant feeding bottles.[39] BPA is linked through ester bonds to most commercial plastic products, but it is subject to hydrolysis at high temperatures or after exposure to an acidic or basic environment. BPA can also leach from polycarbonate containers to contaminate food and beverages. Increased levels of the BPA exposure have been associated with abnormal serum liver enzyme tests in adults. Exposure of children to even low doses of BPA may adversely affect later liver function.[40]

BPA is an endocrine disruptor that may promote obesity and its associated metabolic complications.[41] Micromolar concentrations of BPA enhance adipocyte differentiation and lipid accumulation in vitro. However, studies are needed to determine the relationship between BPA and its derivatives and PPARγ signaling, as has been done for other obesogens.

Conclusion

Future research will likely discover additional hepatotoxicants among the many hundreds of untested chemicals to which children in the modern world are widely exposed. Clinicians should be alert to the possibility of a toxic exposure in any child with unexplained disease or dysfunction of the liver. Researchers in the fields of liver medicine, children's environmental health, and toxicology have the opportunity to discover previously unrecognized mechanisms of action of newly discovered hepatotoxins and also can use hepatotoxicants as probes to unveil previously unknown aspects of liver function.

References

1. Stanca CM, Babar J, Singal V, Ozdenerol E, Odin JA. Pathogenic role of environmental toxins in immune-mediated liver diseases. *J Immunotoxicol*. 2008;5(1):59–68.

2. Kearns GL, Abdel-Rahman SM, Alander SW, Blowey DL, Leeder JS, Kauffman RE. Developmental pharmacology: drug disposition, action, and therapy in infants and children. *N Engl J Med*. 2003;349(12):1157–67.

3. Suchy FJ, Sokol RJ, Balistreri WB, Mack C, Shneider BL, Bezerra J, eds. *Liver Disease in Children*, 5th ed. Cambridge, UK: Cambridge University Press; 2021.

4. Christensen BC, Marsit CJ. Epigenomics in environmental health. *Front Genet*. 2011;2:84.

5. Manibusan MK, Odin M, Eastmond DA. Postulated carbon tetrachloride mode of action: a review. *J Environ Sci Health. Part C, Environ Carcinogen Ecotoxicol Rev*. 2007;25(3):185–209.

6. Sexton K, Adgate JL, Church TR, et al. Children's exposure to volatile organic compounds as determined by longitudinal measurements in blood. *Environ Health Perspect*. 2005;113(3):342–9.

7. Kensler TW, Roebuck BD, Wogan GN, Groopman JD. Aflatoxin: a 50-year odyssey of mechanistic and translational toxicology. *Toxicol Sci*. 2011;120(Suppl 1):S28–48.

8. Gardner R, Hamadani J, Grander M, et al. Persistent exposure to arsenic via drinking water in rural Bangladesh despite major mitigation efforts. *Am J Public Health*. 2011;101 (Suppl 1):S333–8.

9. Jomova K, Jenisova Z, Feszterova M, et al. Arsenic: toxicity, oxidative stress and human disease. *J Appl Toxicol*. 2011;31(2):95–107.

10. Liaw J, Marshall G, Yuan Y, Ferreccio C, Steinmaus C, Smith AH. Increased childhood liver cancer mortality and arsenic in drinking water in northern Chile. *Cancer Epidemiol Biomarkers Prevent*. 2008;17(8):1982–7.

11. Davey JC, Nomikos AP, Wungjiranirun M, et al. Arsenic as an endocrine disruptor: arsenic disrupts retinoic acid receptor-and thyroid hormone receptor-mediated gene regulation and thyroid hormone-mediated amphibian tail metamorphosis. *Environ Health Perspect*. 2008;116(2):165–72.

12. Kelishadi R, Poursafa P. Obesity and air pollution: global risk factors for pediatric nonalcoholic fatty liver disease. *Hepatitis Month*. 2011;11(10):794–802.

13. Garcia E, Stratakis N, Valvi D, et al. Prenatal and childhood exposure to air pollution and traffic and the risk of liver injury in European children. *Environ Epidemiol*. 2021;5(3):e153.

14. Laing S, Wang G, Briazova T, et al. Airborne particulate matter selectively activates endoplasmic reticulum stress response in the lung and liver tissues. *Am J Physiol Cell. Physiol*. 2010;299(4):C736–49.

15. Kelishadi R, Mirghaffari N, Poursafa P, Gidding SS. Lifestyle and environmental factors associated with inflammation, oxidative stress and insulin resistance in children. *Atherosclerosis*. 2009;203(1):311–9.

16. Yuan H, Shyy JY, Martins-Green M. Second-hand smoke stimulates lipid accumulation in the liver by modulating AMPK and SREBP-1. *J Hepatol*. 2009;51(3):535–47.

17. Wu F, Pahkala K, Juonala M, et al. Childhood and adulthood passive smoking and nonalcoholic fatty liver in midlife: a 31-year cohort study. *Am J Gastroenterol.* 2021;116(6):1256–63.

18. Sorahan T, Lancashire RJ. Parental cigarette smoking and childhood risks of hepatoblastoma: OSCC data. *Br J Cancer.* 2004;90(5):1016–8.

19. Diez-Izquierdo A, Cassanello-Penarroya P, Lidon-Moyano C, Matilla-Santander N, Balaguer A, Martinez-Sanchez JM. Update on thirdhand smoke: a comprehensive systematic review. *Environ Res.* 2018;167:341–71.

20. Stratakis N, Conti DV, Borras E, et al. Association of fish consumption and mercury exposure during pregnancy with metabolic health and inflammatory biomarkers in children. *JAMA Netw Open.* 2020;3(3):e201007.

21. Stratakis N, Golden-Mason L, Margetaki K, et al. In utero exposure to mercury is associated with increased susceptibility to liver injury and inflammation in childhood. *Hepatology.* 2021;74(3):1546–59.

22. Chen R, Xu Y, Xu C, et al. Associations between mercury exposure and the risk of nonalcoholic fatty liver disease (NAFLD) in US adolescents. *Environ Sci Pollut Res Int.* 2019;26(30):31384–91.

23. Satarug S. Long-term exposure to cadmium in food and cigarette smoke, liver effects and hepatocellular carcinoma. *Curr Drug Metab.* 2012;13(3):257–71.

24. Mouro VGS, Ladeira LCM, Lozi AA, et al. Different routes of administration lead to different oxidative damage and tissue disorganization levels on the subacute cadmium toxicity in the liver. *Biol Trace Elem Res.* 2021;199(12):4624–34.

25. Mnif W, Hassine AI, Bouaziz A, Bartegi A, Thomas O, Roig B. Effect of endocrine disruptor pesticides: a review. *Int J Environ Res Public Health.* 2011;8(6):2265–303.

26. Badr AM. Organophosphate toxicity: updates of malathion potential toxic effects in mammals and potential treatments. *Environ Sci Pollut Res Int.* 2020;27(21):26036–57.

27. Adigun AA, Wrench N, Seidler FJ, Slotkin TA. Neonatal organophosphorus pesticide exposure alters the developmental trajectory of cell-signaling cascades controlling metabolism: differential effects of diazinon and parathion. *Environ Health Perspect.* 2010;118(2):210–5.

28. Tabb MM, Blumberg B. New modes of action for endocrine-disrupting chemicals. *Mol Endocrinol.* 2006;20(3):475–82.

29. Kahn LG, Philippat C, Nakayama SF, Slama R, Trasande L. Endocrine-disrupting chemicals: implications for human health. *Lancet Diabetes Endocrinol.* 2020;8(8):703–18.

30. Younossi ZM. Non-alcoholic fatty liver disease: a global public health perspective. *J Hepatol.* 2019;70(3):531–44.

31. Heindel JJ, vom Saal FS. Role of nutrition and environmental endocrine disrupting chemicals during the perinatal period on the aetiology of obesity. *Mol Cell Endocrinol.* 2009;304(1–2):90–6.

32. Heindel JJ, Blumberg B. Environmental obesogens: mechanisms and controversies. *Annu Rev Pharmacol Toxicol.* 2019;59:89–106.

33. Stratakis N, Conti DV, Jin R, et al. Prenatal exposure to perfluoroalkyl substances associated with increased susceptibility to liver injury in children. *Hepatology.* 2020;725:1758–70.

34. Jin R, McConnell R, Catherine C, et al. Perfluoroalkyl substances and severity of nonalcoholic fatty liver in children: an untargeted metabolomics approach. *Environ Int.* 2020;134:105220.

35. Grun F, Watanabe H, Zamanian Z, et al. Endocrine-disrupting organotin compounds are potent inducers of adipogenesis in vertebrates. *Mol Endocrinol.* 2006;209:2141–55.

36. Koch HM, Preuss R, Drexler H, Angerer J. Exposure of nursery school children and their parents and teachers to di-n-butylphthalate and butylbenzylphthalate. *Int Arch Occupat Environ Health*. 2005;783:223–9.

37. Wittassek M, Angerer J. Phthalates: metabolism and exposure. *Int J Androl*. 2008;312:131–8.

38. Ghosh J, Das J, Manna P, Sil PC. Hepatotoxicity of di-(2-ethylhexyl)phthalate is attributed to calcium aggravation, ROS-mediated mitochondrial depolarization, and ERK/NF-kappaB pathway activation. *Free Radical Biol Med*. 2010;4911:1779–91.

39. Rubin BS. Bisphenol A: an endocrine disruptor with widespread exposure and multiple effects. *J Steroid Biochem Mol Biol*. 2011;127(1–2):27–34.

40. Lee S, Lee HA, Park B et al. A prospective cohort study of the association between bisphenol A exposure and the serum levels of liver enzymes in children. *Environ Res*. 2018;161:195–201.

41. Rubin BS, Soto AM. Bisphenol A: perinatal exposure and body weight. *Mol Cell Endocrinol*. 2009;304(1–2):55–62.

53

The Environment and Kidney Disease in Children

Darcy K. Weidemann, Jeffrey J. Fadrowski, and Virginia M. Weaver

Introduction

The kidneys may be particularly vulnerable to environmental pollutants secondary to their role in the excretion of metabolic waste and many bioactive foreign substances. Although the kidneys are approximately 1% of body mass, they receive 20–25% of cardiac output. Mounting evidence suggests that environmental exposures may play an important role in the development of acute and chronic kidney disease in children. Chronic kidney disease (CKD) is an increasing public health problem, with an estimated global prevalence of 9.1% (697.5 million cases) in 2017.[1] CKD is the 12th leading cause of death worldwide, causing 1.2 million deaths and an additional 1.4 million deaths from cardiovascular disease attributed to impaired kidney function.[1] CKD is a devastating illness in children, with profound long-term effects on their health and development. Children with end-stage kidney disease (ESKD) experience mortality rates that are 30–150 times higher than the general pediatric population and substantially shortened life spans compared to their peers.[2] In this setting, identification and control of CKD risk factors, particularly those that are potentially preventable, are critical.

The Developing Kidney

The nephron is the basic functional unit of the kidney, composed of a glomerulus that filters the blood and kidney tubules that process waste products and nutrients (Figure 53.1).

Each kidney contains approximately 1,000,000 nephrons that work in tandem to excrete waste products, maintain neutral blood pH and fluid balance in the body, and produce hormones (e.g., for promotion of red blood cell production). Nephrogenesis is complete in humans by 36 weeks of gestation. Genetic and environmental factors associated with decreased nephron endowment include maternal malnutrition, gestational diabetes, prematurity, uteroplacental insufficiency, and maternal–fetal drug exposure. Toxicant exposures to the developing kidney also can reduce the number of nephrons and disrupt nephron structure and/or function.[3] A given nephron can increase its filtration in response to damage to other nephrons, but this adaptive response is limited to the extent that, once a critical mass of nephrons has been damaged, the kidney's ability to compensate is overcome, eventually leading to a decrease in kidney function. Thus any

Darcy K. Weidemann, Jeffrey J. Fadrowski, and Virginia M. Weaver, *The Environment and Kidney Disease in Children*
In: *Textbook of Children's Environmental Health*, Second Edition. Edited by: Ruth A. Etzel and Philip J. Landrigan,
Oxford University Press. © Oxford University Press 2024. DOI: 10.1093/oso/9780197662526.003.0053

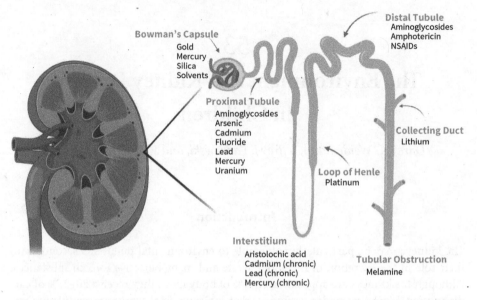

Figure 53.1. The structure of the nephron, the functional unit of the kidney, along with putative sites of injury by selected environmental chemicals
Source: Image created with BioRender.com.

insult to the developing kidney that impacts nephron development may subsequently increase the risk of CKD.

Kidney Outcomes

Assessment of toxicant injury to the kidney requires familiarity with tests that are used to assess kidney function and structure.

Glomerular Filtration Measures

The gold standard to assess kidney function is the glomerular filtration rate (GFR). However, since GFR measurement involves intravenous administration of an exogenous radionuclide or radiocontrast marker followed by collection of timed sequential blood and/or urine samples to measure clearance by the kidneys, less invasive and time-consuming methods to estimate GFR (eGFR) are generally preferred despite their more limited precision and accuracy. Such methods utilize endogenous markers of glomerular filtration, including serum creatinine and cystatin C. The CKiD U25 equations, developed in the prospective Chronic Kidney Disease in Children (CKiD) cohort, are applicable for children and young adults 1–25 years of age.[4]

Kidney Early Biological Effect Markers

Kidney damage involves changes in the structure or function of the kidney. Using kidney filtration measures such as creatinine and GFR estimation equations to assess damage to the kidney is challenging owing to the kidney's remarkable adaptive abilities. Thus, significant injury to the kidneys may occur before a decrease in GFR becomes detectable.

Urinary early biological effect (EBE) markers may provide more sensitive means to detect early kidney damage and represent a burgeoning area of research interest. Validated kidney EBE markers include albumin, a high-molecular-weight protein whose presence in urine primarily reflects glomerular damage, and β2-microglobulin, a low-molecular-weight protein that is normally reabsorbed in the proximal tubules but is excreted in urine after tubular injury. More novel EBE markers such as neutrophil gelatinase-associated lipocalin (NGAL) and kidney injury molecule-1 (KIM-1) are being actively studied. There is also growing interest in identifying biomarkers for CKD progression in children.[5] In 665 children with CKD in the CKiD study, KIM-1, α1 microglobulin, and other biomarkers were associated with CKD progression over a 6.5-year period.[6]

Environmental Toxicants as Risk Factors for Kidney Disease

Owing to their role in blood filtration, the kidneys are exposed to myriad toxicants, which can be removed by glomerular filtration, passive diffusion, and/or or active transport by the kidney tubules. Time is a key factor in distinguishing between acute kidney injury (AKI) and CKD. AKI is defined by change in serum creatinine and/or urine output within 7 days or less.[7] CKD involves kidney injury based on structural and/or functional abnormalities lasting 3 months or longer. High-level exposure to many of the agents discussed in this chapter increases risk for AKI. Fortunately, such exposure is increasingly uncommon, especially in high-income countries. However, chronic exposures to lower levels of nephrotoxicants remain widespread and can also cause kidney injury. Accordingly, this chapter focuses primarily on effects from chronic, lower-level exposures. Research adjusting for common kidney confounders is also prioritized. Additionally, prospective data are emphasized when available due to the potential for reverse causality, in which higher levels of a toxicant in the body result from decreased excretion in CKD rather than causing CKD. Even in the setting of normal kidney function, GFR may potentially impact low levels of chemicals in the body.[8] As noted below, inconsistency in the published literature is apparent; reverse causality and other challenges in nephrotoxicant research may be involved.[8] Kidney reserve and lack of sensitive outcome measures may be additional factors contributing to discrepant research findings. Figure 53.1 highlights a number of established nephrotoxicants and the primary site of action in the nephron.

Metals

Arsenic

A systematic review of arsenic and CKD observed dose-response associations between arsenic and prevalent albuminuria and proteinuria in the few studies with three or more exposure categories.[9] A positive association with kidney disease mortality based on ecological studies was also noted. In Taiwan, standardized mortality ratios for kidney disease declined in the decades following efforts to eliminate arsenic exposure in the drinking water supply.[9]

Studies assessing associations between arsenic exposure and eGFR have been less consistent. One study provides a potential explanation. Cross-sectional analyses of baseline data observed that higher urine inorganic arsenic levels were unexpectedly associated with a *lower* odds ratio of prevalent CKD.[10] However, methylated arsenic metabolites were associated with increased risk of CKD after adjustment for inorganic arsenic. Prospectively, the combined arsenic exposure measure was associated with increased risk of incident CKD over the 6.9-year follow-up period. The authors suggested that kidney function may affect excretion of inorganic arsenic.

The MINIMat children's study in Bangladesh ($n = 1,887$), a country with endemic high arsenic exposure, reported a borderline inverse association ($p = 0.08$) between urine arsenic (at age 18 months) and eGFR at 4.5 years.[11] Maternal urinary arsenic during pregnancy was not associated with eGFR. Arsenic measures were associated with higher blood pressure but not with kidney volume in 4.5-year-olds.[11] Earlier arsenic measures were not associated with eGFR in 540 children at age 9 in MINIMat.[12] A *positive* association between arsenic and eGFR was observed in 1,253 participants ages 12–30 in the US National Health and Nutrition Examination Survey (NHANES) but only when results were corrected for urinary concentration by urine creatinine, not osmolality.[13] Inorganic arsenic was not associated with various markers of GFR and tubular function in a cross-sectional study of more than 300 children living near European smelters, compared to children living in less polluted areas.[14]

Cadmium

Cadmium is a well-known nephrotoxicant that targets the proximal tubule in the kidney resulting in tubulointerstitial fibrosis. Longitudinal studies of populations with chronic, high-level exposures have reported increased risk of CKD or CKD mortality.[15–17]

In longitudinal data from 480 9-year-old children, cumulative dietary cadmium was *inversely* associated with blood urea nitrogen (BUN); no significant associations with serum cystatin C, creatinine, or eGFR were found.[18] In the MINIMat study, no prospective cadmium associations were observed in children at age 4.5.[11] However, in 9-year-olds, urinary cadmium at age 4.5 was inversely associated with cystatin C-based eGFR (although not creatinine-based eGFR).[12] In cross-sectional MINIMat analyses at 4.5 years of age, the inverse urine cadmium association with cystatin C-based eGFR was attenuated by urinary selenium, which was postulated as a potential protective factor.[19] In European adolescents, cadmium levels were not associated with serum cystatin C or urinary β2-microglobulin levels.[20] A study of 594 Thai children reported significant positive associations between urinary cadmium and urinary β2-microglobulin.[21] Another cross-sectional study, in 512 Mexican adolescents, reported a *positive* association between urine cadmium and eGFR but only when urine concentration adjustment used urine creatinine rather than osmolality.[22] Similarly, in European children living near smelters, urine cadmium was associated with *lower* serum creatinine but not with serum β2-microglobulin or cystatin C.[14] Blood cadmium was not associated with any of the three GFR measures although both blood and urine cadmium were positively associated with urinary EBE markers. Among 159 US children 6–17 years of age residing near a zinc smelter, urine cadmium was not associated with albuminuria, *N*-acetyl-beta-*D*-glucosaminidase (NAG), or β2-microglobulin.[23]

Lead

High-dose lead poisoning (blood lead levels >80–100 µg/dL) initially injures the proximal tubules.[24] *Fanconi syndrome*, in which small molecules, such as glucose, are excreted rather than reabsorbed, may occur, especially in children. This has been shown to be reversible with timely chelation therapy.[24] Chronic exposure (blood lead levels > 60 µg/dL) can lead to lead nephropathy, which is characterized by contracted, fibrotic kidneys, decreased GFR, bland urinary sediment, and hypertension.[24]

Higher ESKD mortality occurred in young adults who survived a Queensland, Australia, epidemic of childhood lead paint poisoning.[25] Bone lead levels in affected patients were four times higher than in patients with ESKD from other causes. In subsequent studies of adolescents and adults who sustained childhood lead poisoning, a few cases of kidney disease consistent with lead nephropathy were observed[26,27] as was persistent partial Fanconi syndrome.[28] However, increased mortality and statistically significant differences in ESKD have not been observed. Possible explanations for this discrepancy include lack of chelation therapy in the Queensland children. Unfortunately, severe lead poisoning is not just a historic event. An epidemic occurred in Zamfara, Nigeria, in 2010, resulting in the deaths of more than 400 children.[29] Survivors were chelated, and follow-up of these children could provide relevant information on the adverse kidney impact of extremely high lead exposure and extent of mitigation from chelation.

At lower levels of exposure, a prospective study of 453 children, 8–12 years of age, observed an inverse association between prenatal lead exposure and eGFR but only in overweight children.[30] In Bangladeshi children, prenatal erythrocyte lead was not associated with blood pressure or eGFR at 4.5 years of age.[31] However, in the subset of children who had kidney volume measured, an inverse association with late-gestation lead was observed. Also prospectively, an adverse effect of childhood lead exposure (mean blood lead of 20–40 µg/dL) on erythropoietin production was observed to be reversible by young adulthood as lead exposure declined.[32] In NHANES data from 769 adolescents, higher blood lead (median of 1.5 µg/dL) was associated with lower cystatin C- but not serum creatinine-based GFR.[33] In a cohort of children with CKD, blood lead was inversely associated with measured GFR in the subgroup with glomerular disease underlying their CKD (N = 73) but not the entire cohort (N = 391). Blood lead was positively associated with serum cystatin C and urine β2-microglobulin in 200 Belgian adolescents (mean blood lead of 1.5–2.7 µg/dl).[20] In contrast, *negative* associations with serum creatinine and cystatin C were reported in European children.[14] The authors implicated hyperfiltration, an adaptive increase in GFR in response to nephron injury that ultimately leads to subsequent decline in GFR, as a factor in these contradictory findings.

Male rats exposed to lead in utero through puberty and post-puberty, depending on group, had increased serum creatinine levels at both time points.[34] Blood lead levels were low for an animal model (mean of 7.5 µg/dL). This experimental approach provides additional support for lead nephrotoxicity and suggests increased susceptibility during early development.

Additional longitudinal research in children with a range of outcome assessment methods is needed to determine the impact of current lower lead exposures.

Mercury

Chronic mercury exposure has two distinct effects on the kidney: proximal tubule toxicity and, infrequently, an immune-complex mediated glomerular injury often characterized by massive urinary losses of protein and resultant nephrotic syndrome.[35] AKI has also been reported in cases of mercury poisoning. Age-related differences in kidney function may impact mercury toxicity. Animal data suggest that decreased tubular transport in the young, resulting in less mercury excretion, increases toxicity.[3] In European children residing near smelters, urine mercury was positively associated with urinary NAG but *negatively* associated with serum β2-microglobulin levels.[14]

Mercury amalgam dental fillings are a common source of exposure in the general population. The impact of these fillings on kidney function in children has been examined in two randomized trials, a powerful research design that can rarely be utilized in environmental research. The New England Children's Amalgam Trial enrolled 534 children, 6–10 years of age, who had no prior mercury amalgam fillings.[36] No significant differences between treatment groups (mercury amalgam or resin composite) in average levels of urinary EBE markers were observed over 5 years although a significantly increased prevalence of microalbuminuria occurred in the amalgam group in years 3–5. Results in the Casa Pia trial of 507 children followed for 7 years revealed no significant differences in albuminuria and other EBE markers by treatment group.[37] Casa Pia researchers discounted the statistical approach in an analysis of their data by another group who reported an association between amalgams and a proximal tubule EBE marker.[38] Despite generally reassuring results, continued follow-up of these children would be of interest, as would examination of GFR measures.

Other Metals

The kidney is considered a target organ for uranium toxicity based on animal and limited human data, the latter relying primarily on EBE markers.[39] An evaluation of a family exposed to naturally uranium-contaminated well water suggests that children may be especially susceptible.[40] Water uranium levels were measured at 866 and 1,160 µg/L, much higher than the US Environmental Protection Agency maximum contaminant level of 30 µg/L in public water. Urinary β2-microglobulin was elevated only in the youngest child (3 years old) and declined with cessation of exposure. Exposure from uranium mining was associated with kidney disease and hypertension in members of the Navajo Nation.[41] However, some studies reported *higher* GFR in participants with greater uranium exposure.[39] Age-related differences in uranium kidney toxicity have also been reported in animal models.[3] A cross-sectional study of 512 Mexican children reported an *increase* in eGFR with urine thallium but not urine uranium.[22] Among 83 Mexican children (5–12 years of age) urinary chromium was associated with KIM-1 but not with eGFR or albuminuria.[42]

Melamine

Melamine was patented as a less expensive protein alternative for livestock due to its high nitrogen content but fell out of favor following reports of kidney disease in animals ingesting it. Although not intended for human consumption, it has been used to adulterate food by unscrupulous manufacturers seeking to lower production costs since it falsely elevates the protein content.

Melamine-tainted milk powder was implicated in an outbreak of urinary stones in Chinese infants and toddlers in 2008. The Chinese Ministry of Health reported that, by December of 2008, 294,000 children had been affected.[43] Nephrotoxicity was related to urinary obstruction from precipitation of melamine in the lower urinary tract. Complications related to urinary stones, including AKI, led to almost 52,000 hospitalizations and at least six confirmed deaths.[43] Among 198 children treated for urolithiasis, 17 had residual stones and 10 and 6 had proteinuria and microscopic hematuria, respectively, at a 5-year follow-up.[44] Serum creatinine was nonsignificantly higher in the 49 who had required surgical intervention. Among 45 infants with persistent urolithiasis at hospital discharge, stones and/or hydronephrosis persisted in 11 and 5, respectively, at 4 years.[45]

In response, allowable limits for melamine in food products have decreased worldwide. However, a recent investigation of urinary melamine and cyanuric acid (a related compound) in US children observed relatively higher levels compared to the limited published data (children in Taiwan), with levels above the limit of detection for 78% and 95% of samples, respectively.[46] Cyanuric acid but not melamine demonstrated a borderline positive association with KIM-1. In Taiwanese children, urine melamine was associated with albuminuria.[47]

Other Exposures

Data also are accumulating regarding the nephrotoxicity of other chemicals in children. In prospective data, eGFR was inversely associated with four urinary phthalate metabolites in 103 children (4–13 years) across three seasons.[48] Phthalate metabolites were associated with albuminuria in 667 children in NHANES.[49] In the prospective CKiD study ($n = 618$), bisphenol A (BPA) and phthalates demonstrated consistent patterns of oxidative stress and tubular injury over time, although only phthalic acid was associated with a traditional clinical kidney endpoint (negative eGFR association).[50] However, cross-sectional analyses of baseline CKiD data revealed null and positive BPA and phthalates associations with eGFR, respectively.[51] BPA was associated with albuminuria in 710 children and adolescents in NHANES.[52] Prospectively in CKiD, organophosphate pesticide metabolites were associated with increased KIM-1 but paraxoxically also associated with *higher* eGFR.[53]

In the PROGRESS study of 427 mother–child pairs, exposure to air pollution, assessed as particulate matter 2.5 μm or smaller, between weeks 1–18 of gestation was associated with increased eGFR in children at ages 8–10 years, whereas exposure in the first 14 months after birth was associated with decreased eGFR.[54] Among 5,622 Dutch children, fetal exposure to maternal and paternal tobacco smoke was associated with reduced kidney volume at age 6, and maternal smoking was associated with lower eGFR.[55] Exposure to secondhand tobacco smoke was associated with lower eGFR in 7516 adolescents in NHANES.[56]

A longitudinal study of 483 children from the PROGRESS cohort reported a nonsignificant, negative association between urinary fluoride at age 4 and subsequent eGFR at follow-up 4–8 years later; borderline significance (p = 0.08) was reported in obese children.[57] Cross-sectional data in NHANES adolescents[58] and in Mexican schoolchildren[59] revealed inconsistent effects on kidney injury with plasma and water fluoride, although the latter investigation found significant associations with carotid intima media thickness.

A study of measured and predicted perfluorooctanoic acid (PFOA) levels and eGFR in 9,660 children aged 1 to less than 18 years illustrates the challenges of nephrotoxicant research.[60] The authors hypothesized that predicted serum PFOA (available since this study had extensive exposure information) would be less susceptible to reverse causation than measured levels. Indeed, measured PFOA was inversely associated with eGFR, whereas no associations were observed with predicted levels.

Mixed Exposures

Humans are exposed to a wide range of chemicals in the modern environment. The risk of kidney toxicity related to low-level exposure to a toxicant in isolation may be very different from the risk associated with the same exposure in a setting of multiple toxicants. Despite recognition of this reality, studies examining the potentially synergistic impacts on the kidney of multiple simultaneous exposures have been rare. A cross-sectional examination of 2,709 adolescents 12–19 years in NHANES analyzed associations between co-exposure to four metals (lead, cadmium, mercury, and arsenic) using both blood and urine levels, as applicable, and various kidney outcomes.[61] The investigators found significant positive associations between deciles of urine metal mixtures and BUN, albuminuria, and, paradoxically, eGFR.

Chronic Kidney Disease of Uncertain Etiology

An epidemic of CKD in young and middle-aged adults has been reported in Central America, Sri Lanka, and other equatorial regions. The cause remains unknown but men employed in agricultural jobs are at highest risk, and commonly implicated causes of CKD, such as diabetes and hypertension, are absent.[62] Although primarily diagnosed in adults, increased EBE markers and decreased eGFR have been reported in adolescents and children. A study of participants less than 18 years old in rural Salvador observed CKD (GFR <60 mL/min/1.73 m^2 or albuminuria) in 3.4–5.1% depending on age ($n = 1,623$).[63] Glomerular hyperfiltration was noted as well.

Clinical Impact: Diagnosis and Treatment

The diagnosis of an environmental toxicant as a potential cause of or co-factor for kidney disease in a child involves identification of possible exposure through an environmental history, as discussed in Chapter 56. In the case of bioaccumulative metals (i.e., lead or cadmium), history of past exposure is important. The history also should include nutrition, with assessment of iron and calcium intake, and whether the patient exhibits a history of pica. Family occupational history and residential exposures to potential environmental pollutants, home renovations, age of the home, water source, and history of foreign birthplace and/or potential refugee status also may be relevant.

This information is useful in the evaluation not only of children with kidney disease but also in the clinical assessment of children with CKD risk factors such as obesity, hypertension, or diabetes, in which a chemical exposure may interact with the metabolic risk factor.

Appropriate biological monitoring tests should be obtained. Depending on the metal, levels can be measured in blood or urine, as discussed in the relevant chapters (Chapters 31–34). Assessment for urinary stones with urinalysis and ultrasound is required for melamine; B-mode ultrasonography is recommended since the stones lack calcium and are generally radiolucent.

Treatment depends on the kidney diagnosis; however, minimizing further nephrotoxicant exposure is essential to prevent additional kidney injury. If biomonitoring results are elevated or substantial exposure is indicated by history, follow-up testing and/ or referral to a clinician with expertise in management of patients with the relevant exposure (e.g., occupational and environmental medicine or toxicology) or a pediatric environmental health specialty unit (Chapter 56) is strongly recommended.

References

1. Bikbov B, Purcell CA, Levey AS, et al. Global, regional, and national burden of chronic kidney disease, 1990–2017: a systematic analysis for the Global Burden of Disease Study 2017. *Lancet.* 2020;395(10225):709–33.

2. Warady BA, Chadha V. Chronic kidney disease in children: the global perspective. *Pediatr Nephrol.* 2007;22(12):1999–2009.

3. Solhaug MJ, Bolger PM, Jose PA. The developing kidney and environmental toxins. *Pediatrics.* 2004;113(4 Suppl):1084–91.

4. Pierce CB, Munoz A, Ng DK, Warady BA, Furth SL, Schwartz GJ. Age- and sex-dependent clinical equations to estimate glomerular filtration rates in children and young adults with chronic kidney disease. *Kidney Int.* 2021;99(4):948–56.

5. Greenberg JH, Kakajiwala A, Parikh CR, Furth S. Emerging biomarkers of chronic kidney disease in children. *Pediatr Nephrol.* 2018;33(6):925–33.

6. Greenberg JH, Abraham AG, Xu Y, et al. Urine biomarkers of kidney tubule health, injury, and inflammation are associated with progression of CKD in children. *J Am Soc Nephrol.* 2021;32(10):2664–77.

7. Khwaja A. KDIGO clinical practice guidelines for acute kidney injury. *Nephron Clin Pract.* 2012;120(4):c179–84.

8. Weaver VM, Kotchmar DJ, Fadrowski JJ, Silbergeld EK. Challenges for environmental epidemiology research: are biomarker concentrations altered by kidney function or urine concentration adjustment? *J Expo Sci Environ Epidemiol.* 2016;26(1):1–8.

9. Zheng L, Kuo CC, Fadrowski J, Agnew J, Weaver VM, Navas-Acien A. Arsenic and chronic kidney disease: a systematic review. *Curr Environ Health Rep.* 2014;1(3):192–207.

10. Zheng LY, Umans JG, Yeh F, et al. The association of urine arsenic with prevalent and incident chronic kidney disease: evidence from the strong heart study. *Epidemiology.* 2015;26(4):601–12.

11. Hawkesworth S, Wagatsuma Y, Kippler M, et al. Early exposure to toxic metals has a limited effect on blood pressure or kidney function in later childhood, rural Bangladesh. *Int J Epidemiol.* 2013;42(1):176–85.

12. Akhtar E, Roy AK, Haq MA, et al. A longitudinal study of rural Bangladeshi children with long-term arsenic and cadmium exposures and biomarkers of cardiometabolic diseases. *Environ Pollut*. 2021;271:116333.

13. Weidemann D, Kuo CC, Navas-Acien A, Abraham AG, Weaver V, Fadrowski J. Association of arsenic with kidney function in adolescents and young adults: results from the National Health and Nutrition Examination Survey 2009–2012. *Environ Res*. 2015;140:317–24.

14. de Burbure C, Buchet JP, Leroyer A, et al. Renal and neurologic effects of cadmium, lead, mercury, and arsenic in children: evidence of early effects and multiple interactions at environmental exposure levels. *Environ Health Perspect*. 2006;114(4):584–90.

15. Nishijo M, Nogawa K, Suwazono Y, Kido T, Sakurai M, Nakagawa H. Lifetime cadmium exposure and mortality for renal diseases in residents of the cadmium-polluted Kakehashi River Basin in Japan. *Toxics*. 2020;8():81).

16. Roels HA, Van Assche FJ, Oversteyns M, De Groof M, Lauwerys RR, Lison D. Reversibility of microproteinuria in cadmium workers with incipient tubular dysfunction after reduction of exposure. *Am J Ind Med*. 1997;31(5):645–52.

17. Swaddiwudhipong W, Limpatanachote P, Mahasakpan P, Krintratun S, Punta B, Funkhiew T. Progress in cadmium-related health effects in persons with high environmental exposure in northwestern Thailand: a five-year follow-up. *Environ Res*. 2012;112:194–8.

18. Rodriguez-Lopez E, Tamayo-Ortiz M, Ariza AC, et al. Early-life dietary cadmium exposure and kidney function in 9-year-old children from the PROGRESS Cohort. *Toxics*. 2020;8(4)83.

19. Skröder H, Hawkesworth S, Kippler M, et al. Kidney function and blood pressure in preschool-aged children exposed to cadmium and arsenic: potential alleviation by selenium. *Environ Res*. 2015;140:205–13.

20. Staessen JA, Nawrot T, Hond ED, et al. Renal function, cytogenetic measurements, and sexual development in adolescents in relation to environmental pollutants: a feasibility study of biomarkers. *Lancet*. 2001;357(9269):1660–9.

21. Swaddiwudhipong W, Mahasakpan P, Jeekeeree W, et al. Renal and blood pressure effects from environmental cadmium exposure in Thai children. *Environ Res*. 2015;136:82–7.

22. Weaver VM, Vargas GG, Silbergeld EK, et al. Impact of urine concentration adjustment method on associations between urine metals and estimated glomerular filtration rates (eGFR) in adolescents. *Environ Res*. 2014;132C:226–32.

23. Noonan CW, Sarasua SM, Campagna D, Kathman SJ, Lybarger JA, Mueller PW. Effects of exposure to low levels of environmental cadmium on renal biomarkers. *Environ Health Perspect*. 2002;110(2):151–5.

24. Weaver V, Jaar B. *Lead Nephropathy and Lead-Related Nephrotoxicity*. Waltham, MA: UpToDate; 2022.

25. Inglis JA, Henderson DA, Emmerson BT. The pathology and pathogenesis of chronic lead nephropathy occurring in Queensland. *J Pathol*. 1978;124(2):65–76.

26. Hu H. A 50-year follow-up of childhood plumbism: hypertension, renal function, and hemoglobin levels among survivors. *Am J Dis Child*. 1991;145(6):681–7.

27. Moel DI, Sachs HK. Renal function 17 to 23 years after chelation therapy for childhood plumbism. *Kidney Int*. 1992;42(5):1226–31.

28. Loghman-Adham M. Aminoaciduria and glycosuria following severe childhood lead poisoning. *Pediatr Nephrol*. 1998;12(3):218–21.

29. Bashir M, Umar-Tsafe N, Getso K, et al. Assessment of blood lead levels among children aged ≤ 5 years: Zamfara State, Nigeria, June-July 2012. *MMWR Morb Mortal Wkly Rep.* 2014;63(15):325–7.

30. Saylor C, Tamayo-Ortiz M, Pantic I, et al. Prenatal blood lead levels and reduced preadolescent glomerular filtration rate: modification by body mass index. *Environ Int.* 2021:106414.

31. Skroder H, Hawkesworth S, Moore SE, Wagatsuma Y, Kippler M, Vahter M. Prenatal lead exposure and childhood blood pressure and kidney function. *Environ Res.* 2016;151:628–34.

32. Camaj PR, Graziano JH, Preteni E, et al. Long-term effects of environmental lead on erythropoietin production in young adults: a follow-up study of a prospective cohort in Kosovo. *J Environ Public Health.* 2020;2020:3646252.

33. Fadrowski JJ, Navas-Acien A, Tellez-Plaza M, Guallar E, Weaver VM, Furth SL. Blood lead level and kidney function in US adolescents: The Third National Health and Nutrition Examination Survey. *Arch Intern Med.* 2010;170(1):75–82.

34. Berrahal AA, Lasram M, El Elj N, Kerkeni A, Gharbi N, El-Fazaa S. Effect of age-dependent exposure to lead on hepatotoxicity and nephrotoxicity in male rats. *Environ Toxicol.* 2011;26(1):68–78.

35. Li SJ, Zhang SH, Chen HP, et al. Mercury-induced membranous nephropathy: clinical and pathological features. *Clin J Am Soc Nephrol.* 2010;5(3):439–44.

36. Barregard L, Trachtenberg F, McKinlay S. Renal effects of dental amalgam in children: the New England children's amalgam trial. *Environ Health Perspect.* 2008;116(3):394–9.

37. Woods JS, Martin MD, Leroux BG, et al. Biomarkers of kidney integrity in children and adolescents with dental amalgam mercury exposure: findings from the Casa Pia children's amalgam trial. *Environ Res.* 2008;108(3):393–9.

38. DeRouen TA, Woods JS, Leroux BG, Martin MD. Critique of reanalysis of Casa Pia data on associations of porphyrins and glutathione-S-transferases with dental amalgam exposure. *Hum Exp Toxicol.* 2015;34(3):330–2.

39. Arzuaga X, Rieth SH, Bathija A, Cooper GS. Renal effects of exposure to natural and depleted uranium: a review of the epidemiologic and experimental data. *J Toxicol Environ Health B Crit Rev.* 2010;13(7–8):527–45.

40. Magdo HS, Forman J, Graber N, et al. Grand rounds: nephrotoxicity in a young child exposed to uranium from contaminated well water. *Environ Health Perspect.* 2007;115(8):1237–41.

41. Hund L, Bedrick E, Miller C, et al. A Bayesian framework for estimating disease risk due to exposure to uranium mine and mill waste on the Navajo Nation. *J R Stat Soc: Series A (Statistics in Society).* 2015;178(4):1069–91.

42. Cardenas-Gonzalez M, Osorio-Yanez C, Gaspar-Ramirez O, et al. Environmental exposure to arsenic and chromium in children is associated with kidney injury molecule-1. *Environ Res.* 2016;150:653–62.

43. Gossner CM, Schlundt J, Ben Embarek P, et al. The melamine incident: implications for international food and feed safety. *Environ Health Perspect.* 2009;117(12):1803–8.

44. Chang H, Wu G, Yue Z, Ma J, Qin Z. Melamine poisoning pediatric urolithiasis treatment in Gansu, China 5-year follow-up analysis. *Urology.* 2017;109:153–8.

45. Yang L, Wen JG, Wen JJ, et al. Four years follow-up of 101 children with melamine-related urinary stones. *Urolithiasis.* 2013;41(3):265–6.

46. Sathyanarayana S, Flynn JT, Messito MJ, et al. Melamine and cyanuric acid exposure and kidney injury in US children. *Environ Res.* 2019;171:18–23.

47. Tsai HJ, Wu CF, Hsiung CA, et al. Longitudinal changes in oxidative stress and early renal injury in children exposed to DEHP and melamine in the 2011 Taiwan food scandal. *Environ Int*. 2022;158:107018.

48. Liu M, Zhao L, Liu L, et al. Urinary phthalate metabolites mixture, serum cytokines and renal function in children: A panel study. *J Hazard Mater*. 2022;422:126963.

49. Trasande L, Sathyanarayana S, Trachtman H. Dietary phthalates and low-grade albuminuria in US children and adolescents. *Clin J Am Soc Nephrol*. 2014;9(1):100–9.

50. Jacobson MH, Wu Y, Liu M, et al. Serially assessed bisphenol A and phthalate exposure and association with kidney function in children with chronic kidney disease in the US and Canada: A longitudinal cohort study. *PLoS Med*. 2020;17(10):e1003384.

51. Malits J, Attina TM, Karthikraj R, et al. Renal function and exposure to bisphenol A and phthalates in children with chronic kidney disease. *Environ Res*. 2018;167:575–82.

52. Trasande L, Attina TM, Trachtman H. Bisphenol A exposure is associated with low-grade urinary albumin excretion in children of the United States. *Kidney Int*. 2013;83(4):741–8.

53. Jacobson MH, Wu Y, Liu M, et al. Organophosphate pesticides and progression of chronic kidney disease among children: a prospective cohort study. *Environ Int*. 2021;155:106597.

54. Rosa MJ, Politis MD, Tamayo-Ortiz M, et al. Critical windows of perinatal particulate matter (PM2.5) exposure and preadolescent kidney function. *Environ Res*. 2022;204(Pt B):112062.

55. Kooijman MN, Bakker H, Franco OH, Hofman A, Taal HR, Jaddoe VW. Fetal smoke exposure and kidney outcomes in school-aged children. *Am J Kidney Dis*. 2015;66(3):412–20.

56. Garcia-Esquinas E, Loeffler LF, Weaver VM, Fadrowski JJ, Navas-Acien A. Kidney function and tobacco smoke exposure in US adolescents. *Pediatrics*. 2013;131(5):e1415–23.

57. Saylor C, Malin AJ, Tamayo-Ortiz M, et al. Early childhood fluoride exposure and preadolescent kidney function. *Environ Res*. 2022;204(Pt A):112014.

58. Malin AJ, Lesseur C, Busgang SA, Curtin P, Wright RO, Sanders AP. Fluoride exposure and kidney and liver function among adolescents in the United States: NHANES, 2013–2016. *Environ Int*. 2019;132:105012.

59. Jimenez-Cordova MI, Gonzalez-Horta C, Ayllon-Vergara JC, et al. Evaluation of vascular and kidney injury biomarkers in Mexican children exposed to inorganic fluoride. *Environ Res*. 2019;169:220–8.

60. Watkins DJ, Josson J, Elston B, et al. Exposure to perfluoroalkyl acids and markers of kidney function among children and adolescents living near a chemical plant. *Environ Health Perspect*. 2013;121:625–30.

61. Sanders AP, Mazzella MJ, Malin AJ, et al. Combined exposure to lead, cadmium, mercury, and arsenic and kidney health in adolescents age 12–19 in NHANES 2009–2014. *Environ Int*. 2019;131:104993.

62. Johnson RJ, Wesseling C, Newman LS. Chronic kidney disease of unknown cause in agricultural communities. *N Engl J Med*. 2019;380(19):1843–52.

63. Orantes-Navarro CM, Herrera-Valdes R, Almaguer-Lopez M, et al. Chronic kidney disease in children and adolescents in Salvadoran farming communities: NefroSalva Pediatric Study (2009–2011). *MEDICC Rev*. 2016;18:15–21.

54

Injuries, Trauma, and the Environment

Laura Schwab-Reese, Cara J. Hamann, and Amy A. Hunter

Injuries are defined as damage to the body resulting from "acute exposure to thermal, mechanical, electrical, or chemical energy, or the absence of such essentials as heat or oxygen."[1] Although virtually everyone could describe a time that they were injured, this formal definition can be challenging to understand. When describing injuries, people rarely talk about energy exposure. For example, we may say that a child was "burned" rather than "exposed to thermal energy." Nonetheless, the definition can provide a basic foundation for the numerous environmental exposures that cause injuries.

Injuries are often divided into two categories: unintentional and intentional (violence). Unintentional injuries are a leading cause of death for children (0–17 years old) in the United States.[2] These injuries are often referred to as "accidents" because there is no purposeful intent. Injury prevention professionals do not use the term "accidents" because it implies that these incidents are unavoidable, unpredictable events.[3] However, research demonstrates that unintentional injuries—and also violence—are predictable, exhibit repetitive patterns, and can be prevented.[3] Motor vehicle crashes, drowning, suffocation, and falls are common unintentional injuries. Violence occurs due to purposeful intent and includes homicide, suicide, and other forms of violence. Violence is also a leading cause of death for children, particularly after the first year of life.[2] Although intent is often used to describe injuries, the mechanism (e.g., poisoning), perpetrator (e.g., family violence), or environment (e.g., sports injury) are also common ways to classify injuries (Table 54.1).

Environmental exposures, particularly in the psychosocial and built environment, contribute to risk for injury. Children spend much of their time at home, on roadways, and in the workplace, and each of these has a unique combination of environmental hazards. There are also other factors that contribute to disparate risk for injuries, such as gender, race/ethnicity, and age. For example, the rate of unintentional injury deaths is about three times higher among non-Hispanic Black boys and American Indian/Alaska Native boys compared to Asian/Pacific Islander boys (Figure 54.1). The disparities are even more prominent for violent deaths among boys. Girls have much lower rates of unintentional injury and violent deaths, although there are still substantial differences across the racial/ethnic groups. The complexities of social determinants of health are beyond the scope of this chapter, but Healthy People 2030 contains extensive information.[4]

This chapter provides a brief overview of the most common types of injuries experienced by children, including a summary of the incidence, burden, and physical and social environments that influence the risk of injury. Because other chapters in this volume address injury prevention at school (Chapter 17); during sports, play, and recreation (Chapter 19); on farms (Chapter 21); and in workplaces (Chapter 22), this chapter does not focus on injuries in those environments.

Laura Schwab-Reese, Cara J. Hamann, and Amy A. Hunter, *Injuries, Trauma, and the Environment* In: *Textbook of Children's Environmental Health*, Second Edition. Edited by: Ruth A. Etzel and Philip J. Landrigan, Oxford University Press.
© Oxford University Press 2024. DOI: 10.1093/oso/9780197662526.003.0054

Table 54.1 Leading causes of unintentional injury and violent deaths among 11- to 17-year-olds, United States 2010-2019

Rank	Unintentional injury	Number of deaths	Violence	Number of deaths
1	Motor vehicle crashes	21,322	Firearm	13,625
2	Drowning	7,701	Suffocation	7,457
3	Fire/Flame	2,569	Drug poisoning	783
4	Suffocation	2,252	Cut/Pierce	760
5	Drug poisoning	1,884	Fall	257

Source: Centers for Disease Control and Prevention. Web-based injury statistics query and reporting system (WISQARS). Available at: http://www.cdc.gov/injury/wisqars/index.html.

Unintentional Injury

Unintentional injury is the leading cause of death and disability and children and young people in the United States.[2] From 2010 to 2019, almost 44,000 people aged 1–17 years died from an unintentional injury.[2] The following three most common causes of death—cancer, suicide, and homicide—had a combined total lower than unintentional injuries. Nearly half of unintentional injury deaths were caused by motor vehicle crashes, followed by drowning (18%), fire/flame (6%), and suffocation (5%).[2] There also were more than 65 million emergency department visits among this age group for unintentional injuries from 2010 to 2019.[2] Serious nonfatal injuries, as indicated by visits to the emergency department, are generally caused by different mechanisms than are fatal injuries.[2] Falls and "struck by/against" injuries, which involve an unintentional collision with another person or an inanimate object, account for more than half of emergency department visits.

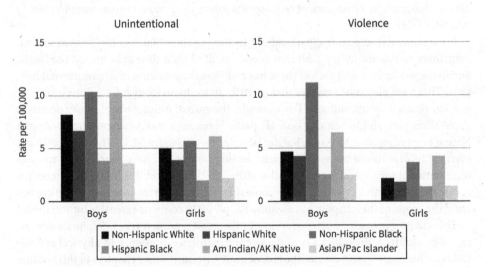

Source: Centers for Disease Control and Prevention. Web-based injury statistics query and reporting system (WISQARS). Available at: http://www.cdc.gov/injury/wisqars/index.html.

Figure 54.1. Rates of unintentional injury and violent deaths among 1- to 17-year-olds, United States 2010–2019

Injuries in the Home Environment

More than 2,000 children die each year, and 3.5 million visit the emergency department related to an injury that occurred in the home. The rate of home injuries among children varies by age, with infants having the highest rate of home injury deaths (17.7 per 100,000 persons).[5] Many of these deaths are related to unsafe sleep environments. The rate of death drops substantially for children ages 1–4 years (4.3 per 100,000), decreases again for children ages 5–14 years (approximately 1 per 100,000), and increases for adolescents (2.8 per 100,000 persons).[5] Across all age groups, boys had higher rates of home injury deaths than girls.[5] Less information is available about nonfatal injuries in the home. In 2012, there were about 3 million of these injuries requiring medical care among children under the age of 12 (58 per 100,000 persons) and about 1.2 million among adolescents ages 12–17 (48.5 per 100,000 persons).[5] There are three types of injuries in the home that are particularly relevant to children: sleep-related deaths, falls, and poisoning.

Sleep-Related Deaths

Unsafe sleep environments, such as sleeping in an adult bed, soft or excess bedding, non-supine sleep positions, and co-sleeping with an adult, increase the risk for these types of deaths. Other risk factors include low birth weight, premature birth, parental smoking, and maternal alcohol or substance use.[5] The exact mechanisms that lead to these deaths are not fully understood, although some have hypothesized that death is more likely to occur when a vulnerable infant (e.g., premature) experiences outside stressors (e.g., prone position) during a critical period of development (e.g., <4 months of age;[5]). Others argue that the underlying causes are more complex.[6]

A recent systematic review identified four promising approaches to prevent sleep-related deaths: a national communication campaign about safe sleep (Back to Sleep), caregiver education, modeling safe sleep behaviors in healthcare settings, and crib distribution programs. Educating parents through communication campaigns, targeted education programs, or modeling behaviors may support parents to make better choices about how they position their infants to sleep.[5] However, education is insufficient if the parents lack access to cribs or other safe sleep equipment, such as bassinets or playpens.[5]

Falls

During a fall, an individual comes to rest unintentionally on the ground, floor, or other lower level. Unintentional falls are a widespread cause of unintentional injury in the home among children.[5] Falling down stairs, falling from windows, and falling from furniture are common types of falls in the home. Of these, falls on stairs are a significant contributor to falls in the home, and infants falling down stairs while in a baby walker are particularly dangerous.[5] Although less common, falling from windows is also particularly dangerous due to the height of the fall.

Many effective fall prevention strategies among children focus on educating parents about ways to reduce children's access to the potential fall hazard. For example, installing gates in the home, particularly fitted stair gates, and/or reducing the use of baby walkers may prevent young children from falling down stairs.[5] Policy interventions, such as mandating a redesign of baby walkers or the installation of window guards, may also reduce falls. For example, the New York City Board of Health mandated window guards for all multistory dwellings with children under the age of 10, which resulted in a 96% decrease in hospital admissions for falls from windows.[5]

Poisonings

Poisoning occurs when an individual ingests or contacts a substance that causes physical harm.[5] About 90% of poisonings among children occur within the home, where children are often exposed to personal care products, cleaners, pesticides, and medications. Beyond these common exposures, children can also be exposed to carbon monoxide, which is produced from the incomplete combustion of carbon-containing substances.[5] Common sources of carbon monoxide include house fires, improperly vented fireplaces, car exhaust, and malfunctioning stoves, space heaters, and furnaces.[5]

Preventing access to potentially harmful substances is key to preventing poisonings. Installing childproof locks on cabinets or drawers is one way to prevent access, as is using lockboxes for medication.[5] However, these approaches depend on individuals to use or install the devices. Child-resistant packaging and sublethal dose packaging (e.g., reducing the number of pills per bottle) are also effective ways of reducing risk and do not depend on the caregiver's behavior.[5]

Injuries in the Road Environment

Road traffic crashes are a leading cause of death and injury among children. Many established factors associated with road traffic crashes and injuries can be categorized into pre-crash (e.g., exposure/miles traveled, driver errors, roadway conditions), during crash (e.g., restraint use), and post-crash (e.g., emergency response, trauma care availability).[7] These factors impact crash occurrence, crash severity, and injury severity. Most of the road safety risk factors that increase crashes and injuries in the general road user population are also risk factors for children. However, some road safety factors are specific to children, such as child passenger restraint use, smaller physical size (e.g., not as easily seen by motorists), and ongoing cognitive and motor development (e.g., increased risk-taking behavior among adolescents).[8]

Children and adolescents are at risk for road traffic injuries across the multiple ways they interact with the road environment. The following sections describe childhood road traffic injury risk factors by these different road use categories.

Passengers

Risk of road traffic injury among the youngest motor vehicle occupants largely depends on adult driving safety behaviors (e.g., following speed limits, driving unimpaired) and proper child restraint (e.g., safety seats, seat belts) use and positioning within the vehicle.[9] For example, best practice indicates that children under 13 years should ride in the rear seat instead of the front seat of a vehicle.[10] In 2021, 863 child passengers were killed in traffic crashes, and, among those whose restraint use was known, 40% were unrestrained.[11] Child restraint decreases as age increases. In 2021, 31% of children younger than 1 year killed in crashes were unrestrained compared to 54% among 13- to 14-year-olds.[11] Child occupants also are subject to universal motor vehicle occupant risks, such as unsafe vehicles and dangerous roadway conditions.[9]

Drivers

Motor vehicle crashes pose a high risk of injury and death to adolescents, and this risk is increased when they are the driver or are riding with other teenage passengers.[9] In 2021, 2,116 drivers aged 15–20 were killed, and 203,256 were injured in traffic crashes in the United States.[12] The high motor vehicle crash injury and fatality risk among adolescents is partly attributable to the young age of licensure (age 15 or 16) possible within most states throughout the United States, when the frontal lobes of their brains are still maturing.[13,14] The frontal lobe is associated with executive functions that are critical for safe driving, including information monitoring, inhibitory control, and task switching and does not fully mature until early adulthood.[13] This frontal lobe immaturity contributes to common teen driving errors such as driving too fast for the road conditions, distraction (e.g., electronic device use, interaction with other teen occupants in the vehicle), and failure to scan for hazards in the road environment.[13] Inexperience also contributes to young driver crash and injury risk. Compared to more experienced drivers, young drivers have less practice in hazard identification and anticipation and navigation of complex driving tasks.[9] Young drivers more frequently speed, brake too late, and follow other vehicles too closely, which reduces the time and space they have to identify and respond to hazards.[9] Alcohol-impaired driving among adolescents frequently exacerbates these age-related risk factors. In 2019, 22% of all 15- to 20-year-old drivers involved in fatal crashes in the United States were under the influence of alcohol.[12]

Pedestrians

Risk of pedestrian injury to children and adolescents is present during many of their common daily activities, such as playing outside, walking to school, or crossing streets. Ongoing cognitive development and small stature make children more vulnerable to pedestrian injury compared to adults.[15] Young children often do not have the cognitive skills to manage complex traffic tasks such as scanning for oncoming traffic and judging vehicle speeds.[9]

Between 2004 and 2018 there was a 32% reduction in pedestrian fatalities and 40% reduction in nonfatal injures among children ages 0–19.[16] However, the reductions have been much larger for those under age 12, as compared to adolescents ages 12–19.[16] Pedestrian injury risk factors vary by age. Pedestrian injuries among younger children (0–11 years old) occur after school (3–6 PM) and at mid-block, while for adolescents (ages 12 and up) they occur most often at night (9 PM–12 AM) and are equally likely to occur mid-block or at an intersection with a crosswalk.[16]

Across all child ages, pedestrian injuries frequently occur in urban and suburban areas, on local neighborhood roads, during the warmer months, and more frequently among boys compared to girls.[9,11,16] Child pedestrian-involved crashes frequently involve large, high-clearance vehicles (SUVs, vans, pickup trucks).[15,16] There are also stark inequities in child pedestrian fatality and injury rates. Black children younger than age 12 have pedestrian fatality rates that are three times higher than White children, and Black adolescents ages 12–19 have pedestrian fatality rates more than twice that of White adolescents.[16] Children from households and neighborhoods of low socioeconomic status also have higher rates of pedestrian injury, partly attributable to having fewer sidewalks and marked crosswalks, poorer road designs and street lighting, higher traffic volumes, and fewer parks and playgrounds than high-income neighborhoods.[16]

Cyclists and Other Micromobility

Child and adolescent cycling includes bicycles, tricycles, and unicycles. These pedal-powered modes are part of the larger field of *micromobility*, which includes "any small, low-speed, human or electric-powered transportation device," such as scooters, skateboards, electric-assisted devices (e.g., e-bikes, e-scooters), and hoverboards.[17] Most child and adolescent micromobility-related injuries occur while bicycling and often on the road.[18,19] Other notable sources of micromobility injuries include e-scooters among adolescents (ages 15 and up) and hoverboards among 5- to 14-year-olds.[20]

Factors associated with increased bicyclist injury include rider inexperience, improper or lack of helmet use, collisions with motor vehicles or stationery objects (e.g., parked vehicles), riding on the sidewalk, unpaved surfaces, and rural areas.[7,18,19,21] Bicycle crashes commonly result in head and musculoskeletal injuries, although head injuries are most concerning and severe.[19] Bicycle helmets are an effective injury prevention measure but remain underutilized.[11,19]

Prevention Approaches

There are several notable strategies for the prevention and reduction in severity of road traffic crashes and injuries among children and adolescents. The majority of prevention approaches are aimed toward improving safety behaviors or road design, some of which include policy components.[7,9,14,15] Examples of impactful interventions relevant to youth road traffic safety include the following:

- Reducing motor vehicle exposure by promoting alternative travel modes (e.g., walking, bicycling)[7]
- Programs to reduce risky behaviors (e.g., underage drinking prevention programs to reduce alcohol-related teen crashes)[9]
- State-level graduated driver licensing (GDL) laws designed to allow young drivers to gain driving experience under conditions that minimize their crash risk (e.g., minimum permit age of 16, 70 supervised practice hours, no teen passengers and night driving restriction starting at 8 PM)[4]
- Laws to require and enforce use of child safety seats, booster seats, and seat belts to reduce vehicle occupant injuries.[21] Large improvements have been made over the past decade, especially related to booster seat use.[10] All 50 US states, the District of Columbia, and Puerto Rico require child safety seats for infants and young children and booster seats for children too large for a child safety seat but too small for a seat belt, though age, weight, and height criteria vary from state to state.[21]

- Improving the built environment (e.g., play areas, green space, infrastructure to separate non-motorists from motorists, such as sidewalks, bicycle lanes and paths, and traffic calming interventions, such as roundabouts, speed bumps)[15,19]
- Post-crash interventions aimed to improve injury outcomes (e.g., on-scene life-saving measures, emergency and trauma services, rehabilitation programs)[7]

Violence

Violence, including homicide and suicide, also is a leading cause of death and disability among children and adolescents in the United States.[2] From 2010 to 2019, approximately 27,000 people ages 1–17 years died from suicide (52%) or homicide (48%; 2). Almost two-thirds of homicide deaths are attributable to firearms.[2] The remaining third is divided between cut/pierce, suffocation, and other causes. Racial disparities among adolescent homicide deaths are particularly pronounced.[22] Firearm homicide in Black adolescents occurred at a rate of 10.4 per 100,000 population, more than eight times the rate of White adolescents. A closer look reveals that 56% of all firearm homicides in adolescents involve Black males.[22] These fatalities were largely concentrated in the South and lower Midwest regions of the United States (Figure 54.2).

Suffocation (50%) and firearms (40%) are the most common causes of suicide death. Almost 70% of suicide deaths among adolescents were in males. Suicide deaths are most prominent in the Mountain and West Central regions of the United States. Characteristics of adolescent suicide decedents varied by mechanism. For example, rates of suffocation, the most common mechanism of suicide among adolescents, were highest in adolescents who were Native American (6.15 per 100,000).

Beyond these deaths, there were more than 3 million emergency department visits among this age group for intentional injuries from 2010 to 2019.[2] Like unintentional injuries, serious nonfatal injuries tend to be caused by different mechanisms than fatal

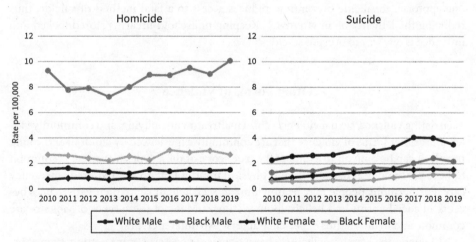

Source: Centers for Disease Control and Prevention. Web-based injury statistics query and reporting system (WISQARS). Available at: http://www.cdc.gov/injury/wisqars/index.html.

Figure 54.2. Rates of homicide and suicide deaths among 1- to 17-year-olds, United States 2010–2019

injuries.[2] Nearly 90% of visits for assault are due to being struck, and only 3% are related to a firearm injury.[2] Approximately 40% of self-harm–related visits are related to poisoning and 35% to being cut/pierced.[2]

Because firearms are used in a substantial number of homicide and suicide deaths, much of this section is devoted to firearms.

Firearm Injury, Homicide, and Suicide

From 2010 to 2019, more than 71,000 US children ages 0–17 years were injured by firearms.[22] During this same period, the rate of firearm injury deaths in this age group increased 31.2%. Fifty-four percent of pediatric firearm deaths were attributed to homicide and 38% to suicide; just 5% of deaths resulted from unintentional discharge.[22] Characteristics of firearm deaths vary greatly by level of intent and child age. Evidence shows that Black children die more often by firearm homicide,[23-25] and White children die more often by firearm suicide.[26,27] Furthermore, rates of firearm death increase with age, from a rate of 0.43 per 100,000 in ages 1–9 years to 13.68 in ages 15–19 years.[22] Across age groups, the rate of firearm death in males is nearly 6 times that of females, and the rate in Black children is 4 times that of White children.

Community-level prevention strategies include hospital-based violence interruption programs,[28] strengthening economic supports through programs like job training, mentoring, and comprehensive background checks. Higher rates of firearm suicide have been observed in states with lenient gun laws.[27]

More than 70% of firearm-related suicides were among White males aged 10–17 years, a significant disparity compared to suicides among their Black counterparts (3%). The majority of adolescent firearm suicide is completed using a weapon from within their own home.[26] Mental illness, such as depression, and bullying are known risk factors for adolescent suicide. Recent evidence suggests that more than a quarter of adolescent suicides now identify the use of technology as a precipitating issue,[29] underscoring the need to examine the online and virtual environment for prevention. Lethal means restriction, one approach to suicide prevention, reduces access to a fatal method of suicide, thus reducing the lethality of an attempt.[30] Keeping household firearms stored locked and unloaded is one way to prevent suicide.[31]

Other Forms of Violence

Teen dating violence, sexual violence, child maltreatment, bullying, and community violence are other forms of violence that are commonly experienced by children and young people. It can be difficult to estimate the true scope of these types because it is common to have psychological harm and mild physical injuries that do not required emergency care. Furthermore, people may be unwilling to seek help or share their experiences because of social norms and beliefs about victims.[32] Many of these types of violence are common, and many victims experience at least two different types.[33]

Although each has some distinct characteristics (e.g., perpetrator, setting, type of injury), there is increasing recognition that many forms of violence are interconnected and share the same root causes.[34] These risk factors fall across the socioecological spectrum, ranging from individual and interpersonal to community factors.[35] Adolescents have an increased risk for perpetrating violence, which tends to decrease as they age. Individuals who have a history of violence, antisocial personality traits, or are experiencing substantial stress have an increased risk of perpetrating violence. Individuals with low parental involvement, family functioning, or emotional attachment to their caregivers also have increased risk. Living in communities with limited economic opportunities, low levels of community participation, and socially disorganized neighborhoods also increases the risk of perpetration.

Generally, violence against children and youth can be prevented through interventions focused on the family or community.[35] Helping parents and communities create healthy, supportive environments, including access to quality education, skill-development programs, and supportive adults, is one approach to preventing youth violence. Additionally, intervening to support youth who have experienced violence may lessen the harm and prevent future problems.

Conclusion

Interventions focused on the individual, such as education, are important, but interventions to improve the social and physical environment also are critical. Unintentional injuries and violence may be prevented most effectively by creating safer physical and social environments through education, community-level interventions, and policies.

References

1. Sleet DA, Ballesteros MF, Baldwin GT. Injuries: an underrecognized lifestyle problem. *Am J Lifestyle Med*. 2010;4(1):8–15.

2. Centers for Disease Control and Prevention. WISQARS™ — Web-based Injury Statistics Query and Reporting System. Updated 2021. Available at: https://www.cdc.gov/injury/wisqars/index.html.

3. Davis RM, Pless B. BMJ bans "accidents": accidents are not unpredictable. *Br Med J*. 2001;322(7298):1320–1.

4. US Department of Health and Human Services. Healthy people 2030: Social determinants of health. Washington, DC: US Department of Health and Human Services. Updated 2021. Available at: https://health.gov/healthypeople/objectives-and-data/social-determinants-health.

5. Gielen AC, McDonald EM, Shields W. Unintentional home injuries across the life span: Problems and solutions. *Annu Rev Public Health*. 2015;36(1):231–53.

6. Guntheroth WG, Spiers PS. The triple risk hypotheses in sudden infant death syndrome. *Pediatrics*. 2002;110(5):e64.

7. Bachani AM, Peden M, Gururaj G, Norton R, Hyder A. Road traffic injuries. In: Mock C, Nugent R, Kobusingye O, et al., eds., *Injury Prevention and Environmental Health*. 3rd ed. Washington, DC: International Bank for Reconstruction and Development/World Bank; 2017. Available at: https://www.ncbi.nlm.nih.gov/books/NBK525212/

8. Peden M, Oyegbite K, Ozanne-Smith J, et al. *World Report on Child Injury Prevention*. Geneva: World Health Organization; 2008.

9. Simons-Morton BG, Caccavale LJ. Motor vehicle and pedestrian injuries among children and adolescents: risk and prevention. In: DeSafey LK, ed. *Injury Prevention for Children and Adolescents*. Washington DC: American Public Health Association; 2012. Available at: https://ajph.aphapublications.org/doi/abs/10.2105/9780875530055ch04

10. Durbin DR, Hoffman BD, Council on Injury, Violence, and Poison Prevention, Agran PF, Denny SA, et al. Child passenger safety. *Pediatrics*. 2018;142(5):e20182460. doi:10.1542/peds.2018-2460.

11. US Department of Transportation. *Traffic Safety Facts 2021 Data: Children*. Washington, DC: National Center for Statistics and Analysis; 2023. Available at: https://crashstats.nhtsa.dot.gov/Api/Public/ViewPublication/813456

12. US Department of Transportation. *Traffic Safety Facts 2021 Data: Young Drivers*. Washington, DC: National Center for Statistics and Analysis; 2023. Available at: https://crashstats.nhtsa.dot.gov/Api/Public/ViewPublication/813492

13. Walshe E, Ward Mcintosh C, Romer D, Winston F. Executive function capacities, negative driving behavior and crashes in young drivers. *Int J Environ Res Public Health*. 2017;14(11):1314.

14. Insurance Institute for Highway Safety. Graduated licensing laws by state. IIHS. Updated 2021. Available at: https://www.iihs.org/topics/teenagers/graduated-licensing-laws-table.

15. Cloutier M-S, Beaulieu E, Fridman L, et al. State-of-the-art review: preventing child and youth pedestrian motor vehicle collisions: critical issues and future directions. *Injury Prevention*. 2021;27(1):77–84.

16. Chandler M, MacKay JM. *Child Pedestrian Safety in the U.S.: Trends and Implications for Prevention*. Washington, DC: SafeKids; 2020.

17. Federal Highway Administration. Micromobility. FHWA. Updated 2020. Available at: https://www.fhwa.dot.gov/livability/fact_sheets/mm_fact_sheet.cfm.

18. Embree TE, Romanow NTR, Djerboua MS, Morgunov NJ, Bourdeaux JJ, Hagel BE. Risk factors for bicycling injuries in children and adolescents: A systematic review. *Pediatrics*. 2016;138(5):e20160282.

19. Marshall S, Gilchrist J, Taneja G, Liller K. Sports and recreational injuries. In: Mock C, Nugent R, Kobusingye O, et al., eds. *Injury Prevention and Environmental Health*. 3rd ed. Washington, DC: International Bank for Reconstruction and Development/World Bank; 2017. Available at: https://www.ncbi.nlm.nih.gov/books/NBK525218/

20. Tark J. *Micromobility Products-Related Deaths, Injuries, and Hazard Patterns: 2017–2019*. Bethesda, MD: US Consumer Product Safety Commission; 2020.

21. Governors Highway Safety Association. Child passenger safety. GHSA. Updated 2020. Available at: https://www.ghsa.org/state-laws/issues/child%20passenger%20safety.

22. Centers for Disease Control and Prevention. Fatal injury reports. CDC. Updated 2021. Available at: https://wisqars.cdc.gov/fatal-reports.

23. Bachier-Rodriguez M, Freeman J, Feliz A. Firearm injuries in a pediatric population: African-American adolescents continue to carry the heavy burden. *Am J Surg*. 2017;213(4):785–9.

24. Kalesan B, Dabic S, Vasan S, Stylianos S, Galea S. Racial/ethnic specific trends in pediatric firearm-related hospitalizations in the United States, 1998–2011. *Matern Child* Health J. 2016;20(5):1082–90.

25. Kalesan B, Vyliparambil MA, Bogue E, et al. Race and ethnicity, neighborhood poverty and pediatric firearm hospitalizations in the United States. *Ann Epidemiol*. 2016;26(1):1–6.

26. Johnson RM, Barber C, Azrael D, Clark DE, Hemenway D. Who are the owners of firearms used in adolescent suicides? *Suicide Life Threat Behav.* 2010;40(6):609–11.

27. Fowler KA, Dahlberg LL, Haileyesus T, Gutierrez C, Bacon S. Childhood firearm injuries in the United States. *Pediatrics.* 2017;140(1):e20163486. doi:10.1542/peds.2016-3486.

28. Kramer EJ, Dodington J, Hunt A, et al. Violent reinjury risk assessment instrument (VRRAI) for hospital-based violence intervention programs. *J Surg Res.* 2017;217:177–86.

29. Orlins E, DeBois K, Chatfield SL. Characteristics of interpersonal conflicts preceding youth suicide: analysis of data from the 2017 National Violent Death Reporting System. *Child Adolesc Ment Health.* 2021;26(3):204–10.

30. Hunter AA, DiVietro S, Boyer M, Burnham K, Chenard D, Rogers SC. The practice of lethal means restriction counseling in US emergency departments to reduce suicide risk: a systematic review of the literature. *Inj Epidemiol.* 2021;8(Suppl 1):54.

31. Johnson RM, Coyne-Beasley T. Lethal means reduction: what have we learned? *Curr Opin Pediatr.* 2009;21(5):635–40.

32. Stubbs-Richardson M, Rader NE, Cosby AG. Tweeting rape culture: examining portrayals of victim blaming in discussions of sexual assault cases on Twitter. *Fem Psychol.* 2018;28(1):90–108.

33. Schwab-Reese LM, Currie D, Mishra AA, Peek-Asa C. A comparison of violence victimization and polyvictimization experiences among sexual minority and heterosexual adolescents and young adults. *J Inter Violence.* 2021;36(11–12):NP5874–NP5891. doi:10.1177/0886260518808853.

34. Wilkins N, Myers L, Kuehl T, Bauman A, Hertz M. Connecting the dots: State health department approaches to addressing shared risk and protective factors across multiple forms of violence. *J Public Health Manag Pract.* 2018;24:S32–S41.

35. Centers for Disease Control and Prevention. Violence prevention. CDC. Updated 2021. Available at: https://www.cdc.gov/violenceprevention/index.html.

55

Acute Pediatric Poisoning

Jennifer Sample

Acute poisoning by toxic chemicals occurs commonly among infants and children in countries around the world. Acute poisonings have resulted in disasters and mass casualties. They cause acute illness that requires urgent medical intervention, but they also cause significant persisting morbidity. Poisoning episodes have repeatedly provided the initial warning of important environmental threats to children's health. Acute chemical intoxications are an important public health problem.

Astute physicians play a pivotal role in recognizing episodes of acute poisoning, in documenting sources of toxicity, and in leading efforts to treat patients and control exposures. Clinical recognition of poisoning requires that physicians be alert to the possibility that an acute environmental exposure can be the cause of disease in a child. It requires that poison always be included in the differential diagnosis. Medical vigilance is the key to early recognition of environmental health disasters.

Although the discipline of children's environmental health deals principally with chronic exposures and their impacts on children's health, it is important that pediatricians, public health officials, and others who care for children have basic knowledge of medical toxicology and understand how the fields of children's environmental health and medical toxicology interface with each other and have each influenced the other's growth.

Scope and Nature of the Problem of Chemical Poisoning

Chemical releases and acute pediatric poisonings are common in countries around the world. Children present daily to clinics and emergency rooms with acute illnesses caused by toxic environmental exposures that range from asthma caused by air pollution, pesticide poisoning, and neurologic symptoms following mercury spills in schools.

Although the true incidence of acute pediatric poisoning is not known and there are no complete databases that track all cases of acute toxicity, the World Health Organization (WHO) estimates that more than 2 million deaths were attributed to chemicals alone in 2019. The WHO tracks these events and responds to them through the International Programme on Chemical Safety (IPCS) http://www.who.int/ipcs/en/.

In the United States, it is estimated that 3–4 million children live within 1 mile of at least one hazardous waste site.[1] The US Environmental Protection Agency (EPA) reports that, in 2019, of the 30 billion pounds of waste produced, American industries released 3.5 billion pounds of chemical agents into the environment, a decrease of almost 11% since 2010.[2] These chemicals were disposed on the land, released to air and water, and injected underground. While overall releases in the second decade of the 21st century

Jennifer Sample, *Acute Pediatric Poisoning* In: *Textbook of Children's Environmental Health*, Second Edition. Edited by: Ruth A. Etzel and Philip J. Landrigan, Oxford University Press. © Oxford University Press 2024. DOI: 10.1093/oso/9780197662526.003.0055

decreased by more than 50% compared to the preceding 10-year period in the United States, this is not evident globally. Efforts are being made to expand Pollutant Release and Transfer Registers (PRTRs) throughout the world. The Toxics Release Inventory (TRI) is the PRTR in the United States and is managed by the EPA. Currently, more than 50 countries have PRTRs with more to be developed in Asia and South America in the coming years.

Some of the chemicals released to the environment are persistent, bioaccumulative, and toxic. These chemicals can remain in the environment for years or even decades after their release, and they can accumulate to high levels in plants and animals. Approximately 180 of the chemicals released to the environment in the United States are known carcinogens. Between 2007 and 2019, environmental releases of carcinogens in air decreased by 30%. Of the carcinogens released into air, styrene (47%), acetaldehyde (12%), and formaldehyde (8%) were the most prevalent.

The chemicals released to the environment can cause acute toxicity in infants and children. Of the persistent, bioaccumulative and toxic chemicals released by industry each year, 97% is comprised of lead and lead compounds. Between 2007 and 2019, the amount of lead released into the environment increased by 26% and is accounted for by the increase in metal mining. The remainder includes mercury and mercury compounds, dioxins, and other chemicals. Exposures to these materials during pregnancy and early childhood are especially dangerous (see Chapter 2). For example, acute exposure during pregnancy to glycol ethers (used in paints and varnishes), pesticides, solvents, and lead may lead to fetal loss.[3] The newborn infant and child have three to two times larger, respectively, surface-to-mass ratio than adults. In addition, the neonate and young infant have decreased keratinization of the skin. Both of these factors may result in increased dermal absorption of toxic chemicals. Case reports of seizures in infants occurring after the use of insect repellants have been attributed to a combination of increased exposure relative to adults and greater biological vulnerability.[4]

Epidemiology of Acute Pediatric Poisoning

In the United States, two major databases collect data on acute pediatric poisonings and exposures to toxic chemicals. They are operated by the Poison Control Centers and Pediatric Environmental Health Specialty Units (PEHSUs) and are described in detail in the sections below (Figure 55.1).

A third relevant database maintained by the National Center for Environmental Health within the Centers for Disease Control and Prevention (CDC) assesses the US population's exposures to environmental chemicals by measuring levels of these chemicals in blood and urine in a nationally representative sample of the population. CDC's most recent *National Report on Human Exposure to Environmental Chemicals*[5] presents data from this source and provides information on levels in blood and urine of more than 200 chemicals. These data do not distinguish acute from chronic exposures nor do they provide information on toxicity or health outcomes.

In Europe, the European Association of Poisons Centres and Clinical Toxicologists facilitates the collection, exchange, and dissemination of information on pediatric acute poisonings among individual physicians, poison control centers, and organizations interested in clinical toxicology (http://www.eapcct.org/index.php?page=home).

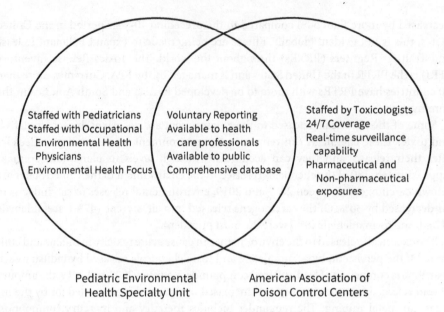

Staffed with Pediatricians
Staffed with Occupational
Environmental Health
Physicians
Environmental Health Focus

Voluntary Reporting
Available to health
care professionals
Available to public
Comprehensive database

Staffed by Toxicologists
24/7 Coverage
Real-time surveillance
capability
Pharmaceutical and
Non-pharmaceutical
exposures

Pediatric Environmental
Health Specialty Unit

American Association of
Poison Control Centers

Figure 55.1. Acute poisoning surveillance programs of the Pediatric Environmental Health Specialty Units and the American Association of Poison Control Centers

The American Association of Poison Control Centers

The best known database on acute poisoning episodes in the United States is the National Poison Data System (NPDS) maintained by the American Association of Poison Control Centers (AAPCC). This database relies on voluntary reporting by Poison Control Centers, and, from January to December 2019, 55 participating Centers submitted data on pediatric poisonings to NPDS.[6] The NPDS database covers the entire population of the 50 US states, American Samoa, the District of Columbia, Federated States of Micronesia, Guam, Puerto Rico, and US Virgin Islands. US Poison Control Centers are staffed by physicians, pharmacists, and nurses. Certified specialists in poison information document each call to a Center. While there are guidelines on case documentation, the information coded largely depends on the judgment of the specialist.

In 2019, US Poison Control Centers logged 2,573,180 total encounters, including 2,148,141 calls reporting human poisoning events. More than 57% of poisonings occurred in children under the age of 20 years with more than 45% in children under the age of 6 years. Ingestion was the route of exposure in 83.6% of poisoning cases followed by dermal (7.2%), inhalation (6.8%), and ocular exposures (4.7%). More than 50% of human poisonings reported to NPDS involved nonpharmaceutical substances. The nonpharmaceutical exposures most frequently involved in pediatric exposures (age <5 years) were cosmetics/personal care products (21.7%), cleaning/household substances (19.7%), foreign bodies (13.6%), pesticides (6.8%), plants (5.2%), arts/crafts/office supplies (4%), chemicals (1.9%), and hydrocarbons (1.5%).

In 2019, NPDS received 351,163 information calls, defined as calls seeking information on exposures rather than reporting poisoning events. This is a marked decrease from years past, with environmental calls only accounting for 3.6% of information calls.

US Poison Control Centers receive many more environmental consultations than the PEHSUs. Thus, in 2019, Poison Control Centers were notified of 1,336 cases of lead exposures in children younger than 20 years, accounting for the most common metal exposure in children. Pediatric cases of mercury thermometers and mercury from other sources only accounted for 608 cases, a third of the cases from 10 years ago. The number of calls to US Poison Control Centers regarding pesticides also far surpasses the number of pesticide calls to PEHSUs. In 2019, more than 33,000 pediatric exposures to all forms of pesticides were reported to US Poison Control Centers.

Despite its many strengths, the Poison Control Center database misses many episodes of acute poisoning that are undiagnosed and unreported, and it also contains little information on more chronic environmental exposures.

Pediatric Environmental Health Specialty Unit Network

The network of PEHSUs, is a second major source of public health surveillance data on acute environmental exposures to children in the United States. Detailed information on the PEHSU network and contact information for PEHSUs is provided in Chapter 56.

Development of the PEHSU network was catalyzed by two major environmental incidents, both of which occurred in 1996.

- In the first of these episodes, methyl parathion, an extremely toxic organophosphate insecticide intended solely for outdoor use, was applied illegally in homes in Illinois, Mississippi, and Ohio. Cases of acute poisoning resulted.[7]
- In the second incident, acute metallic mercury poisoning occurred among children living in apartments in a building that had served previously as a factory for the manufacture of fluorescent lights. The manufacturing process had included the use of liquid metallic mercury. The mercury had not been properly controlled in the manufacturing process, and mercury had permeated walls and floors throughout the building. Families, including small children, were exposed to mercury vapor that came out of the walls and floors and entered their apartments. Similar events have been documented in the literature.[8]

Limitations exist in using PEHSU data for evaluating acute pediatric environmental intoxications. First, the PEHSU program is not well known and is therefore not a primary source for reporting data on pediatric poisoning. A second problem is that the PEHSU database includes both acute and chronic exposure information and also acute and chronic health effects from potential toxicants. A third shortcoming is that the category definitions are broad and may result in misclassification (e.g., "lead" vs. "metals" vs. "heavy metals"). Last, the information is not published or publicly available, leaving many questions about environmental exposures unanswered.

In summary, the Poison Control Center and the PEHSU databases do not capture every poisoning episode that occurs in US children, and they are limited by the fact that both rely on voluntary reporting. But, despite those shortcomings, the AAPCC and PEHSU databases remain the best information resources available today in the United States for healthcare professionals and the public on acute and chronic environmental threats to children's health. These data sources are used by state and local health departments and by the CDC to track outbreaks of acute poisoning and to monitor environmental exposures that may have public health consequences. For example, during the week after Hurricane Sandy in 2012, US Poison Control Centers received

263 carbon monoxide exposure calls and served as a vital public health resource during that crisis.[9]

Role of the Pediatrician in Identifying Acute Environmental Intoxication

Astute physicians play a pivotal role in recognizing episodes of chemical poisoning, documenting sources of toxicity, and leading efforts to treat patients and control exposures.

Clinical recognition of acute and chronic poisoning requires that physicians be alert to the possibility that an environmental exposure can be the cause of disease in a child.

Any pediatrician who is confronted by a child with unexplained illness should always consider the environment as a cause of the illness. Moreover, the pediatrician needs not only to treat any intoxication that is clinically recognized but also to take steps to identify and control the source of the exposure and to become an advocate for the care of that child.[10,11]

Pediatricians need always to be mindful of the fact the most common clinical presentation in environmental medicine is the complete absence of any clinical signs and symptoms of intoxication. Most harmful environmental exposures in children result in subclinical or "silent" toxicity. Many patients with low-level environmental exposures may all too easily be diagnosed as "worried well." Exposures to metals such as lead, methylmercury, and arsenic can all have adverse effects at low levels that are associated with minimal clinical signs and symptoms and can be diagnosed only through special testing. Proper diagnoses of these conditions require obtaining a careful history of environmental exposure coupled with screening of blood or urine in a certified laboratory for suspected toxic chemicals.

Physicians have repeatedly been instrumental in discovering outbreaks of toxic chemical exposure that have had great public health impact. For example, observation by a gynecologist of a cluster of vaginal cancer in young women led to the discovery that all of these women were the daughters of mothers who had been prescribed the synthetic estrogen diethylstilbestrol (DES) during pregnancy to prevent miscarriage. This sentinel episode led to widespread realization that DES causes reproductive tract anomalies and cancer after exposure to the chemical in utero.[12]

However, poisonings that present to the practitioner acutely often occur after significant exposures and result in symptoms that present similarly to other diagnoses. Environmental exposures should always be on the differential diagnosis list to accurately diagnose and treat the patient.

Case Studies in Clinical Recognition of Acute Pediatric Poisoning

The previous edition of this chapter documented sentinel cases of pediatric exposures to toxic chemicals in the environment. While being the first is important, astute physicians since the time of Dr. Robert W. Miller (refer to Chapter 1) have highlighted the importance of recognizing environmental exposures as acute presentations.

Box 55.1 Case Studies

Fever and Rash

Three children, ages 14, 11, and 9 years old, presented to the emergency department with fever and rash. Patient 1 was given ibuprofen and discharged with a diagnosis of viral syndrome. He returned the next day with persistent fever, dry cough, malaise, sore throat, and a generalized headache. Patient 3 presented with fever, rash, headache, and bilateral foot pain. She was given ibuprofen and sent home with a diagnosis of viral syndrome. She returned the next day with symptoms similar to those of Patient 1. On exam, both were found to have a diffuse confluent erythematous maculopapular pruritic rash, injected conjunctiva, and erythematous oropharynx. Patient 2 presented two days after the others with similar symptoms and physical findings.

At the time of the first visit, the differential diagnosis included infectious etiology as the most likely cause, including adenovirus and streptococcus. In this day and age, influenza and coronavirus would also be considered. Kawasaki disease was less likely although it was entertained as a possibility. At the time of the second visit, however, Patient 1 brought in a container of 99.9% elemental mercury which had been "found" 10 days prior to the initial emergency department visit. The patient had read about the toxicity of mercury after the discharge from the first emergency department visit. All three patients had dermal and inhalation exposures to the product.

Mercury is a toxic metal found in three forms: elemental, inorganic, and organic. Elemental mercury can be found in older thermometers and dental amalgams, fluorescent light bulbs, and some industrial products. Elemental mercury is primarily absorbed through the lungs, with some dermal toxicity with prolonged contact. The primary target organs are the brain and kidney. Patients may also develop a rash or acrodynia ("pink disease").[13]

Altered Level of Consciousness

A 7-year-old girl presented to the emergency department with acute onset of vomiting, dizziness, and altered level of consciousness. She presented by ambulance and without parents. Upon arrival, she was found to be lethargic, with pinpoint pupils and dry mouth. Vitals were stable with heart rate of 105/minute, respiratory rate of 19/minute, and blood pressure of 95/69 mmHg. She had normal muscle tone, strength, and reflexes. Visualization of her scalp showed lice nits and eggs as well as a sticky solution on her scalp.

The parents arrived and told the emergency department providers that they had applied Diazinon-60 (an organophosporus compound used for animals) on her hair to rid her of the lice. Given the symptoms and presentation, she was given high-flow oxygen and attached to a cardiac monitor. She was given one dose of atropine and pralidoxime. She was transferred to the pediatric intensive care unit and started on a continuous infusion of pralidoxime and scheduled atropine. Her symptoms improved, and she was discharged after 7 days.

Pesticide misuse is not uncommon and should be considered when patients present to the emergency department with symptoms of cholinergic poisoning. While she did not have the classic symptoms with which an adult would present, children are more likely to present with nicotinic symptoms compared to muscarinic symptoms.

Diazinon is an organophosphate used occupationally where the applicators where protective clothing.[14]

Seizures

A 7-year-old presented to the emergency department after falling unconscious on the playground followed by tonic episodes of the extremities for 5 minutes. He subsequently had vomiting and altered level of consciousness. He had been previously well with no significant medical history. On examination, he was afebrile, heart rate of 92/minute, normal blood pressure, pale, Glasgow coma score of 8, increased tone, brisk deep tendon reflexes, and multiple episodes of tonic-clonic seizures. He was treated per the hospitals "seizure protocol," was given 3% hypertonic saline for increased intracranial pressure, and supportive treatment. Non-contrast computed tomography (CT) of the brain was normal. Initial laboratory assessment was unremarkable except for a microcytic, hypochromic anemia.

Further history obtained on Day 2 revealed that the child spent 8 hours a day during his summer at his father's workplace, which was a car battery recycling factory. The patient did not have a history of pica, but did have poor hygiene. Upon assessment of the father, it was noticed that he had bluish discoloration of his gums and teeth (Bruton's line) and white lines of discoloration on his fingernails (Mees lines). The patient's lead level was found to be 140 µg/dL (reference range: <3.5 µg/dL).

The CDC recently decreased the reference level (97th percentile) for lead to 3.5 µg/dL. In their guidance, severe toxicity from lead can occur at levels above 70 µg/dL. This patient presented to the emergency department with lead encephalopathy, with the correct diagnosis not discerned until Day 2 when the parental occupation was discovered. While the most common source of lead is from paint used prior to 1978, other sources include the parents' workplace. The father's occupation and physical findings led the physician to make the correct diagnosis.[15]

Conclusion

Although less common than low-level, chronic exposures, acute environmental exposures occur and present similarly to other diagnoses. Children may present to the emergency department with acute, nonspecific symptoms that rely on an astute physician to ask the correct question in order to make the correct diagnosis.

The AAPCC and the PEHSUs have played leading roles in documenting poisoning episodes and raising awareness of the importance of poisoning. The development of PRTRs throughout the world will aid in understanding the extent of environmental exposures. These may bring additional awareness to healthcare professionals. In addition, the WHO's International Programme on Chemical Safety promotes chemical safety through strengthening poison centers globally. The work of these organizations has greatly increased knowledge of the problem of acute pediatric toxicity and contributed to medical education around the world.

A major challenge for the future is ongoing enhanced education in medical schools and residencies regarding medical toxicology and pediatric environmental health. This added training will greatly improve clinical recognition of acute pediatric poisoning and help avert environmental disasters.

References

1. Veal K, Lowry JA, Belmont JM. The epidemiology of pediatric environmental exposures. *Pediatr Clin N Am.* 2007;54:15–31.

2. US Environmental Protection Agency. *2019 TRI National Analysis.* Washington DC: Environmental Protection Agency; 2021.

3. Chalupka S, Shalupka AN. The impact of environmental and occupational exposures on reproductive health. *JOGNN.* 2010;39 84–102.

4. Mazur LJ. Pediatric environmental health. *Curr Prob Pediatr Adolesc Health Care.* 2003;33:1–25.

5. Centers for Disease Control and Prevention. National report on human exposure to environmental chemicals. 2023. https://www.cdc.gov/exposurereport/index.html.

6. Gummin DD, Mowry JB, Beuhler MC, et al. 2019 Annual report of the American Association of Poison Control Centers' National Poison Data System (NPDS): 37th annual report. *Clin Toxicol.* 2020;58:1360–541.

7. Rubin C, Esteban E, Hill RH, Pearce R. Introduction. The methyl parathion story: a chronicle of misuse and preventable human exposure. *Environ Health Perspec.* 2002;Suppl 6(Suppl 6):1037–40.

8. Agency for Toxic Substances and Disease Registry. Health consultation: mercury exposure investigation using serial urine testing and medical records review. Kiddie Kollege. 2007 Available at: https://healthy.nj.gov/health/ceohs/documents/eohap/haz_sites/gloucester/franklin_township/kiddie_kollege/kiddiekollegehc.pdf.

9. Centers for Disease Control and Prevention. Notes from the field: carbon monoxide exposures reported to poison centers and related to Hurricane Sandy – Northeastern United States, 2012. *Morb Mort Wk.* 2012;61:905.

10. McClafferty H. Environmental health: children's health, a clinician's dilemma. *Curr Probl Pediatr Adolesc Health Care.* 2016;46:184–9.

11. Duderstadt KG. Environmental health policy and children's health. *J Pediatr Health Care.* 2006;20:411–3.

12. Herbst AL. The effects in the human of diethylstilbestrol (DES) use during pregnancy. *Princess Takamatsu Symp.* 1987;8:67–75.

13. Young AC, Wax PM, Feng SY, Kleinschmidt KC, Ordonez JE. Acute elemental mercury poisoning masquerading as fever and rash. *JMT.* 2020;16:470–6.

14. Hamd MH, Adeel AA, Alhaboob AAN, Ashri AM, Salih MA. Acute poisoning in a child following topical treatment of head lice (Pediculosis capitis) with an organophosphate pesticide. *Sudanese J Paediatr.* 2016;16:63–6.

15. Keshri S, Goel AK, Garg AK. Reversal of acute lead encephalopathy in a child. *Cureus.* 2021;13:e15155.

PART V

PREVENTION AND CONTROL
OF DISEASES OF ENVIRONMENTAL
ORIGIN IN CHILDREN

56

The Environmental History and Examination

The Key to Diagnosis of Environmental Diseases

Jerome A. Paulson and Sandra H. Jee

Child health professionals are increasingly confronted with environmental health issues.[1] These can range from commonly encountered problems such as the impact of lead on a developing child to more complex issues such as the health effects of mold and chronic pesticide exposure,[2] to the appropriate sunblock to use to prevent the harmful effects of ultraviolet radiation, or to environmental health inequities that disproportionately expose some children to air or noise pollution by living near highways, which are associated with increased rates of asthma and respiratory disease.[3-5] The complex intersection of pediatrics, climate change, and structural racism has compounded the disproportionate negative impact on low-income communities and communities of color.[4] Current environmental-related concerns also include the climate crisis and the linkage between climate change and child health harms. Extreme weather patterns and human activity contributing to a warming planet have manifest in increasing prevalence rates of health conditions such as asthma, allergies, eco-anxiety, and heat-related illness that are pressing pediatric health issues.[5]

The provision of evidence-based advice in response to such questions and the diagnosis of disease of environmental origin in children require basic knowledge of environmental medicine, diagnostic openness to the possibility of environmental causation, and a systematic approach to obtaining a history of environmental exposure.

A major problem in this area for the majority of child health professionals is that most have received very little education in environmental medicine during medical school or residency training. For example, a survey of US medical schools found that the average medical student receives only 6 hours of exposure to topics in environmental medicine during the 4 years of medical school.[6] The majority of pediatric practitioners report in surveys that they feel reasonably well prepared to offer advice to parents on prevention of lead poisoning and on the environmental causes of childhood asthma, but that they are not comfortable in dealing with environmental problems beyond those relatively common issues. Recent evidence shows that parents are receptive to discussing climate change in clinical settings.[7-9] Pediatric providers need knowledge on environmental health and climate change to engage in such discussions and answer questions.

This chapter presents a systematic approach to the diagnostic recognition of disease of environmental origin in children. It is important to recognize that children do not exist in isolation from the world around them. The physical, social, and cultural environments within which a child lives all influence health, growth, and development. In gathering information about a child's medical history, one should keep in mind the image of the classical Russian nesting doll, the *matryoshka doll*. In this doll, the child is the smallest

Jerome A. Paulson and Sandra H. Jee, *The Environmental History and Examination* In: *Textbook of Children's Environmental Health*, Second Edition. Edited by: Ruth A. Etzel and Philip J. Landrigan, Oxford University Press. © Oxford University Press 2024.
DOI: 10.1093/oso/9780197662526.003.0056

and nested within its family, which is nested within the extended family, which, in turn, is nested within the community (schools, playgrounds, child care centers, etc.). The converse is also true. The identification of an environmentally related health problem should be a sentinel event indicating the need to evaluate others (family, classmates, playmates, fellow workers) who have been exposed to the same toxicant.

The Environmental History

A comprehensive medical, social, and developmental history is the basis for providing a framework for addressing patient needs. By asking a series of open-ended questions, the healthcare provider can identify the most pressing health and social needs of the family during a clinical visit. For patients who have continuity of care with their provider, interim history can be more focused.

The history also gathers medical and social information about the patient's family, information about the developmental trajectory of the child (when did he or she walk, talk, and reach other developmental milestones), a work history (for an adolescent and for both parents), information about school progress for school-aged children, and a long list of questions about symptoms related to all of the organ systems in the body (the heart, lungs, eyes, skin, liver, brain, etc.), which is known as a *review of systems*.

The history is the single most important instrument for obtaining information on the role of environmental factors in causing disease in children.

For a child who is being brought to a clinician for the evaluation of a complaint that may be related to an environmental factor, it is very important that all of the steps outlined in the paragraph above be undertaken. In addition, there is need to collect information in a systematic and standardized fashion about potentially hazardous exposures in the child's environment with emphasis on the following[10]:

- Places where the child spends substantial amounts of time: home, child care centers, schools, parents' workplaces, grandparents' or neighbors' homes, playgrounds, and fields
- Chemical, physical, and biological hazards such as metals, persistent pollutants, pesticides, toxic gases or vapors, noise, electromagnetic or ionizing radiation, solar radiation in or near the child's home or school
- Nearby hazardous waste sites or industrial "hotspots," nearby agricultural activities (pesticide drift)
- Indoor sources of contamination: building materials, molds, paints, pesticides, home industries or hobbies, "take-home" chemical exposures from a parent's workplaces, secondhand tobacco smoke, poorly ventilated fireplaces or cook stoves
- Exposures on the job, if the child works

Information about current and past environmental exposures should routinely be sought at several logical points while taking every patient's history. At each juncture, a few brief screening questions need to be asked systematically. Then, if relevant information is elicited, more detailed follow-up questions are needed. An efficient, routine screening checklist for environmental disease consists of the following items:

1. *History of the present illness.* Pay attention to any temporal relationship between onset of illness and introduction of toxic exposures in the environment. For example, did symptoms begin shortly after the child's parent started a new job? Did they abate during vacation and then recur after the parent resumed work? Were they related to the introduction of a new factory or landfill in the community? Did they correlate with episodes of pollution or extreme weather patterns or natural disasters? Were there similar illnesses among siblings, neighbors, or other children in the school or community? Did the children who were more heavily exposed suffer greater severity?

   A possible environmental etiology should be sought in every case of acute trauma in children and adolescents because many children work at dangerous jobs, and in every case of repetitive trauma, as for example carpal tunnel syndrome.

2. *Past medical history.* Obtain a list of current and principal past exposures such as secondhand smoke, lead, molds, pesticides, cockroaches and pests, air pollution, asbestos, or radon. The parent of every child with illness of unexplained origin should be asked whether they believe that environmental exposures have caused or contributed to their child's illness.

3. *Review of systems.* Routinely ask the parent of every child, "Does your child now or previously have environmental exposures to asbestos, molds, lead, fumes, chemicals, dust, loud noise, radiation, or other toxic factors?" Also ask the parent of every child whether he or she believes that any of these factors may have caused or contributed to their child's illness. Even if a postulated connection between exposure and disease initially appears tenuous, such suspicions always need to be carefully considered.

4. *Detailed exposure history.* If information from the routine interview suggests an environmental etiology, the physician should obtain a more detailed history of toxic exposures. Duration and intensity of exposures are particularly important. It is necessary to learn how the patient may have come into contact with the suspected toxicant and to consider how he or she may have absorbed the material.

A practical and systematic approach and framework is considered essential to facilitate the task of taking a detailed environmental history and to increase the possibility of detecting potentially harmful environmental exposures. Although many environmental risks to children's health are global, there may also be social, territorial, or cultural characteristics that shape a specific environmental risk profile for a particular region. Therefore, the environmental history needs to include common core questions but at the same time be tailored to the local environment of the practicing health professional.

Different pediatric environmental histories have been developed for use in countries around the world.[11,12]

A pediatric environmental history form that records a core set of basic environmental information has been developed by the World Health Organization/Pan American Health Organization and has been tested in selected low- and middle-income countries.[13] The data collection form is proposed to be in the form of a green sheet, to make it clearly visible within the clinical record. The form has about 50 fields for recording information about children's environments, characteristics of the community, and complementary data. It can be customized according to the local situation, needs, and capacities of highly industrialized and less industrialized regions. It has been developed to be used by health care professionals dealing with infants, children, and adolescents or to follow up pregnant women. Some parts of the environmental history may be completed by a social worker or environmental officer who can visit the places where children spend their time and talk to parents, teachers, or community representatives.

The Pediatric Environmental History of the US National Environmental Education Foundation proposes a set of 14 screening questions and also a more extended questionnaire with seven categories (housing characteristics, indoor and outdoor environment, food and water, toxic chemical exposures, occupation and hobbies) to complement the screening questionnaire.[14]

An environmental health history typically needs to be adapted to capture information on the most severe or prevalent environmental risks locally or regionally. Short questionnaires have been developed locally to screen potential exposures to selected contaminants that represent previously recognized local problems, such as lead from wire burning or metal recycling or carbon monoxide from indoor combustion. A more expanded environmental history (taken by a specialized staff member) is used among children considered at risk.[15]

Environmental exposure history mnemonics have been proposed for primary care physicians and nurses practice. The I PREPARE environmental exposure mnemonic is a quick reference tool to investigate exposure by asking questions about present work, residence, environmental concerns, past work, and activities. The questionnaire provides data about local referrals and resources.[16]

The CH20PD2 questionnaire is aimed at detecting environmental risks by introducing questions about the community, home, hobbies, occupation, personal habits, and drugs.[17]

Once a potential exposure has been suspected or identified, any screening questionnaire needs to be followed by a set of detailed questions to further characterize the exposure, particularly regarding the time and duration of the exposure and its correlation with any symptoms.[18,19]

The Case Studies in Environmental Medicine series of the US Agency for Toxic Substances and Disease Registry provides a summary of questions focused on the details of particular toxic exposures.[20] A paradigmatic set of questions regarding lead exposure was created by the Centers for Disease Control and Prevention (CDC).[21] It includes questions exploring the main sources of lead exposure recognized in the United States and it has been the basis of questionnaires developed around the world.

New evidence may make it necessary to update or expand a questionnaire in order to capture new risk factors or sources of exposure. As an example, the CDC list of risk factors has been expanded in recent years, and therefore new questions may need to be incorporated into the environmental history.[22]

The pediatric environmental history may need to be adapted to determine or document the exposure to risk factors associated with specific health problems such as respiratory diseases or cancer. In some cases it may include specific questions to contribute to understanding the etiology of rare pediatric diseases.[18]

The pediatric environmental history is not only a tool for clinical practice but also a basis for generating new knowledge about pediatric environmental health hazards. The information collected in the history is crucial for increasing knowledge about children's environments. Harmonized data collection in countries around the world facilitates the analysis, comparison, and interpretation of data, enabling epidemiological and other studies.[23] In some cases, the pediatric environmental history also may be an important tool for addressing public health problems, such as a cluster of pediatric disease, uncommon cases or symptoms, epidemics of unknown origin, or changes in epidemiological trends in a given area.[24]

The pediatric (or obstetric) environmental history can be partially completed during a home visit (or in other places where the mother or the child lives or spends his or her time). Moreover, the contents of the pediatric environmental history may be adapted and extended to organize a visual assessment form. Although this is not widely available or easily accessible to people working in the health sector, in some countries primary health workers make visits to the patient's home and surroundings to detect social and environmental risks.[25]

Physical Examination

During a physical examination, the clinician uses her or his senses of seeing, hearing, touching, and occasionally smelling, to gather information about the patient. This is a process with which all who have been to a health practitioner are familiar. The clinician looks in eyes and ears, listens to the heart, and palpates the abdomen along with other activities. For environmentally related exposures, the clinician may be looking for a skin rash or for evidence of irritation of the eyes or the nose. In a newborn infant, there may be very subtle changes in the shape of the patient's ears or the anatomy of the genitalia that may be indicative of maternal exposure to an environmental hazard during pregnancy. In general, there are no specific stigmata or pathognomonic findings associated with most environmentally related problems. For example, an individual who has asthma related to exposure to molds at school may have the same physical findings as an individual who has asthma related to ragweed pollen. The famous "Burton's line" of lead poisoning, a bluish pigmentation of the gingiva, while significant if present, is variable between patients with similar blood lead levels and nonexistent at the blood lead concentrations that are common in the United States today.[26]

Laboratory Evaluation

The use of the laboratory in evaluating a child with a known or suspected environmental health problem has been extensively reviewed.[27,28] Laboratory evaluation of an environmental health issue, as for other clinical problems, should be directed by clinical symptoms or known exposures. Biological samples sent for large numbers of tests (e.g., environmental health screening panels) are more likely to result in false positives than true positives. Tests should be done to confirm suspected exposures, not as screening tests to look for unknown exposures. In particular, it has been known for a long time that hair testing, except in research settings, is unreliable and should not be used for the clinical evaluation of patients with suspected environmentally related problems.[29-35]

Reporting and Referral

If the diagnostic interview indicates or raises the strong suspicion that disease is due to toxic environmental exposures, it is imperative that the physician report the case to state or local public health authorities. Many episodes of these diseases are in essence common-source outbreaks of highly preventable illness. Prompt reporting can lead to identifying additional cases earlier and to prevention by abating a common exposure source. The physician may require access to specialized referral sources in environmental medicine.

Environmental Health Evaluation

If an environmental evaluation of a building or play space is warranted, an expert in performing those types of evaluation should be consulted.[36–38] Samples should only be collected and tests run for specific indications. Random testing is likely to be unrevealing or misleading. Testing for lead dust or mercury vapor can be useful in some settings. Antigen testing for cockroach or mouse antigen is a research tool and not likely to be clinically useful. Measuring air pollutants can be useful in a situation in which very significant air contamination is thought to exist. Indoor mold sampling is not likely to be clinically useful. A good visual inspection of a building can show signs of dampness, water damage, and mold growth. The nose is very sensitive to the odor of mold and is an extremely reliable indicator of mold growth. Encouraging families to take photographs and document evidence of exposure can be helpful in alerting healthcare professionals and accessing available local resources to remediate exposures. Measuring indoor humidity with a humidistat is easy and may be useful.[39–41] Public water in the United States is extremely safe and needs only be tested in special circumstances. Keep in mind that lead water pipes were installed in many communities irrespective of wealth level or racial makeup. Therefore, water should be checked as a potential source in all children with an elevated blood lead level. Well water should routinely be tested for possible contamination.[42]

Consultation for Environmental Health Issues

Pediatric Environmental Health Specialty Units

The clinician evaluating a child may need a consultation to help in the evaluation or management of a child with a known or suspected environmentally related health problem, as is frequently the case with cardiac, pulmonary, developmental, or other specialty problems. To provide such consultation, a network of Pediatric Environmental Health Specialty Units has been created.[43] These units are staffed by pediatricians and nurses with expertise in children's environmental health, occupational-environmental medicine physicians, and other areas. A list of their activities is shown in Table 56.1. In the United States, the units are located at academic medical centers and have a mission to provide evidence-based answers to clinical questions. The units are organized to deal with long-term, low-dose exposures or the long-term sequelae of acute exposures. Consultations by phone or e-mail are free, and, in most instances, the staff can advise the local provider regarding the evaluation and management of the patient without having to send the patient to the unit to be evaluated. A list of units and contact information for those in countries around the world is provided in Table 56.2. Poison control centers can assist in management of acute intoxications (see also Chapter 55). Clinical ecologists and individuals certified by the American Board of Environmental Medicine are not appropriate sources of consultation for children with possible environmentally related health problems.[44–47]

Table 56.1 Activities of Pediatric Environmental Health Speciality Units

Educate the public about the impact of the environment in children's health	Alert the public about existing or potential risks and the appropriate responses Develop educational materials adapted to local needs Organize workshops, lectures, and other events
Train health providers and engage with the health community	Develop and make training materials available Conduct training on the prevention, diagnosis, and management of environmentally related exposures and diseases in children
Develop networks on children's health and environmental health to gain knowledge and share experiences	Conduct campaigns to promote children's environmental health
Provide consultative medical services	Provide advice and referrals on cases involving either children or groups of children Provide guidance on laboratory services Maintain standard data about all cases handled by the unit
Consult with government agencies	Alert government officials about existing or potential environmental hazards and steps to address them Provide technical and policy advice to decision makers and agencies related to children's environmental health

Source: World Health Organization, 2010 (23).

Management of Environmentally Related Health Problems

The management of environmentally related health problems depends on the diagnosis. In many situations, removal of the child from the source is all that is necessary. For example, the family of a child with a blood lead concentration in the 20–45 μg/dL range who lives in a home with flaking and peeling lead paint must be provided with a lead-safe residence until the home can be remediated and cleared for reoccupancy. Alternatively, interrupting the pathway between the child and the source can be effective. For example, the family of a child whose well water is contaminated with organic chemicals migrating underground can be supplied with bottled water or with water piped in from a noncontaminated source.

Removal from the source or interruption of the pathway may not, however, be possible, at least in the short term, or sufficient. Access to legal resources can sometimes be useful or even essential. There are more than 200 sites in the country where health professionals and lawyers come together in medical-legal partnerships (see Chapter 15). Accessing a medical-legal partnership can be quite useful in trying to manage housing problems of individual patients.[48] Moreover, if there is a whole community exposed to high levels of air or water pollution or to a toxic waste site, long-term advocacy with politicians and corporations may be necessary.[49–52]

Chelation is indicated for children with markedly elevated lead levels.[53–55] Chelation has not, however, been shown to reduce the long-term neurocognitive problems associated with lead exposure.[56,57] Chelation may have utility in instances of high-dose mercury exposure, but this treatment should be undertaken in conjunction with a medical toxicologist, a poison control specialist, or a physician experienced with chelation.[58,59] Chelation should never be used to treat an ill-defined environmental problem or autism spectrum disorders.[60,61]

Using Evidence-Based Materials

Evidence-based information is crucial to medical and public health decision-making. Evidence-based medicine and evidence-based public health require the use of the best available peer-reviewed data to create the most appropriate health intervention.[62]

Pediatric Environmental Health

The most readily available resource for pediatric clinicians is the American Academy of Pediatrics publication *Pediatric Environmental Health*, 4th edition.[11] The book contains guidance on taking an environmental history, as well as chapters on a number of chemical-, media-, and place-specific topics.

Pediatric Environmental Health Specialty Units

The Pediatric Environmental Health Specialty Units provide education and training to health professionals through seminars, conferences, and online educational programs and case studies on environmental health issues.[63] A list of resources is available at http://www.pehsu.net under the Resources for Health Professionals tab.

Environmental Health Journals

Environmental Health Perspectives (http://www.ehponline.org) is a peer-reviewed open-access journal that publishes articles from various disciplines within environmental health such as epidemiology, risk assessment, environmental justice, environmental policy, longitudinal studies, environmental medicine case reports, and children's environmental health issues.[64]

Web Sites

Pediatric Environmental Health contains an extensive list of environmental health websites.[11] A useful source for environmental health information is the National Library of Medicine (NLM). On the home page click the drop-down menu. Then click Products and Services and then All Products and Services. On that page, checking Environmental Health & Toxicology and then clicking Search brings up access to six specific NLM databases or services.

Agency for Toxic Substances and Disease Registry Toxic Substances Portal

The Agency for Toxic Substances and Disease Registry (ATSDR) has a toxic substances portal housed on their website that lists toxicological information by substance, routes of exposure, and the health effects associated with exposure to the substance. In addition, toxicological information is tailored by audience.

The general public can access substance information through the Community Members tab on the Web site in which the user can choose from Frequently Asked Questions about a given substance and/or the Community Environmental Health Education Presentations, 20- to 30-minute detailed presentations through various PowerPoint and/or Web seminars. Under the Toxicological and Health Professionals tab, environmental and health professionals can access the ATSDR Tox Guide, which is a quick reference guide with information such as chemical and physical properties, children's health, and health effects.[65] The Health Care Provider and Education tab leads users to a pediatric environmental health toolkit and grand rounds presentation in environmental medicine for continuing education for healthcare providers.

The ATSDR Toxic substances portal can be found at http://www.atsdr.cdc.gov/substances/index.asp.

Office of Children's Health Protection Web Site

The US Environmental Protection Agency created the Office of Child Health Protection (OCHP) as a result of the implementation of Executive Order 13045, Protection of Children from Environmental Health Risks and Safety Risks. The order directs agencies to identify and assess environmental health and safety risks that may disproportionately affect children and ensure that its policies, programs, activities, and standards address disproportionate risks to children that result from environmental health risks or safety risks.[66-68]

On the OCHP Web site (https://www.epa.gov/aboutepa/about-office-childrens-health-protection-ochp) there are fact sheets on various environmental topics, and other information for health professionals and parents.

Conclusion

Health professionals are increasingly called upon to deal with environmentally related questions from parents and patients. There is a burgeoning base of evidence documenting the problems, showing the links between the problems and health outcomes, and providing guidance for pediatric health professionals in the management of individual health effects and community-wide problems.

Table 56.2 Pediatric Environmental Health Specialty Units around the World

Argentina

Unidad Pediátrica Ambiental

Hospital de Pediatría "Prof. Dr. Juan P. Garrahan"

Ciudad de Buenos Aires

Tel.: (54–11) 4308–4300

http://www.garrahan.gov.ar/index.php/hospital/comisiones/
32-comite-de-salud-ambiental-infantil

Unidad Sanitaria Ambiental de la Ciudad

Cuenca Matanza Riachuelo

Osvaldo Cruz 2045.

Ciudad de Buenos Aires

www.buenosaires.gob.ar/noticias/nueva-unidad-sanitaria-ambiental-de-la-ciudad

Chile

Centro de Salud Ambiental de Arica

Calle San Martín 253, Arica

58-2583690

Centro de Salud Ambiental - Servicio Salud Arica

https://www.saludarica.cl/programas-de-salud/centro-salud-ambiental

Colombia

Unidad Pediátrica Ambiental

Hospital Regional de Villavicencio

Cra. 42 #34-28 Villavicencio, Meta-

321-2377777 6717202 – 6626433 – 6610039

Email: recepcion@upacolombia.com

Canada

Canada Pediatric Environmental Health Specialty Unit
Academic Affiliation: University of Alberta and Stollery Children's Hospital *Hospital Affiliation:*
Misericordia Community Hospital *Location: Edmonton, Alberta, Canada* http://www.ChEHC.
ca (780) 735-2443 (local and outside of Canada) Toll free: 1-877-214-2331 (inside Canada only)
E-mail: lorie.grundy@albertahealthservices.ca

Spain

Unidad Pediátrica Especializada en Salud Ambiental
Academic Affiliation: Children's University Hospital La Fe (Hospital Infantil Universitario La Fe)
Location: Valencia
Unidad Pediátrica Especializada en Salud Ambiental
Academic Affiliation: University Hospital Virgen de Arrixaca
Location: Murcia
http://www.pehsu.org/organization/present.htm

Korea

Children's Environmental health Clinic in Korea (CHECK)
Academic Affiliation: Ewha Womans University College of Medicine

Location: Seoul

Mexico

Mexico Pediatric Environmental Health Specialty Unit *(Unidad Pediátrica Ambiental) Academic
Affiliation:* National Institute of Public Health *(Instituto Nacional de Salud Pública) Hospital
Affiliation:* Morelos Children's Hospital *(Hospital del Nino Morelense) Location:* Cuernavaca,
Mexico E-mail: ecifuent@correo.insp.mx

United States

Region 1 Pediatric Environmental Health Specialty Unit
Service area: Connecticut, Maine, Massachusetts, New Hampshire, Rhode Island, Vermont
Academic Affiliation: Harvard Medical School and Harvard T.H. Chan School of Public Health
Hospital Affiliation: Boston Children's Hospital and Cambridge Hospital
http://www.childrenshospital.org/pehc/
Phone: (617) 355-8177
Toll free: (888) CHILD14 or (888) 244-5314

Region 2 Pediatric Environmental Health Specialty Unit
Service area: New Jersey, New York, Puerto Rico, and US Virgin Islands
Academic Affiliation Icahn School of Medicine at Mount Sinai: Department of Environmental Medicine and Public Health
Hospital Affiliation: The Mount Sinai Hospital *Location:* New York, New York
http://www.icahn.mssm.edu/research/pehsu
Toll free: (866) 265-6201

Region 3 Mid-Atlantic Center for Children's Health & the Environment Pediatric Environmental Health Specialty Unit
Service area: Delaware, Maryland, Pennsylvania, Virginia, Washington, DC, West Virginia
Academic Affiliation: Villanova University
Phone: (610) 519-3478
Toll free: (833) 362-2243
https://www1.villanova.edu/university/nursing/macche.html

Region 4 Southeast Pediatric Environmental Health Specialty Unit
Service area: Alabama, Florida, Georgia, Kentucky, Mississippi,
North Carolina, South Carolina, Tennessee
Academic Affiliation: Emory University School of Nursing
Hospital Affiliation: Children's Healthcare of Atlanta—Egleston Children's Hospital and Hughes Spalding Children's Hospital, Atlanta, Georgia
http://www.pediatrics.emory.edu/centers/pehsu/index.html
(404) 727-9428
Toll free (877) 33PEHSU or (877) 337-3478

Region 5 Great Lakes Center for Children's Environmental Health
Service area: Illinois, Indiana, Michigan, Minnesota, Ohio, Wisconsin
Academic Affiliation: University of Illinois at Chicago, School of Public Health
Hospital Affiliation: University of Illinois Hospital and Health Sciences System
https://childrensenviro.uic.edu
IL: (312) 355-0597
Satellite location
Academic Affiliation: University of Cincinnati
Hospital Affiliation: Cincinnati Children's Hospital & Medical Center, Cincinnati, Ohio
Phone: (513) 803-3688
Toll free: (866) 967-7337
http://www.cincinnatichildrens.org/service/e/environmental-health/default/

Region 6 Southwest Center for Pediatric Environmental Health
Service area: Arkansas, Louisiana, New Mexico, Oklahoma, Texas
Academic Affiliation: Texas Tech University Health Sciences Center – Paul L. Foster School of Medicine
Hospital Affiliation: University Medical Center of El Paso & El Paso Children's Hospital
http://www.swcpeh.org
(915) 534-3807
Toll free: (888) 901-5665

Region 7 Mid-America Pediatric Environmental Health Specialty Unit
Service area: Iowa, Kansas, Missouri, Nebraska
Academic Affiliation: University of Missouri-Kansas City School of Medicine
Hospital Affiliation: Children's Mercy Hospitals and Clinics, Kansas City, Missouri
http://www.childrensmercy.org/mapehsu
(913) 588-6638
Toll free: (800) 421-9916

Region 8 Rocky Mountain Pediatric Environmental Health Specialty Unit
Service area: Colorado, Montana, North Dakota, South Dakota, Utah, Wyoming
Academic Affiliation: University of Colorado Health Sciences Center
Hospital Affiliation: Denver Health and Hospitals Authority and the Rocky Mountain Poison
and Drug Center, Denver, Colorado
http://www.rmrpehsu.org
Toll free: (877) 800-5554

Region 9 University California Pediatric Environmental Health Specialty Unit
Service area: Arizona, California, Hawaii, Nevada
Academic Affiliation: University California at San Francisco
Hospital Affiliation: University of California San Francisco Medical Center
(415) 514-0878
Toll free: 866-UC-PEHSU or (866) 827-3478
https://wspehsu.ucsf.edu/

Region 10 Northwest Pediatric Environmental Health Specialty Unit
Academic Affiliation: University of Washington: Occupational and Environmental Medicine
Program, Department of Pediatrics
Hospital Affiliation: Harborview Medical Center, University of Washington Medical Center,
Children's Hospital and Regional Medical Center, Seattle, Washington
http://www.deohs.washington.edu/pehsu
(206) 221-8671

Uruguay

Unidad Pediátrica Ambiental
Enrique Claveaux Health Center. Joint Program. Primary Attention Network. ASSE / Toxicology
Department. Clinical Hospital. Faculty of Medicine.
Location: Montevideo, Uruguay.
598-2-22038679–598-24804000
E-mail: udaupa@hc.edu.uy. Hcciat@hc.edu.uy
http://Toxicologia.hc.edu.uy

References

1. Galvez MP, Balk, SJ. Environmental risks to children: prioritizing health messages in pediatric practice. *Pediatr Rev.* 2017;38:6:263–79.

2. Roberts JR, Karr CJ; Council on Environmental Health. Policy statement. *Pesticide Expos Child.* 130;6:e1765–88. doi:10.1542/peds.2012-2758.

3. Balbus JM, Harvey CE, McCurdy LE. Educational needs assessment for pediatric health care providers on pesticide toxicity. *J Agromedicine.* 2006;11:27–38.

4. Gutshow B, Gray B, Ragavan MI, Sheffield PE, Philipsborn RP, Jee SH. The intersection of pediatrics, climate change, and structural racism: ensuring health equity through climate justice. *Curr Prob Pediatr Adolesc Health Care.* 2021;51:101028.

5. Philipsborn RP, Cowenhoven J, Bole A, Balk SJ, Bernstein A. A pediatrician's guide to climate change-informed primary care. *Curr Prob Pediatr Adolesc Health Care.* 2021;51:101027.

6. Etzel R. Foreward: climate change and children. *Curr Prob Pediatr Adolesc Health Care.* 2021;51:1–2.

7. Ragavan MI, Marcil LE, Philipsborn R, Garg A. Parents' perspectives about discussing climate change during well-child visits. *J Climate Change Health.* 2021;4:100048.

8. Lewandowski AA, Sheffield PE, Ahdoot S, Maibach EW. Patients value climate change counseling provided by their pediatrician: the experience in one Wisconsin pediatric clinic. *J Climate Change Health.* 2021;4:100053.

9. Schenk M, Popp SM, Neale AV, Demers RY. Environmental medicine content in medical school curricula *Acad Med.* 1996;71:499–501.

10. Pollack SH. Adolescent occupational exposures and pediatric-adolescent take-home exposures. *Pediatr Clin North Am.* 2001;48:1267–89.

11. American Academy of Pediatrics Council on Environmental Health. Taking an environmental history and giving anticipatory guidance. In Etzel RA, Balk SJ, eds. *Pediatric Environmental Health.* 4th ed. Itasca, IL: American Academy of Pediatrics; 2019:37–56.

12. Pronczuk J. Pediatric environmental history taking in developing countries. In Pronczuk de Garbino J. *Children's Health and the Environment: A Global Perspective.* Geneva, Switzerland: World Health Organization; 2005:227–31.

13. World Health Organization. The pediatric environmental history. Recording children's exposure to environmental health threats: a "green page" in the medical record. Available at: https://www.who.int/news-room/questions-and-answers/item/q-a-the-paediatric-environmental-history

14. National Environmental Education Foundation. PEHI History Forms. 2015. Available at: https://www.neefusa.org/resource/pediatric-environmental-history-form

15. Facultad de Medicina, Universidad de la República. Urban Environmental Risk Short Questionnaire. Unidad Docente Asistencial- Unidad Pediátrica Ambiental. Available at: https://www.toxicologia.hc.edu.uy/index.php?Itemid=76

16. Paranzino GK, Bitterfield P, Nastoff T, Ranger C. I PREPARE: development and clinical utility of an environmental exposure history mnemonic. *AAOHN J.* 2005;53:37–42.

17. Marshall N, Weir E, Abelsohn A, Sanborn M. Identifying and managing adverse environmental health effects:1. Taking an exposure history. *CMAJ.* 2002;166:1049.

18. Ortega-García JA, Soldin OP, López-Hernández FA, Trasande L, Ferrís-Tortajada J. Congenital fibrosarcoma and history of prenatal exposure to petroleum derivatives. *Pediatrics.* 2012;130:e1019–25.

19. Laumbach R, Kipen HM. Health Care Services. In *PAHO Environmental Health.* 1113–33.

20. Agency for Toxic Substances and Disease Registry. Case studies in environmental medicine. Available at: https://www.atsdr.cdc.gov/csem/csem.html.

21. Managing Elevated Blood Lead Levels Among Young Children: Recommendations from the Advisory Committee on Childhood Lead Poisoning Prevention. Atlanta, GA: Centers for Disease Control and Prevention; 2002.

22. Advisory Committee on Childhood Lead Poisoning Prevention. *Low Level Lead Exposure Harms Children: A Renewed Call for Primary Prevention.* Atlanta, GA: Centers for Disease Control and Prevention; 2012.

23. World Health Organization. Children's environmental health units. 2010. Available at: https://www.who.int/publications/i/item/9789241500425

24. Ortega García JA, Ferris Tortajada J, Claudio Morales L, Berber Tornero O. Pediatric environmental health specialty units in Europe: from theory to practice. *An Pediatr (Barc)*. 2005;63:143–5.

25. US Department of Housing and Human Development. *Healthy Housing Inspection Manual*. Atlanta, GA: Centers for Disease Control and Prevention, US Department of Health and Human Services; 2008.

26. Pearce JM. Burton's line in lead poisoning. *Eur Neurol*. 2007;57:118–9.

27. Hoffman HE, Buka I, Phillips S. Medical laboratory investigation of children's environmental health. *Pediatr Clin North Am*. 2007;54:399–415.

28. Zajac L, Johnson SA, Hauptman M. Doc, can you test me for "toxic metals"? challenges of testing for toxicants in patients with environmental concerns. *Curr Prob Pediatr Adolesc Health Care*. 2020;50:100762. https://dx.doi.org/10.1016/j.cppeds.2020.100762

29. Klevay LM, Bistrian BR, Fleming CR, et al. Hair analysis in clinical and experimental medicine. *Am J Clin Nutr*. 1987;46:233–6.

30. Yoshinaga J, Imai H, Nakazawa M, et al. Lack of significantly positive correlations between elemental concentrations in hair and in organs. *Sci Total Environ*. 1990;99:125–35.

31. Wenning R. Potential problems with the interpretation of hair analysis results. *Forensic Sci Int*. 2000;107:5–12.

32. Eastern Research Group. Hair analysis panel discussion: exploring the state of the science. Available at: http://www.atsdr.cdc.gov/HAC/hair_analysis. Date of publication December 2001.

33. Seidel S, Kreutzer R, Smith D, et al. Assessment of commercial laboratories performing hair mineral analysis. *JAMA*. 2001;285:67–72.

34. Frisch M, Schwartz BS. The pitfalls of hair analysis for toxicants in clinical practice: three case reports. *Environ Health Perspect*. 2002;110:433–6.

35. Harkins DK, Susten AS. Hair analysis: exploring the state of the science. *Environ Health Perspect*. 2003;111:576–8.

36. Portnoy JM, Kennedy K, Barnes C. Sampling for indoor fungi: what the clinician needs to know. *Curr Opin Otolaryngol Head Neck Surg*. 2005;13:165–70.

37. Johnson L, Ciaccio C, Barnes CS, et al. Low-cost interventions improve indoor air quality and children's health. *Allergy Asthma Proc*. 2009;30:377–85.

38. Portnoy J, Kennedy K, Sublett J. Environmental assessment and exposure control: a practice parameter: furry animals. *Ann Allergy Asthma Immunol*. 2012;108:223.e1–15.

39. Institute of Medicine: Committee on Damp Indoor Spaces and Health. *Damp Indoor Spaces and Health*. Washington, DC: National Academies Press; 2004.

40. World Health Organization. *WHO Guidelines for Indoor Air Quality, Dampness and Mould*. Heseltine E, Rosen, J, eds. Geneva: World Health Organization; 2009.

41. Mazur LJ, Kim J. Spectrum of noninfectious health effects from molds. *Pediatrics*. 2006;118:e1909–26.

42. Rogan WJ, Brady MT. Drinking water from private wells and risks to children. *Pediatrics*. 2009;123:e1123–37.

43. Paulson JA, Karr CJ, Seltzer JM, et al. Development of the pediatric environmental health specialty unit network in North America. *Am J Public Health*. 2009;99:S511–6.

44. American College of Physicians. Clinical ecology. *Ann Intern Med*. 1989;111:168–78.

45. American Medical Association Council of Scientific Affairs. Clinical ecology. *JAMA.* 1992;268:3465–7.

46. American Academy of Allergy, Asthma and Immunology Board of Directors. Idiopathic environmental intolerances. *J Allergy Clin Immunol.* 1999;103:36–40.

47. Terr AI. Environmental sensitivity. *Immunol Allergy Clin North Am.* 2003;23:311–28.

48. Weintraub D, Rodgers MA, Botcheva L, et al. Pilot study of medical-legal partnership to address social and legal needs of patients. *J Health Care Poor Underserved.* 2010;21:157–68.

49. Rubin IL, Nodvin JT, Geller RJ, et al. Environmental health disparities: environmental and social impact of industrial pollution in a community—the model of Anniston, AL. *Pediatr Clin North Am.* 2007;54:375–98.

50. Paulson JA. Pediatric advocacy. *Pediatr Clin North Am.* 2001;48:1307–18.

51. Krieger JK, Takaro TK, Allen C, et al. The Seattle-King County healthy homes project: implementation of a comprehensive approach to improving indoor environmental quality for low-income children with asthma. *Environ Health Perspect.* 2002;110:311–22.

52. van den Hazel P, Zuurbier M, Bistrup ML. Policy interpretation network in children's health and the environment. *Acta Paediatrica.* 2006;95:6–12.

53. Ennis JM, Harrison HE. Treatment of lead encephalopathy with PAL (2.3-dimercaptopropanol). *Pediatrics.* 1950;5:853–68.

54. Coffin R, Phillips JL, Staples WI, et al. Treatment of lead encephalopathy in children. *J Pediatr.* 1966;69:198–206.

55. Chisolm JJ Jr. The use of chelating agents in the treatment of acute and chronic lead intoxication in childhood. *J Pediatr.* 1968;73:1–38.

56. Rogan WJ, Dietrich KN, Ware JH, et al. The effect of chelation therapy with succimer on neuropsychological development in children exposed to lead. *N Engl J Med.* 2001;344:1421–6.

57. Dietrich KN, Ware JH, Salganik M, et al. Effect of chelation therapy on the neuropsychological and behavioral development of lead-exposed children after school entry. *Pediatrics.* 2004;114:19–26.

58. Forman J, Moline J, Cernichiari E, et al. A cluster of pediatric metallic mercury exposure cases treated with meso-2,3-dimercaptosuccinic acid (DMSA). *Environ Health Perspect.* 2000;108:575–7.

59. Cao Y, Chen A, Jones RL, et al. Efficacy of succimer chelation of mercury at background exposures in toddlers: a randomized trial. *J Pediatr.* 2011;158:480–5.

60. Senel HG. Parents' views and experiences about complementary and alternative medicine treatments for their children with autistic spectrum disorder. *J Autism Dev Disord.* 2010;40:494–503.

61. Golnik AE, Ireland M. Complementary alternative medicine for children with autism: a physician survey. *J Autism Dev Disord.* 2009;39:996–1005.

62. Brownson RC, Fielding JE, Green LW. Building capacity for evidence-based public health: reconciling the pulls of practice and the push of research. *Annu Rev Public Health.* 2018;39:27–53.

63. Agency for Toxic Substances and Disease Registry. Pediatric environmental health toolkit training module. 2018. Available at: http://www.atsdr.cdc.gov//emes/health_professionals/pediatrics.html.

64. Environmental Health Perspectives. About EHP. Available at: https://ehp.niehs.nih.gov/about-ehp.

65. ATSDR Toxic Substances Portal. Toxicological and health professionals—Tox Guide. 2021. Available at: http://www.atsdr.cdc.gov/toxguides/index.asp.

66. Clinton WJ. Executive Order 13045: Protection of children from environmental health risks and safety risks. *Fed Reg.* 1997;62:19883–8.

67. Payne-Sturges D, Kemp D. Ten years of addressing children's health through regulatory policy at the U.S. Environmental Protection Agency. *Environ Health Perspect.* 2008;116:1720–4.

68. Firestone M, Berger M, Foos B, Etzel R. Two decades of enhancing children's environmental health protection at the U.S. Environmental Protection Agency. *Environ Health Perspect.* 2016;124(12):A214–A218. doi:10.1289/EHP1040.

57

Public Policy and Clinical Practice of Environmental Pediatrics in Latin America and the Caribbean

Amalia Laborde

The Americas are considered one of the world's most inequitable regions in many aspects of development and well-being, both between and within countries. Addressing environmental hazards as cross-cutting themes of social and environmental determinants of health is essential to identify more vulnerable regions and population groups. Understanding the sources of risk and the processes by which vulnerabilities are created in the region is essential to taking evidence-based actions to reduce exposures, improve health, and guarantee the right to health.[1]

Limited research has been conducted to understand how environmental hazards are affecting children's health in South America. Most data come from the region of the Americas or more specifically from Latin America and the Caribbean as a group. Although there have been advances in policies and knowledge on environmental health in Latin America, López-Carrillo et al. found that Mexico and Brazil produced almost 70% of the scientific publications on children's health and the environment, so most countries of Latin America are underrepresented in the scientific literature. The publications addressing environmental exposures showed a decreasing trend in the past decade.[2]

In Latin America and the Caribbean, both traditional and newer environmental risks to children's health are present. The traditional environmental problems, such as indoor air pollution and contaminated drinking water, remain, while new risks such as outdoor air pollution, exposures to toxic pollutants, and climate change are emerging or increasing problems.[3]

The Pan American Health Organization (PAHO) *Atlas of Children's Environmental Health* stated that, in 2011, nearly 100,000 children younger than 5 years died annually due to environmental hazards. Respiratory and diarrheal diseases represented around 60% of deaths associated with the environment, followed by injuries (26%), cancer (7%), and vector-borne diseases (2%). In 2016, the WHO/PAHO estimated that 13% of premature deaths in high-income countries and 19% in low- and middle-income countries of the Americas were attributable to environmental risks. This proportion represents about 1,016,000 annual deaths.[4,5]

Air pollution is one of the leading environmental risks globally: more than 300,000 deaths due to cardiovascular, respiratory diseases, and cancer are attributed to air pollution in the region.[6] The impact of air pollution may be underestimated since a small proportion (<10%) of Latin American and Caribbean (LAC) cities that monitored

Amalia Laborde, *Public Policy and Clinical Practice of Environmental Pediatrics in Latin America and the Caribbean*
In: *Textbook of Children's Environmental Health*, Second Edition. Edited by: Ruth A. Etzel and Philip J. Landrigan,
Oxford University Press. © Oxford University Press 2024. DOI: 10.1093/oso/9780197662526.003.0057

air pollution complied with WHO air quality guidelines.[7] Indoor air pollution is less studied, although household combustion must be a relevant risk factor because 80 million people still depend on polluting fuels for cooking and heating.[6]

Health-related impacts from climate change are expected to increase; the annual average temperature increased more than 1.0°C in 2017 in Latin America and the Caribbean. Latin American countries generate less greenhouse gas than other regions but are disproportionately suffering from the effects through droughts, floods, hurricanes, and seasonal shifts. Although almost all countries of the Americas recognize the importance of health in national climate change commitments, only 17 implement health and climate surveillance activities and more than 30 still lack health national adaptation plans.[8,9]

Access to drinking water systems has been extended in the last several decades. On average around 3% of the LAC population still do not have access to drinking water services, mainly the poor populations in suburban and rural areas. Water quality also is threatened by emerging or increasing risks from cyanobacterial (algal) blooms, unsafe agricultural practices, and informal, artisanal, or uncontrolled industrial and mining activities. Sanitation services have improved in the past several decades as well, but 13% of the population still have inadequate sanitation services.[5,10]

There is limited regional information about exposure to hazardous chemicals and acute poisonings. Data about environmental sources of lead, mercury, arsenic, and pesticides suggest that exposure to these chemicals may be important contributors to noncommunicable diseases, particularly among children. A high risk of adverse health effects among children and adults due to exposure to heavy metals in contaminated sites from mine tailings or industrial emissions has been found.[11]

Small-scale gold mining close to the Amazonian region causes high occupational and dietary exposure to mercury in local communities. South America appears to play a major role in the global burden of mercury pollution while at the same time suffering from the local and global impact of mercury pollution.[12-15]

Soil contamination from landfills and hotspots of pollution linked to informal battery and electronic waste recycling (as well as lead-glazed pottery) are the most frequently reported sources of children's lead exposure in South America. High blood lead levels were reported in the past decade in most LAC countries.[16-18] The Rio Birth Cohort study of environmental exposure and childhood development is starting to show the importance of prenatal exposure to various metals and its relation to neurodevelopment alterations.[19,20]

Arsenic is present in the environment due to mining and metal smelting, particularly as a secondary pollutant from other metals production. Moreover, arsenic in groundwater may be present due to natural geological characteristics in many countries along the Andes region. Data from 2012 estimate that 4.5 million people are exposed to high levels of arsenic in water.[21]

Electronic waste (e-waste) is a combined source of a mixture of heavy metal and a complex mixture of new plastic and persistent organic pollutants. E-waste, electric and electronic waste residues, or digital waste has increased 49% in the past 10 years, and only 3% was safely managed or recycled. This proportion is well under the world average of 17%. Many hotspots of informal manipulation or metal recovering are expected to be identified, particularly in suburban areas.[17,22]

Plastics and microplastics are one of the most complex emerging global environmental risks. A small number of studies describe the South American fate of microplastics in sandy beaches and marine waters of the Atlantic and Pacific oceans.[23] Recent studies have uncovered the presence of antibiotics in water bodies, revealing that this is another emerging environmental problem that needs to be addressed.[24]

Acute pesticide poisoning is widespread in Latin America and the Caribbean, but it is estimated that cases are underreported by more than 50%. Relatively little is known about the extent of children's chronic exposure to pesticides; however, respiratory, neurological, and developmental alterations associated with pesticide exposure have been found in almost all LAC countries.[3,25] Population studies using biomarkers in biological samples are scarce in LAC countries because of lack of laboratories with capabilities to study pesticides or pesticide metabolites in the population (cholinesterase activity is the biological biomarker mostly used to identify significant exposure to organophosphate pesticides in intensive agricultural areas of various LAC countries).[26]

Although accurate statistics are not available in South America, child labor is an important concern in both rural and urban areas. As occurs globally, agricultural areas concentrate working boys and girls older than 10 years. Three countries of South America report 20% of girls working in rural mining regions and have identified 13,000–50,000 children working in small-scale gold mining. [27]

Public Policies in Latin America and the Caribbean

Most LAC countries are signatories of international conventions such as the Stockholm Convention, the Montreal Protocol, the Rotterdam Convention, the Minamata Convention, and the No Lead Paint initiative. There is a wide range of regulations regarding almost all the above-mentioned global environmental issues, although implementation at community levels is uncertain or the results are not publicly available.

The 2021–2024 PAHO Agenda for LAC countries recognizes the vital need to "strengthen environmental public health programs and institutions." Recommended actions include producing technical guidelines for environmental health risk management, developing the analytical competences of environmental laboratories, implementing international conventions at the national level, and creating regional networks to deal with environment exposures and health effects prevention and mitigation.[28] The Toxicology Network of Latin America and the Caribbean as well as regional scientific societies such as the Iberoamerican Society of Environmental Health represent professional groups that address children's environmental health as part of their objectives.[5] Children's environmental health committees in pediatric professional societies are among the groups that continue to sponsor educational activities and advocate for children's environmental risk prevention and management.

Although there have been advances made in policies and knowledge on environmental health in LAC countries there is still a great need to expand and reinforce research on environmental exposures and children's diseases and translate global scientific evidence into regional policies and clinical practice.

Environmental Health in Clinical Practice

Health professionals from LAC have played an essential role in identifying, diagnosing, preventing, and managing environmental exposures and environmentally related diseases in children. However, institutionalized environmental clinical practice has not expanded in LAC countries. One reason for significant gaps in knowledge about environmental hazards across the health sector is the limited educational programs on environmental public health in both graduate- and postgraduate-level training programs. Although a collaborative training manual aimed at medical students was updated in 2010, environmental health—including climate change education—has been insufficiently introduced into the curriculum of health professionals in Latin American universities.[29,30]

Pediatricians, primary care physicians, nurses, and other health professionals are well-positioned to be strategically involved in children's environmental health. A widespread problem is that they may not be able to spend time taking a detailed environmental history to find out about suspected environmental risks, particularly if the environmental history-taking tool is not practical and adapted to the local environment. Surveys of pediatricians have shown that although they are interested in children's environmental health, they perceive barriers such as time, effort, and cost to incorporating a detailed environmental history into their clinical visits.[3] A specific Pediatric Environmental History, adapted to local or regional environmental risks, has been developed by the Pediatric Environmental Units in Argentina and Uruguay. A small set of questions selected from the pediatric environmental history has been developed by the Pediatric Environmental Unit in Uruguay to screen for indoor air pollution and lead exposure among the urban population. This short questionnaire can be completed by nurses, medical students, or other health professionals during or before the clinical visit.[31,32]

Encouraging pediatricians to think about environmental causes of diseases and conditions must be supported by reference centers or specialists. The Pediatric Environmental Health Specialty Units were launched in the United States in 1998. The essential goal of these units is to address gaps in children's environmental health knowledge by enhancing the fundamental knowledge and skills of pediatricians, primary care physicians, and other health professionals. See chapter 56.

These units have been established in Argentina, Chile, Colombia, Mexico and Uruguay as Unidades Pediátricas Ambientales (UPAs). In addition, Clinical Centers dedicated to specific populations exposed to toxic chemicals or metals were developed to serve contaminated sites in a mining region of Chile and an urban area of Argentina in response to community demand. All of them work with multidisciplinary groups: pediatricians, family doctors, nurses, toxicologists, and social workers, among others. Poison Centers are part of or work in close coordination with these Units.

The list of health clinics in South America offering pediatric environmental clinical practice is found in Chapter 56.

More work remains to be done, and new challenges need to be addressed to successfully strengthen and expand clinical practice in children's environmental health in LAC countries. International exchanges and South–South collaboration lay the foundation for advancing the clinical practice of children's environmental health in LAC countries.

References

1. Pan American Health Organization (PAHO). *Equity in Health Policy Assessment: Region of the Americas*. 2020. ISBN978-92-75-12291-4 CC BY-NC Sa 3.0 IGO https://iris.paho.org/handle/10665.2/52931

2. López-Carrillo L, González-González L, Piña-Pozas M, et al. State of children environmental health research in Latin America. *Ann Glob Health*. 2018 Jul 27;84(2):204–11. doi:10.29024/aogh.908.

3. Laborde A, Tomasina F, Bianchi F, et al. 2015. Children's health in Latin America: the influence of environmental exposures. *Environ Health Perspect*. 123, 201–9.

4. Froes Asmus CIR, Camara VM, Landrigan PJ, Claudio L. A systematic review of children's environmental health in Brazil. *Ann Global Health*. 2016;82(1):132–48. doi:http://doi.org/10.1016/j.aogh.2016.02.007.

5. Kroc M, Hochman F. Advancing environmental public health in Latin America and the Caribbean. *Rev Panam Salud Publica*. 2021;45: e118. https://doi.org/10.26633/RPSP.2021.118.

6. Pan American Health Organization (PAHO). Environmental determinants of health. 2022. Available at: https://www.paho.org/en/topics/environmental.

7. Riojas-Rodríguez H, Soares Da Silva A, Texcales-Sangrador JL, Moreno-Banda GL. Air pollution management and control in Latin America and the Caribbean: implications for climate change. *Rev Panam Salud Publica*. 2016;40(3):150–9. Available at: https://iris. paho. org/handle/10665.2/31229.

8. Aracena S, Barboza M, Zamora V, Salaberry O, Montag D. Health system adaptation to climate change: a Peruvian case study. *Health Policy Plan*. 2021;36(1):45–83. doi:10.1093/heapol/czaa072.

9. Pan American Health Organization (PAHO). Snapshot: health and climate change in the Americas. 2021. Available at: https:// www.paho.org/en/snapshot-health-and-climate-change-americas.

10. Juanena C, Negrin A, Laborde A. Cianobacterias en las playas: riesgos toxicológicos y vulnerabilidad infantil. *Revista Médica del Uruguay*. 2020;36(3):157–82. Epub 01 de septiembre de 2020.https://doi.org/10.29193/rmu.36.3.7.

11. Cáceres DD, Jiménez F, Hernández P, et al. Health risk due to heavy metal(loid)s exposure through fine particulate matter and sedimented dust in people living next to a beach contaminated by mine tailings. *Rev Int Contam Ambie*. 2021;37:211–26.

12. Olivero-Verbel J, Alvarez-Ortega N, Alcala-Orozco M, Caballero-Gallardo K. Population exposure to lead and mercury in Latin America. *Curr Opin Toxicol*. 2021;27:27–37.

13. Piñeiro XF, Ae MT, Mallah N, et al. Heavy metal contamination in Perú: implication on children´s health. *Sci Rep*. 2021;11:22729. https://doi.org/10.1038/s41598-021-02163-9.

14. Méndez M, Pose D, Laborde A, Noria A, Gil J; Lindner C. Nivel medio de mercurio en mujeres embarazadas y recién nacidos en Uruguay 2016–2018: Avance de resultados. *Rev Salud Ambient*. 2020;20(1):30–6.

15. Palma-Parra M, Muñoz-Guerrero MN, Pacheco-Garcia O, Ortiz-Gomez Y, Díaz CSM. Niños y adolescentes expuestos ambientalmente a mercurio, en diferentes municipios de Colombia Revista de la Universidad Industrial de Santander. *Salud Mar*. 2019;51(1):43.

16. Juanena C, Pose-Román DA, Sosa A, et al. Identificación de la contaminación con plomo en comunidades vulnerables: experiencia de una Unidad Ambiental en atención primaria en Uruguay. *Rev Salud Ambient*. 2021;21(1):16–22.

17. Pascale A, Sosa A, Bares C, et al. E-waste informal recycling: an emerging source of lead exposure in South America. *Ann Global Health*. 2016;82(1):197–8.

18. Olympio KP, Goncalves CG, Salles FJ, Ferreira AP, Soares AS, Buzalaf MA, Cardozo MR, Bechara EJ. What are the blood lead levels of children living in Latin America and the Caribbean? *Environ Int*. 2017;101:46–58.

19. Assis Araujo MS, Froes-Asmus CIR, de Figueiredo ND, et al. Prenatal exposure to metals and neurodevelopment in infants at six months: Rio birth cohort study of environmental exposure and childhood development (PIPA Project). *Int J Environ Res Public Health*. 2022 Apr 3;19(7):4295.

20. Caravanos J, Carrelli J, Dowling R, Pavilonis B, Ericson B, Fuller R. Burden of disease resulting from lead exposure at toxic waste sites in Argentina, Mexico and Uruguay. *Environ Health*. 2016;15(1):72.

21. Bundschuh J, Armienta MA, Morales-Simfors N, et al. Arsenic in Latin America: new findings on source, mobilization and mobility in human environments in 20 countries based on decadal research 2010–2020. *Crit Rev Environ Sci Technol*. 2021;51(16):1727.

22. Wagner M, Baldé CP, Luda V, Nnorom JC, Kuehr R Iattoni GR. Regional E-waste monitor for Latin America: results for the 13 countries participating in project. *UNIDO-GEF 5554, Bonn (Germany)*. Available at: https://collections.unu.edu/view/UNU:8704

23. Zarate M, Yanacona J. Microplásticos en tres playas arenosas de la costa central del Perú. *Rev Salud Ambient*. 2021;21(2):123–13.

24. Meléndez-Marmolejo J, García-Saavedra Y, Galván-Romero V, Díaz de León-Martínez L, Vargas-Berrones K, Mejía-Saavedra J, Flores-Ramírez R. Contaminantes emergentes. Problemática ambiental asociada al uso de antibióticos. Nuevas técnicas de detección, remediación y perspectivas de legislación en América Latina. *Rev Salud Ambient*. 2020;20(1):53–61.

25. Barcelos da Rocha C, Pinheiro CostaNacimento, Candido Da Silva AM, Botelho. Asma no contrlado en niños y adolescentes expuestos a pesticidas en una región de intensa actividad agroindustrial. *Cada Saude Pública* 2021;37(5).

26. Zúñiga-Venegas L, Saracinoi C, Pancetti F, Muñoz-Quezada MT, Lucero B, Foerster C, Cortés S. Exposición a plaguicidas en Chile y salud poblacional: urgencia para la toma de decisiones [Pesticida exposure in Chile and population health: urgency for decision making.] *Gac Sanit*. 2021 Sep-Oct;3l5(5):480–7.

27. International Labour Office/United Nations Children´s Fund (ILO/UNICEF). *Child Labour: Global Estimates 2020. Trends and the Road Forward*. License CC BY 4.0. New York: UNICEF; 2021.

28. Pan American Health Organization (PAHO). Agenda para las Américas sobre salud, medioambiente y cambio climático 2021–2030. 2021. Available at: https://iris.paho.org/handle/10665.2/55212

29. Palmeiro-Silva YK, Ferrada MT, Flores JR, Cruz ISS. Climate change and environmental health in undergraduate health degrees in Latin America. *Rev Saude Publica*. 2021;23(55):17.

30. Quiroga D, Fernández R, Paris E. *Salud Ambiental Infantil: manual para enseñanza de grado en escuelas de medicina*. 1a ed. Buenos Aires: Ministerio de Salud de la Nación; Organización Panamericana de la Salud; 2010. ISBN 978-950-38-0097-3. https://bancos.salud.gob.ar/sites/default/files/2018-10/0000000271cnt-s12-manual-universitario-salud-ambiental-infantil.pdf.

31. Moll MJ, Laborde A. Briefing on children environmental health in Uruguay. In: Triage J, ed., *Encyclopedia of Environmental Health*. Vol. 1. Amsterdam: Elsevier; 2019: 460–4.

32. Giulia M, Amedo D, González D. Clinic of environmental pediatric care in a high complexity hospital. *Arch Argent Pediatr*. 2014;112:(6):562–6.

58

Public Policy on Children's Environmental Health in the United States

Philip J. Landrigan

In the past twenty five years, countries around the world have established policies to protect children against environmental threats to health. This progress is based on a shared recognition that infants and children are very different from adults in their exposures and vulnerabilities to toxic chemicals and other hazards in the environment and that they therefore require special protections in risk assessment, regulation, and law.

The United States was an early leader in this effort. In recent years, however, the European Union has surpassed the United States and moved to the forefront in developing policies to assess and control risks from chemical exposures. Policies are also under development in Canada and in the more highly industrialized countries of Asia.

This chapter examines the key policies that the United States has put into place to protect children against environmental health hazards. It complements Chapter 57, which discusses children's environmental health policies in Latin America and the Caribbean, Chapter 59, which discusses children's environmental health policies in Europe; Chapter 60, which discusses children's environmental health policies in Asia; Chapter 61, which discusses children's environmental health policies in Africa; and Chapter 62, which examines the international treaties that have been established to protect children around the world and especially children in low- and middle-income countries against toxic chemicals and other hazardous materials in global trade.

US Federal Laws Relevant to Children's Environmental Health

In the United States, there is no single overarching law to protect human health against environmental threats to health. Instead, the United States has enacted a series of federal laws, each designed to protect people against a particular class of chemicals or against pollutants in a particular sector of the environment (Table 58.1). There also are laws in many of the 50 states to protect children against environmental hazards, as well as laws passed in certain US cities.

Most of these laws protect human health by establishing legally enforceable standards for toxic chemicals. These standards regulate the levels of toxic chemicals or pesticides that are legally permitted to be present in air, water, wastes, food, or consumer products. Examples include the level of particulates allowed in air, the level of lead permitted in drinking water, or the level of a particular insecticide allowed on fresh fruit. Most environmental standards in the United States are established through *quantitative risk*

Philip J. Landrigan, *Public Policy on Children's Environmental Health in the United States* In: *Textbook of Children's Environmental Health*, Second Edition. Edited by: Ruth A. Etzel and Philip J. Landrigan, Oxford University Press. © Oxford University Press 2024.
DOI: 10.1093/oso/9780197662526.003.0058

Table 58.1 US federal legislation relevant to children's environmental health

Chemical-specific laws	Regulates
• Food Quality Protection Act	Pesticides
• Toxic Substances Control Act	Industrial and consumer chemicals
• Frank R. Lautenberg Chemical Safety for the 21st Century Act	Industrial and consumer chemicals
• Federal Food, Drug & Cosmetic Act	Food additives, pharmaceuticals, cosmetics
Laws Regulating Environmental Media	
• Clean Air Act	Air pollution
• Safe Drinking Water Act	Drinking water safety
• Clean Water Act	Water pollution
• Resource Conservation and Recovery Act	Waste generation and disposal
• Superfund Authorization and Recovery Act	Cleanup of hazardous wastes
• Consumer Product Safety Improvement Act	Health and safety hazards in consumer products (e.g., toys)
• Residential Lead-Based Paint Hazard Reduction Act	Lead paint in homes

assessment, a process that evaluates the hazardous properties of a chemical and then determines the risks to human health that result from exposure to that chemical.[1] Risk assessments are typically based on extrapolations of data from toxicological and epidemiological studies that examine the occurrence of disease and dysfunction at various levels of chemical exposure.

Risk assessment has four major components: hazard identification, exposure quantification, dose-response analysis, and risk characterization.

Hazard identification is the first step in risk assessment. It defines the diseases and dysfunctions that result from exposure to a toxic chemical or other environmental hazard. Child-centered hazard identification is based on the recognition that the diseases and dysfunctions that result from exposures in early life may be qualitatively and quantitatively different from the effects of exposures in adult life.

Exposure quantification determines the level of exposure to a toxic chemical or other environmental hazard in individuals and populations. Child-centered quantification of exposure recognizes that children have routes and patterns of exposure that are profoundly different from those of adults, such as transplacental exposures for the fetus *in utero* and exposures via breast milk for the nursing infant.

Dose-response analysis is a quantitative process that correlates the frequency or severity of disease and dysfunction with level of exposure to an environmental hazard. Child-centered dose-response analysis understands that the quantitative relationship between exposure and outcome may be very different between children and adults. It also recognizes that during unique windows of vulnerability in the earliest stages of development, timing of exposure may be more important than exposure dose in determining health effects (see Chapter 2). Dose-response typically is thought to be linear and monotonic. However, exposures in early life, and especially early exposures to endocrine-disrupting compounds, may involve nonmonotonic dose-response relationships.[2]

Risk characterization is the final stage of risk assessment. It synthesizes the findings from the first three components of a risk assessment and then characterizes the resulting risk. This characterization typically results in the development of a table depicting the estimated number of cases of disease or dysfunction that will result in a population at given levels of exposure to a toxic chemical or other environmental hazard. Child-centered risk characterization considers the effects of exposures on the health of infants and children separately from effects on adult health.

Risk management is a legal and regulatory process that is separate from and follows risk assessment. It takes the findings from the risk assessment on a particular chemical or other environmental hazard and converts these findings into legally binding regulations and standards or into other governmental enforcement actions, such as requirements for pre-market testing, recalls of toxic products, and outright banning of very hazardous materials.

The distinction between risk assessment and risk management was formulated by an expert committee convened by the US National Academy of Sciences (NAS).[1]

Some standard-setting processes are solely health-based. They set standards at levels that will protect population health and take no cognizance of other factors. Other standard-setting processes balance population health against the cost that a standard will impose on the affected industry and set the standard through a process of compromise, termed "cost-benefit analysis." Under the Food Quality Protection Act (FQPA), pesticide standards ("tolerances") are required to be based solely on protection of public health and specifically on protection of the health of infants and children. By contrast, standards set under the Toxic Substances Control Act (TSCA) are required to consider cost-benefit analyses.

Child-Centered Risk Assessment

A fundamental problem that afflicts many risk assessments is that they are based on a great lack of information about the effects of chemicals on infants and children. This lack of information stems from the fact that most chemicals are tested for toxicity by administering them to experimental animals during adolescence and early adulthood and then sacrificing the animals and examining them for toxic effects 12–24 months later, a life stage equivalent to about age 60–65 years in humans. This truncated approach, which omits both ends of the life span, precludes examination of the possible effects of exposures in utero or in early postnatal life, and it also precludes examination of the late consequences of early exposures. In short, most toxicity testing of chemicals—and in consequence, most risk assessment—fails to consider the special vulnerability of infants and children (see Chapter 2).

Prior to the 1990s, virtually all risk assessment in the United States focused solely on protection of the "average adult." It paid little attention to the unique risks of infants, children, or other vulnerable groups within the population. Thus, prior to the 1990s, risk assessment did not work well to protect children from environmental threats to health.

The National Academy of Sciences Report on Pesticides in the Diets of Infants and Children

Recognition among policymakers of the unique vulnerability of children to environmental hazards had its origins in the United States with the publication in 1993 of the National

Academy of Sciences (NAS) report, "Pesticides in the Diets of Infants and Children."[3] This report found striking differences between children and adults in their patterns of exposure as well as in their susceptibility to toxic chemicals and other environmental hazards (see Chapter 2). This report also identified large gaps in regulation. It found that the federal laws and regulations then in force did not adequately protect infants and children against pesticides and other toxic chemicals. It called for expansion of toxicological testing to assess threats to early human development. It urged reform of risk assessment and regulation to enhance the protection of infants in the womb and young children.

The key consequence of the NAS report was that it elevated consideration of the vulnerability of children from the specialized area of pediatrics to the broad realm of national policy. This report catalyzed a paradigm shift in children's environmental health policy in the United States. It led to new legislative and regulatory initiatives to better protect infants and children against environmental health threats. It catalyzed new approaches to risk assessment that, for the first time, considered the special vulnerabilities of infants and children. It has been especially influential in changing the regulation of pesticides and pharmaceutical chemicals.

The remaining sections of this chapter focuses on two US federal laws that are especially important for children's environmental health, the FQPA of 1996 and the TSCA of 1976. Although these two laws have been selected as examples, many of the conclusions are applicable to other laws governing products that can cause childhood exposure, as described below.

Food Quality Protection Act

The recommendations of the NAS report "Pesticides in the Diets of Infants and Children" were incorporated into federal policy in the United States in 1996 through enactment of the FQPA.[4] FQPA is the principal US federal law governing pesticide use. It was passed unanimously by both houses of Congress in August 1996. It was the first federal environmental law in the United States to affirm the unique vulnerability of children and to contain explicit provisions for protecting children's health. It requires explicit consideration of children in risk assessment and mandates child-protective safety factors in regulation. The passage of FQPA has had a series of far-reaching consequences.

The Food Quality Protection Act and Risk Assessment

FQPA changed risk assessment, especially for pesticides and pharmaceutical chemicals.[5] It forced development of child-protective approaches to risk assessment that explicitly consider children's exposures and susceptibilities.[6] For the first time, it mandated realistic consideration of exposures of multiple pesticides via multiple routes, including diet and drinking water, to assess potentially synergistic effects.[7] It mandates consideration of exposures to chemicals that impact the endocrine system.[8]

The Food Quality Protection Act and Pesticide Regulation

In the regulatory arena, FQPA forced reexamination of pesticide standards (called "tolerances")—the levels of residual pesticides permitted to be present on or in commercially sold fruits and vegetables. It required for the first time that tolerances be based entirely on the protection of human health and that they specifically consider the effects on children. These requirements represented a dramatic change from the previous regulatory regime, in which the health risks of pesticides were balanced against the costs of regulation to the affected industry.

As a result of these regulatory changes, many uses of pesticides were reduced or dropped altogether.[7] For example, agricultural uses of the organophosphate insecticides, a class of pesticide chemicals toxic to brain development, were sharply reduced. Also bans were imposed on residential applications of two organophosphate insecticides—chlorpyrifos and diazinon—that had been widely used for household pest control. These bans were triggered by recognition of these products' neurodevelopmental toxicity coupled with documentation of their long residence time in indoor environments (see Chapter 35).

The Food Quality Protection Act and Biomedical Research

Recognition of children's susceptibility to toxic chemicals in the environment stimulated substantial increases in research in children's environmental health in the United States.[9,10] These initiatives included

- Creation of a national network of federally funded Centers for Children's Environmental Health and Disease Prevention Research supported by the National Institute of Environmental Health Sciences (NIEHS) and the US Environmental Protection Agency (EPA). Research in these Centers has made important contributions to understanding of the environmental causes of asthma, neurobehavioral disorders, endocrine dysfunction, autism, and lead toxicity. Findings from this work are guiding disease prevention efforts.[11]
- Establishment of a network of Pediatric Environmental Health Specialty Units.[12] This network is described in Chapter 56.
- Establishment of fellowship training programs in environmental pediatrics to train the next generation of leaders in children's environmental health.[13]
- Launch of the National Children's Study,[14] a prospective epidemiologic study intended to follow 100,000 children—a statistically representative sample of all children in the United States—from (or before) conception to 21 years of age. The goal of the study was to identify those factors in the environment—chemical, biologic, physical, and psychosocial—that alone or in combination influence children's health, growth, development, and risk of disease. In December 2014, the National Institutes of Health terminated the National Children's Study.[15]

The Food Quality Protection Act and Children's Environmental Health Policy

FQPA has had consequences that extend beyond reform of risk assessment and improved regulation of pesticides. It led to the establishment, in 1996, of the Office of Children's Health Protection within the EPA.[10] It catalyzed a 1997 Presidential Executive Order requiring all US federal agencies to consider children's special susceptibilities in all policy and rule making,[16] and it led to the creation of the President's Task Force on Environmental Health Risks and Safety Risks to Children.

The US experience in addressing the unique vulnerability of infants and children to toxic chemicals in the environment was shared internationally and led to promulgation, in May 1997, of the Miami Declaration on Children's Environment Health, a declaration approved unanimously by the environmental ministers of the G-8 countries.[17]

Toxic Substances Control Act

The TSCA is a federal law intended to regulate "industrial and consumer chemicals,"[18] those many thousands of manufactured chemicals not intended for use as drugs, food additives, cosmetics, or pesticides. Industrial and consumer chemicals are found today in billions of products ranging from construction materials to motor fuels, cleaning products, clothing, consumer electronics, food wrapping, toys, blankets, and baby bottles. American children are exposed to these chemicals on a daily basis, as demonstrated in national biomonitoring surveys conducted by the Centers for Disease Control and Prevention (CDC). These surveys find measurable quantities of several hundred manufactured chemicals in the bodies of virtually all Americans of all ages.[19]

The TSCA was pioneering legislation that was intended to reform chemical policy in the United States. At the time of its passage in 1976, its centerpiece was supposed to have been a requirement for mandatory premarket toxicity testing of all new chemicals as well as a requirement for toxicity testing of high-priority industrial chemicals already in commerce. TSCA never fulfilled these noble intentions.

A grievous error was that TSCA "grandfathered in" the 62,000 chemicals that were already on the market at the time of the law's passage. These chemicals were simply presumed to be safe and were allowed to remain in commerce with no requirement for further assessment. The great majority of these untested chemicals are still in use today.

Two additional provisions of TSCA that have undermined efforts to reduce children's exposures to toxic chemicals are the provision that the EPA cannot ban or restrict a chemical unless the chemical is found to pose an "unreasonable risk" to health—wording that has required the EPA to balance the public health benefits of any proposed regulation against its costs to industry through cost-benefit analysis and a provision instructing the EPA to determine that any proposed regulation is "least burdensome" to the affected industry.[9]

The "unreasonable risk" standard in the TSCA has posed an almost insurmountable barrier to the regulation of industrial chemicals. It has been so difficult to meet that the EPA has not been able to require testing nor remove industrial chemicals from the market except in the very small number of cases where the evidence of potential harm is overwhelming. The result is that only five chemicals have been banned under the TSCA in the more than 45 years since its passage.[9]

The US federal courts' interpretation of the "least burdensome" standard has created still further barriers to enforcement of the TSCA. Thus, in 1991 a federal court found that the EPA had failed to show that a proposed ban on asbestos was the "least burdensome" approach to asbestos control, even though the EPA had analyzed a multiple risk mitigation strategies and concluded that only a ban could effectively protect the public's health. The court thus overturned the EPA's proposed ban on this most thoroughly studied of all human carcinogens, a ruling that has made it virtually impossible for the EPA to regulate dangerous chemicals for the past 30 years.[9]

In summary, the TSCA is failed legislation. The consequence is that, for many decades, US federal policy has presumed chemicals to be safe and has allowed them to enter and remain in commerce with little or no evaluation of their potential toxicity. This policy is neither protective of health nor consistent with current scientific understanding of children's unique vulnerabilities to manufactured chemicals.

Frank R. Lautenberg Chemical Safety for the 21st Century Act

In 2016, to strengthen the TSCA and make it more protective of public health, the US Congress passed the Frank R. Lautenberg Chemical Safety for the 21st Century Act.[20] Two factors that led to passage of the Lautenberg Act were rising concern in the United States among legislators and the public about the hazards of toxic chemicals and the recognition that US chemical policy was falling behind chemical policies other countries, notably Canada and the European Union, that had passed stricter chemical laws. This policy imbalance made it increasingly difficult for US-based chemical corporations to sell their products overseas.[21]

The Lautenberg Act gave the EPA enhanced authority to require toxicity testing of both new and existing chemicals. It explicitly requires that chemical standards protect the health of children and pregnant women. Key provisions of the Act are that it

- Requires that new chemicals be shown to be safe before they can enter the market,
- Requires safety reviews for chemicals already on the market,
- Replaces the TSCA's "unreasonable risk"/cost-benefit safety standard with a solely health-based standard, and
- Limits chemical companies' ability to withhold information on chemicals from governmental agencies and the public by claiming that the information is "confidential."

Enforcement of the Lautenberg Act stalled under the Trump administration between 2017 and 2021,[22] but is now beginning to move forward, albeit imperfectly.[23]

State Laws Relevant to Children's Environmental Health

Several US states have enacted legislation to protect children against environmental threats to health. These developments are important in their own right and additionally because laws passed at the state level are often precursors to national legislation in the United States.

Toxic Chemicals in Consumer Products

California established regulations in 2008 that went into force in 2012 to regulate 1,200 chemicals found in consumer products. The chemicals listed under this regulation are compounds identified as hazardous by the US EPA and the International Agency for Research on Cancer. Manufacturers are required to test products they sell in California for those ingredients to assess their potential hazards. They must then replace hazardous chemicals with safer alternatives or face the threat of strict regulation.[24]

Connecticut enacted legislation in 2011 that bans use of bisphenol A (BPA) in carbonless receipts at banks and retail stores. This action was triggered by the finding that BPA can be absorbed transdermally by persons who regularly handle such receipts such as cashiers and bank tellers, many of whom are women of child-bearing age.

Pesticide Reporting Laws

Three states—California, Massachusetts, and New York—have enacted legislation that requires commercial pesticide applicators and farmers to report all pesticide applications to agencies of state government. Applicators are required to report the name of the pesticide product, the date of application, the amount used, and the specific location (street address or GIS coordinate) of each application. The three states compile this information and maintain it in publicly accessible databases.

The information in these databases has proved very useful to children's environmental health researchers. In California, data from the pesticide registry identified the Salinas Valley as an area of particularly heavy pesticide use, and this information catalyzed the establishment of a prospective birth cohort study led by investigators from the University of California at Berkeley.[25] In New York, pesticide registry data indicated that heaviest pesticide use in the entire state occurred in New York City and its immediately surrounding suburbs. Greatest use was for control of cockroaches and other urban vermin. This finding was quite surprising because the central and western regions of New York State are highly agricultural, and it had been assumed that heaviest pesticide use would occur there. Documentation of heavy pesticide use in New York City led to studies of the effects of the most widely used organophosphate pesticides—chlorpyrifos and diazinon—on children's health.[26,27] These studies documented adverse effects on early brain development and led to a federally imposed ban on residential use of these products. These findings led also to widespread adoption of the principles of integrated pest management by New York City government agencies.

Conclusion

Because of the failure of the TSCA, American children today are placed daily at risk of exposure to hundreds of chemicals that have never been tested for toxicity. The likelihood is high that among these untested chemicals are chemicals that are causing harm to children and whose toxicity has not yet been discovered.

The passage of the Frank R. Lautenberg Chemical Safety for the 21st Century Act in 2016 offers a path to resolution of the current crisis in US chemical policy. Key elements for success of this legislation will be courageous EPA leadership, appropriation of sufficient funds to enable the EPA to implement the law, and recognition by the federal courts that protection of children's health is a sacred national responsibility that transcends the impacts of chemical regulation on the short-term profits of the chemical manufacturing industry.[28]

References

1. National Research Council, Commission on Life Sciences. *Science and Judgment in Risk Assessment*. Washington. DC: National Academies Press, 1994.

2. Vandenberg LN, Colborn T, Hayes TB, et al. Hormones and endocrine-disrupting chemicals: low-dose effects and nonmonotonic dose responses. *Endocr Rev.* 2012;33:378–455.

3. National Academy of Sciences. *Pesticides in the Diets of Infants and Children*. Washington, DC: National Academies Press; 1993.

4. FQPA. Food Quality Protection Act of 1996. Public Law 104–170. 1996. Available at: https://www.epa.gov/laws-regulations/summary-food-quality-protection-act.

5. Landrigan PJ, Kimmel CA, Correa A, Eskenazi B. Children's health and the environment: Public health issues and challenges for risk assessment. *Environ Health Perspect.* 2004;112:257–65.

6. Raffaele KC, Rowland J, May B, Makris SL, Schumacher K, Scaranod LJ. The use of developmental neurotoxicity data in pesticide risk assessments. *Neurotoxicol Teratol.* 2010;32:563–72.

7. Goldman LR. Preventing pollution? US toxic chemicals and pesticides policies and sustainable development. *Environ Law Rep News Analysis*. 2002; 32: 11018–41.

8. Diamanti-Kandarakis E, Bourguignon JP, et al. Endocrine- disrupting chemicals: an Endocrine Society scientific statement. *Endocr Rev.* 2009;30:293–342.

9. Landrigan PJ, Goldman L. Children's vulnerability to toxic chemicals: a challenge and opportunity to strengthen health and environmental policy. *Health Aff (Milwood)*. 2011;30:842–50.

10. Goldman L, Falk H, Landrigan PJ, Balk SJ, Reigart JR, Etzel RA. Environmental pediatrics and its impact on government health policy. *Pediatrics*. 2004;113:1146–57.

11. National Institute of Environmental Health Sciences. Collaborative Centers in Children's Environmental Health Research and Translation. 2022. Available at: https://www.niehs.nih.gov/research/supported/centers/collaborative/index.cfm

12. Wilborne-Davis P, Kirkland KH, Mulloy KB. A model for physician education and consultation in pediatric environmental health: the Pediatric Environmental Health Specialty Units (PEHSU) program. *Pediatr Clin North Am.* 2007;54:1–13.

13. Landrigan PJ, Braun JM, Crain EF, et al. The Academic Pediatric Association retreat for scholars in pediatric environmental health: fifteen-year report on a cross-institutional collaboration in professional education. *Acad Pediatr.* 2019;19(4):421–7.

14. Landrigan PJ, Trasande L, Thorpe LE, et al. The National Children's Study: a 21-year prospective study of 100,000 American children. *Pediatrics* 2006;118:2173–86.

15. Landrigan PJ, Baker DB. The National Children's Study: end or new beginning? *N Engl J Med.* 2015;372:1486–7.

16. Clinton WJ, Gore A. Executive Order 13045: protection of children from environmental health and safety risks. *Fed Reg.* 1997;62:19883–8. http://yosemite.epa.gov/ochp/ochpweb.nsf/content/whatwe_executiv.htm.

17. G-8. 1997 Declaration of the Environment Leaders of the Eight on Children's Environmental Health. Environment Leaders' Summit of the Eight, Miami, Florida. 1997. Available at: http://www.g8.utoronto.ca/environment/1997miami/children.html.

18. Toxic Substances Control Act. US Environmental Protection Agency. 2016. Available at: https://www.epa.gov/laws-regulations/summary-toxic-substances-control-act.

19. Centers for Disease Control and Prevention. National Biomonitoring Program. 2022. Available at: https://www.cdc.gov/biomonitoring/index.html.

20. U.S. Envirronmental Protection Agency. Assessing and managing chemicals under TSCA. 2023. Available at: https://www.epa.gov/assessing-and-managing-chemicals-under-tsca.

21. Denison RA. A Primer on the new Toxic Substances Control Act (TSCA) and what led to it. Environmental Defense Fund. 2017. Available at: https://www.edf.org/sites/default/files/denison-primer-on-lautenberg-act.pdf

22. Woolhandler S, Himmelstein D, Ahmed S, et al. *Lancet* Commission on Public Policy and Health in the Trump Era. *Lancet* 2021;397:705–53.

23. McPartland J, Shaffer RM, Fox MA, Nachman KE, Burke TA, Denison RA. Charting a path forward: assessing the science of chemical risk evaluations under the Toxic Substances Control Act in the context of recent national academies recommendations. *Environ Health Perspect*. 2022 Feb;130^2:25003.

24. Department of Toxic Substances Control. Green chemistry. Department of Toxic Substances Control, California Environmental Protection Agency. 2010. Available at: http://www.dtsc.ca.gov/pollutionpreventi on/greenchemistryinitiative/index.cfm.

25. Vernet C, Johnson M, Kogut K, Hyland C, Deardorff J, Bradman A, Eskenazi B. Organophosphate pesticide exposure during pregnancy and childhood and onset of juvenile delinquency by age 16 years: the CHAMACOS cohort. *Environ Res*. 2021 Jun;197:111055.

26. Berkowitz GS, Wetmur JG, Birman-Deych E, et al. In utero pesticide exposure, maternal paraoxonase activity, and head circumference. *Environ Health Perspect* 2004;112:388–91.

27. Williams MK, Barr DB, Camann DE, et al. An intervention to reduce residential insecticide exposure during pregnancy among an inner-city cohort. *Environ Health Perspect*. 2006;114:1684–9.

28. Woodruff TJ, Rayasam SDG, Axelrad DA, et al. A science-based agenda for health- chemical assessments and decisions: overview and consensus statement. *Environ Health*. 2023 Jan 12;21(Suppl 1):132.

59

Public Policy on Children's Environmental Health in Europe

Peter van den Hazel

The protection of children's health against environmental threats has become an important component of the European policy agenda. The World Health Organization's (WHO) Regional Office for Europe, currently representing 53 countries, and the European Union, which creates law and policy in 27 member countries, are two key policymaking bodies that have been instrumental in getting children's environmental health on the national policy agendas.

The two policy developments in Europe that have probably been most significant for children's environmental health are the WHO-led Children's Environment and Health Action Plan for Europe (CEHAPE) and the European Union's ambitious chemicals legislation, which has a strong focus on protecting the health of vulnerable groups, including children.

The past two decades saw advances in regulatory standards in Europe, particularly on chemical legislation, although these policy advances may not always translate directly into reduced exposures for children (see Table 59.1).

In 1989, the WHO and European nations initiated the first ever Environment and Health process with the aim to eliminate the most significant environmental threats to health. At that time there was no specific mention of children as a vulnerable group. Children's environmental health first emerged as a policy issue in Europe in 1999, at the third Ministerial Conference on Environment and Health, held in London.[1] This Conference built on the 1997 declaration of environmental leaders from the G-8 on children's environmental health,[2] recognized children's unique biological vulnerabilities to hazards in the environment, emphasized the importance of protecting children from toxic environmental exposures, identified priority areas for action, and started a children's environmental health process for Europe.

The fourth Ministerial Conference, held in Budapest in 2004, was a watershed event for children's environmental health. The goal was to define the rationale, structure, and objectives of the new Children's Environment and Health Action Plan for Europe (CEHAPE).[3] Two major products of this process were (a) a thorough review of the scientific evidence on children's environmental health, published by the WHO Regional Office for Europe and the European Environment Agency,[4] and (b) a study that quantified for the first time the burden of disease in children and adolescents in Europe related to environmental exposures.[5]

At the 2004 Budapest Conference, CEHAPE was approved at the highest political level,[6] thus affirming the commitments of the then 52 Member States of the WHO European region to the mitigation of environmental threats to children's health. An important feature of CEHAPE is its recognition that children in particularly adverse

Peter van den Hazel, *Public Policy on Children's Environmental Health in Europe* In: *Textbook of Children's Environmental Health*, Second Edition. Edited by: Ruth A. Etzel and Philip J. Landrigan, Oxford University Press. © Oxford University Press 2024.
DOI: 10.1093/oso/9780197662526.003.0059

Table 59.1 Selected European policies relevant to children's environmental health

The World Health Organization
Children's Environment and Health Action Plan for Europe (CEHAPE)

European Union
Environment and Health Strategy (SCALE)
Environment and Health Action Plan (EHAP)
EU Strategy on the rights of the child

Legislation
Chemical-specific laws
- REACH Regulation
- Plant Protection Products Regulation
- Sustainable Use of Pesticides Directive
- Biocides Regulation

Key law regulating environmental media
- Ambient air quality and cleaner air for Europe Directive

conditions, such as in war or extreme poverty, are at extraordinarily high risk of injuries, psychological trauma, acute and chronic infections, chronic diseases, disability, and death (see Table 59.2).

CEHAPE also led to the adoption of national Children's Environment and Health Action Plans in various European countries. Most countries presented their plans at the Fifth WHO Ministerial Conference in Parma in 2010[7] (e.g., Austria and the United Kingdom published their Children's Environment and Health Action Plans in December 2010).

Although the CEHAPE targets are not legally binding, they give great strength to arguments for allocating funds to children's environmental health because they come from a United Nations body, are agreed upon by many countries, and are based on good evidence.[8] In 2015, in Haifa, a midterm Ministerial Conference assessed how well European countries were meeting the targets set in Parma in 2010[9] (CEHAPE targets were not discussed). Some of the National CEHAPE plans existed but were not high on the Environment and Health agenda. Most plans were focused on classical child health themes like sanitation, hygiene, access to safe drinking water, traffic accidents, and smoking. This was more or less repeated at the sixth Ministerial Conference on Environment and Health in Ostrava in 2017.

European Union Joins the Efforts

At the same time as CEHAPE was being adopted by the WHO Regional Office for Europe and the European Environment Agency, the European Union was also developing a strategy and plan on environment and health. The European Union had a key role to play in environmental health: first because the Treaty of the European Union (Amsterdam Treaty of 1997) introduced a requirement that health protection be incorporated into

Table 59.2 Children's Environment and Health Action Plan: Linking children's diseases and health conditions to environmental factors

---

The Children's Environment and Health Action Plan (CEHAPE) in 2004 [3] calls on European governments to act on four Regional Priority Goals.

Goal 1: Ensuring public health by improving access to safe water and sanitation in Europe

Goal 2: Addressing obesity and injuries through safe environments, physical activity, and healthy diet

Goal 3: Preventing respiratory disease and asthma through improved outdoor and indoor air quality

Goal 4: Preventing disease and disability arising from exposure to chemical, biological, and physical environments during pregnancy, childhood, and adolescence

In 2010, the Parma Declaration on Environment and Health [7], signed by European Union member states and the current 53 World Health Organization (WHO) member countries at the Fifth Ministerial conference, updated the pledges and set several time-bound targets, namely to:

- Strive to provide each child with access to safe water and sanitation by 2020, and revitalize children's hygiene
- Provide safe environments and green spaces to children by 2020
- Provide children with clean indoor air in child care facilities, kindergartens, schools, and public recreational settings, satisfying WHO's air quality guidelines by 2015
- Protect children from harmful substances, including focusing on pregnant and breast-feeding women and places children live, learn, and play, by 2015
- Act on identified risks of exposure to carcinogens, mutagens, and reproductive toxicants, including radon, ultraviolet radiation, asbestos, and endocrine disrupters, and develop by 2015, national programs for the elimination of asbestos-related diseases
- Strive to provide each child with access to safe water and sanitation, and revitalize children's hygiene

---

all European legislation at its inception and, second, because more than 90% of environmental legislation is made at the EU level, rather than at the national level.

Pressure on the European Union to focus on children's environmental health came from the WHO, from the leadership of then Environment Commissioner Margot Wallström, and from the growing scientific evidence from EU-funded research that children are a particularly vulnerable group, especially during the prenatal period and in the first years after birth.

By the early 2000s, the European Union had regulated a number of environmental areas to the benefit of children. For example, legislation to eliminate polychlorinated biphenyls (PCBs) was passed in 1996,[10] all forms of asbestos were banned in 1998–1999,[10] and several carcinogenic and neurotoxic pesticides, including DDT and atrazine, were removed from the market.

Then, in 2004, the European Union introduced a far-reaching strategy and an action plan in children's environmental health.

The European Union Environment and Health Strategy: SCALE

The overall aim of the European Environment and Health Strategy introduced in June 2004 was to reduce diseases caused by environmental factors. The strategy, which is referred to as SCALE (Science, Children, Awareness, Legislation, and Evaluation), gives special emphasis to protecting the most vulnerable groups in society, particularly children, against environmental hazards.[11]

The Environment and Health Action Plan

The first cycle of SCALE addressed environment and health issues in an integrated way through the European Environment and Health Action Plan (EHAP) 2004–2010.[12] The Plan was launched to coincide with the launch of the WHO CEHAPE plan and was designed to support collaboration between the two processes. This led to the development of a Strategic Research and Innovation Agenda by the HERE coordination action (2019–2021).[13] However, this does not have a specific focus on children.

The central focus of SCALE on children is clear. The European Commission's description of EHAP states that

Environmental effects on vulnerable groups are of particular concern.

[and]

The concerns of children are integrated throughout the Action Plan. A number of major child health issues will be covered in the monitoring, as will exposure to the environmental stressors to which children are particularly sensitive. Research on susceptibility is particularly important, so that policy responses can be adjusted to the needs of children in those cases where they are particularly vulnerable. The proposals in the Action Plan on indoor air pollution are a case in point, as the scientific evidence shows that the health impacts of, for instance, Environmental Tobacco Smoke (ETS) are particularly evident for children.[12]

The development of a concerted EU action plan for human biomonitoring was an important outcome of the Environment and Health Action Plan.[14] To launch this biomonitoring plan, a pilot project was initiated in which 4,000 urine and hair samples were collected from mothers and children in 17 countries across Europe. The goal was to measure levels of mercury, cadmium, cotinine, and some phthalates.[15] These findings contributed to the development of a more comprehensive health impact assessment across the European Union with larger biomonitoring studies across Europe such as HBM4EU (2017–2021)[16] and the Partnership for the Assessment of Risk from Chemicals 2022 (PARC).[17] In both projects, cohorts and projects included children in a study of a range of exposures.

Biomonitoring has already made an impact in Europe over the past two decades. For example, in Germany, the findings of the Environmental Survey for Children, which measured levels of 20 chemicals in blood and urine, prompted a lowering of reference values for lead and mercury and the establishment of a new reference value for arsenic.[18]

Additional consequences of the launch of the EU EHAP and the WHO CEHAPE were a revision of pesticide regulations, a new pesticide directive, and a new ambient air quality directive.

The following sections describe a series of EU policies introduced in different areas that further strengthen the protection of Europe's children against environmental threats to health.

Legislation on Chemicals

REACH

After many years of intensive debate, the European regulatory framework for chemicals management known as REACH (Registration, Evaluation, and Authorization of CHemicals) was adopted in 2006. It is considered the most ambitious of the European Union's proposals on environment and health.[19] A stated aim of REACH is "to improve the protection of human health and the environment through the better and earlier identification of the intrinsic properties of chemical substances."[20]

As a consequence of REACH, thousands of synthetic chemicals on the market in Europe are for the first time being registered. Unlike earlier policy, the regulation requires manufacturers and importers to centrally register all substances they use; to put health and safety information about the chemicals in a public database; and to conduct analyses and reduce risks for their own protection, that of other users, the public, and the environment. Crucially, the policy requires the European Chemicals Agency (ECHA) together with EU Member States to carefully scrutinize use of the most harmful substances, aiming at their gradual phase-out.

Under REACH, the burden of proof in establishing the safety of a chemical is on industry. REACH encourages innovation of safer substances through its evaluation and control procedures for chemicals with harmful properties and through its phase-out goals.

A key component of REACH is an initiative to identify "substances of very high concern." Countries can propose a substance for this list if it is carcinogenic, mutagenic, toxic to reproduction; environmentally persistent and bioaccumulative (builds up in the body); or has properties that generate "equivalent concern."[16] Currently, there are 59 categories of restricted substances in REACH Annex XVII, involving more than 1,000 substances.[21] These hazardous substances have specific restrictions, and, as such, certain chemical substances in the specific product are not allowed to be used. The Commission has begun work on a revision of the REACH Regulation as announced in the Chemicals Strategy for Sustainability.[22]

The Council endorsed the strategy on 15 March 2021. In its conclusions, it expressly supported the five main areas of action. It welcomed the aim of the "one substance, one assessment" approach and stressed the importance of clear legal provisions in EU product law and in the sustainable products initiative ensuring that chemicals, materials and products are safe and sustainable-by-design. It underlined the need to ensure that PFAS are eliminated, unless their use is proven essential to society; and to strengthen the EU legal framework in the area of endocrine disruptors. It supported the announced amendment of REACH in a targeted manner, accompanied by a comprehensive impact assessment.

On 16 May 2022, ENVI (European Parliament Committee on Environment) discussed progress in the strategy implementation with the Commission and ECHA. Key steps taken so far include the publication of the "restrictions roadmap," which contains a rolling list of substances—the most harmful to human health and the environment—that will be prioritised for (group) restrictions under REACH. The Commission also adopted a recommendation on a European assessment framework for 'safe and sustainable by design' chemicals and materials; and proposed a revision of the regulation on classification, labelling and packaging of chemicals.

A broad PFAS restriction proposal under REACH, prepared by five European countries, is currently under evaluation by ECHA. The revision of REACH is expected in the last quarter of 2023.

If fully implemented, REACH will contribute to ending the vast ongoing toxicological experiment in which chemicals are being tested on children worldwide instead of in the laboratory.

However, a number of weaknesses hinder the implementation and threaten to reduce the effectiveness of REACH.[23] For example, REACH does not systematically deal with hazardous chemicals produced in small quantities. Also it has only limited means to deal with hazardous substances in imported manufactured products. A third problem is that the chemical departments of many EU member governments lack the resources to properly play their role in nominating "most harmful" chemicals and enforcing company compliance. As a result, the list of harmful substances is growing very slowly.

To date a number of chemicals have been officially identified as endocrine-disrupting chemicals (EDCs) under the existing European chemicals regulation REACH. This shows that EDCs can be identified on a case-by-case basis even in the absence of criteria.[24] However, inadequate control of EDCs under REACH is one of the crucial areas of concern. The multiple adverse health outcomes linked to exposure to EDCs have huge economic and social costs (see results of the EU Horizon 2020 EDC-MixRisk project.[25,26])

In 2012, a scientific consensus statement launched in Paris at an international scientific conference on early-life exposure and health outcomes highlighted those developmental periods, such as prenatal and early postnatal life, that are most sensitive to environmental factors such as nutrients, environmental chemicals, drugs, infections, and other stressors.[27]

The Commission published a new chemicals strategy for sustainability[28] on 14 October 2020. It is part of the EU's zero pollution ambition, which is a key commitment of the European Green Deal. This strategy includes a revision of the REACH Regulation prohibiting the use of the most harmful chemicals in consumer products such as toys, childcare articles, cosmetics, detergents, food contact materials, and textiles unless documented to be essential for society and ensuring that all chemicals are used more safely and sustainably.[29]

Pesticides

By 2005, many studies had pointed to the special vulnerability of children to pesticides, including elevated risks of certain cancers, including childhood leukemia. Legislation on pesticides—the EU Plant Protection Products Regulation (EC No 1107/2009)—was developed in 2009.[30] This legislation sought to protect human health and wildlife against pesticides. It states that particular attention should be paid to the protection of vulnerable groups of the population, including pregnant women, infants, and children.[31]

Another development in the EU Plant Protection Products Regulation is that certain hazardous pesticides—specifically those that are persistent, bioaccumulative, and toxic (PBT); very persistent and very bioaccumulative (vPvB); persistent organic pollutants (POPs); carcinogenic, mutagenic, toxic to reproduction (CMRs); or have endocrine-disrupting properties—will no longer be authorized for use in Europe.[32] This may help end a long discussion about the "safety levels" of pesticides and their residues. It is also of great significance for children. This development is based on the agreement that, for certain pesticides, it is not "the dose that makes the poison" but rather that these chemicals are inherently toxic at any level of exposure[32] (see also Chapter 2). The European pesticide regulation also requires that potential health impacts from mixtures of pesticides be considered (the so-called cocktail effect). This was a very important innovation.

The regulation on Plant Protection Products (PPP) was developed in cooperation with other EU Regulations and Directives (e.g., the regulation on maximum residue levels in food).[32] To protect human health and the environment, monitoring of PPP residues in food is an important activity, and the European Union can evaluate the prediction of the safe use of respective PPPs. However, pesticides that are toxic to neurodevelopment remain a problem under the EU Regulation because the regulation proposes only an additional safety margin on neurotoxic pesticides when, in fact, no safety margin for fetal exposure may exist. This provision of the regulation thus flies in the face of recommendations on protection of fetal neurodevelopment made by medical and scientific experts.

While one piece of the EU PPP Regulation deals with the authorization of pesticides, a second piece of this legislation is aimed at reducing dependency on pesticide use. Under the Directive on Sustainable Use,[33] EU Member States are asked to establish national action plans with pesticide reduction targets. They also must ensure that pesticide use is minimized or even stopped in public spaces (e.g., public parks or schools) frequented by children and other vulnerable groups. Pesticide-free areas are an important way to reduce children's exposure, but the voluntary nature of this law means that its adoption in European cities and towns is not yet widespread.

Under the European Green Deal and particularly the new "chemicals for sustainability," "farm to fork," and "biodiversity" strategies, EU legislation on these issues will be revised in the near future.

Biocides

The European Union classifies biocides (pest control products, wood preservatives, certain disinfectants, rodenticides, anti-fouling paints, and other pesticides used in nonagricultural settings) separately from pesticides. The health of certain populations, such as children and pregnant women, is threatened by the wide and improper use of hazardous biocides.

When the Biocidal Products Regulation[34] was applied (on September 1, 2013), its stated purpose was

> To improve the functioning of the internal market through the harmonization of the rules on the making available on the market and the use of biocidal products, whilst ensuring a high level of protection of both human and animal health and the environment. The provisions of this Regulation are underpinned by the precautionary principle, the aim of which is to safeguard the health of humans, the health of animals and the environment. Particular attention shall be paid to the protection of vulnerable groups.[34]

Like the pesticides legislation, certain hazardous substances will no longer be authorized for use, such as those that are carcinogenic, mutagenic, and toxic to reproduction, or those that disrupt the endocrine system.

However, as in the case of the pesticides regulation, concerns remain in relation to stipulations on those biocides with developmental neurotoxic properties. Another problem in the regulation on biocides is that there are many exemptions on "nonauthorized" biocides. Perhaps even more serious—and quite unlike the pesticides regulation—is that no attention is given in the biocides legislation to strategies for reducing their use. Another issue is that a biocidal product authorized in one Member State also is authorized upon application in other Member States unless there are specific grounds to deviate from this principle of mutual recognition.

Although REACH and the pesticides and biocides legislation are all important steps toward protecting children's environmental health, none adequately addresses concerns about "low-dose" exposure and they only address to a lesser degree exposures to EDCs. Current science suggests that pregnant women and developing children are highly vulnerable to the effects of exposure and that consequences include reproductive abnormalities, cancer, diabetes, obesity, and early puberty. Many products used by children, such as toys, hygiene care products, and plastic products, contain EDCs.

Toys and Other Consumer Products

Toys are of particular concern for policymakers since children come into close contact with toys. The current EU policy on toys dates back to the 1998 Toys Directive, and an update was carried out in 2009 after a wave of toy recalls due to chemical contamination. A revision of the Toys Directive is foreseen. The directive lays down safety criteria for toys, such as mechanical properties, chemical properties, and flammability.

Many toys in the European Union still contain high concentrations of hazardous chemicals, such as brominated flame retardants and heavy metals. Even low-dose exposure to these hazardous substances may put children in Europe at risk of problems in their brain development or cancer later in life.

The lack of a horizontal framework to cover chemicals in all consumer products is a particular shortcoming of EU chemicals legislation. Piecemeal policies cover some products but not others. For example, a certain chemical may be banned in toys but still allowed in fabrics, cosmetics, or other products. This patchwork legislation prevents coherence in incentives to develop and use safer chemicals and it makes enforcement more difficult.

The European Union has cited the precautionary principle as a rationale for protecting children's environmental health. An early precautionary action related to children's exposure to phthalates was passed in 1999. A legally binding EU decision adopted that year forbade the use of phthalates in soft PVC toys and child care articles that are intended to be placed in the mouth by children under 3 years of age.[35] Since then, the decision has been extended more than 20 times in the name of the precautionary principle, and a directive was finally adopted in 2005.[36] The Commission Regulation EU 2020/1245 A rules on plastic materials and articles intended to come into contact with food, but the European Food Safety Authority (EFSA) has warned that there is still a potential risk for children to be in contact with levels of cadmium and lead that are dangerous for their health and that levels in certain products should be reduced.[37] Additionally, allergenic fragrances are forbidden and potential allergens must be labeled.[38]

Most recently, the precautionary principle was used as a basis for the decision to ban bisphenol A in baby bottles under the Food Contact Materials legislation.[39]

Mercury

In 2004, the EFSA recommended "that vulnerable groups in particular select fish from a wide range of species without giving undue preference to large predatory fish likely to contain higher levels of methylmercury, such as swordfish and tuna."[40]

The European Commission adopted in 2005 the Community Strategy Concerning Mercury (European Commission, 2005), which includes a comprehensive plan to address mercury use and pollution and has resulted in the enhancement of Union law on mercury, including restrictions on the inclusion of mercury or mercury compounds in products, a ban of exports of mercury from the EU, and the inclusion of provisions on mercury emissions in EU legislation to protect people against exposure.[41,42]

In view of its risks to health, the European Union adopted a range of regulations, directives, limits, and restrictions in the fields of food safety, application of mercury in products, the environment, and occupational settings.[40] These are meant to reduce or eliminate the use of mercury, including its use in medical measuring devices and vaccines. The strategy also prioritized better education and measures to protect those groups most vulnerable to health damage from mercury, including children[42]. At present, most of the efforts of the European Union on mercury are related to the global treaty, the The Minamata Convention on Mercury of August 2017, which aims to protect human health and the environment from anthropogenic emissions and releases of mercury and mercury compounds[43] (see Chapter 62).

European Union Legislation on Environmental Media

Air Quality

Outdoor Air

The European Union has a long history of developing policies to reduce outdoor air pollution and of setting standards on air quality. The 2008 Directive on Ambient Air Quality and Cleaner Air for Europe[44] is the latest in this set of laws. It defines concentration limits for a number of pollutants that are dangerous to health, including particulate matter (PM_{10}), nitrogen dioxide (NO_2), and ozone. The Ambient Air Quality Directive, adopted as 2008/50/EC, additionally set objectives for fine particulate matter ($PM_{2.5}$). The law recognizes that children and other groups are particularly sensitive to harmful air pollutants. It introduces information requirements and encourages Member States to include specific measures aimed at the protection of sensitive population groups, including children, in air quality plans.

As part of the European Green Deal, the EU is revising these standards to align them more closely with the recommendations of the WHO (the latest Air Quality Guidelines of the WHO were published in 2021[45]). The EU also aims to improve overall EU legislation for clean air, building on the lessons learned from the 2019 evaluation ("fitness check") of the Ambient Air Quality Directives.[46]

EU air quality standards are still above the levels recommended by the WHO, so the current law does not guarantee adequate protection of children. In many places in Europe (especially in cities), air pollution levels are well above the legally binding standards. In addition, although countries are required to inform vulnerable groups such as children and people with asthma of when and where pollution levels are being exceeded, this has not been adequately monitored and therefore little up-to-date information exists on whether any countries have fulfilled these requirements.

Examples of EU-funded research projects show that children are specifically vulnerable to air pollution (e.g., living near busy roads could be responsible for up to 30% of asthma in children).[47]

Indoor Air

The European Union's EHAP placed a specific emphasis on indoor air quality as well as a focus on protecting vulnerable groups, including children. All EU institutions have at one time or other called for action, particularly to achieve smoke-free indoor environments.

Major EU research projects linking indoor air quality to diseases have been funded. For example, one EU-funded research project, SINPHONIE,[48] provided comprehensive data on air quality in Europe's schools and its impact on children's health. It enhanced understanding on of the sources of poor-quality indoor air in schools and allowed comparison of what different countries in Europe have achieved. Currently, unlike the situation for outdoor air, no strategic approach or law exists on for indoor air quality.

Noise

Compared to chemicals and air quality, noise is a much neglected field within EU environmental policy. The focus on of noise policy to date has been on developing a joint approach to measuring noise levels from transport and other sources. Laws currently exist to regulate noise in the transport sector, but this is not connected to an overall strategy. Legislation does not specifically cover the impact of noise on children's health.

The Environmental Noise Directive (Directive 2002/49/EC)[49] relates to the assessment and management of environmental noise. This is the main EU instrument to identify noise pollution levels and to trigger the necessary action both at Member State and EU levels.

To pursue its stated aims, the Environmental Noise Directive focuses on three action areas: (1) the determination of exposure to environmental noise, (2) ensuring that information on environmental noise and its effects is made available to the public, and (3) preventing and reducing environmental noise where necessary and preserving environmental noise quality where it is good.

The EU Environment Action Program to 2020, "Living well, within the limits of our planet," is committed to significantly decrease noise pollution in the Union, moving closer to levels recommended by the WHO by 2020.[50] The Commission published a third implementation report in 2023, setting out how noise can be further reduced.[51] The report shows progress achieved since the second implementation report, which includes a more systematic assessment of noise levels and the adoption of noise management action plans by Member States. However, it warns that the current number and intensity of actions must be increased if the number of people affected by transport noise is to be reduced by 30% by 2030, as set out in the Zero Pollution Action Plan (see next section).[52]

The Way Forward

Over the past 25 years, European policymakers have taken important and positive steps to help make the environment more protective of children's health. The top priority remains to substantially reduce the current long time lag between the publication of scientific findings and their translation into change in policy.

On 12 May 2021, the European Commission adopted the EU Action Plan: "Towards a Zero Pollution for Air, Water and Soil"—a key deliverable of the European Green Deal.[52]

The **zero pollution vision for 2050** is for air, water and soil pollution to be reduced to levels no longer considered harmful to health and natural ecosystems, that respect the boundaries with which our planet can cope, thereby creating a toxic-free environment.

This is translated into key 2030 targets to speed up reducing pollution at source. These targets include:

- improving air quality to reduce the number of premature deaths caused by air pollution by 55%;
- improving water quality by reducing waste, plastic litter at sea (by 50%) and microplastics released into the environment (by 30%);
- improving soil quality by reducing nutrient losses and chemical pesticides use by 50%;
- reducing by 25% the EU ecosystems where air pollution threatens biodiversity;
- reducing the share of people chronically disturbed by transport noise by 30%, and
- significantly reducing waste generation and by 50% residual municipal waste.

The action plan aims to strengthen the EU green, digital and economic leadership, while creating a healthier, socially fairer Europe and planet. It provides a compass to mainstream pollution prevention in all relevant EU policies, to step up implementation of the relevant EU legislation and to identify possible gaps.

A big step forward is the EU strategy on the rights of the child.[53] The protection and promotion of the rights of the child is a core objective of the European Union. This strategy's overarching ambition is to build the best possible life for children in the European Union and across the globe.[53] The strategy contains several actions by the European Commission. Among these are (1) establishing, jointly with the European Parliament and child rights organisations, an EU Children's Participation Platform to connect existing child participation mechanisms at local, national, and EU levels and involve children in the decision-making processes at the EU level; and (2) creating space for children to become active participants of the European Climate Pact through pledges or by becoming Pact Ambassadors. By involving schools in sustainable climate, energy, and environment education, the Education for Climate Coalition will help children become agents of change in the implementation of the Climate Pact and the European Green Deal.[54]

A particularly important area of implementing policy is in regard to endocrine disruptors. In this area, science highlights particular dangers from exposures to EDC mixtures and low-dose exposure in early life. Although there has been an EU strategy on endocrine disruptors in place for 20 years, criteria for determining which substances are agreed EDCs have not yet been properly adopted.

Another focus is on children's mental health. Children's mental health issues are widespread and can be linked to isolation, social exclusion, environmental exposures, poverty, and prolonged use of digital tools. Up to 20% of children worldwide experience mental health issues.[53]

More effective policy implementation will require a more structured and coordinated framework. Continuously, children's health is being significantly compromised by a lack of harmonized policies and by numerous deviations and slowness in policy implementation. For example, deviations on air quality regulations are resulting in unnecessarily high rates of asthma and other respiratory problems. Similarly, slow implementation of EDC strategy in relation to pesticides and biocidal products may be resulting in a range of unnecessary developmental problems and chronic conditions in children.

More political pressure could help push European policymakers to take necessary action more rapidly to protect children. Health experts and advocates, including medical professionals, civil society groups, and young people, have contributed to the policy advances in Europe. These groups should now help create an increased sense of urgency on children's environmental health.

The well-being of future generations depends on taking action to protect the young today. Research makes it clear that harmful environmental exposures during the prenatal period, early life, and childhood have disproportionate effects on future health. Policymakers should seize opportunities to protect children from early environmental exposures and thus prevent unnecessary chronic diseases and disabilities—such as developmental problems, mental health problems, cancer, obesity, and diabetes—that blight the lives of children, undermine the health of adults, shorten human life, and unnecessarily burden future generations.

References

1. Landrigan PJ, Tamburlini G. Children's health and the environment: a transatlantic dialogue. *Environ Health Perspect.* 2005;113:A646–7.

2. 1997 Declaration of the environment leaders of the eight on children's environmental health. Environment Leaders' Summit of the Eight, Miami, FL. Available at: http://www.g8.utoronto.ca/environment/1997miami/children.html.

3. World Health Organization. Declaration of the Fourth Ministerial Conference on Environment and Health. Report No. EUR/04/5046267/6. Available at: http://www.euro.who.int/__data/assets/pdf_file/0008/88577/E83335.pdf.

4. Tamburlini G, von Ehrenstein O, Bertollini R, eds. *Children's Health and Environment: A Review of Evidence.* Copenhagen: World Health Organization Regional Office for Europe and European Environment Agency; 2001.

5. Valent F, Little D, Bertollini R, Nemer L, Barbone F, Tamburlini G. Burden of disease attributable to selected environmental factors and injury among children and adolescents in Europe. *Lancet.* 2004;363:2032–9.

6. World Health Organization. Children's environment and health action plan for Europe. Report No. EUR/04/5046267/7. Available at: http://www.euro.who.int/__data/assets/pdf_file/0006/78639/E83338.pdf.

7. World Health Organization. Parma Declaration on Environment and Health and Commitment to Act. Available at: http://www.euro.who.int/__data/assets/pdf_file/0011/78608/E93618.pdf.

8. Walgate R. European nations agree to improve environmental health. *Lancet.* 2010;375:969.

9. WHO. High-level mid-term review meeting of the European environment and health process. Available at: https://www.euro.who.int/en/health-topics/environment-and-health/pages/european-environment-and-health-process-ehp/governance/european-environment-and-health-task-force/european-environment-and-health-task-force-meetings/high-level-mid-term-review-meeting-of-the-european-environment-and-health-process.

10. European Environment Agency. Late lessons from early warnings: the precautionary principle 1896–2000. Report No. 22/2001. 2001. Available at: http://www.eea.europa.eu/publications/environmental_issue_report_2001_22.

11. SCALE. Environment and health strategy. Available at: http://europa.eu/legislation_summaries/environment/general_provisions/l28133_en.htm.

12. European Commission. Communication from the Commission to the Council, the European Parliament, the European Economic and Social Committee – the European environment and health action plan 2004–2010. Available at: http://eur-lex.europa.eu/smartapi/cgi/sga_doc?smartapi!celexplus!prod!DocNumber&lg=en&type_doc=COMfinal&an_doc=2004&nu_doc=416.

13. European Commission. Environment and Health. Available at: https://ec.europa.eu/info/research-and-innovation/research-area/health-research-and-innovation/environment-and-health_en.

14. Health & Environment Alliance. The EU environment and health action plan (EHAP): assessment and outlook for future action. Available at: http://www.env-health.org/IMG/pdf/ehap_final_report_final.pdf.

15. DEMOCOPHES: European pilot study for assessing chemical exposure. Available at: http://www.eu-hbm.info/democophes.

16. HBM4 Eu project. Available at: https://www.hbm4eu.eu/.

17. Europa. European partnership for the assessment of risks from chemicals (PARC). Available at: https://www.efsa.europa.eu/en/funding-calls/european-partnership-assessment-risks-chemicals-parc.

18. Schulz C, Wilhelm M, Heudorf U, Kolossa-Gehring M; Human Biomonitoring Commission of the German Federal Environment Agency. Update of the reference and HBM values derived by the German Human Biomonitoring Commission: *Int J Hygiene Environ Health*.2012 Feb; 215(2):149. http://www.ncbi.nlm.nih.gov/pubmed/21820957.

19. European Chemical Agency. Homepage. Available at: http://echa.europa.eu.

20. European Commission. REACH. Available at: http://ec.europa.eu/environment/chemicals/reach/reach_intro.htm.

21. CIRS-REACH. REACH restriction list. Available at: https://www.cirs-reach.com/REACH/REACH_Restriction.html.

22. European Commission. Chemicals strategy for sustainability. 2020 Oct 14. Available at: https://ec.europa.eu/environment/strategy/chemicals-strategy_en.

23. Hamrén H. Major shortcomings in the EU regulation for chemicals in goods. Baltic Sea Centre. Stockholm university. 2018. Available at: https://balticeye.org/en/pollutants/reach-report-2018/.

24. Chem Sec. The 32 to leave behind. 2015. Available at: https://chemsec.org/publication/endocrine-disruptors,reach,sin-list/the-32-to-leave-behind-edcs-relevant-for-reach-2015/.

25. European Commission. Feedback from: EDC-Free Europe coalition. 2021 Jun. Available at: https://ec.europa.eu/info/law/better-regulation/have-your-say/initiatives/12959-Chemicals-legislation-revision-of-REACH-Regulation-to-help-achieve-a-toxic-free-environment/F2333093_en.

26. EDC-MixRisk project. Project reference: 634880. Funded under: H2020-EU.3.1.1. 2015 May 1 to 2019 May 1. Available at: https://edcmixrisk.ki.se/.

27. Barouki R, Gluckman PD, Grandjean P, Hanson M, Heindel JJ. Developmental origins of non-communicable disease: implications for research and public health. *Environ Health*. 2012;11:11–42.

28. European Commission. Chemicals strategy for sustainability: towards a toxic-free environment. 2020 Oct. Available at: https://ec.europa.eu/environment/strategy/chemicals-strategy_en.

29. European Parliament. Fact sheets on the European Union. Chemicals and pesticides. Available at: https://www.europarl.europa.eu/factsheets/en/sheet/78/chemicals-and-pesticides.

30. EurLex. Regulation (EC) No. 1107/2009 of the European Parliament and of the Council concerning the placing of plant protection products on the market and repealing Council Directives 79/117/EEC and 91/414/EEC. Available at: https://eur-lex.europa.eu/LexUriServ/LexUriServ.do?uri=OJ:L:2009:309:0001:0050:EN:PDF.

31. EurLex. Regulation (EC) No 1107/2009 of the European Parliament of the Council of October 21, 2009 concerning the placing of plant protection products on the market and repealing Council Directives 79/117/EEC and 91/414/EEC; OJ L 309/1. 2009. Available at: http://eur-lex.europa.eu/LexUriServ/LexUriServ.do?uri=OJ:L:2009:309:0001:0050:EN:PDF.

32. Regulation (EC) No 396/2005 of the European Parliament and of the Council of 23 February 2005 on maximum residue levels of pesticides in or on food and feed of plant and animal origin and amending Council Directive 91/414/EEC. Regulation No. 396/2005 of 16 March 2005. https://eur-lex.europa.eu/eli/reg/2005/396/oj

33. EurLex. Directive 2009/128/EC of the European Parliament and of the Council of October 21, 2009 establishing a framework for Community action to achieve the sustainable use of pesticides. Directive on the sustainable use of pesticides; OJ L 309/71. 2009. Available at: http://eur-lex.europa.eu/LexUriServ/LexUriServ.do?uri=OJ:L:2009:309:0071:0086:EN:PDF.

34. Europa. Biocidal Products Regulation (BPR, Regulation (EU) 528/2012) Available at: https://echa.europa.eu/regulations/biocidal-products-regulation/understanding-bpr.

35. Europa. Phthalate-containing soft PVC toys and childcare articles. Available at: http://europa.eu/legislation_summaries/consumers/consumer_safety/l32033_en.htm.

36. EurLex. Directive 2005/84/EC of the European Parliament and of the Council of December 14, 2005 amending for the 22nd time Council Directive 76/769/EEC on the approximation of the laws, regulations and administrative provisions of the Member States relating to restrictions on the marketing and use of certain dangerous substances and preparations (phthalates in toys and childcare articles); OJ L 26/11. 2005. Available at: http://eur-lex.europa.eu/LexUriServ/LexUriServ.do?uri=OJ:L:2005:344:0040:0043:en:PDF.

37. EurLex. Commission Regulation (EU) 2020/1245 of 2 September 2020 amending and correcting Regulation (EU) No 10/2011 on plastic materials and articles intended to come into contact with food. Available at: https://eur-lex.europa.eu/legal-content/EN/ALL/?uri=CELEX:32020R1245.

38. European Commission. Directive 2009/48/EC, toy safety directives. Available at: https://eur-lex.europa.eu/legal-content/EN/TXT/?uri=CELEX:02009L0048-20191118.

39. Europa. Commission Directive 2011/8/EU of January 28, 2011 amending Directive 2002/72/EC as regards the restriction of use of bisphenol A in plastic infant feeding bottles; OJ L 26/11. Available at: http://eur-lex.europa.eu/LexUriServ/LexUriServ.do?uri=OJ:L:2011:026:0011:0014:EN:PDF.

40. European Food Safety Authority. EFSA provides risk assessment on mercury in fish: precautionary advice given to vulnerable groups. Available at: http://www.efsa.europa.eu/fr/press/news/contam040318.htm.

41. Katsonouri A. Prioritised substance group: Mercury and its organic compounds Available at: https://www.hbm4eu.eu/wp-content/uploads/2019/03/HBM4EU_D4.9_Scoping_Documents_HBM4EU_priority_substances_v1.0-Mercury.pdf.

42. Health & Environment Alliance. Health care without harm Europe. Halting the child brain drain. Available at: http://www.env-health.org/IMG/pdf/mercury_full_report.pdf.

43. Minamata Convention on Mercury. Available at: https://www.mercuryconvention.org/en.

44. EurLex. Directive 2008/50/EC of the European Parliament and of the Council of 21 May 2008 on ambient air quality and cleaner air for Europe; OJ L 152/1. Available at: http://eur-lex.europa.eu/LexUriServ/LexUriServ.do?uri=OJ:L:2008:152:0001:0044:EN:PDF.

45. WHO. Global air quality guidelines. Available at: https://apps.who.int/iris/handle/10665/345334.

46. Europa. Revision of the Ambient Air Quality Directives. Available at: https://ec.europa.eu/environment/air/quality/revision_of_the_aaq_directives.htm.

47. Moshammer H, Forsberg B, Künzli N, Medina S; APHEKOM Project. Improving knowledge and communication for decision making on air pollution and health in Europe. 2009. Available at: https://journals.lww.com/epidem/fulltext/2009/11001/Improving_Knowledge_and_Communication_for_Decision.702.aspx.

48. SINPHONIE. Schools indoor pollution and health: observatory network in Europe Available at: https://publications.jrc.ec.europa.eu/repository/handle/JRC91160.

49. EU Environmental Noise Directive. Available at: https://ec.europa.eu/environment/noise/directive_en.htm.

50. EurLex. Environment Action Programme to 2020 "Living well, within the limits of our planet." Available at: https://eur-lex.europa.eu/legal-content/EN/TXT/?uri=CELEX:32013D1386.

51. Europea. Third implementation report of the Noise Directive. 2023. https://environment.ec.europa.eu/publications/noise-directive-implementation-report_en

52 Europea. Zero pollution action plan. Available at: https://environment.ec.europa.eu/strategy/zero-pollution-action-plan_en

53. EurLex. EU strategy on the rights of the child.2021. Available at: https://eur-lex.europa.eu/legal-content/en/TXT/?uri=CELEX%3A52021DC0142.

54. Europa. A European Green Deal. Available at: https://ec.europa.eu/info/strategy/priorities-2019-2024/european-green-deal_en.

60

Public Policy on Children's Environmental Health in Asia

Eunhee Ha

Asia contains half of the world's children, and the countries of Asia are the most rapidly industrializing nations of the globe. Environmental threats to the health of children are myriad in Asia including the classic infectious disease hazards: pneumonia, dysentery, measles, AIDS, and tuberculosis. As industrial development proceeds, nations pass through the epidemiologic transition and experience rapid urbanization, children confront a multiplying array of new threats to health posed by exposures to toxic chemicals.

In response to these new threats, national governments in Asian countries have taken action since the 1970s to protect children against environmental threats to health and have developed new approaches to the evaluation and management of toxic chemicals in the past decades. This chapter reviews these policy actions.

New Developments in Children's Environmental Health Policy in Korea

Korea has played a leading role in the effort to protect children against environmental threats to health and established a 10-year action plan for children's environmental health policy under the National Environmental Health Plan (2011–2020).

The Korean National Environmental Health Plan is a long-term plan to create and implement practical environmental health policy. It was established in recognition of the fact that a rapidly growing number of Korean children suffer from environmental diseases. The National Environmental Health Plan stresses the importance of environmental policy for vulnerable populations and therefore includes special protections for children's environmental health.[1]

The National Environmental Health Plan includes provisions to prevent and manage health problems caused by hazardous chemicals and pollutants. In particular, clauses 6, 15, 23, and 24 in the Environmental Health Act of 2009 state that the health of vulnerable populations such as children and pregnant women must be continuously investigated and evaluated to perform risk and hazardous chemical management.[2]

Under the National Environmental Health Plan, exposure to environmentally hazardous agents in children's daily living places such as kindergartens and schools will be evaluated. Based on these evaluations, a safety control standard for the environment will be developed and phased in. Environmentally hazardous chemicals in goods for children (e.g., toys) will be listed and strictly controlled. Risk assessments of these goods will help to understand the adverse health effects caused by exposure. Finally, educational

Eunhee Ha, *Public Policy on Children's Environmental Health in Asia* In: *Textbook of Children's Environmental Health*, Second Edition. Edited by: Ruth A. Etzel and Philip J. Landrigan, Oxford University Press. © Oxford University Press 2024.
DOI: 10.1093/oso/9780197662526.003.0060

programs for children's environmental health (precautionary management of environmental diseases, school environment safety checking, and provision of information on hazardous products around schools) will be implemented.[3]

An Environmental Health Policy Division was created in 2004 to implement the National Environmental Health Plan. As the tasks of the environmental health division expanded to control indoor air quality, asbestos, and others in 2009, the government organized the office of environmental health into the four divisions (Environmental Health Policy, Environmental Health Management, Chemicals Management, and Indoor Air and Noise Management) (see Figure 60.1).

The second National Environmental Health Action Plan will be implemented from 2021 to 2030. This plan expands from prevention and management of environmentally harmful factors and moves toward the comprehensive prevention and management of community environments that includes damage response and recovery. It pursues continuous operation and improvement of the children's environmental health birth cohort. The purpose of this plan is two-fold: (1) to establish an integrated risk assessment system for children's activity spaces and supplies (by calculating exposure coefficients according to exposure scenarios and usage time); and (2) to host a "Children's Environmental Health Forum" to initiate a preventive child health protection system with the participation of academia and kindergarten principals. It also highlights reinforcement of environmental safety management of children's activity spaces by conducting a joint inspection of these spaces by local governments and education offices (around 5,000 locations every year).[4] In June 2019, the Ministry of Environment expanded environmental safety management standards to "kids' cafés" along with play areas and nursery, kindergarten, elementary school, and special school classes. The survey of 1,894 sites showed that 75.5% sites exceeded the criteria for detecting heavy metals in paints and finishing materials, and 27.1% sites exceeded indoor air quality standards. Although kids' cafés are primary places for kid's activities, they remain a blind spot for management, which should follow the environmental safety standards for no peeling paint, no rust, good indoor air quality, and no heavy metal exposure. With the amendment, exposure to heavy metals will be reduced in kids' cafés, creating a cleaner and safer environment for children's health.[5]

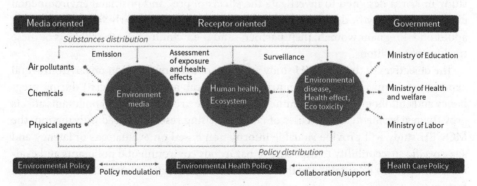

Figure 60.1. Precautionary principle based on the Environmental Health Policy of the Ministry of Environment in the Republic of Korea.

Korean National Studies on Children's Environmental Health

The Children's Health and Environmental Research (CHEER) study was launched in 2005,[6] and the birth cohort study of Mothers and Children's Environmental Health (MOCEH) was launched in 2006.[7] Their purpose is to collect baseline data on vulnerable populations and then use this information to establish a policy that will continuously protect the environment and the health of children in Korea in order to establish a sustainable society.

Children's Health Environmental Research Study

The CHEER study was one of the largest in the world investigating the effects on children's health of exposures to hazardous agents in the environment. In total, 7,059 children from 33 primary schools in 10 cities were enrolled from 2005 to 2010, and one-third of them ($n = 2,770$) were followed-up 2 and 4 years later.

The first purpose of this study is to investigate patterns of exposure of children to hazardous chemicals and establish a baseline. The second goal is to study the effects of these exposures on health conditions such as neurodevelopment (attention deficit hyperactivity disorder [ADHD] and autism spectrum disorders), asthma, and atopic dermatitis and, specifically, to study the impacts of time-related variations in environmental exposures on the prevalence of these diseases. The third objective of the study is to provide evidence on associations between environmental exposures and disease in children that will guide the development of children's environmental health policy.

The study documented the exposure levels of heavy metals such as lead and mercury and endocrine-disrupting chemicals such as phthalates and bisphenol A among Korean children. This longitudinal study showed changes in disease and exposure prevalence during the follow-up period.[8]

Mothers and Children's Environmental Health Study

The MOCEH study is the first prospective hospital- and community-based birth cohort study in Korea designed to investigate the effects of pre- and postnatal environmental exposures on growth, development, and health ranging from early fetal life to young adulthood. Pregnant women, their partners, and their children are the subjects of these tracking observations.

The objectives of the MOCEH study are to (a) collect information on environmental exposures (chemical, biological, nutritional, physical, and psychosocial) during pregnancy and childhood and (b) examine how exposure to environmental pollutants affects growth, development, and diseases of children. Using research-based information, the MOCEH study will provide valuable information based on evaluations of latency and age-specific susceptibility to exposure to hazardous environmental pollutants and evaluations of growth retardation, with a focus on environmental influence and genetic risk factors.[9]

In this study, 1,751 mothers were enrolled from Seoul (metropolitan), Cheonan (a clean city), and Ulsan (an industrial city). The health status of the mothers and their spouses was evaluated and their environmental exposures were measured. The health outcomes of the children include a neurodevelopmental test, growth measures, and collection of data on health conditions such as allergic diseases at birth and at 6, 12, 24, 36, 48, and 60 months. The study showed that heavy metals in a mother's blood can cross the placental barrier and reach the fetus. In fact, higher levels of heavy metals were found in cord blood than in the mother's blood, and exposure levels were related to adverse growth development and neurodevelopment in children. The concentrations of endocrine-disrupting compounds were higher in samples of urine from 24- and 36-month-old children than in the samples of urine from their mothers, and these concentrations were higher than the concentrations reported in studies in other countries.

It is expected that this study will provide new information to support the hypothesis that the gestational environment affects the development of diseases during childhood and adult life. It is also expected that the results of this study will enable an evaluation of latency and age-specific susceptibility to exposure to hazardous environmental pollutants, permit evaluation of growth retardation with a focus on environmental and genetic risk factors, and facilitate the selection of target environmental diseases in children. The study's data will guide development of an environmental health index and the establishment of a national policy for improving the health of pregnant women and their children.

Based on these results of the CHEER and MOCEH studies, the Korean government planned a national large-scale children's cohort study. The objectives of this nationwide cohort study are to (1) provide a framework to develop recommended exposure levels of hazardous environmental factors for the susceptible population, (2) gather scientific evidence on the association of harmful environmental pollutants and health outcomes, and (3) develop guidelines for environmental hazardous factors in terms of environmental health management system.

Thus, a national long-term birth cohort Korean CHildren's ENvironmental health Study (Ko-CHENS) was launched in 2015.[10] Ko-CHENS plans to recruit a total of 70,000 expectant mothers through its two enrollment tracks (65,000 in the Main Cohort and 5,000 in the Core Cohort). The children in the core cohort will be followed up at 6 months, every year before their admission into the elementary school, and every 3 years from the first year after this admission. The children in the main cohort will be followed up through data links (Statistics Korea, National Health Insurance Service and Ministry of Education). The Ko-CHENS data can be linked with the nationally registered health-related database that includes medical utilization, periodic health screening, and the birth/mortality database. Thus, Ko-CHENS aims to conduct a follow-up of 97% of its participants through data linkage between big data systems (see Figure 60.2).

National Network of Environmental Health Centers to Manage Children's Environmental Diseases

In 2007, the Ministry of Environment of the Republic of Korea established a national network of Environmental Health Centers for preventing, managing, and studying environmental diseases in children. There are currently 13 environmental health centers in Korea that monitor exposure to environmentally hazardous agents and study the causal relationships between environmental factors and outcomes.[11] Each center has a specialized focus such as atopic dermatitis, asthma, children's diseases, and adverse health effects from asbestos, radon, and oil exposure. Hospitals and universities host the

Figure 60.2. Korean CHildren's ENvironmental health Study (Ko-CHENS) and its birth cohort activities

designated centers. Eight of these centers focus on children's environmental health and perform relevant education and prevention publicity. As an example of the educational programs, the environmental health centers annually conduct an environmental health class for education and publicity entitled "Atopy and Green Camp." Its purpose is to consult with and educate children suffering from atopic dermatitis. The environmental health centers also run an annual program called "What Is Environmental Disease?" and they aim to provide better health services by exploring consumer demands[12,13] (see Figure 60.3).

To improve reproductive health and children's health and increase awareness of hazards in the pediatric environment, such as particulate matter pollution and immunological disease, South Korea initiated the first Korean Pediatric Environment Health Specialty Units (KPEHSUs) in Ewha Womans University Seoul Hospital. The Pediatric Environmental Health Specialty Units are a national network of experts in the prevention, diagnosis, management, and treatment of health issues that arise from environmental exposures from preconception through adolescence.[14,15] The aim of the KPEHSUs is to improve reproductive health and children's health by integrating environmental health

Figure 60.3. Health education and counseling program in Environmental Health Centers of the Republic of Korea

into clinical care and public health. In the future, more KPEHSU clinics will open in Korea with the collaboration of the Ministry of Environment.

Current Management Policy for Children's Products

The Korean Ministry of Environment conducted studies on exposure levels and risk assessments in 403 children's products between 2007 and 2009 to support better management of hazardous chemicals in these products. One hundred thirty-five environmental factors were identified. To control these chemicals, the Ministry of Environment established a mid- and long-term plan for risk assessment of children's products (2010–2014).[16]

The Ministry of Environment has investigated the use and management of products which most likely contain chemicals and indicate high risk based on risk assessments of the product. In 2010, investigations on six types of phthalates, including di(2-ethylhexyl) phthalate (DEHP), were conducted. Heavy metals with high risk levels have also been investigated. In 2011, the Environmental Health Act was amended, allowing the Ministry of Environment to order withdrawal of products containing hazardous chemicals.

Current Management Policy for Activity Spaces for Children

Since 2009, the Ministry of Environment has prepared regulations for environmental safety management of children's activity spaces such as kindergartens, child-care centers, school classrooms, and playgrounds according to Article 23 of the Environmental Health Act. Standards for heavy metals, wood preservatives, and indoor pollutants were established under these regulations to manage materials at children's play facilities. Paint, lumber, and flooring materials at child care centers and school classrooms were also regulated.[16,17]

In addition, the Ministry of Environment categorizes environmental features of children's activity spaces and reviews the scope of the spaces and the regulated environmental pollutants in them. These projects will help mid- and long-term implementation measures to improve environmental safety and introduce environmental safety standards for unregulated hazards (see Figure 60.4).

Children's Environmental Health Indicators

The "Children's Environmental Health Indicators" were first introduced at the World Summit on Sustainable Development in South Africa in 2002, and the indicators have subsequently been developed by the World Health Organization (WHO). In the United States, Canada, and the European Union, children's environmental health programs have been implemented by governments and/or in cooperation with international organizations.[18]

Based on the Environmental Health Act in Korea, 27 environmental health indicators have been presented to monitor the environmental health of children in seven areas, including ambient atmosphere, indoor air, water quality, chemicals, noise, radiation,

Figure 60.4. Signs to protect the health and safety of children in the Republic of Korea

and safety accidents. A pilot study is being conducted on five of these indicators that have been judged feasible such as infant mortality rate due to respiratory diseases. New indicators will be continuously developed and utilized depending on the developmental process of environmental health indicators.

Action Plan on Children's Environmental Health Monitoring and Survey Project for Life Cycle Environmental Health: Korean National Environmental Health Survey

The Ministry of Environment has been conducting the Korea National Environmental Health Survey every 3 years since 2009 to track the environmental health of the public. A total of 6,311 adults older than 19 years are surveyed using questionnaires and their blood and urine are analyzed to identify four metals (lead, mercury, manganese, and arsenic) and 11 types of organic compounds (e.g., polycyclic aromatic hydrocarbons, phthalates, volatile organic compounds, bisphenol A, pesticides, and secondhand smoke) in urine. Children under 19 years of age are excluded from this survey.[19]

The Ministry of Environment developed a comprehensive national survey plan including children and teenagers in 2012 and toddlers and preschoolers in 2013 and completed a preliminary survey by 2014. The Korean National Environmental Health Survey (KoNEHS) will then provide data that are representative across all ages.

The KoNEHS has completed three phases of survey evaluating the exposure of 33 hazardous chemicals in children, adolescents, and adults.[20] In KoNEHS, it was found that phthalate exposure was associated with early menarche in Korean girls[21] and atopic dermatitis in adolescents.[22] Mercury concentrations were associated with lipid profile and liver enzymes in Korean adults.[23] An association between use of personal care products and concentrations of phthalates, parabens, and triclosan was seen in various age groups.[24]

New Developments in Children's Environmental Health in China

China has launched a prospective birth cohort study in Shanghai. The Shanghai Birth Cohort Study evaluates the effects of genetic, environmental, and behavioral factors on women's reproductive health, pregnancy outcome, child growth, development, and risk of diseases. The study is led by investigators at the Shanghai Key Lab of Children's Environmental Health, Shanghai Jiao Tong University School of Public Health.

The cohort study included a general study of 10,000 women planning pregnancy or in early gestation and a more detailed study of 2,000–3,000 cooperative volunteers. The study started at preconception. To be included in the study, women had to be 20 years of age or older, planning to be pregnant and to receive prenatal care and give birth in a participating hospital, be registered residents of Shanghai municipality, have lived in Shanghai the past 2 years, and not plan to move out of the catchment area in the next 2 years. Women who had tried to conceive spontaneously for more than 12 months were excluded. Children participating in the study were followed up until 2 years of age, possibly longer. Recruitment took place in participating hospitals. There were three tertiary maternity hospitals in Shanghai and 20 maternity hospitals at the district level. This recruitment strategy will give maximum efficiency, although it will miss mothers who do not come to the hospital for preconception care.

The Shanghai study used interviewer-administered questionnaires. Women planning pregnancy and recruited into the study were asked to spend about 1 hour while in the hospital for preconception care answering the interviewer's questions. Sample collection included venous blood, cord blood, saliva, urine, nails, breast milk, semen, placenta, buccal swab, hair, and meconium. A smaller number of cooperative volunteers from among the 10,000 women who agreed to a home visit were visited by a team in the home, and household dust and indoor air measurements were performed on some fraction.

The Shanghai study enrolled 1,180 couples in the preconception cohort and 3,426 pregnant women in the pregnancy cohort. In the preconception cohort, increased exposure of some perfluoroalkyl and polyfluoroalkyl substances (PFAS)—PFOA, perfluorooctyl sulfonate (PFOS), PFNA, and PFHxS—in women was associated with irregular and long menstrual cycles.[25] A study in the pregnancy cohort showed that increased exposure to PFAS in pregnant women disrupted thyroid hormone homeostasis by increasing free T_3 and T_4 and decreasing thyroid stimulating hormone (TSH). Maternal thyroid hormones are important for fetal brain development.[26] The increased exposure to PFAS in pregnant women was found to disrupt fetal thyroid function.[27] Prenatal exposure to PFAS also was found to be associated with childhood adiposity in girls.[28] The Shanghai study identified a critical window of particulate matter ($PM_{2.5}$) exposure affecting fetal growth and birth weight. The Shanghai study intends to study mixture exposures, using exposome and -omics approaches to evaluate the effect of prenatal environmental pollutants on fetal and child growth.

New Developments in Children's Environmental Health in Japan

Japan launched a major new nationwide birth cohort study in 2011. The Japan Environment and Children's Study (JECS) planned to enroll 100,000 participants across all of Japan, recruited over a 3-year period and followed for up to 13 years.[29] The study is led by the Ministry of Environment, among other stakeholders, including 15 regional centers.

The study investigates chemical exposures during the fetal and infant stages and examines how they affect children's health. The main emphasis is given to environmental chemicals such as persistent organic pollutants (POPs) (dioxins, PCBs, organofluorine compounds, and flame retardants), heavy metals (mercury, lead, arsenic, cadmium, etc.), endocrine disruptors (phthalates and bisphenol A), agrichemicals, and volatile organic compounds (benzene and other solvents), as well as genetics, socioeconomics, and lifestyle.

The main outcomes of the study include physical development (low birth weight, anomalies of postnatal development), congenital anomalies (hypospadias, cryptorchidism, cleft lip, cleft palate, spina bifida, digestive tract obstruction, ventricular septal defects, Down syndrome), abnormalities in sexual differentiation (sex ratio, genital development impairment, sexual differentiation of the brain), disorders of neurodevelopment (autism spectrum disorders, learning disorders, ADHD), immune system disorders (pediatric allergies, atopic dermatitis, asthma), and endocrine/metabolic abnormalities (lowered glucose tolerance, obesity).

Anticipated benefits of JECS include (a) identifying environmental factors impacting children's health, (b) developing risk management systems that address vulnerabilities in children, (c) ensuring a sound environment where future generations are able to grow up in good health, and (d) establishing infrastructures for children's studies.

The JECS study enrolled 103,099 pregnant women in the main study, and 5,017 children were enrolled in a subcohort study. The JECS study evaluated how environmental factors affect reproduction, pregnancy complications, congenital anomalies, developmental disorders, and immune and metabolic system dysfunction.[30] Maternal smoking was found to be associated with reduced birth weight.[31] The JECS study found that cadmium levels during pregnancy were associated with early preterm birth in Japanese women, and manganese levels were associated with birth size. The JECS study provides data that allows a population approach toward policy for better health of children in Japan.

New Developments in Children's Environmental Health in Taiwan

Taiwan Birth Panel Study

Taiwan launched a prospective birth cohort study and their first nationwide representative study. The Taiwan Birth Panel Study I (TBPS I) is a prospective cohort study that recruited 486 mother–infant pairs in northern Taiwan since 2004.[32] The primary goals were to examine the low-level, pre- or postnatal exposure and genetic modification effect on the initiation and progress of "environmentally related childhood diseases." Children were followed at 4 and 6 months and at 1, 2, 3, 5, 7, and 9 years of age. Each visit was conducted using standardized questionnaires with sociodemographic information, and biological samples were collected to measure biomarkers for environmental toxicants. In addition, child growth, neurodevelopment, and physical examination were assessed by trained professionals.

Adverse effects related to birth outcomes and child growth have been reported. Cord blood cadmium was associated with decreased head circumference in newborns and the growth of children up to 3 years of age.[33] Higher prenatal PFOS exposure was associated with decreased fetal growth, but the effects were diminished as children reached 8 years of age, and a modest gender-specific effect was observed.[34] Cord blood cotinine adversely affected birth outcomes through the modification effect of GSTT1 and GSTM1 null genotype.[35] Several hazards related to children's neurodevelopment or behavior problems have been reported, such as a gene–environment interaction between the CYP1A1 gene and maternal exposure to secondhand smoke[36] as well as the apolipoprotein E gene and cord blood mercury.[37] The research team also investigated the distribution, correlation, and exposure sources of various heavy metals.

The Taiwan Birth Panel Study II (TBPS II) adopted a similar research protocol since 2010, with approximately half of the children followed-up completely until the age of 7 in 2018. New concerns include chemicals, urban exposome such as air pollution and greenness, puberty development, childhood obesity, and behavior problems. State-of-the-art methods such as nontargeted screening technology, omics technology, and machine learning have been applied to elucidate the critical periods in developmental biology susceptible to environmental pollutants.

Taiwan Birth Cohort Study

The Taiwan Birth Cohort Study (TBCS) is the first nationwide representative birth cohort study in Taiwan with 24,200 mother–infant pairs recruited since 2005.[37,38] By 2019, nine waves of surveys were completed, and the response rates ranged from 87.8% to 94.9%. Structured questionnaires were used to collect information on children's health and behavioral outcomes and personal, family, and social determinants of health. This study focuses on the impact of social environments on children's wellbeing over the life course and documents the health trajectories of Taiwanese children in the 21st century.

These research efforts contribute to significant social impacts on corporate social responsibility for the elimination of environmentally hazardous pollutants and official regulation of food safety. Taiwan is facing the challenges of a declining birth rate, the substitution of hazardous chemicals, and new pediatric morbidities. A collaboration effort between interested parties studying birth cohorts in Taiwan, including active surveillance and an administrative database, is necessary to identify influential environmental hazards while taking into account regional variation and temporal change.

Beginning in 2023, the TBPS, together with the Taiwan Early Life Cohort Study (TEC) and Taiwan Mother–Infant Cohort Study (TMICS) will implement standardized family-based follow-up protocols. This aims to identify influential and important early-life factors that affect children's health and develop predictive models of childhood diseases. The ultimate goal is to translate scientific knowledge into policymaking or strategy implementation to promote the environmental health of Taiwanese children.

Conclusion

The prevalence of environmentally related diseases such as allergic diseases, ADHD, and autism appears to be increasing dramatically in Asian countries. This has generated great public concern about the adverse effect of environmental exposures on children's health.

In response to this widespread concern, a series of rapid changes in children's environmental health policy have been instituted. The government of Japan initiated an International Working Group for Harmonization of the Next Generation of Large-Scale Birth Cohort Studies that brings together researchers and scientists from many countries launching new, large birth cohort studies on child health and the environment.[39] Their leadership on this issue has helped to establish strong international collaboration for the generation of new knowledge about environmental threats to the health of children and for the protection of children's environmental health.

In 2009, the Korean Ministry of Environment held the Third International Conference for Children's Health and Environment in Busan, Korea, in partnership with the WHO.[40]

In July 2010, in Jeju, Korea, the Government of Korea hosted the Second Ministerial Regional Forum on Environment and Health in South-East and East Asian Countries, jointly organized by the United Nations Environment Programme Regional Office for Asia and the Pacific and the WHO Regional Offices for the Western Pacific and South-East Asia.[41]

And, in 2010, the Korean Ministry of Environment convened the International Society for Environmental Epidemiology and the International Society of Exposure Science with the goal of increasing the number of studies on children's environmental health.[42]

These positive actions regarding children's environmental health in China, Japan, Taiwan, and Korea will likely extend to other Asian countries.

References

1. Korean Government. National *Environmental Health Action Plan (NEHAP)*. Seoul, Korea: The Ministry of Environment; 2005.

2. Korean Government. *Environmental Health Act of Korea*. Seoul, Korea: The Ministry of Environment; 2009.

3. Korean Government. *National Environmental Health Action Plan (NEHAP)*. Seoul, Korea: The Ministry of Environment; 2011.

4. Korean Government. *The 2nd National Environmental Health Action Plan (NEHAP)*. Sejong, Korea: The Ministry of Environment; 2020.

5. Korean Government. Ministry of Environment. *The Application of Environmental Safety Management Standards for Kids Cafes*. Sejong: Republic of Korea. 2019.

6. Ha M, Kwon HJ, Lim MH, Jee YK, et al. Low blood levels of lead and mercury and symptoms of attention deficit hyperactivity in children: a report of the children's health and environment research (CHEER). *Neurotoxicology*. 2009;30:31–6.

7. Kim BM, Ha M, Park HS, et al. The Mothers and Children's Environmental Health (MOCEH) study. *Eur J Epidemiol*. 2009;24:573–83.

8. Korean Government. *Annual Report on CHEER (Child Health Environmental Research) Study*. Seoul, Korea: Ministry of Environment, National Institute of Environmental Research; 2011.

9. Korean Government. *Annual Report on MOCEH (Mothers and Children's Environmental Health) Study*. Seoul, Korea: Ministry of Environment, National Institute of Environmental Research; 2011.

10. Jeong KS, Kim S, Kim WJ, et al. Cohort profile: beyond birth cohort study–The Korean CHildren's ENvironmental health Study (Ko-CHENS). *Environmental Res*. 2019 May 1;172:358–66.

11. Korean Government. *The 2nd Forum on Environment and Children's Health-Risk Factors for the Environmental Disease in Korea*. Seoul, Korea: The Association of Environmental Health Centers in Korea; 2011: 1–22. In Korean.

12. Korean Government. *A Guideline of Prevention and Management for Atopy*. Seoul, Korea: Environmental Health Center of Children's Environmental Diseases/Ministry of Environment; 2011: 17–24. In Korean.

13. Gyeonggi Research Institute. *Strategies to Active Atopy Free*. Seoul, Korea: Gyeonggi Research Institute; 2012.

14. Pediatric Environmental Health Specialty Units. PEHSU. 2023. Available at: https://www.pehsu.net.

15. World Health Organization. Children's Environmental Health Units. 2010. Available at: http://who.int/publications/i/items/children's-environmental-health-units.

16. Korean Government. *Annual Report on Children's Products and Space Management Policy*. Seoul, Korea: Ministry of Environment, National Institute of Environmental Research; 2011.

17. Korean Government. *Environmental Review*. Seoul, Korea: Ministry of Environment; 2011.

18. Park CH, Chang JY, Lee YM, Lee BE, Jung SW, Yu SD, Choi K. *A Study on Environmental Health Assessment of National and Local Scale (II)*. Seoul, Korea: National Institute of Environmental Research (NIER-RP2012-307); 2012: 26–31. In Korean.

19. Korean Government. Ministry of Environment. *Concept to Introduce Guidance Value of Mercury Toward Susceptible and Vulnerable Population*. Seoul, Korea: Ministry of Environment; 2011.

20. Lee S, Kim JH, Choi YH, Kim S, Park JB. A review of the literature using the Korean National Environmental Health Survey (cycle 1–3). *J Environ Health Sci*. 2021;47(3):227–44.

21. Park O, Park JT, Chi Y, Kwak K. Association of phthalates and early menarche in Korean adolescent girls from Korean National Environmental Health Survey (KoNEHS) 2015–2017. *Ann Occup Environ Med*. 2021;33(e4):1–12.

22. Kim SW, Lee J, Kwon SC, Lee JH. Association between urinary phthalate metabolite concentration and atopic dermatitis in Korean adolescents participating in the third Korean National Environmental Health Survey, 2015–2017. *Int J Environ Res Public Health*. 2021 Jan;18(5):2261.

23. Lee S, Cho SR, Jeong I, Park JB, Shin MY, Kim S, Kim JH. Mercury exposure and associations with hyperlipidemia and elevated liver enzymes: a nationwide cross-sectional survey. *Toxics*. 2020 Sep;8(3):47.

24. Lim S. The associations between personal care products use and urinary concentrations of phthalates, parabens, and triclosan in various age groups: the Korean National Environmental Health Survey Cycle 3 2015–2017. *Sci Total Environ*. 2020 Nov 10;742:140640.

25. Zhou W, Zhang L, Tong C, et al.; Shanghai Birth Cohort Study. Plasma perfluoroalkyl and polyfluoroalkyl substances concentration and menstrual cycle characteristics in preconception women. *Environ Health Perspect*. 2017 Jun 22;125(6):067012.

26. Aimuzi R, Luo K, Huang R, et al. Perfluoroalkyl and polyfluroalkyl substances and maternal thyroid hormones in early pregnancy. *Environ Pollution*. 2020 Sep 1;264:114557.

27. Aimuzi R, Luo K, Chen Q, Wang H, Feng L, Ouyang F, Zhang J. Perfluoroalkyl and polyfluoroalkyl substances and fetal thyroid hormone levels in umbilical cord blood among newborns by prelabor caesarean delivery. *Environ Int*. 2019 Sep 1;130:104929.

28. Chen Q, Zhang X, Zhao Y, et al. Prenatal exposure to perfluorobutanesulfonic acid and child-hood adiposity: a prospective birth cohort study in Shanghai, China. *Chemosphere*. 2019 Jul 1;226:17–23.

29. Ministry of Environment, Government of Japan. Japan environment and children's study. 2011. Available at: http://www.env.go.jp/en/chemi/hs/jecs.

30. Sekiyama M, Yamazaki S, Michikawa T, et al. Study design and participants' profile in the Sub-Cohort Study in the Japan Environment and Children's Study (JECS). *J Epidemiol*. 2022;32(5):228–36.

31. Suzuki K, Shinohara R, Sato M, Otawa S, Yamagata Z. Association between maternal smoking during pregnancy and birth weight: an appropriately adjusted model from the Japan Environment and Children's Study. *J Epidemiol*. 2016 Jul 5;26(7):371–7.

32. Hsieh CJ, Hsieh WS, Su YN, et al. The Taiwan Birth Panel Study: a prospective cohort study for environmentally-related child health. *BMC Res Notes*. 2011 Dec;4(1):1–8.

33. Lin CM, Doyle P, Wang D, Hwang YH, Chen PC. Does prenatal cadmium exposure affect fetal and child growth? *Occup Environ Med*. 2011 Sep;68(9):641–6. doi:10.1136/oem.2010.059758

34. Chen MH, Ng S, Hsieh CJ, Lin CC, Hsieh WS, Chen PC. The impact of prenatal perfluoroalkyl substances exposure on neonatal and child growth. *Sci Total Environ*. 2017 Dec 31;607–8, 669–75. doi:10.1016/j.scitotenv.2017.06.273.

35. Huang KH, Chou AK, Jeng SF, Ng S, Hsieh CJ, Chen MH, Chen PC, Hsieh WS. The impacts of cord blood cotinine and glutathione-S-transferase gene polymorphisms on birth outcome. *Pediatr Neonatol*. 2017 Aug; 58(4): 362–9.

36. Hsieh CJ, Liao HF, Wu KY, et al. CYP1A1 Ile462Val and GSTT1 modify the effect of cord blood cotinine on neurodevelopment at 2 years of age. *Neurotoxicology*. 2008 Sep;29(5):839–45. doi:10.1016/j.neuro.2008.05.006.

37. Ng S, Lin CC, Hwang YH, Hsieh WS, Liao HF, Chen PC. Mercury, APOE, and children's neurodevelopment. *Neurotoxicology*. 2013 Jul;37:85–92. doi:10.1016/j.neuro.2013.03.012.

38. Chang LY, Lin YH, Lin SJ, Chiang TL. Cohort profile: Taiwan Birth Cohort Study (TBCS). *Int J Epidemiol*. 2021;50(5):1430–1.

39. Etzel R, Charles M-A, Dellarco M, et al. Harmonizing biomarker measurements in longitudinal studies of children's health and the environment. *Biomonitoring*. 2014;1:50–62.

40. World Health Organization. The Busan Pledge for Action on Children's Environmental Health. 2013 July 27. Available at: https://www.who.int/publications/m/item/busan-pledge-for-action-on-ceh-2009

41. Jeju Declaration on Environment and Health, Second Ministerial Regional Forum on Environment and Health in South-East and East Asian Countries, Jeju, Korea. 2010. Available at: http://www.environment-health.asia/userfiles/file/JEJU%20DECLARAT ION%20ON%20ENVIRONMENT%20AND%20HEALTH_final%20unedited.pdf.

42. Gavidia T, Brune MN, McCarty KM, et al. Children's environmental health: from knowledge to action. *Lancet*. 2011; 377:1134–6.

61

Public Policy on Children's Environmental Health in Africa

A. Kofi Amegah and Christian Sewor

Overview

A child's future is determined by its environment, with early-life exposures impacting on adult health owing to fetal programming and early growth potentially being altered by environmental risk factors. A number of multicountry studies conducted in Africa and global systematic reviews have associated household and outdoor environmental conditions with child undernutrition, illnesses, and survival. While epidemiological evidence on the adverse child health effects of deplorable environmental conditions in Africa is increasing, very little progress has been made in the area of development and implementation of national policies to address children's environmental health problems on the continent. South Africa appears to be the only country in the Southern Africa region with a policy intervention aimed at addressing children's environmental health problems. Ghana, Guinea, Liberia, and Mali are the only countries in the Western Africa region to have specific policies targeted at addressing hygiene challenges at the community, school, and health facility levels, which ultimately benefits child health. Ghana has a School Health Education Programme (SHEP) with objectives that link with children's environmental health. In East Africa, Rwanda and Kenya have rolled out child-centered policy interventions such as the School Health Policy to specifically address children's environmental health challenges. These school-based interventions in Ghana, Rwanda, and Kenya could serve as models for improving children's environmental health in Africa. However, they need to target the full spectrum of environmental exposures, particularly air pollution. In North Africa, the Moroccan government recognizes the effects of poor environmental conditions on children's health, with the country's National Health policy facilitating hygiene and safety in schools through a number of health promotion approaches. Children's environmental health challenges are mounting in Africa owing largely to the widespread use of solid fuels for cooking at home; poor water, sanitation, and hygiene (WASH) practices at home and school; and e-waste recycling. In spite of the worrying situation on the continent, there are very limited national policies targeted at children's environmental health in countries.

Background

The highest number of deaths per capita attributable to environmental exposures has been recorded in sub-Saharan Africa, with the highest disease burden (36%) reported among children.[1] Amegah[2] indicated that poor household environmental conditions

A. Kofi Amegah and Christian Sewor, *Public Policy on Children's Environmental Health in Africa* In: *Textbook of Children's Environmental Health*, Second Edition. Edited by: Ruth A. Etzel and Philip J. Landrigan, Oxford University Press.

together with child undernutrition are responsible for the large proportion of under-5 deaths in sub-Saharan Africa and other developing regions. Childhood illnesses with high mortality burden such as malaria, gastroenteritis, pneumonia, and acute lower respiratory infections have all been linked to household environmental exposures.[3] A number of multicountry studies conducted in Africa[4–6] and global systematic reviews[7–10] have associated household and outdoor environmental conditions with child undernutrition, illnesses, and survival. According to Bradshaw et al.,[4] poor child health is a major concern in areas with poor water quality, sanitation, and air quality among other deplorable environmental conditions, all of which are pervasive in Africa.

According to the World Bank, children's environmental health, together with undernutrition, have been neglected by policymakers over the years in the formulation of strategies for improving child survival.[11] This chapter documents children's environmental health problems in Africa and identifies policy actions by governments for addressing these challenges. Figure 61.1 presents an overview of children's environmental health problems in Africa and the existing policy actions.

Children's Environmental Health Problems in Africa

The Demographic and Health Surveys (DHS) program pioneered by the US Agency for International Development more than three decades ago works with governments of low- and middle-income countries to collect nationally representative cross-sectional data every 5 years at most about people, their health, and their health systems. The DHS surveys conducted in African countries provide cross-sectional evidence on the effect on environmental exposures (solid fuel use [SFU] and poor WASH practices) on child health (i.e. low birth weight; stillbirth; undernutrition; acute respiratory infections; and neonatal, infant and child mortality). Multicountry studies conducted in Africa using DHS data have found associations between adverse child health outcomes and SFU[4–6,12,13] and poor WASH practices.[14–16]

Very limited prospective studies have been conducted in Africa on children's environmental health problems. Evidence from the Ghana Randomized Air Pollution and Health Study (GRAPHS), a cluster-randomized trial that evaluated the efficacy of clean

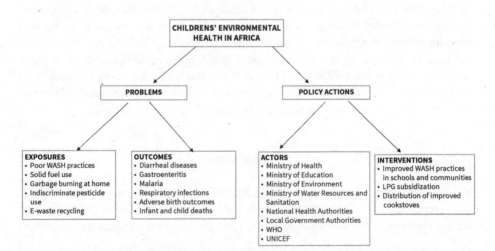

Figure 61.1. Children's environmental health problems and policy actions in Africa.

fuels (liquefied petroleum gas [LPG]) and efficient biomass cookstoves for addressing household air pollution (HAP) found prenatal carbon monoxide (CO) exposure from biomass fuel use in rural Ghanaian households to be associated with impaired lung function and high risk of pneumonia and severe pneumonia in the first year of life.[17–19] This study also found household biomass fuel use to have a negative effect on birth outcomes such as birth weight, birth length, and gestation, albeit the risk estimates were not statistically significant.[17]

Evidence from the Cooking and Pneumonia Study (CAPS), an open cluster-randomized controlled trial conducted in Malawi that sought to establish whether use of a cleaner-burning biomass cookstove could reduce pneumonia incidence in children under 5 years of age living in rural Malawi also found CO exposure due to biomass fuel use to be significantly high among young children and associated with adverse chronic respiratory symptoms.[20–22] This study also found HAP exposure from biomass fuel use to be significantly associated with the prevalence of nasopharyngeal *Streptococcus pneumoniae* carriage among children[23] thereby emphasizing the key role HAP plays in the incidence of infectious diseases among children.

The Venda Health Examination of Mothers, Babies and Their Environment (VHEMBE) study conducted in South Africa also provides longitudinal evidence on child health effects associated with exposure to dichlorodiphenyltrichloroethane (DDT) and dichlorodiphenyldichloroethylene (DDE) due to indoor insecticide spraying to combat malaria. VHEMBE found maternal exposure to DDT and DDE to have negative effects on the anogenital distance among children, albeit the associations reported were relatively weak.[24] VHEMBE also found maternal exposure to DDT and DDE to be associated with adverse respiratory health effects among children.[25] Additionally, VHEMBE found in utero insecticide exposure to have the potential to increase childhood infection rates[26] and also contribute to poor social-emotional development of children.[27]

A cluster-randomized controlled trial conducted in Ethiopia, however, did not find any significant evidence on whether use of improved biomass cookstoves reduces the risk of childhood acute lower respiratory infection compared with cooking on traditional "open fire" stoves.[28] A randomized controlled trial conducted in Nigeria that assessed the impact of HAP from cooking on fetal growth found no significant differences in fetal growth trajectories between the intervention and control groups.[29] The Sanitation Hygiene Infant Nutrition Efficacy (SHINE) trial, a cluster-randomized community-based trial conducted in two contiguous rural districts of Zimbabwe with 15% antenatal HIV prevalence to evaluate independent and combined effects of improved infant and young child feeding (IYCF) and improved WASH on linear growth and hemoglobin levels of children at 18 months of age found improved WASH practices to be associated with clinically important improvements in child development at 2 years of age.[30] A cluster-randomized controlled trial conducted in Kenya found improved WASH practices to significantly reduce *Ascaris* prevalence among children.[31] A similar finding was also noted in a clustered randomized trial conducted in Ethiopia.[32,33] The study found the practice of personal hygiene (nailing clipping and handwashing) to significantly reduce childhood prevalence of anemia and the risk of contracting intestinal parasites.

E-waste burning, which exposes individuals to a mixture of pollutants, is another major environmental health challenge in Africa, especially in the West Africa subregion. Ghana and Nigeria, for instance, have some of the biggest e-waste dumpsites in the developing world. E-waste recycling is dominated by adult males. However, a recent World Health Organization (WHO) report indicated that millions of young children and adolescents also are working in growing e-waste dumpsites in those countries where the problem is pervasive.[34] Nationwide environmental health assessments conducted in Ghana and Nigeria found male children and adolescents to be heavily involved in e-waste recycling and to be most vulnerable to the harmful effects of the pollutants released from e-waste burning.[35,36] Children as young as 5 years of age have been observed to be engaging in e-waste activities in Ghana.[37]

Policy Initiatives

Environmental risk threatens the health, development, and survival of children in Africa and it is important that governments formulate national policies to address environmental health challenges such as household and ambient air pollution, indiscriminate use of pesticides and insecticides, and unsafe water and sanitation to help protect the health and well-being of children.

Southern Africa

South Africa appears to be the only country in the Southern Africa region with a policy intervention aimed at addressing children's environmental health problems. The policy is called Healthy Environments for Children Initiative (HECI-SA) and was initiated in 2002. South Africa, like many other countries in Africa, continues to be saddled with issues of air pollution, unsafe water and sanitation, and indiscriminate pesticide and insecticide use. The main thrust of the HECI-SA policy is to bring to the fore environmental health hazards faced by children in urban and rural settings of South Africa.[38] Through this policy initiative, a special study was commissioned to assess the situation of children with regard to environmental risks from housing, water, sanitation, cooking fuels, transport, and recreation.[38] The findings of this special study have, however, not been reported. In addition, the South African government has implemented a number of laws to protect and enhance environmental quality and reduce environmental pollution exposure within the population, which should ultimately benefit the health of children. These laws include the National Environmental Management and Air Quality Acts, a ban on use of asbestos in homes, the phasing out of leaded petrol, and regulations on the use of lead in paint, among others.[39] The South African government also has adopted the WHO Children's Educational Toolkit, which provides information about environmental health hazards for primary school children.[40,41] Field reports have highlighted that the tool kit is a useful vehicle for improving knowledge of environmental health hazards among primary school children.[41]

West Africa

Ghana, Guinea, Liberia, and Mali are the only countries in the Western Africa region to have specific policies targeted at addressing hygiene challenges in communities, schools, and health facilities,[42] which ultimately benefits child health. In Niger, an Economic and Social Development Plan (ESDP, 2017–2021) was implemented in 2017, with the objective of improving the living environment and addressing climate change. The plan has a subprogram that seeks to strengthen hygiene and sanitation. The policy document, however, fails to identify the key implementers of the policy and also provides no specific linkages with child health and survival.[43]

In Ghana, promotion of proper environmental conditions is backed by several government policies. These policies include the revised National Health Policy, National Medium-Term Development Policy Framework 2022–2025, and Education Strategic Plan 2018–2030. An objective of the revised National Health Policy, implemented in 2020, is to improve the physical environment; to achieve this, the policy outlines the need to improve access to potable water, sanitation, and hygiene and reduce the harmful effects of air, noise, and hazardous substances.[44] Implementation of this policy is expected to be undertaken by the Metropolitan, Municipal and District Assemblies (MMDAs) and nongovernmental organizations (NGOs) engaged in WASH programs.[44] Again, although implementation of this policy will certainly help to improve child health and survival in Ghana, it does not mention the children's environmental health impact of the policy. The National Medium-Term Development Policy Framework 2022–2025 does, however, outline specific initiatives to promote WASH practices among children.[45] Under the objective aimed at enhancing access to and improving sustainable environmental sanitation services, there is a renewed government commitment to scale-up sensitization campaigns to promote proper handwashing and hygiene practices, especially among children. The policy also seeks to leverage the health, education, and nutrition systems to promote optimal WASH practices and also accelerate implementation of the "Toilet for All" program across all MMDAs.[45] The Education Strategic Plan (ESP) 2018–2030 provides directives for promoting children's environmental health. Under the objective that seeks to improve quality of teaching and learning and science, technology, engineering, and mathematics (STEM) at all levels, the policy aims to improve the learning environment (including health and sanitation) and child protection in basic schools.[46] Also, in line with the National Medium-Term Development Policy Framework objective, which seeks to provide life skills training for managing personal hygiene, fire safety, environment, sanitation and climate change, the Ministry of Education, through the ESP, intends to develop and implement a comprehensive curriculum for basic and senior high schools to help meet the objective.[46] In 2020, the Government of Ghana together with UNICEF and Pure Earth developed a 3-year plan to mobilize action to prevent children's exposure to lead.[47]

Ghana has a School Health Education Programme (SHEP) with the goal of ensuring the provision of comprehensive health and nutrition education and related support services in schools to help equip children with basic life skills for healthy living and lead to improvement in child survival and educational outcomes. Among the several objectives of the program are the following that have linkage with children's environmental health: (a) inculcate into schoolchildren health-promoting habits and values of good hygiene and sanitation practices, including hand washing with soap; (b) promote the provision of adequate, safe, and sustainable water and sanitation facilities in schools to reinforce the practice of learned skills for hygiene; and (c) promote good environmental sanitation and hygiene practices in schools that are gender, child, and disability friendly.

In Nigeria, a 13-year strategic National Action Plan (NAP) for the revitalization of the country's water supply, sanitation, and hygiene sector was launched in 2018, with the aim of ensuring universal access to sustainable and safely managed WASH services by 2030. This policy action adopts a population-based approach to improve WASH practices within urban and rural areas.[48] There is, however, no specific mention of children's environmental health issues in the policy document, although the policy is expected to address such challenges. Nigeria's Nation Child Health Policy, which focuses on improving child health indicators such as neonatal mortality, infant mortality rate, under-5 mortality, and incidence of infectious diseases, also does not provide information on how environmental factors associated with these targets can be addressed.[49]

Sierra Leone has a National Policy on WASH and prioritizes the sanitation and hygiene concerns of vulnerable groups, particularly women and children. An objective of the policy is to include WASH issues in the curricula of primary and secondary schools and professional training courses, with implementation of the policy to be led by local Councils with support from government ministries, development partners, civil society, and NGOs.[50] The Liberian government implemented a Community-Led Total Sanitation policy in 2015, recognizing in the document the role that children play in facilitating behavior change in sanitation among their peers and families. This is informed by the policy encouraging the formation of Sanitation Clubs in schools.[51]

According to Lebbie et al.,[52] few African countries have specific legislation to control or manage the growing e-waste problem, and, in countries where such legislation exists it is often ineffective. These policies and legislation also are not targeted at addressing the health needs of those children who have been recognized as being heavily involved in e-waste recycling and are most vulnerable to pollutants from the activity. Policies and legislation implemented in Ghana and Nigeria, for instance, focus on transboundary importation of electronic waste, regulating the work of the recyclers, and revenue generation through taxation.[53,54] In both countries, bills have been passed that mandate producers and importers to register with appropriate authorities to facilitate the collection of tax for government revenue.[53,54] These bills also ban the indiscriminate burning of e-waste, but none of these policy directives emphasizes the protection of child health.[53,54] The WHO is, however, working with international experts and its network of collaborating centers on children's environmental health to compile existing research and knowledge on e-waste and child health to help build capacity within the health sector to protect children from e-waste exposure. As a part of this WHO initiative, pilot e-waste interventions are being implemented in Latin America with plans to commence similar pilot projects in Africa.[34]

Several countries in Africa have implemented LPG subsidization programs with the main objective of addressing socioeconomic inequalities and climate change, as evident in the Rural LPG promotion program implemented in Ghana in 2013 and the LPG scale-up rolled out in Cameroon in 2016.[55,56] Health is often regarded as a downstream benefit of LPG subsidization programs. The accrued health benefits from the policy are the reduction in HAP exposure, which predominantly affects women and children. Improved cookstoves have also been promoted in several African countries with the objective of combating deforestation but with benefits for child health through the curtailing of HAP exposure. The Ghana Action Plan for Clean Cooking, which was rolled out in 2013, seeks to facilitate the adoption of clean cooking solutions (LPG and efficient cookstoves) in low- and middle-income households in urban and rural areas.[57] This policy action also emphasizes the need to adopt measures to raise awareness at antenatal and postnatal clinics on the adverse health effects of HAP exposure.[57]

East Africa

Rwanda National Community Health Policy, which was developed in 2008, seeks to address, among other things, environmental health challenges, mainly by improving sanitation and combatting deforestation.[58] This policy mandates community health workers to facilitate clean up of the environment and to ensure that safe water is distributed to every family. The policy also advocates for the incorporation of community health education into school curricula to train schoolchildren to serve as key proponents of healthy practices at home.

To specifically address children's environmental health challenges, Rwanda and Kenya have rolled out child-centered policy interventions such as the School Health Policy, which aims at providing a safe environment for schoolchildren, one in which adequate water supplies, sanitation, and appropriate hygiene promotion is readily available to the entire school population.[59,60] The Kenyan Environmental Sanitation and Hygiene policy that was implemented in 2016 recognizes the need to give special attention to the promotion of proper sanitary conditions among vulnerable groups such as children, women, the elderly, and persons with disabilities. This policy emphasizes the need for schoolchildren to have a healthy learning environment that includes access to and use of clean, child-friendly environmental sanitation facilities, handwashing, and water supply, and the need for them to have the skills, knowledge, and attitudes to practice effective hygiene at school and home. Per the policy, the Ministry of Education is to include environmental health and hygiene education as a compulsory subject in the school curricula from nursery schools through primary and secondary schools to tertiary institutions, especially in teacher training institutions.[61]

North Africa

In Morocco, the government recognizes the effects of poor environmental conditions on children's health. A health impact assessment conducted in 2003 in the country found key pollutants such as nitrogen oxides (NOx), ozone (O_3), and sulfur dioxide (SO_2) to be associated with asthma attacks among children.[62] In addressing children's environmental health challenges, the country's National Health Policy seeks to facilitate hygiene and safety in schools through a number of health promotion approaches. They include the dissemination of health and safety standards to be observed at all levels of the education system, including universities, and the strengthening of a regulatory framework aimed at ensuring the cleanliness of school environments.[63]

In Egypt, considerable steps have been taken over the past decades to address environmental health problems. In 2000, the Healthy Egyptians 2010 Initiative was implemented to foster health promotion practices by identifying significant and preventable threats to health. Priority areas of this initiative included Environmental Health and Tobacco Control, Maternal and Child Health, and Injury Control.[64] In 2020, the government of Egypt in collaboration with the World Bank launched the Greater Cairo Air Pollution Management and Climate Change Project to help address issues of environmental pollution.[65] As part of this project, a solid waste educational center was to be established to promote improved waste handling, waste segregation at source, reuse, and recycling while also providing special educational materials and activities for children and adults.[66] The project also is expected to enhance public awareness of air pollution, lung health, and epidemics that target the respiratory system such as severe, acute respiratory syndrome (SARS), Middle Eastern respiratory syndrome (MERS), and COVID-19 especially among women, youth, and children.[66]

In Libya, environmental health policy initiatives have focused mainly on improving WASH practices among children. According to a UNICEF report, various interventions have focused on improving access to safe drinking water and sanitation services.[67] Sudan has a similar policy direction that is also concerned with ensuring safe water and sanitation services and the adoption of hygiene practices, as well as strengthening systems for a clean and safe environment for all children, women, girls, and boys.[68]

Conclusion

Children's environmental health challenges are mounting in Africa owing largely to the widespread use of solid fuels for cooking at home, poor WASH practices at home and school, and e-waste recycling. In spite of the worrying situation on the continent, very limited national policies target at children's environmental health in countries. These limited policies focus on improvement in WASH practices in schools and the community in an attempt to reduce the high burden of sanitation- and hygiene-related diseases among children, such as gastroenteritis, diarrheal diseases, and malaria, and they give very little attention to air pollution and other environmental exposures. To comprehensively address children's environmental health problems in Africa, it is important for countries to leverage the growing evidence, the majority of which derives from randomized controlled trials and multicountry cross-sectional studies, to develop effective policy interventions that target all environmental exposures of children at school, home, and the in outdoor environment. In most countries, UNICEF has traditionally led the development of policy initiatives targeted at improved WASH practices for children, and these programs need to be sensitized to incorporate other environmental exposures, especially HAP, in their work as well. The school-based interventions in Ghana, Rwanda, and Kenya can serve as models for improving children's environmental health in Africa. However, they need to target the full spectrum of environmental exposures, particularly air pollution.

References

1. Joubert BR, Mantooth SN, McAllister KA. Environmental health research in Africa: important progress and promising opportunities. *Front Genet*. 2019;10:1166. doi:10.3389/fgene.2019.01166.

2. Amegah AK. Improving child survival in sub-Saharan Africa: key environmental and nutritional interventions. *Ann Glob Health*. 2020;86(1):73. doi:10.5334/aogh.2908.

3. Rutstein SO. Factors associated with trends in infant and child mortality in developing countries during the 1990s. *Bull World Health Organ*. 2000;78(10):1256–70.

4. Bradshaw CJA, Otto SP, Mehrabi Z, Annamalay AA, Heft-Neal S, Wagner Z, Le Souef PN. Testing the socioeconomic and environmental determinants of better child-health outcomes in Africa: a cross-sectional study among nations. *BMJ Open*. 2019;9(9):e029968. doi:10.1136/bmjopen-2019-029968.

5. Fayehun OA. Household environmental health hazards and child survival in Sub-Saharan Africa. 2010. Available at: http://dhsprogram.com/pubs/pdf/WP74/WP74.pdf.

6. Tusting LS, Gething PW, Gibson HS, Greenwood B, Knudsen J, Lindsay SW, Bhatt S. Housing and child health in sub-Saharan Africa: a cross-sectional analysis. *PLoS Med*. 2020;17(3):e1003055. doi:10.1371/journal.pmed.1003055.

7. Amegah AK, Quansah R, Jaakkola JJ. Household air pollution from solid fuel use and risk of adverse pregnancy outcomes: a systematic review and meta-analysis of the empirical evidence. *PloS One.* 2014; 9(12):e113920. doi:10.1371/journal.pone.0113920.

8. Bruce NG, Dherani MK, Das JK, Balakrishnan K, Adair-Rohani H, Bhutta ZA, Pope D. Control of household air pollution for child survival: estimates for intervention impacts. *BMC Public Health.* 2013;13(Suppl 3):S8. doi:10.1186/1471-2458-13-S3-S8.

9. Momberg DJ, Ngandu BC, Voth-Gaeddert LE, Cardoso Ribeiro K, May J, Norris SA, Said-Mohamed R. Water, sanitation and hygiene (WASH) in sub-Saharan Africa and associations with undernutrition, and governance in children under five years of age: a systematic review. *J Dev Orig Health Dis.* 2021;12(1):6–33. doi:10.1017/S2040174419000898.

10. Odo DB, Yang IA, Knibbs LD. A systematic review and appraisal of epidemiological studies on household fuel use and its health effects using demographic and health surveys. *Int J Environ Res Public Health.* 2021;18(4):1411. doi:10.3390/ijerph18041411.

11. World Bank. *Environmental Health and Child Survival: Epidemiology, Economics, Experiences, Environment and Development.* Washington, DC: World Bank; 2008.

12. Akinyemi JO, Adedini SA, Wandera SO, Odimegwu CO. Independent and combined effects of maternal smoking and solid fuel on infant and child mortality in sub-Saharan Africa. *Trop Med Int Health.* 2016;21(12):1572–82. doi:10.1111/tmi.12779.

13. Anand A, Roy N. Transitioning toward Sustainable Development Goals: the role of household environment in influencing child health in sub-Saharan Africa and South Asia using recent demographic health surveys. *Front Public Health.* 2016;4:87. doi:10.3389/fpubh.2016.00087.

14. Headey D, Palloni G. Water, sanitation, and child health: evidence from subnational panel data in 59 countries. *Demography.* 2019;56(2):729–52. doi:10.1007/s13524-019-00760-y.

15. Rakotomanana H, Komakech JJ, Walters CN, Stoecker BJ. The WHO and UNICEF Joint Monitoring Programme (JMP) indicators for water supply, sanitation and hygiene and their association with linear growth in children 6 to 23 months in East Africa. *Int J Environ Res Public Health.* 2020;17(17):6262. doi:10.3390/ijerph17176262.

16. Saaka M, Saapiire FN, Dogoli RN. Independent and joint contribution of inappropriate complementary feeding and poor water, sanitation and hygiene (WASH) practices to stunted child growth. *J Nutr Sci.* 2021;10:e109. doi:10.1017/jns.2021.103.

17. Jack DW, Asante KP, Wylie BJ, et al. Ghana randomized air pollution and health study (GRAPHS): study protocol for a randomized controlled trial. *Trials.* 2015;16(1):420. doi:10.1186/s13063-015-0930-8.

18. Kinney PL, Asante KP, Lee AG, et al. Prenatal and postnatal household air pollution exposures and pneumonia risk: evidence from the Ghana randomized air pollution and health study. *Chest.* 2021;160(5):1634–44. doi:10.1016/j.chest.2021.06.080.

19. Lee AG, Kaali S, Quinn A, et al. Prenatal household air pollution is associated with impaired infant lung function with sex-specific effects. evidence from GRAPHS: a cluster randomized cookstove intervention trial. *Am J Respir Crit Care Med.* 2019;199(6):738–46. doi:10.1164/rccm.201804-0694OC.

20. Havens D, Wang D, Grigg J, Gordon SB, Balmes J, Mortimer K. The Cooking and Pneumonia Study (CAPS) in Malawi: a cross-sectional assessment of carbon monoxide exposure and carboxyhemoglobin levels in children under 5 years old. *Int J Environ Res Public Health.* 2018;15(9):1936. doi:10.3390/ijerph15091936.

21. Mortimer K, Ndamala CB, Naunje A, et al. A cleaner burning biomass-fuelled cookstove intervention to prevent pneumonia in children under 5 years old in rural Malawi (the Cooking and Pneumonia Study): a cluster randomised controlled trial. *Lancet*. 2017;389(10065):167–75. doi:10.1016/S0140-6736(16)32507-7.

22. Rylance S, Nightingale R, Naunje A, et al. Lung health and exposure to air pollution in Malawian children (CAPS): a cross-sectional study. *Thorax*. 2019;74(11):1070–7. doi:10.1136/thoraxjnl-2018-212945.

23. Dherani MK, Pope D, Tafatatha T, et al. Association between household air pollution and nasopharyngeal pneumococcal carriage in Malawian infants (MSCAPE): a nested, prospective, observational study. *Lancet Global Health*. 2022;10(2):e246–56. doi:10.1016/s2214-109x(21)00405-8.

24. Bornman MS, Chevrier J, Rauch S, et al. Dichlorodiphenyltrichloroethane exposure and anogenital distance in the Venda Health Examination of Mothers, Babies and their Environment (VHEMBE) birth cohort study, South Africa. *Andrology*. 2016;4(4):608–15. doi:10.1111/andr.12235.

25. Huq F, Obida M, Bornman R, Di Lenardo T, Chevrier J. Associations between prenatal exposure to DDT and DDE and allergy symptoms and diagnoses in the Venda Health Examination of Mothers, Babies and their Environment (VHEMBE): South Africa. *Environ Res*. 2020;185:109366. doi:10.1016/j.envres.2020.109366.

26. Huang J, Eskenazi B, Bornman R, Rauch S, Chevrier J. Maternal peripartum serum DDT/E and urinary pyrethroid metabolite concentrations and child infections at 2 years in the VHEMBE birth cohort. *Environ Health Perspect*. 2018;126(6):067006. doi:10.1289/EHP2657.

27. Eskenazi, B An S, Rauch SA, et al. Prenatal exposure to DDT and pyrethroids for malaria control and child neurodevelopment: the VHEMBE cohort, South Africa. *Environ Health Perspect*. 2018;126(4):047004. doi:10.1289/EHP2129.

28. Adane MM, Alene GD, Mereta ST, Wanyonyi KL. Effect of improved cookstove intervention on childhood acute lower respiratory infection in Northwest Ethiopia: a cluster-randomized controlled trial. *BMC Pediatr*. 2021;21(1):4. doi:10.1186/s12887-020-02459-1.

29. Dutta A, Alexander D, Karrison T, et al. Household air pollution, ultrasound measurement, fetal biometric parameters and intrauterine growth restriction. *Environ Health*. 2021;20(1):74. doi:10.1186/s12940-021-00756-5.

30. Gladstone MJ, Chandna J, Kandawasvika G, et al. Independent and combined effects of improved water, sanitation, and hygiene (WASH) and improved complementary feeding on early neurodevelopment among children born to HIV-negative mothers in rural Zimbabwe: substudy of a cluster-randomized trial. *PLoS Med*. 2019;16(3):e1002766. doi:10.1371/journal.pmed.1002766.

31. Pickering AJ, Njenga SM, Steinbaum L, et al. Effects of single and integrated water, sanitation, handwashing, and nutrition interventions on child soil-transmitted helminth and Giardia infections: a cluster-randomized controlled trial in rural Kenya. *PLoS Med*. 2019;16(6):e1002841. doi:10.1371/journal.pmed.1002841.

32. Mahmud MA, Spigt M, Bezabih AM, Dinant GJ, Velasco RB. Associations between intestinal parasitic infections, anaemia, and diarrhoea among school aged children, and the impact of hand-washing and nail clipping. *BMC Res Notes*. 2020;13(1):1. doi:10.1186/s13104-019-4871-2.

33. Mahmud MA, Spigt M, Bezabih AM, Pavon IL, Dinant GJ, Velasco, RB. Efficacy of hand-washing with soap and nail clipping on intestinal parasitic infections in school-aged

children: a factorial cluster randomized controlled trial. *PLoS Med.* 2015;12(6):e1001837. doi:10.1371/journal.pmed.1001837.

34. WHO. *Children and Digital Dumpsites: E-Waste Exposure and Child Health.* Geneva: World Health Organization; 2021.

35. Amoyaw-Osei Y, Agyekum OO, Pawmang JA, Mueller E, Fasko R, Schluep M. Ghana e-waste country assessment. SBC e-waste Africa Project. 2011 Mar:1–123. Available at: https://www.basel.int/Portals/4/Basel%20Convention/docs/eWaste/E-wasteAssessmentGhana.pdf

36. Ogungbuyi O, Nnorom IC, Osibanjo O, Schluep M. e-Waste Country Assessment Nigeria. Swiss Federal Laboratories for Materials Science and Technology (Empa). 2012 May:1–97. Available at: http://www.basel.int/Portals/4/Basel%20Convention/docs/eWaste/EwasteAfrica_Nigeria-Assessment.pdf

37. Prakash S, Manhart A, Amoyaw-Osie Y, Agyekum OO. Socio-economic assessment and feasibility study on sustainable e-waste management in Ghana. Commissioned by Inspectorate of the Ministry of Housing, Spatial Planning and the Environment of the Netherlands and the Dutch Association for the Disposal of Metal and Electrical Products. 2010. Available at: https://www.oeko.de/oekodoc/1057/2010-105-en.pdf.

38. World Health Organization Healthy Environments for Children Alliance (HECA). HECA side event at the fourth Ministerial Meeting on Environment And Health. Budapest. 2004. Available at: https://www.who.int/heca/infomaterials/budapest.pdf

39. Mathee A. Environment and health in South Africa: gains, losses, and opportunities. *J Public Health Policy.* 2011;32(Suppl 1):S37–43. doi:10.1057/jphp.2011.21.

40. SAMRC. Children's Environmental Health Educational Toolkit. 2014. Available at: https://www.samrc.ac.za/other/children%E2%80%99s-environmental-health-educational-toolkit.

41. WHO. Public health and environment in the African region: Report on the work of WHO, 2008–2009. 2010. Available at: https://www.afro.who.int/sites/default/files/2017-06/phe2008-2009-fin.pdf.

42. WaterAid. Regional state of hygiene – West Africa. 2021. Available at: https://washmatters.wateraid.org/publications/regional-state-of-hygiene-west-africa

43. The Republic of Niger. Economic and Social Development Plan (ESDP) 2017–2021: a resurgent Niger for a prosperous people. 2017. Available at: https://www.nigerrenaissant.org/sites/default/files/pdf/pdes-summary.pdf.

44. Ministry of Health. National Health Policy: Ensuring healthy lives for all. Ministry of Health, Republic of Ghana; 2020. Available at: https://www.moh.gov.gh/wp-content/uploads/2020/07/NHP_12.07.2020.pdf-13072020-FINAL.pdf

45. National Development Planning Commission. National Medium-Term Development Policy Framework 2022–2025. National Development Planning Commission, Republic of Ghana; 2021: 1–343. Available at: https://ndpc.gov.gh/media/MTNDPF_2022-2025_Dec-2021.pdf

46. Ministry of Education. *Education Strategic Plan 2018–2030.* Ministry of Education, Republic of Ghana; 2018. Available at: https://www.globalpartnership.org/sites/default/files/2019-05-education-strategic-plan-2018-2030.pdf

47. UNICEF. Ghana takes significant strides to protect every child's potential. 2021a. Available at: https://www.unicef.org/ghana/press-releases/ghana-takes-significant-strides-protect-every-childs-potential

48. Federal Ministry of Water Resources. National Action Plan for Revitalization of the Nigeria's WASH Sector. Federal Ministry of Water Resources, Abuja- Nigeria; 2021a. Available at: https://www.wateraid.org/ng/sites/g/files/jkxoof381/files/nigerias-national-action-plan-for-the-revitalization-of-the-wash-sector.pdf

49. Federal Ministry of Health. Draft of National Child Health Policy. Federal Ministry of Health, Abuja-Nigeria; 2006. Available at: http://www.policyproject.com/pubs/countryreports/NIG_RHStrat.pdf

50. Ministry of Energy and Water Resources. *The National Water and Sanitation Policy*. Sierra Leone: Ministry of Energy and Water Resources; 2010. Available at: http://interaide.org/watsan/sl/wp-content/uploads/2015/07/National-WASH-Policy-Final-2010.pdf

51. National Technical Coordinating Unit. *Guidelines for Community-Led Total Sanitation-Implementation in Liberia*. Sierra Leone: National Technical Coordinating Unit; 2015. Available at: https://wash-liberia.org/wp-content/uploads/sites/54/2013/06/Final_CLTS_guidlines2.pdf

52. Lebbie TS, Moyebi OD, Asante KA, et al. E-waste in Africa: a serious threat to the health of children. *Int J Environ Res Public Health*. 2021;18(16):8488. doi:10.3390/ijerph18168488.

53. Amachree M. Update on e-waste management in Nigeria. San Franciso. 2013. Available at: https://www.epa.gov/sites/default/files/2014-05/documents/nigeria.pdf

54. Pwamang JA. Government policy and initiatives on e-waste in Ghana. California; 2013. Available at: https://www.epa.gov/sites/default/files/2014-05/documents/ghana_2.pdf

55. Asante KP, Afari-Asiedu S, Abdulai MA, et al. Ghana's rural liquefied petroleum gas program scale up: a case study. *Energy Sustain Dev*. 2018;46:94–102. doi:10.1016/j.esd.2018.06.010.

56. Bruce N, de Cuevas, RA, Cooper J, et al. The government-led initiative for LPG scale-up in Cameroon: programme development and initial evaluation. *Energy Sustain Dev*. 2018;46:103–10. doi:10.1016/j.esd.2018.05.010.

57. Ghana Energy Commission, Global Alliance for Clean Cook Stoves. Ghana country action plan for clean cooking. 2013: 1–23. Available at: https://cleancooking.org/wp-content/uploads/2021/07/334-1.pdf

58. Ministry of Health. *National Community Health Policy*. Ministry of Health, Republic of Rwanda; 2008. Available at: https://www.advancingpartners.org/sites/default/files/projects/policies/chp_rwanda_2008.pdf

59. Ministry of Health, Ministry of Education. *Kenya: School Health Policy*. Nairobi, Kenya: Ministry of Health; 2018. Available at: https://ncdak.org/wp-content/uploads/2021/08/School-Health-Policy-DFH-MOH-26.06.18.pdf

60. Ministry of Education. *National School Health Policy*. Kigali: Ministry of Education, Kigali-Rwanda; 2014. Available at: https://healtheducationresources.unesco.org/sites/default/files/resources/rwanda_school_health_policy.pdf

61. Ministry of Health. *Kenya Environmental Sanitation and Hygiene Policy*. 2016–2030. Ministry of Health, Republic of Kenya; 2016. Available at: https://faolex.fao.org/docs/pdf/ken179039.pdf

62. UN. *Morocco Environmental Performance Reviews*. New York: United Nations; 2016: 4–4. Available at: https://unece.org/sites/default/files/2021-08/ECE_CEP_170_En.pdf

63. Ministère de la Santé. Stratégie Sectorielle de Santé 2012–2016. Ministère de la Santé, Morocco; 2012. Available at: https://www.sante.gov.ma/Docs/Documents/secteur%20santé.pdf

64. Anwar WA. Environmental health in Egypt. *Int J Hygiene Environ Health*. 2003;206(4):339–50. doi:https://doi.org/10.1078/1438-4639-00230.

65. World Bank. Supporting pollution reduction efforts to protect the health of Egyptians and spur economic recovery. World Bank. 2021. Available at: https://www.worldbank.org/en/

news/feature/2021/04/22/supporting-pollution-reduction-efforts-to-protect-the-health-of-egyptians.

66. World Bank. Greater Cairo air pollution management and climate change project. World Bank. 2020. Available at: https://documents1.worldbank.org/curated/en/78869160177 6867710/pdf/Egypt-Greater-Cairo-Air-Pollution-Management-and-Climate-Change-Proj ect.pdf.

67. UNICEF. Humanitarian end-year situation report January–December 2020. 2021b. Available at: https://www.unicef.org/media/94231/file/Libya-Humanitarian-Situation-Rep ort-End-of-Year-2020.pdf.

68. Federal Ministry of Health. Sudan National Sanitation and Hygiene Strategic Framework. Federal Ministry of Health Khartum; 2016. Available at: https://www.unicef.org/sudan/ media/1026/file/National-Sanitation-Hygiene-Strategic-Framework-2016.pdf

62

Global Treaties and Children's Environmental Health

Ruth A. Etzel

Since the 1980s, the world has become ever more closely interconnected. Economics, politics, and culture are now highly globalized. International trade has grown exponentially. Chemical contamination that affects fundamental natural resources such as water, air, and food is widespread and moves without hindrance across national borders. The global commons—the climate, the ozone layer, and biodiversity—are changing rapidly.

Children today are exposed to multiple environmental hazards that are global in their scope and transcend national boundaries. Three examples are climate change[1]; hazardous chemicals such as mercury that cross borders and contaminate fish worldwide[2]; and depletion of stratospheric ozone that affects children's immune response and increases risk of skin cancer.[3]

To confront the global nature of these hazards, international regulatory mechanisms are needed that involve all countries and allow for consistent global control. In the absence of strong and legally binding international mechanisms, hazardous products can cross borders without hindrance and toxic materials that are forbidden in one country can be exported to other countries that have weaker regulations. The consequences are inequitable environmental degradation and injury to the health of children, especially those in poorer countries.

International conventions are legally binding treaties that the global community has put in place to support coherent, shared regulation of issues that transcend national boundaries, such as arms control, trade, and the environment. This chapter examines a select number of international conventions that are intended to protect human health and the environment and that are designed especially to safeguard the health and well-being of children.

Convention on the Rights of the Child

Over the past century, a number of international treaties and documents have been developed to protect the rights of children. In 1924, the League of Nations adopted the Geneva Declaration of the Rights of the Child. The United Nations took its first step toward declaring the importance of child rights by establishing the United Nations International Children's Emergency Fund in 1946. (The name was shortened to United Nations Children's Fund in 1953 but kept the popular acronym UNICEF.) Two years later, in 1948, the UN General Assembly adopted the Universal Declaration of Human Rights, making it the first UN document to recognize children's need for protection.[4]

Ruth A. Etzel, *Global Treaties and Children's Environmental Health* In: *Textbook of Children's Environmental Health*, Second Edition. Edited by: Ruth A. Etzel and Philip J. Landrigan, Oxford University Press. © Oxford University Press 2024.
DOI: 10.1093/oso/9780197662526.003.0062

It was not until 1989, however, that the global community adopted the United Nations Convention on the Rights of the Child, making it the first international legally binding document concerning child rights. The convention consists of 54 articles covering all four major categories of child rights: right to life, right to development, right to protection, and right to participation.[5]

The Convention on the Rights of the Child became legally binding on September 2, 1990, after 20 countries had ratified it. A number of countries ratified this Convention very soon after it was adopted and others continued to ratify or accede to it in subsequent years, making it the most widely ratified treaty on human rights. Today nearly all 193 member states of the United Nations are parties to the Convention. Somalia and the United States have not yet ratified the Convention but have signed it, indicating their support.

The Convention provides a framework for the defense and promotion of the rights of children around the world, irrespective of race, gender, religion, or social status. It is a legally binding instrument. It defines the responsibilities of governments toward children and also defines the duties of other actors. It contains an ethical statement on core human values.[6]

The United Nations Committee on the Rights of the Child is the body that monitors how well states are meeting their obligations under the Convention on the Rights of the Child. When a country ratifies the Convention, it assumes a legal obligation to implement the rights recognized in the treaty. But signing up is only the first step, because recognition of rights on paper is not sufficient to guarantee that they will be protected in practice. Each country that ratifies the Convention thus incurs an additional obligation to submit regular reports to the Committee on how children's rights are being implemented. This system of human rights monitoring is common to all UN human rights treaties.

To meet their reporting obligation, countries must submit their first report 2 years after joining the Convention and then every 5 years thereafter. In addition to the government report, the Committee receives information on a country's human rights situation from other sources, including nongovernmental organizations, UN agencies, other intergovernmental organizations, academic institutions, and the press. In the light of all the information available, the Committee examines the report together with government representatives. Based on this dialogue, the Committee publishes its concerns and recommendations, referred to as Concluding Observations.

Convention on Long-Range Transboundary Air Pollution

In 1979, 32 countries signed the United Nations Economic Commission for Europe (UNECE) Convention on Long-Range Transboundary Air Pollution, the first international treaty to deal with air pollution. The Convention was designed to protect humans and their environment against air pollution and to limit, reduce, and prevent air pollution including long-range transboundary air pollution.[7] The Convention and its protocols cover numerous pollutants including sulfur dioxide, ground-level ozone precursors—nitrogen oxides and volatile organic compounds—ammonia, persistent organic pollutants, heavy metals, and particulate matter, including black carbon. Fifty-one

UNECE member states are now parties to the Convention.[8] The emissions of sulfur have been reduced by 70% and nitrogen oxides have been reduced by 40% since 1990.[9]

Montreal Protocol on Substances That Deplete the Stratospheric Ozone Layer

In the early 1970s, two young chemists, Mario Molina of Mexico and F. Sherwood Rowland of the United States, theorized that chlorofluorocarbon (CFC) compounds released into the environment from refrigerators and air conditioners would evaporate, rise to the stratosphere, and break down there under the influence of sunlight to release free radicals that destroy stratospheric ozone.[10]

The Stratospheric Ozone Layer

The stratospheric ozone layer is essential for life on Earth. It protects the Earth's surface against solar ultraviolet radiation. The health effects of stratospheric ozone depletion range from malignant melanoma and nonmelanocytic skin cancer to effects on the eye, including photoconjunctivitis and cataract[11] (see also Chapter 44). Children as a group are particularly vulnerable to ultraviolet radiation.[3]

Molina and Rowland's research, first published in 1974,[10] initiated a federal investigation in the United States. Their findings were confirmed by the US National Academy of Sciences in 1976. In 1978, CFC-based aerosols were banned in the United States, Norway, Sweden, and Canada. Further validation of Molina and Rowland's work came in 1985, with the discovery by the British Antarctic Survey of a "hole" in the ozone shield over Antarctica.[11] This hole was recognized to have been caused by CFC compounds.

In 1985, just as the findings on the "ozone hole" were about to be published, representatives from 28 countries met at the Vienna Convention for the Protection of the Ozone Layer. The Montreal Protocol on Substances That Deplete the Ozone Layer[12] is a protocol to the Vienna Convention for the Protection of the Ozone Layer. It was signed in 1987 by 24 parties and went into effect in 1989. In subsequent years, it has been ratified by 197 parties (all 193 UN Member states as well as Niue, the Cook Islands, the Holy See, and the European Union), making it the most widely ratified treaty in UN history. The initial agreement was designed to reduce the production and consumption of several types of CFCs and halons to 80% of 1986 levels by 1994 and to 50% of 1986 levels by 1999. Since then the agreement has been amended to further reduce and completely phase-out CFCs and halons, as well as the manufacture and use of carbon tetrachloride, trichloroethane, hydrofluorocarbons (HFCs), hydrochlorofluorocarbons (HCFCs), hydrobromofluorocarbons (HBFCs), methyl bromide, and other ozone-depleting compounds. These changes were included in five Amendments—London 1990, Copenhagen 1992, Vienna 1995, Montreal 1997, and Beijing 1999. The Montreal Protocol sets out a mandatory timetable for the phase-out of ozone-depleting substances. This timetable has been regularly reviewed, with phase-out dates accelerated in accordance with scientific understanding and technological advances.

The Montreal Protocol has, to date, enabled reductions of more than 97% of all global consumption of controlled ozone-depleting substances.[13] The Montreal

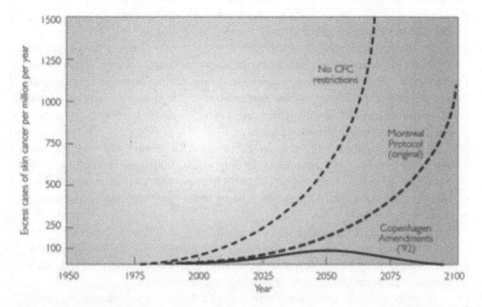

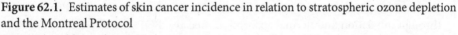

Figure 62.1. Estimates of skin cancer incidence in relation to stratospheric ozone depletion and the Montreal Protocol

Source: Adapted from reference 13.

Protocol is widely considered the most successful environmental protection agreement. Figure 62.1 summarizes the impact of the Montreal Protocol and its amendments on the frequency of skin cancer,[14] confirming that the implementation of this protocol is a significant success story for public health and the international community. In recognition of the importance of the discovery of stratospheric ozone depletion by CFCs, in 1995, Drs. Molina and Rowland were awarded the Nobel Prize in Chemistry.

To support implementation of the Montreal Protocol in all countries, a Multilateral Fund, the first financial mechanism to be created under an international treaty, was set up in 1990, under the Protocol. This Fund provides financial assistance to developing countries to help them achieve their phase-out obligations. The Fund has provided more than US$2.5 billion in financial assistance to developing countries to phase out production and consumption of ozone-depleting substances since the Protocol's inception in 1987.

The Montreal Protocol also has produced other significant environmental benefits. Most notably, the phase-out of ozone-depleting substances is responsible for delaying climate forcing by up to 12 years.

Stockholm Convention on Persistent Organic Pollutants

The Stockholm Convention on Persistent Organic Pollutants (POPs)[15] is a global treaty designed to protect human health and the environment from highly dangerous, long-lasting chemical pollutants by restricting and ultimately eliminating their production, use, trade, release, and storage.

Persistent Organic Pollutants

POPs are organic (carbon-based) chemicals, most based on a carbon-halogen bond, that persist in living organisms and in the environment for long periods of time—years, decades, and even centuries. POPs include pesticides, industrial chemicals, chemicals used in consumer products, and by-products of certain manufacturing and combustion processes.

POPs were produced in vast quantities in the 20th century and were used for beneficial purposes such as increasing crop yields, killing unwanted pests and other vectors, and thermal insulation. POPs have low solubility in water and high solubility in fat and thus accumulate in fatty tissues. They biomagnify in the food chain.[16]

POPs are semivolatile and thus can move long distances in the atmosphere, cross national boundaries, and cause widespread environmental contamination and adverse effects on health and on ecosystems even in regions of the Earth, such as the Antarctic and high Arctic, where they were never used.

The principal route of human exposure to POPs is through the ingestion of contaminated food such as fish, meat, and dairy products. Exposure can also occur through inhalation and dermal absorption. Because POPs are ubiquitous, exposure starts before conception.

Concerns for the effects of POPs on children's health include the possibility of effects on sperm and ova before children are conceived and effects resulting from pregnancy when maternal fat stores are mobilized, leading to exposure of the embryo and then to the fetus through the placenta. Postnatal exposure occurs via breast milk. Exposure to POPs during early life stages may result in effects not only in utero and in childhood but also at later stages, after a latency period, during adolescence or adulthood. The timing of exposure and whether it occurs during a developmental "critical window of vulnerability" is considered a crucial factor in determining the nature of the health effect (see Chapter 2).

The health effects of POPs are discussed in Chapter 36.

The Stockholm Convention on POPs was signed by 92 countries plus the European Commission in 2001, and it entered into force in 2004. Through subsequent signatures and ratifications, the Convention now has 184 parties, including the European Union. The Convention initially included 12 substances: aldrin, chlordane, DDT, dieldrin, endrin, heptachlor, hexachlorobenzene, mirex, polychlorinated biphenyls (PCBs), polychlorinated dibenzo-*p*-dioxins, polychlorinated dibenzofurans, and toxaphene.

In 2009, nine additional substances were added to the Convention through amendments to the original text—alpha hexachlorocyclohexane, beta hexachlorocyclohexane, chlordecone, commercial octabromodiphenyl ether (hexabromodiphenyl ether and heptabromodiphenyl ether), commercial pentabromodiphenyl ether (tetrabromodiphenyl ether and pentabromodiphenyl ether), hexabromobiphenyl, lindane, pentachlorobenzene, perfluorooctane sulfonic acid (PFOS) and its salts, and perfluorooctane sulfonyl fluoride (PFOS-F). These amendments entered into force in 2010.

To achieve the objectives of the Stockholm Convention, all countries that have ratified this treaty have committed to undertake a number of activities to control and reduce exposure to POPs. These activities include strengthening national capacities to address POP-contaminated wastes, including PCBs; evaluating the continued need to use DDT for malaria vector control and promoting the development and deployment of alternatives to DDT; and encouraging reduction of unintentional release of POPs into the environment from anthropogenic activities, such as incinerators and open burning.

Global progress has been made in the 23 years since the signing of the Stockholm Convention.[17] Achievements include developing regional and global monitoring reports assessing baseline levels of POPs in ambient air, human milk, and blood for comparative purposes in future evaluations and the establishment of the Global Alliance for the development and deployment of products, methods, and strategies as alternatives to DDT for disease vector control.

Basel Convention on the Control of Transboundary Movements of Hazardous Wastes and Their Disposal

The Basel Convention on the Control of Transboundary Movements of Hazardous Wastes and Their Disposal[18] was adopted in 1989. The overarching objective of the Convention is to protect human health and the environment against the adverse effects of hazardous wastes. The Convention covers a wide range of "hazardous wastes" as well as household waste and incinerator ash.

The Basel Convention was adopted in response to the international outcry that followed the discovery in the 1980s that toxic wastes from industrialized countries were being dumped in low- and middle-income countries (LMICs) in Africa, Eastern Europe, and other parts of the world. This disposal of hazardous wastes was an unintended consequence of the tightening of environmental regulations that had occurred in the 1970s and 1980s in the industrialized world and that had led to an escalation of disposal costs. This escalation in costs led some operators to seek cheap disposal options in LMICs where environmental awareness was less developed, and regulations and enforcement mechanisms were weak or nonexistent.

It was against this background and to combat the "toxic trade," as it was then termed, that the Basel Convention was negotiated in the late 1980s. The Convention entered into force in 1992. At present, 179 countries are parties to the Basel Convention, and significant progress has been achieved in controlling the global movement of hazardous wastes.[19]

The provisions of the Basel Convention center around the following three principal aims:

1. Reduce the generation of hazardous wastes in countries around the world and promote the environmentally sound management of hazardous wastes, wherever they are disposed
2. Restrict transboundary movements of hazardous wastes except where such movement is perceived to be in accordance with the principles of environmentally sound waste management
3. Create a regulatory system that governs transboundary movements of hazardous wastes in cases where such movements are permissible.

The first of the three aims of the Basel Convention is addressed through a number of general provisions requiring member states to observe the fundamental principles of environmentally sound waste management (Article 4).

The second aim is addressed through a series of prohibitions. For example, hazardous wastes may not be exported to Antarctica, to a member state not party to the Basel Convention, or to a party that has banned the import of hazardous wastes (Article 4). In all cases, transboundary movement of hazardous wastes may take place only if it represents an environmentally sound solution, only if the principles of environmentally sound management and nondiscrimination are observed, and only if it is carried out in accordance with the Convention's regulatory system.

The third aim, the regulatory system, is the cornerstone of the Basel Convention as originally adopted. The regulatory system is based on the concept of prior informed consent. It requires that, before an export of hazardous waste may take place, the authorities of the exporting state must notify the authorities of the prospective importing states, providing them with detailed information on the intended movement of hazardous waste. The movement may proceed only if and when all states concerned have given their written consent (Articles 6 and 7).

Additionally, the Basel Convention provides for cooperation between parties, ranging from exchange of information on issues relevant to the implementation of the Convention to technical assistance, particularly to developing countries (Articles 10 and 13). If a transboundary movement of hazardous wastes is carried out illegally—that is, in contravention of the provisions of Articles 6 and 7—or cannot be completed as foreseen, the Convention assigns responsibility to one or more of the states involved and imposes a duty to these states to ensure safe disposal, either by reimport into the state of generation or otherwise (Articles 8 and 9).

The Basel Convention provides for the establishment of regional or subregional centers for training and technology transfers regarding the management of hazardous wastes and other wastes and the minimization of their generation to cater to the specific needs of different regions and subregions (Article 14). Fourteen such regional centers have been established around the world. They carry out training and capacity-building activities.

The Basel Convention regulates the export and management of hazardous wastes containing lead. Lead exposure continues today to result from inappropriate waste management practices and can be especially hazardous for children.[20] Lead in the hazardous waste stream can originate from industrial activities. It can also derive, particularly in LMICs, from the informal recovery of lead from car batteries, which are often illegally exported from industrialized countries, and from the open burning or incorrect incineration of lead-containing waste. Deaths and cases of severe mental retardation among children in LMICs have resulted from the informal recycling of lead from imported car batteries.[21]

Rotterdam Convention on the Prior Informed Consent Procedure for Certain Hazardous Chemicals and Pesticides in International Trade

In 1998, the world's governments adopted an international treaty called the Rotterdam Convention on the Prior Informed Consent Procedure for Certain Hazardous Chemicals and Pesticides in International Trade.[22]

The objective of the Rotterdam Convention is to promote shared responsibility and cooperative efforts between exporting and importing countries for managing chemicals that pose significant risks to human health and the environment. The Convention also seeks to encourage the environmentally sound management of hazardous chemicals when their use is permitted and to provide and share accurate information on their characteristics, potential dangers, and safe handling and use. At present, 150 countries are parties to the Convention.

The principle of prior informed consent (PIC) is the key to the Rotterdam Convention. PIC is a procedure intended specifically to help developing countries, many of which can find it difficult to monitor and control hazardous imports. If a member country decides not to import a particular chemical or pesticide covered by the PIC procedure, other member states agree under the Rotterdam Convention not to export this product to that country. If a country decides to limit the import of a chemical to certain uses, exporting countries must agree to respect those limits. PIC promises not only to reduce accidents but also to prevent the accumulation of stocks of obsolete or unwanted pesticides and other toxic substances.

Five industrial chemicals and 22 pesticides were included on the initial list of chemicals covered by the PIC procedure when the Rotterdam Convention was adopted. Others have subsequently been added, and the current PIC list can be expected to expand on a regular basis in the years to come. The Rotterdam Convention's PIC list is not a "black list," but rather a "watch list" of industrial chemicals, pesticides, and "severely hazardous pesticide formulations" (which contain a specific percentage of one or more particular active ingredients) whose use should be carefully weighed and whose import needs to be carefully considered.

The chemicals covered by the Rotterdam convention are listed in Box 62.1. A number of these chemicals are carcinogens, neurotoxicants, and endocrine disruptors.[23]

The Rotterdam Convention enables member countries to alert each other to potential dangers. Whenever a member government anywhere in the world takes an action to ban or severely restrict any chemical for health or environmental reasons, this action is reported through the "PIC circular" that the Convention Secretariat distributes to all member countries every 6 months. By ensuring that information is exchanged in this way, the Convention provides an initial warning to governments that a particular chemical may merit a second look.

The Rotterdam Convention is designed also to respond to emerging risks. If two countries from two different regions of the world decide to ban or severely restrict a particular chemical, this chemical automatically becomes a candidate for being placed on the list of substances covered by the Convention's PIC procedure and is referred to the Convention's Chemical Review Committee for investigation.

The Rotterdam Convention's Chemical Review Committee, a panel of experts, is responsible for scrutinizing the potential dangers of the chemicals brought to its attention. Countries that lack expertise thus have a watchdog group on their side. The Committee may recommend adding a chemical to the list covered by the PIC procedure, and the Conference of Parties may then add the chemical to the list.

The Rotterdam Convention calls on member countries to exchange scientific, technical, economic, and legal information concerning chemicals within the scope of the treaty, thus enabling nations with less extensive knowledge to benefit from more advanced techniques for monitoring and analysis.

Box 62.1 List of Chemicals Subject to the Prior Informed Consent Procedure

Hazardous Pesticides
2,4,5-T
Aldrin
Captafol
Chlordane
Chlordimeform
Chlorobenzilate
DDT
Dieldrin
Dinoseb and dinoseb salts
1,2-dibromoethane (EDB)
Fluoroacetamide
HCH (mixed isomers)
Heptachlor
Hexachlorobenzene
Lindane
Mercury compounds, including inorganic mercury compounds, alkyl mercury
   compounds, and alkyloxyalkyl and aryl mercury compounds
Pentachlorophenol

Severely Hazardous Pesticide Formulations
Monocrotophos (soluble liquid formulations of the substance that exceed 600 g active ingredient/L)
Methamidophos (soluble liquid formulations of the substance that exceed 600 g active ingredient/L)
Phosphamidon (soluble liquid formulations of the substance that exceed 1,000 g active ingredient/L)
Methyl-parathion (emulsifiable concentrates [ECs] with 19.5%, 40%, 50%, 60% active ingredient and dusts containing 1.5%, 2%, and 3% active ingredient)
Parathion (all formulations—aerosols, dustable powder [DP], emulsifiable concentrate [EC],
granules [GRs], and wettable powders [WPs]—of this substance are included, except
   capsule suspensions [CSs])

Industrial Chemicals
Crocidolite asbestos
Polybrominated biphenyls (PBBs)
Polychlorinated biphenyls (PCBs)
Polychlorinated terphenyls (PCTs)
Tris (2,3-dibromopropyl) phosphate

The Rotterdam Convention obliges member countries to ensure that their citizens have access to information on hazardous chemicals and pesticides. Since there are many chemicals and pesticides, and new products are developed every day, farmers in LMICs need to know if the materials that they are using are dangerous. They can benefit greatly from being told that affordable alternatives may be available.

The Synergies Process

The Basel, Rotterdam, and Stockholm Conventions described here are multilateral agreements that share the common objective of protecting human health and the environment from hazardous chemicals and wastes. Through a complex process started in 2008, the respective Conferences of the Parties of the three conventions have taken a series of decisions to enhance cooperation and coordination among the conventions. This process, termed the "synergies process" aims at strengthening the implementation of the three conventions at the national, regional, and global levels by providing coherent policy guidance, enhancing efficiency in the provision of support to parties to the conventions, reducing their administrative burden, and maximizing the effective and efficient use of resources at all levels while maintaining the legal autonomy of these three multilateral environmental agreements. An extraordinary simultaneous conference of the parties took place in Bali, Indonesia in 2011, to start this process. It will eventually lead to a coherent and common global administrative and operational structure.

The synergies process is changing and streamlining the way in which implementation of the conventions is undertaken. It constitutes an important example within the UN system for improving coherence and cost-effectiveness of international conventions at the national and regional levels.

Framework Convention on Tobacco Control

In 2003, the World Health Assembly adopted the Framework Convention on Tobacco Control, the first international treaty negotiated under the auspices of the World Health Organization (WHO).[24] It entered into force in 2005. It was developed in response to the globalization of the tobacco epidemic. The Convention provides new legal dimensions for international health cooperation on tobacco control.

The "Minamata" Convention on Mercury

In 2013, representatives of countries from around the world meeting in Geneva, Switzerland agreed to a global, legally binding treaty to prevent emissions and environmental releases of mercury. This Convention is known as the "Minamata" Convention on Mercury—named after the city in Japan where serious health damage occurred as a result of mercury pollution in the mid-20th century (see Chapter 1). Agreement on the treaty followed extensive analysis of evidence and a series of high-level intergovernmental negotiations involving more than 140 countries.

Mercury is recognized as a chemical of global concern because of its ability to travel long distances in the atmosphere; its persistence in the environment; its ability to accumulate in ecosystems, especially in fish; and its significant negative effect on human health and the environment.

Mercury can produce a range of adverse human health effects, including permanent damage to the developing nervous system (see Chapter 32).[2] Because of these effects—and also because mercury can be passed from a mother to an unborn child—infants, children, and women of childbearing age are considered vulnerable groups.

The Minamata Convention mandates controls and reductions on mercury emissions across a range of products, processes, and industries where mercury is used, released, or emitted. These include emissions from coal-fired power plants and from industry, as well as restrictions on the use of mercury in artisanal and small-scale gold mining. The treaty also includes an article specifically dedicated to healthcare, and it sets a phase-out date of 2020 for all mercury-containing thermometers and blood pressure measuring devices used in healthcare.

The Minamata Convention also foresees a "phasing-down" in the use of dental amalgam (a compound of mercury and silver-based alloys). This action will contribute to a reduction of mercury use and the risk of release to the environment.

An important exception to the Minamata Convention that was strongly supported in the negotiations by the public health community was the use of thiomersal (ethylmercury) as a preservative in human and animal vaccines. This followed the recommendation of the WHO Global Advisory Committee on Vaccine Safety, which concluded that "no additional studies of the safety of thiomersal in vaccines are warranted and that available evidence strongly supports the safety of the use of thiomersal as a preservative for inactivated vaccines."[25]

United Nations Framework Convention on Climate Change

In 1992, representatives from 154 nations meeting at the Earth Summit in Rio de Janeiro, Brazil, agreed to a Framework Convention on Climate Change. The objective of this treaty is the stabilization of greenhouse gas concentrations in the atmosphere at a level that would prevent dangerous anthropogenic interference with the climate system. The leaders stated that "Such a level should be achieved within a time frame sufficient to allow ecosystems to adapt naturally to climate change, to ensure that food production is not threatened and to enable economic development to proceed in a sustainable manner."[26]

It was many years before any significant progress was made in addressing the aims of the Framework Convention on Climate Change. In 2015, at the 21st UN Climate Change Conference in Paris, an agreement was reached (the Paris Agreement).[27] The Paris Agreement includes the following long-term goals to guide all nations:

- substantially reduce global greenhouse gas emissions to limit the global temperature increase in this century to 2°C while pursuing efforts to limit the increase even further to 1.5°C;
- review countries' commitments every 5 years;
- provide financing to developing countries to mitigate climate change, strengthen resilience, and enhance abilities to adapt to climate impacts.

The Agreement is a legally binding international treaty. It entered into force on November 4, 2016. Today, 192 Parties (191 countries plus the European Union) joined the Paris Agreement.

The Agreement includes commitments from all countries to reduce their emissions and work together to adapt to the impacts of climate change, and it calls on countries to strengthen their commitments over time. The Agreement provides a pathway for developed nations to assist developing nations in their climate mitigation and adaptation efforts while creating a framework for the transparent monitoring and reporting of countries' climate goals.

In 2022, at the 27th UN Climate Change Conference in Egypt, governments agreed to establish new funding arrangements as well as a dedicated fund to assist developing countries in responding to loss and damage associated with the adverse effects of climate change.[28] In 2023, at the 28th UN Climate Conference in the United Arab Emirates, governments agreed on the need to speed up the transition away from fossil fuels to renewables such as wind and solar power in their next round of climate commitments.[29]

Development of Future Conventions

Scientists from around the world have called for a treaty on plastic, which causes disease, disability, and premature death at every stage of its life cycle.[30] At the 2022 meeting of the United Nations Environmental Assembly, concrete steps were taken to develop such a Convention.[31]

Conclusion

A number of international conventions have been developed under the auspices of the United Nations to protect the environment and human health from hazardous chemicals and other toxic exposures. International conventions are complex and time-consuming mechanisms. Negotiations can take many years, and their implementation may imply long discussions, compromises, and financial resources.

Different countries have differing capacities to implement these conventions. International solidarity is required under these conventions to support action by LMICs through technology transfer and capacity building. Notwithstanding these complexities, international conventions have been essential to addressing major global problems that transcend national boundaries, implementing consistent solutions across countries, and strengthening public health and environmental action against vested interests. The Montreal Protocol, for example, has allowed for an effective action against stratospheric ozone depletion through the elimination of dangerous CFCs and their substitution with safer substances. It has successfully initiated a process that is progressively restoring stratospheric ozone and leading to the closure of the ozone holes over the North and South Poles.

In conclusion, international environmental conventions can be highly effective mechanisms for international environmental governance aiming at health and environment protection. An excellent resource that includes the large number of international treaties is the Health and Environment Interplay Database (HEIDI) available at https://www.chaire-epi.ulaval.ca/en/data/heidi.[32] Of course, the power and the effectiveness of these conventions depends on appropriate implementation, monitoring, and evaluation. Nations must make these a high priority on the global agenda for environmental action.

References

1. Watts N, Amann M, Arnell N, et al. The 2019 report of The Lancet Countdown on health and climate change: ensuring that the health of a child born today is not defined by a changing climate. *Lancet*. 2019;394(10211):1836–78. doi:10.1016/S0140-6736(19)32596-6.

2. Grandjean P. *Only One Chance. How Environmental Pollution Impairs Brain Development and How to Protect the Brains of the Next Generation.* Oxford, UK: Oxford University Press; 2013.

3. American Academy of Pediatrics Council on Environmental Health. *Pediatric Environmental Health*, 4th ed. Etzel, RA, Balk SJ, eds. Itasca, IL: American Academy of Pediatrics; 2019.

4. United Nations. The Universal Declaration of Human Rights. 1948. Available at: http://www.un.org/en/documents/udhr/.

5. Office of the United Nations High Commissioner for Human Rights. Convention on the Rights of the Child. 1989. Available at: http://www2.ohchr.org/english/law/pdf/crc.pdf.

6. Seymour D. 1989–2009: convention brings progress on child rights, but challenges remain. 2009. Available at: http://www.unicef.org/rightsite/237.htm.

7. United National Economic Commission for Europe. UNECE. Convention on Long-Range Transboundary Air Pollution. 1979. Available at: https://unece.org/sites/default/files/2021-05/1979%20CLRTAP.e.pdf.

8. Bull K, Johansson M, Krzyzanowski M. Impacts of the Convention on Long-range Transboundary Air Pollution on air quality in Europe. *J Toxicol Environ Health A*. 2008;71(1):51–5.

9. United National Economic Commission for Europe. UNECE. 2016. Available at: https://unece.org/DAM/env/documents/2016/AIR/Publications/Post-card-for_Air_Convention.pdf. (Gives reductions in pollution from Transboundary Convention.)

10. Molina MJ, Rowland FS. Stratospheric sink for chlorofluoromethanes: chlorine atom-catalysed destruction of ozone. *Nature*. 1974;249:810–2.

11. Farman JC, Gardiner BG, Shanklin JD. Large losses of total ozone in Antarctica reveal seasonal Cl0x/NOx interaction. *Nature*. 1985;315:207–10.

12. United Nations. Montreal Protocol on Substances that Deplete the Ozone Layer (with annex). 2020. Available at: https://ozone.unep.org/treaties/montreal-protocol.

13. Barnes PW, Robson TM, Neale PJ, et al. Environmental effects of stratospheric ozone depletion, UV radiation, and interactions with climate change: UNEP Environmental Effects Assessment Panel, Update 2021. *Photochem Photobiol Sci*. 2022;21(3):275–301. doi:10.1007/s43630-022-00176-5.

14. Slaper H, Velders GJ, Daniel JS, de Gruijl FR, van der Leun JC. Estimates of ozone depletion and skin cancer incidence to examine the Vienna Convention achievements. *Nature*. 1996;384(6606):256–8. doi:10.1038/384256a0.

15. Stockholm Convention on Persistent Organic Pollutants (POPs). 2001. Available at: http://chm.pops.int/Convention/ConventionText/tabid/2232/Default.aspx.

16. World Health Organization. *Persistent Organic Pollutants: Impact on Child Health*. Geneva, Switzerland: World Health Organization; 2010.

17. 20th Anniversary of the Adoption of the Stockholm Convention. 2021.Available at: https://chm.pops.int/TheConvention/Overview/20thAnniversary/tabid/8966/Default.aspx.

18. Basel Convention on the Control of Transboundary Movements of Hazardous Wastes and their Disposal. 1992. Available at: http://www.basel.int/Portals/4/Basel%20Convention/docs/text/BaselConventionText-e.pdf.

19. Basel Convention on the Control of Transboundary Movements of Hazardous Wastes and their Disposal. Milestones. 2009. Available at: http://www.basel.int/TheConvention/Overv iew/Milestones/tabid/2270/Default.aspx.

20. World Health Organization. *Childhood Lead Poisoning*. Geneva, Switzerland: World Health Organization; 2010.

21. Haefliger P, Mathieu-Nolf M, Lociciro S, et al. Mass lead intoxication from informal used lead-acid battery recycling in Dakar, Senegal. *Environ Health Perspect*. 2009;117:1535–40.

22. Rotterdam Convention on the Prior Informed Consent Procedure for Certain Hazardous Chemicals and Pesticides in International Trade. 1998. Available at: http://www.pic.int/TheConvention/Overview/TextoftheConvention/tabid/1048/language/en-US/Defa ult.aspx.

23. Bergman A, Heindel JJ, Jobling S, Kidd KA, Zoeller T. *State of the Science of Endocrine Disrupting Chemicals—2012*. Geneva, Switzerland: World Health Organization and United Nations Environment Programme; 2013.

24. World Health Organization. Framework Convention on Tobacco Control. WHO Framework Convention on Tobacco Control overview. 2003. Available at: https://fctc.who.int/who-fctc/overview

25. World Health Organization. The Global Advisory Committee on Vaccine Safety. Meeting report. 2012 Jun. Available at: https://www.who.int/publications/i/item/WER8730.

26. United Nations Framework Convention on Climate Change. 1992. Available at: https://unf ccc.int/resource/docs/convkp/conveng.pdf.

27. United Nations Framework Convention on Climate Change. Adoption of the Paris Agreement. 2015. Available at: unfccc.int/files/essential_background/convention/applica-tion/pdf.Paris Agreement.

28. Sharm el-Sheikh Implementation Plan. 2022 Nov. Available at: https://unfccc.int/docume nts/624444.

29. United Nations. COP28: What was achieved and what happens next? December 2023. Available at: https://unfccc.int/cop28/5-key-takeaways

30. Symeonides C, Brunner M, Mulders Y, et al. Buy-now-pay-later: Hazards to human and planetary health from plastics production, use and waste. *J Paediatr Child Health*. 2021;57(11):1795–804. https://doi.org/10.1111/jpc.15777.

31. United Nations Environment Assembly. Draft Resolution: End plastic pollution: Towards an international legally binding instrument. UNEP. February 23–26 and February 28–March 2, 2022. Available at: https://www.unep.org/news-and-stories/press-release/historic-day-campaign-beat-plastic-pollution-nations-commit-develop.

32. Morin JF, Blouin C. How environmental treaties contribute to global health governance. *Global Health*. 2019;15:47. https://doi.org/10.1186/s12992-019-0493-7.

Glossary of Key Terms Used in United Nations Treaties

Accede/Accession: An act by which a state signifies its agreement to be legally bound by the terms of a particular treaty. It has the same legal effect as ratification, but it is not preceded by an act of signature. The formal procedure for accession varies according to the national legislative requirements of the state. To accede to a human rights treaty, the appropriate national organ of a state—Parliament, Senate, the Crown, Head of State, or Government, or a combination of these—follows its domestic approval procedures and makes a formal decision to be a party to the treaty. Then, the *instrument of accession*, a formal sealed letter referring to the decision and signed by the state's responsible authority, is prepared and deposited with the United Nations Secretary-General in New York.

Adoption: The formal act by which the form and content of a proposed treaty text are established. Treaties negotiated within an international organization like the United Nations are usually adopted by a resolution of a representative organ of the organization whose membership more or less corresponds to the potential participation in the treaty in question (the United Nations General Assembly, for example).

Convention: A formal agreement between states. the generic term "convention" is thus synonymous with the generic term "treaty." Conventions are normally open for participation by the international community as a whole or by a large number of states. Usually the instruments negotiated under the auspices of an international organization are entitled "conventions" (e.g., The Convention on the Rights of the Child, adopted by the General Assembly of the United Nations in 1989).

Entry into Force: A treaty does not enter into force when it is adopted. Typically, the provisions of the treaty determine the date on which the treaty enters into force, often at a specified time following its ratification or accession by a fixed number of states. For example, the Convention on the Rights of the Child entered into force on September 2, 1990, the 30th day following the deposit of the 20th state's instrument of ratification or accession. A treaty enters into force for those states which gave the required consent.

Ratify/Ratification: An act by which a State signifies an agreement to be legally bound by the terms of a particular treaty. To ratify a treaty, the State first signs it and then fulfills its own national legislative requirements. Once the appropriate national organ of the country—Parliament, Senate, the Crown, Head of State, or Government, or a combination of these—follows domestic constitutional procedures and makes a formal decision to be a party to the treaty. The instrument of ratification, a formal sealed letter referring to the decision and signed by the state's responsible authority, is then prepared and deposited with the United Nations Secretary-General in New York.

State Party or Party: A "state party" to a treaty is a country that has ratified or acceded to that particular treaty and is therefore legally bound by the provisions in the instrument.

Treaty: A formally concluded and ratified agreement between states. The term is used generically to refer to instruments binding at international law, concluded between international entities (states or organizations). Under the Vienna Conventions on the Law of Treaties, a treaty must be (1) a binding instrument, which means that the contracting parties intended to create legal rights and duties; (2) concluded by states or international organizations with treaty-making power; (3) governed by international law; and (4) in writing.

These definitions are adapted from *The Concise Oxford Dictionary of Current English* (8th edition), Clarendon Press, Oxford, 1990 and United Nations Treaty Collection, Treaty Reference Guide, 1999, available at http://untreaty.un.org/English/guide.asp.

63

The Impacts of War on Children's Health

Barry S. Levy

Introduction

War has profound impacts on children's health and well-being.[1-3] During war, children are often killed, severely injured, and permanently disabled. They are at increased risk of abuse, neglect, and exposure to secondary violence. They suffer from malnutrition and communicable diseases, which are often fatal. They develop mental disorders, which may persist throughout their lives. They are exposed to environmental hazards, which can cause both acute disorders as well as cancer and other noncommunicable diseases many years later. They receive inadequate medical care and preventive health services. They are uprooted from their homes and separated from their parents. Some become child soldiers. And children are adversely affected by loss of education and opportunities for social development.

War and Military Expenditures

In 2022, there were 55 armed conflicts, most of which were intrastate conflicts in low- and middle-income countries (LMICs), often with active participation by other countries. Most were armed conflicts over government control or territory, including conflicts over petroleum, water, and other resources.

Over the past several decades, the percentage of people killed or injured during war who have been noncombatant civilians has remained high. Civilians have increasingly been targets of violence. But the vast majority of civilian deaths—including deaths of children—during war have been caused indirectly by damage to health-supporting infrastructure, including healthcare facilities, food supply systems, water treatment plants, and electrical grids.

Military expenditures by the major powers and international arms sales have increased substantially in recent years. In 2022, global military expenditures were estimated at US$2.24 trillion—more than 700 times the entire United Nations budget (US$3.12 billion). In that year, US military expenditures were estimated at US$877 billion, accounting for more than 39% of global military expenditures—greater than those of the next 10 countries combined.[4]

War and the preparation for war divert huge amounts of human and financial resources to military uses and away from health, education, and other programs and services that support the health and well-being of children. As an index of this diversion, the United States ranks first among countries in both military expenditures and export of conventional weapons, but ranks 35th in life expectancy and 46th in infant mortality

Barry S. Levy, *The Impacts of War on Children's Health* In: *Textbook of Children's Environmental Health*, Second Edition.
Edited by: Ruth A. Etzel and Philip J. Landrigan, Oxford University Press. © Oxford University Press 2024.
DOI: 10.1093/oso/9780197662526.003.0063

rate.[5] As another example, the first US$204 billion spent for the Iraq War could have achieved globally for 3 years *all* of the following: a 50% decrease in hunger, provision of needed medication for HIV/AIDS, provision of clean water and sanitation, and immunization for all children in the so-called "developing countries."[6]

Displacement

Many children are uprooted from their homes during armed conflict. At the end of 2022, there were 108.4 million people globally who had been forcibly displaced, most as a result of conflict and violence.[7] These displaced people included 62.5 million internally displaced persons, 35.3 million refugees, 5.4 million asylum-seekers, and 5.2 million other people in need of international protection.[7] The vast majority of displaced people are mothers and children. They may become internally displaced within their own countries or cross international borders to become refugees in other countries, where they often are placed in refugee camps. Children who are internally displaced typically have less access to safe food and water, healthcare, and safety and security than children who become refugees.

During war and its aftermath, children are often separated from their parents who may have been killed or seriously injured. Ideally, children who have been separated from their parents are cared for by other relatives or adults whom they know and trust. Every effort should be made to reunite children with their parents and other family members (Figure 63.1). It is not unusual for separated children to be caring for younger

Figure 63.1. A field worker with UNHCR: The UN Refugee Agency helps a young Rwandan Hutu girl find her parents while evacuating refugees from the Biaro refugee camp in Zaire (now the Democratic Republic of the Congo) in 1997. UN and humanitarian aid agencies were trying to reunite thousands of unaccompanied children with their parents before repatriation to Rwanda. (AP Photo/John Moore.)

siblings who are also displaced, or for separated children to band together. Children in these circumstances are vulnerable to exploitation and violence. They may be recruited into armed groups or sexually exploited.

Assaults and Injuries

During war, children are often wounded or otherwise injured, resulting in many deaths and disabilities.[8] In 1996, UNICEF estimated that, for each child who dies during armed conflict, three more children are injured or permanently disabled.[9] Only a small percentage of disabled children in LMICs have access to needed medical, rehabilitative, and social services and to education.

Antipersonnel landmines and unexploded ordnance have accounted for many injuries and deaths of children during war and its aftermath.[10–12] In 2020, landmines killed 2,492 people and wounded 4,561—a substantial number of them children. Many landmine injuries lead to limb amputations, often in countries where there is limited capacity to produce and repair artificial limbs (Figure 63.2). Landmines and unexploded ordnance also lead to substantial social and economic disruptions that adversely affect children's health, such as making residential areas uninhabitable or preventing cultivation of agricultural land.

Despite the Mine Ban Treaty ("Ottawa Treaty") of 1997, to which 164 countries are states parties, vast areas of land, typically in rural areas of LMICs, remain heavily mined. Antipersonnel landmines, including those that are brightly colored, may be tempting for children to play with. Unexploded ordnance, which includes grenades, bombs, mortar

Figure 63.2. A disabled boy maimed by a landmine stands in a courtyard of a UNICEF-assisted rehabilitation center located in the Wat Tan Temple in Phnom Penh, Cambodia. (Source/Photographer: UNICEF/5907/Roger Lemoyne.)

shells, and cluster munitions that lie on the ground, account for more fatal and nonfatal injuries to children than landmines. Risk behaviors for children include playing in areas where there is unexploded ordnance, tending animals in these areas, and tampering with explosives.

Malnutrition and Communicable Diseases

Acute and chronic malnutrition among children occurs frequently during war and its aftermath.[13,14] Malnutrition adversely affects children in many ways, including impairing neurobehavioral development and suppressing the immune system, making children more vulnerable to acquiring—and dying from—bacterial and viral gastroenteritis, acute respiratory infection, measles, malaria, and other communicable diseases. The children who are most vulnerable to malnutrition are those who are under age 5, but many children 5 years of age and older are also at increased risk. Children under age 5 in LMICs are twice as likely to die if they are mildly malnourished, more than four times likely to die if they are moderately malnourished, and more than eight times likely to die if they are severely malnourished, as compared with children who are adequately nourished.[15]

Refugee and internally displaced mothers and children are especially vulnerable to malnutrition and its adverse effects. In camps for displaced persons, malnourished children have often been severely affected by measles, cholera, and other communicable diseases. For example, during the Yemeni Civil War, an outbreak of cholera occurred with more than 1.2 million cases (58% of which were in children) and more than 3,000 deaths in the first 6 months.

Causes of malnutrition include disruption of families, breakdown of agriculture (due to limited access to seeds and fertilizers, destruction of irrigation and flood control systems, and damage to crops and systems for storage and distribution of food), disruption of medical care and public health services, and damage to the overall economy.[14,16] Childhood malnutrition has often occurred during war when restricting the availability of food has been used as a weapon of war.[17]

Additional causes of malnutrition of children in rural areas include death of farm animals and displacement and death of family members who otherwise raise animals or grow crops. Inadequate food for adults decreases their energy for agricultural work. Deployment of landmines may restrict access to farmland and water, reducing the ability of livestock to graze, and, in turn, reducing family income and access to food.

Breast milk is the ideal food for infants; use of artificial feeding with infant formula can pose a significant risk of diarrheal disease to children in countries where water is likely to be contaminated. Infants and young children who are displaced are less likely to be breastfed—or to be breastfed for shorter periods of time.[18] This is especially true when mothers are themselves malnourished or when mothers need to assume greater responsibilities when fathers are absent.

When parents or other caregivers are killed or physically or mentally disabled during war or its aftermath, children are often left to take care of themselves and therefore are less able to get adequate nutrition. In some situations, children and their families may need to hunt animals or harvest wild plants, activities which may involve additional risks. Even when children have seemingly adequate amounts of food, they may not be ingesting sufficient amounts of calories, proteins, vitamins, and minerals. Malnourished

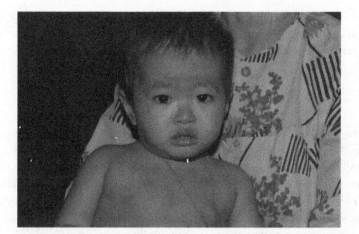

Figure 63.3. Young child with measles in a Cambodian refugee camp. The measles virus is highly contagious and unimmunized children affected by war are at high risk of serious illness and death from measles. (Photograph by Barry S. Levy.)

children are more likely to become seriously ill or die as a result of diarrheal disease, which is frequent during war and its aftermath. And diarrheal disease may exacerbate malnutrition.

It is difficult to prevent communicable diseases in children during war and its aftermath. For example, armed conflict typically reduces childhood vaccination rates and often leads to the reemergence and spread of vaccine-preventable diseases, such as measles (Figure 63.3). In 2017, about two-thirds of the almost 21 million children who did not receive any measles vaccine were living in countries that were affected by armed conflict. Polio has almost been eliminated, except in a few countries where conflict has made polio endemic.[19]

Mental Disorders

Children in war zones have very high rates of mental disorders, including posttraumatic stress disorder and depression.[20,21] They may demonstrate a wide range of behavioral problems, including fear and anxiety, difficulty concentrating, sleeping problems, eating disorders, detachment, and committing violent acts. Mental and behavioral disorders can have long-term adverse effects on school and job performance, ability to form and maintain relationships, family life, coping skills, and overall health and well-being.

Risk factors for mental disorders include death or serious injury to parents, other relatives, and friends; witnessing violent acts of war; and being displaced from their homes. Deployment of a parent to a war zone increases the rate of mental disorders in children.[22,23]

After deployment, soldiers often return home with a "battlefield mentality," impeding their reintegration into their families and communities and predisposing them to committing domestic violence. Children of returning soldiers may develop mental disorders after witnessing this domestic violence. Children may become afraid of their returning parents in such situations. Most psychosocial support and mental health services for children and adolescents during armed conflict are provided in schools and specialized centers. Inadequate security, language and cultural barriers, and inadequate infrastructure may restrict access to mental health services.[24]

Abnormal Birth Outcomes

During war and its aftermath, the percentages of prematurely born infants and low-birth-weight infants often increase. For example, a study in 53 LMICs in which there was armed conflict between 1990 and 2018, found that intrauterine exposure to armed conflict decreased birthweight by 2.8% and increased the occurrence of low birthweight by 3.2%. In addition, war has been associated with increases in neonatal mortality (deaths during the first 28 days of life); in 2015, 14 of the 15 countries with the highest rates of neonatal mortality were experiencing chronic conflict or political instability.

There is some evidence of increased occurrence of congenital anomalies during war. For example, women who had been deployed in the U.S. military during the Gulf War reported a 180% increase in the rate of moderate to severe birth defects among their liveborn children. Civilians exposed to teratogens may also experience increased occurrence of congenital malformations during war. For example, a systematic review and meta-analysis found a significant association between Agent Orange exposure and congenital malformations.[25] Children exposed in utero between the 8th and the 15th gestational week to atomic bomb explosions in 1945 suffered a high frequency of severe mental retardation, often accompanied by small head size and decreased intelligence quotient (IQ).

Noncommunicable Diseases

Children exposed to environmental hazards or nutritional deprivation during war and its aftermath may develop noncommunicable diseases later in life. Children exposed to ionizing radiation or carcinogenic chemicals during war and its aftermath may be at increased risk of malignancies later in life. Children who are undernourished during war may be at increased risk of cardiovascular disease and other noncommunicable diseases as adults. For example, children and adolescents during the German occupation of the Channel Islands during World War II had a 152% increase in cardiovascular disease in adulthood.[26]

Child Soldiers

Child soldiers bear many physical and psychological burdens of war.[27-29] According to the Paris Principles, a child soldier is "any person below 18 years of age who is, or has been, recruited or used by an armed force or armed group in any capacity. . . . It does not only refer to a child who is taking, or has taken, a direct part in hostilities."[30] Child soldiers participate in combat; lay mines and explosives; scout; spy; act as decoys, couriers, and guards; provide logistics and support functions, including portering, cooking, and performing domestic labor; and are often forced to perform sexual acts. They also can be used as human shields and for political advantage or propaganda (Figure 63.4).

Figure 63.4. Child soldiers of the rebel Sudan People's Liberation Army wait for their commander at a demobilization ceremony at their barracks in Malou, southern Sudan, in February 2001. Under an agreement with the United Nations Children's Fund (UNICEF), the SPLA has demobilized 2,500 child soldiers between 8 and 18 years of age. (Source: AP Photo/Sayyid Azim.)

During the past several decades, child soldiers have been deployed in at least 87 countries. Since 1998, there has been armed conflict with child soldiers in at least 36 countries. It is not known how many child soldiers there are today; in 2015, it was estimated that there were 300,000 child soldiers, with the largest number in Africa. Most often, child soldiers are male orphans between 8 and 18 years old; they are often addicted to alcohol or drugs, merciless, amoral, and illiterate.[31] Child soldiers usually come from poor and disadvantaged families. Children may choose to become child soldiers after experiencing the deaths of relatives, destruction of their home environments, displacement (forced migration), economic difficulties, political oppression, and/ or harassment. And they may be kidnapped or enticed into becoming child soldiers by propaganda, a strong belief in the goal of the conflict, the thrill of adventure, entrapment, and/or attraction to identify with a military group. Since children are easily manipulated, they can be directed to perform crimes and atrocities—often without questioning their superiors.

Child soldiers suffer from many physical, psychological, and social consequences of war. In addition to experiencing high rates of injury and illness, they frequently suffer from posttraumatic stress disorder, depression, and other mental disorders.[32] Almost all them experience severe traumatic events, such as witnessing someone being killed or being forced to kill someone. Poor outcomes for child soldiers are associated with trauma and torture, death of a relative, and physical abuse in the household. Resilience of child soldiers is associated with less domestic violence, lower cognition of guilt, less motivation to seek revenge, better socioeconomic status of the family, and more perceived spiritual support. In the aftermath of war, former child soldiers suffer from difficulties with psychosocial adjustment and social reintegration.[33]

The United Nations Convention on the Rights of the Child (CRC) proclaims: "States Parties shall take all feasible measures to ensure that persons who have not attained the age of 15 years do not take a direct part in hostilities." However, the CRC does not prevent children 15 through 17 years of age from voluntarily participating in combat as soldiers. The Optional Protocol to the Convention on the Rights of the Child on the Involvement of Children in Armed Conflict declares: "States Parties shall take all feasible measures to ensure that members of their armed forces who have not attained the age of 18 years do not take a direct part in hostilities" and that they are not compulsorily recruited into their armed forces. It also obligates nations to "take all feasible measures to prevent such recruitment and use, including the adoption of legal measures necessary to prohibit and criminalize such practices."[34]

Economic Sanctions

Economic sanctions have often been brought against countries in lieu of, or before or after, outbreaks of armed conflict. These sanctions have often had profoundly adverse effects on children. For example, from 1991 to 1998, there were many excess deaths among children in Iraq, many of which were due to sanctions on the import of food and medicine during that period. One estimate placed the number between 400,000 and 500,000[35]; another concluded that the likely number was 227,000.[36] In addition, economic sanctions against Cuba and other countries not involved in war have adversely affected children in these countries.

Recommendations Concerning War-Affected Children

A variety of integrated measures need to be implemented to protect and address the needs of war-affected children. Programs and services need to consider socioeconomic, educational, and other inequalities that adversely affect the health and safety of children during war and its aftermath.[37]

Assaults and injuries: Measures need to be taken to protect children from assaults and injuries during war and its aftermath and to improve the availability of specialized expertise, medical supplies, and equipment for treating pediatric patients.[38] Governments of war-affected countries, international agencies, nongovernmental organizations (NGOs), and donor countries should provide greater resources to protect women and girls from gender-based sexual violence. Measures should be taken to ensure that those responsible for assaults and injuries to children during war and its aftermath are held accountable.

Malnutrition and communicable diseases: Addressing childhood malnutrition during war includes

- Adequately assessing the nutritional status of children;
- Providing food for affected children, including supplemental feeding and necessary medical care for severely malnourished children;
- Addressing the adverse effects of malnutrition on child growth and development;
- Maintaining household integrity; and
- Making households socially and economically sustainable.

Feeding programs are most effective when they are part of a broader relief or rehabilitation program that attends to a wider range of needs, including medical care and preventive health services. Factors that may impede relief include political and military impediments, damage to roads and other logistical issues, destruction of storage facilities, and floods, droughts, and other extreme weather-related events. Addressing childhood malnutrition during war and its aftermath requires long-term commitment by relief agencies, which are challenged to maintain public attention and financial support after an acute humanitarian crisis has peaked.

Implementation of immunization programs needs to be a high priority during war and its aftermath. Of highest priority in most situations is immunizing children against measles, given that many children in war zones are not immunized against this disease, which is highly contagious and often fatal in malnourished children.

Mental disorders: Psychosocial support must be a central part of child protection in all phases of emergency and reconstruction. Governments, UN agencies, and NGOs should prevent the institutionalization of children and prioritize their reunification with their families and communities. Children with special needs should receive support within the broader context of reintegration programs for all war-affected children.

Abnormal birth outcomes: Abnormal birth outcomes can be reduced by ensuring that pregnant women receive adequate antenatal care, providing trained birth attendants during labor and delivery, and preventing pregnant women and infants from having hazardous exposures.

Child soldiers: Countries must ratify, incorporate into their national laws, and implement the Optional Protocol to the Convention on the Rights of the Child on the Involvement of Children in Armed Conflict. Programs to disarm, demobilize, and reintegrate child soldiers must be made a priority. Governments and armed groups must prevent recruitment of child soldiers and ensure their demobilization and integration. Child soldiers must be protected from punitive measures.

Roles of Health Professionals

Health professionals can play important roles by (a) diagnosing and treating children who have been adversely affected by war and its aftermath, (b) documenting the adverse health effects of war on children, (c) educating others and raising public awareness about these issues, (d) advocating for policies and programs to protect children during war and its aftermath, and (e) working to prevent war and promote peace. Health professionals can generally be most effective by participating in professional organizations that address these issues, such as the International Physicians for the Prevention of Nuclear War and its national affiliates (such as Physicians for Social Responsibility in the United States and Medact in the United Kingdom) and the Peace Caucus and various sections of the American Public Health Association.

Preventing War and Promoting Peace

The only way to totally eliminate the impacts of war on children is to abolish war and promote peace. A comprehensive approach includes

- Resolving disputes nonviolently by diplomacy, arms control, and interrupting the transmission of violence;

- Reducing the root causes of war by policies and actions to reduce extreme poverty, socioeconomic inequities, militarism, the availability of weapons, poor governance, intergroup animosity, and environmental stress; and
- Strengthening the infrastructure for peace by rehabilitating nations and reintegrating people after a war has ended; supporting post-conflict reconciliation; promoting democracy and the rule of law; and respecting, protecting, and fulfilling human rights.[1]

References

1. Levy BS. *From Horror to Hope: Recognizing and Preventing the Health Impacts of War*. New York: Oxford University Press; 2022.

2. Kadir A, Shenoda S, Goldhagen J. Effects of armed conflict on child health and development: a systematic review. *PloS ONE*. 2019;e0210071.https://doi.org/10.1371/journal.pone.0210071.

3. Machel G. *The Impact of War on Children*. New York: Palgrave; 2001.

4. Stockholm International Peace Research Institute. *SIPRI Yearbook 2023: Armaments, Disarmament, and International Security*. New York: Oxford University Press; 2023.

5. Central Intelligence Agency. *The World Factbook*. 2023. Available at: https://www.cia.gov/the-world-factbook. Accessed September 19, 2023.

6. Bennis P, Leaver E, IPS Iraq Task Force. *The Iraq Quagmire: The Mounting Costs of War and the Case for Bringing Home the Troops*. Washington, DC: Foreign Policy in Focus; 2005.

7. UNHCR: The UN Refugee Agency. Refugee data finder. June 14, 2023. Available at: https://www.unhcr.org/refugee-statistics/. Accessed September 19, 2023.

8. Matos RI, Holcomb JB, Callahan C, Spinella PC. Increased mortality rates of young children with traumatic injuries at a US Army combat support hospital in Baghdad, Iraq, 2004. *Pediatrics*. 2008;122: e959–66.

9. UNICEF. *State of the World's Children 1996*. Oxford: Oxford University Press, 1996, p. 13.

10. Surrency AB, Graitcer PL, Henderson AK. Key factors for civilian injuries and deaths from exploding landmines and ordnance. *Injury Prevent*. 2007;13:197–201.

11. Bilukha OO, Brennan M, Woodruff BA. Death and injury from landmines and unexploded ordnance in Afghanistan. *JAMA*. 2003;290:650–3.

12. Kopjar B, Wiik J, Wickizer TM, et al. Access to war weapons and injury prevention activities among children in Croatia. *Am J Public Health*. 1996;86:397–400.

13. Bendavid E, Boerma T, Akseer N, et al. The effects of armed conflict on the health of women and children. *Lancet*. 2021;397:522–32.

14. Food and Agriculture Organization. Child nutrition and food security during armed conflicts. No date. Available at: http://www.fao.org/DOCREP/W5849T/W5849T07.HTM. Accessed September 19, 2023.

15. Caulfield LE, de Onis M, Blossner M, Black RE. Undernutrition as an underlying cause of child deaths associated with diarrhea, pneumonia, malaria, and measles. *Am J Clin Nutrit*. 2004;80:193–8.

16. Mashal T, Takano T, Nakamura K, et al. Factors associated with the health and nutritional status of children under 5 years of age in Afghanistan: family behaviour related to women and past experience of war-related hardships. *BMC Public Health*. 2008;8:301.

17. Mohareb AM, Ivers LC. Disease and famine as weapons of war in Yemen. *N Engl J Med*. 2019;380:109–11.

18. Andersson N, Paredes-Solis S, Legorreta-Soberanis J, et al. Breast-feeding in a complex emergency: four linked cross-sectional studies during the Bosnian conflict. *Public Health Nutrit*. 2010;13:2097–2104.

19. Ngo NV, Pemunta NV, Muluh NE, et al. Armed conflict, a neglected determinant of child-hood vaccination: some children are left behind. *Hum Vacc Immunother*. 2020;16:1454–63.

20. Panter-Brick C, Eggerman M, Gonzalez V, Safdar S. Violence, suffering, and mental health in Afghanistan: a school-based survey. *Lancet*. 2009;374:807–16.

21. Lustig SL, Kia-Keating M, Knight WG, et al. Review of child and adolescent refugee mental health. *J Am Acad Child Adolesc Psychiatry*. 2004;43:24–36.

22. Klarić M, Frančišković T, Klarić B, et al. Psychological problems in children of war veter-ans with posttraumatic stress disorder in Bosnia and Herzegovina: cross-sectional study. *Croatian Med J*. 2008;49:491–8.

23. Lester P, Peterson K, Reeves J, et al. The long war and parental combat deployment: effects on military children and at-home spouses. *J Am Acad Child Adolesc Psychiatry*. 2010;49:310–20.

24. Kamali M, Munyuzangabo M, Siddiqui F, et al. Delivering mental health and psychosocial support interventions to women and children in conflict settings: a systematic review. *BMJ Global Health*. 2020;5:e002014. doi:10.1136/bmjgh-2019-002014.

25. Ngo AD, Taylor R, Roberts CL, Nguyen TV. Association between Agent Orange and birth defects: systematic review and meta-analysis. *Int J Epidemiol*. 2006;35:1220–30.

26. Head RF, Gilthrope MS, Byrom A, Ellison GTH. Cardiovascular disease in a cohort exposed to the 1940–45 Channel Islands occupation. *BMC Public Health*. 2008;8:303. doi:10.1186/1471-2458-8-303.

27. Etzel RA. Use of children as soldiers. *Pediatr Clin N Am*. 2021;68:437–47.

28. Kohrt BA, Jordans MJ, Tol WA, et al. Social ecology of child soldiers: child, family, and com-munity determinants of mental health, psychosocial well-being, and reintegration in Nepal. *Transcultural Psychiatry*. 2010;47:727–53.

29. Stark L, Boothby N, Ager A. Children and fighting forces: 10 years on from Cape Town. *Disasters*. 2009; 33:522–47.

30. This reference does not have an author.. *The Paris Principles: Principles and Guidelines on Children Associated with Armed Forces or Armed Groups*. New York: UNICEF; 2007.

31. Pearn J. Children and war. *J Pediatr Child Health*. 2003;39:166–72.

32. Song SJ, de Jong J. Child soldiers: children associated with fighting forces. *Child Adolesc Psychiatric Clin N Am*. 2015;24:765–75.

33. Betancourt TS, Khan KT. The mental health of children affected by armed conflict: protec-tive processes and pathways to resilience. *Int Rev Psychiatry*. 2008;20:317–28.

34. Office of the United Nations High Commissioner for Human Rights. Optional Protocol to the Convention on the Rights of the Child on the Involvement of Children in Armed Conflict. United Nations, 2000. Available at: https://www.ohchr.org/en/professionalinterest/pages/opaccrc.aspx.

35. Ali MM, Blacker J, Jones G. Annual mortality rates and excess deaths of children under five in Iraq, 1991–98. *Population Study (Camb)*. 2003;57:217–26.

36. Garfield R. Morbidity and mortality among Iraqi children from 1990 through 1998: assessing the impact of the Gulf War and economic sanctions. Unpublished paper. 1999. Available at: https://reliefweb.int/sites/reliefweb.int/files/resources/A2E2603E5DC88A46852568250 05F211D-garfie17.pdf.

37. Akseer N, Wright J, Tasic H, et al. Women, children and adolescents in conflict countries: an assessment of inequalities in intervention coverage and survival. *BMJ Global Health*. 2020;5:e002214. doi:10.1136/bmjgh-2019-002214.

38. Jain RP, Meteke S, Gaffey MF, et al. Delivering trauma and rehabilitation interventions to women and children in conflict settings: a systematic review. *BMJ Global Health*. 2020;5:e001980. doi:10.1136/bmjgh-2019-001980.

64

Natural Disasters, Environmental Emergencies, and Children's Health

Henry Falk

Children have special needs during natural disasters and environmental emergencies; they are physiologically and anatomically different from adults. They are developmentally immature. They require specialized equipment for treatment. Many emergency drugs and antidotes have not been sufficiently tested in pediatric populations. Emergency personnel and facilities may not be well trained for handling small children (e.g., unfamiliarity of many care providers with pediatric dosages). Communicating with small children and addressing their psychological needs (especially when they are separated from their parents or caregivers) can be very challenging. Beyond the innate vulnerabilities of all children, health disparities, poverty, and socioeconomic factors leave many populations of children at greater vulnerability to the impact of natural disasters and environmental emergencies.[1]

This chapter has two parts. The first summarizes general pediatric issues related to natural disasters and environmental emergencies. The second describes specific categories of disasters, including several of the most striking disasters of the past two decades (Hurricane Katrina, the Haiti earthquake, and the Tsunami of 2004) and, also, the rapidly increasing extreme weather events such as severe storms, wildfires, drought, and heat waves.

The American Academy of Pediatrics (AAP) has released multiple policy, clinical, and technical guidance documents for pediatricians related to disasters (Table 64.1). Terrorism is not dealt with here, but there is considerable overlap in the guidance for disasters, environmental emergencies, and terrorism. There also is a very clear connection between climate change and natural disasters and environmental emergencies; a more complete discussion of climate change is found in Chapter 4.

Disaster Preparedness and Response

In the United States, following the attacks on the World Trade Center and the anthrax events of 2001, disaster preparedness and response became a critical issue; addressing the concerns for children became a central issue for pediatricians[2–4] (Table 64.1).

Gausche-Hill[5] posed the question "Pediatric Disaster Preparedness: Are We Really Prepared?" Her response to this rhetorical question was negative. She emphasized the need to bring more pediatric expertise into disaster planning, obtain appropriate medications and equipment, improve pediatric focus in emergency medical services (EMS) and hospital systems, and better address the special needs of children. A 2005 survey of 3,748 EMS

Henry Falk, *Natural Disasters, Environmental Emergencies, and Children's Health* In: *Textbook of Children's Environmental Health*, Second Edition. Edited by: Ruth A. Etzel and Philip J. Landrigan, Oxford University Press. © Oxford University Press 2024.
DOI: 10.1093/oso/9780197662526.003.0064

agencies noted, for example, that 69% had no plans for a mass casualty event at a school, and only 12.3% had a pediatrician as part of their medical team.[6] A survey of thousands of school districts identified deficiencies in disaster planning[7]; for example, while almost all districts had an evacuation plan, fewer than one-third had ever conducted a drill.

This led to multiple national and local expert groups offering recommendations on how to improve pediatric disaster preparedness and response. Key issues included (a) increased training for EMS in caring for and transporting children, (b) increased engagement of the pediatric community in disaster preparedness and response, (c) the necessity of community-wide or regional planning for the distribution of children so that individual hospitals are not overwhelmed and to ensure that pediatric patients requiring critical care get to hospitals with both the capacity and the expertise to deal with pediatric emergencies, (d) relaxing standards of care during the emergency (e.g., nurses or other professionals would perform certain tasks that normally would be performed only by a physician) so that the needs of large numbers of critically ill children could be addressed, (e) addressing pediatric mental health concerns, and (f) focusing on many practical concerns such as preparedness in schools, drills, and exercises. Incorporating pediatric hospitals, trauma centers, surgeons, and pediatric critical care transport systems (both ground and air) within community-wide or regional surge planning was deemed essential[8] (Table 64.1).

In 2011, the Pediatric Emergency Mass Critical Care Task Force released a set of eight reports on mass critical care for children that provided a wealth of guidance from a series of 44 experts.[9-11] Also, an expert panel convened by the New York City Department of Health and Mental Hygiene developed guidelines for hospitals and emergency planners for the proper decontamination of children exposed to chemical agents in disaster settings.[12]

Recent reports suggest definite improvement: a national assessment of pediatric readiness in 4,149 US emergency departments, released in 2015, shows enhanced readiness associated with improved compliance with published guidelines.[13] However, there is still more to do: a 2015 National Report Card from Save the Children shows that only 21% of recommendations from the National Commission on Children and Disasters have been met; 25% were not met and 54% were partially met.[8,14]

The National Advisory Committee on Children and Disasters (NACCD), established in 2014 to provide expert advice to the Secretary of the US Department of Health and Human Services (DHHS) and the Assistant Secretary for Preparedness and Response (ASPR), provided recommendations in three key areas: coalition building, workforce development, and medical countermeasures readiness.[15] Recent articles provide further guidance on best practices and enhanced coordination and care for pediatric disaster response[16-18] and on caring for children with disabilities in disaster settings.[19]

Concurrent with the efforts in the pediatric health care community, public health programs established since 9/11 have supported state and local government disaster preparedness and response,[20-22] including addressing pediatric needs in public health emergencies.[23]

Mental Health

In the past several decades, it has become clear that major disasters, violence, and personal trauma to children can all lead to an increased risk for mental illness, including posttraumatic stress disorder (PTSD), symptoms of anxiety and depression, and developmental and emotional disorders. Children, when significantly exposed to these events, are clearly vulnerable to such outcomes in virtually all disaster settings. The degree to which such outcomes occur depends to a great deal on multiple factors such as the intensity of the disaster and the child's experience of it; the level of support from family, schools, society, and relief agencies; the general socioeconomic conditions; the child's prior history, personality, and experiences with a variety of traumatic events; the availability of mental health services in the immediate aftermath of the disaster as well as during the stages of recovery; and other factors. Literature searches on children and disasters show an intense focus on studies of PTSD and other mental health outcomes in relation to a wide variety of disasters[24] (Table 64.1), including for related mass trauma events covered in other chapters such as mass shootings, terrorism, pandemics, and climate change.[25-27]

Invariably, these studies demonstrate increases in adverse mental health outcomes post disaster, but the absolute numbers are variable depending on all of the factors noted earlier as well as on the background rates of PTSD, the particular instruments used for the study measurements, the timing of a study post disaster, the experience of the researchers, the numbers of children included in the studies, the quality of translating survey instruments into a variety of languages, and undoubtedly other factors. This chapter will not provide a quantitative summary of this literature. However, all evidence indicates that mental health issues are a very real problem among children and that the magnitude varies by event and location.

In an introduction to a Special Section in *Child Development*,[28] the authors detail challenges to conducting mental health research post-disaster. They note the great rarity of having pre-disaster health assessments available for comparison, the difficulties of obtaining representative samples of populations after mass dispersion and evacuation, the challenge of conducting research in remote areas where there is no culture of research and where trained and experienced researchers are uncommon, the lack of databases on normative development in many disaster-prone areas, and the fact that most such studies are cross-sectional rather than longitudinal.

Several articles summarize the clinical picture of PTSD and related disorders among children of different ages and also review the survey instruments that have been used in disaster settings.[29-31] Recent reports highlight the effect of disaster-related prenatal maternal stress on child development,[32] the emotional and behavioral effects of displacement,[33] and children's cognitive functioning during and after disasters.[34]

Multiple systematic reviews and meta-analyses have evaluated a variety of psychosocial interventions and support programs that appear promising and effective.[35-38]

Violence, War, and Civil Disturbances

Globally, millions of children suffer in complex humanitarian emergencies created by war, civil conflict, and natural disasters, with limited resources for mass critical care; an AAP Technical Report in 2018 notes that more than 1 in 10 children worldwide are affected by armed conflict (Table 64.1). Lautze[39] has reviewed the evolving humanitarian and health practices over the past decades in settings such as Afghanistan, Ethiopia, the Congo, Haiti, and Somalia as they relate to issues ranging from feeding practices to violence prevention. Wars and societal upheavals can be devastating to children in many

Table 64.1 Key American Academy of Pediatrics (AAP) guidance documents on natural disasters and environmental emergencies, 2005–2022

Chemical–Biological Terrorism and Its Impact on Children
Policy Statement: *Pediatrics* 2020;145:e20193749
Technical Report: *Pediatrics* 2020;145:e20193750

Pediatric Readiness in Emergency Medical Services Systems
Policy Statement: *Pediatrics* 2020;145:e20193307
Technical Report: *Pediatrics* 2020;145:e20193308

Understanding Liability Risks and Protections for Pediatric Providers During Disasters
Policy Statement: *Pediatrics* 2019;143:e20183892
Technical Report: *Pediatrics* 2019;143:e20183893

Pediatric Considerations Before, During, and After Radiological or Nuclear Emergencies
Policy Statement: *Pediatrics* 2018;142:e20183000
Technical Report: *Pediatrics* 2018;142:e20183001

Pediatric Readiness in the Emergency Department
Joint Policy Statement: *Pediatrics* 2018;142:e20182459

The Effects of Armed Conflict on Children
Technical Report: *Pediatrics* 2018:142:e20182586

Disaster Preparedness in Neonatal Intensive Care Units
Clinical Report: *Pediatrics* 2017;139:e20170507

Management of Pediatric Trauma
Joint Policy Statement: *Pediatrics* 2016;138:e20161569

Medical Countermeasures for Children in Public Health Emergencies, Disasters, or Terrorism
Policy Statement: *Pediatrics* 2016;137:e20154273

Global Climate Change and Children's Health
Policy Statement: *Pediatrics* 2015;136:e2015–3232
Technical Report: *Pediatrics* 2015;136:e2015–3233

Ensuring the Health of Children in Disasters
Policy Statement: *Pediatrics* 2015;136:e2015–3112

Providing Psychosocial Support to Children and Families in the Aftermath of Disasters and Crises
Clinical Report: *Pediatrics* 2015;136:e2015–2861

Equipment for Ground Ambulances
Joint Policy Statement: *Prehosp Emerg Care* 2014;19:92–97

Consent for Emergency Medical Services for Children and Adolescents
Policy Statement: *Pediatrics* 2011;128:e2011–1166

Guidelines for Care of Children in the Emergency Department
Joint Policy Statement: *Pediatrics* 2009;124:e2009–1807

Disaster Planning for Schools
Policy Statement: *Pediatrics* 2008;122:895–901

Medical Emergencies Occurring at School
Policy Statement: *Pediatrics* 2008;122:887–894

Preparing for Pediatric Emergencies:Drugs to Consider
Clinical Report: *Pediatrics* 2008;121;433–443

Preparation for Emergencies in the Offices of Pediatricians and Pediatric Primary Care Providers
Policy Statement; *Pediatrics* 2007;120:200–212

The Pediatrician and Disaster Preparedness
Policy Statement: *Pediatrics* 2006;117:560–565
Technical Report: *Pediatrics* 2006;117:340–362

Pediatric Mental Health Emergencies in the Emergency Medical Services System
Joint Policy Statement: *Pediatrics* 2006;118:1764–1767

Psychosocial Implications of Disasters or Terrorism in Children:a Guide for the Pediatrician
Clinical Report: *Pediatrics* 2005;116:787–795

ways. Williams[40] notes that in the decade 1993–2003, approximately 2 million children were killed and 6 million injured or disabled in war zones, 1 million children were orphaned, and 20 million were displaced to refugee or other camps. See Chapter 63.

The Syrian civil war has lowered life expectancy by as much as 20 years; Guha-Sapir reports that between 2011 and 2016 almost 14,000 children were killed largely by indiscriminate air bombardments including by barrel bombs.[41] Protecting children in such circumstances is a huge challenge to relief agencies and others, and resettlement efforts are increasingly challenging.[42] As expected, in humanitarian crises quality public health data are very limited,[43,44] particularly for young children.[45]

Post-war, there are enormous difficulties in reintegrating traumatized children and child soldiers into productive lives in civil society.[46,47]

Extreme Weather Events

The website of the National Oceanic and Atmospheric Administration (NOAA) (https://www.ncdc.noaa.gov/billions/overview) contains a wealth of data on natural disasters related to weather and climate. During 2021 (Figure 64.1), there were 20 separate billion-dollar weather and climate disaster events across the United States, the third most costly year on record, behind 2017 and 2005. Figure 64.1 also shows the geographic distribution of these major weather-related disasters within the United States; note that the categories of disasters typically occur in specific regions of the country. Similar records are available going back decades, and these enable trends to be identified. Figure 64.2 shows the increasing trends in the number of major disasters: the inexorable rise in recent decades is partially due to population and economic growth and partially related to climate change. The two costliest years, 2005 and 2017, were dominated respectively by Hurricane Katrina and by the trio of Hurricanes Harvey, Irma, and Maria (Table 64.2). This chapter focuses primarily on the health impact of disasters for children; a more complete discussion of climate change is found in Chapter 4.

The five costliest hurricanes have all occurred within the past decade (Table 64.2); the first three are discussed in more detail below, with a focus on their impact for children.

Hurricane Katrina

In recent years, Hurricane Katrina has been the most devastating natural disaster in the United States. From a public health perspective, it caused widespread destruction in the city of New Orleans, leading to evacuation of the city and to the disruption of almost all

Table 64.2 The five costliest hurricanes in the United States

Hurricane Katrina: $160 billion

Hurricane Harvey: $125 billion

Hurricane Maria: $90 billion

Hurricane Sandy: $70 billion

Hurricane Irma: $50 billion

Data from NOAA website: https://www.ncdc.noaa.gov/billions/overview

U.S. 2021 Billion-Dollar Weather and Climate Disasters

Drought/Heat Wave　Flooding　Hail　Hurricane　Tornado Outbreak　Severe Weather　Wildfire　Winter Storm/Cold Wave

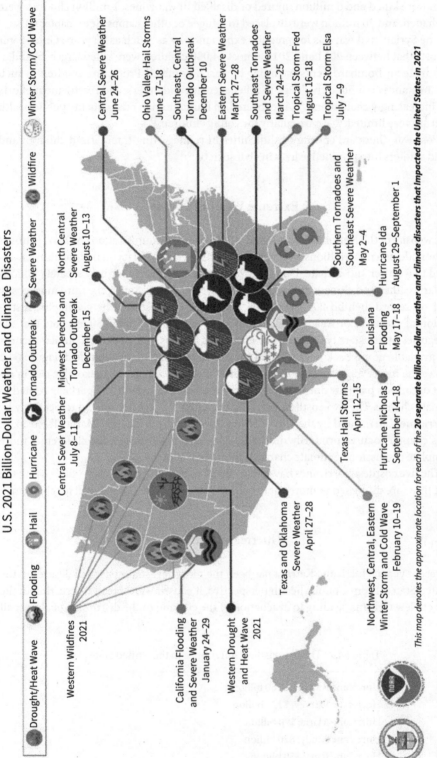

Western Wildfires
2021

California Flooding
and Severe Weather
January 24–29

Western Drought
and Heat Wave
2021

Texas and Oklahoma
Severe Weather
April 27–28

Northwest, Central, Eastern
Winter Storm and Cold Wave
February 10–19

Central Sever Weather
July 8–11

Midwest Derecho and
Tornado Outbreak
December 15

North Central
Severe Weather
August 10–13

Central Severe Weather
June 24–26

Ohio Valley Hail Storms
June 17–18

Southeast, Central
Tornado Outbreak
December 10

Eastern Severe Weather
March 27–28

Southeast Tornadoes
and Severe Weather
March 24–25

Tropical Storm Fred
August 16–18

Tropical Storm Elsa
July 7–9

Southern Tornadoes and
Southeast Severe Weather
May 2–4

Hurricane Ida
August 29–September 1

Louisiana
Flooding
May 17–18

Texas Hail Storms
April 12–15

Hurricane Nicholas
September 14–18

This map denotes the approximate location for each of the **20 separate billion-dollar weather and climate disasters that impacted the United States in 2021**

Figure 64.1. From NOAA.Gov. NOAA map by the National Centers for Environmental Information (NCEI/NOAA)

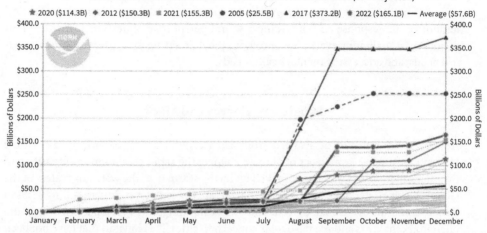

1980–2022 United States Billion-Dollar Disaster Event Cost (CPI-Adjusted)

Figure 64.2. From NOAA.Gov. NOAA map by the National Centers for Environmental Information (NCEI/NOAA)

major municipal services (water, sanitation, natural gas, electricity, transportation, food, housing, etc.).[48]

The destruction caused by Katrina extended also across multiple southern states, causing widespread damage (Figure 64.3). There were urgent needs to provide shelter, subsistence, and medications for an extended period to 1 million evacuees, as well as to reestablish clinical services in a wide area and restore basic public health functions.[49,50] Hurricane Katrina also caused an environmental emergency requiring decision-making about removal of mountains of debris, safe disposal of solid and hazardous wastes, assessment of exposure to hazardous chemicals from damaged industrial facilities and waste sites, protection of vulnerable populations including children from exposure to the ubiquitous mold,[51] and the safety of cleanup workers.

A special supplement in *Pediatrics* vividly describes the heroic efforts of pediatricians to safely transfer very sick children, including neonates, from New Orleans hospitals that could no longer function to out-of-state facilities. It is a primer that provides excellent guidance for the many challenges and lessons associated with such mass transport of pediatric patients.[52–55] Beyond the immediate phase, the unfolding disaster created enormous ongoing challenges to pediatric practices in the area, not all of which could be successfully reestablished, and to the delivery of basic pediatric services such as newborn screening for metabolic diseases.[56,57]

Drury et al.[58] described the pressing mental health threats to the several hundred thousand children who were evacuated during Hurricane Katrina. These threats included the experience of the flooding, the disruption caused by the evacuation, the fragility of the shelters (particularly those in the Superdome and the Convention Center), separation from parents and caregivers for thousands of children, and the shock of returning to destroyed or heavily damaged homes. They also analyzed the limited mental health resources provided as part of the emergency relief operations and the overwhelming challenge of providing services to the many relocated and/or returning children. There have been additional excellent efforts to distill and disseminate the lessons learned from Katrina regarding the reunification of children and families[59] and the optimal provision of medical services to children in shelters.[60] Garrett et al.[61] produced a comprehensive overview of this "mega-disaster" from the perspective of children, covering both the limitations of the response and providing a sound perspective on lessons learned in four major areas: facilitating evacuation, providing shelter, caring for those with special medical needs, and addressing mental health needs.

Hurricanes Harvey and Maria

Additional lessons can be learned from Hurricanes Harvey and Maria. Hurricane Harvey illustrates a key aspect of climate change: the rapid escalation in intensity and the huge amounts of water vapor picked up by the storm over the warming waters of the Gulf of Mexico, later deposited over the flat terrain of Houston and surrounding areas in days of unremitting, drenching rain and severe flooding.[62] This is a pattern seen more often in recent years and is a particular concern to cities like Houston that have not been designed to efficiently and rapidly remove such torrential rainfalls.[63] It was a problem seen as well with Hurricane Sandy and other recent hurricanes. Save the Children has documented the lingering impact on children.[64]

Hurricane Maria did severe damage to Puerto Rico and the US Virgin Islands. Puerto Rico, with decaying and aged infrastructure (e.g., power plants, electric grid, roads, hospitals, housing, food distribution) had prolonged outages of services.[65] Distant from the US mainland, and with Maria coming soon after Hurricanes Harvey and Irma, Puerto Rico suffered greatly from delays in receipt of emergency services and longer-term support for recovery.[66] Maria also illustrated a well-known challenge in disaster settings: getting accurate data on mortality and other key parameters when the public health infrastructure has been decimated. Initial estimates of mortality were severe undercounts, and it took over a year for the most detailed assessments of mortality.[67] Pediatric surveys showed elevated rates for PTSD and other mental health issues among the multiple issues faced by children in this disaster, including loss or separation from family, friends, and neighbors as many departed for the mainland United States.[68,69] There were also major shortages of IV fluids throughout the United States as Puerto Rico produced 44% of the IV fluid bags used in the United States, and production and distribution were severely disrupted.[70]

Other Extreme Weather Events

Figure 64.2 also notes other major weather-related events in 2021: heat waves, drought, winter storms, and wildfires. Shockingly high, record-setting temperature extremes were seen in the Pacific Northwest heat wave in what many are seeing as the early stages of climate change impact; children, of course, are particularly vulnerable in heat waves. Floods also are becoming of greater concern with climate change (Figure 64.1 and Figure 64.2).[71]

Wildfires are a rapidly escalating concern globally and are a dangerous phenomenon for drought-affected forests in the US West (most notably California) (Figures 64.1 and 64.2); given the expanding frequency, size, and intensity of wildfires, as well as the wider geographic and seasonal spread, the link to climate change is very real. Children face many of the same issues as in other disasters (e.g., evacuation, displacement, mental health) and the potential for burn injuries. An emerging concern is the respiratory effects in children due to acute spikes in and wide geographic dispersion of air pollutants.[72-74]

Droughts are another extreme weather event that appears to be increasing under climate change. The impact can vary drastically depending on severity and circumstances. Among marginalized and very-low-income populations, as in parts of Africa, hunger and malnutrition are severe threats, often on a very large scale, and humanitarian relief efforts are crucial for survival, particularly among young children.[75] In the past few years, several very large cities such as Cape Town, South Africa, and Chennai, India, have almost run out of water[76]; this has ominous portents for the future. And, as noted above, in many parts of the world, including the US West, but also in Australia, Europe, and Asia, drought has been the fuel for causing and exacerbating devastating wildfires.[77]

Earthquakes

In the 21st century to date, 17 major earthquakes (defined as 1,000 or more fatalities) have occurred; seven (in China, Haiti, India, Indonesia, Iran, Japan, and Pakistan) were associated with greater than 20,000 fatalities each, and two (the coastal Sumatra, Indonesia tsunami, and the Haiti earthquake) exceeded 200,000 fatalities.[78,79] Although circumstances vary significantly among major earthquakes based on location, building practices, economic conditions, terrain, and population density, in all cases, the two most vulnerable groups are the elderly and young children.[80] Providing shelter, food, water, and public health and other services for these vulnerable groups is a major challenge. The Sichuan, China, earthquake occurred in the afternoon and was particularly devastating to schoolchildren because of the collapse of thousands of school buildings.[81]

Trauma to limbs, head, chest, and abdomen are frequent and often life-threatening. These injuries often can lead to amputations, spinal injuries, and disability. Within a short time of the earthquake, complications from preexisting chronic medical conditions come to the fore due to dislocations, lack of medication, and stresses of the disaster. Infections and other complications may appear during recovery. Particularly characteristic of earthquakes (although it may be seen in other traumatic settings such as war and terrorist events) is *crush syndrome*, with rhabdomyolysis and acute renal failure, which carries a high fatality rate—although with more rapid evacuation and the better availability of dialysis treatment in recent years survival from crush injuries has substantially improved (in the 1976 Tangshan, China, earthquake with more than 240,000 fatalities, as many as 20% may have been related to crush syndrome).[82]

Most striking in terms of the challenges to the well-being of children was the Haiti earthquake. Even before the earthquake, Haiti had the highest mortality rates among infants and children under 5 years of age in the Western Hemisphere. The earthquake greatly exacerbated the harsh lives of large numbers of children in Haiti.[83] Long-term follow-up studies among children in Haiti have been a challenge.

The Haiti earthquake was also notable for what seemed to be a quantum leap forward in the remarkably rapid deployment of sophisticated field hospital facilities that provided critical care services to severely ill and traumatized children.[84]

An Israeli field hospital, which was on site and fully operational within 89 hours of the earthquake, included a 16-bed pediatric ward, an 8-bed emergency department, and a neonatal unit with two incubators and a respirator; 363 pediatric patients were admitted to this facility during its 10 days of operation, including 57 who required surgery. The pediatric team included seven pediatricians, one pediatric surgeon, and six registered nurses.[85]

Other major efforts in Haiti were conducted by Project Medishare, a nongovernmental organization (NGO) focused on community health and development. In collaboration with the University of Miami Miller School of Medicine.[86] The US Naval Ship *Comfort* treated 237 pediatric surgical patients starting 1 week after the earthquake, representing yet another extraordinary relief effort.[87]

2004 Asian Tsunami

The tsunami resulting from a 9.2 Richter scale earthquake off the coast of Sumatra, Indonesia, caused at least 230,000 deaths in Indonesia, Sri Lanka, India, Thailand, and elsewhere. With modern technology, increasingly rapid health assessments are undertaken in the immediate aftermath of natural disasters such as the Asian Tsunami to identify the needs of the devastated communities and help guide humanitarian relief efforts. A rapid health assessment conducted several weeks after the tsunami in three communities of Aceh Jaya district, Indonesia, showed that almost 100% of the dwellings were destroyed in all three communities and that in the most severely affected of the three communities 70% of the population had died at the time of impact. Significant threats to health resulted from the almost complete lack of safe water and sanitation facilities, with rates of diarrhea in children younger than 5 years as high as 85% in the most severely affected community. Twenty-two percent of surviving households were hosting one or more orphans.[88] Highest mortality rates occurred in children younger than 10 years and in the elderly.[89]

Recent reports suggest that a prolonged recovery effort by international and national groups has led to a surprising degree of resiliency among surviving children as measured by growth rates and mortality rates starting after the immediate year of the disaster.[90,91]

Burn Disasters

Fire disasters, both human-made and natural, have plagued humankind for millennia. Mass casualty fires that can affect children, often causing several dozen deaths and hundreds of injured and hospitalized, occur from multiple causes:

1. Discotheque or club fires such as the one in Gothenburg, Sweden, in 1998, which killed 63 and wounded 213 teenagers[92]
2. Fireworks disasters, which can affect neighborhood children
3. Forest fires, wildfires, or bushfires (Figure 64.1)
4. Industrial explosions/fires that extend to nearby residents
5. Explosions and fires resulting from civil strife, war, or sabotage, and often caused by damaged pipelines or leaking tanker trucks[93]
6. Fires in refugee and displaced persons settlements[94]
7. Fires in crowded ferries, trains, and other transportation accidents
8. During the COVID-19 pandemic, multiple explosions of oxygen tanks in hospital wards causing scores of deaths[95]
9. Residential fires in high-rise apartment buildings due in some key instances to faulty building materials[96] or malfunctioning safeguards[97]; in schools or other

buildings with large numbers of children, flames and smoke can spread rapidly and extensively before residents and especially children have an opportunity to evacuate.

The primary health effects of fire disasters are severe burns, often leading to death or long-term disfigurement and disability; respiratory insufficiency from smoke inhalation; and carbon monoxide poisoning from incomplete products of combustion in enclosed spaces such as disco fires. In addition, and particularly in children, there are the ever-present mental health and developmental sequelae of living through and surviving such horrific events.[98]

Radiation/Nuclear Events

The AAP has issued technical and policy statements on Radiation Disasters and Children (Table 64.1) covering background information on radiation science; known radiation effects on children; evacuation, sheltering, and treatment of children post disaster with a detailed discussion of the use of potassium iodide; long-term effects focusing on cancer and psychological impact; and, finally, on disaster preparedness, public health actions, and prevention approaches. There have been three major radiation/nuclear disasters in recent decades.

1. The Three Mile Island nuclear accident highlighted the extraordinary potential effects of a nuclear incident, one that did not release high levels of radiation to the general population outside the facility.[99]
2. The Chernobyl nuclear plant explosion resulted in more than 100 cases of acute radiation sickness and led to a cleanup and containment operation involving hundreds of thousands of workers and a remediation effort which will be ongoing at least till the year 2065, a 30-kilometer exclusion zone where the population was evacuated and has not returned, severe psychological trauma in many strata of the population, and, finally, the occurrence of thousands of cases of thyroid carcinoma resulting from exposure of children to radioactive iodine released from the explosion.[100–102]
3. The Fukushima, Japan, nuclear reactor accident occurred on March 2011, following the earthquake and tsunami. Estimated thyroid radiation doses for children are apparently much lower than in Chernobyl and make long-term effects such as thyroid cancer in children very unlikely. Nevertheless, a massive effort to screen more than 250,000 children using very sensitive detection methods identified significant numbers requiring follow-up and, unfortunately, did little to assuage public concerns over radiation exposure to children.[103,104] As with Chernobyl, clean-up and remediation efforts will take decades.

Chemical Events

The prototype for a chemical disaster is the massive release of methyl isocyanate (MIC) that occurred in 1984, at a pesticide production facility in Bhopal, India. This event led to approximately 3,000 deaths, 100,000 acute illnesses, and the displacement of about 250,000 people in the adjacent community.[105,106] Because the release was sudden and

Figure 64.3. Lead poisoning epidemic associated with artisanal mining in 2010 in Nigeria. Horizontal entrance to gold mine, Zamfara, Anka LGA, Nigeria.
One of the entrances to one of the gold mines in Nigeria. Photos provided by James T. Durant, Agency for Toxic Substances and Disease Registry.

Figure 64.4. Lead poisoning epidemic associated with artisanal mining in Nigeria. Vertical entrance to gold mine, Zamfara, Anka LGA, Nigeria.
Alternative entrance to same mine, which descended vertically.

Figure 64.5. Lead poisoning epidemic in 2010. Ore grinding, Zamfara, Bukkuyum LGA, Nigeria.
Mined rock is first pulverized by hand to gravel, then feed multiple times through grain grinding machines. Between grinding, the ore is spread out to dry in the sun.

Figure 64.6. Lead poisoning epidemic in 2010. Ore grinding, Zamfara, Anka LGA, Nigeria. Ore grinding machines, and ground ore being bagged.

occurred without warning in the middle of the night, essentially the entire neighboring population, including many children, was affected. There were concerns for pulmonary, ocular, and mental health long-term sequelae in children. As a result of the many ongoing health concerns that continue even to the present, the Indian Council for Medical Research has recently created a Center in Bhopal focused on investigating long-term health effects in this population.

Not all chemical disasters are acute events as in Bhopal. Perhaps the world's most severe lead poisoning disaster occurred in 2010 Zamfara, Nigeria, due to widespread and prolonged contamination of the environment with very high lead levels from artisanal mining using crude equipment and techniques for processing the ore. Children in the involved villages were very heavily exposed; initial publications identified approximately 400 deaths among the children, and several thousand who developed severe

Figure 64.7. Lead poisoning epidemic in 2010. Ore washing, Zamfara, Anka LGA, Nigeria
Gold is gravimetrically concentrated using pans and water. Mercury is also used to create
amalgams of gold to aid in extraction. Air concentrations near washing operations were
measured at approximately 10 micrograms per cubic meter.

Figure 64.8. Lead poisoning epidemic in 2010. Mercury-gold amalgam melting, Zamfara,
Anka LGA, Nigeria
Ashed from a small fire (left) is used to heat the amalgams and vaporize the mercury,
leaving extracted gold. Child is holding small grain of extracted gold.

symptoms such as encephalopathy and coma along with long-term disabilities[107–109]
(See Figures 64.3–64.8)).

Decaying Infrastructure Events

The extensive scale of the infrastructure to support modern society (particularly urban
and industrial) requires complex structures whose failure can lead to major disasters;
many of these will heavily impact children along with others in society. Intensive efforts
at monitoring and maintaining all aspects of our infrastructure will be essential for the
safety of children, among others. This section offers three recent examples:

1. The collapse of a 12-story beachfront condo complex in Surfside, Florida: a Washington Post article depicts how the structural support for the building was damaged over time leading to its collapse[110]; in total, 97 people were confirmed dead, including entire families.[111] Similar high-rise residential building collapses have occurred elsewhere worldwide in recent years, and, just in the Miami, Florida, environs there are many more such buildings from the same era that will need to be carefully maintained and evaluated over time.

2. The 2019 Brumadinho tailings dam collapse in Brazil, which killed more than 250 people in its path and devastated huge forest and agricultural areas: the company that owned the dam has 133 mines in Brazil and this was not the first collapse[112]; globally, there are many more such dams.

3. The Merrimack Valley gas explosion centered around the town of Lawrence, Massachusetts, resulting from over-pressurization of the area's natural gas distribution system: an 18-year-old male died, 22 were hospitalized, 50,000 evacuated, and 131 structures damaged.[113] Many natural gas leaks leading to fires and explosions have been reported previously, usually on a smaller scale.

Conclusion

This chapter has highlighted the pediatric aspects of natural disasters and environmental emergencies. It emphasizes the great vulnerability of children in such settings and the critically important role of pediatricians and other healthcare providers to children. The chapter also notes worrisome trends related to global risk factors such as climate change, health disparities, poverty, decaying infrastructure, misuse of technology, and uncivil societies. The impact of natural disasters and environmental emergencies on children is not likely to disappear soon.

Additional excellent sources of information include the websites of key US government agencies, such as the Centers for Disease Control and Prevention (CDC), the Environmental Protection Agency (EPA), the Federal Emergency Management Agency (FEMA), the DHHS, the National Oceanic and Atmospheric Administration (NOAA) and many state and local health departments; key global agencies such as the World Health Organization (WHO), the United Nations Environment Program (UNEP), UNICEF, and many country agencies; news outlets, periodicals, non-governmental organizations, and foundations; and the published literature.

References

1. Rubin IL, Falk H, Mutic AD. Natural disasters and vulnerable populations: a commentary. *Int J Child Health Hum Dev.* 2019;12:303–18.

2. Redlener I, Markenson D. Disaster and terrorism preparedness: what pediatricians need to know. *Dis Mon.* 2004;50:6–40.

3. Markenson D, Redlener I. Pediatric terrorism preparedness national guidelines and recommendations: findings of an evidenced-based consensus process. *Biosecur Bioterror.* 2004;2:301–19.

 4. Bartenfeld MT, Peacock G, Griese S. Public health emergency planning for children in chemical, biological, radiological, and nuclear (CBRN) disasters. *Biosecur Bioterror.* 2004; 12:201–7.

 5. Gausche-Hill M. Pediatric disaster preparedness: are we really prepared? *J Trauma.* 2009;67:S73–6.

 6. Shirm S, Liggin R, Dick R, Graham J. Prehospital preparedness for pediatric mass-casualty events. *Pediatrics.* 2007;120:e756–61.

 7. Graham J, Shirm S, Liggin R, et al. Mass-casualty events at schools: a national preparedness survey. *Pediatrics.* 2006;117:e8–15.

 8. Agency for Healthcare Research and Quality (AHRQ). *National Commission on Children and Disasters: 2010 Report to the President and Congress.* AHRQ Publication No. 10-M037. Rockville, MD: AHRQ; 2010 Oct.

 9. Mace SE, Sharieff G, Bern A, et al. Pediatric issues in disaster management, Part 1: the emergency medical system and surge capacity. *Am J Disaster Med.* 2010;5:83–93.

10. Markovitz BP. Pediatric critical care surge capacity. *J Trauma.* 2009;67:S140–2.

11. Kanter RK. Strategies to improve pediatric disaster surge response: potential mortality reduction and tradeoffs. *Crit Care Med.* 2007;35:2837–42.

12. Kissoon N; Task Force for Pediatric Emergency Mass Critical Care. Deliberations and recommendations of the Pediatric Emergency Mass Critical Care Task Force: executive summary. *Pediatr Crit Care Med.* 2011;12:S103–8.

13. Gausche-Hill M, Ely M, Schmuhl P, et al. A national assessment of pediatric readiness of emergency departments. *JAMA Pediatr.* 2015;169:527–34.

14. 2015 National Report Card on Protecting Children in Disasters. Save the children. 2015. Available at: https://www.savethechildren.org/content/dam/usa/reports/emergency-prep/disaster-report-2015.pdf.

15. Healthcare preparedness for children in disasters: a report of the National Advisory Committee on Children and Disasters. *ASPR/DHHS.* 2015 Nov 13. Available at: https://aspr.hhs.gov/_catalogs/masterpage/ASPR/Documents/Boards%20and%20Committees%20Docs/healthcare-prep-wg-20151311.pdf.

16. Hamele M, Gist RE, Kissoon N. Provision of care for critically ill children in disasters. *Crit Care Clin.* 2019;35:659–75.

17. Gilchrist N, Simpson JN. Pediatric disaster preparedness: identifying challenges and opportunities for emergency department planning. *Curr Opin Pediatr.* 2019;31:306–11.

18. Remick K, Gross T, Adelgais K, et al. Resource document: coordination of pediatric emergency care in EMS systems. *Prehospital Emerg Care.* 2017;21:399–407.

19. Mann M, McMillan JE, Silver EJ, Stein REK. Children and adolescents with disabilities and exposure to disasters, terrorism, and the COVID-19 pandemic: a scoping review. *Curr Psychiatry Rep.* 2021;23:80. https://doi.org/10.1007/s11920-021-01295-z.

20. Rose DA, Murthy S, Brooks J, Bryant J. The evolution of public health emergency management as a field of practice. *AJPH.* 2017;107(Suppl2);S126–33.

21. Katz R, Attal-Juncqua A, Fischer JE. Funding public health emergency preparedness in the United States. *AJPH.* 2017;107(Suppl2):S148–52.

22. Watson CR, Watson M, Sell TK. Public health preparedness funding: key programs and trends from 2001 to 2017. *AJPH.* 2017;107(Suppl2): S165–7.

23. Dziuban EJ, Peacock G, Frogel M. A child's health is the public's health: progress and gaps in addressing pediatric needs in public health emergencies. *AJPH*. 2017;107(Suppl2):S134–7.

24. Mollica RF, Lopes Cardozo B, Osofsky HJ, et al. Mental health in complex emergencies. *Lancet*. 2004;364:2058–67.

25. Comer JS, Bry LJ, Poznanski B, Golik AM. Children's mental health in the context of terrorist attacks, ongoing threats, and possibilities of future terrorism. *Curr Psychiatry Rep*. 2016;18:79. doi10.1007/s11920-016-0722-1.

26. Lowe SR, Galea S. The mental health consequences of mass shootings. *Trauma Violence Abuse*. 2017;18:62–82.

27. Shultz JM, Thoresen S, Flynn BW, et al. Multiple vantage points on the mental health effects of mass shootings. *Curr Psychiatry Rep*. 2014;16:469. doi:10.1007/s11920-014-0469-5.

28. Masten AS and Osofsky JD. Disasters and their impact on child development: introduction to the special section. *Child Dev*. 2010;81:1029–39.

29. Brown EJ. Clinical characteristics and efficacious treatment of posttraumatic stress disorder in children and adolescents. *Pediatr Ann*. 2005;34:138–46.

30. Balaban V. Psychological assessment of children in disasters and emergencies. *Disasters*. 2006;30:178–98.

31. Pfefferbaum B, North CS. Research with children exposed to disasters. *Int J Methods Psychiatr Res*. 2008;17(S2):S49–56.

32. Lafortune S, Laplante DP, Elgbeili G, et al. Effect of natural disaster-related prenatal maternal stress on child development and health: a meta-analytic review. *Int J Environ Res Public Health*. 2021;18:8332. https://doi.org/10.3390/ijerph18168332.

33. Pfefferbaum B, Jacobs AK, Van Horn RL, Houston JB. Effects of displacement in children exposed to disasters. *Curr Psychiatry Rep*. 2016;18:71. doi:10.1007/s11920-016-0714-1.

34. Pfefferbaum B, Noffsinger MA, Jacobs AK, Varma V. Children's cognitive functioning in disasters and terrorism. *Curr Psychiatry Rep*. 2016;18:48. doi:10.1007/s11920-016-0685-2.

35. Gibbs L, Marinkovic K, Nursey J, et al. Child and adolescent psychosocial support programs following natural disasters: a scoping review of emerging evidence. *Curr Psychiatry Rep*. 2021;23:82. https://doi.org/10.1007/s11920-021-01293-1.

36. Pfefferbaum B, Nitiema P, Newman E, Patel A. The benefit of interventions to reduce posttraumatic stress in youth exposed to mass trauma: a review and meta-analysis. *Prehosp Disaster Med*. 2019;34:540–51.

37. Purgato M, Gross AL, Betancourt T, et al. Focused psychosocial interventions for children in low-resource humanitarian settings: a systematic review and individual participant data meta-analysis. *Lancet Glob Health*. 2018;6:e390–400.

38. Pfefferbaum B, North CS. Child disaster mental health services: a review of the system of care, assessment approaches, and evidence base for intervention. *Curr Psychiatry Rep*. 2016;18:5. https://doi.org/10.1007/s11920-015-0647-0.

39. Lautze S, Leaning J, Raven-Roberts A, et al. Assistance, protection, and governance networks in complex emergencies. *Lancet*. 2004;364:2134–41.

40. Williams R. The psychosocial consequences for children and young people who are exposed to terrorism, war, conflict and natural disasters. *Curr Opin Psychiatry*. 2006;19:337–49.

41. Guha-Sapir D, Schluter B, Rodriguez-Llanes JM, et al. Patterns of civilian and child deaths due to war-related violence in Syria: a comparative analysis from the Violation Documentation Center dataset, 2011–16. *Lancet Glob Health*. 2018;6:103–10.

42. Alipui N, Gerke N. The refugee crisis and the rights of children: Perspectives on community-based resettlement programs. In: Leckman JF, Britto PR, eds., *Towards a More Peaceful World: The Promise of Early Child Development Programmes*. New Directions for Child and Adolescent Development. 2018;159:91–8.

43. Checchi F, Warsame A, Treacy-Wong V, et al. Public health information in crisis-affected populations: a review of methods and their use for advocacy and action. *Lancet*. 2017;390:2297–2313.

44. Pyone T, Dickinson F, Kerr R, et al. Data collection tools for maternal and child health in humanitarian emergencies: a systematic review. *Bull World Health Organ*. 2015;93: 648–58A.

45. Slone M, Mann S. Effects of war, terrorism and armed conflict on young children: a systematic review. *Child Psychiatry Hum Dev*. 2016;47:950–65.

46. Stark L, Boothby N, Ager A. Children and fighting forces: 10 years on from Cape Town. *Disasters* 2009;33:522–47.

47. Robjant K, Koebach A, Schmitt S, et al. The treatment of posttraumatic stress symptoms and aggression in female former child soldiers using adapted Narrative Exposure therapy: a RCT in Eastern Democratic Republic of Congo. *Behav Res Ther*. 2019;123:103482. https://doi.org/10.1016/j.brat.2019.103482.

48. Falk H, Baldwin G. Environmental health and Hurricane Katrina. *Environ Health Perspect*. 2006;114:A12–13.

49. Centers for Disease Control and Prevention. Public health response to Hurricanes Katrina and Rita – Louisiana, 2005. *Morb Mortal Wkly Rep*. 2006;55:29–30.

50. Centers for Disease Control and Prevention. Public health response to Hurricanes Katrina and Rita – United States, 2005. *Morb Mortal Wkly Rep*. 2006;55:229–31.

51. Rath B, Young EA, Harris A, et al. Adverse respiratory symptoms and environmental exposures among children and adolescents following Hurricane Katrina. *Public Health Rep*. 2011;126:853–60.

52. Barkemeyer BM. Practicing neonatology in a blackout: the University Hospital NICU in the midst of Hurricane Katrina: caring for children without power or water. *Pediatrics*. 2006;117:S369–74.

53. Perrin K. A first for this century: closing and reopening of a children's hospital during a disaster. *Pediatrics*. 2006;117:s381–5.

54. Baldwin S, Robinson A, Barlow P, Fargason CA Jr. Moving hospitalized children all over the southeast: interstate transfer of pediatric patients during Hurricane Katrina. *Pediatrics*. 2006; 117:S416–20.

55. Distefano SM, Graf JM, Lowry AW, Sitler GC. Getting kids from the Big Easy hospitals to our place (not easy): preparing, improvising, and caring for children during mass transport after a disaster. *Pediatrics*. 2006;117:S421–7.

56. Needle S. Pediatric private practice after Hurricane Katrina: proposal for recovery. *Pediatrics*. 2008;122:836–42.

57. Lobato MN, Yanni E, Hagar A, et al. Impact of Hurricane Katrina on newborn screening in Louisiana. *Pediatrics.* 2007;120:e749–55.

58. Drury SS, Scheeringa MS, Zeanah CH. The traumatic impact of Hurricane Katrina on children in New Orleans. *Child Adolesc Psychiatr Clin N Am.* 2008;17:685–702.

59. Blake N, Stevenson K. Reunification: keeping families together in crisis. *J Trauma.* 2009;67:S147–51.

60. Jenkins JL, McCarthy M, Kelen G, et al. Changes needed in the care for sheltered persons: a multistate analysis from Hurricane Katrina. *Am J Disaster Med.* 2009;4:101–6.

61. Garrett AL, Grant R, Madrid P, et al. Children and megadisasters: lessons learned in the new millennium. *Adv Pediatr.* 2007;54:189–214.

62. Taylor A. The unprecedented flooding in Houston, in photos. *The Atlantic.* 2017 Aug 28. Available at: https://www.theatlantic.com/photo/2017/08/hurricane-harvey-leaves-houston-under-water/538215/.

63. Bogost I. Houston's flood is a design problem. *The Atlantic.* 2017 Aug 28. Available at: https://www.theatlantic.com/technology/archive/2017/08/why-cities-flood/538251/.

64. Kimball M, O'Quinn K, Rahlmian E, Sanborn B. Still at risk: Children one year after Hurricane Harvey. Children at Risk and Save the Children. 2018. Available at: https://catriskprod.wpengine.com/wp-content/uploads/2019/05/Still-At-Risk-Children-One-Year-After-Hurricane-Harvey.pdf.

65. Newkirk VR II. Puerto Rico's environmental catastrophe. *The Atlantic.* 2017 Oct 18. Available at: https://www.theatlantic.com/politics/archive/2017/10/an-unsustainable-island/543207/.

66. Mazzei P. Hunger and an "abandoned" hospital: Puerto Rico waits as Washington bickers. *New York Times.* 2019 Apr 7. Available at: https://www.nytimes.com/2019/04/07/us/puerto-rico-trump-vieques.html?referringSource=articleShare.

67. Santos-Burgoa C, Sandberg J, Suarez E, et al. Differential and persistent risk of excess mortality from Hurricane Maria in Puerto Rico: a time-series analysis. *Lancet Planet Health.* 2018;2:e478–88.

68. Orengo-Aguayo R, Stewart RW, de Arellano MA, et al. Disaster exposure and mental health among Puerto Rican youths after Hurricane Maria. *JAMA Network Open.* 2019;2:e192619. https://jamanetwork.com/journals/jamanetworkopen/articlepdf/2731679/orengoaguayo_2019_oi_190117.pdf.

69. Schonfeld DJ. A call for a better response to major disasters. *JAMA Network Open.* 2019;2:e192628. https://jamanetwork.com/journals/jamanetworkopen/articlepdf/2731675/schonfeld_2019_ic_190030.pdf.

70. Patino AM, Marsh RH, Niles EJ, et al. Facing the shortage of IV fluids: a hospital-based oral rehydration strategy. *N Engl J Med.* 2018;378:1475–7.

71. Mallett LH, Etzel RA. Flooding: what is the impact on pregnancy and child health? *Disasters.* 2018;42:432–58.

72. Henry S, Ospina MB, Dennett L, Hicks A. Assessing the risk of respiratory-related healthcare visits associated with wildfire smoke exposure in children 0–18 years old: a systematic review. *Int J Environ Res Public Health.* 2021;18:8799. https://doi.org/10.3390/ijerph18168799.

73. Holm SM, Miller MD, Balmes JR. Health effects of wildfire smoke in children and public health tools: a narrative review. *J Exposure Sci Environ Epidemiol.* 2021;31:1–20.

74. Cascio WE. Wildland fire smoke and human health. *Sci Total Environ.* 2018;624:586–95.

75. Asmall T, Abrams A, Roosli M, et al. The adverse health effects associated with drought in Africa. *Sci Total Environ.* 2021;793:148500. https://doi.org/10.1016/j.scitotenv.2021.148500.

76. Le Page M. A city without water. *New Scientist.* 2018 Feb 17;237(3165):20–1. doi:10.1016/S0262-4079(18)30301-4

77. Merzdorf J. A drier future sets the stage for more wildfires. NASA Global Climate Change-Vital Signs of the Planet. 2019 Jul 9. Available at: https://climate.nasa.gov/news/2891/a-drier-future-sets-the-stage-for-more-wildfires/.

78. Centers for Disease Control and Prevention. Assessment of health-related needs after tsunami and earthquake: three districts, Aceh Province, Indonesia, July–August 2005. *Morb Mortal Wkly Rep.* 2006;55:93–7.

79. Centers for Disease Control and Prevention. Post-earthquake injuries treated at a field hospital – Haiti, 2010. *Morb Mortal Wkly Rep.* 2011;59:1673–7.

80. Sullivan KM, Hossain SM. Earthquake mortality in Pakistan. *Disasters.* 2010;34:176–83.

81. Chan EY. The untold stories of the Sichuan earthquake. *Lancet.* 2008;372:359–62.

82. Gonzalez D. Crush syndrome. *Crit Care Med.* 2005;33:S34–41.

83. Balsari S, Lemery J, Williams TP, Nelson BD. Protecting the children of Haiti. *N Engl J Med.* 2010;362:e25.

84. Burnweit Stylianos S. Disaster response in a pediatric field hospital: lessons learned in Haiti. *J Pediatr Surg.* 2011;46:1131–9.

85. Farfel A, Assa A, Amir I, et al. Haiti earthquake 2010: a field hospital pediatric perspective. *Eur J Pediatr.* 2011;170:519–25.

86. von Saint Andre-von Arnim A, Brogan TV, Hertzig J, et al. Intensive care for infants and children in Haiti in April 2010. *Pediatr Crit Care Med.* 2011;12:393–7.

87. Walk RM, Donahue TF, Sharpe RP, Safford SD. Three phases of disaster relief in Haiti: pediatric surgical care on board the United States Naval Ship Comfort. *J Pediatr Surg.* 2011;46:1978–84.

88. Brennan RJ, Rimba K. Rapid health assessment in Aceh Jaya District, Indonesia, following the December 26 tsunami. *Emerg Med Australas.* 2005;17:341–50.

89. Doocy S, Rofi A, Moodie C, et al. Tsunami mortality in Aceh Province, Indonesia. *Bull World Health Org.* 2007;85:273–8.

90. Frankenberg E, Friedman J, Ingwersen N, Thomas D. Linear child growth after a natural disaster: a longitudinal study of the effects of the 2004 Indian Ocean tsunami (meeting abstract). *Lancet.* 2017;389:S21. https://www.thelancet.com/action/showPdf?pii=S0140-6736%2817%2931133-9.

91. Lepine A, Restuccio M, Strobl E. Can we mitigate the effect of natural disasters on child health? Evidence from the Indian Ocean tsunami in Indonesia. *Health Econ.* 2021;30:432–52.

92. Cassuto J, Tarnow P. The discotheque fire in Gothenburg 1998: a tragedy among teenagers. *Burn.* 2003;29:405–16.

93. Onuoha FC. Why the poor pay with their lives: oil pipeline vandalisation, fires and human security in Nigeria. *Disasters.* 2009;33:369–89.

94. Kazerooni Y, Gyedu A, Burnham G, et al. Fires in refugee and displaced persons settlements: the current situation and opportunities to improve fire prevention and control. *Burns.* 2016;42:1036–46.

873 NATURAL DISASTERS AND EMERGENCIES

95. Arraf J. Death toll rises to 92 in fire that gutted Iraq hospital coronavirus ward. *New York Times*. 2021 Jul 13; updated 2021 Jul 15. Available at: https://www.nytimes.com/2021/07/13/world/middleeast/iraq-hospital-fire.html?referringSource=articleShare.

96. Schaverien A. Days before Grenfell anniversary, giant warnings on fire safety appear in U.K. *New York Times*. 2019 Jun 13. Available at: https://www.nytimes.com/2019/06/13/world/europe/grenfell-tower-fire-safety.html?referringSource=articleShare.

97. Madani D, Romero D. Malfunctioning space heater blamed in Bronx fire that killed 10 adults, 9 children. *NBC News*. 2022 Jan 9. Available at: https://www.nbcnews.com/news/us-news/numerous-fatalities-anticipated-5-alarm-bronx-fire-fdny-commissioner-s-rcna11524.

98. Sheridan RL, Friedstat J, Votta K. Lessons learned from burn disasters in the post-9/11 era. *Clin Plastic Surg*. 2017. https://www.plasticsurgery.theclinics.com/article/S0094-1298(17)30017-2/pdf.

99. Talbott EO, Youk AO, McHugh-Pemu KP, Zborowski JV. Long-term follow-up of the residents of the Three Mile Island accident area: 1979–1998. *Environ Health Perspect*. 2003;111:341–8.

100. Peplow M. Chernobyl's legacy. *Nature*. 2011;471:562–5.

101. Rahu M. Health effects of the Chernobyl accident: fears, rumours and the truth. *Eur J Cancer*. 2003;39:295–9.

102. Baverstock K, Williams D. The Chernobyl accident 20 years on: an assessment of the health consequences and the international response. *Environ Health Perspect*. 2006;114:1312–7.

103. Nagataki S, Takamura N. A review of the Fukushima nuclear reactor accident: radiation effects on the thyroid and strategies for prevention. *Curr Opin Endocrinol Diabetes Obes*. 2014;21:384–93.

104. Yamashita S, Takamura N, Ohtsuru A, Suzuki S. Radiation exposure and thyroid cancer risk after the Fukushima nuclear power plant accident in comparison with the Chernobyl accident. *Radiation Protection Dosimetry*. 2016;171:41–6.

105. Dhara VR, Dhara R. The Union Carbide disaster in Bhopal: a review of health effects. *Arch Environ Health*. 2002;57:391–404.

106. Koplan JP, Falk H, Green G. Public health lessons from the Bhopal chemical disaster. *JAMA*. 1990;264:2795–6.

107. Dooyema CA, Neri A, Lo Y-C, et al. Outbreak of fatal childhood lead poisoning related to artisanal gold mining in Northwestern Nigeria, 2010. *Environ Health Perspect*. 2012;120:601–7.

108. Burki TK. Nigeria's lead poisoning crisis could have a long legacy. *Lancet*. 2012;379:792.

109. Kaufman JA, Brown MJ, Umar-Tsafe NT, et al. Prevalence and risk factors of elevated blood lead in children in gold ore processing communities, Zamfara, Nigeria, 2012. *J Health Pollut*. 2016;6(11):2–8.

110. Swaine J, Brown E, Lee JS, Mirza A, Kelly M. How a collapsed pool deck could have caused a Florida condo building to fall. *Washington Post*. 2021 Aug 12. Available at: https://www.washingtonpost.com/investigations/interactive/2021/pool-deck-condo-collapse/.

111. Firozi P, Bells T, Paul ML. Search for Florida condo collapse victims nears end as more bodies are identified. *Washington Post*. 2021 Jul 15. Available at: https://www.washingtonpost.com/nation/2021/07/15/surfside-condo-collapse-victims-recovered/.

112. Silva Rotta LH, Alcantara E, Park E, et al. The 2019 Brumadinho tailings dam collapse: possible cause and impacts of the worst human and environmental disaster in Brazil.

Intl J Appl Earth Observ Geoinfor. 2020;90:102119. https://reader.elsevier.com/reader/sd/pii/S0303243420300192?token=3691274A5DE6A41880212BD0042313D8BCD9B024C47DA16AC4E3A0C4FF4512AC47CBFB2E807517BF1A4D86C58D0F2383&originRegion=us-east-1&originCreation=20220214184611.

113. Ly L. Merrimack Valley gas explosions were caused by weak management, poor oversight, NTSB says. *CNN*. 2019 Sep 24. Available at: https://www.cnn.com/2019/09/24/us/ma-gas-explosions-cause/index.html.

65

New Frontiers in Children's Environmental Health

Philip J. Landrigan and Ruth A. Etzel

Children's environmental health has made great progress in the past three decades. Centers of Excellence focusing on key issues in children's environmental health have been formed in major universities in both the Global North and the Global South. New research in these Centers has discovered previously unrecognized environmental causes of disease in children. Findings from this research have been translated into new approaches to disease treatment and prevention. Prospective birth cohort epidemiological studies have been launched to study the short- and longer-term impacts of early-life environmental exposures on children's health and development. Training programs have been established to educate the next generation of pediatric practitioners, epidemiologists, and basic biological scientists. The American Academy of Pediatrics has elevated the former Committee on Environmental Health to Council status and added "Climate Change" to the title; it is now designated the Council on Environmental Health and Climate Change. National and international conferences have been convened. The number of scientific publications in the field has increased sharply. The World Health Organization (WHO) has created a global initiative for children's environmental health. All of these advances have built credibility for children's environmental health, improved children's health and well-being, and saved lives.

Future progress in this still young science will require continuing close study of both the global and local factors in the environment that influence children's health. It will require further advances in interdisciplinary research using new tools from epidemiology, genetics, epigenetics, genomics, exposure science, economics, and urban planning. It will require evidence-based advocacy at the state, national, and international levels to undo the mistakes of the past and to prevent their perpetuation into the future. It will require education of a new generation of leaders. And, last, continuing progress in children's environmental health will require increased international collaboration in research, education, and advocacy to protect the health of children in countries around the world.

Major areas in which children's environmental health can contribute in the years ahead to improving children's health and protecting children from environmental threats include an increased emphasis on the intersection between planetary health and children's health; basic and translational research; disease and hazard tracking; enhanced population biomonitoring for toxic chemicals; development of new approaches to prevention of chemical toxicity; new approaches to risk assessment/risk management; education of students, physicians, parents, and policymakers; advances in clinical practice; social justice and environmental equity; and advocacy (Table 65.1).

Philip J. Landrigan and Ruth A. Etzel, *New Frontiers in Children's Environmental Health* In: *Textbook of Children's Environmental Health*, Second Edition. Edited by: Ruth A. Etzel and Philip J. Landrigan, Oxford University Press. © Oxford University Press 2024. DOI: 10.1093/oso/9780197662526.003.0065

Table 65.1 Areas of focus in children's environmental health in the 21st century

---

Threats to planetary health: climate change, pollution, and biodiversity loss
Research: Basic and translational
Disease and hazard tracking
Enhanced population biomonitoring for toxic chemicals
New approaches to prevention of chemical toxicity
New approaches to risk assessment
Education of students, residents, fellows, parents, and policymakers
Advances in clinical practice
Social justice and environmental equity
Advocacy

Threats to Planetary Health That Affect Children's Health: Climate Change, Pollution, and Biodiversity Loss

These three interlinked planetary-scale threats pose clear and present dangers to the health of all persons, and they put children at disproportionately high risk.[1,2] These factors interact with one another in complex ways, and they are all driven by the massive, unprecedented, rapid growth in the world's population and by increases in global consumption. They have become major foci of research in children's environmental health.

Climate change is the most obvious and urgent of these dangers. It is already causing disease and death in children through heat waves, increased frequency of violent storms, coastal flooding, wildfires, droughts, and famine. Climate change also is influencing the spread of vector-borne diseases as higher temperatures favor the wider extension of habitat for certain disease vectors.[3] Thus diseases such as malaria and dengue are expected to occupy more extensive ranges in future years than they do today. All of these effects are most severe among poor and marginalized populations.[4] Children in small island nations face particular hazards. (Chapter 4).

Pollution—including pollution of air, water, and soil as well as chemical pollution—kills an estimated 9 million persons each year, of whom at least 1 million are children[5] (Chapters 25 and 26). The WHO finds that approximately one death in four among children worldwide is due to unhealthy, polluted environments.[6]

Biodiversity loss results in reduced food yields and undernutrition, increased risk of new infectious diseases, depletion of natural medicine, aesthetic and cultural impoverishment, and reduced mental health in children (Chapter 24).

Research

Interdisciplinary research will be the engine that drives children's environmental health forward in the 21st century. The discovery through research of new environmental threats to children's health will stimulate advocacy, guide pediatric practice and catalyze the enactment of child-protective policies in countries around the world.

Two platforms that have greatly enhanced research in children's environmental health are specialized research centers and large-scale, prospective epidemiological studies.

Collaborative Centers in Children's Environmental Health Research and Translation

National networks of research centers in children's environmental health have been established over the past two decades in the United States and in Korea. In the United States, Collaborative Centers in Children's Environmental Health Research and Translation are supported by the National Institute of Environmental Health Sciences. These Centers conduct multidisciplinary basic and applied research in combination with community-based projects to study the causes and mechanisms of children's developmental disorders, with special emphasis on discovery of environmental exposures that may put children at risk.

These Centers have made important discoveries of environmental causes of disease in children. For example, three studies from the Centers published in 2011 presented compelling new data linking in utero exposure to organophosphate pesticides with decrements in early cognitive development in children.[8-10] Deficits in full-scale IQ, working memory, and perceptual reasoning were consistently observed in these three studies among children exposed in utero to organophosphate insecticides. Other reports from the Centers have linked prenatal exposures to endocrine-disrupting chemicals such as phthalates and bisphenol A to problems in development of the brain and reproductive organs.[11,12]

It will be very important in the years ahead to continue to support independent research centers in children's environmental health. These Centers are critical generators of new knowledge and also incubators for the next generation of leaders in children's environmental health.

Prospective Epidemiological Studies: Critical Engines of Discovery

Large-scale, prospective epidemiological studies, such as the the Norwegian Mother, Father and Child Cohort Study[13] and the Japan Environment and Children's Study[14] are critical engines for discovering the short- and longer-term health consequences of children's early-life exposures to toxic chemicals and other emerging technologies. These studies enrol tens of thousands of infants and their families during pregnancy, measure multiple environmental exposures during pregnancy and in early postnatal life, and then prospectively follow children born to study mothers until adolescence or even adult life. Typically they enrol mothers during (or even before) pregnancy, and they assess exposures during pregnancy in real time, as those exposures are actually occurring.

The great strength of the early and unbiased approach to exposure assessment embodied in prospective studies is that it facilitates the discovery of links between early exposures and disease or dysfunction and minimizes recall bias. The broad-scale discovery research undertaken through these large studies complements and informs the more focused, hypothesis-based research of the Centers, and research findings from the Centers have the potential to generate hypotheses that can guide prospective studies.

Large-scale epidemiological studies have the potential, if they are carefully designed and undertaken, to be one of our generation's most enduring legacies in child health. These investigations have the power to discover new hazards that result from emerging technologies and to identify critical windows of vulnerability in early development.

These studies will not only make scientific breakthroughs of great importance, but they will generate enormous economic benefits through the diseases and premature deaths that they will avert and the increases in children's achievements that they will foster. It is critical that such studies maintain a balanced focus that examines environmental as well as genetic influences on early development (Chapter 10).

Disease and Hazard Tracking

In the years ahead it will be very important to strengthen disease surveillance systems in countries around the world to better track diseases of environmental origin in children. It will also be important to strengthen systems for tracking environmental hazards that threaten children's health (Chapter 7). Hazard tracking systems include air and water monitoring networks using satellite-based networks and international control systems and treaties to monitor the global movement of hazardous chemicals and pesticides (Chapter 62). Disease and hazard tracking systems are often supported by geographic information systems (GIS), which make possible the visual display of quantitative information (Chapter 8).

Enhanced Biomonitoring of Populations for Toxic Chemicals

Biomonitoring programs, such as the US Centers for Disease Control and Prevention (CDC)'s National Biomonitoring Program, that measure levels of chemicals in the bodies of human populations including pregnant and lactating women and young children are critical for assessing population-wide exposures to toxic chemicals.[15] They also play an important role in public education, raising awareness of the ubiquity of chemical threats to children's health.

New Approaches to Prevention of Chemical Toxicity

To better defend children against the unforeseen consequences of emerging technologies and to avoid repetition of tragedies past, countries around the world need urgently to adopt a new framework for risk assessment/risk management and new strategies for responsible stewardship of new chemicals and technologies. A legally mandated, strictly enforced requirement that all new technologies be examined for potential toxicity before commercial introduction must be a cornerstone of this new strategy.

In 2007, the European Union began implementation of a new chemical regulatory system—Registration, Evaluation, Authorisation and Restriction of Chemicals (REACH). REACH is based on the *precautionary principle*: it assumes that chemicals are harmful until shown to be safe, and it requires premarket assessment of all new chemicals and technologies before they are allowed to enter markets[16] (Chapter 59). Under REACH, the chemical industry is required to generate substantial amounts of data on potential risks of commercial chemicals and to register this information in a central database. The extent of required testing depends on the volume of chemical production; high-volume chemicals intended for consumer products will require closer scrutiny than low-volume materials.

Echoing REACH, a bipartisan coalition of US lawmakers, led by the late Senator Frank Lautenberg of New Jersey enacted the Senator Frank R. Lautenberg Chemical Safety for the 21st century Act in 2016.[17] The Lautenberg Act gave the EPA enhanced authority to require toxicity testing of both new and existing chemicals. It explicitly requires that chemical standards protect the health of children and pregnant women. Key provisions of the Act

- Require that new chemicals be shown to be safe before they can enter the market;
- Require safety reviews for chemicals already on the market;
- Replace the Toxic Substances Control Act (TSCA)'s "unreasonable risk"/cost-benefit safety standard with a solely health-based standard; and
- Limit chemical companies' ability to withhold information on chemicals from governmental agencies and the public by claiming that the information is "confidential."

New Approaches to Risk Assessment

Risk assessment is a structured process that evaluates the hazardous properties of a chemical and then determines the magnitude of the resulting risks to human health.[18] Risk assessments are typically based on extrapolations of data from toxicological and epidemiological studies. Risk assessment provides the quantitative basis for most public health and environmental regulation and rule-making (Chapter 58).

Risk assessment has generally not worked well to protect children from environmental threats to health because it has focused on protecting the health of the "average adult" and has had little access to information about the effects of chemicals on children's health. This lack of information on hazards to the young precludes assessment of developmental toxicity and makes it impossible to study late consequences of early exposures.[19] To better protect children in the future, a new, more explicitly child-protective approach to risk assessment will be required.[20–21] This approach is discussed in detail in the concluding section of this chapter.

Education

Education is a powerful strategy for disease prevention that complements and extends legal and regulatory efforts to protect children against disease of environmental origin. Educational efforts in children's environmental health need to be undertaken on several levels.

- General medical education to increase the knowledge of all physicians and health care providers about the basic concepts and principles of children's environmental health;
- Specialty training to prepare tomorrow's leaders in children's environmental health;
- Education of parents about how to protect their children against environmental threats to health; and
- Education of policymakers and elected officials about children's vulnerabilities and the need to include consideration of these unique vulnerabilities in all public policies.

General Medical Education in Children's Environmental Health

There is great need to educate all physicians, nurses, and other health professionals about basic concepts in children's environmental health. Medical and nursing education in most countries around the world has paid scant attention to these issues until recently, and this lack of training is reflected in most providers' inability to recognize environmental health problems. In the United States, for example, the average medical student receives only 6 hours of training in environmental medicine in the 4 years of medical school.[22] Even pediatric residency programs provide little education on topics in environmental health except perhaps on the most fundamental and widely acknowledged problems such as asthma and childhood lead poisoning. Not surprisingly, most physicians and other primary healthcare providers are not knowledgeable about even the most common problems in environmental health. Because of this lack of education in children's environmental health, it is likely that many illnesses of environmental origin in children are not diagnosed, that proper treatments are not provided, and that opportunities for prevention are lost.

Attempts are being made to improve the state of environmental medical education. In the United States, the National Academy of Medicine has convened expert committees to increase the dissemination of information on the teaching of environmental and occupational medicine to medical students, residents, and physicians. The Agency for Toxic Substances and Disease Registry (ATSDR) supports the national network of Pediatric Environmental Health Specialty Units (PEHSUs). The PEHSUs have been very effective in educating a wide range of health professionals in fundamental concepts of children's environmental health (Chapter 56). The state of New York has recently expanded the PEHSU network and created a statewide network of Centers of Excellence in Children's Environmental Health.[23] Canada, Mexico, Chile, Colombia, Argentina, Uruguay, Spain, Korea, and Thailand are among the countries around the world that have adopted this concept and established Pediatric Environmental Health Units modeled on the PEHSUs.

Specialty Training in Children's Environmental Health

Advanced training programs in children's environmental health have been launched to educate pediatricians and research scientists who wish to focus their professional careers in this discipline. The graduates of these advanced training programs will be the next generation of leaders in children's environmental health. The first of these programs was launched in the United States in 2001, by the Ambulatory Pediatric Association, now the Academic Pediatric Association.[24-25] Programs exist today in several major universities in North America and are supported by the National Institutes of Health, major foundations, and philanthropic donors. They provide trainees with formal didactic curricula, clinical experience in children's environmental health, and opportunities for mentored research. These programs are typically of either 2 or 3 years in duration. They train a mix of clinically oriented pediatricians and research-trained (PhD) scientists.

Parental Education

Education of parents, children, and the public about environmental threats to health and about basic strategies for the protection of children are of great importance not only for safeguarding children's health here and now, but also for building broad-based under-standing and support for children's environmental health.

Full disclosure of all the information on toxicity of chemicals that is now hidden be-hind claims of trade secrecy is a powerful tool for the education of parents and com-munities. Information about the release of chemicals into the environment is also very important. Such information becomes especially powerful when it is plotted on maps using GIS technology and patterns of pollution are portrayed in relation to local com-munities. Parents who are informed about the risks of exposure to a contaminant for their children or in their communities can be powerful and well-informed advocates on their children's behalf.

Education of Policymakers

Educating policymakers about children's unique vulnerabilities and about the need to protect children from environmental threats to health is very important. Policymakers include elected officials at the federal, state, and local levels as well as senior officials in government agencies. Policymakers of potentially great importance in children's envi-ronmental health also include religious leaders and members of the press. Time spent in educating and inspiring these persons of great influence is time well spent.

Advances in Clinical Practice

The importance of the pediatric practitioner as a critical first line of defense against the unanticipated hazards of environmental hazards cannot be overemphasized. Regardless of how extensively a chemical or a technology has been tested for possible hazard, the possibility will always remain that unforeseen, previously undetected harmful effects will become evident only after the technology or product has been widely disseminated and millions of children exposed to it. In the case of the many chemicals that have never been tested for toxicity, it is sadly almost inevitable that the first recognition of toxicity will be made by an alert physician or a nurse who is caring for a sick child with unex-plained illness. It therefore falls to the pediatric practitioner to remain always alert to the possibility of environmental causation of disease, to discover novel outcomes, to recog-nize their significance as "sentinel health events," and to act on that information[26] This has happened time and again in the past and undoubtedly will happen again in the future (Chapter 3 and Chapter 56).

Social Justice and Environmental Equity

Environmental injustice—the unequal distribution of environmental hazards among children of different racial, ethnic, or socioeconomic groups, and between the Global North and the Global South—is one of the great evils of our time.[27] Poor children, children of color, and Indigenous children are heavily and often disproportionately exposed to a multitude of toxic environmental hazards. These inequitably distributed exposures include lead, industrial and automotive air pollution, and effluvia from toxic waste disposal sites. They result in disproportionately high rates of disease, disability, and premature death (Chapter 11). Environmental injustice needs to be a matter of great concern for children's environmental health in the 21st century.

To address environmental injustice for children on a global scale, the UN Special Rapporteur on Human Rights and the Environment and UNICEF have elaborated the concept of a Child's Right to a Healthy Environment.[28,29] The Rights concept is a powerful tool to protect children from the impacts of environmental degradation and climate change. It deserves the full attention of all who care for the health of children.

Advocacy

Driven in part by widespread frustration over the shortcomings of governmental approaches to risk assessment and regulation of toxic chemicals, an extensive grassroots advocacy movement in children's environmental health has developed over the past three decades in countries around the world. The goals of this movement are to educate parents and families about environmental hazards to children, support research, and bring about changes in public policy. This movement has become increasingly effective in successfully making changes. Advocacy groups for environmental health have had great success in communicating their concerns to policymakers. Among these groups are the Natural Resources Defense Council, the Children's Environmental Health Network, and Physicians for Social Responsibility.

Pediatricians and pediatric researchers have played important roles in catalyzing evidence-based advocacy in children's environmental health. A great contribution that pediatricians and other trained healthcare professionals can offer to advocacy campaigns is that they are deeply trusted and well equipped to help develop evidence-based strategies for effective advocacy. Pediatricians can work with community groups to help identify the correct targets of advocacy, delineate the likely risks to children's health from environmental exposures, and develop feasible strategies for risk reduction and the protection of children's health. The great respect and trust that pediatricians enjoy in most communities makes them exceptionally powerful and credible advocates for children's health.

Global Challenges to Children's Environmental Health in the 21st Century

The environment in which children live is changing rapidly and these changes will accelerate in the years ahead (Table 65.2). Some of the changes in the global environment that will greatly influence children's health are described in the following paragraphs.

Table 65.2 Challenges in children's environmental health in the 21st century

---
Threats to planetary health: Climate change, pollution, and biodiversity loss
Rapid growth in the world's population
Rapid urbanization
Emergence of new and largely untested synthetic chemicals and other new technologies
Globalization
Widespread poverty
Disparities in educational attainment
Global spread of "Western" values and the "Western" lifestyle
---

Rapid Population Growth

Population growth is a major determinant of child health and will become even more important as we move deeper into the 21st century. World population today is 8 billion, including 2.4 billion children under age 18; 2.0 billion of these children live in low- and middle-income countries (LMICs).[30,31] By 2050, the world population is projected to reach 9.8 billion. This is the most rapid and massive growth in the human population ever recorded. Population growth is a major driver of global environmental change.

Despite overall growth in the world's population, many industrially developed nations such as Japan and Italy are today or will be soon be in negative population growth, with more deaths each year than births. Thus virtually all of the anticipated growth in the world's population will take place in LMICs.[31] When population growth in these countries outstrips basic resources such as food, water, sanitation, and access to healthcare, ill health results.

Increasing Urbanization

In 2008, for the first time in human history, more than half of the world's population lived in cities.[32] The trend toward urbanization continues to accelerate. The greatest concentration of large cities in the world today is in China where there are now 160 cities with populations of greater than 1 million. Africa is rapidly urbanizing and will have 13 megacities with populations of over 30 million by 2100.[33]

Urbanization creates unparalleled opportunities for human development and for the generation of new ideas, industries, and technologies, but it also carries significant risks to health, especially in the new mega-cities of the developing world. These rapidly expanding conurbations often concentrate populations in coastal areas and along tectonic fault lines, places subject to devastating natural disasters and rising sea levels. Another problem is that the growth of cities often comes at the loss of prime agricultural land and may lead to water source vulnerability. The density of city populations places communities at risk for infectious disease transmission and massive energy constraints.

Emergence of New Chemicals and New Technologies

More than 350,000 new chemicals and chemical mixtures—novel materials that never before existed on earth—have been invented since 1950.[34] These new chemicals and other new technologies have time and again shaped and reshaped the lives of children. In the 20th century, mass-produced automobiles and commercial aviation revolutionized transport. Refrigeration brought fresh fruits in winter and led to sharp reductions in stomach cancer. New building materials and motor fuels made possible modern cities and their suburbs, but brought with them the unintended consequences of sedentary lifestyles, obesity, and diabetes. Breakthroughs in microelectronics and physics produced desktop computing and the internet.

Some of these new chemicals and technologies have greatly benefitted children's health. Disinfectants have sterilized drinking water, antibiotics have controlled infectious diseases, and chemotherapeutic agents have made possible the successful treatment of many cancers.

But too many new chemicals and technologies have led to widespread disease and environmental degradation. Classic examples of technologies initially hailed as beneficial but later found to cause great harm include the addition of lead to paint and later to gasoline, asbestos, DDT, thalidomide, polychlorinated biphenyls (PCBs), diethylstilbestrol (DES), and the ozone-destroying chlorofluorocarbons (CFCs). More recent examples include phthalates, bisphenol A, brominated flame retardants, neonicotinoid insecticides, and the per- and polyfluoroalkyl substances (PFAS).[20,21]

A root cause of many of the negative consequences of new chemicals and technologies has been systematic failure to conduct premarket evaluations of their potential toxicity before their introduction to commerce. New chemicals and new technologies are presumed safe and allowed to enter markets until the evidence that they cause harm is overwhelming. Untested new chemicals and new technologies will continue to pose threats to children's health in the years ahead until laws and regulations are established internationally and in individual countries that ensure their proper stewardship (Chapter 3).

Globalization

Steadily increasing global trade results in accelerated movement of toxic materials—older, well-known toxic chemicals as well as newer synthetic chemicals and pesticides—from high-income to LMICs. Toxic chemicals, hazardous industrial processes, and toxic waste no longer tolerated in North America, Japan, or Western Europe are exported to LMICs, where populations are already exposed to lead,[35,36] highly toxic pesticides,[37,38] and products made of asbestos[39] in circumstances where protective controls are weak or nonexistent.[40] In India, for example, consumption of asbestos, much of it destined for use in residential building materials where children are at risk of exposure, is increasing at a rate of 10% per year. The consequences for health of exposures to highly toxic chemicals in LMICs include cancer, birth defects, and sterility.[41]

To protect children against the harmful consequences of exposure to toxic substances in international commerce, a series of treaties and conventions among countries have been established. These treaties are discussed in detail in Chapter 62. In the future, these treaties will need to be extended and enforced.

Widespread Poverty

The relationship between poverty, ill health, and shortened life expectancy has been recognized for at least 200 years[42] (Chapter 11). Stark economic inequities continue to plague the world today.[43] The richest 12% of the world's population owns 85% of the world's assets, while 1.4 billion people live in extreme poverty, with incomes under US$1 per day.[44]

Great inequities in health result from these economic disparities. These inequities exist both between and within countries. For example, infant mortality in the world's poorest countries is approximately 100 per 1,000 live births compared to 6 per 1,000 in the United States.[45] In the world's poorest countries, 16% of the population is undernourished. Life expectancy in the poorest countries is only about half that of the industrialized world. Thus life expectancy at birth in the Central African Republic is 55 years, compared with 77 years in the United States and 81 years in the European Union.[45] In

2018, the average White American was expected to live nearly 6 years longer than the average African American (77.6 vs. 71 years).[46]

Disparities in Educational Attainment

Lack of education is a powerful driver of population growth and thus a major influence on child health. And, on the positive side, education of girls and young women has repeatedly been shown to be one of the most effective means for slowing population growth and thus improving child health.[47,48] Though there has been some improvement in global primary school enrolment over the past decade, there remain regions of the world, such as sub-Saharan Africa and parts of the Middle East, where nearly one-quarter of school-age children do not receive even minimal school-based education.

Global Spread of "Western" Values and the "Western" Lifestyle

The global spread of "Western" values and the "Western" lifestyle is another powerful determinant of health and disease.[49] Multinational corporations target children living in low- and middle-income countries with advertising campaigns, persuading young Cambodians, Bolivians and Malawians that they must have a hamburger, a beer, and a cigarette. Centuries-old diets and traditional behavioral norms and values erode. The result is that noncommunicable diseases once seen only in high-income countries are now epidemic in LMICs. Today nearly 20% of urban schoolchildren in China are overweight or obese. Heart disease, diabetes, and cancer have become major killers in Asia and sub-Saharan Africa.[50] And with over 50% of the world's population now living in cities, this spread of "Western diseases" to the LMICs will likely continue to accelerate.

Conclusion

The protection of children against environmental threats to health is one of the grand challenges of the 21st century. The academic discipline of children's environmental health must be at the forefront of the effort to protect children against these hazards.

The continuing development of new chemicals and new technologies poses a major challenge to children's environmental health. Hundreds of new chemicals are developed every year and are released into the environment. Under current regulatory regimes, too many of these chemicals are untested for their toxic effects. Thus, children's exposure to these untested materials will almost certainly continue to increase.

Global environmental change, climate change in particular, is a second major challenge. The threat of environmental degradation looms large in many countries and is especially grave in LMICs with rapidly expanding populations and weak infrastructure.

To meet these challenges and protect the health of children, we suggest that a new paradigm in health and environment health policy is urgently needed in countries around the world—a paradigm that is explicitly designed to protect the health and well-being of children.[51] Responsible stewardship should be its guiding principle. This new paradigm needs to be based on prudence and precaution. This new framework must overturn the dangerous and outdated assumption that new chemicals and new technologies pose no risks to human health or the environment until they are demonstrated beyond all doubt to cause harm.[20-21] It must overcome the industry tactic of weaponizing scientific uncertainty to delay preventive action. The burden of proof needs to be shifted. Under this new paradigm, safety must be documented and not merely presumed before a new product or technology can be brought to market.

This new paradigm of responsible stewardship calls for a new way of thinking. It will require a retooling of risk assessment to take into account the unique exposures and increased vulnerability of children.[51] It will recast environmental policy. It will prevent recapitulation of the tragedies of the past. It will preserve the earth, our Common Home.[52] It will safeguard the future for our children.[7]

References

1. Perera F. *Children's Health and the Peril of Climate Change*. New York: Oxford University Press; 2022.

2. Nadeau K, Perera F, Salas RN, Solomon CG. Climate, pollution, and children's health. *N Engl J Med*. 2022 Nov 3;387(18):e45.

3. Campbell-Lendrum D, Manga L, Bagayoko M, Sommerfeld J. Climate change and vector-borne diseases: what are the implications for public health research and policy? *Philos Trans R Soc Lond B Biol Sci*. 2015 Apr 5;370(1665):20130552.

4. Levy BS, Patz JA. Climate change, human rights, and social justice. *Ann Glob Health*. 2015 May-Jun;81(3):310–22.

5. Fuller R, Landrigan PJ, Balakrishnan K. Pollution and health: a progress update. *Lancet Planetary Health*. 2022;6(6):e535–e547. doi:10.1016/S2542-5196(22)00090-0.

6. World Health Organization. Children's environmental health. 2023. Available at: https://www.who.int/health-topics/children-environmental-health#tab=tab_1.

7. Whitmee S, Haines A, Beyrer C, et al. Safeguarding human health in the Anthropocene epoch: report of the Rockefeller Foundation-Lancet Commission on planetary health. *Lancet*. 2015 Nov 14;386(10007):1973–2028.

8. Bouchard MF, Chevrier J, Harley KG, et al. Prenatal exposure to organophosphate pesticides and IQ in 7-year old children. *Environ Health Pespect*. 2011;119:1189–95.

9. Engel SM, Wetmur J, Chen J, et al. Prenatal exposure to organophosphates, paraoxonase 1, and cognitive development in childhood. *Environ Health Perspect*. 2011;119:1182–8.

10. Rauh V, Arundjadai S, Horton M, et al. Seven-year neurodevelopmental scores and prenatal exposure to chlorpyrifos, a common agricultural pesticide. *Environ Health Perspect*. 2011;119:1196–201.

11. Engel SM, Miodovnik A, Canfield RL, et al. Prenatal phthalate exposure is associated with childhood behavior and executive functioning. *Environ Health Perspect*. 2010;118:565–71.

12. Swan SH. Environmental phthalate exposure in relation to reproductive outcomes and other health endpoints in humans. *Environ Res*. 2008;108:177–84.

13. Magnus P, Birke C, Vejrup K, et al. Cohort profile update: the Norwegian Mother and Child Cohort Study (MoBa). *Int J Epidemiol*. 2016 Apr;45(2):382–8.

14. Kawamoto T, Nitta H, Murata K, et al. Rationale and study design of the Japan environment and children's study (JECS). *BMC Public Health*. 2014;14:25.

15. Centers for Disease Control and Prevention. National biomonitoring program. 2023. Available at: https://www.cdc.gov/biomonitoring/index.html.

16. European Chemicals Agency. Understanding REACH. European Commission. 2011. Available at: https://echa.europa.eu/regulations/reach/understanding-reach.

17. US Environmental Protection Agency. The Frank R. Lautenberg Chemical Safety for the 21st Century Act. 2023. Available at: https://www.epa.gov/assessing-and-managing-chemicals-under-tsca/frank-r-lautenberg-chemical-safety-21st-century-act.

18. National Research Council. *Science and Judgment in Risk Assessment.* Washington, DC: National Academies Press; 1994.

19. Landrigan PJ. Risk assessment for children and other sensitive populations. *Ann NY Acad Sci.* 1999;895:1–9.

20. Harremoës P, Gee D, MacGarvin M, et al. *Late Lessons from Early Warnings: The Precautionary Principle 1896–2000.* Copenhagen: European Environmental Agency; 2001.

21. European Environment Agency. Late lessons from early warnings: science, precaution, innovation. EEA Report No 1/2013. European Environment Agency, Copenhagen, Denmark. 2013. Available at: https://www.eea.europa.eu/publications/late-lessons-2.

22. Burstein JM, Levy B. The teaching of occupational health in U.S. medical schools: little improvement in 9 years. *Am J Public Health.* 1994;84:846–9.

23. Galvez M, Collins G, Amler RW, et al. Building New York State Centers of Excellence in Children's Environmental Health: a replicable model in a time of uncertainty. *Am J Public Health.* 2019 Jan;109(1):108–12.

24. Etzel RA, Crain EF, Gitterman BA, Oberg C, Scheidt P, Landrigan PJ. Pediatric environmental health competencies for specialists. *Ambul Pediatr.* 2003;3:60–3.

25. Landrigan PJ, Braun JM, Crain EF, et al. The Academic Pediatric Association Retreat for Scholars in Pediatric Environmental Health: fifteen year report on a cross-institutional collaboration in professional education. *Acad Pediatr.* 2019;19(4):421–7.

26. Miller RW. How environmental hazards in childhood have been discovered: carcinogens, teratogens, neurotoxicants, and others. *Pediatrics.* 2004;113:945–51.

27. Bullard RD. *Dumping in Dixie: Race, Class, and Environmental Quality.* 3rd ed. New York: Routledge;1990.

28. United Nations. Office of the High Commissioner on Human Rights. Children's rights and the environment. 2018. Available at: https://www.ohchr.org/en/special-procedures/sr-environment/childrens-rights-and-environment.

29. UNICEF. Every child has the right to a healthy environment. 2022. Available at: www.unicef.org/media/124656/file/Childhood_Right_To_Healthy_Environments.

30. US Census Bureau. U.S. and World Population Clock. 2023. Available at: https://www.census.gov/popclock/.

31. United Nations. Division of Economic and Social Affairs. World population projected to reach 9.8 billion in 2050, and 11.2 billion in 2100. 2017. Available at: https://www.un.org/en/desa/world-population-projected-reach-98-billion-2050-and-112-billion-2100.

32. United Nations. Division of Economic and Social Affairs. World Urbanization Prospects. 2018. Available at: https://population.un.org/wup/.

33. Desjardins J. World's largest megacities by 2100. *Visual Capitalist.* 2018. Available at: https://www.visualcapitalist.com/worlds-20-largest-megacities-2100/.

34. Wang Z, Walker GW, Muir DCG, Nagatani-Yoshida K. Toward a global understanding of chemical pollution: a first comprehensive analysis of national and regional chemical inventories. *Environ Sci Technol*. 2020 Mar 3;54(5):2575–84.

35. Dooyema CA, Neri A, Lo YC, et al. Outbreak of fatal childhood lead poisoning related to artisanal gold mining in northwestern Nigeria, 2010. *Environ Health Perspect*. 2012 Apr;120(4):601–7. doi: 10.1289/ehp.1103965.

36. Lo YC, Dooyema CA, Neri A, et al. Childhood lead poisoning associated with gold ore processing: a village-level investigation-Zamfara State, Nigeria, October-November 2010. *Environ Health Perspect*. 2012;120(10):1450–5. doi:10.1289/ehp.1104793.

37. Diggory HJ, Landrigan PJ, Latimer KP, et al. Fatal parathion poisoning caused by contamination of flour in international commerce. *Am J Epidemiol*. 1977;106:145–53.

38. Etzel RA, Forthal DN, Hill RH, Demby A. Fatal parathion poisoning in Sierra Leone. *Bull World Health Organ*. 1987;65:645–9.

39. Collegium Ramazzini. Asbestos is still with us: repeat call for a universal ban. 2010. Available at: www.collegiumramazzini.org.

40. Chatham-Stephens K, Caravanos J, Ericson B, et al. Burden of disease from toxic waste sites in India, Indonesia, and the Philippines in 2010. *Environ Health Perspect*. 2013;121:791–6.

41. Levy BS, Levin JL, Teitelbaum DT. DBCP-induced sterility and reduced fertility among men in developing countries: a case study of the export of a known hazard. *Int J Occup Environ Health*. 1999;5:115.

42. Krieger N, Smith GD. "Bodies count" and body count: social epidemiology and embodying inequality. *Epidemiol Rev*. 2004;26:92–103.

43. Wilkinson RG, Pickett KE. Why the world cannot afford the rich. *Nature*. 2024 Mar;627(8003):268–270.

44. Visual Capitalist. Global wealth distribution. 2021. Available at: https://www.visualcapitalist.com/distribution-of-global-wealth-chart/.

45. World Bank. Life expectancy at birth. 2022. Available at: https://data.worldbank.org/indicator/SP.DYN.LE00.IN.

46. Centers for Disease Control and Prevention. National Center for Health Statistics. Life expectancy in the U.S. dropped for the second year in a row in 2021. 2022. Available at: https://www.cdc.gov/nchs/pressroom/nchs_press_releases/2022/20220831.htm.

47. Cohen J. Make secondary education universal. *Nature*. 2008;456:572–3.

48. Herz B. The importance of educating girls. *Science*. 2004;305:1910–1.

49. Vedanthan R, Fuster V. Urgent need for human resources to promote global cardiovascular health. *Nat Rev Cardiol*. 2011;8:114–7.

50. GBD 2019, Diseases and Injuries Collaborators. Global burden of 369 diseases and injuries in 204 countries and territories, 1990–2019: a systematic analysis for the Global Burden of Disease Study 2019. *Lancet*. 2020 Oct 17;396(10258):1204–22.

51. Landrigan PJ, Carlson JE. Environmental policy and children's health. *Future Child*. 1995;5:34–51.

52. Pope Francis. Laudato Si'. Encyclical letter on care for our common home. 2015. Available at: https://www.vatican.va/content/francesco/en/encyclicals/documents/papa-francesco_20150524_enciclica-laudato-si.html.

Index

For the benefit of digital users, indexed terms that span two pages (e.g., 52–53) may, on occasion, appear on only one of those pages.

Tables, figures, and boxes are indicated by *t*, *f*, and *b* following the page number